Millionen Jahre	Zeitalter Ära/Árathem		System/Periode	Serie/Epoche	System/Period	Series/Epoch	Era		Million Years
0	Känozoikum	Neophytikum	Quartär	Holozän Pleistozän	Quarternary	Holocene Pleistocene	Neophytic	Cenozoic	0
			Tertiär	Pliozän Miozän Oligozän Eozän Paläozän	Tertiary	Pliocene Miocene Oligocene Eocene			1.6 5.3 23 37 53
50									
100	Mesozoikum	Mesophytikum	Kreide				Me	M	
150			Jura		Jurassic				
200			Trias		Triassic				205
250			Perm		Permian				250
300	Paläozoikum	Paläophytikum	Karbon		Carboniferous		Paleophytic	Paleozoic	290
350			Devon		Devonian				355
400			Silur		Silurian				410
450		Eophytikum	Ordovizium		Ordovician		Eophytic		438
500			Kambrium		Cambrian				510
550									570
2500	Präkambrium		Proterozoikum		Proterozoic		Precambrian		
4000			Archaikum		Archean				

Paläogen Neogen

Paleogene Neogene

Erdzeitalter **Geological Ages**

W0031503

Wörterbuch der Biologie
Dictionary of Biology

Theodor C. H. Cole

Wörterbuch der Biologie
Dictionary of Biology

Englisch ↔ Deutsch
German ↔ English

2. Auflage

unter Mitarbeit von Ingrid Haußer-Siller

Zuschriften und Kritik an:
Elsevier GmbH, Spektrum Akademischer Verlag, Lektorat Naturwissenschaften
Merlet Behncke-Braunbeck (m.braunbeck@elsevier.com)
Jutta Liebau, (j.liebau@elsevier.com), Slevogtstr. 3-5, 69126 Heidelberg

Autoren:
Dipl.Biol. Theodor C.H. Cole, Heidelberg, tchcole@gmx.de;
Dr.rer.nat. Ingrid Haußer-Siller, Heidelberg, ingrid_hausser@med.uni-heidelberg.de

Wichtiger Hinweis für den Benutzer
Der Verlag und der Autor haben alle Sorgfalt walten lassen, um vollständige und akkurate Informationen in diesem Buch zu publizieren. Der Verlag übernimmt weder Garantie noch die juristische Verantwortung oder irgendeine Haftung für die Nutzung dieser Informationen, für deren Wirtschaftlichkeit oder fehlerfreie Funktion für einen bestimmten Zweck. Der Verlag übernimmt keine Gewähr dafür, dass die beschriebenen Verfahren, Programme usw. frei von Schutzrechten Dritter sind. Der Verlag hat sich bemüht, sämtliche Rechteinhaber von Abbildungen zu ermitteln. Sollte dem Verlag gegenüber dennoch der Nachweis der Rechtsinhaberschaft geführt werden, wird das branchenübliche Honorar gezahlt.

Bibliografische Information Der Deutschen Bibliothek
Die Deutsche Bibliothek verzeichnet diese Publikation in der Deutschen Nationalbibliografie; detaillierte bibliografische Daten sind im Internet über http://dnb.ddb.de abrufbar.

Alle Rechte vorbehalten
2. Auflage 2005
© Elsevier GmbH, München
Spektrum Akademischer Verlag ist ein Imprint der Elsevier GmbH.

05 06 07 08   5 4 3 2 1 0

Das Werk einschließlich aller seiner Teile ist urheberrechtlich geschützt. Jede Verwertung außerhalb der engen Grenzen des Urheberrechtsgesetzes ist ohne Zustimmung des Verlages unzulässig und strafbar. Das gilt insbesondere für Vervielfältigungen, Übersetzungen, Mikroverfilmungen und die Einspeicherung und Verarbeitung in elektronischen Systemen.

Planung und Lektorat: Merlet Behncke-Braunbeck, Jutta Liebau
Herstellung: Elke Littmann
Satz: ProSatz Unger, Weinheim
Druck und Bindung: LegoPrint S.p.A., I-Lavis
Umschlaggestaltung: WSP Design, Heidelberg
Gedruckt auf 90 g Tauro-Offset, TCF, chlorfrei gebleicht

Printed in Italy
ISBN 3-8274-1628-0

Aktuelle Informationen finden Sie im Internet unter www.elsevier.de

... to our children!
... believing in their future,
  as we do in our past!

Vorwort zur 2. Auflage

Unser Wörterbuch hat bei Studenten und Forschern eine breite Akzeptanz gefunden – vor allem wegen seiner Übersichtlichkeit, wegen des ansprechenden Formats und nicht zuletzt wegen des günstigen Preis-Leistungsverhältnisses. Diese positive Resonanz hat den Verlag (nun Elsevier Deutschland) veranlasst, eine zweite Auflage anzubieten. Aktualisiert und in neuem Gewand stellt sich das Wörterbuch den Herausforderungen der modernen Biowissenschaften. Für die Autoren bedeutet dieser Erfolg einen Beitrag zum Gelingen internationaler Kommunikation.

Im Frühjahr 2005
Theodor C.H. Cole
Ingrid Haußer-Siller

Preface to the 2nd Edition

Our dictionary has been well received and widely accepted by students and researchers, especially for its clear layout, its appealing format as much as for its attractive price. The overall positive reactions have convinced the publisher, now Elsevier Germany, to launch a second edition. With updated content and appealing new cover, the dictionary is fit to meet the challenges of modern life sciences. For us authors this success signifies a contribution to an efficient international communication.

In the spring of 2005
Theodor C.H. Cole
Ingrid Haußer-Siller

Vorwort zur 1. Auflage

Das *Wörterbuch der Biologie* enthält ca. 50.000 Begriffe der klassischen und modernen Biowissenschaften aus allen Bereichen der Botanik, Zoologie und Mikrobiologie:

- Anatomie/Morphologie
- Biochemie
- Biogeographie
- Biostatistik/Biometrie
- Biotechnologie
- Bodenkunde
- Evolution
- Forstwirtschaft
- Genetik
- Histologie
- Immunologie
- Klimatologie

- Labormethoden/Laborgerät
- Landwirtschaft/Gartenbau
- Mikroskopie
- Molekularbiologie
- Neurobiologie
- Ökologie
- Paläontologie/Erdgeschichte
- Parasitologie
- Physiologie
- Systematik
- Verhaltenslehre
- Zellbiologie

Die Naturwissenschaften erforderten seit jeher einheitliche Kommunikationssysteme. Das Englische hat sich vor allem nach dem zweiten Weltkrieg als internationale Wissenschaftssprache nicht zuletzt wegen seines unkomplizierten Aufbaus – trotz vergleichsweise reichhaltigen Wortschatzes – durchgesetzt.

In letzter Zeit beklagt man oft die zunehmende Anglifizierung der deutschen Sprache. Im Labor-Slang wird: „geblottet, gepoolt, gepatcht und gepottert". Befürworter dieser Entwicklung sehen darin einen Trend zur Internationalisierung der Wissenschaftssprache. Ob eine solche Sprachdynamik wünschenswert ist, bleibt allerdings umstritten. Der deutschen Sprache ist zu wünschen, dass ihre Eigenständigkeit und Ästhetik erhalten bleibt. Sprache ist „lebendig"! Dies bedeutet für deutschsprachige WissenschaftlerInnen die Möglichkeit, im wissenschaftlichen Sprachgebrauch auch aus der deutschen Sprache zu schöpfen.

Viele deutsche Autoren bevorzugen heute die „c"-Schreibung, wo in älterer Literatur traditionell ein „k" oder „z" steht, wie Cuticula statt Kutikula, Caruncula statt Karunkula, Clitoris statt Klitoris, Coccen statt Kokken, Glucose statt Glukose, Citratzyklus statt Zitratzyklus. Diese Übergänge sind aber derzeit noch nicht einheitlich akzeptiert. Im vorliegenden Wörterbuch finden Sie sowohl neue als auch alte Schreibweisen.

Klassische Endungen werden heute vielfach zugunsten der angloamerikanischen Schreibweise vereinfacht, zum Beispiel Thoracopod statt Thorakopodium; der entsprechende Plural lautet dann „Thoracopoden", was den Eindruck einer Tiergruppe (analog zu Decapoden) vermitteln könnte – deshalb wird im Plural meist noch die klassische Endung, beispielsweise Thoracopodien, beibehalten.

Auch in Amerika gibt es interessante sprachliche Entwicklungen. Einige amerikanische Wissenschaftler beginnen heute Schreibweise und Wortendungen der modernen Sprache anzupassen. Der Plural angepasster Fremdwörter wird dann nach den üblichen Regeln der Pluralbildung im Englischen gebildet: flagellums statt flagella, algas statt algae, funguses statt fungi, hyphas statt hyphae, myceliums statt mycelia, larvas statt larvae, antennas statt antennae, vertebras statt vertebrae, phenomenas statt phenomena (von phenomenon), taxons statt taxa, tracheas statt tracheae, und mitochondrions statt mitochondria [Pearse/Buchsbaum, 1992]. Bei einigen Wörtern wird die Pluralform im allgemeinen Sprachgebrauch auch für den Singular verwendet: criteria, agenda, bacteria entsprechen im Plural dann criterias, agendas, bacterias. Dies wird sich sicherlich aus praktischen Gründen durchsetzen. Beim Abfassen von Publikationen sei dem Fremdsprachler jedoch empfohlen, zunächst noch die traditionellen Endungen zu verwenden.

Die deutsche Rechtschreibreform wurde berücksichtigt.

Danksagungen. Für sachdienliche Hinweise bedanken wir uns bei Prof. Dr. Gerd Alberti (Universität Greifswald), Prof. Dr. Ingrun Anton-Lamprecht (Universitäts-Hautklinik, Heidelberg), Prof. Dr. Rolf Beiderbeck (Universität Heidelberg), Dan Bennette, M.A. (University of Maryland), Dr. Rainer Foelix (Aarau), Prof. Dr. Hartmut Hilger (Freie Universität, Berlin), Prof. Dr. Wolfgang H. Kirchner (Universität Konstanz), Prof. Dr. Werner Nachtigall (Saarbrücken), Dipl.Ing. Wolfgang Sittig (Frankfurt), Prof. Dr. Werner A. Stein (Niederjosbach), Prof. Dr. Volker Storch (Universität Heidelberg) sowie Dr. Bernhard Ziegler (Englisches Institut, Heidelberg). Dr. Beatrix Spreier danken wir für ihre wertvolle Hilfe in der Endphase der Manuskriptbearbeitung. Immer zur Seite standen uns Dr. Dietrich Schulz (Umweltbundesamt, Berlin) und Dr. Willi Siller (Universität Heidelberg). Die Universitätsbibliotheken Heidelberg und Berlin sowie die Bestände der Library System of the University of Maryland waren eine Quelle von Inspiration und unerschöpflichem Wissen. Frau Merlet Behncke-Braunbeck und den MitarbeiterInnen des Spektrum Akademischen Verlages gebührt unser Dank für die effiziente und angenehme Zusammenarbeit.

Erika Siebert-Cole, M.A., gilt besondere Wertschätzung. Ihre Kenntnis der deutschen Sprache und ihre Assistenz waren maßgeblich für diese Arbeit.

Allen Kollegen, Studenten und Freunden, die uns auf dem langen Weg durch dieses Projekt beraten und unterstützt haben, gebührt unser ganz herzlicher Dank.

Heidelberg, im Sommer 1998

Theodor C.H. Cole
Ingrid Haußer-Siller

Preface to the 1st Edition

This biological dictionary contains some 50,000 terms from the classical and modern life sciences including all major fields of botany, zoology, and microbiology:

- Anatomy/Morphology
- Biochemistry
- Biogeography
- Biometry
- Biotechnology
- Cell biology
- Climatology/Meteorology
- Earth history/Paleontology
- Ecology
- Ethiology
- Evolution
- Forestry
- Genetics
- Histology
- Horticulture/Agriculture
- Immunology
- Lab equipment & techniques
- Microscopy
- Molecular biology
- Neurobiology
- Parasitology
- Physiology
- Soil science
- Systematics

The *Dictionary of Biology* is intended to help German scholars in the life sciences and related fields in gaining access to the highly specialized terminology of the modern biosciences. This is particularly important as many leading textbooks and the majority of scientific research articles are now written in English. A broad coverage of basic terminology is supplemented with translations of highly specialized terms – thus serving laymen, students, and specialists alike.

The English and German languages have experienced some interesting developments in the spelling of scientific terms: among German authors there is a tendency of changing the traditional "k" and "z" spelling to the English "c" spelling: e. g. Oocyt vs. Oozyt, Cuticula vs. Kutikula, Coccen vs. Kokken, coccal vs. kokkal, or Glucose vs. Glukose.

Particularly some American authors are now suggesting the adaptation of the plural endings in scientific terms to the normal English language endings [Pearse/Buchsbaum], e. g. algas vs. algae, larvas vs. larvae, tracheas vs. tracheae, vertebras vs. vertebrae, taxons vs. taxa, mitochondrions vs. mitochondria, setas vs. setae, antennas vs antennae. The present authors favor this development, while being aware of controversial standpoints on this matter.

The new German orthography rules have been taken into account.

We hope this dictionary may serve you as a useful tool in your research, in writing publications and in translating biological literature.

XI

Acknowledgements. Important help has been granted by Prof. Dr. Gerd Alberti, Prof. Dr. Ingrun Anton-Lamprecht, Prof. Dr. Rolf Beiderbeck, Dan Bennett M.A., Dr. Rainer Foelix, Prof. Dr. Hartmut H. Hilger, Prof. Dr. Wolfgang H. Kirchner, Prof. Dr. Werner Nachtigall, Dr. Dietrich Schulz, Dr. Willi Siller, Dipl. Ing. Wolfgang Sittig, Dr. Beatrix Spreier, Prof. Dr. Werner A. Stein, Prof. Dr. Volker Storch, and Dr. Bernhard Ziegler to all of whom we are particularly grateful. The libraries of the University of Heidelberg, Free University of Berlin, and University of Maryland have been most valuable in researching the contents of this book. The comments of various of our colleagues, students, and of several unnamed reviewers are sincerely appreciated. Ms. Merlet Behncke-Braunbeck, editor at Spektrum Akademischer Verlag, has been a brillant partner and we commend her for the encouragement and energy she devoted to implementing this project.

Erika Siebert-Cole, M.A. has shared her knowledge, time, inspiration, and an intricate sense of the German language to this book. Without her this project would never have been accomplished.

Our gratitude be wholeheartedly expressed to our families and friends who helped and encouraged us throughout this project.

Heidelberg, in the summer of 1998   *Theodor C.H. Cole*
                                                     Ingrid Haußer-Siller

Aufbau und Konzeption

Wortfelder: Um eine zusammenhängende Themenbearbeitung zu ermöglichen, die „Trefferwahrscheinlichkeit" bei der Wortsuche zu erhöhen und somit die Arbeit zu erleichtern, verwenden wir zusätzlich zur gewöhnlichen alphabetischen Ordnung ein auch in amerikanischen Wörterbüchern verwendetes Konzept der thematischen Begriffssammlung (*clusters*) unter den jeweiligen übergeordneten Hauptstichwörtern. So erscheint bei zusammengesetzten Wörtern das angehängte Substantiv als übergeordnetes Stichwort (Unter*art* unter *Art*, Keim*drüse* unter *Drüse* ...). Thematisch verwandte Begriffe werden in Wortfeldern zusammengefasst, auch wenn die einzelnen Begriffe das Hauptstichwort selbst gar nicht enthalten. Beispielsweise findet sich unter dem Hauptstichwort *Hormone* eine Reihe wichtiger Hormone; Larvenformen erscheinen unter *Larve*, Infloreszenzen-Typen im Wortfeld *Infloreszenz*. Dies verschafft Überblick und Arbeitskomfort. In anderen Wörterbüchern müsste jeder Begriff einzeln aufgesucht werden.

Definitionen: Wo zum Verständnis notwendig und zur Unterscheidung ähnlicher Begriffe hilfreich, erscheinen Definitionen oder einschränkende Kommentare in Klammern hinter dem jeweiligen Wort. Allerdings wird der Benutzer für Details gelegentlich auf Definitionswörterbücher und Fachmonographien zurückgreifen müssen. Die Literaturliste am Schluss des Buches verweist auf entsprechende Quellen.

Tier-/Pflanzennamen: Tiere wurden bis zur Ebene der Ordnung, Pflanzen/ Pilze bis zur Ebene der Familie berücksichtigt, das heißt, es finden sich keine Gattungs- oder Artnamen (kompetente, umfassende Listen existieren hierzu in entsprechenden Tier-und Pflanzen-Wörterbüchern).

American English: Dieses Biologie-Wörterbuch folgt in Orthographie und Definitionen der amerikanischen Schreibweise nach *Merriam Webster's Third New International Dictionary* sowie den Autoren der im Anhang aufgeführten Fachliteratur. Besonders berücksichtigt werden Begriffe speziell amerikanischen Sinngehalts, wie bayou, hominy, chitlins, chiggers, snood ...

Concept and Design

Word Clusters: Words are grouped by headwords and topics according to a useful and efficient clustering concept employed in various American biology dictionaries – in addition to regular alphabetical order. This applies to compounds (*underleaf* under *leaf*, *underhair* under *hair*, *subspecies* under *species*) as much as to terms related by meaning, not necessarily by spelling (larva types, such as *maggot*, *grub*, *tadpole* ..., are clustered under the headword *larva*).

Until now, German biology dictionaries have not made use of this concept. Usually German compound nouns are listed only in regular alphabetical order; however, it also may be useful to list the noun as the headword of all related compounds (*Kennart*, *Überart*, *Unterart* ... under *Art*). This helpful concept has proven to allow a much more rapid access and opens the chance for topic-related study and research.

Definitions/Comments: Brief definitions and clarifying comments have been included where considered necessary for understanding the meaning of a term and for referring to the context of usage; for detailed definitions and finer nuances, the reader may need to consult other specialty dictionaries (see list of references on page 763).

Animal/Plant Names: This dictionary has been conceptualized as a tool for accessing biological terms, not as a listing of plant/animal names. Thus, animals have been included only through the category levels of phyla, classes, and orders; plants and fungi through divisions, classes, orders, and families. Species names are not listed. (for this purpose please consult a specialty dictionary of plant names or animal names).

American English: Definitions and orthography are in accordance with *Merriam Webster's Third New International Dictionary* and sources of specialized literature listed in the back of this book.

Autoren

Dipl. Biol. **Theodor C. H. Cole** ist amerikanischer Staatsbürger und arbeitet als Dozent, Übersetzer und Autor in Heidelberg. Während des Studiums der Biologie, Chemie und Physik an der Universität Heidelberg studierte er Allgemeine Biologie und Zoologie bei Prof. Dr. Franz Duspiva und arbeitete am Institut für Systematische Botanik bei Prof. Dr. Werner Rauh. Biochemisch-analytische Forschung betrieb er im Rahmen seiner Diplomarbeit am Institut für Pharmazeutische Biologie bei Prof. Dr. Hans Becker. Als Autor ist er bereits bekannt durch seine Wörterbücher, wie das *Taschenwörterbuch der Botanik* (1994), *Taschenwörterbuch der Zoologie* (1995), *Wörterbuch der Tiernamen* (2000) und *Wörterbuch Labor* (2005).

Dr. rer. nat. **Ingrid Haußer-Siller** studierte Biologie und Chemie an der Universität Heidelberg und arbeitete am Lehrstuhl für Zellenlehre (Prof. Dr. Eberhard Schnepf). Seit ihrer Promotion bei Prof. Dr. Werner Herth arbeitet sie in der medizinisch-genetischen Grundlagenforschung an der Universitäts-Hautklinik in Heidelberg und leitet dort das EM-Labor. Ihre Erfahrungen aus Übersetzertätigkeiten im Teilgebiet Genetik und Molekularbiologie (u.a. Berg/Singer: *Gene und Genome*) bereichern dieses Wörterbuch.

Authors

Dipl. Biol. **Theodor C.H. Cole**, born in 1954 in Plainfield, New Jersey, studied biology, chemistry, and physics mainly at the University of Heidelberg, in particular with Prof. Dr. Franz Duspiva (zoology) and Prof. Dr. Werner Rauh (botany) and wrote his thesis in biochemical analysis of medicinal plants at the Institute of Pharmaceutical Biology under Prof. Dr. Hans Becker. Successful dictionaries by TCH Cole include *Pocket Dictionary of Botany* (1994), *Pocket Dictionary of Zoology* (1995), *Dictionary of Animal Names* (2000), and *Laboratory Dictionary* (2005).

Dr. rer. nat. **Ingrid Haußer-Siller**, born in Heilbronn, Germany, studied biology and chemistry at the University of Heidelberg, especially at the Insititute of Cell Biology (Prof. Dr. Eberhard Schnepf). Since earning her Ph.D. with Prof. Dr. Werner Herth she has been involved in medical-genetic research at the Dermatology Clinic of the University of Heidelberg, now as Head of the EM-Laboratory. This dictionary has greatly benefited from her experience in translating various texts in the field of genetics and molecular biology (e.g., Berg/Singer: *Genes and Genomes*).

Abkürzungen / Abbreviations

sg	Singular – singular
pl	Plural – plural
adv/adj	Adverb/Adjektiv – adverb/adjective
n	Substantiv – noun
vb	Verb – verb
f	weiblich – female
m	männlich – male
nt	sächlich – neuter
allg / general	allgemein – general
aer	Aerodynamik – aerodynamics
agr	Landwirtschaft – agriculture
anat	Anatomie – anatomy
arach	Spinnenkunde – arachnology
biogeo	Biogeographie – biogeography
biot	Biotechnologie – biotechnology
bot	Botanik – botany
cardio	Kardiologie – cardiology
centrif	Zentrifugation – centrifugation
chem	Chemie – chemistry
chromat	Chromatographie – chromatography
ecol	Ökologie – ecology
electroph	Elektrophorese – electrophoresis
embr	Embryologie – embryology
entom	Entomologie – entomology
ethol	Ethologie – ethology
evol	Evolution – evolution
for	Forstwirtschaft – forestry
gen	Genetik – genetics
geol	Geologie – geology
hort	Gartenbau – horticulture
hunt	Jagd – hunting
ichth	Ichthyologie – ichthyology
immun	Immunologie – immunology
lab	Labor – laboratory
limn	Limnologie – limnology
mar	Meereskunde – marine sciences
math	Mathematik – mathematics
med	Medizin – medicine
metabol	Metabolismus/Stoffwechsel – metabolism
meteo	Meteorologie – meteorology
micb	Mikrobiologie – microbiology
micros	Mikroskopie – microscopy
neuro	Neurobiologie – neurobiology
ophthal	Ophthalmologie – ophthalmology
opt	Optik – optics
orn	Ornithologie – ornithology
paleo	Paläontologie – paleontology
photo	Photographie – photography
phys	Physik – physics
physio	Physiologie – physiology
stat	Statistik/Biostatistik – statistics/biostatistics
techn	Technik – engineering
vir	Virologie – virology
zool	Zoologie – zoology

**Englisch
Deutsch**

**English
German**

1   absorption spectrum A

A band (muscle: *anisotropic*)
A-Bande
A-form (of DNA) A-Form,
A-Konformation (der DNA)
A-site (aminoacyl site) A-Stelle
(Aminoacyl-Stelle)
aapa mire/string bog Aapamoor
(Mischmoor), Strangmoor
aardvark (of Africa)/Tubulidentata
Erdferkel (Röhrchenzähner)
abattoir/slaughterhouse
Schlachthaus, Schlachthof
abdomen Abdomen, Unterleib,
Hinterleib
abdomen/opisthosoma
Hinterkörper, Hinterleib, Abdomen,
Opisthosoma
**abdominal breathing/
diaphragmatic respiration**
Bauchatmung, Zwerchfellatmung
abdominal cavity Bauchhöhle
abdominal pregnancy
Abdominalschwangerschaft,
Bauchhöhlenträchtigkeit,
Leibeshöhlenschwangerschaft,
Leibeshöhlenträchtigkeit
abdominal ribs/gastralia (reptiles)
Bauchrippen, Gastralia
abdominal somite/pleomere
Abdominalsegment, Pleomer
abductor muscle Abzieher,
Abduktor, Abductor (Muskel)
aberration Aberration, Abirrung,
Abweichung
 • **autosomal aberration**
  Autosomenaberration,
  autosomale Aberration
 • **chromatic aberration** *opt/micros*
  chromatische Aberration,
  chromatischer Abbildungsfehler,
  Farbabweichung, Farbfehler
 • **chromosomal aberration/
  chromosome aberration**
  chromosomale Aberration,
  Chromosomenaberration
 • **sex-chromosome aberration**
  Heterosomenaberration,
  gonosomale
  Chromosomenaberration
abietic acid Abietinsäure
abiotic abiotisch
**ABM paper (aminobenzyloxymethyl
paper)** ABM-Papier
**abomasum/reed/rennet-stomach/
fourth stomach** Labmagen,
Abomasus
**abort (degenerate/remain
rudimentary)** *vb bot* verkümmern,
in der Entwicklung zurückbleiben
**abort (terminate pregnancy before
term; induced)** *vb* abtreiben,
eine Fehlgeburt herbeiführen

abort/reject abortieren, abstoßen
**abortion/termination of pregnancy
(with expulsion of embryo/fetus;
popularly often *sensu:* induced a.)**
Abort, Abortus, Fehlgeburt, Abgang
 • **early abortion (up to 12th week)**
  Frühabort
 • **induced abortion (termination
  of pregnancy)** Abtreibung,
  Schwangerschaftsabbruch,
  Abortinduktion,
  Beendigung der Schwangerschaft
 • **miscarriage (after 12th week)**
  Spätabort
 • **missed abortion**
  verhaltener Abort
 • **spontaneous abortion/
  miscarriage** Spontanabort,
  Fehlgeburt
abortive abortiv, verkümmert,
(zu)rückgebildet, rudimentär,
unvollkommen/unvollständig
entwickelt; steril, taub, unfruchtbar;
med abgekürzt verlaufend, vorzeitig,
verfrüht, gemildert
abortive infection *med/vet* abortive
Infektion, im Frühstadium
unterdrückte Infektion
abortive transduction
abortive Transduktion
abortive transfection
abortive Transfektion
**aboveground/overground/
superterranean** oberirdisch
 • **underground/belowground/
  subterranean** unterirdisch
**abscission/falling off/dropping off/
shedding** Abszission, Abwerfen,
Abwurf
 • **leaf abscission** Blattabwurf
abscission layer/separation layer
Abszissionsschicht,
Ablösungsschicht, Trennschicht,
Trennungsschicht
abscission zone/separation zone
Abszissionszone, Trennungszone
absorb absorbieren
**absorbance/absorbancy/
absorbency/extinction (optical
density)** Absorbanz, Extinktion
absorbance index/absorptivity
Absorptionsindex
absorbency Absorptionsvermögen,
Absorptionsfähigkeit,
Aufnahmefähigkeit
absorbent absorbierend,
absorptionsfähig
absorption Absorption
absorption coefficient
Absorptionskoeffizient
absorption spectrum
Absorptionsspektrum

absorptivity

absorptivity/absorbance index Absorptionsindex
abundance (frequency of occurrence) Abundanz, Individuenzahl, Häufigkeit
abundance index/cover abundance index/species importance value Artmächtigkeit
abundant mRNA abundante mRNA
abyssal/deep-sea floor/ocean floor Abyssal, Meeresgrund, Tiefseeboden
abyssal plain Tiefseeebene, Tiefsee-Ebene
abyssal zone/deep-sea zone Abyssal (unterste Wasserschicht im Meer), Tiefseebereich, Tiefseezone
abyssobenthic/deep-sea floor/ocean floor Abyssobenthal, Tiefseeboden (auf dem Tiefseeboden lebend/den Tiefseeboden bewohnend)
abyssopelagic zone Abyssopelagial, unterster Tiefseebereich (Wasserschicht), Meerestiefenbereich, unterste Tiefseezone
abzyme Abzym
acanthor larva Acanthor-Larve, Hakenlarve
acanthosome Akanthosom
acanthus family/Acanthaceae Akanthusgewächse
acaricide Acarizid, Akarizid, Milbenbekämpfungsmittel
acarology Acarologie, Milbenforschung
acaulescent/stemless akauleszent, stammlos
accelerating voltage (EM) *micros* Beschleunigungsspannung
acceleration of gravity Erdbeschleunigung
acceleration phase Beschleunigungsphase, Anfahrphase
acceptor Akzeptor
acceptor stem *gen* **(protein synthesis)** Akzeptorstamm
accessory akzessorisch, zusätzlich
accessory agent Hilfsstoff
accessory bud akzessorische Knospe, Beiknospe
accessory cell/auxiliary cell/subsidiary cell Nebenzelle (Spaltöffnung)
accessory chromosome akzessorisches Chromosom, zusätzliches Chromosom
accessory gland akzessorische Drüse, Anhangsdrüse
accessory pigment akzessorisches Pigment
accessory teat Afterzitze

accidental host/wrong host Fehlwirt, Irrwirt
acclimate (artificial conditions)/acclimatize (climate/seasons)/adapt akklimatisieren, anpassen
acclimation/acclimatization/adaptation Akklimatisation, Akklimatisierung, Eingewöhnung, Anpassung, Adaptation; (climate/seasons) Klimaanpassung
acclimatize (climate/seasons)/acclimate (artificial conditions)/adapt akklimatisieren, anpassen
accommodation (eye) Akkommodation, Scharfstellung
accrescence Zuwachs (nach Blüte)
accrescent fortwachsend, weiterwachsend
accretion/accrescence/additional growth/new growth/enlargement Zuwachs
accretion (cell wall growth) Auflagerung (Zellwandwachstum)
accrustation/adcrustation Akkrustierung, Adkrustierung
accumbent (lying along/against another body) anliegend (entlang anderem Gegenstand)
accumulation Akkumulation, Anhäufung
accuracy of measurement Messgenauigkeit
acellular/noncellular azellulär, nicht zellig
acellular slime molds/plasmodial slime molds/Myxomycetes echte Schleimpilze (plasmodial)
acelous/acoelous acöl, acoel
acentric chromosome azentrisches Chromosom
Aceraceae/maple family Ahorngewächse
aceriform ahornblattartig
acerose/needle-shaped nadelförmig
acetate (acetic acid/ethanoic acid) Acetat, Azetat (Essigsäure/Ethansäure)
acetic acid/ethanoic acid (acetate) Essigsäure, Ethansäure (Acetat)
acetic anhydride/ethanoic anhydride/acetic acid anhydride Essigsäureanhydrid
acetoacetic acid (acetoacetate)/β-ketobutyric acid Acetessigsäure (Acetacetat), β-Ketobuttersäure
acetyl CoA/acetyl coenzyme A "aktivierte Essigsäure", Acetyl-CoA
acetylcholine Acetylcholin (ACh)
acetylene Acetylen
N-acetylmuramic acid N-Acetylmuraminsäure
achene/akene Achäne, Achaene

acid A

Achilles' tendon/tendon of the heel
Achillessehne
achlamydeous achlamydeisch
achromatic achromatisch, unbunt
achromatic lens *micros*
achromatische Linse
achromatic objective/achromat
micros achromatisches Objektiv,
Achromat
achromatic substage/condenser
micros achromatischer Kondensor
aciculate/acicular/needle-shaped
nadelförmig
acid/acidic *adv/adj* azid, sauer
acid *n* Säure *(the following is a list of some important bio-organic acids)*
- **abietic acid** Abietinsäure
- **acetic acid/ethanoic acid (acetate)** Essigsäure,
  Ethansäure (Acetat)
- **acetoacetic acid (acetoacetate)/ β-ketobutyric acid**
  Acetessigsäure (Acetacetat),
  β-Ketobuttersäure
- **acetyl CoA/acetyl coenzyme A**
  "aktivierte Essigsäure",
  Acetyl-CoA
- **N-acetylmuramic acid (NAM)**
  N-Acetylmuraminsäure
- **N-acetylneuraminic acid (NAN)**
  N-Acetylneuraminsäure
- **aconitic acid (aconitate)**
  Aconitsäure (Aconitat)
- **adenylic acid (adenylate)**
  Adenylsäure (Adenylat)
- **adipic acid (adipate)**
  Adipinsäure (Adipat)
- **alginic acid (alginate)**
  Alginsäure (Alginat)
- **allantoic acid** Allantoinsäure
- **amino acid** Aminosäure
- **amygdalic acid/mandelic acid/ phenylglycolic acid** Mandelsäure,
  Phenylglykolsäure
- **anthranilic acid/2-aminobenzoic acid** Anthranilsäure
- **arachic acid/arachidic acid/ icosanic acid** Arachinsäure,
  Arachidinsäure, Eicosansäure
- **arachidonic acid/icosatetraenoic acid** Arachidonsäure
- **ascorbic acid (ascorbate)**
  Ascorbinsäure (Ascorbat)
- **asparagic acid/aspartic acid (aspartate)**
  Asparaginsäure (Aspartat)
- **azelaic acid/nonanedioic acid**
  Azelainsäure, Nonandisäure
- **behenic acid/docosanoic acid**
  Behensäure, Docosansäure
- **benzoic acid (benzoate)**
  Benzoesäure (Benzoat)

- **butyric acid/butanoic acid (butyrate)** Buttersäure,
  Butansäure (Butyrat)
- **caffeic acid** Kaffeesäure
- **capric acid/decanoic acid (caprate/decanoate)**
  Caprinsäure, Decansäure
  (Caprinat/Decanat)
- **caproic acid/capronic acid/ hexanoic acid (caproate/ hexanoate)** Capronsäure,
  Hexansäure (Capronat/Hexanat)
- **caprylic acid/octanoic acid (caprylate/octanoate)**
  Caprylsäure, Octansäure
  (Caprylat/Octanat)
- **carbonic acid (carbonate)**
  Kohlensäure (Karbonat/Carbonat)
- **carboxylic acids (carbonates)**
  Carbonsäuren, Karbonsäuren
  (Carbonate/Karbonate)
- **cerotic acid/hexacosanoic acid**
  Cerotinsäure, Hexacosansäure
- **chinic acid/kinic acid/quinic acid (quinate)** Chinasäure
- **chinolic acid** Chinolsäure
- **chlorogenic acid** Chlorogensäure
- **cholic acid (cholate)**
  Cholsäure (Cholat)
- **chorismic acid (chorismate)**
  Chorisminsäure (Chorismat)
- **cinnamic acid (cinnamate)**
  Cinnamonsäure,
  Zimtsäure (Cinnamat)
- **citric acid (citrate)** Citronensäure,
  Zitronensäure (Citrat/Zitrat)
- **crotonic acid/α-butenic acid**
  Crotonsäure, Transbutensäure
- **cysteic acid** Cysteinsäure
- **deoxyribonucleic acid (DNA)**
  Desoxyribonucleinsäure (DNS),
  Desoxyribonukleinsäure
- **diprotic acid** zweiwertige Säure,
  zweiprotonige Säure
- **ellagic acid/gallogen** Ellagsäure
- **erucic acid/(Z)-13-docosenoic acid**
  Erucasäure, Δ^{13}-Docosensäure
- **fatty acid** Fettsäure
- **ferulic acid** Ferulasäure
- **folic acid (folate)/ pteroylglutamic acid** Folsäure
  (Folat), Pteroylglutaminsäure
- **formic acid (formate/formiate)**
  Ameisensäure (Format/Formiat)
- **fumaric acid (fumarate)**
  Fumarsäure (Fumarat)
- **galacturonic acid**
  Galakturonsäure
- **gallic acid (gallate)**
  Gallussäure (Gallat)
- **gamma-aminobutyric acid (GABA)** γ-Aminobuttersäure

acid

- **gentisic acid** Gentisinsäure
- **geranic acid** Geraniumsäure
- **gibberellic acid** Gibberellinsäure
- **glacial acetic acid** Eisessig
- **glucaric acid/saccharic acid** Glucarsäure, Zuckersäure
- **gluconic acid (gluconate)** Gluconsäure (Gluconat)
- **glucuronic acid (glucuronate)** Glucuronsäure (Glukuronat)
- **glutamic acid (glutamate)/ 2-aminoglutaric acid** Glutaminsäure (Glutamat), 2-Aminoglutarsäure
- **glutaric acid (glutarate)** Glutarsäure (Glutarat)
- **glycolic acid (glycolate)** Glykolsäure (Glykolat)
- **glycyrrhetinic acid** Glycyrrhetinsäure
- **glyoxalic acid (glyoxalate)** Glyoxalsäure (Glyoxalat)
- **glyoxylic acid (glyoxylate)** Glyoxylsäure (Glyoxylat)
- **guanylic acid (guanylate)** Guanylsäure (Guanylat)
- **gulonic acid (gulonate)** Gulonsäure (Gulonat)
- **homogentisic acid** Homogentisinsäure
- **humic acid** Huminsäure
- **hyaluronic acid** Hyaluronsäure
- **ibotenic acid** Ibotensäure
- **imino acid** Iminosäure
- **indolyl acetic acid/indoleacetic acid (IAA)** Indolessigsäure
- **inosinic acid/inosine monophosphate (IMP)/inosine 5-phosphate** Inosinmonophosphat
- **iodoacetic acid** Jodessigsäure
- **isovaleric acid** Isovaleriansäure
- **jasmonic acid** Jasmonsäure
- **keto acid** Ketosäure
- **kojic acid** Kojisäure
- **lactic acid (lactate)** Milchsäure (Laktat)
- **lauric acid/decylacetic acid/ dodecanoic acid (laurate/ dodecanate)** Laurinsäure, Dodecansäure (Laurat/Dodecanat)
- **levulinic acid** Lävulinsäure
- **lichen acid** Flechtensäure
- **lignoceric acid/tetracosanoic acid** Lignocerinsäure, Tetracosansäure
- **linolenic acid** Linolensäure
- **linolic acid/linoleic acid** Linolsäure
- **lipoic acid (lipoate)/thioctic acid** Liponsäure, Thioctsäure (Liponat)
- **lipoteichoic acid** Lipoteichonsäure
- **litocholic acid** Litocholsäure
- **lysergic acid** Lysergsäure
- **maleic acid (maleate)** Maleinsäure (Maleat)
- **malic acid (malate)** Äpfelsäure (Malat)
- **malonic acid (malonate)** Malonsäure (Malonat)
- **mandelic acid/phenylglycolic acid/amygdalic acid** Mandelsäure, Phenylglykolsäure
- **mannuronic acid** Mannuronsäure
- **mevalonic acid (mevalonate)** Mevalonsäure (Mevalonat)
- **monoprotic acid** einwertige, einprotonige Säure
- **mucic acid** Schleimsäure, Mucinsäure
- **muramic acid** Muraminsäure
- **myristic acid/tetradecanoic acid (myristate/tetradecanate)** Myristinsäure, Tetradecansäure (Myristat)
- **nervonic acid/ (Z)-15-tetracosenoic acid/ selacholeic acid** Nervonsäure, Δ^{15}-Tetracosensäure
- **neuraminic acid** Neuraminsäure
- **nicotinic acid (nicotinate)/niacin** Nikotinsäure (Nikotinat)
- **nucleic acid** Nucleinsäure, Nukleinsäure
- **oleic acid/(Z)-9-octadecenoic acid (oleate)** Ölsäure, Δ^9-Octadecensäure (Oleat)
- **orotic acid** Orotsäure
- **orsellic acid/orsellinic acid** Orsellinsäure
- **osmic acid** Osmiumsäure
- **oxalic acid (oxalate)** Oxalsäure (Oxalat)
- **oxalosuccinic acid (oxalosuccinate)** Oxalbernsteinsäure (Oxalsuccinat)
- **oxoglutaric acid (oxoglutarate)** Oxoglutarsäure (Oxoglutarat)
- **palmitic acid/hexadecanoic acid (palmate/hexadecanate)** Palmitinsäure, Hexadecansäure (Palmat/Hexadecanat)
- **palmitoleic acid/ (Z)-9-hexadecenoic acid** Palmitoleinsäure, Δ^9-Hexadecensäure
- **pantoic acid** Pantoinsäure
- **pantothenic acid (pantothenate)** Pantothensäure (Pantothenat)
- **pectic acid (pectate)** Pektinsäure (Pektat)
- **penicillanic acid** Penicillansäure
- **performic acid** Perameisensäure

- **phosphatidic acid** Phosphatidsäure
- **phosphoric acid (phosphate)** Phosphorsäure (Phosphat)
- **phthalic acid (phthalate)** Phthalsäure (Phthalat)
- **phytanic acid** Phytansäure
- **phytic acid** Phytinsäure
- **picric acid (picrate)** Pikrinsäure (Pikrat)
- **pimelic acid** Pimelinsäure
- **plasmenic acid** Plasmensäure
- **prephenic acid (prephenate)** Prephensäure (Prephenat)
- **propionic acid (propionate)** Propionsäure (Propionat)
- **prostanoic acid** Prostansäure
- **pyrethric acid** Pyrethrinsäure
- **pyroligneous acid/wood vinegar** Holzessig
- **pyruvic acid (pyruvate)** Brenztraubensäure (Pyruvat)
- **resin acids** Harzsäure
- **retinic acid** Retinsäure
- **ribonucleic acid (RNA)** Ribonucleinsäure, Ribonukleinsäure (RNS/RNA)
- **saccharic acid/aldaric acid (glucaric acid)** Zuckersäure, Aldarsäure (Glucarsäure)
- **saccharinic acid** Saccharinsäure
- **saccharonic acid** Saccharonsäure
- **salicic acid (salicylate)** Salicylsäure (Salicylat)
- **shikimic acid (shikimate)** Shikimisäure (Shikimat)
- **sialic acid (sialate)** Sialinsäure (Sialat)
- **silicic acid (silicate)** Kieselsäure (Silikat)
- **sinapic acid** Sinapinsäure
- **sorbic acid (sorbate)** Sorbinsäure (Sorbat)
- **stearic acid/octadecanoic acid (stearate/octadecanate)** Stearinsäure, Octadecansäure (Stearat/Octadecanat)
- **stomach acid/gastric acid** Magensäure
- **suberic acid/octanedioic acid** Suberinsäure, Korksäure, Octandisäure
- **succinic acid (succinate)** Bernsteinsäure (Succinat)
- **tannic acid (tannate)** Gerbsäure (Tannat)
- **tartaric acid (tartrate)** Weinsäure (Tartrat)
- **teichoic acid** Teichonsäure
- **teichuronic acid** Teichuronsäure
- **uric acid (urate)** Harnsäure (Urat)
- **uridylic acid** Uridylsäure

- **urocanic acid (urocaninate)** Urocaninsäure (Urocaninat), Imidazol-4-acrylsäure
- **uronic acid (urate)** Uronsäure (Urat)
- **usnic acid** Usninsäure
- **vaccenic acid/11-octadecenoic acid** Vaccensäure
- **valeric acid/pentanoic acid (valeriate/pentanoate)** Valeriansäure, Pentansäure (Valeriat/Pentanat)
- **vanillic acid** Vanillinsäure
- **xanthic acid/xanthonic acid/ xanthogenic acid/ ethoxydithiocarbonic acid** Xanthogensäure

acid amide Säureamid
acid ester Säureester
acid rain saurer Regen
acid-base balance Säure-Basen-Gleichgewicht
acid-fast säurefest
acid-fastness Säurefestigkeit
acidic azid, sauer; säuerlich; säurebildend, säurehaltig
acidification Säuerung, Säurebildung; Versauerung
acidifier (food) Säuerungsmittel
acidify ansäuern; versauern
acidity Azidität, Säuregrad, Säuregehalt
acidosis Acidose, Azidose
acinus cell Acinuszelle
acoelous/acelous acöl, acoel
aconitic acid (aconitate) Aconitsäure (Aconitat)
acontium (anthozoans) Acontium
acorn Eichel
acorn worms/enteropneusts/ Enteropneusta Eichelwürmer, Enteropneusten
acoustical hair/trichobothrium/ vibratory sensory hair Hörhaar, Becherhaar, Trichobothrium
acquire erwerben, aneignen, annehmen; erlernen
acquired behavior/learned behavior erlerntes Verhalten, Lernverhalten
acquired characteristic erworbenes Merkmal
acquired immune deficiency syndrome (AIDS) erworbenes Immunschwächesyndrom, Immunmangel-Syndrom (AIDS)
acquired immunity erworbene Immunität
acquisition time *vir* Aufnahmezeit
ACR (ancient conserved region) *gen* vorgeschichtliche konservierte Region

acranians

acranians/Acrania Schädellose
acrid/irritating/sharp/pungent beißend, scharf, stechend
acridine dye Acridinfarbstoff
acritarchs/Acritarcha Acritarchen (*sg* Acritarch)
acrocarpic/acrocarpous akrokarp, gipfelfrüchtig
acrocentric chromosome akrozentrisches Chromosom
acrodont acrodont
acron/prostomium Acron, Kopflappen, Prostomium
acronematic flagellum/whiplash flagellum Peitschengeißel
acropetal/basifugal akropetal, basifugal
acropetal development Akropetalie
acrorhagus (tubercle with stinging cells) Acrorhagus, Nesselsack
acrosome Akrosom, Acrosom
acrostichal acrostich
acrostichoid acrostichisch
acrotony Akrotonie
act/work/be effective/causing an effect/take effect wirken
actin Aktin, Actin
actin cable Aktinkabel
actin filament/microfilament Aktinfilament, Actinfilament, Mikrofilament
actinidia family/Chinese gooseberry family/Actinidiaceae Strahlengriffelgewächse
actinomorphic/actinomorphous/ star-shaped/radial/radially symmetrical/regular/cyclic aktinomorph, strahlenförmig, sternförmig, radiär, radiärsymmetrisch, zyklisch
action potential/impulse/spike Aktionspotential, Spitzenpotential, Impuls
activated carbon Aktivkohle
activated sludge Belebtschlamm, Rücklaufschlamm
activated state aktivierter Zustand
activation energy/energy of activation Aktivierungenergie
activator protein Aktivatorprotein
active ingredient/active component Wirkstoff
active metabolic rate Arbeitsumsatz, Leistungsumsatz
active metabolism Leistungsstoffwechsel, Arbeitsstoffwechsel
active principle Wirkstoff, Wirksubstanz, aktiver Bestandteil
active site/catalytic site aktives Zentrum, katalytisches Zentrum
active transport/uphill transport aktiver Transport
activity curve Aktivitätskurve
actual value/effective value Istwert
actualism Aktualismus
acuity, visual Sehschärfe
acuminate/taper-pointed lang zugespitzt (konkav zulaufend: z.B. Blattspitze)
acute/sharp/pointed/sharp-pointed spitz
acute phase protein Akutphasenprotein
acute transforming retrovirus akut transformierendes Retrovirus
Adam's apple/laryngeal prominence/prominentia laryngea (largest cartilage of larynx) Adamsapfel
adamantoblast/ameloblast/ enamel cell Adamantoblast
adapt/adjust to/acclimate/ acclimatize anpassen, akklimatisieren
adaptability Adaptabilität, Anpassungsfähigkeit
adaptation/acclimation/ acclimatization Adaptation, Adaption, Anpassung
adaptation/accustomation/ habituation Adaptation, Adaption, Anpassung, Gewöhnung, Eingewöhnung
adapted angepasst, adaptiert
adaptive adaptiv, anpassungsfähig
adaptive landscape/adaptive surface adaptive Landschaft
adaptive peak Anpassungsgipfel
adaptive radiation adaptive Radiation
adaptive value/selective value Anpassungswert, Selektionswert
adcrustation/accrustation Adkrustierung, Akkrustierung
adder's-tongue family/grape fern family/Ophioglossaceae Natternzungengewächse, Rautenfarngewächse
addicted süchtig
• **become addicted** süchtig werden
addicted to drugs drogenabhängig, drogensüchtig
addiction/dependance Sucht, Süchtigkeit, Abhängigkeit
addictive suchterzeugend, süchtig machend, abhängig machend
additive *n* Zusatzstoff, Additiv
additive genetic variance additive genetische Varianz
addorsed/addossed adossiert, rückseitig

adductor muscle Anzieher, Schließmuskel, Adduktor, Adductor(-Muskel) (auch: Schalenschließmuskel)
adecticous pupa Pupa adectica
adelphogamy Adelphogamie, Geschwisterbestäubung
adelphotaxon/sister taxon Adelphotaxon, Schwestertaxon
adenine Adenin
adenohypophysis/anterior lobe of pituitary gland Adenohypophyse, Hypophysenvorderlappen
adenosine Adenosin
adenosine diphosphate (ADP) Adenosindiphosphat (ADP)
adenosine monophosphate (AMP)/ adenylic acid Adenosinmonophosphat (AMP), Adenylsäure
• **cyclic AMP (cAMP)** zyklisches/ cyclisches AMP (cAMP)
adenosine triphosphate (ATP) Adenosintriphosphat (ATP)
adenovirus Adenovirus
adenylate cyclase Adenylylcyclase, Adenylatcyclase
adenylic acid (adenylate)/ adenosine monophosphate Adenylsäure (Adenylat), Adenosinmonophosphat
adequate stimulus adäquater Reiz
adherence/adhesion/attachment Adhärenz, Adhäsion, Anheftung
adherence factor Adhärenzfaktor
adhesin Adhäsin, Adhesin
adhesion/adherence/attachment Adhäsion, Adhärenz, Anheftung
adhesion root/anchorage root Ankerwurzel
adhesive body Klebkörper, Klemmkörper
adhesive cell/colloblast/lasso cell Klebzelle, Kolloblast, Colloblast
adhesive disk Haftplatte, Haftscheibe, Saugnapf; Kletterorgan
adhesive gland/colleterial gland/ cement gland (insects) Haftdrüse, Klebdrüse, Kittdrüse, Zementdrüse
• **pedal gland** Fußdrüse
adhesive pad/pulvillus Haftlappen, Pulvillus, Lobulus lateralis
adhesive sac/metasomal sac/ internal sac (bryozoans) Anheftungsorgan
adhesive thread (spiderweb: viscid, sticky line) Klebfaden
adhesive trap/flypaper trap Klebfalle
adiantum family/maidenhair fern family/Adiantaceae Frauenhaarfarngewächse

adipic acid (adipate) Adipinsäure (Adipat)
adipocyte/adipose cell/fat cell Adipozyt, Adipocyt, Fettzelle
adipose/fatty (animal fat) fettartig, fetthaltig, fettig, Fett...
adipose capsule of kidney/fatty capsule of kidney Nierenfettkapsel, Nierenkapsel
adipose cell/adipocyte/fat cell Adipozyt, Adipocyt, Fettzelle
adipose fin Fettflosse
adipose tissue/fatty tissue Fettgewebe
adiposity Fettsucht
adjacent benachbart, angrenzend
adjust/focus micros **(fine/coarse)** justieren, fokussieren (Scharfeinstellung des Mikroskops: fein/grob)
adjustable variable Stellgröße
adjusted mean stat bereinigter Mittelwert, korrigierter Mittelwert
adjustment stat Bereinigung
adjustment/focus adjustment/ focus micros **(fine/coarse)** Justierung, Fokussierung (Scharfeinstellung des Mikroskops: fein, grob)
adjustment knob/focus adjustment knob micros Justierschraube, Justierknopf, Triebknopf
adjuvant Adjuvans (pl Adjuvantien), Hilfsstoff
admixture Beimischung
adnate (unequal parts) angewachsen, verwachsen (der Länge nach/ungleiche Organe)
adnation (unequal parts) Verwachsung (unterschiedlicher Organe)
adolescence/juvenile stage/juvenile phase Jugendstadium, Jugendphase, Jugendzeit
adoral zone membranelles (AZM) (ciliates) adorales Membranellenband
Adoxaceae/moschatel family Moschuskrautgewächse
adrenal gland Nebenniere
adrenaline/epinephrine Adrenalin, Epinephrin
adrenergic adrenerg
adrenocorticotropic hormone/ corticotropin (ACTH) adrenocorticotropes Hormon, Kortikotropin, Adrenokortikotropin
adsorb adsorbieren
adsorption chem/phys Adsorption
adsorption/apposition Anlagerung, Apposition
adult plant Adultpflanze

advanced fortgeschritten, höher entwickelt, weiterentwickelt, abgeleitet
advantage Vorteil
• **selective advantage** Selektionsvorteil
adventitious adventiv
adventitious plantlet (e.g. Kalanchoe) Brutpflänzchen
adventitious root Adventivwurzel, Beiwurzel, Nebenwurzel
• **stem-borne adventitious root** sprossbürtige Wurzel
adventitious shoot Adventivspross
aecium/aecidium/cluster cup Aecidium
aedeagus/intromittent organ/penis *entom* Aedeagus, Aedoeagus, Penis
aeolian *geol* äolisch
aeolidacean snails/aeolidaceans/ Aeolidiacea/Eolidiacea Fadenschnecken
aerate belüften, durchlüften
aeration Belüftung, Durchlüftung
aeration tank/aerator Belüftungsbecken (Belebungsbecken)
aerenchyma Aerenchym, Durchlüftungsgewebe
aerial courtship Flugbalz
aerial root/air root Luftwurzel
aerial survey *ecol/stat* Lufterhebung
aerie/eyrie (bird's nest on cliff/ mountain) Horst, Adlerhorst, Raubvogelnest, Greifvogelnest
aerobic aerob, sauerstoffbedürftig
aerofoil/airfoil (wing) Tragfläche, Flügel
aerofoil section/airfoil section Flügelprofil
aerophyte/air plant/aerial plant/ epiphyte Luftpflanze, Epiphyt, Aufsitzerpflanze
aestival (appearing in summer) sommerlich
aestivate/estivate/pass summer in dormant stage übersommern
aestivation/estivation Ästivation, Aestivation, Knospendeckung; Sommerschlaf
aethalium (slime molds) Aethalium, Aethalie (*pl* Aethalien), Sammelfruchtkörper
afferent hinführend, zuführend, zuleitend; (rising) aufsteigend
affination (sugar filtration) Affination
affinated sugar Affinade
affinity Affinität
affinity blotting Affinitäts-Blotting
affinity constant Affinitätskonstante
affinity labeling Affinitätsmarkierung
affinity maturation *immun* Affinitätsreifung
affinity partitioning Affinitätsverteilung
affix/attach fixieren (befestigen, fest machen)
affixment/attachment Anheftung
afforestation/reforestation/ reafforestation Aufforstung, Wiederbewaldung
African elephant shrews/ Macroscelidea Rüsselspringer
African eyeworm disease/loa (Loa loa) Kamerunbeule
African violet family/gesneria family/gesneriad family/gloxinia family/Gesneriaceae Gesneriengewächse
after-hyperpolarization Nach-/ Hyperpolarisation
afterbirth Nachgeburt
afterbrain/metencephalon (pons + cerebellum) Nachhirn, Metencephalon
afterdischarge *n neuro* Nachfeuerung, Nachentladung
afterdischarge *vb neuro* nachfeuern
afterfeather/accessory plume/ hypoptile/hypoptilum/hypopenna Afterfeder, Nebenfeder
afterripening/after-ripening Nachreifen
aftershaft/hyporachis/hyporhachis (median shaft of hypopenna) Afterschaft
aftervane/hypovexillum Nebenfahne
agamous/asexual/neuter ungeschlechtig
agar Agar
• **blood agar** Blutagar
• **chocolate agar** Kochblutagar, Schokoladenagar
agar diffusion test Agardiffusionstest, Diffusionstest
agar plate Agarplatte
agar slant culture Schrägagarkultur, Schrägkultur
agaric Blätterpilz, Lamellenpilz
agarics/Agaricales Lamellenpilze
Agaricus family/Agaricaceae Egerlinge
agava family/century plant family/ Agavaceae Agavengewächse
age *n* Alter; (geological era) Zeitalter, Ära
age/become old/senesce *vb* altern
age cohort Alterskohorte, altersspezifische Gruppe

aldehyde

age composition Alterszusammensetzung
age distribution Altersverteilung
age group/age class Altersgruppe
age structure Altersaufbau, Ätilität (Population)
ageing/aging/senescence Alterung, Altern, Seneszenz
agent Agenz, Agens (*pl* Agenzien)
agglutinate/clump together agglutinieren, sich zusammenballen, verklumpen
aggradation *ecol* Anlandung
aggregate *n* Aggregat, Anhäufung, Ansammlung, Masse
aggregate *vb* anhäufen, ansammeln, sich ansammeln, vereinigen
aggregate fruit Sammelfrucht
aggregate structure (soil) Aggregatgefüge
aggregated species/species aggregate (agg.) Sammelart
aggregation Aggregation, Zusammenschluss, Zusammenlagerung, Vereinigung, Ansammlung
aggression Aggression
aggressive inhibition Angriffshemmung, Aggressionshemmung
aggressive mimicry/Peckhammian mimicry Angriffs-Mimikry, Angriffsmimikry, Peckhammsche Mimikry
aggressiveness Aggressivität
aging/ageing/senescence Alterung, Altern, Seneszenz
aging mutant Alterungsmutante
agitator vessel Rührbehälter
aglossate/without tongue zungenlos, ohne Zunge
aglycone Aglycon
agranulocyte/agranular leukocyte Agranulozyt, agranulärer Leukozyt
agriculture/farming Landwirtschaft
agroinfection Agroinfektion
agroinoculation Agroinokulation
agronomist Agronom, diplomierter Landwirt
agronomy Agronomie, Ackerbaukunde, Ackerbaulehre
AIDS (Acquired Immune Deficiency Syndrome) AIDS (erworbenes Immunschwächesyndrom)
air bladder/float/pneumatophore *bot* (algas) Schwimmblase (Algen)
air bladder/swimbladder *zool* Schwimmblase
air capillaries Luftkapillaren
air chamber *(Marchantia)* Luftkammer

air curtain Luftvorhang (Vertikalflow-Biobench)
air embolism/cavitation Luftembolie
air humidity Luftfeuchtigkeit
air layering Luftablegerverfahren
air plant/aerial plant/aerophyte/ epiphyte Luftpflanze, Epiphyt, Aufsitzerpflanze
air pocket/air bag/air sac/vesiculum Luftsack (Pollen)
air pollutant Luftschadstoff
air pore *(Marchantia)* Atemöffnung
air root/aerial root Luftwurzel
air sac Luftsack
air-breathing luftatmend
airborne windgetragen
airfoil/aerofoil (wing) Tragflügel, Flügel
airfoil section/aerofoil section Flügelprofil
airlift loop reactor Mammutschlaufenreaktor
airlift reactor/pneumatic reactor Airliftreaktor, pneumatischer Reaktor (Mammutpumpenreaktor)
airproof/airtight luftdicht, hermetisch verschlossen
airspace (leaf parenchyma) Hohlraum
Aizoaceae/fig marigold family/ carpetweed family/ mesembryanthemum family Mittagsblumengewächse, Eiskrautgewächse
akebia family/lardizabala family/ Lardizabalaceae Fingerfruchtgewächse
akin/related verwandt
akinesis (absence/arrest of motion) (see also: catalepsy) Akinese (reflektorische Bewegungslosigkeit)
akinete/resting cell Akinet, (unbewegliche) Dauerzelle
alanine Alanin
alar cell (mosses) Alarzelle, Blattflügelzelle
alarm substance/alarm pheromone Alarmstoff, Schreckstoff, Alarm-Pheromon
alary/alar/winglike flügelartig, schwingenartig
alate/winged geflügelt
albedo/reflective power (as a percentage) Albedo, Rückstrahlung
albumen gland Eiweißdrüse
albumin Albumin
alcohol Alkohol
aldehyde/acetic aldehyde/ acetaldehyde Aldehyd, Acetaldehyd

aldosterone

aldosterone Aldosteron
alembic (retort) (historischer) Distillierapparat (Retorte)
aleurone layer Aleuronschicht
alevin/sacfry/yolk fry (salmon larvae) Setzling, junge Fischbrut, Dottersackbrut (v. a. Lachs)
alga (pl algae/algas) Alge
- **bluegreen algae/cyanobacteria/ Cyanophyceae** Blaualgen, Cyanobakterien
- **brown algae/phaeophytes/ Phaeophyceae** Braunalgen
- **calcareous alga** Kalkalge
- **floridean algae/florideans/ Florideophyceae (red algae)** Florideen
- **golden algae/golden-brown algae/Chrysophyceae** Goldalgen, Chrysophyceen
- **green algae/Chlorophyceae/ Isokontae** Grünalgen
- **red algae/Rhodophyceae** Rotalgen
- **terrestrial alga** Landalge, Luftalge
- **yellow-green algae/ Xanthophyta** Gelbgrünalgen, Xanthophyten
algal bloom Algenblüte
algal fungi/lower fungi/ Phycomycetes Algenpilze, niedere Pilze
algal mat/algal layer Algenteppich, Matte
algesia Schmerzempfindlichkeit
algicide Algenbekämpfungsmittel, Algizid
alginic acid (alginate) Alginsäure (Alginat)
algology Algologie, Algenkunde
alicyclic alizyklisch
aliform/wing-shaped flügelförmig
alimentary zur Nahrung/Unterhalt dienend, Nahrungs.., Ernährungs.., Speise.., Verdauungs..
alimentary canal/alimentary tract/ digestive canal/digestive tract Verdauungskanal, Verdauungstrakt
aliphatic aliphatisch
aliquot aliquoter Teil
Alismataceae/water-plantain family/arrowhead family Froschlöffelgewächse
alisphenoid bone Keilbeinflügelknochen, großer Keilbeinflügel
alive/living/biological/biotic lebend, lebendig
alive and well (idiomatic) gesund und munter; wohlauf
alkali blotting Alkali-Blotting

alkali-heath family/sea heath family/Frankeniaceae Nelkenheidegewächse, Frankeniengewächse
alkaline/basic alkalisch, basisch
alkalosis Alkalose
alkaptonuria Alkaptonurie
all-or-none response Alles-oder-Nichts Antwort, Alles-oder-Nichts Reaktion
allantoic acid Allantoinsäure
allantoic placenta Allantoisplazenta
allantoin Allantoin
allantois Allantois, Harnsack, Harnhaut
allele Allel
- **multiple alleles** multiple Allele
- **null allele** Nullallel
- **wild-type allele** Wildallel
allele frequency Allelenfrequenz
allele specific oligonucleotide (ASO) allelspezifisches Oligonucleotid (ASO)
allelic exclusion Allelausschluss, allele Exclusion, allele Exklusion
allelic replacement Allelen-Austauschtechnik
Allen's law/Allen's rule/ proportion rule Allen'sche Regel, Proportionsregel
allergen Allergen
allergenic allergen
allergic allergisch
allergy Allergie, Überempfindlichkeitsreaktion
Alliaceae/onion family Zwiebelgewächse, Lauchgewächse
alliaceous (smelling/tasting like garlic/onions) zwiebelartig (Geruch/Geschmack)
alliance Allianz, Verband, Assoziationsgruppe
allochory Allochorie, Fremdausbreitung
allochthonous biotopfremd, bodenfremd, eingeführt
allofeeding füttern
allogamy/xenogamy/cross- fertilization Allogamie, Fremdbefruchtung
allograft/homograft/syngraft Allotransplantat, Homotransplantat
allogrooming/social grooming Allogrooming, Fremdputzen (z.B. Fell bei Säugern)
allopatric ("other country") evol allopatrisch (in getrennten Arealen)
allopreening Fremdputzen, Putzen an fremdem Körper (Vögel: gegenseitiges Gefiederkraulen/ Gefiederputzen)

11   **amine** A

allosteric interaction allosterische
Wechselwirkung
allosteric transition allosterische
Transition, allosterischer
Übergang
allotype Allotyp, Allotypus
alluvial fan Schwemmkegel,
Schwemmfächer
alluvial plain/floodland/floodplain
Schwemmland
aloe family/Aloeaceae
Aloegewächse
alpha DNA/α DNA Alpha-DNA,
α-DNA
alpine alpin
alpine grassland Matte, Mattenstufe
alpine marsh Hochgebirgsmoor
alpine meadow alpine Bergwiese,
Matte
**alpine mountains/alpine mountain
chain** Hochgebirge
alpine region Hochgebirgsregion
alpine zone Hochgebirgsstufe
alternate *adv/adj bot* **(leaf
arrangement)** alternierend,
abwechselnd, wechselständig,
zerstreut, schraubig
alternate *vb* alternieren, wechseln,
abwechseln zwischen zweien
**alternate disjunction (of
chromosomes)** alternierende
Verteilung (von Chromosomen)
alternate host Wechselwirt
alternating field gel electrophoresis
Wechselfeld-Gelelektrophorese
alternation Alternanz, Wechsel
alternation of generations
Generationswechsel
alternation of hosts Wirtswechsel
alternation of nuclear phase
Kernphasenwechsel
alternation rule Alternanzregel
alternative splicing *gen* alternatives
Spleißen
altitude (above sea level) Höhe
(über dem Meeresspiegel)
altitude/elevation/higher location
Höhe (über dem Meeresspiegel),
Höhenlage
altitudinal zonation vertikale
Stufung
altitudinal zone/region/belt
Höhenstufe (Vegetationsstufe *see*
vegetational zone)
**altricial animal/nidicolous animal
(e.g. birds)** Nesthocker
altruism/self-sacrifice Altruismus,
Selbstlosigkeit, Selbstaufopferung,
Uneigennützigkeit, Gemeinnutz
Alu **family** *gen Alu*-Familie
alula (insects wing: small lobe)
Alula (Insekten: Flügellappen)

alula/spurious wing/bastard wing
orn Alula, Daumenfittich,
Afterflügel, Nebenflügel, Ala spuria
(Federngruppe am 1. Finger)
alular quill Daumenfeder
aluminum Aluminium
alveolar alveolär
**alveolate/deeply pitted/
honeycombed/favose/faveolate**
kleingrubig; wabig
amacrine cell amakrine Zelle
Amanita family/Amanitaceae
Freiblättler, Wulstlinge
**amaranth family/cockscomb family/
pigweed family/Amaranthaceae**
Amarantgewächse,
Fuchsschwanzgewächse
**Amaryllidaceae/amaryllis family/
daffodil family**
Amaryllisgewächse,
Narzissengewächse
amber Bernstein
ambient pressure Umgebungsdruck
ambient temperature
Umgebungstemperatur
amble/pace (gait) Passgang
ambrosia cell (fungus cell)
Ambrosiazelle
ambulacral foot/tube foot/podium
Ambulakralfüßchen, Saugfüßchen
ambulacral groove
Ambulakralfurche
ambulacral plate Ambulakralplatte
**ambulacral system/water-vascular
system** Ambulakralsystem,
Wassergefäßsystem
ambush auflauern, aus dem
Hinterhalt angreifen
ambush predator Laurer
amebas/amoebas/Amoebozoa
Amöben, Wechseltierchen,
Wurzeltierchen, Rhizopoden
• **testate amebas/Testacea**
Thekamöben, beschalte Amöben
**amebic dysentery/amebiasis
(Entamoeba histolytica)**
Amöbenruhr, Amöbiasis
amebocyte Amöbozyt, Amöbocyt
ameboid amöboid
amenity forest/recreational forest
Erholungswald
amensalism Amensalismus
**American Society for the
Prevention of Cruelty to Animals
(New York)** U.S. nationale
Tierschutzvereinigung
Ames test Ames-Test
amictic amiktisch
amidation Amidierung
amide Amid
amination Aminierung
amine Amin

amino acid

amino acid Aminosäure
amino sugar Aminozucker
aminoacyl site (A-site)
　Aminoacyl-Stelle (A-Stelle)
aminoacyl-tRNA synthetase
　Aminoacyl-tRNA-Synthetase
aminoacylation Aminoacylierung
γ-aminobutyric acid (GABA)
　γ-Aminobuttersäure
Ammon's horn/anterior
　hippocampus/pes hippocampi
　Ammonshorn
ammonia Ammoniak
ammoniotelic/ammonotelic
　ammoniotelisch, ammonotelisch
ammonites/Ammonoidea
　Ammoniten
ammonitic suture lines
　ammonitische Lobenlinien
amniocentesis Amniozentese,
　Amnionpunktion,
　Fruchtwasserpunktion
amnion/"bag of waters" Amnion,
　Schafhaut, innere Eihaut, innere
　Keimhülle, inneres Eihüllepithel
　• pleuramnion Pleuramnion,
　　Faltamnion
　• schizamnion Schizamnion,
　　Spaltamnion
amniote egg amniotisches Ei,
　Ei amniotischer Tiere
amniotic band amniotischer Strang,
　Simonart'-Band
amniotic cavity Amnionhöhle
amniotic fluid/"water"
　Amnionflüssigkeit, Amnionwasser,
　Fruchtwasser
amniotic fold Amnionfalte
amniotic sac Amnionsack,
　Fruchtblase, Fruchtwassersack,
　Fruchtsack
amorphous amorph
amphiarthrosis Amphiarthrose
amphibian/amphibious adv/adj
　amphibisch
amphibians/Amphibia Amphibien,
　Lurche
amphibious/amphibian adv/adj
　amphibisch
amphibolic amphibol
amphibolic pathway/central
　metabolic pathway amphiboler
　Stoffwechselweg
amphicarpy bot Amphikarpie
amphicelous amphicöl, amphicoel
amphid (nematodes) Amphid
amphidiploid/allopolyploid
　amphidiploid, allopolyploid
amphidromous amphidrom
amphiesmal vesicle
　Amphiesmalvesikel,
　Amphiesmalbläschen
amphigastrium/ventral leaf/
　underleaf (liverworts)
　Amphigastrium, Bauchblatt,
　"Unterblatt"
amphiphilic amphiphil
amphipods (beach hoppers/sand
　hoppers & sand fleas)/Amphipoda
　Flohkrebse, Flachkrebse
amphisbenids/amphisbenians/
　worm lizards/Amphisbaenia
　Wurmschleichen, Doppelschleichen
amphitoky/amphitokous
　parthenogenesis Amphitokie
amphitrichous amphitrich
amphitropous (ovule) amphitrop,
　hufeisenförmig gekrümmt
　(Samenanlage)
amphoteric amphoter, amphoterisch
amphotropic amphotrop,
　amphotropisch
amplexicaul/stem-clasping
　amplexikaul, stengelumfassend
amplexus (mating embrace of male
　amphibians) Umklammerung
　(Amphibien)
amplification Amplifikation,
　Vervielfältigung, Vermehrung
amplification cascade
　Verstärkungskaskade
amplification refractory mutation
　system (ARMS) System der
　amplifizierungsresistenten Mutation
　(ARMS)
amplified gene amplifiziertes Gen
amplifier Verstärker
amplify amplifizieren, verstärken
amplimer Amplimer
ampoule/ampulla Ampulle
ampulla/bladder bot (Utricularia)
　Ampulle, Blase
ampulla/reservoir (Mastigophora)
　Ampulle, Geißelsäckchen, Schlund
ampullary gland (reproductive
　gland) Ampullardrüse
ampullary organ ichth
　Ampullenorgan
amygdaloid nucleus/amygdaloid
　nuclear complex/nucleus
　amygdalae/corpus
　amygdaloideum Mandelkern,
　Mandelkörper, Mandelkernkomplex
amylopectin Amylopektin
amyocerate antenna (muscles
　only in base segment)
　Geißelantenne, Ringelantenne,
　amyocerate Antenne
anabatic wind Hangaufwind
anabiosis/suspended animation
　Anabiose, latentes Leben
anabolic (synthetic reactions of
　metabolism) anabol, anabolisch,
　aufbauend

13   **angiocarpic** **A**

**anabolic pathway/biosynthetic
pathway**
anaboler Stoffwechselweg,
Biosynthese-Stoffwechselweg
**anabolism/synthetic reactions/
synthetic metabolism**
Anabolismus, Aufbau
(Stoffwechsel), Synthesestoffwechsel
anabolite
Stoffwechselsyntheseprodukt,
Anabolit
**Anacardiaceae/cashew family/
sumac family** Sumachgewächse
anadromous anadrom
anaerobic anaerob
anaerobic fermentation anaerobe
Dissimilation, anaerobe Gärung
anagenesis Anagenese
**anal callosity/ischial callosity/
sitting pad** Gesäßschwiele,
Sitzschwiele, Analkallosität
anal cell Analzelle
anal cone/anal tube (crinoids)
Afterhügel, Afterröhre, Analtubus
anal fan Analfächer
**anal field/anal area/vannal area/
vannus** Analfeld, Vannus
anal fin Afterflosse, Analflosse
anal fold *embr* Analfalte
**anal fold/vannal fold/anal fold-
line/vannal fold-line** *entom*
Analfalte, Vannalfalte, Plica analis,
Plica vannalis
anal gland/rectal gland (shark)
Analdrüse (Hai)
anal loop (draggonflies)
Analschleife (Libellen)
anal margin Analrand
anal pit *embr* Analgrube,
Analgrübchen
anal plate *embr* Analplatte
anal proleg/anal leg (caterpillars)
letzter Afterfuß, Nachschieber,
Postpes, Propodium anale (Raupen)
**anal sac/paranal sinus/sinus
paranalis** Analbeutel
anal tubercle/anal papilla
Analhöcker, Analhügel
analeptic amine Weckamin
analog/analogue Analogon
(*pl* Analoga)
analogize analogisieren
analogous analog, funktionsgleich
analogy Analogie
analysis (*pl* analyses) Analyse
**analysis of amniotic fluid (for
prenatal diagnosis)**
Fruchtwasseruntersuchung
analytic(al) analytisch
analytical balance Analysenwaage
analytical centrifugation
analytische Zentrifugation

analyze analysieren
analyzer Analysator
**anamorphic stage/imperfect stage/
asexual stage** *fung*
Nebenfruchtform
anamorphosis Anamorphose
ananthous/flowerless blütenlos
anaphylaxis Anaphylaxe
anaplerotic reaction anaplerotische
Reaktion, Auffüllungsreaktion
anapophysis Anapophyse
anastomize (blood vessels)
ineinandermünden; verästeln
anastomosis Anastomose,
Einmündung, Querverbindung,
Ineinandermünden
anatomy (internal structure)
Anatomie (Morphologie der inneren
Gestalt)
anatropous (ovule) anatrop,
umgewendet (Samenanlage)
ancestor/forebear/progenitor
Vorfahre, Ahne
ancestral Ahnen..., angestammt,
Ur..., Erb..., ererbt
**ancestral birds/"lizard birds"/
Archaeornithes** Urvögel
ancestral chromosome
Urchromosom
ancestrula (larva) Ancestrula
ancestry Abstammung
anchor ice *limn* Grundeis
anchor impeller *biot* (bioreactor)
Ankerrührer
anchorage Verankerung
anchorage root/adhesion root
Ankerwurzel
ancient DNA vorgeschichtliche DNA
**androconium (*pl* androconia)/scent
scale(s) (♂ Lepidoptera wings)**
Duftschuppe, Duftfeld, Androkonie
androecium Staubblattkreis
androgen Androgen
androgenesis Androgenese
androgynophore/gynoandrophore
Androgynophor
androsterone Androsteron
anellus Anellus
anemia "Blutarmut", Anämie
 • **pernicious anemia** bösartige
Blutarmut, perniziöse Anämie
anemochore *bot* Windstreuer
anemophilous/wind-pollinated
anemophil, windblütig
aner (♂ insect) männliches Insekt
(speziell: Ameisen)
anerobiosis/anoxybiosis
Anerobiose
anestrus Anestrus, Anöstrus
aneuploid aneuploid
aneuploidy Aneuploidie
angiocarpic/angiocarpous angiokarp

angiosperm 14

angiosperm/flowering plant/ anthophyte/Anthophyta Angiosperme, Bedecktsamer, Decksamer, Blütenpflanze
angle of attack *ethol* Angriffswinkel; *aer* Anstellwinkel
angle rotor/angle head rotor *centrif* Winkelrotor
anglerfishes/Lophiiformes Anglerfische, Armflosser
angular kantig
angular aperture Öffnungswinkel
anhydrobiosis Anhydrobiose, Trockenstarre
animal abuse Tierquälerei
animal asylum/animal shelter Tierheim
animal breeding Tierzucht, Tiere züchten *(sensu lato)*
animal cage Tierkäfig
animal community/zoocoenosis Tiergemeinschaft, Zoozönose
animal feces/animal droppings/ dung Tierexkremente, Tierkot; Losung
animal feed Viehfutter
animal husbandry/ranching/ ranging Viehwirtschaft, Viehzucht
animal keeper Tierpfleger, Tierwärter
animal kingdom Tierreich
animal lover Tierliebhaber
animal model Tiermodell
animal parasite tierischer Parasit, Zooparasit
animal physiology/zoophysiology Tierphysiologie
animal pole (cell) animaler Pol
animal psychology/zoopsychology Tierpsychologie
animal rightist/animal rights advocate Tierrechtler, Tierschützer
animal shelter Tierheim
animal society Tiergesellschaft (essentielle Vergesellschaftung, Sozietät), Tierstaat
animal virus Tiervirus
animal-dispersal/zoochory Tierausbreitung
animalcule (little/small animal) (mikroskopisch kleines) Tierchen
animalism Animalismus
animate/alive belebt, lebendig
• **inaminate** unbelebt
anion exchanger Anionenaustauscher
anisic aldehyde/anisaldehyde Anisaldehyd
anisophylly/heterophylly Anisophyllie, Heterophyllie, Verschiedenblättrigkeit, Ungleichblättrigkeit

anisotropic anisotrop, doppelbrechend
ankle (foot)/malleolus Knöchel (>Fußknöchel)
ankle joint Fußgelenk
anklebone/talus/astragalus Sprungbein, Talus
anklebone/tarsal bone/tarsal/os tarsi Fußwurzelknochen, Fußgelenksknochen, Tarsal
ankylosis Ankylose
anlage/precursor/preformation/ early form/primordium Anlage, Keim, Ansatz, Primordium
annatto family/bixa family/ Bixaceae Annattogewächse
annealing/reannealing/ reassociation/renaturation (of DNA) Doppelstrangbildung, Annealing, Reannealing, Reassoziation
annelids/segmented worms/ Annelida Anneliden, Ringelwürmer, Gliederwürmer, Borstenfüßer
Annonaceae/cherimoya family/ custard apple family Schuppenapfelgewächse
annual annuell, jährlich; *bot* einjährig
annual growth Jahreswachstum, Jahreszuwachs
annual plant/therophyte Therophyt, Annuelle, Einjährige
annual ring/growth ring Jahresring
annual shoot/one-year shoot Jahrestrieb
annular/cyclic ringförmig, zyklisch
annular cartilage/cricoid cartilage Ringknorpel, Cartilago crinoidea
annular placenta/zonary placenta/ placenta zonaria Gürtelplazenta
annulate lamellae annulierte Lamellen
annulus Annulus, Anulus
• **inferior annulus/inferior ring (remains of velum partiale)** Annulus inferus, Ring, Kragen
• **superior annulus/armilla/ manchette (remains of velum universale)** Annulus superus, Armilla, Manchette
anogenital licking *ethol* Anogenitalmassage
anomalous/irregular anomal, irregulär, unregelmäßig
anomalous rectification anomale Gleichrichtung
anomalous secondary thickening anomales sekundäres Dickenwachstum

anthophilous

anomaly/irregularity Anomalie, Unregelmäßigkeit
anorexia Anorexie, Fressunlust, Appetitlosigkeit
anoxia Anoxie, Sauerstoffmangel
anoxic anoxisch, sauerstofffrei
anserine Gänse betreffend, gänseartig, Gänse ...
ant *siehe:* ants
ant eaters/Vermilingua (Xenarthra) Ameisenbären
antedating/anticipation Antizipation
antenatal diagnosis/prenatal diagnosis pränatale Diagnose
antenna (*pl* antennas/antennae) (feeler) Antenne (Fühler)
• **aristate antenna** grannenartiger Fühler
• **brachycerous antennae/ with short antennae** mit kurzen Fühlern
• **bristlelike/setaceous antenna** borstenartiger Fühler
• **capitate antenna (knoblike)** kolbenförmiger Fühler
• **clubbed/clavate antenna** keulenförmiger, gekeulter Fühler
• **comblike/pectinate antenna** gekämmter Fühler
• **elbowed/geniculate antenna** geknieter Fühler
• **fan-shaped/flabelliform antenna/flabellate antenna** fächriger Fühler, gefächerter Fühler, Fächerfühler
• **filiform/threadlike antenna/ hairlike antenna** fadenförmiger Fühler, Fadenfühler
• **flabelliform/flabellate antenna/ fan-shaped antenna** fächriger Fühler, gefächerter Fühler, Fächerfühler
• **geniculate/elbowed antenna** geknieter Fühler
• **lamellate antenna** blätterförmiger Fühler, lamellenartiger Fühler
• **moniliform antenna** Perlschnurfühler, rosenkranzförmiger Fühler
• **pectinate/comblike antenna** gekämmter Fühler
• **sawlike/serrate antenna** gesägter Fühler
• **setaceous/bristlelike antenna** borstenartiger Fühler
• **stylate antenna** stilettförmiger Fühler
• **threadlike/hairlike antenna/ filiform antenna** fadenförmiger Fühler, Fadenfühler

antenna complex (of chlorophyll molecules) Antennenkomplex (von Chlorophyllmolekülen)
antenna pigment Antennenpigment
antennal aperture Fühleröffnung, Antennalforamen
antennal club Fühlerkeule
antennal furrow/antennal pit Antennengrube, Fühlergrube, Fossa antennalis
antennal gland/antennary gland/ green gland Antennendrüse, Antennennephridium, grüne Drüse
antennal organ Antennalorgan
antennal pedicel Pedicellus, Wendeglied
antennal scale/scaphocerite Antennenschuppe, Scaphocerit
antennal scape Fühlerschaft, Scapus
antennal segment/antennomer Antennensegment, Antennomer
antennal socket Fühlerpfanne
antennal sulcus Fühlerfurche, Antennalfurche, Fühlerrinne
antennal suture Antennennaht, Antennalnaht, Fühlerringnaht
antennary gland/antennal gland/ green gland Antennendrüse, Antennennephridium, grüne Drüse
antennate Fühler besitzend
antennation (insects) Antennenweisung, Antennenkommunikation, Kommunikation durch Antennenberührung
antennifer/socket of antenna Antennifer, Antennenträger
antennomer/antennal segment Antennomer, Fühlersegment, Antennensegment, Antennalsegment
antennule (first antenna: crustaceans) Antennula, erste Antenne
antenodal antenodal
antenodal cross vein Antenodalquerader
anthecology/pollination ecology Blütenökologie
anthela Anthela, Spirre, Trichterrispe
anther Anthere, Staubbeutel
antheridium Antheridium
anthesis/florescence/flowering period Anthese, Floreszenz, Blütezeit, Blütenentfaltung
anthill/ant mound Ameisenhaufen
anthocarp/infructescence/multiple fruit Fruchtstand, Fruchtverband
anthoclade Anthocladium
anthophilous/flower-frequenting Blüten besuchend

anthracene

anthracene Anthrazen
anthracite/hard coal Anthrazit, Kohlenblende
anthranilic acid/2-aminobenzoic acid Anthranilsäure
anthropogenic (resulting from influence of humans) anthropogen (durch den Menschen bedingt/vom Menschen geschaffen)
anthropoid anthropoid, menschenartig, menschenähnlich
anthropoid apes/"apes"/ Hominoidea Menschenaffen, Menschenartige
anthropology/science of human beings Anthropologie, Menschenkunde, Lehre/Wissenschaft vom Menschen
anthropomorphic anthropomarph, von menschlicher Gestalt
anthropophagous/feeding on human flesh anthropophag, den Menschen befallend
antibiosis Antibiose, Widersachertum
antibiotic Antibiotikum (*pl* Antibiotika)
antibiotic resistance Resistenz gegen Antibiotika
antibody Antikörper
- **autoantibody** Autoantikörper
- **bispecific antibody/hybrid antibody** bispezifischer Antikörper
- **catalytic antibody** katalytischer Antikörper
- **intrabody (intracellular antibody)** Intrakörper (intrazellulärer Antikörper)
- **monoclonal antibody** monoklonaler Antikörper
- **polyclonal antibody** polyklonaler Antikörper

anticipation/antedating Antizipation
anticoding strand/antisense strand anticodierender Strang, Nichtsinnstrang
anticodon Anticodon
antidiuretic hormone (ADH)/ vasopressin antidiuretisches Hormon (ADH), Adiuretin, Vasopressin
antidote/antitoxin/antivenin (tierische Gifte) Gegengift
antidromic antidrom
antifeeding compound/feeding deterrent Fraßhemmer
antifoam/antifoaming agent Entschäumer, Schaumhemmer, Antischaummittel

antigen Antigen
- **autoantigen/self-antigen** Autoantigen
- **cross reacting antigen** kreuzreagierendes Antigen
- **differentiation antigen** Differenzierungsantigen
- **group-specific antigen (gag)** gruppenspezifisches Antigen
- **histocompatibility antigen** Histokompatibilitätsantigen
- **type-specific antigen** typenspezifisches Antigen
- **very late antigen (VLA)** "sehr spätes Antigen" (bildet sich spät in der Entwicklung)

antigen combining site/antigen binding site/paratope Antigenbindungsstelle, Antigenbindestelle, Paratop
antigen drift/antigenic drift Antigendrift
antigen presentation Antigenpräsentation
antigen processing Antigen-Processing, Antigenweiterverarbeitung
antigen receptor Antigenrezeptor
antigen shift/antigenic shift Antigenshift
antigen variation Antigenvariation
antigen-presenting cell (APC) antigenpräsentierende Zelle
- **professional a.** professionelle antigenpräsentierende Zelle

antigenic determinant/epitope antigene Determinante, Antigendeterminante, Epitop
antigenic drift/antigen drift Antigendrift
antigenic shift/antigen shift Antigenshift
antigenic variance Antigenvarianz
antigenic variation antigene Variation
antigenicity Antigenität
antimetabolite Antimetabolit
antimicrobial index antimikrobieller Index
antimony Antimon
anting *orn* einemsen
antiparallel antiparallel
antipetalous antipetal, vor den Kelchblättern stehend
antiphlogistic/anti-inflammatory antiphlogistisch, entzündungshemmend
antiphonal singing Wechselgesang
antipode/antipodal cell Antipode
antisense oligonucleotide Antisense-Oligonucleotid/ Anti-Sinn-Oligonucleotid/ Gegensinn-Oligonucleotid

apical meristem

antisense RNA Antisense-RNA, Gegensinn-RNA
antisense strand/anticoding strand Nicht-Sinnstrang, anticodierender Strang
antisense technique Antisense-Technik, Anti-Sinn-Technik, Gegensinntechnik
antisepalous antisepal, vor den Kronblättern stehend
antiserum Antiserum
antitermination protein Antiterminationsprotein, Antiterminator
antithetic theory/interpolation theory of alternation of generations antithetischer, heterophasischer Generationswechsel (Heterogenese)
antlers Geweih (Gefänge)
• **bez tine** Eisspross
• **brow tine** Augspross
• **burr (base of antler)** Geweihbasis, Hornbasis
• **main beam** Stange
• **pedicel** Geweihrose
• **prong/spike** Geweihzacke, Geweihspross
• **royal antler** Mittelspross
• **rubbing** *vb* fegen
• **surroyal antler** Wolfsspross
• **velvet** Bast
antlion Ameisenjungfer (Ameisenlöwe *siehe* doodlebug)
antorbital gland/preorbital gland Voraugendrüse, Antorbitaldrüse
antrorse vorwärts gerichtet, aufwärts gerichtet
ants/Formicidae Ameisen
• **dinergate/soldier** Dinergat, Soldat
• **ergate/worker** Ergat, Arbeiterin
• **gamergate (fertilized, ovipositing worker)** Gamergat
• **macraner (large-size ♂ ant)** Makraner (große ♂ Ameise)
• **macrergate** Makroergat
• **macrogyne (large ♀ ant/queen)** Makrogyne (große Ameisenkönigin)
• **micraner (dwarf ♂ ant)** Mikraner (kleine ♂ Ameise)
• **micrergate/microergate/dwarf worker** Mikrergat, Zwergarbeiterin
• **microgyne (dwarf ♀ ant)** Mikrogyne (kleine ♀ Ameise)
• **pterergate (with rudiments of wings)** Pterergat (Arbeiterin mit Stummelflügeln)

anucleate kernlos
anus Anus, After
anus vent/anal aperture/vent After, Analöffnung
anvil (bone)/incus (ear) Amboss, Incus
aortic arch Aortenbogen
aortic bodies/Zuckerkandl's bodies Aortenkörper, Zuckerkandl'-Organ
aortic valve/valva aortae (heart) Aortenklappe
AP site (apurinic or apyrimidinic site) AP-Stelle (purin- oder pyrimidinlose Stelle)
aperture/opening/orifice Apertur, Öffnung, Mündung
aperture *bot* **(pollen)** Keimpore, Keimstelle
aperture *micros* Apertur, Blende
apes/anthropoid apes/Hominoidea Menschenartige
• **great apes/pongids/Pongidae** Menschenaffen
• **lesser apes/gibbons/Hylobatidae** Gibbons
• **old-world monkeys & apes/Catarrhina** Altweltaffen, Schmalnasenaffen
apetalous apetal
apex (*pl* **apices**) **(summit/point/spike/fastigium)** Spitze, Gipfel, Höhepunkt; (peak: highest among other high points) Scheitelpunkt
aphids/Aphidina Blattläuse
aphrodisiac *n* Aphrodisiakum (*pl* Aphrodisiaka)
aphthous fever/foot-and-mouth disease (Aphthovirus) Maul-und-Klauenseuche
aphyllous/without leaves blattlos
aphylly/absence of leaves Blattlosigkeit, Aphyllie
Apiaceae/parsley family/carrot family/umbellifer family/Umbelliferae Umbelliferen, Doldenblütler, Doldengewächse
apiary/bee yard Bienenhaus, Bienenstand
apical apikal, endständig, gipfelständig, an der Spitze gelegen
apical bud/terminal bud Gipfelknospe
apical cell Scheitelzelle
apical complex (Apicomplexa) Apikalkomplex
apical furrow Scheitelfurche, Scheitelgrube
apical grafting Veredeln auf den Kopf
apical growth Spitzenwachstum
apical meristem Apicalmeristem, Apikalmeristem, Scheitelmeristem, Spitzenmeristem, Vegetationspunkt

apical sense organ

apical sense organ/statocyst (ctenophores) Scheitelorgan, Apikalorgan, Statocyste
apical shoot/terminal shoot Gipfeltrieb
apiculate/pointleted (leaf tip) zugespitzt (mit feiner Spitze endend)
apiculture/beekeeping Bienenzucht
apiculus *bot* Spitzchen
apiology Bienenzüchterei
aplacophorans/Aplacophora Aplacophoren, Wurmmollusken
apnea Apnoe
apocarpous apokarp, chorikarp, freiblättriges Fruchtblatt
apochlamydeous apochlamydeisch
apocrine gland apokrine Drüse
Apocynaceae/periwinkle family/ dogbane family Hundsgiftgewächse, Immergrüngewächse
apod apod, fußlos, beinlos
apoenzyme Apoenzym
apolar unpolar
apomorphic apomorph, apomorphisch
apomorphism Apomorphie
Aponogetonaceae/water hawthorn family/cape-pondweed family Wasserährengewächse
apopetalous apopetal
apophysis Apophyse
apoplast pathway apoplastischer Wasserstrom
apoptosis/programmed cell death Apoptose, programmierter Zelltod
apopyle Apopyle
appearance Erscheinungsbild, Erscheinungsform
appeasement gesture Befriedungsgebärde
appeasement ritual Befriedungsritual
appendage/appendix Anhang, Anhängsel, Anhangsgebilde
appendages *zool* Gliedmaßen
appendicular musculature Extremitätenmuskulatur
appendicular skeleton Extremitätenskelett, Gliedmaßenskelett
appendicularians/Appendicularia/ Larvacea Appendicularien, geschwänzte Schwimm-Manteltiere
appendix/vermiform appendix Appendix, Wurmfortsatz des Blinddarms, Appendix vermiformis
appetence/appetency (fixed/strong desire) Appetenz, natürliches Verlangen, Trieb

appetitive behavior Appetenzverhalten
application *chromat* Auftragung, Applikation
applied angewandt
apply *chromat* auftragen, applizieren
apposition/accretion (Zellwandwachstum durch) Auflagerung
apposition eye Appositionsauge
appressed angedrückt
appressorium/holdfast Appressorium, Haftscheibe, Haftorgan
approach/method *n* Ansatz, Methode
approach *vb math/stat* (e.g. a value) sich annähern, näherkommen, annähern, erreichen (z. B. einen Wert)
approach grafting *bot* Ablaktieren, Ablaktion
approximation *math* Näherung
apterial/featherless federlos
apterium (*pl* **apteria)/featherless space** Federrain, federlose Stelle, Apterium
apterous/unwinged/exalate/ wingless ungeflügelt, flügellos
aptery Flügellosigkeit
apterygial (without fins) flossenlos
apterygotes/Apterygota Flügellose, Urinsekten
apurinic or apyrimidinic site (AP site) purin- oder pyrimidinlose Stelle (AP-Stelle)
apyrene (sperm) apyren
aquarium/fishtank Aquarium
aquatic aquatisch, wasserlebend, wasserbewohnend, im Wasser lebend
aquatic animals/hydrocoles Wassertiere
aquatic life/life in the water Wasserleben
aquatic plant/water plant/ hydrophyte Wasserpflanze, Hydrophyt
aqueous wässrig
aqueous channel Wasserkanal
aqueous humor (eye) Kammerwasser
Aquifoliaceae/holly family Stechpalmengewächse
arable urbar
arable land/tillable land kultivierbares Land, anbaufähiger Boden, urbarer Boden
arachic acid/arachidic acid/icosanic acid Arachinsäure, Arachidinsäure, Eicosansäure

aristulate

arachidonic acid/icosatetraenoic acid Arachidonsäure
arachnid/arachnoid *n* spinnenartiges Tier
arachnids/Arachnida Arachniden, Spinnentiere
arachnoid/arachnoidea *neuro* **(membrane of brain)** Spinnwebenhaut
arachnoid/spiderlike/spidery spinnenartig
arachnology/araneology *sensu lato* Spinnenkunde
Araliaceae/ginseng family/ivy family Efeugewächse
araneologist/arachnologist *sensu lato* Spinnenforscher, Spinnenkundler
araneology/arachnology *sensu lato* Spinnenkunde
araucaria family/monkey-puzzle tree family/Araucariaceae Araukariengewächse
arbitrary/random willkürlich, zufällig
arbor/bowery Gartenlaube, Laube
arboreal/treelike baumartig, Baum...
arboreal/living in trees baumbewohnend
arboreous/forested/wooded bewaldet
arborescence/arborization baumartiger Wuchs
arborescent baumartig verwachsen, baumartig verzweigt, sich baumartig ausbreitend
arboretum Arboretum, Baumgarten
arboricole/arboricolous/arboreal/tree-dwelling/living in trees arborikol, baumbewohnend
arboriculture Baumzucht (speziell Ziersträucher/Zierbäume)
arborist/arboriculturist Baumzüchter
arboroid/dendroid/dentritic baumartig
arbuscule (tree-like small shrub/dwarf tree) Bäumchen, baumartiger Strauch, niedriger Baum
arch of the aorta Aortenbogen
archaeocyte/archeocyte Archäozyt, Archäocyt, Archaeozyt, Archaeocyt
archaeornithes/ancestral birds/"lizard birds"/Archaeornithes Urvögel
Archean Eon/Archeozoic Eon (early Precambrian) Archaikum (Altpräkambrium/frühes Präkambrium)
archegonium Archegonium

archencephalon/primitive brain Archencephalon, Archenzephalon, Urhirn
archeopulmonates/Archaeopulmonata Altlungenschnecken
archespore/sporoblast Archespor
archetype/prototype (also in the sense of: stock) Urtyp, Urform
archicoel/archicoele (blastocoel/blastocoele) Archicöl, Archicoel (Blastocöl/Blastocoel)
archicoelomates/Archicoelomata Archicoelomaten, Ur-Leibeshöhlentiere, Ur-Coelomaten
archinephric duct Vornierengang
archinephros Vorniere
arciform/arcuate/arched bogenförmig, gebogen
arctic/polar arktisch, polar
arcualium Arcualium, Wirbelbogen
area Gegend, Gebiet
area of distribution/geographic range Areal, Verbreitunggebiet
area of expansion Ausbreitungsgebiet
area sampling Flächenstichprobe(nverfahren)
Arecaceae/palm family/Palmae Palmen
arenaceous/sandy/sandlike sandhaltig; sandartig; in sandigem Boden wachsend
arenicolous/sabulicolous/psammobiontic im Sand lebend, im Sand wohnend
areola/areole Areole; kleiner Hof, kleiner (Haut-)Bezirk; Gewebsspalte
areola/areola mammae (of mammary gland) Warzenvorhof, Warzenhof
areolar glands Warzenvorhofdrüsen, Montgomery-Knötchen
areolar tissue lockeres Bindegewebe
areole/areola Areole; kleiner Hof, kleiner (Haut-)Bezirk; Gewebsspalte
areometer (hydrometer) Aräometer, Senk-, Tauch-, Flüssigkeitswaage
argillaceous tonhaltig, tonartig, Ton...
arginine Arginin
arid/dry arid, trocken
arid land/arid region/dryland Trockengebiet
aril Arillus, Samenmantel; Samenwarze
aristate/awned begrannt, Grannen tragend
aristulate mit kurzer Granne

Aristotle's lantern

Aristotle's lantern Laterne des Aristoteles
- **auricle** Aurikel
- **comminator muscle/interpyramid muscle** Interpyramidalmuskel, interpyramidaler Muskel
- **compass** Kompass
- **epiphysis** Epiphyse
- **pyramid** Pyramide
- **rotule** Rotula
- **tooth guide** Zahnführung

arithmetic growth arithmetisches Wachstum
arithmetic mean *stat* arithmetisches Mittel
arks/Arcidae Archenmuscheln
arm/limb *micros* Trägerarm
arm/pinnule (crinoids) Arm
arm bone/humerus Oberarmknochen, Oberarmbein, Humerus
arm canal/brachial canal Mundarmgefäß
arm disk Mundarmscheibe
arm-palisade (conifers) *bot* Armpalisade
armadillos (Xenarthra: Cingulata/Loricata) Gürteltiere
armilla/manchette/superior annulus (remains of velum universale) Armilla, Manschette, Annulus superus
armor/test/theca/lorica/shell/case Panzer
armored/thecate gepanzert
armpit/axilla Achsel, Achselhöhle (Arm)
ARMS (amplification refractory mutation system) ARMS (System der amplifizierungsresistenten Mutation)
aroid family/arum family/calla family/Araceae Aronstabgewächse
arolium (pretarsus) Arolium, Afterkralle, Haftläppchen
aromatic aromatisch
arousal/exitement Erregung, Aufregung
arousal reaction *ethol* Weckreaktion
arrangement/set-up (experiment) Anordnung, Ansatz (Versuchsansatz/Versuchsaufbau)
arrest *vb gen* **(chromosomes in metaphase)** arretieren
arrhenotokous arrhenotok
arrhenotoky/arrhenotokous parthenogenesis Arrhenotokie, arrhenotoke Parthenogenese
arrhizous/arrhizal/rootless wurzellos
arrow poison Pfeilgift
arrow worms/chaetognathans Borstenkiefer, Pfeilwürmer, Chaetognathen
arrow-grass family/scheuchzeria family/Scheuchzeriaceae Blumenbinsengewächse, Blasenbinsengewächse
arrowgrass family/Juncaginaceae Dreizackgewächse
arrowhead family/water-plantain family/Alismataceae Froschlöffelgewächse
arrowhead-shaped/sagittate/sagittiform pfeilförmig
arrowroot family/prayer plant family/Marantaceae Pfeilwurzelgewächse
ARS (autonomously replicating sequence) autonom replizierende Sequenz (ARS)
arsenic Arsen
arteriole (small artery) Arteriole, Äderchen
artery Arterie *(for different arteries please consult a medical dictionary)*
arthral Gelenk...
arthrobranch/arthrobranchia/joint gill Arthrobranchie
arthropods/Arthropoda Arthropoden, Gliederfüßer
arthrospore/Oidie arthrospore, oidium
articular gegliedert, Gelenk..., Glieder...
articular bristle Gliederborste
articular capsule/joint capsule Gelenkkapsel
articular cartilage Gelenksknorpel
articular disk (meniscus) Gelenkzwischenscheibe (Meniskus)
articular pivot of antenna Fühlergelenk
articular process Gelenkfortsatz
articular socket/acetabulum Gelenkpfanne, Gelenkgrube
articulate(d)/jointed gegliedert, mit Gelenk, gelenkig, gelenkartig verbunden, Glied..., Glieder.., Gelenk...
articulates/articulated animals/Articulata Articulaten, Gliedertiere
articulation Artikulation, Gelenkverbindung, Gelenk
artifact/artefact Artefakt
artificial color(s)/coloring künstliche(r) Farbstoff(e)
artificial flavor(s)/flavoring künstliche(r) Geschmackstoff(e)
artificial insemination künstliche Befruchtung

assortative mating

artificial light/lighting künstliche Beleuchtung
artificial selection künstliche Selektion, künstliche Auslese
arum family/calla family/aroid family/Araceae Aronstabgewächse
arundinaceous/reedlike schilfrohrartig, riedförmig
arytenoid cartilage/Cartilago arytaenoidea Aryknorpel, Stellknorpel
ascarid worms (Ascaris spp.) Spulwürmer
ascendent/ascendant aufsteigend
aschelminths/nemathelminths/ pseudocoelomates Aschelminthen, Schlauchwürmer, Rundwürmer *(sensu lato)*, Nemathelminthen (Pseudocölomaten)
ascidial/ascidiform/sac-like/ bag-like/pitcher-shaped/ flask-shaped schlauchförmig, kannenförmig, krugförmig, sackförmig
ascidiate kannenartig, krugartig, sackartig, schlauchartig
ascidiate leaf/pitcher leaf Schlauchblatt, Kannenblatt
ascidiform kannenförmig, krugförmig, schlauchförmig
Asclepiadaceae/milkweed family Schwalbenwurzgewächse, Seidenpflanzengewächse
ascorbic acid/vitamin C (ascorbate) Ascorbinsäure (Ascorbat)
asexual/agamous/neuter ungeschlechtig
asexual reproduction/vegetative reproduction ungeschlechtliche Vermehrung/Fortpflanzung, vegetative Vermehrung/ Fortpflanzung
ASO (allele specific oligonucleotide) ASO (allelspezifisches Oligonucleotid)
asparagic acid/aspartic acid (aspartate) Asparaginsäure (Aspartat)
asparagine/aspartamic acid Asparagin
asparagus family/Asparagaceae Spargelgewächse
ASPCA (American Society for the Prevention of Cruelty to Animals, New York) U.S. nationale Tierschutzvereinigung
aspect ratio *aer* Flügelstreckung
aspidium family/sword fern family/ Aspidiaceae/Dryopteridaceae Schildfarngewächse, Wurmfarngewächse

Aspleniaceae/spleenwort family Streifenfarngewächse
ass/*Equus asinus* (domestic ass: donkey/burro) Esel
• **hinny** (♀ ass/donkey x ♂ horse stallion) Maulesel (Eselstute x Pferdehengst)
• **jack/jackass** (♂ uncastrated ass/ donkey) männlicher Esel
• **jennet/jenny** (♀ ass/donkey) Eselin
• **mule** (♀ horse x ♂ ass/donkey) Maultier (Pferdestute x Eselhengst)
assay/test/trial/examination/exam/ investigation Probe, Versuch, Untersuchung, Test, Prüfung
assay material/test material/ examination material Probe, Probensubstanz, Untersuchungsmaterial
assay medium Untersuchungsmedium, Prüfmedium, Testmedium
assemblage/group Gruppe, Gruppierung
assembly Assemblierung, Zusammenbau
• **disassembly** Disassemblierung, Zerlegen, Zerlegung, Auseinandernehmen
• **reassembly** Reassemblierung, Wiederzusammenbau
• **self-assembly** Selbst- Assemblierung, Selbstassoziierung, Selbstzusammenbau, spontaner Zusammenbau (molekulare Epigenese), Spontanzusammenbau
assembly initiation complex *vir* Initiationskomplex der Assemblierung
assess *vb* bewerten, erfassen
assessment Bewertung, Erfassung
assimilate *n* Assimilat
assimilate *vb* assimilieren
assimilate stream Assimilatstrom
assimilation (anabolism) Assimilation
assimilative root Assimilationswurzel
assimilatory assimilatorisch
associated species Begleitart
association area *neuro* Assoziationsfeld
associative learning/learning by experience Assoziationslernen, Erfahrungslernen
assortative mating assortative Paarung, bewusste Paarung, übereinstimmende Paarung, sortengleiche Paarung, Gattenwahl

assortment Sortieren, Ordnen; Auswahl; Sammlung, Zusammenstellung
- **independent a.** *gen* unabhängige Verteilung, freie Verteilung
- **law of random assortment/ principle of random assortment/ principle of independent assortment (Mendel)** Kombinationsregel, Unabhängigkeitsregel

aster *cyt* Aster, Polstrahl (*pl* Polstrahlen)

Asteraceae/aster family/composite family/daisy family/sunflower family/Compositae Köpfchenblütler, Korbblütler

asteronome/star mine Sternmine

astichous astich

astigmatism Astigmatismus

astomatous *bot* ohne Spaltöffnungen

astomous/astomatous/mouthless astom, mundlos, ohne Mund, ohne Öffnung

astomy/mouthlessness Astomie, Mundlosigkeit

astringency Adstringenz, zusammenziehende Wirkung

astringent/astringent agent/ astringent substance Adstringens, adstringierender Stoff

astringent/styptic *adv/adj* adstringent, zusammenziehend

astrocyte Astrozyt

atavism/throwback (primitive characteristic) Atavismus, Rückschlag (ursprüngliches Merkmal)

Athyriaceae/lady fern family Frauenfarngewächse

atoke atok

atoll Atoll, Atollriff, Lagunenriff

atomic atomar, Atom...

atomic bond Atombindung

atomic force microscopy (AFM) Rasterkraftmikroskopie

atomic number Atomzahl

atomic weight Atomgewicht

atomize zerstäuben

atomizer Zerstäuber, Sprühgerät

atrial cavity/peribranchial cavity (ascidians/lancelets) Peribranchialraum

atrial natriuretic peptide (ANP)/ atrial natriuretic factor (ANF)/ atriopeptin atriales natriuretisches Peptid (ANP), Atriopeptin

atrichous haarlos, unbehaart

atrioventricular bundle/bundle of His (heart) Hissches Bündel (tert. Autonomiezentrum)

atrioventricular node/Aschoff and Tawara node/A-T node (heart) Atrioventrikularknoten, Aschoff-Tawara-Knoten (sek. Autonomiezentrum)

atrium/auricle (heart) Atrium, Aurikel, Vorhof, Herzvorhof, Herzvorkammer

atropine Atropin

atropous/orthotropous/orthotropic (ovule) atrop, aufrecht, geradläufig (Samenanlage)

attached befestigt, angeheftet
- **firmly attached (permanently)/ sessile** festsitzend, festgewachsen, festgeheftet, aufsitzend, sessil

attached X-chromosomes "Attached" X-Chromosomen, verbundene X-Chromosomen, verklebte X-Chromosomen

attachment/affixment Anheftung, Befestigung

attachment site Anheftungsstelle, Ansatzstelle

attack *n* Attacke, Angriff, Übergriff

attack *vb* attackieren, angreifen

attack and rend (prey) reißen (Beute)

attempt *n* Ansatz, Versuch

attenuate/tapering/pointed *adj/adv bot* verjüngt, spitz zulaufend, (keilförmig) zugespitzt

attenuate *vb micb/vir* attenuieren, abschwächen (die Virulenz vermindern)

attenuation Attenuation, Attenuierung, Abschwächung

attract/lure locken, anlocken, anziehen

attractant Attraktans (*pl* Attraktantien), Lockmittel, Lockstoff

attracting call *orn* Lockruf

attracting song *orn* Lockgesang

attraction Lockung, Anlockung

audition Hörvermögen, Gehör

auditory das Gehör/die Ohren betreffend, Gehör..., Hör...

auditory acuity Hörschärfe

auditory canal/auditory meatus Gehörgang

auditory center Hörzentrum

auditory nerve Hörnerv

auditory ossicle/ear ossicle/ ossiculum Ohrenknöchelchen, Gehörknöchelchen

auditory tube/eustachian tube Ohrtrompete, Eustachische Röhre

auks/Alcidae (Charadriiformes) Alken

aulopiform fishes/Aulopiformes Fadensegelfische

avicularium

aural Ohr(en) betreffend, Ohr..., Ohren...
auricle (leaf) Blattöhrchen (Gräser)
auricle/atrium Aurikel, Atrium, Herzvorhof, Herzvorkammer
• **pinna/ear conch/external ear/ outer ear** Ohrmuschel, äußeres Ohr, Pinna
• **pollen press/pollen packer (bees)** Pollenschieber, Aurikel
auriculars (ear coverts) Ohrenfedern
auriculate/eared/ear-like geöhrt
austral/southern südlich
Australian oak family/silk-oak family/protea family/Proteaceae Proteagewächse, Silberbaumgewächse
Australian pitcher-plant family/ Cephalotaceae Krugblattgewächse
autapomorphy Autapomorphie
autecology Autökologie
autoalloploidy Autoalloploidie
autoantibody Autoantikörper
autoantigen/self-antigen Autoantigen
autocatalysis Autokatalyse
autochthonous bodenständig, einheimisch, alteingesessen
autoclave n Autoklav
autoclave vb autoklavieren
autocrine autokrin, autocrin
autoecious autözisch, wirtstreu
autogamous autogam, selbstbefruchtend
autogamy/orthogamy/ self-fertilization/selfing Autogamie, Selbstbefruchtung
autogeneous control autogene Kontrolle
autograft (autologous graft) Autotransplantat
autoignitable selbstentzündlich
autoignition Selbstentzündung
autoimmune autoimmun
autoimmune disease Autoimmunkrankheit
autoimmunity Autoimmunität
autologous autolog
automaticity center (sinus node) Automatiezentrum (Sinusknoten)
automictic automiktisch
autonomic/autonomous/self- governing/self-controlling/ spontaneous/independent autonom, unabhängig, selbstständig funktionierend, spontan, vegetativ
autonomic center vegetatives Zentrum
autonomic nervous system (ANS)/ involuntary nervous system autonomes Nervensystem, vegetatives Nervensystem

autonomic reflex vegetativer Reflex
autonomous/autonomic/self- governing//self-controlling/ spontaneous/independent autonom, selbstständig funktionierend, spontan, unabhängig, vegetativ
autonomous control element autonomes Kontrollelement
autonomous(ly) replicating sequence (ARS) autonom replizierende Sequenz (ARS)
autoparasitism Autoparasitismus
autophagosome/autosome Autophagosom
autophagy Autophagie
autophily/self-pollination Selbstbestäubung
autopolyploid autopolyploid
autopreening/autogrooming/self- grooming (mammals) Putzen am eigenen Körper, Selbstputzen
autopreening orn Gefiederputzen, Gefiederkrauen (selbst)
autoradiography/radioautography Autoradiographie, Autoradiografie
autosomal-dominant autosomal- dominant
autosomal-recessive autosomal- rezessiv
autosome Autosom
autotherapy/self-healing Selbstheilung
autotomize autotomieren, selbst verstümmeln
autotomy Autotomie, Selbstverstümmelung
autotroph/autotrophic ("self-feeding") autotroph
autotrophy Autotrophie
autozooid/feeding polyp (Octocorallia) Autozooid, Zooid, Fangpolyp, Fresspolyp
autozygosity Autozygotie
autumn/fall Herbst
autumn coloration/fall coloration Herbstfärbung
autumn foliage Herbstlaub
autumnal herbstlich, Herbst...
auxillary heart Nebenherz, auxilläres Herz
auxins Auxine
auxotrophic auxotroph
auxotrophy Auxotrophie
availability Verfügbarkeit
average/mean Durchschnitt (Mittelmaß)
avian Vögel betreffend, Vogel...
aviary/birdhouse Voliere, Vogelhaus
avicularium (pl avicularia) Avicularie, "Vogelköpfchen"

aviculture

aviculture Vogelzucht
aviculturist Vogelzüchter
avidity Avidität
avirulent nicht virulent
avoidance Vermeidung, Meidung
avoidance behavior Meideverhalten
avoidance reaction/avoiding
  reaction Meidereaktion,
  Vermeidungsreaktion
avoidance strategy
  Vermeidestrategie
awl-shaped/subulate pfriemlich
awn Granne
awned/aristate begrannt,
  Grannen tragend
axenic culture/pure culture
  Reinkultur
axial/axile achsenständig
axial bundle/cauline bundle
  Stammbündel, Stammleitbündel
axial cell (*Dicyemida*) Axialzelle,
  Zylinderzelle
axial filament Axialfilament
axial gland (echinoderms)
  Axialorgan (Axialdrüse, braune
  Drüse)
axial rod/axoneme Achsenfaden,
  Axonema, Axonem
axial sinus (echinoderms)
  Axialsinus, "Fortsatzsinus",
  Fortsatz des Axialorgans
axial skeleton Axialskelett,
  Achsenskelett, Stammskelett,
  Rumpfskelett, Skelett des Stammes
axial spur Achsensporn
axil Achsel

axillary achselständig
axillary bud/lateral bud
  Achselknospe, Seitenknospe
axillary bulbil Achsenbulbille
axillary feathers/axillars
  Achselfedern
axillary meristem Achselmeristem
axillary region *entom* (wing)
  Axillarregion, Gelenkfeld
axis Achse
• principal axis
  Abstammungsachse, Hauptachse
axis/epistropheus/second cervical
  vertebra Axis, zweiter Halswirbel
axolemma Axolemm, Axolemma,
  Mauthnersche Scheide
axon Axon
• giant axon Riesenaxon,
  Riesenfaser, Kolossalfaser
axon hillock/growth cone
  Axonhügel, Axonkegel,
  Wachstumskegel
axoneme/axial rod/axial complex
  Axonema, Axonem, Achsenfaden
axopodium Axopodium
axosome Axosom, Axialkorn
azelaic acid/nonanedioic acid
  Azelainsäure, Nonandisäure
azeotropic azeotrop
azeotropic mixture azeotropes
  Gemisch
**Azollaceae/mosquito fern family/
  duckweed fern family**
  Algenfarngewächse
AZT (azidothymidine) AZT
  (Azidothymidin)

25 balance **B**

B cell/B lymphocyte (B = Bursa)
B-Zelle, B-Lymphozyt,
B-Lymphocyt
• **virgin B cell** unreife B-Zelle
B form (of DNA) B-Form,
B-Konformation (der DNA)
baa (bleat of sheep) mäh
**BAC (bacterial artificial
chromosome)** künstliches
Bakterienchromosom
**bacca/berry (esp. from interior
ovary)** Beere (unterständig) (*see
also:* nuculane)
baccate/pulpy/fleshy/as a berry
beerig, Beeren..., fleischig
**bacciferous/berry-producing/
bearing berries** beerentragend
bacciform/berry-shaped
beerenförmig
bachelor Junggeselle
bacillary bazillär, Bazillen...,
bazillenförmig, stäbchenförmig
back *n* Rücken; Rückseite
back mutation Rückmutation
back of the head Hinterkopf
back of the knee/bend of the knee
Kniekehle
back-mutation/reverse mutation
Rückmutation
backcrossing/backcross
Rückkreuzung
background Hintergrund
• **genetic background**
genetischer Hintergrund,
genotypischer Hintergrund
background radiation
Hintergrundsstrahlung
backshore *mar* Hochstrand,
Sturmstrand
bacteremia Bakteriämie
bacteria (*pl* but also used as *sg*)
(*sg* actually: bacterium) Bakterien
• **bacilli (*sg* bacillus) (rods)**
Bacillen, Bazillen (Stäbchen)
• **cocci (*sg* coccus) (spherical
forms)** Coccen, Kokken (kuglig)
• **denitrifying bacteria**
denitrifizierende Bakterien,
Denitrifikanten
• **hydrogen bacteria (aerobic
hydrogen-oxidizing bacteria)**
Knallgasbakterien,
Wasserstoffbakterien
• **luminescent bacteria**
Leuchtbakterien
• **myxobacteria** Myxobakterien,
Schleimbakterien
• **nitrogen-fixing bacteria**
stickstofffixierende/
stickstoffixierende Bakterien
• **nodule bacteria**
Knöllchenbakterien

• **putrefactive bacteria**
Fäulnisbakterien
• **rickettsias/rickettsiae
(*sg* rickettsia) (rod-shaped to
coccoid)** Rickettsien (Stäbchen-
oder Kugelbakterien)
• **spirilla (*sg* spirrilum) (spiraled
forms)** Spirillen (schraubig
gewunden)
• **sulfur bacteria** Schwefelbakterien
• **thermophilic bacteria**
wärmesuchende Bakterien,
thermophile Bakterien
• **vibrios (mostly comma-shaped)**
Vibrionen (meist gekrümmt)
bacterial bakteriell
**bacterial artificial chromosome
(BAC)** künstliches
Bakterienchromosom
bacterial culture Bakterienkultur
bacterial flora Bakterienflora
bacterial infection bakterielle
Infektion
bacterial lawn Bakterienrasen
bacterial nodule Bakterienknöllchen
bacterial strain Bakterienstamm
bactericidal/bacteriocidal
bakterizid, keimtötend (*see
germicidal)
bacterioids Bakterioide
bacteriologic/bacteriological
bakteriologisch
bacteriology Bakteriologie,
Bakterienkunde
**bacteriophage/phage/bacterial
virus** Bakteriophage, Phage
**bacterium (*pl* bacteria:
also used as *sg*)** Bakterie,
Bakterium (*pl* Bakterien)
bacterivorous/bactivorous
bakterienfressend
badlands erodierte und wüstenartige
Landschaft (zerrachelte
Erosionslandschaft)
baffle *biot* **(bioreactor impellers)**
Strombrecher
baffle plate Prallblech, Prallplatte,
Ablenkplatte (Strombrecher z.B. am
Rührer von Bioreaktoren)
"bag of waters"/amnion Amnion,
Schafhaut, innere Eihaut, innere
Keimhülle, inneres Eihüllepithel
bailer/gill bailer/scaphognathite
Atemplatte, Scaphognathide,
Scaphognathit
bait Köder
baker's yeast Backhefe, Bäckerhefe
baking/heat treatment Backen,
Hitzebehandlung
balance *vb* wiegen, ausbalancieren;
abgleichen
balance/equilibrium Gleichgewicht

B balance

balance *metabol/math* Bilanz
balance *phys* (for measuring mass/
weight) Waage
• analytical balance
Analysenwaage
• laboratory balance Laborwaage
• precision balance Feinwaage
balanced growth ausgewogenes
Wachstum
balanced lethal balanciert letal
balanced polymorphism
balancierter Polymorphismus
balanced translocation balancierte
Translokation
balancer/haltere (dipterans)
Schwingkölbchen, Haltere
balancer/spring (ctenophores)
Wimperfeder
balanophora family/
Balanophoraceae
Kolbenträgergewächse,
Kolbenschmarotzer
Balbiani rings Balbiani-Ringe
bald/bare/barren/glabrous kahl
bale (hay/straw/cotton) Ballen
(Heuballen/Strohballen)
baleen/whalebone Fischbein,
Walbein
ball-and-socket joint/spheroid joint
Kugelgelenk
ball-and-stick model/stick-and-ball
model *chem* Kugel-Stab-Model,
Stab-Kugel-Model
ballistic capsule Schleuderkapsel
ballistic dispersal
Schleuderausbreitung
ballistic fruit Schleuderfrucht,
ballistische Frucht
ballooning Fadenflug,
"Luftschiffen", "Ballooning"
(Spinnen: Altweibersommer)
balsam (a plant exudate) Balsam
balsam family/jewelweed family/
touch-me-not family/
Balsaminaceae
Balsaminengewächse,
Springkrautgewächse
BALT (bronchus/bronchial/bursa-
associated lymphoid tissue)
bronchusassoziiertes lymphatisches
Gewebe, bronchienassoziiertes
lymphatisches Gewebe
banana family/Musaceae
Bananengewächse
band *electrophor/chromat* Bande
• main band Hauptbande
• satellite band Satellitenbande
band shift assay Gelretentionsanalyse
band-shaped/fascial/fasciate
bandförmig
banded/fasciate gebändert, breit
gestreift

banding/ringing (e.g. birds)
Beringung
banding pattern *gen* (of
chromosomes) Bänderungsmuster,
Bandenmuster (von Chromosomen)
banding technique *gen*
Bänderungstechnik
bank(s)/shore/coast Ufer
bank/clone bank/DNA-library
Genbank, DNA-Bibliothek
bank reef/patch reef Fleckenriff
banking *orn/aer* Kurvenschräglage,
Querneigung, Querlage
banner/banner petal/standard
petal/vexillum Fahne
bar diagram/bar graph/bar chart
Stabdiagramm
barb Bart; Widerhaken
• main branch of feather Federast,
Ramus
Barbados cherry family/malpighia
family/Malpighiaceae
Malpighiengewächse,
Barbadoskirschengewächse
barbate/bearded gebärtet
barbellate (e.g. pappus) mit
borstigen Widerhaken versehen
barbels/barbs/beard Barteln
barberry family/Berberidaceae
Berberitzengewächse,
Sauerdorngewächse
barbs/barbels/whiskers Bartfäden,
Barthaare (Fische)
barbule (notched/hooked barbule)
orn Federstrahl, Radius
(Bogenstrahl, Hakenstrahl)
bare/barren/bald/glabrous kahl
bark *vb* (e.g. dog) bellen
bark/cortex *n bot* Rinde
• tertiary bark/dead outer bark/
rhytidome Borke, Rhytidom
bark grafting/rind grafting
Rindenpfropfung, Pfropfen hinter
die Rinde
barn/shed Stall
barnacles (cirripedes/Cirripedia)
Seepocken (Rankenfüß(l)er/
Cirripeden)
barophile *n* barophiler Organismus
barophilic/barophilous barophil
Barr body Barrkörperchen
barrel *biochem* (protein structure)
Fass, Fass-Struktur
barrel (trunk of a quadruped)
Rumpf
barren *adv/adj* (unproductive
wasteland) öde, dürr, kahl,
unproduktiv
barren *n*/barren land Ödland
barren/empty (seed/fruit) taub, leer
barren/incapable of producing
offspring unfruchtbar, steril

**barrier filter/selective filter/
stopping filter/selection filter**
Sperrfilter
barrier function Barrierefunktion
barrier reef Barriereriff, Wallriff
barrier web *arach* dreidimensionales
Schutzgewebe
**Bartholin's gland/greater vestibular
gland/gl. vestibularis major**
Bartholin-Drüse
basal body Basalkörper
basal cell Basalzelle
basal disk/pedal disk (anthozoa)
Fußscheibe
basal fold/plica basalis Basalfalte
**basal ganglion/cerebral nucleus/
corpus striatum** Basalganglion,
Stammganglion
basal lamina/basement membrane
Basallamina, Basalmembran
basal layer/basal zone Basalschicht
basal medium Basisnährmedium,
Basisnährboden
**basal metabolic rate/base
metabolic rate (BMR)**
Basalumsatz, basale
Stoffwechselrate,
Grundstoffwechselrate, Grundumsatz
basal metabolism
Grundstoffwechsel,
Ruhestoffwechsel
basal placentation basale
Plazentation, basiläre Plazentation,
grundständige Plazentation
basal plate Basalplatte
**basal rate/metabolic basal rate/
metabolic base rate**
Grundstoffwechselrate, Grundumsatz
basapophysis Basapophyse
base *bot/hort* Stock, Grundlage
base *chem* Base
base/foot *micros* **(supporting
stand)** Fuß, Stativfuß
base/nitrogenous base *gen/
biochem* stickstoffhaltige Base
base analogue/base analog *gen*
Basenanalogon (*pl* Basenanaloga)
base composition/base ratio *gen*
Basenzusammensetzung
base deficit *gen* Basendefizit
base excess *gen* Basenüberschuss
**base material/ground substance/
matrix** Grundsubstanz,
Grundgerüst, Matrix
**base material/starting material/raw
material** Grundstoff, Rohstoff
**base metabolic rate/basal
metabolic rate** Grundumsatz,
Basalumsatz, Grundstoffwechselrate,
basale Stoffwechselrate
base of leaf blade Blattspreitengrund,
Blattspreitenbasis

base pair *gen* Basenpaar
base pairing *gen* Basenpaarung
base substitution *gen*
Basensubstitution, Basenaustausch
base-pairing rules *gen*
Basenpaarungsregeln
base-stacking Basenstapelung
**basella family/Madeira vine family/
Basellaceae**
Schlingmeldengewächse
basement membrane/basal lamina
Basalmembran, Basallamina
basic/alkaline basisch, alkalisch
basic building block Grundbaustein
basic research Grundlagenforschung
basicity Basizität, basischer Zustand
basidium (*pl* **basidia)** Basidie
(*pl* Basidien)
basifixed basifix
basify basisch machen
basilar membrane Basilarmembran
basin Bassin, Becken; *geol*
Senkungsmulde, Kessel; (kleine)
Bucht
• **deflation basin/blowout** *geol*
Deflationskessel, Windmulde
• **depression** Vertiefung, Mulde
• **drainage basin/drainage area/
catchment basin/catchment
area/watershed**
Wassereinzugsgebiet,
Grundwassereinzugsgebiet,
Flusseinzugsgebiet, Sammelbecken
basipetal basipetal
basipetal development Basipetalie
basipodite Basipodit
basitony *bot* Basitonie
basket cell *neuro* Korbzelle
**bast/secondary phloem/secondary
bark** Bast, sekundäres Phloem
bast fiber Bastfaser
bast ray Baststrahl
bastard/hybrid Bastard, Hybride
bastardize/hybridize bastardieren,
hybridisieren
bastardization/hybridization
Bastardisierung, Hybridisierung
bat pollination/chiropterophily
Fledermausbestäubung,
Fledermausblütigkeit,
Chiropterophilie
**bat-pollinated flower/
chiropterophile** Fledermausblume
batch Ansatz, Charge
batch culture Satzkultur,
diskontinuierliche Kultur, Batch-
Kultur
batch process Satzverfahren
Batesian mimicry Batessche Mimikry
bathyal zone (*upper***: continental
slope;** *lower***: continental rise)**
Bathyal (Meeresboden)

B bathypelagic zone

bathypelagic zone Bathypelagial, mittlerer Tiefseebereich, mittlere Tiefseezone

batis family/saltwort family/ Bataceae Batisgewächse

bats/chiropterans/Chiroptera Fledermäuse, Flattertiere

battery of nematocysts/cnidophore Nesselbatterie, Cnidophore

bay/bight/gulf Bucht, Meerbusen, Golf

baymouth bar/bay bar/bay barrier Nehrung (vor einem Haff)

bayou Altwasser, sumpfiger Flussnebenarm

BDNF (brain-derived neurotrophic factor) *used as such (not translated!)*

beach hoppers/sand hoppers & sand fleas and relatives/ amphipods/Amphipoda Flohkrebse, Flachkrebse

bead-bed reactor Kugelbettreaktor

beads-on-a-string structure (chromatin) Perlschnurstruktur

beak (strong/short/broad bill) Schnabel
- **seed-cracking beak** Kernbeißerschnabel, Kernbeißer
- **water-straining beak** Seihschnabel

beak/rostrum (e. g. bugs) Schnabel, Rüssel (Stechrüssel), Rostrum (Wanzen)

beaker *chem* Becherglas, Zylinderglas

beaker tongs *lab* Becherglaszange

beam Balken, Holzbalken

beam *zool* **(antlers)** Stange (Geweih)

beam of light Lichtstrahl, Lichtbündel

beam of rays Strahlenbündel

bean family/pea family/legume family/pulse family/Fabaceae/ Papilionaceae (Leguminosae) Hülsenfruchtgewächse, Hülsenfrüchtler, Schmetterlingsblütler

bear young/give birth/bear offspring gebären, niederkommen, Junge bekommen
- **litter** Junge werfen

bear's-foot fern family/davallia family/Davalliaceae Hasenfußfarngewächse

beard worms/beard bearers/ pogonophorans Bartwürmer, Bartträger

bearded/barbate (having hair tufts) gebärtet, bärtig

beast (wildes) Tier, Bestie (*bes.* vierbeinige Säuger)

beast of burden/pack animal Lasttier

beaver lodge Bieberburg

bed load *soil/mar* Bodenfracht, Geschiebe

bedrock/rock base/parent material/ parent rock (unmodified) Grundgestein, Muttergestein, Ausgangsgestein, fester Untergrund, Felsuntergrund

bedstraw family/madder family/ Rubiaceae Rötegewächse, Labkrautgewächse, Krappgewächse

bee Biene, Imme
- **drone** Drohne, Drohn
- **follower bee** Nachläuferin
- **forager/field bee/flying bee** Flugbiene, Trachtbiene, Sammelbiene, Sammlerin
- **guard bee** Wächterbiene, Wehrbiene
- **house bee** Stockbiene
- **nurse bee** Ammenbiene
- **queen bee** Bienenkönigin, Königin, Weisel
- **scout bee** Spurbiene, Kundschaftler, Pfadfinder
- **worker bee** Arbeitsbiene, Arbeiterin

bee bread/cerago Bienenbrot, Cerago

bee colony Bienenvolk

bee dance Bienentanz
- **bumping run** Rumpellauf
- **buzzing run/breaking dance** Schwirrlauf
- **fanning with lifted abdomen (exposing Nasanov organ)** Sterzeln
- **figure-eight dance/waggle dance/wagging dance/tail-wagging dance** Schwänzeltanz
- **jerk dance** Rucktanz
- **jostling dance/run** Drängeln
- **night dance** Nachttanz
- **persistent dance** Dauertanz
- **quiver dance/tremble dance/ trembling dance** Zittertanz
- **round dance** Rundtanz
- **shaking dance** Rütteltanz
- **sickle dance (bowed figure 8)** Sicheltanz
- **spasmodic dance** Trippeln
- **tremble dance/trembling dance/ quiver dance** Zittertanz
- **vibrating dance/dorsoventral abdominal vibrating dance (DVAV)** Schütteltanz, Schüttelbewegung
- **waggle dance/wagging dance/ tail-wagging dance/figure-eight dance** Schwänzeltanz

29   behavior **B**

bee language Bienensprache
bee milk/royal jelly Königin-
Futtersaft, Gelée Royale
bee pollination/melittophily
Bienenbestäubung
bee venom Bienengift
bee yard/apiary Bienenhaus,
Bienenstand
beech family/Fagaceae
Buchengewächse, Becherfrüchtler
beechnut Buchecker
beef Rindfleisch
beef-steak fungi/Fistulinaceae
Reischlinge
beefwood family/she-oak family/
Casuarinaceae
Streitkolbengewächse
beehive (artificial nest)
Bienenstock, Bienenkorb (künstliche
Behausung)
beekeeper/apiarist Bienenzüchter,
Imker
beekeeping/apiculture
Bienenzucht
beeswax Bienenwachs
beet Rübe
beet sugar (cane sugar)/table
sugar/sucrose Rübenzucker
(Rohrzucker), Sukrose, Sucrose
beetle pollination/cantharophily
Käferbestäubung
beetle-pollinated flower/
coleopterophile Käferblume
beetles/Coleoptera Käfer
beg betteln
begging behavior Bettelverhalten
begonia family/Begoniaceae
Schiefblattgewächse
behavior Verhalten
• acquired behavior/learned
behavior erlerntes Verhalten,
Lernverhalten
• alarm behavior/warning
behavior Warnverhalten
• aposematic behavior/warning
behavior Warnverhalten
• appetitive behavior
Appetenzverhalten
• avoidance behavior
Meideverhalten
• begging behavior
Bettelverhalten
• conditioned behavior
konditioniertes Verhalten,
angepasstes Verhalten,
beeinflusstes Verhalten
• consummatory behavior
Konsumverhalten
• courting behavior
Werbeverhalten
• curiosity/inquisitiveness
Neugier, Neugierverhalten

• display behavior
Imponierverhalten,
Imponiergehabe, Imponiergebaren
• epideictic behavior
epideiktisches Verhalten
• epimeletic behavior
epimeletisches Verhalten
• exploratory behavior
Erkundungsverhalten
• expressive behavior
Ausdrucksverhalten
• fighting behavior
Kampfverhalten
• following behavior
Nachfolgeverhalten
• foraging behavior
Weideverhalten
• herding (guarding behavior)
Herden (Hüteverhalten)
• innate behavior angeborenes
Verhalten
• instinct behavior/instinctive
behavior Instinktverhalten,
Triebverhalten
• learned behavior Lernverhalten,
erlerntes Verhalten
• marking behavior
Markierverhalten
• mating behavior
Paarungsverhalten
• migratory behavior
Wanderverhalten
• mobbing behavior Hassen,
Hassverhalten
• orientational behavior/
orientation
Orientierungsverhalten,
Orientierung
• play behavior/play
Spielverhalten
• postcopulatory behavior Nachbalz
• precopulatory behavior/
precopulatory rite
Begattungsvorspiel
• preening behavior Putzverhalten
• reproductive behavior
Fortpflanzungsverhalten
• ritualized behavior
Ritualverhalten
• rutting behavior Brunstverhalten
• selfish behavior egoistisches
Verhalten
• sexual behavior Sexualverhalten,
Geschlechtsverhalten
• social behavior Sozialverhalten
• soliciting behavior
Begattungsaufforderung
• startle behavior
Schreckverhalten
• territorial behavior
Territorialverhalten
• threat behavior Drohverhalten

B behavior genetics

behavior genetics Verhaltensgenetik
behavioral barrier/ethological barrier Verhaltensbarriere
behavioral disorder/behavioral anomaly/deviant behavior Verhaltensstörung, Verhaltensanomalie
behavioral ecology Verhaltensökologie, Ethökologie
behavioral pattern Verhaltensmuster
beheading/pollarding (of trees) köpfen
behenic acid/docosanoic acid Behensäure, Docosansäure
being/creature Wesen, Kreatur
belemnites/Belemnitida Belemniten
belemnoid pfeilförmig
bell *n* **(medusa)** Glocke
bell *vb* **(deer/stag)** röhren (Hirsch)
bell-shaped/campanular/campaniform glockenförmig
bell-shaped curve (Gaussian curve) Glockenkurve (Gauß'sche Kurve)
bellflower family/bluebell family/Campanulaceae Glockenblumengewächse
bellows/lung Lunge
belly/undersurface of animal's body Unterseite, Bauchseite (Vorderseite)
belly/venter/abdomen Bauch, Abdomen
belt desmosome Gürteldesmosom
belt transect Gürteltransekt
Beltian bodies/Belt's bodies/food bodies (Acacia) Beltsche Körperchen, Futterkörper
ben oil/benne oil Behenöl
bench (lab bench) Werkbank (Labor-Werkbank)
• **sterile bench** sterile Werkbank
bench grafting *hort* Tischveredelung
bend of the arm/bend of elbow/crook of the arm/inside of the elbow Armbeuge, Ellenbeuge
bend of the knee/back of the knee/hollow behind knee/poples/popliteal fossa/popliteal space Kniekehle
bending resistance (wood) Biegesteifigkeit
bending strength (wood) Biegefestigkeit, Tragfähigkeit
beneficial/beneficient/useful nützlich
beneficial insect/beneficient insect Nutzinsekt
beneficial species/beneficient species Nutzart, Nützling

benefit/advantage *n* Nutzen, Vorteil, Gewinn
benefit *vb* nützen
benign benigne, gutartig
benignity/benign nature Benignität, Gutartigkeit
benne family/sesame family/Pedaliaceae Sesamgewächse
benthic/benthonic den Meeresboden bewohnend, bodenbewohnend, untergrundbewohnend (Ozean)
benthic zone (floor) Benthal, Meeresbodenbereich
benthon/benthos/benthos community Benthos, Gewässergrundbewohner, Bodenbewohner, Meeresbodenorganismen (Organismen des Benthal)
benzofuran/coumarone Benzofuran, Cumaron
benzoic acid (benzoate) Benzoesäure (Benzoat)
Berlese funnel Berlese-Apparat
berm *mar* Berm, Strandwall
berries *general/culinary* Beerenobst
berry *bot* Beere
berry with hard rind (hesperidium and pepo/gourd) Panzerbeere (Hesperidium und Kürbisfrucht, *siehe dort*)
bet-hedging *ethol* Risikoversicherung, auf Nummer sicher gehen
beta barrel/β barrel beta-Fass, β-Fass, beta-Rohr
beta meander/β meander beta-Mäander, β-Mäander
beta sheet/β sheet/beta pleated sheet/β pleated sheet beta-Faltblatt, β-Faltblatt
beta turn/-turn/bend/hairpin loop/reverse turn (DNA/protein) beta-Schleife, β-Schleife, Haarnadelschleife, Winkelschleife, Umkehrschleife, beta-Drehung, β-Drehung
betaine/lycine/oxyneurine/trimethylglycine Betain
Betulaceae/birch family Birkengewächse
betweenbrain/interbrain/diencephalon Zwischenhirn
bez tine (antlers) Eisspross (Geweih)
bezoar (stomach ball/hair ball) Bezoar (Magenkugel)
Bi Bi reaction *biochem* Bi-Bi-Reaktion (zwei Substrate-zwei Produkte)

bias/systematic error
Voreingenommenheit; *stat* Bias,
systematischer Fehler
● **unbiasedness**
Unvoreingenommenheit; *stat*
Treffgenauigkeit
biased voreingenommen; *stat*:
verfälscht, verzerrt, mit einem
systematischen Fehler behaftet
● **unbiased** unvoreingenommen;
stat unverfälscht, unverzerrt,
frei von systematischen Fehlern
bicarpellate *bot* aus zwei
Fruchtblättern bestehend
**bichirs & reedfishes and allies/
Polypteriformes** Flösselhechte,
Flösselhechtverwandte
bicistronic bizistronisch,
bicistronisch
bicuspid *adv/adj* zweihöckerig,
bikuspidal
bicuspid *n* Prämolar, vorderer
Backenzahn
bicuspid valve/mitral valve (heart)
Mitralklappe
Bidder's organ (toads) Bidder'sches
Organ
bidentate zweizähnig
bidirectional replication
bidirektionale Replikation
bidiscoidal placenta Placenta
bidiscoidalis
biennial zweijährig
bifacial/dorsiventral/zygomorph
bifazial, zweiseitig, dorsiventral,
zygomorph
bifid zweispaltig
biflagellated zweigeißlig
**bifurcate/Y-shaped/forked/
dichotomous** zweigabelig,
einfach gegabelt, dichotom
bifurcation/forking/dichotomy
Gabelung, Gabelteilung,
Dichotomie
big-bang reproduction/semelparity
Big-Bang-Fortpflanzung,
Semelparitie
bight Bucht, Einbuchtung
**bignonia family/trumpet-creeper
family/trumpet-vine family/
Bignoniaceae** Bignoniengewächse,
Trompetenbaumgewächse
bilabiate zweilippig
bilateral bilateral, zweiseitig; beide
Seiten betreffend
bilateral cleavage Bilateralfurchung,
bilaterale Furchung
**bilaterally symmetrical/
monosymmetrical/zygomorphic/
irregular** bilateralsymmetrisch,
monosymmetrisch, zygomorph,
unregelmäßig

bilaterians/Bilateria Zweiseitentiere
bile Galle, Gallflüssigkeit
bile acid Gallensäure
bile duct Gallengang
bile salts Gallensalze
**bilharziosis/schistosomiasis/blood
fluke disease (*Schistosoma spp.*)**
Bilharziose, Schistosomiasis
bill *zool* (*see:* beak) Schnabel,
schnabelähnliche Schnauze
● **decurved bill** Hakenschnabel
● **probing bill** Stocherschnabel,
Sondenschnabel
● **water-straining bill/
filter-feeding bill (flamingo)**
Seihschnabel
billing schnäbeln
billygoat/male goat Ziegenbock
bilocular zweifächerig
bimanous zweihändig
**bimodal distribution/two-mode
distribution** bimodale Verteilung
binary fission/bipartition binäre
Zellteilung, Zweiteilung
binding curve Bindungskurve
binding energy/bond energy
Bindungsenergie
**bindweed family/morning glory
family/convolvulus family/
Convolvulaceae** Windengewächse
binoculars Binokular
binomial distribution *stat/math*
Binomialverteilung
binomial formula binomische
Formel
**binomial nomenclature/binary
nomenclature** binominale/binäre
Nomenklatur, zweigliedrige
Benennung/Bezeichnung
bioassay/biological assay
biologischer Test
bioavailability Bioverfügbarkeit
**biocenosis/biocoenose/biotic
community** Biocönose, Biozönose,
Biozön, Organismengemeinschaft,
Lebensgemeinschaft
biochemical *n* Biochemikalie
**biochemical oxygen demand/
biological oxygen demand (BOD)**
biochemischer Sauerstoffbedarf,
biologischer Sauerstoffbedarf (BSB)
biochemistry Biochemie
biodegradability biologische
Abbaubarkeit
biodegradable biologisch abbaubar
biodegradation Biodegradation,
biologischer Abbau
biodeterioration biologische
Zersetzung
**biodiversity/biological diversity/
biological variability** biologische
Vielfalt, Lebensvielfalt

B bioenergetics

bioenergetics Bioenergetik
bioengineering Biotechnik,
biologische Verfahrenstechnik,
Bioingenieurwesen
bioequivalence Bioäquivalenz
bioethics Bioethik
**biogenetic law/biogenetic
principle/law of recapitulation/
Haeckel's law** biogenetische Regel,
biogenetisches Grundgesetz
biogenic biogen
**biogeochemical cycle (nutrient
cycle)** biogeochemischer Zyklus/
Kreislauf (Stoffkreislauf/
Nährstoffkreislauf)
biohazard biologische Gefahr,
biologisches Risiko
bioherm Bioherm
bioindicator/biological indicator
Bioindikator
bioinorganic bioanorganisch
**biolistics/microprojectile
bombardment** Biolistik
biologic(al)/biotic biologisch,
biotisch
biological clock biologische Uhr
biological containment biologische
Sicherheit(smaßnahmen)
**biological oxygen demand/
biochemical oxygen demand
(BOD)** biologischer
Sauerstoffbedarf, biochemischer
Sauerstoffbedarf (BSB)
biological pest control biologische
Schädlingsbekämpfung
**biological response mediator/
biological response modifier**
"biological response" Mediator,
biological response modifier (*not
translated!*)
biological rhythm/biorhythm
Biorhythmus
biological species biologische Art
biological warfare/biowarfare
biologische Kriegsführung
biological warfare agent
biologischer Kampfstoff
biological weapons biologische
Waffen
biologist/bioscientist/life scientist
Biologe *m*, Biologin *f*
biology/bioscience/life sciences
Biologie, Biowissenschaften
**biology lab technician/biological
lab assistant** BTA (biologisch-
technischeR AssistentIn)
bioluminescence Biolumineszenz
biomass Biomasse
• **total biomass** Gesamtbiomasse
biomathematics Biomathematik
biome/vegetational zone
Vegetationszone

biomechanics Biomechanik
biomedicine Biomedizin
biomembrane biologische Membran
biometry/biometrics Biometrie
biomolecule Biomolekül
bionics Bionik
biophysics Biophysik
bioreactor Bioreaktor (Reaktortypen
siehe Reaktor)
bioremediation biologische
Sanierung
biorhythm/biological rhythm
Biorhythmus
biorhythmicity Biorhythmik
biosafety biologische Sicherheit
biosafety committee Komitee für
biologische Sicherheit
bioscience/life science/biology
Biowissenschaft, Biologie
biosphere Biosphäre
biostatics Biostatik
biostatistics Biostatistik
biosynthesis Biosynthese
biosynthesize biosynthetisieren
**biosynthetic reaction (anabolic
reaction)** Biosynthesereaktion
biosynthetic(al) biosynthetisch
biotechnology Biotechnologie
biotic/biological biotisch, biologisch
biotic community/biocenosis
Lebensgemeinschaft,
Organismengemeinschaft, Biozön,
Biozönose, Biocönose
biotic pyramid Nahrungspyramide
biotin (vitamin H) Biotin
biotin labelling/biotinylation
Biotin-Markierung, Biotinylierung
biotope/life zone Biotop,
Lebensraum
• **humid biotope** Feuchtbiotop
biotransformation/bioconversion
Biotransformation, Biokonversion
biparental biparental
bipartite/dimeric/in two parts
zweiteilig
bipectinate zweikammreihig
biped *n* Zweifüß(l)er
bipedal biped, bipedisch, zweibeinig,
zweifüßig
bipedalism/bipedality Bipedie,
Bipedität, Zweibeinigkeit,
Zweifüßigkeit
bipennate/bipinnate zweifach
pennat/pinnat, zweifach gefiedert
bipolar cell bipolare Zelle,
Bipolarzelle
biradial symmetry
Biradialsymmetrie
biramous *bot* **(having two
branches)** biram, einfach
verzweigt/gegabelt (in zwei
Hauptäste)

33   **bladderworm** **B**

biramous appendage (schizopodal)
Spaltbein, Schizopodium
birch family/Betulaceae
Birkengewächse
bird cage Vogelkäfig
bird flower/ornithophile/bird-pollinated flower Vogelblume
bird migration Vogelzug
bird of passage Strichvogel
bird of prey/predatory bird/raptorial bird Raubvogel,
Greifvogel
bird pollination/ornithophily
Vogelbestäubung, Vogelblütigkeit,
Ornithophilie
bird-dispersal/ornithochory
Vogelausbreitung
bird-hipped dinosaurs/ornithischian reptiles/Ornithischia
Vogelbecken-Dinosaurier
bird-pollinated/ornithophilous
vogelblütig
bird's eye (wood texture)
Vogelaugenholz
bird's-nest fungi/bird's-nest family/Nidulariaceae Nestpilze, Nestlinge,
Vogelnestpilze, Teuerlinge
birdhouse/aviary Vogelhaus, Voliere
birding/bird watching
Vogelbeobachtung
birds/Aves Vögel
birefringence/double refraction
Doppelbrechung
birth Geburt
● **to give birth** gebären
birth control Geburtenkontrolle
birth defect Geburtsfehler
birth weight Geburtsgewicht
birthmark Muttermal
birthrate/natality Geburtenrate,
Geburtenzahl, Natalität
birthwort family/Dutchman's-pipe family/Aristolochiaceae
Osterluzeigewächse
biseriate/in two rows/two-row
zweireihig
bisexual/hermaphroditic
zweigeschlechtig, zwittrig
bisexual flower/hermaphroditic flower/perfect flower
zwittrige Blüte, Zwitterblüte
bisubstrate reaction
Bisubstratreaktion,
Zweisubstratreaktion
bisymmetrical/bilateral
disymmetrisch, bilateral
bitch Weibchen (speziell: Hündin)
bite Biss
biting beißend
biting lice/chewing lice/Mallophaga Haarlinge &
Federlinge

biting-chewing/mandibulate (mouthparts) beißend-kauend
(Mundwerkzeuge)
bitter bitter
bitterness Bitterkeit
bitters Bitterstoffe
bittersweet family/staff-tree family/spindle-tree family/Celastraceae
Spindelbaumgewächse,
Baumwürgergewächse
bituminous coal/soft coal (siehe
unter: Kohle) Steinkohle,
bituminöse Kohle
bivalence/divalence
Zweiwertigkeit
bivalent/divalent *chem* bivalent,
divalent, zweiwertig
bivalent *gen* bivalent,
doppelchromosomig
bivalent *n* Bivalent,
Chromosomenpaar
bivalve/bivalved *adj/adv*
zweiklappig, doppelklappig
bivalves/pelecypods/"hatchet-footed animals"/Pelecypoda/Bivalvia/Lamellibranchiata (clams: sedimentary/mussels: freely exposed) Muscheln
("Zweischaler")
bivium (holothurians) Bivium,
Interradius
bivouac (ants) Biwak
bixa family/annatto family/Bixaceae Annattogewächse
black body schwarzer Körper
black corals/thorny corals/antipatharians/Antipatharia
Dornkorallen, Dörnchenkorallen
black horse Rappe
black mildews/Meliolales
schwarze Mehltaupilze
black smoker (white smoker)
schwarzer Raucher (weißer
Raucher)
blackboy family/grass tree family/Xanthorrhoeaceae
Grasbaumgewächse
blackwater fever
Schwarzwasserfieber
bladder Blase
bladder cell Blasenzelle
bladder hair Blasenhaar
bladder trap/utriculus/utricle
Fangblase
bladderlike/bladdery/utriculate/utricular blasenartig, blasenförmig
bladdernut family/Staphyleaceae
Pimpernussgewächse
bladderworm/cysticercus (tapeworm larva) Blasenwurm,
Finne, Cysticercus

B bladderwort family 34

bladderwort family/butterwort family/Lentibulariaceae Wasserschlauchgewächse
blade/lamina (phyllid: certain algas) Spreite; Phylloid
blade/stalk (grass) Halm
blanch (bleach by excluding light) bleichen
blank *n stat/math* Blindwert
• **to leave blank** *math* frei lassen
blanket bog/climbing bog Deckenmoor
blastocoel/blastocoele Blastula-Höhle, Blastocöl, Furchungshöhle, primäre Leibeshöhle, primäre Körperhöhle
blastoderm Blastoderm
blastodisc/germinal disk/germ disk (>cicatricle/"eye") Keimscheibe, Embryonalschild, Diskus, Discus (> Hahnentritt)
blastogenesis Blastogenese, Furchungsteilung
blastokinesis Blastokinese, Keim(es)bewegung
blastomere Blastomere, Furchungszelle
blastopore/protostoma Blastoporus, Protostom, Urmund
blastozooid Blastozooid
blastula Blastula, Blasenkeim, Keimblase
blaze (horse: white stripe) Blesse
bleach *n* Bleiche
bleach *vb* bleichen (*activ*: weiss machen/aufhellen), ausbleichen
bleak (goat) meckern
bleat blöken
blechnum family/deer fern family/Blechnaceae Rippenfarngewächse
bleed bluten
bleeding *zool/bot* (e.g. plant wound) Bluten
bleeding/hemorrhage *med/vet* Bluten, Blutung, Hämorrhagie
bleeding heart family/fumitory family/Fumariaceae Erdrauchgewächse
blemish *n zool* Fehler, Makel
blending inheritance Mischvererbung
blepharoplast/blepharoblast/mastigosome/kinetosome (flagellar basal granule/corpuscle/body) Blepharoplast, Kinetosom, Basalkörper
blet *vb* (internal decay of fruit) teigig werden, überreif werden (Innenfäule)
blind spot (optic disk) blinder Fleck
blizzard heftiger Schneesturm

bloat blähen, aufblähen
bloated (auf)gebläht, (auf)geschwollen, aufgedunsen
bloating Blähung
block holder *micros* Blockhalter
block synthesis Blockverfahren
blocking reagent Blockierungsreagenz
blood Blut
blood agar Blutagar
blood cell/blood corpuscle/blood corpuscule Blutkörperchen
• **band granulocyte/stab cell/band cell/rod neutrophil** stabkerniger Granulocyt
• **basophil granulocyt** basophiler Granulocyt
• **eosinophil granulocyte** eosinophiler Granulocyt
• **granulocyte (polymorphonuclear)** Granulocyt, Granulozyt (polymorphkerniger Leukozyt)
• **lymphocyte** Lymphozyt, Lymphocyt
• **monocyte** Monozyt, Monocyt
• **neutrophil granulocyte** neutrophiler Granulocyt
• **red blood cell (RBC)/erythrocyte** rotes Blutkörperchen, Erythrocyt, Erythrozyt
• **reticulocyte/proerythrocyte (immature RBC)** Retikulocyt, Reticulozyt, Proerythrozyt
• **segmented granulocyte/filamented neutrophil** segmentkerniger Granulocyt
• **thrombocyte/blood platelet** Thrombozyt, Thrombocyt, Blutplättchen
• **white blood cell (WBC)/leukocyte** weißes Blutkörperchen, Leukocyt, Leukozyt
blood clot Blutgerinnsel, Blutkoagulum
blood clotting Blutgerinnung
blood clotting factor Blutgerinnungsfaktor, Gerinnungsfaktor
blood coagulation Blutgerinnung
blood corpuscle/blood corpuscule/blood cell Blutkörperchen
blood count/hematogram Blutbild, Blutkörperchenzählung, Blutzellzahlbestimmung, Hämatogramm
blood culture Blutkultur
blood donation Blutspende
blood dust/hemoconia/hemokonia Blutstäubchen, Hämokonia
blood factor Blutfaktor
blood flow Blutfluss, Durchblutung

35   **body cavity** **B**

**blood fluke/schistosome
(*Schistosoma spp.*)** Pärchenegel
blood group Blutgruppe
blood group incompatibility
Blutgruppenunverträglichkeit
blood grouping/blood typing
Blutgruppenbestimmung
blood island Blutinsel
blood meal (feeding on blood)
Blutmahlzeit
blood meal (ground dried blood)
Blutmehl
blood plasma Blutplasma
blood platelet/thrombocyte
Blutplättchen, Thrombozyt,
Thrombocyt
blood poisoning Blutvergiftung,
Sepsis
blood pressure Blutdruck
blood sample Blutprobe
blood sedimentation Blutsenkung
blood smear Blutausstrich
blood substitute Blut-Ersatz
blood sugar Blutzucker
blood sugar level
Blutzuckerspiegel
blood supply/circulation
Blutzufuhr, Blutversorgung,
Durchblutung
blood test Bluttest,
Blutuntersuchung
blood typing/blood grouping
Blutgruppenbestimmung
blood vessel Blutgefäß (Ader)
blood-brain barrier Blut-Hirn-
Schranke
**blood-sucking/sanguivorous/
hematophagous** blutsaugend, sich
von Blut ernährend
blood-typing
Blutgruppenbestimmung
bloodstream Blutstrom,
Blutkreislauf
**bloodwort family/redroot family/
kangaroo paw family/
haemodorum family/
Haemodoraceae**
Haemodorumgewächse
bloom *vb* blühen
• **in bloom** in Blüte (stehen)
bloom/flower *n* Blüte, Blume
bloom/the flowering state Blüte
(blühender Zustand)
blossom/flower Blüte
blot *vb* blotten; klecksen, Flecken
machen, beflecken
blot *n* Blot; Klecks, Fleck
• **enzyme-linked immunotransfer
blot (EITB)** enzymgekoppelter
Immunoelektrotransfer
blot hybridization
Blothybridisierung

blotch mine Platzmine
blotting/blot transfer *n* Blotten
• **affinity blotting**
Affinitäts-Blotting
• **alkali blotting** Alkali-Blotting
• **capillary blotting**
Diffusionsblotting
• **direct blotting electrophoresis/
direct transfer electrophoresis**
Blotting-Elektrophorese,
Direkttransfer-Elektrophorese
• **dry blotting** Trockenblotten
• **genomic blotting** genomisches
Blotting
• **ligand blotting**
Liganden-Blotting
• **wet blotting** Nassblotten
blowhole/vent/spiracle (whales)
Blasloch, Spritzloch, Spiraculum
blowout/deflation basin *geol*
Deflationskessel, Windmulde
blubber Speck, Tran, Blubber
(Walspeck: Unterhautfettschicht)
**blue corals/Coenothecalia/
Helioporida** Blaukorallen
**bluebell family/bellflower family/
Campanulaceae**
Glockenblumengewächse
**bluegreen algae/cyanobacteria/
Cyanophyceae** Blaualgen,
Cyanobakterien
bluff (e.g. river bluff) *geol*
Felsufer, Steilufer, Steilküste,
steile Felsküste, Steilwand,
Felswand, Klippe
**bluff/grove (clump of trees on
open plain)** Baumgruppe
blunt/obtuse stumpf
blunt end *gen* glattes Ende,
bündiges Ende
blunt end ligation *gen* Ligation
glatter Enden
**blurred/out of focus/not in focus
(picture)** unscharf
**blurredness/blur/obscurity/
unsharpness** *photo* Unschärfe
boar (male pig: not castrated) Eber
(männliches Schwein)
• **male wild boar** Keiler
board/plank Brett
boat conformation (cycloalkanes)
chem Wannenform
body/soma Körper
body cavity Körperhöhle,
Leibeshöhle
• **primary body cavity/blastocoel/
blastocoele** primäre Körperhöhle,
primäre Leibeshöhle, Blastocöl
• **secondary body cavity/coelom/
perigastrium** sekundäre
Körperhöhle, sekundäre
Leibeshöhle, Cölom, Coelom

B body cell 36

body cell/somatic cell Körperzelle,
Somazelle
body covering/integument (skin)
Körperdecke, Integument
body covering/vesture/vestiture
Körperhülle, Hülle, Mantel
body fluid Körperflüssigkeit
body heat Körperwärme
body of water/water body
Gewässer
body plan/construction/structure
Bauplan
body stalk *embr* Bauchstiel
body surface Körperoberfläche
(generell)
body surface area Körperoberfläche
(spez. Maß)
**bog (ombrogenic/ombrotrophic
peatland)** Moor (ombrogen/
oligotroph), Torfmoor; Luch
● **blanket bog/moor**
Deckenmoor,
terrainbedeckendes Moor
● **half-bog/early bog** Anmoor
● **palsa bog** Palsenmoor,
Torfhügelmoor
● **peat bog** Torfmoor, Sphagnum-
Moor
● **quaking bog/quagmire/
schwingmoor** Schwingrasen
● **raised bog/(upland/high) moor**
Hochmoor
● **string bog/aapa mire**
Strangmoor, Aapamoor
● **transitory bog** Übergangsmoor,
Zwischenmoor
bog drainage Moorentwässerung
bog drainage rill Rülle
bog forest (upland)
Hochmoorwald
**bog hollow/ditch/rivulet (in raised
bog)** Schlenke
bog moss/peat moss (*Sphagnum*)
Torfmoos
**bog myrtle family/wax-myrtle
family/sweet gale family/
Myricaceae** Gagelgewächse
bog plant/marsh plant/helophyte
Sumpfpflanze, Moorpflanze
bogbean family/Menyanthaceae
Bitterkleegewächse,
Fieberkleegewächse
boggy/swampy sumpfig, torfig,
moorig
**bogmoss family/mayaca family/
Mayacaceae** Moosblümchen
boil *vb* sieden, kochen
boil/furuncle *n med* Furunkel
("Blutschwär")
boil disease (*Myxobolus pfeifferi*)
Beulenkrankheit
boiling flask Siedegefäß

boiling flask with round bottom
Rundkolben (Siedegefäß)
boiling point Siedepunkt
boiling stone/boiling chip
Siedestein, Siedesteinchen
Bolbitius family/Bolbitiaceae
Mistpilze
**boldo family/monimia family/
Monimiaceae** Monimiengewächse
bole Baumstamm
**boletes/bolete mushroom family/
Boletaceae** Röhrenpilze,
Röhrlinge
**boletus mushroom/pore
mushroom/pore fungus**
Röhrenpilz, Röhrling, Porling
boll (cotton) Baumwollkapsel
**Bollinger body/Bollinger's granule
(inclusion body)** Bollinger Körper
(*viraler Einschlusskörper*)
bolting/shooting *bot/hort* schießen
(früh in Blüte)
**Bombacaceae/kapok-tree family/
silk-cotton tree family/cotton-tree
family** Wollbaumgewächse
bond/link *vb chem* binden
bond/linkage Bindung
● **atomic bond** Atombindung
● **carbon bond**
Kohlenstoffbindung
● **chemical bond** chemische
Bindung
● **conjugated bond** konjugierte
Bindung
● **covalent bond** kovalente
Bindung
● **disulfide bond/disulfide bridge**
Disulfidbindung (Disulfidbrücke)
● **double bond** Doppelbindung
● **glycosidic bond/glycosidic
linkage** glykosidische Bindung
● **heteropolar bond** heteropolare
Bindung
● **high energy bond** energiereiche
Bindung
● **homopolar bond/nonpolar bond**
homopolare Bindung, unpolare
Bindung
● **hydrogen bond**
Wasserstoffbrücke,
Wasserstoffbrückenbindung
● **hydrophilic bond** hydrophile
Bindung
● **hydrophobic bond** hydrophobe
Bindung
● **ionic bond** Ionenbindung
● **multiple bond** Mehrfachbindung
● **peptide bond** Peptidbindung
● **peptide bond/peptide linkage**
Peptidbindung
● **triple bond** Dreifachbindung
bond angle Bindungswinkel

bone Knochen

- **alisphenoid bone**
Keilbeinflügelknochen, großer
Keilbeinflügel
- **anklebone/talus/astragalus**
Sprungbein, Talus
- **anklebone/tarsal bone/tarsal**
Fußwurzelknochen,
Fußgelenksknochen, Tarsal
- **anvil/anvil bone/incus (ear)**
Amboss, Incus
- **arm bone/humerus**
Oberarmknochen, Oberarmbein,
Humerus
- **breastbone/sternum** Brustbein,
Sternum
- **calcaneus/calcaneum/heelbone**
Fersenbein
- **cancellous bone/spongy bone**
spongiöser Knochen
- **cannon bone (from hock to
fetlock)** Kanonenbein, Sprungbein
(Hauptmittelfußknochen der
Huftiere), *Pferd:* Röhrbein
- **capitate bone/capitate/
capitatum/os capitatum**
Kapitatum, Kopfbein
- **cardiac bone/heart ossicle/os
cordis** Herzknochen
- **carpal bones** Handwurzelknochen
- **cheekbone/malar bone/
zygomatic bone/os
zygomaticum** Backenknochen,
Wangenbein, Jochbein
- **chevron (bone)/hemal arch**
Sparrknochen, Haemapophyse,
ventraler Wirbelbogen, Chevron
- **coccyx/os coccygis** Steiß,
Steißbein, Schwanzfortsatz
- **coffin bone/pedal bone (horses:
distal phalanx)** Hufbein
- **collarbone/clavicle**
Schlüsselbein, Clavicula
- **compact bone/dense bone**
kompakter Knochen
- **coronary bone (horse: small
pastern bone)** Kronbein
- **cranial bone** Schädelknochen
- **cuboid bone** Würfelbein
- **cuneiform bone/os cuneiforme**
Keilbein
- **dermal bone** Hautknochen,
Deckknochen, Belegknochen
- **elbow bone/ulna** Ulna, Elle
- **ethmoid bone/os ethmoidale**
Siebbein
- **femoral bone/femur/thighbone/
os femoris** Femur,
Oberschenkelknochen
- **fibrous bone** Faserknochen,
Geflechtknochen
- **fishbone** Gräte

- **flank bone/ilium/os ilium**
Darmbein, Ilium
- **flat bone/os planum** platter
Knochen
- **frontal bone/forehead bone/
os frontale** Stirnbein
- **hamate bone/unciform bone**
Hakenbein
- **hammer/malleus (ear)** Hammer,
Malleus
- **heelbone/calcaneum/calcaneus/
os calcis** Fersenbein
- **hipbone/innominate bone/os
coxae (lateral half of pelvis)**
Hüftbein, Hüftknochen
(Beckenhälfte: Darmbein, Sitzbein,
Schambein)
- **hollow bone/pneumatic bone/os
pneumaticum** Hohlknochen,
pneumatischer Knochen
- **hyoid bone/lingual bone/os
hyoideum** Hyoideum,
Zungenbein
- **ilium/flank bone/os ilium** Ilium,
Darmbein
- **incisive bone/os incisivum**
Zwischenkieferknochen
- **innominate bone/hipbone/os
coxae (lateral half of pelvis)**
Hüftbein, Hüftknochen
(Beckenhälfte: Darmbein, Sitzbein,
Schambein)
- **interparietal bone/os
interparietale**
Zwischenscheitelbein
- **jawbone** Kieferknochen
- >**lower jawbone/lower jaw/
submaxilla/submaxillary bone/
mandible** Unterkiefer,
Unterkieferknochen, Mandibel
- **jugal/jugal bone (see:
cheekbone)** Jugale, Jochbein
- **kneecap/knee bone/patella**
Kniescheibe, Patella
- **lacrimal bone/os lacrimale**
Tränenbein
- **laminar bone** lamellärer
Knochen, Lamellenknochen
- **lingual bone** see hyoid bone
- **long bone/os longum** langer
Knochen, Röhrenknochen
- **lunate bone/semilunar bone/os
lunatum** Mondbein
- **marsupial bone/os marsupialis**
Beutelknochen
- **mastoid bone/mastoid/
processus mastoideus**
Warzenfortsatz (des Schläfenbeins)
- **maxilla/upper jawbone
(vertebrates)** Oberkiefer
- **maxillary bone/os maxillare**
Oberkieferbein

bone

- **metacarpal bone** Mittelhandknochen
- **metatarsal bone** Mittelfußknochen
- **nasal bone/os nasale** Nasenbein
- **navicular bone/distal sesamoid/ os sesamoideum distale (horse)** Strahlbein, distale Sesambein
- **navicular bone/os naviculare** Kahnbein, Fußwurzelknochen
- **occipital bone/os occipitale** Hinterhauptbein
- **ossicle/small bone** Knöchelchen
- **palatine bone/os palatinum** Palatinum, Gaumenbein
- **parietal bone/os parietale** Scheitelbein
- **pastern bone (long pastern)** Fesselknochen, Fesselbein
- **>short p.b./small p.b./middle phalanx (horse)** Kronbein
- **patella/knee bone/genu** Patella, Kniescheibe
- **pedal bone/coffin bone (horse)** Hufbein
- **pelvic bone/pelvis/pubis (ilium & ischium)** Beckenknochen (Darmbein & Sitzbein)
- **penis bone/baculum/os penis** Penisknochen
- **periotic bone/periotic/os perioticum** Felsenbein
- **petrous bone/petrosal bone (of temporal bone)** Felsenbein (des Schläfenbeins)
- **pinbone/ischium/os ischii (hipbone esp. in quadrupeds)** Sitzbein
- **pisiform bone/pisiform/os pisiforme** Erbsenbein
- **ploughshare bone/vomer** Pflugscharbein, Vomer
- **pneumatic bone/hollow bone/os pneumaticum** pneumatischer Knochen, Hohlknochen, Knochen mit lufthaltigen Zellen
- **pterygoid bone/pterygoid process/os pterygoides** Pterygoid, Flügelbein
- **pubic bone/pubis/os pubis** Schambein
- **pygostyle (ploughshare bone/ vomer of birds)** Pygostyl, Schwanzstiel (Pflugscharbein/ Vomer der Vögel)
- **quadrate bone/quadratum** Quadratbein
- **radius (bone of forearm)** Speiche
- **rib/costa** Rippe, Costa
- **rostral bone/os rostrale** Rüsselbein
- **scaphoid bone/os scaphoideum** Kahnbein, Handwurzelknochen
- **semilunar bone/lunate bone/ os lunatum** Mondbein
- **sesamoid bone/os sesamoideum** Sesambein, Sesamknöchelchen; Gleichbein
- **sesamoid bone, distal (horse)** Strahlbein
- **shinbone/tibia** Schienbein(knochen), Schiene, Tibia
- **short bone/os breve** kurzer Knochen
- **shoulder blade/scapula** Schulterblatt
- **shoulder girdle/pectoral girdle** Schultergürtel, Brustgürtel
- **sphenethmoid bone/os sphenethmoideum/os en ceinture** Gürtelbein
- **sphenoid bone/os sphenoidale (skull)** Keilbein
- **splint bone/fibula** Wadenbein, Fibula
- **spongy bone/cancellous bone** spongiöser Knochen
- **sternum/breastbone (vertebrates)** Sternum, Brustbein
- **stifle bone (horse)** Kniescheibe
- **stirrup/stapes (ear)** Steigbügel, Stapes
- **sutural bone/epactal bone/ wormian bone** Schaltknochen, Nahtknochen
- **talus/os tali (astragalus)** Rollbein, Sprungbein
- **tarsal bone** Fußwurzelknochen
- **temporal bone/os temporale** Schläfenbein
- **thigh bone/femur/os femoris** Oberschenkelknochen, Femur
- **tibia** *see* shinbone
- **trapezium bone/greater multangular bone** großes Vieleckbein
- **trapezoid bone/lesser multangular bone** kleines Vieleckbein
- **triangular bone/triquestral bone/os triquestrum** Dreiecksbein
- **tubular bone (long/hollow)** Röhrenknochen
- **turbinals/turbinate bones/ conchae nasalis** Nasenmuscheln (schleimhautüberzogene Knorpelplatten)
- **tympanic bone/tympanic** Paukenbein, Tympanicum
- **unciform bone/hamate bone** Hakenbein

Bowman's capsule

- **ungual bone/unguicular bone/os ungulare** Klauenbein
- **upper jawbone/upper jaw/ maxilla** Oberkiefer, Oberkieferknochen, Maxilla
- **whalebone/baleen** Walbein, Fischbein
- **wishbone/furcula/fourchette (birds: united clavicles)** Gabelbein, Furcula
- **wormian bone/sutural bone/ epactal bone** Schaltknochen, Nahtknochen
- **wrist bone/carpal bone** Handwurzelknochen, Handgelenksknochen, Carpalia (Ossa carpalia)

bone cell/osteocyte Knochenzelle, Osteocyt, Osteozyt
- **bone-forming cell/osteoblast** Osteoblast, knochenbildende Zelle

bone crest Knochenleiste, Knochenkamm
bone marrow/medulla of bone Knochenmark
- **red bone marrow** rotes Knochenmark, blutbildendes Knochenmark
- **yellow bone marrow/fatty b.m.** Fettmark, gelbes fetthaltiges Knochenmark

bone marrow cell/myelocyte Knochenmarkzelle, Myelozyt, Myelocyt
bone marrow grafting/marrow grafting Knochenmarkstransplantation
bone meal Knochenmehl
bone tissue/osseous tissue Knochengewebe
bone-forming cell/osteoblast Osteoblast, knochenbildende Zelle
bonitation Bonitierung
bony Knochen..., knöchern
book gills (gill book) Buchkiemen, Kiemenbeine, Blattbeine, Blattfüße (dichtstehende Kiemenlamellen: Xiphosuriden)
book lice/psocids/Psocoptera/ Copeognatha Bücherläuse (Staubläuse & Flechtlinge)
book lung/lung book Buchlunge, Fächerlunge, Fächertrachee
booster vaccination (booster shot) Auffrischimpfung (Auffrischinjektion)
borage family/Boraginaceae Bor(r)etschgewächse, Rau(h)blattgewächse
bordered flowerbed/border Rabatte
bordered pit Hoftüpfel

boreal/northern nördlich
boreal forest/temperate coniferous forest borealer Wald, borealer Nadelwald, Nadelwald der gemäßigten Zone
borer (insects) Bohrer (gang-/ lochgrabendes Insekt)
boron Bor
boss/protuberance *med/vet* Schwellung, Beule, Höcker
bostryx (helicoid cyme) Schraubel
botanical garden/botanic garden Botanischer Garten
botanist Botaniker
botany/plant science Botanik, Pflanzenkunde
botfly & cattle grub infestation (*Hypoderma bovis* et spp.) Dasselbeule
bothrium Bothrium, Sauggrube, Haftgrube
bothrosome/sagenogen/ sagenogenetosome Bothrosom, Sagenogen, Sagenogenetosom
bothryoidal tissue (hirudinids) Bothryoidgewebe
botryose/racemose traubig, razemös, racemös
bottle with faucet (carboy with spigot) Ballon (für Flüssigkeiten)
bottleneck *stat* Flaschenhals, Engpass
bottom/floor (of sea/ocean/lake) Boden (Meeresgrund/ Gewässergrund)
bottom (basis) Untergrund
bottom dweller Grundbewohner
bottom fermenting (beer) untergärig
- **top fermenting (beer)** obergärig
bottomland (river floodplain wetland) Tiefland (Schwemmland)
boulder(s) *geol* Block (*pl* Blöcke)
boulder layer/loose rock layer (soil/ ground) Untergrund (C-Horizont)
bound/leap/jump springen, hochhüpfen
boundary layer *mar/limn* Grenzschicht
bourse/knob/cluster base *hort/bot* Fruchtkuchen
bouton/bouton terminal/end-foot *neuro* Endknöpfchen, Bouton
bovine Rinder betreffend, Rinder...
bovine spongiform encephalopathy (BSE) bovine spongiforme Enzephalopathie (BSE)
bowel movement Stuhlgang, Entleerung
bowl-shaped flower Napfblume
Bowman's capsule Bowman-Kapsel

B   Bowman's gland

Bowman's gland/olfactory gland
Bowman-Spüldrüse, Bowman'sche
Spüldrüse, Drüse der
Riechschleimhaut
box family/Buxaceae
Buchsbaumgewächse
box jellies/sea wasps/
cubomedusas/Cubomedusae/
Cubozoa Würfelquallen
brace root/prop root (e. g. corn)
Stützwurzel
brachial/arm-like brachial, Arm...,
armartig
brachial canal/arm canal
Mundarmgefäß
brachiate/brachiferous/having arms
armtragend, mit Armen
brachiate (swing using arms:
gibbons) (mit Armschwung von
Halt zu Halt) hangeln
brachiation Brachiation, Hangeln,
Schwingklettern
brachiator Hangler, Schwingkletterer
brachidium (brachiopods)
Brachidium, Armgerüst
brachiomeric musculature
Brachialmuskulatur
brachiopods/lampshells/
Brachiopoda Brachiopoden,
Armfüßer
brachycerous/short-horned
kurzgehörnt
brachydont/brachyodont (with low
crowns) brachyodont,
niedrigkronig
brachypodous/with short legs
kurzfüßig, mit kurzem Fuß
brachypterous/with short wings
kurzflüglig, stummelflüglig
brachyptery Kurzflügligkeit,
Kurzflügeligkeit
brachystomatous/with a short
proboscis (insects) mit kurzem
Rüssel
brachyural/brachyurous
kurzschwänzig (Krabben: Abdomen
unter den Thorax geklappt)
bracken ferns (hypolepis family/
Hypolepidaceae) Adlerfarne
bracket/conk (shelf-like
sporophyte) Konsole
(Fruchtkörper von Baumpilzen,
z. B. *Fomes*)
bracket fungi/polypore family/
Poriales/Polyporaceae Porlinge,
Echte Porlinge
bracket fungus/shelf fungus/tree
fungus konsolenförmiger
Baumschwamm, Baumpilz
brackish marsh Brackmarsch
brackish water (somewhat salty)
Brackwasser

bract/subtending bract *bot*
Braktee, Tragblatt, Deckblatt
bract *zool* **(siphonophorans)**
Deckstück
bract-scale/secondary bract/
bracteole/bractlet Tragschuppe,
Deckschuppe, Brakteole, Tragblatt
zweiter Ordnung/zweiten Grades
bracteate mit Tragblättern/
Deckblättern, Brakteen...
bracteolate mit Tragschuppen/
Deckschuppen
bracteole/bract-scale/bractlet/
secondary bract Brakteole,
Tragschuppe, Deckschuppe,
Tragblatt zweiter Ordnung/zweiten
Grades
bradytelic bradytelisch
bradytely Bradytelie
brae/steep bank/slope/hillside
Abhang
brain (encephalon) Gehirn, Hirn
(Encephalon/Enzephalon)
• **afterbrain/metencephalon (pons**
+ cerebellum) Nachhirn,
Metencephalon
• **archencephalon/primitive brain**
Archencephalon, Urhirn
• **betweenbrain/interbrain/**
diencephalon Zwischenhirn,
Diencephalon
• **cerebellum/epencephalon**
Cerebellum, Kleinhirn,
Hinterhirn
• **cerebrum/endbrain/**
telencephalon Großhirn,
Endhirn, Telencephalon
• **deutocerebrum (insects)**
deutocerebrum, Mittelhirn
• **diencephalon/interbrain/**
betweenbrain Diencephalon,
Zwischenhirn
• **endbrain/cerebrum/**
telencephalon Endhirn,
Großhirn, Telencephalon
• **forebrain/prosencephalon**
(telencephalon + diencephalon)
(vertebrates) Vorderhirn,
Prosencephalon
• **hindbrain/rhombencephalon**
Rautenhirn, Rhombencephalon
• **interbrain/betweenbrain/**
diencephalon Zwischenhirn,
Diencephalon
• **marrow brain/medullary brain/**
myelencephalon Markhirn,
Myelencephalon
• **medulla/medulla oblongata**
verlängertes Rückenmark
• **mesencephalon/midbrain**
(vertebrates) Mesencephalon,
Mittelhirn

- **metencephalon/afterbrain (pons + cerebellum)** Nachhirn, Metencephalon (Hinterhirn)
- **midbrain/mesencephalon (vertebrates)** Mittelhirn, Mesencephalon
- **myelencephalon/marrow brain/ medullary brain** Myelencephalon, Markhirn (*sometimes:* Nachhirn)
- **neocerebellum (lateral lobes of cerebellum)** Neocerebellum, Neukleinhirn
- **neoencephalon/neencephalon** Neoencephalon, Neencephalon, Neuhirn
- **olfactory brain/"nose brain"/ rhinencephalon** Riechhirn, Rhinencephalon
- **paleoencephalon/paleencephalon** Paläoencephalon, Paleencephalon, Althirn
- **primitive brain/archencephalon** Urhirn, Archencephalon
- **prosencephalon/forebrain (telencephalon + diencephalon) (vertebrates)** Prosencephalon, Vorderhirn
- **protocerebrum (insects)** Protocerebrum, Vorderhirn
- **rhinencephalon/olfactory brain/ "nose brain"** Rhinencephalon, Riechhirn
- **rhombencephalon/hindbrain** Rhombencephalon, Rautenhirn
- **tritocerebrum (insects)** Tritocerebrum, Hinterhirn

brain stem/truncus cerebri Hirnstamm
brain-heart infusion agar Hirn-Herz-Infusionsagar
braincase/skull/head capsule/ cranium Hirnkapsel, Schädel, Cranium, Kranium
bramble Dornenstrauch
bramble/bush berries (*Rubus* species) Himbeersträucher, Brombeersträucher
bramble sharks & dogfishes sharks and allies/Squaliformes Stachelhaie
bran Kleie
branch/limb Ast
branch collar Astwulst, Astring (am Astansatz)
branch migration (DNA) *gen* Schenkelwanderung
branch out/ramify sich verzweigen
branch site Verzweigungsstelle
branched *chem* verzweigt
branched/ramified *bot* verzweigt
branched appendage/branched leg (arthropods) Spaltfuß, Spaltbein

branched-chained *chem* verzweigtkettig
branchial branchial, die Kiemenbögen betreffend
branchial arch/gill arch/visceral arch/gill bar Branchialbogen, Kiemenbogen
branchial gut/branchial basket/ pharynx Kiemendarm, Pharynx
branchial pump Kiemenpumpe
branchial skeleton Branchialskelett, Kiemenskelett
branchiferous kiementragend
branching *chem* Verzweigung
branching/ramification Verästelung, Verzweigung, Ramifikation
branching system (monopodial/ sympodial) (monopodiales/ sympodiales) Verzweigungssystem
branchiomeric musculature Branchialmuskulatur
branchiostegal membrane/ branchiostegous membrane/gill membrane Branchiostagal-membran, Kiemenhaut
branchiostegal ray/radius branchiostegus Branchiostegalstrahl, Kiemenhautstrahl
branny kleiig, grobmehlig
Brassicaceae/mustard family/ cabbage family/Cruciferae Kreuzblütler, Kreuzblütlergewächse
brawn (full strong muscles) Muskeln, muskulöser Teil
brawny/muscular muskulös, fleischig
bray (cry of donkey/mule) schreien (Esel)
brazil-nut family/lecythis family/ Lecythidaceae Deckeltopfbäume
bread molds/zygospore fungi/ zygomycetes (coenocytic fungi) Jochpilze, Zygomyceten
break through *n* **(surviving carrier of a lethal mutation)** Durchbrenner (überlebender Träger einer Letalmutation)
breakage-fusion/breakage and reunion *gen* Bruch-Fusion, Bruch und Wiedervereinigung
breakage-fusion-bridge Bruch-Fusions-Brücke
breakdown Abbau, Zerfall, Zusammenbruch
breaker zone Wellenschlagzone
breakers/surf Brandung, Meeresbrandung
breakwater/jetty/mole Mole
breast/mamma/bosom weibliche Brust, Busen
breast/thorax Brust, Thorax

B breastbone

breastbone/sternum Brustbein, Sternum

breastbone keel/breastbone ridge/ carina Brustbeinkamm, Carina

breath *n* Atem; Atemzug
- **to take a breath** Atem holen, Atem schöpfen, Luft holen, einen Atemzug machen

breathe/respire *vb* atmen
- **breathe in/inhale** einatmen, Luft holen, Atem schöpfen
- **breathe out/exhale** ausatmen

breathing/respiration Atmung, Respiration

breech position (birth) Steißlage (Geburt)

breed/cultivate/grow züchten, kultivieren

breed/breeding/cultivation/ growing Züchtung, Kultivierung

breed *n* **(form with new characteristics)** Zuchtform, Zucht, Brut

breed in hineinzüchten, hineinkreuzen

breed out wegzüchten, herauskreuzen

breed true/breed pure *vb* reinerbig sein
- **true-bred/pure-bred** *adj/adv* reinerbig

breeding experiment Züchtungsexperiment

breeding period/incubation period Brutdauer, Inkubationszeit

breeding place/breeding ground Brutstätte

breeding season Brutzeit, Brützeit (Jahreszeit)

breeze *meteo* Brise
- **land breeze** Landbrise
- **sea breeze/ocean breeze** Meeresbrise

brew *n* Gebräu, Bräu (Gebrautes)

brew *vb* brauen

brewers' grains Treber, Biertreber

brewers' yeast/brewer's yeast Brauereihefe, Bierhefe

brewery Brauerei

bridge grafting/repair grafting Überbrückung, Wundüberbrückung

bridge line/bridging thread (spiderweb) Brückenfaden

bridge of the nose Nasenrücken

bridging cross Überbrückungskreuzung

bright field *micros* Hellfeld

brightener/clearant/clearing agent (optical brightener) Aufheller, Aufhellungsmittel (optischer A.)

brightfield microscopy Hellfeld-Mikroskopie

brisket (cattle: breast or lower chest) Bruststück, Vorderbrust (bei Schlachttieren)

bristle/seta Borste

bristle worms/polychetes/polychete worms Borstenwürmer, Vielborster, Polychaeten

bristle-like/setaceous borstenartig

bristle-like coat (cell surface: clathrin) Stachelsaum

bristle-pointed borstig spitz

bristle-shaped/setiform borstenförmig

bristletails/thysanurans Borstenschwänze, Thysanuren (Felsenspringer, Fischchen)

bristly/setose/setaceous/ chaetigerous borstig

brittle stars/serpent stars & basket stars (*Gorgonocephalus/ Astrophyton*) Schlangensterne, Ophiuroiden

broad heritability (H²) allgemeine Erblichkeit (H^2)

broadleaf tree (*pl* broadleaves)/ hardwood Laubbaum (*pl* Laubbäume)

bromatium/gongylidium (*pl* gongylidia) Bromatium, Gongylidie, "Kohlrabi"

bromeliads/bromelia family/ pineapple family/Bromeliaceae Bromelien, Bromeliengewächse, Ananasgewächse

bronchiole Bronchiole, Bronchiolus, Bronchulus

bronchus (*pl* bronchi) Bronchus (*pl* Bronchien), Ast der Luftröhre

brood/breed/incubate *vb* bebrüten, brüten, inkubieren

brood/hatch *n* Brut (Nachkommen)

brood bud/bulbil Brutknöllchen, Brutknospe, Bulbille

brood capsule Brutkapsel

brood chamber (*Daphnia*) Brutraum

brood parasite Brutparasit, Brutschmarotzer

brood parasitism Brutparasitismus

brood pouch/marsupium Brutbeutel, Marsupium

brood provisioning/brood care/ brooding/parental care Brutfürsorge, Brutpflege

brood spot/brood patch *orn* Brutfleck

brooding/incubating Bebrüten, Brüten, Inkubieren

broodmare Zuchtstute

brook/creek Bach

broom *bot* Ginster

broomrape family/Orobanchaceae Sommerwurzgewächse

brotulas & cusk-eels & pearlfishes/
Ophidiiformes Eingeweidefische
brow/eyebrow Braue, Augenbraue
brow/forehead Stirn
brow ridge/brow crest
Überaugenwulst, Augenbrauenwulst
brow tine (antlers) Augspross
(Geweih)
brown algae/phaeophytes/
Phaeophyceae Braunalgen
brown body (bryozoans) brauner
Körper
brown coal/lignite (see also: coal)
Braunkohle, Lignit
brown fat braunes Fett
brown rot Braunfäule
(Destruktionsfäule)
browse (twigs/leaves of shrubs/
bark) abfressen (see: graze)
browser (woody shoots/leaves/
bark) junge Sprösslinge
abfressendes Tier
browsing damage (damage caused
by game) Verbiss, Wildverbiss,
Fraßschaden
bruise n Quetschung, Prellung;
Bluterguss, blauer Fleck
Brunner's gland/duodenal gland
Brunnersche Drüse,
Duodenaldrüse
brush/scrubland Buschland
brush/thicket/scrub/thick
shrubbery Gestrüpp, Dickicht
brush/underwood/undergrowth
Unterholz
brush border cyt Bürstensaum,
Stäbchensaum, Mikrovillisaum,
Rhabdorium
brush fire Buschfeuer
brush hair bot Fegehaar
brushwood/spray Reisig, Gestrüpp,
Gesträuch
bryology Bryologie, Mooskunde
bryozoans/moss animals/
Ectoprocta/Polyzoa/Bryozoa
Bryozoen, Moostierchen
• cheilostomates/Cheilostomata
Lappenmünder, Lippenmünder
• ctenostomates/Ctenostomata
Kammünder
• gymnolaemates/"naked throat"
bryozoans/Stelmatopoda/
Gymnolaemata Kreiswirbler
• phylactolaemates/"covered
throat" bryozoans/freshwater
bryozoans/Lophopoda/
Phylactolaemata Armwirbler,
Süßwasserbryozoen
• stenostomates/stenolaemates/
"narrow-throat" bryozoans/
Stenostomata/Stenolaemata
Engmünder

BSE (bovine spongiform
encephalopathy) BSE (bovine
spongiforme Enzephalopathie)
bubble linker PCR Blasen-Linker
PCR
bubble shells/Cephalaspidea
Kopfschildschnecken,
Kopfschildträger
bubonic plague (Yersinia pestis)
Beulenpest, Bubonenpest
buccal buccal, den Mund betreffend,
Mund...
buccal cavity/peristome/
peristomium (protozoans)
Buccalhöhle, Zellmundhöhlung,
Peristom
buccal field/mouth field
Buccalfeld, Mundfeld
buccal nerve ring Schlundring
(Nervenring)
buccal sucker Mundsaugnapf
(Prohaptor bei Trematoden)
buck (adult male goat/deer/
antilope) adultes Säugermännchen:
Bock (Ziegenbock/Rehbock),
Männchen (male rabbit > Rammler)
buckeye family/horse chestnut
family/Hippocastanaceae
Rosskastaniengewächse
bucking (cutting felled tree into
specific lengths) Ausformung,
Aushaltung
buckling strength/buckling
resistance/folding strength
(wood) Knickfestigkeit
buckthorn family/coffeeberry
family/Rhamnaceae
Kreuzdorngewächse
buckwheat family/dock family/
knotweed family/smartweed
family/Polygonaceae
Knöterichgewächse
bud vb Knospen treiben, knospen;
sprießen, ausschlagen;
okulieren
bud/eye bot Knospe, Auge
• accessory bud akzessorische
Knospe, Beiknospe
• adventitious bud
Adventivknospe
• apical bud Gipfelknospe
• axillary bud Achselknospe,
Seitenknospe
• dormant bud/resting bud/
quiescent bud schlafende
Knospe, ruhende Knospe
• flower bud/floral bud
Blütenknospe
• foliage bud Blattknospe,
Laubknospe
• latent bud Ersatzknospe,
Proventivknospe

bud

- **naked bud** nackte Knospe, offene Knospe (ohne Knospenschuppen)
- **protected bud** geschlossene Knospe (mit Knospenschuppen)
- **renewal bud** Erneuerungsknospe
- **terminal bud** Endknospe, Terminalknospe
- **winter bud/hibernaculum/turio/ turion** Winterknospe, Überwinterungsknospe, Hibernakel, Turio, Turione

bud/frustule *zool* **(polyps)** Knospe, Frustel, Frustula
bud bracts/bud envelope Knospenhülle
bud cluster/eye cluster Beiknospengruppe
bud cutting/eye cutting/single eye cutting/leaf bud cutting Augensteckling
bud dormancy Knospenruhe
bud gap Knospenlücke
bud grafting/budding Okulieren, Okulation, Augenveredlung, Äugeln
bud primordium Knospenanlage
budding *adj/adv* knospend
budding *n* Knospung
budding/bud grafting Okulieren, Okulation, Augenveredlung, Äugeln

- **chip budding** Chipveredelung, Chipveredelung, Span-Okulation
- **flute budding/ring budding** Ring-Okulation, Ringveredlung
- **patch budding/plate budding** Platten-Okulation
- **ring budding/annular budding** Ring-Okulation
- **shield budding** Schild-Okulation (Augenschild/Schildchen)
- **T-budding (shield budding)** Okulieren mit T-Schnitt

budding/frustulation *zool* **(polyps)** Knospung, Frustulation
budding/sprouting knospend, sprießend; Knospung, Sprossung
budding potential/budding rate Ausschlagvermögen
buddleja family/Buddlejaceae Sommerfliedergewächse
buffer *n* Puffer
buffer *vb* puffern
buffer well Pufferwanne
buffering Pufferung
buffering capacity Pufferkapazität
bufonin Bufonin
bufotenine Bufotenin
bufotoxin Bufotoxin
bugs/hemipterans/Rhynchota/ Hemiptera (Heteroptera & Homoptera) Schnabelkerfe, Halbflügler

"bugs" *populär für:* "Insekten"
building block Baustein, Bauelement
bulb Blumenzwiebel, Zwiebelknolle
bulb "plate" with short internodes Zwiebelkuchen
bulb vegetable Zwiebelgemüse
bulb-shaped/bulbous/tuberous knollenförmig, zwiebelförmig
bulbil/brood bud Bulbille, Brutknöllchen, Brutknospe, Brutspross; Zehe, Brutzwiebel
- **axillary bulbil (aboveground)** Achsenbulbille

bulblet Brutkörper, unterirdischer Zwiebelbrutkörper, Tochterzwiebel
bulbotuber/corm/"solid bulb" (swollen shoot base) unterirdische Hypokotylknolle, "Knollenzwiebel"
bulbourethral gland/Cowper's gland Bulbourethraldrüse, Cowpersche Drüse
bulbous/bulbose/bulb-shaped/ tuberous zwiebelförmig, knollenförmig, knollig
bulbous perennial herb Zwiebelstaude
bulge/collar/protuberant seam Wulst, Kragen
bulk flow/mass flow (water) Massenströmung
bulking sludge Blähschlamm
bull (adult male mammal: cattle/ elephant/whale/seal) Bulle (erwachsenes männliches Tier)
bullate/bubblelike/blistered blasig, bläschenartig
bulliform/bubble-shaped/ bubblelike bläschenförmig
bulliform cell/motor cell Motorzelle, motorische Zelle, Gelenkzelle (im Schwellkörper des Blattes)
bunch/cluster/tuft Büschel
bundle/fascicle Bündel, Faszikulus
- **closed bundle (vascular bundle)** *bot* geschlossenes Leitbündel
- **open bundle (vascular bundle)** *bot* offenes Leitbündel

bundle of His/atrioventricular bundle (heart) Hissches Bündel (tert. Autonomiezentrum)
bundle sheath Bündelscheide, Leitbündelscheide
bundle-sheath extension erweiterte Bündelscheide
bundled/fasciculate gebündelt
bunny Häschen
bunodont (with low crowns and cusps) bunodont, stumpfhöckrig, rundhöckrig

45

by-product

buoyancy Auftrieb
buoyant schwebend
buoyant density Schwimmdichte, Schwebedichte
bur/burr/burry fruit Klette, Klettfrucht, Klettenfrucht
bur-reed family/Sparganiaceae Igelkolbengewächse
burden Last (auch: Ausmaß eines Parasitenbefalls)
• **viral burden** Virenlast
buret/burette Bürette
Burgess shale Burgess-Schiefer
burl/lignotuber/wood knot (woody outgrowth with wavy grain) Maserknolle, Wurzelhalsknolle (Auswuchs an bestimmten Bäumen)
burmannia family/Burmanniaceae Burmanniagewächse
burn/combust/incinerate verbrennen
burn *n med* Verbrennung
burr (base of antler) Geweihbasis, Hornbasis
burr cell/echinocyte/crenocyte Stechapfelform, Echinozyt (>Erythrozyt)
burr knot *bot/hort* Wurzelfeld
burrow/earth hole/cave Bau, Loch, Erdloch, Höhle, Erdhöhle (auch: Fraßgang)
burrow *vb* einen Gang/Höhle graben
bursa copulatrix/genital pouch Begattungstasche
bursa of Fabricius Bursa Fabricii
Burseraceae/incense tree family/ torchwood family/frankincense tree family Balsambaumgewächse
burst *vb general* aufplatzen, zum Platzen bringen
burst *n neuro* Burst, Salve, Aktivitätsschub
burst *n vir* Wurf
burst period Zeitpunkt der Virusfreisetzung
burst size *vir* Wurfgröße (Anzahl freigesetzter Viren)
burst-forming unit (BFU) *not translated*: im Knochenmark gebildete Vorläuferzelle bzw. Stammzelle (pre-CFU)
burster neuron Bursterneuron
bush Busch
bush fruit/bush berries (*Ribes*: currents/gooseberries, etc.) Strauchbeeren, Strauchbeerenobst

bushes/shrubbery/thicket/ underbrush (in forest) Gebüsch
bushy/shrubby/fruticose buschig
Butomaceae/flowering rush family Wasserlieschgewächse, Schwanenblumengewächse
butt *n zool* Stoß (mit Hörnern)
butt *vb zool* stoßen (mit Hörnern)
butt (base of plant stem from which roots arise) *bot* Stammbasis
buttercup family/crowfoot family/ Ranunculaceae Hahnenfußgewächse
buttercup tree family (silk-cotton tree family)/cochlospermum family/Cochlospermaceae Nierensamengewächse
butterfly pollination/psychophily Schmetterlingsbestäubung
butterfly-pollinated flower/ psychophile Schmetterlingsblume
butterwort family/bladderwort family/Lentibulariaceae Wasserschlauchgewächse
butting stoßen (mit dem Kopf)
buttocks/posterior/behind/rump Gesäß
buttocks display *ethol* Gesäßweisen
button/knob/key Taste
button *fung* **(immature stage of mushroom)** junger Pilz
button gall/spangle gall (oak) große Linsengalle, Bechergalle
buttress (supportive ridge at base of tree trunk) Wurzelanlauf, Stammanlauf
• **leaf buttress** Blatthöcker, frühe Blattanlage
• **plank buttress/buttress root (*esp.* tropical trees)** Brettwurzel
buttress root (*esp.* tropical trees) Brettwurzel
buttress zone/spur-and-groove zone (reef) Grat-Rinnen-System
butyric acid/butanoic acid (butyrate) Buttersäure, Butansäure (Butyrat)
Buxaceae/box family Buchsbaumgewächse
buzzing run (bees) Schwirrlauf
buzzing sound (bees during waggle dance) Schwirrgeräusch
by-product/residual product Nebenprodukt

C form (of DNA) C-Form, C-Konformation (der DNA)
C-banding (centromere-banding) C-Banding, Centromer-Banding
c-onc (cellular oncogene) c-*onc* (zelluläres Onkogen)
C₀t-analysis/value (*pronounce:* cot; product of total concentration of DNA at time 0 and hybridization time t) C₀t-Analyse/Wert (*sprich:* kott; Produkt aus DNA-Gesamtkonzentration zur Zeit 0 und Hybridisierungszeit t)
CAAT box (a component of the nucleotide sequence that makes up the eukaryotic promoter) CAAT-Box (Bestandteil der Nucleotidsequenz des eukaryotischen Promotors)
cabbage family/mustard family/ Cruciferae/Brassicaceae Kreuzblütler, Kreuzblütlergewächse
cable theory Kabeltheorie
cacao family/cocoa family/ Sterculiaceae Sterkuliengewächse, Kakaogewächse
cackle (geese) schnattern
cactus family/Cactaceae Kaktusgewächse, Kakteen
cadaver/carcass/corpse Kadaver, Leiche
cadaverine Cadaverin
CADD (computer-aided drug design) computerunterstützte Planung von Wirkstoffen
caddis flies/Trichoptera Haarflügler, Köcherfliegen
cadence (rhythm of horse beat) Kadenz
caduceus *med* Merkurstab
caducous (falling off prematurely) frühzeitig abfallend/absterbend
caesalpinia family/Caesalpiniaceae Caesalpinogewächse, Johannisbrotgewächse
caespitose/cespitose (growing densely in tufts) rasig, rasenartig, grasbüschelartig
caffeic acid Kaffeesäure
caffeine/theine (1,3,7-trimethylxanthine) Koffein, Thein
cage Käfig
Cainozoic Era/Caenozoic Era/ Cenozoic Era/Cenozoic/Neozoic Era Känozoikum, Kaenozoikum, Erdneuzeit, Neozoikum (erdgeschichtliches Zeitalter)
Cala-Azar/kala azar (*Leishmania donovani*) Cala-Azar, Kala-Azar, schwarzes Fieber, viszerale Leishmaniasis

calamistrum Calamistrum, Kräuselkamm
calamites/calamite family/giant horsetail family/Calamitaceae Calamiten, Schachtelhalmbäume
calcaneal tendon/Achilles' tendon/ tendon of the heel Achillessehne
calcaneal tuber/tuberosity of calcaneus/tuber calcanei Fersenbeinhöcker
calcaneus/calcaneum/heelbone Fersenbein
calcaneus/heel Ferse
calcar/spur (bony/spur-like process) *zool* Sporn (Knochen-/ Knorpelspange)
calcarate/spurred gespornt
calcareous kalkig, kalkartig, kalkhaltig
calcareous alga Kalkalge
calcareous corpuscle/calcareous body Kalkkörper
calcareous shell Kalkschale
calcareous sponges/Calcarea Kalkschwämme
calcariform/spur-shaped sporenförmig
calcicole/calciphile calcicol, kalzikol, kalkhold, kalkliebend
calciferol/ergocalciferol Calciferol, Ergocalciferol (Vit. D₂)
calciferous gland Kalkdrüse
calcification Calcifikation, Kalzifizierung, Verkalkung, Kalkeinlagerung
calcified verkalkt
calcified cartilage verkalkter Knorpel
calcifuge/basifuge/calciphobe calcifug, kalzifug, kalkfliehend, kalkmeidend; Kalkflieher
calcify verkalken
calciphile/calcicole kalkliebend, kalziphil, kalzikol, kalkhold
calcitonin Calcitonin, Kalzitonin
calcium Kalzium, Calcium
caldarium/heated greenhouse/ hot-house Caldarium, Warmhaus
calf *vb* kalben
calf *n* Kalb, Jungtier
calf (of the leg) Wade
calibrate kalibrieren
calibration Kalibrierung
caliche/lime pan Caliche, Kalkanreicherungshorizont
call/note/call note *n* Ruf
  • **attracting call** *orn* Lockruf
  • **mating call/courtship call** Paarungsruf, Werberuf
  • **mobbing call** *orn* Sammelruf
call *vb* rufen

canker

calla family/arum family/aroid family/Araceae Aronstabgewächse
Callitrichaceae/water starwort family/starwort family Wassersterngewächse
callosity Kallosität, Schwiele
callosum/corpus callosum Balken
callus *bot* **(wound)** Kallus, Wundholz, Wundcallus
callus *zool* Kallus, Hornhaut (verhornte Haut)
callus culture Callus-Kultur, Kallus-Kultur
calm *n meteo* Windstille
calmodulin Calmodulin
caloric value Brennwert
calorie Kalorie
calorimetry Kalorimetrie
calotte/polar cap (*Dicyemida*) Kalotte, Polarzellen-Kappe
calotte cell/polar cell (*Dicyemida*) Polzelle, Polarzelle
caltrop family/creosote bush family/Zygophyllaceae Jochblattgewächse
calvarium/calvaria/dome of skull Schädeldach, Kalotte
Calvin cycle Calvin-Zyklus, Calvin-Cyclus
calving *n* Kalben
Calycanthaceae/strawberry-shrub family/spicebush family Gewürzsträucher
calycera family/Calyceraceae Kelchhorngewächse
calyciform/cup-shaped kelchförmig
calyptra Calyptra, Kalyptra, Haube; Mooshaube
calyx (sepals) Kelch, Blütenkelch (Sepalen)
cambium Cambium, Kambium
 • **cork cambium/phellogen** Korkcambium, Phellogen
 • **fascicular cambium** Faszikularcambium
 • **interfascicular cambium** Zwischenbündelcambium
 • **nonstoried cambium/nonstratified cambium** nichtetagiertes Cambium
 • **storied/stratified cambium** etagiertes Cambium, Stockwerk-Cambium
 • **wound cambium** Wundcambium
Cambrian Period/Cambrian (geological time) Cambrium, Kambrium
camellia family/tea family/Theaceae Teegewächse, Kamel(l)iengewächse, Teestrauchgewächse

camellia gall Weidenrose (Galle)
cameral fluid (nautilus) Kammerflüssigkeit
camouflage *n* Tarnung
camouflage *vb* tarnen
campaniform/campanular/campanulate/bell-shaped glockenförmig, glockig
campaniform sensilla Sinneskuppel
Campanulaceae/bluebell family/bellflower family Glockenblumengewächse
campylokinesis Krümmungsbewegung
campylotropous/bent (ovule) kampylotrop, campylotrop, gekrümmt, krummläufig (Samenanlage)
canalized character kanalisiertes Merkmal
cancellous bone/spongy bone spongiöser Knochen
cancer (malignant neoplasm) Krebs
cancer therapy Krebstherapie
cancerogenic/cancer-causing kanzerogen, canzerogen, krebserzeugend
cancerous/malignant kanzerös, krebsartig, krebsbefallen, Krebs betreffend; bösartig
cancroid cancroid, krabbenartig
candelabra-shaped kandelaberförmig
candidate gene Kandidatengen
cane (thin shoot) dünner Schaft
cane (woody shoot of brambles/shrubs) Rute, Beerenrute
cane/stick (hollow/pithy reed or sugarcane) Rohr (Schilfrohr/Zuckerrohr), Rohrstock
cane sugar Rohrzucker
canella family/white cinnamon family/wild cinnamon family/Canellaceae Kaneelgewächse
canescent *bot* **(hoary/greyish-white/fine/short white hair)** fein grau-weißlich behaart
canine *adj/adv* **(relating to dogs)** Hunde betreffend, Hunde…, Hunds…
canine *adj/adv* **(relating to cuspid/canine tooth)** Eckzahn/Eckzähne betreffend, Eckzahn…
canine/dogs and relatives *n* Hund (Hundeverwandte)
canine/canine tooth/dog tooth/cuspid (tooth with one point) Eckzahn
canine ascarid (*Toxocara canis*) Hundespulwurm
canker *bot* Baumkrebs

cankerworm

cankerworm/inchworm/measuring worm Spannerraupe
canna family/Queensland arrowroot family/Cannaceae Blumenrohrgewächse
Cannabaceae/hemp family Hanfgewächse
cannibalism Kannibalismus
cannibalistic kannibalistisch
cannon bone (from hock to fetlock) Kanonenbein, Sprungbein (Hauptmittelfußknochen der Huftiere), *Pferd:* Röhrbein
cannula Kanüle
canopy (cover of foliage in forest) Baumkronenbereich, Baumwipfelzone, Blätterdach, Laubdach (Wald)
canter/Canterbury gallop (slow gallop) Kanter (kurzer/leichter Galopp)
• **disunited canter** Kreuzgalopp
canterelle family/cantarelle family/ chanterelle family/chantarelle family/Cantharellaceae Pfifferlinge
cantharophily/beetle pollination Käferbestäubung
canyon/gorge Schlucht
caoutchouc/rubber/india rubber Kautschuk
cap/pileus *fung* Hut, Pilzhut
cap cell (scolopidium) Kappenzelle
cap structure (modified 5' end of eukaryotic mRNA molecule) Cap-Struktur (modifiziertes 5'-Ende eines eukaryotischen mRNA-Moleküls)
CAP (catabolite activator protein) Katabolitaktivatorprotein
CAP site (attachment point for the catabolite activator protein) CAP-Stelle (Anheftungspunkt des Katabolitaktivatorproteins)
capacitance (C) elektrische Kapazität
capacitive current (I_c) *neuro* kapazitiver Strom
capacitor Kondensator
capacity Kapazität
cape-pondweed family/water hawthorn family/ Aponogetonaceae Wasserährengewächse
caper family/Capparidaceae/ Capparaceae Kaperngewächse
capillary *n* Kapillare, Haargefäß
capillary blotting Diffusions-Blotting
capillary pipet Kapillarpipette
capitate (with knob-like head or tip) kopfig, köpfchenartig

capitate sorelium/capitiform sorelium (lichens) Kopfsoral
capitulum/cephalium/flower head *bot* Capitulum, Cephalium, Blütenköpfchen, Korb, Körbchen
capitulum/rounded articular eminence/rounded articular extremity *zool* Capitulum, Gelenkkopf
capnophilic kapnophil, kohlendioxidliebend
capon Kapaun
capping *gen* Capping (Anlagerung der Cap-Struktur)
capreolate Ranken..., rankentragend, mit Ranken versehen
capreolus/shoot-tendril Ranke (speziell: Sprossranke)
capric acid/decanoic acid (caprate/ decanoate) Caprinsäure, Decansäure (Caprinat/Decanat)
Caprifoliaceae/honeysuckle family Geißblattgewächse
caprine Ziegen betreffend, ziegenähnlich, Ziegen...
caprinized vaccine durch Ziegen erzeugter Impfstoff
caproic acid/capronic acid/hexanoic acid (caproate/hexanoate) Capronsäure, Hexansäure (Capronat/Hexanat)
caprylic acid/octanoic acid (caprylate/octanoate) Caprylsäure, Octansäure (Caprylat/Octanat)
capsid (viral shell) Capsid, Kapsid
capsomere (virion: morphological unit) Capsomer, Kapsomer
capsule Kapsel
• **ballistic capsule** Schleuderkapsel
• **circumscissile capsule/lid capsule/pyxis/pyxidium** Deckelkapsel
• **dorsicidal capsule** dorsizide/dorsicide Spaltkapsel
• **explosive capsule** Explodierkapsel, Explosionskapsel (Springkapsel)
• **lid capsule/circumscissile capsule/pyxis/pyxidium** Deckelkapsel
• **loculicidal capsule** lokulizide/loculicide Spaltkapsel, fachspaltige Kapsel
• **longitudinally dehiscent capsule** Spaltkapsel
• **poricidal/porose capsule** Lochkapsel, Löcherkapsel, Porenkapsel, porizide Kapsel
• **septicidal capsule** septizide/septicide Spaltkapsel, wandspaltige Kapsel

carpophagous

captacula (Scaphopoda) Captacula, Fangfäden(büschel)
captive gefangen
captivity Gefangenschaft
capturing zone (web) Fangzone, Fangnetz (innerhalb Radnetz)
carapace (insects/turtles) Carapax
carbohydrate Kohlenhydrat
carbon Kohlenstoff
carbon bond Kohlenstoffbindung
carbon compound Kohlenstoffverbindung
carbon source Kohlenstoffquelle
carbonic acid (carbonate) Kohlensäure (Karbonat/Carbonat)
carboniferous swamp forests Steinkohlenwälder
Carboniferous Period/ Carboniferous/"Carbon Age" (geological time) Karbon, Steinkohlenzeit
carbonization/coalification *paleo* Karbonisation, Inkohlung
carboxylic acids (carbonates) Carbonsäuren, Karbonsäuren (Carbonate/Karbonate)
carboxysome/polyhedral body Carboxysom
carcass/corpse/cadaver Leiche, Kadaver
carcinoembryonic antigen carcinoembryonales Antigen
carcinogen Karzinogen
carcinogenic karzinogen, carcinogen, krebserregend, krebserzeugend
carcinoma Karzinom
cardboard/paperboard/fiberboard Karton
cardia (opening between esophagus and stomach) Magenmund
cardia/cardiac stomach/gizzard (insects) Kaumagen, Cardia
cardiac (*see also:* coronary) kardial, das Herz betreffend, Herz...
cardiac activity Herzaktivität
cardiac arrest Herzstillstand
cardiac jelly Herzgallerte
cardiac muscle Herzmuskel
cardiac output Herzausstoß, Herzminutenvolumen
cardiac output per minute Herzminutenvolumen (HMV)
cardiac skeleton/skeleton of the heart Herzskelett
cardiac stomach/cardia/gizzard (insects) Kaumagen, Cardia
cardiac valve/coronary valve/heart valve Herzklappe
cardinal temperature Vorzugstemperatur

cardinal tooth (bivalves) Hauptzahn
cardinal vein Kardinalvene
cardo (basal segment of maxilla) Cardo (*pl* Cardines), Angelstück
care/provide for *vb* pflegen, versorgen
care/provisioning *n* Pflege, Versorgung, Fürsorge
caretaker/nurse (e.g. asexual individual in social insects) Amme
Caricaceae/papaya family Melonenbaumgewächse
caridoid caridoid, garnelenartig
carinate/keeled/having a keel gekielt
carinate birds Carinaten, Kielbrustvögel
carnassial Fleischfresser..., Reiß...
carnassial tooth/fang (carnivores) Reißzahn, Fangzahn, Fang
carnation family/pink family/ Caryophyllaceae Nelkengewächse
carnitine (vitamin B$_T$) Carnitin (Vitamin T)
carnivore/flesh-eater/meat-eater Carnivor, Karnivor, Fleischfresser
carnivores/Carnivora Carnivore, Karnivore, Raubtiere
• **terrestrial carnivores/Fissipedia** Landraubtiere
carnivorous/flesh-eating/meat- eating carnivor, karnivor, fleischfressend
carnose fleischig
carob gum/carob seed gum/locust gum/locust bean gum Karobgummi, Johannisbrotkernmehl
carotid/carotid artery Halsschlagader, Kopfschlagader
carotid body/Glomus caroticum Carotidenkörper
carotin/carotene (vitamin A precursor) Carotin, Caroten, Karotin (Vitamin A Vorläufer)
carpal bones Handwurzelknochen
carpel Carpell, Karpel, Fruchtblatt
carpellate/pistillate/female weiblich
carpellate flower/pistillate flower Stempelblüte, weibliche Blüte
carpellode *bot* Karpellodium
carpetweed family/fig marigold family/mesembryanthemum family/Aizoaceae Mittagsblumengewächse, Eiskrautgewächse
carpogonium (algae) Karpogon, Carpogon
carpophagous/feeding on fruit/ frugivorous karpophag, fruchtfressend, frugivor, fruktivor

C carpophore 50

carpophore/fruit bearer (receptacle)
bot Karpophor, Fruchtträger,
Fruchthalter
carpopodite Carpopodit
carposis Karpose
carpospore (algae) Karpospore,
Carpospore, Carpogonidie
**carps & characins & minnows &
suckers & loaches/Cypriniformes**
Karpfenfische, Karpfenartige
carpus Carpus, Karpus, Handwurzel
carr (fen woodland) Bruchmoor,
Bruchwald, Übergangs-Waldmoor
**carrageenan/carrageenin (*Irish
moss* extract)** Carrageenan
carrier Träger; Trägersubstanz;
chromat Träger
● **gene carrier** Genträger
carrier electrophoresis
Trägerelektrophorese
carrier molecule Trägermolekül
carrier protein Trägerprotein,
Schlepperprotein
carrion/decaying carcass Aas
carrion feeder/scavenger
Aasfresser, Unratfresser
carrion flower Aasblume
**carrot family/parsley family/
umbellifer family/Apiaceae/
Umbelliferae** Umbelliferen,
Doldenblütler, Doldengewächse
carrying capacity Kapazitätsgrenze,
Umweltkapazität,
Belastungsfähigkeit, Grenze der
ökologischen Belastbarkeit,
Tragfähigkeit (Ökosystem)
cartilage Knorpel
● **articular cartilage**
Gelenksknorpel
● **arytenoid cartilage/cartilago
arytaenoidea** Aryknorpel,
Stellknorpel
● **calcified cartilage** verkalkter
Knorpel
● **costal cartilage** Rippenknorpel
● **elastic cartilage** elastischer
Knorpel
● **epiglottic cartilage**
Kehldeckelknorpel, Schließknorpel
● **fibrous cartilage/fibrocartilage**
Faserknorpel
● **hyaline cartilage** hyaliner
Knorpel
● **Meckel's cartilage** Meckelscher
Knorpel
● **thyroid cartilage/cartilago
thyreoidea** Schildknorpel
● **tracheal cartilage/tracheal ring**
Trachealknorpel, Knorpelspange
der Luftröhre, Tracheálring
● **xiphoid cartilage/cartilago
xiphoidea** Schaufelknorpel

cartilage cell/chondrocyte
Knorpelzelle, Chondrozyt
cartilaginous knorpelig
**cartilaginous fishes/
chondrichthians/Chondrichthyes**
Knorpelfische
cartridge/cassette Kassette
caruncle *bot* Caruncula, Karunkula,
Mikropylenwulst, Samenwarze (an
Mikropyle)
caruncle *zool/orn* Fleischlappen,
Fleischauswuchs
**Caryophyllaceae/pink family/
carnation family** Nelkengewächse
caryopsis/grain Caryopse, Karyopse,
"Kernfrucht", Kornfrucht
(Grasfrucht)
cascade Kaskade, Kascade
cascade system (e.g. enzymes)
Kaskadensystem
case/casing/shell Gehäuse, Panzer
case/chamber/valve Kammer, Fach
case *med* Fall
case *zool* (***Trichoptera*: in some
species> tube)** Köcher
casein Casein
caseous käsig, käseartig
cash crop (leicht verkäufliches)
Landbauprodukt, Kultur mit
"garantiertem" Ertrag
**cashew family/sumac family/
Anacardiaceae** Sumachgewächse
casing Gehäuse
Casparian strip Casparischer
Streifen
casque (bill) Schnabelaufsatz
(Nashornvögel)
cassowaries & emus/Casuariiformes
Kasuare & Emus
cast *vb* werfen, gebären;
abwerfen, verlieren
cast *zool* (**insect exuvia)** Exuvie,
abgestoßene Haut
cast *zool* (**worm excrement)** von
Würmern aufgeworfenes
Erdhäufchen (>Kotwurst)
cast/mold *paleo* Abguss
caste Kaste
casting (lugworm: *Arenicola*)
Kotwurst
castor/castoreum (beaver oil)
Bibergeil, Castoreum
**castor gland/preputial gland
(beaver)** Präputialdrüse
castor oil/ricinus oil Kastoröl,
Rizinusöl
castrate/geld kastrieren
castration Kastration
casual host Gelegenheitswirt
**Casuarinaceae/she-oak family/
beefwood family**
Streitkolbengewächse

51   **cauliflory** **C**

cat's-tail/cattail/reedmace
Rohrkolben
catabolic (degradative reactions)
catabol, katabol, katabolisch,
abbauend
catabolism Stoffwechselabbau
catabolite Katabolit,
Stoffwechselabbauprodukt
catabolite activator protein (CAP)
Katabolitaktivatorprotein
catabolite repression Katabolit-
Repression, Katabolitrepression
(Hemmung), katabolische
Repression
catadromous katadrom
**catalepsy/catalepsis/shamming
dead reflex** Katalepsie, *med*
Starrsucht; Totstellreaktion,
Totstellreflex
catalysis Katalyse
catalyst Katalysator
catalytic(al) katalytisch
**catalytical unit/unit of enzyme
activity (*katal*)** katalytische
Einheit, Einheit der Enzymaktivität
(*katal*)
catalyze katalysieren
cataphyll Niederblatt
catapult fruit/catapult capsule
Katapultfrucht, Katapultkapsel
catastrophism Katastrophentheorie
catbrier family/Smilacaceae
Stechwindengewächse
catch tentacle/tentacular arm
Fangarm, Tentakelarm
**catchment area/catchment basin/
drainage area/drainage basin/
watershed** Wassereinzugsgebiet,
Grundwassereinzugsgebiet,
Flusseinzugsgebiet, Sammelbecken
catecholamine Katecholamin
categorization Kategorisierung,
Einstufung
categorize kategorisieren, einstufen,
einteilen, einer Rangstufe zuordnen
category Kategorie, Rangstufe
catenane/concatenate Catenan,
Concatenat
catenation Catenation, Ringbildung
caterpillar Raupe
**caterpillar movement/rectilinear
movement (snakes)**
Raupenbewegung,
Integumentbewegung
catfishes/Siluriformes Welse,
Welsartige
catgut (from sheep intestines)
Katgut, Darmsaite (Nähfaden aus
Schafgedärmen)
cation exchanger
Kationenaustauscher
catkin/ament/amentum Kätzchen

**catkin-mistletoe family/
Eremolepidaceae** Eremolepidaceae
**cattail family/cat's-tail family/
reedmace family/Typhaceae**
Rohrkolbengewächse
cattle Vieh, Rindvieh, Rinder
**cattle grub & botfly infestation
(*Hypoderma bovis* et spp.)**
Dasselbeule
cattle production/cattle breeding
Viehzucht, Rinderzucht
cattle ranching Rinderwirtschaft
cattle shed/barn Viehstall
caudal den Schwanz betreffend,
Schwanz…
caudal furca Schwanzgabel
caudal gland (rectal gland)
Schwanzdrüse (Rektaldrüse)
caudal plate/pygidium (trilobites)
Schwanzschild, Pygidium
caudal shield/pygidium (bugs)
Analschild, Pygidium (Wanzen)
caudal vertebra/coccygial vertebra
Schwanzwirbel
**caudate/tail-pointed (ending with
tail-like appendage)** geschwänzt,
geschwanzt (Blatt: mit
Träufelspitze)
caudate nucleus/nucleus caudatus
Schweifkern
caudex/stalk/stem Strunk, Schaft,
Achse
**caudex/stump/stub/stool/stem
base** Caudex, Stumpf, Strunk
**caudex/trunk of tree (palms/
treeferns)** Stamm (Palmen/
Baumfarne)
caudicle (stalk of pollinium)
Kaudikula, Caudicula, Stielchen
caudofoveates/Caudofoveata
Schildfüßer
caulescence *bot* Kauleszenz,
Cauleszenz
● **concaulescence** *bot*
Konkauleszenz, Concauleszenz
**caulescent (with stem above
ground)** *bot* cauleszent,
kauleszent, stammbildend,
stengeltreibend
● **acaulescent/stemless**
akauleszent, stammlos
● **concaulescent** konkauleszent
(Seitensprosse auf/an Hauptspross)
**caulescent perennial herb/giant
rosette plant/giant leaf-rosette
plant** Schopfrosettenpflanze
caulid/stemlet/stipe (algas/mosses)
Cauloid, Kauloid, Stämmchen
(Algen/Moose)
cauliflorous kauliflor, stammblütig
cauliflory Kauliflorie,
Stammblütigkeit

cauline (arising from the stem) stammbürtig
cauline bundle/axial bundle Stammbündel, Stammleitbündel
causal morphology Entwicklungsmechanik
caustic/corrosive/mordant *chem* ätzend, korrosiv
cauterization *med* Ätzen, Ätzung
cauterize *med* ätzen, ausbrennen, kauterisieren
caution! (careful!) Vorsicht!
caution/cautiousness/care/ carefulness/precaution Vorsicht
cautious/careful vorsichtig
cave/crypt/cavity Höhle, Höhlung
cave-dwelling/cavernicolous (troglophilic) höhlenbewohnend
cavernous bodies/erectile tissue Corpora cavernosa, Schwellkörper
cavitate/break (water column) reißen (Wassersäule)
cavitation (xylem/phloem; rupture of water column) Kavitation
cavity/chamber/ventricle Höhle, Kammer, Ventrikel (kleine Körperhöhle)
cavity/lumen Hohlraum, Höhlung, Lumen
CBA-paper (cyanogen bromide activated paper) CBA-Papier
CD (cluster of differentiation) CD (Differenzierungscluster)
CDE (centromere DNA sequence elements) CDE-Elemente (DNA-Sequenzelemente am Centromer)
cDNA (complementary DNA) cDNA (komplementäre DNA)
● **cDNA library** cDNA-Bibliothek
CDR (complementarity determining region) *gen* CDR (komplementaritätsbestimmende Region)
cecal blind endend
cecidiology Cecidologie, Gallenkunde, Lehre von den Gallen
cecidium/gall Cecidium, Galle, Pflanzengalle
cecidology Cecidologie, Gallenkunde, Lehre von den Gallen
cecidozoa Cecidozoen, gallerzeugende Tiere
Celastraceae/staff-tree family/ spindle-tree family/bittersweet family Spindelbaumgewächse, Baumwürgergewächse
celiac/coeliac (pertaining to the abdominal cavity) die Bauchhöhle betreffend, Bauchhöhlen...
celiac plexus/solar plexus Sonnengeflecht

cell *cyt* Zelle
● **body cell/somatic cell** Körperzelle, somatische Zelle
● **enucleate cell** kernlose Zelle
● **feeder cell** Feeder-Zelle
● **non-permissive cell** nicht-permissive Zelle
● **permanent cell** Dauerzelle
● **permissive cell** permissive Zelle
● **transformed cell** transformierte Zelle
● **vegetative cell** vegetative Zelle
cell (mericarpic nutlet/segment of loment) *bot* Klause
cell adhesion molecule Zelladhäsionsmolekül
cell aggregate Zellverband
cell biology/cytology Zellbiologie, Zellenlehre, Cytologie, Zytologie
cell body/soma Zellkörper, Soma
cell colony Zellkolonie
cell constancy/eutely Zellkonstanz, Eutelie
cell content Zellinhalt
cell count (number of cells) Zellzahl
cell count/germ count Keimzahl (Anzahl von Mikroorganismen)
cell culture Zellkultur
cell cycle Zellzyklus, Zellcyclus
cell death Zelltod
cell density Zelldichte
cell division/cytokinesis Zellteilung, Cytokinese, Zytokinese
cell envelope Zellhülle
cell extract Zellextrakt
cell fate Zellschicksal
cell fractionation Zellaufschluss, Zellfraktionierung
cell fusion Zellfusion, Zellverschmelzung
cell homogenization Zellhomogenisation, Zellhomogenisierung, Zellaufschluss
cell hybridization Zellhybridisierung
cell inclusion/cellular inclusion Zelleinschluss (Inklusion)
cell junction Zellkontakt
cell line Zellinie
● **cloned cell line** klonierte Zellinie
● **continuous cell line** kontinuierliche Zellinie
● **established cell line** etablierte Zellinie
cell lineage/cell line/celline Zellinie
cell lysis Zellaufschluss (Öffnen der Zellmembran)
cell membrane (outer)/unit membrane/plasmalemma/ biological membrane Zellmembran, Plasmamembran, Plasmalemma, biologische Membran

53   **centromere** **C**

cell process Zellfortsatz
cell proliferation Zellproliferation
cell sap Zellsaft
cell separation Zelltrennung,
Zellseparation
cell sorter Zellsorter, Zellsortierer,
Zellsortiergerät (Zellfraktionator)
cell sorting Zellsortierung
cell surface Zelloberfläche
cell surface marker
Zelloberflächenmarker
cell theory Zelltheorie
cell transformation
Zelltransformation
cell wall Zellwand
cell-anus/cytopyge/cytoproct
Zellafter, Zytopyge, Cytopyge,
Zytoproct, Cytoproct
cell-attached patch zellulär-
befestigter *Patch*/Membranflicken
cell-bound zellgebunden
cell-free zellfrei
cell-free extract zellfreier Extrakt
**cell-free protein synthesizing
system** zellfreies
Proteinsynthesesystem
cell-free system zellfreies System
cell-mediated zellvermittelt
cell-mediated immune response
zellvermittelte Immunantwort
cell-mouth/cytostome/cytostoma
Zellmund, Zellmundöffnung,
Zytostom, Cytostom
cell-wall defective *adv/adj* mit
defekter Zellwand, mit schadhafter
Zellwand
cellobiose Zellobiose, Cellobiose
cellular zellig, zellulär
cellular fraction Zellfraktion
cellular metabolism
Zellstoffwechsel
cellular respiration Zellatmung
**cellular slime molds/
Acrasiomyycetes/
Dictyosteliomycetes
(Myxomycota)** Acrasiomyceten,
zelluläre Schleimpilze
cellularity Zellularität, zellulärer
Aufbau
  • **multicellularity** Vielzelligkeit,
  vielzellige Organisationsstufe
  • **unicellularity** Einzelligkeit,
  einzellige Organisationsstufe
cellulose Zellulose, Cellulose
celom/coelom Cölom
cement gland/adhesive gland
Zementdrüse, Kittdrüse, Klebdrüse
**Cenozoic/Cenozoic Era/Caenozoic
Era/Cainozoic Era/Neozoic Era**
Känozoikum, Kaenozoikum,
Erdneuzeit, Neozoikum
(erdgeschichtliches Zeitalter)

census Zensus, Erhebung
(*auch:* Volkszählung)
center *vb* zentrieren
centile/percentile Zentil, Perzentil,
Prozentil
**centimorgan (unit of genetic
recombination)** Centimorgan
(Einheit für genetische
Rekombination)
centipedes/chilopodians
Hundertfüßer, Chilopoden
central axis Mittelachse
central body (insect brain)
Zentralkörper
central cylinder/stele (of stem)
Zentralzylinder des Sprosses,
Sprossstele; (of root)
Zentralzylinder der Wurzel,
Wurzelstele
central dogma zentrales Dogma
**central fibril/axial filament/
axoneme** Achsenfaden, Axonema
**central granule/axoplast/
centroplast** Zentralkorn,
Centroplast
central leader *bot/hort*
Mittelleittrieb
central nervous system
Zentralnervensystem
central placentation
Zentralplazentation
centrifugal zentrifugal
centrifugal force Zentrifugalkraft,
Fliehkraft
centrifugation Zentrifugation
  • **analytical centrifugation**
  analytische Zentrifugation
  • **density gradient centrifugation**
  Dichtegradientenzentrifugation
  • **differential centrifugation**
  Differentialzentrifugation,
  differentielle Zentrifugation
  • **isopycnic centrifugation**
  isopycnische Zentrifugation,
  isopyknische Zentrifugation
  • **preparative centrifugation**
  präparative Zentrifugation
  • **ultracentrifugation**
  Ultrazentrifugation
  • **zonal centrifugation**
  Zonenzentrifugation
centrifuge *n* Zentrifuge
centrifuge *vb* zentrifugieren
centrifuge cell Zentrifugenzelle
centrifuge rotor/centrifuge head
Zentrifugenrotor
centrifuge tube Zentrifugenröhrchen
centriole Centriol, Zentriol
centripetal zentripetal
centro-acinar cell centroacinäre
Zelle
centromere Centromer, Zentromer

centromere DNA sequence elements   54

centromere DNA sequence elements (CDE) CDE-Elemente (DNA-Sequenzelemente am Centromer)
centroplast/central granule/axoplast Centroplast, Zentralkorn
centrum (of vertebra) Wirbelkörper, Centrum
century plant family/agava family/Agavaceae Agavengewächse
cepaceous zwiebelartig (Geruch/Geschmack)
cephalic den Kopf betreffend, Kopf...
cephalic flexure/cranial flexure *embr* Scheitelbeuge
cephalic gland/frontal gland Kopfdrüse, Stirndrüse, Frontaldrüse
cephalic vesicles *embr* Hirnbläschen
cephalization/head development Cephalisation, Kopfbildung
cephalopods Cephalopoden, Kopffüßer
cephalotaxus family/plum yew family/Cephalotaxaceae Kopfeibengewächse
cephalothorax (fused head and thorax) Cephalothorax, Kopfbruststück
ceraceous/waxy/wax-like wachsartig
cerago/bee bread Cerago, Bienenbrot
ceramide Ceramid
cerata (nudibranchs) Cerata, Rückenanhänge
ceratitic suture line ceratitische Lobenlinie/Nahtlinie
Ceratophyllaceae/hornwort family Hornblattgewächse
cercaria Cercarie, Zerkarie, Schwanzlarve
Cercidiphyllaceae/katsura-tree family Katsuragewächse
cercus/cercopod (clasping organs) Cercus, Aftergriffel, Afterraife, Afterfühler, Schwanzborsten (Schwanzanhang)
cere (on bill of birds) Cera, Wachshaut (am Schnabel)
cereal(s) (grain) Getreide, Getreidepflanze(n); (foodstuff made from grain) Getreidekost (Frühstücksgetreidekost); (flakes) Getreideflocken
cerebellum/epencephalon Kleinhirn, Hinterhirn, Cerebellum
cerebral cerebral, zerebral, das Hirn betreffend, Hirn...
cerebral commissure Cerebralkommissur, Zerebralkommissur
cerebral cortex Großhirnrinde
• **auditory cortex** Hörrinde
• **neocortex** Neocortex, Neokortex
• **optic cortex** Sehrinde
cerebral ganglion Cerebralganglion, Zerebralganglion
cerebral hemisphere Hirnhemisphäre, Großhirnhälfte
cerebral membrane/cerebral meninx (*pl* meninga) Gehirnhaut, Hirnhaut, Meninx (*pl* Meninga)
cerebral meninx (*pl* meninga/meninges)/cerebral membrane Hirnhaut, Gehirnhaut, Meninx (*pl* Meninga)
cerebral peduncle/crura cerebri Hirnschenkel, Hirnstiel
cerebroside Cerebrosid
cerebrospinal fluid (CSF) Gehirn-Rückenmark-Flüssigkeit, Liquor cerebrospinalis
cerebrum/endbrain/telencephalon Großhirn, Endhirn, Telencephalon
cerianthids/tube anemones/Ceriantharia Zylinderrosen
cerotic acid/hexacosanoic acid Cerotinsäure, Hexacosansäure
ceruminous gland/wax gland Wachsdrüse; Ohrenschmalzdrüse
cervical cervikal, zervikal, Hals/Nacken/Gebärmutterhals betreffend
cervical flexure *embr* Nackenbeuge
cervical plexus Halsgeflecht
cervical sclerite Halsschild
cervical spine Halswirbelsäule (HWS)
cervical vertebra Halswirbel (HW), Cervikalwirbel
cervix Cervix, Zervix, Gebärmutterhals
cesarean section/cesarean *med/vet* Kaiserschnitt
cesium chloride gradient Cäsiumchloridgradient
cespitose/caespitose/caespitulose (growing densely in tufts) grasbüschelartig, rasig, rasenartig
cesspool/cesspit Klärgrube
cetaceans (whales & porpoises & dolphins)/Cetacea Wale und Delphine
CFCs (chlorofluorocarbons/chlorofluorinated hydrocarbons) FCKW (Fluorchlorkohlenwasserstoffe)
CGH (comparative genome hybridization) CGH (vergleichende Genomhybridisierung)
chaetiferous/chaetiphorous mit Borsten versehen, Borsten...

chelate

chaetoblast (annelids) Chaetoblast, Borstenbildungszelle
chaetotaxy Chaetotaxie
chaff/bracts (small dry scales) Spreu, Kaff
chaffy/paleaceous spreuartig, voller Spreu
Chagas disease (*Trypanosoma cruzi*) Chagas Krankheit
chain (branched/unbranched) Kette (verzweigte/unverzweigte)
 • **heavy chain (H-chain)** schwere Kette (H-Kette)
 • **light chain (L-chain)** leichte Kette (L-Kette)
chain form/open-chain form Kettenform
chain formula/open-chain formula Kettenformel
chain length Kettenlänge
chain of cells/filament Zellfaden, Filament
chain reaction Kettenreaktion
chain-terminating technique Kettenabbruchverfahren
chair conformation (cycloalkanes) *chem* Sesselform
chalaza/treadle Chalaze, Hagelschnur
chamaephyte Chamaephyt
 • **woody chamaephyte/dwarf-shrub** holziger Chamaephyt, Zwergstrauch
chamber/valve/case Kammer, Fach
chamber *electrophor* Kammer
chambered/valvate gekammert, gefächert, fächerig
chance Zufall
change/modification/variation Veränderung, Variation
channel (*see also:* membrane channel) Kanal
 • **ligand-gated channel** ligandenregulierter/ligandengesteuerter Kanal
 • **mechanically gated channel** mechanisch gesteuerter Kanal
 • **resting channel/leakage channel** Ruhemembrankanal, Leckkanal
 • **voltage-sensitive channel/voltage-gated channel** spannungsregulierter/spannungsgesteuerter Kanal
channel current Kanalstrom
channel gate Kanaltor
channel protein Kanalprotein, Tunnelprotein
chanterelle family/chantarelle family/canterelle family/cantarelle family/Cantharellaceae Pfifferlinge
chaos theory Chaostheorie

chaotropic agent chaotrope Substanz
chaotropic series chaotrope Reihe
chaperone protein/chaperone/molecular chaperone Chaperon, molekulares Chaperon, Begleitprotein
chaperonin Chaperonin
characins (tetras & piranhas)/Characiformes Salmler
character/characteristic Charakter, Eigenschaft, Merkmal
 • **acquired c.** erworbene Eigenschaft
 • **canalized c.** kanalisiertes Merkmal
 • **derived c.** abgeleitetes Merkmal
 • **inborn/innate c./genetically determined c.** Erbmerkmal, angeborenes Merkmal
character difference Merkmalsunterschied
character displacement *evol* Konstrastbetonung, Merkmalsverschiebung (Merkmalsdifferenz-Regel)
character divergence Merkmalsdivergenz
character phylogeny Merkmalsphylogenetik
character species Charakterart, Leitart
characteristic/character Eigenschaft, Merkmal; Charakter
 • **acquired characteristic** erworbenes Merkmal
 • **derived characteristic** abgeleitetes Merkmal
characteristic value Kennwert
charcoal Holzkohle
charge/feed beschicken
charge separation Ladungstrennung
chase *n hunt* Hatz, Hetzjagd
chase *vb* jagen, hetzen
chatter/jabber (monkeys/apes) schnattern
cheek/gena Backe, Wange, Gena
cheek pouch (e.g. hamster) Backentasche
 • **cheek gland** Backendrüse
cheek tooth Backenzahn
cheekbone/zygomatic bone/malar bone/os zygomaticum Backenknochen
cheese-skipper (Piophilidae larva) Käsefliegenlarve
cheilostomates/Cheilostomata (bryozoans) Lappenmünder, Lippenmünder
chelate *n chem* Chelat, Komplex
chelate *vb chem* komplexieren

chelate

chelate/claw-bearing mit Scheren versehen
chelating agent/chelator *chem* Chelatbildner, Komplexbildner
chelation/chelate formation *chem* Chelatbildung, Komplexbildung
chelicera/fang/cheliceral fang Chelicere, Chelizere, Kieferfühler, Scherenkiefer, Klaue, Fresszange, Greifzange
cheliceral fang Chelicerenklaue, Scherenfinger (Unguis)
chelicerates/Chelicerata Cheliceraten, Chelizeraten
cheliferous mit Scheren versehen, scherentragend, Scheren...
chelifore (pycnogonids) Chelifore
cheliform/chelate/pincerlike/ clawlike scherenartig, zangenartig
cheliped Scherenfuß
chemical bond (*see also:* bond) chemische Bindung
chemical complexity chemische Komplexität
chemical oxygen demand (COD) chemischer Sauerstoffbedarf (CSB)
chemical warfare chemische Kriegsführung
chemical warfare agent chemischer Kampfstoff
chemical weapons chemische Waffen
chemiosmosis Chemiosmose
chemiosmotic hypothesis/theory chemiosmotische Hypothese/Theorie
chemisorption Chemisorption, chemische Adsorption
chemoaffinity hypothesis Chemoaffinitäts-Hypothese
chemoheterotroph(ic) chemoheterotroph
chemoheterotrophy Chemoheterotrophie
chemolithotroph(ic)/ chemoautotroph(ic) chemolithotroph, chemoautotroph
chemolithotrophy/ chemoautotrophy Chemolithotrophie, Chemoautotrophie
chemomorphosis Chemomorphose (Gestaltentwicklung entsprechend chemischer Umweltreize)
chemoorganotroph(ic) chemoorganotroph
chemoorganotrophy Chemoorganotrophie
chemostat Chemostat
chemosynthesis Chemosynthese
chemotaxis Chemotaxis (*pl* Chemotaxien)
chemotherapy Chemotherapie
Chenopodiaceae/goosefoot family Gänsefußgewächse
cherimoya family/custard apple family/Annonaceae Schuppenapfelgewächse
chest/thorax Oberkörper, Brustkorb, Brustkasten, Thorax
chestnut (horse: above knee/lower hock on medial side of leg) Kastanie
chevron (bone)/hemal arch Sparrknochen, Haemapophyse, ventraler Wirbelbogen, Chevron
chew/masticate kauen, zerkauen
• **chew the cud (regurgitate)** wiederkäuen
chewing/mastication *n* Kauen, Zerkauen
chewing/masticatory *adj* kauend
chewing lice/biting lice/ Mallophaga Haarlinge & Federlinge
chewing-biting (mouthparts) kauend-beißend
Chi form Chi-Form
chi-square test Chi-Quadrat-Test
chiasma (*pl* **chiasmata)** Chiasma (Chiasmata), Überkreuzung
• **optic chiasma/chiasma opticum** Sehnervenkreuzung
chick Küken
chicken Huhn (hen > Henne); Hühnerfleisch
chicken coop Hühnerstall
chicken embryo culture Eikultur (Hühnerei)
chicle (chicle gum) Chicle, Chiclegummi (Kaugummirohstoff)
chief association Hauptassoziation
chief cell (stomach) Hauptzelle
chigger/"red bug"/harvest mite (parasitic red mite larva) Erntemilbenlarve, parasitische Larve von Trombidiidae
childbed fever/puerperal fever (bacterial) Kindbettfieber, Wochenbettfieber, Puerperalfieber
childbirth Kindesgeburt, Niederkunft
childhood Kindheit
chilidium/chilidial plate Chilidium, Verschlussplatte
chilling Abkühlung, Kühlen, Gefrieren
chilling damage/chilling injury Kälteschaden, Kälteschädigung, Erkältung, Unterkühlungsschaden
chimaeras/ratfishes/rabbit fishes/ Holocephali Chimären, Seedrachen, Seekatzen
chimera Chimäre, Pfropfhybride, Zellhybride
chine (backbone/spine) Rückgrat, Kreuz

chine (cattle: a cut with all or part of the backbone) Kamm, Kammstück

Chinese gooseberry family/ actinidia family/Actinidiaceae Strahlengriffelgewächse

chinic acid/kinic acid/quinic acid (quinate) Chinasäure

chinine/quinine Chinin

chinolic acid Chinolsäure

chinoline/quinoline Chinolin

chinone Chinon

chip budding Chipveredelung, Chipveredlung, Span-Okulation

chipboard Spanplatte

chiral chiral

chirality Chiralität, Drehsinn

chiropatagium (bats) Chiropatagium (Flughaut der Fledermäuse)

chiropterochory/bat-dispersal Fledermausausbreitung

chiropterophile/bat-pollinated flower Fledermausblume

chiropterophilous/bat-pollinated fledermausblütig

chiropterophily/bat-pollination/ pollination by bats Fledermausbestäubung

chirp/cheep *orn* piepen, piepsen

chirp/stridulate *entom* zirpen, schrillen, stridulieren

chirping/stridulation Zirpen, Schrillen, Stridulation

chitin Chitin

chitinous chitinös, Chitin…

chitinous shell Chitinschale

chitterlings/chitlins (hog intestines) Gedärme, Gekröse (Schweine)

chlamydospore Chlamydospore (Gemme)

chloragogen cell/chloragogue cell (oligochetes) Chloragozelle

chlorenchyma Chlorenchym, Assimilationsparenchym

chloride cell *ichth* Chloridzelle, Ionocyt

chlorinate chlorieren; chloren

chlorinated hydrocarbon chlorierter Kohlenwasserstoff

chlorination Chlorierung

chlorine Chlor

chlorogenic acid Chlorogensäure

chlorophyll Chlorophyll

chloroplast Chloroplast

chlorosome Chlorosom (Chlorobium-Vesikel)

chocolate agar Schokoladenagar, Kochblutagar

choke *vb* würgen, erwürgen, erdrosseln, ersticken

cholecalciferol (vitamin D₃) Cholecalciferol, Calciol (Vitamin D_3)

cholecystokinin-pancreozymin (CCK-PZ) Cholecystokinin-Pankreozymin (CCK-PZ)

cholesterol Cholesterin, Cholesterol

cholic acid (cholate) Cholsäure (Cholat)

cholinergic cholinerg

chondroblast Chondroblast, knorpelbildende Zelle

chondrocyte Chondrozyt, Knorpelzelle

chondroid tissue chondroides Gewebe, Knorpelgewebe (Parenchymknorpel)

chondrotonal sensilla/scolopidium/ scolophore stiftführende Sensille, Scolopidium

chordal plate/notochordal plate Chordaplatte

chordates Chordaten, Chordatiere, Rückgrattiere

chordotonal chordotonal

chordotonal organ Chordotonalorgan, Saitenorgan

chorioallantoic placenta Chorioallantoisplazenta, Zottenplazenta

chorion (external extraembryonic membrane) Chorion, äußere Eihaut

chorion/eggshell (insect egg) Chorion, Eischale

chorion frondosum Zottenhaut

chorion laeve (nonvillous chorion) Chorion laeve, Zottenglatze

chorionic placenta Chorionplazenta

chorionic villi Chorionzotten

chorionic villus biopsy/chorionic villus sampling/chorion villi biopsy Chorionzotten-Biopsie

choripetalous (having separate petals) choripetal

chorismic acid (chorismate) Chorisminsäure (Chorismat)

choroid/chorioid/chorioidea Aderhaut

chorology/biogeography Chorologie, Arealkunde, Verbreitungslehre

christmas mistletoe family/ Viscaceae Mistelgewächse

chromaffin/chromaffine/ chromaffinic chromaffin

chromatid Chromatid

• **UESCE (unequal exchange of sister chromatids)** UESCE (ungleicher Austausch von Schwesterchromatiden)

chromatid conversion Chromatidenkonversion

chromatin Chromatin

chromatin thread Chromatinfaden

chromatogram Chromatogramm

chromatograph Chromatograph
chromatography Chromatographie, Chromatografie
- **affinity chromatography** Affinitätschromatographie
- **bonded-phase chromatography** Festphasenchromatographie
- **capillary chromatography** Kapillarchromatographie
- **chiral chromatography** enantioselektive Chromatographie
- **circular chromatography** Zirkularchromatographie, Rundfilterchromatographie
- **column chromatography** Säulenchromatographie
- **gas chromatography** Gaschromatographie (GC)
- **gel permeation chromatography/molecular sieving chromatography** Gelpermeationschromatographie, Molekularsiebchromatographie
- **high-pressure liquid chromatography/ high performance liquid chromatography (HPLC)** Hochdruckflüssigkeits-/Hochleistungschromatographie
- **immunoaffinity chromatography** Immunaffinitätschromatographie
- **ion-exchange chromatography** Ionenaustauschchromatographie
- **liquid chromatography (LC)** Flüssigkeitschromatographie
- **molecular sieve/molecular sieving chromatography/ gel permeation chromatography/ gel filtration** Gelfiltration, Molekularsiebchromatographie, Gelpermeationschromatographie
- **partition chromatography** Verteilungschromatographie
- **preparative chromatography** präparative Chromatographie
- **recognition site affinity chromatography** Erkennungssequenz-Affinitätschromatographie
- **reversed phase/reverse-phase chromatography** Umkehrphasenchromatographie
- **salting-out chromatography** Aussalzchromatographie
- **size exclusion chromatography (SEC)** Ausschlusschromatographie, Größenausschlusschromatographie
- **supercritical fluid chromatography (SFC)** überkritische Fluidchromatographie
- **thin-layer chromatography** Dünnschichtchromatographie

chromatophore/pigment cell Chromatophore, Pigmentzelle, Farbzelle
chromium Chrom
chromocenter Chromozentrum
chromomere Chromomer
chromoplast Chromoplast
chromosomal aberration/ chromosome aberration chromosomale Aberration, Chromosomenaberration
chromosomal breakage syndrome Syndrom mit erhöhter Chromosomeninstabilität
chromosome Chromosom
- **accessory chromosome** akzessorisches Chromosom, zusätzliches Chromosom
- **acentric chromosome** azentrisches Chromosom
- **acrocentric chromosome** akrozentrisches Chromosom
- **ancestral chromosome** Urchromosom
- **artificial chromosome** künstliches Chromosom
- **dicentric chromosome** dizentrisches Chromosom
- **giant chromosome** Riesenchromosom
- **harlequin chromosomes** Harlekin-Chromosomen
- **homologous chromosome** homologes Chromosom
- **human artificial chromosome (HAC)** künstliches Humanchromosom, künstliches menschliches Chromosom, menschliches Minichromosom
- **isodicentric chromosome** isodizentrisches Chromosom
- **lampbrush chromosome** Lampenbürstenchromosom
- **metacentric chromosome** metazentrisches Chromosom
- **metaphase chromosome** Metaphasenchromosom
- **multiforked chromosome** Chromosom mit mehreren Replikationsgabeln
- **polytene chromosomes** polytäne Chromosomen
- **ring chromosome** Ringchromosom
- **satellite chromosome** Satellitenchromosom
- **submetacentric chromosome** submetazentrisches Chromosom
- **telocentric chromosome** telozentrisches Chromosom

chromosome complement Chromosomensatz, Chromosomenbestand

chromosome hopping/jumping/walking Chromosomenhopsen, -springen, -wandern

chromosome instability Chromosomeninstabilität

chromosome painting Fluoreszenzmarkierung ganzer Chromosomen

chromosome puff Chromosomenpuff

chromosome set Chromosomensatz

chromosome theory (of inheritance) Chromosomentheorie (der Vererbung)

chromosome-mediated gene transfer Chromosomen-vermittelter Gentransfer

chronic/chronical chronisch

chronospecies Chronospezies

chronotropic chronotrop

chrysalis (*pl* chrysalids/chrysalides) (pupa of holometabolic insects) Chrysalis (Puppe holometaboler Insekten)

chryseous goldgelb

Chrysobalanaceae/cocoa-plum family/coco-plum family Goldpflaumengewächse

chuck (cut of beef: parts of neck/shoulder & area of first three ribs) Schulterstück, Bugstück

chyle Chylus, Darmlymphe

chylific ventricle/chylific cecum Chylusblindsack

chyme Chymus, Speisebrei, Magenbrei

chymosin/lab ferment/rennin Chymosin, Labferment, Rennin

chymotrypsine Chymotrypsin

chytrids/Chytridiomycetes/Archimycetes (Chytridiales) Urpilze

cicadas/Auchenorrhyncha (Homoptera) Zikaden, Zirpen

cicatricle/"eye"/blastodisc/germinal disk Hahnentritt, Keimscheibe

cicatrix/cicatrice/scar *med* Cicatricula, Narbe, Wundnarbe

cicatrization Vernarbung

cidaroids/*Cidaroida* Lanzenseeigel

ciliary band Wimpernkranz, Wimperkranz

ciliary body/corpus ciliare Ciliarkörper, Ziliarkörper

ciliary feeder/ciliary suspension feeder ciliärer Suspensionsfresser, Nahrungsstrudler

ciliary feeding Nahrung herbei strudeln

ciliary loop/corona ciliata (*Chaetognatha*) Corona ciliata

ciliary pit (*Gnathostomulida*) Ciliengrube

ciliary processes Ziliarfortsätze

ciliary tuft Wimperschopf

ciliated/bearing cilia/cilium-bearing/ciliferous bewimpert, gewimpert, zilientragend, cilientragend

ciliated crown/ciliated organ/corona (rotifers) Räderorgan, Krone (Rotatorien)

ciliated epithelium Flimmerepithel, Wimperepithel, Geißelepithel

ciliated groove (ctenophores) Wimperfurche

ciliated larva Wimperlarve

ciliated pit Wimpergrube

ciliates/Ciliata Ciliaten, Wimpertierchen

ciliation Bewimperung

cilium (*pl* cilia) Cilie, Zilie, Wimper, Flimmerhaar, Flimmerhärchen, Kinozilie, Kinozilium (Haarzelle)

cincinnal bract wickelartige Schuppe/Deckschuppe

cincinnus (scorpioid cyme) Wickel

cinereous/ash-colored aschenfarbig, aschfarben

cinnamic acid Cinnamonsäure, Zimtsäure (Cinnamat)

cinnamic alcohol/cinnamyl alcohol Zimtalkohol

cinnamic aldehyde/cinnamaldehyde Zimtaldehyd

cinnamon fern family/flowering fern family/royal fern family/Osmundaceae Königsfarngewächse, Rispenfarngewächse

circadian rhythm circadianer Rhythmus, Tagesrhythmus (-rhythmik)

circannual rhythm circannueller Rhythmus, Jahresrhythmus (-rhythmik)

circinate/coiled/volute aufgerollt (schneckenförmig)

circinate vernation (ferns) Blattentwicklung aus aufgerollter Knospenlage (Farne)

circle Kreis

circuit/electric circuit Stromkreis

circuit/neural circuit *neuro* Schaltkreis, Schaltsystem

• **reverberating circuit** zurückwirkender Schaltkreis

circuitry *neuro* Verschaltung

circular/orbicular kreisförmig, kreisrund

circular chromatography

circular chromatography Zirkularchromatographie, Rundfilterchromatographie
circular dichroism Circulardichroismus, Zirkulardichroismus
circular shaker/rotary shaker Rundschüttler
circularization Zirkularisierung, Ringschluss
circulate zirkulieren
circulating/circulatory zirkulierend, Zirkulations...
circulation Zirkulation, Zirkulieren; Kreislauf
- **pulmonary circulation** kleiner Blutkreislauf, Lungenkreislauf
circulation/blood supply/blood circulation Durchblutung
circulation/bloodstream Kreislauf, Blutkreislauf
circulatory shock Kreislaufschock, Kreislaufkollaps
circulatory system (open/closed) Zirkulationssystem, Blutkreislaufsystem, Kreislaufsystem (offenes/geschlossenes)
circumapical band (rotifers) Circumapicalband
circumcise/circumcising (prepuce/clitoris) beschneiden (Präputium/Clitoris)
circumcision (prepuce/clitoris) Beschneiden (Präputium/Clitoris)
circumduction Zirkumduktion, Kreisbewegung
circumnutation Circumnutation
circumscissile (e.g. capsule) rundherum aufreißend
cirrate mit Ranken
cirrose/cirrous (leaf with prolonged midrib) mit Ranken, rankenartige Blattspitze
cirrus (*pl* cirri) *bot/zool* Cirre (*pl* Cirren)
cirrus/tendril *bot* Ranke
cirrus/feeding leg (thoracopod of barnacles) Rankenfuß
cirrus pouch/cirrus sac (flatworms) Cirrusbeutel
***cis*-acting locus** *gen* *cis*-aktiver Lokus
***cis-trans* complementation test** *cis-trans*-Test
CISS (chromosomal in situ suppression hybridization) chromosomale in-situ Suppressionshybridisierung
Cistaceae/rockrose family Cistrosengewächse, Zistrosengewächse, Sonnenröschengewächse

cisterna (*pl* cisternae/cisternas) (of ER) Zisterne (des ER)
- **paired cisternae/cisternas** paarweise liegende Zisternen (des ER)
cistron Cistron
CITES (Convention on International Trade in Endangered Species): Washington 1975 Washingtoner Artenschutzabkommen/Artenschutzübereinkommen
citric acid (citrate) Citronensäure, Zitronensäure (Citrat/Zitrat)
citric acid cycle/tricarboxylic acid cycle (TCA cycle)/Krebs cycle Citrat-Zyklus, Citratcyclus, Zitronensäurezyklus, Tricarbonsäure-Zyklus, Krebs-Cyclus
citrulline Citrullin, Zitrullin
clade (branch of phylogenetic tree) Clade, Klade, Kladus
cladistics/phylogenetic classification/phylogenetic taxonomy Cladistik, Kladistik
cladode/cladophyll Cladodium, Kladodium (Flachspross eines Langtriebs)
cladogenesis Cladogenese, Kladogenese
cladogram Cladogramm, Kladogramm
clam shrimps/Conchostraca Muschelschaler
clam shrimps & water fleas/Diplostraca/Onychura Doppelschaler, Krallenschwänze
clamp/clip Klammer, Klemme
clamp *fung* **(Basidiomycetes)** Schnalle (Basidiomyceten)
clamp/corpuscle/corpusculum *bot* **(milkweeds)** Klemmkörper (Asclepiadaceen)
clamp connection *fung* **(Basidiomycetes)** Schnallenverbindung (Basidiomyceten)
clamp holder *lab* Muffe
clan Klan, Sippe, Großfamilie
clarification/purification Klärung (z.B. absetzen, entfernen von Schwebstoffen aus einer Flüssigkeit)
clasp reflex Klammerreflex
clasper *zool* **(sharks/rays/skates)** Haftorgan
clasper/tendril/climbing shoot *bot* Ranke
claspered/bearing tendrils mit Ranken versehen
class Klasse
class frequency/cell frequency *stat* Klassenhäufigkeit, Besetzungszahl, absolute Häufigkeit

class switch *immun* Klassenwechsel
class trait Klassenmerkmal
class-switching *immun*
  Klassenwechsel, Klassensprung
classical conditioning (Pavlovian c.)
  klassische Konditionierung
  (Pawlowsche K.)
classification/classifying
  Klassifizierung, Klassifikation,
  Einteilung, Gliederung (taxonomic
  c.>Gruppeneinteilung)
classify klassifizieren, gliedern
classifying/classification
  Klassifizierung, Klassifikation,
  Einteilung, Gliederung
claustral cell (sealed queen ant cell)
  Kessel, Brutkammer
clavate/club-shaped/club-like
  keulenartig
claw (orn talon) Klaue, Kralle;
  (nail) Nagel
  ● **sole of claw** Krallensohle
  ● **wall of claw** Krallenwall
claw/unguis *bot* Nagel (des
  Kronblattes), Unguis
claw gland Klauendrüse
clay Ton
  ● **heavy clay soil (marsh)** Klei,
    schwerer Kleiboden
  ● **modeling clay** Knete, Knetmasse,
    Plastilin
clean bench Sicherheitswerkbank
clean room Reinraum bzw.
  Reinstraum (je nach Partikelanzahl/
  m3)
cleaning symbiosis Putzsymbiose
clear/clarify/purify klären (z. B.
  absetzen, entfernen von
  Schwebstoffen aus einer Flüssigkeit)
clear/fell (of a forest) abholzen,
  fällen, roden, kahlschlagen
clear plaque klarer Plaque
**clear-cut/clear cutting/clearance/
  land clearance** Abholzung,
  Kahlschlag
clear-cutting/land clearing
  abholzen, abforsten, kahlschlagen
**clear-zone eye/optical
  superposition eye** optisches
  Superpositionsauge
**clearance (not translated: used as
  such!)** Clearance, Klärung,
  Klärfaktor, Clearance-Wert,
  Filtrierung
clearance rate Filtrierrate
clearant/clearing agent *micros*
  Aufheller, Aufhellungsmittel
clearing (felling) *for* Abholzung,
  Kahlschlag, Rodung
clearing/aisle Schlag, Waldschlag;
  Lichtung; Schneise
cleavage Spaltung, Furchung

**cleavage/breakage/opening/
  cracking/splitting/breakdown**
  chem Spaltung; Spaltbarkeit
cleavage/segmentation *embr* **(egg
  cleavage)** Furchung,
  Furchungsteilung (Eifurchung)
  ● **bilateral c.** bilaterale Furchung,
    Bilateralfurchung
  ● **complete/holoblastic c.** totale
    Furchung, vollständige Furchung,
    holoblastische Furchung
  ● **determinate c.** determinative
    Furchung, determinierte Furchung
    (nichtregulative)
  ● **equal c.** äquale Furchung,
    gleichmäßige Furchung
  ● **holoblastic/complete c.**
    holoblastische/vollständige/totale
    Furchung
  ● **irregular c.** unregelmäßige
    Furchung
  ● **meridional c.** Meridionalfurchung
  ● **regulative c.** regulative Furchung
    (nichtdeterminativ)
  ● **spiral c.** Spiralfurchung
  ● **superficial c.** superfizielle
    Furchung, oberflächliche
    Furchung, Oberflächenfurchung
  ● **total c.** totale Furchung
  ● **unequal c.** inäquale Furchung,
    ungleichmäßige Furchung
cleavage fusion Spaltfusion
cleavage site *gen* Schnittstelle
**cleave (groove/striate/furrow/
  fissure)** furchen
**cleave/break/open/crack/split/
  break down** *chem* spalten
cleft/crack/slit/crevice Spalt, Spalte
cleft grafting/wedge grafting
  Spaltpfropfung, Pfropfen in den
  Spalt
cleft lip Lippenspalte
cleft palate Gaumenspalte
**cleidoic egg/shelled egg/"land egg"
  (reptiles/birds)** kleidoisches Ei,
  beschaltes Ei
cleistocarpous kleistokarp
cleistogamous kleistogam
cleistogamy Kleistogamie
cleistothecium/cleistocarp
  Kleistothecium
cleptobiosis Cleptobiose,
  Kleptobiose
**clethra family/white-alder family/
  Clethraceae** Scheinellergewächse
clicking sound (whales) Klicklaut
cliff Fels, Klippe
climacteric *n* Klimakterium,
  kritische/entscheidende Phase
climacteric/menopause
  Klimakterium, Klimax,
  Wechseljahre, Menopause

C climate 62

climate Klima
- **continental climate**
Kontinentalklima, Binnenklima,
Landklima
- **oceanic climate/marine climate/
maritime climate/coastal climate**
Meeresklima, Küstenklima
climatic klimatisch, Klima...
climatic belt Klimagürtel
climatology Klimatologie,
Klimakunde
climax/culmination Klimax,
Höhepunkt
climax/orgasm Orgasmus
climax/to have an orgasm *vb* den
Höhepunkt erreichen,
einen Orgasmus haben
climax community
Klimaxgesellschaft
climax formation Klimaxformation
climax vegetation Klimaxvegetation
climber/vine/scandent plant
Kletterpflanze, Rankengewächs
climbing/scandent klimmend,
kletternd
climbing fern family/curly-grass
family/Schizaeaceae
Spaltfarngewächse
climbing fiber Kletterfaser
climbing foot Kletterfuß
cline (phenotypic/character
gradient) Cline, Kline, Klin,
Merkmalsgefälle, Merkmalsgradient,
Merkmalsprogression
- **chemocline** *limn* Chemokline
(chemische Sprungschicht)
- **ecocline (gradient of vegetation
and biotopes)** Ökocline,
Ökokline, Ökoklin
- **step cline** Stufen-Kline
- **thermocline** *limn* Thermokline,
Sprungschicht
clingfishes/Gobiescociformes
Saugfischverwandte, Schildfische,
Spinnenfischartige
clinical feature klinisches Merkmal
clinical symptom klinisches
Symptom
clip *n* Klammer
- **stage clip** *micros* Objekttisch-
Klammer
clip *vb* abschneiden, beschneiden,
stutzen; scheren (z. B. Schafe)
clip/shears Schere; Schermaschine
(Schur der Schafwolle)
clipping Scheren, Stutzen,
Beschneiden
clitellates/Clitellata Clitellaten,
Gürtelwürmer
clitellum Clitellum, Drüsengürtel
clitoris Clitoris, Klitoris, Kitzler
cloaca Kloake

cloacal opening/cloacal aperture/
vent Kloakenöffnung
cloaked adult/coarctate pupa
coarctate Puppe
clock, biological biologische Uhr
clod (of soil) Scholle, Klumpen,
Erdklumpen
clonal selection theory
Klonselektionstheorie, klonale
Selektionstheorie
clone *n* Klon
clone *vb* klonieren
clone bank/bank/DNA-library
Genbank, DNA-Bibliothek
cloning Klonierung
- **positional cloning**
Positionsklonierung
- **somatic cell nuclear transfer
cloning** Kerntransferklonierung
einer somatischen Zelle
- **subcloning** Subklonierung
- **subtractive cloning** subtraktive
Klonierung,
Subtraktionsklonierung
cloning vector Klonierungsvektor
close-set stand/dense stand (of
trees) dichter Baumbestand
clot *n* Gerinnsel; *vb* gerinnen
clotting Gerinnung
clotting factor (blood)
Gerinnungsfaktor
cloud forest/fog forest/humid
forest/perhumid forest Nebelwald
cloud seeding (silver iodide) *meteo/
ecol* Wolkenimpfung
clove (of garlic) Zehe
(Knoblauchzehe)
cloven-hoofed paarzehig
cloven-hoofed animals/even-toed
ungulates/artiodactyls/
Artiodactyla Paarhufer
cloverleaf *gen* Kleeblatt
club fungi/Basidiomycetes
Ständerpilze
club fungus Keulenpilz
club mosses/lycopods
Bärlappgewächse
club-like/clubbed/clavate
keulenartig
club-moss trees/lepidophyte trees/
Lepidodendrales Lepidophyten,
Bärlappbäume
club-root/hernia Hernie
club-shaped/club-like/clavate
keulenartig
clubmoss family/Lycopodiaceae
Bärlappgewächse
clusia family/mamey family/St.
John's wort family/mangosteen
family/Clusiaceae/Guttiferae/
Hypericaceae Hartheugewächse,
Johanniskrautgewächse

cluster Gruppe
- **gene cluster** Gengruppe, Gencluster

cluster/bunch *bot* Büschel
cluster analysis *gen* Clusteranalyse
cluster base/knob/bourse *hort/bot* Fruchtkuchen
cluster cup/aecium/aecidium *fung* Aecidium
cluster of differentiation (CD) Differenzierungscluster (CD)
clustered in Gruppen
clutch (nest of eggs) Gelege, Eigelege, Brut, Nest mit Eiern
clutch size Brutgröße
Cneoraceae/spurge olive family Zeilandgewächse, Zwergölbaumgewächse
cnida/thread capsule/nematocyst (urticator) Cnide, Nesselkapsel, Nematocyste
- **glutinant cnida** Glutinant, Klebkapsel, Haftkapsel
- **penetrant cnida** Penetrant, Durchschlagskapsel
- **tube cnida/ptychocyst (Ceriantharia)** Ptychozyste, Ptychonema (eine Astomocnide)
- **volvent cnida** Volvent, Wickelkapsel

cnidarians/coelenterates/Cnidaria/ Coelenterata Hohltiere, Nesseltiere, Coelenteraten
cnidocyte/cnidoblast/nematocyte/ stinging cell Cnidocyt, Cnidozyt, Nematocyt, Ncmatozyt, Nesselzelle
coacervate Koazervat
coacervation Koazervierung, Entmischung
coagulate/set/curdle koagulieren, gerinnen
coal Kohle
- **anthracite/hard coal** Anthrazit, Kohlenblende
- **bituminous coal/soft coal** bituminöse Kohle, Steinkohle
- **charcoal** Holzkohle
- **lignite/brown coal** Lignit, Braunkohle
- **subbituminous coal** subbituminöse Kohle, Glanzbraunkohle

coalescence/symphysis Verwachsung (allgemein)
coalescent/fused by growth verwachsen
coalification/carbonization Inkohlung
coarctate/compressed/constricted zusammengepresst, aneinandergedrückt, verengt

coarctate larva/larva coarctata/ pseudocrysalis Scheinpuppe, Pseudocrysalis
coarctate pupa/pupa coarctata Tönnchenpuppe
coarctation/stricture Verengung, Striktur, Koarktation
coarse adjustment/coarse focus adjustment *micros* Grobjustierung, Grobeinstellung (Grobtrieb)
coarse adjustment knob *micros* Grobjustierschraube, Grobtrieb
coarse/sturdy/rough/robust/tough/ hard derb
coarse-grained grobfaserig
coast/seaboard/shore (*see also:* shoreline) Küste (Ufer/Gestade)
- **alluvial coast/shoreline of progradation** Anschwemmungsküste, Anwachsküste
- **cliffed coast** Kliffküste
- **fjord(ed) coast/fjord shoreline** Fjordküste
- **low coast** Flachküste
- **ria coast/ria shoreline** Riasküste
- **skerry coast/schären-type coastline (rocky isle)** Schärenküste
- **steep coast** Steilküste

coastal desert Küstenwüste
coastal dune Küstendüne
coastal flat/tidal flat Watt
coastal marsh Küstenmarsch, Seemarsch
coastal plain Küstenebene
coastal strip Küstenstreifen, Küstenstrich
coastal swamp/marsh Küstensumpf
coastal upwelling (water) aufwärtsstrebende vertikale Küstenströmung (Auftrieb/ Auftriebswasser)
coastal vegetation/maritime vegetation Küstenvegetation
coastal waters Küstengewässer
coastal zone/littoral zone Küstenzone, Uferzone
coastline/shoreline (waterline) (for different types see: shoreline) Küstenlinie, Küstenstrich
coat *techn* Schutzschicht, Schutzfilm
coat/plumage Tracht, Kleid (Fell/ Gefieder)
coat protein Hüllprotein
coated pit "coated pit", Stachelsaumgrübchen
coated vesicle Korbvesikel, Stachelsaumbläschen, Stachelsaumvesikel
cob/corn cob/ear Maiskolben

cobalt 64

cobalt Kobalt, Cobalt
coca family/Erythroxylaceae
Kokastrauchgewächse
cocaine Kokain
cocarcinogen Kokarzinogen
coccal coccal, kokkal
coccoid coccoid, kokkoid
coccolith Coccolit, Kokkolit
Kalkplättchen, Kalkkörperchen
coccoliths/Coccolithales/
Coccolithophoridae
Coccolithophoriden, Kalkflagellaten
coccus (pl cocci) Kokkus
(pl Kokken), Kugelbakterium
coccygial vertebra/caudal vertebra
Steißwirbel, Steißbeinwirbel,
Schwanzwirbel
coccyx/os coccygis Steiß, Steißbein,
Schwanzfortsatz
cochlea (ear) Schnecke
cochlear/spoon-like cochlear,
löffelartig
cochlear duct/scala media
Cortischer Kanal, Schneckengang
cochleate/cochleiform/irregularly
helical/coiled like a snail's shell
cochlear, schraubenartig gewunden,
schneckenhausförmig eingerollt,
schneckenhausartig gewunden
cochlospermum family/buttercup
tree family (silk-cotton tree
family)/Cochlospermaceae
Nierensamengewächse
cock/rooster Hahn
cock's comb/crista galli (bony ridge
in skull) Hahnenkamm, Crista
galli
cockerel (under 1 year) Hähnchen,
junger Hahn
cockroaches/Blattodea Schaben
cockscomb family/amaranth family/
Amaranthaceae Amarantgewächse,
Fuchsschwanzgewächse
cocoa-plum family/coco-plum
family/Chrysobalanaceae
Goldpflaumengewächse
coconversion Cokonversion
cocoon Kokon, Puppenhülle,
Gespinst (Raupenkokon)
cod-liver oil Lebertran
code/encode vb codieren, kodieren
code n Code
 ● **one-letter-code** Ein-Buchstaben-
Code
codfishes & haddock & hakes/
Gadiformes Dorschfische
coding capacity
Codierungskapazität,
Kodierungskapazität
coding strand/sense strand
codierender Strang, kodierender
Strang, Sinnstrang

codominance Codominanz,
Kodominanz
codominant codominant,
kodominant
codon Codon, Kodon
 ● **initiation codon** Startcodon,
Initiationscodon
 ● **nonsense codon** Nichtsinncodon,
Nonsense-Codon
 ● **PTC (premature termination**
codon) vorzeitiges Stoppcodon
 ● **punctuation codon**
Satzzeichencodon
 ● **stop codon/termination codon/**
terminator codon Abbruchcodon,
Stoppcodon
codon preference
Codon-Präferenz
codon usage Codon-Nutzung
coefficient of association stat
Assoziationskoeffizient
coefficient of coincidence
Coinzidenzfaktor,
Koinzidenzfaktor
coefficient of contingency stat
Kontingenzkoeffizient
coefficient of relatedness
Verwandtschaftskoeffizient
coefficient of variation stat
Variationskoeffizient
coeliac/celiac (pertaining to the
abdominal cavity) die Bauchhöhle
betreffend, Bauchhöhlen...
coeliac plexus/celiac plexus/solar
plexus Sonnengeflecht
coelom/celom Cölom
coelomic fluid Cölomflüssigkeit
coenenchyme Coenenchym
coenobium (pl coenobia)/cell family
Coenobium, Cönobium, Zönobium
(pl Coenobien)
coenocytic coenocytisch
coenosarc Coenosark
coenzyme Coenzym, Koenzym
coevolution Coevolution,
Koevolution
coexistance Koexistenz
cofactor Cofaktor
coffeeberry family/buckthorn
family/Rhamnaceae
Kreuzdorngewächse
coffin bone/pedal bone (horses:
distal phalanx) Hufbein
coffin joint (horses) Hufgelenk
cognate (nucleotide/tRNA)
zugehörig, verwandt (Nucleotid/
tRNA)
cohabitation Beischlaf
cohesion Kohäsion
cohesion theory (cohesion-tension
theory) Kohäsionstheorie
cohesive kohäsiv

colony

cohesive end *gen* überhängendes Ende, klebriges Ende, kohäsives Ende
cohesiveness Kohäsivität
cohort Kohorte
coil *n* Knäuel, Spule
coil *vb* aufwinden, wickeln, in Ringen übereinanderlegen
coil conformation/loop conformation Knäuelkonformation, Schleifenkonformation
coiled/twisted/wound aufgewickelt
coiled coil (superspiraled helices/ helixes) Doppelwendel-Dimer (superspiralisierte Helices)
cointegrate structure (fusion of two replicons) Fusionsprodukt (Fusion zweier Replikons)
coir (coconut fiber) Coir (Kokosfaser)
coitus/coition/copulation/sexual intercourse Koitus, Kopulation, Kopulationsakt, Begattungsakt, Geschlechtsverkehr
Colchicaceae/crocus family Krokusgewächse, Zeitlosengewächse
colchicine Colchicin, Kolchizin
cold *n* Kälte; *med* Erkältung, "Schnupfen"
cold desert Kältewüste
cold frame *hort* glasgedeckter Pflanzkasten/Anzuchtkasten/ Frühbeetkasten
cold house/cold storage Eishaus
cold house/cool greenhouse/ orangery Kalthaus, Frigidarium, Orangerie
cold room/cold-storage room Kühlraum
cold shock Kälteschock
cold storage Kühllagerung
  • **cold-storage room/cold room** Kühlraum
cold store Kühlhaus, Kühllager, Eishaus
cold-blooded/poikilothermal/ poikilothermic wechselwarm, wechselblütig, poikilotherm
cold-sensitive kälteempfindlich, kältesensitiv
cole/cabbage Kohl
coleoids/Coleoidea/Dibranchiata Tintenfische
coleopterophile/cantharophile/ beetle-pollinated flower Coleopterophile, Cantharophile, Käferblume
coleoptile/plumule sheath Coleoptile, Koleoptile, Keimscheide, Keimblattscheide

coleorhiza/root sheath/radicle sheath Koleorhiza, Coleorhiza, Wurzelscheide
colies/mousebirds/Coliiformes Mausvögel
colinearity Colinearität, Kolinearität
collagen Kollagen
collapse *n* Kollaps
collapse/deflate (lung) kollabieren
collar Kragen
collar cell/choanocyte Kragengeißelzelle, Kragenzelle, Choanozyt
collarbone/clavicle Schlüsselbein, Clavicula
collateral kollateral
collection Kollektion, Sammlung
collector lens/collecting lens Kollektorlinse
collenchyma Collenchym, Kollenchym
  • **angular c.** Kantenkollenchym, Eckenkollenchym
  • **lacunar c.** Lückenkollenchym
  • **lamellar c./tangential c.** Plattenkollenchym
collencyte Collenzyt, Collencyt
colleter (multicellular glandular trichome; sticky/viscous secretions) Colletere, Kolletere (Leimzotte/Drüsenzotte)
colleterial gland/adhesive gland/ cement gland (insects) Kittdrüse, Klebdrüse, Zementdrüse (Lepidoptera: Glandula sebacea)
collide kollidieren
colligative property *stat* kolligative Eigenschaft (Teilchenzahl)
collimate *vb* Lichtstrahlen parallel ausrichten
collimating lens parallel-richtende Sammellinse
collimating slit Kollimationsblende, Spaltblende
collimator Kollimator
collision (enzyme kinetics) Kollision
colloblast/lasso cell/adhesive cell Colloblast, Kolloblast, Klebzelle
colloidal gold kolloidales Gold
coloboma Kolobom
colon Colon, Kolon, Grimmdarm
colonial/colony-forming kolonial, koloniebildend (forming a corm > stockbildend)
colonization Kolonisation, Kolonisierung, Besiedlung
colonize kolonisieren, besiedeln
colony (ants/bees) Kolonie, Volk, Staat (Bienenvolk/Ameisenstaat)
colony/cormus (corals/bryozoans) Stock, Tierstock (Korallenstock/ Bryozoenstock)

C

colony bank 66

colony bank Koloniebank
colony-forming/colonial
koloniebildend, kolonial
colony-forming unit (CFU)
koloniebildende Einheit (KBE) (im
Knochenmark gebildete
Vorläuferzelle/Stammzelle)
colophony/rosin Kolophonium
color/shade/tint/tone/pigmentation
Färbung, Farbton, Pigmentation
color blindness Farbenblindheit
color vision Farbensehen
color-matching Farbanpassung
colostrum/foremilk Colostrum,
Kolostralmilch, Vormilch, Biestmilch
**colt (male horse/pony under 4
years)** männliches Fohlen
**colugos/flying lemurs/
dermopterans/Dermoptera**
Pelzflatterer, Riesengleitflieger
colulus Colulus
columella Columella, Gewebesäule
columella *zool* **(snail shell)**
Columella, Spindel
(Schneckenschale)
column *biot* **(bioreactor)** Kolonne,
Turm
column/gynostemium (orchids)
Säule, Säulchen
column/pillar Säule
column chromatography
Säulenchromatographie,
Säulenchromatografie
column reactor Säulenreaktor,
Turmreaktor
columnar säulenartig, säulenförmig
columnar epithelium
Zylinderepithel, Säulenepithel
columnicidal columnicid,
columnizid, säulenspaltig
**coma/comal tuft/hair-tuft/head of
hairs** Haarschopf, Haarbüschel,
Haarkranz; Blattschopf
**comatulids/feather stars/
Comatulida** Haarsterne,
Federsterne, Comatuliden
comb/crest/ridge/pecten Kamm,
Crista (Hahn > Hahnenkamm)
comb cell Kammzelle
**comb jellies/sea combs/sea
gooseberries/sea walnuts/
ctenophores/Ctenophora**
Rippenquallen, Kammquallen,
Ctenophoren
comb honey Scheibenhonig
**comb plate/swimming plate/ciliary
comb/ctene** Wimperkamm,
Wimperplatte, Wimperplättchen,
Schwimmplatte, Ruderplatte,
Ruderplättchen, "Kamm" comb row/
costa Wimpermeridian,
Wimperrippe, Rippe,

comb row/costa (ctenophores)
Wimpermeridian, Wimperrippe,
Rippe, Pleurostiche
comb-shaped/cteniform
kammförmig
**comblike/rakelike/ctenoid/ctenose/
pectinate/pectiniform** kammartig,
kammförmig, gekämmt
**Combretaceae/white mangrove
family/Indian almond family**
Strandmandelgewächse
combustible brennbar
combustion/incineration
Verbrennung
**combustion heat/heat of
combustion** Verbrennungswärme
comestible/edible genießbar, essbar
comma-less (DNA-code) kommalos
(DNA-Code)
command function
Kommandofunktion
command neuron Kommandoneuron
Commelinaceae/spiderwort family
Commelinengewächse
commensal *n* Kommensale, Mitesser
commensalism Kommensalismus
● **hostile commensalism/
synechthry/synecthry**
Raubgastgesellschaft, Synechthrie
**comminator muscle/interpyramid
muscle (Aristotle's lantern)**
Interpyramidalmuskel
commissure Kommissur,
Verbindungsstelle; (Knochen-) Fuge,
Naht
common name/vernacular name
volkstümlicher Name,
Vernakularname
**communal breeding system/
cooperative breeding** kooperatives
Brutpflegesystem
communal courtship Gruppenbalz
communicating ramus Ramus
communicans
● **gray ramus** Ramus communicans
griseus
● **white ramus/visceral ramus**
Ramus communicans albus
communication Kommunikation,
Verständigung
community/association
Gemeinschaft, Gesellschaft
● **plant community**
Pflanzengesellschaft
**comose/tufted/having a tuft of
hairs** dichthaarig, schopfig,
haarschopfig, mit Haarbüschel,
haarbüschelig
compact bone/dense bone
kompakter Knochen
companion cell Begleitzelle,
Geleitzelle

company Gesellschaft, Beisammensein

comparative embryology vergleichende Embryologie

comparative genome hybridization (CGH) vergleichende Genomhybridisierung (CGH)

comparative morphology vergleichende Morphologie

comparative substance Vergleichssubstanz

compartimentalize kompartimentieren

compartmentation/ compartmentalization/ sectionalization/division Kompartimentierung, Unterteilung, Fächerung

compass plant/heliotropic plant Kompasspflanze, Medianpflanze

compatibility/tolerance Kompatibilität, Verträglichkeit, Toleranz

compatible/tolerant kompatibel, verträglich, tolerant
● **incompatible/intolerant** inkompatibel, unverträglich, intolerant

compensation depth *mar* Kompensationstiefe

compensation point Kompensationspunkt

compensation sac/ascus (bryozoans) Wassersack, Ascus

compete konkurrieren
● **outcompete** im Widerstreit/in der Konkurrenz überlegen sein, durch Konkurrenz "ausschalten"

competence (for evocation) Kompetenz (zur Blühinduktion)

competent (cell/culture) kompetent

competition Kompetition, Wettbewerb, Konkurrenz, Existenzkampf
● **evasion of competition** Konkurrenzvermeidung
● **sperm competition (*now:* sperm precedence)** Spermienkonkurrenz

competitive kompetitiv

competitive exclusion, principle of/ exclusion principle (Gause's rule/ principle) Konkurrenz-Ausschluss-Prinzip, Konkurrenz-Exklusions-Prinzip

competitive inhibition kompetitive Hemmung, Konkurrenzhemmung

competitor Konkurrent

complement *immun* Komplement

complement binding reaction/ complement fixation reaction Komplementbindungsreaktion (KBR)

complement cascade Kaskade des Komplementsystems

complement fixation (CF) *immun* Komplementbindung

complement fixation test (CFT) *immun* Komplementbindungsreaktion (KBR)

complementarity Komplementarität

complementarity determining region (CDR) komplementaritätsbestimmende Region (CDR)

complementary komplementär

complementary DNA (cDNA) komplementäre DNA (cDNA)

complementation Komplementation
● **alpha c.** Alpha-Komplementation
● ***cis-trans* c.** *cis-trans* Test
● **intraallelic c.** intraallele Komplementation
● **intragenic c.** intragene Komplementation

complementation groups *gen* Komplementationsgruppen

complete flower vollständige Blüte

complete medium Vollmedium

complexity (chemical/kinetic) Komplexität (chemische/kinetische)

compliance *med* Befolgung (von Medikationsvorschriften)

compliance/capacitance (vessel wall) Weitbarkeit

composite family/daisy family/ sunflower family/aster family/ Compositae/Asteraceae Köpfchenblütler, Korbblütler

compost Kompost

compost heap Komposthaufen

composter/compost bin Kompostbehälter

compound *adv/adj* **(e. g. leaf)** zusammengesetzt (z. B. Blatt)

compound *n chem* Verbindung
● **chemical compound** chemische Verbindung
● **high energy compound** energiereiche Verbindung

compound crossing over Mehrfachaustausch

compound eye/facet eye Komplexauge, Facettenauge, Netzauge, Seitenauge
● **stalked compound eye (Ephemeroptera)** Turbanauge

compound heterozygote zusammengesetzt-heterozygot, compound-heterozygot

compound leaf/divided leaf Fiederblatt (ganzes!), gefiedertes Blatt

compound locus *gen* Locus aus mehreren eng gekoppelten Genen
compound microscope zusammengesetztes Mikroskop
compress/compressed stauchen, gestaucht
compressed/contracted gestaucht, zusammengezogen
compression Kompression, Stauchung
compression resistance (wood) Druckfestigkeit
compression wood Druckholz, Rotholz
computed tomography (CT) Computertomographie, Computertomografie
computer-aided drug design (CADD) computerunterstützte Planung von Wirkstoffen
concatemer Concatemer, Konkatamer
concatenate/catenane *n* Concatenat, Catenan
concatenate *vb* verketten
concatenation Verkettung
concaulescence *bot* Konkauleszenz
concave mirror Hohlspiegel
concentrate konzentrieren, einengen
concentration gradient Konzentrationsgradient, Konzentrationsgefälle
concentricycloids/ concentricycloideans/sea daisies/ Concentricycloidea Concentricycloidea, Seegänseblümchen
conceptacle (*Fucus*) Konzeptakel
conception (fertilization of egg cell by sperm) Konzeption, Empfängnis
conceptive (capable of conceiving) empfängnisfähig
concertina movement (snakes) Harmonikabewegung, Regenwurmbewegung
conch (mollusk shell) Muschel, Muschelschale von Gastropoden (speziell: große gewundene Meeresschnecken-Muschelschale)
conch/ear conch/external ear/outer ear/auricle/pinna Ohrmuschel, äußeres Ohr, Pinna
conchiform/conchoid/shell-shaped schalenförmig, muschelförmig
conchology Konchyliologie, Muschelkunde
concomitant immunity/premunition begleitende Immunität, Prämunität
concrement vacuole (Placozoa) Konkrementvakuole
concrescence Verwachsung
condensation *chem* Kondensation

condensation/compaction/ compression *gen* Verdichtung, Komprimierung, Verkürzung
condensation reaction/dehydration reaction Kondensationsreaktion, Dehydrierungsreaktion
condense/liquify *chem/lab* **(from gaseous state)** kondensieren, flüssig werden
condense (make denser or more compact)/shorten (e.g. chromosomes) verdichten, komprimieren, kürzen, verkürzen
condenser *chem/lab* Kühler
condenser *micros* Kondensor
 • **Liebig condenser** Liebigkühler
 • **reflux condenser** Rückflusskühler
condenser adjustment knob/ substage adjustment knob *micros* Kondensortrieb
condenser diaphragm (iris diaphragm) *micros* Aperturblende, Kondensorblende (Irisblende)
condensing point *chem* Kondensationspunkt
condiment Würze
condition *n math* **(sufficient/ necessary)** (hinreichende/ notwendige) Bedingung
condition *n med* Leiden
condition/prerequisite *n* Vorraussetzung
condition *vb chromat* konditionieren
conditional/conditioned konditional, eingeschränkt; *ethol* bedingt
conditional reinforcement *ethol* bedingte Verstärkung
conditional reflex (CR) bedingter Reflex
conditional-lethal mutant konditional letale Mutante, bedingt letale Mutante
conditional-lethal mutation konditional letale Mutation, bedingt letale Mutation
conditioned/conditional *ethol* angepasst, beeinflusst; bedingt
conditioned/conditional reflex (CR) (classical conditioning) bedingter Reflex
conditioned/conditional response bedingte Reaktion
conditioned/conditional stimulus bedingter Reiz
conditioning/training Konditionierung, Dressur; *chromat* Konditionierung
 • **classical c./Pavlovian c.** *ethol* klassische Konditionierung, Pawlowsche Konditionierung

connection

- **operant c./instrumental c. (trial-and-error learning)** *ethol* operante Konditionierung, instrumentelles Lernen
- **conduct/transport/translocate/lead** leiten (Elektrizität, Flüssigkeiten)
- **conductance (G)** Leitfähigkeit
- **conducting tissue/vascular tissue** Leitgewebe (Gefäße)
- **conduction/conductance/transport/translocation** Leitung, Fortleitung, Weiterleitung, Transport
- **conductivity** Leitfähigkeit
- **conductor** *electr* Leiter
- **condyle** Condylus, Gelenkhöcker, Gelenkfortsatz
- **condyloid joint/ellipsoidal joint** Eigelenk, Ellipsoidgelenk
- **cone** Kegel, Zapfen, Zäpfchen
- **cone (e. g. vegetative cone)** Kegel
- **cone/strobile/strobilus** Zapfen, Blütenzapfen
- **cone cell** Zapfenzelle, Zapfen, Zäpfchen
- **cone gall** *bot* Kegelgalle
- **cone scale/cone bract** Zapfenschuppe
- **cone-shaped/conical** kegelförmig, konisch
- **conelet** kleiner Zapfen
- **confidence interval** *stat* Konfidenzintervall, Vertrauensbereich, Mutungsintervall
- **confidence level** *stat* Konfidenzniveau, Konfidenzwahrscheinlichkeit
- **confidence limit** *stat* Konfidenzgrenze, Vertrauensgrenze, Mutungsgrenze
- **confirmatory data analysis** konfirmatorische Datenanalyse
- **confluent (cells)** konfluent
- **confocal laser scanning microscopy** konfokale Laser-Scanning Mikroskopie
- **conformation** Konformation
  - **boat conformation (cycloalkanes)** *chem* Wannenform
  - **chair conformation (cycloalkanes)** *chem* Sesselform
  - **coil conformation/loop conformation** *gen* Knäuelkonformation, Schleifenkonformation
  - **relaxed (conformation)** relaxiert, entspannt
  - **repulsion conformation** *gen* **(DNA)** Repulsionskonformation
  - **ring conformation/ring form** Ringform
- **conformation polymorphism** Konformationspolymorphismus
- **conformational epitope/ discontinuous epitope** Konformationsepitop
- **congenial** kongenial, verwandt, gleichartig
- **congenic strains (mice)** kongene/ congene Stämme (Mäuse)
- **congenital/innate/inborn** kongenital, angeboren, ererbt
- **congruence** Kongruenz, Übereinstimmung; *math* Deckungsgleichheit
- **conical/cone-shaped** konisch, kegelförmig
- **conidiophore** Conidienträger, Konidienträger
- **conidium** Conidie, Konidie, Knospenspore
- **conies/Hyracoidea** Schliefer
- **coniferous** Zapfen tragend
- **coniferous forest** Nadelwald
- **coniferous tree/conifer/softwood tree** Nadelbaum
- **coniferyl alcohol/coniferol** Coniferylalkohol
- **conispiral** turmförmig gewunden
- **conjugate** konjugieren, zusammenfügen, verschmelzen; sich paaren
- **conjugated bond** konjugierte Bindung
- **conjugation** Konjugation
- **conjugation mapping** Konjugationskartierung
- **conjunctival sac (eye)** *zool* Konjunktivalsack
- **conjunctive tissue (eye/monocots)** *zool/bot* Bindegewebe
- **conk** konsolenförmiger Pilz-Fruchtkörper; (dadurch bedingte) Holzfäule
- **connard family/zebra wood family/Connaraceae** Connaragewächse
- **connate/coalescent (firmly united)** fest verwachsen/angewachsen, zusammengewachsen, fest vereinigt (gleiche Teile)
- **connation/cohesion** Verwachsung (gleicher Organe)
- **connect/bond/link** verbinden
- **connectance** Konnektanz, Beziehungsgefüge, Verknüpfungsgrad
- **connecting link** Bindeglied (Brückentier)
- **connecting strand (sieve pore)** *bot* Verbindungsstrang (Siebpore)
- **connection/bond/linkage** Verbindung

connective

connective (part of anther) Konnektiv, Mittelband
connective tissue Bindegewebe
conodonts/Conodontophorida ("fascinating little whatzids") Conodonten
coreceptor Korezeptor
corepressor Korepressor
consanguineal/consanguineous konsanguin, blutsverwandt
consanguineous marriage konsanguine Ehe, Ehe unter Blutsverwandten, Verwandtenehe
consanguinity Konsanguinität, Blutsverwandtschaft
conscious bewusst, bei Bewusstsein
- **unconscious** unbewusst; *med* bewusstlos

consciousness Bewusstsein; Bewusstseinszustand
- **unconsciousness** Bewusstlosigkeit

consensus sequence *gen* Consensussequenz, Konsensussequenz
consent Einverständnis
- **informed consent (medical/genetic counseling)** Einverständniserklärung nach ausführlicher Aufklärung

conservation *ecol* Konservierung, Erhaltung, Bewahrung, Schutz (Naturschutz), Schonung; Einsparung
- **energy conservation** Energieeinsparung
- **environmental conservation** Erhaltung der Umwelt, Umweltschonung
- **resource conservation** Ressourcenschonung

conservationist *ecol* Naturschützer (Anhänger des Naturschutzgedankens)
conservator (amtlicher) Konservator, Museumsdirektor
conservatory Konservatorium
conserve/preserve (keep from spoiling) konservieren, präservieren, haltbar machen (vor Fäulnis schützen)
consist konsistieren, beschaffen sein
consistency Konsistenz, Beschaffenheit
conspecific *adj/adv* konspezifisch (von der gleichen Art)
conspecific/peer *n* **(member of same species)** Artgenosse
conspecificity Konspezifität
constancy/presence degree Konstanz, Stetigkeit

constant region *immun* konstante Region
constant truncation konstanter Schwellenwert
constitution Konstitution
constitutive heterochromatin konstitutives Heterochromatin
constitutive mutant konstitutive Mutante
constriction Verengung, Enge, Einschnürung
construction/structure/body plan/anatomy Aufbau, Struktur, Bauplan, Anatomie
consume konsumieren, verbrauchen
consumer (primary/secondary/tertiary) *ecol* Konsument, Verbraucher (1./2./3. Ordnung)
consumer protection Verbraucherschutz
consummatory act Konsumhandlung
consummatory behavior Konsumverhalten
contact cell (wood parenchyma) Belegzelle (Holzparenchym)
contact inhibition Kontaktinhibition, Kontakthemmung
contagion/infection Ansteckung, Infektion
contagion index Kontagionsindex, Infektionsindex
contagious/infectious ansteckend, ansteckungsfähig, infektiös
contagious disease/infectious disease ansteckende Krankheit, infektiöse Krankheit
contagiousness Kontagiosität
containment (biological/physical) Sicherheit(smaßnahmen) (biologische/physikalische)
containment host Sicherheitswirt
containment vector Sicherheitsvektor
contaminate kontaminieren, verunreinigen
contaminated kontaminiert, belastet (verschmutzt)
contamination Kontamination, Verunreinigung; Belastung (Verschmutzung)
contemporary/extant zeitgenössisch, heute lebend, existierend, bestehend
contiguous/adjoining/boardering/touching angrenzend, anliegend, anstoßend, berührend
continental climate Kontinentalklima, Binnenklima, Landklima
continental drift Kontinentaldrift, Kontinentalverschiebung

71   **coordination**

continental edge/continental fringe
Kontinentalrand
continental location
Kontinentallage
continental rise Kontinentalfuß,
Kontinentalfußregion
continental shelf Festlandsockel,
Kontinentalsockel,
Kontinentalschelf, Schelf
continental slope
Kontinentalböschung
contingency table Kontingenztafel
**contingent negative variation
(CNV)** Erwartungspotential
**continuous culture/maintenance
culture** kontinuierliche Kultur
contort drehen, verdrehen, krümmen
contorted gedreht, verdreht,
verkrümmt, eingewunden
contortion Verdrehung,
Verkrümmung
contour/outline Umriss
contour farming/contour planting
Anbau/Anpflanzung entlang der
Höhenlinien
contour feather *orn* Konturfeder,
Umrissfeder
contour plowing Pflügen entlang
der Terrainkonturen
contraception Kontrazeption,
Empfängnisverhütung
contraceptive *adv/adj* kontrazeptiv,
empfängnisverhütend
contraceptive *n* Kontrazeptivum,
empfängnisverhütendes Mittel,
Verhütungsmittel
**contracted cymoid/cymose umbel/
pseudosciadioid** Scheindolde,
Pseudosciadioid
contractile fiber kontraktile Faser
contractile protein/motile protein
kontraktiles Protein, motiles Protein
contractile root Zugwurzel
**contractile vacuole/water expulsion
vesicle** kontraktile Vakuole,
pulsierende Vakuole
contraction period
Kontraktionsphase
contracture Kontraktur
contrast *n* Kontrast
- **negative contrast/negative
contrasting** *micros* negativer
Kontrast, Negativkontrastierung
contrast *vb* kontrastieren
**contrast staining/differential
staining** Kontrastfärbung,
Differentialfärbung
control/regualtion *n* **(e.g.
metabolism)** Kontrolle,
Regulierung; Steuerung
- **autogeneous control** autogene
Kontrolle

control/regulate steuern, regulieren
control element/control unit
Regelglied
control system of a process
Regelstrecke
**controlled variable/controlled
condition** Regelgröße
**controlling element/adjuster/
actuator** Stellglied
**Convallariaceae/lily-of-the-valley
family** Maiglöckchengewächse
**Convention on International Trade
in Endangered Species (CITES):
Washington 1975** Washingtoner
Artenschutzabkommen/
Artenschutzübereinkommen
conventional pseudogene
konventionelles Pseudogen
converge konvergieren,
zusammenlaufen, sich (einander)
annähern
convergence Konvergenz,
Zusammenlaufen, Annäherung
convergent konvergent,
zusammenlaufend, sich (einander)
annähernd
convergent circuit konvergenter
Schaltkreis
convergent evolution konvergente
Evolution
conversion Konversion, Umordnung,
Übergang; Genkonversion
- **chromatid conversion**
Chromatidenkonversion
- **half-chromatid conversion**
Halbchromatidenkonversion
convolute/convoluted/rolled up
zusammengerollt, gewickelt,
gewunden, seitlich eingewickelt,
übereinandergerollt
convolution Einrollung (seitlich
eingewickelt/zusammengerollt)
**convolvulus family/morning glory
family/bindweed family/
Convolvulaceae** Windengewächse
cook/boil kochen
cooked-meat broth Fleischbrühe,
Kochfleischbouillon,
Siedfleischbouillon
cool *vb* kühlen, abkühlen
cooling Abkühlung
cooling coil Kühlschlange
Coomassie Blue Coomassie-Blau
coop/pen/hutch (e.g. chickens)
kleiner Tierkäfig, kleiner Verschlag,
Auslauf (z.B. Geflügelstall)
cooperate *vb* kooperieren
cooperative binding kooperative
Bindung
cooperativity Kooperativität
coordinate *vb* koordinieren
coordination Koordination

C copepods 72

copepods/Copepoda Ruderfußkrebse, Ruderfüßer
copper Kupfer
copper grid *micros* Kupfernetz
coppice/coppicing Rückschnitt bis auf den Stumpf für Neuaustrieb
coppice forest/sprout forest Ausschlagswald, Niederwald (durch Rückschnitt; kleines, niedriges Wäldchen)
copra (dried coconut meat) Kopra
coprodeum Coprodeum, Kotdarm, Kotraum
coprolith Koprolith, Kotstein
coprophagist Coprophage, Koprophage, Kotfresser, Dungfresser
coprophagous/coprophagic coprophag, koprophag, kotfressend, dungfressend
coprophagy Coprophagie, Koprophagie, Dungfressen, Kotfressen
coprophilic/coprophilous *fung/bact* coprophil, koprophil, mistbewohnend, dungbewohnend
copulate/mate kopulieren, begatten, paaren
copulation/coitus/sexual union/ mating/sexual intercourse Kopulation, Kopulationsakt, Koitus, Begattungsakt, Geschlechtsverkehr, Paarung
copulatory organ/intromittent organ/penis Begattungsorgan (männliches), Penis
copy number *gen* Kopienzahl
coracidium (cestoda larva) Coracidium-Larve, Korazidium (eine Schwimmlarve)
coracine rabenschwarz
coracoid Coracoid, Rabenbein, Rabenschnabelbein
coral (bright-red ovary/roe of lobster) unbefruchteter Hummerrogen (auch gekocht)
coral fungus family/Clavariaceae Korallenpilze, Keulenpilze
coral reef Korallenriff
coral reef pinnacle Korallenpfeiler
corals Korallen
- **black corals/thorny corals/ antipatharians/Antipatharia** Dornkorallen, Dörnchenkorallen, schwarze Edelkorallen
- **blue corals/Coenothecalia/ Helioporida** Blaukorallen
- **fire corals/stinging corals/ milleporine hydrocorals/ Milleporina** Feuerkorallen
- **hexacorals/hexacorallians/ Hexacorallia** Hexacorallia

- **horny corals/gorgonians/ Gorgonaria** Rindenkorallen, Hornkorallen
- **milleporine hydrocorals/fire corals/stinging corals/ Milleporina** Feuerkorallen
- **octocorals/octocorallians/ Octocorallia** Octocorallia
- **soft corals/alcyonaceans/ Alcyonaria/Alcyonacea** Lederkorallen
- **stony corals/scleractinians/ Madreporaria/Scleractinia** Steinkorallen, Riffkorallen
- **stylasterine hydrocorals/ Stylasterida** Stylasteriden
- **thorny corals/black corals/ antipatharians/Antipatharia** Dornkorallen, Dörnchenkorallen, schwarze Edelkorallen
corbiculum/pollen basket Corbiculum, Pollenkörbchen, Körbchen
cordate/cordiform/heart-shaped herzförmig
cordon *bot/hort* Kordon, Schnurbaum, Schnurspalierbaum
cordwood Klafterholz
core Core, Kern, Mark
core *bot* **(fruit)** Kerngehäuse (einer Frucht)
core *vir* zentrale Virionstruktur, Kernstruktur, Zentrum
core/drill core *geol* Kern, Bohrkern
core enzyme Core-Enzym, Kernenzym (RNA-Polymerase)
core octamer Core-Octamer
core particle Kernpartikel
core species Kernart
coremium Coremium, Koremium, Koremie
corepressor Corepressor, Korepressor
Cori cycle Cori-Zyklus, Cori-Cyclus
coriaceous/leathery ledrig, lederartig
coriaria family/Coriariaceae Gerbersträucher, Gerberstrauchgewächse
corium/dermis/cutis vera Corium, Korium, Lederhaut, Dermis
cork/phellem/secondary bark Kork, Phellem
cork cambium/phellogen Korkkambium, Phellogen
corkwood family/Leitneriaceae Korkholzgewächse
corky/suberous verkorkt
corm *bot* **(swollen shoot base)/ bulbotuber/"solid bulb"** unterirdische Hypokotylknolle/ Stengelknolle, "Knollenzwiebel"

corm/cormus/colony *zool* Kormus, Stock, Tierstock (corals: Korallenstock)
cormel/cormlet (*Gladiolus*) Brutknolle
cormidium (siphonophores) Cormidium
cormophyte Kormophyt, Achsenpflanze, Sprosspflanze
cormus *bot* Kormus
cormus/corm/colony Tierstock (corals: Korallenstock)
corn/grain/kernel Korn
corn grits/hominy grits Maisgrütze, Maisgrieß
Cornaceae/dogwood family Hartriegelgewächse, Hornstrauchgewächse
cornea (eye) Cornea, Hornhaut
corned gepökelt, eingesalzen (corned beef: gepökeltes Rindfleisch)
corneous/hornlike (horny texture) hornartig (hornartige Beschaffenheit)
corner frequency Grenzfrequenz
cornification Verhornung
cornified/horny verhornt
cornified cell envelope (CE) (epidermis) zellplasma-umhüllende Schicht bei Hautverhornung
cornified sheath/horn sheath Hornscheide
cornify (converting/changing into horn) verhornen
cornigerous/horned gehörnt, mit Hörnern
cornstalk Maisstengel
cornsteep liquor Maisquellwasser
cornu (*pl* **cornua**) Horn, hornförmige Struktur
• **cornu Ammonis** Ammonshorn
cornute gehörnt, hornförmig, hornartig
corolla Blumenkrone, Blütenkrone, Krone
corolla tube/tubular corolla Blütenröhre, Röhrenblüte (mit verwachsenen Kronblätter)
corona/crown Krone, Kranz; (teeth) Zahnkrone
coronal/wreath-shaped kranzförmig
coronal groove *zool* Coronalfurche, Ringfurche
coronal scale *bot* Schlundschuppe
coronal stomach/coronal sinus Kranzdarm
coronal suture (skull) Kranznaht
coronary band/coronary ring/ coronary cushion (horse) Kronsaum, Kronband

coronary bone (horse: small pastern bone) Kronbein
coronary groove/coronary sulcus (heart) Herzkranzfurche
coronary valve/cardiac valve/heart valve Herzklappe
• **aortic valve** Aortenklappe
• **atrioventricular valve** Atrioventrikularklappe, Segelklappe
• **mitral valve/bicuspid valve (with two cusps/flaps)** Mitralklappe, Bikuspidalklappe, Zweisegelklappe, zweizipflige Segelklappe
• **pulmonary valve/pulmonic valve** Pulmonalklappe
• **semilunar valve (consisting of three semilunar cusps/flaps)** Taschenklappe (aus drei halbmondförmigen Falten = Semilunarklappen)
• **tricuspid valve (with three cusps/flaps)** Trikuspidalklappe, Dreisegelklappe, dreizipflige Segelklappe
coronary vessel (arteries/veins) Herzkranzgefäß (Arterien/Venen)
coronate medusas/Coronatae Kranzquallen, Tiefseequallen
coronet (at hoof) Krone, Haarkranz (am Huf)
corpora allata Corpora allata
corpora cardiaca Corpora cardiaca
corpora pedunculata/pedunculate bodies/mushroom bodies Corpora pedunculata, Pilzkörper
corpse Gebeine, sterbliche Hülle
corpse/carcass/cadaver Leiche, Kadaver
corpus luteum Gelbkörper
corpus striatum Corpus striatum, Basalkern, Basalkörper
corrasion (mechanical erosion by wind/water/snow) Korrasion
• **sand blasting/wind carving** Sandschliff, Windschliff, Windkorrasion
correctness/exactness/accuracy *stat* Richtigkeit, Genauigkeit
correlation coefficient *stat* Korrelationskoeffizient
• **partial correlation coefficient** Teilkorrelationskoeffizient
• **product-moment correlation coefficient** Produkt-Moment-Korrelationskoeffizient, Maßkorrelationskoeffizient
• **rank correlation coefficient** Rangkorrelationskoeffizient
corresponding entsprechend; *math* einander zugeordnet

corresponding member
korrespondierendes Mitglied
**corroborative growth (*Troll*)/
establishment growth
(*Zimmermann/Tomlinson*)** *bot*
Erstarkungswachstum
corrode korrodieren, ätzen, beizen
corrosion Korrosion, Ätzen, Ätzung
corrosive korrodierend, ätzend, beizend
corrugated/corrugative/crumpled irregularly/in folds gerunzelt, runzelig, gewellt, geriffelt
corrugation irrigation Furchenberieselung
corrugative/corrugated/rugose/wrinkled gewellt, geriffelt, runzelig, gerunzelt
corset bearers/loriciferans/Loricifera Korsetttierchen, Panzertierchen, Loriciferen
cortex Rinde
- **auditory cortex** *neuro* Hörrinde
- **cerebral cortex** Großhirnrinde
- **hair cortex** Haarrinde
- **neocortex/neopallium** *neuro* Neocortex, Neokortex, Neopallium, Neuhirnrinde
- **optic cortex** *neuro* Sehrinde
- **paleocortex** *neuro* Palaeocortex, Palaeokortex, Archicortex, Althirnrinde
- **renal cortex** Nierenrinde
- **secondary cortex/phelloderm** *bot* Korkrinde, Phelloderm
cortical layer/cortical zone Cortexschicht
cortical parenchyma Rindenparenchym
cortication Kortikation, Berindung; Rindenbildung
corticoliberin/corticotropin-releasing hormone (CRH)/corticotropin-releasing factor (CRF) Corticoliberin, Corticotropinfreisetzendes Hormon, corticotropes Releasing-Hormon (CRH)
corticotropin/adrenocorticotropic hormone (ACTH) Corticotropin, Kortikotropin, Adrenokortikotropin, adrenocorticotropes Hormon, adrenokortikotropes Hormon (ACTH)
cortina *fung* Cortina, Schleier
cortinal zone Schleierzone
Cortinarius family/Cortinariaceae Schleierlinge, Haarschleierpilze
cortisol/hydrocortisone Cortisol, Hydrocortison
cortisone Cortison, Kortison
Corylaceae/hazel family Haselnussgewächse

corymb (inflorescence) Corymbus, Ebenstrauß
- **umbel-like panicle** Doldenrispe, Schirmrispe
- **umbel-like raceme** Doldentraube, Schirmtraube
cos **site** *cos*-Stelle
cosmid Cosmid
cosmine Cosmin
cosmoid scale *ichth* Cosmoidschuppe
cosmopolitan/cosmopolite *n* Kosmopolit
cosmopolitan/occurring worldwide *adj/adv* kosmopolitisch, weltweit verbreitet
costa/rib *zool* Costa (große Längsader), Rippe
costa/rib/vein *bot* **(leaf)** Costa, Rippe (Blattrippe)
costal kostal, Rippen..
costal arch/arcus costalis Rippenbogen
costal cartilage (vertebrates) Rippenknorpel
costal field/costal area *entom* Costalfeld, Remigium
costal fold *entom* Costalfalte
costal plate (crinoids/tortoises) Costalplatte
costal pleura/pleura costalis Rippenfell
costal process/processus costalis Lendenwirbelquerfortsatz
costate gerippt, mit Rippen
cotransduction Cotransduktion, Kotransduktion
cotransfection Cotransfektion, Kotransfektion
cotransformation Cotransformation, Kotransformation
cotranslational cotranslational
cotton swab/swab Wattebausch, Tupfer
cotton-tree family/silk-cotton tree family/kapok-tree family/Bombacaceae Wollbaumgewächse
cotyledon/seminal leaf Kotyledone, Cotyledone, Keimblatt
cotyledonary placenta Placenta cotyledonaria, Placenta multiplex
cotyloid/cotyliform/cup-shaped schalenförmig, tassenförmig, becherförmig
cotyloid cavity (joint) Hüftgelenkpfanne, Hüftpfanne, Acetabulum
cotyloid ligament Hüftband
couch (otter) (höhlenartiger) Bau, Lager
cough *n* Husten
cough *vb* husten

Coulter counter Coulter-Zellzählgerät
counterattack Gegenangriff
countercurrent Gegenstrom
countercurrent distribution Gegenstromverteilung
countercurrent electrophoresis Gegenstromelektrophorese, Überwanderungselektrophorese
countercurrent extraction Gegenstromextraktion
counterevolution Evolutionsumkehr
counterselection Gegenselektion, Gegenauslese
countershading (e.g. fish) Gegenschattierung
counterstain/counterstaining *micros* Gegenfärbung
counterstain *vb micros* gegenfärben
counting chamber Zählkammer
counting plate Zählplatte
country Land
  • **developed countries/ industrialized nations/core countries/more-developed countries (MDCs)** Industrieländer
  • **developing countries/peripheral countries/less-developed countries (LDCs)** Entwicklungsländer
  • **semi-peripheral countries** Schwellenländer
countryside/landscape Landschaft
couple *vb* koppeln
coupled (couple)/linked (link) gekoppelt (koppeln)
coupled reaction gekoppelte Reaktion
coupled transport/co-transport gekoppelter Transport
coupling potential Kopplungspotential
course microscope Kursmikroskop
course of development Entwicklungsgang
courting behavior Werbeverhalten
courtship/mating behavior/display Liebeswerbung, Balz
courtship song/mating song Balzgesang
cousin (first/second cousin) Cousin(e)/Kousin(e) (ersten/zweiten Grades)
covalent bond kovalente Bindung
covalently closed circles DNA (cccDNA) DNA aus kovalent geschlossenen Ringen (cccDNA)
covariance analysis Kovarianzanalyse
covariance Kovarianz
cove enge Schlucht; kleine Bucht (am Meer mit kleiner Mündung)
cove forest Schluchtwald
cover *n* Decke, Abdeckung, Schutz
cover/stand/growth *n bot* Bewuchs
cover/protect *vb* abdecken, zudecken, schützen
cover/serve/leap *vb zool* decken, begatten, bespringen
cover abundance index/species importance value Artmächtigkeit
cover cell Deckzelle
cover value Deckungswert
coverage percentage/coverage level Deckungsgrad
covering cell/supporting cell Deckzelle, Stützzelle
covering gall Umwallungsgalle
coverslip/coverglass *micros* Deckglas
covert/shelter/hiding place Deckung, Schutz, Versteck, Schlupfwinkel
covert/wing covert/protective feather/tectrix (*pl*** tectrices)** Decke, Deckfeder, Tectrix
  • **lesser/minor covert** kleine Decke (Flügeldecke/Flügelfeder)
  • **marginal/marginal tectrix** Randdecke
  • **median covert** mittlere Decke
covey (quails) Schwarm
cow dropping/cow pat/cow dung Kuhfladen
Cowper's gland/bulbourethral gland Cowpersche Drüse, Cowper-Drüse, Bulbourethraldrüse
coxal gland Coxaldrüse
coxal joint/hip joint Hüftgelenk
coxopodite Coxopodit
crab apple Holzapfel
crab pincers/chela Krebsschere, Klaue
crabs/Brachyura Krebse, echte Krabben
crack/break down/open *vb chem* aufspalten, spalten, öffnen
cracking/opening *chem* Aufspaltung, Öffnen; *Erdöl:* kracken
cram/stuff (fowl: geese) mästen (Geflügel)
cranes & rails and allies/Gruiformes Kranichvögel, Kranichverwandte
cranesbill family/geranium family/ Geraniaceae Geraniengewächse, Storchschnabelgewächse
cranial kranial, Schädel..., den Schädel betreffend
cranial/cephalic/superior Kopf..., am Kopfende stehend/befindlich (oben/apikal/terminal)
cranial base Schädelbasis
cranial bone Schädelknochen

cranial flexure

cranial flexure/cephalic flexure *embr* Scheitelbeuge, Mittelhirnbeuge
cranial floor/skull base/base of skull (interne) Schädelbasis
cranial fossa Schädelgrube
cranial index Schädelindex
cranial roof/skull roof/clavarium Schädeldach, Schädeldecke
cranial suture Schädelnaht
- **coronal suture** Kranznaht
- **frontal suture** Stirnnaht
- **lambdoid suture** Lambdanaht
- **plane suture** ebenflächige Naht (Schädelknochennaht mit ebenen Flächen)
- **sagittal suture** Pfeilnaht
- **serrate suture** Sägenaht
- **squamosal suture** Schuppennaht

cranium/braincase/skull/head capsule Cranium, Kranium, Hirnkapsel, Schädel
- **branchiocranium/branchial cranium** Branchiokranium, Kiemenschädel, Kiemenskelett
- **chondrocranium/cartilaginous cranium** Chondrokranium, Knorpelschädel
- **dermatocranium** Dermatokranium, Hautknochenschädel
- **desmocranium (presursor of chondrocranium)** *embr* Desmokranium, Bindegewebsschädel
- **neurocranium/cerebral cranium** Neurokranium, Neuralkranium, Hirnschädel
- **osteocranium** Osteokranium, Knochenschädel
- **viscerocranium/visceral cranium/facial skeleton** Viscerokranium, Viszerokranium, Gesichtsschädel

Crassulaceae/stonecrop family/sedum family/orpine family Dickblattgewächse
crate planks/crate boards Kistenbretter
crater teat/crater tit/crater nipple Stülpzitze
crawl kriechen
crawling traces/repichnia *paleo* Kriechspuren
crayfishes/crawdads Flusskrebse
creatine Kreatin
creature/being Kreatur, Wesen, Geschöpf
crèche/nesting colony Nistkolonie
creek/brook Bach; *Brit* kleine Bucht
creep (turgor steady state) Turgor-Fließgleichgewicht
creeper/trailing plant Kriechpflanze
creeping/crawling/repent kriechend (am Boden entlang kriechend)
creeping foot (mollusks) Kriechfuß
creeping frustule (asexual polyp bud) Kriechfrustel (Polypenknospe)
cremaster Cremaster
cremocarp (schizocarp of Umbelliferae) Doppelachäne
crenate/with rounded teeth/scalloped kerbzähnig, zackig gekerbt
crenulate/finely notched fein kerbzähnig, feingekerbt, feinkerbig
creodonts/Creodonta Ur-Raubtiere
creosote bush family/caltrop family/Zygophyllaceae Jochblattgewächse
Crepidotus family/crep fungus family/Crepidotaceae Stummelfüßchen, Krüppelfüße
crepuscular dämmerungsaktiv, im Zwielicht erscheinend
crescent *adj/adv* sichelförmig, halbmondförmig; zunehmend, wachsend
crescent *n* Halbmond, Sichel
- **gray crescent (amphibian egg)** grauer Halbmond
crescentic/crescent-shaped halbmondförmig
- **with crescent-shaped ridges/selenodont** halbmondhöckrig, selenodont (Zahnhöcker)
Cretaceous Period/Cretaceous Kreidezeit, Kreide
crevice/fissure/crack Spalt, Spalte, Riss; Felsspalte
cri-du-chat syndrome Katzenschrei-Syndrom
crib biting/cribbing (horses: nervous habit) Krippenbeißen
cribellum *arach* Cribellum, Spinnplatte, Spinnsieb
cribriform (pierced with small holes like a sieve) siebartig, siebförmig, kribriform
cribriform plate (horizontal plate of ethmoid bone) Siebbeinplatte
crinkle/leaf curl Blattkräuselkrankheit
crinophagy Crinophagie
crippled/stunted krüppelig, krüppelhaft, verkrüppelt
criss-cross inheritance Überkreuzvererbung
cristate/crested scheitelartig, kammartig, schopfartig
criteria *sg&pl* (*sg actually* criterion) Kriterium, Kennzeichen, unterscheidendes Merkmal
critical point kritischer Punkt

crossing over

critical point drying (CPD) Kritisch-Punkt-Trocknung
critical temperature kritische Temperatur
• **lower critical temperature (LCT)** untere kritische Temperatur (UKT)
• **upper critical temperature (UCT)** obere kritische Temperatur (OKT)
critter *dial*/**domestic animal/farm animal** Haustier
critter *dial*/**lower animal** Getier (e. g. Insekten/Schnecken)
CRM⁺ (positive for cross-reacting material) positiv für kreuzreagierendes Material
Cro repressor Cro-Repressor
croak (frog) quaken
crochet (any small hooklike structure) Häkchen; Hakenfortsatz
crocodiles/Crocodilia Krokodile, Panzerechsen
crocus family/Colchicaceae Krokusgewächse, Zeitlosengewächse
crook of the arm/bend of the arm/ inside of the elbow Ellenbeuge, Armbeuge
crop *agr* **(plant crop)** Kultur, Pflanzenkultur
crop/produce *agr* **(plant/animal product grown and harvested)** Feldfrucht, Bodenprodukt, Landerzeugnis, Naturerzeugnis, Landbauprodukt
• **heavy crop** reiche Ernte
crop (pouched enlargement of gullet) *zool* Vormagen, Vorderdarm, Kropf, Ingluvies (Insekten, Vögel)
crop milk/pigeon milk (milky secretion from crop lining) Kropfmilch, Kropfsekret (Tauben)
crop plant/cultivated plant Kulturpflanze
crop rotation Anbaurotation, Fruchtwechsel, Fruchtfolge
crop yield/harvest/crop Ernteertrag
cropping/plant production Ackerbau
cropping method/technique/ procedure Anbaumethode/ verfahren
cross/breed/crossbreed/interbreed kreuzen, züchten
cross *vb* **(legs) (>arms: fold)** kreuzen (>Arme: schränken/ verschränken)
cross *n* **(e. g. between different species)** Kreuzung, Kreuzungsprodukt
• **bridging cross** Überbrückungskreuzung
• **dihybrid cross** Dihybridkreuzung
• **double cross** Doppelkreuzung
• **monohybrid cross** Monohybridkreuzung
• **outcrossing** Herauskreuzen, Auskreuzen
• **single cross** Einfachkreuzung
• **testcross** Testkreuzung
• **three point testcross** Drei-Faktor-Kreuzung
cross bridge (myosin filament) Querbrücke
cross flow filtration Kreuzstrom-Filtration
cross hybridization Kreuzhybridisierung
cross linker/crosslinking agent quervernetzendes Agens
• **CRM⁺ (positive for cross-reacting material)** positiv für kreuzreagierendes Material
cross section/transverse section Hirnschnitt, Querschnitt
cross vein Querader
cross wall Querwand
cross-fertilization/allogamy/ xenogamy Kreuzbefruchtung, Fremdbefruchtung, Allogamie, Xenogamie
cross-field electrophoresis (CEP) Kreuzelektrophorese
cross-flow filtration Querstromfiltration
cross-fostering Fremdaufzucht
cross-hair disk *micro:* **eyepiece** Zwischenlegscheibe mit Fadenkreuz (Okular)
cross-linked quervernetzt
cross-linking Quervernetzung
cross-matching *immun* Kreuzprobe
cross-pollination Kreuzbestäubung
cross-protection Kreuzimmunität, übergreifender Schutz
cross-reaction Kreuzreaktion
cross-reactive kreuzreaktiv
cross-reactivity Kreuzreaktivität
crossgrain Querfaserung
crossgrained querfaserig, widerspänig
crossgrained timber/crosscut wood Hirnholz
crossing/cross/crossbre(e)d/breed/ crossbreeding/interbreeding Kreuzung, Züchtung, Kreuzzüchtung
crossing over/crossover Crossing over, Überkreuzungsaustausch (homologer Chromatidenabschnitte), Überkreuzungsstelle
• **compound crossing over** Mehrfachaustausch
• **unequal crossing over** ungleiches Crossing-over

crosswall Querwand
crotch/crutch Schritt (zwischen den Beinen)
crotonic acid/α-butenic acid Crotonsäure, Transbutensäure
croup (rump of horse) Kruppe, Hinterteil
crow vb krähen
crowberry family/Empetraceae Krähenbeerengewächse
crowded/tufted (leaves) gedrängt (Blätter)
crowfoot family/buttercup family/ Ranunculaceae Hahnenfußgewächse
crown/treetop/apex/tip bot Krone, Gipfel, Spitze, Baumkrone, Baumgipfel, Stammkrone
crown/calyx zool **(crinoids)** Calyx (kelchförmiger Körper der Crinoiden)
crown gall tumor (by *Agrobacterium tumefaciens*) "Wurzelhalstumor" (Infektionstumor an Stamm und Wurzel)
crown grafting hort Kronenveredlung
crown layer/upper canopy Kronenregion, Kronenschicht, obere Baumschicht
crown rosette plant/tree (terminally tufted leaves) Schopfpflanze/baum
crozier (Ascomycetes) Askushaken, Askushakenzelle
crozier/fiddlehead (ferns) eingerolltes junges Farnblatt
cruciate kreuzförmig
cruciate ligament Kreuzband
crucible lab Schmelztiegel
crucible tongs lab Tiegelzange
Cruciferae/mustard family/cabbage family/Brassicaceae Kreuzblütler, Kreuzblütlergewächse
cruciform kreuzförmig
cruciform DNA kreuzförmige DNA
cruciform structure kreuzförmige Struktur
crude/coarse/rough/tough derb, grob
crude extract Rohextrakt
crumb structure (soil) Krümelstruktur
crural/femoral krural, Schenkel..., den Schenkel betreffend, zum Schenkel gehörig
crural feathers Beinfedern, Schenkelfedern
crust fungus family/Corticiaceae Rindenpilze
crustaceans Crustaceen, Krebstiere, "Krebse"

crustose/crustaceous krustig
crustose fungus Krustenpilz
crustose lichens Krustenflechten
cryo-electron microscopy Kryoelektronenmikroskopie
cryofracture/freeze-fracture Gefrierbruch
cryophyte (plant preferring low temperatures) Kryophyt, Kältepflanze
cryoprotectant Gefrierschutzmittel
cryostat section Kryostatschnitt
cryoultramicrotomy Kryoultramikrotomie
crypt/cave/cavity Höhle, Höhlung, Grube, Vertiefung
crypt of Lieberkühn/intestinal gland Lieberkühnsche Krypte
cryptic verborgen, geheim
cryptic coloration/concealing coloration Tarnfärbung, Schutzfärbung
cryptic dress/camouflage dress (plumage/pelage/coat) Tarntracht, Tarnkleid
cryptic splice site verborgene Spleißstelle
cryptocarpous verborgenfrüchtig
cryptogam Kryptogame, blütenlose Pflanze
Cryptogrammaceae/rock-brake fern family/parsley fern family Rollfarngewächse
cryptomonads/Cryptophyceae Kryptomonaden
cryptophyte/geophyte/ geocryptophyte (*sensu lato*) Cryptophyt, Kryptophyt, Geophyt, Erdpflanze, Staudengewächs
• **hemicryptophyte** Hemikryptophyt
cryptorchid n Kryptorchide
cryptorchid horse/ridgeling Spitzhengst, Klopphengst
cryptorchid pig Spitzeber, Binneneber
cryptorchism Kryptorchismus
crystal cell Kristallzelle, Cristallogenzelle, Sempersche Zelle
crystalline cone Kristallkegel, Kristallkörper, Linsenzylinder, Conus
crystalline style (bivalves) Kristallstiel
crystallization Kristallisation
crystallize/crystalize kristallisieren
crystallography Kristallographie, Kristallografie
• **X-ray crystallography** Röntgenkristallographie, Röntgenkristallografie

ctene/swimming plate/ciliary comb Schwimmplatte, Ruderplatte, Ruderplättchen, "Kamm", Wimperplättchen
cteniform/comb-shaped kammförmig
ctenoid scale *ichth* Ctenoidschuppe, Kammschuppe
ctenose/ctenoid/pectinate/ pectiniform/comblike/rakelike kammartig, kammförmig, gekämmt
ctenostomates/Ctenostomata (bryozoans) Kammmünder
cuboid bone Würfelbein
cuboidal epithelium kubisches Epithel, Pflasterepithel
cubomedusas/box jellies/sea wasps/Cubomedusae/Cubozoa Würfelquallen
cuckoos & turacos and allies/ Cuculiformes Kuckucksvögel
cucullate/hooded/hood-like (e.g. petals) helmartig, haubenförmig
cucumiform gurkenförmig
Cucurbitaceae/cucumber family/ gourd family/pumpkin family Kürbisgewächse
cucurbitaceous gurkenartig, kürbisartig, zu den Kürbisgewächsen gehörend
cud/bolus wiedergekäutes Futter (Klumpen)
• **chew the cud** wiederkäuen
cue Signal, Bedeutung, Hinweis, Orientierungshinweis, Schlüssel, Auslöser
culinary kulinarisch
culinary mushroom/edible mushroom Speisepilz
culinary herbs Küchenkräuter
cull *n* Aussortierte, Ausschuss; Merzvieh; Ausschussware; Ausschussholz
cull *vb* aussortieren, auslesen, auswählen; Merzvieh aussondern
culm *bot* Grashalm, Grasstengel
culmen *orn* Culmen, Schnabelfirst
culmiferous halmtragend
cultivability Anbaueignung
cultivar/cultivated variety/domestic variety Kulturform
cultivate *micb* kultivieren
cultivate/till/crop/grow kultivieren, anbauen
cultivated land landwirtschaftliche Nutzfläche
cultivated variety/domestic variety/ cultivar Kulturform
cultivation/breeding/growing Kultivierung, Zucht, Züchtung
cultivation/cropping/growing Anbau

culture Kultur
• **batch culture** Satzkultur, Batch-Kultur, diskontinuierliche Kultur
• **blood culture** Blutkultur
• **cell culture** Zellkultur
• **chicken embryo culture** Eikultur
• **continuous culture/maintenance culture** kontinuierliche Kultur
• **dilution shake culture** Verdünnungs-Schüttelkultur
• **enrichment culture** Anreicherungskultur
• **maintenance culture** Erhaltungskultur
• **mixed culture** Mischkultur
• **perfusion culture** Perfusionskultur
• **pure culture/axenic culture** Reinkultur
• **roller tube culture** Rollerflaschenkultur
• **shake culture** Schüttelkultur
• **slant culture/slope culture** Schrägkultur (Schrägagar)
• **smear culture** Abstrichkultur
• **stab culture** Einstichkultur, Stichkultur (Stichagar)
• **static culture** statische Kultur
• **stem culture/stock culture** Stammkultur
• **streak culture** Ausstrichkultur
• **submerged culture** Submerskultur, Eintauchkultur
• **surface culture** Oberflächenkultur
• **synchronous culture** Synchronkultur
• **tissue culture** Gewebekultur
culture dish Kulturschale, Petrischale
culture flask/vessel Kulturgefäß
culture medium/medium Kulturmedium, Nährmedium, Medium
• **complete medium** Komplettmedium, Vollmedium
• **complex medium** komplexes Medium
• **deficiency medium** Mangelmedium
• **defined medium** synthetisches Medium (chem. definiertes Medium)
• **differential medium** Differenzierungsmedium
• **enrichment medium** Anreicherungsmedium
• **minimal medium** Minimalmedium
• **selective medium** Selektivmedium, Elektivmedium
cumaceans/Cumacea Kumazeen

cumulative frequency

cumulative frequency *stat* Summenhäufigkeit, kumulative Häufigkeit
cuneate/cuneiform/sphenoid/ wedge-shaped keilförmig
cuneiform bone/os cuneiforme Keilbein
Cunoniaceae/lightwood family Cunoniaceae
cup animals/scyphozoans/ Scyphozoa Schirmquallen, Scheibenquallen, Echte Quallen, Scyphozoen
cup fern family/Dennstaedtiaceae Schüsselfarngewächse
Cupressaceae/cypress family Zypressengewächse
cupule/cupula Cupula
- **flower cup/floral cup** *bot* Blütenbecher

cupule/cupulasmall sucker *zool* kleine Haftscheibe/Saugnapf; Kuppel; Gallertkuppe(l)
curable heilbar
curare Curare
curd (milk) geronnene/dicke Milch
curd/coagulate *n* Gerinnsel, Koagulat
curdle/set/coagulate *vb* gerinnen, koagulieren
cure *vb* **(meat)** pökeln
cure *vb polym* härten, aushärten
cure/drug/medication *n* Heilmittel
cure/heal *vb* heilen
cure/healing *n* **(recovery/relief from disease)** Heilung
cure/treatment *n* Behandlung
cure-all/panacea Allheilmittel
curing (meat) Pökeln
curing *polym* Härten, Aushärten
curing agent *polym* Härter, Aushärtungskatalysator
curing period *polym* Härtezeit, Abbindezeit
curiosity/inquisitiveness (behavior) Neugier, Neugierverhalten
curious neugierig
curly-grass family/climbing fern family/Schizaeaceae Spaltfarngewächse
currant family/gooseberry family/ Grossulariaceae Stachelbeergewächse
current *phys/electro/neuro* **(charge per time)** Strom (*pl* Ströme)
- **capacitive current (I_c)** *neuro* kapazitiver Strom
- **channel current** *neuro* Kanalstrom
- **end-plate current/synaptic current** *neuro* Endplattenstrom
- **gating current** *neuro* Torstrom
- **ionic current** *neuro* Ionenstrom
- **leak current/leakage current (I_l)** *neuro* Leckstrom
- **membrane current** Membranstrom
- **threshold current** *neuro* Schwellenstrom
- **unitary current** Einheitsstrom

current/flow (liquid) Strömung (Flüssigkeit)
- **convection current** Konvektionsstrom, Konvektionsströmung
- **density current** Konzentrationsströmung
- **eddy current** Wirbelstrom (Vortex-Bewegung)
- **inshore current** *mar* auf die Küste zufließende Strömung
- **longshore current** *mar* Brandungslängsstrom, Längsströmung (am Strand)
- **ocean current** Meeresströmung
- **rip current** *mar* Brandungsrückströmung, Rippstrom, Reißstrom
- **tidal current** *mar* Gezeitenströmung, Gezeitenstrom
- **turbidity current** Trübungsstrom, Trübungsströmung

current flow *neuro* Stromfluss
curry/currycomb *vb* **(clean coat of a horse)** striegeln
currycomb *n* Striegel
Cuscutaceae/dodder family Seidengewächse
cushion plant Polsterpflanze
cushion-shaped/pulvinate kissenförmig, polsterförmig
cuspid/canine/canine tooth/dog tooth (tooth with one point) Eckzahn
cuspidate stachelspitzig
custard apple family/cherimoya family/Annonaceae Schuppenapfelgewächse
cut/incised schnittig, geschnitten, eingeschnitten
cut/prune/trim beschneiden, zurückschneiden
cut flower Schnittblume
cut grass/hay/mowing Mahd
cutaneous kutan, Haut...
cutaneous respiration/cutaneous breathing/integumentary respiration Hautatmung
cutaway drawing Ausschnittszeichnung
cuticle/cuticula Cuticula, Kutikula
cuticle of the hair Oberflächenhäutchen

81   **cynthia** C

cuticular skeleton Cuticularskelett, Kutikularskelett

cuticularization Cutikularisierung, Cutin/Kutin-Auflagerung, Cutin/Kutin-Anlagerung

cutinization Cutinisierung, Cutin Einlagerung

cutis/skin Cutis, Haut, eigentliche Haut

cutis vera/dermis/corium Lederhaut, Korium, Corium, Dermis

cuttage *hort* Stecklingsvermehrung

cutting/pruning *hort* Beschneiden, Zurückschneiden

cutting/slip *hort* Steckling
- **bud cutting/eye cutting/single eye cutting/leaf bud cutting** Augensteckling
- **heel cutting** Steckling mit Astring (Stammsteckling)
- **root cutting** Wurzelsteckling

cutting face/cutting plane *bot* Schnittfläche, Schnittebene
- **apical cell with one/two/three cutting faces** einschneidige/zweischneidige/dreischneidige Scheitelzelle

cuttlebone Schulp

cuttlefish & sepiolas/Sepioidea (Sepiida) eigentliche Tintenschnecken

cuttlefish & squids/Decabrachia/Decapoda zehnarmige Tintenschnecken, Zehnarmer

cutworm Raupe bestimmter Eulenfalter

cuvette/spectrophotometer tube Küvette

Cuvierian tubules/tubules of Cuvier Cuvier'sche Schläuche

cyanastrum family/Cyanastraceae Cyanastrumgewächse

cyanelle Cyanelle

cyanogen bromide activated paper (CBA-paper) Bromcyan-aktiviertes Papier (CBA-Papier)

cyanogen bromide cleavage Bromcyanspaltung

cyanthium Cyanthium, Zyanthium

cyathea family/tree fern family/Cyatheaceae Becherfarne, Baumfarne

cyathiform/cup-shaped becherförmig

cycads/cycad family/Cycas family/Cycadaceae Palmfarngewächse

cyclanthus family/panama-hat family/jipijapa family/Cyclanthaceae Scheinpalmen

cycle Kreis, Kreislauf

cyclic zyklisch, cyclisch, ringförmig

cyclic AMP/cAMP (adenosine monophosphate) cyclisches AMP, zyklisches AMP, Cyclo-AMP, Zyklo-AMP, cAMP (Adenosinmonophosphat)

cyclic electron transport zyklischer/cyclischer Elektronentransport

cyclic phosphorylation zyklische/cyclische Phosphorylierung

cyclization *chem* Zyklisierung, Ringschluss

cyclobutyl dimer Cyclobutyldimer

cycloid scale *ichth* Cycloidschuppe, Rundschuppe

cyclomorphosis *ecol* Zyklomorphose, Temporalvariation

cyclone (rotating low-pressure wind system or storm) Zyklone (Tiefdruckgebiet)

cyclone (tropical whirlwind/tornado) Zyklon (trop. Wirbelsturm)

cyclostomes/Cyclostomata Rundmäuler, Kreismünder

cydippid larva Cydippe-Larve

cygotene Zygotän

cylinder *chem* Zylinder

cylinder cell/ciliated cell (Trichoplax) Zylinderzelle, Wimperzelle

cylindric/cylindrical cylindrisch, zylindrisch, walzenförmig

cyme/cymose inflorescence Cyme, Cyma, Cymus, Zyme, cymöser/zymöser Blütenstand, Trugdolde, Scheindolde
- **dichasial cyme/dichasium** Dichasium, zweigablige Trugdolde
- **fan-shaped cyme/rhipidium** Fächel
- **helicoid cyme>bostryx** Schraubel
- **helicoid cyme>drepanium** Drepanium, Sichel
- **monochasial cyme/simple cyme/monochasium** Monochasium, eingablige Trugdolde
- **pleiochasial cyme/pleiochasium** Pleiochasium, vielgablige Trugdolde
- **scorpioid cyme/cincinnus** Wickel

cyme with sessile flowers Knäuel

cyme with very short pedicles Büschel

Cymodoceaceae/manatee-grass family Tanggrasgewächse

cymoid/cymose (sympodially branched) cymös, cymos, zymös, trugdoldig (sympodial verzweigt)

cymule verkürzte Trugdolde (als Teilblütenstand), Scheinquirl (bei Tubifloren)

cynthia Zynthie, Cynthia

Cyperaceae

Cyperaceae/sedge family Riedgräser, Riedgrasgewächse, Sauergräser
cypress family/Cupressaceae Zypressengewächse
cypris larva Cypris-Larve
cypsela/inferior bicarpellary achene unterständige Achäne (Asteraceen)
cyrilla family/leatherwood family/ Cyrillaceae Lederholzgewächse
cyst Cyste, Zyste
cyst wall Cystenwand, Zystenwand
cystacanth Cystacanthus, Hakencyste
cysteamine Cysteamin
cysteic acid Cysteinsäure
cysteine Cystein
cystic fibrosis/mucoviscidosis zystische Fibrose, Mukoviszidose, Mucoviszidose
cystid/zooecium (*sensu lato*) Cystid
cystidium Cystidium, Cystidie, Zystid
cystine Cystin
cystozygote/oospore Cystozygote, Zygotenfrucht, Zygokarp (Oospore)
cytidine Cytidin, Zytidin
 • **deoxycytidine** Desoxycytidin
cytidine triphosphate Cytidintriphosphat
cytochemistry Cytochemie, Zytochemie, Zellchemie
cytochrome Cytochrom, Zytochrom
cytocidal zelltötend, zytozid
cytogenetics Cytogenetik, Zytogenetik
cytohet Cytohet
cytokeratin Zytokeratin, Cytokeratin
cytokine (biological response mediator) Cytokin, Zytokin
cytokinesis/cell division Cytokinese, Zytokinese, Zellteilung
cytology/cell biology Cytologie, Zytologie, Zellenlehre, Zellbiologie
cytolytic cytolytisch, zytolytisch
cytometry Zytometrie, Cytometrie

cytopathic (cytotoxic) cytopathisch, zytopathisch, zellschädigend (zytotoxisch)
cytopathic effect zytopathischer Effekt, zytopathogener Effekt
cytopempsis Cytopempsis, Zytopempsis, Vesikulartransport
cytopenia Cytopenie, Zytopenie
cytopharynx/gullet Cytopharynx, Zytopharynx, Zellschlund
cytoplasm Cytoplasma, Zytoplasma, Zellplasma
cytoplasmic cytoplasmatisch, zytoplasmatisch
cytoplasmic inheritance cytoplasmatische Vererbung, zytoplasmatische Vererbung
cytoplasmic plaque cytoplasmatischer Plaque, zytoplasmatischer Plaque
cytoplasmic streaming/plasma streaming/cyclosis Cytoplasmaströmung, Zytoplasmaströmung, Plasmaströmung, Dinese
cytoproct/cytopyge/cell-anus Cytoproct, Zytoproct, Cytopyge, Zytopyge, Zellafter
cytosine Cytosin
cytoskeletal Cytoskelett..., Zytoskelett...
cytoskeleton Cytoskelett, Zytoskelett, Zellskelett
cytosol Cytosol, Zytosol
cytosome/microbody Cytosom
cytostatic cytostatisch, zytostatisch
cytostatic agent/cytostatic Cytostatikum, Zytostatikum (meist *pl* Zytostatika/Cytostatika)
cytotoxic cytotoxisch, zytotoxisch
cytotoxic T cell/killer T cell/T-killer cell (T_K or T_c) cytotoxische T-Zelle
cytotoxicity Zytotoxizität, Cytotoxizität
cytotoxin Zellgift, Zytotoxin, Cytotoxin

83   **decay** **D**

dabble (waterfowl) gründeln
(Wasservögel)
dactylopodite Dactylopodit
daffodil family/amaryllis family/
Amaryllidaceae
Amaryllisgewächse,
Narzissengewächse
dairy Molkerei
dairy cattle Milchvieh
dairy cow Milchkuh
dairy husbandry Milchwirtschaft
dairy product Milchprodukt,
Molkereiprodukt
daisy family/sunflower family/aster
family/composite family/
Compositae/Asteraceae
Köpfchenblütler, Korbblütler
dam (riverine/fluvial/coastal)
Damm
dam (mother animal: horses etc.)
Muttertier
damage caused by game/browsing
damage Wildverbiss (z. B. an
Baumrinde)
dance Tanz
• **bee dance** Bienentanz
dander Kopf-/Haar-/Hautschüppchen
(von Tieren; evtl. allergene Wirkung)
dandruff Schuppen (Kopfschuppen/
Haarschuppen/Hautschuppen)
danger/hazard/risk/chance Gefahr,
Risiko
danger area Gefahrenbereich
danger zone Gefahrenzone
dangerous/hazardous/risky
gefährlich, riskant
dangerous goods/hazardous
materials Gefahrgut
dangerous substance/hazardous
material Gefahrstoff
dansylation Dansylierung
daphne family/mezereum family/
Thymelaeaceae
Spatzenzungengewächse,
Seidelbastgewächse
dapple-grey horse Apfelschimmel
dark reaction Dunkelreaktion
dark repair/light-independent DNA
repair lichtunabhängige DNA-
Reparatur
darkfield *micros* Dunkelfeld
darkfield microscopy Dunkelfeld-
Mikroskopie
darners/darning needles
(dragonflies) große Libellen
(einige Arten speziell in
Nordamerika)
dart/love dart (gastropods) Pfeil,
Liebespfeil
dart sac (mollusks) Pfeilsack
Darwinian fitness Darwinsche
Fitness

Darwinian selection/natural
selection natürliche Selektion/
Auslese
data (*pl;* **used as** *sg* **&** *pl;* **often**
attrib)/fact Daten, Tatsache,
Angabe
date *n* Datum
date *vb* datieren
dating *n* Datierung
datisca family/durango root family/
Datiscaceae Scheinhanfgewächse
dauer larva (temporarily dormant
larva) Dauerlarve
daughter cell Tochterzelle
daughter strand Tochterstrang
davallia family/bear's-foot fern
family/Davalliaceae
Hasenfußfarngewächse
day-neutral plant tagneutrale
Pflanze
daylily family/Hemerocallidaceae
Tagliliengewächse
deacidification Entsäuerung
dead *adv/adj* tot
dead ripeness (fruit/grain) Totreife
dead volume Totvolumen
deadly/lethal tödlich, letal
deadnettle family/mint family/
Lamiaceae/Labiatae
Lippenblütengewächse,
Lippenblütler
deadspace Totraum
deaf taub, gehörlos
deafness Taubheit, Gehörlosigkeit
dealate (having shed the wings)
flügellos, mit abgeworfenen
Flügeln
deamidation/deamidization/
desamidization Desamidierung
deamination/desamination
Desaminierung
death Tod
death phase/decline phase/phase
of decline *micb* Absterbe-Phase
decanter Abklärflasche,
Dekantiergefäß
decapitate/crown/top köpfen,
kappen, abwerfen
decapitation of tree/beheading of
tree/topping/pollarding Köpfen,
Kappen (von Bäumen), Abwerfen
(Krone)
decapods/Decapoda
Zehnfußkrebse
decay/decompose/disintegrate/fall
apart *vb* zerfallen, sich zersetzen
decay/decomposition/
disintegration *n* Zersetzung,
Zerfall, Verrottung, Verfaulen
• **radioactive decay/radioactive**
disintegration radioaktiver
Zerfall

decay

decay/rot/foul/putrefy *vb* faulen, verfaulen, verwesen, modern, vermodern
decay/rot/fouling/putrefaction *n* Fäulnis, Verwesung, Moder, Vermoderung
decay of variability Variabilitätsrückgang
decaying/rotting verfaulend, modernd, moderig
deceleration phase/retardation phase Verlangsamungsphase, Bremsphase, Verzögerungsphase
deception/delusion/illusion Täuschung
deceptive flower Täuschblume
decerating agent *micros* **(for removing paraffin)** Entparaffinierungsmittel
decidua Decidua, Dezidua, Hinfallhaut, Siebhaut
deciduous/falling/shedding abfallend
deciduous (dropping of leaves/ leaf-dropping) laubwerfend, blattwerfend
deciduous (summergreen) sommergrün
deciduous dentition/lacteal dentition/primary dentition Milchgebiss
deciduous forest/broadleaf forest Laubwald, Fallaubwald
deciduousness/dropping of leaves/ leaf-dropping Laubwerfen
decline (physical) Verfall (körperlicher)
decline phase Absterbe-Phase
decompose/disintegrate/decay/fall apart zersetzen, zerfallen
decomposer Zersetzer, Destruent, Reduzent
decomposing wood holzzersetzend
decomposition/disintegration/ decay Zersetzung, Zerfall
decontaminate dekontaminieren, entseuchen, reinigen, säubern, entgiften
decontamination Dekontaminierung, Dekontamination, Entseuchung, Entgiftung, Reinigen (Beseitigung von Verunreinigungen)
decorticate/debark entrinden, schälen (Rinde)
decortication Entrindung
decouple/uncouple entkoppeln
decoupling/uncoupling Entkopplung
decticous pupa Pupa dectica
decumbent/lodged/prostrate with tips rising up (cereals) niederliegend, niedergedrückt

decurrent herablaufend, herabhängend
decurrent/deliquescent *bot* **(tree form)** (nach oben) ausladend (sympodiale Wuchsform)
decussate/crossed dekussiert, gekreuzt, kreuzgegenständig
decussation Dekussation, Wirtelung
dedifferentiation Dedifferenzierung, Entdifferenzierung
deep etching Tiefenätzung
deep sea (*see also:* **abyssal)** Tiefsee
deep-rooted plant tiefwurzelnde Pflanze
deep-sea basin Tiefseebecken
deep-sea floor Tiefseeboden
deep-sea trench Tiefseegraben, Tiefseerinne
deep-sea trough Tiefseetrog
deer/fallow deer Damwild
deer/venison *culinary* Rotwild
deer fern family/blechnum family/ Blechnaceae Rippenfarngewächse
deer path/run/runway Wildwechsel
defecate/egest den Darm entleeren, Stuhlgang haben (>Mensch)
defecation/egestion Defäkation, Darmentleerung, Klärung, Koten; (Mensch) Stuhlgang
defecation ceremony Defäkationszeremonie
defecation disturbance Defäkationsstörung
defecation reflex Defäkationsreflex
defective gene defektes Gen, Defektgen
defective interfering particle (DI particle)/von Magnus particle DI-Partikel, Von-Magnus-Partikel
defective mutant Defektmutante
defective virus defektes Virus
defense Verteidigung, Abwehr
defense protein Abwehrprotein
defensive gland (*Peripatus:* **slime gland)** Wehrdrüse (*Peripatus*: Schleimdrüse)
defensive medicine Defensivmedizin
defervescence/delay in boiling Siedeverzug
deficiency Defizienz, Mangel
deficiency medium Mangelmedium
deficiency symptom Defizienzerscheinung, Mangelerscheinung, Mangelsymptom
deficient/lacking mangelnd, Mangel...
definite/restricted (growth) beschränkt, begrenzt, bestimmt
deflagrate verpuffen
deflagration Verpuffung

dendritic cell

deflation *geol* (*see also:* **corrasion**) Deflation, Ausblasung
deflation basin/blowout *geol* Deflationskessel, Windmulde
deflect ablenken
deflection Ablenkung
deflexed/abruptly bent or turned downward umgeknickt, umgebogen, zurückgebogen, heruntergebogen
deflorate *adv bot* abgeblüht, verblüht
deflorate/deflower *vb zool* deflorieren, entjungfern
defloration *bot* Abblühen, Verblühen
defloration *zool* Defloration, Entjungferung
defoliate/denude entblättern, entlauben
defoliated/denuded entblättert, entlaubt
defoliation/denudation/stripping of leaves Entlaubung
 • **complete d. (by pests)** Kahlfraß (durch Schädlinge)
defoliation by pests Kahlfraß durch Schädlinge
deforestation Abholzung, Entwaldung, Waldzerstörung
deformation Deformation, Verformung, Formänderung
degas/outgas entgasen
degassing/gassing-out Entgasen, Entgasung
degeneracy Degeneration
degenerate degenerieren, entarten
degenerate/regress rückbilden
degeneration/regression Degeneration, Rückbildung
deglutition/swallowing Schlucken, Schluckakt
degradability Abbaubarkeit
degradation/decomposition/ breakdown Abbau, Zersetzung
degrade/decompose/break down abbauen, zersetzen
degree of freedom (df) *stat* Freiheitsgrad
degree of latitude/parallel Breitengrad
dehisce/break open aufplatzen, aufspringen; sich öffnen
dehiscence/breaking open Dehiszenz, Aufspringen, Aufplatzen
dehiscent aufspringend, aufplatzend
dehiscent fruit Streufrucht, Springfrucht, Öffnungsfrucht
dehydrate dehydratisieren, entwässern
dehydration Dehydratation, Entwässerung
dehydrogenate dehydrieren

dehydrogenation Dehydrierung
delamination (endoderm) Delamination, Abblätterung (Entodermbildung)
delay/retard *vb* verzögern
delay/retardation *n* Verzögerung
delayed effect Verzögerungseffekt
delayed rectification verzögerte Gleichrichtung
delayed-type hypersensitivity reaction (T_{DTH}) Überempfindlichkeitsreaktion vom Spättyp, verzögerter Typ
deletion *gen* Deletion (Mutation unter Verlust von Basenpaaren)
deletion analysis Deletionsanalyse
deletion mapping Deletionskartierung
deletion mutation Deletionsmutation
deliberate release experiment Freisetzungsexperiment
deliquescence Zerfließen, Zerschmelzen, Zergehen
deliquescent *chem* zerfließend, zerschmelzend, zergehend
deliquescent *bot* **(branching)** zerfließend, sich fein verästelnd, reich verzweigt
delouse/delousing lausen, Lausen
deltoid deltaförmig, dreieckig; breiteiförmig (Blattform)
deltoid muscle Deltamuskel
demanding/having high requirements/having high demands anspruchsvoll
deme Dem
demersal auf den Meeresboden sinkend, nahe dem Meeresboden lebend
demister Entfeuchter
demography (study of populations: growth rates/age structure) Demographie
demosponges/Demospongiae Gemeinschwämme
den (bear: often a hollow or cavern) Bau; (lions) Lager, Rastplatz
denaturation/denaturing Denaturierung
denature denaturieren
denatured (DNA/protein/egg white) denaturiert (DNS/Proteine/Eiweiß)
denatured egg white denaturiertes Eiweiß
denaturing/denaturation Denaturierung
denaturing gel denaturierendes Gel
dendrite Dendrit, Markfortsatz
dendritic cell Dendritenzelle, dendritische Zelle

dendritic sheath
Dendritenscheidezelle
dendritic spine Dendritenspine
dendrogram (phylogenetic relationships) Dendrogramm
dendroid/dentritic/arboroid baumartig
dendrologist Dendrologe
dendrology Dendrologie, Gehölzkunde, Baumkunde
dendronotacean snails/ dendronotaceans/Dendronotacea Bäumchenschnecken
denitrification Denitrifikation, Denitrifizierung (Nitrat-Atmung)
denitrify denitrifizieren
Dennstaedtiaceae/cup fern family Schüsselfarngewächse
dense (mass/vol) dicht
dense body dense body (*not translated!*)
density (mass/volume) Dichte
density dependent dichteabhängig
density gradient Dichtegradient
density gradient centrifugation Dichtegradientenzentrifugation
density independent dichteunabhängig
dental alveolus/alveolar cavity/ tooth socket Zahnfach, Zahnalveole
dental cavity/pulp cavity Zahnhöhle, Pulpahöhle
dental cusp/dental ridge Zahnhöcker
dental enamel Zahnschmelz
dental formula Zahnformel
dental lamina Zahnleiste
dental pulp/pulpa Zahnmark
dental replacement Zahnersatz
dental ridge/cusp Zahnhöcker
dental root Zahnwurzel
dentate/toothed gezähnt
dentate gyrus *neuro/anat* Gyrus dentatus
denticle Zähnchen
denticulate/finely dentate gezähnelt, fein gezähnt
dentin/dentine/substantia eburnea Dentin, Zahnbein
dentition/teeth Gebiss
dentition (development/cutting of teeth) Dentition, Zahndurchbruch
dentition (type/number/ arrangement of teeth) Zahntyp, Zahnsystem, Zahnformel, Zahnstruktur
- **acrodont/attached to outer surface of bone/summit of jaws (teleosts/lizards)** akrodont, auf der Kieferkante stehend (Teleostei/Echsen)
- **brachydont/brachyodont/with low crowns** brachyodont, niedrigkronig
- **bunodont/with low crowns and cusps** bunodont, rundhöckrig, stumpfhöckrig
- **deciduous dentition/milk dentition/lacteal dentition/ primary dentition** Milchgebiss
- **diphyodont (with two sets of teeth)** diphyodont (einmaliger Zahnwechsel)
- **heterodont/anisodont** heterodont, ungleichzähnig
- **homodont/isodont** homodont, gleichartig bezahnt
- **hypsodont/hypselodont (high crowns/short roots)** hypsodont, hypselodont, hochkronig
- **isodont/homodont** homodont, gleichartig bezahnt
- **labyrinthodont (with complicated arrangement of dentine)** labyrinthodont (mit komplexer Struktur)
- **lophodont/with transverse ridges** lophodont, mit Querjochen
- **monophyodont (only one set of teeth)** monophyodont (einfaches Gebiss/ohne Zahnwechsel)
- **permanent dentition/permanent teeth** Dauergebiss
- **pleurodont/attached to inside surface of jaws** pleurodont, an der Kieferinnenseite
- **plicodont/with folded cusps (elephants)** plicodont (mit gefalteten Höckern)
- **polyphyodont** polyphyodont (mehrfacher Zahnwechsel)
- **selenodont/crescentic/with crescent-shaped ridges** selenodont, halbmondhöckrig (Zahnhöcker)
- **tetralophodont/with four transverse ridges** tetralophodont, mit vier Querjochen
- **thecodont/teeth in sockets** thekodont, in Zahnfächern verankert
- **triconodont (three crown prominences in a row)** triconodont, dreihöckrig (in einer Reihe)

denuded/stripped of leaves *bot* entlaubt
deoxycytidine Desoxycytidin
deoxyribonucleic acid (DNA) Desoxyribonucleinsäure, Desoxyribonukleinsäure (DNS)
depauperate/starved/reduced/ underdeveloped/impoverished verarmt, verkümmert

87 desolation **D**

dephosphorylation
Dephosphorylierung
depolarization Depolarisation
depopulate entvölkern
depopulation Entvölkerung
deposit/sediment *vb* ablagern,
sedimentieren, sich niederschlagen
deposit/sediment (precipitate) *n*
Ablagerung, Sediment, Niederschlag
(Präzipitat)
deposit feeder Depositfresser
deposition/deposit/sedimentation
Ablagerung, Sedimentation
depression/basin Vertiefung, Mulde
depressor muscle Depressor, Senker
depth of focus/depth of field
Tiefenschärfe, Schärfentiefe
depurination Depurinisierung
derivation *math/theor* Ableitung
derivative Derivat, Abkömmling
(von etwas abgeleitet)
derivatization Derivatisation,
Derivatisierung
derivatize derivatisieren
derive *math/theor* ableiten
derived abgeleitet
derived characteristic abgeleitetes
Merkmal
dermal/dermic/dermatic dermal,
Haut…
dermal bone/membrane bone
Hautknochen, Deckknochen,
Belegknochen
dermal branchiae/skin gills/papulae
Papulae
dermal denticle/placoid scale
Hautzahn, Zahnschuppe,
Placoidschuppe, Dentikel
dermal gland Dermaldrüse,
Hautdrüse
dermal musculature Hautmuskulatur
dermal papilla Dermispapille,
Hautpapille
dermal papula Papula
dermal plate Hautplatte
**dermal skeleton/dermatoskeleton/
dermoskeleton/exoskeleton**
Hautskelett, Dermalskelett
**dermal tissue/boundary tissue/
exodermis** Abschlussgewebe
dermatome Dermatom
dermatophyte Dermatophyt,
Hautpilz
**dermatoskeleton/dermal skeleton/
exoskeleton** Hautskelett,
Außenskelett, Hautpanzer,
Exoskelett
dermis/corium/true skin/cutis vera
Dermis, Korium, Lederhaut
dermo-epidermal junction zone
dermo-epidermale Junktionszone
dermomyotome Dermomyotom

**dermopterans/colugos/flying
lemurs/Dermoptera** Pelzflatterer,
Riesengleitflieger
dermotrichium Dermotrichium,
Flossenstrahl (aus Hautknochen)
desalinate entsalzen
desalination Entsalzung
**descend from/originate from/
derive from** abstammen von
descendant/descendent *adv/adj*
mech/phys absteigend; (derived)
gen/evol abstammend von
descendant/offspring/progeny *n*
Deszendent, Abkömmling,
Nachkomme
descending *mech/phys* absteigend
descent/origin Abstammung
desert *n* Wüste
• **cold desert** Kältewüste
• **fog desert** Nebelwüste
• **gravel desert (serir)** Kieswüste,
Geröllwüste (Serir)
• **hot desert** Wärmewüste
• **sand desert** Sandwüste
• **semidesert** Halbwüste
• **stone desert/stony desert/rock
desert (hammada)** Steinwüste
(Hamada)
desert bloom Wüstenblüte
desert inky cap fungi/Podaxales
Podaxales
desert pavement/stone pavement
Wüstenpflaster, Steinpflaster
desert plant/eremophyte/eremad
Wüstenpflanze, Eremiaphyt
desert varnish Wüstenlack
desertification/desert expansion
Wüstenausbreitung
desiccate/dry up/dry out trocknen,
austrocknen
desiccation Austrocknung, Trocknis
desiccation avoidance
Austrocknungsvermeidung
desiccation tolerance
Austrocknungstoleranz
desiccator *chem/lab* Exsikkator
desire/craving starkes Verlangen,
Bedürfnis
desmids/Desmidiaceae Zieralgen
**desmosome/bridge corpuscle/
bridge corpuscle/macula
adherens** Desmosom, Macula
adhaerens
• **belt desmosome**
Gürteldesmosom, Banddesmosom
• **hemidesmosome** Halbdesmosom
• **spot desmosome**
Plaquedesmosom
desolated verödet, verwüstet;
verlassen
desolation/obliteration Verödung,
Verwüstung

detassel *vb* **(of corn)** entfernen der männlichen Blütenstände des Mais
detect/prove nachweisen
detection/proof Nachweis
detection limit Nachweisgrenze
detection method Nachweismethode
detergent Detergens, Reinigungsmittel
determinate/restricted beschränkt, begrenzt, bestimmt; *bot* endständig
determinate cleavage determinative Furchung, determinierte Furchung (nichtregulative)
determinate growth/limited growth begrenztes Wachstum, beschränktes Wachstum
determinate inflorescence geschlossene Infloreszenz
determination Determination, Determinierung, Bestimmung
determine/elucidate bestimmen, feststellen, aufklären
deterrent/repellent Abschreckstoff, Schreckstoff, Repellens
detorsion (gastropods: nerve cords) Detorsion, Rückdrehung
detoxification Entgiftung
detoxify entgiften
detritivore/detritus-feeder Abfallfresser, Detritusernährer, Detritivor
detritivory Detritivorie
detritus Detritus
detritus food chain Detritusnahrungskette
detritus-feeder/detritivore Detritusernährer, Detritusfresser, Detritivor
deuter cell/pointer cell/eurycyst Deuter
deuterostomes/Deuterostomia Zweitmünder, Neumundtiere, Neumünder
deuterotoky Deuterotokie
deutocerebrum (insects) Deutocerebrum, Mittelhirn
deutoplasm Deutoplasma, Nahrungsdotter
develop/emerge/unfold entwickeln, entstehen
developing chamber (TLC) Trennkammer (DC)
development Entwicklung
- **course of development** Entwicklungsgang
- **embryonal development/ embryonic development/ embryogenesis/embryogeny** Embryonalentwicklung, Embryogenese, Embryogenie, Keimesentwicklung
- **head development/cephalization** Kopfbildung, Cephalisation
- **hemimetabolic development/ hemimetabolous development** hemimetabole Entwicklung
- **holometabolic development** holometabole Entwicklung
- **mosaic development** Mosaikentwicklung
- **regulative development** regulative Entwicklung
- **retrogressive development/ retrogressive evolution** Rückentwicklung
- **sustainable development** dauerhaft-umweltgerechte Entwicklung, nachhaltige Entwicklung

developmental biology Entwicklungsbiologie
developmental cycle Entwicklungszyklus
developmental genetics Entwicklungsgenetik
developmental level Entwicklungsstufe
developmental noise Entwicklungsschwankung
developmental stage/ developmental phase Entwicklungsstadium (*pl* Entwicklungsstadien), Entwicklungsphase
deviate from abweichen von
deviation Abweichung
- **random deviation** *stat* Zufallsabweichung
- **standard deviation/root-mean-square deviation** *stat* Standardabweichung
- **statistical deviation** statistische Abweichung

device *n* Vorrichtung, Einrichtung, Gerät
devil's-claw family/unicorn plant family/martynia family/ Martyniaceae Gemsbockgewächse
devoid of ohne, bar, ...los
devoid of life leblos, ohne Leben
devoid of plants pflanzenlos
Devonian Period/Devonian (geological time) Devon
devour/gulp down verschlingen, verzehren, herunterschlingen
dew Tau
dewclaw/false foot Afterklaue
dewdrop Tautropfen
dewlap (cattle) Wamme; (birds/ reptiles: wattle) Kehllappen
dextrorse dextrors, rechtsdrehend, rechtswindend

89   **dictyosome** **D**

diabetes mellitus Zuckerkrankheit,
Diabetes mellitus
diadelphous diadelphisch,
zweibrüderig
diadromous diadrom
diagnosis Diagnose
- **antenatal diagnosis/prenatal
  diagnosis** pränatale Diagnose,
  pränatale Diagnostik
- **differential diagnosis**
  Differentialdiagnose
- **prenatal diagnosis/antenatal
  diagnosis** pränatale Diagnose,
  pränatale Diagnostik
- **presymptomatic diagnosis**
  präsymptomatische Diagnose,
  präsymptomatische Diagnostik
diagnostic diagnostisch
diagnostic approach diagnostischer
Ansatz
diagnostic species Kennart
diagonal gait Diagonalgang,
Kreuzgang
diagram/plot *math/graph/stat*
Diagramm, Kurve
- **bar diagram/bar graph/bar chart**
  Stabdiagramm
- **dot diagram** Punktdiagramm
- **floral diagram/flower diagram**
  Blütendiagramm
- **frequency diagram**
  Häufigkeitsdiagramm,
  Häufigkeitskurve
- **hist(i)ogram/strip diagram**
  Hist(i)ogramm, Streifendiagramm
- **line diagram/line graph**
  Strichdiagramm
- **Lineweaver-Burk plot/double-
  reciprocal plot** Lineweaver-Burk-
  Diagramm
- **phase diagram**
  Phasendiagramm
- **pie chart** Kreisdiagramm
- **Ramachandran plot**
  Ramachandran-Diagramm
- **Scatchard plot** Scatchard-
  Diagramm
- **scatter diagram/scattergram/
  scattergraph/scatterplot**
  Streudiagramm
diakinesis Diakinese
**dialypetalous/choripetalous/with
separate petals** frei/
getrenntkronblättrig, frei/
getrenntblumenblättrig, choripetal
dialysis Dialyse
dialyze dialysieren
diameter at breast height (dbh)
Brusthöhendurchmesser (BHD)
diapause Diapause
diapensia family/Diapensiaceae
Diapensiagewächse

diaphragm/diaphragma
zool Diaphragma, Zwerchfell;
micros Blende
diaphragm aperture *micros*
Blendenöffnung
**diaphragmatic respiration/
abdominal breathing**
Zwerchfellatmung, Bauchatmung
**diapophysis (transverse process of
neural arch for rib attachment))**
Diapophyse, Rippenfortsatz
diarrhea Diarrhö
**diarthrodial joint/diarthrosis/
synovial joint** Diarthrose, Gelenk,
Articulatio
diarthrosis Diarthrose, echtes
Gelenk
diaspore/propagule/disseminule
Diaspore, Ausbreitungseinheit,
Disseminule
diastema (toothless space)
Diastemma, Zwischenraum:
Zahnlücke, Lücke in Zahnreihe
diatom Diatomee, Kieselalge
diatomaceous earth Diatomeenerde,
Kieselerde
diatoms/Bacillariophyceae
Diatomeen, Kieselalgen
diatropism Diatropismus
dibber/dibble *agr/hort* Dibbelstock,
Setzholz, Pflanzholz
dicentric chromosome dizentrisches
Chromosom
**dichasium/dichasial cyme
(inflorescence)** Dichasium,
zweigablige Trugdolde
**dichlorodiphenyltrichloroethane
(DDT)** Dichlordiphenyltrichlorethan
(DDT)
**dichlorodiphenyltrichloroethylene
(DDE)**
Dichlordiphenyldichlorethylen
(DDE)
dichogamy/heteracmy Dichogamie
dichotomous/forked dichotom,
gabelig verzweigt
dichotomous branching gabelige
Verzweigung
dichotomous venation
Gabeladerung, Gabelnervatur,
Fächeraderung
**dichotomy/(repeated) forking/
bifurcation** Dichotomie, Gabelung,
Gabelteilung
dicksonia family/Dicksoniaceae
Baumfarngewächse
dicotyledon/dicot Dikotyle,
Dikotyledone, Zweikeimblättrige
dicotyledonous dikotyl,
zweikeimblättrig
dictyosome/Golgi body Diktyosom,
Dictyosom (des Golgi-Apparates)

dictyotene Dictyotän
didelphic zweischeidig
dideoxy sequencing Didesoxy-Sequenzierung
dideoxynucleotide Didesoxynucleotid, Didesoxynukleotid
didierea family/Didieraceae Armleuchterbäume
diductor muscle Klaffmuskel
didymous/twinlike/occurring in pairs doppelt, gepaart
die sterben
die off absterben
dieback teilweise absterben
diecious/dioecious diözisch, zweihäusig, getrenntgeschlechtig
diel pattern/diel rhythm/diel periodicity 24-Stunden Rhythmus/Takt, Tag-Nacht Rhythmus, Tag-Nacht-Periodizität
dielectric constant Dielektrizitätskonstante
diencephalon/interbrain/betweenbrain Zwischenhirn
diestrus Diöstrus, Dioestrus
diet/food/feed/nutrition Diät, Kost, Speise, Nahrung
• **balanced diet** ausgewogene Diät
• **to be on a diet** eine Diät machen
dietary Diät..., diät, die Diät betreffend
dietary fiber Ballaststoffe
dietetic diätetisch
dietetics Diätetik
difference/differing/variability Unterschied; Verschiedenartigkeit, Unterschiedlichkeit, Variabilität
differential centrifugation Differentialzentrifugation, differentielle Zentrifugation
differential diagnosis Differentialdiagnose
differential display (form of RT-PCR) differentieller Display (Form der RT-PCR)
differential interference (Nomarski) Differential-Interferenz
differential species Differentialart, Trennart
differential staining/contrast staining Differentialfärbung, Kontrastfärbung
differentiate differenzieren
differentiating characteristic Unterscheidungsmerkmal
differentiation Differenzierung
differentiation antigen Differenzierungsantigen
diffract *opt* beugen
diffraction pattern Beugungsmuster

diffuse diffundieren
diffuse light diffuses Licht
diffuse placenta Placenta diffusa
diffuse porous (wood) zerstreutporig
diffuse secondary thickening (certain monocots) "anomales" sekundäres Dickenwachstum
diffusion coefficient Diffusionskoeffizient
dig *vb* graben
dig/excavation *geol/paleo* Ausgrabung
digest *vb metabol* verdauen
digest *vb (sewage)* faulen (im Faulturm der Kläranlage)
digest *n (enzymatic)* Verdau (enzymatischer)
• **double digest** Doppelverdau
• **partial digest** Partialverdau
digester/digestor/sludge digester/sludge digestor Faulturm
digestibility Verdaulichkeit, Bekömmlichkeit
digestible verdaulich
digestion *general* Verdauung
digestion/degradative reactions/degradative metabolism/catabolism Abbau, Stoffwechselabbau
digestive canal/digestive tract/alimentary canal/alimentary tract Verdauungskanal, Verdauungstrakt
digestive cavity/gastrovascular cavity/enteron Verdauungshohlraum
digestive enzyme Verdauungsenzym
digestive gland Verdauungsdrüse
digestive gland/"liver" (mollusks/echinoderms) Mitteldarmdrüse, Darmdivertikel
digestive gland duct (echinoderms) Darmkanal
digestive system Verdauungssystem
digestive tract Verdauungstrakt
• **one-way digestive tract** durchgängiger Verdauungstrakt
digging/fossorial/burrowing grabend, Grab...
digital cushion/cuneal cushion/pulvinus digitalis (horse) Hufkissen (Strahlkissen + Ballenkissen)
digital pad *zool (e.g. frogs)* Fingerballen
digitate/fingered digitat, gefingert
digitate venation fingerförmige Nervatur/Aderung
digitiform/fingershaped/fingerlike fingerförmig
digitigrade Digitigrade, Zehengänger

discoidal

digitigrade gait Digitigradie, Zehengang
digitoxin Digitoxin
digoxin Digoxin
dihedral symmetry diedrische Symmetrie
dihybrid cross Dihybridkreuzung
dikaryotic mycelium Paarkernmyzel
dikaryotic phase Dikaryophase, Paarkernphase
dike Damm, Deich (am Meer)
dilation/dilatation/expansion Dilatation, Ausweitung, Erweiterung
dilation growth/dilatation growth/ expansion growth/extension growth Dilatationswachstum, Erweiterungswachstum
dilator muscle Dilator, Erweiterer
dillenia family/silver-vine family/ Dilleniaceae Rosenapfelgewächse
dilute verdünnen
dilution Verdünnung
dilution shake culture Verdünnungs-Schüttelkultur
dilution streak/dilution streaking *micb* Verdünnungsausstrich
dimegaly Dimegetismus, sexueller Größenunterschied
dimer Dimer
  • **cyclobutyl dimer** Cyclobutyldimer
  • **thymine dimer** Thymindimer
dimerization Dimerisierung
dimerize dimerisieren
dimerous dimer, zweizählig
dimictic *limn* dimiktisch
dimitic dimitisch
dimorphism Dimorphismus
dinergate/soldier Dinergat, Soldat
dinoflagellates/Pyrrhophyceae Dinoflagellaten, Panzergeißler
dinosaur(s) Dinosaurier
  • **dinosaur ancestors/thecodonts/ Thecodontia** Urwurzelzähner
dioecious/diecious (postnatal: gonochoric/gonochoristic) diözisch, zweihäusig, getrenntgeschlechtig (speziell postnatal: gonochor)
dioecy/dioecism (postnatal: gonochory/gonochorism) Diözie, Zweihäusigkeit, Getrenntgeschlechtigkeit (speziell postnatal: Gonochorismus)
dioestrus/diestrus Diöstrus, Dioestrus
diopter (D) (unit) Dioptrie
dioptric dioptrisch
Dioscoreaceae/yam family Yamswurzelgewächse, Schmerwurzgewächse
diphasic diphasisch

diphycercal diphycerk, diphyzerk, protocerk, protozerk
diphyodont (single dentition) diphyodont (einmaliger Zahnwechsel)
diploid diploid
diplosome Diplosom
diplostemonous diplostemon
diplotene Diplotän
dipole moment Dipolmoment
diprotic acid zweiwertige/ zweiprotonige Säure
Dipsacaceae/teasel family/scabious family Kardengewächse
dipterocarpus family/meranti family/Dipterocarpaceae Zweiflügelfruchtgewächse, Flügelnussgewächse
direct genetics direkte Genetik
direct repeats *gen* direkte Sequenzwiederholungen
direct transfer electrophoresis/ direct blotting electrophoresis Direkttransfer-Elektrophorese, Blottingelektrophorese
directional orientation Richtungsorientierung
directional selection gerichtete Selektion/Auslese
disadvantage Nachteil
disappearing layer (anther) *bot* Schwundschicht
disassortative mating Fremdpaarung
disc *see also:* disk
disc/meniscus (articular disc) Meniskus, Diskus (Gelenkmeniskus/ Gelenkzwischenscheibe)
disc diaphragm (annular aperture) *micros* Ringblende
disc electrophoresis Diskelektrophorese, diskontinuierliche Elektrophorese
disc flower/disc floret/tubular flower Scheibenblüte, Röhrenblüte (Asterales)
disc turbine impeller Scheibenturbinenrührer
disc-shaped scheibenförmig
discal cell/discoidal cell Diskalzelle, Discalzelle, Discoidalzelle
discharge *n neuro* Entladung
discharge/outflow/draining off *n* Ausfluss, Abfluss
discharge/drain/lead out/lead away/carry away *vb* ausführen, wegführen, ableiten (Flüssigkeit)
disclimax Disklimax (Störungsklimax)
discoidal/disk-like/disc-like discoidal, diskoidal

discoidal cleavage

discoidal cleavage discoidale/ diskoidale/scheibenförmige Furchung
discoidal placenta Placenta discoidalis
discontinuity Diskontinuität
discontinuous gene gestückeltes (unterbrochenes) Gen, Mosaikgen
discontinuous replication diskontinuierliche Replikation
disease/illness Krankheit
- **contagious disease/infectious disease** ansteckende Krankheit, infektiöse Krankheit
- **inheritable disease** Erbkrankheit
- **inherited/hereditary/genetic disease** Erbkrankheit, erbliche Erkrankung
- **monogenic disease** monogene (Erb-)Krankheit
- **polygenic disease** polygene (Erb-)Krankheit
- **sexually transmitted disease (STD)** Geschlechtskrankheit
- **transmissible disease/ communicable disease** übertragbare Krankheit

disease-causing/pathogenic krankheitserregend, pathogen
disease-causing agent/pathogen Krankheitserreger
diseases of civilization Zivilisationskrankheiten
disembowel ausweiden, ausnehmen
disequilibrium Ungleichgewicht
disguise n Verschleierung, Verkleidung
disinfect desinfizieren
disinfectant Desinfektionsmittel
disinfection Desinfizierung, Desinfektion
disinfest von Ungeziefer befreien, entseuchen
disinfestation Befreiung von Ungeziefer, Entseuchung
disinhibition Enthemmung, Disinhibition
disintegrate/decay/decompose zersetzen, zerfallen
disintegration/decay/ decomposition Zersetzung, Zerfall
disjunct/disjunctive disjunkt, zerstückelt, voneinander isoliert
disjunction Trennung, Verteilung; gen Disjunktion (der Tochterchromosomen)
- **alternate disjunction** alternierende Verteilung
- **nondisjunction** Non-Disjunktion, Chromosomenfehlverteilung
- **nonrandom disjunction** nicht- zufallsgemäße Verteilung

disjunction/discontinuity/isolation Disjunktion, Isolierung, Isolation
disk (disc) Scheibe
- **adhesive disk** Haftplatte, Haftscheibe, Saugnapf; Kletterorgan
- **articular disk/meniscus/disk** Gelenkmeniskus, Gelenkzwischenscheibe, Meniskus, Diskus
- **basal disk/pedal disk (anthozoa)** Fußscheibe
- **cross-hair disk** *micros* **(eyepiece)** Zwischenlegscheibe mit Fadenkreuz (Okular)
- **intervertebral disk** Bandscheibe
- **oral disk/peristome/peristomium** Oralscheibe, Mundscheibe, Peristom
- **suction disk** Saugscheibe, Saugnapf

disk/meniscus (articular disk) Meniskus, Diskus (Gelenkmeniskus/ Gelenkzwischenscheibe)
disk diaphragm (annular aperture) *micros* Ringblende
disk electrophoresis Diskelektrophorese, diskontinuierliche Elektrophorese
disk flower/disk floret/tubular flower Scheibenblüte, Röhrenblüte (Asterales)
disk turbine impeller Scheibenturbinenrührer
disk-shaped scheibenförmig
dislocate dislozieren, verlagern
dislocation *gen* Dislokation, Verlagerung (von Chromosomenabschnitten)
disomic disom
disomy Disomie
disorder Ordnungslosigkeit
disorder/disease *med* Störung, Krankheit, Erkrankung
- **heritable disorder** erbliche Erkrankung, Erbkrankheit

dispersal/dissemination/ propagation Ausbreitung, Streuung, Propagation
- **animal-dispersal/zoochory** Tierausbreitung
- **ant-dispersal/myrmecochory** Ameisenausbreitung
- **ballistic dispersal** Schleuderausbreitung
- **bat-dispersal/chiropterochory** Fledermausausbreitung
- **bird-dispersal/ornithochory** Vogelausbreitung
- **passive dispersal** passive Ausbreitung, Verdriftung

distribution

- **self-dispersal/autochory**
  Selbstausbreitung
- **sweepstake dispersal**
  Zufallsverbreitung
- **water-dispersal/hydrochory**
  Wasserausbreitung
- **wind-dispersal/anemochory**
  Windausbreitung

dispersal unit/propagule/diaspore/ disseminule Ausbreitungseinheit, Propagationseinheit, Fortpflanzungseinheit, Diaspore

disperse/scatter zerstreuen, dispergieren

dispersion/colloid *chem* Dispersion, Kolloid

dispersion/scattering/spreading Dispersion, Zerstreuung, Dispergierung, Verteilung

dispersive replication *gen* disperse Replikation

displaced loop/displacement loop Verdrängungsschlaufe, Verdrängungsschleife

displacement Verdrängung

displacement activity *ethol* Übersprungshandlung

displacement reaction *biochem* Verdrängungsreaktion

- **double displacement reaction (ping-pong reaction)** doppelte Verdrängungsreaktion, Doppel-Verdrängung (Pingpong-Reaktion)
- **ordered displacement reaction** geordnete Verdrängungsreaktion
- **random displacement reaction** zufällige Verdrängungsreaktion, nicht-determinierte Verdrängungsreaktion
- **single displacement reaction** einfache Verdrängungsreaktion, Einzel-Verdrängung

display *vb* zurschaustellen, zeigen

display *n ethol* Schaustellung (protzig)

display *n* **(apparatus)** Anzeige (*Gerät*)

display behavior Imponierverhalten, Imponiergehabe, Imponiergebaren

disposable gloves Einweghandschuhe

disposable syringe Einwegspritze

dispose of (e.g. trash/waste/ chemicals) wegwerfen, wegschaffen, beseitigen, entsorgen

disposition Disposition, Veranlagung, Anfälligkeit

disrupt (e.g. tissue) zerreißen, zertrümmern (z.B. Gewebe)

disruptive selection/diversifying selection disruptive Selektion/Auslese

dissect *anat* präparieren, sezieren

dissected *bot* zerschnitten

dissecting dish/dissecting pan Präparierschale

dissecting instruments (dissecting set) Präparierbesteck

dissecting microscope Präpariermikroskop

dissecting needle/probe Präpariernadel

dissection *anat* Präparation, Sezierung

disseminate/disperse/spread/ release ausstreuen

dissemination/dispersal/spreading/ releasing Ausstreuung

disseminule/propagule/diaspore Diaspore

dissepiment/partition/cross-wall/ dividing wall/septum Scheidewand, Septe, Septum

dissilient *bot* aufspringend, -platzend

dissimilation/catabolism Dissimilation, Katabolismus, Stoffwechselabbau

dissimilatory dissimilatorisch

dissipate/scatter streuen

dissipation/scattering Streuung

dissociate dissoziieren

dissociation Dissoziation

dissociation constant Dissoziationskonstante (K_i)

dissociation rate Dissoziationsgeschwindigkeit

dissolution/disintegration *chem* Auflösung, Aufschluss

dissolve lösen (*chem*: in einem Lösungsmittel), auflösen

dissolve/disintegrate/break up *chem* aufschließen

dissolved gelöst (lösen)

dissymmetrical/asymmetrical dissymmetrisch, asymmetrisch, unsymmetrisch

distichous/distichate/two-ranked distich, zweizeilig

distichy Distichie

distil/distill/still destillieren

distillate Destillat

distillation Destillation

distillation receiver *chem* Vorlage

distillers' grains/stillage Schlempe (Nassschlempe)

distilling apparatus/still Destilliergerät, -apparatur

distilling flask/retort Destillierkolben

distribute aufspalten, verteilen

distribution/expansion/spread/ spreading Verbreitung (Ausbreitung *see* dispersal)

distribution

distribution *stat/chem* Verteilung
- **bimodal d./two-mode d.**
  bimodale Verteilung
- **statistical d.** statistische
  Verteilung
distribution function *stat*
  Verteilungsfunktion
distribution map *biogeo*
  Verteilungskarte
disturbance value/interference factor Störgröße
distyly/dimorphic heterostyly Distylie
disulfide bond/disulfhydryl bridge/ disulfide bridge Disulfidbindung, Disulfidbrücke
disymmetrical/bilateral/biradial/ bilaterally symmetrical/radially symmetrical disymmetrisch, bilateral
ditch Graben
ditch-grass family/Ruppiaceae Saldengewächse
dithalamous/dithalamic/with two chambers/dithecal dithalam, zweikammerig, zweikämmrig, dithekal
diuresis Diurese, Harnfluss, Harnausscheidung
diurnal tagaktiv
diurnal birds of prey (falcons and others)/Falconiformes Greifvögel
diurnal plant Tagblüher, Tagpflanze
diurnal rhythm Tagesrhythmus (*as opposed to*: Nachtrhythmus)
divaricate (widely divergent) ausgespreizt, sperrig
dive *n* (in air) Sturzflug; (in water) Tauchgang
dive *vb* (in air) im Sturzflug fliegen; (in water) tauchen
diverge divergieren, auseinandergehen, auseinanderstreben
divergence/divergency Divergenz, Auseinanderstreben
divergent circuit *neuro* divergenter Schaltkreis
divergent transcription *gen* divergente Transkription
divers/loons/Gaviiformes Seetaucher
diverse divers, vielfältig
diversiflorous verschiedenblütig
diversity/variability Diversität, Vielfalt, Variabilität, Mannigfaltigkeit
diverticulum/cecum (blind-ended) Divertikulum, Divertikel, Aussackung, Blindsack, Blinddarm, Darmblindsack, Darmdivertikulum, Caecum

divide *vb* gliedern, einteilen
divide/fission/separate teilen
divided unterteilt, gegliedert
divided/parted/partite (divided into parts) geteilt, gegliedert, unterteilt
dividing wall/cross-wall/partition/ dissepiment/septum Scheidewand, Septe, Septum
diving bell (water spiders) *arach* Tauchglocke
division Division, Teilung, Unterteilung, Gliederung, Einteilung
- **subdivision** Untergliederung, Unterteilung
division/fission/separation *cyto* Teilung
- **binary fission/bipartition** binäre Zellteilung, Zweiteilung
- **cell division/cytokinesis** Zellteilung, Cytokinese, Zytokinese
- **equatorial division** Äquatorialteilung
- **longitudinal division/fission** Längsteilung
- **multiple fission** Vielfachteilung, Mehrfachteilung
- **nuclear division/mitosis (karyokinesis)** Kernteilung, Mitose
- **reduction division/meiosis** Reduktionsteilung, Reifeteilung, Meiose
division/phylum Abteilung, Phylum
division phase Teilungsphase
dixenous/dixenic dixen, zweiwirtig
dizygosity Dizygotie, Zweieiigkeit
dizygous/dizygotic dizygot, zweieiig
DNA (deoxyribonucleic acid) DNS (Desoxyribonucleinsäure/ Desoxyribonukleinsäure), DNA
- **$3' \rightarrow 5'$ (three prime five prime/ three prime to five prime)** $3' \rightarrow 5'$ (drei Strich-fünf Strich/drei Strich nach fünf Strich)
- **A form** A-Form, A-Konformation
- **alpha-DNA** alpha-DNA
- **ancient DNA** vorgeschichtliche DNA
- **anonymous DNA** anonyme DNA
- **B form** B-Form, B-Konformation
- **C form** C-Form, C-Konformation
- **cccDNA (covalently closed circles DNA)** cccDNA (DNA aus kovalent geschlossenen Ringen)
- **cDNA (complementary DNA)** cDNA (komplementäre DNA)
- **cruciform DNA** kreuzförmige DNA
- **DNA footprint** DNA-Fußabdruck, DNA-Footprint

dominance

- **extragenic DNA** extragene DNA
- **figure eight** Achterform
- **fold-back DNA/snap-back DNA** in sich gefaltete DNA, zurückgebogene DNA
- **foreign DNA** Fremd-DNA
- **junk DNA** unnütze DNA, überflüssige DNA, wertlose DNA
- **linker DNA** Linker-DNA
- **minisatellite DNA** Minisatelliten-DNA
- **native DNA** native DNA
- **oc-DNA (open circle DNA)** oc-DNA (offene ringförmige DNA)
- **passenger DNA** passagere DNA, Passagier-DNA
- **promiscuitive DNA** promiskuitive DNA
- **repetitive DNA** repetitive DNA
- **satellite DNA** Satelliten-DNA
- **selfish DNA** egoistische DNA
- **single copy DNA** Einzelkopie-DNA, nichtrepetitive DNA
- **stuffer DNA** Stuffer-DNA
- **Z-form** Z-Form, Z-Konformation

DNA bending DNA-Biegung, DNA-Verbiegung
DNA fingerprinting/DNA profiling DNA-Fingerprinting, genetischer Fingerabdruck
DNA library/DNA bank DNA-Bibliothek, DNA-Bank
DNA polymerase DNA-Polymerase
DNA repair DNA-Reparatur
- **dark repair/light independent DNA repair** lichtunabhängige DNA-Reparatur
- **excision repair** Exzisionsreparatur
- **light repair** Lichtreparatur
- **mismatch repair** Fehlpaarungsreparatur

DNA replication (see also: replication) DNA-Replikation
DNA sequencer DNA-Sequenzierungsautomat
DNA sequencing DNA-Sequenzierung
DNA strand DNA-Strang
DNA synthesis DNA-Synthese
- **unscheduled DNA synthesis** außerplanmäßige DNA-Synthese

DNA tumor virus DNA-Tumorvirus
DNA-binding protein DNA-bindendes Protein
DNA-dependent-DNA polymerase DNA-abhängige-DNA Polymerase
DNA-world DNA-Welt
DOC (dissolved organic carbon) gelöster organischer Kohlenstoff
docile gelehrig, folgsam, gefügig, fromm (Pferd)

docility Gelehrigkeit, Folgsamkeit, Fügsamkeit, Gefügigkeit
dock *n* **(horses)** Schwanzstumpf, Schwanzstummel, Stummelschwanz
dock/docking *vb* **(horses)** Schwanz stutzen, anglisieren
dock family/buckwheat family/ knotweed family/smartweed family/Polygonaceae Knöterichgewächse
docking protein Docking-Protein, Andockprotein
docoglossate radula docoglosse Radula, Balkenzunge
dodder family/Cuscutaceae Seidengewächse
doe adultes Säugerweibchen, Geiß (Rehgeiß/Hirschkuh; auch: Ziege/ Hase/Känguruh etc.)
dog Hund
- **female dog** Hündin
- **male dog** Rüde

dogbane family/periwinkle family/ Apocynaceae Hundsgiftgewächse, Immergrüngewächse
dogwood family/Cornaceae Hartriegelgewächse, Hornstrauchgewächse
doldrums Kalmen, Kalmengürtel
dolioform/barrel-shaped tonnenförmig
doliolaria larva Doliolaria, Tönnchenlarve
doliolids/Doliolida Tonnensalpen
dolipore Doliporus
DOM (dissolved organic matter) gelöste organische Substanz
domain (tertiary structure) Domäne
domatium (lodging for insects/ mites) Domatium (*pl* Domatien)
dome (ctenophores) Kuppel
dome web *arach* Haubennetz
domestic häuslich, Haus.., heimisch; einheimisch; inländisch, Inlands.., im Inland erzeugt; Kultur…
domestic animal/domesticated animal domestiziertes Tier, Haustier
domestic fowl Haushuhn
domestic variety/cultivated variety/ cultivar Kulturform
domesticate (to make domestic) domestizieren, zu Haustieren/ Kulturpflanzen machen, zähmen, züchten
domesticated animal/domestic animal domestiziertes Tier, Haustier
domestication Domestikation; Zähmung; Kultivierung
dominance Dominanz
- **codominance** Kodominanz

dominance index

- **delayed dominance** verzögerte Dominanz
- **incomplete dominance** Semidominanz, Partialdominanz, unvollständige Dominanz
- **shifting dominance** variable Dominanz

dominance index Dominanzindex
dominance variance Dominanzvarianz
dominant dominant
dominant negative dominant negativ
dominate dominieren, beherrschen, vorherrschen
DON (dissolved organic nitrogen) gelöster organische(r) Stickstoff(verbindungen)
donor Donor, Spender
donor cell Donorzelle, Spenderzelle
doodlebug (antlion larva) Ameisenlöwe (Larve der Ameisenjungfer)
DOP (dioctyl phthalate) smoke DOP (Dioctylphthalat)-Vernebelung
DOP-PCR (degenerate oligonucleotide primer PCR) DOP-PCR (PCR mit degeneriertem Oligonucleotidprimer)
dopamine Dopamin
doridacean snails/doridaceans/Doridacea/Holohepatica Warzenschnecken, Sternschnecken
dories (John Dory) and others/Zeiformes Petersfischartige: Petersfische und Eberfische
dormancy/inactive state *general* Ruhezustand; (endogenous) Dormanz (*see*: quiescence)
dormancy period Ruhephase, Ruheperiode
dormant/resting/quiescent schlafend, ruhend, im Ruhezustand
dormant egg/resting egg (winter egg) Latenzei, Dauerei
dorsal dorsal, rückseitig
dorsal fin Rückenflosse
dorsal horn Flügelplatte (Neuralrohr)
dorsal ocellus Stirnauge, Scheitelauge (Stirn-Ocelle)
dorsal root/posterior root *neuro* Dorsalwurzel
dorsal root ganglion/spinal ganglion/posterior root ganglion Spinalganglion
dorsal shield/carapace (turtles) Rückenschild, Carapax
dorsal suture/dorsal seam Dorsalnaht, Rückennaht

dorsicidal dorsizid, dorsicid, rückenspaltig
dorsifixed dorsifix
dorsiventral/dorsoventral/bifacial dorsiventral, dorsoventral, bifazial, zweiseitig
dorsoventral abdominal vibrating dance (DVAV)/vibrating dance (bees) Schütteltanz, Schüttelbewegung
dosage/dose Dosis
dosage compensation Dosiskompensation
dosage effect Dosiseffekt
dose *vb* dosieren
dose *n* Dosis; Gabe, Portion
- **lethal dose** Letaldosis, letale Dosis, tödliche Dosis
- **median lethal dose (LD$_{50}$)** mittlere Letaldosis, mittlere letale Dosis
- **overdose** Überdosis

dose equivalent *rad* Dosisäquivalent
dose-response curve/dose-effect curve *stat* Dosis-Wirkungskurve
dot blot/spot blot Rundlochplatte
dot diagram *stat* Punktdiagramm
double blind assay Doppelblindversuch
double bond *chem* Doppelbindung
double diffusion/double immunodiffusion (Ouchterlony technique) Doppeldiffusion, Doppelimmundiffusion
double digest Doppelverdau
double displacement reaction (ping-pong reaction) *biochem* doppelte Verdrängungsreaktion, Doppel-Verdrängung (Pingpong-Reaktion)
double fertilization doppelte Befruchtung
double helix (DNA) Doppelhelix
double heterozygote doppelt-heterozygot
double infection Doppelinfektion
double layer/bilayer (membrane) Doppelschicht
double membrane Doppelmembran
double raceme Doppeltraube
double recombination *gen* doppelte Rekombination
double refraction/birefringence Doppelbrechung
double spike Doppelähre
double strand *gen* Doppelstrang
double sugar/disaccharide Doppelzucker, Zweifachzucker, Disaccharid
double umbel (inflorescence) Doppeldolde

97   **driftwood** **D**

double-headed intermediate (enzymatic reaction) *biochem* doppelköpfiges Zwischenprodukt, janusköpfiges Zwischenprodukt
double-strand sequencing *gen* Doppelstrangsequenzierung
double-stranded/two-stranded *gen* zweisträngig
double-working (grafting with interstock) Zwischenveredlung
doubling time (generation time) Verdopplungszeit (Generationszeit)
dough Teig
dough stage (grain) Teigreife
dourine (*Trypanosoma equiperdum*) Beschälseuche
doves & pigeons and allies/ Columbiformes Taubenvögel
dovetail connection *micros* Schwalbenschwanzverbindung
down *orn* Flaum
- **natal down/neossoptile/neoptile (a down feather)** Nestdune, Neossoptile, Neoptile
- **powder-down feather/ pulviplume** Puderdune, Pulvipluma
down feather/down/plumule Daune, Dune, Dunenfeder, Flaumfeder, dunenartige Feder
down-mutation "Down-Mutation"
down-regulation/downregulation *metabol* Herunterregulierung, Herabregulation, Runterregulierung
- **receptor down-regulation** Rezeptor-Ausdünnungsregulation
- **up-regulation/upregulation** *metabol* Hochregulierung, Heraufregulation
downstream abwärts (Richtung 3′-Ende eines Polynucleotids)
- **upstream** stromaufwärts, aufwärts (Richtung 5′-Ende eines Polynucleotids)
downward classification Herunterstufung
downward stroke of wing Flügelabschlag
downy mildews/Peronosporaceae falsche Mehltaupilze
downy/pubescent flaumig, feinstflaumig
Dracaenaceae/dragon-blood tree family Drachenbaumgewächse
draff (malting residue) Treber, Trester (hier speziell: Malzrückstand)
draft animal Zugtier
drag *n aer/orn* Luftwiderstand, Strömungswiderstand
drag *vb* schleppen, schleifen, ziehen
drag effect *physio* Schleppeffekt

dragline *arach* Schleppfaden, Schleppleine, Zugleine
dragon-blood tree family/ Dracaenaceae Drachenbaumgewächse
dragonflies (anisopterans) and damselflies (zygopterans)/ Odonata Libellen
drain *vb* entwässern, drainieren, abfließen/ablaufen lassen; (bog/ swamp) trockenlegen (Moor/Sumpf)
drainage/draining Drainage, Abfluss, Ablauf, Entwässerung, Trockenlegung
drainage basin/drainage area/ catchment basin/catchment area/ watershed Wassereinzugsgebiet, Grundwassereinzugsgebiet, Flusseinzugsgebiet, Sammelbecken
drainage channel Entwässerungskanal, Ablaufrinne
drainage ditch Entwässerungsgraben
drainage water/leachate/soakage/ seepage/gravitational water Sickerwasser
drake (♂ duck) Enterich, Erpel
drepanium (a helicoid cyme) *bot* Drepanium, Sichel
drepanoid/sickle-shaped/crescent/ falcate/falciform sichelförmig
dress (coat or treat with fungicides/pesticides) beizen (Saatgut)
dressing/fertilizing material *n agr* Dünger, Düngmittel, Düngung
dressing (removing feathers and blood from birds) Geflügel ausbluten lassen und rupfen (küchenfertig machen)
dressing agent (pesticides/ fungicides) Saatgutbeizmittel
drift Drift; Verschiebung; *meteo* Verwehung; Fluktuation
- **antigen drift/antigenic drift** Antigendrift
- **continental drift** *geol* Kontinentaldrift, Kontinentalverschiebung
- **genetic drift/Sewall Wright effect** Gendrift, genetische Drift
- **glacial drift** *geol* Glazialgeschiebe
- **random drift (Sewall-Wright)** zufallsbedingte Drift, Zufallsdrift, ungerichtete Fluktuation
- **snow drift** *meteo* Schneewehe, Schneeverwehung
- **steady drift** gerichtete Fluktuation
drift line/intertidal fringe (on shore) Spülsaum
driftwood Treibholz

drill

drill *n agr* Saatrille; Drillreihe (drill row)
drill core *geol/paleo* Bohrkern
drill furrow *agr* Saatrille, Drillfurche
drink *vb* trinken
drinkability/potability Trinkbarkeit
drinkable/potable trinkbar
drinking water Trinkwasser
drip *vb* tropfen
drip irrigation/trickle irrigation Tropfbewässerung, Tröpfchenbewässerung
drip tip (leaf) *bot* Träufelspitze
drive *n ethol* Antrieb, Trieb
• sex drive Sexualtrieb
driving potential Antriebspotential
drone (bee) Drohne, Drohn
drooping schlaff herabhängend
drooping funnel *lab* Tropftrichter
drooping leaf herunterhängendes Blatt
drop *n* (of a liquid) Tropfen
droplet infection Tröpfcheninfektion
dropper Tropfglas, Tropfpipette
dropping bottle Tropfflasche
• dropper vial Pipettenflasche
dropping funnel *lab* Tropftrichter
droppings Tierexkremente, Dung, Tiermist
Droseraceae/sundew family Sonnentaugewächse
drought Dürre
drought avoidance Dürrevermeidung
drought hardiness/drought tolerance Dürrehärte, Dürrefestigkeit, Dürrebeständigkeit
drought resistance Dürreresistenz
drought tolerance Austrocknungstoleranz
drought-avoiding dürremeidend
drought-enduring dürreertragend, dürreüberdauernd
drought-evading dürremeidend
drought-resistant (xerophytic) dürreresistent, dürrefest, trockenresistent
drought-tolerant dürretolerant, dürreduldend
drug Droge
• addicted to drugs drogenabhängig, drogensüchtig
• herbal drug Pflanzendroge
• to be drugged unter Drogen stehen
drug abuse Drogenmissbrauch
drug addiction Drogenabhängigkeit, Sucht
drug design zielgerichtete "Konstruktion" neuer Medikamente am Computer

• computer-aided drug design (CADD) computerunterstützte Planung von Wirkstoffen
drumlin (elongate/oval hill of glacial drift) Drumlin, Drummel, langgestreckter Moränenhügel (Rückenberg/Schildberg)
drumstick/leg (fowl) Keule, Schlegel
drupaceous fruit Steinobst
drupe/drupaceous fruit/stone Steinfrucht
druse/granule Druse
dry/arid *adv/adj* trocken
dry *vb* trocknen
dry farming/dryland farming Trockenkultur, Trockenlandwirtschaft
dry fruit Trockenfrucht
dry mass/dry matter Trockenmasse, Trockensubstanz
dry matter Trockensubstanz
dry meadow/arid grassland Trockenrasen
dry rot Trockenfäule
dry rot family/Coniophoraceae Kellerschwämme, Warzenschwämme
dry spell/drought Trockenperiode
dry wash/dry valley (wadi) Trockental (Wadi)
dry weight (*sensu stricto*: dry mass) Trockengewicht (*sensu stricto*: Trockenmasse)
drying bed Trockenbeet (Kläranlage)
drying cabinet (plant-drying cabinet) Trockenschrank
dryness/drought Trockenheit, Dürre
Dryopteridaceae/dryopteris family/ male fern family/Aspidiaceae/ aspidium family Wurmfarngewächse
duckweed family/Lemnaceae Wasserlinsengewächse
duckweed fern family/mosquito fern family/Azollaceae Algenfarngewächse
duct *anat* Gang, Kanal
duct/passageway Ausführgang, Ausführkanal
ductility Duktilität, Dehnbarkeit, Streckbarkeit
duetting *orn* Duettgesang, Paargesang
duff (raw humus) humöser/ humusartiger Waldboden (Rohhumus)
Dufour's gland/alkaline gland (hymenopterans) Dufour-Drüse
dulosis (ants) Dulosis, Sklavenhaltung, Sklavenhalterei
dun (horse: black points/dorsal stripe) Falbe

dun (subadult/sub-imago of mayflies) Subimago der Eintagsfliegen; künstl. Angelfliege

dune Düne
- **barchan/crescentic dune/ crescent-shaped dune** Sicheldüne, Bogendüne, Barchan
- **blowout dune** Deflationsdüne, Haldendüne
- **brown dune** Braundüne
- **coastal dune** Küstendüne
- **dome dune** Kuppeldüne, Haufendüne
- **foredune** Vordüne
- **inland dune** Binnendüne, Inlandsdüne, Innendüne, Festlandsdüne, Kontinentaldüne
- **lineal dune** Strichdüne, Silk-Düne
- **linguoid dune** Zungendüne
- **parabolic dune** Paraboldüne, Parabeldüne
- **primary dune** Primärdüne
- **secondary dune (yellow dune/ white dune)** Sekundärdüne (Gelbdüne/Weißdüne)
- **seif dune/longitudinal dune** Seif, Längsdüne, Longitudinaldüne
- **shifting dune/mobile dune/ migratory dune** Wanderdüne
- **shore dune** Stranddüne
- **shrub-coppice dune/nebkha** Kupste, Kupstendüne
- **star dune** Sterndüne, Pyramidendüne
- **tertiary dune (grey dune)** Tertiärdüne (Graudüne)
- **transverse dune** Tranversaldüne, Querdüne

dune field Dünenfeld

dung/manure Dung (tierische Exkremente)

dung-fly flower/sapromyophile Aasfliegenblume, Sapromyiophile

duodenal gland/Brunner's gland Duodenaldrüse, Brunnersche Drüse

duodenum Duodenum, Zwölffingerdarm

dura mater/pachymeninx (tough membrane around brain) Dura mater, Pachymeninx, harte Gehirnhaut, harte Hirnhaut

durability/shelf-life Haltbarkeit

durability (wood) Verwitterungsbeständigkeit

duramen/heartwood Kernholz, Hartholz

durango root family/datisca family/ Datiscaceae Scheinhanfgewächse

dust cell (large alveolar macrophage: a pulmonary histiocyte) Staubzelle, Körnchenzelle, Rußzelle (Alveolarmakrophage)

dust plug *micros* **(nosepiece)** Schutzkappe

Dutchman's-pipe family/birthwort family/Aristolochiaceae Osterluzeigewächse

dwarf *n* Zwerg

dwarf male Zwergmännchen

dwarf mutant Zwergmutante

dwarf vegetation Zwergvegetation

dwarf-shrub/chamaephyte Zwergstrauch

dwarfed growth/dwarfishness/ dwarfism/stunted appearance/ nanism/microsomia Zwergwuchs, Nanismus, Kümmerwuchs

dwell *vb* sich aufhalten, leben, wohnen

dwelling Wohnquartier, Behausung

dwelling structures/domichnia *paleo* Wohnbauten

dy/gel mud Dy, Torfschlamm

dyability/stainability Anfärbbarkeit

dyable/stainable anfärbbar

dyad Dyade

dye/add color/add pigment *vb* färben, einfärben; (stain) anfärben

dye/colorant/pigment Farbstoff, Pigment
- **vital dye/vital stain** Vitalfarbstoff

dyeable/stainable anfärbbar

dyeing/staining Anfärbung

dynamic soaring dynamischer Segelflug

dynein Dynein

dynorphin Dynorphin

dysentery Ruhr

dysmorphic dysmorph

dysmorphy Dysmorphie

dysodont dysodont

dysphotic zone *limn* dysphotische Zone, Dämmerzone

dysplasia Dysplasie

dyspnea Dyspnoe

dystrophic *ecol* dystroph (nährstoffarm und humusreich)

dystrophic/wrongly nourished/ inadequately nourished *physio* dystroph, schlecht ernährt, mangelhaft ernährt

ear

ear Ohr
ear/cob (corn) Getreideähre, Fruchtstand des Getreides; Kolben (Mais)
"ear bone"/"ear stone"/otolith Hörsteinchen, Gehörstein, Otolith
ear conch/auricle/external ear/ outer ear/pinna Ohrmuschel, äußeres Ohr, Pinna
ear opening/auditory meatus Ohrenöffnung
eardrum/tympanic membrane/ tympanum Trommelfell, Ohrtrommel, Tympanalmembran, Tympanum
earlobe Ohrläppchen
early bloomer Frühblüher
early gene frühes Gen
early protein *vir* Frühprotein
earlywood/springwood Frühholz, Weitholz, Frühlingsholz
earth/ground/soil Erde, Boden
Earth/World Erde, Welt
earth balls (Geastraceae) Erdsterne; (Sclerodermataceae) Hartboviste
earth history/history of the Earth/ geologic history Erdgeschichte
earth tongues/Geoglossaceae Erdzungen
Earth history/history of the Earth/ geologic history Erdgeschichte
Earth science/geology Geologie
earwax/cerumen Ohrenschmalz, Cerumen
earwigs/Dermaptera Ohrwürmer
East Indian pitcher plant family/ nepenthes family/Nepenthaceae Kannenpflanzengewächse
easterlies Ostwinde
eat essen
eat into/corrode *chem* ätzen, korrodieren
eat through *chem* hindurchfressen
eatable/edible essbar, genießbar
- **uneatable/inedible** nicht essbar, ungenießbar
ebb/low tide/ebb tide Ebbe
ebony family/Ebenaceae Ebenholzgewächse
eccrine ekkrin
ecdysis/molt/molting Ekdyse, Ecdysis, Häutung, Federverlust, Haarverlust
- **loosing feathers** Federverlust
- **loosing/shedding hair** Haarverlust
- **shedding skin** Häutung
ecdysone Ecdyson
ecesis (pioneer stage of dispersal to a new habitat) *ecol* Neubesiedlung
echinate igelborstig

echinocyte/crenocyte/burr cell Echinozyt, Stechapfelform (>*Erythrozyt*)
echinoderms/Echinodermata Echinodermen, Stachelhäuter
echinulate/with small bristles kleinborstig, kleindornig
echiuroid worms/spoon worms/ Echiura Echiuriden, Igelwürmer, Stachelschwänze
echolocation Echolotpeilung, Echoortung
eclipse period/eclipse *vir* Eklipse
eclipse plumage/inconspicuous plumage (drake/duck) Schlichtkleid
eclosion (insects: hatching from egg/larva/pupa) Schlüpfen (Insekt aus Ei/Larve/Puppe); Entpuppung
ecobalance (life cycle assessment/ analysis) Ökobilanz
ecocline (gradient of vegetation and biotopes) Ökocline, Ökokline, Ökoklin
ecogenetics Ökogenetik
ecogram Ökogramm
ecological ökologisch
ecological balance ökologisches Gleichgewicht
ecological diversity/biodiversity ökologische Vielfalt
ecological efficiency ökologische Effizienz, ökologischer Wirkungsgrad
ecological niche ökologische Nische
ecological potency ökologische Potenz
ecological pyramid ökologische Pyramide
ecological succession ökologische Sukzession
ecological valency/valence ökologische Valenz
ecologist Ökologe
ecology Ökologie
- **anthecology/pollination ecology** Blütenökologie
- **autecology** Autökologie
- **behavioral ecology** Verhaltensökologie, Ethökologie
- **geoecology/environmental geology** Geoökologie
- **habitat ecology** Standortlehre
- **human ecology** Humanökologie
- **landscape ecology** Landschaftsökologie
- **paleoecology** Paläoökologie, Palökologie
- **phytoecology/plant ecology** Pflanzenökologie, Vegetationskunde, Vegetationsökologie

101   **egg** **E**

- **population ecology**
  Populationsökologie, Demökologie
- **predictive ecology**
  vorausschauende Ökologie,
  voraussagende Ökologie
- **synecology** Synökologie
- **systems ecology**
  Systemökologie
- **terrestrial ecology** terrestrische
  Ökologie, Festlandsökologie,
  Epeirologie
- **urban ecology** Urbanökologie,
  Stadtökologie
**economic plant/useful plant/crop
plant** Nutzpflanze,
Weltwirtschaftspflanze,
Wirtschaftspflanze
ecophene Ökophän *nt*
ecophenotypy Ökophänotypie
ecospecies Ökospezies
ecosphere Ökosphäre
ecosystem Ökosystem
- **agroecosystem/agricultural
  ecosystem** Agrarökosystem,
  Agroökosystem
ecotone Ökoton,
Übergangsgesellschaft
ecotope Ökotop
ecotropic ecotropisch
ecotype Ökotyp
ecozone Ökozone
ectocarp/epicarp/exocarp Ektokarp
ectocochlea/external shell
Ectocochlea, Außenschale
ectoderm/outer germ layer
Ectoderm, Ektoderm, primäres
Keimblatt, äußeres Keimblatt
**ectoparasite/exoparasite/epizoon
(*see also*: skin parasite)**
Ektoparasit, Exoparasit,
Außenparasit (*siehe*: Hautparasit)
ectopic ektopisch, verlagert (an
unüblicher Stelle liegend/auf
unübliche Weise)
ectopic pairing (of chromosomes)
ektopische Paarung (unspezifische
Paarung von Chromosomen)
ectopy Ektopie (an unüblicher
Stelle/auf unübliche Weise)
edaphic edaphisch
eddy/swirl Strudel
eddy current Wirbelstrom (Vortex-
Bewegung)
edentate/toothless zahnlos
**edentates/"toothless" mammals/
xenarthrans/Edentata/Xenarthra**
Zahnarme, Nebengelenktiere
edge/margin Rand
edge effect *ecol* Randeffekt
edible/eatable essbar, genießbar
- **inedible/uneatable** nicht essbar,
  ungenießbar

editing Editieren, Redigieren
- **RNA editing** Redigieren von RNA
Edman degradation Edmanscher
Abbau
eel-grass family/Zosteraceae
Seegrasgewächse
eel-like/anguilliform aalartig,
anguilliform
eels/Anguilliformes Aalfische,
Aalartige
effect *n* Wirkung
effect *vb* bewirken, verursachen,
veranlassen
- **affect** *vb* betreffen, sich
  auswirken auf; *med* angreifen,
  befallen
effective stroke/power stroke
Wirkungsschlag, Kraftschlag
effector organ Erfolgsorgan
effector T cell T-Effektorzelle
efferent ausführend, wegführend,
ableitend (Flüssigkeit); absteigend
efficiency Effizienz, Wirkungsgrad
efficiency of plating
Plattierungseffizienz
effluent Ablauf, Ausfluss
(herausfließende Flüssigkeit)
efflux Ausstrom
egest/excrete ausscheiden (Exkrete/
Exkremente)
egestion/excretion Ausscheidung,
Exkretion
egg/egg cell/ovum (female gamete)
Ei, Eizelle (weibliche
Geschlechtszelle)
- **dormant egg/resting egg
  (winter egg)** Latenzei, Dauerei
- **mosaic egg** Mosaikei
- **nurse egg/trophic egg** Nährei
- **parthenogenetic egg** Subitanei,
  Jungfernei
- **regulative egg** Regulationsei
- **resting egg/dormant egg
  (winter egg)** Latenzei, Dauerei
- **roe (fish eggs esp. enclosed in
  ovarian membrane)** Rogen
  (Fischeier innerhalb der
  Eierstöcke), Fischlaich
- **trophic egg/nurse egg** Nährei
**egg (reproductive body: embryo &
nutrients & hard shell)** Ei
(Fortpflanzungseinheit: Embryo &
Nährstoffe & Schale)
- **amniote egg** amniotisches Ei,
  Ei amniotischer Tiere
- **cleidoic egg/shelled egg/"land
  egg" (reptiles/birds)** kleidoisches
  Ei, beschaltes Ei
- **clutch (nest of eggs)** Gelege,
  Eigelege, Brut, Nest mit Eiern
- **fish eggs (roe)** Fischeier,
  Fischlaich (Rogen)

egg burster

- hatch eggs/brood eggs ausbrüten
- lay eggs/deposit eggs Eier legen, ablegen
- roe (crustaceans: lobster eggs) Eier
- spawn n (many small eggs of aquatic animals: esp. fish/mollusks) Laich

egg burster/hatching spine (insects) Eizahn, Oviruptor
egg capsule/ovicapsule/ootheca Eikapsel, Oothek
egg case Eiertasche, Eierbeutel
egg cell/egg/ovum (female gamete) Eizelle, Ei (weibliche Geschlechtszelle)
egg cell/ovocyte/oocyte (before and during meiosis) Eizelle, Ovozyt, Oozyt, Ovocyt, Oocyt (vor und während Meiose)
egg culture medium Eiernährboden
egg glue Eierleim
egg guide/gonapophysis Gonapophyse
egg jelly (amphibians) gallertige Eihülle
egg medium Eiernährmedium
egg membrane Eihaut, Eihülle, "Eimembran", Oolemma
egg raft (gastropods/culicids) Eischiffchen, Eiplatte
egg sac (copepods) Eisäckchen
egg sac/"cocoon" arach Eisack, Eipaket, Eikokon
egg string Laichschnur, Laichkette
egg tooth (reptiles) Eizahn, Eischwiele
egg tube/ovarian tube/ovariole Eiröhre, Eischlauch, Ovariole, Ovariolschlauch (Insekten)
egg white/egg albumen Eiweiß
- denatured egg white denaturiertes Eiweiß
- native egg white natives Eiweiß, Eiklar

egg yolk/yolk/vitellum Eidotter, Dotter, Eigelb
- centrolecithal (yolk aggregated in center) zentrolezithal, centrolecithal, Dotter im Zentrum
- isolecithal (yolk distributed nearly equally) isolezithal, isolecithal, Dotter gleichmäßig verteilt
- mesolecithal (with moderate yolk content) mesolezithal, mesolecithal, mäßig dotterreich
- oligolecithal (with little yolk) oligolezithal, oligolecithal, mikrolecithal, dotterarm
- polylecithal (with large amount of yolk) polylezithal, polylecithal, makrolecithal, dotterreich
- telolecithal (yolk in one hemisphere) telolezithal, telolecithal, Dotter an einem Pol

egg-laying/oviparous adj/adv eierlegend, ovipar
egg-laying/egg deposition/deposit of eggs/oviposition n Eiablage, Oviposition
egg-laying apparatus/egg-laying organ/egg depositor/ovipositor Legeapparat, Legeorgan, Ovipositor (Insekten)
egg-rolling ethol Eirollbewegung
egg-shaped/ovate eiförmig
egg-string Eischnur
eggshell Eischale; (insect egg: chorion) Chorion
eglandulous/eglandular drüsenlos
ejaculate/discharge sperm vb ejakulieren, Samen ausspritzen, sich entsamen
ejaculate/discharged sperm n Ejakulat, ausgespritzte Samen(flüssigkeit)
ejaculation/seminal discharge Ejakulation, Samenerguss, Samenausstoß
ejaculatory duct/ductus ejaculatorius Ausspritzungsgang, Samenausführgang, Samengang
ejection device/ballistic device Schleudervorrichtung
ejectisome/ejectosome (an extrusome) Ejectisom
Elaeagnaceae/oleaster family Ölweidengewächse
Elaeocarpaceae/makomako family Elaeocarpusgewächse
elaiosome Elaiosom, Ölkörper (Samen)
Elaphoglossaceae/elephant's-ear fern family Zungenfarngewächse
elastic elastisch
elastic cartilage elastischer Knorpel
elastic fiber elastische Faser
elasticity Elastizität
elastin Elastin
elastotubule Elastotubulus (pl Elastotubuli)
elater Elatere, Schleuderzelle
Elatinaceae/waterwort family Tännelgewächse
elbow/cubitus Ellenbogen, Cubitus
- bend of the elbow/bend of the arm/crook of the arm/inside of the elbow Armbeuge, Ellenbeuge

elbow bone/ulna Elle, Ulna
electric rays/Torpediniformes elektrische Rochen, Zitterrochen

elm family **E**

electricity (*colloquial:* power/juice)
Elektrizität, Strom
electrocardiogram
Elektrokardiogramm (EKG)
electroencephalogram
Elektroencephalogramm (EEG)
electrogenic elektrogen
**electroimmunodiffusion/counter
immunoelectrophoresis**
Elektroimmunodiffusion
electromotive force (emf/E.M.F.)
elektromotorische Kraft (EMK)
electron acceptor
Elektronenakzeptor,
Elektronenraffer
electron carrier
Elektronenüberträger
electron donor Elektronendonor,
Elektronenspender
**electron energy loss spectroscopy
(EELS)** Elektronen-Energieverlust-
Spektroskopie
electron micrograph
elektronenmikroskopisches Bild,
elektronenmikroskopische
Aufnahme
**electron spin resonance (ESR)/
electron paramagnetic resonance
(EPR)** Elektronenspinresonanz
(ESR)
electron transfer
Elektronenübertragung
electron transport
Elektronentransport
• **cyclic e.t.** zyklischer, cyclischer
Elektronentransport
• **noncyclic e.t.**
Elektronentransportnichtzyklischer,
nichtcyclischer, linearer
Elektronentransport
electron-transport chain
Elektronentransportkette
electroneutral (electrically silent)
elektoneutral
electronic elektronisch
electrophilic attack electrophiler
Angriff
electrophorese *vb* elektrophoretisch
auftrennen
electrophoresis Elektrophorese
• **alternating field gel
electrophoresis** Wechselfeld-
Gelelektrophorese
• **capillary electrophoresis**
Kapillarelektrophorese
• **carrier electrophoresis**
Trägerelektrophorese
• **countercurrent electrophoresis**
Gegenstromelektrophorese,
Überwanderungselektrophorese
• **cross field electrophoresis (CEP)**
Kreuzelektrophorese

• **disk electrophoresis/
discontinuous electrophoresis**
Diskelektrophorese,
diskontinuierliche Elektrophorese
• **direct transfer electrophoresis**
Direkttransfer-Elektrophorese,
Blotting-Elektrophorese
• **free electrophoresis (carrier-free
electrophoresis)** freie
Elektrophorese
• **gel electrophoresis**
Gelelektrophorese
• **multilocus enzyme
electrophoresis (MLEE)**
Multilokus-Enzymelektrophorese
• **paper electrophoresis**
Papierelektrophorese
• **pulsed field gel electrophoresis
(PFGE)** Puls-Feld-
Gelelektrophorese
• **zone electrophoresis**
Zonenelektrophorese
electrophoretic elektrophoretisch
electrophoretic mobility
elektrophoretische Mobilität
electroplaque Elektroplaque (*pl*
Elektroplaques/*slang:* Elektroplaxe)
electroporation Elektroporation
electroretinogram
Elektroretinogramm (ERG)
electrotonic potential
elektrotonisches Potential
elementary body
Elementarkörperchen
**elephants and relatives/
Proboscidea** Rüsseltiere
elephant birds/Aepyornithiformes
Elefantenvögel, Madagaskarstrauße
**elephant shrews, African/
Macroscelidea** Rüsselspringer
**elephant's-ear fern family/
Elaphoglossaceae**
Zungenfarngewächse
elfin forest/elfin woodland
Zwergwald, Zwergwaldstufe
elicitation (of a reaction) Auslösung
(einer Reaktion)
elicitor Elicitor, Auslöser
eliminate/eradicate/extirpate
eliminieren, ausrotten, ausmerzen
**ELISA (enzyme-linked
immunosorbent assay)** ELISA
(enzymgekoppelter
Immunadsorptionstest,
enzymgekoppelter
Immunnachweis)
ellagic acid/gallogen Ellagsäure
ellipsoidal joint/condyloid joint
Ellipsoidgelenk, Eigelenk
elliptic/elliptical elliptisch
elm family/Ulmaceae
Ulmengewächse

elodea family

elodea family/tape grass family/ frog-bit family/Hydrocharitaceae Froschbissgewächse
elongation/extension Elongation, Streckung, Verlängerung
- **region of elongation (growth)** Streckungszone

elongation factor *gen* Elongationsfaktor
elongational growth/extension growth Streckungswachstum
eluate *n* Eluat
eluate *vb* eluieren
elucidate (interrelationships/ chemical structures) aufklären (Zusammenhänge/Strukturen)
elucidation Aufklärung (Strukturen, Zusammenhänge)
eluent/eluant Elutionsmittel, Eluens (Laufmittel)
eluotropic series eluotrope Reihe (Lösungsmittelreihe)
eluting strength (eluent strength) Elutionskraft
elution Elution, Auswaschen, Auswaschung, Herausspülen
elutriate auswaschen, schlämmen, reinigen
elutriation *ecol* Auswaschen, Auswaschung, Schlämmung
eluviation Auswaschung
elver junger Aal
elytron/elytrum (*pl* **elytra)/wing sheath/wing cover/wing case (insects)** Elytre, Deckflügel, Flügeldecke
emanate hervorquellen
emarginate/shallowly notched ausgerandet
emasculation (Blüte) Emaskulation, Entmannung, Kastrierung
embankment künstliche Böschung/ Damm
Embden-Meyerhof pathway/ Embden-Meyerhof-Parnas pathway (EMP pathway)/ hexosediphosphate pathway/ glycolysis Embden-Meyerhof-Weg, Glykolyse
embed einbetten
embedded specimen eingebettetes Präparat
embedding Einbettung
embedding machine/embedding center Einbettautomat, Einbettungsautomat
embolism (obstruction) Embolie
embolium (insect wing) Embolium
embolus Embolus
emboly/invagination Embolie, Invagination, Einfaltung, Einstülpung

embryo Embryo, Keimling
embryo sac (gametophyte) *bot* Embryosack, Keimsack
embryo transfer Embryotransfer
embryoid/somatic embryo Embryoid, somatischer Embryo
embryonal/embryonic embryonal
embryonal development/embryonic development/embryogenesis/ embryogeny Embryonalentwicklung, Embryogenese, Embryogenie, Keimesentwicklung
embryonation Embryonenbildung
embryonic shell Embryonalschale, Larvenschale, Primärschale
- **prodissoconch (mollusks: bivalves)** Prodissoconch
- **protoconch (mollusks: gastropods)** Protoconch

embryonic stalk/body stalk/ connecting stalk Bauchstiel
emerge/develop/unfold entwickeln, entstehen
emerge (e. g. rise from a fluid) herausragen, herauskommen, hervorkommen, hervortreten, auftauchen (>Wasser)
emergence Emergenz, Auswuchs
emergency Notfall
emergency response Notfalleinsatz
emigration Emigration, Abwanderung, Auswanderung
emission Emission, Ausstoß, Ausstrahlung
emissivity Strahlungsvermögen, Emissionsvermögen (Wärmeabstrahlvermögen)
emissivity coefficient (absorptivity coefficient) Emissionskoeffizient
emit ausstrahlen, abstrahlen, aussenden, emittieren; absondern, ausscheiden; ausströmen, verströmen
emollient erweichendes Mittel
Empetraceae/crowberry family Krähenbeerengewächse
empiric(al) empirisch
empirical formula empirische Formel
emulsification Emulgieren, Emulgierung
emulsifier Emulgator, Emulgierungsmittel
emulsify emulgieren
emulsion Emulsion
enamel (tooth) Zahnschmelz
enamel organ Schmelzorgan
enantiomere Enantiomer
enation (*Lycophyta***)** Auswuchs
enbalm einbalsamieren
encapsulation Einkapselung
encapsule einkapseln

energy source

enclosed pasture/fenced pasture Koppel (Weide)
enclosure (e.g. within zoos) Gehege (Tiergehege)
- **game preserve/game reserve** Wildgehege
- **outdoor enclosure** Freigehege

encode/code *vb* codieren
encrustation/incrustation Inkrustierung
encrusting krustenbildend
encyst encystieren, enzystieren, zystieren
encystment Enzystierung, Encystierung
end labelling Endmarkierung
end moraine Endmoräne
end-bulb/Krause's bulb/Krause's corpuscle Krause'-Endkolben
end-foot (astrocyte) Endfüßchen
end-group analysis/terminal residue analysis Endgruppenanalyse, Endgruppenbestimmung
end-plate current/synaptic current Endplattenstrom
end-plate potential (epp) Endplattenpotential
end-point dilution technique Endpunktverdünnungsmethode (Virustitration)
end-product inhibition/feedback inhibition Endprodukthemmung, Rückkopplungshemmung
endanger (threaten) *ethol* gefährden (bedrohen)
endangered *ethol* gefährdet; *ecol* bedroht
endangered species *ecol* bedrohte Art
endangerment (threat) Gefährdung (Bedrohung)
endbrain/cerebrum/telencephalon Endhirn, Großhirn, Telencephalon
endemic *adv/adj* **(disease/species)** endemisch, auf ein bestimmtes Gebiet beschränkt, lokal begrenzt auftretend
endemic/native *adv/adj* **(e.g. plant)** einheimisch
endemic *n***/endemic species/ endemic organism/endemic lifeform** Endemit
- **neoendemic** Neoendemit, primärer Endemit
- **paleoendemic** Paläoendemit, Reliktendemit

endemic (occurrance of an endemic disease) *n* Endemie
endemism/endemicity Endemismus
endergonic endergon, energieverbrauchend

endite Endit
endocarp Endokarp
endocrine endokrin
endocrine gland endokrine Drüse
endocuticle Endocuticula, Endokutikula
endocytic vesicle/endosome Endozytosevesikel, Endosom
endocytosis Endocytose, Endozytose
- **receptor-mediated endocytosis** rezeptorvermittelte Endozytose, rezeptorgekoppelte Endozytose

endoderm/entoderm Endoderm, Entoderm, inneres Keimblatt, primäres Keimblatt
endodermis Endoderm, Endodermis, Innenhaut
endogamy/inbreeding Inzucht
endolymph Endolymphe
endomitosis Endomitose
endoparasite Endoparasit, Innenparasit
endoplasmic reticulum (ER) (smooth/rough ER) endoplasmatisches Retikulum (ER) (glattes/raues ER)
endopod/endopodite (inner branch) Endopodit (Innenast)
endopolyploidy Endopolyploidie
endoreic/endorheic *limn* endorheisch (Entwässerung im Inland)
endorphin Endorphin
endoskeleton Endoskelett, Innenskelett
endosome/endocytic vesicle Endosom, Endozytosevesikel
endosperm Endosperm, Nährgewebe
endosymbiont Endosymbiont
endosymbiont theory Endosymbiontentheorie
endotheliochorial placenta endothelio-choriale Plazenta
endothelium Endothel
endothermic endotherm
endurance/persistence/hardiness/ perseverance Ausdauer, Dauerhaftigkeit
endure ausdauern
endysis Endysis, Federneubildung (feathers), Fellneubildung (fur/ pelage)
enemy/predator Feind, Fressfeind
energetics Energetik
energy barrier Energiebarriere
energy charge Energieladung
energy flux Energiefluss
energy metabolism Energiestoffwechsel
energy profile Energieprofil
energy requirements Energiebedarf
energy source Energiequelle

energy transfer

energy transfer Energieübergang, Energietransfer
energy-rich energiereich
engorge (e. g. ticks/leeches) mit Blut vollsaugen
engulf einverleiben, verschlingen
enhance *metabol* verstärken
enhancer *metabol* Verstärker (Substanz)
enhancer *gen* Enhancer, Verstärker
enhancer sequence *gen* Verstärkersequenz
enkephalin Enkephalin
enlarge/magnify *micros* vergrößern
enlarged/thickened verdickt
enlargement Erweiterung, Ausdehnung (dicker werden)
enlargement/magnification *micros* Vergrößerung
enology Önologie, Weinbaukunde
enrich anreichern
enrichment Anreicherung
  • **filter enrichment** Filteranreicherung
enrichment culture Anreicherungskultur
enrichment zone/paracladial zone Bereicherungszone
ensiform/gladiate/xiphoid/sword-shaped schwertförmig
ensilage Silospeicherung von Grünfutter
ensile einsilieren, einmieten, Grünfutter in Silo aufbewahren
entelechy Entelechie
enteral/enteric enterisch
enterocoel/"intestine coelom" Enterocöl
enthalpy Enthalpie
entire/simple (leaf margin) ganzrandig (Blatt)
entocodon Glockenkern
entoderm/endoderm/inner germ layer Entoderm, Endoderm, inneres Keimblatt, primäres Keimblatt
entoecism Entökie, Einmietung, Schutzeinmietung
Entoloma family/Entolomataceae Rotblättler
entomology/study of insects Entomologie, Insektenkunde
entomophile/endophile-pollinated flower Insektenblume
entomophilous/insect-pollinated entomophil, insektenblütig
entomophily/insect-pollination Entomophilie, Insektenbestäubung
entrails/innards/viscera/guts (fish viscera etc.) Innereien, Eingeweide
entrainment (rhythm adjustment) Rhythmus-Anpassung (circadian)

entropy Entropie
enucleate (remove nucleus) *vb* entkernen, den Kern entfernen
enucleate cell kernlose Zelle
enurination/urine spraying Harnspritzen
envelope/hull Hülle (*auch:* Viren/ Bakterien); (water: jacket) Wasserhülle
envenom vergiften (Tiergift)
envenomation/envenomization Vergiftung (Tiergift)
environment Umwelt, Milieu
environmental die Umwelt betreffend, Umwelt.., Milieu...
environmental analysis Umweltanalyse
environmental analytics Umweltanalytik
environmental audit Öko-Audit, Umweltaudit
environmental chemistry Umweltchemie
environmental compatibility Umweltverträglichkeit
environmental conditions Umweltverhältnisse, Umweltbedingungen
environmental contamination Umweltverschmutzung
environmental crime Umweltkriminalität
environmental degradation Umweltzerstörung
environmental factor Umweltfaktor, Milieufaktor
environmental geology/geoecology Geoökologie
environmental hazard Umweltgefahr, Umweltgefährdung
environmental impact assessment (EIA) Umweltverträglichkeitsprüfung (UVP)
environmental insult "Umweltschmähung", Angriff auf die Umwelt
environmental medicine Umweltmedizin
environmental monitoring Umweltmessung(en)
environmental politics Umweltpolitik
environmental pollution Umweltverschmutzung
environmental protection/nature protection/environmental conservation/nature conservation/nature preservation Umweltschutz, Naturschutz
environmental requirements Umweltansprüche

107   epicuticle **E**

environmental resistance
Umweltwiderstand
environmental science
Umweltwissenschaft
environmental variance
Umweltvarianz
environmental warden
Landschaftspfleger
**environmentalism (doctrine
emphasizing environmental
factors over hereditary traits)**
"Milieutheorie"
environmentalist Umweltschützer
environmentally compatible
umweltgerecht, umweltverträglich
enzymatic enzymatisch, Enzym...
enzymatic catalysis Enzymkatalyse
enzymatic coupling
Enzymkopplung
enzymatic degradation
enzymatischer Abbau
**enzymatic inhibition/enzymatic
repression/inhibition of enzyme**
Enzymhemmung
enzymatic pathway enzymatische
Reaktionskette
enzymatic reaction Enzymreaktion
**enzymatic specificity/enzyme
specificity** Enzymspezifität
enzyme Enzym, Ferment
● **apoenzyme** Apoenzym
● **coenzyme** Coenzym, Koenzym
● **core enzyme** Kernenzym
(RNA-Polymerase)
● **digestive enzyme**
Verdauungsenzym
● **holoenzyme** Holoenzym
● **isozyme/isoenzyme** Isozym,
Isoenzym
● **key enzyme** Schlüsselenzym,
Leitenzym
● **multienzyme complex/
multienzyme system**
Multienzymkomplex,
Multienzymsystem, Enzymkette
● **processive enzyme** progressiv
arbeitendes Enzym
● **proenzyme/zymogen** Proenzym,
Zymogen
● **repair enzyme** Reparaturenzym
● **restriction enzyme**
Restriktionsenzym
● **tracer enzyme** Leitenzym
**enzyme activation/activation of
enzyme** Enzymaktivierung
enzyme activity (*katal*)
Enzymaktivität (*katal*)
enzyme cascade Enzymkaskade
**enzyme-immunoassay/enzyme
immunassay (EIA)**
Enzymimmunoassay,
Enzymimmuntest (EMIT-Test)

**enzyme-linked immuno-
electrotransfer blot (EITB)**
enzymgekoppelter
Immunoelektrotransfer
**enzyme-linked immunosorbent
assay (ELISA)** enzymgekoppelter
Immunadsorptionstest,
enzymgekoppelter Immunnachweis
(ELISA)
enzyme-substrate complex Enzym-
Substrat-Komplex, Enzym-Substrat-
Zwischenverbindung
**Eocene/Eocene Epoch (geological
time)** Eozän
eon (*pl* eons) (geological time)
Äon *m* (*pl* Äonen), Weltalter
● **Archean Eon/Archeozoic Eon
(early Precambrian)** Archaikum
(Altpräkambrium/frühes
Präcambrium)
● **Phanerozoic Eon/Phanerozoic**
Phanerozoikum
● **Proterozoic Eon/Proterozoic (late
Precambrian)** Proterozoikum
(Jungpräkambrium/spätes
Präcambrium/Eozoikum)
Eophytic Era (geobotanical age)
Eophytikum
**eosuchians (ancient two-arched
reptiles)/Eosuchia/Younginiformes**
Urschuppensaurier
epacris family/Epacridaceae
Australheidegewächse
epaxial epaxionisch
(Rumpfmuskulatur)
ependymal cell Ependymzelle
**ephedra family/mormon tea family/
joint-pine family/Ephedraceae**
Meerträubelgewächse
**ephemeral/taking place or
occurring once only/short-lived**
ephemer, ephemerisch, flüchtig,
vergänglich, vorübergehend; (e. g.
insect/plant) kurzlebig; nur einen
Tag dauernd
ephemere Ephemere
ephippium Ephippium
epiblast Hypoblast
epiblem(a)/rhizodermis
Rhizodermis, Wurzelepidermis
epiboly Epibolie (Umwachsung)
epibranchial furrow
Epibranchialrinne
epicalyx *bot* Außenkelch
epicardium Epikard
epicarp/exocarp/ectocarp
Ektokarp
epicondyle Epikondyle,
Gelenkhöcker
epicotyl *bot* Epikotyl, Epicotyl
epicuticle Epicuticula, Epikutikula,
Grenzlamelle

epideictic behavior epideiktisches Verhalten
epidemic Epidemie, Seuche
epidemiologic(al) epidemiologisch
epidemiology Epidemiologie
epidermal/cutaneous epidermal, Haut.., die Haut betreffend
epidermal growth factor (EGF)/ urogastrone epidermaler Wachstumsfaktor, Epidermiswachstumsfaktor
epidermis Epidermis, primäres Abschlussgewebe; Oberhaut; (certain invertebrates) Hypodermis
epididymis Epididymis, Nebenhoden
epifauna Epifauna
epigean/epigeal/epigeous (insects living on surface) auf der Oberfläche lebend, auf dem Erdboden lebend, epigäisch
epigean germination/epigeal germination epigäische Keimung
epigenetic epigenetisch
epigenetic factors epigenetische Faktoren
epigeous *bot* epigäisch
epiglottic cartilage Kehldeckelknorpel, Schließknorpel
epiglottis Epiglottis, Kehldeckel
epigynous epigyn, unterständig
epigynum Epigyne
epiillumination/epi-illumination/ incident illumination *micros* Auflicht, Auflichtbeleuchtung
epimeletic behavior epimeletisches Verhalten
epimerization Epimerisierung
epimerize epimerisieren
epinephrine/adrenaline Epinephrin, Adrenalin
epipetalous epipetal
epiphyll Epiphyll
epiphyllous (attached/growing on leaves) epiphyll (auf Blättern wachsend)
epiphysis (pineal body) Epiphyse, Zirbeldrüse, Pinealorgan
epiphyte/air plant/aerial plant/ aerophyte Epiphyt, Aufsitzerpflanze, Luftpflanze
epipod/epipodite Epipodit
episepalous episepal
episome *gen* Episom (integrationsfähiges Plasmid)
epistasis Epistase (Unterdrückung des Phänotyps eines nichtallelen Gens)
epitheca/epicone (dinoflagellate frustule) Epitheka
epithelial tissue Epithelgewebe
epitheliochorial placenta epitheliochoriale Plazenta
epitheliomuscular cell/ epitheliomuscle cell Hautmuskelzelle
epitheliomuscular tube Hautmuskelschlauch
epithelium (*pl* **epithelia)** Epithel (*pl* Epithelien)
● **ciliated e.** Flimmerepithel, Wimperepithel, Geißelepithel
● **columnar e.** Zylinderepithel, Säulenepithel
● **cuboidal e.** kubisches Epithel, Pflasterepithel
● **germinal e.** Keimepithel
● **glandular e.** Drüsenepithel
● **olfactory e./nasal mucosa** Nasenschleimhaut, Riechepithel
● **pseudostratified columnar e.** hohes-mehrreihiges Epithel
● **sensory e.** Sinnesepithel
● **simple columnar e.** hochprismatisches Epithel, hohes Zylinderepithel
● **squamous e.** Plattenepithel, Säulenepithel, Zylinderepithel
● **stratified e.** zweischichtiges Epithel, mehrschichtiges Epithel
● **transitional e.** Übergangsepithel
epithet (subordinate unit within genus) Epitheton, Beiname, Zusatzbezeichnung, zusätzliche Bezeichnung
● **specific epithet** Artname, Artbezeichnung (zweiter, kleingeschriebener Teil des Artnames)
epitoke epitok
epitope/antigenic determinant Epitop, Antigendeterminante
● **conformational epitope/ discontinuous epitope** Konformationsepitop
● **continuous epitope/linear epitope** kontinuierliches, lineares Epitop
epizoic (living attached to body of an animal) aufsiedelnd
epizoic/epizoochorous epizoochor (epizoochore Verbreitung)
epizooic/epizootic eine Tierseuche betreffend
epizooic disease/epizooic pest/ livestock epidemic Tierseuche, Viehseuche
epizoon/ectoparasite Ektoparasit
epizoon/epizoan/epizoite Epizoon (*pl* Epizoen), Aufsiedler
epoch (lower/middle/upper *or* **early/middle/late) (see also: eon/ era/period)** Epoche (frühe/mittlere/ späte)
● **Eocene Epoch/Eocene** Eozän

109   **Erlenmeyer flask**   **E**

- **Holocene Epoch/Holocene/ Recent Epoch/Recent** Holozän, Jetztzeit, Alluvium
- **Ice Age/Glacial Epoch/ Pleistocene Epoch/Diluvial** Eiszeit, Glazialzeit, Pleistozän, Diluvium
- **Miocene Epoch/Miocene** Miozän
- **Oligocene Epoch/Oligocene** Oligozän
- **Paleocene Epoch/Paleocene** Paläozän
- **Pleistocene Epoch/Glacial Epoch/ Diluvial/Ice Age** Pleistozän, Diluvium, Glazialzeit, Eiszeit
- **Pliocene Epoch/Pliocene** Pliozän
- **Triassic, Lower** Buntsandstein
- **Triassic, Middle** Muschelkalk
- **Triassic, Upper** Keuper

epoecism Epökie, Aufsiedlung
equal cleavage äquale Furchung, gleichmäßige Furchung
equally pinnate/equally pennate/ paripinnate paarig gefiedert, paarig pinnat, paarig pennat
equate *math* gleichen
equation *math* Gleichung
equation of the xth order Gleichung *x*ten Grades
equatorial cleavage Äquatorialfurchung, äquatoriale Furchung
equatorial division Äquatorialteilung
equidistance Äquidistanz
equilibrium centrifugation/ equilibrium centrifuging Gleichgewichtszentrifugation
equilibrium constant Gleichgewichtskonstante
equilibrium dialysis Gleichgewichtsdialyse
equilibrium organ (static/dynamic) Gleichgewichtsorgan (statisches/ dynamisches)
equilibrium potential Gleichgewichtspotential
equilibrium state Gleichgewichtszustand
equine Pferde betreffend, Perde…
equinox Äquinotikum, Tag-Nacht-Gleiche, Tagundnachtgleiche
Equisetaceae/horsetail family Schachtelhalmgewächse
era (*pl*** eras)/geological era (***see also:*** **eon/epoch/period)** Ära (*pl* Ären), Erdzeitalter

- **Cenozoic Era/Cenozoic/Neozoic Era/Neozoic (Cainozoic Era/ Caenozoic Era)** Känozoikum, Kaenozoikum, Neozoikum, Erdneuzeit

- **Eophytic Era/Eophytic** Eophytikum
- **Mesophytic Era/Mesophytic** Mesophytikum
- **Mesozoic Era/Mesozoic** Mesozoikum, Erdmittelalter
- **Neozoic Era/Neozoic/Cenozoic Era/Cenozoic (Cainozoic Era/ Caenozoic Era)** Neozoikum, Känozoikum, Kaenozoikum, Erdneuzeit
- **Paleophytic Era/Paleophytic** Paläophytikum, Florenaltertum
- **Paleozoic Era/Paleozoic** Paläozoikum, Erdaltertum
- **Precambrian Era/Precambrian** Präcambrium, Präkambrium

eradicate/extirpate/eliminate ausrotten, ausmerzen, eliminieren
eradication/extirpation (pests) Ausrottung, Ausmerzung
erect/strict/upright/straight aufrecht
erect *vb* **(state of penis/clitoris)** erigieren, aufrichten, aufstellen, steif sein
erectile tissue/cavernous tissue/ cavernous bodies Schwellkörper, Corpora cavernosa
erection Erektion, Erigieren, Aufrichten, Aufrichtung
eremophyte/eremad/desert plant Eremiaphyt, Wüstenpflanze
ergastic substance ergastische Substanz
ergate/worker (ants) Ergat, Arbeiterin

- **dinergate/soldier** Dinergat, Soldat
- **gamergate (fertilized, ovipositing worker)** Gamergat
- **macrergate** Makroergat
- **micrergate/microergate/dwarf worker** Mikrergat, Zwergarbeiterin
- **pterergate (with rudiments of wings)** Pterergat (Arbeiterin mit Stummelflügeln)

ergot/calcar metacarpeum *zool* **(horses: horny stub behind fetlock)** Sporn
ergot fungi/ergot family/ Clavicipitaceae Kernkeulen
ergotamine Ergotamin
Erica Erika
Ericaceae/heath family Heidekrautgewächse
Eriocaulaceae/pipewort family Eriocaulongewächse
eriophyllous wollblättrig
Erlenmeyer flask Erlenmeyer Kolben

erosion Erosion
- **gully erosion** Grabenerosion, rinnenartige Erosion (Schluchterosion)
- **pluvial erosion** Regenerosion
- **sheet erosion** Schichterosion, Schichtfluterosion, Flächenerosion
- **soil erosion** Bodenerosion

errant (straying outside proper path/bounds) abweichend, umherirrend
erratic/eccentric/strange abartig, seltsam, unberechenbar, launenhaft, unzuverläßig, exzentrisch, ausgefallen
erratic/free-moving/unattached (often: moved by other agent) frei beweglich, wandernd
error/mistake (defect) Fehler (Defekt)
- **inborn error** *gen* Erbleiden, angeborener Fehler
- **random error** *stat* zufälliger Fehler, Zufallsfehler
- **standard error (standard error of the mean = SEM)** *stat* Standardfehler (des Mittelwerts), mittlerer Fehler
- **statistical error** statistischer Fehler
- **systematic error/bias** *stat* systematischer Fehler, Bias

error in measurement/measuring mistake Messfehler
error of estimation *stat* Schätzfehler
erucic acid/(Z)-13-docosenoic acid Erucasäure, Δ^{13}-Docosensäure
eruciform larva (with more than 5 pairs of abdominal prolegs: Tenthredinidae) Afterraupe (Blattwespenlarven)
erythroblast/normoblast Erythroblast, Normoblast
erythrocyte/red blood cell (RBC) Erythrocyt, Erythrozyt, rotes Blutkörperchen
erythrocyte ghost Erythrozytenschatten, Schatten (ausgelaugtes rotes Blutkörperchen)
erythropoiesis Erythrozytenreifung, Erythropoese
erythropoietin/erythropoiesis-stimulating factor (ESF) Erythropoetin
Erythroxylaceae/coca family Kokastrauchgewächse
escape *vb* entkommen; verwildern
escape traces/fugichnia (trace fossil) Fluchtspuren (Spurenfossil)
escarpment Böschung, steiler Abhang; Steilabbruch

eserine/physostigmine Eserin, Physostigmin
esker (long/narrow ridge/mound of deposited debris along stagnant glacier) Esker, langgestreckter Geschiebehügel (am Gletscher), fluvioglazialer Wallberg
esophageal gland (gastropods) Schlunddrüse
esophagus/oesophagus/gullet Ösophagus, Speiseröhre, Kehle
espalier/trellis Spalier
espalier fruit Spalierobst
essential essentiell, wesentlich, unentbehrlich
essential for life/vital lebenswichtig, lebensnotwendig, vital
essential oil/ethereal oil ätherisches Öl
EST (expressed sequence tag) *gen* exprimierte sequenzmarkierte Stelle (EST)
establish/naturalize/acclimate einbürgern
establish/start (a culture) *micb* etablieren, anzüchten (einer Kultur)
establish an hypothesis/a theory eine Hypothese/Theorie aufstellen
established cell line etablierte Zellinie
establishment/settlement/naturalization/acclimatization Einbürgerung
establishment growth/corroborative growth Erstarkungswachstum
establishment phase Eingewöhnungsphase
esterification Veresterung
esterify verestern
esthetasc/aesthetasc Ästhetask, Riechschlauch
esthete/aesthete (photosensitive structure in chitons) Ästhet (*pl* Ästheten)
estimate *n stat* Schätzwert
estimate/assume *vb* schätzen, annehmen
estimation/estimate/assumption Schätzung, Annahme
estival/aestival frühsommerlich
estivate/aestivate *vb* (pass summer in dormant stage) übersommern
estivation/aestivation Ästivation, Aestivation, Knospendeckung; Sommerschlaf, Übersommerung
estradiol/progynon Östradiol
estrogen Östrogen
estrone Östron, Estron
estrous *adj/adv* östrisch, östral, Brunst..., die Brunst betreffend

estrous cycle/estrus cycle/estral cycle Östruszyklus, Brunstzyklus
estrus Östrus, Brunst (nicht: Brunft! *siehe dort*)
- **anestrus** Anöstrus, Anestrus
- **diestrus** Diöstrus, Dioestrus
- **metestrus** Metöstrus, Metoestrus, Nachbrunst
- **proestrus** Proöstrus, Prooestrus, Vorbrunst

estuarine estuarin, Flussdelta betreffend, Ästuar...
estuarine marsh Flussmarsch (an der Flussmündung/im Flussdelta)
estuary Ästuar, Ästuarium (*pl* Ästuarien) (trichterförmige Flussmündung), Flussdelta
etch *vb metal/techn/micros* ätzen (*see:* freeze etching)
etchant *metal/techn/micros* Ätzmittel
etching *metal/techn/micros* Ätzen, Ätzung, Ätzverfahren (*see:* freeze etching)
ethanol/ethyl alcohol/alcohol Äthanol, Ethanol, Äthylalkohol, Ethylalkohol, Alkohol
- **graded ethanol series** aufsteigende Alkoholreihe

ether Äther, Ether
ethereal oil/essential oil ätherisches Öl
ethmoid bone/os ethmoidale Siebbein
ethogram/behavioral inventory/behavioral retertoire Ethogramm, Verhaltensinventar, Verhaltensrepertoire
ethological/behavioral ethologisch, Verhaltens...
ethological barrier/behavioral barrier Verhaltensbarriere
ethological isolation/behavioral isolation ethologische Isolation
ethology/study of behavior Ethologie, Verhaltensforschung, Verhaltensbiologie
ethylene Äthylen, Ethylen
etiolation Etiolement, Vergeilung
etioplast Etioplast
eucarpic eukarp
euchromatin Euchromatin
eucommia family/Eucommiaceae Guttaperchagewächse
eugenics/eugenetics Eugenik, Eugenetik, Erbhygiene, Rassenhygiene
euglenoids/euglenids/Euglenophyta Euglenen, Euglenophyta, Augenflagellaten
eukaryote (eucaryote) Eukaryont, Eukaryot, Eucaryont, Eucaryot

eukaryotic (eucaryotic) eukaryontisch, eukaryotisch, eucaryontisch, eucaryotisch
eukaryotic cell Eucyt, Eucyte
eulamellibranch bivalves/Eulamellibranchia Lamellenkiemer, Blattkiemer
eulittoral/eulittoral zone Eulitoral
Euphorbiaceae/spurge family Wolfsmilchgewächse
euphotic zone euphotische Zone
eupnea Eupnoe
eupyrene (sperm) eupyren
euryhaline euryhalin
euryoecious/euryecious/euryoecic euryök
eurypterids/sea scorpions/Eurypterida Seeskorpione
eurytele Eurytele
eustachian tube/auditory tube Eustachische Röhre
eutely/cell constancy Eutelie, Zellkonstanz
euthyneural nerve pattern Euthyneurie, sekundäre Orthoneurie
eutrophic/nutrient-rich eutroph, nährstoffreich
eutrophicate eutrophieren
eutrophication Eutrophierung
evaluate (e. g. results) auswerten (z. B. Ergebnisse)
evaluation Beurteilung, Bewertung, Abschätzung; Auswertung; *math* Bestimmung, Berechnung;
evanescent schwindend, verwelkend
evaporate verdunsten
evaporating dish Abdampfschale
evaporation Verdunstung
evaporative cooling Verdunstungskälte
evapotranspiration Evapotranspiration
even-toed ungulates/cloven-hoofed animals/artiodactyls/Artiodactyla Paarhufer
evening-primrose family/willowherb family/Oenotheraceae/Onagraceae Nachtkerzengewächse
evenness/equitability *ecol* Äquität, Äquitabilität
evergreen immergrün
eversed evers
eviction vector Apportiervektor
eviscerate (sea cucumber) auswerfen der Eingeweide (Cuviersche Schläuche)
evocation Evocation, Blühinduktion
evoke hervorrufen, wachrufen, herbeirufen
evoked potential evoziertes Potential

evolution

evolution/phylogeny/phylogenesis Evolution, Phylogenie, Phylogenese, Abstammungsgeschichte, Stammesgeschichte, Stammesentwicklung
- **coevolution** Koevolution
- **convergent evolution** konvergente Evolution
- **counterevolution** Evolutionsumkehr
- **determinate evolution** Orthoevolution
- **iterative evolution** iterative Evolution, sich wiederholende Evolution
- **phyletic evolution** phyletische Evolution
- **quantum evolution** Quantenevolution
- **reticulate evolution** netzartige Evolution
- **retrogressive evolution/ retrogressive development** Rückentwicklung

evolutionarily stable strategy (ESS) evolutionär stabile Strategie, evolutionsstabile Strategie

evolutionary/phylogenetic/phyletic evolutionär, abstammungsgeschichtlich, phylogenetisch, phyletisch

evolutionary genetics Evolutionsgenetik

evolutionary studies Abstammungslehre

evolutionary theory/theory of evolution Evolutionstheorie, Deszendenztheorie

ewe/mother sheep Mutterschaf

exalbuminous eiweisslos

examination under a microscope/ usage of a microscope Mikroskopieren

examine under a microscope/use a microscope mikroskopieren

exarate pupa/pupa exarata (free appendages) gemeißelte Puppe

excavate *geol* ausgraben

excavation *geol* Ausgrabung

exchange *n* Austausch

exchange *vb* austauschen

exchange reaction Austauschreaktion

exciple/excipulum (lichens) Excipulum, Rand (Flechten)
- **ectal exciple** Ektalexcipulum
- **excipulum proprium/proper margin (without algae)** Eigenrand
- **medullary exciple** Entalexcipulum, inneres Excipulum

- **thalline exciple/excipulum thallium** Lagerrand

excise *gen* herausschneiden, aussschneiden

excise *med* exzidieren, herausschneiden, abschneiden

excision Excision, Exzision, Herausschneiden

excision repair *gen* Excisionsreparatur, Exzisionsreparatur

excitability/irritability/sensitivity Erregbarkeit

excitable erregbar

excitation/irritation Erregung, Irritation

excitatory exzitatorisch, erregend

excitatory postsynaptic potential (EPSP) exzitatorisches postsynaptisches Potential

excite/irritate erregen

excite/stimulate reizen, anregen, stimulieren

excited state erregter Zustand, angeregter Zustand

exciter *micros* Erreger

exciter filter Erregerfilter

exclusion Exklusion, Ausschluss
- **allelic exclusion** Allelausschluss, allele Exclusion, allele Exklusion

exclusion principle Ausschlussprinzip
- **competitive exclusion principle/ principle of competitive exclusion (Gause's rule/ principle)** Konkurrenz-Ausschlussprinzip, Konkurrenz-Exklusionsprinzip

exclusive exklusiv, ausschließlich; anspruchsvoll

exclusive (fidelity/sociality) *ethol/ ecol* treu, fest

excreta/excretions Ausscheidungen, Exkrete, Exkremente

excretion Exkret, Exkretion, Ausscheidung

excretions/excreta *zool* Exkretion, Exkrete, Exkremente, Ausscheidungen

excretory canal Exkretionskanal

excretory cell Exkretzelle

excretory organ Exkretionsorgan, Ausscheidungsorgan

excretory system Exkretionssystem, Ausscheidungssystem

excretory tissue/secretory tissue Ausscheidungsgewebe

excurrent überragend, heraustretend; (Blatt) auslaufend

excurrent (tree form: main stem reaching top) *bot* geradstämmig, monopodiale Wuchsform, astlos in die Spitze auslaufend; überragend/ heraustretend (schlanker Wipfel/ unten ausladend)

excurrent aperture/exhalant aperture (egestive aperture) Ausströmöffnung (Egestionsöffnung)

excursion/field trip Exkursion

exergonic exergon, energiefreisetzend

exhalation Exhalation, Ausatmung, Expiration

exhale/breathe out ausatmen

exhaust/deplete *vb* **(soil)** ausmergeln, auslaugen

exhaust/tire *vb phys* erschöpfen, ermüden, entkräften

exhaustion Erschöpfung, Ermüdung, Entkräftung

exhaustion hybridization Erschöpfungshybridisierung

exhibit/show/display *vb* ausstellen, zurschaustellen, zeigen

exhibition/show/display Ausstellung, Zurschaustellung

exine (pollen/spore) Exine, Außenschicht

existing/extant existierend, bestehend

exit Ausgang

exit *techn* Austritt (z. B. Austrittsöffnung)

exite (arthropods) Exit

exoconjugant Exokonjugant

exocytosis Exozytose

exodermis/dermal tissue Exodermis, Außenhaut, Abschlussgewebe

exogenous exogen

exogenous cyst Tochtercyste

exon *gen* **(encoding sequence)** Exon

exon cloning Exonklonierung

exon shuffling Hin-und Herschieben von Exons, Exonmischung, Exonshuffling

exoparasite/ectoparasite/epizoon (*see also***: skin parasite)** Exoparasit, Ektoparasit, Außenparasit (*siehe*: Hautparasit)

exopod/exopodite (outer branch) Exopodit (Außenast)

exoreic/exorheic *limn* exorheisch (Entwässerung in den Ozean)

exoskeleton Exoskelett, Außenskelett, Hautpanzer

exothermic exotherm

expansion/dilation/dilatation Expansion, Dilatation, Ausweitung; *bot* (growth) Erweiterungswachstum, Dilatationswachstum

expansion tissue mechanisches Gewebe, Expansionsgewebe

expansivity Dehnbarkeit

experiment *vb* experimentieren, einen Versuch machen

experiment/test/trial Versuch
- **deliberate release experiment/ environmental release experiment** Freisetzungsexperiment
- **perform an experiment** einen Versuch durchführen
- **performance of an experiment** Versuchsdurchführung
- **preliminary experiment/pretrial** Vorversuch

experimental experimentell, Versuchs…

experimental procedure/ experimental protocol/ experimental method Versuchsverfahren

experimental series Versuchsreihe

experimental setup Versuchsanordnung

expert Experte (referee>Gutachter)

expertise Expertise, Gutachten

expiration/exhalation Ausatmen, Ausatmung, Expiration, Exhalation

expiration date Verfallsdatum

expire/exhale ausatmen

explant *n* Explantat

explant *vb* explantieren, auspflanzen

explode/blow up/detonate explodieren, in die Luft fliegen, detonieren

explorative data analysis explorative Datenanalyse

exploratory behavior Erkundungsverhalten

explosion/detonation Explosion, Detonation

explosive fruit/explosive capsule Explodierfrucht (Springfrucht), Explodierkapsel (Springkapsel)

explosives Explosivstoffe

exponential growth phase exponentielle Wachstumsphase, exponentielle Entwicklungsphase

expose *vb* **(film)** *opt* belichten

expose *vb* **(e. g. to a hazardous chemical/radiation)** aussetzen (einem Schadstoff/einer Strahlung aussetzen)

exposed hymenium fungi/ Hymenomycetes Hautpilze

exposure Ausgesetztsein, Gefährdung

exposure (to light: film/plant) Belichtung

express *vb* ausdrücken; *gen* exprimieren

expressed sequence tag (EST) exprimierte sequenzmarkierte Stelle (EST)

E expression 114

expression Expression, Ausdruck
- **high level expression** *gen*
Überexpression
expression cassette/cartridge
Expressionskassette
expression cloning
Expressionsklonierung
expression library
Expressionsbibliothek
expression vector Expressionsvektor
expressivity Expressivität,
Ausprägungsgrad
exsert/protrude vorstehen,
herausstehen
exserted/protruding hervorgestreckt
exstipulate/astipulate/estipulate
nebenblattlos, ohne Stipeln
extant/contemporary/recent heute
lebend, zeitgenössisch, gegenwärtig
existierend, derzeit bestehend, rezent
extend strecken (in die Länge
ziehen)
extension Dehnung, Streckung;
Ausdehnung, Verlängerung
- **primer extension** *gen* Primer-
Extension, Primer-Verlängerung
**extension growth/elongational
growth** *bot* Streckungswachstum
**extensive farming/extensive
agriculture** Extensivwirtschaft,
extensive Wirtschaft
extensor Strecker, Extensor
exterior layer/outer layer
Außenschicht
external/extrinsic äußerlich, von
außen, extern
external shell/ectocochlea
Außenschale, Ectocochlea
external stimulus Außenreiz
extinct/died out ausgestorben
- **become extinct/die out**
aussterben
extinction *opt* Extinktion,
Auslöschung
extinction/dying out Aussterben
extinction coefficient/absorptivity
Extinktionskoeffizient
extirpate/eradicate ausrotten,
ausmerzen
extirpation/eradication Ausrottung,
Ausmerzung
extracellular extrazellulär,
außerzellulär
extrachromosomal gene
extrachromosomales Gen
extract *vb* extrahieren, herauslösen
extract *n* Extrakt, Auszug
- **cell extract** Zellextrakt
- **cell-free extract** zellfreier Extrakt
- **crude extract** Rohextrakt
- **meat extract** *micb* Fleischextrakt
- **yeast extract** Hefeextrakt

extraction Extraktion
extraembryonic membrane
Embryonalhülle, Keimhülle,
extraembryonale Membran
extragenic DNA extragene DNA
extranuclear gene extranukleäres/
extranucleäres Gen
extravasation Extravasation,
Flüssigkeitsaustritt aus einem Gefäß
extremity/limb Extremität
extrinsic/extrinsical extrinsisch
extrorse extrors
extrusome/extrusive organelle
Extrusom, Ausschleuderorganelle
- **discobolocyst** Discobolocyste
- **ejectisome/ejectosome/ejectile
body** Ejectisome
- **haptocyst** Haptocyste
- **kinetocyst** Kinetocyste
- **muciferous body** Schleimsack
- **mucocyst** Mucocyste
- **nematocyst** Nematocyste
- **rhabdocyst** Rhabdocyste
- **spindle trichocyst**
Spindeltrichocyste
- **toxicyst** Toxicyste
**exudate/exudation/discharge/
secretion** Exsudat, Absonderung,
Abscheidung
exude/secrete/discharge absondern,
abscheiden (Flüssigkeiten)
exumbrella (medusa) Exumbrella,
Schirmoberseite, Schirmaußenwand
exuvia (cast-off skin/shell etc.)
Exuvie
exuvial fluid/molting fluid
Exuvialflüssigkeit,
Häutungsflüssigkeit,
Ecdysialflüssigkeit
eyas (unfledged bird) Nestling
(bes. Nestfalke/-habicht)
eye Auge
- **apposition eye** Appositionsauge
- **clear-zone eye/optical
superposition eye** optisches
Superpositionsauge
- **compound eye/facet eye**
Komplexauge, Facettenauge,
Netzauge, Seitenauge
- **compound eye, stalked
(*Ephemeroptera*)** Turbanauge
- **facet eye/compound eye**
Facettenauge, Komplexauge,
Netzauge, Seitenauge
- **lateral eye/lateral ocellus/
stamma** Lateralauge, Seitenauge,
Lateralocelle, Stemma
- **lens eye/lenticular eye** Linsenauge
- **main eye** Hauptauge
- **median eye/midline eye
(a dorsal ocellus)** Mittelauge,
Medianauge (Stirnocelle)

eyrie

- **naupliar eye (median eye)** Naupliusauge
- **neural superposition eye** neurales Superpositionsauge
- **ocellus/simple eye (dorsal and lateral)** Ocellus, Ocelle, Einzelauge, Punktauge, Nebenauge (*siehe:* Scheitelauge/Stirnauge)
- **optic superposition eye/clear-zone eye** optisches Superpositionsauge
- **parietal eye (*Sphenodon*)** Parietalauge
- **pigment cup eye/inverted eye** Pigmentbecherauge, Becherauge
- **pineal eye/epiphyseal eye (median eye)** Pinealauge (bei Neunauge: *Petromyxon*)
- **retinal cup eye/everted eye** Blasenauge
- **simple eye/ocellus (dorsal and lateral)** Einzelauge, Punktauge, Nebenauge, Ocelle, Ocellus (*siehe:* Scheitelauge/Stirnauge)
- **stalked compound eye (*Ephemeroptera*)** Turbanauge
- **stalked eye** Stielauge
- **stemma (insect larvas)/lateral eye/lateral ocellus** Lateralauge, Lateralocellus, Seitenauge (Punktauge/Einzelauge)
- **stemma (*pl* stemmata/stemmas) (dorsal/lateral ocellus)** Stemma, Punktauge (Einzelauge/Ocelle)
- **superposition eye** Superpositionsauge
- **>neural s.e.** neurales Superpositionsauge
- **>optical s.e./clear-zone eye** optisches Superpositionsauge

eye *bot* **(node/bud, e.g. potato)** Auge (z.B. Kartoffel)
eye chamber Augenkammer
eye cluster/bud cluster *bot* Beiknospengruppe
eye cutting/single eye cutting/bud cutting/leaf bud cutting *hort* Augensteckling
eye facet Augenfacette
eye lens/ocular lens Augenlinse, Okularlinse
eye socket/eyepit/orbit Augenhöhle, Orbita
eyeball Augapfel, Bulbus
eyebrow Augenbraue
eyebrow flash/eyebrow raise *ethol* Augengruß
eyecup Augenbecher
eyelash Augenwimper, Wimper
eyelid/palpebra Augenlid, Lid, Augendeckel
- **third eyelid** Blinzelhaut, Nickhaut

eyepiece/ocular Okular
- **binoculars** Binokular
- **pointer eyepiece** Zeigerokular
- **spectacle eyepiece/ high-eyepoint ocular** Brillenträgerokular
- **trinocular head** Trinokularaufsatz, Tritubus, Dreiertubus

eyepit/eye socket/orbit Augenhöhle
eyesight Sehkraft, Sehvermögen, Sehleistung
eyespot/stigma Augenfleck, Stigma
eyestalk Augenstiel
eyetooth (canine tooth of upper jaw) Augenzahn (oberer Eckzahn/ Reißzahn)
eyeworm disease, African/loa (*Loa loa*) Kamerunbeule
eyrie/aerie (bird's nest on cliff/ mountain) Horst, Adlerhorst, Raubvogelnest, Greifvogelnest

F factor (fertility factor) F-Faktor (Fertilitäts-Faktor)
F plasmid F-Plasmid
F⁺ cell F⁺-Zelle
F-distribution/Fisher distribution/ variance ratio distribution *stat* F-Verteilung, Fisher-Verteilung, Varianzquotientenverteilung
F-duction F-duktion
***Fab* (antigen-binding fragment of an immunoglobulin)** *Fab* Fragment
Fabaceae/pea family/bean family/ legume family/pulse family/ Papilionaceae (Leguminosae) Hülsenfruchtgewächse, Hülsenfrüchtler, Schmetterlingsblütler
fabric/mesh/network (e.g. siders) Gewebe (z. B. Spinngewebe)
face *n* Gesicht
facet eye/compound eye Facettenauge, Komplexauge, Netzauge, Seitenauge
facial expression Miene, Gesichtsausdruck
facial features Gesichtszüge
facies Fazies
facilitate erleichtern, fördern
facilitated transport erleichterter Transport
facilitating *neuro* bahnend
facilitation *neuro* Facilitation, Bahnung
facilitator neuron Bahnungsneuron
facultative/optional fakultativ, optional, freigestellt
facultative heterochromatin fakultatives Heterochromatin
FAD/FADH₂ (flavin adenine dinucleotide) FAD/FADH₂ (Flavin-Adenin-Dinukleotid)
fade (see: bleichen) ausbleichen (*passiv*/z. B. Fluoreszenzfarbstoffe)
fading (see: bleichen) Ausbleichen (*passiv*/z. B. Fluoreszenzfarbstoffe)
failure to thrive Gedeihstörung
fairy ring *fung* Hexenring
fairy shrimps/anostracans/ Anostraca Schalenlose, Kiemenfüße
falcate/falciform/sickle-shaped sichelförmig
falconer/hawker Falkner
falconry Falknerei
fall/autumn Herbst
fall ill/get sick/sicken/contract a disease erkranken
falling liquid film Rieselfilm
Fallopian tube/uterine tube (oviduct of mammals) Eileiter

fallow Brache, Brachfeld
• **to lie fallow** brachliegen
fallow deer Damwild
false/spurious falsch
false mermaid family/meadowfoam family/Limnanthaceae Sumpfblumengewächse
false negative falsch negativ
false positive falsch positiv
false scorpions/pseudoscorpions/ Pseudoscorpiones/Chelonethi Afterskorpione, Pseudoskorpione
false spiders/sun spiders/ windscorpions/solifuges/ solpugids/Solifugae/Solpugida Walzenspinnen
false tissue/paraplectenchyma/ pseudoparenchyma Scheingewebe, Pseudoparenchym
false whorl/pseudowhorl Scheinquirl, Scheinwirtel, Doppelwickel
falx cerebri (sickle-shaped fold in dura mater) Großhirnsichel, Falx cerebri
family Familie
• **multiplex family** Familie mit mehreren befallenen Mitgliedern
family trait Familienmerkmal
family tree/genealogical diagram/ dendrogram (pedigree) Stammbaum
fan *n* Fächer
fan *vb* **(bees)** fächeln
fan/fanning with lifted abdomen (bees: exposing Nasanov organ) sterzeln
fan palm Fächerpalme
fan-shaped/flabellate fächerförmig
fang/carnassial tooth (carnivores) Reißzahn, Fangzahn, Fang
• **poison fang/venomous fang/ unguis (chelicerates)** Giftklaue
• **poison tooth/venom tooth (snakes)** Giftzahn
fanning (bees) Fächeln
fanning with lifted abdomen (bees: exposing Nasanov organ) sterzeln
farm *n* Farm, Bauernhof
farm *vb* Landwirtschaft betreiben
farmer Bauer, Landwirt
farming/agriculture Landwirtschaft, Ackerwirtschaft
farming/tillage/cultivation Bodenbestellung, Ackern, Ackerbau; Ackerwirtschaft
farmland/tillage/tilth/cultivated land/arable land Ackerland
farrow (bring forth young pig litter) ferkeln, abferkeln (Wurf kleiner Schweine hervorbringen)

feather

fascia (ensheating band of connective tissue) Bindegewebshülle, Faszie
fascial/fasciate/band-shaped bandförmig
fasciate/banded/broadely striped breit gestreift, breit streifig, gebändert
fasciate *bot* **(stems teratologically grown together)** verbändert, zusammengewachsen (flächig verwachsen)
fasciation Fasziation, Verbänderung
fascicle (bundle) *zool* Faserbündel, Strang, Faszikel
fascicle *bot* **(inflorescence: cyme with very short pedicles)** Faszikel, Fasciculus, Büschel
fascicled *bot* büschelig, in Büscheln
fascicular büschelförmig
fasciculate/clustered/bundled/ growing in bundles gebündelt, bündelartig, büschelartig wachsend
fast *vb* fasten
fast-growing/rapid-growing schnellwachsend
fast-twitch fiber schnell-kontrahierende Faser
fasten befestigen
fastigiate gebüschelt, zur Spitze zu gedrängt, zugespitzt
fastigium/spike Spitze
fasting Fasten
fat Fett
- **brown fat** braunes Fett
fat body Fettkörper, Corpus adiposum
fat cell/adipocyte/adipose cell Fettzelle, Adipozyt, Adipocyt
fat droplet Fettröpfchen
fat storage/fat reserve Fettspeicher, Fettreserve
fat-soluble fettlöslich
fate map *embr* Determinationskarte, Schicksalskarte
fate mapping *gen* Schicksalskartierung
fatigue/tiring *n* Ermüdung
fatigue/tire *vb* ermüden
fatten mästen
fatty/adipose fettartig, fetthaltig, fettig, Fett...
fatty acid Fettsäure
- **monounsaturated fatty acid** einfach ungesättigte Fettsäure
- **polyunsaturated fatty acid** mehrfach ungesättigte Fettsäure
- **saturated fatty acid** gesättigte Fettsäure
fatty capsule of kidney/adipose capsule of kidney Nierenfettkapsel, Nierenkapsel
fatty tissue/adipose tissue Fettgewebe
faucet gland (of bucket orchid) Pleuridium
fauna/animal life Fauna, Tierwelt
fauna (faunal work/manual: with key) Fauna, Tierbestimmungsbuch
faunal faunistisch
faunal break Faunenschnitt
faunal complex Faunenkomplex
faunal element Faunenelement
faunal province Faunenprovinz
faunal realm Faunenreich
faunal region/zoogeographical region Faunenregion, Tierregion, tiergeographische Region
faunistics Faunistik
faveolate/favose/honeycombed/ alveolate/deeply pitted wabig; kleingrubig
fawn Rehkitz
Fc **(crystallizable fragment of an immunoglobulin)** *Fc* Fragment
FCS (fetal calf serum) fetales Kälberserum
feather Feder
- **auriculars (ear coverts)** Ohrenfedern
- **axillary feather/axillar** Achselfeder
- **contour feather** Konturfeder, Umrissfeder
- **covert/wing covert/protective feather/tectrix (***pl*** tectrices)** Decke, Deckfeder, Tectrix
- **down feather/down/plumule** Daune, Dune, Dunenfeder, Flaumfeder, dunenartige Feder
- **flight feather/remex (***pl*** remiges)** Schwungfeder, Remex
- **humeral feather/humeral/tertial feather (tertiaries)** Humeralfeder (Humeralflügel)
- **lesser covert/minor covert** kleine Decke (Flügeldecke/ Flügelfeder)
- **marginal covert/marginal tectrix** Randdecke
- **median covert/median tectrix** mittlere Decke
- **natal down/neossoptile/neoptile (a down feather)** Nestdune, Neossoptile, Neoptile
- **powder-down feather/ pulviplume** Puderdune, Pulviplume
- **primary feather (primaries/ primary remiges)** Handschwinge, Hautschwinge, Handschwungfeder
- **primary tectrix** Handdecke
- **quill feather** Schwanzfeder
- **rectrix (***pl*** rectrices)** Steuerfeder

feather

- remex (*pl* remiges)/flight feather Schwungfeder
- scapular feather Schulterfeder, Schulterblattfeder
- secondary feather (secondaries/secondary remiges) Armschwinge, Armschwungfeder, Unterarmschwungfeder
- secondary tectrix Armdecke
- tectrix (*pl* tectrices)/covert/wing covert/protective feather/deck feather Decke, Deckfeder, Tectrix

feather papilla Federpapille, Federbalg
feather parasite (bird louse/body louse: Mallophaga) Federling
feather ruffling Fiedersträuben
feather stars/comatulids/Comatulida Haarsterne, Federsterne, Comatuliden
feather tract/pteryla Federflur, Pteryla
featherless/apterial federlos
featherless space/apterium (*pl* apteria) Federrain, federlose Stelle, Apterium
feathery/plumose fedrig, federig
feathery mistletoe family/Misodendraceae Federmistelgewächse
feces Fäkalien, Kot, Stuhl
fecund/prolific fruchtbar, produktiv
fecundate fruchtbar machen, befruchten
fecundation Fekundation, Befruchtung
fecundity Fekundität, Fruchtbarkeit
fed-batch culture Zulaufkultur, Fedbatch-Kultur (semi-diskontinuierlich)
fed-batch process/fed-batch procedure Zulaufverfahren, Fedbatch-Verfahren (semi-diskontinuierlich)
fed-batch reactor/fed-batch reactor Fedbatch-Reaktor, Fed-Batch-Reaktor, Zulaufreaktor
feed *vb* füttern
feed *n* Futter, Nahrung
feed on something/ingest fressen, sich ernähren, etwas zu sich nehmen, sich von etwas ernähren
feed-forward inhibition/reciprocal inhibition *neuro* Vorwärtshemmung
feedback Rückkopplung
feedback inhibition/end-product inhibition Rückkopplungshemmung, Endprodukthemmung, negative Rückkopplung
feedback loop Rückkopplungsschleife

feedback system/feedback control system Regelkreis
feeder cell Feeder-Zelle
feeder root Nährwurzel
feeding Füttern, Fütterung
feeding attractant fraßauslösender Stoff, Fraßstimulans
feeding burrows/fodinichnia *paleo* Fressbauten
feeding deterrent Fraßhemmer, Phagodeterrens
feeding efficiency/profitability Profitabilität
feeding grounds Futterplatz, Futterstelle
feeding habits Fressgewohnheiten
feeding polyp/nutritive polyp/gastrozooid/trophozooid Fresspolyp, Nährtier, Gasterozoid, Trophozoid
feeding value Nährwert (des Futters)
feedlot/feed yard Viehkoppel
feel/sense/perceive empfinden, fühlen, spüren
feel/touch/palpate tasten
feeler/antenna (*see also under:* antenna) Fühler, Antenne
feeling/sensation Gefühl
Fehling's solution Fehlingsche Lösung
feign/sham vortäuschen, simulieren, sich verstellen, heucheln
feign death/play dead totstellen
feigning death Sich-Totstellen
feline katzenartig, Katzen betreffend, Katzen... (Felidae)
fell *vb for* fällen
fellfield Felsrasen, Felssteppe (Hochland)
felling (clearing) Rodung
felling (logging) Fällen, Baumfällen
felty/felt-like/tomentose filzig
female *adj/adv* weiblich
female/carpellate/pistillate *bot* weiblich
female *n* Weibchen
femoral femoral, zum Oberschenkel gehörend, den Oberschenkel betreffend
femoral artery Oberschenkelschlagader, Oberschenkelarterie
femoral gland (follicular gland: lizards) Schenkeldrüse (Follikulärorgan: Eidechsen)
femur (*pl* femora) (arthropods) Femur, Schenkel
femur/femoral bone/thighbone/os femoris Femur, Oberschenkelknochen

119   **fiber** **F**

fen/fenland (minerotrophic peatland: fed by underground water or interior drainage) Fehn, Fenn (minerotropher vererdeter Flachmoor/Niedermoor)
fence post Zaunpfosten
fenestrated gefenstert
fenestrated flame cell (protonephridia) Reusengeißelzelle, Cyrtocyte (*siehe*: Flammenzelle)
fenestrated leaf gefenstertes Blatt
fenistiform pit Fenstertüpfel
feral verwildert
ferment *vb* fermentieren, gären, vergären
fermentation Fermentation, Gärung, Vergärung
fermentation chamber reactor/ compartment reactor/cascade reactor/stirred tray reactor Rührkammerreaktor
fermentation layer (soil) Fermentationsschicht, Vermoderungshorizont
fermentation tube Gärröhrchen, Einhorn-Kölbchen
fermenter Fermenter, Gärtank (*siehe auch:* Reaktor)
fern Farn
fern family/Polypodiaceae Tüpfelfarngewächse
ferredoxin Ferredoxin
ferrunginous rost-rot
fertile fertil, fruchtbar, fortpflanzungsfähig
• **infertile/sterile** infertil, steril, unfruchtbar, nicht fortpflanzungsfähig
fertility Fertilität, Fruchtbarkeit
fertilization Fertilisation, *sensu stricto:* Befruchtung (Bestäubung see pollination)
• **cross-fertilization/xenogamy/ allogamy** Kreuzbefruchtung (Xenogamy), Fremdbefruchtung (Allogamie)
• **double fertilization** *bot* doppelte Befruchtung
• **in-vitro fertilization (IVF)** In-vitro-Fertilisation, Reagenzglasbefruchtung
• **self-fertilization/autogamy** Selbstbefruchtung, Autogamie
fertilization/application of fertilizer *agr* Düngung
fertilization cone Empfängnishügel
fertilization membrane Befruchtungsmembran
fertilize (gametes) befruchten
fertilize (supply soil with plant nutrients) düngen

fertilizer/manure Dünger, Düngemittel
fertilizer/plant food/manure Dünger, Düngemittel
ferulic acid Ferulasäure
fetal fetal, fötal
fetal calf serum (FCS) fetales Kälberserum
fetid/foetid (adverse odor) stinkend (übelriechend)
fetlock (horse: metatarso-phalangeal articulation) Köte, Fesselkopf, Fesselgelenk (Pferd)
fetlock hair Kötenbehang, Kötenhaare, Fesselhaare
fetus Fötus *m*, Fetus *m*, Fet *m*
fiber Faser
• **bast fiber** *bot* Bastfaser
• **climbing fiber** Kletterfaser
• **coir (coconut fiber)** Coir (Kokosfaser)
• **contractile fiber** kontraktile Faser
• **dietary fiber** Ballaststoffe
• **elastic fiber** elastische Faser
• **fast-twitch fiber** schnell-kontrahierende Faser
• **giant fiber/Mauthner's cell** Riesenfaser, Mauthnersche Zelle, Mauthner-Zelle (Fische)
• **intrafusal fiber** Intrafusalfaser
• **libriform fiber (wood)** Libriformfaser, Holzfaser
• **lint (cotton fiber)** Lint, Lintbaumwolle
• **linters (short cotton fibers)** Linters (kurze Baumwollfasern)
• **mantle fiber** Zugfaser
• **mossy fiber** Moosfaser
• **Muellerian fiber/fiber of Müller** *ophth* Müller-Faser
• **muscle fiber/myofiber** Muskelfaser
• **nerve fiber** Nervenfaser
• **nuclear bag fiber** Kernhaufenfaser
• **nuclear chain fiber** Kernkettenfaser
• **parallel fiber** Parallelfaser
• **phasic fiber** phasische Faser
• **polar fiber (microtubule)** Polfaden (Mikrotubulus)
• **Purkinje fiber/conduction myofiber** Purkinje-Faser
• **sclerenchymatous fiber** *bot* Sklerenchymfaser
• **Sharpey's fiber** Sharpeysche Faser
• **slow-twitch fiber** langsam-kontrahierende Faser
• **spindle fiber** *cyt* Spindelfaser
• **stress fiber** Stressfaser
• **zonule fibers** Zonulafasern

F | **fiber cell** | 120

fiber cell Faserzelle
fiber plant/fiber crop Faserpflanze
fiber tracheid Fasertracheide
fibril Fibrille
fibrillar fibrillär
fibrin Fibrin (Blutfaserstoff)
fibrinogen Fibrinogen
fibroblast Fibroblast
fibroin Fibroin
fibrous/stringy faserig
fibrous bone Faserknochen,
Geflechtknochen
fibrous cartilage/fibrocartilage
Faserknorpel
fibrous layer (anther) Faserschicht
fibrous proteins fibrilläre Proteine,
Faserproteine
fibrous root system
Büschelwurzelsystem
fibula Fibula, Wadenbein
Fick diffusion equation Ficksche
Diffusionsgleichung
fiddle-shaped/panduriform
geigenförmig
fiddlehead/crozier schneckenförmig
eingerolltes junges Farnblatt
fidelity (community) Treue,
Gesellschaftstreue
field/land/farmland Acker;
(meadow) Acker
field/plain/open fields/
meadowland Feld, Flur
field bee/forager Sammelbiene,
Sammlerin
field boundary strip/balk
Ackerrain, Feldrain
field capacity/field moisture
capacity/capillary capacity (soil
moisture) Feldkapazität
field diaphragm Feldblende,
Leuchtfeldblende, Kollektorblende
field exercise Geländeübung
field guide Feldführer
field inversion gel electrophoresis
(FIGE) Feldinversions-
Gelelektrophorese
field layer ecol Krautschicht
field lens micros Feldlinse
field of view/scope of view/field
of vision/range of vision/visual
field Sehfeld, Blickfeld,
Gesichtsfeld
field stop (a field diaphragm in
eyepiece: ocular aperture) micros
Sehfeldblende, Gesichtsfeldblende
field study/field investigation/field
trial Feldversuch,
Freilanduntersuchung,
Freilandversuch
field trip/excursion Exkursion
fig family/mulberry family/
Moraceae Maulbeergewächse

fig marigold family/carpetweed
family/mesembryanthemum
family/Aizoaceae
Mittagsblumengewächse,
Eiskrautgewächse
fight-or-flight reaction Kampf-oder-
Flucht Reaktion
fighting behavior Kampfverhalten
figure/design (wood) Maserung,
Masertextur, Fladerung, Figur,
Zeichnung (Holz)
figure eight (of DNA) Achterform
(der DNA)
figwort family/foxglove family/
snapdragon family/
Scrophulariaceae
Braunwurzgewächse, Rachenblütler
filament bot (stamen) Filament,
Staubfaden (Staubblatt)
filament/chain of cells Filament,
Zellfaden
filament/thread Filament, Faden
• intermediate filament
intermediäres Filament
filamentous/filliform/thread-
shaped/threadlike trichal
(haarförmig), fadenförmig
filamentous lichens/hairlike lichens
Fadenflechten, Haarflechten
filarial worms Filarien
filbert Haselnussfrucht
file meristem/rib meristem
Rippenmeristem
filial generation (first/second)
(erste/zweite) Tochtergeneration,
Filialgeneration
filibranch bivalves/Filibranchia
Fadenkiemer
filibranch gill Filibranchie,
Fadenkieme
filiciform farnartig
filiform/filamentous/threadlike/
hairlike fadenförmig, fadenartig,
haarförmig, trichal
fill-in reaction/filling in reaction
Auffüllreaktion
filling (of palps) Laden (der Palpen)
filly (female horse/pony under 4
years) weibliches Fohlen/Füllen,
Stutenfohlen
film reactor (bioreactor)
Filmreaktor
film water Haftwasser
filmy fern family/
Hymenophyllaceae
Hautfarngewächse,
Schleierfarngewächse
filoplume Filopluma, Fadendune,
Fadenfeder
filopodium Filopodium
filter/pass through vb filtrieren,
passieren

121   "first animals"   **F**

filter *n* Filter
- **barrier filter/selective filter/ stopping filter/selection filter** Sperrfilter
- **exciter filter** Erregerfilter
- **folded filter** Faltenfilter
- **HEPA filter (high efficiency particulate air filter)** HOSCH-Filter (Hochleistungsschwebstofffilter)
- **membrane filter** Membranfilter
- **noise filter** Rauschfilter
- **polarizing filter/polarizer** Polarisationsfilter, "Pol-Filter", Polarisator
- **round filter (filter paper disk)** Rundfilter
- **stopping filter/barrier filter/ selective filter/selection filter** Sperrfilter
- **suction filter/vacuum filter** Nutsche, Filternutsche
- **syringe filter** Spritzenvorsatzfilter, Spritzenfilter
- **trickling filter** Tropfkörper (Tropfkörperreaktor/ Rieselfilmreaktor)

filter disk method Filterblättchenmethode
filter enrichment Anreicherung durch Filter
filter feeder Filtrierer, Filterer
filter flask/vacuum flask Filtrierflasche, Filtrierkolben, Saugflasche
filter holder *micros* Filterträger
filter network/filtering network Filternetzwerk
filter pump Filterpumpe
filter-feeding Filtern, Nahrungsfiltern
filtrate *n* Filtrat
filtrate *vb* klären, filtrieren
filtration Klärung, Filtrierung, Filtration
filz gall Filzgalle
fimbriate/fimbriated/fringed fransenartig, gefranst, befranst
fimbriated funnel of oviduct/ infundibulum Flimmertrichter, Wimperntrichter, Eileitertrichter, Infundibulum (mit Ostium tubae)
fimbricidal fimbricid, fimbrizid, fransenspaltig
fin Flosse
- **adipose fin** Fettflosse
- **anal fin** Afterflosse, Analflosse
- **caudal fin/tail fin** Schwanzflosse
- **dorsal fin** Rückenflosse
- **lobe fin** Quastenflosse
- **pectoral fin** Brustflosse

- **pelvic fin/ventral fin** Bauchflosse
- **ray fin** Strahlenflosse
- **tail fin/caudal fin** Schwanzflosse
- **ventral fin/pelvic fin** Bauchflosse

fin rot Flossenfäule
final host Endwirt
final image *micros* Endbild
finalism Finalismus
findings Befund
fine adjustment/fine focus adjustment *micros* Feinjustierung, Feineinstellung
fine structure Feinstruktur, Feinbau
fine-adjustment knob *micros* Feinjustierschraube, Feintrieb, Mikrometerschraube
finely notched/crenulate feingekerbt
finely serrate/finely saw-edged/ serrulate/serratulate feingesägt, kleingesägt
finely striped/striated feingestreift, feinstreifig
finger/digit Finger, Digitus
fingered/digitate gefingert
fingerlike/fingershaped/digitiform fingerförmig, handförmig, digitat
fingernail Fingernagel
fingerprint Fingerabdruck
fingerprinting/genetic fingerprinting/DNA fingerprinting Fingerprinting, genetischer Fingerabdruck
fingertip Fingerspitze, Fingerkuppe
- **volar, soft portion of fingertip/ torulus tactilis** Fingerbeere, Fingerballen

fir clubmoss family/Huperziaceae Teufelsklauengewächse
fir family/pine family/Pinaceae Tannenfamilie, Kieferngewächse, Föhrengewächse
fire/firing *vb neuro* feuern
fire brigade/fire department Feuerwehr
fire corals/stinging corals/ milleporine hydrocorals/ Milleporina Feuerkorallen
fire extinguisher Feuerlöscher, Feuerlöschgerät, Löschgerät
firefighter/fireman Feuerwehrmann
firewood/fuelwood Brennholz, Feuerholz
firmly attached (permanently)/ sessile festsitzend, festgewachsen, festgeheftet, aufsitzend, sessil
firn/névé Firn, Gletschereis
firn region/firn zone Firnregion
"first animals"/protozoans Urtierchen, Urtiere, "Einzeller", Protozoen

F first cousin

first cousin Cousin ersten Grades
first-degree relative Verwandter ersten Grades
first-order kinetics Kinetik erster Ordnung
first-order reaction Reaktion erster Ordnung (Reaktionskinetik)
Fischer projection/Fischer formula/ Fischer projection formula Fischer-Projektion, Fischer-Formel, Fischer-Projektionsformel
fish *vb* fischen, angeln
fish *n* Fisch; *pl* **fish** Fische ein und derselben Art; *pl* **fishes** Fische verschiedener Arten)
fish *n* (culinary) Fisch (kulinarisch)
FISH (fluorescence *in situ* **hybridization)** FISH (*in situ* Hybridisierung mit Fluoreszenzfarbstoffen)
fish birds/Ichthyornithiformes Fischvögel
fish eggs (roe) Fischeier, Fischlaich (Rogen)
fish hatchery Fischzuchtanlage, Fischzuchtstation
fish ladder Fischleiter
fish lice/Branchiura/Argulida Fischläuse, Karpfenläuse, Kiemenschwänze
fish meal Fischmehl
fish-eater/piscivore Fischfresser
fish-eating/piscivorous fischfressend
fishery Fischerei
fishes/Pisces Fische (Fische verschiedener Arten)
fissiparity Fissiparie
fist Faust
fist-walking Faustgang (Handknöchel)
fitness/suitability Fitness, Eignung
• **Darwinian fitness** Darwinsche Fitness
• **frequency-dependent fitness** frequenzabhängige Fitness
• **inclusive fitness** Gesamteignung, Gesamtfitness
fix fixieren (mit Fixativ härten)
fixation Fixierung, Fixieren
fixative Fixativ, Fixiermittel
fixed action pattern Erbkoordination (formkonstante Verhaltenselemente)
fixed bed reactor/solid bed reactor (bioreactor) Festbettreaktor
fixed-angle rotor *centrif* Festwinkelrotor
flabellate antenna/flabelliform antenna/fan-shaped antenna Fächerfühler
flabellate/fan-shaped fächerförmig

flaccid/limp/weak (wilting/deficient in turgor) schlaff, schlapp, erschlaffend (welkend)
flacourtia family/Indian plum family/Flacourtiaceae Flacourtiagewächse
flag leaf Fahnenblatt, Fähnchenblatt
flagellar geißelartig, begeißelt, Geißel...
flagellar basal body/flagellar corpuscle/flagellar granule/ kinetosome/mastigosome/ blepharoplast/blepharoblast Basalkörper, Kinetosom, Blepharoplast
flagellar pocket/reservoir/anterior pocket Geißelsäckchen
flagellar swelling/paraflagellar body Paraflagellarkörper
flagellaria family/Flagellariaceae Peitschenklimmer
flagellate(d)/bearing flagella flagellat, begeißelt
flagellates/mastigophorans/ Flagellata/Mastigophora Geißeltierchen, Geißelträger, Flagellaten
flagellation Begeißelung
flagelliform geißelförmig, peitschenförmig
flagellomer Flagellomer
flagellum (*pl* **flagella/flagellums**) Flagelle, Geißel (Antenne: Fühlergeißel)
• **acronematic flagellum/whiplash flagellum** Peitschengeißel
• **pleuronematic flagellum/tinsel flagellum/flimmer flagellum** Flimmergeißel
• **pulling flagellum** Zuggeißel
• **pushing flagellum** Schubgeißel
• **tinsel flagellum/flimmer flagellum/pleuronematic flagellum** Flimmergeißel
• **trailing flagellum** Schleppgeißel
• **whiplash flagellum/acronematic flagellum** Peitschengeißel
flagellum (mosses) *bot* peitschenähnlicher, kleinblättriger oberirdischer Spross
flame bulb Wimperkölbchen, Wimperkolben
flame cell/flame bulb (terminal flame bulb) Flammenzelle, Wimperflammenzelle
• **fenestrated flame cell (protonephridia)** Reusengeißelzelle, Cyrtocyte (*siehe*: Flammenzelle)
flame ionization detector (FID) Flammenionisationsdetektor

123   flexuose   **F**

flame retardant/flame retarder
Flammschutzmittel,
Flammenverzögerungsmittel
flame-resistant nicht entflammbar
flame-retardant feuerhemmend,
flammenhemmend, nicht leicht
entflammbar
flameproof flammsicher,
flammensicher, feuerbeständig,
flammfest, feuerfest, nicht/schwer
entflammbar
**flamingoes and allies/
Phoenicopteriformes** Flamingos
flammability Entflammbarkeit,
Entzündbarkeit, Brennbarkeit
**flammable/inflammable/
combustible** entflammbar,
entzündlich, brennbar
flank/side (e.g. of horse) Flanke
flank bone/ilium/os ilium
Darmbein, Ilium
**flank meristem/peripheral
meristem** Flankenmeristem
flanking region flankierende
Region
flap/flutter (e.g. with wings) *vb*
flattern; (mit den Flügeln) schlagen
flap grafting *hort* seitliches
Anplatten mit langer Gegenzunge
flap of skin Hautfalte, Hautlappen
flapping flight Schlagflug,
Flatterflug
**flark (open water pool in bog: esp.
in aapa mire)** Flarke
(wassergefüllte Risse im Hochmoor)
flash destillation Kurzweg-
Destillation
flash point Flammpunkt
flask Kolben, Gefäß, Flasche
- **boiling flask with round bottom**
Rundkolben, Siedegefäß
- **distilling flask/retort**
Destillierkolben, Retorte
- **Erlenmeyer flask** Erlenmeyer
Kolben
- **filter flask/vacuum flask**
Filtrierkolben, Filtrierflasche,
Saugflasche
- **Florence boiling flask/Florence
flask (boiling flask with flat
bottom)** Stehkolben, Siedegefäß
- **shake flask** Schüttelkolben
- **swan-necked flask/S-necked
flask/gooseneck flask**
Schwanenhalskolben
- **volumetric flask** Messkolben
flat bed gel/horizontal gel
horizontal angeordnetes Plattengel
flat bone/os planum platter
Knochen
flat-blade impeller (bioreactor)
Scheibenrührer, Impellerrührer

flatfishes/Pleuronectiformes
Plattfische
flatsawn (wood) flach-aufgesägt
**flatworms/platyhelminths/
Platyhelminthes** Plattwürmer,
Plathelminthen, Plathelminthes
- **free-living flatworms/
turbellarians/Turbellaria**
Strudelwürmer, Turbellarien
flavine mononucleotide (FMN)
Flavinmononukleotid (FMN)
flavonoid Flavonoid
flavor/flavoring/aromatic substance
Geschmacksstoff, Aromastoff
- **off-flavor (spontaneous food
constituent alteration)**
Aromafehler
flax family/Linaceae Leingewächse
flax lily family/Phormiaceae
Phormiumgewächse
flay *bot* abschälen (Rinde etc)
flay (remove skin from carcass) *zool*
die Haut abziehen
**flea-bitten grey horse/flea-bitten
white horse** Fliegenschimmel
**fleas/Siphonaptera/Aphaniptera/
Suctoria** Flöhe
fledge flügge werden
- **fully fledged/full-fledged (able
to fly)** flügge (flugfähig)
fledgling eben flügge gewordener
Vogel/Jungvogel
fleece (coat of wool) Wolle,
Wollkleid
**fleece (wool of a sheep from one
shearing)** Schur (Schurwolle)
flense flensen (abhäuten/Walspeck
abziehen)
flesh/meat Fleisch;
fung Pilzfleisch, Fleisch
flesh-eater/meat-eater/carnivore
Fleischfresser, Karnivor, Carnivor
**flesh-eating/meat-eating/
carnivorous** fleischfressend,
karnivor, carnivor
fleshy fleischig
fleshy cone/"berry" Beerenzapfen
fleshy fruit Saftfrucht
fleshy taproot Rübe
**fleshy-finned fishes/
sarcopterygians/Sarcopterygii/
Choanichthyes** Fleischflosser
flews (dogs) Lefze(n)
**flex/bend/curve back (arm/leg/
muscles)** beugen (Arm/Bein/
Muskeln)
flexible/pliable biegsam
flexion (muscle) Beugung
flexor (muscle) Flexor, Beuger
**flexuose/flexuous (curved in a
zig-zag manner)** wellig gebogen,
schlängelnd gebogen

F flexure

flexure Biegung, Beugung, Krümmung, Flexur
- **cephalic flexure** *embr* Scheitelbeuge, Mittelhirnbeuge
- **cervical flexure** *embr* Nackenbeuge
- **pontine flexure** *embr* Brückenbeuge

flight Flug
- **bounding flight** Bolzenflug, Bogenflug
- **circling/circling flight** Kreisen
- **climb** Steigflug
- **dive** Sturzflug
- **drag** Luftwiderstand, Strömungswiderstand
- **dynamic soaring** dynamischer Segelflug
- **flapping flight** Schlagflug (Flatterflug)
- **glide/gliding/gliding flight** Gleitflug
- **homing** Zielflug
- **hovering flight (hummingbirds)** Schwirrflug (Kolibris)
- **hovering** Schwebeflug (in der Luft stehen)
- **level flight** Geradeausflug
- **lift** Auftrieb
- **migratory flight** Wanderflug
- **nuptial flight** Brautflug
- **propulsive flight/powered flight** Kraftflug
- **slope soaring** Hangsegeln
- **soaring** *(see there)* Segelflug
- **stall** *n* Sackflug
- **stall** *vb* absacken, abrutschen
- **static soaring** statischer Segelflug
- **sustained flight** Dauerflug
- **tethered flight** fixierter Flug
- **thermal soaring** Thermiksegelflug, Thermiksegeln
- **thrust** Vortrieb, Anschub, Schub, Schubkraft
- **windhovering** Rütteln im Wind

flight/flock (birds) Schwarm, Schar (Vogelschar)
flight distance Flugentfernung, Flugweite
flight feather/remex (*pl*** remiges)** Schwungfeder, Remex
flight reaction/escape reaction Fluchtreaktion
flightless/unable to fly flugunfähig
flightlessness/unableness to fly Flugunfähigkeit
flimmer/tinsel Flimmerhärchen
flimmer flagellum/tinsel flagellum/ pleuronematic flagellum Flimmergeißel
flip-flop Drehung um 180°, Handstandüberschlag

flip-flop mechanism (membrane lipids, gene expression) Flip-Flop-Mechanismus (Membranlipide, Genexpression)
flipper/fluke Schwimmflosse (groß/ fleischig), paddelartig Flosse (z.B. bei Delphinen), Paddel
float/air sac/pneumatophore Schwebeorgan, Gasbehälter, Pneumatophor
float/suspend *vb* schweben
floating/suspended schwebend
floating fern family/water fern family/Parkeriaceae Hornfarngewächse
floating leaf Schwimmblatt
floating ribs/costae fluitantes frei endende Rippen
floatoblast (bryozoans) Flottoblast
floccose flockig
flocculation Flockulation
flock *n* Schar, Schwarm (Vögel), Herde (Schafe)
flock *vb* **(e.g. birds)** sammeln, versammeln, zusammenscharen, zu Scharen zusammenkommen
flock *vb chem* flocken, ausflocken
flocking Flockung
flood/inundate *vb* überschwemmen, überfluten
flood/flooding/inundation Überschwemmung, Überflutung
flood irrigation Bewässerung durch Überflutung
floodplain/alluvial plain/floodland/ alluvial land Überschwemmungsebene, Schwemmland (Flussaue)
floodplain forest Auwald, Überschwemmungswald
floodplain meadow Überschwemmungswiese
floor plate/subplate *neuro* Bodenplatte
flora Flora, Pflanzenwelt
floral Blumen.., Blüten.., geblümt
floral biology Blütenbiologie
floral bract (hypsophyll) Braktee (Hochblatt)
floral bud/flower bud Blütenknospe
floral diagram/flower diagram Blütendiagramm
floral envelope Blütenhülle
floral guide Blütenmal
floral induction Blühinduktion, Blüteninduktion
floral leaves Blütenblätter
floral realm Florenreich
Florence boiling flask/Florence flask (boiling flask with flat bottom) Stehkolben, Siedegefäß

**florescence/flowering period/
anthesis** Floreszenz, Blütezeit,
Anthese
floret Blütchen, Blümchen, kleine
Blume; Einzelblüte (z. B. Grasblüte/
Compositenblüte)
floricane (second-year cane)
Fruchtrute (Beerensträucher)
floriculture Blumenzucht
floriculturist Blumenzüchter,
Blumengärtner
floridean algae/florideans
Florideen
floridean starch Florideenstärke
florist Florist, Blumenzüchter;
Blumenhändler
florist shop Blumengeschäft
floristic/floristics *n* Floristik
floristic *adv/adj* floristisch
floristic composition
Florenzusammensetzung
floristic element Florenelement
floristic region Florengebiet
floristic unit Floreneinheit
flour Mehl
  • **whole-grain flour** Vollkornmehl
flourish/thrive florieren, gedeihen,
sprießen (gut wachsen)
flow *n* Fluss, Fließen
flow *vb* fließen
flow chart Fließschema
flow cytometry
Durchflusszytometrie,
Durchflusscytometrie
flow pattern Strömungsmuster
flow rate (volume per time) Strom
(Volumen pro Zeit),
Durchflussgeschwindigkeit
flow reactor (bioreactor)
Durchflussreaktor
flow resistance/resistance to flow
Strömungswiderstand
flower *vb* blühen, in Blüte stehen
flower *n* (blossom) Blüte; (plant)
Blume
  • **bat-pollinated flower/
chiropterophile**
Fledermausblume
  • **bee-pollinated flower/
melittophile** Bienenblume
  • **beetle-pollinated flower/
cantharophile/coleopterophile**
Käferblume
  • **bird-pollinated flower/bird
flower/ornithophile** Vogelblume
  • **bisexual flower/hermaphrodite
flower/perfect flower** zwittrige
Blüte, Zwitterblüte,
zweigeschlechtige Blüte
  • **bowl-shaped flower** Napfblume
  • **butterfly-pollinated flower/
psychophile** Schmetterlingsblume

  • **carpellate flower/pistillate
flower** Stempelblüte, weibliche
Blüte
  • **carrion flower** Aasblume
  • **complete flower** vollständige
Blüte
  • **cut flower** Schnittblume
  • **deceptive flower** Täuschblume
  • **disk flower/disk floret/tubular
flower** Scheibenblüte,
Röhrenblüte (Asterales)
  • **dung-fly flower/sapromyophile**
Aasfliegenblume
  • **fly-pollinated flower/myiophile**
Fliegenblume
  • **funnel-shaped flower**
Trichterblüte
  • **gall flower (figs)** Gallenblüte
  • **incomplete flower**
unvollständige Blüte
  • **insect-pollinated flower/
entomophile** Insektenblume
  • **moss "flower"** Moosblüte
  • **moth-pollinated flower
(geometers)/phalaenophile**
Mottenblume (Spanner)
  • **moth-pollinated flower (hawkmoths)/sphingophile**
Nachtschwärmerblume
  • **perfect flower/bisexual flower/
hermaphroditic flower** zwittrige
Blüte, Zwitterblüte,
zweigeschlechtliche Blüte
  • **pinch-trap flower**
Klemmfallenblume,
Klemmfallenblüte
  • **pistillate flower/carpellate
flower/female flower**
Stempelblüte, weibliche Blüte
  • **pitfall trap/slippery-slide trap
(flower)** Kesselfallenblume,
Gleitfallenblume
  • **ray flower/ray floret/ligulate
flower** Strahlenblüte, Zungenblüte
  • **showy flower** auffällige Blüte,
prachtvolle Blüte, prächtige Blüte
  • **solitary flower/single flower**
Solitärblüte, Einzelblüte
  • **staminate flower** Staubblüte,
männliche Blüte
  • **strawflower** Strohblume,
Trockenblume
  • **trap flower/trap blossom/prison
flower** Fallenblume, Fallenblüte
  • **tubular flower/disk flower/disk
floret** Röhrenblüte, Scheibenblüte
(Asterales)
  • **unisexual flower/imperfect
flower** eingeschlechtige Blüte
  • **wildflower** Wildpflanze,
wildwachsende Pflanze
flower abscission Blütenfall

flower animals

**flower animals/anthozoans/
Anthozoa** Blumentiere,
Blumenpolypen, Anthozoen
flower bud/floral bud Blütenknospe
**flower cup/floral cup/cupule/
cupula** Blütenbecher, Cupula
flower diagram/floral diagram
Blütendiagramm
flower funnel Blütenschlund
flower head/capitulum/cephalium
Blütenköpfchen, Köpfchen, Korb,
Körbchen, Capitulum, Cephalium
flower organ Blütenorgan
flower pot Blumentopf
**"flower pot" leaf/urn-shaped leaf/
pouch leaf (*Dischidia*)** Urnenblatt
flower scent/flower perfume
Blütenduft
flower stalk/peduncle
Blütenstengel, Blütenstiel
flower structure Blütenbau
flower tuft Blütenschopf
flowerbed/patch Blumenbeet, Beet
flowering blühend, in Blüte stehend
- **free-flowering** mit freien
Blütenblättern (radialsymmetrisch/
nicht verwachsen)
**flowering fern family/cinnamon
fern family/royal fern family/
Osmundaceae**
Königsfarngewächse,
Rispenfarngewächse
**flowering period/anthesis/
florescence** Blütezeit, Anthese,
Blütenentfaltung, Floreszenz
**flowering plant/angiosperm/
anthophyte** Blütenpflanze,
Angiosperme
flowering rush family/Butomaceae
Wasserlieschgewächse,
Schwanenblumengewächse
flowering sequence Aufblühfolge
flowing water (river/stream)
Fließgewässer (Fluss/Strom)
fluctuate schwanken, fluktuieren
fluctuation Fluktuation,
Schwankung
fluctuation analysis/noise analysis
Fluktuationsanalyse, Rauschanalyse
fluctuation of population
Populationsschwankung,
Bevölkerungsschwankung
fluctuation test Fluktuationstest
flue gases *ecol* Rauchgase
fluence Flussrate
fluid/liquid *adv/adj* flüssig
fluid/liquid *n* Flüssigkeit
fluid bed reactor (bioreactor)
Fließbettreaktor
fluid-mosaic model
Flüssigmosaikmodell
fluidity Fluidität, Fließfähigkeit

fluidized bed reactor
Wirbelschichtreaktor,
Wirbelbettreaktor
**fluke/tail fluke (lobe on whale's
tail)** Fluke, Schwanzruder,
Schwanzflosse
flukes/trematodes/Trematoda
Saugwürmer, Egel, Trematoden
fluoresce fluoreszieren
fluorescence Fluoreszenz
**fluorescence photobleaching
recovery/fluorescence recovery
after photobleaching (FRAP)**
Fluoreszenzerholung nach
Lichtbleichung
fluorescence quenching
Fluoreszenzlöschung
**fluorescence-activated cell sorter
(FACS)** fluoreszenzaktivierter
Zellsorter/Zellsortierer
**fluorescence-activated cell sorting
(FACS)** fluoreszenzaktivierte
Zelltrennung
**fluorescence-*in-situ*-hybridization
(FISH)** Fluoreszenz-*in-situ*-
Hybridisierung (FISH)
fluorescent fluoreszierend
fluoridate fluoridieren
fluoridation Fluoridierung
fluorinate fluorieren
fluorinated hydrocarbon
Fluorkohlenwasserstoff
fluorine Fluor
flush end *gen* glattes Ende,
bündiges Ende
flush irrigation Berieselung
flute budding/ring budding *hort*
Ring-Okulation, Ringveredlung
flutter/flap (the wings) flattern (mit
den Flügeln schlagen)
fluttering leaves flatternde Blätter
fluvial plain Flussebene
flux Strömung
**flux (volume per time per transect;
light/energy)** Fluss
fly *vb* fliegen
fly-pollinated flower/myiophile
Fliegenblume, Myiophile
**flying lemurs/colugos/
dermopterans/Dermoptera**
Pelzflatterer, Riesengleitflieger
fMet (N-formyl methionine) fMet
(N-Formylmethionin)
FMN (flavin mononucleotide) FMN
(Flavinmononucleotid)
foal Fohlen, Füllen
- **colt (male horse/pony under 4
years)** männliches Fohlen/Füllen
- **filly (female horse/pony under 4
years)** weibliches Fohlen/Füllen
foaling Gebären eines Fohlens,
Fohlengebären

foam *vb* schäumen
foam *n* Schaum
foam/froth/sea spray/ocean spray
mar Gischt
focal length Brennweite
focal plane Brennebene
focal point/focus Brennpunkt
focus/adjustment *micros* **(fine/**
coarse) Justierung
(Scharfeinstellung des Mikroskops:
fein/grob)
focus/focal point Brennpunkt
focus (focusing or focussing) *vb*
fokussieren, scharf einstellen
• **in focus/focussed (picture)** scharf
• **not in focus/blurred/out of**
focus (picture) unscharf
focus formation Fokusbildung
focus map Fokuskarte
focus-forming unit (ffu)
fokusbildende Einheit
focussing Scharfeinstellung
fodder/forage Futterpflanze
fog Nebel
fog desert Nebelwüste
fold/plication/wrinkle Falte
fold gall (leaf margin)
Blattrandgalle
fold-back DNA/snap-back DNA
in sich gefaltete DNA,
zurückgebogene DNA
folded/pleated/plicate gefaltet,
faltig
folded filter Falterfilter
foliaceous/foliose/phylloid/leaf-like
blattartig, laubblattartig
foliage/leafage/leaves Belaubung,
Blattwerk, Laubwerk, Laub
foliage eruption/leafing
Laubausbruch
foliage leaf Laubblatt, Folgeblatt
• **primary foliage leaves/first**
foliage leaves Primärblätter,
Erstlingsblätter
foliage plant/leafy plant
Grünpflanze, Blattpflanze
foliaged/foliate/provided with
leaves/leaf-bearing/leaved
beblättert
foliar Blatt..., blättrig
foliar gap/leaf gap Blattlücke
foliar plantlet/adventitious plantlet
(*Kalanchoe*) Brutpflänzchen
foliar trace/leaf trace Blattspur
foliate/provided with leaves/leaf-
bearing/foliaged beblättert
foliation/leafing (leaf
development/ontogeny)
Blattbildung, Blattentwicklung
foliation/prefoliation/vernation
Blattfolge in der Knospe, Vernation,
Knospenlage

folic acid (folate)/pteroylglutamic
acid Folsäure (Folat),
Pteroylglutaminsäure
foliferous/foliating/producing
leaves sich belauben
foliiform/leaf-shaped/leaf-like
blattförmig, blattartig
foliolate/leafletted blättchenartig,
kleinblättrig, fiederblättrig
foliole/leaflet/pinna Blättchen,
Fieder, Blattfieder, Fiederblättchen,
Teilblatt
foliose/folious/leafy/resembling
a leaf Blatt..., Laub..., blattartig,
blättrig, laubartig; vielblättrig
foliose lichens Blattflechten,
Laubflechten
follicle *zool* Follikel
• **Graafian follicle/vesicular**
ovarian follicle Graafscher
Follikel, Graaf-Follikel,
Tertiärfollikel
• **hair follicle** Haarfollikel,
Haarbalg
• **lymph follicle/lymph nodule**
Lymphfollikel, Lymphknötchen
• **primary follicle** Primärfollikel
• **secondary follicle**
Sekundärfollikel
• **splenic follicle/splenic node/**
splenic nodule/splenic
corpuscle/Malpighian body/
Malpighian corpuscle
Milzfollikel, Milzknötchen,
Malpighi-Körperchen,
Milzkörperchen
follicle *bot* **(fruit)** Follikel, Balg,
Balgfrucht
follicle-stimulating hormone (FSH)
Follitropin, follikelstimulierendes
Hormon (FSH)
follicular gland (femoral gland)
Follikulärorgan, Follikulärdrüse
(Schenkeldrüse: Eidechsen)
follow-up *med* Nachsorge
follower *zool* Laufsäugling
following behavior
Nachfolgeverhalten
following substrate Folgesubstrat
fontanel Fontanelle
food Essen, Futter, Nahrung
• **plant food/fertilizer/manure**
Dünger, Düngemittel
• **staple food/basic food/main**
food source
Grundnahrungsmittel,
Hauptnahrung,
Hauptnahrungsquelle
food/diet Kost, Essen, Diät
food additive
Lebensmittelzusatzstoff
food begging Futterbetteln

F food chain 128

food chain Nahrungskette
- **detritus food chain**
  Detritusnahrungskette
- **grazing food chain**
  Fraßnahrungskette,
  Abweidenahrungskette,
  Weidenahrungskette
food chemistry
  Lebensmittelchemie
food crop/forage plant/food plant
  Nahrungspflanze
food crop production
  Nahrungspflanzenanbau
food hoarding Futterhorten
food inspection
  Lebensmittelüberwachung,
  Lebensmittelkontrolle
food poisoning
  Nahrungsmittelvergiftung
food preservation
  Nahrungsmittelkonservierung
food preservative
  Lebensmittelkonservierungsstoff
food requirements
  Nahrungsbedürfnisse
food source Nahrungsquelle
food vacuole/gastriole
  Nahrungsvakuole
food value/nutritive value
  Nährwert
food web Nahrungsnetz,
  Nahrungsgefüge
foodstuff/nutrients Lebensmittel
foot (*pl* **feet**) Fuß
- **ambulacral foot/tube foot/
  podium** Ambulakralfüßchen,
  Saugfüßchen
- **climbing foot** Kletterfuß
- **creeping foot** Kriechfuß
- **false foot/dewclaw** Afterklaue
- **grasping foot/prehensile foot**
  Greiffuß
- **side-foot/parapod** Parapodium
- **swimming foot** Schwimmfuß
- **wading foot** *orn* Schreitfuß
- **webbed foot/swimming foot
  (e.g. birds)** Schwimmfuß
foot/haustorium Fuß, Haustorium
foot sole/pedal sole/planta
  Fußsohle, Planta (*also*: Kriechsohle/
  Kriechfußsohle)
**foot-and-mouth disease/aphthous
  fever (*Aphthovirus*)** Maul-und-
  Klauenseuche
footfall Schritt, Tritt
foothills/foothill zone kolline Stufe,
  Hügelstufe, Hügellandstufe,
  Vorgebirge
footpad/torus Fußballen
footprint (DNA footprint) DNA-
  Fußabdruck, DNA-Footprint
footprinting Fußabdruckmethode

forage *n* **(animal food, *esp*. by
  browsing/grazing)** Futter,
  Viehfutter
forage *n* **(the act of foraging)**
  Futtersuche, Nahrungssuche
forage *vb* auf Nahrungssuche gehen,
  Nahrung suchen, Futter suchen
forager Sammler
forager/field bee Sammelbiene,
  Sammlerin
foraging behavior Weideverhalten
foraminicidal foraminicid,
  foraminizid, fensterspaltig
foraminiferans/forams Lochträger,
  Foraminiferen
**forb (nongraminoid herbaceous
  plant)** Krautpflanze, krautige
  Pflanze (nicht Gräser)
force/forcing *bot* **(fast growing/
  early flowering)** rasch
  hochzüchten, früh zur Reife bringen
force microscopy Kraftmikroskopie
- **atomic force microscopy (AFM)**
  Rasterkraftmikroskopie
forcing bed/hotbed *hort* Frühbeet,
  Mistbeet
forcipate/forked like foreceps
  gegabelt, scherenförmig
forcipulatids/Forcipulatida
  Zangensterne
**forcipule/prehensor/poison claw
  (*Chilopoda*)** Giftklaue
fore reef Vorriff
fore-kidney/pronephros Vorniere,
  Pronephros
forearm Unterarm
forebear/ancestor/progenitor
  Vorfahre, Ahne
**forebrain/prosencephalon
  (telencephalon & diencephalon)**
  Vorderhirn, Prosencephalon
forefoot (*pl* forefeet)/front foot
  Vorderfuß (*pl* Vorderfüße)
foregut *embr* Vorderdarm
foregut/stomodaeum/stomodeum
  Vorderdarm, Munddarm,
  Mundbucht, Stomodaeum
forehand (horse) Vorhand
forehead/frons Stirn, Frons
forehoof Vorderhuf
foreign DNA Fremd-DNA
foreleg/front leg Vorderbein
forelimb Vorderextremität
forelock (e.g. horse) Stirnhaare,
  Schopf
foremilk/colostrum Vormilch,
  Biestmilch, Kolostralmilch,
  Colostrum
forensic *adv/adj* forensisch,
  Gerichts..., gerichtlich
forensic medicine Gerichtsmedizin
forensics Forensik

129   **four-o'clock family**

forequarter(s) Vorderviertel
(Pferd: Vor(der)hand)
forerun Vorlauf
foreshore *mar* Vorstrand,
Gezeitenstrand
foreskin/preputium/prepuce/sheath
Vorhaut, Präputium, Scheide
forest (*see:* woods) Wald größerer
Ausdehnung
• **cultivated forest/tree farm/tree
plantation** Forst, Wirtschaftswald
• **urban forest/community**
Stadtwald
• **young forest** Jungwald
forest administration/forest service
Forstverwaltung
forest canopy Blätterdach,
Kronendach (Wald)
forest damage Waldschaden
forest deterioration/forest decline
Waldsterben
forest edge/fringe Waldrand
forest fire Waldbrand
forest floor Waldboden
forest line/timberline Waldgrenze
forest litter Waldstreu
forest plantation Forstkultur
(Pflanzung)
• **young and protected forest
plantation** Schonung
forest ranger/forest warden
Forstwart
forest science/forestry
Forstwissenschaft, Forstkunde
forest tree Forstbaum
forest warden/forest ranger
Forstwart
forested/wooded/arboreous
bewaldet
forester Förster
forestry Forstwesen, Forstwirtschaft
forewing/front wing/tegmina
Vorderflügel (Oberflügel/
Deckflügel/Flügeldecke)
forficulate/forficiform
scherenförmig
fork *n* Gabel, Gabelung, Abzweig
fork *vb* **(bifurcate)** sich gabeln,
sich zwieseln
forked/furcate gegabelt
• **bifurcate/Y-shaped/dichotomous**
gegabelt, dichotom
forking/bifurcation/dichotomy
Gabelung, Gabelteilung, Dichotomie
**forking of trunk at base/lower
trunk** Zwieselung
(Stammverzweigung nah am Boden)
forking of trunk at midhight
Gabelung (Forstbaum)
form pruning/shape pruning
Erziehungsschnitt
formal genetics formale Genetik

formation Formation
formic acid (formate) Ameisensäure
(Format)
formyl methionine Formylmethionin
forniciform sorelium (lichens)
Helmsoral, Gewölbesoral
fornix Fornix, Gewölbe
fortified milk mit Vitaminen
(Mineralien) angereicherte Milch
fortify/enrich anreichern
forward mutation Vorwärtsmutation
fossil *adj/adv* fossil, versteinert
fossil *n* Fossil (*pl* Fossilien),
Versteinerung
• **index fossil/zone fossil/zonal
fossil** Leitfossil, Faziesfossil
• **living fossil** lebendes Fossil
• **trace fossil/ichnofossil**
Spurenfossil, Ichnofossil
• **transitional fossil**
Übergangsfossil
• **zone fossil/zonal fossil/index
fossil** Leitfossil, Faziesfossil
fossil fuel fossiler Brennstoff
fossil record fossiles Zeugnis,
Fossilieninventar
fossil remains fossile Überreste
fossiliferous (strata)
fossilienführend, Fossilien
enthaltend
fossilization Fossilisierung
fossilized fossilisiert
fossorial zum Graben geeignet,
Grab...
fossorial leg Grabbein
foster/nurture/rear/bring up
aufziehen, erziehen
foster child Pflegekind
foster mother Pflegemutter
foster parents Pflegeeltern
foster raising Ammenaufzucht
foul/rot/decompose/decay
verfaulen, zersetzen
foulbrood (bees) Faulbrut
fouling/rotting verfaulen
founder effect Gründereffekt
founder mouse Gründermaus
founder polyp/primary polyp
Gründerpolyp, Primärpolyp
founder principle/founder effect
Gründerprinzip, Gründereffekt
Fouquieriaceae/ocotillo family
Ocotillogewächse
**four flat-blade paddle impeller
(bioreactor)** Kreuzblattrührer
four-angled/quadrangular
vierkantig
four-handed/quadrumanous
vierhändig
four-locular vierfächerig
four-o'clock family/Nyctaginaceae
Wunderblumengewächse

fovea/small pit/small depression
kleine Grube, Grübchen
foveate/pitted grubig
foveolate (having small pits or depressions) kleingrubig
fowl Federvieh, Geflügel
• **poultry (domestic)** Geflügel (Hausgeflügel)
foxglove family/figwort family/ snapdragon family/ Scrophulariaceae
Braunwurzgewächse, Rachenblütler
fraction Fraktion
fraction collector Fraktionssammler
fractional precipitation fraktionierte Fällung
fractionate fraktionieren
fractionating column Fraktioniersäule
fractionation Fraktionierung
fragile X chromosome (syndrome) fragiles X-Chromosome (Syndrom)
fragrance/perfume/pleasant scent angenehmer Duft/Geruch/ Geruchsstoff
fragrant/pleasantly smelling (angenehm) duftend, wohlriechend
frame line (spiderweb) *arach* Rahmenfaden (Spinnennetz)
frameshift *gen* Rasterverschiebung, Leserasterverschiebung
frameshift mutation Rasterschub-Mutation, Rastermutation, Rasterverschiebungsmutation
framework region (of immunoglobulins) Gerüstregion (von Immunglobulinen)
Frankeniaceae/sea heath family/ alkali-heath family
Frankeniengewächse, Nelkenheidegewächse
frankincense tree family/incense tree family/torchwood family/ Burseraceae Balsambaumgewächse
frass (debris produced by insects) Fraßmehl
frass (feces of insect larvas) Kot von Insektenlarven
fray *vb* ausfransen
frayed/fringed/fimbriate(d) fransig
freckle/macula solaris Sommersprosse
free pupa/pupa libra freie Puppe
free zone (web) freie Zone
free-floating/pendulous frei schwebend
free-flowering mit freien Blütenblättern (radialsymmetrisch/ nicht verwachsen)
free-living freilebend
free-living flatworms/turbellarians/ Turbellaria Strudelwürmer, Turbellarien
free-running rhythm freilaufender Rhythmus
freemartin/martin heifer (sterile ♀ calf: twin of ♂ calf) Zwicke, Zwitterrind (steriles Kuhkalb: Zwillings-Geschwister eines ♂ Kalbs)
freeze einfrieren, gefrieren
freeze preservation/cryopreservation Gefrierkonservierung, Kryokonservierung
freeze storage Gefrierlagerung
freeze substitution Gefriersubstitution
freeze-dry/lypophilize gefriertrocknen, lyophilisieren
freeze-drying/lyophilization Gefriertrocknung, Lyophilisierung
freeze-etch *vb* gefrierätzen
freeze-etching Gefrierätzung
freeze-fracture/freeze-fracturing/ cryofracture Gefrierbruch
freezer Kühltruhe, Tiefkühltruhe, Gefrierschrank, Tiefkühlschrank
freezer compartment Kühlfach (eines Kühlschranks)
freezing microtome/cryomicrotome Gefriermikrotom
freezing point Gefrierpunkt
freezing-point depression Gefrierpunktserniedrigung
French layering/continuous layering *hort* Ablegen (mehrere Jungpflanzen pro Trieb)
frenulum Frenulum, Zügel, Stützleiste
frequence-dependent selection frequenzabhängige Selektion/ Auslese
frequency Frequenz, Häufigkeit
• **gen frequency** Genfrequenz, Genhäufigkeit
frequency (of occurrence)/ abundance Häufigkeit
frequency diagram *stat* Häufigkeitsdiagramm, Häufigkeitskurve
frequency distribution *stat* Häufigkeitsverteilung
frequency histogram Häufigkeitshistogramm
frequency of occurrence/abundance Häufigkeit
frequency ratio *stat* relative Häufigkeit
frequency-dependent fitness frequenzabhängige Fitness
frequency-dependent selection frequenzabhängige Selektion

131   **fruit** **F**

frequent/abundant häufig
"fresh mass" (fresh weight) "Frischmasse" (Frischgewicht)
fresh weight (*sensu stricto*: fresh mass) Frischgewicht (*sensu stricto*: Frischmasse)
freshwater Süßwasser
freshwater snails/*Basommatophora* (*Pulmonata*) Wasserlungenschnecken
Freund's adjuvant Freundsches Adjuvans
fright Schreck, Angst, Ängstigung
fright coloration Schreckfärbung, Schrecktracht
frighten *vb* ängstigen, erschrecken
fringe/seam/border/edge Saum
fringe community/gallery community Saumgesellschaft
fringed/fimbriate/fimbriated fransenartig, befranst, gefranst
fringing reef Saumriff, Küstenriff, Strandriff
frit *lab* Fritte
frog (triangular horny pad inside horse hoof) (*see also*: hoof) Hufstrahl, Strahl (am Pferdehuf)
frog-bit family/tape grass family/ elodea family/Hydrocharitaceae Froschbissgewächse
frog-stay/spine of frog/spina cunei (horse hoof) Hahnenkamm
frogs and toads/anurans/Salientia/ Anura Froschlurche (Frösche und Kröten)
frond (ferns/palms/kelp) Blattwedel, Wedel; Braunalgenspreite
front *meteo/mar* Front
● **cold front** Kaltfront
● **occluded front/occlusion** Okklusion
● **stationary front** stationäre Front
● **warm front** Warmfront
front side/ventral vorderseitig (bauchseitig), ventral
frontal bone/os frontale Stirnbein
frontal carina Frontalleiste, Stirnleiste
frontal gland/cephalic gland Frontaldrüse, Stirndrüse, Kopfdrüse
frontal heart/frontal sac Stirnherz
frontal lobe/lobus frontalis Frontallobus, Stirnlappen
frontal membrane (bryozoans) Frontalmembran
frontal organ Stirnorgan
frontal plate Stirnplatte
frontal sinus/sinus frontalis Stirnhöhle (Nebenhöhle: Schädel)
frontal suture (skull) Stirnnaht
frontal tuber/frontal tuberosity/ tuber frontale Stirnhöcker

frost Frost
● **ground frost** Bodenfrost
● **hoarfrost/white frost (fine/ feathery)** feinflockiger Reif, Raureif
● **permafrost** Permafrost, Dauereis, Dauerfrost
● **rime/rime frost** Raufrost, Raureif
frost blight/nip Frostbrand
frost damage/frost injury/freezing injury Frostschaden, Frostschädigung
frost desiccation damage/damage by winter drought Frosttrocknis
frost drought damage/frost desiccation damage/winter desiccation damage Frosttrocknis
frost hardening Frosthärtung
frost hardiness Frosthärte, Frostbeständigkeit
frost protection irrigation Frostschutzberegnung
frost tolerance Frostverträglichkeit
frost-resistant/frost hardy frostbeständig, frostresistent
frost-tender/susceptible to frost frostempfindlich
frostbite Erfrierung (lokal begrenzt), Congelatio
frozen section *micros* Gefrierschnitt
fructiferous/bearing fruit/fruiting fruchtend, fruchttragend
fructification/fruit formation Fruchtbildung
fructification/fruitbody/fruiting body/carposoma Fruchtkörper, Karposom
fructose (fruit sugar) Fruktose, Fructose (Fruchtzucker)
frugivore/fructivore Frugivor, Fruktivor, Fruchtfresser
frugivorous/fruit-eating/ carpophagous/feeding on fruit frugivor, fruktivor, fruchtfressend, karpophag
fruit Frucht; (culinary) Obst
● **achene/akene** Achäne, Achaene
● **>inferior bicarpellary achene/ cypsela** unterständige Achäne (Asteraceen)
● **acorn** Eichel
● **aggregate fruit** Sammelfrucht
● **apocarpous fruit/unicarpellary fruit/monocarpellate fruit/ simple fruit** Einblattfrucht
● **ballistic fruit/ballist** ballistische Frucht, Schleuderfrucht
● **beechnut** Buchecker
● **berry** Beere
● **bur/burr/burry fruit** Klette, Klettenfrucht, Klettenfrucht

F fruit

132

- **capsule** Kapsel
- **>ballistic c.** Schleuderkapsel
- **>catapult c.** Katapultkapsel
- **>circumscissile c./lid capsule/ pyxis/pyxidium** Deckelkapsel
- **>dorsicidal c.** dorsizide/dorsicide Spaltkapsel
- **>explosive c.** Explodierkapsel, Explosionskapsel (Springkapsel)
- **>lid c./circumscissile c./pyxis/ pyxidium** Deckelkapsel
- **>loculicidal c.** lokulizide/ loculicide Spaltkapsel, fachspaltige Kapsel
- **>longitudinally dehiscent c.** Spaltkapsel
- **>poricidal c./porose c.** Lochkapsel, Löcherkapsel, Porenkapsel, porizide Kapsel
- **>septicidal c.** septizide/septicide Spaltkapsel, wandspaltige Kapsel
- **caryopsis/grain** Karyopse, Caryopse, "Kernfrucht", Kornfrucht (Grasfrucht)
- **catapult fruit/catapult capsule** Katapultfrucht, Katapultkapsel
- **cell/mericarpic nutlet (one-seeded segment/fruitlet of loment)** Klause
- **cereal** Halmfrucht (Getreide)
- **cocklebur** Klette (Spitzklette)
- **cypsela/inferior bicarpellary achene** unterständige Achäne (Asteraceen)
- **cystocarp/cystocarpium** Hüllfrucht, Cystokarp
- **dehiscent fruit** Springfrucht, Streufrucht, Öffnungsfrucht
- **drupe/drupaceous fruit/stone** Steinfrucht
- **dry fruit** Trockenfrucht
- **explosive fruit** Explodierfrucht
- **false fruit/spurious fruit/ pseudocarp/pseudofruit** Scheinfrucht
- **filbert** Haselnussfrucht
- **fissile fruit** Zerfallfrucht
- **fleshy fruit** Saftfrucht
- **follicle** Follikel, Balg, Balgfrucht
- **gourd/pepo** Gurkenfrucht, Kürbisfrucht, Panzerbeere
- **grain/caryopsis** "Kernfrucht", Kornfrucht, Karyopse, Caryopse
- **hesperidium** Hesperidium, Citrusfrucht, Zitrusfrucht (eine Panzerbeere)
- **indehiscent fruit** Schließfrucht
- **key/samara** Flügelnuss
- **legume/pod** Hülse
- **lid capsule/circumscissile capsule/pyxis/pyxidium** Deckelkapsel

- **loment/lomentum/lomentaceous fruit/jointed fruit** Bruchfrucht, Gliederhülse, Gliederfrucht, Klausenfrucht
- **lomentose siliqua** Gliederschote
- **mericarp** Merikarp, Teilfrucht
- **mericarpic nutlet/cell (one-seeded segment/fruitlet of loment)** Klause
- **nuculane/nuculanium** (*Henderson*: berry from superior ovary: medlar/grape) oberständige Beere (z. B. Traube); (*Spjut*: dry pericarp/hard endocarp/outer layer fibrous: coconut/almond/walnut) Nussfrucht mit trocken-faserigem Perikarp
- **nut** Nuss
- **nutlet/nucule** Nüsschen
- **pepo/gourd** Gurkenfrucht, Kürbisfrucht, Panzerbeere
- **pod/legume** Hülse
- **pome/core-fruit** Pomum, Apfelfrucht
- **poricidal capsule/porose capsule** Lochkapsel, Löcherkapsel, Porenkapsel, porizide Kapsel
- **pseudocarp/pseudofruit/false fruit/spurious fruit** Scheinfrucht
- **pyxis/pyxidium/lid capsule/ circumscissile capsule** Deckelkapsel
- **samara/key** Flügelnuss
- **schizocarp/schizocarpium** Schizokarp, Spaltfrucht
- **septicidal capsule** Bruchkapsel, septicide/septizide Spaltkapsel
- **silicle** Schötchen
- **silique/siliqua** Schote
- **>lomentose s.** Gliederschote
- **simple fruit** Einzelfrucht
- **simple fruit/apocarpous fruit/ unicarpellary fruit/ monocarpellate fruit** Einblattfrucht
- **sorosis/fleshy multiple fruit** Beerenverband, Beerenfruchtstand
- **spurious fruit/false fruit/ pseudocarp/pseudofruit** Scheinfrucht
- **stone/drupe/drupaceous fruit** Steinfrucht
- **syconium/sycone** Sykonium, Steinfruchtverband, Feigenfrucht
- **utricle/utriculus** Utriculus, Schlauchfrucht
- **winged fruit** Flügelfrucht

fruit abscission Fruchtfall
fruit body/fruitbody/fruiting body/ fructification Fruchtkörper, Karposom
fruit core Kerngehäuse (Frucht)

fruit growing Obstbau
fruit orchard Obstplantage
fruit pulp Fruchtfleisch, Fruchtmus
fruit skin/peel Haut einer Frucht,
  Fruchtschale
fruit stalk Fruchtstiel
fruit sugar/fructose Fruchtzucker,
  Fruktose
fruit tree/fruit-bearing tree
  Obstbaum
**fruit wall/ovary wall/seed vessel/
  pericarp** Fruchtwand, Perikarp
fruit-bearing shrubs
  Beerensträucher
fruiting fruchtend
fruiting body/fruitbody
  Fruchtkörper
fruitlet Früchtchen; Einzelfrucht,
  Teilfrucht, Karpid, Karpidium
  (entire carpel)
frustule (diatoms) Schale
**frutescent/fruticose/shrub-like/
  shrubby/bushy** sich strauchartig
  entwickeln; strauchartig, strauchig,
  buschig
fruticose lichens/shrublike lichens
  Strauchflechten
fruticulose (etwas) strauchartig
fry *ichth* Brut, Fischbrut
fucose/6-deoxygalactose Fukose,
  Fucose, 6-Desoxygalaktose
**fugacious/short-living/short-lived/
  soon disappearing** kurzlebig,
  hinfällig, flüchtig; früh abfallend,
  früh verblühend
fulcrum Stützorgan
full song Vollgesang
full-grown ausgewachsen
fulvous/tawny gelbbraun, rötlich-
  gelb, lohfarben
fumaric acid (fumarate) Fumarsäure
  (Fumarat)
fume hood/hood Rauchabzug,
  Dunstabzugshaube, Abzug
fumigate begasen
fumigation Begasung
**fumitory family/bleeding heart
  family/Fumariaceae**
  Erdrauchgewächse
functional group *chem* funktionelle
  Gruppe
functional system/behavior system
  Funktionskreis
fundus gland Fundusdrüse
fungicide Pilzbekämpfungsmittel,
  Fungizid
**fungicide treatment/pesticide
  treatment (of seeds)** Beizmittel
  (zur Saatgutbehandlung)
fungus (*pl*** funguses/fungi)** Pilz

fungus garden Pilzgarten
**funicle/funiculus/ovule stalk/seed
  stalk** Funiculus, Nabelstrang,
  Samenstiel
funnel/siphon/infundibulum
  Trichter, Sipho, Infundibulum
funnel trap, unidirectional
  Trichterfalle, Reusenfalle
funnel web/tube web *arach*
  Trichternetz, Röhrennetz
**funnel-leaf/ascidiate leaf
  (***Nepenthes***)** Trichterblatt,
  Schlauchblatt
**funnel-leaved plant/infundibulate
  plant** Trichterpflanze
funnel-shaped/infundibulate
  trichterförmig
funnel-shaped flower Trichterblüte
funnelform trichterförmig
fur/coat Fell, Pelz
furan Furan
furca Furca, Sprunggabel
furcal retinaculum Retinaculum,
  Sprunggabelhalter
furcate/forked gegabelt
furcula (insects) Furcula,
  Sprunggabel
**furcula/fourchette/wishbone (birds:
  united clavicles)** Furcula,
  Gabelbein
furfuraceous/scurfy schorfig,
  Schorf…, kleinschuppig;
  kleiig, mehlig
furrow/groove/sulcus Furche,
  Graben, Rinne
furrow irrigation
  Grabenbewässerung,
  Furchenbewässerung
furrowed/grooved/fissured/sulcate
  gefurcht, furchig, gerieft
furuncular furunkulös, Furunkel…
fused/coalescent/connate
  verwachsen, angewachsen
fusiform/spindle-shaped
  spindelförmig
fusiform initial *bot* Fusiforminitiale
fusion Fusion, Verschmelzung;
  Verwachsung
  • **transcription fusion**
    Transkriptionsfusion
  • **translation fusion**
    Translationsfusion
fusion/coalescence/symphysis
  Verwachsung
fusion gene Fusionsgen
fusion of nuclei/caryogamy
  Kernverschmelzung, Karyogamie
fusion protein Fusionsprotein
futile cycle Leerlauf-Zyklus,
  Leerlaufzyklus

G

G-banding 134

G-banding G-Banding, Giemsa-Banding
G1 phase G1-Phase (von "gap = Lücke")
G2 phase G2-Phase
gaggle *n* **(flock of geese: not in flight)** Schar Gänse
Gaia hypothesis Gaia-Hypothese
gain/increase *n* Zunahme
gain/increase *vb* zunehmen
gain of function mutation Funktionsgewinnmutation
gait/pace Gang, Gangart
- **canter/Canterbury gallop/lope (slow gallop)** Kanter (leichter Galopp)
- **diagonal gait** Diagonalgang, Kreuzgang
- **digitigrade gait** Digitigradie, Zehengang
- **disunited canter** Kreuzgalopp
- **gallop/run (fast three-beat gait)** Galopp (Sprunglauf)
- **lope (horse: easy natural gait resembling a canter)** leichter Kanter
- **orthograde gait/erect gait/ upright gait** aufrechter Gang, aufrechte Gangart
- **pace** Passgang
- **paso** Pasos
- **plantigrade gait** Plantigradie, Sohlengang
- **rack/single foot** schneller Passgang
- **running walk** Tölt
- **trot/trotting gait** Trab, Trott (schnelle Gangart)
- **unguligrade gait** Unguligradie, Zehenspitzengang, Hufgang
- **walk** Schritt
galactosamine Galaktosamin
galactose Galaktose
galactosemia Galaktosämie
galacturonic acid Galakturonsäure
gale/strong wind (51–101 km/h) *meteo* Sturmwind
galea/outer lobe of maxilla Galea, Außenlade
gall/cecidium Galle, Cecidium, Pflanzengalle
- **ball gall** Galle durch *Eurosta solidaginis* an *Solidago* (USA)
- **button gall (oak)** Große Linsengalle, Bechergalle
- **camellia gall** Weidenrose (durch *Rhabdophaga rosaria*)
- **cone gall** Kegelgalle
- **covering gall** Umwallungsgalle
- **filz gall** Filzgalle
- **fold gall** Blattrandgalle
- **knopper gall** Knoppergalle

- **leaf gall** Blattgalle
- **marble gall** Schwammkugelgalle
- **mark gall/medullar gall** Markgalle
- **oak apple** Eichenschwammgalle
- **petiolar gall** Blattstielgalle
- **pin cushion gall/bedeguar** Schlafapfel, Bedeguar
- **pineapple gall** Ananasgalle
- **pit gall** Zweiggalle durch *Asterolecanium variolosum* an *Quercus*
- **pouch gall** Beutelgalle
- **purse gall** Blattstielgalle durch *Pemphigus bursarius* an *Populus*
- **roll gall** Rollgalle
- **root gall** Wurzelgalle
- **twig gall** Stengelgalle, Zweiggalle
gall apple *bot* Gallapfel
gall bladder *zool* Gallenblase
gall flower (figs) Gallenblüte
galler(s)/gallmaker(s) Gallerreger (*sg & pl*)
gallery Gallerie, unterirdischer Gang, Stollen, Laufgang
gallery forest/fringing forest Galeriewald
gallic acid (gallate) Gallussäure (Gallat)
gallicolous gallicol, gallenbewohnend, Gall...
gallinaceous hühnerartig
gallinaceous birds/fowl-like birds/ Galliformes Hühnervögel
gallop/run (fast three-beat gait) Galopp (Sprunglauf)
GALT (gut-associated lymphatic tissue) darmassoziiertes lymphatisches Gewebe
game/play Spiel
game/hunted animals *hunt* Wild (jagbare Tiere), Wildbret
game birds (legally hunted) Jagdgeflügel
game fish Sportfisch
game population/stock of game Wildbestand
game preserve/game reserve/game enclosure Wildgehege
game theory *ethol* Spieltheorie
gamergate (fertilized, ovipositing worker ant) Gamergat
gametangiogamy Gametangiogamie
gametangiophore Gametangiophor
gamete/sex cell Gamet, Keimzelle, Geschlechtszelle
gametocyst Gametocyst, Gametozyst
gametocyte Gametocyt, Gametozyt
gametogamy/syngamy Gametogamie, Syngamie

135 gastrotrichs G

gametogony/gamogony
Gametogonie, Gamogonie
gametophore Gametangienträger
gametophyte Gametophyt
gamma particle (chytrids)
Gammakörper
gander (adult male goose) Ganter,
Gänserich
ganglion Ganglion, Nervenknoten
(*siehe auch Wörterbücher der
Human- und Veterinärmedizin*)
- **basal ganglion/cerebral nucleus/
  corpus striatum** Basalganglion,
  Stammganglion
- **cerebral ganglion**
  Cerebralganglion,
  Zerebralganglion
- **pedal ganglion** Pedalganglion
- **spinal ganglion/dorsal root
  ganglion/posterior root
  ganglion** Spinalganglion
- **subesophageal ganglion/
  suboesophageal ganglion**
  Subösophagealganglion,
  Unterschlundganglion
- **supraesophageal ganglion/
  supraoesophageal ganglion/
  "brain"** Oberschlundganglion,
  Supraösophagealganglion,
  "Gehirn"
- **ventral ganglion**
  Ventralganglion,
  Bauchnervenknoten
- **visceral ganglion**
  Visceralganglion, Viszeralganglion
ganglionic ganglionär
ganglioside Gangliosid
ganoid scale *ichth* Ganoidschuppe,
Schmelzschuppe
ganoine Ganoin
gap Lücke
gap junction
Kommunikationskontakt, Macula
communicans, Nexus, Gap junction
(Zellkontakte)
gape/gaping *vb* klaffen, offen
stehen (Muschel); sperren,
aufsperren (Schnabel)
gape *n* **(beak/bill)** Schnabelspalt
garden *n* Garten
garden/gardening *vb* Gartenbau
betreiben, im Garten arbeiten,
gärtnern
**garden market/gardening market/
horticulture shop** Gärtnerei
garden peat/granulated peat
Torfmull
garden plant Gartenpflanze
gardener/horticulturist Gärtner
gardening/horticulture Gärtnerei,
Gärtnern; Gartenbau
gardening supplies Gärtnereibedarf

**Garryaceae/silk-tassel tree family/
silktassel-bush family**
Becherkätzchengewächse
gars/Lepisosteiformes
Knochenhechte
gas chamber (nautilus) Gaskammer
gas constant Gaskonstante
**gas exchange/gaseous interchange/
exchange of gases** Gasaustausch
gas gland Gasdrüse
gas mask Gasmaske
gasket Dichtungsring,
Dichtungsmanschette
**gaskin (horse: lower thigh
between stifle and hock)**
Unterschenkel (Hinterschenkel),
Hose
gasohol Treibstoffalkohol, Gasohol
gasp *vb* **(for air)** schnappen (nach
Luft)
gaster Gaster
gastralia/abdominal ribs (reptiles)
Gastralia, Bauchrippen
**gastric cecum/digestive cecum/
gastric diverticulum/digestive
diverticulum** Magenblindsack,
Magendivertikel
gastric filament Gastralfilament
gastric gland (rotifers) Magendrüse
**gastric inhibitory peptide (GIP)/
glucose-dependent insulin-release
peptide** gastrointestinal-
inhibitorisches Peptid, gastrisches
Inhibitor-Peptid (GIP),
glucoseabhängiges Insulin-releasing-
Peptid
gastric juice Magensaft
gastric mill/triturating mill
Magenmühle
**gastric mucosa/mucous tunic
(mucosal layer of stomach)**
Magenschleimhaut, Tunica mucosa
gastric pit/foveola gastrica
Magengrübchen
gastric pouch Gastraltasche,
Darmsack
gastricsin/pepsin C Gastricsin
(Pepsin C)
gastrin Gastrin
**gastrocoel/archenteron/primitive
gut** *embr* Gastrocöl, Archenteron,
Urdarm
gastrodermal tube/solenia
Gastrodermis-Kanal, Solenie
gastrointestinal tract
Gastrointestinaltrakt, Magen-Darm-
Trakt
gastrolith Gastrolith, Magenstein,
Magensteinchen, Hummerstein
gastrotrichs Gastrotrichen,
Bauchhaarlinge, Bauchhärlinge,
Flaschentierchen

G gastrovascular system 136

gastrovascular system
Gastrovaskularsystem
gastrula Gastrula, Becherkeim,
Becherlarve
gate *neuro* Tor
gate impeller Gitterrührer
gate neuron steuerndes Neuron,
regulierendes Neuron
gated ion channel Ionenschleuse
gatherer/collector Sammler
gating current *neuro* Torstrom
(*pl* Torströme)
gating mechanism
Schleusenmechanismus
gauge/calibrate/adjust *vb* eichen,
kalibrieren, justieren
• **measure precisely** *vb* genau
messen, abmessen, ausmessen
**gauge (instrument for measuring/
testing)** Messgerät, Messfühler,
Anzeiger, Messer (auch: Zollstab,
Lehre)
gauge/diameter (e. g. needle)
Durchmesser, Stärke, Dicke
gauge/dimensions/size
Dimensionen, Größe
gauge/standard Maß, Maßstab,
Norm, Normmaß, Standard,
Standardmaß
Gaussian curve Gauß'sche Kurve
**Gaussian distribution (Gaussian
curve/normal probability curve)**
Gauß-Verteilung, Normalverteilung,
Gauß'sche Normalverteilung
gauze Gaze
GC (gas chromatography) GC
(Gaschromatographie)
GC box *gen* GC-Box
Geiger counter Geiger-Zähler
**geitonogamy (*sensu stricto*:
geitonophily)** Geitonogamie,
Nachbarbestäubung
gel *vb* gelieren
gel *n* Gel, Gallerte
• **denaturing gel** denaturierendes
Gel
• **flat bed gel/horizontal gel**
horizontal angeordnetes Plattengel
• **native gel** natives Gel
• **running gel/separating gel**
Trenngel
• **slab gel** hochkant angeordnetes
Plattengel
• **stacking gel** Sammelgel
gel electrophoresis
Gelelektrophorese
• **alternating field gel
electrophoresis** Wechselfeld-
Gelelektrophorese
• **field inversion gel
electrophoresis (FIGE)**
Feldinversions-Gelelektrophorese

• **gradient gel electrophoresis**
Gradienten-Gelelektrophorese
• **pulsed field gel electrophoresis
(PFGE)** Pulsfeld-Gelelektrophorese
• **SDS gel electrophoresis
(sodium dodecyl sulfate)**
SDS-Gelelektrophorese,
Natriumdodecylsulfat-
Gelelektrophorese
• **temperature gradient gel
electrophoresis**
Temperaturgradienten-
Gelelektrophorese
gel filtration Gelfiltration (*see:* gel
permeation)
**gel permeation chromatography/
molecular sieve chromatography**
Gelpermeations-Chromatographie,
Molekularsiebchromatographie
gel retardation analysis
Gelretentionsanalyse
**gel retention assay/electrophoretic
mobility shift assay (EMSA)**
Gelretentionstest
gel well *electrophor* Geltasche
gel-retention analysis
Gelretentionsanalyse
gel-sol-transition Gel-Sol-Übergang
gelatin Gelatine
gelatinous/gel-like gallertartig
gelatinous lichens Gallertflechten
gelation Gelieren
geld/castrate (stallion) reißen
(Hengst), kastrieren
gelding kastriertes ♂ Tier
(Pferd: Wallach)
gelling agent Geliermittel
gelling point Gelierpunkt
gemma (*pl* gemmae or gemmas)
Gemma, Gemme (*pl* Gemmen),
Brutkörper, Brutkörperchen
gemma cup Brutbecher
gemmation/budding Knospung,
Knospenbildung; Knospenanordnung
gemmiform/bud-shaped
knospenförmig
gender/sex Geschlecht
gene Gen, Erbfaktor
• **amplified gene** amplifiziertes
Gen
• **candidate gene** Kandidatengen
• **cell-specific gene** zellspezifisches
Gen
• **discontinuous gene**
diskontinuierliches Gen,
gestückeltes Gen, Mosaikgen
• **early gene** frühes Gen
• **extrachromosomal gene**
extrachromosomales Gen
• **extranuclear gene**
extranukleäres/extranucleäres Gen
• **fusion gene** Fusionsgen

137   **generic name** **G**

- **heterologous gene** Fremdgen
- **homeotic gene** homöotisches Gen
- **housekeeping gene** Haushaltsgen, Haushaltungsgen, konstitutives Gen
- **jumping gene** springendes Gen
- **late gene** spätes Gen
- **luxury gene** Luxusgen
- **master gene** Meistergen
- **mimic genes** mimische Gene
- **modifier gene** Modifikationsgen
- **nested genes** ineinandergesetzte Gene, ineinandergeschachtelte Gene
- **overlapping genes** überlappende Gene
- **regulatory gene** Regulationsgen
- **reporter gene** Reportergen
- **resistance gene** Resistenzgen
- **sex-influenced gene** geschlechtsbeeinflusstes Gen
- **sex-limited gene** geschlechtsbeschränktes Gen
- **silent gene** stummes Gen
- **single copy gene** Einzelkopie-Gen
- **split gene** gestückeltes Gen, Mosaikgen
- **structural gene** Strukturgen
- **suppressor gene** Suppressorgen
- **switch gene** Schaltergen, Schlüsselgen
- **syngenic genes** syngene Gene (Gene auf *einem* Chromosom)
- **tissue-specific gene** gewebespezifisches Gen

gene activation Genaktivierung
gene amplification Gen-Verstärkung, Genamplifikation
gene carrier Genträger
gene cloning Genklonierung
gene cluster Gengruppe, Gencluster
gene complex Genkomplex
gene conversion Genkonversion
gene disruption/gene replacement/ gene targeting Allelen-Austauschtechnik
gene dosage Gendosis
gene dosage effect Gendosiseffekt
gene egoism Genegoismus
gene eviction/gene rescue Genrückgewinnung
gene exchange Genaustausch
gene expression Genexpression
- **control of g.e.** Kontrolle der Genexpression, Genexpressionskontrolle
- **differential g.e.** differentielle Genexpression

gene family Genfamilie
gene farming Gen-Farming

gene flow Genfluss, Genwanderung
gene frequency Genfrequenz, Genhäufigkeit
gene knockout Gen-Knockout (Ausschaltung von Genen durch homologe Rekombination)
gene linkage Genkopplung
gene linkage map Genkopplungskarte
gene locus Genlocus
gene map/genetic map (see also: map) Genkarte
gene mapping/genetic mapping Genkartierung
gene pool Genpool
gene product Genprodukt
gene replacement/gene disruption/ gene targeting Allelen-Austauschtechnik
gene superfamily Gensuperfamilie
gene surgery/gene therapy Gentherapie
gene targeting/gene disruption/ gene replacement Gen-Targeting, Allelen-Austauschtechnik
gene technology/genetic engineering Gentechnologie, Gentechnik, Genmanipulation
gene therapy Gentherapie
- **germ line g.t.** Keimbahngentherapie
- **somatic g.t.** somatische Gentherapie

gene tracking Bestimmung von Vererbungslinien
gene transfer Gentransfer, Genübertragung
gene transplacement Allelen-Austauschtechnik
genealogy Genealogie, Stammbaumforschung, Ahnenforschung, Familienforschung
generalized recombination allgemeine Rekombination
generate/develop (gases) bilden, entwickeln
generation Generation
- **filial generation (F_1=first/ F_2=second)** (erste/zweite) Tochtergeneration, Filialgeneration

generation/development *chem* (gases) Bildung, Entwicklung
generation period Generationsdauer
generation time (doubling time) Generationszeit (Verdopplungszeit)
generator potential Generatorpotential
generic name (nonproprietary name) Sammelname, allgemeingültiger Name, allgemeingültige Bezeichnung

G generic name   138

generic name/genus name
Gattungsname
genesis Genese, Genesis,
Entstehung, Entwicklung
genet Genet (Gesamtheit eines
Klons)
genetic analysis Erbanalyse
genetic code genetischer Code
genetic colonization genetische
Kolonisierung
genetic counsel(l)ing genetische
Beratung
genetic diagnostics Gendiagnostik
genetic disorder Erbkrankheit
genetic dissection genetische
Dissektion
genetic distance genetische Distanz
genetic drift/Sewall Wright effect
Gendrift, genetische Drift
genetic engineering Gentechnik
(*sensu lato:* Gentechnologie > gene
technology)
**genetic fingerprinting/DNA
fingerprinting** genetischer
Fingerabdruck, DNA-Fingerprinting
genetic fixation genetische
Fixierung
genetic hazard Erbschaden,
genetischer Schaden
genetic immunization genetische
Immunisierung
**genetic load/genetic burden/
genetic bond/mutational bond**
Erblast, genetische Last, genetische
Bürde, genetische Belastung
genetic mapping Genkartierung
genetic marker genetischer Marker
genetic predisposition genetische
Prädisposition
genetic risk genetisches Risiko
genetic screening genetischer
Suchtest
genetic susceptibility genetische
Anfälligkeit
genetic variation genetische Varianz
genetically engineered
gentechnisch verändert
genetically engineered organism
gentechnisch veränderter
Organismus (GVO)
genetics (study of inheritance)
Genetik (Vererbungslehre)
• **behavior genetics**
Verhaltensgenetik
• **biochemical genetics/molecular
genetics** Molekulargenetik
• **clinical genetics** klinische
Genetik
• **developmental genetics**
Entwicklungsgenetik
• **direct genetics** direkte Genetik
• **ecogenetics** Ökogenetik

• **eugenics/eugenetics** Eugenik,
"Erbhygiene"
• **formal genetics** formale Genetik
• **human genetics** Humangenetik
• **molecular genetics**
Molekulargenetik
• **pharmacogenetics**
Pharmakogenetik
• **phenogenetics** Phänogenetik
• **population genetics**
Populationsgenetik
• **reverse genetics** reverse Genetik
• **transmission genetics**
Vererbungslehre
**genicular (pertaining to region of
the knee)** das Knie(gelenk)
betreffend, Knie...; knieartig
geniculate/bent like a knee
knieförmig gebogen
geniculate body *neuro* Kniehöcker
geniculate nucleus *neuro/anat*
Kern des Kniehöckers
genital display *ethol*
Genitalpräsentieren, Präsentierung
des Genitals
genital fold *embr* Genitalfalte
**genital opening/genital aperture/
genital pore/gonopore**
Genitalöffnung, Begattungsöffnung,
Genitalporus, Geschlechtsöffnung,
Gonopore
genital plate *embr* Genitalplatte
genital pouch/bursa copulatrix
Begattungstasche
genital presentation *ethol*
Genitalpräsentieren
genital primordium *embr*
Genitalanlage
genital ridge/gonadal ridge *embr*
Genitalleiste, Keimdrüsenleiste
**genital tubercle/tuberculum
genitale** *embr* Genitalhöcker,
Geschlechtshöcker
**genitals/genitalia/genital organs/
sexual organs** Genitalien
genome Genom
• **nuclear genome** Kerngenom
genome analysis Genomanalyse
genomic blotting genomisches
Blotting
genomic imprinting genomische
Prägung
genomic library genomische
Bibliothek
genotype Genotyp, Genotypus
genotyping Gendiagnostik,
Bestimmung des Genotyps
gentian family/Gentianaceae
Enziangewächse
gentisic acid Gentisinsäure
genu/patella (arthropods) Knie,
Patella

ghost

genus (*pl* genera) Gattung
geo-ecology/environmental geology Geoökologie
geobotany/plant geography/phytogeography Geobotanik, Pflanzengeographie, Pflanzengeografie
geocole/geodyte/terricole/soil-dwelling organism Bodenorganismus
geoecology/environmental geology Geoökologie
geogenous geogen
geographic range/area of distribution Verbreitungsgebiet, Areal
geographical geographisch, geografisch, erdkundlich
geography Geographie, Geografie, Erdkunde
geological epoch erdgeschichtliche Epoche
geological era Erdzeitalter
geological period erdgeschichtliche Periode
geology/Earth science Geologie
geonasty Geonastie
geophagous geophag, erdeessend
geophagy/geophagism Geophagie, Erdeessen
geophilomorphs/Geophilomorpha Erdläufer
geophilous geophil
geophyte/geocryptophyte/cryptophyte (*sensu lato*) Geophyt, Erdpflanze, Kryptophyt, Cryptophyt, Staudengewächs
geotaxis Geotaxis (*pl* **Geotaxien**)
geranic acid Geraniumsäure
geranium family/cranesbill family/Geraniaceae Geraniengewächse, Storchschnabelgewächse
geranyl acetate Geranylacetat
germ *micb* Keim
germ/embryo Keim, Keimling, Embryo
germ band (insect egg) Keimstreifen
germ cell/embryonic cell Keimzelle, embryonale Zelle, Embryonalzelle
germ count/cell count Keimzahl (Anzahl von Mikroorganismen)
germ disk/germinal disk/blastodisc Keimscheibe, Embryonalschild, Diskus, Discus
germ layer *embr* Keimschicht, Keimblatt, Blatt
  • **inner germ layer/endoderm/entoderm** Endoderm, Entoderm, inneres Keimblatt, primäres Keimblatt
  • **mesoderm** Mesoderm, sekundäres Keimblatt
  • **outer germ layer/ectoderm** Ectoderm, Ektoderm, primäres Keimblatt, äußeres Keimblatt
germ line/germline Keimbahn
germ plasm/idioplasm/gonoplasm Keimplasma, Idioplasma
germ spot/macula germinativa Keimfleck
germ-free/sterile keimfrei, steril
germ-tube (hypha from spore) Keimschlauch
germicidal keimtötend
germinability Keimfähigkeit
germinal cell Keimzelle
germinal center Keimzentrum
germinal disk/germ disk/blastodisc Keimscheibe, Embryonalschild, Diskus, Discus
germinal epithelium Keimepithel
germinal layer (*Echinococcus*) Keimschicht
germinal streak/primitive streak *embr* Keimstreifen, Primitivstreifen
germinal vesicle/Purkinje's vesicle Keimbläschen (großer Oocytenkern)
germinate/sprout keimen, sprießen
germinating after frost (frost germinator) Frostkeimer
germinating in darkness (dark germinator) Dunkelkeimer
germination Keimung
  • **light-induced germination (photodormancy)** Hellkeimung
germination aperture Keimpore
germination percentage Keimzahl, Keimunganteil
germination period Keimzeit
germline/germ line Keimbahn
germline hypothesis/germline theory Keimbahnhypothese, Keimbahntheorie
germline mosaic/germinal mosaic/gonadal mosaic/gonosomal mosaic Keimbahnmosaik
gesneria family/gesneriad family/gloxinia family/African violet family/Gesneriaceae Gesneriengewächse
gestagen/progestin Gestagen, Progestin, Corpus-luteum-Hormon, "Schwangerschaftshormon"
gestation/pregnancy/gravidity Gestation, Schwangerschaft, Trächtigkeit, Gravidität
gestational period/period of gestation Tragzeit, Tragezeit, Schwangerschaftsperiode
gesture Gestik, Geste, Gebärde
ghost/cell ghost *Ghost*, leere Zellhülle (*see:* erythrocyte ghost)

G

giant axon 140

giant axon Riesenaxon, Riesenfaser, Kolossalfaser
giant cell Riesenzelle, Kolossalzelle
giant chromosome Riesenchromosom
giant fiber/Mauthner's cell Riesenfaser, Mauthnersche Zelle, Mauthner-Zelle (Fische)
giant horsetail family/calamites/ calamite family/Calamitaceae Calamiten, Schachtelhalmbäume
giardiasis (*Giardia lamblia*) Giardiasis, Lamblienruhr
gibberellic acid Gibberellinsäure
gibberellins Gibberelline
gibbose/gibbous aufgetrieben, angeschwollen; *orn* (on bill of birds:) höckerig, buckelig)
gibbosity Auftreibubg, Schwellung; *orn* (on bill of bird:) Höcker, Buckel, Wölbung
giblets (edible viscera of fowl) Innereien (essbare Organe des Geflügels)
gill Kieme
● **book gills (gill book)** Buchkiemen, Kiemenbeine, Blattbeine, Blattfüße (dichtstehende Kiemenlamellen: Xiphosuriden)
● **compressible gill** kompressible Gaskieme
● **dendrobranchiate gill** Dendrobranchie
● **external gill/ectobranch** Außenkieme, äußere Kieme
● **filibranch gill** Fadenkieme, Filibranchie
● **foot gill/podobranch** Podobranchie
● **gaseous gill/gaseous plastron/ air-bubble gill** Gaskieme
● **gill plume/gill comb/ctenidium** Kammkieme, Fiederkieme, Ctenidie, Ctenidium
● **hemibranch** Hemibranchie
● **holobranch** Holobranchie
● **internal gill/entobranch** Innenkieme, innere Kieme
● **joint gill/arthrobranch** Arthrobranchie
● **lamellar gill/sheet gill/ lamellibranch/eulamellibranch** Blattkieme, Lamellibranchie, Eulamellibranchie
● **phyllobranchiate gill** Phyllobranchie
● **physical gill** physikalische Kieme
● **pseudobranch (accessory/ spurious gill in some fish)** Pseudobranchie

● **pseudolamellar gill/ pseudolamellibranch** Scheinblattkieme, Pseudolamellibranchie
● **side gill/pleurobranch** Pleurobranchie
● **tracheal gill** Tracheenkieme
● **trichobranchiate gill** Trichobranchie
gill/lamella *fung* Lamelle, Pilzlamelle, "Blatt"
gill arch/branchial arch/visceral arch Kiemenbogen, Branchialbogen, Viszeralbogen (Gesamtheit der Teile)
gill bailer/bailer/scaphognathite Atemplatte, Scaphognathide, Scaphognathit
gill bar/branchial bar/visceral bar (skeleton only) Kiemenbogen, Branchialbogen, Viszeralbogen (nur Knorpelspange)
gill basket Kiemenkorb
gill cavity/gill chamber Kiemenhöhle, Kiemenkammer
gill cleaner Kiemenbürste (Flabellum)
gill comb/gill plume/ctenidium Kammkieme, Fiederkieme, Ctenidie
gill cover/operculum Kiemendeckel, Operkulum
gill filament Kiemenfilament, Kiemenfaden
gill fungus/gill mushroom Lamellenpilz, Blätterpilz
gill heart/branchial heart (cephalopods) Kiemenherz
gill lamella (fish/bivalves) Kiemenblatt, Kiemenblättchen, Kiemenlamelle, Hemibranchie (Fische, Muscheln)
gill opening/gill aperture Kiemenöffnung
gill plume/gill comb/ctenidium Fiederkieme, Kammkieme, Ctenidie
gill pouch/branchial sac/ pharyngeal pouch Kiementasche, Kiemensack
gill raker (bristle-like process on gill arch) Kiemendorn (*pl* gill rakers Kiemenreuse)
gill ray Kiemenstrahl
gill rod (cephalochordates) Kiemenbalken
gill slit/pharyngeal slit/gill cleft/ branchial cleft/pharyngeal cleft Kiemenspalte, Viszeralspalte
gill trama/dissepiment Lamellentrama
gilled puffballs/Hymenogastrales (Hymenogastraceae) Erdnußartige (Pilze)

gland

gilt (youngpig before becoming a sow) junge Sau, Jungsau
gin/ginning (cotton) *vb* (Baumwolle) entkörnen, egrenieren
ginger family/Zingiberaceae Ingwergewächse
ginglymus joint/hinge joint Scharniergelenk
ginkgo family/Ginkgoaceae Ginkgogewächse
ginseng family/ivy family/Araliaceae Efeugewächse
girdle *vb* umgürten
girdle/ring *vb* **(tree bark)** for ringeln
girdle/cingulum *n* Gürtel, Gurt, Cingulum
girdled pupa/pupa cingulata Gürtelpuppe
girdling/ringing (tree bark) for Gürteln, Ringelung
girth Umfang
give birth/bear young/bear offspring gebären, niederkommen, Junge bekommen
giving birth/parturition Gebären, Niederkunft
gizzard *orn* Muskelmagen
gizzard/proventriculus (insects/crustaceans) Kaumagen, Proventriculus
glabrous/hairless (smooth) haarlos, unbehaart
glacial acetic acid Eisessig
glacial drift Glazialgeschiebe
glacial lake Gletschersee
glacial till/glacial detritus/moraine/till Moräne, Gletscherschutt, Glazialschutt, Gletschergeröll
glacier Gletscher
glade Lichtung, Schneise
gladiate/xiphoid/ensiform/sword-shaped schwertförmig
gladius/pen (chitinous internal shell) Gladius, Rückenfeder
gland Drüse
- **accessory gland** akzessorische Drüse, Anhangsdrüse
- **adhesive gland/colleterial gland/cement gland (insects)** Haftdrüse, Klebdrüse, Kittdrüse, Zementdrüse
- **adrenal gland** Nebenniere
- **albumen gland** Eiweißdrüse
- **ampullary gland (reproductive gland)** Ampullardrüse
- **anal gland/rectal gland (shark)** Analdrüse (Hai)
- **antennal gland/antennary gland/green gland** Antennendrüse, Antennennephridium, grüne Drüse
- **antorbital gland/preorbital gland** Voraugendrüse, Antorbitaldrüse
- **apocrine gland** apokrine Drüse
- **Bartholin's gland/greater vestibular gland/gl. vestibularis major** Bartholin-Drüse
- **Bowman's gland/olfactory gland** Bowman'-Spüldrüse, Drüse der Riechschleimhaut
- **Brunner's gland/duodenal gland** Brunnersche Drüse, Duodenaldrüse
- **bulbourethral gland/Cowper's gland** Bulbourethraldrüse, Cowpersche Drüse
- **calciferous gland** Kalkdrüse
- **castor gland/preputial gland (beaver)** Präputialdrüse
- **caudal gland (rectal gland)** Schwanzdrüse (Rektaldrüse)
- **cement gland/adhesive gland** Zementdrüse, Kittdrüse, Klebdrüse
- **cephalic gland/frontal gland** Kopfdrüse, Stirndrüse, Frontaldrüse
- **ceruminous gland/wax gland** Wachsdrüse; Ohrenschmalzdrüse
- **cheek gland** Backendrüse
- **circumanal gland** Zirkumanaldrüse
- **claw gland** Klauendrüse
- **colleterial gland/adhesive gland/cement gland (insects)** Kittdrüse, Klebdrüse, Zementdrüse (Lepidoptera: Glandula sebacea)
- **Cowper's gland/bulbourethral gland** Cowpersche Drüse, Cowper-Drüse, Bulbourethraldrüse
- **coxal gland** Coxaldrüse
- **defensive gland (*Peripatus*: slime gland)** Wehrdrüse (Schleimdrüse)
- **dermal gland** Dermaldrüse, Hautdrüse
- **digestive gland** *sensu lato* Verdauungsdrüse
- **digestive gland/"liver" (mollusks/echinoderms)** Mitteldarmdrüse, Darmdivertikel
- **Dufour gland/alkaline gland (hymenopterans)** Dufour-Drüse
- **duodenal gland/Brunner's gland** Duodenaldrüse, Brunnersche Drüse
- **endocrine gland** endokrine Drüse
- **esophageal gland (gastropods)** Schlunddrüse
- **exocrine gland/eccrine gland** exokrine Drüse, ekkrine Drüse
- **femoral gland (follicular gland: lizards)** Schenkeldrüse (Follikulärorgan: Eidechsen)

gland

- **follicular gland (femoral gland)** Follikulärorgan, Follikulärdrüse (Schenkeldrüse: Eidechsen)
- **frontal gland/cephalic gland** Frontaldrüse, Stirndrüse, Kopfdrüse
- **fundus gland** Fundusdrüse
- **gas gland** Gasdrüse, Schwimmblasendrüse (Roter Körper)
- **gastric gland (rotifers)** Magendrüse
- **gland of Zeis/sebaceous ciliary gland** Zeis-Drüse
- **granular gland/poison gland (amphibians)** Körnerdrüse
- **green gland/antennal gland/ antennary gland** grüne Drüse, Antennendrüse, Antennennephridium
- **gustatory gland (mammals)** Spüldrüse, von Ebnersche Drüse, Ebner' Drüse
- **Harderian gland/Harder's gland** Hardersche Drüse
- **hedonic gland (amphibians)** hedonische Drüse
- **hermaphroditic gland/ hermaphroditic gonad/ovotestis** Zwitterdrüse
- **holocrine gland** holokrine Drüse
- **hypopharyngeal gland (bees)** Hypopharynxdrüse, Futtersaftdrüse
- **inguinal gland** Leistendrüse
- **ink gland/ink sac** Tintendrüse, Tintensack, Tintenbeutel
- **interdigital gland** interdigitale Drüse, Interdigitaldrüse, Zwischenzehendrüse
- **labial gland** Labialdrüse
- **lacrimal gland** Tränendrüse
- **lingual gland** Zungendrüse
- **lymphatic gland** Lymphdrüse
- **mammary gland** Brustdrüse, Milchdrüse
- **mandibular gland** Mandibulardrüse, Unterkieferdrüse
- **maxillary gland** Schalendrüse, Maxillendrüse, Maxillennephridium
- **Mehlis' gland/shell gland (cement gland)** Mehlissche Drüse, Schalendrüse
- **Meibomian gland** Meibom-Drüse
- **merocrine gland** merokrine Drüse
- **metatarsal gland** Metatarsaldrüse
- **midgut gland/digestive gland/ "liver"** Mitteldarmdrüse, "Leber"
- **midgut gland/hepatopancreas** Mitteldarmdrüse, Hepatopankreas
- **molting gland/Y organ** Carapaxdrüse, Y-Organ
- **mucous gland** Schleimdrüse
- **musk gland (scent gland)** Moschusdrüse (Duftdrüse)
- **Nassanov gland/Nassanov's gland** Nassanov Drüse, Nassanoffsche Drüse
- **nidamental gland/shell gland** Nidamentaldrüse, Schalendrüse, Eischalendrüse
- **odoriferous gland/scent gland** Duftdrüse, Brunftdrüse, Brunftfeige (Gemse)
- **oil gland** Öldrüse, Schmierdrüse
- **olfactory gland/Bowman's gland** Bowman-Drüse, Bowmansche Drüse, Drüse der Riechschleimhaut
- **parathyroid gland/parathyroidea** Nebenschilddrüse, Beischilddrüse, Epithelkörperchen
- **parotid gland/parotis/parotid (mammals: salivary gland)** Parotis, Ohrspeicheldrüse
- **parotoid gland (amphibians)** Parotoiddrüse, Parotisdrüse, Ohrdrüse, Duvernoysche Drüse
- **pedal gland/adhesive gland/ cement gland (rotifers)** Fußdrüse, Klebdrüse, Kittdrüse
- **perineal gland** Perinealdrüse, Dammdrüse
- **pineal gland/pineal body/ conarium/epiphysis** Pinealorgan, Epiphyse, Zirbeldrüse
- **pituitary gland/hypophysis** Hirnanhangsdrüse, Hypophyse
- **poison gland** Giftdrüse
- **preorbital gland/antorbital gland** Voraugendrüse, Antorbitaldrüse
- **preputial gland/castor gland (beaver)** Präputialdrüse
- **prostate gland/prostate** Prostata, Prostatadrüse, Vorsteherdrüse
- **prostatic gland (annelids: spermiducal gland)** Kornsekretdrüse
- **prothoracic gland (an ecdysial, molting gland)** Prothoraxdrüse
- **purple gland** Purpurdrüse
- **pygidial gland/anal gland** Pygidialdrüse, Analdrüse
- **rectal gland** Rectaldrüse, Rektaldrüse (*see*: Klebdrüse, Kittdrüse, Zementdrüse)
- **rectal gland/anal gland (shark)** Analdrüse (Hai)
- **renette gland/ventral gland (nematodes)** Ventraldrüse

globose soralium

- **repugnatorial gland** Stinkdrüse (*siehe:* Wehrdrüse)
- **salivary gland** Speicheldrüse
- **salt gland** Salzdrüse
- **scent gland/odoriferous gland** Duftdrüse, Brunftdrüse, Brunftfeige (Gemse)
- **sebaceous gland** Talgdrüse, Haartalgdrüse
- **secretory gland** Sekretdrüse
- **seminal gland/vesicular gland/ seminal vesicle (♂ accessory reproductive gland)** Bläschendrüse, Samenblase, Samenbläschen
- **sex gland/germ gland/gonad** Geschlechtsdrüse, Keimdrüse, Gonade
- **shell gland/Mehlis' gland (cement gland)** Schalendrüse, Mehlissche Drüse
- **shell gland/nidamental gland** Schalendrüse, Nidamentaldrüse
- **silk gland/spinning gland/ sericterium (caterpillars: labial gland)** Seidendrüse, Spinndrüse, Sericterium (Labialdrüse: Raupen)
- **sinus gland** Sinusdrüse
- **slime gland/mucous gland** Schleimdrüse
- **sublingual gland** Unterzungendrüse
- **suborbital gland** Unteraugendrüse
- **sugar gland/subradular organ** Zuckerdrüse
- **supracaudal gland** Schwanzwurzeldrüse
- **sweat gland/sudoriferous gland/ sudoriparous gland** Schweißdrüse
- **tarsal gland** Tarsaldrüse
- **tarsal gland/Meibomian gland** Meibom'-Drüse
- **temporal gland (elephant)** Schläfendrüse
- **thymus (gland)** Thymus, Thymusdrüse, Bries (Halsthymus/ Brustthymus)
- **thyroid gland/thyreoidea** Schilddrüse
- **urethral gland** Urethraldrüse
- **uropygial gland/preen gland/oil gland** Bürzeldrüse
- **ventral gland/renette gland (nematodes)** Ventraldrüse
- **vesicular gland/seminal gland/ seminal vesicle (♂ accessory reproductive gland)** Bläschendrüse, Samenblase, Samenbläschen
- **vestibular gland (♀ vaginal gland)** Vorhofdrüse
- **violet gland/supracaudal gland** Violdrüse (Schwanzwurzeldrüse des Fuchses)
- **wax gland/ceruminous gland** Wachsdrüse
- **wool fat gland** Wollfettdrüse
- **yolk gland/vitellarian gland/ vitelline gland/vitellarium/ vitellogen** Dotterstock, Dotterdrüse, Vitellar, Vitellarium

gland cell Drüsenzelle
glandular drüsig
glandular epithelium Drüsenepithel
glandular hair Drüsenhaar
glandular portion of stomach Drüsenmagen
glandular secretion (secreted substance matter) Drüsensekret; (process/phenomenon) Drüsensekretion
glandular tissue Drüsengewebe
glass homogenizer (Potter-Elvehjem homogenizer; Dounce homogenizer) Glashomogenisator ("Potter"; Dounce)
glass pestle *lab* Glasstößel, Glaspistill (Homogenisator)
glass rod *lab* Glasstab
glass sponges/Hexactinellida Glasschwämme, Hexactinelliden
glasshouse see greenhouse
glaucous/grey-green (with a bloom) blaugrün, bläulich-grün, graugrün, wachsartig schimmernd, weißlich reflektierend (Blattoberfläche)
glaze n Glatteis, dünne Eisschicht
GLC (gas-liquid chromatography) GFC (Gas-Flüssig-Chromatographie)
gleichneria family/Gleichneriaceae Gleichneriaceae
gleization (soil) Gleybildung, Vergleyung
glenoid glenoid, flachschalig
glenoid cavity/glenoid fossa/ cavitas glenoidalis/fossa glenoidalis Gelenkpfanne der Skapula, Schultergelenkpfanne
glial cell Gliazelle
glide vb gleiten
glide/gliding n **(flight)** Gleitflug
glide angle/gliding angle Gleitwinkel
gliding joint/plane joint (arthrodia) Gleitgelenk, ebenes Gelenk
globe daisy family/globularia family/Globulariaceae Kugelblumengewächse
globose soralium (lichens) Kugelsoral

G globular protein 144

globular protein globuläres Protein, Sphäroprotein
globularia family/globe daisy family/Globulariaceae Kugelblumengewächse
globulin Globulin
glochidiate/provided with barbed hairs widerhakig, mit widerhakigen Borsten
glochidium (larva) Glochidium
glomerular filtration Glomerulusfiltration
glomerular filtration rate (GFR) glomeruläre Filtrationsrate
glomerular ultrafiltrate Primärharn, Glomerulusfiltrat
glomerule/flower cluster Blütenknäuel
glomerulus/network of blood capillaries Glomerulus, Gefäßknäuel
glossiness Lackglanz
glossy glänzend
glottis Glottis
glove *lab* Handschuh (Laborschutzhandschuh)
glove box Handschuhkasten, Handschuhschutzkammer
gloxinia family/gesneria family/ gesneriad family/African violet family/Gesneriaceae Gesneriengewächse
glucaric acid/saccharic acid Glucarsäure, Zuckersäure
glucocorticoid Glucocorticoid
gluconeogenesis Gluconeogenese
gluconic acid (gluconate) Gluconsäure (Gluconat)
glucosamine Glukosamin, Glucosamin
glucose (grape sugar) Glukose, Glucose (Traubenzucker)
glucosuria/glycosuria Glukosurie, Glycosurie
glucuronic acid (glucuronate) Glucuronsäure (Glukuronat)
glumaceous spelzblütig, spelzig
glume *bot* Hüllspelze
glumella/palea/pale/inner glume Vorspelze
glumellule/lodicule/paleola Schüppchen, Lodicula, Schwellkörper (Grasblüte)
glumose spelzig, Spelzen...
glutamic acid (glutamate)/ 2-aminoglutaric acid Glutaminsäure (Glutamat), 2-Aminoglutarsäure
glutamine Glutamin
glutaric acid (glutarate) Glutarsäure (Glutarat)
glutathione Glutathion

gluteal fold Gesäßfalte
gluten (glutelin & gliadin) Gluten
glutinant Glutinant, Klebkapsel, Haftkapsel
glutine (glue from animals) Glutin (Knochenleim)
glutinous/mucilaginous/viscid/ slimy (sticky) glutinös, schleimig (klebrig)
glycemia Glykämie
glyceraldehyde/dihydroxypropanal Glyzerinaldehyd, Glycerinaldehyd
glycerol Glyzerin, Glycerin, Propantriol
glycine/glycocoll Glycin, Glyzin, Glykokoll
glycocalyx (cell coat) Glykokalyx
glycocoll/glycine Glykokoll, Glycin, Glyzin
glycogen Glykogen
glycol aldehyde/glycolal/ hydroxyaldehyde Glykolaldehyd, Hydroxyacetaldehyd
glycolic acid (glycolate) Glykolsäure (Glykolat)
glycolysis Glykolyse
glycometabolism Zuckerstoffwechsel
glycosaminoglycan/ mucopolysaccharide Glykosaminoglykan
glycosidic bond/glycosidic linkage glykosidische Bindung
glycosuria/glucosuria Glykosurie, Glukosurie
glycyrrhetinic acid Glycyrrhetinsäure
glyoxalic acid (glyoxalate) Glyoxalsäure (Glyoxalat)
glyoxylate cycle Glyoxalatzyklus
glyoxylic acid (glyoxylate) Glyoxylsäure (Glyoxylat)
glyoxysome Glyoxysom
glyphosate Glyphosat
GM-CSF (granulocyte-macrophage stimulating factor) GM-CSF (Granulocyten-Makrophagen-stimulierender Faktor)
gnarl/burl/burr *bot* Knorren (an Baum), Holzmaser, Maser, Maserknolle
gnarled knorrig
gnathobase/blade (crustaceans) Kaulade
gnathochilarium Gnathochilarium
gnathopod Gnathopod
gnathos Gnathos
gnathosoma/capitulum Gnathosoma, Capitulum
gnathostomulids/Gnathostomulida Kiefermäuler, Kiefermündchen, Gnathostomuliden

gnaw nagen (an etwas nagen)
gnawer/rodent Nager, Nagetier
gnawing nagend
Gnetaceae/joint-fir family
Gnetumgewächse,
Gnemonbaumgewächse
Goblet cell (mucus-producing)
Becherzelle, Schleimzelle
goggles/safety goggles/safety
spectacles Schutzbrille
gold-labelling Goldmarkierung
golden algae/golden-brown algae/
Chrysophyceae Goldalgen,
Chrysophyceen
Golgi apparatus/Golgi complex
Golgi-Apparat
Golgi body/dictyosome Diktyosom,
Dictyosom
Golgi staining method Golgi-
Anfärbemethode
Golgi vesicle Golgi-Vesikel
Gomphidius family/Gomphidiaceae
Schmierlinge
gomphosis Gomphose, Einzapfung
gonad/sex gland Gonade,
Keimdrüse, Geschlechtsdrüse
• **ovary** Ovar, Ovarium, Eierstock
• **ovotestis/hermaphroditic**
gonad/hermaphroditic gland
Ovotestis, Zwitterdrüse
• **testis** (*pl* **testes)/testicle** Hoden,
Samendrüse
gonadal mosaic/gonadic mosaic/
germline mosaic/germinal mosaic/
gonosomal mosaic
Keimbahnmosaik, gonadales
Mosaik
gonadoliberin/gonadotropin
releasing hormone, factor (GnRH,
GnRF) Gonadoliberin,
Gonadotropin-Freisetzungshormon
gonadotropin Gonadotropin
gonadotropin releasing hormone,
factor (GnRH/GnRF)/
gonadoliberin Gonadotropin-
Releasing Hormon, Gonadoliberin
gongylidium (*pl* **gongylidia)/**
bromatium Gongylidie,
Bromatium, "Kohlrabi"
goniatitic suture line goniatische
Lobenlinie/Nahtlinie
gonochoric/gonochoristic
gonochor
gonochory/gonochorism
Gonochorismus
gonocyte Gonocyt, Gonozyt
gonopalpon Genitaltaster
gonopod/gonopodium
Gonopodium, Genitalfuß,
Begattungsfuß
goodness of fit *stat* Güte der
Anpassung

gooseberry family/currant family/
Grossulariaceae
Stachelbeergewächse
gooseflesh/goose pimples/goose
bumps Gänsehaut
goosefoot family/Chenopodiaceae
Gänsefußgewächse
gordian worms/horsehair worms/
hairworms/threadworms/
nematomorphans/nematomorphs/
Nematomorpha Saitenwürmer
gorge/canyon Schlucht
gorger (animal which gulps down
entire prey) Schlinger
gorgonians/horny corals/
Gorgonaria Rindenkorallen,
Hornkorallen
gosling junge Gans, Gänschen,
Gössel (Gänseküken)
gossamer (film of cobwebs floating
in air) Altweibersommer
(Spinnengewebe)
gourd/pepo Kürbisfrucht,
Gurkenfrucht, Panzerbeere
gourd family/pumpkin family/
cucumber family/Cucurbitaceae
Kürbisgewächse
GPP (gross primary production)
Bruttoprimärproduktion
Graafian follicle/vesicular ovarian
follicle Graafscher Follikel, Graaf-
Follikel, Tertiärfollikel
gradation Abstufung, Staffelung,
Stufenfolge
grade (group at same
organizational level) Gruppe
(derselben Organisationsstufe)
graded ethanol series Alkoholreihe,
aufsteigende Äthanolreihe
graded potential graduiertes
Potential
gradient Gradient, Gefälle
gradient gel electrophoresis
Gradienten-Gelelektrophorese
gradient hypothesis Gradienten-
Hypothese
graduate *vb* graduieren, in Grade
einteilen/unterteilen
graduated graduiert, mit einer
Gradeinteilung versehen
graduated cylinder Messzylinder
graduated pipette/measuring pipet
Messpipette
graft *vb bot/hort* pfropfen
graft/slip/scion/cion *bot/hort*
Pfropfreis (*pl* Pfropfreiser), Edelreis,
Reis, Pfröpfling
graft/transplant *n* **(tissue/skin)**
Transplantat
graft/transplant *vb* **(tissue/skin)**
transplantieren, verpflanzen
graft rejection Transplantatabstoßung

G graft union 146

graft union *bot* Pfropfstelle
graft-versus-host reaction (GVH) Transplantat-anti-Wirt-Reaktion
grafting *med* Transplantation, Implantation
grafting *bot* Pfropfung, Veredelung, Veredlung
- **apical grafting** Veredeln auf den Kopf
- **approach grafting** Ablaktieren, Ablaktion
- **bark grafting/rind grafting** Rindenpfropfung, Pfropfen hinter die Rinde
- **bench grafting** Tischveredelung
- **bridge grafting/repair grafting** Überbrückung, Überbrücken, Wundüberbrückung
- **bud grafting/budding** Augenveredlung, Äugeln, Okulieren, Okulation
- **cleft grafting/wedge grafting** Spaltpfropfung, Pfropfen in den Spalt
- **double working/intergrafting** Zwischenpfropfung, Zwischenveredlung
- **flap grafting** seitliches Anplatten mit langer Gegenzunge
- **inarching** Ammenveredlung, Anhängen, Vorspann geben
- **inlay grafting** Geißfußpfropfung, Geißfußveredelung (Triangulation)
- **intergrafting/double working** Zwischenpfropfung, Zwischenveredlung
- **nurse grafting (nurse-root grafting)** Ammenveredelung
- **rind grafting/bark grafting** Rindenpfropfung, Pfropfen hinter die Rinde
- **root grafting** Wurzelpfropfung, Wurzelveredlung
- **saddle grafting** Sattelschäften
- **shield grafting/sprig grafting** seitliches Einspitzen
- **side grafting** Seitenpfropfung, Seitenveredelung, Veredeln an die Seite
- **side-tongue grafting** seitliches Anplatten mit Gegenzunge
- **side-veneer grafting/veneer side grafting/spliced side grafting** Anplatten, seitliches Anplatten
- **splice grafting/whip grafting** Kopulation, Kopulieren, Schäften (Pfropfung)
- **sprig grafting/shield grafting** seitliches Einspitzen
- **top grafting/top working** Astpfropfung, Astveredlung

- **wedge grafting/cleft grafting** Spaltpfropfung, Pfropfen in den Spalt
- **whip grafting/splice grafting** Kopulation, Kopulieren, Schäften (Pfropfung)
grain (form of wood texture) Faser, Faserung, Faserorientierung, Struktur, Fibrillenanordnung (Schnittholz)
grain (particle size) Körnung (Korngröße)
grain/kernel (cereal) Korn (Getreide)
grain filling (poorly or well-filled) *bot/agr* Ährenfüllung, Kornfüllung
gram equivalent Grammäquivalent
Gram stain Gram-Färbung
gram-negative gramnegativ
gram-positive grampositiv
Gramineae/grass family/Poaceae Süßgräser, Gräser
graminifoliose grasblättrig
graminoid/graminaceous/grassy grasartig
Grandry's corpuscle (duck bill) Grandrysches Körperchen
granivorous granivor, samenfressend, körnerfressend
granivorous animal Körnerfresser
granular granulär
granular gland/poison gland (amphibians) Körnerdrüse
granule cell Körnerzelle (Cerebellum)
granule cell layer Körnerschicht
granulocyte Granulozyt, Granulozyt (polymorphonuklearer Leukozyt)
- **band granulocyte/stab cell** stabkerniger Granulozyt
- **basophil(ic) granulocyte** basophiler Granulozyt
- **eosinophil(ic) granulocyte** eosinophiler Granulozyt
- **neutrophil(ic) granulocyte** neutrophiler Granulozyt
- **polymorphonuclear** polymorphkerniger Granulozyt, polymorphonuklearer Leukozyt
- **segmented granulocyte/ filamented granulocyte** segmentkerniger Granulozyt
granulocyte-macrophage stimulating factor (GM-CSF) Granulocyten-Makrophagen-stimulierender Faktor (GM-CSF)
granulopoesis Granulopoese
granulosis viruses Granulaviren
granum (*pl* grana) Granum (*pl* Grana)

gregariousness

grape family/vine family/Vitaceae Weinrebengewächse
grape fern family/adder's-tongue family/Ophioglossaceae Natternzungengewächse, Rautenfarngewächse
grape sugar/glucose Traubenzucker, Glukose, Glucose
grapevine Weinrebe, Weinstock
graph/diagram (*math* curve) Grafik, Graphik, Diagramm, Schaubild (*math* Kurve)
graph paper Millimeterpapier
 • **semi-log graph paper** halblogarithmisches Millimeterpapier
graphic representation grafische/graphische Darstellung
graptolites/Graptolithina Graptolithen
grasp *n* Klammergriff
grasp *vb* ergreifen, zupacken, festhalten
grasping/prehensile/able to grasp/raptorial Greif..., zum Greifen geeignet, zupackend, ergreifend
grasping claws/clasper(s)/clasps Greifzange, Haltezange, Klasper
grasping foot/prehensile foot Greiffuß
grass/lawn Gras, Rasen
grass cover/sod/turf (nonforage grass) Rasendecke
grass family/Gramineae/Poaceae Süßgräser, Gräser
grass heath (a tussock community) Grasheidenstufe
grass of Parnassus family/Parnassiaceae Herzblattgewächse
grass tree family/blackboy family/Xanthorrhoeaceae Grasbaumgewächse
grasses (Poaceae) echte Gräser, Süßgräser (Spelzenblütler)
grassy/graminoid/graminaceous grasartig
grate/bar screen (sewage treatment plant) Rechen (Kläranlage)
gratuitous inducer freiwilliger Induktor
graupel/sleet/soft hail Graupel, Schneeregen
gravel Kies, Schotter
gravel bar Schotterbank
gravel pit Kiesgrube
gravid/pregnant trächtig, schwanger
gravidity/pregnancy Trächtigkeit, Schwangerschaft, Gravidität
gravitation/gravity/gravitational force Gravitation, Schwerkraft
gravitational field Schwerefeld
gravitational sense Schweresinn

gravitational water/seepage water Senkwasser, Sickerwasser
gravity/gravitation/gravitational force Gravitation, Schwerkraft
gray crescent (amphibian egg) *embr* grauer Halbmond
gray horse Schimmel
gray matter/substantia grisea *neuro* graue Substanz (Hirn- u. Rückenmark)
gray ramus/gray communicating ramus Ramus communicans griseus
gray-green/glaucous graugrün, blaugrün
graze/pasture (herbaceous plants) grasen, abgrasen, abfressen, weiden (Wild: äsen)
grazer (grazing on herbaceous plants) grasendes Tier
grazer (invertebrates) Weidegänger
grazing/browsing Weiden (Wild>Nahrungsaufnahme: Äsung/Geäse)
grazing animals Weidevieh
grazing food chain Fraßnahrungskette, Abweidenahrungskette, Weidenahrungskette, Weidekette
grazing traces/pascichnia *paleo* Weidespuren
grease *n* Fett, Schmalz; Schmierfett
grease *vb* fetten, einfetten, schmieren, einschmieren
grebes/Podicipediformes Lappentaucher
green algae/Chlorophyceae/Isokontae Grünalgen
green density Rohdichte
green forage/greenstuff/soilage Grünfutter, Grünzeug
green gland/antennal gland/antennary gland grüne Drüse, Antennendrüse, Antennennephridium
green manure Gründünger
green revolution grüne Revolution
greenery/green (floristics) Grün
greenhouse Treibhaus, Gewächshaus (Glashaus); (open to the public) Pflanzenschauhaus
greenhouse effect *ecol* Treibhauseffekt
greens/potherbs Suppenkraut, Blattgemüse (gekochtes)
greenstuff/green forage/soilage Grünfutter, Grünzeug
gregarious gregär, Herden..., in Herden lebend, Gruppen..., in Gruppen lebend, gesellig (Herdentiere/Insekten)
gregariousness/sociability Geselligkeitsgrad, Soziabilität

G gressorial

gressorial/gressorious/adapted for walking zum Laufen geeignet, Lauf…

gressorial leg/walking leg Laufbein

grey *see* **gray** grau

grid *micros* **(for EM)** Gitter, Netz, Gitternetz, Trägernetz, Probenträger(netz) für Elektronenmikroskopie

grilse Jackobslachs, Bartolomäuslachs

grind (with teeth/jaws) zermalmen, mahlen

grinding tooth Mahlzahn

grip *n* Griff

gristle/cartilage Knorpel

grit *n geol* Kies, Grus, grober Sand

grit *n agr* Korn, Schrot, Grütze

• **grits/hominy grits (U.S.: coarse cornmeal)** Maisgrütze, Maisgrieß, grobes Maismehl

grit *vb* mahlen, knirschen

grit cell (in fruit) Steinzelle (Frucht)

grit chamber (sewage treatment plant) Sandfang (Kläranlage)

grits (U.S.: coarse cornmeal) Maisgrütze, Maisgrieß, grobes Maismehl

grizzly (grizzled coat/fur) grauhaarig (mit gräulichem Fell)

groats/grits (hulled/ground grain) Grütze, Grieß

groin/inguinal zone/regio inguinalis Hüftbeuge, Leiste, Leistenbeuge, Leistengegend, Inguinalgegend

groom putzen, säubern

groom/brush/currycomb (horses) pflegen, striegeln

grooming Putzen, Grooming

• **allogrooming/social grooming** Allogrooming, Fremdputzen (z. B. Fell bei Säugern)

• **autogrooming/autopreening/ self-grooming (mammals)** Putzen am eigenen Körper, Selbstputzen

• **pseudogrooming** Scheinputzen

groove/furrow/sulcus Furche, Rinne, Grube, Sulcus

• **major groove** *gen* **(DNA)** große Furche, große Rinne, tiefe Rinne (DNA-Struktur)

• **minor groove** *gen* **(DNA)** kleine Furche, kleine Rinne, flache Rinne (DNA Struktur)

grooved/furrowed/sulcate gefurcht, furchig, gerieft

gross potential Summenpotential

gross production Bruttoproduktion, Gesamtproduktion

gross productivity Bruttoproduktivität

gross weight Bruttogewicht

Grossulariaceae/currant family/ gooseberry family Stachelbeergewächse

ground (soil) Boden, Erde, Erdoberfläche

ground cover/herbaceous soil cover Bodendecker

ground frost Bodenfrost

ground level Erdoberfläche

ground meristem Grundmeristem

ground moraine/basal moraine Grundmoräne, Untermoräne

ground pecking *ethol* Bodenpicken

ground state Grundzustand

ground stratum/ground layer Bodenschicht

ground tissue/fundamental tissue/ parenchyma Grundgewebe, Parenchym

ground water Grundwasser

group/assemblage Gruppe

group importance value Gruppenmächtigkeit

group transfer Gruppenübertragung

group value Gruppenwert

group-specific antigen (gag) gruppenspezifisches Antigen

grove Hain, Baumhain, Plantage; Gehölz, Waldung, Wäldchen, kleines Waldstück

grow wild/overgrow verwildern

growing point/apical meristem *bot* Wachstumspunkt

growl/snarl (wütend) knurren; (bear) brummen

growth Wachstum; Wuchs

• **acroplastic growth** akroplastes Wachstum

• **annual growth** Jahreswachstum

• **apical growth** Spitzenwachstum

• **arithmetic growth** arithmetisches Wachstum

• **balanced growth** ausgewogenes Wachstum

• **basiplastic growth** basiplastes Wachstum

• **cell growth** Zellwachstum

• **corroborative growth (*Troll*)/ establishment growth (*Zimmermann/Tomlinson*)** Erstarkungswachstum

• **determinate growth/restricted growth** begrenztes/beschränktes Wachstum

• **diffuse growth (animals)** diffuses Wachstum, zerstreutes/ verteiltes Wachstum

• **dilation growth/dilatation growth** Dilatationswachstum, Erweiterungswachstum

149   **Guarnieri body**   **G**

- **direction of growth**
  Wuchsrichtung
- **elongational growth/extension growth** Streckungswachstum
- **establishment growth (*Zimmermann/Tomlinson*)/ corroborative growth (*Troll*)**
  Erstarkungswachstum
- **extension growth/elongational growth** Streckungswachstum
- **head growth** Kopfwachstum, kopfseitiges Wachstum
- **indeterminate grwth/ unrestricted growth**
  unbegrenztes/unbeschränktes Wachstum
- **intrusive growth** intrusives Wachstum
- **localized growth (plants)**
  lokalisiertes Wachstum, örtlich begrenztes Wachstum
- **longitudinal growth**
  Längenwachstum
- **polar growth** polares Wachstum
- **primary growth** Primärwachstum
- **secondary growth**
  Sekundärwachstum
- **symplastic growth** symplastes Wachstum
- **tail growth** Schwanzwachstum, endständiges Wachstum
- **thickening growth**
  Dickenwachstum
growth cone/axon hillock *neuro*
Wachstumskegel, Axonhügel, Axonkegel
growth curve Wachstumskurve
growth flush (leaves of trop. plants) Blattausschüttung
growth form/habit/external appearance Wuchsform, Habitus, äußere Gestalt, äußeres Erscheinungsbild
growth hormone (GH)/ somatotropin Wachstumshormon, Somatotropin, somatotropes Hormon
growth inhibitor
Wachstumshemmer, Wuchshemmer, Wuchshemmstoff
growth layer Zuwachszone
growth period Wachstumsperiode
growth phase Wachstumsphase
- **acceleration phase**
  Beschleunigungsphase, Anfahrphase
- **death phase/decline phase/ phase of decline** Absterbephase
- **deceleration phase/retardation phase** Verlangsamungsphase, Bremsphase, Verzögerungsphase
- **decline phase/phase of decline/ death phase** Absterbephase

- **division phase** Teilungsphase
- **dormancy period** Ruhephase, Ruheperiode
- **establishment phase**
  Eingewöhnungsphase
- **exponential growth phase**
  exponentielle Wachstumsphase, exponentielle Entwicklungsphase
- **lag phase/incubation phase/ latent phase/establishment phase** lag-Phase, Adaptationsphase, Anlaufphase, Latenzphase, Inkubationsphase
- **logarithmic phase (log-phase)**
  logarithmische Phase
- **stationary phase/stabilization phase** stationäre Phase
growth rate Zuwachsrate, Wachstumsrate (Wachstumsgeschwindigkeit)
- **specific growth rate** spezifische Wachstumsrate
growth ring/annual ring (wood)
Jahresring
growth substance Wuchsstoff
growth-retarding/growth-inhibiting wachstumshemmend
growth-stimulating
wachstumsfördernd
grub/scarabeiform larva/thick wormlike larva (Coleoptera/ Hymenoptera/certain Diptera)
Engerling (im Boden lebende Käferlarve), Bienenlarve u. a.
grub *vb* **(for food)** graben, wühlen
grunt grunzen
GT-AC rule GT-AC-Regel
guaiazulene Guajazulen
guanidine Guanidin
guanine Guanin
guanosine Guanosin
guanosine triphosphate (GTP)
Guanosintriphosphat (GTP)
guanylic acid (guanylate)
Guanylsäure (Guanylat)
guar gum/guar flour Guargummi, Guarmehl
guar meal/guar seed meal
Guar-Samen-Mehl
guard *n* Wächter, Bewachung
guard *vb* bewachen
guard/rostrum (thunderbolt)
Rostrum (Donnerkeil)
guard bee Wächterbiene, Wehrbiene
guard cell *bot* Schließzelle
guard hair *zool* Deckhaar (Grannenhaare und Leithaare)
- **long & smooth guard hair**
  Leithaar
- **short guard hair** Grannenhaar
Guarnieri body (an inclusion body)
Guarnierischer Einschlusskörper

G gubernaculum testis 150

gubernaculum testis Leitband
guest Gast
guide (guiding information: pamphlet/brochure) Führer (Broschüre/Informationsschrift)
guide/tour guide Führer (Führungsperson)
guide RNA Guide-RNA
guided tour Führung
guidepost cell Wegweiserzelle
guidepost neuron Wegweiserneuron
guild Gilde, Lebensgemeinschaft (Pflanzen)
guinea worm/medina worm (*Dracunculus medinensis*) Drachenwurm, Medinawurm, Guineawurm
guitarfishes/Rhinobatoidei Geigenrochen
guitarfishes & skates/Rajiformes Rochenartige
gular den Kehlbereich betreffend
gular plate/gula (fish/prognathous insects) Gularplatte, Kehlplatte, Schlundplatte
gular pouch Kehlsack (Pelikan, Frosch)
gullet/cytopharynx Zellschlund, Zytopharynx, Cytopharynx
gullet/pharynx/hypostome/oral cone/manubrium Mundrohr, Magenstiel, Manubrium
gulls & shorebirds & auks/Charadriiformes Möwenvögel & Watvögel & Alken
gully erosion *geol* Grabenerosion, rinnenartige Erosion (Schluchterosion)
gulonic acid (gulonate) Gulonsäure (Gulonat)
gum (plant gum) Gummi (Pflanzengummi)
gum (eye) Augenbutter, Augenschmalz
gum(s)/gingiva Zahnfleisch
gum arabic (acacia gum) Gummi arabicum, Acacia-Gummi
gust Windstoß, Bö
gustatory den Geschmack betreffend, Geschmacks...
gustatory bud/taste bud Geschmacksknospe
gustatory gland (mammals) Spüldrüse, von Ebnersche Drüse, Ebner' Drüse
gustatory nerve Geschmacksnerv
gustatory organ Geschmacksorgan
gustatory papilla/lingual papilla Geschmackspapille
● **filiform papilla/threadlike papilla** fadenförmige Papille

● **foliate papilla** Blattpapille, blättrige Papille
● **fungiform papilla** Pilzpapille
● **lentiform papilla** linsenförmige Papille
● **vallate papilla** Wallpapille
gusty böig (stürmisch)
gut/alimentary canal/digestive tract *sensu lato* (stomach & intestines) Verdauungskanal
gut/intestines Darmkanal
● **branchial gut** Kiemendarm
● **foregut** Vorderdarm
● **head gut** Kopfdarm
● **hindgut** Hinterdarm, Enddarm
● **midgut** Mitteldarm, Rumpfdarm
● **primitive gut/archenteron/gastrocoel** Urdarm, Archenteron, Gastrocöl
● **tail gut/postanal gut** Schwanzdarm
gut/eviscerate/disembowel *vb* **(e.g. fish)** ausweiden, ausnehmen
gut-associated lymphatic tissue (GALT) darmassoziiertes lymphatisches Gewebe
guts/bowels/entrails Gedärme, Eingeweide
guttation/droplet secretion/exudation Guttation, Tropfenabscheidung, Exsudation
Guttiferae/St. John's wort family/mamey family/mangosteen family/clusia family/Clusiaceae/Hypericaceae Hartheugewächse, Johanniskrautgewächse
guyline (e.g. spiral guyline) *arach* Spannfaden (beim Netzbau der Spinnen)
guyot/flat-topped seamount (tablemount) *mar* Tiefseekuppe, Tiefseetafelberg
gymnocarps (coral fungi & pore fungi and allies) Aphyllophorales
gymnolaemates/"naked throat" bryozoans/Stelmatopoda/Gymnolaemata Kreiswirbler
gymnophionas/caecilians/wormlike amphibians (legless)/Gymnophiona/Caecilia/Apoda Blindwühlen
gymnosperm *adj/adv* nacktsamig
gymnosperms/naked-seed plants Gymnospermen, Nacktsamer
gynandromorph/gyander/sex mosaic Gynandromorph, Gynander
gynandromorphism Gynandromorphismus
gynecoid/gamergate (ants) eierlegende Arbeiterin
gynoandrophore/androgynophore Gynoandrophor, Androgynophor

gyttja G

gynogenesis Gynogenese
gynophore Gynophor
gynostegium (Asclepiadaceae)
Gynostegium
gynostemium/column (orchids)
Gynostemium, Griffelsäule, Säule,
Säulchen (Orchideen)

**gyrencephalous/gyrencephalic
(convoluted surface)** gyrencephal,
gefurcht (Gehirn)
gyroconic gyrocon (Cephalopoden:
Gehäuse)
gyttja/necron mud Gyttia, Gyttja,
Grauschlamm, Halbfaulschlamm

H zone

H zone (muscle) H-Zone
habenular body Habenula, Zirbeldrüsenstiel, Epiphysenstiel
habit (regular performance) Gewohnheit
habit/external appearance/aspect/ growth form Habitus, Wuchsform, äußeres Erscheinungsbild, Wuchsform
habit/growth form/appearance Habitus, Wuchsform, Erscheinung, Wuchs
habitat/place of living/place of growth (*sensu stricto*) Habitat, Standort, Lebensraum
habitat assessment (*sensu lato*: site assessment) Standortbewertung
habitat ecology Standortlehre
habitat imprinting Biotopprägung
habitat requirements Standortansprüche
habituate/get used to/adapt gewöhnen, anpassen
habituation/habit-formation/ adaptation Habituation, Gewöhnung, Anpassung (>Gewöhnungslernen)
hackle (dog: erectile hairs along neck and back) aufstellbare Rücken- bzw. Nackenhaare
hackle (domestic fowl: neck plumage>long narrow feathers) lange Nackenfedern/Halsfedern
hackled band/zig-zag silk *arach* zickzackförmiges Stabiliment, Stabilimentum
hadal zone (slopes) hadische Zone, Hadal (Böden/Hänge der Tiefseegrabenzone)
hadopelagic zone Tiefseegrabenbereich (Wasser)
Haeckel's law/biogenetic law/ biogenetic principle biogenetische Regel, biogenetisches Grundgesetz
Haemodoraceae/haemodorum family/bloodwort family/redroot family/kangaroo paw family Haemodorumgewächse
Hageman factor (blood clotting factor XII) Hageman-Faktor (Blutgerinnungsfaktor XII)
hagfishes/Myxiniformes (Myxinida) Schleimaale, Ingerartige, Inger
hail Hagel
hair Haar
- **acoustical hair/trichobothrium/ vibratory sensory hair** Hörhaar, Becherhaar, Trichobothrium
- **bladder hair** Blasenhaar
- **brush hair** Fegehaar
- **fetlock hair** Kötenbehang, Kötenhaare, Fesselhaare
- **glandular hair** Drüsenhaar
- **guard hair** Deckhaar (Grannenhaare & Leithaare)
- **>long & smooth guard hair** Leithaar
- **>short guard hair** Grannenhaar
- **olfactory hair** Riechhärchen (Sinneshaar)
- **pelage/furcoat (mammals: hairy covering/thick coat of hair)** dichte Haarbedeckung, Körperbedeckung der Säugetiere
- **pile (coat of short/fine furry hairs)** *zool* Flaum, Wolle, Pelz, Haar (Fell)
- **root hair** Wurzelhaar, Wurzelhärchen
- **sensitive hair/trigger hair** Fühlhaar, Reizhaar
- **sensory hair** Sinneshaar
- **stellate hair** Sternhaar
- **stinging hair/urticating hair/ urticating trichome** Brennhaar
- **tactile hair/vibrissa (***p****l vibrissae)** Vibrissa, Spürhaar, Sinushaar, Tasthaar (ein Sinneshaar)
- **trigger hair/sensitive hair** Reizhaar, Fühlhaar
- **underhair** Unterhaar, Wollhaar
- **urticating hair/urticating trichome/stinging hair** Brennhaar
- **wooly hair** Wollhaar

hair/hairiness/pilosity Behaarung
hair/trichome Haar
hair bulb Haarbulbus, Haarzwiebel
hair cell Haarzelle
hair cortex Haarrinde
hair erector muscle/arrector pili muscle/musculus arrector pili Haarmuskel
hair follicle Haarfollikel, Haarbalg
hair medulla Haarmark
hair loss Haarausfall
hair pencil/tibial tuft (butterflies) Haarpinsel, Duftpinsel
hair root Haarwurzel
hair root sheath Haarwurzelscheide
hair sensilla Haarsensille
hair shaft Haarschaft
hair sheath Haarwurzelscheide
hair-covering/indumentum/coat of hair/furcoat Haarkleid
hair-tuft Haarschopf, Haarbüschel, Haarkranz
hairiness/hair Behaarung
hairless/glabrous/bald haarlos, unbehaart, kahl
hairpin loop/reverse turn/beta turn/β bend Haarnadelschleife, Winkelschleife, Umkehrschleife, Haarnadelstruktur, beta-Schleife, β-Schleife

Hassall's corpuscle | 153

hairworms/horsehair worms/ threadworms/nematomorphans/ nematomorphs/Nematomorpha Saitenwürmer

hairy *zool* haarig (*siehe*: behaart)

hairy/pilose *bot* behaart

half-bog/early bog Anmoor

half-bog soil anmooriger Boden

half-chromatid conversion Halbchromatidenkonversion

half-life *rad* Halbwertszeit; (enzyme) Halblebenszeit

half-shrub/suffrutecsent plant Halbstrauch

half-sibs Halbgeschwister

Haller's organ (ticks) Hallersches Organ

halophyte Salzpflanze

Haloragaceae/water milfoil family/ milfoil family Seebeerengewächse, Tausendblattgewächse

haltere/balancer Haltere, Schwingkölbchen

Hamamelidaceae/witch-hazel family Zaubernussgewächse

hamate/hamulose/hooked hakig, hakenartig, Haken…

hamate bone/unciform bone Hakenbein

hamburger (lean & fat ground beef) (mager und fettes) Rinderhack(fleisch) (ohne Herz/ Leber/Niere…)

hamiform/hook-shaped hakenförmig

hammer/malleus (ear) Hammer, Malleus (Ohr)

hamstring Kniesehne (Mensch); Sehne der ischiokruralen Muskeln; Achillessehne (Vierfüßer)

hamstring muscles (quadrupeds: caudal muscles) ischiokrurale Muskeln, Oberschenkelbeuger

hamulate/with small hook-like processes mit kleinen Häkchen versehen

hamulus/hooklet Haken, Häkchen, hakenartiger Fortsatz

hand-shaped/palmate handförmig

hapaxanthic/hapaxanthous/ hapanthous/monocarp/ monocarpic hapaxanth, monokarp(isch)

haplochlamydeous/ monochlamydeous haplochlamydeisch, monochlamydeisch, einfachblumen-blättrig, mit einfacher Blütenhülle

haploid haploid

haploidization Haploidisierung

haploinsufficiency Haploinsuffizienz

haplostemonous haplostemon

haplotype Haplotype

haplotyping Bestimmung des Haplotyps, Haplotypanalyse

haptocyst (an extrusome) Haptocyste

haptor (trematodes: attachment organ) Haptor (Haftorgan)

hard bast Hartbast

hard coal/anthracite Glanzkohle, Anthrazit

hard-leaf/hard-leaved plant/ sclerophyll/sclerophyllous plant Hartlaub, Hartlaubgewächs, Sklerophyll

harden härten

hardening *n* Abhärtung

● **hardening off** *n* Abhärten

Harderian gland/Harder's gland Hardersche Drüse

hardiness/persistence/perseverance Ausdauer

hardness/toughness Härte

hardpan (soil) *geol* verhärtete Bodenschicht

hardwood (tree) Laubbaum (*speziell:* Angiospermen)

hardwood (wood of hardwood trees) Hartholz

hardy/persistent/enduring abgehärtet, ausdauernd (widerstandsfähig); winterfest, winterhart

Hardy-Weinberg equilibrium Hardy-Weinberg-Gleichgewicht

Hardy-Weinberg law Hardy-Weinberg-Gesetz

hare (großer) Hase

hare lip/cleft lip Hasenscharte, Lippenspalte

harem Harem

harlequin chromosomes Harlekin-Chromosomen

harmful/causing damage schädlich

harmful organism/harmful lifeform Schadorganismus

harpagone/harpe (insects: male claspers) Harpagon, Harpe (Valven bei Insekten)

harrow *n* Egge

harrow *vb* eggen

Hartig net *fung* Hartig'sches Netz

harvest *n* Ernte; *vb* ernten

harvest mite/chigger/red bug (parasitic red mite larva) Erntemilbenlarve, parasitische Larve von Trombidiidae

harvestmen/"daddy longlegs"/ Opiliones/Phalangida Weberknechte

Hassall's corpuscle/thymic corpuscle Hassall-Körperchen

hastate

hastate/hastiform/spear-shaped
(e.g. leaf) spießförmig
hatch/brood (eggs/young)
ausbrüten
hatch/emerge (from egg/chrysalis/
pupa) schlüpfen, ausschlüpfen
hatchery *zool* Züchterei,
Zuchtanlage
"hatchet-footed animals"/bivalves/
pelecypods/Pelecypoda/Bivalvia/
Lamellibranchiata (clams:
sedimentary/mussels: freely
exposed) Muscheln
hatchetfishes and relatives (deep-
sea)/Stomiiformes Stomiiformes
(Tiefseefische)
hatchling frisch ausgebrütetes
Junges
Hatschek's pit/Hatschek's groove
Hatscheksche Grube
haunch/hindquarter Hinterbacke,
Gesäß; Keule, Lendenstück
haunch/hip Hüfte, Lende
haunch bone/ilium/iliac/os ilium
Darmbein
haustellate mit saugenden
Mundwerkzeugen
haustorium/foot Haustorium, Fuß
haustorium/holdfast Haustorium,
Senker
haustorium/sucker Haustorium,
Saugorgan
Haversian canal/haversian canal
(central canal) Haversscher Kanal
Haversian system/haversian
system/osteon Haverssches
System, Osteon
hawker (Anisoptera) große Libelle
Haworth projection/Haworth
formula *chem* Haworth-Projektion,
Haworth-Formel
hay Heu
hay/mowing/cut grass Mahd
hay infusion Heuaufguß
hay meadow/mowed meadow
Mähwiese
haylage (hay silage) Halbheu,
Gärheu, Abwelkheu
hazard/source of danger
Gefahrenquelle
hazard class Gefahrenstufe,
Gefahrenklasse, Risikostufe
hazard code Gefahrencode,
Gefahrenkennziffer
hazard icon Gefahrensymbol
hazard potential Gefahrenpotential
hazardous gefährlich; riskant
hazardous materials regulations
Gefahrgutbestimmungen
hazardous waste Sondermüll
hazel family/Corylaceae
Haselnussgewächse

head (cephalon/caput) Kopf
head/front part Vorderteil,
Kopfende
head *chem* (fat molecule) Kopf
head/flower head/capitulum/
cephalium *bot* Korb, Körbchen,
Köpfchen, Capitulum, Cephalium
head growth Kopfwachstum,
kopfseitiges Wachstum
head gut/foregut Kopfdarm
head process/notochordal process
Kopffortsatz, Chordafortsatz
head wind *meteo/aer* Gegenwind
head-body length Kopf-Rumpf-
Länge
head-foot (mollusks) Kopffuß,
Cephalopodium
head-space gas chromatography
Dampfraum-Gaschromatographie
head-to-head ramming *ethol*
Kopf-an-Kopf-Stoßen
head-to-head repeats *gen*
Kopf-an-Kopf-Wiederholungen
heading *aer/orn* Kurs, Flugrichtung
headwater(s) Quellbereich,
Quellgebiet, Oberlauf (der Flüsse)
heal heilen
healing Heilung
• self-healing/autotherapy
Selbstheilung
health Gesundheit
healthy gesund
hear hören (vernehmen)
hearing/sense of hearing Gehör,
Gehörsinn
hearing threshold/auditory
threshold Hörschwelle, Hörgrenze
heart Herz
• auxillary heart Nebenherz,
auxilläres Herz
• frontal heart/frontal sac
Stirnherz
• gill heart/branchial heart
(cephalopods) Kiemenherz
• lateral heart Lateralherz
• lymph heart Lymphherz
• tubular heart Röhrenherz,
Herzschlauch
heart beat (simple contraction)
Herzschlag (einfache Kontraktion)
heart rate Herzfrequenz
heart rot (wood) Kernfäule
heart urchins/Spatangoida
Herzigel, Herzseeigel
heart valve/cardiac valve/coronary
valve (*see:* "coronary valve" for
different types) Herzklappe
heart-shaped/cordate/cordiform
herzförmig
heart-weight rule/Hesse's rule *evol*
Herzgewichtsregel, Reihenregel,
Hessesche Regel

155   hematopoiesis **H**

heartburn/acid indigestion
Sodbrennen
heartwood/duramen Kernholz
heartworm (*Dirofilaria spp.*)
Herzwurm
heat *n phys* Hitze
heat *vb phys* heizen, erhitzen
heat (female) *zool* Brunst
(weibliche)
  • **in heat (female)/sexually
aroused** brünstig, in der Brunst,
geschlechtlich erregt; rossig
(Stute); läufig (Hündin)
heat cramps Hitzekrämpfe
heat exchanger Wärmetauscher
heat exhaustion Hitzeerschöpfung
heat of vaporization
Verdunstungswärme
heat shock Hitzeschock
heat shock gene Hitzeschockgen
heat shock protein
Hitzeschockprotein
**heat shock reaction/heat shock
response** Hitzeschockreaktion
heat transfer Wärmeübergang
heat transport Wärmetransport
heat treatment/baking
Hitzebehandlung, Backen
heat-resistant/heat-stable
hitzebeständig
heat-tolerant hitzeverträglich
heath/heathland Heide,
Heideland
heath family/Ericaceae
Heidekrautgewächse
heath forest Heidewald
heath sedge Heidegras
heather (*Calluna vulg.*)/heath
Heidekraut
heating coil Heizschlange
heatstroke Hitzschlag
heavy chain (H chain) *immun*
schwere Kette (H-Kette)
heavy metal contamination
Schwermetallverunreinigung,
Schwermetallbelastung
**hectocotylus/hectocotylized arm/
heterocotylus** Geschlechtstentakel,
Geschlechtsarm, Hectocotylus
hedge Hecke
hedge clippers Heckenschere
hedge plant Heckenpflanze
hedonic gland (amphibians)
hedonische Drüse
heel (mammals) Ferse
**heel/calcaneum/calcaneus/
hypotarsus (birds)** Fersenbein,
Hypotarsus
heel cutting *bot* Steckling mit
Astring (Stammstecking)
**heelbone/calcaneum/calcaneus/
os calcis** Fersenbein

heifer Färse
(junge Kuh: noch nicht gekalbt)
**height equivalent to theoretical
plate (HETP)** Trennstufenhöhe
helical/spiraled helical, helikal,
treppenhausförmig gewunden,
schraubig, spiralig
  • **irregularly helical (like a snail
shell)/cochleate** schraubenartig
gewunden, schneckenhausartig
gewunden, cochlear
helical ribbon impeller (bioreactor)
Wendelrührer
helicone (gastropod shell)
Kegelspirale
heliconome/serpentine mine
Spiralmine
heliotropism/solar tracking
Heliotropismus, Lichtwendigkeit,
Sonnenwendigkeit
helix (*pl* helices or helixes)/spiral
Helix, Spirale (*pl* Helices)
helix-loop-helix Helix-Loop-Helix
(Strukturmotiv)
helix-turn-helix Helix-Turn-Helix
(Strukturmotiv)
**hellgrammite/dobson (dobsonfly
larva: esp. *Corydalis cornutus*)**
Schlammfliegenlarve (Megaloptera)
helophyte/bog plant/marsh plant
Sumpfpflanze, Moorpflanze
helotism Helotismus
Helotium family/Helotiaceae
Becherchen
helper cell Helferzelle
helper T cell/T-helper cell (T$_H$)
T-Helferzelle, Helfer T-Zelle
helper virus Helfervirus
**hemadsorption inhibition test (HAI
test)** Hämadsorptionshemmtest
(HADH)
**hemagglutination inhibition test
(HI test)**
Hämagglutinationshemmtest (HHT)
hemal arch Hämalbogen,
Haemalbogen
hemal canal/hemal duct
Hämalkanal, Haemalkanal
hemal system Hämalsystem,
Haemalsystem
hemapophysis Hämapophyse,
unterer Dornfortsatz
hematocrit Hämatokrit
hematocyte/hemocyte Hämatocyt,
Hämatozyt, Hämocyt, Hämozyt,
Haematozyt, Blutzelle
**hematophagous/sanguivorous/
blood-sucking** blutsaugend, sich
von Blut ernährend
hematopoiesis Haematopoese,
Hämatopoese, Blutbildung,
Blutzellbildung

H heme

heme Häm
Hemerocallidaceae/daylily family
Tagliliengewächse
hemiacetal Halbacetal
hemiangiocarpic/hemiangiocarpous
hemiangiokarp
hemibranch (gill) Hemibranchie
hemichordates/Branchiotremata/
Hemichordata Hemichordaten,
Kragentiere
hemicryptophyte Hemikryptophyt
hemicyclic hemizyklisch
hemielytron/hemelytron
Hemielytre, Halbdecke
hemihomocercal hemihomocerk,
hemihomozerk
hemimetabolic development/
hemimetabolous development
hemimetabole Entwicklung
Hemionitidaceae/strawberry fern
family Nacktfarngewächse
hemiparasite/semiparasite
Hemiparasit, Halbparasit,
Halbschmarotzer
hemipenis Hemipenis
hemisphere Hemisphäre; Halbkugel
• **cerebral hemisphere** *anat*
Hirnhemisphäre, Großhirnhälfte
• **Eastern Hemisphere** *geol* **(Earth)**
Osthemisphäre
• **global hemisphere** *geol* **(Earth)**
Erdhemisphäre, Erdhalbkugel,
Erdhälfte
• **Northern Hemisphere** *geol*
(Earth) Nordhemisphäre,
Nordhalbkugel
• **Southern Hemisphere** *geol*
(Earth) Südhemisphäre,
Südhalbkugel
• **Western Hemisphere** *geol*
(Earth) Westhemisphäre
hemizygosity Hemizygotie
hemizygous hemizygot
hemochorial placenta haemo-
choriale Plazenta
hemoendothelial placenta
Labyrinthplazenta
hemoglobin Hämoglobin
hemolymph Hämolymphe
hemorrhage/profuse bleeding
Hämorrhagie, Blutung, Blutsturz
hemorrhagic hämorrhagisch,
Blutung betreffend, durch Blutung
gekennzeichnet
hemp family/Cannabaceae
Hanfgewächse
hen Henne
• **young hen/pullet** Hühnchen
Henderson-Hasselbalch equation
Henderson-Hasselbalch Gleichung,
Henderson-Hasselbalchsche
Gleichung

HEPA filter (high efficiency
particulate air filter)
HOSCH-Filter
(Hochleistungsschwebstofffilter)
heparin Heparin
hepatic die Leber betreffend,
Leber...; *bot* Lebermoose
betreffend
hepatic duct/ductus hepaticus
Lebergang; Gallengang; *med*
(Mensch) gemeinsamer Gallengang
an der Leberpforte (nach
Vereinigung des r. u. l. Gallengangs)
hepatic lobe Leberlappen
hepatic lobules Leberläppchen
hepatic portal Leberpforte
hepatic sacculation Lebersack,
Lebersäckchen, Leberblindsack
(*siehe*: Mitteldarmdrüse)
hepatopancreas (decapods)
Hepatopankreas, Mitteldarmdrüse
hepatotoxic leberschädigend,
hepatotoxisch
heptamer Heptamer
heptamerous siebenteilig
herb/herbaceous plant (annual and
biennial)/wort/weed Kraut,
Krautpflanze, krautige Pflanze
herb garden Kräutergarten
herbaceous krautig
herbaceous plant/herb
Krautpflanze, krautige Pflanze
herbaceous plant layer
Krautschicht
herbal *n* Kräuterbuch
herbal drug Pflanzendroge
herbarium Herbar
herbicide/weed killer Herbizid,
Unkrautvernichtungsmittel,
Unkrautbekämpfungsmittel
herbivore Herbivore, Pflanzenfresser,
pflanzenfressendes Tier
herbivorous pflanzenfressend
herbs/vegetables for soup making
Suppengrün
• **culinary herbs** Küchenkräuter
Herbst corpuscle *orn* Herbstsches
Körperchen
herd/flock *n* Herde
herd instinct/herding instinct
Herdentrieb, Herdeninstinkt
herding zu einer Herde sammeln
herding (guarding behavior)
Herden (Hüteverhalten)
hereditary/heritable heredität,
ererbt, vererbt, erblich, Erb...
hereditary disease/genetic disease/
genetic defect/inherited disease/
heritable disorder Erbkrankheit,
erbliche Erkrankung
hereditary information/genetic
information Erbinformation

157   heterologous probing   **H**

hereditary material/genome
Erbgut, Erbträger, Erbmaterial,
Erbsubstanz, Genom
hereditary relationship
Verwandtschaft, verwandtschaftliche
Beziehung
**hereditary trait/hereditary
characteristic** Erbmerkmal
**heredity/inheritance/transmission
(of hereditary traits)** Vererbung
heritability Heritabilität,
Erblichkeitsgrad
● **broad h. (H^2)** allgemeine
Erblichkeit (H^2)
● **h. in the narrow sense (h^2)**
Erblichkeit im engeren Sinne (h^2)
hermaphrodism Hermaphroditismus,
Zwittertum, Zwittrigkeit
hermaphrodite Hermaphrodit,
Zwitter
● **sequential h.**
Sukzedanhermaphrodit
● **simultaneous/synchronous
hermaphrodite**
Simultanhermaphrodit
hermaphroditic (bisexual)
hermaphroditisch, zwittrig
(zweigeschlechtig)
hermaphroditic duct Zwittergang
**hermaphroditic gland/
hermaphroditic gonad/ovotestis**
Zwitterdrüse
hermaphroditism/hermaphrodism
Zwittertum, Zwittrigkeit,
Hermaphroditismus
hermatypic/reef-building
hermatypisch, riffbildend
Hernandiaceae/Hernadia family
Eierfruchtbaumgewächse
**herons & storks & ibises and allies/
Ciconiiformes** Stelzvögel,
Schreitvögel
herpetology Herpetologie
(Amphibien- und Kriechtierkunde/
Reptilienkunde)
herrings and relatives/Clupeiformes
Heringsfische, Heringsverwandte
**HERV-family (HERV=human
endogeneous retrovirus) (DNA
element)** HERV-Familie
hesperidium Hesperidium,
Zitrusfrucht, Citrusfrucht (eine
Panzerbeere)
Hesse's rule/heart-weight rule *evol*
Hessesche Regel,
Herzgewichtsregel, Reihenregel
heterobasidium (*pl* heterobasidia)
Heterobasidie, Phragmobasidie
heterocarpous heterokarp,
verschiedenfrüchtig
heterocelous/heterocoelous
heterocöl, heterocoel

heterocercal heterocerk, heterozerk
heterochlamydeous
heterochlamydeisch
heterochromatin Heterochromatin
● **constitutive h.** konstitutives
Heterochromatin
● **facultative h.** fakultatives
Heterochromatin
heterochronous heterochron
heterochrony/heterochronism
Heterochronie
heterocyclic heterozyklisch
heterodont/anisodont heterodont,
ungleichzähnig
heteroduplex *gen* Heteroduplex
heteroduplex mapping
Heteroduplex-Kartierung
**heteroecious/heterecious/
heteroxenous** heterözisch
heteroecy/heteroecism Heteröcie,
Heterözie
heterogametic sex
heterogametisches Geschlecht
heterogamous heterogam
heterogamy Heterogamie,
Heterogonie, zyklische
Parthenogenese
heterogeneity/heterogenous state
Heterogenität,
Ungleichartigkeit,
Verschiedenartigkeit,
Andersartigkeit
**heterogeneous/consisting of
dissimilar parts/mixed** heterogen,
ungleichartig, verschiedenartig,
andersartig (*antonym:* homogen)
**heterogeneous nuclear RNA
(hnRNA)** heterogene Kern-RNA
heterogenesis Heterogenese
heterogenetic heterogenetisch,
genetisch unterschiedlichen
Ursprungs
heterogenote Heterogenote *f*
heterogenous (of different origin)
heterogen, unterschiedlicher
Herkunft
heterogeny Heterogenie,
unterschiedlicher Herkunft
heterogony/heterogamy
Heterogonie, zyklische
Parthenogenese
heterograft *bot* Fremdpfropfen,
Fremdpfropfung
heterolactic fermentation
Milchsäuregärung/
heterofermentative, unreine
heterologous gene/foreign gene
Fremdgen
heterologous probe *gen* heterologe
Sonde
heterologous probing *gen* mit
Hilfe einer heterologen Sonde

heteromorphous

heteromorphous heteromorph, anders gestaltet, verschiedengestaltig
heterophylly/anisophylly Heterophyllie, Anisophyllie, Verschiedenblättrigkeit, Ungleichblättrigkeit
heteropolar bond heteropolare Bindung
heteropolymer Heteropolymer
heteroscedasticity *stat* Varianzheterogenität, Heteroskedastizität
heterosis/hybrid vigor Heterosis, Bastardwüchsigkeit
heterostyly Heterostylie, Verschiedengriffeligkeit
heterothermic heterothermisch
heterothermy Heterothermie
heterotroph/heterotrophic ("other-feeding") heterotroph
heterotrophy Heterotrophie
heterotypic heterotypisch
heterozygosity Heterozygotie, Mischerbigkeit
• **loss of heterozygosity (LoH)** Verlust der Heterozygotie, Heterozygotieverlust
heterozygote Heterozygote
heterozygote advantage Heterozygotenvorteil
heterozygous heterozygot, mischerbig
• **compound heterozygous** zusammengesetzt-heterozygot, compound-heterozygot
• **double heterozygous** doppelt-heterozygot
heuristic heuristisch
hexacanth larva/hooked larva/ oncosphere (cestodes) Sechshakenlarve, Oncosphaera-Larve
hexacorallians/hexacorals/ Hexacorallia Hexacorallia
hexamer/hexon *vir* Hexamer, Hexon
hexose monophosphate shunt (HMS)/pentose phosphate pathway/pentose shunt/ phosphogluconate oxidative pathway Hexosemonophosphatweg, Pentosephosphatweg, Phosphogluconatweg
Hfr cell (*from:* high frequency of recombination) Hfr-Zelle (hohe Rekombinationshäufigkeit)
hibernaculum/winter bud Hibernaculum, Hibernakel, Dauerknospe, Winterknospe
hibernate/overwinter überwintern
hibernation/overwintering Überwinterung, Winterschlaf

hide/conceal *vb* verstecken
hide *n zool* (esp. large/heavy skins: cowhide) Fell, Haut
hideout/hideaway/hiding place/ retreat/refuge Versteck, Unterschlupf
hiding/concealment Verstecken
hierarchical hierarchisch
hierarchy Hierarchie, Rangfolge, Rangordnung
high density lipoprotein (HDL) Lipoprotein hoher Dichte
high fructose corn syrup Isomeratzucker, Isomerose
high mobility group (HMG-box) Gruppe von hoher Beweglichkeit (HMG-Box)
high tide/flood Tide, Flut
high voltage electron microscopy (HVEM) Höchstspannungselektronenmikroskopie, Hochspannungselektronenmikroskopie
high-energy bond energiereiche Bindung
high-energy compound energiereiche Verbindung
high-level expression *gen* Überexpression
high-molecular hochmolekular
high-pressure liquid chromatography/high performance liquid chromatography (HPLC) Hochdruckflüssigkeitschromatographie, Hochleistungschromatographie
higher plants höhere Pflanzen
highland Hochland
highmoor peat/sphagnum peat/ moss peat Hochmoortorf
hill Hügel
hill country/rolling countryside Hügelland
Hill coefficient/Hill constant Hill-Koeffizient, Kooperativitätskoeffizient
Hill equation Hill-Gleichung
Hill plot Hill-Auftragung
Hill reaction Hill-Reaktion
hillside/hill slope/mountainside/ mountain slope Hang, Abhang (Hügel/Berg)
hillside location/slope location Hanglage
hilly hügelig
hilly terrain hügeliges Gelände, Hügellandschaft
hilum/funiculus scar Hilum, Nabel, Samennabel
hindbrain/rhombencephalon Rautenhirn, Rhombencephalon

159   **hollow bone** H

hindgut *embr* Hinterdarm
hindgut/proctodeum Hinterdarm,
Enddarm (Colon & Rectum),
Proctodaeum
hindlegs/posterior legs Hinterbeine
hindlimb Hinterextremität
hindpaw Hinterpfote
hindquarter(s)/haunch Hinterteil,
Hinterleib, Hinterviertel (Pferd:
Hinterhand)
hindwing Hinterflügel (Unterflügel)
hinge Scharnier, Schlossleiste
(Muscheln)
**hinge (immunglobulin molecule/
collagen type VII molecule)**
Gelenk (Immunglobulinmolekül/
Kollagen Typ VII-Molekül)
hinge joint/ginglymus joint
Scharniergelenk
hinge ligament Schlossligament,
Schlossband
hinge socket (bivalves) Zahngrube
(an Schloss der Muschelschale)
hinge teeth (bivalves)
Schlosszähne
**hinny (♀ ass/donkey x ♂ horse
stallion)** Maulesel
(Eselstute x Pferdehengst)
hip Hüfte
  ● **point of hip/coxal tuber/tuber
coxae** Hüfthöcker
hip/rose hip *bot* Hagebutte
hip joint/coxal joint/coxa *vert/
entom* Hüftgelenk, Coxa
**hipbone/innominate bone/os
coxae (lateral half of pelvis)**
Hüftbein, Hüftknochen
(Beckenhälfte: Darmbein, Sitzbein,
Schambein)
**Hippocastanaceae/buckeye family/
horse chestnut family**
Rosskastaniengewächse
**Hippuridaceae/marestail family/
mare's-tail family**
Tannenwedelgewächse
hirsute (with coarse/stiff hairs)
rauhaarig, borstig
hirtellous (finely hirsute)
feinhaarig, feinbehaart
**hispid (with stiff hairs/spines/
bristles)** kurzborstig, steifhaarig
hiss (e.g. snake) fauchen, zischen
histamine Histamin
histidine Histidin
histiocyte (*actually:* macrophage)
Histiozyt, Gewebswanderzelle,
Gewebs-Makrophage (*eigentlich:*
Makrophage)
histiogram *see* histogram
histocompatibility
Histokompatibilität,
Gewebeverträglichkeit

histocompatibility antigen
Histokompatibilitätsantigen
  ● **major histocompatibility
antigens**
Haupthistokompatibilitätsantigene
  ● **minor histocompatibility
antigens**
Nebenhistokompatibilitätsantigene
histocompatibility complex
Histokompatibilitätskomplex
  ● **major histocompatibility
complex (MHC)**
Haupthistokompatibilitätskomplex
histogram *stat* Histogramm
  ● **frequency histogram**
Häufigkeitshistogramm
histoincompatibility
Histoinkompatibilität,
Gewebeunverträglichkeit
histology Histologie, Gewebelehre
histone Histon
**HLA complex (human leucocyte
antigen complex)** menschlicher
Leukozytenantigen-Komplex (HLA-
Komplex)
HMG-box (*high mobility group*)
gen HMB-Box (Gruppe von hoher
Beweglichkeit)
hoard/hoard up *vb* hamstern, raffen,
horden
hoarding *n* Hamstern
hoarding of food Vorratshaltung
**hoarfrost/white frost (fine/
feathery)** feinflockiger Reif,
Raureif
hock (quadrupeds: horse/cattle)
Sprunggelenk (Knöchel)
hock (slaughter animals) Hachse
(Sprunggelenk der Schlachttiere)
hock *orn* Mittelfußgelenk
hoe culture/cultivation/agriculture
Hackkultur, Hackbau
hog/swine/pig Schwein
hog cholera/swine fever (viral)
Schweinefieber
Hogness box *gen* Hogness-Box
holandric holandrisch
holandry Holandrie
hold-up time *biot* Rückhaltezeit,
Verweildauer, Aufenthaltsdauer
holdfast/appressorium Haftscheibe,
Haftorgan, Senker; Rhizoid (kelp/
mosses); Appressorium
holdfast root Haftwurzel
hole/burrow Erdhöhle, Erdloch,
Bau
holistic holistisch
Holliday structure *gen* Holliday-
Struktur
hollow *adj/adv* hohl
hollow bone/pneumatic bone
Hohlknochen

hollow impeller shaft

hollow impeller shaft *biot*
(bioreactors) Hohlwelle (Rührer in
Bioreaktoren)
hollow stirrer *biot* Hohlrührer
holly family/Aquifoliaceae
Stechpalmengewächse
**holobasidium (*pl* holobasidia)/
homobasidium** Holobasidie
(*pl* Holobasidien), Homobasidie
**holoblastic cleavage/complete
cleavage** holoblastische/
vollständige/totale Furchung
holobranch (gill) Holobranchie
(Kieme)
holocarpic holokarp
**Holocene/Recent/Holocene Epoch/
Recent Epoch** Holozän, Jetztzeit,
Alluvium (erdgeschichtliche
Epoche)
holocrine holokrin
holoenzyme Holoenzym
hologynic hologyn
hologyny Hologynie
holometabolic/holometabolous
holometabol
holometabolic development
holometabole Entwicklung
holometabolism Holometabolie
holometabolous/holometabolic
holometabol
holomictic holomiktisch
holonephros Holonephros
holoparasite/obligate parasite
Holoparasit, Vollschmarotzer,
Vollparasit
holotype Holotypus, Holotyp,
Holostandard
home range *ecol* Aktionsraum,
Streifgebiet
homeobox Homöobox
homeobox gene/*Hox* gene
Homöobox-Gen, *Hox* Gen
homeodomain *gen* Homöodomäne
homeostasis Homöostase,
Homöostasie
homeotic gene homöotisches Gen
homeotic mutation homöotische
Mutation
homing/philopatry *ethol* Ortstreue
homing instinct
Heimkehrvermögen,
Heimfindevermögen, Zielflug
**homing receptor (lymphocyte
surface protein)** *homing*-
Rezeptor
**hominy (soaked/washed/hulled
corn kernels)** eingeweichte/
gewaschene/geschälte Maiskörner
**hominy corn/corn grits (U.S.:
coarse cornmeal)** Maisgrütze,
Maisgrieß, grobes Maismehl
homocelous homocöl, homocoel

homocercal homocerk, homozerk
homodont/isodont homodont,
gleichartig bezahnt
homoduplex *gen* Homoduplex
homeosmotic/homoiosmotic
homoiosmotisch
**homeotherm/homoiotherm/warm-
blooded animal** Homoiotherme,
Warmblütler
**homeothermic/homoiothermic/
endothermic/warm-blooded**
homoiotherm, gleichwarm,
endotherm, warmblütig
**homeothermy/homoiothermy/
homoiothermism/warm-
bloodedness** Homoiothermie,
Warmblütigkeit
homogametic sex homogametisches
Geschlecht
homogamy Homogamie
**homogeneity (with same kind of
constituents)** Homogenität,
Einheitlichkeit, Gleichartigkeit
**homogeneous (having same kind
of constituents)** homogen,
einheitlich, gleichartig
homogenization Homogenisation,
Homogenisierung
homogenize homogenisieren
homogenizer Homogenisator
homogenote Homogenote *f*
homogenous (of same origin)
homogen, gleicher Herkunft
homogentisic acid
Homogentisinsäure
homogeny/homogeneity
Homogenität
homograft/syngraft/allograft
Homotransplantat, Allotransplantat
**homoiochlamydeous/
homochlamydeous**
homoiochlamydeisch, gleichartige
Hüllblätter
homoiosmotic/homeosmotic
homoiosmotisch
**homoiotherm/homeotherm/warm-
blooded animal** Homoiotherme,
Warmblütler
**homoiothermic/homeothermic/
endothermic/warm-blooded**
homoiotherm, gleichwarm,
endotherm, warmblütig
**homoiothermy/homeothermy/
homoiothermism/warm-
bloodedness** Homoiothermie,
Warmblütigkeit
homolactic fermentation
homofermentative
Milchsäuregärung, reine
Milchsäuregärung
homologization Homologisierung
homologize homologisieren

161   **hoofed game** **H**

homologous homolog, ursprungsgleich
homologous chromosome homologes Chromosom
homologous recombination homologe Rekombination
homologous theory/transformation theory of alternation of generations homologer Generationswechsel
homology Homologie, Ursprungsgleichheit
homonomous homonom
homonym Homonym
homonymous/homonymic homonym
homonymy Homonymie
homopolar bond/nonpolar bond homopolare Bindung
homopolymer Homopolymer
homopterans (cicadas & aphids & scale insects)/Homoptera Pflanzensauger
homoscedasticity *stat* Varianzhomogenität, Varianzgleichheit, Homoskedastizität
homoserine Homoserin
homotype Homotyp
homotypic homotypisch
homozygosity Homozygotie (Reinerbigkeit/Reinrassigkeit)
homozygous (true-bred/pure-bred) homozygot (reinerbig/reinrassig)
honey Honig
 • **comb honey** Scheibenhonig
honey crop/honey stomach/honey sac (bees) Kropf, Honigmagen
honey guide *bot* Honigmal
honeybush family/Melianthaceae Honigstrauchgewächse
honeycomb Wabe, Honigwabe
honeycomb stomach/honeycomb bag/second stomach/reticulum Netzmagen, Haube, Retikulum
honeycombed/favose/faveolate/ alveolate/deeply pitted wabig, wabenförmig; kleingrubig
honeydew Honigtau
honeysuckle family/Caprifoliaceae Geißblattgewächse
hood *zool* (**Nautilus**) Kopfkappe
hood/fume hood *chem/lab* Abzug, Rauchabzug, Dunstabzugshaube
hoof (*pl* **hoofs/hooves**)/**ungula** Huf, Lauf (Huftier)
 • **bar/pars inflexa lateralis** Eckstrebe
 • **bulb/pad** Hufballen
 • **buttress of heel/angle of heel/ angle of wall** Trachte
 • **coffin bone/pedal bone** Hufbein

 • **collateral groove/commissure/ sulcus paracunealis** seitliche Strahlfurche
 • **cuneal cushion/pulvinus cunealis** Strahlkissen, Strahlpolster
 • **digital cushion/plantar cushion/ pulvinus digitalis** Hufkissen (Strahlkissen + Ballenkissen)
 • **forehoof/toe** Vorderhuf, Zeh
 • **frog/cuneus ungulae** Hufstrahl
 • >**cleft of frog/median cleft/ central cleft/central groove/ central sulcus/sulcus cunealis centralis** mittlere Strahlfurche
 • >**crus of frog/crus cunei** Strahlschenkel
 • >**horny frog/cuneus corneus** Hornstrahl
 • >**point of frog/apex cunei** Strahlspitze
 • **heel/buttress** Trachte
 • **heel wall/wall of heel** Trachtenwand
 • **limbus** Saum
 • **navicular bone/distal sesamoid/ os sesamoideum distale** Strahlbein, distale Sesambein
 • **navicular bursa/bursa podotrochlearis (hoof)** Hufrollenschleimbeutel
 • **pastern bone>long pastern/first phalanx** Fesselbein, Fesselknochen,
 • **pastern bone>short pastern/ small pastern bone/second phalanx/middle phalanx** Kronbein
 • **quarter (pars lateralis: side of wall between toe and heel)** Seitenwand, Seitenteil
 • **seat of corn*** Eckstrebenwinkel, Sohlenwinkel (*corn >cornu >horn = hardening/thickening of epidermis)
 • **semilunar zone (podotrochlea)** Hufrolle
 • **toe** Zehe
 • **toe wall** Zehenwand, Zehenteil, Rückenteil
hoof bar Eckstrebe
hoof capsule Hufkapsel
hoof dermis (corium) Huflederhaut
hoof plate Hufplatte
hoof sole (horny sole) Hufsohle (Hornsohle)
hoof wall (paries corneus) Hufwand, Hufwall
hoof-like/ungulate hufartig
hoof-shaped/unguliform hufförmig
hoofed/hooved/ungulate behuft, mit Hufen, Huf…
hoofed game Schalenwild

hoofed mammals

hoofed mammals/ungulates
Huftiere
hoofprint Hufabdruck (Spur)
hook-shaped/unciform/hamiform
hakenförmig
hooked/hook-like/uncinate/hamate
hakig
hooklet/barbicel/hamulus
(*pl* **hamuli**) Häkchen, Hamulus
hookworms (*Ancylostoma/Necator spp.*) Hakenwürmer
hoot (owl) heulen, schreien
hop (e.g. rabbit) hoppeln
hop/jump/skip/leap hüpfen
horizon Horizont
- **boulder layer/loose rock layer (C-horizon)** Untergrund (C-Horizont)
- **fermentation layer (soil)** Fermentationsschicht, Vermoderungshorizont
- **litter layer** *bot* **(forest)** Streuschicht, Streuhorizont, Förna
- **soil horizon** Bodenhorizont
- **subsoil (zone of accumulation/illuviation)** Unterboden, unterer Mineralhorizont (B-Horizont)
- **time horizon** Zeithorizont
- **topsoil (zone of leaching/eluviation)** Oberboden, Krume, Bodenkrume, Bodendeckschicht, oberer Mineralhorizont
- **zone of accumulation/zone of illuviation (B-horizon)** Einwaschungshorizont

horizontal gel/flat bed gel
horizontal angeordnetes Plattengel
horizontal transmission *gen/med*
horizontale Transmission, horizontale Übertragung
hormocyst *fung/lich* Hormocyste
hormocystangium
Hormocystangium
hormonal hormonal, hormonell
hormone Hormon
- **adrenaline/epinephrine** Adrenalin, Epinephrin
- **adrenocorticotropic hormone/corticotropin (ACTH)** adrenocorticotropes Hormon, Kortikotropin, Adrenokortikotropin
- **aldosterone** Aldosteron
- **androsterone** Androsteron
- **antidiuretic hormone (ADH)/vasopressin** antidiuretisches Hormon (ADH), Adiuretin, Vasopressin
- **atrial natriuretic peptide (ANP)/atrial natriuretic factor (ANF)/atriopeptin** atriales natriuretisches Peptid (ANP), Atriopeptin
- **calcitonin** Calcitonin, Kalzitonin
- **cholecystokinin-pancreozymin (CCK-PZ)** Cholecystokinin-Pankreozymin (CCK-PZ)
- **corticoliberin/corticotropin-releasing hormone (CRH)/corticotropin-releasing factor (CRF)** Corticoliberin, Corticotropin-freisetzendes Hormon, corticotropes Releasing-Hormon (CRH)
- **corticosterone** Corticosteron, Kortikosteron
- **corticotropin/adrenocorticotropic hormone (ACTH)** Corticotropin, Kortikotropin, Adrenokortikotropin, adrenocorticotropes Hormon, adrenokortikotropes Hormon (ACTH)
- **cortisol/hydrocortisone** Cortisol, Hydrocortison
- **cortisone** Cortison, Kortison
- **endorphin** Endorphin, Endomorphin
- **epinephrine/adrenaline** Epinephrin, Adrenalin
- **erythropoietin/erythropoiesis-stimulating factor (ESF)** Erythropoetin
- **estrogen** Östrogen
- **estrone** Östron, Estron
- **follicle-stimulating hormone (FSH)** Follitropin, follikelstimulierendes Hormon (FSH)
- **gastric inhibitory peptide (GIP)/glucose-dependent insulin-release peptide** gastrointestinal-inhibitorisches Peptid, gastrisches Inhibitor-Peptid (GIP), glucoseabhängiges Insulin-releasing-Peptid
- **gastricsin/pepsin C** Gastricsin (Pepsin C)
- **gastrin** Gastrin
- **gestagen/progestin** Gestagen, Progestin, Corpus-luteum-Hormon, "Schwangerschaftshormon"
- **glucagon** Glucagon, Glukagon
- **glucocorticoids** Glukokortikoide
- **gonadoliberin/gonadotropin releasing hormone/gonadotropin releasing factor (GnRH, GnRF)** Gonadoliberin, Gonadotropin-Freisetzungshormon
- **gonadotropin** Gonadotropin
- **gonadotropin releasing hormone/gonadotropin releasing factor (GnRH, GnRF)/gonadoliberin** Gonadotropin-Releasing Hormon, Gonadoliberin

163   **horn sharks** **H**

- **growth hormone (GH)/ somatotropin** Wachstumshormon, Somatotropin, somatotropes Hormon
- **human chorionic gonadotropin (hCG)** Choriongonadotropin (hCG)
- **human growth hormone (hGH)/ human somatotropin** menschliches Wachstumshormon (Somatotropin, somatotropes Hormon)
- **human placental lactogen (HPL)/ human chorionic somatomammotropin (HCS)** Plazentalaktogen
- **inhibin** Inhibin
- **insulin** Insulin
- **interstitial-cell stimulating hormone (ICSH)/luteinizing hormone (LH)** zwischenzellstimulierendes Hormon, Lutropin, Luteotropin, luteinisierendes Hormon (LH)
- **juvenile hormone (JH)** Juvenilhormon
- **luteinizing hormone (LH)/ interstitial-cell stimulating hormone (ICSH)** Lutropin, Luteotropin, luteinisierendes Hormon (LH), zwischenzellstimulierendes Hormon
- **melanocyte-stimulating hormone (MSH)** Melanotropin, melanozytenstimulierendes Hormon (MSH)
- **melanoliberin/melanotropin releasing hormone/ melanotropin releasing factor (MRH, MRF)** Melanoliberin, Melanotropin-Freisetzungshormon
- **melatonin** Melatonin
- **molt-inhibiting hormone (MIH)** häutungshemmendes Hormon
- **Mullerian inhibiting hormone (MIH)** Anti-Müller-Hormon (AMH)
- **norepinephrine/noradrenaline** Norepinephrin, Noradrenalin
- **oxytocin** Oxytocin, Oxytozin
- **parathyroid hormone/ parathyrin/parathormone (PTH)** Nebenschilddrüsenhormon, Parathyrin, Parathormon (PTH)
- **phytohormone** Phytohormon, Pflanzenwuchsstoff
- **progesterone** Progesteron
- **prolactin (PRL)/luteotropic hormone (LTH)** Prolaktin, Prolactin (PRL), Mammatropin, mammotropes Hormon, lactotropes Hormon, luteotropes Hormon (LTH)

- **prolactoliberin/prolactin releasing hormone/prolactin releasing factor (PRH/PRF)** Prolaktoliberin, Prolaktin-Freisetzungshormon
- **prostaglandin** Prostaglandin
- **relaxin** Relaxin
- **releasing hormone/release hormone/releasing factor/ release factor** Freisetzungshormon, Freisetzungsfaktor, freisetzendes Hormon, freisetzender Faktor
- **secretin** Secretin, Sekretin
- **sex hormone** Sexualhormon
- **somatoliberin/somatotropin release-hormone/somatotropin releasing factor (SRF)/growth hormone release hormone/ factor (GRH/GRF)** Somatoliberin, Somatotropin-Freisetzungshormon
- **somatomedin/insulin-like growth factor (IGF) (sulfation factor/serum sulfation factor)** Somatomedin
- **somatostatin/somatotropin release-inhibiting factor/growth hormone release-inhibiting hormone (GRIH)** Somatostatin
- **somatotropin (STH)/growth hormone (GH)** Somatotropin, somatotropes Hormon, Wachstumshormon
- **testis-determining factor (TDF)** Testis-Determinationsfaktor
- **testosterone** Testosteron
- **thyroliberin/thyreotropin releasing hormone/factor (TRH/ TRF)** Thyroliberin, Thyreotropin-Freisetzungshormon (TRH/TRF)
- **thyrotropin/thyroid-stimulating hormone (TSH)** Thyr(e)otropin, Tyrotropin, thyreotropes Hormon, thyreoideastimulierendes Hormon (TSH)
- **thyroxine (*also:* thyroxin)/ tetraiodothyronine (T$_4$)** Thyroxin
- **triiodothyronine (T$_3$)** Triiodthyronin
- **vasoactive intestinal polypeptide (VIP)** vasoaktives intestinales Peptid (VIP)
- **vasopressin/antidiuretic hormone (ADH)** Vasopressin, antidiuretisches Hormon (ADH), Adiuretin
- **vasotocin** Vasotocin

hormone release Hormonausschüttung
hormonopoiesis Hormonbildung
horn sharks/Heterodontiformes Doggenhaiartige

horn tubule

horn tubule (hoof) Hornröhrchen
horned/cornigerous gehörnt
horned pondweed family/Zannichelliaceae Teichfadengewächse
horns Gehörn, Hörner (Geweih > antlers)
hornworm (hawk moth/sphinx moth larva) Schwärmerlarve, Schwärmerraupe (mit Afterhorn)
hornwort Hornmoos
hornwort family/Ceratophyllaceae Hornblattgewächse
horny Horn…, aus Horn, hornig, schwielig
horny/excited sexually sexuell erregt, geschlechtlich sein, geil sein
horny cell Hornzelle
horny corals/gorgonian corals/gorgonians/Gorgonaria Hornkorallen, Rindenkorallen
horny frog/foot pad (hoof) Hornstrahl
horny hoof/hoof capsule (horse) Hornschuh, Hufkapsel
horny layer Hornschicht
horny sole (hoof) Hornsohle
horny sponges/Cornacuspongiae Hornschwämme, Netzfaserschwämme
horny wall (hoof) Hornschuhwand, Hornwand, Hufwand
horotelic horotelisch
horotely Horotelie
horse Pferd
- **black horse** Rappe
- **broodmare** Zuchtstute
- **colt (male horse/pony under 4 years)** männliches Fohlen/Füllen
- **dam (mother animal)** Muttertier
- **dapple-grey horse** Apfelschimmel
- **filly (female horse/pony under 4 years)** weibliches Fohlen/Füllen, Stutenfohlen
- **flea-bitten grey horse/flea-bitten white horse** Fliegenschimmel
- **grey horse** Schimmel
- **hinny (♀ ass/donkey x ♂ horse stallion)** Maulesel (Eselstute x Pferdehengst)
- **mare** Stute
- **mule (♀ horse x ♂ ass/donkey)** Maultier (Pferdestute x Eselhengst)
- **mustang (small naturalized horse of western plains)** Mustang (verwildertes Präriepferd)
- **roan (red roan/strawberry roan)** Rotschimmel
- **sire (male parent)** Vatertier, männliches Stammtier (Pferd: Beschäler/Zuchthengst)
- **skewbald horse** Schecke
- **stallion** Hengst (Zuchthengst: stud/studhorse)
- **stock horse** Zuchtpferd
- **stud (group of horses bred and kept by one owner)** Gestüt
- **studhorse/stud (see: stallion)** Zuchthengst, Schälhengst, Deckhengst, Beschäler
- **thoroughbred** Vollblut
- **white horse** weißer Schimmel

horse chestnut family/buckeye family/Hippocastanaceae Rosskastaniengewächse
horse latitudes Rossbreiten
horse radish peroxidase Meerrettichperoxidase
horsehair worms/hairworms/threadworms/gordian worms/nematomorphans/nematomorphs/Nematomorpha Saitenwürmer
horseradish tree family/Moringaceae Moringagewächse, Bennussgewächse, Behennussgewächse, Pferderettichgewächse
horseshoe crabs/Xiphosura Pfeilschwanzkrebse
horsetail/scouring rush Schachtelhalm
horsetail family/Equisetaceae Schachtelhalmgewächse
horticultural show/horticultural exhibit Blumenschau, Gartenschau, Gartenbauausstellung
horticulture/gardening Gartenbau; Gärtnerei, Gärtnern
host Wirt (allgemein: Wirtsorganismus/Wirtstier)
- **alternate host** Wechselwirt
- **containment host** Sicherheitswirt
- **final host** Endwirt
- **intermediary host** Zwischenwirt
- **main host/primary host/definitive host** Hauptwirt
- **non-permissive host** nicht-permissiver Wirt
- **paratenic host/transfer host** paratenischer Wirt, Sammelwirt, Stapelwirt, Transportwirt
- **permissive host** permissiver Wirt
- **reservoir host** Reservoir-Wirt
- **secondary host** Nebenwirt

host animal Wirtstier
host cell Wirtszelle
host organism Wirtsorganismus
host plant Wirtspflanze
host race Wirtsrasse
host range Wirtsspektrum, Wirtsbereich
host specificity Wirtsspezifität

hostile commensalism/synechthry/ synecthry Raubgastgesellschaft, Synechthrie
hot plate *lab* Heizplatte
hot spot Hot-Spot, sensible Position (Stelle in einem Gen mit hoher Mutabilität)
hot spring heiße Quelle
hotbed/forcing bed *hort* (cold frame & heating) beheizter glasgedeckter Pflanzkasten (Frühbeet/Treibbeet); (heated with fermenting manure) Mistbeet
hothouse (greenhouse) Warmhaus (Gewächshaus)
hourglass (*Latrodectus*) *arach* Sanduhr
house bee Stockbiene
house plant Zimmerpflanze
housekeeping gene konstitutives Gen, Haushaltungsgen, Haushaltsgen
hovering *orn* Schweben, in der Luft stehen
hovering flight (hummingbirds) Schwirrflug (Kolibris)
howl heulen
hub/nub (web) *arach* Netznabe
huddling *ethol* Kontaktverhalten
HUGO (Human Genome Project) Menschliches Genomprojekt (HUGO)
hull (e.g. cereal seed husk/grain husk/outer covering) Schale, Hülle, Hülse (Samenschale); Außenkelch
hum/buzz (insects/hummingbirds etc.) summen
human *adj/adv* menschlich, den Menschen betreffend, human, Human…
human/human being Mensch
human artificial chromosome (HAC) künstliches Humanchromosom, künstliches menschliches Chromosom
human chorionic gonadotropin (hCG) Choriongonadotropin (hCG)
human genetics Humangenetik, Anthropogenetik
human growth hormone (hGH)/ human somatotropin menschliches Wachstumshormon (Somatotropin/somatotropes Hormon)
human leucocyte antigen complex (HLA complex) menschlicher Leukozytenantigen-Komplex (HLA-Komplex)
human placental lactogen (HPL)/ human chorionic somatomammotropin (HCS) Plazentalaktogen

Human Genome Project (HUGO) Menschliches Genomprojekt (HUGO)
human race *sensu* human species Mensch, Menschen, Menschheit
humanity Menschheit
humanize vermenschlichen; der menschlichen Natur anpassen
humankind/humans Menschheit, Menschengeschlecht
humanlike menschenähnlich
humeral cross-vein Humeralquerader
humeral feathers/humerals/ tertiaries/tertial feathers Humeralflügel
humic acid Huminsäure
humid/moist feucht
humid biotope Feuchtbiotop
humidifier/mist blower/sprayer Zerstäuber, Wasserzerstäuber
humidify befeuchten
humidity/moisture Feuchtigkeit, Feuchte
humification Humifizierung, Humifikation, Humusbildung
humify humifizieren
hummingbirds/Trochiliformes Kolibris
hummock/hillock/tussock Bult, Bülte
humor/body fluid Humor, Körperflüssigkeit
• **aqueous humor (eye)** Kammerwasser
hump (cattle/camels) Höcker, Fetthöcker
hump/bulge/knoll/mound Buckel, Erhebung
humped höckerig, buckelig
humpless ohne Höcker
humus Humus
• **duff (raw humus)** humöser/ humusartiger Waldboden (Rohhumus)
• **moder (humus layer)** Moder
• **mor humus (acid pH)** Rohhumus, saurer Auflagehumus; Trockentorf
• **mull humus/mull (near neutral pH)** Mull (milder Dauerhumus)
• **raw humus/skeletal humus** Rohhumus
• **unstable humus/friable humus/ crustable humus** Nährhumus
humus layer Humusschicht, Humusauflage
hunch/guess/assumption Vermutung, Annahme
hunger Hunger
hungry hungrig
hunt *vb* jagen
hunt/hunting *n* Jagd, Jägerei

hunter Jäger
hunter-gatherer Jäger-Sammler
hunting ground Wildbahn
**hunting range/hunting grounds/
hunting territory** Jagdgründe
Huperziaceae/fir clubmoss family
Teufelsklauengewächse
hurricane (≥ 115 km/h) Hurrikan,
Orkan
hurt/be painful schmerzen
husk (corn) Liesche, Maishülse
husk/coat/cover Schale,
Schutzschicht, Hülle
husk/glume (small bract) Spelze
husk/pod Hülse, Schote, Schale
hutch/coop/pen (e.g. chickens)
kleiner Tierkäfig, kleiner Verschlag,
Auslauf (z.B. Geflügelstall)
hyacinth family/Hyacinthaceae
Hyazinthengewächse
hyaline/clear/transparent hyalin,
glasartig, klar, glasklar,
durchsichtig, transparent
hyaline cartilage hyaliner Knorpel
hyaline cell Hyalinzelle, Hyalocyt
hyaluronic acid Hyaluronsäure
hybrid/crossbred *adj/adv* hybrid,
durch Kreuzung erzeugt
hybrid/crossbreed *n* Hybride,
Hybrid, Mischling, Bastard
hybrid antibody bispezifischer
Antikörper
hybrid cell Hybridzelle
hybrid DNA/chimeric DNA Hybrid-
DNA
hybrid sterility Hybridensterilität,
Bastardsterilität
hybrid swarm Hybridschwarm,
Bastardschwarm (Bastardpopulation)
hybrid vigor/heterosis
Bastardwüchsigkeit, Heterosis
hybrid zone Hybridisierungzone,
Bastardisierungszone
hybrid-arrest translation (HART)
hybridarretierte Translation
hybrid-release translation (HRT)
Hybrid-Freisetzungstranslation
hybridization/bastardization
Hybridisierung, Bastardisierung
- **CISS (chromosomal** *in situ*
  suppression hybridization)
  chromosomale *in-situ*
  Suppressionshybridisierung
- **comparative genome
  hybridization (CGH)**
  vergleichende
  Genomhybridisierung (CGH)
- **competition h.**
  Kompetitionshybridisierung
- **cross h.** Kreuzhybridisierung
- **DNA-driven h.** DNA-getriebene
  Hybridisierung

- **exhaustion hybridization**
  Erschöpfungshybridisierung
- **fluorescence-*in-situ*-
  hybridization (FISH)**
  Fluoreszenz-*in-situ*-Hybridisierung
  (FISH)
- ***in situ* h.** *in situ* Hybridisierung
- **RNA-driven h.** RNA-getriebene
  Hybridisierung
- **sandwich h.** Sandwich-
  Hybridisierung
- **saturation h.**
  Sättigungshybridisierung
hybridize hybridisieren
hybridoma Hybridom
**hydathode/water stoma/water
pore** Hydathode, Wasserspalte
hydatid Hydatide
hydnora family/Hydnoraceae
Lederblumengewächse
hydrangea family/Hygrangeaceae
Hortensiengewächse
hydranth (Cnidaria) Hydranth
hydrate Hydrat
hydration/solvation Hydratation,
Hydratisierung, Solvation
(Wassereinlagerung,
Wasseranlagerung)
hydration shell Hydratationsschale,
Hydrathülle, Wasserhülle
hydric hydrisch
hydrocarbon Kohlenwasserstoff
- **chlorinated hydrocarbon**
  chlorierter Kohlenwasserstoff
- **CFCs (chlorofluorocarbons/
  chlorofluorinated hydrocarbons)**
  FCKW
  (Fluorchlorkohlenwasserstoffe)
- **fluorinated hydrocarbon**
  Fluorkohlenwasserstoff
**Hydrocharitaceae/frog-bit family/
tape grass family/elodea family**
Froschbissgewächse
hydrochory/water-dispersal
Hydrochorie, Wasserausbreitung
hydrocoel/hydrocoele Hydrocoel
hydrocoles/aquatic animals
Wassertiere
Hydrocotylaceae/pennywort family
Wassernabelgewächse
hydrogen Wasserstoff
**hydrogen bacteria (aerobic
hydrogen-oxidizing bacteria)**
Knallgasbakterien,
Wasserstoffbakterien
hydrogen bond Wasserstoffbrücke,
Wasserstoffbrückenbindung
**hydrogen cyanide/hydrocyanic
acid/prussic acid** Blausäure,
Zyanwasserstoff
hydrogen electrode Wasserstoff-
Elektrode

167   **hypertrophic** **H**

hydrogen peroxide
Wasserstoffperoxid
hydrogenate hydrieren
hydrogenation Hydrierung
(Wasserstoffanlagerung)
hydrologic cycle/water cycle
Wasserkreislauf
hydrolysis Hydrolyse,
Wasserspaltung
hydrolytic hydrolytisch,
wasserspaltend
hydrophilic (water-attracting/
water-soluble) hydrophil
(wasseranziehend/wasserlöslich)
hydrophilicity (water-attraction/
water-solubility) Hydrophilie
(wasseranziehend/wasserlöslich)
hydrophobic (water-repelling/
water-insoluble) hydrophob
(wasserabweisend/wasserabstoßend/
wasserunlöslich)
hydrophobicity (water-insolubility)
Hydrophobie (wasserabweisend/
wasserabstoßend/wasserunlöslich)
Hydrophyllaceae/waterleaf family
Wasserblattgewächse
hydrophyte/aquatic plant
Hydrophyt, Wasserpflanze
hydroponics (soil-less culture/
solution culture) Hydrokultur
hydrosere *eco* Hydroserie
hydroskeleton/hydrostatic skeleton
Hydroskelett, hydrostatisches Skelett
hydrosphere Hydrosphäre,
Wasserhülle
hydrostachys family/
Hydrostachyaceae
Wasserröhrengewächse
hydrothermal vent hydrothermaler
Schlot
hydrous *chem* wasserhaltig
hydroxyapatite Hydroxyapatit
hydroxylation Hydroxylierung
hydroxyproline Hydroxyprolin
hydrozoans/hydra-like animals/
hydroids/Hydroidea Hydrozoen
hygiene Hygiene
hygienic hygienisch
hygrophorus family/
Hygrophoraceae Wachsblättler
hygrophyte (thriving in moist
habitats) Hygrophyt
hygroscopic hygroskopisch
hymen Hymen, Jungfernhäutchen
hymenochaete family/
Hymenochaetaceae
Borstenporlinge
Hymenophyllaceae/filmy fern
family Hautfarngewächse,
Schleierfarngewächse
hymenopterans/Hymenoptera
Hautflügler

hymenopterous wing Hautflügel
hyoid arch Hyoidbogen,
Zungenbeinbogen (Gesamtheit der
Teile)
hyoid bar (skeleton only)
Hyoidbogen, Zungenbeinbogen (nur
Knorpelspange)
hyoid bone/lingual bone/os
hyoideum Hyoideum, Zungenbein
hyolithids/Hyolithida Hyolithen
hypanthium/floral tube
Hypanthium, Achsenbecher,
Blütenröhre, vergrößerter/
scheibenförmiger Blütenboden/
Blütenachse
hypaxial hypaxionisch
(Rumpfmuskulatur)
hypercalcemia Hyperkalzämie
hypercapnia Hyperkapnie
hyperchromasia/hyperchromia/
hyperchromatism Hyperchromasie
hyperchromicity/hyperchromic
effect/hyperchromic shift
Hyperchromizität
hyperchromicity/hyperchromism
Hyperchromie
hyperemia Hyperämie
hyperglycemia Hyperglykämie
hyperglycemic hyperglykämisch
Hypericaceae/St. John's wort
family/mamey family/mangosteen
family/clusia family/*Clusiaceae/*
Guttiferae Hartheugewächse,
Johanniskrautgewächse
hypernatremia *hema*
Hypernatriämie (erhöhter
Natriumgehalt)
hyperparasite/superparasite
Hyperparasit, Überparasit
hyperphagia Hyperphagie,
Fresssucht, Esssucht, Gefräßigkeit
hyperploid hyperploid
hyperpnea Hyperpnoe
hyperpolarization
Hyperpolarisierung
hypersensitivity/allergy
Hypersensibilität,
Überempfindlichkeit, Allergie
hypersensitivity reaction
Überempfindlichkeitsreaktion
● **delayed-type h.r. (T$_{DTH}$)**
Überempfindlichkeitsreaktion vom
Spättyp, verzögerter Typ
● **immediate-type h.r. (T$_{ITH}$)**
Überempfindlichkeitsreaktion vom
Soforttyp, anaphylaktischer Typ
hypertension *med* Hochdruck,
Bluthochdruck
hypertonic hyperton(isch)
hypertonicity/hypertonia
Hypertonie
hypertrophic hypertroph

hypertrophy

hypertrophy Hypertrophie
hypervariable region *immun*
 hypervariable Region
hypha (*pl* **hyphas/hyphae)** Hyphe,
 Pilzfaden
 • **raquet hypha (raquet mycelium)**
 Raquettehyphe, Keulenhyphe
 (Raquettemyzel/Keulenmyzel)
**hypnospore/resting spore/
 persistant spore/dormant spore**
 Hypnospore, Dauerspore
hypobranchial furrow/endostyle
 Hypobranchialrinne, Endostyl
hypocercal hypocerk, hypozerk
hypocotyl Hypokotyl,
 Keimsprossachse
hypodermic needle Nadel, Kanüle,
 Hohlnadel (Spritze)
**hypogean germination/hypogeal
 germination** hypogäische Keimung
hypogeous hypogäisch
hypoglycemia Hypoglykämie
hypoglycemic hypoglykämisch
hypognathous hypognath
hypogynous hypogyn, oberständig
**hypolepis family/Hypolepidaceae
 (incl. bracken ferns)**
 Buchtenfarngewächse
 (inkl. Adlerfarne)
**hypophyseal pouch/hypophyseal
 sac/Rathke's pouch**
 Hypophysentasche, Rathkesche
 Tasche
hypoploid hypoploid

hypopneustic hypopneustisch
**hypoptile/afterfeather/accessory
 plume/hypopenna** *orn*
 Nebenfeder, Afterfeder
hyporheic hyporheisch
**hypostome/oral cone/peduncle/
 gullet/manubrium** Magenstiel,
 Mundrohr, Manubrium
**hypotheca/hypocone
 (dinoflagellate frustule)**
 Hypotheka
hypothermic hypothermisch
hypothesis Hypothese
hypothetic/hypothetical
 hypothetisch
hypotonic hypotonisch
hypotonicity/hypotonia Hypotonie
hypotrophic hypotroph
hypotrophy Hypotrophie
hypoxia Hypoxie
hypoxic hypoxisch
hypsodont (with high crowns)
 hypsodont, hypselodont,
 hochkronig
hypsophyll Hochblatt
hypural fan *ichth* Hypurale,
 Schwanzfächer
hysteresis Hysterese
hysterotely *entom* Hysterotelie
hysterothecium *fung*
 Hysterothecium
hystrichoglossate radula
 hystrichoglosse Radula,
 Bürstenzunge

169

immobile **I**

I band (muscle: *isotropic*) I-Bande
ibotenic acid Ibotensäure
ice cap/ice sheet (glacial ice cover)
Eisdecke; Gletscher
ice nucleating activity *micb*
Eiskernaktivität
ice scouring *meteo/bot*
Eisabscheuerung
Ice Age/Glacial Epoch/Pleistocene
Epoch/Diluvial Eiszeit, Glazialzeit,
Pleistozän, Diluvium
ice-bath Eisbad
ichnofossil/trace fossil Ichnofossil,
Spurenfossil
ichnology Ichnologie, Spurenkunde
ichthyology Ichthyologie, Fischkunde
ichthyosaurs/fish-reptiles (ocean-
living reptiles)/Ichthyosauria/
Ichthyoptergia Fischsaurier
identification Identifizierung,
Bestimmung
identify identifizieren, bestimmen
identity Identität
• **nonidentity** *immun*
Verschiedenheit (Nicht-Identität)
• **partial identity** *immun*
Teilidentität, partielle
Übereinstimmung
identity by descent (IBD) identisch
aufgrund gemeinsamer Abstammung
identity by state (IBS) identisch
aufgrund von Zufällen
idioblast Idioblast
idiophase Idiophase
(Produktionsphase)
idioplasm/germ plasm/gonoplasm
Idioplasma, Keimplasma
idiotope Idiotop
idiotype Idiotyp
idling reaction *gen* Leerlaufreaktion
igneous glutflüssig, magmatisch
igneous rock *geol*
Erstarrungsgestein, Eruptivgestein
ignitable entzündbar
ignite entzünden
ignition Entzündung, Zündung
ileum Ileum, Hüftdarm
ilium/flank bone/os ilium Ilium,
Darmbein
ill/sick krank
illegitimate recombination *gen*
illegitime Rekombination
illicium family/star-anise family/
Illiciaceae Sternanisgewächse
illness/sickness/disease/disorder
Erkrankung, Krankheit, Störung
illuminance Beleuchtungsstärke
illuminate beleuchten
illumination Beleuchtung
• **epiillumination/incident**
illumination *micros* Auflicht,
Auflichtbeleuchtung

• **Koehler illumination**
Köhlersche Beleuchtung
• **oblique illumination**
Schräglichtbeleuchtung
• **transillumination/transmitted**
light illumination Durchlicht,
Durchlichtbeleuchtung
illuminator Leuchte
illuviation Einwaschung
image *vb* abbilden
• **intermediate image** *micros*
Zwischenbild
• **virtual image** virtuelles Bild
image point Bildpunkt
imaginal anlage Imaginalanlage
imaginal disk/imaginal bud
Imaginalscheibe
imaginal ring Imaginalring
imago (pl **imagoes/imagines)/adult**
insect Imago (*pl* Imagines),
Vollinsekt, Adultinsekt
imbalance/disequilibrium
Ungleichgewicht
imbibe/hydrate imbibieren,
hydratieren
imbibition/hydration Imbibition,
Hydratation
imbricate/overlapping dachig,
dachziegelartig, schuppenartig,
schindelartig überlappend
imidazole Imidazol
imino acid Iminosäure
imitate/mimic nachahmen, mimen
imitating/mimetic nachahmend,
fremde Formen nachbildend,
mimetisch
imitation Imitation, Nachahmung
immature unreif
immaturity/immatureness Unreife
immediate-type hypersensitivity
reaction
Überempfindlichkeitsreaktion vom
Soforttyp, anaphylaktischer Typ
immerse eintauchen, untertauchen;
einbetten
immersed *bot* ganz unter Wasser
immersed slot reactor (bioreactor)
Tauchkanalreaktor
immersing surface reactor
(bioreactor) Tauchflächenreaktor
immersion Immersion, Eintauchen,
Untertauchen
immersion heater *lab* Tauchsieder
immigrate einwandern (Zellen)
immigration Immigration,
Einwanderung, Zuwanderung
immiscibility Unvermischbarkeit
immiscible unvermischbar,
nicht mischbar
immobile/fixed/motionless
immobil, fixiert, bewegungslos,
unbeweglich

I immobility

immobility/motionlessness
Immobilität, Bewegungslosigkeit
immobilization Immobilisation,
Immobilisierung
immobilize (to make immobile)
immobilisieren
immortal unsterblich
immortality Immortalität,
Unsterblichkeit
immortalized cell immortalisierte
Zelle
immortalized celline/cell line
immortalisierte Zelllinie
immotile/fixed unbeweglich, fixiert
immune immun
immune adherence Immunadhärenz
immune complex Immunkomplex
immune defect Immundefekt
immune deficiency/
immunodeficiency
Immunschwäche
● **acquired immune deficiency**
syndrome (AIDS) erworbenes
Immunschwächesyndrom,
erworbenes Immunmangel-
Syndrom (AIDS)
● **severe combined immune**
deficiency (SCID) schwerer
kombinierter Immundefekt
immune electron microscopy (IEM)
Immun-Elektronenmikroskopie
(IEM)
immune reaction Immunreaktion
immune recognition
Immunerkennung
immune response Immunantwort
immune tolerance/immunological
tolerance Immuntoleranz
immunity Immunität
● **acquired immunity/adaptive**
immunity (active/passive)
erworbene Immunität (aktive/
passive)
● **artificial immunity**
künstliche Immunität
● **cellular immunity**
zelluläre Immunität
● **concomitant immunity/**
premunition begleitende
Immunität, Prämunität
● **natural immunity**
natürliche Immunität
● **passive immunity**
passive Immunität
immunization/vaccination
Immunisierung, Impfung
● **protective immunization**
Schutzimpfung
● **genetic immunization**
genetische Immunisierung
immunize/vaccinate immunisieren,
impfen

immunoaffinity chromatography
Immunaffinitätschromatographie
immunoassay Immunoassay
immunoblot/Western blot
Immunoblot, Western-Blot
immunocompetence/immunologic
competence Immunkompetenz
immunocompromized
abwehrgeschwächt
immunodiffusion Immundiffusion,
Immunodiffusionstest;
Gelpräzipitationstest
● **double diffusion/double**
immunodiffusion
Doppelimmundiffusion
● **double radial immunodiffusion**
(DRI) (Ouchterlony technique)
doppelte radiale Immundiffusion
(Ouchterlony-Methode)
● **radial immunodiffusion (RID)**
radiale Immundiffusion
● **single immunodiffusion (Oudin**
test) einfache Immundiffusion
(Oudin-Methode)
● **single radial immunodiffusion**
(SRI) (Mancini technique)
einfache radiale Immundiffusion
(Mancini-Methode)
immunoelectrophoresis
Immunelektrophorese
● **countercurrent**
immunoelectrophoresis/
counterelectrophoresis
Überwanderungs-
immunelektrophorese,
Überwanderungselektrophorese
● **rocket immunoelectrophoresis**
Raketenimmunelektrophorese
● **charge-shift**
immunoelectrophoresis Tandem-
Kreuzimmunelektrophorese
● **crossed immunoelectrophoresis/**
two-dimensional
immunoelectrophoresis
Kreuzimmunelektrophorese
immunofluorescence
Immunfluoreszenz
immunofluorescence
chromatography
Immunfluoreszenzchromatographie
immunofluorescence microscopy
Immunfluoreszenzmikroskopie
immunogen Immunogen
immunogenetics Immungenetik
immunogenic immunogen
immunogenicity Immunogenität,
Immunisierungsstärke
immunoglobulin fold
Immunglobulinfaltung
immunogold-silver staining
Immunogold-Silberfärbung
immunologic(al) immunologisch

impulse

immunological memory
immunologisches Gedächtnis
immunology Immunologie
immunopathology
Immun(o)pathologie
immunopathy Immunkrankheit,
Immunopathie
immunoprecipitation
Immunpräzipitation
immunoprophylaxis
Immunprophylaxe
immunoradiometric assay (IRMA)
immunoradiometrischer Assay
(IRMA)
immunoscreening Immunscreening
**immunosuppression/immune
suppression** Immunsuppression
**immunosurveillance/immunological
surveillance** immunologische
Überwachung, Immunüberwachung
impaling (shrikes) ethol/orn
Aufspießen
**imparipinnate/odd-pinnate/
unequally pinnate (pinnate with
an odd terminal leaflet)** unpaarig
gefiedert
impeller/stirrer/agitator biot
(bioreactors) Rührer, Rührwerk
● **anchor impeller** Ankerrührer
● **crossbeam impeller**
Kreuzbalkenrührer
● **disk turbine impeller**
Scheibenturbinenrührer
● **flat-blade impeller**
Scheibenrührer, Impellerrührer
● **four flat-blade paddle impeller**
Kreuzblattrührer
● **gate impeller** Gitterrührer
● **helical ribbon impeller**
Wendelrührer
● **hollow stirrer** Hohlrührer
● **marine screw impeller**
Schraubenrührer
● **multistage impulse
countercurrent impeller**
Mehrstufen-Impuls-Gegenstrom
(MIG) Rührer
● **off-center impeller** exzentrisch
angeordneter Rührer
● **paddle stirrer/paddle impeller**
Schaufelrührer, Paddelrührer
● **pitch screw impeller**
Schraubenspindelrührer
● **pitched-blade fan impeller/
pitched-blade paddle impeller/
inclined paddle impeller**
Schrägblattrührer
● **profiled axial flow impeller**
Axialrührer mit profilierten
Blättern
● **propeller impeller**
Propellerrührer

● **rotor-stator impeller/Rushton-
turbine impeller** Rotor-Stator-
Rührsystem
● **screw impeller** Schneckenrührer
● **self-inducting impeller with
hollow impeller shaft**
selbstansaugender Rührer mit
Hohlwelle
● **stator-rotor impeller/Rushton-
turbine impeller** Stator-Rotor-
Rührsystem
● **turbine impeller** Turbinenrührer
● **two flat-blade paddle impeller**
Blattrührer
● **two-stage impeller** zweistufiger
Rührer
● **variable pitch screw impeller**
Schraubenspindelrührer mit
unterschiedlicher Steigung
impeller shaft biot **(biorectors)**
Rührerwelle
**imperfect fungi/deuteromycetes/
Deuteromycetes** Deuteromyceten,
unvollständige Pilze, Fungi
imperfecti
**imperfect stage/anamorphic stage/
asexual stage** fung
Nebenfruchtform
impermeability/imperviousness
Impermeabilität, Undurchlässigkeit
impermeable/impervious
impermeabel, undurchlässig
impervious/impermeable
undurchlässig
implant n **(organs)** Implantat
implant vb **(organs)** einpflanzen
implant vb embr einnisten,
implantieren
implantation (organ) Einpflanzung
implantation/nidation embr
Implantation, Nidation, Einnistung
impoundment (of water)
Wassersammelbecken
impregnate befruchten, schwängern;
chem durchdringen, sättigen;
phys imprägnieren, durchtränken
impregnation Befruchtung,
Schwängerung; chem
Durchdringung, Sättigung;
phys Imprägnation, Durchtränkung
impress general beeindrucken;
paleo abdrücken (einen Abdruck
hinterlassen)
impression general Eindruck;
paleo Addruck
imprint vb ethol/gen prägen
imprinting ethol/gen Prägung
● **habitat imprinting**
Biotopprägung
● **genomic imprinting** genomische
Prägung
impulse Impuls, Erregung

impulse propagation

impulse propagation Nervenleitung
impurity/contamination
Verunreinigung, Kontamination
in bloom in Blüte
in focus/sharp *micros* scharf, im
Fokus
in leaf/leaved beblättert
in silk (corn) blühend, in Blüte (Mais)
in situ **hybridization** *in situ*
Hybridisierung
• **FISH (fluorescence** *in situ*
hybridization) FISH (*in situ*
Hybridisierung mit
Fluoreszenzfarbstoffen)
in-vitro fertilization (IVF) In-vitro-
Fertilisation,
Reagenzglasbefruchtung
inactive inaktiv
inactive state/dormant state/
dormancy Ruhezustand
inanimate/lifeless/nonliving/dead
unbelebt, leblos, tot
inarching *hort* Ammenveredelung,
Anhängen (Vorspann geben)
inborn angeboren
inborn error Erbleiden, angeborener
Fehler
inbred line Inzuchtlinie
inbred strain Inzuchtstamm
inbreed *vb* Inzucht betreiben
inbreeding/endogamy Inzucht,
Reinzucht
incense tree family/torchwood
family/frankincense tree family/
Burseraceae Balsambaumgewächse
incest Inzest
inchworm/measuring worm/looper/
spanworm (geometer moth larva)
Spannerraupe
incidence Vorkommen, Auftreten,
Häufigkeit, Verbreitung,
Ausdehnung; Einfallen (Licht)
incident light einfallendes Licht,
auftreffendes Licht
incipient anfangend, anfänglich,
beginnend
incipient plasmolysis
Grenzplasmolyse
incised/evenly notched/evenly cut
gleichmäßig geschlitzt/zerschlitzt/
eingeschnitten
incision/indentation/cut Einschnitt
incisive bone
Zwischenkieferknochen
incisor/front tooth Schneidezahn,
Vorderzahn, Beißzahn
incite/chase (inciting/chasing)
hetzen
inclination Neigung,
Neigungswinkel
incline/slope Hang, Abhang,
Schräglage, Neigung

inclusion (intracellular) Einschluss
inclusion/intercalation Einlagerung
inclusion body
Einschlusskörperchen
• **Bollinger body/Bollinger's**
granule Bollinger Körper (viraler
Einschlusskörper)
• **cell inclusion/cellular inclusion**
Zelleinschluss (Inklusion)
• **Guarnieri body** Guarnierischer
Einschlusskörper
• **Negri body** Negrisches
Körperchen, Negri-Körper
• **nuclear inclusion body**
Kerneinschlusskörper
• **x body** X-Körper
inclusive fitness Gesamtfitness,
Gesamteignung
incompatibility Inkompatibilität,
Unverträglichkeit
incompatibility group
Inkompatibilitätsgruppe
incompatibility reaction
Inkompatibilitätsreaktion,
Unverträglichkeitsreaktion
incompatible inkompatibel,
unverträglich
incomplete dominance
unvollständige Dominanz
incomplete metamorphosis/gradual
metamorphosis unvollkommene/
unvollständige Metamorphose/
Verwandlung
incomplete penetrance *gen*
unvollständige Penetranz
incompletely linked genes
unvollständig gekoppelte Gene
incrustation/encrustation
Inkrustierung
incubate (brood/breed) inkubieren
(brüten/bebrüten)
incubation (brooding/breeding)
Inkubation (Brüten/Bebrütung/
Bebrüten)
incubation period/breeding period
Inkubationszeit, Brutdauer
incubator Brutschrank
incubous incub, oberschlächtig
incurrent aperture/inhalant
aperture/ingestive aperture
(e. g. sponges) Einströmöffnung,
Ingestionsöffnung
incus/anvil (bone) Amboss
indefinite/unrestricted (growth)
unbeschränkt, unbegrenzt,
unbestimmt
indehiscent fruit Schließfrucht
indentation/indenture/notch/
crenation/cut Einbuchtung, Kerbe,
Einkerbung, Einschnitt
indentation/projection/spike/
notch/serration Zacke

infection

indented eingedrückt
independent assortment
unabhängige Verteilung, freie
Verteilung
indeterminate cleavage
nichtdeterminative Furchung
(regulative Furchung)
indeterminate growth unbegrenztes
Wachstum, unbeschränktes
Wachstum
indeterminate inflorescence offene
Infloreszenz
index fossil Leitfossil,
Faziesfossil
index number/indicator *stat*
Kennzahl, Kennziffer
index plant/indicator plant
Zeigerpflanze, Anzeigerpflanze,
Leitpflanze
**index species/guide species
(indicator species)** Leitart
(Zeigerart)
**Indian almond family/white
mangrove family/Combretaceae**
Strandmandelgewächse
**Indian lotus family/lotus lily
family/Nelumbonaceae**
Lotusblumengewächse
Indian pipe family/Monotropaceae
Fichtenspargelgewächse
**Indian plum family/flacourtia
family/Flacourtiaceae**
Flacourtiagewächse
indican/indoxyl sulfate Indikan,
Indoxylsulfat
indicator Indikator, Anzeiger
indicator plant/index plant
Indikatorpflanze, Zeigerpflanze,
Anzeigerpflanze, Leitpflanze
**indicator species (index species/
guide species)** Zeigerart (Leitart)
indifferent (soil/fidelity)
indifferent, vag, vage (Bodentreue/
Gesellschaftstreue)
indifferent species indifferente Art
indigenous/native/endemic
einheimisch
**indigenous species/native species/
native organism/native lifeform**
Indigen, einheimische Art
individual *n* Individuum
individual(ly) *adv/adj* individuell
**indolyl acetic acid/indoleacetic acid
(IAA)** Indolessigsäure
induce induzieren
induced fit (enzymes) induzierte
Anpassung, induzierte Passform
inducer Induktor
  ● **gratuitous inducer** freiwilliger
Induktor
inducible induzierbar
induction Induktion

indusium/episporangium (fern)
Indusium, Schleier (Farn)
industrial melanism
Industriemelanismus
inedible/uneatable nicht essbar,
ungenießbar
inert *chem* träg, träge
inert gas/rare gas Edelgas
inertia Trägheit
inertial force Trägheitskraft
infancy Säuglingsalter, frühe
Kindheit
infant (child under age of 2 years)
Säugling, Kleinkind
(unter 2 Jahren)
infant mortality
Säuglingssterblichkeit
infanticide Kindermord
infauna Infauna
infect *vb* infizieren
infection/contagion Infektion,
Ansteckung
  ● **abortive infection** abortive
Infektion
  ● **agroinfection** Agroinfektion
  ● **airborne/aerial infection**
aerogene Infektion
  ● **chronic infection**
chronische Infektion
  ● **concurrent/complex infection**
Mehrfachinfektion
  ● **contact infection**
Kontaktinfektion
  ● **covert/silent/inapparent/
subclinical infection**
stumme Infektion, stille Feiung
  ● **droplet infection**
Tröpfcheninfektion
  ● **double infection** Doppelinfektion
  ● **incomplete infection**
unvollständige Infektion
  ● **latent infection** latente Infektion
  ● **local infection** lokale Infektion,
örtliche Infektion
  ● **lytic infection** lytische Infektion
  ● **multiplicity of infection**
Infektionsmultiplizität
  ● **nosocomial infection/hospital-
acquired infection** nosokomiale
Infektion, Nosokomialinfektion,
Krankenhausinfektion
  ● **opportunistic infection**
opportunistische Infektion
  ● **persisting infection** anhaltende
Infektion, persistente Infektion
  ● **productive infection**
produktive Infektion
  ● **secondary infection**
Sekundärinfektion
  ● **silent/covert/inapparent/
subclinical infection**
stumme Infektion, stille Feiung

infectiosity

infectiosity Infektiosität, Ansteckungsfähigkeit
infectious infektiös, ansteckend, ansteckungsfähig
infectious disease Infektionskrankheit
infectious dose (ID$_{50}$ = 50 % infectious dose) Infektionsdosis
infectious waste infektiöser Abfall
infective infektiös; übertragbar
infectivity Infektionsvermögen, Ansteckungsfähigkeit
inference Schlussfolgerung; *stat* Schlussweise
inferior tieferstehend, tiefer, unten, Unter...; untergeordnet; (defeated) unterlegen; (inadequate) minderwertig; *bot* unterständig
inferiority Untergeordnetheit; Unterlegenheit; (inadequacy) Minderwertigkeit
infertile/sterile infertil, steril, unfruchtbar, nicht fortpflanzungsfähig
infertility/sterility Unfruchtbarkeit, Sterilität
infest (pests/parasites) befallen (Schädlingsbefall)
infestation (with pests/parasites) Befall (Parasitenbefall)
infiltrate *vb* infiltrieren, eindringen; (seep into) einsickern
infiltration Infiltration; (seepage) Versickerung
inflame entzünden
inflammation Entzündung
inflammed/inflammatory entzündlich
inflate (inflated) aufblasen (aufgeblasen)
inflected/inflexed (nach innen) geknickt
inflorescence/flower cluster Infloreszenz, Blütenstand
- **bostryx (a helicoid cyme)** Schraubel
- **botrys/raceme** Traube, Botrys
- **capitulum/cephalium/flower head** Capitulum, Cephalium, Blütenköpfchen, Köpfchen, Korb, Körbchen
- **cincinnus/scorpioid cyme** Wickel
- **corymb** Corymbus, Ebenstrauß
- **>umbel-like panicle** Doldenrispe, Schirmrispe
- **>umbel-like raceme** Doldentraube, Schirmtraube
- **cyme/cymose inflorescence** Cyme, Cyma, Cymus, Zyme, cymöser/zymöser Blütenstand, Trugdolde, Scheindolde

- **>scorpioid cyme/cincinnus** Wickel
- **>dichasial cyme/dichasium** Dichasium, zweigablige Trugdolde
- **>pleiochasial cyme/pleiochasium** Pleiochasium, vielgablige Trugdolde
- **>fan-shaped cyme/rhipidium** Fächel
- **>monochasial cyme/simple cyme/monochasium** Monochasium, eingablige Trugdolde
- **>helicoid cyme>bostryx** Schraubel
- **>helicoid cyme>drepanium** Drepanium, Sichel
- **>with sessile flowers** Knäuel
- **>with very short pedicles** Büschel
- **cymule** verkürzte Trugdolde (als Teilblütenstand), Scheinquirl (bei Tubifloren)
- **determinate inflorescence** geschlossene Infloreszenz
- **dichasium/dichasial cyme** Dichasium, zweigablige Trugdolde
- **drepanium (a helicoid cyme)** Drepanium, Sichel
- **fascicle (inflorescence: cyme with very short pedicles)** Faszikel, Fasciculus, Büschel
- **indeterminate inflorescence** offene Infloreszenz
- **juba/loose panicle/panicle of grasses** (lockere) Grasrispe
- **monochasium/monochasial cyme/simple cyme** Monochasium, eingablige Trugdolde
- **panicle** Rispe, Blütenrispe
- **>loose panicle/juba/panicle of grasses** (lockere) Grasrispe
- **>umbel-like panicle (a corymb)** Doldenrispe, Schirmrispe
- **pleiochasium/pleiochasial cyme** Pleiochasium, vielgablige Trugdolde
- **raceme/botrys** Traube, Botrys
- **>umbel-like raceme (a corymb)** Doldentraube, Schirmtraube
- **rhipidium/fan-shaped cyme** Fächel
- **spike** Ähre
- **spire** Blütenähre
- **tassel** männliche Infloreszenz des Mais
- **truncate inflorescence** Rumpfinfloreszenz
- **truncate synflorescence** Rumpfsynfloreszenz

175 inhibition

- **umbel/sciadium** Dolde, Umbella, Sciadium
- **>simple umbel** einfache Dolde
- **>compound umbel** zusammengesetzte Dolde
- **umbel-like panicle (a corymb)** Doldenrispe, Schirmrispe
- **umbel-like raceme (a corymb)** Doldentraube, Schirmtraube

influence Influenz, Einfluss
influent *n ecol* Influent (*pl* Influenten)
influx Einstrom
informed consent (medical/genetic counseling) Einverständniserklärung nach ausführlicher Aufklärung
infrared spectroscopy Infrarot-Spektroskopie, IR-Spektroskopie
infructescence/multiple fruit/ anthocarp Fruchtstand, Fruchtverband
infundibulate plant/funnel-leaved plant Trichterpflanze
infundibulum/fimbriated funnel of oviduct Infundibulum, Wimpertrichter, Flimmertrichter, Eileitertrichter (mit Ostium tubae)
infusiform infusiform
infusiform larva infusiforme Larve
infusorigen Infusorigen
ingest einnehmen, etwas zu sich nehmen, Nahrung aufnehmen
ingestion/food intake Einnahme, Nahrungsaufnahme
ingress (cells) einwandern
ingression (cells) Einwanderung
ingrown eingewachsen
inguinal inguinal, Leisten…
inguinal canal/canalis inguinalis Leistenkanal
inguinal gland Leistendrüse
inguinal pouch/inguinal sinus/sinus inguinalis (sheep) Inguinaltasche
inguinal zone/groin/regio inguinalis Hüftbeuge, Leiste, Leistenbeuge, Leistengegend, Inguinalgegend
inhabit/lodge/occupy/dwell/reside bewohnen
inhabitant/dweller Bewohner
inhalant *n* Inhalant, Inhalationsmittel
inhalation Inhalation, Einatmung, Einatmen, Inspiration
inhale/breathe in einatmen, Luft holen, Atem schöpfen
inherent innewohnend, eigen; angeboren
inherit erben, ererben
inheritable erbbar, vererbbar, Erb…
inheritable disease Erbkrankheit

inheritance Vererbung; (mode of i.) Erbgang
- **autosomal i.** autosomale Vererbung
- **blending i.** Mischvererbung
- **criss-cross i.** Überkreuzvererbung
- **cytoplasmic i.** cytoplasmatische Vererbung
- **dominant i.** dominante Vererbung
- **intermediate i.** intermediärer Erbgang, intermediäre Vererbung
- **maternal i.** maternale Vererbung
- **matroclinous i.** matrokline Vererbung
- **mitochondrial inheritance** mitochondriale Vererbung
- **multifactorial i.** multifaktorielle Vererbung
- **mode of inheritance** Vererbungsmodus, Erbgang
- **monogenic inheritance** monogener Erbgang
- **particulate i.** partikuläre Vererbung
- **polygenic i.** polygene Vererbung
- **recessive i.** rezessive Vererbung
- **sex-linked i.** geschlechtsgebundene Vererbung
- **uniparental i.** uniparentale Vererbung
- **X-linked i.** X-chromosomale Vererbung

inherited ererbt, vererbt, Erb…
inhibit hemmen
inhibition Inhibition, Hemmung
- **aggressive inhibition** Angriffshemmung, Aggressionshemmung
- **allosteric inhibition** allosterische Hemmung
- **competitive inhibition** kompetitive Hemmung, Konkurrenzhemmung
- **contact inhibition** *cyt* Kontakthemmung
- **end-product inhibition** Endprodukthemmung
- **feedback inhibition** Rückwärtshemmung, Rückkopplungshemmung
- **feed-forward inhibition/ reciprocal inhibition** *neuro* Vorwärtshemmung
- **irreversible inhibition** irreversible Hemmung
- **noncompetitive inhibition** nichtkompetitive Hemmung
- **reciprocal inhibition** reziproke Hemmung, gegenseitige Hemmung
- **reversible inhibition** reversible Hemmung

inhibition 176

- **substrate inhibition**
Substratinhibition
- **suicide inhibition**
Suizidhemmung
- **uncompetitive inhibition**
unkompetitive Hemmung
inhibition zone Hemmzone
inhibitory hemmend, inhibierend,
inhibitorisch
inhibitory neuron Hemmungsneuron
**inhibitory postsynaptic potential
(IPSP)** inhibitorisches
postsynaptisches Potential
inhomogeneity Inhomogenität
inhomogeneous inhomogen,
ungleichmäßig beschaffen
initial/stem cell (primordial cell)
Initiale, Initialzelle, Stammzelle
(Primordialzelle/Primane)
initial distribution *stat*
Ausgangsverteilung
initial magnification
Primärvergrößerung
initial population
Ausgangspopulation
initial segment Initialsegment
(myelinisierte Fasern)
initial velocity (vector)/initial rate
Anfangsgeschwindigkeit
(v_0: Enzymkinetik)
initiating ring Initialring,
Initialschicht
initiation codon Initiationscodon
initiation complex Initiationskomplex
initiation factor Initiationsfaktor
inject injizieren, spritzen; einspritzen
injection Injektion, Spritze;
Einspritzung
injure verletzen
injury Verletzung
ink Tinte
ink duct Tintengang
ink sac/ink gland Tintensack,
Tintenbeutel, Tintendrüse
inky cap family/Coprinaceae
Tintlinge, Tintenpilze
inland landeinwärts
inland sea Binnenmeer (saltwater),
Binnensee (freshwater)
inland water/inland waterbody
Binnengewässer
inlay grafting Geißfußpfropfung,
Geißfußveredelung (Triangulation)
inlet Zulauf (Eintrittsstelle einer
Flüssigkeit)
innate angeboren; angewachsen,
im Inneren entstanden, endogen
innate behavior angeborenes
Verhalten
innate releasing mechanism (IRM)
ethol angeborener auslösender
Mechanismus (AAM)

inner cell mass innere Zellmasse
inner ear Innenohr
inner layer/interior layer
Innenschicht
innervate innervieren
innervation Innervation,
Innervierung
inoculate/vaccinate inokulieren,
einimpfen, impfen
inoculating loop Impföse
inoculating needle Impfnadel
inoculating wire Impfdraht
**inoculation/vaccination
(immunization)** Impfen, Impfung,
Einimpfung, Beimpfung,
Inokulation, Vakzination
(Immunisierung)
inoculum/vaccine Impfstoff,
Inokulum, Inokulat, Vakzine
inorganic anorganisch
inosine Inosin
**inosine monophosphate (IMP)/
inosinic acid/inosine 5-phosphate**
Inosinmonophosphat
inosine triphosphate (ITP)
Inosintriphosphat
**inosinic acid/inosine
monophosphate (IMP)/inosine
5-phosphate**
Inosinmonophosphat
inositol Inosit, Inositol
inotropic inotrop
input Eingabe, Einsatz, Einbringen
(eingebrachte/zugeführte Menge);
ecol Eintrag; *techn/electr*
Eingangsleistung; *Computer:*
Eingabe
- **output** *agr* Ertrag, Produktion;
ecol Austrag; *techn/electr*
Ausgangsleistung; *Computer:*
Ausgang, Ausgabe
inquiline (e. g. mussels/snails)
mitbewohnend
inquilinism Inquilinismus,
Einmietung, Synökie
insatiable unersättlich
insect Insekt, Kerf, Kerbtier
insect frass (debris from feeding)
Fraßmehl; (feces) Insektenfäkalien/
-kot in Bohrgängen und Minen
insect pollination/entomophily
Insektenbestäubung,
Insektenblütigkeit, Entomophilie
**insect-pollinated flower/
entomophile** Insektenblume,
Entomophile
insect-trap Insektenfalle
insectary Insektarium
insecticide Insektizid,
Insektenvernichtungsmittel
insectivore Insektivore,
Insektenfresser

177   intercistronic region

insectivorous insektivor, insektenfressend
insects Insekten, Kerfe, Kerbtiere
● **wingless insects/Apterygota** ungeflügelte Insekten
inseminate besamen, inseminieren
insemination Insemination, Inseminierung, Besamung, Samenübertragung
● **artificial i.** künstliche Besamung
● **heterologous i.** heterologe Insemination
● **homologous i.** homologe Insemination
insensible/unconscious unsensibel, unbewußt (gefühllos/nicht reagierend)
insensible (without awareness of the senses) unempfindlich, empfindungslos
inserted inseriert, eingefügt
insertion Einfügung
insertion mutation Insertionsmutation
insertion sequence Insertionssequenz
insertional activation Insertionsaktivierung
insertional inactivation Insertionsinaktivierung
inshore current auf die Küste zufliessende Strömung
inside-out patch Inside-out patch (Innenseite nach außen)
inside-out vesicle Inside-out Vesikel (Vesikel mit der Innenseite nach außen)
insidious/developing gradually (disease) *med* schleichend, langsam; tückisch
insolation Sonneneinstrahlung
insolubility Unlöslichkeit
insoluble unlöslich
insoluble in fat fettunlöslich
insoluble in water wasserunlöslich
inspection (on-site inspection) Begehung, Besichtigung (z. B. Geländebegehung)
inspiration/inhalation Einatmung, Einatmen, Inspiration, Inhalation
inspire/inhale inspirieren, einatmen
instar Entwicklungsstadium zwischen Häutungen bei Insekten, Erscheinungsform
instinct Instinkt
● **death instinct/aggressive instinct** Todestrieb
● **herd instinct/herding instinct** Herdentrieb, Herdeninstinkt
● **homing instinct** Heimkehrvermögen, Heimfindevermögen, Zielflug

● **interlocking (instinct)** Verschränkung (Instinkt)
● **sexual instinct/life instinct/eros** Sexualtrieb, Geschlechtstrieb
instinct behavior/instinctive behavior Instinktverhalten, Triebverhalten
instinct interlocking Instinktverschränkung
instinct-training-interlocking Instinkt-Dressur-Verschränkung
instinctive instinktiv
insufficiency/hypofunction Unterfunktion, Insuffizienz
insula (brain) Insel, Inselfeld
integral proteins (intrinsic proteins) integrale Proteine (intrinsische Proteine)
integrated pest management (IPM) integrierte Schädlingsbekämpfung, integrierter Pflanzenschutz
integument/covering (e. g. body covering, skin) Integument, Decke, Hülle (z. B. Körperdecke, Haut)
intensifying screens (autoradiography) Verstärkerfolien (Autoradiographie)
intensive farming/intensive agriculture Intensivwirtschaft, intensive (Land)wirtschaft
interaction Interaktion, Wechselwirkung
interaction variance Interaktionsvarianz
intercalary (inserted between others) intercalar, interkalar, eingeschoben
intercalary meristem *bot* interkalares Meristem, Restmeristem
intercalary vein *entom* Intercalarader, Interkalarader
intercalated disk (muscle/bivalves) Glanzstreifen, Kittlinie
intercalation/inclusion Interkalation, Einlagerung
intercalation agent/intercalating agent interkalierendes Agens
intercellular interzellulär
intercellular junction interzelluläre Verbindung, interzelluläre Junktion
intercellular space Interzellularraum, Interzellulare, Zwischenzellraum
intercentrum/hypocentrum Intercentrum, Zwischenwirbelkörper
interchange/interchromosomal rearrangement *gen* Austausch (zwischen Chromosomen), reziproke Translokation, interchromosomale Umordnung
intercistronic region/intergenic region intergene Region

intercostal field Interkostalfeld,
Zwischenrippenfeld
intercourse/sexual intercourse
Verkehr, Geschlechtsverkehr
intercropping/double cropping
Zwischenkultur
interdigital gland interdigitale
Drüse, Interdigitaldrüse,
Zwischenzehendrüse
interdisciplinary research
interdisziplinäre Forschung
interface Grenzfläche
interfascicular cambium
Zwischenbündelcambium
interference assay Interferenzassay
interference microscopy
Interferenzmikroskopie
interferon Interferon
intergenic region intergene Region
interglacial/interglacial period
Zwischeneiszeit
**intergrade/intermediary form/
transitory form/transient**
Zwischenstufe, Übergangsform
intergrafting/double working *hort*
Zwischenpfropfung,
Zwischenveredlung
interlocking (instinct)
Verschränkung (Instinkt)
intermediary intermediär,
dazwischenliegend
**intermediary form/transitory form/
transient/intergrade**
Übergangsform, Zwischenstufe
intermediary host Zwischenwirt
intermediary metabolism
intermediärer Stoffwechsel,
Zwischenstoffwechsel
intermediate *n* Zwischenprodukt
• **double-headed intermediate**
doppelköpfiges Zwischenprodukt,
janusköpfiges Zwischenprodukt
**intermediate density lipoprotein
(IDL)** Lipoprotein mittlerer Dichte
intermediate filament
intermediäres Filament,
Intermediärfilament
intermediate image *micros*
Zwischenbild
intermediate inheritance
intermediärer Erbgang, intermediäre
Vererbung
**intermediate product/intermediate
form** Zwischenprodukt,
Zwischenform
**intermediate state/intermediate
stage** Zwischenstadium,
Zwischenstufe
intermembrane space
Intermembranraum
internal/intrinsic innerlich,
von innen, intern

internal nostril/choana innere
Nasenöffnung, Choane
**International Unit (IU)/SI unit
(***fr:* **Système Internationale)**
Internationale Maßeinheit,
SI Einheit
**international unit system/SI unit
system (***fr:* **Système
Internationale)** internationales
Maßeinheitensystem,
SI Einheitensystem
interneuron Zwischenneuron,
Interneuron
internode Internodium,
Zwischenknoten
interparietal bone/os interparietale
Zwischenscheitelbein
interphase Interphase
**interpyramid muscle/comminator
muscle (Aristotle's lantern)**
Interpyramidalmuskel
interrelate in Zusammenhang
bringen, in Wechselbeziehung setzen
interrelation/interrelationship
Wechselbeziehung
interrupted mating unterbrochene
Paarung
intersex Intersex, Zwitter
intersexual zwischengeschlechtlich
intersexuality Intersexualität,
Zwischengeschlechtlichkeit
interspecific interspezifisch,
zwischenartlich
intersperse/disperse verstreut
(liegen)
interstitial interstitiell
interstitial cell Interstitialzelle,
Zwischenzelle
interstitial fauna (meiofauna)
Interstitialfauna, Sandlückenfauna
(Meiofauna)
interstitial fluid (ISF)/tissue fluid
Interstitialflüssigkeit, interstitielle
Flüssigkeit
interstitial region interstitielle
Region
**interstitial space/interstice
(***pl* **interstices)** Interstitialraum,
Interstitium, (Gewebs)Zwischenraum
intertidal flats Wattenmeer
**intertidal zone/tidal zone/littoral
zone/eulittoral zone** Tidebereich,
Gezeitenzone, Eulitoral
interval Intervall
interval scale *stat* Intervallskala
intervening sequence (IVS)/intron
intervenierende Sequenz,
dazwischenliegende Sequenz, Intron
**intervertebral disk/discus
intervertebralis**
Zwischenwirbelscheibe,
Bandscheibe

intestinal Darm..., Intestinal...
intestinal loop *embr* Darmbucht
intestine(s) Darm
- **foregut** *embr* Vorderdarm
- **hindgut** *embr* Hinterdarm
- **large intestines** Dickdarm
- **mid intestine/midgut/ mesenteron/ventricle/ ventriculus/"stomach"** Mitteldarm, Mesenteron, "Magen", Ventriculus (Insekten)
- **midgut** *embr* Mitteldarm
- **midgut/intestine** Nahrungsdarm
- **small intestines** Dünndarm
- **spiral intestine** Spiraldarm

intestines/entrails/innards/guts/ viscera (human: bowels/ intestines/guts) Gedärme, Eingeweide, Innereien, Viscera, Splancha
- **chitterlings/chitlins (hog intestines)** Gedärme, Gekröse (Schweine)

intima (innermost layer, esp. of blood vessels) Intima (innerste Schicht, bes. Blutgefäßwand)
intine (pollen/spore) Intine, Innenschicht
intrabody (intracellular antibody) Intrakörper (intrazelluläre Antikörper)
intracellular(ly) *adv/adj* intrazellulär
intrachange/intrachromosomal recombination *gen* intrachromosomale Umordnung
intrafusal fiber (muscle) Intrafusalfaser
intramembrane particle/membrane intercalated particle Intramembran-Partikel
intrasexual innerhalb des gleichen Geschlechts
intraspecific intraspezifisch, innerartlich
intrinsic/intrinsical intrinsisch
intrinsic factor/hemopoietic factor Intrinsic-Faktor, hämopoetischer Faktor
introduce/import einführen, importieren
introduced/imported (allochthonous) eingeführt
introgression Introgression
intron/intervening sequence Intron, intervenierende Sequenz, dazwischenliegende Sequenz
introrse (anthers) *bot* intrors, einwärts gewendet
introvert/turn inward *vb* einstülpen, nach innen richten
introvert (proboscis: Priapula) *n* Introvert

intrude eindringen
intruder Eindringling
intussusception/introsusception/ invagination *med* Intussuszeption, Invagination
intussusception (cell wall growth) Intussuszeption, Einlagerung (Zellwandwachstum)
invagination/emboly Invagination, Einstülpung, Einfaltung, Embolie
invariant residue *math* unveränderter Rest, invarianter Rest
invasion Invasion, Eindringung
invasive invasiv, in die Umgebung hineinwachsend
invasiveness Invasivität
inventive erfinderisch
inventory Inventar; Bestandsaufnahme
- **to make an inventory** eine Bestandsaufnahme/-liste machen

inversion Inversion, Umkehrung
- **chromosome inversion** *gen* chromosomale Inversion
- **paracentric inversion** *gen* parazentrische Inversion
- **pericentric inversion** *gen* perizentrische Inversion
- **temperature inversion** *meteo* Temperaturinversion

inversion mutation Inversionsmutation
invert umkehren, auf den Kopf stellen
invert sugar Invertzucker
invertebrates Invertebraten, Evertebraten, Wirbellose, Wirbellose Tiere
inverted invers, umgekehrt
inverted repeat/inverted repetition/palindrome *gen* invertierte Sequenzwiederholung, gegenläufige Sequenzwiederholung, umgekehrte Sequenzwiederholung, Palindrom
inverted terminal repetitions/ inverted terminal repeats (ITR) *gen* umgekehrte terminale Repetitionen
investigate/examine/test/try/assay/ analyze prüfen, untersuchen, testen, probieren, analysieren
investigation/examination (exam)/ test/trial/assay/analysis Untersuchung, Prüfung, Test, Probe, Analyse
involucral bracts/phyllary *bot* Involukralschuppe, Involukralblätter
involucre/envelope Involukrum, Hülle

I involucre 180

involucre *bot* **(whorl of bracts at base of inflorescence)** Hüllkelch, Hüllblattkreis
involuntary musculature unwillkürliche Muskulatur
involute/rolled inward involutiv, nach oben eingerollt
involution *bot* Involution, Einrollung (Blatt)
iodination Jodierung (mit Jod reagieren/substituieren)
iodine Jod
iodine number/iodine value Jodzahl
iodization Jodierung (mit Jod/Jodsalzen versehen)
iodize jodieren (mit Jod/Jodsalzen versehen)
iodoacetic acid Jodessigsäure
ion channel (membrane channel) Ionenkanal (Membrankanal)
ion equilibrium/ionic steady state Ionengleichgewicht
ion exchange Ionenaustausch
ion exchanger Ionenaustauscher
ion pair Ionenpaar
ion pore Ionenpore
ion product Ionenprodukt
ion pump Ionenpumpe
ion transport Ionentransport
ion-exchange resin Ionenaustauscherharz
ionic ionisch
ionic bond Ionenbindung
ionic conductivity Ionenleitfähigkeit
ionic coupling Ionenkopplung
ionic current Ionenstrom
ionic radius Ionenradius
ionic strength Ionenstärke
ionization Ionisation
ionize ionisieren
ionizing radiation ionisierende Strahlen, ionisierende Strahlung
ionophore Ionophor
ionophoresis Ionophorese, Iontophorese
IPM (integrated pest management) integrierte Schädlingsbekämpfung, integrierter Pflanzenschutz
Iridaceae/iris family Schwertliliengewächse
iridescent schillernd
iridocyte/iridophore/leucophore/guanophore Iridocyt, Iridozyt, Flitterzelle, Leucophor, Guanophor
iris *zool* Iris, Regenbogenhaut
iris diaphragm *opt* Irisblende
iris family/Iridaceae Schwertliliengewächse
IRM (innate releasing mechanism) *ethol* angeborener auslösender Mechanismus (AAM)

iron Eisen
iron-regulating factor (IRF) eisenregulierender Faktor
iron-sulfur protein Eisen-Schwefel-Protein
ironpan/hardpan/ortstein Eisenstein, Ortstein
ironwood Eisenhölzer
IRP (island specific PCR) IRP (inselspezifische PCR)
irradiance/fluence rate/radiation intensity/radiant-flux density Bestrahlungsintensität, Bestrahlungsdichte
irradiate bestrahlen
irradiation Bestrahlung
irregular/non-uniform ungleichmäßig, unregelmäßig
irregular/zygomorphic/bilaterally symmetrical/monosymmetrical unregelmäßig, zygomorph, bilateralsymmetrisch, monosymmetrisch
irregular cleavage unregelmäßige Furchung
irregular grain (wood) Streuungstextur
irregularity Unregelmäßigkeit
irregularly helical (like a snail shell)/cochleate schraubenartig gewunden, schneckenhausartig gewunden, cochlear
irreversible inhibition irreversible Hemmung
irrigate bewässern, berieseln
irrigated crop Bewässerungskultur
irrigation *agr* Bewässerung; Beregnung, Berieselung; *med* Ausspülung
● **furrow irrigation** Grabenbewässerung, Furchenbewässerung
● **overhead irrigation** Beregnung von oben
● **surge irrigation** Schwallbewässerung
irrigation ditch Bewässerungsgraben
irritability/excitability/sensitivity Reizbarkeit, Erregbarkeit
irritable/excitable/sensitive reizempfänglich, reizbar
irritable/sensible empfindlich (reizempfänglich)
irritate *med/physio/chem* reizen, irritieren
irritation/stimulation Irritation, Reizung, Stimulation
irritation/stimulus Reiz, Stimulus
isabelline isabellfarben (gelb-olivbraun)
ischemia Ischämie

ischial callosity/sitting pad/anal callosity Gesäßschwiele, Sitzschwiele, Analkallosität
ischial tuber/ischial tuberosity/ tuber ischiadicum Sitzbeinhöcker
ischiopodite Ischiopodit
ischium/os ischii Sitzbein, Gesäßbein, Sitzknochen
isidium Isidie
island Insel
• **blood island** Blutinsel
isle/island (esp. islet) Insel (*bes.* kleine Insel)
island biogeography Inselbiographie, Inselbiografie
islet/a little island Inselchen, kleine Insel; Zellinsel
islet of Langerhans/pancreatic islet Langerhanssche Insel, Pankreasinsel, Inselorgan
islet organ Inselorgan
isoacceptors Isoakzeptoren
isobar Isobare (*pl* Isobaren)
isocoenosis (*pl* isocoenoses) Isozönose
isodicentric chromosome isodizentrisches Chromosom
isoelectric focusing isoelektrische Fokussierung, Isoelektrofokussierung
isoelectric point isoelektrischer Punkt
Isoetaceae/quillwort family Brachsenkrautgewächse
isogamous isogam
isogamy Isogamie
isogenous/genetically identical isogen, genetisch identisch
isolate isolieren; absondern, abtrennen
isolation Isolation; Absonderung, Abtrennen
• **ecological isolation** ökologische Isolation
• **ethological isolation/behavioral isolation** ethologische Isolation
• **prezygotic isolation** präzygotische Isolation
• **postzygotic isolation** postzygotische Isolation
• **reproductive isolation** reproduktive Isolation
• **seasonal isolation/temporal isolation** saisonale Isolation
• **spatial isolation** räumliche Isolation
isolation medium Isolationsmedium

isolecithal isolecithal, isolezithal (mit gleichmäßig verteiltem Dotter)
isoleucine Isoleucin
isomer Isomer
isomeric isomer
isomerism/isomery Isomerie
isomerization Isomerisation
isomerize isomerisieren
isomerous isomer, gleichzählig
isophene Isophän *nt*
isopods/Isopoda (sea slaters & rock lice & pill bugs etc.) Isopoden, Asseln
isoprene Isopren
isopycnic centrifugation isopycnische Zentrifugation, isopyknische Zentrifugation
isosmotic/iso-osmotic iso(o)smotisch
isotachophoresis Isotachophorese
isotelic isotel, die gleiche Wirkung erzielend
isotomy *bot/zool* Isotomie, Gabelung in gleiche Achsen
isotonic isotonisch
isotonicity Isotonie
isotope Isotop
isotope assay Isotopenversuch
isotropic/isotropous isotrop, einfachbrechend
isotype Isotyp, Isotypus, Isostandard
isotype switching Isotypwechsel, Klassenwechsel
isovaleric acid Isovaleriansäure
isozyme/isoenzyme Isozym, Isoenzym
isthmus (of oviduct) Isthmus, Eileiterenge
iterative evolution iterative Evolution, sich wiederholende Evolution
iteroparity Iteroparitie
iteroparous iteropar
ITR (inverted terminal repetitions/ inverted terminal repeats) *gen* umgekehrte terminale Repetitionen (ITR)
IVF (in-vitro fertilization) In-vitro-Fertilisation, Reagenzglasbefruchtung
ivory Elfenbein
IVS (intervening sequence)/intron *gen* IVS (intervenierende Sequenz/ dazwischenliegende Sequenz), Intron
ivy family/ginseng family/ Araliaceae Efeugewächse

J jack

182

jack/jackass (♂ uncastrated ass/
donkey) männlicher Esel
jacket cell *bot* (mosses)
Mantelzelle
jacket cell *zool* (*Dicyemida*)
Hüllzelle
Jacobson's organ/vomeronasal
organ Jacobsonsches Organ,
vomeronasales Organ
jamming avoidance reaction
elektrische Meidereaktion
japygids/diplurans/Diplura
Doppelschwänze
jasmonic acid Jasmonsäure
jaw Kiefer, Maul; Schlund, Kehle,
Rachen, Mundöffnung
• lower jaw (horses) Ganasche
• lower jaw/lower jawbone/
submaxilla/submaxillary bone/
mandible Unterkiefer,
Unterkieferknochen, Mandibel
• pharyngeal jaw(s)/trophi
(rotifers) Kiefer, Trophi
jaw/beak (squid/cuttlefish) Kiefer,
Schnabel
jawbone Kieferknochen
jawed vertebrates/jaw-mouthed
animals/gnathostomatans/
Gnathostomata Kiefermünder
jawless fishes/agnathans/*Agnatha*
Kieferlose, Agnathen
jejunum Jejunum, Leerdarm
jelly Gelee, Gallerte
jelly fungi/Tremellales Zitterpilze,
Gallertpilze
jellyfishes Quallen
jennet/jenny (♀ ass/donkey)
Eselin
jerk Zuckreflex, ruckartige
Bewegung
jet loop reactor (bioreactor)
Düsenumlaufreaktor, Strahl-
Schlaufenreaktor
jet reactor (bioreactor)
Strahlreaktor
jewelweed family/balsam family/
touch-me-not family/
Balsaminaceae
Balsaminengewächse,
Springkrautgewächse
jimmy (*pl* jimmies) (adultblue crab)
männliche Blaukrabbe (*see also:*
sook)
jipijapa family/panama-hat family/
cyclanthus family/Cyclanthaceae
Scheinpalmen
Joe-wood family/Theophrastaceae
Theophrastaceen
joey (baby kangaroo) *Australian*
junges Känguruh
Johnston's organ Johnstonsches
Organ

joint/articulation/hinge Gelenk,
Verbindung, Angelpunkt
• amphiarthrodial joint/
amphiarthrosis Wackelgelenk,
straffes Gelenk
• ankle joint Fußgelenk
• ball-and-socket joint/spheroid
joint/spheroidal joint/
enarthrodial articulation/
enarthrosis Kugelgelenk,
Nussgelenk, Enarthrose,
Articulatio cotylica
• condylar joint/articulatio
bicondylaris Walzengelenk
• diarthrodial joint/diarthrosis/
synovial joint Diarthrose,
Gelenk, Articulatio
• ellipsoidal joint Ellipsoidgelenk,
Eigelenk
• ginglymus joint/hinge joint
Scharniergelenk
• gliding joint/plane joint/
arthrodial joint (arthrodia)
Gleitgelenk, ebenes Gelenk,
Arthrodialgelenk
• hinge joint/ginglymus joint
Scharniergelenk
• hip joint/coxal joint
Hüftgelenk
• knee joint Kniegelenk
• pastern joint Fesselbeingelenk,
Krongelenk
• pivot joint/trochoid(al) joint/
rotary joint Zapfengelenk
• saddle joint/sellaris joint
Sattelgelenk
• stifle joint (horse) Kniegelenk
• synarthrodial joint/synarthrosis
Synarthrose, Fuge, Haft
• thurl (hip joint in cattle)
Hüftgelenk
• trochoid(al) joint/pivot joint/
rotary joint Zapfengelenk
• wrist/wrist joint Handgelenk
joint capsule/articular capsule
Gelenkkapsel
joint cavity Gelenkspalt
joint-fir family/Gnetaceae
Gnetumgewächse,
Gnemonbaumgewächse
joint-pine family/mormon tea
family/ephedra family/
Ephedraceae Meerträubelgewächse
jointed/articulate/articulately
jointed gelenkig, gelenkartig
verbunden
Jordan's organ/chaetosoma/
chaetosema Jordansches Organ,
Chaetosoma, Chaetosema
jostling run (bees) Drängeln
jowl (a cut of fish: head and
adjacent parts) *ichth* Kopfstück

jowl (wattle/dewlap/pendulous part of double chin) Kehllappen, Kinnlappen (Wamme/Doppelkinnfalte)

jowl/cheek Backe; (cheek meat of a hog) Backenfleisch

jowl/jaw/mandible Kiefer; Unterkiefer

juba/loose panicle/panicle of grasses (lockere) Grasrispe

jugal (bone) Jugale, Jochbein

jugal area/jugal region/jugum/neala Jugalfeld, Jugum, Neala

jugal cell Jugalzelle

jugal fold Jugalfalte

jugal vein Jugalader

Juglandaceae/walnut family Walnussgewächse

jugular vein/vena jugularis Drosselvene

jump/spring/bound/leap springen

jumping gene springendes Gen

Juncaceae/rush family Simsengewächse

juncaceous/rushy/rushlike binsenartig

junciform/rush-shaped binsenförmig

junction zone *cyt* Junktionszone
● **dermo-epidermal j.z.** dermo-epidermale Junktionszone

jungle Dschungel

jungle book Dschungelbuch

junk DNA unnütze DNA, überflüssige DNA, wertlose DNA

Jurassic Period/Jurassic (geological time) Jurazeit, Jura

juvenile jugendlich, Jugend…

juvenile form Jugendform

juvenile hormone (JH) Juvenilhormon

juvenile plant/young plant Jungpflanze

juvenile song/subsong *orn* Jugendgesang, Dichten (Jungvögel)

juvenile stage/juvenile phase/adolescence Jugendstadium, Jugendphase, Jugendzeit

juvenility Jugendlichkeit, Jugend

K selection

K selection K-Selektion
**K strategist/K-selected species
(slow development)** K-Stratege
kairomone Kairomon
**kala azar/Cala-Azar (*Leishmania
donovani*)** Kala-Azar, Cala-Azar,
schwarzes Fieber, viszerale
Leishmaniasis, Leishmaniose
kallikrein Kallikrein
**kame (short ridge/mount at glacial
front formed by meltwater)**
Kame (fluvioglazialer Sand-/
Kieshügel an Gletscherfront)
kamptozoans/Entoprocta
Kelchwürmer, Nicktiere,
Kamptozooen
**kangaroo paw family/bloodwort
family/redroot family/
haemodorum family/
Haemodoraceae**
Haemodorumgewächse
**kapok-tree family/silk-cotton tree
family/cotton-tree family/
Bombacaceae** Wollbaumgewächse
karst lake Karstsee
karyogamy/nuclear fusion
Karyogamie, Kernvereinigung,
Kernverschmelzung
karyogram/karyotype Karyogramm,
Karyotyp
karyoplasm/nucleoplasm
Kernplasma, Karyoplasma,
Nucleoplasma
karyotype Karyotyp
karyotyping Bestimmung des
Karyotyps, Karyotypanalyse
katabatic wind Hangabwind
**katsura-tree family/
Cercidiphyllaceae**
Katsuragewächse
keel/carina Kiel
keel *bot* Kiel, Schiffchen
keeled/having a keel/carinate
gekielt
kelp/brown seaweed (Laminariales)
Brauntang
**kennel (commercial caretaking of
cats/dogs)** öffentlich-kommerzielle
Tierpension
  ● **dog kennel** Hundepension,
Hundeheim
  ● **pack of dogs** Hundemeute
  ● **shelter/container for dogs**
Hundezwinger
keratin Keratin
keratin filament Keratinfilament
keratinization/cornification
Keratinisierung, Verhornung
keratinized/cornified/horny
keratinisiert, verhornt
keratinocyte Keratinozyt,
Keratinocyt

**keratinosome/Odland body/
lamellar body** Keratinosom
kernel/corn/grain Korn
kernel/seed Kern
keto acid Ketosäure
ketoaldehyde/aldehyde ketone
Ketoaldehyd
ketone Keton
ketone body (acetone body)
Ketonkörper
ketonuria/acetonuria Ketonurie
kettle *geol* Kessel, Gletschertopf,
Gletschermühle
kettle lake Muldensee
key (for identification)
Bestimmungsschlüssel
key bed/marker bed *geol/paleo*
Leithorizont
key enzyme Schlüsselenzym,
Leitenzym
key evolutionary innovation (KEI)
evolutionäre Schlüsselinnovation
**key stimulus/sign stimulus (release
stimulus)** Schlüsselreiz,
Auslösereiz
key substance Schlüsselsubstanz
keystone predator Schlüsselräuber
keystone species Schlüsselart
kidding (parturition in goats)
Zicklein gebären
kidney Niere
  ● **holonephros/archinephros**
Holonephros, Archinephros
  ● **mesonephros/middle kidney/
midkidney** Mesonephros,
Urniere, Wolffscher Körper
  ● **metanephros/hind kidney/
definitive kidney** Metanephros,
Nachniere, definitive Niere
  ● **multilobular kidney/
multipyramidal kidney/
polypyramidal kidney**
multipyramidale Niere,
mehrwarzige Niere,
zusammengesetzte Niere, gelappte
Niere
  ● **opisthonephros** Opisthonephros,
Rumpfniere
  ● **pronephros/fore-kidney/
primitive kidney/primordial
kidney/head kidney** Pronephros,
Vorniere, Kopfniere
  ● **unilobular kidney/unipyramidal
kidney/monopyramidal kidney**
unipyramidale Niere, einwarzige
Niere
kidney-shaped/reniform
nierenförmig
kieselguhr (loose/porous diatomite)
Kieselgur
killer cell/K cell Killer-Zelle,
Killerzelle

killifishes/Cyprinodontiformes
Kleinkärpflinge
kiln/kiln oven Darre, Darrofen
kiln-dry (grain/lumber/tobacco)
darren
kilosequencing *gen*
Kilosequenzierung
kin Sippe, Geschlecht,
Verwandtschaft, Familie
kin selection Verwandtenselektion
kind/species Art, Spezies
kindle *n* **(young rabbit)** Häschen,
frischgeborenes Häschen/Kaninchen
kindle *vb* **(bear young rabbits)**
gebären (speziell bei Hasen/
Kaninchen)
kinematic viscosity kinematische
Viskosität
kinesis Kinese
kinetic complexity kinetische
Komplexität
kinetics Kinetik
 • **first-order kinetics**
Kinetik erster Ordnung
 • **nonsaturation kinetics**
Nichtsättigungskinetik
 • **reaction kinetics**
Reaktionskinetik
 • **reassociation kinetics**
Reassoziationskinetik
 • **saturation kinetics**
Sättigungskinetik
 • **second-order kinetics** Kinetik
zweiter Ordnung
 • **zero-order kinetics** Kinetik
nullter Ordnung
kinetin/zeatin Kinetin
kinetlum/kinety Kinet
kinetochore Kinetochor
kinetocyst Kinetocyst
kinetoplast Kinetoplast
kinetosome/basal body Kinetosom,
Basalkörper
**kingfishers & bee-eaters & hoopoes
& rollers & hornbills/
Coraciiformes** Rackenvögel
kinorhynchs/Kinorhyncha
Hakenrüssler
kinship Verwandtschaft,
Blutsverwandtschaft
kinship selection
Verwandtschaftsselektion
kinship theory
Verwandtschaftstheorie
kit (young fur-bearing animal)
Junges (bes. von Felltieren)

kitten (young cat) Kätzchen,
junge Katze, Katzenjunges
kiwis/Apterygiformes Kiwis
kleptoparasite/cleptoparasite
Kleptoparasit
knee Knie
 • **bend of the knee/back of the
knee/hollow behind knee/
poples/popliteal fossa/popliteal
space** Kniekehle
knee/knee-root *bot* Knie,
Wurzelknie
knee joint Kniegelenk
kneecap/knee bone/patella
Kniescheibe, Patella
knockout mutation
Knockout-Mutation
knoll/hummock (rounded knoll)
kleiner Hügel
knoll/mound/bulge/hump Buckel,
Erhebung
knopper gall Knoppergalle
knot *bot* Knoten, Astknoten; Auge,
Knospe (Holz)
**knotweed family/dock family/
buckwheat family/smartweed
family/Polygonaceae**
Knöterichgewächse
knuckle (hand) Knöchel,
Handknöchel
knuckle-walking Knöchelgang
(Fußknöchel)
Koch's postulate Koch's Postulat,
Koch'sches Postulat
Koehler illumination
Köhlersche Beleuchtung
kojic acid Kojisäure
**krameria family/ratany family/
Krameriaceae**
Krameriagewächse
Krause's end bulb/bouton
Krause'sches Körperchen
**Krebs cycle/citric acid cycle/
tricarboxylic acid cycle (TCA cycle)**
Krebs-Cyclus, Citrat-Zyklus,
Citratcyclus, Zitronensäurezyklus,
Tricarbonsäure-Zyklus
krill and allies/Euphausiacea Krill,
Leuchtkrebse
**Kupffer cell/stellate
reticuloendothelial cell**
Kupffer-Zelle, Kupffer-Sternzelle
kurtosis *stat* Häufungsgrad,
Häufigkeitsgrad; Wölbung,
Koeffizient der Wölbung, Exzess,
Steilheit

L lab

lab/laboratory Labor
lab ferment/rennin/chymosin
Labferment, Rennin, Chymosin
lab grade *chem* **(quality
designation)** technisch
(Qualitätsbezeichnung)
label *n* Markierung, Etikett
label *vb* markieren, ein Etikett
aufkleben
labial gland Labialdrüse
**labial palp/labipalp/labial feeler/
palp/palpus/palpus labialis**
Labialpalpus, Labialtaster,
Lippentaster, Taster, Tastfühler,
Palpe
labial suture Labialnaht
**Labiatae/mint family/deadnettle
family/Lamiaceae**
Lippenblütengewächse,
Lippenblütler
labidognathous labidognath
labium/lower lip (vertebrates)
Labium, Unterlippe
labium/second maxilla (insects)
Labium, Unterlippe, 2. Maxille
labium (folds at margin of vulva)
Labium vulvae, Schamlippe
labor/uterine contractions *n*
Wehen (in den Wehen liegen)
labor *vb* in den Wehen liegen,
Wehen haben
laboratory/lab Labor
laboratory apron Laborschürze
laboratory balance Laborwaage
laboratory coat/labcoat
Laborkittel
**laboratory equipment/lab
equipment** Laborgerät, -ausrüstung
laboratory facilities/lab facilities
Laboreinrichtung, Laborausstattung
laboratory findings Laborbefund
laboratory jack
höhenverstellbarer Tisch
laboratory safety Laborsicherheit
laboratory scale/lab scale
Labormaßstab
**laboratory table/lab table/
laboratory bench/lab bench/lab
table** Labortisch, Labor-Werkbank
**laboratory technician/lab
technician/technical lab assistant**
technische(r) Assistent(in),
Laborassistent(in), Laborant(in)
labriform soralium (lichens)
Lippensoral
labrum/upper lip Labrum,
Oberlippe
**labware/laboratory supplies/
lab supplies** Laborbedarf
**labyrinthulids/slime nets/
Labyrinthulomycetes**
Netzschleimpilze

**labyrithine placenta/
hemoendothelial placenta**
Labyrinthplazenta
***lac* operon** *lac*-Operon
lacerate/torn ungleichmäßig
zerschlitzt/geschlitzt, ungleichmäßig
eingeschnitten
laceration Laceration, Riss,
Zerreißung; *med* Fleischwunde,
Risswunde
lachrymal/lacrimal tränenartig,
Tränen...
lacinia/inner lobe of maxilla
Lacinia, Innenlade
laciniate/slashed gefranst, geschlitzt
lacrimal/lachrymal tränenartig,
Tränen...
lacrimal bone/os lacrimale
Tränenbein
lacrimal duct/tear duct Tränengang,
Tränenkanal
lacrimal gland Tränendrüse
lacrimal lake/lacus lacrimalis
Tränensee
lactamide Laktamid, Lactamid,
Milchsäureamid
lactate (lactic acid) Laktat
(Milchsäure)
lactate *vb* laktieren, Milch geben/
produzieren/absondern
lactation Laktation,
Milchabsonderung (aus
Milchdrüsen)
lactic acid (lactate) Milchsäure
(Laktat)
**lactic acid fermentation/lactic
fermentation** Milchsäuregärung
lactifer/laticifer *bot* Milchröhre,
Milchsaftröhre
lactiferous milchführend
lactiferous duct/milk duct
Milchgang
lactiferous sinus/milk cistern
Milchsinus, Milchzisterne
lactose (milk sugar) Laktose,
Lactose (Milchzucker)
lactose repressor (*lac* repressor)
Lactose-Repressor (*lac*-Repressor)
lacuna/space/cavity Lakune, Spalt,
Hohlraum
lacunar system Lakunensystem
lacustrine Seen betreffend
(Binnenseen), an/in Seen wachsend
oder lebend, See...
lacustrine plant Seepflanze
ladder-shaped/scalariform
leiterförmig
**ladder-type nerve system/double-
chain nerve system**
Strickleiternervensystem
lady fern family/Athyriaceae
Frauenfarngewächse

lag *vb* zurückbleiben, nachhinken

lag phase/latent phase/incubation phase/establishment phase Anlaufphase, Latenzphase, Inkubationsphase, Verzögerungsphase, Adaptationsphase, lag-Phase

lagg/fen water trough (drainage channel within a raised bog) Lagg (Entwässerungsgraben im Hochmoor)

lagging nachhängend, zurückbleibend

lagging strand *gen* Folgestrang

lagoon Lagune

lair (resting/living place: game/wild animal) Lager, Rastplatz

lake See
 • **freshwater lake** Süßwassersee
 • **saline lake/salt lake** Salzsee

lake zonation/lacustrine zonation Seenzonierung (Gewässerzonierung)

lakeshore/shore of a lake/ banks of a lake Seeufer

lamb/little sheep Lamm, Schäfchen

lambdoid suture (skull) Lambdanaht

lamb/lambing *vb* Lämmer gebären

lame *adj/adv* lahm

lame *vb* lahmen

lamella Lamelle

lamellar gill/sheet gill/ lamellibranch/eulamellibranch Blattkieme, Lamellibranchie, Eulamellibranchie

lamellate lamellenartig, blattartig

lamellate antenna lamellenartiger Fühler, blätterförmiger Fühler

lamellated corpuscle/Pacinian body/Pacinian corpuscle Lamellenkörperchen, Endkörperchen, Pacinisches Körperchen, Pacini Körperchen

lameness Lähme

Lamiaceae/mint family/deadnettle family/Labiatae Lippenblütengewächse, Lippenblütler

lamina/lamella/blade (thin layer) Lamina, Lamelle (Platte/Spreite/ Blatt)
 • **basal lamina/basement membrane** Basallamina, Basalmembran
 • **blade/frond/phyllid (algas/ mosses)** Algenspreite, Moosblättchen, Phylloid
 • **dental lamina** Zahnleiste
 • **leaf lamina/leaf blade** Blattspreite
 • **nuclear lamina** Kernfaserschicht, Kernlamina

laminar/laminiform/laminous laminal, spreitig, spreitenförmig, blättrig, plättchenartig geschichtet

laminar bone lamellärer Knochen, Lamellenknochen

laminar flow laminare Strömung, Schichtströmung

laminar flow workstation/laminar flow hood/laminar flow unit Laminarstrombank

laminate placentation/ laminar placentation/ lamellate placentation laminale Plazentation, flächenständige Plazentation

laminated/layered laminiert, geschichtet

laminiform plattenförmig, plättchenartig

lammas shoot Johannistrieb

lampbrush chromosome Lampenbürstenchromosom

lampreys/Petromyzontida Neunaugen, Neunaugenartige

lampshells/brachiopods/ Brachiopoda Armfüßer, Brachiopoden ("Lampenmuscheln")

lanate/wooly wollig

lancelet/cephalochordates (Amphioxiformes) Lanzettfischchen, Cephalochordaten

lanceolate lanzettförmig, lanzettlich

land life/life on land/terrestrial life Landleben

land snails/Stylommatophora (Pulmonata) Landlungenschnecken

landfill/sanitary landfill Mülldeponie, Müllgrube (geordnet)

landscape/countryside Landschaft

landscape architect Landschaftsplaner, Landschaftsarchitekt

landscape ecology Landschaftsökologie

landscape planning Landschaftsplanung

Lang's vesicle Bursalorgan

language Sprache
 • **animal language** Tiersprache
 • **body language** Körpersprache
 • **dance language (bees)** Tanzsprache
 • **symbol language** Symbolsprache

lanosterol Lanosterin, Lanosterol

lanternfishes & blackchins/ Myctophiformes Laternenfische

lanthionine Lanthionin

lapinized vaccine durch Hasen erzeugter Impfstoff

lappet Hautlappen, Fleischlappen

lard Schmalz (Schweineschmalz), Schweinefett

lardizabala family

lardizabala family/akebia family/ Lardizabalaceae Fingerfruchtgewächse
lardon Speckstreifen
large intestines Dickdarm
larva (pl larvas/larvae) Larve
- **ancestrula** Ancestrula
- **auricularia larva** Auricularia
- **bladderworm/cysticercus (tapeworm larva)** Blasenwurm, Finne, Cysticercus
- **cercaria** Cercarie, Zerkarie, Schwanzlarve
- **cheese skipper/cheese maggot (Piophilidae larva)** Käsefliegenlarve
- **chigger/"red bug"/harvest mite (parasitic red mite larva)** Erntemilbenlarve, parasitische Larve von Trombidiidae
- **ciliated larva** Wimperlarve
- **coarctate larva/larva coarctata/ pseudocrysalis** Scheinpuppe, Pseudocrysalis
- **coracidium (cestoda larva)** Coracidium-Larve, Korazidium (eine Schwimmlarve)
- **cutworm** Raupe bestimmter Eulenfalter
- **cydippid larva** Cydippe-Larve
- **cypris larva** Cypris-Larve
- **dauer larva (temporarily dormant larva)** Dauerlarve
- **doliolaria larva (vitellaria larva) (Crinoida)** Doliolaria-Larve
- **doodlebug (antlion larva)** Ameisenlöwe (Larve der Ameisenjungfer)
- **eruciform larva (with more than 5 pairs of abdominal prolegs: Tenthredinidae)** Afterraupe (Blattwespenlarven)
- **gastrula** Gastrula, Becherkeim, Becherlarve
- **glochidium** Glochidium
- **grub (thick wormlike larva/ scarabeiform larva: Coleoptera/ Hymenoptera/certain Diptera)** Engerling (im Boden lebende Käferlarve), Bienenlarve u. a. (*see:* maggot)
- **hellgrammite/dobson (dobsonfly larva: esp. *Corydalis cornutus*)** Schlammfliegenlarve (Megaloptera)
- **hexacanth larva/hooked larva/ oncosphere (cestodes)** Sechshakenlarve, Oncosphaera-Larve, Onkosphäre
- **hornworm (hawk moth/sphinx moth larva)** Schwärmerlarve, Schwärmerraupe (mit Afterhorn)
- **inchworm/measuring worm (geometer moth larva)** Spannerraupe
- **infusiform larva** infusiforme Larve
- **lasidium** Lasidium
- **maggot (apodal larva)** Made (apode Larve)
- **miracidium (fluke larva)** Miracidium (*pl* Miracidien), Mirazidium (Digenea-Larve)
- **mitraria larva (a metatrochophore)** Mitraria, Mitraria-Larve
- **Mueller's larva** Müllersche Larve
- **mysis larva** Mysis-Larve
- **naupliar larva/nauplius** Naupliuslarve, Nauplius
- **oncosphere/hexacanth larva/ hooked larva (cestodes)** Oncosphaera-Larve, Onkosphäre, Sechshakenlarve
- **pericalymma/test-cell larva** Pericalymma, Hüllglockenlarve
- **pilidium larva (nemertines)** Pilidium, Pilidium-Larve
- **pluteus larva** Pluteuslarve, Pluteus
- **primary larva** Primärlarve, Junglarve, Eilarve
- **procercoid (cestodes)** Procercoid, Prozerkoid (Cestoda-Postlarve)
- **redia (flukes)** Redie
- **rotiger/pseudotrochophore (larva)** Rotiger, Pseudotrochophora
- **sacfry/yolk fry/alevin (salmonid larvas with yolk sac)** Dottersackbrut (Lachs)
- **silkworm (silkmoth larva)** Seidenraupe
- **tornaria larva** Tornaria-Larve
- **toxophore (butterfly larva)** Toxophorium (Schmetterlingsraupen: Brenn- und Gifthaare)
- **trochophore larva** Trochophora-Larve, Wimperkranzlarve
- **veliger larva** Veligerlarve, Segellarve
- **vitellaria larva/yolk larva** Vitellaria-Larve, Vitellaria
- **wiggler (mosquito larva)** Schnakenlarve
- **wireworm (elaterid larva)** Drahtwurm
- **yolk fry/sacfry/alevin (salmon larvas)** Dottersackbrut
- **yolk larva/vitellaria/lecithotroph larva** Dotterlarve, Vitellaria-Larve
- **zoëa (decapod crustacean larva)** Zoëa

larval proleg/false leg Abdominalbein, Bauchfuß, Propes, Pes spurius (larval)
larviform larvenförmig
larviparous larvipar
larvipary Larviparie
laryngeal laryngeal, Larynx.., Kehlkopf…
laryngeal aditus Kehlritze
laryngeal prominence/Adam's apple Adamsapfel
larynx (Adam's apple) Larynx, Kehlkopf (Adamsapfel)
lasidium (larva) Lasidium
late gene spätes Gen
latency Latenz
latency period Latenzzeit
latent bud Proventivknospe, Ersatzknospe
latent egg Dauerei
latent period Latenzzeit
latent phase/incubation phase/ establishment phase/lag phase Latenzphase, Adaptationsphase, Anlaufphase, Inkubationsphase, lag-Phase
latent shoot Ersatztrieb, Stresstrieb, Proventivtrieb
lateral lateral, seitlich; seitenwendig
lateral axis/lateral branch Seitenachse
lateral branch/offshoot Seitenast
lateral bud/axillary bud Seitenknospe, Achselknospe
lateral cerebral sulcus/fissura lateralis cerebri Sylvische Furche
lateral eye/lateral ocellus/stamma Lateralauge, Seitenauge, Lateralocelle, Stemma
lateral heart Lateralherz
lateral line system/lateralis system/ acoustico-lateralis system Seitenliniensystem
lateral magnification Lateralvergrößerung, Seitenverhältnis, Seitenmaßstab, Abbildungsmaßstab
lateral moraine Seitenmoräne
lateral pore Lateralpore
lateral root Nebenwurzel, Seitenwurzel
lateral shoot/side shoot/offshoot Seitentrieb
lateral undulation/lateral undulatory movement (snakes) seitliche/horizontale Wellenbewegung, Schlängelbewegung, Schlängeln
lateralis organ Seitenlinienorgan, Lateralisorgan
laterite (soil) Laterit

laterization/latosolization (soil) Laterisation, Lateritisierung, Lateritbildung
latewood Spätholz, Engholz
latex Latex, Milchsaft
latex tube Milchröhre
lath/plank Latte
laticifer/lactifer *bot* Milchröhre, Milchsaftröhre
latitude Breitengrad
lattice sampling/grid sampling *stat* Gitterstichprobe(nverfahren)
laurel family/Lauraceae Lorbeergewächse
Laurer's canal (vestigial copulatory canal) Laurerscher Kanal
lauric acid/decylacetic acid/ dodecanoic acid (laurate, dodecanate) Laurinsäure, Dodecansäure (Laurat, Dodecanat)
law of combining ratios Gesetz der konstanten Proportionen (Mischungsverhältnisse)
law of conservation of matter/mass Massenerhaltungssatz, Gesetz von der Erhaltung der Masse
law of conservation of energy Energieerhaltungssatz
law of large numbers *stat* Gesetz der großen Zahlen
law of mass action Massenwirkungsgesetz
law of random assortment/ principle of random assortment/ principle of independent assortment (Mendel) Kombinationsregel, Unabhängigkeitsregel
law of segregation/principle of segregation (Mendel) Spaltungsregel
law of thermodynamics (first/ second) (1./2.) Hauptsatz (der Thermodynamik)
law of uniformity/principle of uniformity (F1 of monohybrid cross) (Mendel) Uniformitätsregel
lawn *bot*/*micb* Rasen
lawn culture Rasenkultur
lay eggs/deposit eggs Eier legen, ablegen
layer/story/stratum/sheet *allg*/*geol* Schicht
layer *bot* Ableger, Absenker
layering *allg*/*geol* Schichtung
layering/layerage *bot*/*hort* Absenkervermehrung, Absenken, Ablegervermehrung
  ● **continuous layering/French layering** Ablegen mehrerer Jungpflanzen pro Trieb

L layering

- **mound layering/stool layering/ stooling** Ablegervermehrung durch Anhäufeln
- **simple layering** Absenken
- **stool layering/mound layering** Ablegervermehrung durch Anhäufeln (Abrisse nach Anhäufeln)
- **tip layering** Absenken von Triebspitzen
- **trench layering** Absenken in Bodenfurchen

LCR (locus control region) LCR (Lokus-Kontrollregion)

LD$_{50}$ (median lethal dose) mittlere Letaldosis, mittlere letale Dosis

LDCs (less-developed countries/ developing countries/peripheral countries) Entwicklungsländer

LDL (low density protein) LDL (Lipoproteinfraktion niedriger Dichte)

leach (soil/minerals) auslaugen

leachate (lixivium)/soakage/ seepage/gravitational water/ drainage water Lauge, Laug(en)lösung (Bodenauslaugung), Sickerwasser

leaching (soil: dissolved minerals) Auslaugung, Auswaschung, Herauslösen

lead *n chem* **(the element)** Blei

lead (of a key) führende(r)/ übergeordnete(r) Stelle/Eintrag (eines Bestimmungsschlüssels)

leader *zool* **(animal in group)** Leittier

leader *bot* **(main shoot)** Haupttrieb, Leittrieb, Höhentrieb

leader segment *gen* Leader-Sequenz

leading shoot/main shoot/primary shoot/main axis/primary axis Hauptspross, Primärspross, Hauptachse

leading strand *gen* Leitstrang

leading substrate Leitsubstrat

leadwort family/sea lavender family/plumbago family/ Plumbaginaceae Bleiwurzgewächse, Grasnelkengewächse

leaf (*pl* **leaves)** Blatt (*pl* Blätter)
- **amphigastrium/ventral leaf/ underleaf (liverworts)** Amphigastrium, Bauchblatt, "Unterblatt"
- **asciidate leaf/pitcher leaf** Schlauchblatt, Kannenblatt
- **compound leaf/divided leaf** zusammengesetztes Blatt, Fiederblatt (ganzes!), gefiedertes Blatt
- **cotyledon/seminal leaf** Kotyledone, Cotyledone, Keimblatt
- **drooping leaf** herunterhängendes Blatt
- **fenestrated leaf** gefenstertes Blatt
- **flag leaf** Fahnenblatt, Fähnchenblatt
- **floating leaf** Schwimmblatt
- **"flower pot" leaf/urn-shaped leaf/pouch leaf (***Dischidia***)** Urnenblatt
- **foliage leaf** Laubblatt, Folgeblatt
- **>primary foliage leaves/first foliage leaves** Primärblätter, Erstlingsblätter
- **funnel-shaped/ascidiate leaf (***Nepenthes***)** Trichterblatt, Schlauchblatt
- **nectar leaf/nectariferous leaf/ honey leaf** Nektarblatt, Honigblatt
- **nest leaf** Nischenblatt, Mantelblatt
- **peltate leaf** peltates Blatt, Schildblatt
- **pitcher leaf/ascidiate leaf** Kannenblatt, (*sensu lato*: Schlauchblatt)
- **primary leaf** Primärblatt
- **producing leaves/coming into leaf** Blätter austreiben
- **production of leaves/coming into leaf** Blattaustrieb
- **seminal leaf/cotyledon** Keimblatt, Kotyledone, Cotyledone
- **shade leaf/sciophyll** Schattenblatt
- **submerged leaf** Wasserblatt
- **sun leaf** Sonnenblatt, Lichtblatt
- **trap leaf** Fallenblatt
- **underleaf/hypophyll** Unterblatt
- **urn-shaped leaf/pouch leaf/ "flower pot" leaf (***Dischidia***)** Urnenblatt
- **ventral leaf/amphigastrium/ underleaf (liverworts)** Bauchblatt, "Unterblatt", Amphigastrium

leaf abscission/shedding of leaves Blattfall, Laubfall

leaf apex/leaf tip Blattspitze
- **acuminate/taper-pointed** lang zugespitzt (konkav zulaufend: z.B. Blattspitze)
- **acute/sharp/pointed/sharp-pointed** spitz
- **apiculate/pointleted (forming small tip)** (fein) zugespitzt (mit feiner Spitze endend)
- **aristate/awned** begrannt, Grannen tragend

- **caudate/tail-pointed (ending with tail-like appendage)** geschwänzt, geschwanzt, mit Schwanz versehen (mit Träufelspitze)
- **cirrose/cirrous (leaf with prolonged midrib)** mit Ranken, rankenartige Blattspitze
- **cuspidate (gradually terminating in sharp/rigid point)** stachelspitzig (lang zugespitzt)
- **emarginate/shallowly notched** ausgerandet
- **mucronate (abruptly terminating in sharp/hard point)** stachelspitz
- **mucronulate** kleinspitzig
- **obtuse (blunt or rounded end of leaf)** stumpf
- **pungent (with stiff/sharp point)** spitzig
- **retuse (obtuse with broad shallow notch)** eingebuchtet
- **rotund/rounded** abgerundet, rundlich
- **setose/bristly/set with bristles** borstig
- **truncate/terminating abruptly** gestutzt
- **uncinate/barbed/hooked** hakig, mit Haken

leaf area index (LAI) Blattflächenindex (BFI)

leaf area ratio (LAR) Blattflächenverhältnis

leaf arrangement/phyllotaxy/ phyllotaxis Blattanordnung, Blattstellung, Beblätterung, Phyllotaxis
- **alternate l.a.** wechselständige Blattstellung
- **crowded l.a.** gedrängte Blattstellung
- **decussate l.a.** kreuzgegenständige (dekussierte) Blattstellung
- **distichous/two-ranked/ two-rowed l.a.** distiche Blattstellung, zweizeilige Blattstellung
- **opposite l.a.** gegenständige Blattstellung
- **scattered l.a.** zerstreute (disperse) Blattstellung
- **spiral l.a.** schraubige Blattstellung
- **whorled l.a.** quirlständige, wirtelige Blattstellung

leaf axil Blattachsel

leaf axis Blattachse

leaf base (see also: leaf blade base) Blattbasis, Blattgrund

leaf blade/leaf lamina (see also: leaf shapes) Blattspreite

leaf blade apex see leaf apex

leaf blade base/base of blade Blattspreitengrund
- **acute (equally curved convexly to the base)** (gleichmäßig) in den Stiel verschmälert, zugespitzt
- **attenuate/tapering/pointed (convex sides/concave towards base)** verschmälert (an der Basis konkave Spreitenränder)
- **auriculate/eared/ear-like** geöhrt
- **cordate/cordiform/heart-shaped** herzförmig
- **cuneate/cuneiform/sphenoid/ wedge-shaped** keilförmig
- **hastate/hastiform/spear-shaped** spießförmig
- **oblique** schief, schräg
- **reniform/kidney-shaped** nierenförmig
- **rotund/rounded** rundlich, abgerundet
- **sagittate/sagittiform/ arrowhead-shaped** pfeilförmig
- **truncate** gestutzt

leaf blade margin/leaf blade edge/ edge of blade/margin of blade (see also: leaf margin) Blattspreitenrand

leaf bundle Blattbündel

leaf buttress Blatthöcker, frühe Blattanlage

leaf cast (caused by frost/dryness/ fungal disease) Schütte, Blattschütte, Nadelschütte; Frostschütte, Trockenschütte

leaf curl/leaf roll Blattkräuselkrankheit

leaf cushion/leaf pulvinus Blattkissen, Blattpolster, Gelenkpolster

leaf cutter Blattschneider

leaf cutting Blattsteckling

leaf drop, early frühzeitiger Blattfall

leaf flushing, rapid Blattausschüttung, Laubausschüttung

leaf flutter Blattflattern

leaf gall Blattgalle

leaf gap/foliar gap Blattlücke

leaf lamina/leaf blade Blattspreite

leaf litter Blattstreu, Laubstreu, Laubschicht

leaf margin/leaf edge Blattrand
- **ciliate/bearing cilia** bewimpert, gewimpert
- **crenate/with rounded teeth/ scalloped** kerbzähnig, zackig gekerbt
- **crenulate/finely notched** fein kerbzähnig, feingekerbt, feinkerbig
- **curled** kräuselig, gekräuselt
- **dentate/toothed** gezähnt

leaf margin

- **denticulate/finely dentate** gezähnelt, fein gezähnt
- **digitate/fingered** digitat, gefingert
- **entire/simple** ganzrandig
- **gnawed** ausgebissen
- **incised/evenly notched/evenly cut** gleichmäßig geschlitzt/zerschlitzt/eingeschnitten
- **lacerate/torn** ungleichmäßig zerschlitzt/geschlitzt, ungleichmäßig eingeschnitten
- **laciniate/slashed** gefranst, geschlitzt
- **lobate/lobed** lappig, gelappt
- **palmate/fingered/hand-shaped** gefingert, handförmig
- **palmatifid (divided to middle)** fingerspaltig, handförmig gespalten
- **palmatilobate/palmately lobed** fingerlappig, handförmig gelappt
- **palmatipartite/palmately partite** fingerteilig, handförmig geteilt
- **palmatisect** fingerschnittig, handförmig (ein)geschnitten
- **repand (slightly uneven and waved margin)** leicht gewellt, geschweift, randwellig
- **runcinate/retroserrate/hook-backed (e.g. dandelion leaf margin)** schrotsägeförmig, rückwärts gesägt
- **serrate/serrated/sawed/saw-edged** sägeförmig gezackt, gesägt
- **serrulate/serratulate/finely serrate/finely saw-edged** feingesägt, kleingesägt
- **sinuate (strongly waved margin)** buchtig, gebuchtet
- **spinose/spinous** (grob)stachelig
- **undulate** gewellt, wellig

leaf primordium Blattprimordium, Blattanlage

leaf pulvinus/leaf cushion Blattpolster, Blattkissen, Gelenkpolster

leaf roll/leaf curl Blattkräuselkrankheit

leaf scar Blattnarbe

leaf shape Blattform

- **acerose/aciculate/acicular/needle-shaped** nadelförmig
- **auriculate/eared/ear-like** geöhrt
- **cordate/cordiform/heart-shaped** herzförmig
- **cuneate/cuneiform/sphenoid/wedge-shaped** keilförmig
- **deltoid** breiteiförmig, deltaförmig, dreieckig
- **elliptic/elliptical** elliptisch
- **ensiform/gladiate/xiphoid/sword-shaped** schwertförmig
- **falcate/falciform/sickle-shaped** sichelförmig
- **hastate/hastiform/spear-shaped** spießförmig
- **lanceolate** lanzettförmig, lanzettlich
- **lineal/linear** linear, linealisch
- **lyrate/lyriform/lyre-shaped** leierförmig, lyraförmig
- **obcordate/obcordiform/inversely heart-shaped** verkehrt herzförmig
- **oblanceolate/inversely lanceolate** verkehrt lanzettförmig
- **oblong** länglich
- **obovate/inversely egg-shaped** verkehrt eiförmig
- **orbicular/circular** kreisförmig, kreisrund
- **orbiculate/nearly round** kreisförmig, fast rund
- **ovate/egg-shaped** eiförmig
- **panduriform/fiddle-shaped** geigenförmig
- **peltate leaf** peltates Blatt, Schildblatt
- **peltate/peltiform/shield-shaped** schildförmig
- **reniform/kidney-shaped** nierenförmig
- **rhombic/rhomboid/diamond-shaped** rhombisch, rautenförmig
- **runcinate/retroserrate/hook-backed (e.g. dandelion leaf)** schrotsägeförmig
- **sagittate/sagittiform/arrowhead-shaped** pfeilförmig, pfeilspitzenförmig
- **spathulate/spatulate/spoon-shaped** spatelförmig
- **subulate/awl-shaped** pfriemlich

leaf sheath Blattscheide

leaf stalk/petiole Blattstiel

leaf stalk vegetable Stengelgemüse

leaf surface Blattoberfläche

- **lower leaf surface/abaxial leaf surface** Blattunterseite
- **upper leaf surface** Blattoberseite

leaf tendril Blattranke

leaf tip/leaf apex Blattspitze, Blattspreitenspitze

leaf trace/foliar trace Blattspur

leaf trace bundle Blattspurstrang, Blattspurbündel

leaf vein/leaf rib Blattader, Blattnerv, Blattrippe

leaf venation Blattaderung, Blattnervatur

leaf-borne blattbürtig

lepidodendron family

leaf-eating/folivorous blattfressend, blätterfressend
leaf-like/phylloid/foliaceous/foliose blattartig, blattförmig
leafage/foliage/leaves Belaubung, Blattwerk, Laubwerk, Laub
leafing/unfolding of leaves Blattentfaltung
leafless/aphyllous blattlos, unbeblättert
leaflet Blättchen, Blattfieder
leafy vegetable/leaf vegetable Blattgemüse
leak *n* Leck
leak *vb* tropfen, herauslaufen, undicht sein
leak current/leakage current (I$_l$) *neuro* Leckstrom
leakage Leckage, Leck, Lecken
leakage channel/resting channel *neuro* Leckkanal, Ruhemembrankanal
leakage conductance (g$_1$) *neuro* Leckleitfähigkeit
leaky mutant durchlässige Mutante
leaky mutation durchlässige Mutation
lean (meat etc.) *adj/adv* mager
learn *vb* lernen
learned behavior Lernverhalten, erlerntes Verhalten
learning Lernen
learning theory Milieutheorie
least significant difference/critical difference *stat* Grenzdifferenz (GD)
least squares method *stat* Methode der kleinsten Quadrate
leatherwood family/cyrilla family/ Cyrillaceae Lederholzgewächse
leathery/coriaceous ledrig, lederartig
leavening Treibmittel, Gärmittel, Gärstoff
lecithin Lecithin
lecithotrophic lecithotroph
lectin Lektin
lectotype Lectotypus, Lectotyp, Lectostandard
lecythis family/brazil-nut family/ Lecythidaceae Deckeltopfbäume
lee/lee side Lee, windgeschützte Seite, Windschattenseite (dem Wind abgekehrte Seite)
leeches/hirudineans/Hirudinea (Annelida) Egel, Blutegel, Hirudineen
leeward (opposite windward) im Windschatten, auf der Windschattenseite
left-handed (helix) *gen/biochem* linksgängig
left-handed/sinistral linkshändig
leg Bein
leg/drumstick (fowl) Keule, Schlegel
leg/foot (birds) Lauf
leg beat Laufschlag
legal regulations on species protection Artenschutzverordnung
legume/leguminous plant Hülsenfrüchtler
legume/pod Hülse
legume family/bean family/pea family/pulse family/Fabaceae/ Papilionaceae (Leguminosae) Hülsenfruchtgewächse, Hülsenfrüchtler, Schmetterlingsblütler
leishmaniasis Leishmaniose
  • **cutaneous l./oriental sore (*Leishmania spp.*)** kutane Leishmaniose, Hautleishmaniose, Orientbeule
  • **visceral l./kala azar/Cala-Azar (*Leishmania donovani*)** viszerale Leishmaniose, Kala-Azar, Cala-Azar, schwarzes Fieber
Leitneriaceae/corkwood family Korkholzgewächse
lek (communal mating ground) *ethol/ecol* Balzarena
lemma/lower palea/outer palea Deckspelze
Lemnaceae/duckweed family Wasserlinsengewächse
lemniscus (*pl* lemnisci) Lemnisk (*pl* Lemnisken)
length constant (membrane) Längskonstante
lennoa family/Lennoaceae Lennoagewächse
lens/lense Linse
  • **collimating lens** parallel-richtende Sammellinse
lens/magnifying glass Lupe, Vergrößerungsglas
lens eye/lenticular eye Linsenauge
lens placode Linsenplakode
lens tissue Linsenpapier
Lentibulariaceae/bladderwort family/butterwort family Wasserschlauchgewächse
lentic/lenitic (of/in standing water) lentisch, lenitisch (in stehendem Gewässer lebend)
lenticel Lentizelle, Korkpore
lenticular nucleus/nucleus lentiformis Linsenkern
lentiform/lenticular/lentil-shaped linsenförmig
lepidodendron family (clubmoss trees)/Lepidodendraceae Schuppenbäume

**lepidophyte trees/club-moss trees/
Lepidodendrales** Lepidophyten,
Bärlappbäume
**lepidopterans (butterflies &
moths)/Lepidoptera**
Schuppenflügler (Schmetterlinge &
Motten)
lepidosaurs/Lepidosauria
Schuppenkriechtiere
leprose leprös
leptocaulous/slender-stemmed
dünnstämmig, schlankstämmig
**leptomeninx/pia-arachnoid
membrane** weiche Hirnhaut,
Leptomeninx (Arachnoidea & Pia
mater)
**leptospirosis/Weil's disease/swamp
fever/infectious anemia
(*Leptospira interrogans*)**
Leptospirose, Weil-Krankheit,
Weilsche Krankheit
leptotene Leptotän
**lerp (Australia/Tasmania: sweet/
waxy secretion/manna on
eucalyptus from jumping plant
lice)** Manna an Eucalyptus
(Schutzsekret von Pflanzenläusen)
• **sugar-lerp (*Australian term*)/
honeydew** Honigtau
lesion/injury/harm Läsion
(Schädigung/Verletzung/Störung),
krankhafte Veränderung (durch
Verletzung/Krankheit)
**less-developed countries (LDCs)/
developing countries/peripheral
countries)** Entwicklungsländer
lesser covert/minor covert *orn*
kleine Decke (Flügeldecke/
Flügelfeder)
lestobiosis Lestobiose
lethal/deadly letal, tödlich
lethal dose Letaldosis, letale Dosis,
tödliche Dosis
• **median lethal dose (LD$_{50}$)**
mittlere Letaldosis,
mittlere letale Dosis
lethal mutant Letalmutante
lethal mutation letale Mutation
lethality Letalität
leucine Leucin
leucine zipper (protein) *gen*
Leucin-Reißverschluss
leucopenia Leukopenie
leucoplast Leukoplast
leukemia Leukämie,
"Weißblütigkeit"
**leukocyte/white blood cell
(WBC) (*see also:* blood cells)**
Leukocyt, Leukozyt, weißes
Blutkörperchen
leukocytosis Leukocytose
levan Lävan

**levator/lifter (muscle: raising an
organ or part)** Levator, Heber
levee/dike Deich (Fluss)
leverage mechanism
Hebelmechanismus
leveret (hare in 1.year) Häschen,
junger Hase
levulinic acid Lävulinsäure
Leydig cell Leydigsche Zwischenzelle
liana/woody climber Liane,
Kletterpflanze, Schlingpflanze
(holzig/verholzt)
library/bank (clone bank)
Bibliothek, Klonbank
• **genomic library** genomische
Bibliothek, genomische Genbank
• **subgenomic library**
subgenomische Bibliothek,
subgenomische Genbank
• **subtractive library** subtraktive
Genbank, Subtraktionsbank,
Subtraktionsbibliothek
libriform fiber Libriformfaser,
Holzfaser
lichen Flechte
• **crustose lichens** Krustenflechten
• **filamentous lichens/hairlike
lichens** Haarflechten,
Fadenflechten
• **foliose lichens** Blattflechten,
Laubflechten
• **fruticose lichens/shrublike
lichens** Strauchflechten
• **gelatinous lichens**
Gallertflechten
• **umbilicate foliose lichens**
Nabelflechten (Blattflechten)
lichen acid Flechtensäure
lichenin Lichenin (Flechtenstärke,
Moosstärke)
lichenization Lichenisierung
lichenized lichenisiert
lick *vb* lecken
**lid/opercle/operculum (e. g. gill
cover)** Deckel, Operculum,
Operkulum (z. B. Kiemendeckel)
**lid capsule/circumscissile capsule/
pyxis/pyxidium** Deckelkapsel
lid-like/operculiform deckelförmig,
deckelartig
**Lieberkühn's organelle/watchglass
organelle (hymenostome ciliates)**
Lieberkühnsches Organell
life Leben
**life community/biotic community/
biocoenose** Lebensgemeinschaft,
Biocönose
life cycle/"life history"
Lebenszyklus, Lebenskreislauf,
Entwicklungs-Zyklus
life cycle assessment (LCA) *ecol*
Ökobilanz

life expectancy Lebenserwartung
life form Lebensform
life process(es) Lebensvorgang
(-vorgänge)
life sciences/biology
Biowissenschaften, Biologie
life scientist/biologist Biologe
life size Lebensgröße
life span Lebensdauer, Lebensspanne
life table/mortality table
Sterbetafel, Sterblichkeitstabelle
life zone/biotope Lebenszone,
Lebensraum, Biotop
lifeform/organism Lebensform,
Lebewesen, Organismus
lifeless/inanimate/dead leblos, tot
**lifestyle/mode of life/way of life/
habits** Lebensweise
lifetime Lebenszeit; *neuro*
Öffnungsdauer (eines
Membrankanals)
ligament Ligament, Band
ligament sac Ligamentsack
ligand Ligand
ligand blotting Liganden Blotting
ligation Ligation, Verknüpfung
• **blunt end ligation** Ligation
glatter Enden
• **self-ligation** Selbst-Ligation
light *n* Licht
• **artificial light/lighting**
künstliche Beleuchtung
• **beam of light** Lichtstrahl,
Lichtbündel
• **diffracted light** gebeugtes Licht
• **diffuse light** diffuses Licht
• **exposure (to light)** Belichtung
• **incident light** einfallendes Licht,
auftreffendes Licht
• **path of light (ray diagram)**
Strahlengang (Strahlendiagramm)
• **plane-polarized light** linear
polarisiertes Licht
• **polarized light** polarisiertes Licht
• **ray/beam (of light)** Strahl
• **reflected light** Reflexlicht,
reflektiertes Licht
• **scattered light** Streulicht
light chain (L-chain) leichte Kette
(L-Kette)
**light independent DNA repair/
dark repair** lichtunabhängige
DNA-Reparatur
light intensity Lichtstärke,
Lichtintensität
**light microscope (compound
microscope)** Lichtmikroskop
(zusammengesetztes Mikroskop)
light microscopy Lichtmikroskopie
light permeability
Lichtdurchlässigkeit
light reaction Lichtreaktion

light repair (DNA) Lichtreparatur
light scattering Lichtstreuung
light sensibility Lichtempfindbarkeit
light sensitivty Lichtempfindlichkeit
(leicht reagierend)
light source Lichtquelle
light stimulus Lichtreiz
light-harvesting complex (LHC)
Lichtsammelkomplex
**light-induced germination of seed/
photodormant seed** Hellkeimer,
Lichtkeimer (Samen)
light-sensitive lichtempfindlich
(leicht reagierend)
lightwood family/Cunoniaceae
Cunoniaceae
ligneous/woody holzartig, holzig
lignicolous/lignicole
holzbewohnend, auf Holz wachsend
lignification/sclerification
Lignifizierung, Verholzung
lignified/sclerified lignifiziert,
verholzt
lignin Lignin
lignite/brown coal Weichbraunkohle
& Mattbraunkohle, Lignit
lignoceric acid/tetracosanoic acid
Lignocerinsäure, Tetracosansäure
lignotuber/burl/woody outgrowth
ebenerdige Maserknolle,
Wurzelhalsknolle, Kropf (Auswuchs
am Wurzelanlauf bestimmter
Bäume)
ligular/tongue-shaped
zungenförmig
ligulate/strap-shaped
streifenförmig
ligule/ligula Ligula, Zünglein
ligule *bot* (grasses) Ligula,
Blatthäutchen
likelihood function
Wahrscheinlichkeitsfunktion
lily family/Liliaceae Liliengewächse
**lily-of-the-valley family/
Convallariaceae**
Maiglöckchengewächse
**limb/extremity/appendage
(articulated)** Gliedmaße,
Extremität (*pl* Gliedmaßen/
Extremitäten)
• **ascending limb (loop of Henle)**
aufsteigender Ast (Henlesche
Schleife)
• **descending limb (loop of Henle)**
absteigender Ast (Henlesche
Schleife)
limb muscle/appendicular muscle
Extremitätenmuskel
limbate gesäumt, andersfarbig
gerändert
lime *n chem* Kalk
lime/calcify *vb* kalken

lime tree family/linden family/ Tiliaceae Lindengewächse
liminal (pertaining to a threshold) einen Grenzwert/Schwellenwert betreffend, Schwellen...
liminal stimulus Schwellenreiz
liming Kalkung
limit of resolution *opt* Auflösungsgrenze
limited capacity control system (LCCS) limitiertes Kapazitätskontrollsystem
limiting factor begrenzender Faktor, limitierender Faktor, Grenzfaktor
Limnanthaceae/false mermaid family/meadowfoam family Sumpfblumengewächse
limnetic/limnal/limnic limnisch, im Süßwasser lebend
Limnocharitaceae/water-poppy family Wassermohngewächse
limnocrene Limnokrene, Tümpelquelle
limnogenous limnogen
limnology Limnologie, Seenkunde (Binnengewässerkunde)
limnomedusas/Limnomedusae/ Limnohydrina Limnomedusen, Limnohydrinen
limonene Limonen
limp schlaff (welk)
limpets & keyhole limpets & abalone/archeogastropods/ Archaeogastropoda/Diotocardia Altschnecken
limy/limey kalkig, kalkartig, kalkhaltig
Linaceae/flax family Leingewächse
linden family/lime tree family/ Tiliaceae Lindengewächse
line diagram Strichdiagramm
line transect *ecol* Linientransekt
line transect method *ecol/stat* Linienstichprobe(nverfahren)
LINE (long interspersed nuclear element) *gen* LINE (langes eingeschobenes nukleäres Element)
lineal/linear linear, linealisch
Lineweaver-Burk plot/double-reciprocal plot Lineweaver-Burk-Diagramm
lingual die Zunge betreffend
lingual papilla/gustatory papilla Zungenpapille, Geschmackspapille
• **filiform papilla/threadlike papilla** fadenförmige Papille
• **foliate papilla** Blattpapille, blättrige Papille
• **fungiform papilla** Pilzpapille
• **lentiform papilla** linsenförmige Papille
• **vallate papilla** Wallpapille

lining Auskleidung, Überzug, Oberflächenschicht
linkage Kopplung (Genkopplung)
• **partial linkage** partielle Kopplung
• **sex linkage** Geschlechtskopplung
linkage analysis Kopplungsanalyse
linkage disequilibrium Kopplungsungleichgewicht
linkage equilibrium Kopplungsgleichgewicht
linkage group Kopplungsgruppe
• **partial l.g.** partielle Kopplungsgruppe
linkage map Genkopplungskarte
linked genes gekoppelte Gene
linker DNA Linker-DNA
• **polylinker** Polylinker
linolenic acid Linolensäure
linolic acid/linoleic acid Linolsäure
lint (cotton fiber) Lint, Lintbaumwolle
linters (short cotton fibers) Linters (kurze Baumwollfasern)
lip/labellum Lippe, Labellum
• **lower lip/labium** Unterlippe, Labium
lip smacking *ethol* Lippenschnalzen
lip-curling/"flehmen" flehmen
lipid Lipid
lipid bilayer Lipiddoppelschicht
lipofection Lipofektion
lipoic acid (lipoate)/thioctic acid Liponsäure, Dithiooctansäure, Thioctsäure, Thioctansäure (Liponat)
lipophilic lipophil
liposome Liposom
lipoteichoic acid Lipoteichonsäure
liquefaction Verflüssigung
liquefy verflüssigen
liquid *n* Flüssigkeit
liquid *adj/adv* flüssig
liquid chromatography (LC) Flüssigkeitschromatographie
liquid manure (total excretions diluted with water) Gülle; (urine) Jauche
lissencephalous (no/few convolutions) lissencephal (ungefurcht/glattes Gehirn)
list (dark stripe on back) *zool* Aalstrich
lithobiomorphs/Lithobiomorpha Steinläufer
lithotroph(ic) lithotroph
lithotrophy Lithotrophie
litocholic acid Litocholsäure
litter/bear young *vb* Junge werfen
litter *n bot* Streu
litter *n zool* Wurf, Tracht (Jungtiere/ Wurf)

litter *vb ecol* verunreinigen, (Müll) herumliegenlassen
• **don't litter!** entsorgen Sie Ihren Müll umweltgerecht!
litter layer *bot* **(forest)** Streuschicht, Streuhorizont, Förna
litter mate Wurfgeschwister, Geschwister eines Wurfes
litter size *zool* Wurfgröße
littoral *adj/adv* küstenbewohnend, uferbewohnend, am Ufer lebend
littoral fringe Küstensaum, Ufersaum
littoral/littoral zone/intertidal zone/eulittoral zone (marine) Litoral, Litoralzone, Litoralbereich, Gezeitenzone, Tidebereich, Eulitoral
littoral/littoral zone (lake) Uferregion, Uferzone (Gewässer)
live *adj/adv* lebend, lebendig
live *vb* leben, lebendig sein
live culture/living culture Lebendkultur
live germ count Lebendkeimzahl
live vaccine Lebendimpfstoff, Lebendvakzine
live weight Lebendgewicht
live-bearing/viviparous *adj/adv* lebendgebärend, vivipar
live-bearing/vivipary/viviparity *n* Lebendgebären, Viviparie
live-birth/vivipary Lebendgeburt, Viviparie
liver Leber
liverwort Lebermoos
livestock landwirtschaftlich genutztes Vieh
livestock breeding Tierzucht (Nutztiere in der Landwirtschaft), Viehzucht (*sensu lato*)
livestock keeping/animal husbandry Viehhaltung
livestock unit Großvieheinheit (500kg Lebendgewicht)
living fossil lebendes Fossil
lizard's tail family/Saururaceae Molchschwanzgewächse
lizard-hipped dinosaurs/saurischian reptiles/saurischians/Saurischia Echsenbecken-Dinosaurier
lizards/Lacertilia (Squamata) Echsen, Eidechsen
load/freight Last, Belastung; Fracht (Flüssigkeit/Abwasser)
• **bed load** *soil/mar* Bodenfracht, Geschiebe
• **genetic load** Erblast, genetische Last, genetische Bürde, genetische Belastung
• **mutational load** Mutationsbelastung, Mutationslast
• **silt load/suspension load** *soil/ mar* Schwebfracht
loaded form beladene Form (z.B. ADP→ATP)
• **unloaded form** entladene Form (z.B. ATP→ADP)
loading Beladung, Belastung
• **phloem loading** Phloembeladung
• **phloem unloading** Phloementladung
• **wing loading** *aer/orn* Flügel-Flächen-Belastung
loam Lehm
loasa family/Loasaceae Blumennesselgewächse
lobe Lappen
• **adenohypophysis/anterior lobe of pituitary gland** Adenohypophyse, Hypophysenvorderlappen
• **earlobe** Ohrläppchen
• **frontal lobe/lobus frontalis** *neuro* Frontallobus, Stirnlappen
• **galea/outer lobe of maxilla** Galea, Außenlade
• **lacinia/inner lobe of maxilla** Lacinia, Innenlade
• **mantle lobe (brachiopods)** Mantellappen, Kragenlappen
• **neurohypophysis/posterior lobe of pituitary gland** Neurohypophyse, Hypophysenhinterlappen
• **occipital lobe/lobus occipitalis** *neuro* Okzipitallappen, Hinterhauptslappen
• **optic lobe/visual lobe/lobus opticus** Sehlappen
• **parietal lobe/lobus parietalis** *neuro* Parietallappen, Scheitellappen
• **polar lobe (cleavage)** Pollappen
• **temporal lobe/lobus temporalis** *neuro* Temporallappen, Schläfenlappen
• **visual lobe/optic lobe/lobus opticus** Sehlappen
lobe fin Quastenflosse
lobe-finned fishes/ crossopterygians/Crossopterygii Quastenflosser
lobed/lobate lappig, gelappt
lobelia family/Lobeliaceae Lobeliengewächse
lobopodium Lobopodium
lobster pot (Genlisea) Schlauchblatt
lobule Läppchen
local/native/endemic *adj/adv* örtlich, heimisch, einheimisch, endemisch
locale *n* Ort, Platz, Standort

localized potential lokales Potential, Lokalpotential
locate orten
location Lage (Ort)
lock-and-key principle Schlüssel-Schloss-Prinzip, Schloss-Schlüssel-Prinzip
locomotion Lokomotion, Fortbewegung, Ortsveränderung (Bewegung)
locule/lock/loculus (chamber/cell) Fach, Loculament, Lokulament (von Ovar/Anthere/Sporangium)
loculicidal loculizid, lokulicid, fachspaltig
loculicidal capsule lokulizide Spaltkapsel
locus Locus, Lokus, Ort, Stelle
• **compound locus** *gen* Locus aus mehreren eng gekoppelten Genen
• **LCR (*locus control region*)** *gen* LCR (Lokus-Kontrollregion)
locust gum/locust bean gum/carob gum/carob seed gum Johannisbrotkernmehl, Karobgummi
lod score ("logarithm of the odds ratio") Lod-Wert
lodge *n* (e.g. beaver) (höhlenartig) Bau, Lager
lodge *vb* **(animals)** lagern
lodge *vb* **(crops)** umlegen (z. B. durch Wind)
lodicule/paleola/glumellule Lodicula, Schwellkörper, Schüppchen (Grasblüte)
loess Löss
log *n* gefällter Holzstamm, gefällter Baumstamm
log/cut trees for lumber (Bäume) fällen
log/log off/clear land in lumbering kahlschlagen
log on/off *vb techn* ein Gerät an-/abschalten, in ein Programm einsteigen/aus einem Programm "aussteigen", draufgehen/runtergehen
logania family/Loganiaceae Strychnosgewächse, Brechnussgewächse
logarithmic normal distribution/lognormal distribution *stat* logarithmische Normalverteilung, Lognormalverteilung
logarithmic phase (log-phase) *ecol/micb* logarithmische Phase
logging/lumbering/felling of trees/timber harvesting Holzfällen
lognormal distribution/logarithmic normal distribution *stat* Lognormalverteilung, logarithmische Normalverteilung
logs Langhölzer
LoH (loss of heterozygosity) Verlust der Heterozygotie, Heterozygotieverlust
loin Lende
lomasome Lomasom
loment/lomentum/lomentaceous fruit/jointed fruit Gliederfrucht, Bruchfrucht, Klausenfrucht, Gliederhülse
London dispersion forces London-Dispersionskräfte
lone wolf/loner *ethol* Einzelgänger
long *vb* **(for something)** nach etwas verlangen, sich nach etwas sehnen
long bone/os longum langer Knochen, Röhrenknochen
long interspersed nuclear element (LINE) langes eingeschobenes nukleäres Element (LINE)
long period interspersion langphasige Einstreuung
long shoot/axis Langtrieb
long terminal repeat (LTR) *gen* lange terminale Sequenzwiederholung (LTR)
long-chain langkettig
long-day plant Langtagspflanze
long-distance transport Ferntransport
long-lived langlebig
long-term potentiation (LTP) Langzeitpotenzierung
longevity Langlebigkeit, lange Lebensdauer
longing/yearning Verlangen, Sehnsucht
longipetalous langkronblättrig
longisection/longitudinal section/long section Längsschnitt
longitude *geogr* Längengrad
longitudinal division/fission Längsteilung
longitudinal muscle Längsmuskel
longitudinal vein Längsader
longshore current *mar* Brandungslängsstrom, Längsströmung (am Strand)
loop Schlaufe
• **displacement loop** Verdrängungsschlaufe
loop conformation/coil conformation *gen* Schleifenkonformation, Knäuelkonformation
loop of Henle Henle-Schleife, Henlesche Schleife
loop reactor/circulating reactor/recycle reactor (bioreactor) Umlaufreaktor, Umwälzreaktor, Schlaufenreaktor

lush vegetation

looper/measuring worm/inchworm/ spanworm *entom* Spannerraupe
looping (movement of loopers) *entom* Spannen (Fortbewegung bei Spannerraupen)
loosestrife family/Lythraceae Weiderichgewächse, Blutweiderichgewächse
lope *n* **(horse: easy natural gait resembling a canter)** leichter Kanter
lophodont (teeth with transverse ridges) lophodont, mit Querjochen
lophotrichous lophotrich
loppers *hort* Astschere
lopseed family/Phrymaceae Phrymagewächse
Loranthaceae/mistletoe family (showy mistletoe family) Mistelgewächse, Riemenblumengewächse
lore *orn* Zügel (Vogel: Bereich zwischen Schnabel und Augen)
loreal (scale: snakes) Loreale, Zügelschild
Lorenzini flask Lorenzinische Ampulle (Ampullenrezeptor)
lorica (a girdle-like skeleton)/case Lorica, Panzer, Panzerplatte
loriciferans/corset bearers/ Loricifera Loriciferen, Korsetttierchen, Panzertierchen
loss *n* Verlust
loss of function mutation Funktionsverlustmutation
loss of heterozygosity (LoH) Verlust der Heterozygotie, Heterozygotieverlust
lotic (of/in actively moving water) lotisch
lotus lily family/Indian lotus family/Nelumbonaceae Lotusblumengewächse
low density lipoprotein (LDL) Lipoprotein niedriger Dichte, Lipoproteinfraktion niedriger Dichte
low tide/ebb tide/ebb Ebbe
low-molecular niedermolekular
lower critical temperature (LCT) untere kritische Temperatur (UKT)
lower plants/primitive plants niedere Pflanzen
lower primates/prosimians/Prosimii Halbaffen
Lower Triassic (epoch) Buntsandstein (Epoche)
lowland Tiefland, Niederung
LSE (least squares estimation) MSQ-Schätzung (Methode der kleinsten Quadrate)

LTR (long terminal repeat) *gen* LTR (lange terminale Sequenzwiederholung)
lubricate gleitfähig machen, schmieren
lubrication Schmieren, Schmierung, Ölen
lucid/luminous glatt und glänzend; leuchtend
lumbar lumbar, Lenden...
lumbar vertebra Lumbalwirbel, Lendenwirbel
lumber/timber/wood *n* Stammholz, Brauchholz, Bauholz, Nutzholz, Schnittholz, Holz
lumber *vb* **(wood)** Holz aufbereiten
lumber industry/timber industry Holzwirtschaft
lumbering/logging/felling of trees Holzfällen
lumberer/lumberjack/woodcutter/ woodchopper Holzfäller
lumen Lumen, Hohlraum, Höhlung
luminescence Lumineszenz
luminescent lumineszent, leuchtend
luminescent bacteria Leuchtbakterien
luminosity Leuchtkraft
luminous leuchtend, Leucht...
luminous organ/light-emitting organ/photophore Leuchtorgan, Photophore
lump/lump of soil Scholle, Erdscholle
lumpy (soil) klumpig, schollig (schwerer Boden)
lunar cycle (28d) Mondzyklus
lunar periodicity Lunarperiodik, Lunarperiodizität, Mondperiodik
lunar rhythm/circamonthly rhythm Lunarrhythmus, Mondrhythmus
lunate bone/semilunar bone Mondbein
lung Lunge
• **book lung/lung book** Buchlunge, Fächerlunge, Fächertrachee
• **suction lung** Sauglunge
lung book/book lung (arachnids) Buchlunge, Fächerlunge, Fächertrachee
lungfishes/Dipnoi Lungenfische
lunule Lunula
lure *n* Köder
lure/attract *vb* anlocken, locken, ködern
luring/attraction Anlockung
luring song/soliciting song Lockgesang
lush üppig
lush vegetation üppige Vegetation

luteinizing hormone (LH)/ interstitial-cell stimulating hormone (ICSH) Lutropin, Luteotropin, luteinisierendes Hormon (LH), zwischenzellstimulierendes Hormon
luv/windward side Luv, Wetterseite, Windseite (dem Wind zugewandte Seite)
luxury gene Luxusgen
lymph Lymphe
lymph heart Lymphherz
lymph node Lymphknoten
lymph nodule/lymph follicle Lymphknötchen, Lymphfollikel
lymph vessel/lymphatic vessel Lymphgefäß
lymphatic lymphatisch
lymphatic gland Lymphdrüse
lymphatic system Lymphsystem, Lymphgefäßsystem
lymphatic tissue lymphatisches Gewebe, Lymphgewebe
lymphatic vessel/lymph vessel Lymphgefäß
lymphocyte (*see also:* blood cells) Lymphocyt, Lymphozyt
lymphokine Lymphokin (lymphozytäres Zytokin/Cytokin)
lyonization Lyonisierung
lyophilization/freeze-drying Lyophilisierung, Gefriertrocknung
lyophilize/freeze-dry lyophilisieren, gefriertrocknen
lyotropic series/Hofmeister series lyotrope Reihe, Hofmeistersche Reihe
lyre-shaped/lyrate/lyriform leierförmig, lyraförmig
lysate Lysat
 • **cleared lysate** geklärtes Lysat
lyse *vb* lysieren, auflösen
lysergic acid Lysergsäure
lysigenic/lysigenous lysigen
lysine Lysin
lysis Lyse, Auflösung, Zerfall
lysogenic (temperate) lysogen (temperent)
lysogenic conversion lysogene Konversion
lysogeny Lysogenie
lysosome Lysosom
 • **secondary lysosome/ phagolysosome** sekundäres Lysosom
lysozyme Lysozym
Lythraceae/loosestrife family Weiderichgewächse, Blutweiderichgewächse
lytic lytisch
lytic cycle lytischer Zyklus
lytic infection lytische Infektion
lytic phage lytischer Phage
lytic plaque lytischer Hof

M line (M disk) (M = mesophragma)
M-Linie, M-Streifen (M-Scheibe)
M phase (mitotic phase of cell cycle) M-Phase (Mitosephase des Zellzyklus)
maar/volcanic lake Maar
macerate mazerieren
maceration Mazeration
mackerel sharks and relatives/ Lamniformes Makrelenhaiverwandte
macrandrous makrandrisch
macrergate (ants) Makroergat
macrocyte Makrocyt, Makrozyt
macroevolution Makroevolution
macrofauna Makrofauna
macrogyne (ant queen) Makrogyne
macromere Makromer
macromolecule Makromolekül
macronucleus/meganucleus Makronukleus, somatischer Zellkern
macronutrients Kernnährelemente
macrophage Makrophage
macroscopic makroskopisch
macrospecies Makrospezies, Großart
macrospore/megaspore Makrospore, Megaspore
macrosporophyll Makrosporophyll, Samenblatt
macula/spot Macula, Fleck, fleckenförmiger Bezirk
 • **macula adherens/macula adherens/desmosome/bridge corpuscle/bridge corpuscle** Desmosom, Macula adhaerens
 • **macula lutea/yellow spot** *ophthal* gelber Fleck
 • **sensory spot/acoustic macula (of membranous labyrinth)** *ichth* Hörfleck, Sinnesfleck, Sinnespolster (im inneren Labyrinth)
macular/maculate/spotted gefleckt
maculiform soralium (lichens) Flecksoral
mad cow disease (bovine spongiform encephalopathy = BSE) Rinderwahnsinn
madder family/bedstraw family/ Rubiaceae Rötegewächse, Labkrautgewächse, Krappgewächse
Madeira vine family/basella family/ Basellaceae Schlingmeldengewächse
madreporian body Madreporenköpfchen
madreporic canal/stone canal/ hydrophoric canal Steinkanal
madreporic plate/madreporite/ sieve plate Madreporenplatte, Siebplatte

maggot (apodal larva) Made (apode Larve)
magnesium Magnesium
magnetic magnetisch
magnetic resonance imaging (MRI)/ nuclear magnetic resonance imaging Magnetresonanztomographie (MRT), Kernspintomographie (KST)
magnetic stirrer Magnetrührer
magnetism Magnetismus
magnetosome Magnetosom
magnification/enlargement Vergrößerung
 • **initial magnification** Primärvergrößerung; Maßstabzahl
 • **lateral magnification** Lateralvergrößerung, Seitenverhältnis, Seitenmaßstab, Abbildungsmaßstab
 • **total magnification/overall magnification** Gesamtvergrößerung
magnification at *x* diameters *x*-fache Vergrößerung
magnify/enlarge vergrößern
magnify at *x* diameters *x*-fach vergrößern
magnifying glass/magnifier/lens Vergrößerungsglas, Lupe
mahogany family/Meliaceae Zedrachgewächse
maiden flight Jungfernflug
main axis/principal axis Hauptachse
main band Hauptbande
main beam (antler) Stange (Geweih)
main eye Hauptauge
main host/primary host/definitive host Hauptwirt
mainland Festland
maintenance coefficient Erhaltungskoeffizient
maintenance culture *micb* Erhaltungskultur (kontinuierliche Kultur)
maintenance energy Erhaltungsenergie
maintenance metabolism Betriebsstoffwechsel
maintenance pruning *hort* Erhaltungsschnitt
major groove (DNA structure) große Furche, große Rinne, tiefe Rinne (DNA-Struktur)
major histocompatibility complex (MHC) Haupthistokompatibilitätskomplex
makomako family/Elaeocarpaceae Elaeocarpusgewächse
Mal de Calderas (*Trypanosoma equinum*) Kreuzlähme
malaceous apfelartig

malacology

malacology/study of mollusks Malakologie, Weichtierkunde
malacophily/snail pollination Schneckenbestäubung
malacostracous/soft-shelled weichschalig
malady/disease/disorder Krankheit, Gebrechen
malar region Backenregion, Jochbeingegend (Vögel)
malaria (*Plasmodium spp.*) Malaria, Sumpffieber
male *adj/adv* männlich, männlichen Geschlechts
male *n* Männchen
male/staminate *bot* männlich, staminat
male chicken/cock/rooster Hahn
male clasper/harpagone/harpe Haltezange (Insektenmännchen), Harpagon, Harpe
male dog Rüde
male fern family/Dryopteridaceae/ drypopteris family/Aspidiaceae/ aspidium family Wurmfarngewächse
male wild boar Keiler
maleic acid (maleate) Maleinsäure (Maleat)
malformation Fehlbildung
malfunction Funktionsstörung
malic acid (malate) Äpfelsäure (Malat)
malignancy Malignität, Bösartigkeit
malignant maligne, bösartig
mallard wilder Enterich
malleate/hammer-shaped hammerförmig
mallee scrub/formation (Australia) Macchie-ähnliche Formation (mehrstämmige Sträucher aus Lignotuber)
malleolus/ankle (hammershaped projection) Malleolus, Knöchel, Fußknöchel
malleus/hammer (an ear bone) Hammer (Knochen im Innenohr)
mallow family/Malvaceae Malvengewächse
malnourished fehlernährt
malnutrition Fehlernährung
malodorous schlecht riechend, unangenehm riechend
malonic acid (malonate) Malonsäure (Malonat)
malpighia family/ barbados cherry family/ Malpighiaceae Malpighiengewächse, Barbadoskirschengewächse

Malpighian body/Malpighian corpuscle (splenic nodule) Malpighi-Körperchen, Milzkörperchen, Milzknötchen, Milzfollikel; (renal corpuscle) Malpighi-Körperchen, Nierenkörperchen
Malpighian layer/germinal layer/ germinative layer/stratum germinativum Keimschicht, Stratum germinativum
Malpighian tubule Malpighi-Gefäß, Malpighisches Gefäß, Malpighischer Schlauch, Malpighi-Schlauch
MALT (mucosa-associated lymphoid tissue) schleimhautassoziiertes lymphatisches Gewebe
malt sugar/maltose Malzzucker, Maltose
malting (process) Mälzung, Mälzen, Vermälzung
maltose (malt sugar) Maltose
Malvaceae/mallow family Malvengewächse
mamey family/St. John's wort family/mangosteen family/clusia family/Clusiaceae/Guttiferae/ Hypericaceae Hartheugewächse, Johanniskrautgewächse
mamilla/mammilla/nipple (multiple ducts)/teat (single duct) Mamille, Brustwarze, Zitze
mamillary body/corpus mammilare *neuro* Mamillarkörper
mamillary process/mamillary tubercle/processus mammillaris (vertebras) Zitzenfortsatz
mamillate/with nipplelike protuberances mit warzenartigen Erhebungen, mit kleinen Warzen
mamilliform/nipple-shaped warzenförmig
mamma/breast (*pl* mammae/ mammas) Brust, Busen
mammallike reptiles (advanced synapsids)/Therapsida säugetierähnliche Reptilien, Therapsiden
mammalogy Mammalogie, Säugerkunde, Säugetierkunde
mammals/Mammalia Säugetiere, Säuger
- **"toothless" mammals/edentates/ xenarthrans/Edentata/Xenarthra** Zahnarme, Nebengelenktiere
- **pouched mammals/ metatherians/Metatheria/ Didelphia** Beutelsäuger (marsupials/Marsupialia Beuteltiere)
- **small mammals** Kleinsäuger, Kleintiere

203 marc **M**

mammary gland Brustdrüse, Milchdrüse

"man"/mankind (*better:* **humankind/humans**) Mensch, Menschheit

manatee-grass family/ Cymodoceaceae Tanggrasgewächse

manchette/armilla/superior annulus (remains of velum universale) Manschette, Armilla, Annulus superus

mandelic acid/phenylglycolic acid/ amygdalic acid Mandelsäure, Phenylglykolsäure

mandible Mandibel

mandibular arch Mandibularbogen, Kieferbogen (Gesamtheit der Teile)

mandibular bar (skeleton only) Kieferbogen, Mandibularbogen (nur Knorpelspange)

mandibular gland Mandibulardrüse, Unterkieferdrüse

mane Mähne

maned gemähnt, mit einer Mähne

mangal/mangrove formation *biogeo* Mangrove(n), Mangrovewald, Gezeitenwald

manganese Mangan

manger Futtertrog, Futterraufe, Krippe

mangosteen family/mamey family/ St. John's wort family/clusia family/Clusiaceae/Guttiferae/ Hypericaceae Hartheugewächse, Johanniskrautgewächse

mangrove(s) (*see also:* mangal) Mangrove(n)

mangrove family/red mangrove family/Rhizophoraceae Mangrovengewächse

mangrove swamp Mangrovensumpf

maniciform soralium (lichens) Manschettensoral

mankind/humankind/"man" Menschen, Menschheit

manner/character/nature Wesen, Wesensart, Charakter, Natur

mannitol Mannit

mannuronic acid Mannuronsäure

manoxylic wood locker gebautes Sekundärholz

mantids/Mantodea/Mantoptera Gottesanbeterinnen & Fangschrecken

mantis shrimps/Hoplocarida/ Stomatopoda Fangschreckenkrebse, Maulfüßer

mantle/pallium (mollusks) Mantel, Pallium

mantle/tunic Mantel, Tunica

mantle cavity/pallial cavity Mantelhöhle

mantle cell Mantelzelle

mantle fiber *cyt* Zugfaser

mantle fold Marginalfalte, Mantelfalte (Mantelrand)

mantle girdle (chitons) Mantelgürtel, Gürtel, Perinotum

mantle layer *embr* Mantelschicht

mantle leaf Mantelblatt; Nischenblatt

mantle lobe (brachiopods) Mantellappen, Kragenlappen

manual (with keys for identification) Bestimmungsbuch

manual/handbook Handbuch

manual/primary feather Handschwinge

manubrium Manubrium, Schlundrohr

manure Mist, Stallmist, Dünger
- **liquid manure** (total excretions diluted with water) Gülle, Flüssigmist; (urine) Jauche

manure/dung/droppings (Tierkot) Mist, Dung

manyplies/third stomach/ psalterium/omasum Blättermagen, Vormagen, Psalter, Omasus

map *n* Karte, Landkarte
- **biological map** biologische Karte
- **fate map** *gen* Determinationskarte, Schicksalskarte
- **focus map** Fokuskarte
- **genetic map** genetische Karte
- **linkage map** Kopplungskarte
- **physical map** physikalische Karte

map/plot kartieren

maple family/Aceraceae Ahorngewächse

mapping/plotting Kartierung
- **deletion mapping** *gen* Deletionskartierung
- **fate mapping** *gen* Schicksalskartierung
- **gene mapping/genetic mapping** Genkartierung
- **positional mapping** Positionskartierung
- **transduction mapping** Transduktionskartierung
- **transformation mapping** Transformationskartierung

mapping function Kartierungsfunktion

maquis/macchie Maquis, Macchia, Macchie, Buschwald

marattia family/Marattiaceae Marattiaceae

marble gall Schwammkugelgalle

marc (fruit/grape press residue) Trester, Treber (hier speziell: Frucht-/Traubenrückstände)

marcescent

marcescent/shrivelling (withered leaves on plant) verwelkend, abtrocknend (an lebender Pflanze)
Marcgraviaceae/shingleplant family Honigbechergewächse
marcotage *hort* Markottage
marcotage using moss *hort* Abmoosen
mare Stute
- **broodmare** Zuchtstute

marestail family/mare's-tail family/ Hippuridaceae Tannenwedelgewächse
marginal marginal, randständig
marginal coverts/marginal tectrices Randdecken
marginal distribution *stat* Randverteilung
marginal meristem Marginalmeristem, Randmeristem
marginal placentation randständige Plazentation
marginal plate (tortoise carapace) Randplatte
marginal veil *fung* Marginalvelum
marginate gerandet, mit Rand
mariculture Meereskultur, marine Aquakultur
marine/maritime marin, Meeres…, das Meer betreffend; im Meer lebend, meeresbewohnend; maritim
marine animal Meerestier
marine biology Meeresbiologie, Marinbiologie
marine carnivores (seals/sealions/ walruses)/Pinnipedia Flossenfüßer, Robben
marine climate/maritime climate/ oceanic climate/coastal climate Meeresklima, Küstenklima
marine phosphorescence Meeresleuchten
marine sciences/oceanography Meereskunde, Ozeanographie, Ozeanografie
maritime/marine maritim; Meeres…, das Meer betreffend; meeresbewohnend
maritime climate/marine climate/ oceanic climate/coastal climate Meeresklima, Küstenklima, ozeanisches Klima
maritime vegetation/coastal vegetation Küstenvegetation, Meeresküstenvegetation
mark/brand/earmark markieren, kennzeichnen
marker (genetic/radioactive) Marker, Markersubstanz (genetischer/radioaktiver)
marker bed/key bed *paleo* Leithorizont

marking *ethol* Markieren
marking behavior Markierverhalten
marking of territory/territorial marking Reviermarkierung
marl *geol* Mergel
marrow Mark
- **bone marrow** Knochenmark

marrow brain/medullary brain/ myelencephalon Markhirn, Myelencephalon
marrow cavity/medullary cavity Markhöhle
marrow grafting/bone marrow grafting Knochenmarktransplantation
marsh (dominated by grasses) Marsch
- **brackish marsh** Brackmarsch
- **coastal marsh** Küstenmarsch, Seemarsch
- **estuarine marsh** Flussmarsch (an der Flussmündung/im Flussdelta)
- **freshwater marsh** Süßwassermarsch
- **high marsh** Hochmarsch
- **peat marsh** Torfmarsch
- **river-mouth marsh** Flussmündungsmarsch
- **riverine marsh** Flussmarsch
- **salt marsh (salt meadow)** Salzmarsch (Salzwiese)
- **shallow marsh/low marsh** Tiefmarsch
- **tidal marsh** Tidenmarsch, Gezeitenmarsch
- **young marsh/juvenile marsh** Koog

marsh fern family/Thelypteridaceae Sumpffarngewächse, Lappenfarngewächse
marsh plant/bog plant/helophyte Moorpflanze, Sumpfpflanze
marshland/marsh Marschland
marshy sumpfig, moorig
marsilea family/water clover family/Marsileaceae Kleefarne, Kleefarngewächse
marsupial bone/os marsupialis Beutelknochen
marsupials/pouched mammals/ Marsupialia Beuteltiere
martynia family/devil's-claw family/ unicorn plant family/ Martyniaceae Gemsbockgewächse
masculin/male maskulin, männlich
masculinity Männlichkeit, männliche Art
masculinization/virilization Maskulinisierung, Virilisierung, Vermännlichung
mash (e.g. for brewing) Maische

205   **mature** **M**

mass Masse
- **dry mass/dry matter**
  Trockenmasse, Trockensubstanz
- **"fresh mass" (fresh weight)**
  "Frischmasse" (Frischgewicht)
- **molar mass ("molar weight")**
  Molmasse, molare Masse
  ("Molgewicht")
- **molecular mass ("molecular weight")** Molekülmasse
  ("Molekulargewicht")
- **relative molecular mass/ molecular weight (M_r)**
  relative Molekülmasse,
  Molekulargewicht (M_r)

mass action constant
  Massenwirkungskonstante
mass exchange/substance exchange Stoffaustausch
mass extinction Massensterben
mass flow/bulk flow (water)
  Massenströmung
mass reproduction/mass spread/ outbreak Massenvermehrung
mass spectroscopy
  Massenspektroskopie (MS)
mass transfer Stoffübergang,
  Massenübergang, Stofftransport,
  Massentransport, Massentransfer
mass transfer coefficient
  Stoffübergangszahl,
  Stofftransportkoeffizient,
  Massentransferkoeffizient
mast/fattening/stuffing (of animals) Mast (Viehmast/
  Tiermast)
mast cell Mastzelle
master gene Meistergen
master sequence Mastersequenz
mastic *n* (resin) Mastix (Harz)
mastication/chewing *n* Kauen,
  Zerkauen
masticatory/chewing *adj* kauend
masticatory/gum/chewing gum *n*
  Kaumittel (Gummiharz), Kaugummi
masticatory muscle/muscle of mastication Kaumuskel
masticatory surface Kaufläche
mastigonema Geißelhärchen
mastoid(al) warzenartig,
  warzenähnlich, brustwarzenförmig
mastoid bone/mastoid/processus mastoideus Warzenfortsatz (des
  Schläfenbeins)
mat Matte, Polster
mat-like vegetation
  Polstervegetation
mate/copulate/pair begatten,
  sich paaren, kopulieren
mate/mating partner
  Geschlechtspartner
mate feeding *ethol* Partnerfüttern

maternal matern, mütterlich,
  mütterlicherseits, Mutter...
maternal effect maternaler Effekt,
  maternale Prädetermination
maternity/motherhood
  Mutterschaft
matgrass Borstgras
mating Paarung
- **assortative mating** assortative
  Paarung, übereinstimmende
  Paarung, sortengleiche Paarung,
  Gattenwahl, bewusste Paarung
- **backcross mating/backcross**
  Rückkreuzung
- **disassortative mating**
  Fremdpaarung, sortenungleiche
  Paarung
- **interrupted mating**
  unterbrochene Paarung
- **random mating** Zufallspaarung,
  zufällige Paarung, Panmixie
- **tripartite mating** *gene*
  Dreifachpaarung

mating affinity Paarungaffinität
mating barrier Paarungsschranke
mating behavior Paarungsverhalten
mating call/courtship call
  Paarungsruf, Werberuf
mating line/mating thread (male spider) *arach* Begattungsfaden
mating partner/mate
  Geschlechtspartner
mating plug/sphragis (*Lepidoptera*)
  Sphragis, Begattungssiegel,
  Kopulationssiegel
mating preference
  Paarungsbevorzugung
mating song/courtship song
  Werbegesang
mating system
  Partnerschaftssystem,
  Paarungssystem
mating type Paarungstyp,
  Kreuzungstyp
matriarchal matriarchalisch
matrilineal *gen* durch die
  mütterliche Linie vererbt
matrix/base material Matrix,
  Grundgerüst, Grundsubstanz
matrix/stroma (chloroplast) Matrix,
  Stroma
matroclinous inheritance
  matrokline Vererbung
matrotrophic matrotroph
matted verflochten, verfilzt
maturation Reifung, Reifen,
  Gedeihen
maturation promoting factor
  Reifungs-Förderfaktor
maturation-development
  Reifungsentwicklung
mature/ripe *adv/adj* reif

mature

mature/ripen *vb* reifen, gedeihen
maturing/ripening Reifen
maturity/ripeness Reife
Mauthner's cells Mauthnersche Zellen, Riesenfasern
maxilla (*pl* **maxillas/maxillae)** Maxille, Kiefer
- **first maxilla/maxilla prima (insects)** erste Maxille, Unterkiefer
- **second maxilla/labium (insects)** zweite Maxille, Labium, Unterlippe

maxilla/upper jawbone (vertebrates) Oberkiefer
maxillary bone/os maxillare Oberkieferbein
maxillary gland Schalendrüse, Maxillendrüse, Maxillennephridium
maxillary palp/palpus maxillaris Maxillarpalpus, Maxillartaster, Kiefertaster
maxillary plate/maxilliped plate (insects: galea & lacinia) Kaulade
maxilliped/maxillipede/ gnathopodite/jaw-foot/foot-jaw/ pes maxilliaris Maxilliped, Maxillarfuß, Kieferfuß
maxillopods Kieferfüßer
maxillule/maxillula (first maxilla: crustaceans) Maxillula, erste Maxille
maximum permissible workplace concentration/maximum permissible exposure MAK-Wert (maximale Arbeitsplatz-Konzentration)
maximum rate Maximalgeschwindigkeit (V_{max}: Enzymkinetik, Wachstum)
may apple family/Podophyllaceae Fußblattgewächse, Maiapfelgewächse
mayaca family/bogmoss family/ Mayacaceae Moosblümchen
mayflies/Ephemeroptera Eintagsfliegen
MCS (multiple cloning site) *gen* Vielzweckklonierungsstelle
MDCs (more-developed countries/ developed countries/ industrialized nations/core countries) Industrieländer
meadow Wiese
- **alpine meadow** alpine Bergwiese, Matte
- **damp meadow/wet meadow/ wetland** Nasswiese
- **dry meadow/arid grassland** Trockenrasen
- **floodplain meadow** Überschwemmungswiese
- **hay meadow/mowed meadow** Mähwiese
- **native meadow** Naturwiese
- **rich meadow/pasture** Fettwiese
- **riverine floodplain meadow** Auwiese, Auenwiese, Flussauenwiese
- **rough meadow/rough pasture/ poor grassland** Magerwiese
- **salt meadow** Salzwiese

meadow-beauty family/melastome family/Melastomataceae Schwarzmundgewächse
meadowfoam family/false mermaid family/Limnanthaceae Sumpfblumengewächse
meadowland Wiesenland
meal (act of eating) Mahlzeit
meal (coarsely ground grain) grobes Mehl
- **whole meal** Getreideschrot

mealworm Mehlwurm
mealy/farinaceous mehlig
mealybugs (Coccoidea) Wollschildläuse
mean/mean value/arithmetic mean/ average *stat* Mittelwert, arithmetisches Mittel, Durchschnittswert
- **adjusted mean** bereinigter Mittelwert, korrigierter Mittelwert
- **harmonic mean** harmonisches Mittel
- **regression to the mean** Regression zum Mittelwert
- **quadratic mean/root mean square (RMS)** quadratisches Mittel, Quadratmittel

mean residence time mittlere Verweilzeit
mean square deviation/variance *stat* mittlere quadratische Abweichung, mittleres Abweichungsquadrat, Varianz
β meander β-Mäander
measle Finne, Blasenwurm-Larve (speziell: im Fleisch des Haustiers)
measles *vir* Masern
- **German measles/rubella** Röteln

measly (infected with measels) mit Masern infiziert
measly (meat: with larval tapeworms) finnig, finnenhaltig
measly (meat: with trichina worms) trichinös, trichinenhaltig
measure *n* Maß
measure *vb* messen, abmessen
measurement/test/testing/reading/ recording Messung
measuring cup Messbecher
measuring unit/measuring device Messglied (Größe)

megaspore

measuring worm/looper/inchworm/ spanworm (geometer moth larva) Spannerraupe
meat/flesh Fleisch
meat extract *micb* Fleischextrakt
meat infusion (meat digest, tryptic digest) Fleischwasser, Fleischbrühe, Fleischsuppe
meat inspection Fleischbeschau
mechanical stage *micros* Kreuztisch
Meckel's cartilage Meckelscher Knorpel (Mandibulare)
Meckel's diverticulum Meckelsches Divertikel
media/median vein (insect wing) Media, Medialader (Medianzelle)
medial moraine Mittelmoräne
median covert mittlere Decke
median eye/midline eye (a dorsal ocellus) Mittelauge, Medianauge (Stirnocelle)
median layer/median zone Mittelschicht
median lethal dose (LD$_{50}$) mittlere Letaldosis, mittlere letale Dosis
median longitudinal plane Sagittalebene (parallel zur Mittellinie)
median value *stat* Medianwert, Zentralwert
median vein/media (insect wing) Media, Medialader (Medianzelle)
mediator Vermittler, Mediator
medical examination/medical exam/physical examination/ physical medizinische (körperliche) Untersuchung
medical lab technician medizinisch-technische(r) LaborassistentIn (MTLA)
medicinal plant Arzneipflanze, Heilpflanze
medicine/drug Medikament, Medizin, Droge
medium/culture medium/nutrient medium Medium, Kulturmedium, Nährmedium
• **complete medium** Komplettmedium, Vollmedium
• **complex medium** komplexes Medium
• **conditioned medium** konditioniertes Medium
• **deficiency medium** Mangelmedium
• **defined medium** synthetisches Medium (chemisch definiertes Medium)
• **differential medium** Differenzierungsmedium
• **egg medium** Eiermedium, Eiernährmedium
• **enrichment medium** Anreicherungsmedium
• **maintenance medium** Erhaltungsmedium
• **minimal medium** Minimalmedium
• **rich medium/complete medium** Vollmedium, Komplettmedium
• **selective medium** Selektivmedium, Elektivmedium
• **test medium (for diagnosis)** Testmedium, Prüfmedium
medulla/pith/core Mark
medulla oblongata verlängertes Rückenmark
medullar/medullary/pithy medullär, markhaltig, markig, Mark...
medullar gall/mark gall Markgalle
medullary cavity/marrow cavity Markhöhle
medullary tube/tubus medullaris *embr* Medullarrohr, Markrohr, Neuralrohr
medullation Verkernung
medusa Meduse, "Qualle"
medusas Quallen
• **box jellies/sea wasps/ cubomedusas/Cubomedusae/ Cubozoa** Würfelquallen
• **comb jellies/sea combs/sea gooseberries/sea walnuts/ ctenophores/Ctenophora** Rippenquallen, Kammquallen, Ctenophoren
• **coronate medusas/Coronatae** Kranzquallen, Tiefseequallen
• **cup animals/scyphozoans/ Scyphozoa** Schirmquallen, Scheibenquallen, Echte Quallen, Scyphozoen
• **rhizostome medusas/ Rhizostomeae** Wurzelmundquallen
• **semeostome medusas/ Semaeostomeae** Fahnenquallen, Fahnenmundquallen
• **siphonophorans/Siphonophora** Siphonophoren, Staatsquallen
• **stauromedusas/Staruomedusae** Stielquallen, Becherquallen
• **tentaculiferans/"tentaculates" (Ctenophora)** Tentaculiferen, tentakeltragende Rippenquallen
megalopterans: dobsonflies, fishflies, alderflies (neuropterans)/Megaloptera Schlammfliegen
megaphyllous großblättrig
megasequencing *gen* Megasequenzierung
megaspore/macrospore Megaspore, Makrospore

megaspore mother cell

megaspore mother cell/macrospore mother cell/megasporocyte Megasporenmutterzelle, Makrosporenmutterzelle; Embryosackmutterzelle
Mehlis' gland/shell gland (cement gland) Mehlissche Drüse, Schalendrüse
meiofauna/mesofauna (0.2–2 mm) Meiofauna, Mesofauna
meiosis/reduction division Meiose, Reifeteilung, Reduktionsteilung
meiotic nondisjunction meiotische Non-Disjunktion, Chromosomenfehlverteilung bei der Meiose
Meissner's corpuscle/corpuscle of touch Meissner Körperchen, Meissner-Tastkörperchen
melanization Melanisierung
melanocyte Melanozyt, Melanocyt
melanocyte-stimulating hormone (MSH) Melanotropin, melanozytenstimulierendes Hormon (MSH)
melanoliberin/melanotropin releasing hormone/melanotropin releasing factor (MRH, MRF) Melanoliberin, Melanotropin-Freisetzungshormon
melanoma Melanom
- **malignant melanoma** malignes Melanom, schwarzer Hautkrebs

melanophage Melanophage
melastome family/meadow-beauty family/Melastomataceae Schwarzmundgewächse
melatonin Melatonin
Meliaceae/mahogany family Zedrachgewächse
Melianthaceae/honeybush family Honigstrauchgewächse
melittophily/bee pollination Bienenbestäubung
mellowness Gare (Boden)
melon (cetaceans) Melone
melt *n* Schmelze
melt *vb* schmelzen, aufschmelzen
melting curve Schmelzkurve
melting point Schmelzpunkt
melting temperature Schmelztemperatur
meltwater Schmelzwasser
membrane Membran
- **basement membrane/basal lamina** Basalmembran, Basallamina
- **basilar membrane** Basilarmembran
- **cell membrane** Zellmembran
- **cerebral membrane/meninx** Hirnhaut, Meninx
- **double membrane** Doppelmembran
- **egg membrane** "Eimembran", Eihaut, Eihülle
- **extraembryonic membrane** Embryonalhülle, Keimhülle
- **fertilization membrane** Befruchtungsmembran
- **fetal membranes** Eihäute
- **frontal membrane (bryozoans)** Frontalmembran
- **mucous membrane/mucosa** Schleimhaut, Schleimhautepithel
- **nuclear membrane** Kernmembran
- **outer membrane** Außenmembran
- **peritrophic membrane** peritrophische Membran
- **plasma membrane/(outer) cell membrane/unit membrane/ ectoplast/plasmalemma** Plasmamembran, Zellmembran, Ektoplast, Plasmalemma
- **stacked membranes** Membranstapel
- **tympanic membrane/eardrum/ tympanum** Trommelfell, Ohrtrommel, Tympanalmembran, Tympanum
- **undulating membrane** undulierende Membran
- **unit membrane/double membrane** Elementarmembran, Doppelmembran
- **vitelline membrane/vitelline layer/membrana vitellina** Vitellinmembran, Dotterhaut, Dottermembran, primäre Eihülle

membrane attack complex *immun* Membran-Angriffskomplex
membrane capacitance Membrankapazität
membrane channel Membrankanal
- **ion channel** Ionenkanal
- **ligand-gated channel** ligandenregulierter/ ligandengesteuerter Kanal
- **mechanically gated channel** mechanisch gesteuerter Kanal
- **resting channel/leakage channel** Ruhemembrankanal, Leckkanal
- **voltage-sensitive channel/ voltage-gated channel** spannungsregulierter/ spannungsgesteuerter Kanal

membrane conductance Membranleitfähigkeit
membrane current Membranstrom
membrane filter Membranfilter
membrane flow Membranfluss
membrane flux Membrandurchfluss
membrane fusion Membranfusion

membrane ghost Membran-Ghost (künstlich hergestellte leere Membran)
membrane length constant (space constant) Membranlängskonstante (Raumkonstante)
membrane potential Membranpotential
membrane protein Membranprotein
membrane reactor Membranreaktor
membrane time constant Membranzeitkonstante
membrane trafficking Transport durch eine Membran hindurch
membrane transport Membrantransport
membrane-bound membrangebunden
membrane-coated membranumgeben
membranelle Membranelle
membranous membranös, membranartig, häutig
memnospore (remaining at place of origin) Memnospore
memory Gedächtnis, Erinnerungsvermögen
  • **long-term memory** Langzeitgedächtnis
  • **short-term memory** Kurzzeitgedächtnis
memory cell Gedächtniszelle
menadione (vitamin K₃) Menadion
menaquinone (vitamin K₂) Menachinon
menarche (first menstruation) Menarche (erste Menstruation)
Mendel's law(s) Mendelsche(s) Gesetz(e)
mendelian *adj/adv* mendelnd, nach den Mendelschen Gesetzen vererbt
Mendelian Inheritance in Man (MIM) Mendelsche Vererbung beim Menschen
mendelize mendeln
meninx (*pl* **meninga/meninges)** Hirnhaut, Gehirnhaut
meniscus (articular disk) Meniskus (Gelenkzwischenscheibe)
Menispermaceae/moonseed family Mondsamengewächse
menopause (cessation of ovulation/ menstruation) Menopause
menstrual cycle Menstruationszyklus
menstruate menstruieren
menstruation/period Menstruation, Periode, Regel, Monatsblutung, Blutung
Menyanthaceae/bogbean family Bitterkleegewächse, Fieberkleegewächse

MER-family (MER=medium reiteration frequency) (DNA element) MER-Familie (MER=mittlere Wiederholungshäufigkeit) (DNA-Element)
meranti family/dipterocarpus family/Dipterocarpaceae Zweiflügelfruchtgewächse, Flügelnussgewächse
mercury Quecksilber
mericarp Merikarp, Teilfrucht
mericlinal chimera Meriklinalchimäre
meridional canal/gastrovascular canal (ctenophores) Meridionalkanal, Rippengefäß
meridional cleavage Meridionalfurchung
meristem Meristem, Bildungsgewebe
  • **apical meristem/growing point** Spitzenmeristem, Scheitelmeristem, Wachstumspunkt, Vegetationspunkt
  • **axillary meristem** Achselmeristem
  • **block meristem** Blockmeristem
  • **flank meristem/peripheral meristem** Flankenmeristem
  • **ground meristem** Grundmeristem
  • **intercalary meristem** interkalares Meristem, Restmeristem
  • **lateral meristem** laterales Meristem
  • **marginal meristem** Randmeristem
  • **plate meristem** Plattenmeristem
  • **primary thickening (PTM) meristem** primärer Meristemmantel
  • **rib meristem/file meristem** Rippenmeristem
  • **secondary meristem** Folgemeristem
  • **terminal meristem** Endmeristem
  • **tiered meristem** Etagenmeristem
Merkel cell Merkelzelle
Merkel's corpuscle/Merkel's disk/ tactile disk Merkelsches Körperchen
mermaid's purse/sea purse Seemaus, Eikapsel der Knorpelfische
meroblastic cleavage/incomplete cleavage meroblastische Furchung, unvollständige Furchung, partielle Furchung
merocrine gland merokrine Drüse
merocyte Merocyt, Merozyt
merogamy Merogamie
merogenesis/segmentation Merogenese, Segmentierung

M merognathite 210

merognathite Merognathit
merogony Merogonie
meromictic meromiktisch
meromyosin Meromyosin
meropodite Meropodit
merospermy Merospermie
**merostomes/merostomates/
Merostomata** Hüftmünder
Mertensian mimicry Mertenssche
Mimikry
**mesembryanthemum family/fig
marigold family/carpetweed
family/Aizoaceae**
Mittagsblumengewächse,
Eiskrautgewächse
mesenchymal mesenchymatisch
mesenchyme Mesenchym
(embryonales Bindegewebe)
mesentery Mesenterium, Gekröse
mesh web *arach* Maschennetz
meshy maschig
mesic/moderately moist
gekennzeichnet durch mittlere
Feuchtigkeitsmenge
mesocarp Mesokarp
mesocoel/mesocoele Mesocöl,
Mesocoel
mesoderm Mesoderm, sekundäres
Keimblatt
**mesogastropods: periwinkles &
cowries/Mesogastropoda/
Taenioglossa** Mittelschnecken
mesogloea/mesoglea Mesoglöa,
Stützschicht
mesohyl Mesohyl
mesolecithal mesolecithal,
mesolezithal, mäßig dotterreich
mesomerism Mesomerie
mesonotum Mesonotum,
Mesothorakalschild (dorsal)
mesophilic mesophil
**mesophyll (spongy + palisade
parenchyma)** Mesophyll
mesophyte Mesophyt
Mesophytic Era Mesophytikum
mesoptile Mesoptile, Zwischenfeder
mesosaprobes Mesosaprobien
mesosaurs/Mesosauria
Rechengebissechsen
mesoscutellum (bugs)
Mesoscutellum
mesosome Mesosom
mesothelium Mesothel
mesothorax Mesothorax, Mittelbrust
**mesotrophic (intermediate levels of
minerals)** mesotroph
**Mesozoic/Mesozoic Era (geological
age)** Mesozoikum, Erdmittelalter
messenger Bote, Botenstoff
messenger RNA/mRNA Messenger-
RNA, Boten-RNA, mRNA
metabiosis Metabiose

metabolic metabolisch,
sich verwandelnd
**metabolic disturbance/metabolic
derangement** Stoffwechselstörung
**metabolic pathway/metabolic
shunt** Stoffwechselweg
metabolic rate Metabolismusrate,
Stoffwechselrate, Energieumsatzrate
• **active metabolic rate**
Arbeitsumsatz, Leistungsumsatz
**metabolic scope/index of metabolic
expansibility** metabolisches
Spektrum, Stoffwechselspektrum
metabolic turnover
Stoffwechselumsatz
metabolism Metabolismus,
Haushalt, Stoffwechsel;
Verwandlung, Formänderung
• **active metabolism**
Arbeitsstoffwechsel,
Leistungsstoffwechsel
• **basal metabolism**
Grundstoffwechsel,
Ruhestoffwechsel
• **cellular metabolism**
Zellstoffwechsel
• **energy metabolism**
Energiestoffwechsel
• **holometabolism** Holometabolie
• **intermediary metabolism**
intermediärer Stoffwechsel,
Zwischenstoffwechsel
• **maintenance metabolism**
Betriebsstoffwechsel
• **primary metabolism**
Primärstoffwechsel
• **secondary metabolism**
Sekundärstoffwechsel
• **synthetic metabolism/synthetic
reactions/anabolism**
Synthesestoffwechsel,
Anabolismus
metabolite Metabolit,
Stoffwechselprodukt
metabolize umwandeln
(Stoffwechsel), "verstoffwechseln"
metacarpal *n* Metacarpus, Mittelhand
metacarpal bone
Mittelhandknochen
metacentric chromosome
metazentrisches Chromosom
metacercaria/adolescaria
Metacercarie
**metachromatic granules/volutin
granules** metachromatische
Granula (*pl*)
**metagenesis (alternation of
generations)** Metagenese
(Generationswechsel)
**metal-ore leaching, microbial/
microbial leaching of metal ores**
mikrobielle Erzlaugung

micromere

metallothionein Metallothionein
metamere/segment Metamer, echtes Segment
metamerism/segmentation Metamerie, Segmentierung
metamorphic metamorph, metamorphisch, die Gestalt verändernd
metamorphose/metamorphize *vb* metamorphosieren (sich verwandeln)
metamorphosis Metamorphose, Umwandlung, Verwandlung (*geol* metamorphism Metamorphismus)
- **complete m.** vollkommene/vollständige Metamorphose, vollkommene/vollständige Verwandlung
- **incomplete m.** unvollkommene/unvollständige Metamorphose, unvollkommene/unvollständige Verwandlung

metamorphotic metamorphotisch
metanephros/hind kidney/ definitive kidney Metanephros, Nachniere, definitive Niere
metanotum Metanotum, Metathorakalschild (dorsal)
metaphase chromosome Metaphasenchromosom
metaphyll Folgeblatt
metaphyte (multicellular plant) Metaphyt
metasaprobity Metasaprobität
metastasis Metastase, Tochtergeschwulst
metastasize metastasieren, Metastasen bilden
metatarsal *n* Metatarsus, Mittelfuß
metatarsal bone Mittelfußknochen
metatarsal gland Metatarsaldrüse
metathorax Metathorax, Hinterbrust
metatroph metatrophic
metaxyphyll Zwischenblatt
metazoans/Metazoa Metazoen, "Vielzeller", Mitteltiere, Gewebetiere
meteorology (study of weather and weather forecasting) Meteorologie, Wetterkunde
metestrus Metöstrus, Metoestrus, Nachbrunst
methanogenic methanbildend, methanogen
methanogenic organism/ methanogen Methanbildner
methanophile methanophil
methionine Methionin
method of estimation *stat* Schätzverfahren

methroxate Methroxat
methylate methylieren
methylation Methylierung, Methylieren
metoestrus *see* metestrus
metric scale metrische Skala
metric unit metrische Einheit
metrological messtechnisch
metrology Messtechnik
mevalonic acid (mevalonate) Mevalonsäure (Mevalonat)
mezereum family/daphne family/ Thymelaeaceae Spatzenzungengewächse, Seidelbastgewächse
micellation Micellierung
micelle Mizelle
Michaelis constant/ Michaelis-Menten constant Michaeliskonstante, Halbsättigungskonstante (K_M)
Michaelis-Menten equation Michaelis-Menten-Gleichung
micrergate/microergate/ dwarf worker (ants) Mikrergat
microanalysis Mikroanalyse
microanatomy/histology Mikroanatomie, Histologie
microbe/microorganism Mikrobe, Mikroorganismus
microbial mikrobiell
microbial metal-ore leaching/ microbial leaching of metal ores mikrobielle Erzlaugung
microbiological mikrobiologisch
microbiologist Mikrobiologe
microbiology Mikrobiologie
microbody Mikrobody, Mikrokörperchen
microcell Mikrozelle
microcephalic/microcephalous mikrocephal, kleinköpfig
microclimate Mikroklima
microcosm Mikrokosmos
microfauna Mikrofauna, Kleintierwelt
microfilament/actin filament Mikrofilament, Aktinfilament, Actinfilament
microglia Mikroglia
microglial cell Mikrogliazelle
micrograph/microscopic image/ microscopic picture mikroskopische Aufnahme, mikroskopisches Bild
microinjection Mikroinjektion
micromanipulation Mikromanipulation
micromanipulator Mikromanipulator
micromere Mikromer

micrometer screw/fine-adjustment knob/fine-adjustment *micros* Mikrometerschraube
micronema (*pl* **micronemas)** Mikronema (*pl* Mikronemen)
micronucleus Mikronukleus, generativer Zellkern
micronutrient/trace element Spurenelement
microorganism/microbe Mikroorganismus (*pl* Mikroorganismen)
microphage (small phagocyte) Mikrophage
micropipet Mikropipette
micropipet tip Mikropipettenspitze
microprobe/probe Sonde
• **proton microprobe** Protonensonde
microprocedure Mikroverfahren
micropropagation Mikrovermehrung
micropyle Mikropyle, Keimmund
microsatellite Mikrosatellit
microscope Mikroskop
• **compound microscope/light microscope** zusammengesetztes Mikroskop, Lichtmikroskop
• **condenser** Kondensor
• **course microscope** Kursmikroskop
• **dissecting microscope** Präpariermikroskop
• **electron microscope** Elektronenmikroskop
• **examination under a microscope/usage of a microscope** Mikroskopieren *n*
• **examine under a microscope/use a microscope** mikroskopieren
• **light microscope/compound microscope** Lichtmikroskop, zusammengesetztes Mikroskop
• **polarizing microscope** Polarisationsmikroskop
• **scanning electron microscope** Rasterelektronenmikroskop
• **stereo microscope** Stereomikroskop
• **transmission electron microscope** Transmissionselektronenmikroskop
microscope accessories Mikroskopzubehör
microscope arm/limb Arm, Trägerarm
microscope clip/stage clip Objekttischklammer
microscope depression slide/concavity slide/cavity slide Objektträger mit Vertiefung
microscope foot/base Fuß, Basis
microscope illuminator Mikroskopierleuchte
microscope pillar Säule
microscope slide Objektträger
• **prepared microscope slide** Mikropräparat
microscope stage Objekttisch
microscopic(al) mikroskopisch
microscopic image/microscopic picture/micrograph mikroskopisches Bild, mikroskopische Aufnahme
microscopic preparation/microscopic mount mikroskopisches Präparat
• **blood smear** Blutausstrich
• **permanent mount/permanent slide** mikroskopisches Dauerpräparat
• **scraping mount/scraping** Schabepräparat
• **squash mount/squash** Quetschpräparat
• **wet mount/wet preparation** Nasspräparat, Frischpräparat, Lebendpräparat, Nativpräparat
• **whole mount** Totalpräparat
microscopic procedure Mikroskopierverfahren
microscopy Mikroskopie
• **atomic force microscopy (AFM)** Rasterkraftmikroskopie
• **brightfield microscopy** Hellfeld-Mikroskopie
• **confocal laser scanning microscopy** konfokale Laser-Scanning Mikroskopie
• **cryo electron microscopy (IEM)** Kryoelektronenmikroskopie
• **darkfield microscopy** Dunkelfeld-Mikroskopie
• **force microscopy** Kraftmikroskopie
• **high voltage electron microscopy (HVEM)** Hochspannungselektronenmikroskopie
• **immune electron microscopy** Immun-Elektronenmikroskopie
• **interference microscopy** Interferenzmikroskopie
• **light microscopy (LM)** Lichtmikroskopie
• **phase contrast microscopy** Phasenkontrastmikroskopie
• **polarizing microscopy** Polarisationsmikroskopie
• **scanning electron microscopy (SEM)** Rasterelektronenmikroskopie (REM)
• **scanning force microscopy (SFM)** Raster-Kraftmikroskopie (RKM)

213 **milk stage** M

- **scanning tunneling microscopy (STM)** Raster-Tunnelmikroskopie (RTM)
- **transmission electron microscopy (TEM)** Transmissionselektronenmikroskopie, Durchstrahlungselektronenmikroskopie

microscopy accessories Mikroskopierzubehör

microspecies Mikrospezies, Kleinart

microsphere Mikrosphäre

microsporangiate cone/ pollen-bearing cone männlicher Zapfen

microspore Mikrospore

microtome Mikrotom
- **freezing microtome/ cryomicrotome** Gefriermikrotom
- **rotary microtome** Rotationsmikrotom
- **sliding microtome** Schlittenmikrotom
- **ultramicrotome** Ultramikrotom

microtome blade Mikrotommesser

microtome chuck Mikrotom-Präparatehalter, Objekthalter (Spannkopf)

microtomy Mikrotomie

microtrabecular network Mikrotrabekulargeflecht

microtubule Mikrotubulus (*pl* Mikrotubuli)

microtubule organizing center (MTOC) Mikrotubulus-Organisationszentrum

microtubule-associated protein (MAP) mikrotubuliassoziertes Protein

microvillus (*pl* microvilli) Mikrovillus (*pl* Mikrovilli); Stereocilien (Lateralisorgan)

microwhipscorpions/palpigrades/ Palpigradi Palpigraden

mictic miktisch

micturate/urinate miktuieren, harnen, urinieren

micturition/urination Miktion, Harnen, Harnlassen, Urinieren

mid intestine Mitteldarm (*sensu stricto*)

midbrain/mesencephalon Mittelhirn

middle ear/midear Mittelohr

middle lamella Mittellamelle

Middle Triassic (epoch) Muschelkalk

middlings/shorts (from wheat milling) Mittelmehl (Weizenfuttermehl)

midgut *embr* Mitteldarm

midgut/intestine Nahrungsdarm

midgut/mesenteron/ventricle/ ventriculus/"stomach" *entom* Mitteldarm, Mesenteron, "Magen", Ventriculus (Insekten)

midgut gland/digestive gland/ "liver" Mitteldarmdrüse, "Leber"

midgut gland/hepatopancreas Mitteldarmdrüse, Hepatopankreas
- **midgut diverticulum/cecum** Mitteldarmdivertikel, Mitteldarmventrikel (Blindsack)

midparent value Elternmittelwert

midrib/midvein/costa Mittelrippe, Costa (*see also:* rachis)

mignonette family/Resedaceae Resedagewächse, Resedengewächse, Waugewächse

migrate wandern

migration *biogeo* Migration, Wanderung, Zug (Vögel)

migration *chromat/electrophor* Wanderung

migratory animal Durchzügler

migratory behavior Wanderverhalten

migratory bird Zugvogel

migratory fish Zugfisch

migratory restlessness Zugunruhe

migratory route Zugstraße

mildew *fung* Mehltaupilz
- **black mildews/Meliolales** schwarze Mehltaupilze
- **downy mildews/Peronosporaceae** falsche Mehltaupilze
- **powdery mildews/Erysiphales** echte Mehltaupilze

milfoil family/water milfoil family/ Haloragaceae Seebeerengewächse, Tausendblattgewächse

milk *n* Milch
- **bee milk/royal jelly** Königin-Futtersaft, Gelée Royale
- **crop milk/pigeon milk (milky secretion from crop lining)** Kropfmilch, Kropfsekret (Tauben)
- **crude milk** Rohmilch
- **curd** geronnene, dicke Milch
- **foremilk/colostrum** Vormilch, Biestmilch, Kolostralmilch
- **treading/kneading (milk elicitation movement)** Milchtritt
- **uterine milk** *ichth/entom* Uterusmilch, Uterinmilch

milk *vb* melken

milk cistern/lactiferous sinus Milchzisterne, Milchsinus

milk duct/lactiferous duct Milchgang

milk line/mammary ridge Milchleiste

milk stage/milk ripeness (grain) Milchreife

milk sugar/lactose Milchzucker, Laktose
milk teeth/deciduous teeth/first teeth/primary teeth Milchzähne
milk vein (e.g. cow) Milchvene
milk well (e.g. cow) Milchgrube
milkfishes and relatives/ Gonorhynchiformes Milchfischverwandte, Sandfische
milkweed family/Asclepiadaceae Schwalbenwurzgewächse, Seidenpflanzengewächse
milkwort family/Polygalaceae Kreuzblümchengewächse, Kreuzblumengewächse
mill/shape vb **(wood)** fräsen (Holz)
milleporine hydrocorals/fire corals/ stinging corals/Milleporina Feuerkorallen
milling (fish school) Kreisen (Fischschwärme)
millipedes/"thousand-leggers"/ diplopods/myriapodians/ Myriapoda Doppelfüßer, Diplopoden, Tausendfüßler, Myriapoden
milt (sperm-containing liquid of male fish) Milch (Spermaflüssigkeit der männlichen Fische)
milt/soft roe (testes of fish) Fischhoden
milter (male fish during spawning season) Milchner, Milcher
mimesis Mimese
mimic n Mimik
mimic/imitate vb mimen, nachahmen, nachmachen ("nachäffen")
mimic genes mimische Gene
mimicry Mimikry, Schutztracht, Warntracht, Angleichung, schützende Nachahmung
• **aggressive mimicry/Peckhamian mimicry** Angriffs-Mimikry, Peckhamsche Mimikry
• **automimicry** Automimikry
• **Batesian mimicry** Batessche Mimikry
• **behavioral mimicry/ethomimicry** Verhaltensmimikry, Ethomimikry
• **Mertensian mimicry** Mertenssche Mimikry
• **Muellerian mimicry** Müllersche Mimikry
• **Peckhamian mimicry/aggressive mimicry** Peckhamsche Mimikry, Angriffs-Mimikry
• **protective mimicry** Verteidigungs-Mimikry
• **vocal mimicry** stimmliche Nachahmung

mimosa family/Mimosaceae Mimosengewächse
mine (damage by insect larvae on leaves) Mine, Blattmine, Fraßgang
• **blotch mine** Platzmine
• **serpentine mine/heliconome** Spiralmine
• **star mine/asteronome** Sternmine
mineral n Mineral (pl Mineralien)
mineral adj Mineral.., anorganisch
mineral oil Mineralöl
mineral soil Mineralboden
mineralization Mineralisation, Mineralisierung
mineralocorticoid Mineralokortikoid, Mineralocorticoid
minerotrophic minerotroph
miniature endplate potential (mepps) neuro Miniaturenplattenpotential, Miniaturenendplattenpotential (MEPP)
minichromosome Minichromosom
minigene Minigen
minimal inhibitory concentration/ minimum inhibitory concentration (MIC) minimale Hemmkonzentration (MHK)
minimal medium Minimalmedium
miniprep/minipreparation Miniprep, Minipräparation
minisatellite gen Minisatellit
minor groove (DNA) kleine Furche, kleine Rinne, flache Rinne (DNA-Struktur)
minor histocompatibility antigens Nebenhistokompatibilitätsantigene
minor histocompatibility complex Nebenhistokompatibilitätskomplex
mint family/deadnettle family/ Lamiaceae/Labiatae Lippenblütengewächse, Lippenblütler
minus strand (noncoding strand) gen Minus-Strang, Negativ-Strang (nichtcodierender Strang)
minute respiratory volume Atemminutenvolumen (AMV)
Miocene/Miocene Epoch Miozän (erdgeschichtliche Epoche)
miracidium (fluke larva) Miracidium (pl Miracidien), Mirazidium (Digenea-Larve)
mire (European: from old Norse term)/peatland (peat-forming wetlands: bogs & fens) n Moor
miscarriage Fehlgeburt, Spontanabort
miscibility Mischbarkeit
miscible mischbar

mismatch (of bases) *gen*
Fehlpaarung, Basenfehlpaarung
mismatch DNA repair
Fehlpaarungsreparatur
Misodendraceae/feathery mistletoe family Federmistelgewächse
mispairing of chromosomes
Fehlpaarung,
Chromosomenfehlpaarung
• **slipped strand mispairing/ slippage replication/replication slippage** Fehlpaarung durch Strangverschiebung
missense mutation Missense-Mutation, Fehlsinnmutation
missing contact analysis
Kontaktpunktanalyse
missing link fehlende Zwischenstufe
Mississippian/Lower Carboniferous
Frühes Karbon
mist/drizzle *n* Sprühregen
mist/slight fog *n* leichter Nebel
mistletoe family (showy mistletoe family)/Loranthaceae
Mistelgewächse,
Riemenblumengewächse
misty dunstig, leicht nebelig
mites & ticks/Acari/Acarina
Milben & Zecken
mitochondrial crista (*pl* **cristae/ cristas)** Crista (*pl* Cristae) (mitochondrial)
mitochondrial inheritance
mitochondriale Vererbung
mitochondrion (*pl* **mitochondria/ mitochondrions)** Mitochondrion, Mitochondrium (*pl* Mitochondrien)
mitosis/nuclear division/duplication division Mitose, Kernteilung
mitotic mitotisch
mitotic cycle Mitosezyklus
mitotic recombination
mitotische Rekombination
mitotic stage Mitosestadium
mitral cell Mitralzelle
mitral valve/bicuspid valve (heart)
Mitralklappe, Bikuspidalklappe
mix *vb* mixen, mischen
mix *n* Mix, Mischung
mixed antiserum Mischantiserum
mixed crop/mixed stand
Mischkultur
mixed culture Mischkultur
mixed forest Mischwald
mixed-function oxidase
mischfunktionelle Oxidase
mixer/blender (vortex) Mixer, Mixette, Küchenmaschine (Vortex)
mixing Vermischung
mixis Mixis
mixoploid mixoploid
mixoploidy Mixoploidie

mixotrophic/mesotrophic
mixotroph
mixotropic series mixotrope Reihe
mixture Mischung, Gemenge
mneme Gedächtnis, Erinnerung
moan stöhnen, ächzen
moas/Dinornithiformes Moas
moat Wassergraben
mobbing behavior Hassen, Hassverhalten
mobbing call *orn* Sammelruf
mobile/vagile/wandering
beweglich, mobil, vagil
(Ortsveränderung des Gesamtorganismus)
mobility/vagility Beweglichkeit, Mobilität, Vagilität (Ortsveränderung des Gesamtorganismus)
mobility shift experiment
Gelretardationsexperiment
mock-hunting Jagdspiel
modal value *stat* Modalwert
mode Modus, Art und Weise, Modalwert
mode of action/mechanism
Wirkungsweise, Mechanismus
mode of inheritance
Vererbungsmodus, Erbgang
model building Modellbau
moder (humus layer) Moder
modern bowfin/Amiiformes
Kahlhechte (Schlammfische)
modification Modifikation, Veränderung
modifier gene Modifikationsgen
modify modifizieren, verändern, abändern
module Modul, Funktionseinheit
moiety/part/section Teil (des Ganzen), Anteil, Hälfte
moist feucht
moistness Feuchte, Feuchtheit
moisture Feuchtigkeit
moisture capacity/water holding capacity of soil Wasserkapazität
molality Molalität
molar Molar, hinterer Backenzahn
molar mass ("molar weight")
molare Masse, Molmasse ("Molgewicht")
molar volume Molvolumen
mold/cast *paleo* Abguss (eines Fossils)
mold/mould/mildew (rot) *general*
Schimmel, Moder
mold/mould *fung* Schimmelpilz
• **acellular slime molds/plasmodial slime molds/Myxomycetes** echte Schleimpilze (plasmodial)
• **bread molds/zygospore fungi/ Zygomycetes (coenocytic fungi)**
Jochpilze, Zygomyceten

M mold 216

- cellular slime molds/
  Acrasiomyycetes/
  Dictyosteliomycetes
  (Myxomycota) Acrasiomyceten,
  zelluläre Schleimpilze
- **slime molds/Myxomycota**
  Schleimpilze
- **water molds/Saprolegniales**
  Wasserschimmel
moldy/mouldy/putrid/musty
  moderig (Geruch)
mole/jetty/breakwater Mole
molecular biology
  Molekularbiologie
molecular cytogenetics molekulare
  Cytogenetik
molecular formula Summenformel
molecular genetics
  Molekulargenetik
molecular ion Molekülion
**molecular mass ("molecular
  weight")** Molekülmasse
  ("Molekulargewicht")
molecular sieve Molekularsieb,
  Molekülsieb
**molecular sieve chromatography/
  gel permeation chromatography
  (gel filtration)**
  Molekularsiebchromatographie,
  Gelpermeationschromatographie
  (Gelfiltration)
**molecular weight/relative
  molecular mass (M_r)**
  Molekulargewicht, relative
  Molekülmasse (M_r)
molecule Molekül
molluscicide
  Schneckenbekämpfungsmittel,
  Molluskizid
mollusks/Mollusca Mollusken,
  Weichtiere
**molt/molting n (shedding
  plumage/feathers)**
  Gefiederwechsel, Mauser
- **shedding skin/shedding
  exoskeleton** Häutung
molt/shed feathers vb mausern
molt/shed skin or exoskeleton vb
  häuten
molting fluid/exuvial fluid
  Häutungsflüssigkeit,
  Ecdysialflüssigkeit,
  Exuvialflüssigkeit
**molting season/molting time/
  molting period/molt/deplumation**
  Mauser, Mauserzeit,
  Gefiederwechsel
molybdenum Molybdän
monascous monask
monecious (monoecious)
  monözisch, einhäusig,
  gemischtgeschlechtig

**monecy/monoecy/monecism/
  monoecism** Monözie,
  Einhäusigkeit,
  Gemischtgeschlechtigkeit
monestrous (monoestrous)
  monöstrisch
moniliform moniliat, perlschnurartig
moniliform antenna
  rosenkranzförmiger Fühler,
  Perlschnurfühler
**monimia family/boldo family/
  Monimiaceae** Monimiengewächse
monitor n Monitor, Bildschirm;
  Überwacher
monitor vb überwachen
monitoring Überwachung
monkey Äffchen
**monkey-puzzle tree family/
  araucaria family/Araucariaceae**
  Araukariengewächse
**monkfishes/angel sharks/
  Squatiniformes** Engelhaie,
  Engelhaiartige
**monocarp/monocarpic/
  hapaxanthic/hapaxanthous/
  hapanthous** monokarpisch,
  hapaxanth
monocarpellate fruit Einblattfrucht
monocentric species
  monozentrische Art
**monochasium/monochasial cyme/
  simple cyme** Monochasium,
  eingablige Trugdolde
**monochlamydeous/
  haplochlamydeous**
  monochlamydeisch,
  haplochlamydeisch, einfachblumen-
  blättrig, mit einfacher Blütenhülle
monocistronic monocistronisch,
  monozistronisch
monoclonal antibody (MAb)
  monoklonaler Antikörper (mAb)
monocolpate monocolpat
monocotyledon/monocot
  Monokotyle, Monokotyledone,
  Einkeimblättrige
monocotyledonous einkeimblättrig
monoculture Monokultur
monocyte/mononuclear leucocyte
  Monocyt, Monozyt
**monodelphous/monodelphic/
  monadelphous (having single
  female genital tract)**
  monodelphisch, einscheidig
monoecious/monecious monözisch,
  einhäusig, gemischtgeschlechtig
monoestrous see monestrous
monogamous monogam
monogamy Monogamie, Einehe
monogenic monogen
monogenic diseases monogene
  Krankheiten

Moringaceae

monogonont monogonont
monogynous monogyn, einweibig
monohybrid cross
  Monohybridkreuzung
monolayer cell culture
  Einschichtzellkultur
monolayer/monomolecular layer
  einlagige Schicht,
  monomolekulare Schicht
monomictic monomiktisch
monomitic monomitisch
monomorphic/monomorphous
  monomorph, gleichgestaltet
monomorphism Monomorphismus
mononuclear mononukleär,
  mononucleär
**monophagous/monotrophic/
univorous** monophag,
  monotroph
monophasic monophasisch,
  einphasisch
monophyletic monophyletisch
monophyodont monophyodont
**monoplacophorans/
Monoplacophora**
  Urmützenschnecken, Einplatter,
  Monoplacophoren
monopodial (indeterminate)
  monopodial
monopodial branching system
  monopodiales Verzweigungssystem
monopodium Monopodium
monoprotic acid einwertige,
  einprotonige Säure
monospecific monospezifisch
monospecificity Monospezifität
monosymmetrical/zygomorphic
  monosymmetrisch, zygomorph
**monothalamous/monothalamic/
single-chambered/monothecal**
  monothalam, einkammerig,
  einkämmrig, monothekal
monothetic monothetisch
monotokous monotok
**monotremes (prototherians)/
Monotremata (Prototheria)**
  Kloakentiere
monotrichous monotrich
Monotropaceae/Indian pipe family
  Fichtenspargelgewächse
monotypic species
  monotypische Art
monounsaturated
  einfach ungesättigt
monounsaturated fatty acid
  einfach ungesättigte Fettsäure
monoxenous/monoxenic einwirtig,
  homoxen
monozygosity Monozygotie,
  Eineiigkeit
monozygous/monozygotic eineiig,
  monozygot

**monozygous twins/monozygotic
twins/identical twins**
  eineiige Zwillinge
monsoon forest Monsunwald
montane/mountain adj/adv
  montan, Berg…, Gebirgs…
montane forest Bergwald
  (immergrüne Coniferenstufe)
montane heathland Bergheide
montane perennial herb
  Hochstaude
montane plant Bergpflanze,
  Gebirgspflanze
montane rain forest Bergregenwald,
  Nebelwald
montane zone/montane region
  Bergstufe, Bergwaldstufe,
  montane Stufe
moo (cattle) muhen
moonfishes/Lampriformes
  Glanzfische, Glanzfischartige,
  Gotteslachsverwandte
moonseed family/Menispermaceae
  Mondsamengewächse
moor/peatland (bogs/fens) Moor,
  Torfmoor; (raised bog) Hochmoor;
  (dry) Bergheide; Heidemoor
mooring thread/guyline arach
  Spannfaden
moorland Moorlandschaft,
  Sumpflandschaft; Heideland
mor humus (acid pH) Rohhumus,
  saurer Auflagehumus; Trockentorf
**Moraceae/fig family/mulberry
family** Maulbeergewächse
moraine/till/glacial till Moräne,
  Gletscherschutt, Gletschergeröll
  • **end moraine** Endmoräne
  • **ground moraine/basal moraine**
    Grundmoräne, Untermoräne
  • **lateral moraine** Seitenmoräne
  • **medial moraine** Mittelmoräne
  • **terminal moraine** Frontalmoräne,
    Stirnmoräne
morbid morbid, erkrankt, krankhaft,
  kränklich
morbidity Morbidität (Häufigkeit
  der Erkrankungen), Erkrankungsrate
**mordant (fixing dye onto
specimen)** Beize,
  Beizenfärbungsmittel
**more-developed countries (MDCs)/
developed countries/
industrialized nations/core
countries**) Industrieländer
morels/morel family/Morchellaceae
  Morcheln
**Moringaceae/horseradish tree
family** Moringagewächse,
  Bennussgewächse,
  Behennussgewächse,
  Pferderettichgewächse

mormon tea family 218

mormon tea family/joint-pine family/ephedra family/Ephedraceae Meerträubelgewächse
mormyrids/Mormyriformes Nilhechte
morning glory family/bindweed family/convolvulus family/Convolvulaceae Windengewächse
morph Morphe
morphogenesis Morphogenese
morphogenetic morphogenetisch
morphology (external/descriptive morphology) Eidonomie (Morphologie der äußeren Gestalt)
- **causal morphology** Entwicklungsmechanik
- **comparative morphology** vergleichende Morphologie
morphometrics Morphometrie
morphopoesis Morphopoese
morphospecies Morphospezies, morphologische Art
mortal sterblich
mortality Mortalität, Sterblichkeit, Sterberate
mortality rate Absterberate
mortar (and pestle) Mörser (und Stößel/Pistill)
morula Morula, Maulbeerkeim
mosaic Mosaik
- **germline mosaic/germinal mosaic/gonadal mosaic/gonosomal mosaic** Keimbahnmosaik
- **gonadic mosaic** gonadales Mosaik
mosaic bilayer model Mosaikdoppelschichtmodel
mosaic development *embr/gen* Mosaikentwicklung
mosaic egg Mosaikei
mosaic gene Mosaikgen, gestückeltes Gen
mosaicism *gen* Vorkommen eines Gens im Mosaik
moschatel family/Adoxaceae Moschuskrautgewächse
mosquito fern family/duckweed fern family/Azollaceae Algenfarngewächse
mosquitoes/Nematocera (Diptera) Mücken & Schnaken
moss Moos, Laubmoos
moss animals/bryozoans/Ectoprocta/Polyzoa/Bryozoa Moostierchen, Bryozoen
moss carpet Moosteppich, Moospolster, Mooskissen
moss "flower" Moosblüte
moss layer Moosschicht
moss mat Moosdecke

moss peat/sphagnum peat/highmoor peat Hochmoortorf
mosses Moose, Laubmoose
mossy fiber Moosfaser
moth pollination/phalaenophily Mottenbestäubung, Phalaenophilie
moth-pollinated flower (geometers)/phalaenophile Mottenblume (Spanner)
moth-pollinated flower (hawkmoths)/sphingophile Nachtschwärmerblume
mother cell Mutterzelle
mother of vinegar/mother Essigmutter
mother plant Mutterpflanze
mother-of-pearl Perlmutter, Perlmutt
moths/Heterocera Motten
motile (capable of moving) motil, beweglich, bewegungsfähig (Bewegung eines Körperteils)
motile/vagile motil, beweglich, vagil
motility (capable of movement) Motilität, Beweglichkeit, Bewegungsvermögen (Bewegung eines Körperteils)
motility/vagility (free to move about) Motilität, Beweglichkeit, Vagilität (frei beweglich)
motion/movement/locomotion Bewegung, Fortbewegung, Lokomotion
- **rotational motion** Rotationsbewegung
- **translational motion** Translationsbewegung
- **vibrational motion** Schwingungsbewegung
motoneuron/motor neuron Motoneuron
motor cell/bulliform cell Motorzelle, motorische Zelle, Gelenkzelle (im Schwellkörper des Blattes)
motor endplate/myoneural junction motorische Endplatte
motor neuron/motoneuron Motoneuron
motor root/anterior root/ventral root *neuro* motorische Wurzel, Ventralwurzel
motor unit motorische Einheit
motoric/motor ... *physiol/neuro* motorisch, Motor...
mottled gefleckt, gesprenkelt
mould/mold *fung* Schimmelpilz
mould/mold/mildew (rot) Schimmel, Moder
mouldy/moldy/putrid/musty moderig (Geruch)
mound Erdhügel, Erdwall, Erddamm, Erhebung, kleiner Hügel

Müllerian inhibiting hormone **M**

**mound layering/stool layering/
stooling** *hort/bot*
Ablegervermehrung durch
Anhäufeln
mount/cover *vb* **(copulate)**
bespringen
mount *vb micros* präparieren
mount *n micros* Präparat
- **blood smear** Blutausstrich
- **microscopic preparation/
microscopic mount**
mikroskopisches Präparat
- **permanent mount/permanent
slide** mikroskopisches
Dauerpräparat
- **scraping mount/scraping**
Schabepräparat
- **squash mount/squash**
Quetschpräparat
- **wet mount/wet preparation**
Nasspräparat, Frischpräparat,
Lebendpräparat, Nativpräparat
- **whole mount** Totalpräparat
mountain Berg
(*pl* mountains > Berge/Gebirge)
**mountain chain/mountain range/
mountain ridge** Bergkette,
Gebirgskette
mountain crest/mountain ridge
Gebirgskamm, Berggrat
mountain forest/montane forest
Bergwald, Gebirgswald
mountain ridge/mountain crest
Gebirgskamm, Berggrat,
Bergrücken
mountain slope/hillslope
Berghang
mountainous gebirgig
mountains Gebirge
mountainside/mountain slope
Berghang
mountant/mounting medium
micros Einbettungsmittel,
Einschlussmittel
mousebirds/colies/Coliiformes
Mausvögel
mouth Mund; Mündung
**mouth cavity/oral cavity/buccal
cavity** Mundhöhle
mouth field/buccal field Mundfeld,
Buccalfeld
**mouth of the uterus/orifice of the
uterus/orificium uteri**
Gebärmuttermund, Muttermund
mouthbreeder Maulbrüter
**mouthbreeding/oral gestation/
buccal incubation** Maulbrüten
mouthpart/oral appendage
Mundgliedmaße
(*pl* Mundgliedmaßen)
mouthparts Mundwerkzeuge
move *vb* bewegen

movement/motion/locomotion
Bewegung, Fortbewegung,
Lokomotion
moving bed reactor (bioreactor)
Fließbettreaktor
MS (mass spectroscopy) MS
(Massenspektroskopie)
mucic acid Schleimsäure,
Mucinsäure
mucilage/slime (plant) Schleim
(pflanzlich)
mucilage cell Schleimzelle
mucilage gland Schleimdrüse
mucilaginous/glutinous/slimy
schleimig
mucilaginous canal
schleimführender Kanal
mucin Mucin
**muck (feces/urine from domestic
animals in wet state)** Mist
(Stallmist), Dung (flüssig)
muck (highly decomposed peat)
Sumpferde
mucous/slimy schleimig
mucous gland Schleimdrüse
mucous membrane/mucosa
Schleimhaut, Schleimhautepithel
mucoviscidosis/cystic fibrosis
Mukoviszidose, Mucoviszidose,
zystische Fibrose
mucro (sharp point) Mucrone
mucronate (sharp/hard pointed)
bot stachelspitz
mucronulate kleinspitzig
mucus/mucilage/slime/ooze
Schleim
mud (alluvial: silt/sludge)
Schlamm, Schlick
mud bottom Schlickgrund
mud flat Watt, Schlickwatt
Mueller... *see* Müller...
**mulberry family/fig family/
Moraceae** Maulbeergewächse
mulch *n* Mulch
mulch *vb* mulchen
mule (♀ horse x ♂ ass/donkey)
Maultier (Pferdestute x Eselhengst)
mull/mull humus (near neutral pH)
Mull (milder Dauerhumus)
Müller cell *ophth* Müller-Stützzelle
**Müller's duct/Mueller's duct/
Müller's canal/Müllerian duct/
paramesonephric duct**
Müllerscher Gang
Müller's larva/Mueller's larva
Müllersche Larve
**Müllerian fiber/Muellerian fiber/
fiber of Müller** *ophth*
Müller-Faser
**Müllerian inhibiting hormone/
Muellerian inhibiting hormone
(MIH)** Anti-Müller-Hormon (AMH)

Müllerian mimicry

Müllerian mimicry/Muellerian mimicry Müllersche Mimikry
multicellular vielzellig, mehrzellig
multicellular lifeform Vielzeller, vielzelliges Lebewesen
multicistronic/polycistronic multizistronisch, multicistronisch, polyzistronisch, polycistronisch
multicomponent virus Multikomponentenvirus
multidentate vielzähnig, mehrzähnig
multienzyme complex/multienzyme system Multienzymkomplex, Multienzymsystem, Enzymkette
multifactorial inheritance/polygenic inheritance multifaktorieller Erbgang, polygener Erbgang
multiforked chromosome Chromosom mit mehreren Replikationsgabeln
multifunctional vector/multipurpose vector Vielzweckvektor, multifunktioneller Vektor
multigene family Multigenfamilie
multilayered vielschichtig, mehrschichtig
multilocus enzyme electrophoresis (MLEE) Multilokus-Enzymelektrophorese
multinucleate(d)/multinuclear/polynucleate(d)/polynuclear vielkernig, mehrkernig
multiparous multipar (mehrmals geboren habend/mehrere Junge gleichzeitig werfen)
multipartite/pluripartite vielteilig
multiple alleles multiple Allele
multiple birth Mehrlingsgeburt
• **progeny of m.b.** Mehrlinge
multiple bond Mehrfachbindung
multiple cloning site (MCS) Vielzweckklonierungsstelle, Polylinker
multiple fission Vielfachteilung, Mehrfachteilung
multiple fruit/infructescence Fruchtstand, Fruchtverband
multiple sugar/polysaccharide Vielfachzucker, Polysaccharid
multiple-factor hypothesis *gen* Mehrfaktortheorie, Polygentheorie
multiplex family *gen* Familie mit mehreren befallenen Mitgliedern
multiplication Multiplikation, Vervielfältigung, Vermehrung
multiplicity of infection (m.o.i.) *vir* Multiplizität der Infektion, Infektionsmultiplizität
multipolar cell Multipolarzelle

multiseriate/multiple rowed/in several rows multiseriat, mehrreihig, vielreihig
multistage *adv/adj* mehrstufig
multistage impulse countercurrent impeller (bioreactor) Mehrstufen-Impuls-Gegenstrom (MIG) Rührer
multivesicular body multivesikulärer Körper
multivoltine/polyvoltine multivoltin, polyvoltin, plurivoltin
mummification Mummifizierung
mummify mumifizieren (ledern werden/trocken werden)
mummy Mumie
muramic acid Muraminsäure
murein Murein
muriform (like a brick wall) mauerförmig, mauerartig
murine Mäuse/Ratten betreffend, zu den Mäusen/Ratten gehörig, von Mäusen/Ratten stammend, Maus.../Ratten...
muscarine Muscarin
muscarinic receptor muscarinischer Rezeptor, muskarinischer Rezeptor
muscle (see *also*: musculature) Muskel
• **abductor muscle** Abzieher, Abduktor, Abductor(-Muskel)
• **adductor muscle** Anzieher, Schließmuskel, Adduktor, Adductor(-Muskel)
• **cardiac muscle** Herzmuskel
• **comminator muscle/interpyramid muscle** Interpyramidalmuskel, interpyramidaler Muskel (Aristotle's lantern)
• **depressor muscle** Depressor, Senker, Niederleger
• **diductor muscle** Klaffmuskel
• **dilator muscle** Dilator, Erweiterer
• **flexor** Flexor, Beuger
• **hair erector muscle/arrector pili muscle/*musculus arrector pili*** Haarmuskel
• **levator/lifter (muscle: raising an organ or part)** Levator, Heber
• **limb muscle/appendicular muscle** Extremitätenmuskel
• **longitudinal muscle** Längsmuskel
• **masticatory muscle/muscle of mastication** Kaumuskel
• **pedal retractor muscle** Fußretraktor
• **protractor muscle** Protraktor, Vorzieher, Vorwärtszieher
• **retractor muscle** Retraktormuskel, Retraktor, Rückzieher, Rückwärtszieher

mutation

- **ring muscle/circular muscle**
  Ringmuskel
- **rotator muscle** Rotator,
  Drehmuskel, Dreher
- **smooth muscle/plain muscle/
  non-striated muscle/unstriped
  muscle** glatter Muskel
  (glatte Muskulatur)
- **sphincter muscle** Sphinkter,
  Schließmuskel
- **strap muscle**
  bandförmiger Muskel
- **striated muscle/striped muscle**
  gestreifte Muskulatur
muscle belly/venter musculi
  Muskelbauch
muscle bundle Muskelbündel
muscle fascia Muskelfaszie,
  Muskelbinde
muscle fascicle Muskelfaserbündel
muscle fiber/myofiber Muskelfaser
muscle fibril/myofibril/myofibrilla
  Myofibrille, Muskelfibrille
muscle insertion Muskelansatz
muscle mass Muskelmasse
muscle origin Muskelursprung
muscle segment/myotome
  Ursegment, Myotom
muscle tone Muskelspannung,
  Muskeltonus
muscle twitching Muskelzucken
muscular muskulär, die Muskeln
  betreffend; muskulös
muscular contraction
  Muskelkontraktion
musculature/muscles Muskulatur
- **appendicular m.**
  Extremitätenmuskulatur
- **brachiomeric m.**
  Brachialmuskulatur
- **dermal m.** Hautmuskulatur
- **involuntary m.**
  unwillkürliche Muskulatur
- **obliquely striated m.**
  schräggestreifte Muskulatur
- **skeletal m.** Skelettmuskulatur
- **smooth m. (plain muscle/non-
  striated muscle/unstriped
  muscle)** glatte Muskulatur
- **striated m./striped m.**
  gestreifte Muskulatur
- **trunk m.** Rumpfmuskulatur
- **visceral m.** viscerale Muskulatur,
  Eingeweidemuskulatur
- **voluntary m.**
  willkürliche Muskulatur
mushroom/fungus Ständerpilz, Pilz
**mushroom bodies/pedunculate
  bodies/corpora pedunculata**
  Pilzkörper *pl*, Corpora pedunculata
musk Moschus
musk bag Moschusbeutel

musk gland (scent gland)
  Moschusdrüse (Duftdrüse)
**muskeg (Canadian term for
  peatlands)** Moor (ombrogen/
  oligotroph), Torfmoor; kanadisches
  Tundramoor
mussels/Mytiloidea Miesmuscheln
must (unfermented/uncleared juice)
  Most
must/musth (elephants) Brunst-Wut
  männlicher Elefanten
**mustang (small naturalized horse
  of western plains)** Mustang
  (verwildertes Präriepferd)
**mustard family/cabbage family/
  Cruciferae/Brassicaceae**
  Kreuzblütler, Kreuzblütlergewächse
mustard oil Senföl
mustard-tree family/Salvadoraceae
  Senfbaumgewächse
mutability Mutabilität,
  Mutierbarkeit, Mutationsfähigkeit
mutable mutabel, mutierbar
mutagen Mutagen, mutagen
mutagenesis Mutagenese
- **directed m.**
  gerichtete Mutagenese
- *in vitro* **m.** *in vitro*-Mutagenese
- **oligonucleotide-directed m.**
  oligonucleotidgesteuerte
  Mutagenese
- **site-directed m.**
  ortsspezifische Mutagenese
- **site-specific m.**
  sequenzspezifische Mutagenese
mutagenic mutagen,
  mutationsauslösend
mutagenicity Mutagenität
mutant Mutante
- **aging mutant** Alterungsmutante
- **breakthrough** "Durchbrenner"
  (überlebende letale Mutation)
- **conditional-lethal**
  konditional-letale Mutante,
  bedingt letale Mutante
- **constitutive mutant**
  konstitutive Mutante
- **cryptic mutant**
  kryptische Mutante
- **leaky mutant**
  durchlässige Mutante
- **lethal mutant** Letalmutante
mutarotation Mutarotation
mutate mutieren
mutation Mutation
- **ARMS (amplification refractory
  mutation system)** ARMS
  (System der
  amplifizierungsresistenten
  Mutation)
- **back mutation/reverse mutation**
  Rückmutation

mutation

- **conditional lethal mutation** konditional letale Mutation, bedingt letale Mutation
- **deletion mutation** Deletionsmutation
- **down mutation** "Down-Mutation"
- **forward mutation** Vorwärtsmutation
- **frameshift mutation** Leserasterverschiebung(smutation)
- **gain of function mutation** Funktionsgewinnmutation
- **germ-line mutation** Keimbahnmutation
- **homeotic mutation** homöotische Mutation
- **induced mutation** induzierte Mutation
- **insertion mutation** Insertionsmutation
- **inversion mutation** Inversionsmutation
- **knockout mutation** Knockout-Mutation
- **leaky mutation** durchlässige Mutation
- **lethal mutation** letale Mutation, Letalmutation
- **loss of function mutation** Funktionsverlustmutation
- **missense mutation** Missense-Mutation, Fehlsinnmutation
- **new mutation** Neumutation
- **nonsense mutation** Nonsense-Mutation, Nichtsinnmutation
- **pleiotropic mutation** pleiotrope Mutation
- **point mutation** Punktmutation
- **polar mutation** polare Mutation
- **pre-mutation** Prämutation
- **reverse mutation/back mutation** Rückmutation
- **sense mutation** Sinnmutation
- **silent mutation/samesense mutation** stumme Mutation
- **somatic mutation** somatische Mutation
- **spontaneous mutation** Spontanmutation
- **suppressor mutation** Suppressormutation
- **temperature-sensitive mutation** temperatursensitive Mutation
- **uniparental mutation** uniparentale Mutation
- **unstable mutation** instabile Mutation

mutation rate Mutationsrate
mutational bond/genetic load/ genetic burden/genetic bond Erblast, genetische Last, genetische Belastung
mutational load Mutationsbelastung, Mutationslast
mute stumm
- **deaf mute** taubstumm

mutilate verstümmeln
mutilation Verstümmelung
mutual/mutualistic gegenseitig, wechselseitig
mutualism/mutualistic symbiosis Mutualismus, Gegenseitigkeit, gemeinnützige Symbiose
mutualist Symbiont (in gegenseitiger/gemeinnütziger Lebensgemeinschaft)
mutualistic symbiotisch (gemeinnützig)
mutualistic symbiosis/mutualism gemeinnützige Symbiose
muzzle/snout Maul, Schnauze
muzzle (glandular muzzle of bovids) Flotzmaul
mycelial cord Myzelstrang
mycelium (*pl* myceliums/mycelia/ mycelias) Myzel (*pl* Myzelien), Pilzgeflecht
- **aerial mycelium** Luftmyzel
- **dikaryotic mycelium** Paarkernmyzel
- **persistent mycelium/mycelium perenne** Dauermyzel
- **primary mycelium** Primärmyzel, Einkernmyzel
- **raquet mycelium (raquet hyphae/raquet hyphas)** Raquettemyzel, Keulenmyzel (Raquettehyphen/Keulenhyphen)
- **secondary mycelium** Sekundärmyzel, Paarkernmyzel
- **spawn/mycelium fecundum** Pilzbrut

mycetism/mushroom poisoning Pilzvergiftung
mycobiont Mykobiont, Mycobiont, Pilzpartner
mycologist Mykologe
mycology Mykologie, Pilzkunde
mycophagy Mykophagie, Mycetophagie
mycoplasma Mykoplasma (*pl* Mykoplasmen)
mycorrhiza Mykorrhiza, "Pilzwurzel"
mycotoxin Mykotoxin
myelin sheath Markscheide, Myelinscheide
myelinated myelinisiert, markhaltig
myelination/myelinization Myelinisierung
myelocyte/bone marrow cell (an early polymorphonuclear leukocyte) Myelozyt, Myelocyt, Knochenmarkzelle

myxomatosis

myeloma Myelom
myiasis (Diptera larva) Myiasis, Madenkrankheit
myoblast/myogenic cell/sarcoblast/ sarcogenic cell Myoblast, Sarkoblast
myocardial infarction Herzschlag, Herzinfarkt
myocerate antenna (muscles in each antennal segment) myocerate Antenne, Gliederantenne
myofibril/muscle fibril/myofibrilla Myofibrille, Muskelfibrille
myogenic myogen
myomere/myotome Myomer, Myotom
myoneme Myonem
myotatic reflex/stretch reflex myotatischer Reflex, Dehnungsreflex
myotome/muscle segment Myotom, Ursegment
myotubule (*pl* myotubules) Myotubulus (*pl* Myotubuli)

Myricaceae/wax-myrtle family/ bayberry family/bog myrtle family/sweet gale family Gagelgewächse
myristic acid/tetradecanoic acid (myristate/tetradecanate) Myristinsäure, Tetradecansäure (Myristat)
Myristicaceae/nutmeg family Muskatnussgewächse
myrmecochory/ant-dispersal Ameisenausbreitung
myrmecophilous myrmekophil
myrmecophyte/myrmecoxenous plant Myrmekophyt, Ameisenpflanze
myrsine family/Myrsinaceae Myrsinaceae
mysis larva Mysis-Larve
myxobacteria Myxobakterien, Schleimbakterien
myxocyte Myxozyt, Schleimzelle
myxomatosis Myxomatose

Nacré wall/nacreous wall Nacréwand
nacreous perlmuttartig glänzend, perlmutterartig glänzend
nacreous layer/hypostracum Nacréschicht, Perlmutt(er)schicht, Hypostracum
nacrine/mother-of-pearl colored permuttfarben, perlmutterfarben
NAD/NADH (nicotinamide adenine dinucleotide) NAD/NADH (Nikotinamid-adenin-dinucleotid)
NADP/NADPH (nicotinamide adenine dinucleotide phosphate) NADP/NADPH (Nikotinamid-adenin-dinucleotid-phosphat)
nagana/nagana disease (*Trypanosoma spp.*) Nagana, Naganaseuche
naiad/aquatic nymph Wassernymphe
nail/unguis/ungula Nagel, Unguis
najas family/water nymph family/Najadaceae Nixenkrautgewächse
naked/nude nackt
naked bud *bot* nackte Knospe, offene Knospe (ohne Knospenschuppen)
nakedness/nudeness/nudity Nacktheit
name/term/designation Namensbezeichnung, Bezeichnung
  • **common name/vernacular name** volkstümlicher Name, volkstümliche Bezeichnung, Vernakularname,
  • **nonproprietary name** ungeschützter Name/Bezeichnung
  • **official name** offizieller Name, amtliche Bezeichnung
  • **proper name** Eigenname
  • **proprietary name** Markenbezeichnung
  • **scientific name** wissenschaftlicher Name
  • **species name** Artname
  • **specific name/specific epithet** Artname, Artbezeichnung, Epitheton
  • **substitute name** Ersatzname
  • **systematic name** systematischer Name
  • **trivial name** Trivialname
  • **vernacular name/common name** Vernakularname, volkstümliche Bezeichnung, volkstümlicher Name
name tag Namensetikett, Namensschildchen
naming/designation (nomenclature) Namensgebung, Benennung, Bezeichnung (Nomenklatur)

nanander/nanandrium (dwarf male plant) Nannandrium, Zwergmännchen
nanandrous nannandrisch
nanism/dwarfishness/dwarfism/microsomia Nanismus, Zwergwuchs, Kümmerwuchs; Verzwergung
nanophanerophyte (shrubs under 2 meters in height) Nanophanerophyt, Strauch
nanous/dwarfish/undersized zwergenhaft
napaceous/turnip-like rübenartig
nape/back of the neck/nucha Nacken, Genick
naphthalene Naphthalin
napiform/turnip-shaped rübenförmig
narcomedusas/Narcomedusae Narkomedusen
narcosis Narkose
narcotic *adj/adv* narkotisch, betäubend; berauschend
narcotic *n* Narkotikum, Betäubungsmittel; Rauschgift
nacotize narkotisieren, betäuben
nare/naris (mostly *pl*: nares)/nostril of vertebrates Nasenloch, Nasenöffnung (Vertebraten)
nasal bone/os nasale Nasenbein
nasal capsule Nasenkapsel
nasal cavity/nasal chamber Nasenhöhle
nasal opening/nasal aperture Nasenloch, Nasenöffnung
nascent (in process of formation) entstehend, werdend, in Entstehung begriffen
nascent *chem* freiwerdend
nasolacrimal duct Tränennasengang, Tränennasenkanal
nasopalatine duct Nasengaumengang
nasopharyngeal duct Nasenrachengang
Nassanov gland/Nassanov's gland (bees) Nassanov Drüse, Nassanoffsche Drüse
nastic nastisch
nastic movement Nastie, nastische Bewegung
nasturtium family/Tropaeolaceae Kapuzinerkressengewächse
nasute (termites) Nasensoldat, Nasutus-Soldat
natal/relating to birth Geburts..., die Geburt betreffend
natal down/neossoptile/neoptile (a down feather) *orn* Nestdune, Neossoptile, Neoptile

225   **nectary** N

natality/birthrate Natalität,
Geburtenrate
**natatorial leg/swimming leg
(insects)** Schwimmbein
natatorial/swim.../swimming ...
zum Schwimmen geeignet,
Schwimm...
national park Nationalpark
native/indigenous heimisch,
einheimisch
native/original im Urzustand,
naturbelassen, ursprünglich
native (not denatured) nativ
(nicht-denaturiert)
native egg white natives Eiweiß,
Eiklar
native meadow Naturwiese
native plant einheimische Pflanze
natural natürlich
natural balance Naturhaushalt
(natürliches Gleichgewicht)
natural catastrophe/disaster
Naturkatastrophe
natural enemy natürlicher Feind,
natürlicher Fressfeind
**natural environment/natural
setting** Naturlandschaft
natural gas Erdgas
natural history Naturgeschichte
natural history museum
Naturkundemuseum
natural law Naturgesetz
natural monument Naturdenkmal
natural product Naturstoff
natural product chemistry
Naturstoffchemie
natural resources natürliche Rohstoffe
natural sciences/science
Naturwissenschaften
natural scientist/scientist
Naturwissenschaftler, Naturforscher
natural selection
natürliche Selektion/Auslese
**naturalization/acclimatization/
settlement/establishment**
Naturalisation, Einbürgerung
naturalize/acclimatize
naturalisieren, einbürgern
**nature conservation league/nature
protection league/
environmentalist group**
Naturschutzbund, Naturschutzverein
nature conservation movement
Naturschutzbewegung
nature guide Naturführer
**nature protection/nature
conservation/nature preservation/
environmental protection**
Naturschutz, Umweltschutz
**nature reserve/wildlife reserve/
wildlife sanctuary/protected area**
Naturschutzgebiet, Naturreservat

nature trail/nature walk
Naturlehrpfad
**nature-nurture/nature-versus-
nurture** Veranlagung contra
Umwelt und Erziehung
naupliar eye (median eye)
Naupliusauge
naupliar larva/nauplius
Naupliuslarve, Nauplius
nausea Übelkeit, Brechreiz
nauseate übel werden
nauseous/nauseating/sickening
Übelkeit/Brechreiz verursachend/
erregend/hervorrufend/auslösend
nautilus (*pl* nautili)/Nautiloidea
Nautilusverwandte
navel/umbilicus/ophamos Nabel
navel-like/umbilicate/omphaloid
nabelartig, omphaloid
navicular/scaphoid/cymbiform
navikular, kahnförmig, bootförmig
navicular bone Kahnbein,
Fußwurzelknochen
**navicular bone/distal sesamoid/
os sesamoideum distale (horse)**
Strahlbein, distale Sesambein
**navicular bursa/bursa
podotrochlearis (hoof)**
Hufrollenschleimbeutel
**navicular zone/semilunar zone
(hoof: navicular bursa/bursa
podotrochlearis)** Hufrolle,
Fußrolle
neap tide Nipptide
nebulin Nebulin
neck Hals
neck *micros* Hals, Tubusträger
neck (tooth) Zahnhals
neck cell/neck canal cell *bot*
Halskanalzelle
necron unzersetztes totes Algen-/
Pflanzenmaterial
necron mud/gyttja Grauschlamm,
Halbfaulschlamm, Gyttia, Gyttja
**necrophilous/growing on dead
tissue** nekrophil, auf totem Gewebe
wachsend
necrosis Nekrose
necrotic nekrotisch
necrotize nekrotisieren; Nekrose
verursachend
necrotroph/necrotrophic nekrotroph
nectar Nektar (Blütensaft)
nectar guide/honey guide Saftmal,
Honigmal
nectar leaf/honey leaf Nektarblatt,
Honigblatt
**nectariferous leaf/nectar leaf/
honey leaf** Honigblatt
nectariferous scale Honigschuppe
nectary/nectar gland Nektarium,
Nektardrüse, Honigdrüse

nectophore

nectophore/nectocalyx/swimming bell (siphonophores) Nectophore, Schwimmglocke
needle Nadel; Kanüle, Hohlnadel
needle arrangement Benadelung
needle litter/needle litter layer *bot* Nadelstreu
needle-shaped/acicular nadelförmig
negative staining/negative contrasting Negativkontrastierung
Negri body (an inclusion body) Negrisches Körperchen, Negri Körper
neigh/whinny (low/gentle) wiehern
nekton (high mobility) Nekton (starke Eigenbewegung)
Nelumbonaceae/lotus lily family/Indian lotus family Lotusblumengewächse
nemathelminths/aschelminths/Nemathelminthes/Aschelminthes Nemathelminthen, Aschelminthen, Schlauchwürmer, Rundwürmer (*sensu lato*)
nematicide Nematizid, Nematodenbekämpfungsmittel
nematocyst/cnida/thread capsule/urticator Nematocyste, Nematozyste, Cnide, Nesselkapsel
• **adherent nematocyst** Glutinant
• **battery of nematocysts/cnidophore** Nesselbatterie, Cnidophore
• **penetrant nematocyst** Penetrant
• **volvent neamtocyst** Volvent
nematocyte/cnidocyte/cnidoblast/stinger cell Nematocyt, Nematozyt, Cnidocyt, Cnidozyt, Cnidoblast, Nesselzelle
nematodes/roundworms/nematoda Nematoden, Fadenwürmer
nematogen (*Dicyemida*) Nematogen
nematophore/nematocalyx Nematophore
nemertines/nemerteans/proboscis worms/rhynchocoelans/ribbon worms (broad/flat)/bootlace worms (long)/Nemertini/Rhynchocoela Schnurwürmer
neocerebellum (lateral lobes of cerebellum) Neukleinhirn, Neocerebellum
neocortex/neopallium Neocortex, Neopallium, Neuhirnrinde
neoencephalon/neencephalon Neuhirn, Neoencephalon, Neencephalon
neogastropods: whelks & cone shells/Neogastropoda/Stenoglossa Neuschnecken, Schmalzüngler

Neogene (geological period) Neogen, Jung-Tertiär, Jungtertiär
neonatal Neugeborene betreffend, neonatal
neonate *n* Neugeborenes (*adj* neugeboren)
neossoptile/neoptile/natal down (a down feather) Neossoptile, Neoptile, Nestdune
neoteny Neotenie
neotype Neotyp, Neotypus, Neostandard
Neozoic Era/Cenozoic Era/Cenozoic/Caenozoic Era/Cainozoic Era Neozoikum, Erdneuzeit, Känozoikum, Kaenozoikum (erdgeschichtliches Zeitalter)
nepenthes family/East Indian pitcher plant family/Nepenthaceae Kannenpflanzengewächse
nepheloid layer *mar* Bodentrübe
nephelometry Nephelometrie, Streulichtmessung
nephric/renal (see also: renal) Nieren..., die Niere betreffend
nephric ridge/nephrogenic ridge Nierenleiste
nephridiopore Nephridialporus, Nephridialöffnung, Nierenöffnung (Exkretionsporus)
nephrocyte Nephrocyt, Nephrozyt
nephron/functional unit of kidney (see also: kidney/renal) Nephron, Nierenelement, "Elementarapparat"
• **collecting duct/collecting tubule/papillary duct/ductus papillaris** Sammelrohr
• **convoluted tubule (distal/proximal)** gewundenes Kanälchen (distal/proximal)
• **loop of Henle/loop of the nephron/nephronic loop** Henlesche Schleife, Henle-Schleife
nephronic loop/loop of the nephron/loop of Henle Henlesche Schleife, Henle-Schleife
nephrostome Nephrostom
nephrotome/renal plate Nephrotom, Nierenplatte
neritic zone/neritic province neritische Region, Flachmeerzone
Nernst equation Nernst-Gleichung, Nernstsche Gleichung
nerve Nerv
nerve bundle Nervenbündel
nerve cell Nervenzelle
nerve cord Nervenstrang, Markstrang; Nervenbahn
nerve ending Nervenendigung
nerve fiber Nervenfaser

nerve growth factor (NGF)
Nervenwachstumsfaktor
nerve impulse Nervenimpuls
nerve net (invertebrates)/neuronal network Nervennetz,
Nervengeflecht
nerve ring Nervenring
● **buccal nerve ring** Schlundring
nerve strand/nerve cord
Nervenstrang
nerve system/nervous system (*see also there*) Nervensystem
● **autonomic/vegetative/visceral/ involuntary nerve system**
autonomes/vegetatives/viszerales/
unwillkürliches Nervensystem
● **double-chain nerve system/ ladder-type nerve system**
Strickleiternervensystem
● **peripheral nerve system (PNS)**
peripheres Nervensystem
● **somatic/voluntary nerve system (SNS)** animales Nervensystem,
animalisches Nervensystem
nervonic acid/(Z)-15-tetracosenoic acid/selacholeic acid Nervonsäure,
Δ^{15}-Tetracosensäure
nervous nervös; Nerven…, Nerven
betreffend
nervous system/nerve system
Nervensystem
● **autonomic nervous system (ANS)/involuntary nervous system** autonomes Nervensystem,
vegetatives Nervensystem
● **central nervous system (CNS)**
Zentralnervensystem (ZNS)
● **peripheral nervous system (PNS)**
peripheres Nervensystem
● **somatic/voluntary nervous system (SNS)**
animales Nervensystem,
animalisches Nervensystem
nervous tissue Nervengewebe
nest *vb* nisten
nest leaf Nischenblatt, Mantelblatt
nest odor Nestgeruch
nest parasitism Nestparasitismus
nest relief Brutablösung
nested genes ineinandergesetzte
Gene, ineinandergeschachtelte Gene
nested primer *gen* verschachtelter
Primer
nesting/nestling/nidulant nistend
(eingebettet in einer Aushöhlung)
nesting box Nistkasten
nesting colony/crèche Nistkolonie
nesting site Nistplatz
nestle *vb* schmiegen,
sich anschmiegen, kuscheln,
sich einnisten,
sich behaglich niederlassen

nestling *n* **(young bird still confined to nest)** Nestling
net/web Netz (web types *see also under:* web)
net production *ecol*
Nettoproduktion
net weight Nettogewicht
net-like/reticulate/reticular
netzartig, netzförmig, retikulär
netted/meshy/reticulate vernetzt,
netzartig
nettle family/Urticaceae
Nesselgewächse
nettle ring Nesselring
network Netzwerk
network of interactions
Beziehungsgefüge
network theory *immun*
Netzwerktheorie, Gittertheorie
neural/neuric neural
neural arch/basidorsale
Neuralbogen, oberer Wirbelbogen
neural crest Neuralleiste,
Ganglienleiste
neural encoding neuronale Kodierung
neural fold Neuralfalte, Neuralwulst,
Medullarwulst
neural groove Neuralrinne,
Medullarrinne
neural map neurale Karte
neural network neuronales Netz,
Netzwerk
neural plate Neuralplatte,
Nervenplatte, Medullarplatte,
Markplatte
neural superposition eye
neurales Superpositionsauge
neural tube/medullary tube/nerve cord/spinal cord Neuralrohr,
Medullarrohr, Markrohr,
Rückenmark
neuraminic acid Neuraminsäure
neurapophysis Neurapophyse,
Spinalfortsatz, oberer Dornfortsatz
neurobiology Neurobiologie
neurocranium Neurocranium,
Hirnschädel, Gehirnschädel
neurofilament Neurofilament
neurogenesis Neurogenese
neuroglial cell Neurogliazelle
neurohemal organ
Neurohämalorgan
neurohypophysis/posterior lobe of pituitary gland Neurohypophyse,
Hypophysenhinterlappen
neurolemma Neurolemm(a),
Neurilemm(a)
neurolemma cell/neurolemmocyte/ Schwann cell Schwannsche Zelle
neurology Neurologie
neuromast Neuromaste *f*,
Sinneshügel, Endhügel

neuromere

neuromere Neuromer
neuromotor network neuromotorisches Netzwerk
neuron/neurone/nerve cell Neuron, Nervenzelle
- **afferent neuron** afferentes Neuron
- **bipolar neuron/bipolar cell** Bipolarzelle
- **burster neuron** Bursterneuron
- **command neuron** Kommandoneuron
- **efferent neuron** efferentes Neuron
- **guidepost neuron** Wegweiserneuron
- **inhibitory neuron** Hemmungsneuron
- **interneuron** Zwischenneuron, Interneuron
- **motoneuron/motor neuron** Motoneuron
- **multipolar neuron/multipolar cell** Multipolarzelle
- **pioneer neuron** Pionierneuron
- **pseudounipolar cell** Pseudounipolarzelle
- **pyramidal neuron/pyramidal cell** Pyramidenzelle
- **relay neuron** Relaisneuron
- **sensory neuron** Sensorneuron, sensorisches Neuron
- **trigger neuron/command interneuron** Triggerneuron, Befehlsinterneuron
- **unipolar neuron/unipolar cell** Unipolarzelle

neuronal neuronal, neuronisch
neuronal circuit neuronaler Schaltkreis
neuropeptide Neuropeptid
neuropil Neuropil
neuropterans (dobson flies&ant lions)/Planipennia/Neuroptera Hafte, echte Netzflügler
neurosecretory neurosekretorisch
neurotoxic neurotoxisch
neurotransmitter Neurotransmitter
neuston Neuston
neuter/agamous ungeschlechtig
neuter/castrate *vb* kastrieren
neutralization test (NT) Neutralisationstest (NT)
neutrophil Neutrophil *n*
- **rod neutrophil/band neutrophil/ stab neutrophil/stab cell** stabkerniger Neutrophil
- **segmented neutrophil/ filamented neutrophil/ polymorphonuclear granulocyte** segmentkerniger Neutrophil

neutrophilic neutrophil

névé/firn Firn, Gletschereis
new mutation Neumutation
new-world monkeys (South American monkeys and marmosets)/Platyrrhina Neuweltaffen, Breitnasenaffen
Newtonian fluid Newton'sche Flüssigkeit
nexin Nexin
nibble (rabbits) knabbern
niche Nische, Wirkungsfeld
- **ecological niche** ökologische Nische
- **fundamental niche** fundamentale Nische
- **realized niche** realisierte Nische

niche overlap Nischenüberlappung
niche shift Nischenverschiebung
niche size Nischengröße
niche width/niche breadth Nischenbreite
nick Kerbe, Schlitz, Bruchstelle; *gen* (in single-strand DNA) Einzelstrangbruch
- **staggered nicks (e.g. in double-strand DNA)** versetzte Einzelstrangbrüche (z.B. in doppelsträngiger DNA)

nick translation *gen* Nick-Translation
nicked/notched eingekerbt, gekerbt, kerbig
nickel Nickel
nicotine Nikotin, Nicotin
nicotinic acid (nicotinate)/niacin Nikotinsäure, Nicotinsäure (Nikotinat)
nicotinic acid amide/nicotinamide/ niacinamide Nikotinsäureamid, Nicotinsäureamid, Niacinamid
nicotinic receptor nikotinischer, nicotinischer Rezeptor
nictitate/blink blinzeln, die Augen zwinkern
nictitating membrane/third eyelid Blinzelhaut, Nickhaut
nictitation/nictation Blinzeln, Augenzwinkern
nidamental gland/shell gland Nidamentaldrüse, Schalendrüse, Eischalendrüse
nidation/implantation Einnistung
nidicolous *orn* nesthockend
nidifugous animal/precocial animal *orn* Nestflüchter
nidulant/nesting/nestling nistend, eingebettet (in kleiner Aushöhlung)
night-active/nocturnal nachtaktiv
nightjars/goatsuckers/oilbirds/ Caprimulgiformes Nachtschwalben
nightshade family/potato family/ Solanaceae Nachtschattengewächse

229 nonviable **N**

NIOSH (National Institute for Occupational Safety and Health) U.S. Institut für Sicherheit und Gesundheit am Arbeitsplatz
nipple/mamilla/mammilla/teat Zitze, Mamille; Brustwarze
Nissl granules (rough ER with ribosomes) Nissl-Schollen, Tigroidschollen (raues ER)
nit (lice) Nisse
nitrate Nitrat
nitrification Nitrifikation, Nitrifizierung
nitrite Nitrit
nitrogen Stickstoff
nitrogen cycle Stickstoffkreislauf
nitrogen deficiency Stickstoffmangel
nitrogen fixation Stickstofffixierung
nitrogen-fixing bacteria stickstofffixierende Bakterien
nitrogenous/nitrogen-containing stickstoffenthaltend, Stickstoff…
nitrogenous base stickstoffhaltige Base
nitrogenous compound/ nitrogen-containing compound Stickstoffverbindung
nival zone nivale Stufe
noble rot Edelfäule
nociception Nozizeption, Noziception, Wahrnehmung von Schmerz, Schmerzempfinden
nociceptive nozizeptiv, noziceptiv, Schmerz empfindend, schmerzempfindlich
nocturnal nachtaktiv; nächtlich, Nacht…
nocturnal animal Nachttier
nocturnal plant Nachtpflanze, Nachtblüher
nod nicken
nodal plexus Knotengeflecht
nodding nickend
node Nodium, Knoten
nodule Knötchen; *bot* Knöllchen
● **bacterial nodule** Bakterienknöllchen
nodule bacteria Knöllchenbakterien
noise *neuro* Rauschen
noise filter Rauschfilter
nolana family/Nolanaceae Glockenwindengewächse
nomenclature/designation/name Nomenklatur, Benennung, Bezeichnung, Name
● **binomial nomenclature/binary nomenclature** binominale/binäre Nomenklatur, zweigliedrige Benennung/Namensgebung

nomenclature/system of terms Nomenklatur, Gesamtheit der Fachausdrücke
nominal scale *stat* Nominalskala
nominal value/rated value/desired value Sollwert
non-Mendelian ratio nicht-Mendelsches Aufspaltungsverhältnis
non-Newtonian fluid nicht-Newton'sche Flüssigkeit
non-parental tetrade (NPD) nicht-parentaler Dityp (NPD)
non-permissive cells nicht-permissive Zellen
non-permissive host nichtpermissiver Wirt
non-point source Flächenquelle
non-polar unpolar
noncompetitive inhibition nichtkompetitive Hemmung
noncyclic electron transport nichtzyklischer/nichtcyclischer/ linearer Elektronentransport
noncyclic phosphorylation nichtzyklische/nichtcyclische/lineare Phosphorylierung
nondisjunction (of chromosomes) Non-Disjunktion, Chromosomenfehlverteilung
nonessential nicht essentiell
nonhomologous recombination nichthomologe Rekombination
nonidentity *immun* Verschiedenheit (Nicht-Identität)
nonmigratory bird/resident bird Standvogel
nonmotile/immotile/immobile/ motionless/fixed unbeweglich, bewegungslos, fixiert
nonmotile unicellular kapsal, capsal, kokkal, coccal
nonmyelinated myelinlos, nicht myelinisiert, marklos, markfrei
nonoverlapping nicht-überlappend
nonpermissive cell nichtpermissive Zelle
nonpersistent transmission *vir* nicht-persistente Übertragung
nonruminant Nichtwiderkäuer
nonsaturation kinetics Nichtsättigungskinetik
nonsense codon Nonsense-Codon, Nichtsinn-Codon
nonsense mutation Nonsense-Mutation
nonspecific unspezifisch
nonstructural protein Nichtstrukturprotein
nontemplate strand (noncoding strand) Nichtmatrizenstrang (nichtcodierender Strang)
nonviable/not viable lebensunfähig

nonvolatile

nonvolatile nicht flüchtig, schwerflüchtig
noosphere Noosphäre
norepinephrine/noradrenaline Norepinephrin, Noradrenalin
norm of reaction Reaktionsnorm
normal distribution *stat* Normalverteilung
northern/boreal nördlich
"nose brain"/rhinencephalon Riechhirn, Rhinencephalon
nosema disease/nosemosis (Nosema apis) Nosemaseuche
nosepiece/nosepiece turret Revolver, Objektivrevolver
- **double nosepiece** Zweifachrevolver
- **quadruple nosepiece** Vierfachrevolver
- **quintuple nosepiece** Fünffachrevolver
- **triple nosepiece** Dreifachrevolver

nosocomial infection/hospital-acquired infection Nosokomialinfektion, nosokomiale Infektion, Krankenhausinfektion
nostril Nüster, Nasenloch
not viable/nonviable lebensunfähig
notacanthiforms/Notacanthiformes Dornrückenaale
notaspideans/Notaspidea Flankenkiemer
notation Notierung, Aufzeichnung; *chem/med* Bezeichnungssystem
notation/scoring *stat* Bonitur
notch Kerbe, Einschnitt
notched/nicked/crenate gekerbt, kerbig, eingeschnitten
- **finely notched/crenulate (small rounded teeth)** feingekerbt

notched zone/strengthening zone (orb web) *arach* Befestigungszone
nothosaurs/Nothosauria Nothosaurier
notochord Notocorda, Rückensaite, Chorda dorsalis
notochordal plate/chordal plate Chordaplatte
notochordal process/head process Chordafortsatz, Kopffortsatz
notochordal sheath Chordascheide
notopodium Notopodium
notoungulates/Notoungulata Süd-Huftiere
notum (*pl* nota) (thoracic tergum) Notum (Thorakalrückenplatte/dorsales Thorakalschild)
nozzle loop reactor/circulating nozzle reactor (bioreactor) Düsenumlaufreaktor, Umlaufdüsen-Reaktor

NPP (net primary production) Nettoprimärproduktion
NSF (National Science Foundation) "Nationale Wissenschaftsstiftung" (U.S. Forschungsgemeinschaft)
nub/hub (web) *arach* Netznabe
nucellus Nuzellus, "Knospenkern"
nucha/nape of the neck Nacken
nuchal nuchal, zum Nacken gehörend, den Nacken betreffend, Nacken...
nuchal crest Nackenkamm
nuchal ligament Nackenband
nuchal organ Nuchalorgan, Nackenorgan
nuchal region Nackengegend, Nackenregion
nuciferous/nut-bearing Nuss..., Nüsse bildend
nuclear nukleär, nucleär
nuclear bag fiber Kernhaufenfaser
nuclear cap Kernkappe
nuclear catastrophe *phys/ecol* Nuklearkatastrophe
nuclear chain fiber Kernkettenfaser
nuclear dimorphism Kerndimorphismus
nuclear disintegration *phys* atomarer Zerfall
nuclear division/mitosis (karyokinesis) Kernteilung, Mitose
nuclear dualism Kerndualismus
nuclear envelope Kernhülle
nuclear genome Kerngenom
nuclear lamina Kernfaserschicht, Kernlamina
nuclear magnetic resonance (NMR) Kernspinresonanz, kernmagnetische Resonanz
nuclear magnetic resonance spectroscopy (NMR spectroscopy) kernmagnetische Resonanzspektroskopie, Kernspinresonanz-Spektroskopie
nuclear matrix Kernmatrix, Kerngrundsubstanz
nuclear medicine/nuclear radiology Nuklearmedizin
nuclear membrane Kernmembran
nuclear phase Kernphase
nuclear polyhedrosis viruses (NPV) Kernpolyederviren
nuclear pore Kernpore
nuclear threat nukleare Bedrohung
nuclear transfer/nuclear transplantation Kerntransfer, Kerntransplantation
nuclear waste *ecol* Atommüll
nuclease Nuclease, Nuklease
nucleic acid Nucleinsäure, Nukleinsäure

nurture

nucleic acid hybridization Nucleinsäurehybridisierung, Nukleinsäurehybridisierung
nucleocapsid Nucleokapsid, Nukleokapsid
nucleoid/nuclear body Nucleoid, Nukleoid, Kernäquivalent, Karyoid, "Bakterienkern"
nucleolar organizer/nucleolus organizer (NOR) Nukleolus-Organisator, Nucleolus-Organisator
nucleolus Nucleolus, Nukleolus, Kernkörperchen
nucleophilic attack *chem* nucleophiler Angriff
nucleoplasm Nucleoplasma, Nukleoplasma, Kernplasma
nucleoside Nucleosid, Nukleosid
nucleoskeleton Kerngerüst
nucleosome Nucleosom, Nukleosom
nucleotide Nucleotid, Nukleotid
nucleotide-pair substitution Nucleotidpaaraustausch, Nukleotidpaaraustausch
nucleus/karyon Nucleus, Nukleus, Zellkern
- **amygdaloid nucleus/amygdaloid nuclear complex/nucleus amygdalae/corpus amygdaloideum** *neuro/anat* Mandelkern, Mandelkörper, Mandelkernkomplex
- **anucleate cell/enucleate cell** kernlose Zelle
- **caudate nucleus/nucleus caudatus** *neuro/anat* Schweifkern
- **cerebral nucleus/basal ganglion/corpus striatum** *neuro/anat* Basalganglion, Stammganglion
- **enucleate/to remove the nucleus** entkernen, den Kern entfernen
- **geniculate nucleus** *neuro/anat* Kern des Kniehöckers
- **lenticular nucleus** *neuro/anat* Linsenkern, Nucleus lentiformis
- **macronucleus/meganucleus** *cyt* Makronukleus, somatischer Zellkern
- **micronucleus** *cyt* Mikronukleus, generativer Zellkern
- **olivary nucleus** *neuro/anat* Olivenkern
- **paranucleus** *cyt* Nebenkern
- **polar nucleus** *cyt* Polkern
- **pronucleus** *cyt* Pronukleus, Pronucleus, Vorkern
- **resting nucleus** *cyt* Ruhekern
- **zygote nucleus/synkaryon** *cyt* Zygotenkern, Synkaryon
nucleus-associated organelle kernassoziiertes Organell

nuculane/nuculanium (*Henderson*: berry from superior ovary: medlar/grape) oberständige Beere (z. B. Traube); (*Spjut*: dry pericarp/hard endocarp/outer layer fibrous: coconut/almond/walnut) Nussfrucht mit trocken-faserigem Perikarp
nucule/nutlet Nüsschen
nude/naked nackt
nude mouse Nacktmaus
nudeness/nudity/nakedness Nacktheit
nudibranchs/sea slugs/ Nudibranchia Nudibranchier, Nacktkiemer, Meeresnacktschnecken
nudity/nudeness/nakedness Nacktheit
null allele Nullallel
null cell Null-Zelle
null hypothesis Nullhypothese
nulliparous nullipar
nullisomic nullisom
nullizygous nullizygot
numb taub, gefühllos
numbness Taubheit, Gefühllosigkeit
numerical taxonomy/taxometrics/ phenetics numerische Taxonomie, Phänetik
nuptial dress/nuptial plumage/ breeding plumage/courtship plumage Brutkleid, Hochzeitskleid (Vögel)
nuptial flight Brautflug
nurse *n* (animal) Amme
nurse *vb* nähren, füttern; versorgen, betreuen
nurse bee/nursery bee (worker bee) Ammenbiene (Arbeitsbiene/Arbeiterin)
nurse cell Nährzelle, Saftzelle
nurse crop *agr* Untersaat; *for* Vorholz
nurse grafting *bot* (nurse-root grafting) Ammenveredelung
nursery *zool* (e. g. in a zoo) Aufzuchtstätte, "Kinderstube"
nursery *hort* Pflanzgarten, Pflanzschule, Pflanzenaufzuchtbetrieb
- **tree nursery** Baumschule
nursery bed *hort* Aufzuchtbeet
nursery web (nursery tent) *arach* Brutgespinst, Eigespinst (Schutzgespinst für Jungspinnen)
nursing mother/foster mother Pflegemutter
nursing position Saugstellung
nurture *n* Nahrung; Pflege, Erziehung
nurture/bring up/rear/foster aufziehen, erziehen

N nurture 232

nurture/feed *vb* ernähren, nähren,
  füttern
nut Nuss
nut clams/*Nuculacea* Nussmuscheln
nut shell Nussschale
nutation Nutation
nutlet/nucule Nüsschen
  • **mericarpic nutlet/cell** Klause
nutlike/nutty nussartig
nutmeg family/Myristicaceae
  Muskatnussgewächse
nutrient Nährstoff
nutrient broth Nährbouillon,
  Nährbrühe
nutrient budget Nährstoffhaushalt
**nutrient cycle (biogeochemical
  cycle)** Stoffkreislauf,
  Nährstoffkreislauf
  (biogeochemischer Zyklus/Kreislauf)
  • **mineral cycle**
    Mineralstoffkreislauf
  • **nitrogen cycle** Stickstoffkreislauf
  • **oxygen cycle** Sauerstoffkreislauf
  • **phosphorus cycle**
    Phosphorkreislauf
  • **sulfur cycle** Schwefelkreislauf
  • **water cycle/hydrologic cycle**
    Wasserkreislauf
nutrient deficiency Nährstoffarmut,
  Nährstoffverknappung
**nutrient demand/nutrient
  requirement** Nährstoffbedarf
**nutrient medium (solid and liquid)/
  culture medium/substrate**
  Nährboden, Nährmedium,
  Kulturmedium, Medium, Substrat
  (*siehe auch:* Medium/
  Kulturmedium)
nutrient protein Nährstoffprotein
**nutrient requirements/nutritional
  requirements** Nahrungsbedürfnisse

nutrient salt Nährsalz
nutrient solution Nährlösung
nutrient tissue Nährgewebe
nutrient uptake Nährstoffaufnahme
**nutrient-deficient/oligotroph/
  oligotrophic** nährstoffarm,
  oligotroph
nutrient-rich/eutroph/eutrophic
  nährstoffreich, eutroph
nutrition Nahrung, Ernährung
nutritional requirements
  Nahrungsbedarf
  (*pl* Nahrungsbedürfnisse)
nutritive/nourishing nahrhaft
nutritive/nutritional Nahrung
  betreffend, Ernährung betreffend,
  Nähr...
**nutritive animal/feeding polyp/
  gasterozooid/trophozooid
  (nutritive polyp)** Nährtier,
  Fresspolyp, Gasterozoid, Trophozoid
  (Fresspolyp)
nutritive ratio/nutrient ratio
  Nährstoffverhältnis
nutritive value/food value
  Nährwert
nutritive-muscular cell
  Nährmuskelzelle
nuzzle/sniff at beschnuppern,
  beschnüffeln
Nyctaginaceae/four-o'clock family
  Wunderblumengewächse
nymph Nymphe
nymph/naiad (water)
  Wassernymphe
Nymphaeaceae/water-lily family
  Seerosengewächse
nymphiparous nymphipar
nymphipary Nymphiparie
Nyssaceae/sourgum family
  Tupelobaumgewächse

ocular

oak apple (gall) Eichenschwammgalle
obcordate/obcordiform/inversely heart-shaped verkehrt herzförmig
obdiplostemonous obdiplostemon
obese/excessively fat/overweight fett, fettleibig, korpulent, übergewichtig
objective *micros* Objektiv
oblanceolate/inversely lanceolate verkehrt lanzettförmig
obligate/restricted obligat, Zwangs...
obligate parasite/holoparasite obligater Parasit, Vollschmarotzer, Vollparasit, Holoparasit
obligatory/obligate obligatorisch, obligat
oblique schief, schräg
oblique illumination Schräglichtbeleuchtung
obliquely striated musculature schräggestreifte Muskulatur
obliterate verwüsten, zerstören; vernichten; *med* obliterieren, veröden
obliteration Verwüstung, Zerstörung; Vernichtung; *med* Verödung
obliterative shading unauffällige Färbung, unauffälliger Farbton (z.B. Süßwasserfische)
oblong länglich
obovate/inversely egg-shaped verkehrt eiförmig
obstetrics Geburtshilfe
obtect pupa/pupa obtecta bedeckte Puppe, Mumienpuppe
obturator (outgrowth) Obturator (Gewebewucherung)
obtuse (blunt or rounded end of leaf) stumpf
oc-DNA/open circle DNA oc-DNA, offene ringförmige DNA
oceanic climate/marine climate ozeanisches Klima, Meeresklima
oceanography/oceanology Ozeanographie, Ozeanografie, Ozeanologie
occipital bone/os occipitale Hinterhauptbein
occipital lobe/lobus occipitalis Hinterhauptslappen, Okzipitallappen
occiput (dorsal/posterior part of head) Occiput, Hinterhaupt
occlude/obstruct/close verschließen, verstopfen
occlude (teeth) schließen (obere u. untere Zähne)
occluded front/occlusion *meteo* Okklusion
occludens junction *see* tight junction

occlusion/obstruction/blockage Okklusion, Verschließung, Verstopfung; Verschluss
occlusion (teeth) Okklusion, Kieferschluss, Zahnreihenschluss
occlusion body *vir* Verschlusskörper
occlusor Occlusor, Okklusor
occupational hazard Gefahr am Arbeitsplatz
occupational hygiene Arbeitsplatzhygiene
occupational safety/workplace safety Arbeitsplatzsicherheit
occurrence/presence Vorkommen, Anwesenheit, Präsenz
ocean Ozean
ocean disposal/ocean dumping Ozeanverklappung
ocean floor/seafloor/seabed Meeresboden, Meeresgrund
ocean spray Sprühwasser
oceanfront Meeresfront, Küstenfront
oceanic ozeanisch
oceanic location/coastal location Meeresküstenlage
oceanic climate/maritime climate/ marine climate/coastal climate Meeresklima, Küstenklima, ozeanisches Klima
oceanic zone/oceanic region/ oceanic province ozeanische Region, Hochsee
oceanography Ozeanographie, Ozeanografie, Meereskunde
ocellar center Ocellenzentrum
ocellar pedicel Ocellenstiel
ocellus/eye spot (Lepidoptera) Ocellus, Augenfleck
ocellus/simple eye (dorsal and lateral) Ocellus, Ocelle, Einzelauge, Punktauge, Nebenauge (*siehe:* Scheitelauge/Stirnauge)
ochna family/Ochnaceae Grätenblattgewächse
ochre ocker
ochrea/ocrea/mantle *bot/fung* Ochrea, Tute
ocotillo family/Fouquieriaceae Ocotillogewächse
octad Oktade
octocorallians/octocorals/ Octocorallia Octocorallia
octopods/octopuses/Octopoda/ Octobrachia achtarmige Tintenschnecken, Kraken
ocular/eyepiece *micros* Okular
 • **binoculars** Binokular
 • **high-eyepoint ocular/spectacle eyepiece** Brillenträgerokular
 • **pointer eyepiece** Zeigerokular
 • **trinocular head** Trinokularaufsatz, Tritubus

ocular aperture

ocular aperture *micros* Sehfeldblende

ocular lens Okularlinse, Augenlinse

ocular micrometer Okularmikrometer

odd-pinnate/unequally pinnate/ imparipinnate (pinnate with an odd terminal leaflet) unpaarig gefiedert

odd-toed ungulates/Perissodactyla Unpaarhufer

odontoblast Odontoblast, Zahnbeinbildner

odontophore/radula support Odontophor, Radulapolster

odor Geruch, Duft

odor trail Geruchsfährte, Geruchsspur, Duftspur

odoriferous gland/scent gland Duftdrüse, Brunftdrüse, Brunftfeige (Gemse)

oenocyte Oenocyt, Oenozyt

oenology/enology Önologie, Weinkunde

Oenotheraceae/willowherb family/ evening-primrose family/ Onagraceae Nachtkerzengewächse

oestrous *see* estrous

oestrus *see* estrus

off-center impeller *biot* exzentrisch angeordneter Rührer

off-flavor (foods) Aromafehler

offal (butchery: parts removed in animal dressing) *agr* Fleischabfall, Innereien

offset *bot/hort* kurzer Seitentrieb (am Wurzelhals); Wurzelspross, Wurzeltrieb

offset bulb/bulblet/bulbil Brutzwiebel

offshoot/lateral shoot Nebentrieb, Seitentrieb

offshoot/offset/slip/sucker Wurzelspross, Wurzeltrieb

offshoot (derived descendant) *evol* Seitenzweig, Seitenlinie

offshore wind Landwind

offspring/descendant/progeny/ young Abkömmling, Deszendent, Nachkomme, Nachwuchs

oidium/oidiospore Oidium, Oidie, Oidiospore
- **arthrospore (arthroconidium)** Arthrospore

oil Öl
- **ben oil/benne oil** Behenöl
- **canola oil (rapeseed/colza oil)** Speise-Rapsöl, Rüböl
- **castor oil/ricinus oil** Rizinusöl
- **coconut oil** Kokosöl
- **cod-liver oil** Lebertran
- **corn oil** Maisöl

- **cotton oil** Baumwollsaatöl
- **crude oil/petroleum** Erdöl
- **essential oil/ethereal oil** ätherisches Öl
- **fusel oil** Fuselöl
- **linseed oil** Leinöl
- **lubricating oil** Schmieröl
- **mineral oil** Mineralöl
- **mustard oil** Senföl
- **olive oil** Olivenöl
- **olive kernel oil** Olivenkernöl
- **palm oil** Palmöl
- **peanut oil** Erdnussöl
- **pumpkinseed oil** Kürbiskernöl
- **safflower oil** Safloröl
- **sesame oil** Sesamöl
- **soybean oil** Sojaöl
- **sperm oil (whale)** Walratöl
- **sunflower seed oil** Sonnenblumenöl
- **vegetable oil** Pflanzenöl
- **virgin oil (olive)** Jungfernöl

oil bath *lab* Ölbad

oil cavity *bot* Ölbehälter

oil crops Ölsaaten (ölliefernde Pflanzen)

oil gland *zool* Öldrüse, Schmierdrüse

oil pollution Ölverschmutzung, Ölpest

oil reservoir *geol* Ölvorkommen, ölführende Schicht

oil shale *geol* Ölschiefer, Brandschiefer

oil slick Ölschlick, Ölteppich

oil spill *ecol* Ölkatastrophe

oil well *geol* Ölquelle

oilbirds/nightjars/goatsuckers/ Caprimulgiformes Nachtschwalben

oilseed Ölsaat, Ölsamen

oily ölig

Okazaki fragment Okazaki-Fragment

olax family/tallowwood family/ Olacaceae Olaxgewächse

Old man's ears and allies/ Auriculariales Ohrlappenpilze

old-growth/old-growth forest/ old-growth stand/mature forest Altbestand, alter Baumbestand (Wald)

old-world monkeys (incl. apes)/ Catarrhina Altweltaffen, Schmalnasenaffen

Oleaceae/olive family Ölbaumgewächse

Oleandraceae/stalwart sword fern family Nierenfarngewächse

oleaster family/Elaeagnaceae Ölweidengewächse

olecranon/point of the elbow Olekranon, Ellenbogenhöcker, Ellenbogenspitze, Ellenbogenfortsatz

oleic acid/(Z)-9-octadecenoic acid (oleate) Ölsäure, Δ^9-Octadecensäure (Oleat)
oleosome Oleosom
olericulture Gemüseanbau
olfaction/process of smelling Riechen
olfaction/sense of smell Geruchssinn
olfactory den Geruchssinn betreffend, Geruchs…, Riech…
olfactory dome (sensory dome)/ olfactory bulb/bulbus olfactorius Riechhügel, Riechkolben
olfactory epithelium/nasal mucosa Nasenschleimhaut, Riechepithel
olfactory gland/Bowman's gland Bowman-Drüse, Bowmansche Drüse, Drüse der Riechschleimhaut
olfactory hair Riechhärchen (Sinneshaar)
olfactory mucosa Riechschleimhaut
olfactory nerve Riechnerv, Geruchsnerv
olfactory organ Riechorgan
olfactory peg/sensory peg (sensilla styloconica/basiconica) Riechkegel, Sinneskegel, Sinnesstäbchen
olfactory pit (a sensory pit) Riechgrube
olfactory plate (sensory plate) Riechplatte, Porenplatte (Sensilla placodea)
olfactory sense olfaktorischer Sinn, Geruchssinn
olfactory threshold Riechschwelle
olfactory tract/tractus olfactorius Riechbahn
olfactory trail/scent trail Duftspur
olfactory tubercle Riechwulst
Oligocene/Oligocene Epoch Oligozän (erdgeschichtliche Epoche)
oligochetes/Oligochaeta Wenigborster, Oligochaeten
oligodendrocyte Oligodendrozyt, Oligodendrocyt
oligolecithal (with little yolk) oligolecithal, oligolezithal, mikrolecithal, dotterarm
oligomer Oligomer
oligomerous oligomer
oligomictic oligomiktisch
oligonucleotide Oligonucleotid, Oligonukleotid
  • **antisense oligonucleotide** Antisense-Oligonucleotid, Anti-Sinn-Oligonucleotid, Gegensinn-Oligonucleotid
  • **ASO (allelspecific oligonucleotide)** allelspezifisches Oligonucleotid

oligonucleotide-directed mutagenesis oligonucleotidgesteuerte Mutagenese, oligonukleotidgesteuerte Mutagenese
oligopod *adj/adv* oligopod, wenigbeinig
oligosaprobic oligosaprob
oligotrophic oligotroph, nährstoffarm
olivary nucleus *neuro* Olivenkern
olive/olivary nucleus *neuro/anat* Olive
olive family/Oleaceae Ölbaumgewächse
omasum/third stomach/manyplies/ psalterium Omasus, Blättermagen, Vormagen, Psalter
ombrogenous ombrogen, niederschlagsbedingt
ombrophilous (thriving under conditions of abundant rain) ombrophil
ombrophobous (intolerant to prolonged rain) ombrophob
ombrophyte Ombrophyt
ombrotrophic ("rainstorm fed") ombrotroph (Nährstoffe aus Niederschlägen)
Omega loop/Ω loop (proteins) Omega-Schleife, Ω-Schleife (Proteine)
ommatidium/facet/stemma Ommatidium, Sehkeil
omnipotence Omnipotenz
omnipotent omnipotent
omnivore Omnivore, Allesfresser
omnivorous omnivor, allesfressend
Onagraceae/willowherb family/ evening-primrose family/ Oenotheraceae Nachtkerzengewächse
oncogene/onc gene Onkogen
oncogenic/oncogenous onkogen, oncogen, krebserzeugend
oncogenic protein Onkoprotein, onkogenes Protein
oncogenicity Onkogenität
oncology Onkologie
oncosphere/hexacanth larva/ hooked larva (cestodes) Oncosphaera-Larve, Onkosphäre, Sechshakenlarve
oncotic pressure onkotischer Druck, kolloidosmotischer Druck
one-enzyme-one-gene theory Ein Enzym-ein Gen-Theorie
one-gene-one-polypeptide hypothesis Ein Gen-ein Polypeptid-Theorie

one-gene-one-protein theory
Ein Gen-ein Protein-Theorie
**one-horned/unicornate/
unicornuate/unicornuous**
einhörnig
one-letter code *gen*
Ein-Buchstaben-Code
one-toed einzehig
one-way digestive tract
durchgängiger Verdauungstrakt
onion family/Alliaceae
Zwiebelgewächse, Lauchgewächse
onshore wind Seewind
**ontogenesis/ontogeny/
development (of the individual)**
Ontogenese, Ontogenie,
Entwicklungsgeschichte
(des Einzelorganismus)
ontogenetic ontogenetisch,
entwicklungsgeschichtlich
oocyst Oocyste
**oocyte/ovocyte/ovicyte/egg cell
(before and during meiosis)**
Oocyt, Oozyt, Ovocyt, Ovozyt,
(unreife) Eizelle (vor und während
Meiose)
ooecium/ovicell Ooecium, Ovicelle
(Brutkammer)
oogamy Oogamie, Eibefruchtung
oogenesis Oogenese, Eibildung
oogonium/ovogonium Oogonium,
Ureizelle; *fung* (oosporangium)
Oogonium, Sporangium
ookinete Ookinet
oolemma/zona pellucida Oolemma,
Eihülle, Eihaut, Zona pellucida
**oomycetes/Oomycota (water molds
& downy mildews)** Eipilze,
Oomyzeten, Oomyceten
**oophagy/egg cannabalism (social
insects)** Oophagie
ooplasm/ovoplasm Ooplasma,
Bildungsplasma, Eiplasma
oospore Oospore
oostegite (crabs) Oostegit,
Brutplatte
ootheca (egg case/egg pod)
Oothek (bei *Blattopteroida*:
Eipaket)
ootype Ootyp
oozooid/oozoite (ascidians)
Oozooid
open promoter complex *gen*
offener Promotorkomplex
open reading frame (ORF) *gen*
offenes Leseraster
**open sea/pelagic zone/oceanic
zone/oceanic province** Hochsee,
offenes Meer, Hochseebereich,
ozeanische Region
open time *neuro* Öffnungszeit,
Offenzeit

**opening/aperture/orifice/mouth/
perforation/entrance** *n* Öffnung,
Mund, Mündung
opening/dehiscent *adv/adj bot*
öffnend
operant conditioning *ethol*
operante Konditionierung,
instrumentelles Lernen
operator *gen* Operator
**operculate/opercular/
operculiferous/bearing a lid**
gedeckelt, mit Deckel versehen,
Deckel...
operculiform/lid-like deckelförmig,
deckelartig
operculum/opercle/lid Operkulum,
Deckel
operon Operon
ophidian/snake-like schlangenartig,
Schlangen...
ophidiophobia Schlangenphobie
**Ophioglossaceae/adder's-tongue
family/grape fern family**
Natternzungengewächse,
Rautenfarngewächse
ophiology/study of snakes
Ophiologie, Schlangenkunde
opiate Opiat
**opisthaptor/opisthohaptor/Baer's
disk** Opisthaptor, Opisthohaptor
**opisthobranch snails/
opisthobranchs/Opisthobranchia**
Hinterkiemer,
Hinterkiemenschnecken
opisthocelous/opisthocoelous
opisthocöl, opisthocoel
opisthognathous opistognath
opisthonephros Opisthonephros,
Rumpfniere
**opossum shrimps/mysids/
Mysidacea** Spaltfüßer, Mysidaceen
opportunistic opportunistisch
opportunistic species/opportunist
Opportunist
opposable opponierbar,
entgegenstellbar
opposite/opposing (position)
gegenständig, gegenüberliegend
opsin/scotopsin Opsin
opsonin Opsonin
opsonization Opsonierung,
Opsonisation, Opsonisierung
optic/optical optisch
optic chiasma/chiasma opticum
Sehnervkreuzung
optic cup *embr* Augenbecher
**optic diffusion/optical diffusion/
dispersion/dissipation/scattering
(light)** Lichtstreuung
optic lobe/visual lobe/lobus opticus
Sehlappen
optic nerve/nervus opticus Sehnerv

optic refraction Refraktion,
optische Brechung, Lichtbrechung
optic stalk *embr* Augenbecherstiel
**optic superposition eye/clear-zone
eye** optisches Superpositionsauge
**optic tectum/optic lobe/tectum
opticum** Mittelhirndach
optic vesicle/vesicula ophthalmica
embr Augenblase, Augenbläschen
optical density/absorbance
optische Dichte, Absorption
**optical diffusion/dispersion/
dissipation/scattering (light)**
Streuung (Lichtstreuung)
optical refraction Refraktion,
optische Brechung, Lichtbrechung
optical resolution
optische Auflösung
optical specificity
optische Spezifität
optics Optik
option/choice/alternative Option,
Wahl, Alternative
optional/facultative optional,
freiwillig, freigestellt, wahlfrei,
fakultativ
oral apparatus (ciliates)
Oralapparat, Buccalapparat
oral arm (polyps/echinoderms)
Mundarm
oral disk/peristome/peristomium
Oralscheibe, Mundscheibe, Peristom
oral vestibule Mundbucht,
Vestibulum
orb web *arach* Radnetz
orbicular/circular kreisförmig,
kreisrund
orbiculate/nearly round
kreisförmig, fast rund
orbicule (pollen) *bot* kleine
kreisförmige Erhebung auf
Pollenexine
orbit/eye socket Augenhöhle,
Orbita
orchard/grove Baumgarten,
Baumhain
**orchids/orchid family/orchis family/
Orchidaceae** Orchideen,
Knabenkrautgewächse
order Ordnung
order of rank/ranking/hierarchy
Rangordnung, Rangfolge,
Stufenfolge, Hierarchie
order statistics Ordnungsstatistik
ordered displacement reaction
geordnete Verdrängungsreaktion
ordinal scale *stat* Ordinalskala
**Ordovician Period/Ordovician
(geological time)** Ordovizium
ORF (open reading frame) *gen*
offenes Leseraster
organ Organ

organ of Corti/spiral organ
Cortisches Organ
organ of Jullien Julliensches Organ
organ of Tömösvary
Tömösvarysches Organ,
Postantennalorgan
organelle Organell *nt*, Organelle *f*
organic organisch
organic debris/organic waste
organischer Abfall,
organische Abfallstoffe
organic farming
organischer Landbau
organic matter
organisches Material
organism/lifeform Organismus,
Lebensform, Lebewesen
organismal organismisch
organizational form
Organisationstyp, Organisationsform
**organizational level/grade of
organization** Organisationsstufe
organizer *embr* Organisator
(dorsale Blastoporenlippe)
organogenesis Organogenese,
Organbildung, Organentwicklung
orgasm/sexual climax Orgasmus,
sexueller Höhepunkt
• **clitoral orgasm** Klitoralorgasmus
oricidal oricid, orizid, rachenspaltig
**oriental sore/cutaneous
leishmaniasis (*Leishmania spp.*)**
Orientbeule, Hautleishmaniose,
kutane Leishmaniose
orientation/orientational behavior
Orientierung, Orientierungsverhalten
orientational movement/taxy/taxis
(*pl* taxes) Orientierungsbewegung,
Taxie, Taxis
orifice/mouth/opening Öffnung,
Mund, Mündung
origin/descent/provenance
Ursprung, Abstammung, Herkunft,
Provenienz
origin of replication
Replikationsursprung,
Replikationsstartpunkt
original/basic/simple/primitive
originär, ursprünglich
ornamental garden/amenity garden
Ziergarten
ornamental grass Ziergras
ornamental plant Zierpflanze
ornamental shrub Zierstrauch
ornithine Ornithin
ornithine-urea cycle
Ornithin-Harnstoff-Zyklus
**ornithischian reptiles/bird-hipped
dinosaurs/Ornithischia**
Vogelbecken-Dinosaurier
ornithochory/bird-dispersal
Vogelausbreitung

ornithology

ornithology/study of birds
Ornithologie, Vogelkunde
**ornithophile/bird-pollinated
flower/bird flower** Vogelblume
ornithophilous/bird-pollinated
vogelblütig
Orobanchaceae/broomrape family
Sommerwurzgewächse
orobranchial cavity
Orobranchialhöhle, Mund-
Kiemenhöhle
orophyte/mountain plant Orophyt,
Bergpflanze
orotic acid Orotsäure
**orpine family/sedum family/
stonecrop family/Crassulaceae**
Dickblattgewächse
orsellic acid/orsellinic acid
Orsellinsäure
ortet Ortet, Klonausgangspflanze,
Klonmutterpflanze
orthodromic *neuro* orthodrom
**orthoevolution/determinate
evolution** Orthoevolution
**orthogenesis (apparently
predetermined development)**
Orthogenese (geradlinige
Entwicklung)
orthognathous orthognath
ortholog *gen* ortholog (Gene)
orthology Orthologie
orthopterans/Orthoptera
Geradflügler
orthostichous orthostich
orthotropism Orthotropismus
orthotropous/orthotropic/atropous
orthotrop
**OSHA (Occupational Safety and
Health Administration)** U.S.
Bundesamt für Sicherheit und
Gesundheit am Arbeitsplatz
osmeterium/osmaterium
Osmaetherium
osmic acid Osmiumsäure
**osmiophilic (staining readily with
osmium stains)** osmiophil
(kontrastierbar mit
Osmiumsäurederivaten)
osmium tetraoxide Osmiumtetroxid
osmoconformer Osmokonformer
osmolality Osmolalität
osmolarity Osmolarität
osmoregulation Osmoregulation
osmoregulator Osmoregulierer
osmosis Osmose
osmotic osmotisch
osmotic potential osmotisches
Potential
osmotic pressure osmotischer
Druck
osmotic shock osmotischer Schock
osmotrophic osmotroph

**Osmundaceae/flowering fern
family/cinnamon fern family/royal
fern family** Königsfarngewächse,
Rispenfarngewächse
osphradium (*pl* **osphradia)**
Osphradium (*pl* Osphradien)
ossicle/small bone Knöchelchen
ossification Ossifikation,
Verknöcherung, Knochenbildung
ossify verknöchern
osteoblast/bone-forming cell
Osteoblast, knochenbildende Zelle
osteoclast/giant cell Osteoklast
osteocyte/bone cell Osteocyt,
Osteozyt, Knochenzelle
osteogenesis Osteogenese,
Knochenbildung,
Knochenentstehung
**osteoglossiforms/
Osteoglossiformes**
Knochenzüngler,
Knochenzünglerartige
ostium Ostium, Öffnung
ostium (of oviduct)/ostium tubae
Eileiteröffnung
ostracoderms/Ostracodermata
Schalenhäuter
ostriches/Struthioniformes
Laufvögel, Strauße, Straußenvögel
**OTA (Office of Technology
Assessment)** US-Büro für
Technikfolgenabschätzung
otic otisch, Ohr...
otic capsule Ohrkapsel
otic placode Ohrplakode
outbreak (of disease) Ausbruch
(einer Krankheit)
outbred durch Kreuzung entfernt
oder nicht verwandter Individuen
gezüchtet
outbreed nicht verwandte Individuen
kreuzen
outbreeding Züchtung durch
Kreuzung entfernt oder nicht
verwandter Individuen
outcompete im Widerstreit/in der
Konkurrenz überlegen sein, durch
Konkurrenz "ausschalten"
outcrop *n geol* zutageliegende
Schicht, Ausstrich, Ausbiss
outcrop *vb geol* zutagetreten,
zutageliegen, anstehen, ausstreichen,
ausbeißen
outcrossing Herauskreuzen,
Auskreuzen
outer layer/exterior layer
Außenschicht
outgroup (cladistics) Außengruppe
outgrowth/protrusion Auswuchs
outlet Ablauf, Ausfluss
(Austrittsstelle einer Flüssigkeit)
outlier *stat* Ausreißer

oxidative

outpocketing/evagination/ protrusion Ausbuchtung, Ausstülpung (z. B. Darmdivertikel)
output *agr* Ertrag, Produktion; *ecol* Austrag; *techn/electr* Ausgangsleistung; *Computer:* Ausgang, Ausgabe
• **input** Eingabe, Einsatz, Einbringen (eingebrachte/ zugeführte Menge); *ecol* Eintrag; *techn/electr* Eingangsleistung; *Computer:* Eingabe
outside-out patch outside-out Patch
outside-out vesicle outside-out Vesikel (Vesikel mit der Außenseite nach außen)
outwash/outwash plain Sander
ovarian ball Ovarialballen
ovarian pregnancy Ovarioträchtigkeit, Eierstockschwangerschaft, Ovarialschwangerschaft
ovariole/egg tube (insects) Ovariole, Ovariolschlauch, Eischlauch, Eiröhre
ovary *bot* Ovar, Ovarium, Fruchtknoten
ovary *zool* Ovar, Ovarium, Eierstock
ovary/germarium Keimstock, Germarium
ovary wall/fruit wall/pericarp Fruchtwand, Perikarp
ovate/egg-shaped eiförmig
overactivity/hyperactivity Überfunktion, Hyperaktivität
overdominance Überdominanz
overdose *n* Überdosis
overexpression *gen* Überexpression
overgraze überweiden
overgrazing Überweidung
overgrow zuwachsen, überwachsen, überwuchern; verwildern
overgrown überwuchert, zugewachsen
overhanging end/protruding end/ protruding extension *gen* überhängendes Ende
overhead irrigation Bewässerung/ Beregnung von oben
overlapping/imbricate überlappend
overlapping genes überlappende Gene
overpopulation Überpopulation; Überbevölkerung, Übervölkerung
overreplication *gen* Überreplikation
overshoot *n neuro/ecol* Überschuss
overshoot *vb neuro/ecol* überschießen (z. B. Kapazitätsgrenze)
overstory/overstory growth Oberholz, Oberstand, Schirmbestand, Überwuchs

overtopping (unilateral dominance) *bot* Übergipfelung
overturn *n* **(lake water)** *limn* Umwälzung (Vollzirkulation)
overturn *vb* **(lake water)** *limn* umwälzen
overwinding *gen* Überdrehung
overwinter/hibernate überwintern
ovicell/ooecium (bryozoans) Ovicelle, Ooecium, Ooecie, Embryosack (Brutkammer)
ovicyte/ovocyte/oocyte/egg cell (before and during meiosis) Ovocyt, Ovozyt, Oocyt, Oozyt, (unreife) Eizelle (vor und während Meiose)
oviduct Ovidukt, Eileiter
oviger (egg-carrying leg: certain arachnids/pycnogonids) Oviger, Eiträger (Brutbein)
ovigerous oviger, eitragend, eiführend
ovine Schafe betreffend, Schafs...
oviparous/egg-laying ovipar, eierlegend
oviposit Eier (ab)legen, eierlegen
ovipositor (insects)/egg-laying apparatus/egg-laying organ/egg depositor Ovipositor, Legeapparat, Legeorgan
ovipositor sheath Legescheide
ovisac/brood pouch/egg case Eisack, Bruttasche
ovoid/egg-shaped eiförmig
ovotestis/hermaphroditic gonad/ hermaphroditic gland Ovotestis, Zwitterdrüse
ovulate *vb* ovulieren, springen (Ei)
ovulate cone/ovuliferous cone *bot* weiblicher Zapfen
ovulation Ovulation, Eisprung, Follikelsprung
ovule *bot* Samenanlage
ovuliferous scale/seed scale Samenschuppe, Fruchtschuppe
owls/Strigiformes Eulen
ox (*pl* **oxen)** Ochse
oxalic acid (oxalate) Oxalsäure (Oxalat)
Oxalidaceae/wood-sorrel family Sauerkleegewächse
oxaloacetic acid (oxaloacetate) Oxalessigsäure (Oxalacetat)
oxalosuccinic acid (oxalosuccinate) Oxalbernsteinsäure (Oxalsuccinat)
oxbow *limn* Altwasserarm
oxidant/oxidizing agent Oxidationsmittel
oxidation Oxidation
oxidation-reduction reaction Redoxreaktion
oxidative oxidativ

oxidative phosphorylation

oxidative phosphorylation/carrier-level phosphorylation oxidative Phosphorylierung
oxidative stress oxidativer Stress
oxidize oxidieren
oxidizing oxidierend
oxidizing agent/oxidant Oxidationsmittel
oxoglutaric acid (oxoglutarate) Oxoglutarsäure (Oxoglutarat)
oxygen Sauerstoff
oxygen debt Sauerstoffschuld, Sauerstoffverlust, Sauerstoffdefizit
oxygen deficiency Sauerstoffmangel
oxygen partial pressure Sauerstoffpartialdruck
oxygen transfer rate (OTR) Sauerstofftransferrate
oxygen-deficient sauerstoffarm
oxytocin (OT) Oxytocin, Oxytozin
oyster spat Austernlaich
ozone Ozon
ozone depletion Ozonabbau
ozone hole Ozonloch
ozone layer Ozonschicht, Ozonosphäre

P-protein body/phloem protein body/slime body/slime plug P-Protein, Proteinkörper, Schleimkörper, Schleimpfropfen (in Siebröhren)
pace/fast amble (gait) Passgang
pace/rate of movement Fortbewegungsgeschwindigkeit, Tempo
pacemaker Schrittmacher (*siehe:* Sinusknoten)
pacemaker potential Schrittmacherpotential
pacer/side-wheeler Passgänger
pachydont pachydont
pachymeninx harte Hirnhaut, Pachymeninx (Dura mater)
pachytene (in meiotic prophase) *gen* Pachytän
Pacinian body/Pacinian corpuscle/ lamellated corpuscle Pacini Körperchen, Pacinisches Körperchen, Lamellenkörperchen, Endkörperchen
pack (dogs/wolves) Rudel, Meute
pack animal Lasttier, Tragtier
packaging (e. g. viral nucleic acid by viral proteins) Verpackung (z. B. Virusnucleinsäure mit Virusproteinen)
● *in vitro* **packaging** *in vitro*-Verpackung
packed bed reactor (bioreactor) Packbettreaktor, Füllkörperreaktor
packing efficiency/packing ratio *biochem/gen* Packungsverhältnis (DNA: Spiralisierungsgrad)
pad (finger pad/foot pad/toe pad/ paw pad/palmar pad) Ballen (Fingerballen/Fußballen/ Zehenballen/Sohlenballen)
paddle stirrer/paddle impeller (bioreactor) Schaufelrührer, Paddelrührer
paddle wheel reactor Schaufelradreaktor
paddock (*dial***: frog/toad)** Kröte, Frosch
paedogenesis/pedogenesis Pädogenese
Paeoniaceae/peony family Pfingstrosengewächse
pain Schmerz
pain sensation Schmerzgefühl
painful schmerzhaft
pair *n* Paar
pair/mate *vb* **(copulate)** paaren (begatten/kopulieren)
paired cisternae (of ER) paarweise liegende Zisternen (des ER)
pairing/mating (copulating) Paarung (Begattung/Kopulation)

pairing season/mating season Paarungssaison, Paarungszeit
palatable genießbar, schmackhaft
palatal arch Gaumenbogen
palate/roof of mouth (vertebrates)/ roof of pharynx (insects) Gaumen
● **bony palate/osseous palate/ palatum osseum** knöcherner Gaumen
● **hard palate** harter Gaumen
● **soft palate/velum palatinum** weicher Gaumen, Gaumensegel, Velum
palatine bone/os palatinum Palatinum, Gaumenbein
palea/pale/palet/glumella/inner glume *bot* Vorspelze
paleobotany/paleophytology Paläobotanik, Paläophytologie, Phytopaläontologie
Paleocene/Paleocene Epoch Paläozän (erdgeschichtliche Epoche)
paleocortex Palaeocortex, Archicortex, Althirnrinde
paleoecology Paläoökologie, Palökologie
paleoencephalon/paleencephalon Althirn, Paläoencephalon, Paleencephalon
paleoendemic Paläoendemit, Reliktendemit
Paleogene (geological period) Paläogen, Alt-Tertiär, Alttertiär
paleola/lodicule/glumellule Schwellkörper, Lodicula, Schüppchen (Grasblüte)
paleontology Paläontologie
Paleophytic Era/"Age of Ferns" Paläophytikum, Florenaltertum, Farnzeitalter
paleospecies (chronospecies) Paläospezies (Chronospezies)
Paleozoic/Paleozoic Era Paläozoikum, Erdaltertum (erdgeschichtliches Zeitalter)
palet/palea/pale/glumella/inner glume *bot* Vorspelze
palindrome/inverted repeat *gen* Palindrom, umgekehrte Repetition, umgekehrte Wiederholung, invertierte Sequenzwiederholung
palingenesis Palingenese
palisade cell Palisadenzelle
palisade parenchyma Palisadenparenchym
palisade worm (*Strongylus equinus et spp.***)** Palisadenwurm
pallial line Palliallinie, Mantellinie
pallial sinus Pallialraum, Mantelhöhle
pallium/mantle Pallium, Mantel

palm (of hand) Handfläche, hohle Hand

palm family/Arecaceae/Palmae Palmen

palmar/volar palmar, volar, die Handinnenfläche betreffend, Handflächen...

palmate/fingered/hand-shaped gefingert, handförmig

palmately veined handnervig

palmatifid (divided to middle) fingerspaltig, handförmig gespalten

palmatilobate/palmately lobed fingerlappig, handförmig gelappt

palmatipartite/palmately partite fingerteilig, handförmig geteilt

palmatisect fingerschnittig, handförmig (ein)geschnitten

palmfern/cycad Palmfarn

palmitic acid/hexadecanoic acid (palmate/hexadecanate) Palmitinsäure, Hexadecansäure (Palmat/Hexadecanat)

palmitoleic acid/(Z-9-hexadecenoic acid Palmitoleinsäure, Δ^9-Hexadecensäure

palp/palpus Palpe, Taster, Tastfühler
- **labial palp/labipalp/labial feeler/ palpus labialis** Labialpalpus, Lippentaster
- **maxillary palp/palpus maxillaris** Maxillarpalpus, Kiefertaster

palp proboscis/palp appendage (bivalves) Mundlappenanhang

palpal endite (with scapula) Pedipalpenlade

palpebra/eyelid/blepharon Augenlid, Lid

palpebral palpebral, das Augenlid betreffend, Augenlid..., Lid...

palpebral sebum/gum/sebum palpebrale (from Meibomian gland) Augenbutter, Augenschmalz

palpiger Palpiger, Tasterträger

palsa bog Palsenmoor, Torfhügelmoor

paludification *limn* Versumpfung

palustrine Sumpf..., sumpfig, im Sumpf wachsend

palynology Pollenkunde

pamphlet/brochure/booklet Broschüre, Informationsschrift

pampiniform rankenförmig

pampiniform body/Rosenmüller's body Nebenovar, Nebeneierstock, Epoophoron, Rosenmüller'-Organ

pan *n geol* Pfanne, Mulde, Becken

panacea Panazee, Universalmittel, Allheilmittel, Wundermittel

panama-hat family/jipijapa family/ cyclanthus family/Cyclanthaceae Scheinpalmen

pancreas Pankreas, Bauchspeicheldrüse

pancreatic islet/islet of Langerhans Pankreasinsel, Inselorgan, Langerhanssche Insel

pancreatic juice Bauchspeichel, Pankreassaft

pancreatic polypeptide (PP) pankreatisches Polypeptid (PP)

Pandanaceae/screw-pine family Schraubenpalmen

pandemic *adv/adj* **(occurring over a wide area)** pandemisch, sich weit ausbreitend, weit verbreitet

pandemic *n* Pandemie

panduriform/fiddle-shaped geigenförmig

pangolins/scaly anteaters/ Pholidota Schuppentiere

panicle (an inflorescence) Rispe, Blütenrispe

paniculate/panicular paniculat, panikulat, rispig

panspermia/panspermatism *evol* Panspermie

pant/panting (e.g. dog) hecheln

pantophagous/pantophagic pantophag

pantothenic acid (pantothenate) Pantothensäure (Pantothenat)

pap/nipple Brustwarze

pap/tit/teat (of cow) Euter

PAP stain/Papanicolaou's stain PAP-Färbung, Papanicolaou-Färbung

Papaveraceae/poppy family Mohngewächse

papaya family/Caricaceae Melonenbaumgewächse

paper chromatography Papierchromatograph(f)ie

paper electrophoresis Papierelektrophorese

Papilionaceae/Fabaceae/pea family/ bean family/legume family/pulse family (Leguminosae) Hülsenfruchtgewächse, Hülsenfrüchtler, Schmetterlingsblütler

papilionaceous/butterfly-like (flower) schmetterlingsartig, schmetterlingsblütig

papilla (see also: lingual papilla) Papille
- **anal papilla/anal tubercle** *arach* Analhügel
- **dermal papilla** Dermispapille, Hautpapille
- **feather papilla** Federpapille, Federbalg
- **gustatory papilla/lingual papilla (see there)** Geschmackspapille
- **mammary papilla** Brustwarze

243 parasitism **P**

pappus *bot* Pappus, Haarkelch, Federkelch (Haarkranz des Blütenkelchs)
parabasal body Parabasalkörper
parabiosis Parabiose
parabronchus (*pl* **parabronchi)** Parabronchus (*pl* Parabronchien), Lungenpfeife
paracarpous coeno-parakarp
paracentric inversion parazentrische Inversion
parachute (patagium) Fallschirm (Spannhaut)
paraclade Parakladium, Wiederholungstrieb (Synfloreszenz)
paracorolla Parakorolle, Nebenkrone
paracrine parakrin, paracrin
paradidymis Paradidymis
paradigm/pattern/example Paradigma, Muster, Beispiel
paraflagellar body/flagellar swelling (euglenids) Paraflagellarkörper
paraglossa Paraglossa, Nebenzunge
parallel evolution/parallelism parallele Evolution, Parallelentwicklung, Parallelismus
parallel veined parallelnervig
parallel venation Paralleladerung, Parallelnervatur
parallelism/parallel evolution Parallelismus, parallele Evolution, Parallelentwicklung
parallely striped parallelgestreift
parallely veined paralleladrig
paralog *gen* paralog
paralogy Paralogie
paralysis Paralyse, Lähmung
paralyze paralysieren, lähmen
paramere Paramer *nt*
paramesonephric duct/Mueller's duct Müllerscher Gang
paranal sinus/anal sac/sinus paranalis Analbeutel
paranemic joint *gen* paranemische Verbindung
paranephros Nebenniere
paranucleus Nebenkern
paraparietal organ/parapineal organ/parietal organ of epiphysis Parietalorgan, Parapinealorgan
parapatric parapatrisch
parapatry Parapatrie, Kontakt-Allopatrie
paraphyletic paraphyletisch
paraphysis Paraphyse ("Saftfaden")
parapod/side-foot Parapodium
parapterum/parapteron *orn* Schulterfittich; *entom* (small sclerite like a shoulder lappet/tegula) Parapterum
pararetrovirus Pararetrovirus

parasite Parasit, Schmarotzer
- **animal parasite/zooparasite/ a parasitic animal** (*see:* zoophagous parasite) tierischer Parasit, parasitierendes Tier (Zooparasit)
- **brood parasite** Brutparasit, Brutschmarotzer
- **cleptoparasite/kleptoparasite** Kleptoparasit
- **endoparasite** Endoparasit, Innenparasit
- **exoparasite/ectoparasite/ epizoon** Außenparasit, Exoparasit, Ektoparasit (Hautparasit)
- **facultative parasite** fakultativer Parasit, Gelegenheitsparasit
- **hemiparasite/semiparasite** Hemiparasit, Halbschmarotzer
- **hyperparasite/superparasite** Hyperparasit, Überparasit
- **kleptoparasite/cleptoparasite** Kleptoparasit
- **obligate parasite/holoparasite** Vollschmarotzer, Vollparasit, Holoparasit
- **phytoparasite/parasitic plant** (*see:* plant parasite) pflanzlicher Parasit, parasitierende Pflanze (Phytoparasit)
- **plant parasite (thriving in/on plants)** Phytoparasit, Pflanzenparasit (Schmarotzer in/auf Pflanzen)
- **skin parasite/dermatozoan** Hautparasit, Hautschmarotzer, Dermatozoe
- **superparasite/hyperparasite** Hyperparasit, Überparasit
- **zooparasite/animal parasite/a parasitic animal (***see:* zoophagous parasite) tierischer Parasit, parasitierendes Tier (Zooparasit)
- **zoophagous parasite (thriving in/on animals)** Zooparasit (Schmarotzer in/auf Tieren)
parasitemia Parasitämie
parasitic parasitär, parasitisch, schmarotzend
parasitism Parasitismus, Schmarotzertum
- **autoparasitism** Autoparasitismus
- **brood parasitism** Brutparasitismus
- **hyperparasitism/superparasitism** Hyperparasitismus, Überparasitismus
- **multiple parasitism/ polyparasitism** Multiparasitismus
- **social parasitism** Sozialparasitismus

P parasitize 244

parasitize parasitieren,
schmarotzen
parasitoid Parasitoide
parasitologist Parasitologe
parasitology Parasitologie,
Parasitenkunde
parasitosis Parasitose
parasomal sac parasomaler Sack
parasome Parasome, Nebenkörper
parasympathetic parasympathisch
(autonomes Nervensystem)
parasympathetic nerve system
parasympathisches Nervensystem,
Parasympathikus (autonomes
Nervensystem)
paratenic host/transfer host
paratenischer Wirt, Sammelwirt,
Stapelwirt, Transportwirt
parathion Parathion (E 605)
parathyrin/parathormone/
parathyroid hormone (PTH)
Parathyrin, Parathormon,
Nebenschilddrüsenhormon (PTH)
parathyroid gland/parathyroid
Nebenschilddrüse, Beischilddrüse,
Epithelkörperchen
parathyroid hormone/parathyrin/
parathormone (PTH)
Nebenschilddrüsenhormon,
Parathyrin, Parathormon (PTH)
paratope/antigen combining site/
antigen binding site Paratop,
Antigenbindestelle,
Antigenbindungsstelle
paratype Paratypus, Parastandard
paraxial rod (Kinetoplastida)
Achsenstab
paraxon Paraxon, Nebenaxon
parchment fungus family/
Stereaceae Rindenschichtpilze
parchment-bark family/pittosporum
family/tobira family/
Pittosporaceae
Klebsamengewächse
parchmentlike (e.g. wings/egg
case) pergamentartig
parenchyma/ground tissue/
fundamental tissue Parenchym,
Grundgewebe
• **boundary parenchyma**
Kontaktparenchym
• **cortical parenchyma**
Rindenparenchym
• **palisade parenchyma**
Palisadenparenchym
• **pseudoparenchyma/**
paraplectenchyma
Pseudoparenchym
• **ray parenchyma**
Markstrahlparenchym
• **spongy parenchyma**
Schwammparenchym

• **stellate parenchyma**
Sternparenchym
• **storage parenchyma**
Speicherparenchym
• **traumatic parenchyma**
Wundparenchym
parenchymatous parenchymatisch
parent Elternteil
• **parents** Eltern
parent compound/parent molecule
(backbone) *chem* Grundkörper
parent rock/bedrock/base
(unmodified) fester Untergrund,
Muttergestein, Grundgestein,
Ausgangsgestein
parent substance Muttersubstanz
parental care *ethol*
elterliche Fürsorge
parental ditype (PD)
parentaler Dityp (PD)
parental investment *ethol*
Elternaufwand
parenthood Elternschaft
parenthosome/parenthesome/
septal pore cap Parenthosom,
Parenthesom, Porenkappe
parents Eltern
parfocal *opt* abgeglichen
parfocality *opt* Abgeglichenheit
paries Wand, Wandung
(eines Hohlraums)
parietal/borne on the wall parietal,
wandbürtig, wandständig
parietal bone/os parietale
Scheitelbein
parietal cell/oxyntic cell
(HCl production) Belegzelle
parietal eye (*Sphenodon*)
Parietalauge
parietal lobe/lobus parietalis *neuro*
Scheitellappen
parietal organ of epiphysis/
paraparietal organ/parapineal
organ Parietalorgan,
Parapinealorgan
parietal placentation
Parietalplazentation,
wandständige Plazentation
parietal pleura/pleura parietalis
Rippenfell
parietal tuber/parietal tuberosity/
tuber parietale Scheitelhöcker
paripinnate/even-pinnate/equally
pinnate paarig gefiedert
park tree Parkbaum
Parkeriaceae/floating fern family/
water fern family
Hornfarngewächse
parkland Parkwald, Parklandschaft
Parnassiaceae/grass of Parnassus
family Herzblattgewächse
paroecism Parökie, Beisiedlung

paroophoron/parovarium
Paroophoron, Beieierstock
parotid gland/parotis/parotid (mammals: salivary gland)
Parotis, Ohrspeicheldrüse
parotoid gland (amphibians)
Parotoiddrüse, Parotisdrüse, Ohrdrüse, Duvernoysche Drüse
parovarium/epoophoron
Nebeneierstock, Epoophoron
paroxysm Paroxysmus, Krampf, Anfall
parr (young salmon: stage between fry and smolt) Sälmling, Lächsling (junge Lachsbrut)
parrots & parakeets/Psittaciformes
Papageien
parsimony Parsimonie
parsley family/carrot family/ umbellifer family/Apiaceae/ Umbelliferae Umbelliferen, Doldenblütler, Doldengewächse
parsley fern family/rock-brake fern family/Cryptogrammaceae
Rollfarngewächse
parted/partite (leaf margin) teilig, geteilt
parthenocarpic parthenokarp
parthenocarpy Parthenokarpie
parthenogenesis Parthenogenese, Jungfernzeugung
parthenogenetic egg
parthenogenetisches Ei, Subitanei, Jungfernei
partial correlation coefficient
Teilkorrelationskoeffizient
partial digest *biochem*
Partialverdau
partial linkage partielle Kopplung
partial pressure Partialdruck
particulate inheritance
partikuläre Vererbung
particulate organic carbon (POC)
partikulärer organischer Kohlenstoff
partite/parted (leaf margin) teilig, geteilt (Blattrand)
partition/cross-wall/dividing wall/ dissepiment/septum Scheidewand, Septe, Septum
parturition/delivery/giving birth
Niederkunft, Gebären
party (wild boar) Rudel, Schar
PAS stain (periodic acid-Schiff stain)
PAS-Anfärbung (Periodsäure-Schiff-Reagens)
passage Gang, Weg; Durchgang; Verbindungsgang;
med/physiol (bowel movement) Stuhlgang, Entleerung
passage/subculture Passage, Subkultivierung
passage cell Durchlasszelle

passageway Ausführgang
passenger DNA passagere DNA, Passagier-DNA
passeriform sperlingsartig
passerines/passeriforms (perching birds)/Passeriformes
Sperlingsvögel
passionfruit family/Passifloraceae
Passionsblumengewächse
passive dispersal
passive Ausbreitung, Verdriftung
pastern (horse) Fessel
pastern bone (long pastern)
Fesselknochen, Fesselbein
● **short/small pastern bone/middle phalanx (horse)** Kronbein
pastern joint Fesselbeingelenk, Krongelenk
Pasteur effect Pasteur-Effekt
Pasteur pipet Pasteurpipette
pasteurize pasteurisieren
pasteurizing/pasteurization
Pasteurisierung, Pasteurisieren
pastoral (relating to shepherds/ herdsmen) Schäfer…, Hirten…, Schäfer/Hirten betreffend
pastoral (devoted to livestock raising) Vieh…, Viehzucht betreffend
pastoral (relating to the countryside) ländlich
pastoral economy/pastoralism/ pasture farming Weidewirtschaft
pasture/pasturage Weide, Weidewiese (Grünland), Trift (Heide)
● **permanent pasture** Dauerweide
pasture farming/pastoral economy/ pastoralism/agropastoralism
Weidewirtschaft
pasturing Beweidung
patagium Patagium, Flughaut, Flatterhaut, Spannhaut, Gleithaut
patch *allg* Flicken; Fleck, Stelle
patch/flowerbed Beet, Blumenbeet
patch budding *hort*
Platten-Okulation
patch clamp
"Membranfleck-Klemme"
patch clamp technique
Patch-Clamp Verfahren
patch reef/bank reef Fleckenriff
patching Patching, Verklumpung, Fleckbildung
patella/knee bone/genu Patella, Kniescheibe
paternal väterlich, väterlicherseits, Vater…
paternity Vaterschaft
paternity exclusion
Ausschluss der Vaterschaft

paternity test

paternity test
Vaterschaftsbestimmung,
Vaterschaftstest,
Vaterschaftsnachweis
path/pathway Bahn, Pfad, Weg;
Leitung
path difference *opt*
Gangunterschied
path of light (ray diagram)
Strahlengang (Strahlendiagramm)
pathogen *n* Krankheitserreger,
Erreger, pathogener
(Mikro)Organismus
pathogenic (causing or capable of causing disease) pathogen,
krankheitserregend
pathogenicity Pathogenität
pathological (altered or caused by disease) pathologisch, krankhaft
pathology Pathologie, Lehre von
den Krankheiten
pathway Bahn, Pfad, Weg; Leitung;
(Reaktions-)Kette
- **amphibolic pathway/central metabolic pathway**
  amphiboler Stoffwechselweg
- **anabolic pathway/biosynthetic pathway** anaboler
  Stoffwechselweg,
  Biosynthese-Stoffwechselweg
- **apoplast pathway**
  apoplastischer Wasserstrom
- **Embden-Meyerhof pathway/ Embden-Meyerhof-Parnas pathway (EMP pathway)/ hexosediphosphate pathway/ glycolysis** Embden-Meyerhof-Weg
- **enzymatic pathway**
  enzymatische Reaktionskette
- **metabolic pathway/metabolic shunt** Stoffwechselweg
- **pentose phosphate pathway/ pentose shunt/ phosphogluconate oxidative pathway/hexose monophosphate shunt (HMS)**
  Pentosephosphatweg,
  Hexosemonophosphatweg,
  Phosphogluconatweg
- **reaction pathway** Reaktionskette
- **salvage pathway**
  Wiederverwertungs-
  stoffwechselwege,
  Wiederverwertungsreaktionen
- **transpiration pathway**
  Transpirationsweg
- **uricolytic pathway** uricolytischer
  Weg, urikolytischer Weg
- **water transport pathway**
  Wassertransportweg
patriarchal patriarchalisch
pattern/design Muster, Musterung

pattern/sample/model; specimen
Muster, Vorlage, Modell
pattern formation Musterbildung
pattern recognition *neuro*
Mustererkennung
paturon/basal segment of chelicera
Paturon, Chelicerengrundglied,
Chelizerenbasalsegment
paunch/rumen/ingluvies/first stomach Pansen, Rumen
pauropods/Pauropoda Wenigfüßer,
Pauropoden
pavement epithelium
einschichtiges Plattenepithel
paw *n* Pfote, Tatze, Pranke (Tatze
großer Raubtiere)
paw/pawing *vb* (mit der Pfote)
scharren
paw/pawing/paw the ground (horse: scraping the ground)
scharren
Paxillus family/Paxillaceae
Kremplinge
PCR (polymerase chain reaction)
PCR (Polymerasekettenreaktion)
- **bubble linker PCR**
  Blasen-Linker PCR
- **differential display (form of RT-PCR)** differentieller Display (Form
  der RT-PCR)
- **DOP-PCR (degenerate oligonucleotide primer PCR)**
  DOP-PCR (PCR mit degeneriertem
  Oligonucleotidprimer)
- **inverse PCR** inverse
  Polymerasekettenreaktion
- **IRP (island specific PCR)** IRP
  (inselspezifische PCR)
- **ligation-mediated PCR**
  ligationsvermittelte
  Polymerasekettenreaktion
- **RACE-PCR (rapid amplification of cDNA ends-PCR)** RACE-PCR
  (schnelle Vervielfältigung von
  cDNA-Enden-PCR)
- **RT-PCR (reverse transcriptase-PCR)** RT-PCR (PCR mit reverser
  Transkriptase)
pea family/bean family/legume family/pulse family/Fabaceae/ Papilionaceae (Leguminosae)
Hülsenfruchtgewächse,
Hülsenfrüchtler,
Schmetterlingsblütler
peak value/maximum (value)
Scheitelwert, Höchstwert,
Maximum
peanut worms/sipunculoids/ sipunculans Spritzwürmer,
Sternwürmer, Sipunculiden
pear-shaped/pyriform birnenförmig
pearl Perle

247   **pelagic organisms** **P**

peat Torf
- **black peat** Schwarztorf
- **granulated peat/garden peat** Torfmull
- **reed peat** Schilftorf
- **sedge peat** Seggentorf
- **white peat** Weißtorf (Hochmoortorf)

peat bank/peatery Torfstich
peat bog Sphagnum-Moor, Torfmoor
peat clay/organic silt Mudde, organogener Schlamm
peat moss/bog moss (*Sphagnum*) Torfmoos
pebble Kiesel, Kieselstein
pebrine (*Nosema bombycis*) Fleckenkrankheit (Seidenraupen)
peck/pick *vb* picken
peck order/pecking order Hackordnung
Peckhamian mimicry Peckhamsche Mimikry
pecten (bird's eye) Pecten
pecten/comb/brush/rake (*pl* pectines) Pecten, Kamm, Bürste, Rechen
pecten/pollen comb Pollenkamm
pectic acid (pectate) Pektinsäure (Pektat)
pectin Pektin
pectinate/pectiniform/ctenose/ ctenoid/comblike/rakelike kammartig, kammförmig, gekämmt
pectinella Pektinelle
pectineus (muscle) Kammmuskel, Pektineus
pectoral brustständig, Brust...
pectoral fin Brustflosse
pectoral girdle/shoulder girdle Brustgürtel, Schultergürtel
ped/soil aggregate Gefügekörper (Boden)
pedal bone/coffin bone (horse) Hufbein
pedal ganglion Pedalganglion
pedal gland/adhesive gland/cement gland (rotifers) Fußdrüse, Klebdrüse, Kittdrüse
pedal laceration (sea anemones: *Actiniaria*) Laceration der Fußscheibe
pedal retractor muscle Fußretraktor
pedal sole/foot sole/planta Fußsohle, Planta (auch: Kriechfußsohle)
Pedaliaceae/sesame family/benne family Sesamgewächse
pedate fußförmig
pedicel Pedicellus, Stiel
pedicel *bot* Blütenstiel (einzelner Infloreszenzblüten)
pedicel *zool* **(antler)** Geweihrose

pedicellaria (*pl* pedicellarias/ pedicellariae) Pedicellarie, Pedizellarie
- **gemmiform pedicallaria/ globiferous pedicellaria/ poison(ous) pedicellaria/toxic pedicellaria** gemmiforme Pedicellarie, globifere Pedicellarie, drüsige Pedicellarie, Giftpedicellarie, Giftzange
- **ophiocephalous pedicallaria** ophiocephale Pedicellarie, gezähnte Beißzange
- **tridentate pedicallaria** tridactyle Pedicellarie, tridentate Pedicellarie, dreiklappige Zange, Klappzange
- **trifoliate pedicallaria/triphyllous pedicellaria** trifoliate Pedicellarie, triphyllate Pedicellarie, Putzzange

pedicellus/pedicel Pedicellus, Pedizellus, Stiel
pedigree Stammbaum
pediveliger Pediveliger, Velichoncha
pedobiology/soil biology Pedobiologie, Bodenbiologie
pedogamy/paedogamy Pädogamie
pedogenesis/paedogenesis Pädogenese
pedologist/soil scientist Pedologe, Bodenkundler
pedology/soil science Pedologie, Bodenkunde
pedomorphosis Pädomorphose
pedonic pedonisch
pedosphere Pedosphäre
peduncle/flower stalk Blütenstiel, Blütenschaft; Blütenstandsstiel
pedunculate/stalked (flower stalk) gestielt
peel/skin *n* Haut, Schale
peel/skin *vb* **(e. g. fruit)** schälen, die Haut abmachen
peep/squeak (mouse) piepen, piepsen
peer *vb* spähen, schauen, starren; hervorschauen, hervorstehen, herausstehen
peer *n* Ebenbürtige, Gleichrangige, Genossen, Kollegen; Kumpels
peer-review Begutachtungsverfahren (Manuskripte)
peg Pflock
peg (sensilla) *zool* Kegel
pelage/furcoat (mammals: hairy covering/thick coat of hair) dichte Haarbedeckung, Körperbedeckung der Säugetiere
pelagic/pelagial/open-sea pelagisch, pelagial
pelagic organisms/pelagic community Pelagos (Organismen des Pelagial)

pelagic zone

pelagic zone/open sea Pelagial, pelagische Zone, Hochseebereich, Freiwasserzone
pellet *centrif* Pellet (Niederschlag in Zentrifugierröhrchen)
pellet(s) Pellet, Kügelchen, Pille
- **feed pellets** Futterkügelchen
- **rabbit feces (dry/round/brown)** Hasenkot (trocken/rund)
- **regurgitated matter (predatory birds)** Gewölle
- **reingested rabbit feces (soft/greenish)** Blinddarmkot, Vitaminkot
pellicula/pellicle Pellicula
pellucid/translucent (not hyaline) durchscheinend, lichtdurchlässig
pelt Fell, Tierhaut, Tierpelz (die ganze abgezogene Tierhaut)
peltate/peltiform/shield-shaped schildförmig
peltate leaf peltates Blatt, Schildblatt
pelvic bone/pelvis/pubis (ilium & ischium) Beckenknochen (Darmbein & Sitzbein)
pelvic cavity Beckenhöhle
pelvic fin Bauchflosse
pelvic floor/pelvic diaphragm Beckenboden
pelvic girdle/pelvic arch/hip girdle Beckengürtel
pelvis Pelvis, Becken
pelycosaurs/pelycosaurians (early synapsids)/Pelycosauria Urraubsaurier
pen/coop/hutch (e.g. chickens) kleiner Tierkäfig, kleiner Verschlag, Auslauf (z.B. Geflügelstall)
pen/female swan weiblicher Schwan
pen/fold (small fenced-in area for sheep/cattle/pigs etc.) Pferch (kleines Gehege)
pen/gladius Rückenfeder, Gladius, pergamentartiger Schulp
pen shells/Pectinidae Kammuscheln
pen tray *lab* Federschale
pendulous/hanging down hängend, herabhängend; frei schwebend
penetrance Eindringen
penetrance (complete/incomplete) *gen* Penetranz (vollständige/ unvollständige)
penetrate penetrieren, eindringen, durchstoßen, durchstechen; (den Penis einführen)
penetrating odor/smell penetranter/ aufdringlicher Geruch
penguins/Sphenisciformes Pinguine

penial spicule/copulatory spicule (nematodes) Spiculum (Kopulationshaken: Nematoden)
penicillanic acid Penicillansäure, Penizillansäure
penicillin Penicillin, Penizillin
peninsula Halbinsel
penis (*pl* penes)/phallus/copulatory organ/intromittent organ Penis, Phallus, männliches Glied, Rute, männliches Begattungsorgan
penis/aedeagus/intromittent organ *entom* Penis, Aedeagus, Aedoeagus
penis bone/baculum/os penis Penisknochen
penis bulb (trematoda) Bursa
penninerved/pinnately nerved/ pinnately veined fiedernervig, fiederadrig
Pennsylvanian/Upper Carboniferous Spätes Karbon
pennywort family/Hydrocotylaceae Wassernabelgewächse
pentadactyl/limb with five digits (five-toed) fünffingerig, fünffingrig, fünfstrahlig, pentadaktyl (Fünfzahl von Fingern, Zehen)
pentadactylism/pentadactyly Pentadaktylie, Fünffingerigkeit, Fünfzahl von Fingern/Zehen
pentamer/penton *vir* Pentamer, Penton
pentameric/pentamerous pentamer, fünfzählig, fünfstrahlig, fünfteilig
pentamerous symmetry/pentameral symmetry/five-sided symmetry pentamere Symmetrie, fünfstrahlige Symmetrie
pentamery Pentamerie, fünfstrahlige Radiärsymmetrie
pentandrous mit 5 Staubblättern
pentavalent fünfwertig
pentose phosphate pathway/ pentose shunt/phosphogluconate oxidative pathway/hexose monophosphate shunt (HMS) Pentosephosphatweg, Hexosemonophosphatweg, Phosphogluconatweg
pentosuria Pentosurie
peony family/Paeoniaceae Pfingstrosengewächse
peperomia family/Peperomiaceae Zwergpfeffergewächse
peplomer Peplomer
pepo/gourd Kürbisfrucht, Gurkenfrucht, Panzerbeere
pepper family/Piperaceae Pfeffergewächse
pepsin (pepsin A) Pepsin (Pepsin A)
peptide Peptid

peptide bond/peptide linkage
Peptidbindung
peptide chain Peptidkette
peptidoglycan/mucopeptide
Peptidoglykan, Mukopeptid
peptidyl transferase
Peptidyltransferase
peptidyl-site/P-site Peptidyl-Stelle,
P-Stelle
peptone Pepton
peptone water Peptonwasser
peptonization Peptonisierung
peptonize peptonisieren
perceive wahrnehmen, empfinden
(Reiz)
percentage Prozentsatz,
prozentualer Anteil
perceptible/sensible wahrnehmbar,
empfindbar
perception Perzeption,
Wahrnehmung, Empfindung
perch/roost *vb* rasten, sich setzen,
sich niederlassen (zur Ruhe/Schlaf)
perch & perchlike fishes/
Perciformes Barschfische,
Barschartige
percolate (flow through)
durchsickern, durchfließen
percolation (flowing through)
Durchsickern, Durchfluss
perennial perennierend, ausdauernd,
mehrjährig
perennial herb (hardy/with woody
base) Staude
pereon/pereion Pereion, Peraeon,
Pereon (Brust/Thorax bei
Crustaceen)
pereopod/pereiopod (walking leg
of pereion) Pereiopode,
Peraeopode (Schreitbein/Brustfuß
des Pereon)
perfect flower/bisexual flower/
hermaphroditic flower
zwittrige Blüte, Zwitterblüte
perfect stage/telomorphic stage
fung Hauptfruchtform
perfoliate durchwachsen,
durchwachsenblättrig
perforated perforiert, löcherig
perforation plate (xylem)
Perforationsplatte,
perforierte Endwand (Xylem)
performance value/performance
coefficient Leistungszahl
performic acid Perameisensäure
perfuse perfundieren, durchströmen,
durchspülen; übergießen,
überströmen
perfusion culture Perfusionskultur
perianth Perianth, differenzierte
Blütenhülle, differenzierter
Blütenhüllkreis/Blütenhüllblattkreis

peribranchial cavity/atrial cavity
(ascidians/lancelets)
Peribranchialraum
pericalymma/test-cell larva
Pericalymma, Hüllglockenlarve
pericambium/pericycle
Perikambium, Perizykel
pericardial cavity/pericardial
chamber/pericardial sac/
pericardial sinus Perikardialhöhle,
Perikardialraum, Perikardialsack,
Perikardialbeutel, Perikardialsinus
pericardial septum/diaphragm
Perikardialseptum, Diaphragma
pericardium Pericard, Perikard,
Herzbeutel
pericarp Fruchtwand, Perikarp
perichondrium Perichondrium,
Knorpelhaut
periclinal chimera
Periklinalchimäre
pericyte Pericyt, Perizyt
peridium Peridie
perigon Perigon, einheitliche
Blütenhülle, einheitlicher
Blütenhüllkreis/Blütenhüllblattkreis
perigonadial cavity/gonocoel
Gonadenhöhle, Gonocoel
perigynium Fruchtsack,
Fruchtknotenhülle (Gräser)
perigynous perigyn, mittelständig
perikaryon/cell body/soma
(neurons) Perikaryon, Zellkörper,
Soma
perilymph Perilymphe
perineal perineal, den Damm
betreffend, Damm…
perineal gland Perinealdrüse,
Dammdrüse
perineal swelling (chimpanzee)
Dammschwiele
perinephric fat/perirenal fat
Nierenfett
perineum Perineum, Damm
perinuclear space/perinuclear
cistern perinukleärer Raum,
perinukleärer Spaltraum,
perinukleäre Zisterne, Cisterna
karyothecae
period (geological time) (see also:
eon/epoch/era) Periode
● **Cambrian Period/Cambrian**
Cambrium, Kambrium
● **Carboniferous Period/**
Carboniferous/'Coal Age'
Karbon, Steinkohlenzeit
● **Cretaceous Period/Cretaceous**
Kreidezeit, Kreide
● **Devonian Period/Devonian**
Devon
● **Jurassic Period/Jurassic** Jurazeit,
Jura

P period

250

- **Neogene** Neogen, Jung-Tertiär, Jungtertiär
- **Ordovician Period/Ordovician** Ordovizium
- **Paleogene** Paläogen, Alt-Tertiär, Alttertiär
- **Permian Period/Permian** Perm
- **Quaternary Period/Quarternary** Quartärzeit, Quartär
- **Silurian Period/Silurian** Silur
- **Tertiary Period/Tertiary** Tertiärzeit, Tertiär, Braunkohlenzeit
- **Triassic Period/Triassic** Triaszeit, Trias

period of gestation/gestational period Tragzeit, Tragezeit, Schwangerschaft(speriode)
periodic(al) periodisch
periodic acid-Schiff stain (PAS stain) Periodsäure-Schiff-Reagens (PAS-Anfärbung)
periodic table (of the elements) Periodensystem (der Elemente)
periodicity Periodizität
periople/epidermis limbi (on hoof of equines) Perioplum
periosteum Periost, Knochenhaut
periostracum Periostracum, Schalenhäutchen
periotic (situated around the ear) um das Innenohr herum liegend (betr. Knochenelemente)
periotic bone/periotic/os perioticum Felsenbein
peripatric peripatrisch
peripheral peripher
peripheral meristem/flank meristem Flankenmeristem
peripheral nervous system peripheres Nervensystem
peripheral protein/extrinsic protein peripheres Protein, extrinsisches Protein
periphyton (attached algae) *limn* Periphyton, Aufwuchs, Bewuchs
periplasmic space periplasmatischer Raum
periproct Periprokt, Afterfeld
perirenal fat/perinephric fat Nierenfett
perish verderben; zugrunde gehen
perishable (foods) verderblich
- **highly perishable (fruit)** leicht verderblich (Früchte)
perisperm Nährgewebe (nucellar)
peristalsis Peristaltik
peristaltic peristaltisch
peristome/peristomium (border of mouth/oral margin) (polyps/sea urchins) Peristom, Mundfeld; (buccal cavity: ciliates) Peristom, Buccalhöhle (Zellmundhöhlung)

peritoneum Peritoneum, Bauchfell
peritrichous peritrich
peritrophic membrane peritrophische Membran
periwinkle family/dogbane family/ Apocynaceae Hundsgiftgewächse, Immergrüngewächse
perlite Perlit, Perlstein
permafrost Permafrost, Dauereis, Dauerfrost
permafrost soil Permafrostboden, Dauerfrostboden
permanent cell Dauerzelle
permanent mount/permanent slide *micros* mikroskopisches Dauerpräparat
permanent pasture Dauerweide
permanent tissue/secondary tissue Dauergewebe
permanent wilting percentage permanenter Welkungsgrad
permeability Permeabilität, Durchlässigkeit
permeable permeabel, durchlässig
- **impermeable/impervious** impermeabel, undurchlässig
permeant(s) *zool* Permeant(en)
Permian Period/Permian (geological time) Perm
permissible workplace exposure zulässige Arbeitsplatzkonzentration
permissive cell permissive Zelle
permissive host permissiver Wirt
permissivity/permissive conditions Permissivität, permissive Bedingungen
pernicious anemia bösartige Blutarmut, perniziöse Anämie
peroxisome Peroxisom
persistence/hardiness/perseverance Persistenz, Ausdauer
persistent persistent, ausdauernd
persistent spore/resting spore Dauerspore
perthotrophic/perthophytic perthotroph, perthophytisch
pest Schädling, Ungeziefer
pest control Schädlingskontrolle, Schädlingsbekämpfung
- **biological pest control** biologische Schädlingsbekämpfung
pest infestation Schädlingsbefall
pest insect Schadinsekt
pesticide/plant-protective agent/ biocide Pestizid, Schädlingsbekämpfungsmittel, Pflanzenschutzmittel, Biozid
pesticide resistance Pestizidresistenz
pestle (and mortar) Stößel, Pistill (und Mörser)
pet zahmes Haustier (Liebhaberei)

petaloid *adv/adj* petaloid, kronblattartig
petaloid *n* Petalodium
petals/corolla Kronblätter, Blütenkronblätter
petasma Petasma
petiolar gall Blattstielgalle
petiolate/stipitate/stalked gestielt
petiole/leaf stalk Blattstiel
petiole/podeon/podeum *zool* (hymenopterans) Petiolus, Hinterleibsstiel, "Taille"
petite mutant Petite-Mutante
Petri dish Petrischale
petrification Versteinerung (*Vorgang*)
petrified versteinert
petrify versteinern
petroleum/crude oil Erdöl
petroleum ether Petroläther
petrophyte/rock plant Felspflanze
petrosal das Felsenbein betreffend, Felsenbein…
petrosal bone/petrous bone (of temporal bone) Felsenbein (des Schläfenbeins)
petrous steinhart, felsig
Peyer's patch Peyerscher Plaque
phaeomelanin Phäomelanin
phage/bacteriophage Phage, Bakteriophage
• **lytic phage** lytischer Phage
• **temperate phage** temperenter Phage
• **virulent phage/lysogenizing phage** virulenter Phage
phagemid Phagemid
phagocyte Phagozyt, Phagocyt
phagocytize phagozytieren
phagocytosis Phagozytose
phagolysosome/secondary lysosome sekundäres Lysosom
phagosome/heterophagosome (*siehe auch:* Autophagosom) Phagosom, Heterophagosom
phagotrophic phagotroph
phalaenophily/moth pollination Mottenbestäubung, Phalaenophilie
phalangeal phalangeal, Finger-/Zehenglieder betreffend
phalanx (*pl* phalanges) Phalanx, Fingerglied, Zehenglied
phallic threat Phallusdrohen
phallus/penis/copulatory organ/ intromittent organ Phallus, Penis, männliches Glied, männliches Begattungsorgan
phanerophyte (woody plant; aerial dormant buds) Phanerophyt, Holzgewächs (Bäume/Sträucher)
Phanerozoic Eon/Phanerozoic Phanerozoikum

pharate (cloaked adult/coarctate pupa) pharat (coarctate Puppe)
pharmacist Pharmazeut(in), Apotheker
pharmaceutic(al) *adj/adv* pharmazeutisch
pharmaceutical *n* Pharmazeutikum, Pharmakon, Medikament, Arznei, Arzneimittel, Drogen
pharmacogenetics Pharmakogenetik
pharmacognosy Pharmakognosie, Drogenkunde
pharmacology Pharmakologie
pharmacopeia Arzneibuch
pharmacy (science/study of drugs) Pharmazie, Pharmazeutik, Arzneikunde, Arzneilehre
pharmacy/drugstore Apotheke (subscription drugs)
pharyngeal pharyngial, den Schlund betreffend; den Rachen betreffend
pharyngeal arch Pharyngialbogen, Schlundbogen (*see also:* branchial arch)
pharyngeal basket (ciliates) Reusenapparat der Mundbucht
pharyngeal jaws/trophi (rotifers) Kiefer, Trophi
pharyngeal plane Schlundebene
pharyngeal pouch Schlundtasche
pharyngeal tonsil Rachentonsille, Rachenmandel
pharynx/air pipe Pharynx, Luftröhre
pharynx/gullet Pharynx, Schlund
pharynx/mastax (rotifers) Pharynx, Mastax, Kaumagen
phase boundary Phasengrenze, Phasentrennlinie
phase contrast *micros* Phasenkontrast
phase contrast microscopy Phasenkontrastmikroskopie
phase diagram Phasendiagramm
phase ring/phase annulus Phasenring
phase transition Phasenübergang
phase transition temperature Phasenübergangstemperatur
phase variation Phasenveränderung
phasic fiber phasische Faser
phasmid Phasmid, Schwanzpapillendrüse
phellem/cork Phellem
phelloderm/secondary cortex Phelloderm, Korkrinde
phellogen/cork cambium Phellogen
phenanthrene Phenanthren
phene Phän *nt*
• **ecophene** Ökophän
• **isophene** Isophän
phenetics/numerical taxonomy/ taxometrics Phänetik, numerische Taxonomie

phenocopy

phenocopy Phänokopie
phenogenesis Phänogenese
phenogenetics Phänogenetik
phenogram (numerical taxonomy) Phänogramm, Ähnlichkeitsdendrogramm
phenol Phenol
phenology Phänologie
phenospermy Leerfrüchtigkeit, Kenokarpie
phenotype Phänotyp, Phaenotypus
phenylalanine Phenylalanin
phenylketonuria Phenylketonurie
pheophytin Phäophytin
pheromone Pheromon
philopatry/homing Philopatrie, Ortstreue
phlebotomy/venesection Phlebotomie, Aderlass, Veneneröffnung
phloem Phloem, Siebteil, Bastteil
• **external phloem** externes Phloem, äußeres Phloem, Außenphloem
• **internal phloem** internes Phloem, inneres Phloem, Innenphloem
• **interxylary phloem** interxylares Phloem
• **intraxylary phloem** intraxylares Phloem
phloem loading Phloembeladung
phloem sap Phloemsaft
phloem unloading Phloementladung
phlox family/Polemoniaceae Sperrkrautgewächse, Himmelsleitergewächse
phonation Phonation, Lautbildung, Stimmbildung
phorbol ester Phorbolester
phoresis/phoresy/phoresia Phoresie
Phormiaceae/flax lily family Phormiumgewächse
phoronids/Phoronidea Phoroniden, Hufeisenwürmer
phorozooid Phorozooid, Pflegetier, Tragtier (Ascidien)
phosphate Phosphat
phosphatidic acid Phosphatidsäure
phosphatidylcholine Phosphatidylcholin
phosphodiester bond Phosphodiesterbindung
phosphogluconate oxidative pathway/pentose phosphate pathway/pentose phosphate shunt/hexose monophosphate shunt (HMS) Phosphogluconatweg, Pentosephosphatweg, Hexosemonophosphatweg
phosphorescence Phosphoreszenz
• **marine phosphorescence** Meeresleuchten

phosphoric acid (phosphate) Phosphorsäure (Phosphat)
phosphorous *adj/adv* phosphorhaltig, phosphorig, Phosphor...
phosphorus *n* Phosphor
phosphorylation Phosphorylierung
• **cyclic phosphorylation** zyklische/cyclische Phosphorylierung
• **noncyclic phosphorylation** nichtzyklische/nichtcyclische/lineare Phosphorylierung
• **oxidative phosphorylation/carrier-level phosphorylation** oxidative Phosphorylierung
• **substrate-level phosphorylation** Substratkettenphosphorylierung
• **transphosphorylation** Transphosphorylierung
photoallergenic photoallergen
photoautotrophic photoautotroph
photobleaching Photobleichung, Lichtbleichung
photoheterotroph(ic) photoheterotroph
photolithotroph(ic)/photoautotroph(ic) photolithotroph, photoautotroph
photomultiplier *opt* Photomultiplier, Photovervielfacher
photoorganotrophic(al) photoorganotroph
photoperception Lichtwahrnehmung
photoperiodism Photoperiodismus
photophore/luminous organ/light-emitting organ Photophore, Leuchtorgan
photoreactivation Photoreaktivierung
photorespiration Photorespiration, Photoatmung, Lichtatmung
photosensibilization Photosensibilisierung
photostability Lichtbeständigkeit
photostable lichtbeständig
photosynthesis Photosynthese
photosynthesize photosynthetisieren
photosynthetic photosynthetisch
photosynthetic photon flux (PPF) Photonenstromdichte
photosynthetic product/photosynthate Photosyntheseprodukt
photosynthetic quotient Photosynthesequotient, Assimilationsquotient
photosynthetic unit Photosynthese-Einheit
photosynthetically active radiation (PAR) photosynthetisch aktive Strahlung

phototroph(ic)/photosynthetic
phototroph, photosynthetisch
phototrophy Phototrophie
phototropic phototrop
phototropism Phototropismus
phragmocone Phragmokon
phratry Phratrie
phreatic (pertaining to groundwater) phreatisch, Grundwasser…
phreatophyte Phreatophyt, Grundwasserpflanze
phrenic Zwerchfell…
phrenic nerve Zwerchfellnerv
phrenology (according to: Gall) Phrenologie
Phrymaceae/lopseed family Phrymagewächse
phthalic acid (phthalate) Phthalsäure (Phthalat)
phthirapterans/Phthiraptera (Mallophaga & Anoplura) Tierläuse
phycology Algenkunde
phyletic phyletisch, Stammes…
phyletic evolution phyletische Evolution
phyllary/involucral bract Involukralblatt, Involukralschuppe
phyllid/leaflet/blade/lamina (algas/mosses) Phylloid
phylloclade Phyllokladium, Phyllocladium (Flachspross eines Kurztriebs)
phyllode Phyllodium, Blattstielblatt
phyllody (floral organ into leaflike structure) Phyllodie, Verlaubung
phylloid/phylloidal/leaf-like/foliaceous blattartig
phylloid *n* **(leaf-like organ)** Phylloid, blattartiges Organ (*pl* Phyllidien)
phyllome Phyllom, Blattorgan
phyllopod (a swimming appendage) Phyllopodium (*pl* Phyllopodien), Blattbein
phyllopods/branchiopods/Phyllopoda/Branchiopoda Blattfußkrebse, Kiemenfüßer
phylloquinone/phytonadione (vitamin K₁) Phyllochinon
phyllotaxis/phyllotaxy/leaf arrangement/leaf position Blattstellung, Blattanordnung, Blattfolge, Phyllotaxis
phylogenesis/phylogeny/evolution Phylogenese, Phylogenie, Stammesgeschichte, Stammesentwicklung, Abstammungsgeschichte, Evolution
phylogenetic/phyletic/evolutionary phylogenetisch, phyletisch, stammesgeschichtlich, evolutionär

phylogenetic tree Stammbaum
phylum (*pl* **phyla/phylums)** Phylum, Stamm
physical physikalisch; *med* körperlich
physical containment (level) Laborsicherheitsstufe
physical map physikalische Karte
physician/doctor Arzt, Doktor
physician assistant (PA) medizinisch-technische(r) AssistentIn (MTA)
physiologic(al) physiologisch
physiologic barrier physiologische Schranke
physiologist Physiologe
physiology Physiologie
physoclist (without pneumatic duct) Physoclist
physogastry Physogastrie
physostome (with open pneumatic duct) Physostom
phytanic acid Phytansäure
phytic acid Phytinsäure
phytoalexin Phytoalexin
phytobezoar (stomach ball) Phytobezoar (Magenkugel)
phytocecidium Phytocecidium, von Pilzen hervorgerufene Pflanzengalle
phytocoenon/community type/nodum/abstract plant community Phytozönon, Pflanzengesellschaft (allgemein/abstrakt)
phytocoenose/concrete plant community Phytozönose, spezifische Pflanzengesellschaft
phytoecology/plant ecology Pflanzenökologie, Vegetationskunde, Vegetationsökologie
phytogeography/plant geography/geobotany Pflanzengeographie, Pflanzengeografie, Geobotanik
phytohormone Phytohormon, Pflanzenwuchsstoff
phytol Phytol
Phytolaccaceae/pokeweed family Kermesbeerengewächse
phytophagous/herbivorous/plant-eating/feeding on plants pflanzenfressend
phytosterol Phytosterin
phytotoxic phytotoxisch, pflanzenschädlich
pia mater Pia mater
PIC (polymorphism information content) PIC (Informationsgehalt eines Polymorphismus)
pickerel-weed family/water hyacinth family/Pontederiaceae Hechtkrautgewächse
pickle pökeln, sauer einlegen (Gurken, Hering etc.)

P pickling

pickling Pökeln, in Salzlake oder Essig einlegen (Gurken, Hering etc.)
picric acid (picrate) Pikrinsäure (Pikrat)
pie chart Kreisdiagramm
pier/quay Pier
pierce *vb* stechen, durchstechen; durchdringen (Laut)
piercing-sucking/stylate-haustellate (mouthparts) stechend-saugend (Mundwerkzeuge)
pig (*see also:* swine) Schwein
piglet/little pig Ferkel
pigment/colorant Pigment, Farbstoff
● **accessory pigment** akzessorisches Pigment
● **antenna pigment** Antennenpigment
● **bile pigments** Gallenfarbstoffe
● **respiratory pigment** Atmungspigment
pigment cell/chromatophore Pigmentzelle, Farbzelle, Chromatophore
pigment cup eye/inverted eye Pigmentbecherauge, Becherauge
pigment layer (eye) Pigmentschicht
pigweed family/cockscomb family/ amaranth family/Amaranthaceae Amarantgewächse, Fuchsschwanzgewächse
pile (coat of short/fine furry hairs) *zool* Flaum, Wolle, Pelz, Haar (Fell)
pile (supportive long/slender column of timber) *bot/techn* Pfahl, Holzpfahl, Stützpfahl, Pfeiler
pileate/pileiform/cap-shaped/ having a pileus pileat, haubenförmig, kappenförmig, hutförmig, konsolenförmig
pileus/cap *fung* Hut, Schirm, Haube, Kappe, Pilzhut
piliferous/piligerous/bearing hairs (hairy) behaart (haarig)
piliform/trichoid haarförmig, haarartig
pill bugs/woodlice/sowbugs/ Isopoda Asseln
pillar/column Säule
piloerection *zool* Haarsträuben, Fellsträuben
pilose/downy/pubescent feinbehaart, flaumig
pilose/piliferous/bearing hairs (hairy) behaart (haarig)
pilot plant *techn* Versuchsanlage
piloting lotsen, lenken, führen
pilus/a hair (*pl* pili) Pilus, Haar
pilus (on bacterial surface) Pilus, Konjugationsrohr, Konjugationsfortsatz (auf Bakterienoberfläche)

pimelic acid Pimelinsäure
pin cushion gall/bedeguar Schlafapfel, Bedeguar
pin feather sich entwickelnde Feder (noch innerhalb Federscheide)
pin-point colony *micb* Kleinstkolonie
Pinaceae/pine family/fir family Tannenfamilie, Kieferngewächse, Föhrengewächse
pinacocyte Pinacocyt, Pinakozyt
pinbone/ischium/os ischii (hipbone esp. in quadrupeds) Sitzbein
pincers Pinzette; Krebsschere
● **tail pincers (e.g. earwigs)** Schwanzzange
pinch clamp *lab* Quetschhahn
pinch trap *bot* Klemmfalle
pinch-trap flower Klemmfallenblume, Klemmfallenblüte
pine Kiefer, Föhre
"pine"/cone Zapfen
pine family/fir family/Pinaceae Tannenfamilie, Kieferngewächse, Föhrengewächse
pineal body/pineal gland/ conarium/epiphysis Pinealorgan, Epiphyse, Zirbeldrüse
pineal eye/epiphyseal eye (median eye) Pinealauge (bei Neunauge: *Petromyxon*)
pineapple family/bromeliads/ bromelia family/Bromeliaceae Bromelien, Bromeliengewächse, Ananasgewächse
pineapple gall Ananasgalle
pinecone/"pine" Kiefernzapfen, Kienapfel
pinewood chip Kienspan
ping-pong reaction/double-displacement reaction Pingpong-Reaktion, doppelte Verdrängungsreaktion
pinion *n orn* **(terminal section of wing: carpus/metacarpus/ phalanges)** Flügelspitze
pinion/flight feather *n orn* Schwungfeder (Armschwinge)
pinion *vb orn* **(restrain flight by cutting pinion of one wing)** die Flügel stutzen
pink family/carnation family/ Caryophyllaceae Nelkengewächse
pinna/leaflet Fieder, Fiederblatt, Fiederblättchen (ersten Grades), Teilblatt
pinnate/pennate pinnat, pennat, gefiedert, fiedrig, fiederig, fiederblättrig, federförmig
● **bipinnate** zweifach gefiedert

255   **pitch** P

- **imparipinnate/odd-pinnate/ unequally pinnate (pinnate with an odd terminal leaflet)** unpaarig gefiedert
- **paripinnate/even-pinnate/ equally pinnate (pinnate with paired terminal leaflets)** paarig gefiedert
- **quadripinnate/divided pinnately four times** vierfach gefiedert
- **tripinnate (three times pinnate)** dreifach gefiedert

pinnate appendage/pinnate leg Fiederfuß
pinnate venation Fiederaderung, Fiedernervatur
pinnately cleft/pinnately split/ pinnatifid fiederspaltig
pinnately incised/pinnatisect fiederschnittig
pinnately lobed/pinnatilobate fiederlappig
pinnately parted/pinnately partite/ pinnatipartite fiederteilig
pinnately veined/pinnately nerved/ penninerved fiederaderig, fiedernervig
pinnately-leaved palm Fiederpalme
pinnatifid/pinnately split/ pinnately cleft fiederspaltig
pinnatilobate/pinnately lobed fiederlappig
pinnation Fiederung
pinnatipartite/pinnately parted/ pinnately partite fiederteilig
pinnatisect/pinnately incised fiederschnittig
pinnule/pinnula Pinnula, Fiederchen, Fiederblättchen (zweiten Grades)
pinocytosis Pinozytose
pinwheel *neuro* Orientierungszentrum, Windmühle, Windrad
pinworm (*Enterobius vermicularis*) Madenwurm
pioneer neuron Pionierneuron
pioneer organism Pionierorganismus
pioneer plant Pionierpflanze
pioneer species Pionierart
pioneer vegetation Pioniervegetation
pip (small seed of several-seeded fleshy fruit) Kern (einer vielsamigen Frucht, z. B. Apfel/ Traube/Zitrus)
pipe/tube Rohr, Röhre
Piperaceae/pepper family Pfeffergewächse
piperazine Piperazin
piperidine Piperidin

piperine Piperin
pipet *vb* pipettieren
pipet/pipette Pipette
- **capillary pipet** Kapillarpipette
- **graduated pipet/measuring pipet** Messpipette
- **micropipet** Mikropipette
- **Pasteur pipet** Pasteurpipette
- **suction pipet/patch pipet** Saugpipette
- **transfer pipet/volumetric pipet** Vollpipette, volumetrische Pipette

pipet bulb/rubber bulb Pipettierball, Pipettierbällchen
pipet helper Pipettierhilfe
pipeting nipple/rubber nipple Pipettierhütchen, Pipettenhütchen, Gummihütchen
pipewort family/Eriocaulaceae Eriocaulongewächse
pirate perch and freshwater relatives/Percopsiformes Barschlachse
piriform/pear-shaped birnenförmig
piriform recess/recessus laryngis Morgagnische Tasche
pisiform/pea-shaped erbsenförmig
pisiform bone/pisiform Erbsenbein
pistil Pistill, Stempel
pistillate/carpellate/female pistillat, weiblich
pistillate flower/carpellate flower/ female flower Stempelblüte, weibliche Blüte
pistillode Pistillodium
pit/crypt (snakes) Grube
pit/fovea Grube, Loch, Vertiefung
pit/stone/putamen/pyrene *bot* Stein, Steinkern, Putamen (Endokarp), Obststein
pit (phloem) Tüpfel
- **fenistriform pit** Fenstertüpfel
- **ramiform pit** verzweigter Tüpfel

pit aperture Tüpfelapertur, Tüpfelöffnung
pit casing (fruit) Steinschale
pit cavity Tüpfelhöhle
pit chamber/pit cavity Tüpfelhof
pit connection (red algas) Tüpfelverbindung (mit Tüpfelkanal)
pit field Tüpfelfeld
pit membrane Tüpfelschließhaut
pit organ (snakes) Grubenorgan
pit plug (red algas) Tüpfelpfropfen
pit-pair Tüpfelpaar
- **aspirated pit-pair** verschlossenes/ aspirates Tüpfelpaar
- **boardered pit-pair** behöftes Tüpfelpaar

pitch *n* **(DNA: helix periodicity)** *gen* Ganghöhe (DNA-Helix: Anzahl Basenpaare pro Windung)

P pitch 256

pitch *n* **(highness/lowness of sound)** Tonhöhe
pitch *n* **(resin from conifers)** Koniferenharz, Terpentinharz
pitch screw impeller (bioreactor) Schraubenspindelrührer
pitched-blade fan impeller/pitched-blade paddle impeller/inclined paddle impeller (bioreactor) Schrägblattrührer
pitcher/ascidium Krug, Kanne, *sensu lato*: Schlauch
pitcher leaf/ascidiate leaf Kannenblatt, *(sensu lato*: Schlauchblatt)
pitcher plant Kannenpflanze
pitcher-plant family/Sarraceniaceae Schlauchpflanzengewächse, Krugpflanzengewächse
pitfall Falle, Fallgrube, Fallstrick
pitfall trap *hunt* Bodenfalle
pitfall trap/slippery-slide trap *bot* **(flower)** Kesselfallenblume, Gleitfallenblume
pith/medulla/core Mark
pith ray/medullary ray Markstrahl
pithy/medullary markig, markhaltig, medullär
pitted (phloem) getüpfelt
pitted (removal of pit/stone) entkernt
pitted/foveate grubig, mit einer Grube versehen, vertieft
pitting Entkernung, Entkernen; (phloem) Tüpfelung
pittosporum family/tobira family/parchment-bark family/Pittosporaceae Klebsamengewächse
pituitary gland/pituitary/hypophysis Hirnanhangdrüse, Hypophyse
pituitary stalk Hypophysenstiel
pivot joint/trochoid(al) joint Walzengelenk
pizzle (penis esp. of bull) Penis, Rute (spez. des Rinderbullen)
placebo Placebo, Plazebo, Scheinarznei
placenta *bot* Plazenta, Samenleiste
placenta *zool* Plazenta, Mutterkuchen
- **allantoic placenta** Allantoisplazenta
- **annular placenta/zonary placenta/Placenta zonaria** Gürtelplazenta
- **bidiscoidal placenta** Placenta bidiscoidalis
- **chorioallantoic placenta** Chorioallantoisplazenta, Zottenplazenta

- **cotyledonary placenta** Placenta cotyledonaria, Placenta multiplex
- **deciduate placenta/placenta vera/placenta deciduata** Vollplazenta
- **diffuse placenta** Placenta diffusa
- **discoidal placenta** Placenta discoidalis
- **hemochorial placenta** haemo-choriale Plazenta
- **hemoendothelial placenta** Labyrinthplazenta
- **labyrinthine placenta/hemoendothelial placenta** Labyrinthplazenta
- **semiplacenta/nondeciduate placenta/placenta adeciduata** Semiplacenta, Halbplazenta
- **syndesmochorial placenta** syndesmo-choriale Plazenta
- **zonary placenta/annular placenta/Placenta zonaria** Gürtelplazenta
placentals/eutherians/Placentalia/Eutheria Plazentatiere
placentation Plazentation
- **axile p.** zentralwinkelständige Plazentation
- **basal p.** basale Plazentation, basiläre Plazentation, grundständige Plazentation
- **free central p.** Zentralplazentation
- **laminar/lamellate p.** laminale Plazentation, flächenständige Plazentation
- **marginal p.** randständige Plazentation
- **parietal p.** Parietalplazentation, wandständige Plazentation
placentome Plazentom
placer *geol* Seife, erzseifenhaltige Stelle
placode Plakode
- **lens placode** Linsenplakode
- **otic placode** Ohrplakode
placoderms/Placodermi Placodermen, Plattenhäuter
placodonts/placodontians (mollusk-eating euryapsids)/Placodontia Pflasterzahnsaurier
placoid/plate-like tellerförmig, schildförmig
placoid scale/dermal denticle Placoidschuppe, Zahnschuppe, Hautzahn, Dentikel
placophorans (incl. chitons)/Placophora Käferschnecken
plagiotropic/plagiotropous/obliquely inclined plagiotrop, schief/waagrecht wachsend
plagiotropism Plagiotropismus

plague Plage, Seuche; Pest
- **bubonic plague (*Yersinia pestis*)** Beulenpest, Bubonenpest

plain (extensive level country) *n geogr* Ebene

plain/ordinary *adv/adj* einfach, schlicht, gewöhnlich

plain stage *micros* Standardtisch

plainsawn/flatsawn/tangential section (wood) Sehnenschnitt, Fladerschnitt (Holz)

planarians/Tricladida Planarien

planation Planation

plane/flat/level *adv/adj math/techn* eben, flach

plane/flat surface/level surface *n math/techn* Ebene, Fläche

plane family/plane tree family/ sycamore family/Platanaceae Platanengewächse

plane mirror/plano-mirror Planspiegel

plane suture (skull) ebenflächige Naht (Schädelknochennaht mit ebenen Flächen)

plane-polarized light linear polarisiertes Licht

planispiral planspiral, flach-scheibenförmig gewunden

plank Planke, Bohle

plank buttress/buttress root *bot* Brettwurzel

plankter/planktonic organism Plankter, Planktont, Planktonorganismus

planktivorous planktivor, planktonfressend

plankton (passive drifters) Plankton (passiv schwebend)
- **femtoplankton** Femtoplankton
- **microplankton** Mikroplankton, Kleinplankton
- **nanoplankton/nannoplankton** Nanoplankton
- **phytoplankton** Phytoplankton, pflanzliches Plankton
- **potamoplankton** Potamoplankton, Flussplankton
- **ultraplankton** Ultraplankton
- **zooplankton** Zooplankton, tierisches Plankton

plankton strainer (a food-strainer) Planktonseiher

planktonic organism/plankter Planktonorganismus, Plankter, Planktont

planktotroph *n* Planktonfresser

plano-concave mirror Plan-Hohlspiegel, Plankonkav

plant/cultivate/grow *vb* anpflanzen

plant *vb* **(to set/put in ground for growth)** pflanzen

plant/flower/growth/wort *n* Pflanze, Blume, Gewächs

plant biogeography/plant geography/phytogeography/ geobotany Pflanzengeographie, Pflanzengeografie, Geobotanik

plant chemical/phytochemical Pflanzeninhaltsstoff

plant community Pflanzengesellschaft, Pflanzengemeinschaft

plant consumer Pflanzenkonsument

plant cover/vegetational cover/ vegetation Pflanzendecke

plant debris sich zersetzendes Pflanzenmaterial

plant disease Pflanzenkrankheit

plant diversity Pflanzenvielfalt

plant ecology/phytoecology Pflanzenökologie, Vegetationskunde, Vegetationsökologie

plant geography/plant biogeography/phytogeography/ geobotany Pflanzengeographie, Pflanzengeografie, Geobotanik

plant kingdom Pflanzenreich

plant parasite (thriving in/on plants) Phytoparasit, Pflanzenparasit (Schmarotzer in/auf Pflanzen)

plant pest Pflanzenschädling

plant physiology Pflanzenphysiologie

plant pigment Pflanzenfarbstoff

plant press Pflanzenpresse

plant production/cropping Feldbau

plant protection Pflanzenschutz

plant show Pflanzenschau

plant sociology/phytosociology Pflanzensoziologie

plant specimen Pflanzenmaterial, Belegexemplar

plant virus Pflanzenvirus

plant waste Pflanzenabfälle

plant-eating/phytophagous/ herbivorous pflanzenfressend

plant-protective agent/pesticide Pflanzenschutzmittel, Pestizid

plantain family/Plantaginaceae Wegerichgewächse

plantation Plantage, Pflanzung, Anpflanzung

plantigrade (animal) Plantigrade, Sohlengänger

plantigrade gait Plantigradie, Sohlengang

plantlet Pflänzchen

plaque Hof, Lysehof, Aufklärungshof, Plaque
- **clear plaque** klarer Plaque

plaque assay Plaque-Test

P plaque-forming unit (PFU) 258

plaque-forming unit (PFU) Plaque-
bildende Einheit (PBE), Kolonie-
bildende Einheit (KBE)
plasma (blood) Blutplasma
plasma/cytoplasm Plasma,
Cytoplasma, Zytoplasma
plasma cell Plasmazelle, Plasmozyt
**plasma membrane/(outer) cell
membrane/unit membrane/
ectoplast/plasmalemma**
Plasmamembran, Zellmembran,
Ektoplast, Plasmalemma
plasma skimming
Plasmaabschöpfung
**plasma streaming/cytoplasmic
streaming/cyclosis**
Plasmaströmung, Dinese
plasmalogen Plasmalogen
plasmatic plasmatisch
plasmenic acid Plasmensäure
plasmid Plasmid
• **broad host range plasmid**
Plasmid mit breitem Wirtsbereich
• **conjugative plasmid/self-
transmissible plasmid/
transferable plasmid**
konjugatives Plasmid
• **cryptic plasmid**
kryptisches Plasmid
• **mobilizable plasmid**
mobilisierbares Plasmid
• **non-conjugative plasmid**
nicht-konjugatives Plasmid
• **relaxed plasmid**
relaxiertes Plasmid,
schwach kontrolliertes Plasmid
• **single copy plasmid**
Einzelkopie-Plasmid
plasmid amplification
Plasmidamplifikation
plasmid curing Plasmidkurierung
(Entfernung eines Plasmid aus einer
Wirtszelle)
plasmid incompatibility
Plasmidinkompatibilität
plasmid instability
Plasmidinstabilität
plasmid mobilization
Plasmidmobilisierung
plasmid promiscuity
Plasmidpromiskuität
plasmin/fibrinolysin Plasmin,
Fibrinolysin
plasmodesm/plasmodesma (*pl***
plasmodesmas/plasmodesmata)**
Plasmodesmos, Plasmodesma
(*pl* Plasmodesmen/Plasmodesmata)
plasmodesmatal frequency
Plasmodesmendichte
plasmodial tapetum
Periplasmodialtapetum
plasmodiocarp Plasmodiokarp

plasmolysis Plasmolyse
• **incipient plasmolysis**
Grenzplasmolyse
plasticine Plastilin
plasticity Plastizität, Formbarkeit
plasticizer Plastifikator,
Weichmacher
plastid *n* Plastide
plastination Plastination
• **whole mount plastination**
Ganzkörperplastination
plastome Plastom
plastron/plastrum Plastron,
Bauchpanzer, Bauchplatte,
Brustschild (Schildkröten/Vögel)
plate *bot/zool/micb* Platte
plate (HPLC) Trennstufe
• **maxilliary plate/maxilliped plate
(insects: galea and lacinia)**
Kaulade
plate assay/plating *micb*
Platten-Test, Plattenverfahren
plate count Plattenzählverfahren
plate meristem Plattenmeristem
plateau *math* Plateau (flache Stelle
in einer Kurve)
plateau/elevated plane/tableland
Hochfläche, Hochebene
plated puffballs/Gautieriales
Morcheltrüffeln
platelet (blood) Plättchen,
Blutplättchen
**platelet-derived growth factor
(PDGF)** Plättchenwachstumsfaktor,
Blutplättchen-Wachstumsfaktor,
Plättchenfaktor
plating (plating out) Plattierung,
Plattieren (Ausplattieren)
• **efficacy of plating**
Plattierungseffizienz
• **replica plating** Replicaplattierung
plating method Plattierungsmethode
platyclade Platycladium, Flachspross
play/play behavior Spielverhalten
play face *ethol* (apes) Spielgesicht
play song Spielgesang
play-fight(ing) *ethol* Kampfspiel
pleated/plicate/folded gefaltet,
faltig
**pleated sheet/beta pleated sheet/
beta-sheet** beta-Faltblatt
plectenchyma Plectenchym,
Flechtgewebe
**plectognath fishes/
Tetraodontiformes/Plectognathi**
Kugelfischverwandte
plectonemic winding
plektonemische Windung
pleiochasium/pleiochasial cyme
Pleiochasium, vielgablige Trugdolde
pleiotropic mutation
pleiotrope Mutation

259   **pluviometer**   **P**

pleiotropy Pleiotropie
**Pleistocene Epoch/Glacial Epoch/
Diluvial/Ice Age** Pleistozän,
Diluvium, Glazialzeit, Eiszeit
(erdgeschichtliche Teilepoche)
pleomere/abdominal somite
Pleomer, Abdominalsegment
pleomorphism/polymorphism
Pleomorphismus, Polymorphismus,
Mehrgestaltigkeit
pleon (abdomen of crustaceans)
Pleon
plerocercoid Plerocercoid
plesiomorphic plesiomorph
plesiomorphism Plesiomorphie
plesiosaurs/Plesiosauria
Plesiosaurier
pleura Pleura
• **parietal pleura/pleura parietalis**
Brustfell
• **pulmonary pleura/visceral
pleura/pleura pulmonalis/pleura
visceralis** Lungenfell
pleural cavity/cavitas pleuralis
Pleurahöhle
pleurapophysis Pleurapophyse
pleurite/lateral sclerite Pleurit,
Seitenplatte, Seitenstück,
lateraler Sklerit
pleurobranch/pleurobranchia
Pleurobranchie
pleurocarpic/pleurocarpous
pleurokarp, seitenfrüchtig
pleuron/lateral plate Pleuron,
Seitenteil
**pleuronematic flagellum/tinsel
flagellum/flimmer flagellum**
Flimmergeißel
pleuston Pleuston
pliability/flexibility Biegsamkeit
pliable/flexible biegsam
plicate/pleated/folded gefaltet,
faltig
plication/fold/wrinkle Falte
Pliocene/Pliocene Epoch Pliozän
(erdgeschichtliche Epoche)
ploidy Ploidie
plot *vb math/geom* auftragen,
"plotten", aufzeichnen
plot (graph/diagram/curve) *n math/
geom* Diagramm, Kurve
• **Lineweaver-Burk plot/double-
reciprocal plot**
Lineweaver-Burk-Diagramm
• **Ramachandran plot**
Ramachandran-Diagramm
• **Scatchard plot**
Scatchard-Diagramm
plotter Plotter, Kurvenzeichner
plotting/mapping Kartierung
ploughshare bone/vomer
Pflugscharbein, Vomer

**plough/ploughing/plow/plowing/
till/tiling** pflügen
pluck *n* **(innards)** Innereien
(Schlachttiere: Schweine/Rinder)
pluck *vb* **(feathers)** ausreißen,
rupfen (Federn)
plug flow Pfropfströmung,
Pfropfenströmung
plug-flow reactor (bioreactor)
Pfropfenströmungsreaktor,
Kolbenströmungsreaktor
**plum yew family/cephalotaxus
family/Cephalotaxaceae**
Kopfeibengewächse
plumage/ptilosis Gefieder,
Federkleid, Ptilosis
• **basic plumage** Schlichtkleid
• **breeding plumage/nuptial
plumage/courtship plumage**
Brutkleid, Hochzeitskleid
• **camouflage plumage/cryptic
plumage (dress/pelage/coat)**
Tarntracht, Tarnkleid
• **display plumage/conspicuous
plumage** Prachtkleid
• **eclipse plumage/inconspicuous
plumage (drake/duck)**
Schlichtkleid
• **juvenile plumage** Jugendkleid
• **nuptial plumage/breeding
plumage/courtship plumage**
Brutkleid, Hochzeitskleid (Vögel)
plumage ruffling/feather ruffling
Gefiedersträuben
**plumbago family/sea lavender
family/leadwort family/
Plumbaginaceae**
Bleiwurzgewächse,
Grasnelkengewächse
plumed seed
Samen mit fedrigen Flughaaren
plumose/feathery fedrig, federig
**plumule/plumula/terminal
embryonic bud** Plumula,
Keimknospe, Stammknospe,
Sprossknospe,
terminale Embryoknospe
plunge *vb* eintauchen
**plunging jet reactor/deep jet
reactor/immersing jet reactor
(bioreactor)** Tauchstrahlreaktor
pluriennal plurienn, mehrjährig
wachsend bis zur Blüte (z. B. *Agave*)
plurilocular/multilocular
plurilokulär, mehrkammerig
plus strand (coding strand) *gen*
Plus-Strang, Positiv-Strang
(codierender Strang)
Pluteus family/Pluteaceae
Dachpilze, Dachpilzartige
pluviometer/rain gauge
Regenmesser

P plywood 260

plywood Sperrholz, Furnierholz
pneumathode *bot* Pneumatode,
 Atemöffnung
pneumatic bone/hollow bone
 pneumatischer Knochen,
 Hohlknochen, Luftknochen
pneumatophore/air root/aerating
 root Pneumatophore, Atemwurzel
pneumonia Pneumonie,
 Lungenentzündung
pneumonic plague (*Yersinia pestis*)
 Lungenpest
Poaceae/grass family/Gramineae
 Süßgräser, Gräser
poach wildern, räubern, stehlen
poacher Wilderer, Wilddieb
POC (particulate organic carbon)
 partikulärer organischer
 Kohlenstoff
pocket (enzyme) Tasche
pocosin/"swamp-on-a-hill" (U.S.
 peatland: SE coastal plains)
 amerikan. Waldmoor
pod (whales/dolphins/seals)
 Koppel, Schwarm, Zug (Gruppe:
 Wale/Delphine/Seehunde)
pod/legume Hülse
podeon/podeum/petiole
 (hymenopterans) Hinterleibsstiel,
 "Taille", Petiolus
podetium Podetium
podium/tube-foot/ambulacral foot
 Ambulakralfüßchen, Saugfüßchen
podobranch/podobranchia/foot-gill
 Podobranchie
podocarpus family/Podocarpaceae
 Steineibengewächse
podocyte Podocyt, Podozyt,
 Füßchenzelle
Podophyllaceae/may apple family
 Fußblattgewächse,
 Maiapfelgewächse
Podostemaceae/riverweed family
 Blütentange
poikilohydrous poikilohydrisch,
 poikilohydr, wechselfeucht
poikilothermic/poikilothermal/
 poikilothermous/cold-blooded/
 ectothermal/heterothermal
 poikilotherm, wechselwarm,
 ektotherm
point mutation Punktmutation
point source Punktquelle
pointed/acute/sharp spitz,
 zugespitzt
poison/intoxicate *vb* vergiften
poison/toxin *n* Gift, Toxin
poison claw Giftklaue
poison control center/poison
 control clinic Entgiftungszentrale,
 Vergiftungszentrale,
 Entgiftungsklinik

poison fang/venomous fang
 (unguis) Giftzahn, Giftklaue
poison gland Giftdrüse
poison information center
 Giftinformationszentrale
poison tooth/venom tooth/fang
 (snakes) Giftzahn
poisoning/intoxication Vergiftung,
 Intoxikation
poisonous/toxic giftig, toxisch
poisonous materials Giftstoffe
poisonous plant Giftpflanze
poisonousness/toxicity Giftigkeit,
 Toxizität
Poisson distribution Poissonsche
 Verteilung, Poisson Verteilung
pokeweed family/Phytolaccaceae
 Kermesbeerengewächse
polar polar
polar body Polkörper,
 Richtungskörper
polar cap (within polar sac)
 Polkappe
polar cap/calotte (*Dicyemida*)
 Polarzellen-Kappe, Kalotte
polar capsule Polkapsel
polar cell/calotte cell (Dicyemida)
 Polzelle, Polarzelle
polar fiber (microtubule) Polfaden
 (Mikrotubulus)
polar field (ctenophores) Polplatte
polar growth polares Wachstum
polar lobe (cleavage) Pollappen
polar mutation polare Mutation
polar nucleus Polkern
polar plasm/pole plasm Polplasma
polar plates (coelenterates)
 Polfelder
polar ring (*Apicomplexa*) Polring
polarimeter Polarimeter
polarity Polarität
polarization Polarisation
polarized light polarisiertes Licht
 ● **plane-polarized light**
 linear polarisiertes Licht
polarizing filter/polarizer
 Polarisationsfilter, "Pol-Filter",
 Polarisator
polarizing microscope
 Polarisationsmikroskop
pole (long/slender length of wood)
 Stange
pole (opposite ends) Pol
pole cell Polzelle
pole granules Polgranula
Polemoniaceae/phlox family
 Sperrkrautgewächse,
 Himmelsleitergewächse
polian vesicle Polische Blase
poll/back of the head/occiput
 (crest/top/apex: *esp.* horses/
 cattle) Hinterkopf, Hinterhaupt

pollakanthic pollakanth
pollarding/beheading of tree/ decapitation of tree Kappen, Köpfen (eines Baumes)
pollen Pollen, Blütenstaub
pollen basket/corbiculum (bees) Pollenkörbchen, Corbiculum
pollen brush (bees) Pollenbürstchen
pollen case/theca Pollensackgruppe, Theca, Theka
pollen chamber Pollenkammer
pollen comb/pecten (bees) Pollenkamm
pollen cone/male cone/ microstrobilus/microsporangiate strobilus Pollenzapfen, männlicher Zapfen
pollen grain Pollenkorn
pollen press/pollen packer/auricle (bees) Pollenschieber, Aurikel
pollen sac (saccus/locule/loculus) Pollensack, Pollenfach (Lokulament)
pollen transfer Pollenübertragung
pollen tube Pollenschlauch
pollen tube cell Pollenschlauchzelle
pollenkitt/pollen coat *bot* Pollenkitt
pollinarium Pollinarium
pollinate bestäuben
pollination *sensu stricto:* Bestäubung
(Befruchtung *see* fertilization)
● **animal pollination/zoophily** Tierbestäubung, Tierblütigkeit, Zoophilie
● **bat pollination/chiropterophily** Fledermausbestäubung, Fledermausblütigkeit, Chiropterophilie
● **bee pollination/melittophily** Bienenbestäubung
● **beetle pollination/cantharophily** Käferbestäubung
● **bird pollination/ornithophily** Vogelbestäubung, Vogelblütigkeit, Ornithophilie
● **butterfly pollination/ psychophily** Schmetterlingsbestäubung
● **cross-pollination** Kreuzbestäubung, Fremdbestäubung
● **fly-pollination/myiophily** Fliegenbestäubung, Myiophilie
● **hawk moth pollination/ sphingophily** Nachtfalterbestäubung, Nachtschwärmerbestäubung, Sphingophilie
● **insect pollination/entomophily** Insektenbestäubung, Insektenblütigkeit, Entomophilie

● **moth pollination/phalaenophily** Mottenbestäubung, Phalaenophilie
● **self-pollination/autophily** Selbstbestäubung, Autophilie
● **snail pollination/malacophily** Schneckenbestäubung
● **wasp pollination/sphecophily** Wespenbestäubung
● **water pollination/hydrophily** Wasserbestäubung, Wasserblütigkeit, Hydrophilie
● **wind pollination/anemophily** Windbestäubung, Windblütigkeit, Anemophilie
pollination drop/pollination droplet Bestäubungstropfen, Befruchtungströpfchen
pollinator Bestäuber
pollinium Pollinium
polliwog/tadpole Kaulquappe
pollutant/contaminant Umweltgift, Schadstoff, Schmutzstoff
pollute/contaminate verschmutzen, kontaminieren
polluter Umweltverschmutzer
pollution/contamination Verschmutzung, Kontamination, Umweltverschmutzung
● **air pollution** Luftverschmutzung
● **environmental pollution** Umweltverschmutzung
● **noise pollution** Lärmverschmutzung
● **water pollution** Wasserverschmutzung
pollution control Umweltschutz
poly(A) tail Poly(A)-Schwanz
polyacrylamide Polyacrylamid
polyadenylation Polyadenylierung
polyandrous polyandrisch
polyandry Polyandrie, Vielmännerei
polyaxenic polyaxenisch
polycentric species polyzentrische Art
polychetes/polychaetes/polychete worms/bristle worms Polychaeten, Vielborster, Borstenwürmer
polycistronic mRNA polycistronische mRNA
polycistronic polycistronisch
polyculture Polykultur, Mischkultur
polyenergid polyenergid
polyestrous (polyoestrous) polyöstrisch
Polygalaceae/milkwort family Kreuzblümchengewächse, Kreuzblumengewächse
polygamous polygam; *bot* polygamisch
polygamy Polygamie, Vielehe
polygenic polygen (von mehreren Genen abhängig)

polygenic diseases
 polygene (Erb-)Krankheiten
Polygonaceae/dock family/ buckwheat family/knotweed family/smartweed family
 Knöterichgewächse
polygynous polygyn, vielweibig
polygyny Polygynie, Vielweiberei
polyhedral symmetry polyedrische Symmetrie, vielflächige Symmetrie
polylecithal polylecithal, polylezithal, makrolecithal, dotterreich
polylinker/multiple cloning site *gen* Polylinker, multiple Klonierungsstelle
polymer Polymer
polymerase chain reaction (see also: PCR) Polymerasekettenreaktion
polymerization Polimerisation
polymerize polymerisieren
polymictic polymiktisch
polymorphic/pleomorphic mehrgestaltig, polymorph, pleomorph
polymorphism/pleomorphism Polymorphismus, Pleomorphismus, Mehrgestaltigkeit
 • **balanced polymorphism** *gen* balancierter Polymorphismus
 • **conformation polymorphism** *gen* Konformationspolymorphismus
 • **STRPs (short tandem repeat polymorphisms)** *gen* Polymorphismen von kurzen direkten Wiederholungen
polymorphism information content (PIC) Informationsgehalt eines Polymorphismus (PIC)
polymorphonuclear leukocyte/ granulocyte polymorphonuklearer Leukozyt, Granulocyt, Granulozyt
polynuclear polynukleär, polynucleär
polynucleotide Polynucleotid, Polynukleotid
polyoestrous *see* polyestrous
polyp (outgrowth of tissue) Polyp (Gewebewucherung)
polyp/hydroid Polyp
 • **autozooid/feeding polyp (*Octocorallia*)** Autozooid, Zooid, Fangpolyp, Fresspolyp
 • **defense polyp/protective polyp/ stinging zooid/dactylozooid** Wehrpolyp, Dactylozoid, Dactylozooid
 • **feeding polyp/nutritive polyp/ gastrozooid/trophozooid** Fresspolyp, Nährtier, Gasterozoid, Gastrozooid, Trophozoid, Trophozooid
 • **founder polyp/primary polyp** Gründerpolyp, Primärpolyp
 • **protective polyp/defense polyp/ stinging zooid/dactylozooid** Wehrpolyp, Dactylozoid, Dactylozooid
 • **reproductive polyp/gonozooid** Geschlechtspolyp, Gonozoid, Gonozooid
 • **siphonozooid** Siphonozooid, Pumppolyp
 • **stinging zooid/protective polyp/ defense polyp/dactylozooid** Wehrpolyp, Dactylozoid, Dactylozooid
polyphage Vielfraß
polyphagous/polyphagic polyphag, allesfressend (begrenzte Nahrungsauswahl)
polyphenism Polyphänismus
polypheny/pleiotropy/pleiotropism Polyphänie, Pleiotropie
polyphyletic polyphyletisch
polyphyodont polyphyodont
polypide Polypid
polyploid polyploid
polyploidy Polyploidie
polypnea Polypnoe
polypod polypod, vielbeinig
Polypodiaceae/fern family Tüpfelfarngewächse
polypoid stage Polypenstadium
polypore family/bracket fungi/ Poriales/Polyporaceae Porlinge, Echte Porlinge
polyprotein Polyprotein
polysome/polyribosome Polysom, Polyribosom
polystemonous polystemon
polytene chromosomes polytäne Chromosomen
polythalamous/polythalamic/with many chambers/polythecal polythalam, vielkammerig, vielkämmrig, mehrkammerig, polythekal
polythetic polythetisch
polytokous polytok
polytypic species polytypische Art
polyunsaturated mehrfach ungesättigt
polyunsaturated fatty acid mehrfach ungesättigte Fettsäure
polyvoltine/multivoltine polyvoltin, plurivoltin, mit mehreren Jahresgenerationen
polyxenous/polyxenic mehrwirtig, heteroxen
pomaceous fruit Kernobst
pome Apfelfrucht
pomegranate family/Punicaceae Granatapfelgewächse

263   **pot** **P**

pomology Obstbaukunde
pond/pool Teich, Tümpel
- **small pond (e.g. fish pond)**
  Weiher (z. B. Fischweiher)
pond scum Kahmhaut,
  Oberflächenhäutchen (auf Teich)
pondweed family/
  Potamogetonaceae
  Laichkrautgewächse
pongids/great apes/Pongidae
  Menschenaffen
pons varolii Varolsbrücke, Brücke
Pontederiaceae/pickerel-weed
  family/water hyacinth family
  Hechtkrautgewächse
pontine flexure *embr* Brückenbeuge
pool *n* (*see also:* pond) Tümpel;
  Lache
pool *n* **(whole quantity of a**
  particular substance: body
  substance, metabolite etc)
  "Pool" (Gesamtheit einer
  Stoffwechselsubstanz)
pool/combine/accumulate poolen,
  vereinigen, zusammenbringen,
  zusammenfassen
pooling of data *stat*
  Zusammenfassung von Daten
poor grassland/rough pasture/
  rough meadow Magerwiese
poples/popliteal fossa/popliteal
  space/bend of knee/back of the
  knee/hollow behind knee
  Kniekehle
poppy family/Papaveraceae
  Mohngewächse
populate bevölkern
population/reproductive group
  Population, Fortpflanzungsgemeinschaft; Bevölkerung
population control
  Populationskontrolle,
  Bevölkerungskontrolle
population crash
  Populationszusammenbruch,
  Bevölkerungszusammenbruch
population curve Populationskurve,
  Bevölkerungskurve
population density
  Populationsdichte,
  Bevölkerungsdichte
population ecology
  Populationsökologie, Demökologie
population genetics
  Populationsgenetik
population growth
  Populationszuwachs,
  Populationswachstum,
  Bevölkerungswachstum
population pressure
  Populationsdruck,
  Bevölkerungsdruck

population pyramid
  Populationspyramide,
  Bevölkerungspyramide
population size Populationsgröße,
  Bevölkerungsgröße
porcine schweineartig, Schweine
  betreffend, Schweine...
pore mushroom/pore fungus/
  boletus mushroom Porling,
  Röhrenpilz, Röhrling
pore of Kohn Kohnsche Pore
poricidal poricid, porizid, lochspaltig
poricidal capsule Porenkapsel,
  Lochkapsel
porin Porin
pork Schweinefleisch
porker Mastferkel, Mastschwein
porosity Porosität, Durchlässigkeit
porous porös, porig, durchlässig
porpoise (no "beak" as in dolphins)
  Kleintümmler
portal vein/vena portae Pfortader
portion/fraction Teilmenge, Portion,
  Fraktion
Portulacaceae/purslane family
  Portulakgewächse
Posidonia family/Posidoniaceae
  Neptungrasgewächse, Neptunsgräser
position effect Positionseffekt,
  Lageeffekt
positional cloning
  Positionsklonierung
positional mapping *gen*
  Positionskartierung
positron emission tomography
  (PET)
  Positronenemissionstomographie
  (PET)
post *n* **(upright piece of wood/**
  metal as a stay/support) Pfosten,
  Pfahl
post-emergence treatment *agr*
  Nachauflaufbehandlung
postabdomen/metasoma
  Postabdomen, Metasoma
postantennal organ
  Postantennalorgan,
  Hinterfühlerorgan
postcopulatory behavior Nachbalz
postfloration Postfloration,
  Nachblüte
posttetanic potentiation (PTP)
  posttetanische Potenzierung
posttranslational posttranslational
posture/stance Haltung, Stellung,
  Lage
pot *n* Topf
- **flower pot** Blumentopf
pot *vb* eintopfen
- **potted plant** Topfpflanze
- **potting soil** Topferde,
  Blumenerde

pot

pot/marijuana/marihuana (dried leaves/flowering tops of pistillate hemp plant) Marihuana
potamic potamal, von Flüssen transportiert, Fluss...
potamodromous/migrating in fresh water potamodrom, in Fließgewässern wandernd
Potamogetonaceae/pondweed family Laichkrautgewächse
potamoplankton Potamoplankton, Flussplankton
potassium Kalium
potato family/nightshade family/ Solanaceae Nachtschattengewächse
potential *adj/adv* potentiell, möglich, eventuell, latent vorhanden
potential *n* Potential; *electr* Spannung
• **action potential** Aktionspotential
• **contingent negative variation (CNV)** Erwartungspotential
• **coupling potential** Kopplungspotential
• **electrotonic potential** elektrotonisches Potential
• **end plate potential (epp)** Endplattenpotential
• **equilibrium potential** Gleichgewichtspotential
• **evoked potential** evoziertes Potential
• **excitatory postsynaptic potential (EPSP)** exzitatorisches postsynaptisches Potential
• **generator potential** Generatorpotential
• **graded potential** graduiertes Potential
• **gross potential** Summenpotential
• **inhibitory postsynaptic potential (IPSP)** inhibitorisches postsynaptisches Potential
• **localized potential** lokales Potential, Lokalpotential
• **membrane potential** Membranpotential
• **osmotic potential** osmotisches Potential
• **pacemaker potential** Schrittmacherpotential
• **readiness potential** Bereitschaftspotential
• **resting potential** Ruhepotential
• **reversal potential** Umkehrpotential
• **solute potential** Löslichkeitspotential
• **threshold potential** Schwellenpotential
potential difference/voltage Potentialdifferenz, Spannung

potherbs/greens Suppenkraut, Blattgemüse (gekochtes)
pothole/deep pool Kolk, Moorauge, Blänke
potted plant Topfpflanze
Potter-Elvehjem homogenizer (glass homogenizer) Potter" (Glashomogenisator)
potting soil (potting mixture: soil & peat a.o.) Topferde, Blumenerde
pouch/sac/sac-like cavity/pocket/ bursa Beutel, Sack, Tasche, Bursa
pouch/marsupium Beutel, Brutbeutel, Marsupium
pouch *orn* Kehlhautsack
pouch gall Beutelgalle
pouched/saccate sackförmig, taschenförmig
pouched mammals/marsupials/ metatherians/Metatheria/ Didelphia Beuteltiere, Marsupialia
poulard/poularde (sterilized hen/or killed before sexual maturity) Poularde, Masthuhn (geschlachtet vor Geschlechtsreife) *(see also:* pullet)
poult (young fowl) Junggeflügel; junger Truthahn, Jungpute
poultry Geflügel, Hausgeflügel, Federvieh
pound (public enclosure for stray dogs/cats) institutionalisierter Zwinger für verwaiste Hunde/ Katzen
• **dog pound** Hundezwinger
pour/water the plants gießen
pour-plate method Plattengussverfahren, Gussplattenmethode
pout *n* Schmollen; Schmollmund, Schnute, Flunsch
pout (pouting) *vb* schmollen; eine Schnute/Flunsch ziehen
powder-down feather/pulviplume *orn* Puderdune, Pulvipluma
powdery mildews/Erysiphales echte Mehltaupilze
power diving/nose diving (from air) *orn* Stoßtauchen
power stroke/effective stroke (forward stroke) Kraftschlag, Wirkungsschlag
power supply Stromquelle, Stromzufuhr
prairie Prärie
prayer plant family/arrowroot family/Marantaceae Pfeilwurzgewächse
pre-emergence treatment *agr* Vorauflaufbehandlung
pre-mutation Prä-Mutation

prehensile hand **P**

pre-proinsulin/preproinsulin
Prä-Proinsulin, Präproinsulin
pre-mRNA (precursor mRNA)
Prä-mRNA, Vorläufer-mRNA
pre-rRNA (precursor rRNA)
Prä-rRNA, Vorläufer-rRNA
pre-tRNA (precursor tRNA)
Prä-tRNA, Vorläufer-tRNA
pre-T cell/T-cell precursor
T-Vorläuferzelle
preadaptation Präadaptation,
Vorangepasstsein
prebiotic(al) präbiotisch
prebiotic soup präbiotische Suppe
prebiotic synthesis präbiotische
Synthese
Precambrian/Precambrian Era
Präkambrium, Präcambrium
(erdgeschichtliches Zeitalter)
precaution/precautionary measure
Vorsichtsmaßnahme,
Vorsichtsmaßregel
precipitate/deposit *n* Präzipitat,
Fällung, Ausfällung, Abscheidung
precipitate/deposit *vb*
präzipitieren, fällen, ausfällen,
abscheiden
precipitation Präzipitation,
Ausfällung, Ausfällen, Fällung,
Fällen; *meteo* Niederschlag
precise/exact präzis, genau, exakt
precision/exactness Präzision,
Genauigkeit
precision balance Präzisionswaage,
Feinwaage
precision grip Präzisionsgriff
precoated plate Fertigplatte
precocial animal/nidifugous animal
orn Nestflüchter
precocious verfrüht
precocious *bot* **(flowering before
leaf formation)** frühblühend (vor
der Beblätterung)
**precopulatory rite/precopulatory
behavior** Begattungsvorspiel
preculture Vorkultur
precursor Präkursor, Vorläufer
precursor cell Vorläuferzelle
predaceous/predatory räuberisch
predaceous instinct Raubinstinkt
predation Raub, Räubertum, Jagd
predator/predatory animal
Prädator, Räuber, Raubfeind,
Raubtier, Fressfeind, Jäger
predator-prey relationship
Räuber-Beute-Verhältnis
predatory/predaceous/raptorial
(greifend) räuberisch, Raub...
predatory animal Raubtier
**predatory bird/bird of prey/
raptorial bird** Raubvogel,
Greifvogel

predatory insect Raubinsekt,
räuberisches Insekt
predictive ecology
vorausschauende Ökologie,
voraussagende Ökologie
predictive medicine
vorhersagende Medizin
predilection site
Prädilektionsstelle
predisposition Prädisposition,
Veranlagung
predominate vorherrschen
preen *vb* (das Gefieder) putzen,
sich putzen
preening Körperpflege
preening *orn* **(plumage)**
Gefiederputzen, Gefiederkraulen
preening behavior Putzverhalten
preferential preferentiell, Vorzugs...,
bevorrechtigt
**preferential/favorably associated
(fidelity)** *biogeo/ecol* preferentiell,
hold, bevorzugt
prefloration (aestivation)
Präfloration, Knospendeckung der
Blütenblätter
**prefoliation/foliation/vernation/
ptyxis** Blattlage in der Knospe,
Vernation, Knospenlage der
Laubblätter
pregerminate vorkeimen
pregnancy/gravidity/gestation
Schwangerschaft, Trächtigkeit,
Gestation
● **abdominal pregnancy**
Abdominalschwangerschaft,
Bauchhöhlenträchtigkeit,
Leibeshöhlenschwangerschaft,
Leibeshöhlenträchtigkeit
● **tubal pregnancy**
Eileiterschwangerschaft,
Oviduktträchtigkeit
● **ectopic pregnancy/extrauterine
pregnancy (EUP)** ektopische
Schwangerschaft,
Extrauteringravidität (EUG)
● **ovarian pregnancy**
Ovarialschwangerschaft,
Ovariolenträchtigkeit,
Ovarialträchtigkeit
pregnant/gravid/gestational
schwanger, trächtig
● **pseudopregnant** pseudopregnant,
scheinschwanger
pregnenolone Pregnenolon
**prehensile/grasping/able to grasp/
raptorial** zum Greifen geeignet,
zupackend, ergreifend, Greif...
prehensile foot/grasping foot
Greiffuß
prehensile hand/grasping hand
Greifhand

prehensile mask (dragonfly larva: retractible prehensile labium) Fangmaske
prehensile organ Greiforgan
prehensile tail Greifschwanz
prehensile tentacle Greifarm (Tentakel)
prehensility Fähigkeit zum Greifen, Greiffähigkeit
preimplantation testing Präimplantationstest (Untersuchung vor Einnistung des Eis)
preinitiation complex Präinitiationskomplex
premature verfrüht, frühzeitig
premature birth Frühgeburt
prematurity Frühzeitigkeit, Vorzeitigkeit; Prämaturität, Frühreife
premaxilla Prämaxille
premolar/bicuspid tooth Prämolar, vorderer Backenzahn
premunition/concomitant immunity Prämunität, Präimmunität, Prämunition, begleitende Immunität
prenatal pränatal, vorgeburtlich
prenatal diagnosis Pränataldiagnose, pränatale Diagnose
prenatal diagnostics Pränataldiagnostik, pränatale Diagnostik
prenylation Prenylierung
preorbital gland/antorbital gland Voraugendrüse, Antorbitaldrüse
preparation Präparation, Vorbereitung; Zubereitung
preparation (*Lebewesen:* preserved specimen) Präparat
preparative centrifugation präparative Zentrifugation
prepare/dissect/mount präparieren
prepatent period Präpatenz, Latenzzeit
prephenic acid (prephenate) Prephensäure (Prephenat)
prepriming complex Prä-Startkomplex
preproinsulin/pre-proinsulin Präproinsulin, Prä-Proinsulin
prepuce/foreskin/sheath Präputium, Vorhaut, Scheide
prepupa/propupa Präpupa, Propupa, Vorpuppe, Semipupa
preputial gland/castor gland (beaver) Präputialdrüse
preputial sac (insects) Präputialsack, Penisblase
prescutum Präscutum, Vorschild
presence Präsenz, Anwesenheit, Vorhandensein, Vorliegen
presence degree Stetigkeit

preservation Preservierung, Bewahrung, Erhaltung
preservation of species Arterhaltung
preservative *n* Konservierungsstoff, Konservierungsmittel, Präservierungsstoff; Präservativ
preserve/conserve preservieren, konservieren, bewahren, erhalten, haltbar machen
preserved specimen Präparat
pressboard Pressspan
pressure cycle reactor (bioreactor) Druckumlaufreaktor
pressure-flow theory/pressure-flow hypothesis Druckstromtheorie, Druckstromhypothese
presymptomatic diagnosis präsymptomatische Diagnose, präsymptomatische Diagnostik
pretarsus (insects) Prätarsus, Klauenglied, Krallenglied, Krallensegment, Krallensockel
pretrial/preliminary experiment Vorversuch
prevalence/prevalency Prävalenz
prevalent prävalent sein, vorherrschen, überwiegen; überhandnehmen
prey *n* Beute, Jagdbeute, Beutetier
prey *vb* erbeuten, auf Beutejagd gehen
priapulans Priapuliden, Priapswürmer
Pribnow box *gen* Pribnow-Box
prick/prickle Stachel (Epidermisauswuchs)
prickle cell (skin) Stachelzelle (der Haut)
prickle-shaped/bristle-shaped/styliform stilettförmig, griffelförmig
prickly stachelig, stachlig
prickly shrub/bramble stacheliger Strauch
pride *n* Stolz
pride *n* (a company of lions) Rudel, Schar
primaries/primary feathers (primary remiges) Handschwingen, Hautschwingen
primary consumer Primärkonsument
primary culture Primärkultur
primary growth Primärwachstum, primäres Dickenwachstum
primary host/main host/definitive host Hauptwirt
primary larva Primärlarve, Junglarve, Eilarve
primary leaf Primärblatt
primary metabolism Primärstoffwechsel

process

primary metabolite Primärmetabolit, Primärstoffwechselprodukt
primary product/initial product Ausgangsprodukt
primary production Primärproduktion
primary response Primärantwort
primary settlement/primary succession Erstbesiedlung, primäre Sukzession
primary settling tank Vorklärbecken
primary structure Primärstruktur
primary tectrices *orn* Handdecken
primary thickening meristem (PTM) primärer Meristemmantel
primary transcript Primärtranskript
primary wall Primärwand
primary xylem Primärxylem
primates Primaten, Herrentiere
• **lower primates/prosimians/ Prosimii** Halbaffen
prime *adj/adv* primär, grundlegend, erster
prime *n* Anfang, Beginn
prime *vb* vorbereiten, präparieren
primer *gen* Primer
• **nested primer** verschachtelte Primer
• **universal primer** Universalprimer
primer effect *physiol* Umstimmungs-Effekt
primer extension *gen* Primer-Extension, Primer-Verlängerung
primer extension analysis Primer-Extensionsanalyse (Verfahren zur Bestimmung des 5′-Endes einer mRNA)
primeval forest/virgin forest/ pristine forest/jungle Urwald
priming *gen* Priming
• **RNA priming** RNA-Priming
• **self-priming** Selbst-Priming
primiparous erstgebärend
primitive/primordial/of earliest origin primitiv, ursprünglich, früh (frühen Ursprungs)
primitive/not differentiated or specialized primitiv, undifferenziert (>omnipotent)
primitive brain/archencephalon Urhirn, Archenzephalon, Archencephalon
primitive form/basic form/parent form Stammform, Urform
primitive groove *embr* Primitivrinne
primitive node/Hensen's node/ primitive knot/Hensen's knot Primitivknoten, Hensenscher Knoten, Hensen'-Knoten

primitive pit Primitivgrube
primitive plate *embr* Primitivplatte
primitive ray-finned bony fishes/ Chondrostei Altfische
primitive streak/germinal streak (gastrulation) *embr* Primitivstreifen, Keimstreifen
primocane Erstjahrestrieb einer zweijährigen Rutenpflanze (z. B. Himbeere)
primordial/primitive/original/first formed primordial, ursprünglich, zuerst angelegt
primordial cell/initial Primordialzelle, Primane, Initiale, Initialzelle
primordial male germ cell/ spermatogonium Ursamenzelle, Spermatogonium
primordial soup Ursuppe
primordium/anlage Primordium, Anlage
primosome Primosom
primrose family/Primulaceae Primelgewächse, Schlüsselblumengewächse
principle of competitive exclusion/ exclusion principle (Gause's rule/ principle) Konkurrenz-Ausschluss-Prinzip, Konkurrenz-Exklusions-Prinzip
prion (protein infectious agent) Prion
priority rule Prioritätsregel
prismatic layer Prismenschicht
pristine ursprünglich, urtümlich
pristine forest/primeval forest/ virgin forest/jungle Urwald
pro-oestrus *see* proestrus
probability Wahrscheinlichkeit
probe (microprobe) Sonde (Mikrosonde)
• **riboprobe** Ribosonde, RNA-Sonde
probe/probing head Tastkopf
probing bill Stocherschnabel
probiosis Probiose, Nutznießung
proboscis Proboscis, Rüssel
proboscis/acorn (prosoma/ protocoel of enteropneusts) Eichel
proboscis receptacle/rhnychocoel Rüsselscheide, Rhynchocoel
proboscis receptacle retractor Rüsselscheidenretraktor
procercoid (cestodes) Procercoid, Prozerkoid (Cestoda-Postlarve)
process *vb biochem/gen/neuro* prozessieren, verarbeiten, weiterverarbeiten, behandeln; weiterleiten
process/metabolize *vb* umsetzen

P process 268

process *n* Prozess, Verfahren, Behandlung
process *n neuro* Fortsatz
process control Prozesskontrolle
process engineering Verfahrenstechnik
processed antigen prozessiertes Antigen, weiterverarbeitetes Antigen
processed pseudogene weiterverarbeitetes Pseudogen
processing Prozessierung, Verarbeitung, Weiterverarbeitung
  • **antigen processing** Antigenprocessing, Antigenweiterverarbeitung
  • **RNA processing** RNA-Processing, RNA-Weiterverarbeitung
processivity Prozessivität
procoel/procoele Procöl, Procoel
procoelous procöl, procoel
procurved procurv, prokurv
prodissoconch (bivalve larva: premetamorphic shell) Prodissoconch, Prodissoconcha
prodronal phase Vorläuferstadium
prodrug Prodrug (Arzneivorstufe)
produce *n* Erzeugnis, Naturerzeugnis, Landbauprodukt
produce/make produzieren, erzeugen, herstellen
producer Produzent, Erzeuger, Hersteller
product Produkt
product inhibition Produkthemmung
product rule Produktregel
product-moment correlation coefficient Produkt-Moment-Korrelationskoeffizient, Maßkorrelationskoeffizient
productive produktiv
productivity Produktivität
proenzyme/zymogen Proenzym, Zymogen
proestrus Proöstrus, Vorbrunst
professional antigen presenting cell professionelle antigenpräsentierende Zelle
profiled axial flow impeller (bioreactor) Axialrührer mit profilierten Blättern
profitability/feeding efficiency Profitabilität
proflavin Proflavin
profundal *n* Tiefe (Meerestiefe)
profundal depth Meerestiefe
profundal zone (inland waterbody) Untergrundbereich (Binnensee)
progenesis (precocious reproduction) Progenese
progenitor/forebear/ancestor Vorfahre, Ahne
progenote Progenot *m*

progeny/descendant/offspring Nachkomme, Deszendent, Abkömmling; Nachkommenschaft
progeny of a multiple birth Mehrlinge
progesterone Progesteron
progestin Progestin
proglottis/proglottid/tape "segment" (tapeworms) Proglottide, "Segment"
prognathism/prognathy Prognathie, Vorkiefrigkeit
prognathous/prognathic prognath, vorkiefrig
programmed cell death (apoptosis) programmierter Zelltod (Apoptose)
progymnosperms Progymnospermen
prohaptor/buccal sucker (flukes) Prohaptor, Mundsaugnapf
proinsulin Proinsulin
projection/spike/serration vorstehender Teil, Vorsprung, Zacke
projection field/projection area *neuro* Projektionsfeld
prokaryote (procaryote) Prokaryont, Prokaryot *m* (Procaryont/Procaryot)
prokaryotic (procaryotic) prokaryontisch, prokaryotisch (procaryontisch/procaryotisch)
prokaryotic cell prokaryotische Zelle, Protocyt, Protocyte
prolactin (PRL)/luteotropic hormone (LTH) Prolaktin, Prolactin (PRL), Mammatropin, mammotropes Hormon, lactotropes Hormon, luteotropes Hormon (LTH)
prolactoliberin/prolactin releasing hormone/prolactin releasing factor (PRH/PRF) Prolaktoliberin, Prolaktin-Freisetzungshormon
prolamellar body Prolamellarkörper
proleg/pes spurius Propes, Bauchfuß (larvales Abdominalbein)
  • **with crochets on planta in a row** Klammerfuß, Pes semicoronatus (Bauchfüße bei Großschmetterlinglarven)
  • **with crochets on planta in a circle** Kranzfuß, Pes coronatus (Bauchfüße bei Kleinschmetterlinglarven)
proliferate proliferieren, wuchern, sich stark vermehren
proliferation Proliferation, Wucherung
proliferative wuchernd, proliferierend
proliferative zone/budding zone (tapeworms) Proliferationszone, Sprossungszone
proline Prolin

promiscuity Promiskuität
(wahllose/ungebundene
Geschlechtsbeziehungen)
promiscuous promiskuitiv, gemischt;
in Promiskuität lebend
promiscuous DNA
promiskuitive DNA
promontory (a bodily prominence)
anat vorstehender/vorspringender
Körperteil
promoter (strong/weak)
Promotor (starker/schwacher)
promoter complex (open/closed)
Promotorkomplex
(offener/geschlossener)
pronation Pronation,
Einwärtsdrehung (um Längsachse)
pronephric duct Pronephros-Gang,
primärer Harnleiter
prong/spike (point of antler)
Geweihzacke, Geweihspross,
Geweihspitze
pronghorn gegabeltes Horn/Geweih
pronotum Pronotum,
Prothorakalschild, dorsaler
Halsschild
pronucleus Pronukleus, Pronucleus,
Vorkern
proof (alcohol grade) 50% des
jeweiligen Alkoholgehaltes
(60 proof = 30% Alkohol)
proofreading Korrekturlesen
prop Stütze
prop root/stilt root Stützwurzel,
Stelzwurzel
propagate/reproduce fortpflanzen,
vermehren, reproduzieren
propagate *neuro* weiterleiten,
fortleiten, propagieren
propagation/reproduction
Vermehrung, Fortpflanzung,
Reproduktion
propagation *neuro* **(nerve impulse)**
Fortleitung, Weiterleitung
propagative transmission *vir*
propagative Übertragung
**propagule/dispersal unit/diaspore/
disseminule** Ausbreitungseinheit,
Propagationseinheit,
Fortpflanzungseinheit, Diaspore
propellants/propellents Treibmittel,
Treibgase
propeller impeller Propellerrührer
proper name Eigenname
prophage Prophage
prophase Prophase
**prophyll (a bracteole/bractlet/
secondary bract)** Prophyll, Vorblatt
propionic acid (propionate)
Propionsäure (Propionat)
**propionic aldehyde/
propionaldehyde** Propionaldehyd

proplastid Proplastide
propodite Propodit
**proportion rule/Allen's law/Allen's
rule** Proportionsregel,
Allen'sche Regel
proportional truncation
proportionaler Schwellenwert
propositus Proband, Propositus
proprioceptor/proprioreceptor
Proprioceptor
propulsion Antrieb, Voranbringen
(Fortbewegung)
propulsive flight *orn* Kraftflug,
Schlagflug
propulsive force Antriebskraft,
Triebkraft
propupa/prepupa Propupa, Präpupa,
Vorpuppe, Semipupa
prosenchymatous prosenchymatisch
prosimians/lower primates/Prosimii
Halbaffen
**prosobranch snails/prosobranchs/
Streptoneura/Prosobranchia**
Vorderkiemer,
Vorderkiemenschnecken
**prosoma/proterosoma/
cephalothorax** Prosoma,
Vorderkörper, Vorderleib,
Cephalothorax, "Kopf"
prosopyle (sponges)
zuführende Kammerpore
prostaglandin Prostaglandin
prostanoic acid Prostansäure
prostate/prostate gland Prostata,
Prostatadrüse, Vorsteherdrüse
**prostatic gland (annelids:
spermiducal gland)**
Kornsekretdrüse
prosthecate prosthekat, prostekat
prosthetic group
prosthetische Gruppe
prostomium Prostomium,
Kopflappen
prostrate/procumbent/trailing/lying
liegend, niederliegend
protandric/protandrous
protandrisch, proterandrisch
protandric hermaphrodite
proterandrischer
Hermaphroditismus,
protandrisches Zwittertum
protandry Protandrie, Proterandrie,
Vormännlichkeit
**protea family/Australian oak
family/silk-oak family/Proteaceae**
Proteagewächse,
Silberbaumgewächse
proteasome Proteasom
protect (protected) schützen
(geschützt)
protected bud geschlossene Knospe
(mit Knospenschuppen)

P protected forest

protected forest Schutzwald (Schonwald/Hegewald/Bannwald/ Naturwaldreservat)
protected forest plantation, young Schonung
protection assay/protection experiment Schutzversuch, Schutzexperiment
protective clothing Schutzkleidung
protective device Schutzvorrichtung
protective gloves Schutzhandschuhe
protective hood Schutzhaube
protective immunization Schutzimpfung
protective measure/precautionary measure Schutzmaßnahme
protective polyp/defensive polyp/ stinging zooid/dactylozooid Wehrpolyp, Dactylozoid
protective resemblance/protective adaptation *ethol* Schutzanpassung
protein Protein, Eiweiß
- **acute phase protein** Akutphasenprotein
- **carrier protein** Trägerprotein, Schlepperprotein
- **contractile protein/motile protein** kontraktiles Protein, motiles Protein
- **defense protein** Abwehrprotein
- **fibrous proteins** fibrilläre Proteine, Faserproteine
- **globular proteins** globuläre Proteine, Sphäroproteine
- **heat shock protein** Hitzeschockprotein
- **integral (intrinsic) proteins** integrale (intrinsische) Proteine
- **integral membrane protein** integrales Membranprotein
- **nonstructural protein** Nichtstrukturprotein
- **nutrient protein** Nährstoffprotein
- **peripheral (extrinsic) proteins** periphere (extrinsische) Proteine
- **protective protein** Schutzprotein
- **regulative protein/regulatory protein** Regulatorprotein, regulatives Protein, regulatorisches Protein
- **scleroprotein** Skleroprotein
- **signal protein** Signalprotein
- **single-cell protein (SCP)** Einzellerprotein
- **storage protein** Speicherprotein, Reserveprotein
- **structural protein** Strukturprotein, Gerüstprotein, Stützprotein
- **transport protein** Transportprotein

protein aggregation Proteinaggregation, Proteinzusammenlagerung
protein body Proteinkörper
protein deficiency Eiweißmangel
protein engineering gezielte Konstruktion von Proteinen
protein folding Proteinfaltung
protein tagging Protein-Tagging
protein targeting Steuerung von Proteinen
protein truncation test (PTT) Nachweis verkürzter Proteine
proteinaceous proteinartig, proteinhaltig, Protein…, aus Eiweiß bestehend, Eiweiß…
proteinoid Proteinoid
proteinuria Proteinurie
proteoglycan Proteoglycan
proteolysis Proteolyse, Eiweißspaltung
proteolytic proteolytisch
proterandry *bot* Vormännlichkeit
proterogyny Vorweiblichkeit
Proterozoic Eon (late Precambrian) Proterozoikum (Jungpräkambrium/ spätes Präcambrium/Eozoikum)
prothallus Prothallus, Prothallium; Vorlager (Vorkeim von Farnen)
prothoracic gland (an ecdysial, molting gland) Prothoraxdrüse
prothoracicotropic hormone (PTTH)/brain hormone prothoracotropes Hormon, Aktivierungshormon
prothorax Prothorax, Vorderbrust
prothrombin/thrombinogen Prothrombin
protists/Protista Protisten
proto-oncogene Protoonkogen
protobranch (gills) Protobranchie, Kammkieme
protobranch bivalves/ Protobranchiata (Bivalvia) Kammkiemer, Fiederkiemer
protocerebrum (insects) Protocerebrum, Vorderhirn
protocoel/protocoele Protocöl, Protocoel, Axocöl, Axocoel
protoconch (nuclear whorls) Protoconch, Larvenschale, Embryonalschale, Embryonalgewinde
protocorm Protokorm
protocorm-like body Protokormus-ähnlicher Körper
protofilament Protofilament
protofrogs/proanurans/Proanura Proanura
protogynous protogyn, proterogyn
protogyny Protogynie, Proterogynie, Vorweiblichkeit

protomer Protomer
proton gradient Protonengradient
proton microprobe Protonensonde
proton motive force
protonenmotorische Kraft
proton pump Protonenpumpe
protonema Protonema (Vorkeim von
Moosen/gewissen Algen)
protoplasm Protoplasma
protoplast Protoplast
protopod/protopodite (basal part)
Protopodit (Sympodit)
protostelids/Protosteliomycetes
Urschleimpilze, haploide
Schleimpilze
protostomes/Protostomia
Urmundtiere, Urmünder,
Erstmünder
protozoans/"first animals"
Protozoen, "Einzeller", Urtierchen,
Urtiere
protractor muscle Protraktor,
Vorzieher, Vorwärtszieher
protrude vorstehen, hervorstehen,
herausstehen
protruding vorstehend,
hervorstehend
**protruding end/overhanging end/
overhanging extension (DNA)**
überhängendes Ende
protrusion/projection/outgrowth
Ausbuchtung, Auswuchs
**protrusive (e. g. pharynx of
turbellarians)** ausstülpbar
protuberance/tubercle Auswuchs,
Wölbung (Höcker/Beule/Warze)
proturans Beintastler, Protura
proventriculus Proventrikulus,
Ventiltrichter, Ventilkropf
(Vormagen der Honigbiene)
provirus Provirus
prowl *vb zool* **(e. g. cats)**
umherstreifen, herumschleichen
proximal proximal, ursprungsnah,
dem Zentrum zu gelegen
pruinose bereift
prune/trim *hort* beschneiden,
zuschneiden, zurückschneiden,
stutzen, einkürzen
**pruners/pruning shears/pruning
snips (secateurs)** Gartenschere,
Trimmschere
pruning *hort* Rückschnitt, Stutzen
(Gehölzschnitt)
• **form pruning/shape pruning/
training** Erziehungsschnitt
• **maintenance pruning**
Erhaltungsschnitt
• **tree pruning** Baumschnitt
pruning back *hort* Rückschnitt
pruning of woody plants
Gehölzschnitt

**psammobiontic/sabulicolous/
arenicolous** im Sand lebend,
im Sand wohnend
psammolittoral habitat
Sandlückensystem
psammon (interstitial flora/fauna)
Psammon
**psammophilous/thriving in sandy
habitats** psammophil, sandliebend
**psammophyte/plant growing in
unconsolidated sand**
Psammophyt, Sandpflanze
pselaphognaths/Pselaphognatha
Pinselfüßer
pseudanthium Scheinblüte
pseudaxis/sympodium Scheinachse,
Sympodium
pseudobranch Pseudobranchie,
Tracheenkieme
pseudobulb Pseudobulbe,
Luftknolle
**pseudocarp/false fruit/spurious
fruit** Scheinfrucht
pseudodominance Pseudodominanz
pseudofruit Scheinfrucht
pseudogamy/pseudomixis
Pseudogamie
pseudogene Pseudogen
• **conventional p.**
konventionelles Pseudogen
• **processed p.**
prozessiertes Pseudogen,
weiterverarbeitetes Pseudogen
pseudogrooming Scheinputzen
pseudohermaphrodism
Pseudohermaphrod(it)ismus,
Scheinzwittertum
pseudohermaphrodite
Pseudohermaphrodit, Scheinzwitter
**pseudolamellar gill/
pseudolamellibranch**
Pseudolamellibranchie,
Scheinblattkieme
**pseudoparenchyma/
paraplectenchyma/false tissue**
Pseudoparenchym, Scheingewebe
pseudopod *cyt* Pseudopodium,
Scheinfüßchen
pseudopregnant pseudopregnant,
scheinschwanger
**pseudoscorpions/false scorpions/
Pseudoscorpiones/Chelonethi**
Pseudoskorpione, Afterskorpione
pseudostem/false stem (banana)
Scheinstamm
pseudostigmatic organ (orbatids)
pseudostigmatisches Organ
**pseudostratified columnar
epithelium**
hohes-mehrreihiges Epithel
pseudounipolar cell
Pseudounipolarzelle

P psilopsids

psilopsids Psilopsida, Urfarne, Nacktfarne
psilotum family/Psilotaceae Gabelfarngewächse
psocids/Psocoptera (booklice and barklice) Staubläuse
psychobiology Psychobiologie
psychogenetic(al) psychogenetisch
psychologic(al)/psychic/ psychogenic psychologisch, psychisch, psychogen
psychology Psychologie
psychotic (e. g. animals in zoos) psychotisch
psychrometer/wet-and-dry-bulb hygrometer Psychrometer (ein Luftfeuchtigkeitsmessgerät)
psychrophyt Psychrophyt (kälteangepasste Pflanze)
psychrotrophic/psychrophilic (thriving at low temperatures) psychrotroph, psychrophil
PTC (premature termination codon) vorzeitiges Stoppcodon (PTC)
ptenoglossate radula ptenoglosse Radula, Federzunge
pterergate Pterergat (Arbeiterin mit Stummelflügeln)
pteridology Farnkunde
pteris family/Pteridaceae Flügelfarngewächse, Schwertfarne
pterobranchs/Pterobranchia Flügelkiemer
pteropods/Pteropoda Flügelschnecken, Flossenfüßer
- **naked pteropods/Gymnosomata** Ruderschnecken, nackte Flossenfüßer (Flügelschnecken)
- **shelled pteropods/sea butterflies/Thecosomata** Seeschmetterlinge, beschalte Flossenfüßer (Flügelschnecken)
pterosaurs (extinct flying reptiles)/ Pterosauria Flugsaurier
pterostigma Pterostigma, Flügelmal, Flügelrandmal, Makel am Flügelrand
pterygiophore Pterygophor, Flossenträger, Radius
pterygoid bone/pterygoid process/ os pterygoides Pterygoid, Flügelbein
pteryla/feather tract Pteryla, Federflur
pterylosis Pterylographie, Pterylografie, Flurenmuster
PTT (protein truncation test) Nachweis verkürzter Proteine
ptychocyst/tube cnida (Ceriantharia) Ptychozyste, Ptychonema (eine Astomocnide)
ptychophyllous faltenblättrig
ptychospermous faltensamig

ptyxis/vernation/prefoliation Blattlage in der Knospe, Knospenlage der Blätter
puberty/sexual maturity Pubertät, Geschlechtsreife
pubescence Flaumbehaarung, Feinbehaarung; Geschlechtsreife
pubescent (arriving at or having reached puberty) geschlechtsreif werdend
pubescent/downy feinbehaart, flaumig
pubic arch Schambogen
pubic bone/pubis/os pubis Schambein
pubic presentation *ethol* Schamweisen
pubic prominence/mons pubis Venushügel, Schamhügel
pubic region/pubic zone Schamgegend, Scham, Schambeingegend
pubic symphysis/symphysis pubica Schambeinfuge, Schamfuge
public park/public gardens öffentliche Grünanlage
puddle Pfütze, Lache
pudendal fissure/rima pudendi Schamspalte
puerperal fever/childbed fever (bacterial) Puerperalfieber, Kindbettfieber, Wochenbettfieber
puffballs/Lycoperdales Stäublinge, Boviste
- **plated puffballs/Gautieriales** Morcheltrüffeln
pullet/young hen Hühnchen, junges Huhn
pulling flagellum Zuggeißel
pulmonary die Lunge betreffend, Lungen…
pulmonary alveolus/alveola Lungenalveole, Lungenbläschen, Alveole
pulmonary artery Lungenarterie
pulmonary cavity/pulmonary sac (pulmonate snails) Lungenhöhle
pulmonary circulation Lungenkreislauf, kleiner Blutkreislauf
pulmonary pleura/pleura pulmonalis Lungenfell
pulmonary valve/pulmonic valve Pulmonalklappe
pulmonate snails (freshwater & land snails and slugs)/Pulmonata Lungenschnecken, Pulmonaten
pulp *general* Pulpe, Brei (breiige Masse)
pulp *bot* **(fruit)** Fruchtpulpe, Fruchtfleisch
pulp *bot* **(stem)** Stengelmark

pygidial gland **P**

pulp *zool* **(tooth)** Pulpa, Pulpe
pulp cavity of quill Pulpa,
Federseele
pulp cavity of tooth Pulpahöhle
pulpwood Faserholz, Papierholz
pulsate/throb/beat pulsieren
pulsatile flow Pulsstrom,
Pulsströmung
pulsation/pulse beat/throb
Pulsschlag
pulse Puls
**pulse family/bean family/pea
family/legume family/Fabaceae/
Papilionaceae (Leguminosae)**
Hülsenfruchtgewächse,
Hülsenfrüchtler,
Schmetterlingsblütler
pulse labeling Pulsmarkierung
pulse rate Pulszahl
**pulsed field gel electrophoresis
(PFGE)**
Puls-Feld-Gelelektrophorese,
Wechselfeld-Gelelektrophorese
pulvinate/cushion-shaped
polsterförmig, kissenförmig
pulvinus Pulvinus, Blattpolster,
Blattkissen, Gelenkpolster
pumice Bims, Bimsstein
pumping stomach Saugmagen
(Vorratsmagen: Kropf der
Culiciden)
**pumpkin family/gourd family/
cucumber family/Cucurbitaceae**
Kürbisgewächse
punctiform punktförmig
punctiform soralium (lichens)
Punktsoral
punctualism Punktualismus
punctuated gepunktet, punktiert
punctuated equilibrium *evol*
durchbrochenes Gleichgewicht
punctuation codon *gen*
Satzzeichencodon
puncture *n* **(needle biopsy)**
Punktion; (Ein)Stich, Loch
puncture/tap *vb* punktieren
pungency scharfer Geruch,
stechender Geruch
pungent (odor) scharf, stechend,
beizend, ätzend
pungent (with stiff/sharp point)
spitzig
Punicaceae/pomegranate family
Granatapfelgewächse
Punnett square Punnett-Schema
pupa Puppe
• **adecticous pupa/pupa adectica**
Pupa adectica
• **decticous pupa/pupa dectica**
Pupa dectica
• **exarate pupa/pupa exarata (free
appendages)** gemeißelte Puppe

• **free pupa/pupa libra** freie Puppe
• **girdled pupa/pupa cingulata**
Gürtelpuppe
• **obtect pupa/pupa obtecta**
bedeckte Puppe, Mumienpuppe
• **suspended pupa/pupa suspensa**
Stürzpuppe
puparium/pupal instar Puparium
pupate sich verpuppen
pupation/pupating Verpuppung
pupil (eye) Pupille
pupiparous pupipar
pupipary Pupiparie
**pure breeding line/pure breeding
strain** reinerbige Linie,
reine Linie, reinerbiger Stamm,
reiner Stamm
pure culture/axenic culture
Reinkultur
pure-bred/true-bred reinrassig
purification Reinigung, Säuberung;
Aufarbeitung; (isolation) Isolation,
in Reinform darstellen
**purification procedure/purification
technique** Reinigungsverfahren
purify reinigen, säubern;
aufarbeiten; (isolate) isolieren,
darstellen
purine Purin
Purkinje cell Purkinje-Zelle
Purkinje fiber/conduction myofiber
Purkinje-Faser
purple gland Purpurdrüse
purple membrane Purpurmembran
purr (cat) schnurren
purse web *arach* Gespinstschlauch,
Röhrennetz, Röhrengespinst
purslane family/Portulacaceae
Portulakgewächse
pus Eiter
pushing flagellum Schubgeißel
pusule Pusule
putamen Putamen
putrefaction/rotting/decomposition
Verwesung, Zersetzung
putrefactive fäulniserregend
putrefactive bacteria
Fäulnisbakterien
putrefy/rot/decompose verwesen,
zersetzen
putrescine Putrescin, Putreszin
pycnium/pycnidium Pyknidium,
Pyknidie (Pyknosporenlager)
**pycnogonids/pantopods/
seaspiders/Pycnogonida/
Pantopoda** Asselspinnen
pycnospore/pycnidiospore
Pyknospore
pycnoxylic wood
dichtfaseriges Holz
pygidial gland/anal gland
Pygidialdrüse, Analdrüse

pygidium 274

pygidium/caudal shield (arthropods) Pygidium, Afterlappen, Afterschild, Analschild

pygmy conifer woodland Zwergnadelwald, Zwergstrauchzone

pygostyle (ploughshare bone/ vomer of birds) Pygostyl, Schwanzstiel (Pflugscharbein/Vomer der Vögel)

pyknosis/pycnosis Pyknose (Kernverdichtung, Karyoplasmaagglutination)

pyloric cecum Pylorus-Anhang

pyloric stomach (echinoderms) Mitteldarm

pylorus (crustaceans: posterior region of gizzard) Pylorus, Filtermagen

pylorus (lower opening of stomach into duodenum) Pylorus, Pförtner

pyramid tract/corticospinal tract Pyramidenbahn (verlängertes Rückenmark)

pyramid-shaped treetop/crown Pyramidenkrone

pyramidal bone/triangular bone/ triquetral bone/os triquetrum Dreiecksknochen

pyramidal cell/pyramidal neuron Pyramidenzelle

pyran Pyran

pyrethric acid Pyrethrinsäure

pyrethrin Pyrethrin

pyridoxine/adermine (vitamin B$_6$) Pyridoxin, Pyridoxol, Adermin

pyriform organ/piriform organ (bryozoans) birnenförmiges Organ

pyriform/piriform/pear-shaped birnenförmig

pyrimidine Pyrimidin

Pyrolaceae/wintergreen family/ shinleaf family Wintergrüngewächse

pyroligneous acid/wood vinegar Holzessig

pyroligneous alcohol/wood alcohol/pyroligneous spirit/ wood spirit (chiefly: methanol) Holzgeist

pyrophyte *ecol* Pyrophyt (stark feuerresistente Pflanze/ durch Brände gefördert)

pyrosomes/Pyrosomida (phosphorescent tunicates) Feuerwalzen

pyrrhophytes Pyrrhophyten, Feueralgen

pyrrole Pyrrol

pyrrolidine Pyrrolidin

pyruvic acid (pyruvate) Brenztraubensäure (Pyruvat)

pyxidate (with lid) gedeckelt, mit Deckel

pyxis/pyxidium/circumscissile capsule/lid capsule Deckelkapsel

quack (duck) quaken
quadrate bone/quadratum
Quadratbein
quadrifid/deeply cleft in four parts
vierspaltig
quadrilobate/four-lobed vierlappig
**quadripartite/divided into four
parts** vierteilig
**quadripinnate/divided pinnately
four times** vierfach gefiedert
quadrumanous/four-handed
vierhändig
quadruped *adj/adv* vierfüßig
quadruped *n* Quadrupede, Vierfüßer
quadrupedal vierfüßig
quadrupedalism Vierfüßigkeit
quadruplets *gen* Vierlinge
**quagmire/quaking bog (swampy/
muddy ground)** Morast,
Sumpfland, Moorboden, Moorgrund,
Schwingrasen (vibrierender/
schwankender Hochmoorboden)
qualitative qualitativ
quality Qualität
quality control Qualitätskontrolle
quantification/quantitation
med/chem Quantifizierung
quantify/quantitate *med/chem*
quantifizieren
quantile/fractile Quantil, Fraktil
quantitative quantitativ
quantitative phenotye quantitativer
Phänotyp
quantity Quantität
quantity (to be measured)
Messgröße
quantum evolution
Quantenevolution
quarantine Quarantäne
quarry *geol* Steinbruch
quarry/prey *zool* verfolgtes Tier/
Wild, Jagdbeute
quarter (hindquarter) *zool*
Viertel (Hinterviertel);
croup (rump of horse) Kruppe

quarter (hoof) Trachte
quarternary structure (proteins)
Quartärstruktur
**quartersawn/radial section
(wood)** vierteilig-aufgesägt
(Holzstamm/Stammholz),
Radialschnitt, Spiegelschnitt
quartile *stat* Quartil, Viertelswert
quasi-equivalence theory
Quasi-Äquivalenz-Theorie
quasispecies Quasispezies
quassia family/Simaroubaceae
Bittereschengewächse
**Quaternary Period/Quaternary
(geological age)** Quartär
quay/wharf Kai
queen Königin
queen bee Bienenkönigin, Weisel
queen substance (bees)
Königin-Substanz
**Queensland arrowroot family/
canna family/Cannaceae**
Blumenrohrgewächse
quick-stain *micros* Schnellfärbung
quiescence (exogenous) Quieszenz,
Ruhe, Stille
quiescent/resting ruhend
quiescent bud/resting bud
ruhende Knospe
quiescent center/quiescent zone
ruhendes Zentrum
quiescent stage Ruhestadium
**quill/calamus (central shaft of
feather)** Federspule, Calamus
**quill/horny spine (hedgehogs/
porcupine)** Stachel, Hornstachel
quill feather *orn* Schwanzfeder
quillwort Brachsenkraut (*Isoetes*)
quillwort family/Isoetaceae
Brachsenkrautgewächse
quinic acid Chinasäure
quinquefoliolate
fünfblättrig gefiedert
quinquepartite fünfteilig
quiver/tremble zittern

R banding

R banding R-Banding, Revers-Banding
R factor/resistance factor R-Faktor, Resistenz-Faktor
r selection (*rapid* development) *r*-Selektion
r-selected species (r strategist) *r*-"Stratege"
r strategy (*actually not a "strategy"*) *r*-Strategie
rabbit ears Löffel (Hasenohren)
rabbits/lagomorphs/Lagomorpha Hasen
rabid *adj/adv* (a rabid animal) tollwütig; wild
rabies (*Lyssavirus*) Rabies, Tollwut, Hundswut, Lyssa
race Rasse
race hygiene/racial hygiene Rassenhygiene (*Nazi term for Aryan eugenics*), Erbhygiene
RACE-PCR (rapid amplification of cDNA ends-PCR) RACE-PCR (schnelle Vervielfältigung von cDNA-Enden-PCR)
raceme/botrys (an inflorescence) Traube, Botrys
racemose/botryoid/botryose/grape-cluster-like razemös, racemös, racemos, botrytisch, traubig, traubenförmig
rachiglossate radula Schmalzunge, rhachiglosse Radula, stenoglosse Radula
rachilla *bot* Blütenstiel einzelner Grasblüte, Ährchenachse, kleine sekundäre Rhachis
rachis (midrib of compound leaf) Rhachis, Blattachse, Blattspindel, Blattstiel (an gefiedertem Blatt), Fiederblattachse, Mittelrippe eines Fiederblattes
rachitic/rickety rachitisch
racial rassisch, Rassen…
racial discrimination Rassendiskriminierung
racism Rassismus
racist *adj* rassistisch
racist *n* Rassist
rack (vial rack/test tube rack) Ständer
racquet organ/malleolus (solifugids) Malleolus, hammerförmiges Organ
radial/cyclic/radially symmetrical/regular/actinomorphic radiär, radiärsymmetrisch, zyklisch, strahlenförmig, aktinomorph
radial canal Radialkanal, Radiärkanal
radial cell/radius Radialzelle, Radius

radial cleavage Radiärfurchung
radial glial cell Radialgliazelle
radial nerve Radiärnerv
radial section/quartersawn (wood) Radialschnitt, Spiegelschnitt (Holz)
radial symmetry Radialsymmetrie, Radiärsymmetrie
radial thread/radius/spoke (spiderweb) *arach* Netzspeiche, Radius
radially symmetrical/actinomorphic radiärsymmetrisch
radiant energy Strahlungsenergie
radiant heat Strahlungswärme
radiate strahlen, ausstrahlen, verbreiten
radiation *evol* Entfaltung, Ausstrahlung
radiation *phys* Strahlung; Ausstrahlung
- **background radiation** Hintergrundstrahlung
- **electromagnetic radiation** elektromagnetische Strahlung
- **harmful radiation** gesundheitsschädliche Strahlung
- **ionizing radiation** ionisierende Strahlung
- **nuclear radiation** Kernstrahlung
- **radioactive radiation/radioactivity** radioaktive Strahlung, Radioaktivität
- **scattered radiation/diffuse radiation** Streustrahlung
- **solar radiation** Sonnenstrahlung
radiation biology Strahlenbiologie
radiation burn Strahlenverbrennung
radiation control/protection from radiation Strahlenschutz
radiation genetics Strahlengenetik
radiation intensity Strahlungsintensität
radiation load Strahlenbelastung
radiation protection Strahlenschutz
radiation sickness/radiation syndrome Strahlenkrankheit, Strahlensyndrom
radiation source Strahlenquelle
radical *chem* Radikal
- **free radical** freies Radikal
radical scavenger Radikalfänger
radication/rootage/rooting Bewurzelung
radiciform/rhizoid/rootlike wurzelförmig, wurzelartig
radicle/radicula/embryonic root Radicula, Keimwurzel, Hauptwurzelanlage
radicle sheath/root sheath/coleorhiza Wurzelscheide, Koleorhiza, Coleorhiza
radioactive radioactiv

random screening

radioactive decay/radioactive disintegration radioaktiver Zerfall
radioactive exposure radioaktive Belastung
radioactive marker radioaktiver Marker
radioactively contaminated radioaktiv verseucht
radioactivity Radioaktivität
radioallergosorbent test (RAST) Radio-Allergo-Sorbent Test
radioautography Autoradiographie, Autoradiografie
radiobiology Radiobiologie, Strahlenbiologie
radiocarbon method Radiokarbonmethode, Radiokohlenstoffmethode
radiodiagnosis Strahlendiagnose
radioimmunoassay Radioimmunassay, Radioimmunoassay
radioimmunoelectrophoresis Radioimmunelektrophorese
radiolabeling radioaktive Markierung
radiolarian ooze Radiolarienschlamm
radiolarians/Radiolaria Radiolarien (*sg* Radiolarie), Strahlentierchen
radiology Radiologie, Strahlenkunde
radionuclide Radionuklid
radiosensitive strahlenempfindlich
radiotherapy Strahlentherapie
radius *math/geom* Radius
radius (bone of forearm) Speiche
radix/root (crinoids) Wurzel
radula Radula, Reibplatte, "Zunge"
- **docoglossate radula** docoglosse Radula, Balkenzunge
- **hystrichoglossate radula** hystrichoglosse Radula, Bürstenzunge
- **ptenoglossate radula** ptenoglosse Radula, Federzunge
- **rachiglossate radula** rhachiglosse Radula, stenoglosse Radula, Schmalzunge
- **rhipidoglossate radula** rhipidoglosse Radula, Fächerzunge
- **taenioglossate radula** taenioglosse Radula, Bandzunge
- **toxiglossate radula (hollow radula teeth)** toxoglosse Radula, Pfeilzunge (hohl)
radula support/odontophore Radulapolster, Odontophor
Rafflesiaceae/rafflesia family Schmarotzerblumengewächse
rafted timber/rafted logs/raft wood Floßholz, Flößholz
rafting (wood) Flößen, Treiben

rain forest Regenwald
rain shadow Regenschatten
rain showers Regenfälle
rain-shadow desert Regenschattenwüste
rainy season Regenzeit, Pluvialzeit
raise/grow/cultivate *hort/agr* ziehen, anbauen, kultivieren
raise/rear *zool* aufziehen, erziehen
raised bog/(upland/high) moor/ peat bog Hochmoor
ram (male sheep) Schafbock, Widder
ram ventilation *ichth* Staudruck-Ventilation
Ramachandran plot Ramachandran-Diagramm
ramate/ramous/ramose verzweigt
ramate (rotifers: grinding-type trophi) ramat
ramentum/chaffy scale/palea/pale Spreuschuppe
rameous/ramal astständig
ramet (individual member of clone) Ramet, Klonindividuum, Klonmitglied, Einzelpflanze eines Klons (>Zweig/Steckling eines Ortets)
ramification/branching Ramifikation, Verästelung, Verzweigung
ramiflorous zweigblütig, astblütig
ramiform zweigförmig
ramify/branch sich verästeln, sich verzweigen
ramous/ramose/ramate verzweigt
rancid ranzig
rancidity Ranzigkeit
rand/slope community of raised bog Randgehänge
random *adj/adv* zufällig, wahllos, ziellos, ungeordnet, verstreut
random coil "Zufallsknäuel", ungeordnetes Knäuel
random deviation Zufallsabweichung
random displacement reaction zufällige Verdrängungsreaktion, nicht-determinierte Verdrängungsreaktion
random drift (Sewall-Wright) zufallsbedingte Drift, Zufallsdrift
random error zufälliger Fehler, Zufallsfehler
random event Zufallsereignis
random mating Zufallspaarung, zufällige Paarung, Panmixie
random number Zufallszahl, beliebige Zahl
random sample Zufallsprobe, Zufallsstichprobe
random screening Zufallsauslese

random variable

random variable Zufallsvariable
random-walk model
"Random-Walk"-Modell
random-walk process "Random-Walk"-Prozess, Irrfahrtprozess
randomization Randomisierung
randomize randomisieren, eine Zufallsauswahl treffen
range vb agr
das Vieh frei weiden lassen
range n ecol Bereich, Gebiet, Raum; Ausdehnung, Umfang; geogr Areal, Verbreitungsgebiet
- **geographic range/area of distribution** Verbreitungsgebiet, Areal
- **home range** ecol Aktionsraum
- **host range (parasites)** Wirtsspektrum
- **hunting range/hunting grounds/ hunting territory** Jagdgründe
- **mountain range/mountain chain/mountain ridge** Bergkette, Gebirgskette
range/field (see also: rangeland) Freiland, Feld; unkultiviertes Weide-/Jagdgebiet
range/distance rad Entfernung, Reichweite
range stat Spannweite; Streuungsbreite, Toleranzbreite, Bereich
range chart/range map Arealkarte
range of distribution/range of variation stat Variationsbreite
range of measurement Messbereich
range of occurrence/geographic range/area of occurrence Verbreitungsgebiet, Areal
range of saturation/zone of saturation Sättigungsbereich, Sättigungszone
range of vision/field of view/scope of view/field of vision/visual field Sehfeld, Blickfeld, Gesichtsfeld
rangeland/grazing land/pasture/ pastureland/pasturage Weideland, Weidefläche
ranger Forstaufseher, Jagdaufseher, Wächter
raniform froschartig, froschförmig
ranine (of or relating to frogs) Frosch..., froschartig, zu den Fröschen gehörig, Frösche betreffend
ranine (referring to undersurface of tongue) Unterzungen...
ranine artery/arteria profunda linguae tiefe Zungenschlagader
rank n **(relative standing/position)** Rang, Stufe

rank/classify vb einordnen, einstufen, klassifizieren
rank correlation coefficient stat Rangkorrelationskoeffizient
rank statistics/rank order statistics Rangmaßzahlen
ranking scale stat ranking scale (used as such: not translated)
Ranunculaceae/buttercup family/ crowfoot family Hahnenfußgewächse
Ranvier's node/node of Ranvier/ neurofibral node Ranvierscher Schnürring
raphe bot Raphe, Samennaht, Samenwulst
rapid freezing Schnellgefrieren
raptorial/raptorious Raub..., räuberisch
raptorial bird/raptor/predatory bird/bird of prey Raubvogel, Greifvogel
raptorial claw (predatory birds) Greiffuß, Fang
raquet hyphae/raquet hyphas (raquet mycelium) Raquettehyphen, Keulenhyphen (Raquettemyzel/Keulenmyzel)
rare/scarce selten
rare species seltene Art
rareripe (fruit/vegetable) früh reifend, frühreif
rarity/scarcity Seltenheit
rasorial (fowl) scharrend
rasp/scraper n **(plectrum)** zool/ entom (>stridulation) Schrillkante
ratany family/krameria family/ Krameriaceae Krameriagewächse
rate constant (enzyme kinetics) Geschwindigkeitskonstante
rate of metabolism/metabolic rate Stoffwechselrate, Stoffwechselintensität, Metabolismusrate
rate-determining step/reaction geschwindigkeitsbestimmende(r) Schritt/Reaktion
rate-limiting step/reaction geschwindigkeitsbegrenzende(r) Schritt/Reaktion
Rathke's pouch/hypophyseal pouch/hypophyseal sac Rathkesche Tasche, Hypophysentasche
rating scale stat rating scale (used as such: not translated)
ratio/quotient/proportion/relation Verhältnis, Quotient, Proportion
ratio scale stat Verhältnisskala, Ratioskala
ratite/having an unkeeled sternum (insects) mit ungekieltem Sternum

279   **reactor** **R**

ratite birds (flightless birds)
Ratiten, Flachbrustvögel
ratoon *bot* Schössling (speziell:
Zuckerrohr); Schösslinge treiben
ravine *n* Schlucht, Bergschlucht,
Klamm, Hohlweg
raw data Rohdaten
raw material Rohstoff
raw sewage Rohabwasser
raw sludge Rohschlamm
**raw sugar/crude sugar (unrefined
sugar)** Rohzucker
ray Strahl; (beam of light)
Lichtstrahl
 • **pith ray/medullary ray**
Markstrahl
 • **wood ray** Holzstrahl
ray fin Strahlenflosse
**ray floret/ray flower/ligulate
flower** Strahlenblüte, Zungenblüte
ray initial Markstrahlinitiale
ray parenchyma
Markstrahlparenchym
**ray-finned bony fishes/
actinopterygians/Actinopterygii**
Strahlenflosser
**rays & skates/Batoidea
(*Superorder!*)** Rochen
razor blade Rasierklinge
re-uptake Wiederaufnahme
reabsorb reabsorbieren,
wiederaufnehmen
("rückresorbieren")
reabsorption Reabsorption,
Wiederaufnahme
react reagieren
reactant Reaktand,
Reaktionsteilnehmer, Ausgangsstoff
reaction Reaktion
 • **first-order reaction** *kinet*
Reaktion erster Ordnung
 • **second-order reaction** *kinet*
Reaktion zweiter Ordnung
 • **third-order reaction** *kinet*
Reaktion dritter Ordnung
 • **zero-order reaction** *kinet*
Reaktion nullter Ordnung
reaction center Reaktionszentrum,
Photosynthesereaktionszentrum
reaction intermediate
Reaktionszwischenprodukt
reaction kinetics Reaktionskinetik
reaction pathway Reaktionskette
reaction rate Reaktions-
geschwindigkeit, Reaktionsrate
**reaction sequence/reaction
pathway** Reaktionsfolge
reaction wood Reaktionsholz
reactive force Gegenkraft,
Rückwirkungskraft
reactivity/reactiveness
Reaktionsfähigkeit

reactor/bioreactor *biot* Reaktor,
Bioreaktor
 • **airlift loop reactor**
Mammutschlaufenreaktor
 • **airlift reactor/pneumatic reactor**
Airliftreaktor, pneumatischer
Reaktor, Mammutpumpenreaktor
 • **bead-bed reactor**
Kugelbettreaktor
 • **bubble column reactor**
Blasensäulen-Reaktor
 • **column reactor** Säulenreaktor,
Turmreaktor
 • **fedbatch reactor/fed-batch
reactor** Fedbatch-Reaktor, Fed-
Batch-Reaktor, Zulaufreaktor
 • **fermentation chamber reactor/
compartment reactor/cascade
reactor/stirred tray reactor**
Rührkammerreaktor
 • **film reactor** Filmreaktor
 • **fixed-bed reactor/solid bed
reactor** Festbettreaktor
 • **flow reactor** Durchflussreaktor
 • **fluidized-bed reactor**
Wirbelschichtreaktor,
Wirbelbettreaktor
 • **immersed slot reactor**
Tauchkanalreaktor
 • **immersing surface reactor**
Tauchflächenreaktor
 • **jet loop reactor**
Strahlschlaufenreaktor,
Strahl-Schlaufenreaktor
 • **jet reactor** Strahlreaktor
 • **loop reactor/circulating
reactor/recycle reactor**
Umlaufreaktor, Umwälzreaktor,
Schlaufenreaktor
 • **membrane reactor**
Membranreaktor
 • **moving-bed reactor**
Fließbettreaktor
 • **nozzle loop reactor/circulating
nozzle reactor**
Düsenumlaufreaktor,
Umlaufdüsen-Reaktor
 • **packed bed reactor**
Packbettreaktor, Füllkörperreaktor
 • **paddle wheel reactor**
Schlaufenradreaktor
 • **plug-flow reactor**
Pfropfenströmungsreaktor,
Kolbenströmungsreaktor
 • **plunging jet reactor/deep jet
reactor/immersing jet reactor**
Tauchstrahlreaktor
 • **pressure cycle reactor**
Druckumlaufreaktor
 • **sieve plate reactor**
Siebbodenkaskadenreaktor,
Lochbodenkaskadenreaktor

reaktor

- **solid phase reactor**
  Festphasenreaktor
- **stirred cascade reactor**
  Rührkaskadenreaktor
- **stirred loop reactor**
  Rührschlaufenreaktor,
  Umwurfreaktor
- **stirred-tank reactor**
  Rührkesselreaktor
- **tray reactor** Gärtassenreaktor
- **trickling filter reactor**
  Tropfkörperreaktor,
  Rieselfilmreaktor
- **tubular loop reactor**
  Rohrschlaufenreaktor

read/record messen, ablesen (z. B. Messdaten)
read through (a stop codon) hinweglesen über (ein Stoppcodon)
readiness potential Bereitschaftspotential
reading/recording (data) Ablesung (z. B. Messdaten)
reading fidelity Lesetreue
reading frame *gen* Leserahmen, Leseraster
- **closed r.f.** geschlossenes Leseraster
- **open r.f.** offenes Leseraster
- **unassigned r.f./unclassified r.f.** nicht zugeordnetes Leseraster
- **unidentified r.f.** unbekanntes Leseraster

readthrough *gen* darüber hinweg lesen, Durchlesen, Überlesen (eines Terminationssignals)
reagent Reagenz, Reagens (*pl* Reagenzien)
reagent grade (chemicals) *lab* analysenrein, zur Analyse (pA-Ware)
reagin/reaginic antibody (IgE antibodies) Reagin, IgE-Antkörper
real image *micros* reelles Bild
reallocation of arable land/ consolidation of arable land Flurbereinigung
reanimate wiederbeleben
reannealing/annealing/ reassociation/renaturation (of DNA) Doppelstrangbildung, Reannealing, Annealing, Reassoziation, Renaturierung
rear/foster/nurture/bring up aufziehen, erziehen
rear reef Rückriff
rearing/fostering/nurturing/ upbringing Aufzucht, Erziehung
rearrange *chem* umlagern, umordnen
rearrangement *chem* Umlagerung, Umordnung

rearrangement/reassortment *gen* **(DNA/genes/genome)**
Rearrangement, Umordnung, Neuordnung
reassembly Wiederzusammenbau, Wiederzusammenfügung
reassociation/annealing/ reannealing/renaturation/ reassociation (DNA)
Doppelstrangbildung, Annealing, Reannealing, Renaturierung, Reassoziation
reassociation kinetics Reassoziationskinetik
reassuring gesture Beschwichtigungsgeste, Beruhigungsgeste
recalcitrant/not responsive to treatment (e. g. seeds) nicht auf Reize reagierend
recalcitrant/resistant beständig, schwer abbaubar, widerstandsfähig
recapitulation theory/ principle of recapitulation Rekapitulations-Theorie
receiving water Vorfluter
recent/contemporary/extant rezent, gegenwärtig, heute lebend
Recent/Holocene/Recent Epoch/ Holocene Epoch Jetztzeit, Holozän (erdgeschichtliche Epoche)
recent publication/paper kürzlich erschienene/veröffentliche Publikation, kürzlich erschiene Veröffentlichung
receptacle/receptaculum Rezeptakel, Rezeptakulum; *bot*: **torus of a flower** Blütenbasis, Blütenachse, Blütenboden, Torus
receptive empfänglich
receptor Rezeptor, Empfänger
receptor-down regulation Rezeptor-Ausdünnungsregulation
receptor-mediated endocytosis rezeptorvermittelte Endozytose, rezeptorgekoppelte Endozytose
recessed end *gen* zurückgesetztes Ende
recessive rezessiv
recipient (also: host) (transplants/ graft) Empfänger, Rezipient (z. B. Transplantate)
recipient cell Empfängerzelle
reciprocal *adj/adv* reziprok
reciprocal *n stat* Kehrwert, reziproker Wert
reciprocal translocation reziproke Translokation
reciprocating shaker Reziprokschüttler
recognition Erkennung
recognition site Erkennungssequenz

recognition site affinity chromatography Erkennungssequenz-Affinitätschromatographie
recognize erkennen
recolonization Wiederbesiedlung
recolonize wiederbesiedeln
recombinant *adv/adj* rekombiniert, rekombinant
recombinant *n* **(cell)** Rekombinante (Zelle)
recombinant DNA molecule rekombiniertes DNA-Molekül, rekombinantes DNA-Molekül
recombinant DNA technology rekombinierte DNA-Technologie (Methoden mit Hilfe rekombinierter DNA), rekombinante DNA-Technologie
recombinant protein rekombiniertes Protein, rekombinantes Protein
recombination Rekombination
• **double r.** doppelte Rekombination
• **general r.** allgemeine Rekombination
• **homologous r.** homologe Rekombination
• **illegitimate r.** illegitime Rekombination
• **mitotic r.** mitotische Rekombination
• **non-homologous r.** nichthomologe Rekombination
• **site-specific r.** sequenzspezifische Rekombination
• **targeted homologous r.** Allelen-Austauschtechnik
recombination frequency Rekombinationsfrequenz
recombination nodule Rekombinationsknoten
recombination signal sequences Rekombinationssignalsequenzen
recombine rekombinieren
recommended daily allowance (RDA) empfohlener täglicher Bedarf
reconstitute rekonstituieren
reconstitution Rekonstitution
record(s)/protocol *n* Aufzeichnung, Protokoll, Akte
• **lab records/lab protocol** Laboraufzeichnungen, Laborprotokoll
record *vb neuro* ableiten
record/read *vb* **(data)** ablesen (z. B. Messdaten)
record/register *vb* aufnehmen, aufschreiben, registrieren
recording *neuro* Ableitung
recording/reading (data) Ablesung (z. B. Messdaten)

recording/registration Aufnahme, Aufschreiben, Registrierung
recover erholen
recovery Erholung
recovery stroke (backstroke) Erholungsschlag
recreational forest/amenity forest Erholungswald
rectal gland Rectaldrüse, Rektaldrüse (*see*: Klebdrüse, Kittdrüse, Zementdrüse)
• **anal gland (shark)** Analdrüse (Hai)
rectification Gleichrichtung
rectifier Gleichrichter
rectify gleichrichten
rectilinear movement/caterpillar movement (snakes) Integumentbewegung, Raupenbewegung
rectum Rectum, Rektum, Mastdarm
recurrence risk Wiederholungsrisiko
recurved/bent backwards recurv, rekurv, zurückgebogen
recycle wiederverwerten
recycling Recycling, Wiederverwertung
red algae/Rhodophyceae Rotalgen
red blood cell (RBC)/erythrocyte rotes Blutkörperchen, Erythrocyt, Erythrozyt
red body (swimbladder) roter Körper
Red Data Book Rote Liste
red mangrove family/mangrove family/Rhizophoraceae Mangrovengewächse
Red Queen's hypothesis *evol* Rote-Königin-Hypothese
red tide Rote Tide (rötliche Wasserblüte)
red-water fever/hemoglobinuric fever/Texas fever (babesiosis) (Babesia ssp.) Texasfieber
redd Laichstelle des Lachses (Vertiefung im Flussschotter)
redia (fluke larva) Redie
redox potential-discontinuity (RPD) *mar* Redoxpotential Diskontinuität
redroot family/bloodwort family/ kangaroo paw family/ haemodorum family/ Haemodoraceae Haemodorumgewächse
reduce *chem* **(vs oxidize)** reduzieren (*vs* oxidieren)
reduce/concentrate einengen, konzentrieren
reducing agent Reduktionsmittel
reduction Reduktion, Verminderung; Verarmung

reduction division/meiosis Reduktionsteilung, Reifeteilung, Meiose
redundance/redundancy Redundanz; Überfluss, Übermaß, Überfülle, Überflüssigkeit; unnötige Wiederholung
redundant redundant, überreichlich, übermäßig, überflüssig
redwood family/swamp-cypress family/taxodium family/ Taxodiaceae Taxodiumgewächse, Sumpfzypressengewächse
reed *bot* Ried, Schilfrohr, Schilfgras, Schilfröhricht
reed/rennet-stomach/fourth stomach/abomasum Labmagen, Abomasus
reed bank/reeds Röhricht
reed swamp Riedsumpf
reedmace/cat's-tail Rohrkolben
reedmace family/cattail family/ Typhaceae Rohrkolbengewächse
reef Riff
- **bank reef/patch reef** Fleckenriff
- **barrier reef** Barriereriff, Wallriff
- **buttress zone/spur-and-groove zone** Grat-Rinnen-System
- **fore reef** Vorriff
- **fringing reef** Saumriff, Küstenriff, Strandriff
- **rear reef** Rückriff
- **table reef** Plattformriff

reef crest Riffkrone
reef edge Riffkante
reef flat Riffdach
reef slope Riffhang
reef-building/hermatypic riffbildend, hermatypisch
- **not reef-building/non-hermatypic** nicht riffbildend, ahermatypisch

reestablish/resettle wiederherstellen, wiederbesieden
reestablishment/resettlement Wiederherstellung, Wiederbesiedlung
refection (rabbits/guinea pigs) Caecotrophie, Coecotrophie, Coecophagie
referee/reviewer Schiedsrichter; Gutachter (wissenschaftliche Manuskripte)
reference strain Referenzstamm
reflex Reflex
- **clasp reflex** Klammerreflex
- **conditioned reflex (CR) (classical conditioning)** bedingter Reflex
- **myotatic reflex/stretch reflex** myotatischer Reflex, Dehnungsreflex
- **snapping reflex** Schnappreflex
- **startle reflex** Schreckreflex
- **stretch reflex/myotatic reflex** Dehnungsreflex, myotatischer Reflex
- **suction reflex** Saugreflex
- **unconditioned reflex (UCR)** unbedingter Reflex

reflex arc Reflexbogen
reflux condenser Rückflusskühler
reforestation/reafforestation/ afforestation Wiederaufforstung, Wiederbewaldung
refractile body (Placozoa) Glanzkugel
refracting angle Brechungswinkel
refraction Refraktion, Brechung
- **optical refraction** optische Brechung, Lichtbrechung

refractive index/index of refraction Brechungsindex, Brechungszahl, Brechungskoeffizient
refractivity Brechungsvermögen
refractometer Refraktometer
refractory period Refraktärperiode, Refraktärzeit
refractory stage Refraktärphase, Refraktärstadium
refrigerant *n techn* Kühlmittel
refrigerate kühlen, in den Kühlschrank stellen
refrigerator Kühlschrank
refuge *ecol* Refugium
regenerate/regrow/reestablish regenerieren, nachwachsen, wiederergänzen
regeneration Regenerierung, Regeneration, Wiederergänzung
regional association *biogeo/ecol* Gebietsassoziation
regression Regression
regression coefficient/coefficient of regression *stat* Regressionskoeffizient
regression to the mean *stat* Regression zum Mittelwert
regressive regressiv, zurückbildend, zurückentwickelnd
regular regulär, "richtig"; regelmäßig
regular *bot* **(corolla)** radiär
regulate/control regeln, kontrollieren
regulation/control Regulierung, Regelung, Kontrolle
regulative cleavage regulative Furchung (nichtdeterminativ)
regulative development regulative Entwicklung
regulative egg Regulationsei
regulative protein/regulatory protein Regulatorprotein, regulatives Protein, regulatorisches Protein

renal cortex

regulatory gene Regulationsgen
regulatory mechanism Steuerungsmechanismus
regulatory procedure Reglungsprozess
regurgitate regurgitieren, hochwürgen, wiederaufstoßen
regurgitation Regurgitation, Hochwürgen, Wiederaufstoßen
rehearsal song *orn* Studiergesang
rehydrate rehydrieren
rehydration Rehydratation, Rehydratisierung
reinfestation Wiederbefall
reinforce/amplify (stimulus) verstärken (Reiz)
reinforcement/amplification (stimulus) Verstärkung (Reiz); Bekräftigung
Reinig's line Reinig-Linie
Reissner's membrane/membrana vestibularis Reißnersche Membran, Reißner-Membran
reject (graft rejection) abstoßen (Transplantat)
rejection (graft rejection) Abstoßung (Transplantat)
rejection reaction Abstoßungsreaktion
rejuvenate/regenerate verjüngen, regenerieren
rejuvenation/regeneration Verjüngung, Regeneration
relapsing fever (*Borrelia recurrentis*) Rückfallfieber
relation/correlation/ interrelationship/connection Zusammenhang, Verhältnis, Verbindung
relationship/connection Verhältnis, Beziehung
relationship/relatedness/kinship Verwandtschaft
relative *n* Verwandter
 • **first-degree relative** Verwandter ersten Grades
 • **second-degree relative** Verwandter zweiten Grades
relative *adj/adv* relativ, verwandt
relative biological effectiveness (RBE) *rad* relative biologische Wirksamkeit (RBW)
relative frequency relative Häufigkeit
relative molecular mass/ molecular weight (M_r) relative Molekülmasse, Molekulargewicht (M_r)
relatives Verwandtschaft, Verwandte
relax *physiol* entspannen, erschlaffen
relaxation *physiol* Relaxation, Erschlaffung, Entspannung
relaxation period *neuro* Erholungsphase
relaxed (conformation) relaxiert, entspannt
relaxed plasmid relaxiertes Plasmid, schwach kontrolliertes Plasmid
relaxin Relaxin
relay cell *neuro* Relaiszelle
relay neuron Relaisneuron, Projektionsneuron, Hauptneuron
release *n* **(e.g. hormones/ neurotransmitter)** Ausschüttung, Freisetzung
release *vb* freisetzen, entlassen, befreien
release factor Freisetzungsfaktor
releaser Auslöser
releasing hormone/release hormone/releasing factor/release factor Freisetzungshormon, Freisetzungsfaktor, freisetzendes Hormon, freisetzender Faktor
releasing mechanism (RM) *ethol* Auslösemechanismus (AM)
 • **acquired releasing mechanism (ARM)** erworbener Auslösemechanismus (EAM)
 • **innate releasing mechanism (IRM)** angeborener Auslösemechanismus (AAM)
reliability Zuverlässigkeit
reliable zuverlässig
relic/relics Überbleibsel, Überrest
relict Relikt (>Reliktart)
relief Relief
remex (*pl* remiges) Flugfeder
 • **primary remex/primary feather (primaries/primary remiges)** Handschwinge, Hautschwinge, Handschwungfeder
 • **secondary remex/secondary feather (secondaries/secondary remiges)** Armschwinge, Armschwungfeder, Unterarmschwungfeder
remote sensing Fernerkundung
removal of side shoots/suckers *hort* Geizen, Ausgeizen
removing side shoots/removing suckers *hort* geizen, ausgeizen
renal die Niere betreffend, Nieren...
renal blood flow (RBF) Nierendurchblutung
renal calix (*pl* calices) Nierenkelch
renal capsule Nierenkapsel
renal column/column of Bertin/ columna renis Bertin Säule, Bertinsche Säule
renal corpuscle Nierenkörperchen, Malpighi-Körperchen, Malpighisches Körperchen
renal cortex Nierenrinde

renal hilus Nierenhilus, Nierenpforte, Nierenstiel
renal lobule Nierenlappen
renal medulla Nierenmark
renal papilla Nierenpapille
renal pelvis/pelvis of the kidney Nierenbecken
renal plasma flow (RPF) Nierenplasmadurchströmung
renal pyramid Nierenpyramide
renal threshold Nierenschwelle
renal tubule/kidney tubule Nierenkanälchen
renaturation/renaturing Renaturierung
- **annealing/reannealing/ reassociation (of DNA)** Renaturierung, Annealing, Reannealing, Reassoziation, Doppelstrangbildung

renature renaturieren
renette cell (nematodes) H-Zelle
renette gland/ventral gland (nematodes) Ventraldrüse
renewal bud Erneuerungsknospe
reniform/kidney-shaped nierenförmig
renin (angiotensinogen >angiotensin) Renin
rennet Labferment
rennet-stomach/fourth stomach/ reed/abomasum Labmagen, Abomasus
rennin/lab ferment/chymosin Rennin, Labferment, Chymosin
Rensch's rule Rensch'sche Haarregel
reorient/reorientate umstimmen
reorientation Reorientierung, Umorientierung; *physiol* Umstimmung
repair *n gen* Reparatur
- **dark repair/light independent DNA repair** lichtunabhängige DNA-Reparatur
- **excision repair** Exzisionsreparatur
- **light repair** Lichtreparatur

repair enzyme Reparaturenzym
repair grafting/bridge grafting *hort* Wundüberbrückung, Überbrückung
repair mechanism *gen* Reparaturmechanismus
repand (slightly uneven and waved margin) leicht gewellt, geschweift, randwellig
repatriation Wiedereinbürgerung
repeat/repetition (of a sequence) *gen* Wiederholung, Sequenzwiederholung
- **direct repeats** direkte Sequenzwiederholungen
- **dispersed/interspersed repeats** verstreut liegende Sequenzwiederholungen
- **head-to-head repeats** Kopf-an-Kopf Wiederholungen
- **indirect repeats** indirekte Sequenzwiederholungen
- **inverted repeat** invertierte/ gegenläufige Sequenzwiederholung
- **long terminal repeats** lange terminale Sequenzwiederholungen
- **tail-to-tail repeats** Schwanz-an-Schwanz Wiederholungen
- **tandem repeats** Tandemwiederholungen

repel/deter abschrecken, abstoßen
repellent/deterrent *n* Repellens Schreckstoff, Abschreckstoff
repellent/deterrent *adj/adv* abschreckend, abstoßend; widerlich
repent/creeping/crawling kriechend (am Boden entlang/an Nodien bewurzelnd)
replacement vector Substitutionsvektor
replant/transplant verpflanzen
replica *micros* Abdruck (Oberflächenabdruck: *EM*)
replica plating *micb* Replikaplattierung, Stempel-Methode
replication Replikation
- **bidirectional r.** bidirektionale Replikation
- **discontinuous r.** diskontinuierliche Replikation
- **dispersive r.** disperse Replikation
- **overreplication** Überreplikation
- **rolling circle r.** Rollender-Ring-Replikation
- **saltatory r.** saltatorische Replikation
- **semiconservative r.** semikonservative Replikation
- **semidiscontinuous r.** semidiskontinuierliche Replikation

replication bubble/replication eye Replikationsblase
replication fork Replikationsgabel
replication origin/origin or replication (ori) Replikationsursprung, Replikationsstartpunkt
replication slippage/slipped strand mispairing/slippage replication Fehlpaarung durch Strangverschiebung
replicative form *gen* replikative Form
replicon/unit of replication Replikon, Replikationseinheit

replisome Replisom, Replikationskomplex
replum *bot* Rahmen (Scheidewand bei Cruciferen)
reporter gene Reportergen
repot (a plant) *hort* umtopfen
repress unterdrücken, hemmen, reprimieren
repression Repression, Unterdrückung, Hemmung; Reprimierung
reproduce reproduzieren, vervielfältigen; vermehren (Nachkommen produzieren)
reproducibility Reproduzierbarkeit
reproducible reproduzierbar, vervielfältigbar; nachvollziehbar (Ergebnisse)
reproduction/propagation Reproduktion, Vermehrung, Fortpflanzung
  • **asexual reproduction** ungeschlechtliche Fortpflanzung
  • **sexual reproduction** geschlechtliche Fortpflanzung
  • **vegetative reproduction** vegetative Fortpflanzung
reproductive cell Fortpflanzungszelle
reproductive cycle Fortpflanzungszyklus
reproductive organ Fortpflanzungsorgan, Vermehrungsorgan
reproductive polyp/gonozooid Geschlechtspolyp, Gonozoid, Gonozooid
reproductive rate Fortpflanzungsrate, Vermehrungsrate
reproductive system Fortpflanzungsorgane
reptile-like dinosaurs/lizard-hipped dinosaurs/saurischian reptiles/saurischians/Saurischia Echsenbecken-Dinosaurier
reptiles/Reptilia Reptilien, Kriechtiere
repugnant substance unangenehmer, abweisender Geruchsstoff
repugnatorial gland Stinkdrüse (*siehe:* Wehrdrüse)
research *vb* forschen, untersuchen, wissenschaftlich arbeiten
research *n* Forschung, Forschungsarbeit, Untersuchung
  • **basic research** Grundlagenforschung
  • **cancer research** Krebsforschung
  • **interdisciplinary research** interdisziplinäre Forschung

research scientist/natural scientist Naturforscher, Naturwissenschaftler
researcher Forscher, Wissenschaftler
Resedaceae/mignonette family Resedagewächse, Resedengewächse, Waugewächse
resemblance/similarity Gleichartigkeit, Ähnlichkeit
resemble/be similar sich gleichen, gleichartig sein, ähneln
reserve/nature reserve Reservat, Naturreservat
reserve material/storage material/food reserve Reservestoff, Nahrungsreserve
reserve volume Reservevolumen
reservoir *electrophor* Reservoir, Wanne (Pufferwanne)
reservoir/ampulla (Mastigophora) Geißelsäckchen, Ampulle, Schlund
reservoir host Reservoir-Wirt
resettlement/reestablishment Wiederbesiedlung
residence time Verweilzeit, Verweildauer
residual *adj/adv* übrig, zurückbleibend, Rest…
residual/residuum *n math* Rest, Restbetrag, Restwert, Differenz
residual body Residualkörper
residual urine Restharn
residual volume Residualvolumen
residue Rest, Überbleibsel; *chem* Rückstand
  • **invariant residue** *math* unveränderter Rest, invarianter Rest
  • **variable residue** *math* variabler Rest
residue/rest (amino acid side chain) Rest
resilience *ecol* Wiederherstellung des biol. Gleichgewichts
resilium (flexible horny hinge) Resilium, Schließknorpel
resin Harz
resin acids Harzsäure
resin canal/resin duct Harzkanal, Harzgang
resin gall Harzgalle
resiniferous harzabsondernd
resinification/becoming resinous Verkienung, Verharzung
resinous/resiny harzig, kienig, harzreich, Harz…
resinous gum Gummiharz
resinous pinewood Kien, Kienholz
resist resistieren, widerstehen, ausdauern
resistance/resistivity/hardiness Resistenz, Beständigkeit, Widerstandsfähigkeit

resistance *electr* Widerstand
resistance factor/R factor Resistenz-Faktor, R-Faktor
resistance gene Resistenzgen
resistance tissue (fruit) Widerlagergewebe
resistant/resistive resistent, beständig, widerstandsfähig, widerstehend; abweisend
resistivity spezifischer Widerstand
resolution *chromat* Trennschärfe
• **optical resolution** optische Auflösung
resolve *opt* auflösen
resolving power *opt* Auflösungsvermögen
resonance pouch (frogs: vocal pouch) Schallblase
resorb resorbieren, aufnehmen, aufsaugen
resorption Resorption
resources Rohstoffe
• **natural resources** natürliche Rohstoffe
• **nonrenewable resources** nichterneuerbare Rohstoffe
• **regenerating resources** nachwachsende Rohstoffe
• **renewable resources** erneuerbare Rohstoffe
respiration/breathing Respiration, Atmung
respiratory center Atemzentrum
respiratory chain/electron transport chain Atmungskette, Elektronentransportkette, Elektronenkaskade
respiratory epithelium Respirationsepithel, respiratorisches Epithel, Atmungsepithel
respiratory pigment Atmungspigment
respiratory poison Atmungsgift
respiratory quotient respiratorischer Quotient, Atmungsquotient
respiratory tree (sea cucumbers) baumartig verzweigte Wasserlunge der Seewalzen
respirometry Spirometrie
respond/react wirken, Wirkung zeigen, reagieren, anschlagen, ansprechen auf
response (to stimulus) *neuro* Antwort
response/reaction *ethol/med/stat* Wirkung, Reaktion
• **conditioned response** *ethol* bedingte Reaktion
• **unconditioned response** *ethol* unbedingte Reaktion
rest/lie dormant *vb* ruhen

rest/residue *n chem/biochem* **(amino acid side chains)** Rest (z.B. Aminosäuren-Seitenkette)
resting/quiescent/dormant ruhend
resting bud/dormant bud ruhende Knospe, schlafende Knospe
resting egg/dormant egg (winter egg) Latenzei, Dauerei
resting nucleus Ruhekern
resting period/quiescent period/ dormancy period Ruhephase, Ruheperiode
resting posture Ruhestellung
resting potential Ruhepotential
resting spore Dauerspore
resting traces/cubichnia *paleo* Ruhespuren
restio family/Restionaceae Restiogewächse
restitute restituieren, wiederherstellen
restitution Restitution, Wiederherstellung
restriction endonuclease Restriktionsendonuclease
restriction enzyme Restriktionsenzym
restriction fragment length polymorphism (RFLP) Restriktionsfragment-Längenpolymorphismus
restriction site Restriktionsschnittstelle
resupinate um 180° gedreht
resupination (inversion) Resupination
resurrection plant Auferstehungspflanze, Wiederauferstehungspflanze
ret/retting *vb* rösten, rötten (Flachsrösten)
retained water Haftwasser
retard retardieren, verzögern, verlangsamen, zurückbleiben (Entwicklung/geistig)
retardation Retardierung, Verzögerung, Verlangsamung, Entwicklungshemmung
retarded/stunted (growth) zurückgeblieben
rete mirabile Wundernetz
retene Reten
retention time Retentionszeit, Verweildauer, Aufenthaltszeit
reticulate/netted/meshy vernetzt, netzförmig; netznervig
reticulate evolution netzartige Evolution
reticulate venation/net venation/ netted venation Netznervatur, Netzaderung
reticulately veined netzadrig

287

rhoptry **R**

reticulocyte/proerythrocyte (immature RBC) Retikulozyt, Reticulocyt, Proerythrozyt
reticulopodium/reticulopod Reticulopodium, Retikulopodium
reticulum/honeycomb stomach/ honeycomb bag/second stomach Retikulum, Netzmagen, Haube
retina Retina, Netzhaut
retinal/retinene Retinal
retinal cup eye/everted eye Blasenauge
retinic acid Retinsäure
retinol (vitamin A) Retinol
retinular cell Retinulazelle
retort Retorte
retractile einziehbar, zurückziehbar
retractor muscle Retraktormuskel, Retraktor, Rückzieher, Rückwärtszieher
retreat/refuge/hideout/hideaway/ hiding place Versteck, Unterschlupf
retreat *arach* Rückzug, Versteck, Schlupfwinkel, Retraite
retrix (*pl*** retrices)** Steuerfeder
retrocerebral organ Retrocerebralorgan, Retrocerebralkomplex
retrocerebral sac Retrocerebralsack
retroelement Retroelement
retrogene Retrogen
retrogressive development/ retrogressive evolution Rückentwicklung
retrorse rückwärts gerichtet, nach unten gerichtet, nach unten gebogen
retrotransposon Retrotransposon
retroviral retroviral
retrovirus Retrovirus
  ● **acute transforming retrovirus** akut transformierendes Retrovirus
retuse (leaf apex) eingebuchtet
reusable wiederverwendbar
reuse *n* Wiederverwendung
reuse *vb* wiederverwenden
reverberating circuit *neuro* zurückwirkender Schaltkreis
reversal potential Umkehrpotential
reverse genetics reverse Genetik
reverse osmosis Reversosmose, Umkehrosmose
reverse phase Reversphase, Umkehrphase
reverse transcriptase reverse Transkriptase, Revertase, Umkehrtranskriptase
reverse transcription reverse Transkription
reverse translation reverse Translation
reverse turn *gen* Umkehrschleife

reversibility Reversibilität, Umkehrbarkeit
reversible reversibel, umkehrbar
reversible inhibition reversible Hemmung
revival/resuscitation Wiederbelebung
revive/resuscitate wiederbeleben
revolute/rolled backward nach hinten eingerollt, zurückgerollt
reward *n* (e.g. nectar) Belohnung
reward *vb* belohnen
Reynolds number Reynold'sche Zahl, Reynolds-Zahl, Reynoldsche Zahl
RFLP (restriction fragment length polymorphism) RFLP (Restriktionsfragment-längenpolymorphismus)
rhabdite (turbellarians) Rhabdit, Epithelstäbchen
rhabdome Rhabdom
rhabdomere Rhabdomer
rhagon (sponges) Rhagon
Rhamnaceae/coffeeberry family/ buckthorn family Kreuzdorngewächse
rheas/Rheiformes Nandus
rhinarium Rhinarium, Riechplatte
rhipidium (fan-shaped cyme) Fächel
rhipidoglossate radula rhipidoglosse Radula, Fächerzunge
rhithral *limn* Rhithral, Salmonidenregion
rhizodermis/epiblem(a) Rhizodermis, Wurzelepidermis
rhizoid/rootlet *n* Rhizoide, Würzelchen
rhizoid/rootlike *adj/adv* wurzelartig
rhizomatous tuber Rhizomknolle
rhizome/rootstock (creeping underground stem) Rhizom, Erdspross, Erdausläufer, Wurzelstock
Rhizophoraceae/mangrove family/ red mangrove family Mangrovengewächse
rhizosphere Rhizosphäre
rhizostome medusas/Rhizostomeae Wurzelmundquallen
rhodopsin/rose-purple Rhodopsin, Sehpurpur
rhombic/rhomboid rhombisch, rautenförmig
rhombogen (mesozoans) Rhombogen
rhopalium (tentaculocyst) Rhopalium, Randsinnesorgan, Randkörper, "Hörkölbchen"
rhopaloneme Rhopalonema
rhoptry (*pl*** rhoptries)** Rhoptrie (*pl* Rhoptrien)

rhynchocephalians/Rhynchocephalia (Sphenodon) Brückenechsen
rhytidome/tertiary bark/dead outer bark Rhytidom, Borke
ria coast/ria shoreline Riasküste
rib/costa zool Rippe, Costa
- **abdominal rib/gastralia (reptiles)** Bauchrippe, Gastralrippe, Gastralia
- **cervical rib/costa cervicalis** Halsrippe
- **false rib/costa spuria** falsche Rippe, unechte Rippe
- **floating ribs/costae fluitantes** frei endende Rippen
- **sacral rib** Sakralrippe
- **thoracic rib/costa thoracalis** Thorakalrippe
- **true rib/costa vera** echte Rippe

rib/vein (leaf) bot Rippe, Ader, Nerv
- **midrib/midvein/costa** Mittelrippe, Costa (mittlere/zentrale Blattrippe)

rib basket/rib cage (chest/thorax) Brustkasten, Brustkorb (Thorax)
rib meristem/file meristem Rippenmeristem
riboflavin/lactoflavin (vitamin B$_2$) Riboflavin, Lactoflavin
ribonuclear protein Ribonukleoprotein
ribonuclease Ribonuclease, Ribonuklease
ribonucleic acid (RNA) Ribonucleinsäure, Ribonukleinsäure (RNS/RNA)
riboprobe Ribosonde, RNA-Sonde
ribosomal RNA/rRNA ribosomale RNA, rRNA
ribosome Ribosom
ribosome binding site Ribosomenbindungsstelle
ribozyme Ribozym
rich meadow/pasture Fettwiese
ricinuleids/tick spiders/Ricinulei/Podogona Kapuzenspinnen
rickettsias Rickettsien
rickety/rachitic rachitisch; schwach, wackelig
rictal die Schnabelinnenseite betreffend, Schnabel...
rictal bristle Schnabelborsten
rictus orn Sperrweite (bzgl. Schnabelinnenseite)
ride vb (e.g. horse) reiten
ridge (range of hills/mountains) Hügelkette, Bergkette, Gebirgskette
ridge hort Hügelbeet
- **to ridge up** anhäufeln

ridge/crest Gebirgskamm, Berggrat, Bergrücken
ridge soaring/slope soaring aer/orn Hangsegeln

ridgeling/ridgling (cryptorchid stallion) Spitzhengst, Klopphengst (Pferd mit Kryptorchismus)
right side-out vesicle Right side-out Vesikel (Vesikel mit der richtigen Seite nach außen)
right-handed/dextral rechtshändig (Spirale: rechtsgängig)
rigor mortis Totenstarre, Leichenstarre
rill/bog drainage Rülle, Bächlein, Rinnsal
rima/cleft/crack/fissure Ritze, Spalt, Spalte, Furche; Stimmritze
- **palpebral fissure/rima palpebrarum** Lidspalte
- **pudendal fissure/rima pudendi** Schamspalte
- **rima glottidis** Stimmritze (zw. Stimmlippen und Aryknorpeln des Kehlkopfs)
- **rima oris/opening of the mouth** Mundspalte

rimate/having fissures rissig
rime/crust/incrustation Kruste
rime meteo fest aufgefrorener Reif, Raureif, Raufrost
rimiform soralium/fissoral soralium (lichens) Spaltensoral
rind bot Rinde, Baumrinde, Borke; Fruchtwand, Fruchtschale
rind grafting/bark grafting hort Rindenpfropfung, Pfropfen hinter die Rinde
ring (for support stand/ring stand) lab Stativring
ring/inferior annulus (remains of velum partiale) fung Ring, Kragen, Annulus inferus
ring budding/annular budding (flute budding) Ring-Okulation
ring canal/radial canal Ringkanal, Radiärkanal, Ambulakralring
ring chromosome Ringchromosom
ring cleavage chem Ringspaltung
ring form/ring conformation Ringform
ring formation/cyclization chem Ringschluss
ring formula chem Ringformel
ring muscle/circular muscle Ringmuskel
ring porous (wood) ringporig (cyclopor)
ring species Ringart
Ringer's solution Ringer-Lösung, Ringerlösung
ringing (by browser) for Rundfraß (durch Wild an Bäumen)
ringing/girdling (tree bark) Ringelung, Gürteln, Rundfraß
ringlike/annular ringartig

rod neutrophil **R**

rip *mar* Kabbelung
rip current *mar*
Brandungsrückströmung, Rippstrom,
Reißstrom
rip tide *mar* Ripptide
riparian/riparious/riparial
uferbewohnend, am Ufer lebend
(Flussufer), Ufer...
riparian forest Uferwald
ripe/mature reif
ripeness/maturity Reife
ripple *geol* Rippel
ripple mark Rippelmarke
risk/danger Risiko (*pl* Risiken),
Gefahr
risk class/security level
Sicherheitsstufe, Risikostufe
risk factor Risikofaktor
rite Ritus
rival Rivale; Nebenbuhler
rivalry Rivalität
river blindness/onchocercosis
(*Onchocerca volvulus*)
Flussblindheit, Onchocercose
river bluff Steilufer, Felsufer/
Felswand am Fluss
river mouth Flussmündung
river plain/river valley
Flussniederung, Flusstal
riverbank/bank/embankment
Flussufer, Uferböschung,
Flussböschung, Böschung
riverbed Flussbett
riverbed filtration Uferfiltration
riverine Flüsse/Fließgewässer
betreffend, Fluss...
riverine floodplain Aue, Flussaue
riverine floodplain meadow
Auwiese, Auenwiese,
Flussauenwiese
riverweed family/Podostemaceae
Blütentange
rivulet/streamlet/rill Rinnsal,
Bächlein; Rülle
RNA (ribonucleic acid) RNA, RNS
(Ribonucleinsäure/Ribonukleinsäure)
● 3′→5′ (three prime five prime/
three prime to five prime)
3′→5′ (drei Strich-fünf Strich/
drei Strich nach fünf Strich)
● antisense RNA Antisense-RNA,
Anti-Sinn-RNA, Gegensinn-RNA
● guide RNA Guide-RNA
● mRNA/messenger RNA mRNA,
Boten-RNA
● pre-mRNA Prä-mRNA
● pre-rRNA Prä-rRNA,
Vorläufer-rRNA
● pre-tRNA Prä-tRNA,
Vorläufer-tRNA
● RNA editing RNA-Editing,
Redigieren von RNA

● rRNA/ribosomal RNA rRNA,
ribosomale RNA
● snRNA/small nuclear RNA
snRNA, kleine nucleäre-RNA
● stable RNA stabile RNA
● tRNA/transfer RNA tRNA,
Transfer-RNA
RNA editing Redigieren von RNA
RNA polymerase RNA-Polymerase
RNA priming RNA-Priming
RNA processing RNA-Processing,
RNA-Weiterverarbeitung
RNA transcript RNA-Transkript
RNA-world RNA-Welt
RNase (ribonuclease) RNase
(Ribonuclease/Ribonuklease)
roar *n* Brüllen, Gebrüll
roar/bellow *vb* brüllen
rock Stein, Gestein, Fels
● bedrock/rock base/parent rock
Ausgangsgestein, Grundgestein,
Muttergestein
● evaporites Evaporite,
Eindampfungsgesteine
● extrusive rocks Extrusivgesteine,
Effusivgesteine, Ergussgesteine,
Ausbruchsgesteine
● gangue rock (dikes etc.)
Ganggestein
● igneous rock Erstarrungsgestein,
Eruptivgestein
● intrusive rocks Intrusivgesteine
● metamorphic rock
Umwandlungsgesteine
● parent rock/bedrock/rock base
Muttergestein, Ausgangsgestein,
Grundgestein
● primary rock/primitive rock
Urgestein
● sedimentary rock
Sedimentgestein, Absetzgestein,
Schichtgestein
rock base/bedrock Grundgestein
rock debris/loose stones Geröll
rock plant/petrophyte Felspflanze
rock-brake fern family/parsley fern
family/Cryptogrammaceae
Rollfarngewächse
rocket immunoelectrophoresis
Raketenimmunelektrophorese
rockpool/lithotelma Felstümpel,
Gesteinstümpel, Lithotelme
rockrose family/Cistaceae
Cistrosengewächse,
Zistrosengewächse,
Sonnenröschengewächse
rocky steinig, felsig
rod/rod cell (eye) Stäbchen,
Stäbchenzelle
rod neutrophil/band neutrophil/
stab neutrophil/stab cell
stabkerniger Neutrophil

rod organ

rod organ (*Peranema*) Staborganell
rod-shaped/rod-like stabförmig, rutenförmig
rodent *adj/adv* nagend
rodents/gnawing mammals (except rabbits)/Rodentia Nagetiere
rods/bacilli Stäbchen, Stäbchenbakterien, Bazillen
roe (crustaceans: lobster eggs) Eier
roe (fish eggs esp. enclosed in ovarian membrane) Rogen (Fischeier innerhalb der Eierstöcke), Fischlaich
• **soft roe/milt (testes of a fish)** Fischhoden (*see also:* milt)
rogue *adj/adv zool* bösartig, zerstörerisch; *Pferd:* ungezogen, unartig, bockend
rogue *n bot/zool* aus der Art schlagendes Individuum; bösartiger Einzelgänger
rogue *vb bot/agr* minderwertige/kranke/missgebildete/schwache Pflanze ausjäten
rogue elephant bösartiger Einzelgänger (Elefant)
roll gall Rollgalle
rolled backward/revolute zurückgerollt
roller tube culture Rollflaschenkultur
rolling-circle replication *gen* Rollender-Ring-Replikation
rookery (breeding ground: herons/penguins/seals; *also:* **colony of such animals)** Nistplatz, Brutplatz; Brutkolonie (Seevögel/Robben)
roost/perch *n* **(resting site/lodging site)** Ruheplatz, Rastplatz, Schlafplatz, Schlafsitz, Unterkunft (Geflügel: Hühnerstange/Hühnerstall)
roost/perch *vb* rasten, sich setzen, sich zur Rast/zum Schlaf niederlassen, auf der Stange sitzen, sich zum Schlafen niederhocken (Hühner)
root/grub (boar: dig up with snout) *vb zool* mit der Schnauze (auf)wühlen (Schwein)
root *vb bot* bewurzeln, Wurzeln treiben/schlagen
root *n* Wurzel
• **adventitious root** Adventivwurzel, Luftabsenker
• **air root/aerating root/pneumatophore** Atemwurzel, Pneumatophore
• **anchorage root/adhesion root** Ankerwurzel

• **anterior root/ventral root/motor root** *neuro* Ventralwurzel, motorische Wurzel
• **assimilative root** Assimilationswurzel
• **brace root/prop root (e.g. corn)** Stützwurzel
• **buttress root (*esp. tropical trees*)** Brettwurzel
• **central cylinder of root** Wurzelstele
• **contractile root** Zugwurzel
• **dorsal root/posterior root** *neuro* Dorsalwurzel
• **feeder root** Nährwurzel
• **fibrous root system** Büschelwurzelsystem
• **hair root** Haarwurzel
• **holdfast root** Haftwurzel
• **lateral root** Seitenwurzel
• **motor root/anterior root/ventral root** *neuro* motorische Wurzel, Ventralwurzel
• **pneumatophore/air root/aerating root** Pneumatophore, Atemwurzel
• **posterior root/dorsal root** *neuro* Dorsalwurzel
• **primary root** Primärwurzel, Hauptwurzel
• **prop root/stilt root** Stützwurzel, Stelzwurzel
• **radicle/radicula/embryonic root** Radicula, Keimwurzel, Hauptwurzelanlage
• **region of maturation of root** Wurzelhaarzone
• **secondary root** Sekundärwurzel, Nebenwurzel, Seitenwurzel
• **seminal root** Keimwurzel
• **shallow root/surface root** Flachwurzel, Oberflächenwurzel
• **storage root** Speicherwurzel
• **suction root/sucking root** Saugwurzel
• **take root** *vb* anwachsen, bewurzeln, anwurzeln
• **ventral root/anterior root/motor root** *neuro* Ventralwurzel, motorische Wurzel
root apex/root tip Wurzelspitze, Wurzelpol (Embryo)
root bud/root sucker/tiller Stockausschlag, Stockreis
root bulbil Wurzelbulbille
root cap/calyptra Wurzelhaube, Wurzelhäubchen, Kalyptra
root climber Wurzelkletterer, Wurzelklimmer
root collar Wurzelhals
root crop/root vegetable Hackfrucht, Wurzelgemüse

rough-leaved

root crown Wurzelhals, Wurzelkrone
root cutting Wurzelsteckling
root gall/root knot (nematodes)
  Wurzelgalle
root graft Wurzelpfropf
root grafting Wurzelpfropfung,
  Wurzelveredlung
root hair Wurzelhaar,
  Wurzelhärchen
root mean square (RMS) *stat*
  quadratisches Mittel, Quadratmittel
root nodules Wurzelknöllchen
root pressure Wurzeldruck
root primordium Wurzelanlage
root rot Wurzelfäule
root sheath/radicle sheath/
  coleorhiza Wurzelscheide,
  Koleorhiza, Coleorhiza
root sucker/offshoot/offset/slip
  wurzelbürtiger Spross,
  Wurzelausschlag, Wurzeltrieb,
  Wurzelschössling, Wurzelreis
root tendril Wurzelranke
root trace Wurzelspur
root water tension
  Wurzelsaugspannung
root-collar rot Wurzelhalsfäule
root-collar shoot/sucker/offshoot
  Wurzelhalsschössling
root-tuber/tuberous root
  Wurzelknolle
rootage (system of roots)
  Wurzelwerk
rootage/rooting/radication
  Bewurzelung
rootless/arrhizous/arrhizal
  wurzellos
rootlet Würzelchen
rootlike/radiciform/rhizoid
  wurzelförmig, wurzelartig
rootstock/caudex Wurzelstock,
  Caudex
rootstock/rhizome (creeping
  underground stem) Erdspross,
  Rhizom, Erdausläufer
rootstock/stock (base for grafting)
  Wurzelpfropfgrundlage,
  Wurzelpfröpfling, Unterlage
rose family/Rosaceae
  Rosengewächse
rose-purple/rhodopsin Sehpurpur,
  Rhodopsin
rosehip/hip Hagebutte
rosette (whorl) Rosette
  (Wirtel/Quirl)
rosette of leaves/whorl of leaves
  Blattrosette, Blattwirtel
rosette plant Rosettenpflanze
rosette plate (bryozoans)
  Rosettenplatte
ross/remove bark/debark/bark/
  decorticate schälen, entrinden

rostellum (hooked prominence on
  head of tapeworm) *zool*
  Rostellum, Hakenkranz
rostellum/adhesive body (part of
  gynostemium) *bot* Rostellum,
  Klebkörper
rostral bone/os rostrale
  Rüsselbein
rostral plate/planum rostrale (pigs)
  Rüsselscheibe
rostrate/beaked geschnäbelt
rot/foul/putrefy/decompose/decay/
  disintegrate *vb* faulen, verfaulen,
  verwesen, modern, vermodern, sich
  zersetzen
rot/mold/mildew/blight Fäule
rotary evaporator
  Rotationsverdampfer
rotary microtome
  Rotationsmikrotom
rotary shadowing
  Rotationsbedampfung
rotary shaker Rundschüttler
rotating stage *micros* Drehtisch
rotational grazing
  Umtriebsbeweidung
rotational motion
  Rotationsbewegung
rotational sense/sense of rotation
  Rotationssinn, Drehsinn
rotator muscle Rotator, Drehmuskel,
  Dreher
rotenone Rotenon
rotifers/Rotifera Rotatorien,
  Rädertiere
rotiger/pseudotrochophore (larva)
  Rotiger, Pseudotrochophora
rotor *centrif* Rotor
  ● **angle rotor/angle head rotor**
    centrif Winkelrotor
  ● **fixed-angle rotor** *centrif*
    Festwinkelrotor
  ● **swinging-bucket rotor** *centrif*
    Ausschwingrotor
  ● **vertical rotor** *centrif*
    Vertikalrotor
rotor-stator impeller/Rushton-
  turbine impeller (bioreactor)
  Rotor-Stator-Rührsystem
rotten/decayed/decomposed
  verwest, vermodert, verfault,
  zersetzt
rotting/decaying/putrefying/
  decomposing moderig, faulend,
  verfaulend
rotting process Fäulnisprozess
rotund/rounded rundlich,
  abgerundet
rotundifolious/with rounded leaves
  rundblättrig
rough-leaved/trachyphyllous
  raublättrig

roughage Raufutter; unverdauliche Nährstoffe (undigestible components of diet)
roughened/scabrid aufgeraut
round (rounded/rotund) rund (abgerundet, rundlich)
round dance (bees) Rundtanz
round window/oval window (ear) rundes Fenster, ovales Fenster
rounded/roundish/rotund rundlich, abgerundet
roundwood/log timber Rundholz
roundworms/nematodes/Nematoda Rundwürmer (*sensu stricto*), Fadenwürmer, Nematoden
royal fern family/flowering fern family/cinnamon fern family/ Osmundaceae Königsfarngewächse, Rispenfarngewächse
royal jelly/bee milk Königin-Futtersaft, Gelée Royale
RPD (redox potential-discontinuity) *mar* Redoxpotential Diskontinuität
RT-PCR (reverse transcriptase-PCR) RT-PCR (PCR mit reverser Transkriptase)
RTLV-family (RTLV=reverse transcriptase-like virus) (DNA-element) RTLV-Familie (retrovirusartiges Element) (DNA-Element)
rubble/debris/detritus Schutt; Gesteinsschutt (= coarse rock debris)
rudder (tail fluke of whales) Ruder, Ruderflosse (Schwanzflosse der Wale)
ruderal/growing among rubbish or debris ruderal, auf Schutt wachsend
ruderal plant Ruderalpflanze, Schuttpflanze
rudiment (*sensu lato*: vestige) Rudiment
rudimentary (*sensu lato*: vestigial) rudimentär
rue family/Rutaceae Rautengewächse
ruff/ruffle *n* (feathers/hair around neck) *orn/mammals* Krause, Halskrause
Ruffini's endings/Ruffini's organ/ corpuscles of Ruffini Ruffini'sches Körperchen
ruffle/ruff *n* (feathers/hair around neck) *orn/mammals* Krause, Halskrause
ruffle *vb zool* sträuben (Federn/Haare), sich aufplustern (Vögel)
ruffle/ruffled *vb bot* (strongly wavy leaf margin) kräuseln, gekräuselt (Blatt)

rugate/rugose/wrinkled/wrinkly (corrugative/corrugated) runzelig, gerunzelt (gewellt/geriffelt), faltig
rugulose/finely wrinkled feinrunzelig
rumen/paunch/ingluvies/ first stomach Rumen, Pansen
• **atrium ruminis** Pansenvorhof, Schleudermagen
ruminal pillar Pansenpfeiler
ruminant/"cud chewers" Wiederkäuer, Retroperistaltiker, Ruminantier
ruminants/Ruminantia Wiederkäuer
rump/hindquarters (quadrupeds) Rumpf, Hinterteil, Steiß
rump/tail/uropygium *orn* Bürzel
rump patch (deer) Spiegel
rumposome Rumposom
run *n* (fish) Laichwanderung
run *vb* rennen, laufen
runaway effect *gen* Selbstläufer-Effekt
runcinate/retroserrate/hook-backed (e.g. dandelion leaf) schrotsägeförmig, rückwärts gesägt
runner/sarment oberirdischer Ausläufer, photophiler Ausläufer, Kriechspross
runner/sucker/offshoot Ableger, Ausläufer
running/cursorial rennend, Renn...
running free/free-running/free- ranging (fowl etc.) freilaufend
running gel/separating gel Trenngel
running-step (gait of horses) Tölt
runoff Abfluss (oberflächlich abfließend), Abschwemmung
runt (puny) Zwergtier, Zwerg... (z.B. Zwergrind/Zwergschwein)
rupestrine/rupicoline/rupicolous (living/growing on/among rocks) auf Steinen/Felsen lebend (zwischen Steinen wachsend)
rupicaprine/chamois-like gemsartig
rupicolous/rupicoline/rupestrine (living/growing on/among rocks) auf Steinen/Felsen lebend (zwischen Steinen wachsend)
Ruppiaceae/ditch-grass family Saldengewächse
rural ländlich
rush *n bot* Simse
rush family/Juncaceae Simsengewächse
rush-shaped/junciform binsenförmig
rushy/rushlike/juncaceous binsenartig

rynchocephalians **R**

Russula family/Russulaceae
Sprödblätterpilze, Sprödblättler
rusts/rust fungi/Uredinales
Rostpilze
rut *n* **(male)** Brunst (männliche)
rut/courting *n* (*spec* **deer>stag)**
Brunft (Hirsch)
rut *vb* **(male)/court** brunsten,
brunften
rut/mate/copulate *vb* rammeln,
kopulieren
Rutaceae/rue family
Rautengewächse

rutting (male)/in heat (female)/
sexually aroused brünstig,
in der Brunst, geschlechtlich erregt
rutting behavior
Brunstverhalten
rutting season/rutting time/
courting season/mating season/
season of heat Brunstzeit,
Paarungssaison; (*spec:* deer>stag)
Brunftzeit, Paarungssaison (Hirsch)
rynchocephalians/sphenodontids/
Rynchocephalia/Sphenodonta
Schnabelköpfe

S phase

S phase (synthesis phase during cell cycle) S-Phase (Synthesephase im Zellzyklus)

sabre tooth/saber tooth Säbelzahn

sabulicolous/arenicolous/ psammobiontic/living in sand im Sand lebend, im Sand wohnend

sac fungi/cup fungi/ascomycetes/ Ascomycetes Schlauchpilze

saccade Sakkade, Blicksakkade, Blickbewegung

saccadic sakkadisch, ruckartig, stoßartig, ruckartig unterbrochen

saccate/pouched sackförmig, taschenförmig

saccharic acid/aldaric acid (glucaric acid) Zuckersäure, Aldarsäure (Glucarsäure)

sacchariferous zuckerhaltig; zuckerbildend

saccharification Verzuckerung

saccharify verzuckern

saccharimeter Saccharimeter

saccharinic acid Saccharinsäure

saccharose/sucrose/table sugar/ beet sugar/cane sugar Saccharose, Rübenzucker, Rohrzucker

sacfry/yolk fry/alevin (salmonid larvae with yolk sac) Dottersackbrut (Lachs)

sacoglossans/Sacoglossa/ Saccoglossa Sackschnecken, Schlauchschnecken, Schlundsackschnecken

sacral sakral, zum Kreuzbein gehörig

sacral rib Sakralrippe

sacral vertebra Sakralwirbel, Kreuzwirbel

sacrum/os sacrum Sakrum, Kreuzbein

saddle (horse etc.) Sattel

saddle (butchery) Rückenstück, Grat

saddle (male fowl) Bürzel

saddle feathers (male fowl) Bürzelfedern

saddle fungi & false morels/ Helvellaceae Lorchelpilze

saddle grafting *hort* Sattelschäften

saddle joint/sellaris joint Sattelgelenk

safe/secure *techn* **(personal protection)** sicher

safety/security *techn* **(personal protection)** Sicherheit
• **lab safety** Laborsicherheit

safety data sheet Sicherheitsdatenblatt

safety device Sicherheitsvorrichtung

safety goggles/safety spectacles Schutzbrille

safety guidelines Sicherheitsbestimmungen, Sicherheitsrichtlinien

safety line *arach* Sicherheitsfaden

safety measure Sicherheitsmaßnahme, Sicherheitsmaßregel

safety pipet filler Peleusball (Pipettierball)

safety precaution Sicherheitsvorkehrung, Sicherheitsvorbeugemaßnahme

safety regulations Sicherheitbestimmungen

safety spectacles/safety goggles Schutzbrille

sagittal/median longitudinal sagittal, in Pfeilrichtung, in Pfeilebene

sagittal crest/crista sagittalis Sagittalkamm, Scheitelkamm

sagittal section/median longisection Sagittalschnitt (parallel zur Mittelebene)

sagittal suture (skull) Pfeilnaht

sagittate/sagittiform/ arrowhead-shaped pfeilförmig, pfeilspitzenförmig

sail *n* **(e.g. Velellina medusas)** Segel

sail *vb* segeln

salamanders & newts and relatives/urodeles/Urodela/ Caudata Schwanzlurche (Salamander & Molche und Verwandte)

Salicaceae/willow family Weidengewächse

salicic acid (salicylate) Salicylsäure (Salicylat)

saline *adv/adj* salzig, salzhaltig, Salz...

saline lake Brackwassersee

saline *n* Salzlösung; physiologische Kochsalzlösung
• **isotonic saline** isotone Kochsalzlösung
• **phosphate buffered saline (PBS)** phosphatgepufferte Salzlösung

salinity/saltiness Salinität, Salzgehalt

salinization Versalzung

saliva Speichel

salivarium Salivarium, Speicheltasche

salivary gland Speicheldrüse

salmon & trout/Salmoniformes Lachse & Lachsverwandte

salps/Salpida (order) eigentliche Salpen

salps/thaliceans/Thaliacea (class) Salpen, Thaliaceen

295 saprophilic **S**

salt bridge (ion pair) Salzbrücke
(Ionenpaar)
salt flat Salzsteppe
salt gland Salzdrüse
salt in *vb* einsalzen
salt lake Salzsee
salt marsh Salzmarsch
salt out *vb* aussalzen
salt wedge *limn* Salzwasserkeil
**saltatory/saltatorial (adapted for or
used in jumping)** saltatorisch,
zum Springen geeignet, Sprung...,
Spring...
saltatory conduction
saltatorische Erregungsleitung
**saltatory leg/saltatorial leg/
jumping leg** Springbein, Sprungbein
saltatory movement
saltatorische Bewegung
saltatory replication
saltatorische Replikation
saltiness Salzigkeit
salting in Einsalzen, Einsalzung
salting out Aussalzen
salting-out chromatography
Aussalzchromatographie,
Aussalzchromatografie
saltmarsh Salzmarsch, Salzsumpf
saltpan/salina Salzpfanne
saltwater Salzwasser
**saltwort family/batis family/
Bataceae** Batisgewächse
salty/saline salzig, salzhaltig
Salvadoraceae/mustard-tree family
Senfbaumgewächse
salvage logging/salvage felling
Kalamitätennutzung (Holzernte)
salvage pathway Wieder-
verwertungsstoffwechselwege,
Wiederverwertungsreaktionen
salverform stieltellerförmig,
präsentiertellerförmig
salvinia family/Salviniaceae
Schwimmfarngewächse,
Schwimmfarne
SAM (S-adenosylmethionine)
S-Adenosylmethionin
samara/key (single-winged nutlet)
Flügelnuss
sample *chem/biochem* Probe
(Teilmenge eines zu untersuchenden
Stoffes)
sample *stat* Stichprobe
 • **random sample** *stat*
Zufallsstichprobe
 • **subsample** *stat* Teilstichprobe
sample function/sample statistic
stat Stichprobenfunktion
sample preparation
Probenvorbereitung
sample size Stichprobenumfang;
stat Fallzahl

sample-taking Probenahme,
Probeentnahme
sampling Stichprobenentnahme
sanctuary/sanctuary area
Schongebiet
 • **wildlife sanctuary/wildlife
refuge** Wildreservat
**sand dollars/true sand dollars/
Clypeasteroida** Sanddollars,
Schildseeigel
sandalwood family/Santalaceae
Sandelholzgewächse,
Leinblattgewächse
sandbank Sandbank
sandbar längliche Sandbank,
Sandriff, Sandbarre
sandflat Watt, Sandwatt
sandy sandig
sandy soil Sandboden,
sandiger Boden
**sanguivorous/hematophagous/
blood-sucking** blutsaugend,
sich von Blut ernährend
sanguineous blutig, Blut...,
Blut betreffend; blutrot
 • **consanguineous** blutsverwandt
Santalaceae/sandalwood family
Sandelholzgewächse,
Leinblattgewächse
sap Saft, Flüssigkeit
 • **plant sap (xylem & phloem fluid)**
Pflanzensaft
Sapindaceae/soapberry family
Seifenbaumgewächse
sapling *bot/hort* Sprössling,
Bäumchen
sapodilla family/Sapotaceae
Sapotegewächse, Breiapfelgewächse
saponification Verseifung
Sapotaceae/sapodilla family
Sapotegewächse,
Breiapfelgewächse
saprobe/saprobiont
Fäulnisbewohner (*pl* Saprobien)
saprobic/saprophilic/saprophytic
saprob, saprophil, saprophytisch,
von faulenden Stoffen lebend
saprobity Saprobie, Saprobität
saprobity system Saprobiensystem
saprogen *n*
fäulniserregendes Lebewesen
saprogenic saprogen,
fäulniserregend
saprophage/saprotroph/saprobiont
Saprophage, Saprovore,
Fäulnisernährer, Fäulnisfresser
saprophagous/saprotrophic
saprophag
saprophagy Saprophagie
saprophilic/saprophytic/saprobic
saprophil, saprophytisch, saprob,
von faulenden Stoffen lebend

saprophyte/saprobiont Saprophyt, Fäulnispflanze, Faulpflanze, Fäulnisbewohner
saprozoic lifeform Saprozoe
sapwood/alburnum/splintwood Splintholz
sarcolemma Sarcolemm, Sarkolemm
sarcoplasmatic reticulum (SR) sarkoplasmatisches Retikulum (SR)
sarcosine Sarcosin
sarcosome/sarcosoma (mitochondrion of striated muscle fiber) Sarcosom, Riesenmitochondrium
sarcotesta Sarkotesta
sarcotubular system sarkotubuläres System
sarment/runner *bot* Ausläufer, Kriechspross (oberirdisch/photophil)
sarmentose *bot* (oberirdische) Ausläufer bildend, kriechend
Sarraceniaceae/pitcher-plant family Schlauchpflanzengewächse, Krugpflanzengewächse
satellite band Satellitenbande
satellite DNA (sat-DNA) Satelliten-DNA
satellite species/marginal species Satellitenart, Randart
satellite virus Satellitenvirus
satiate sättigen
satiation Sättigung
saturate (saturated) sättigen (gesättigt)
saturated fatty acid gesättigte Fettsäure
saturation deficit Sättigungsverlust, Sättigungsdefizit
saturation kinetics Sättigungskinetik
saurischian reptiles/saurischians/ lizard-hipped dinosaurs/reptile-like dinosaurs/Saurischia Echsenbecken-Dinosaurier
sauropteryians/Sauropterygia Paddelechsenartige
Saururaceae/lizard's tail family Molchschwanzgewächse
savanna Savanne
Savi vesicle (*Torpedo:* around electric organ) Savisches Bläschen
sawfishes/Pristiformes Sägerochen, Sägefische
sawmill/timber mill Sägewerk
sawsharks/Pristiophoriformes Sägehaie
saxicolous auf oder zwischen Steinen wachsend, Stein besiedelnd, felsbewohnend
saxifrage family/Saxifragaceae Steinbrechgewächse

scab Schorf, Wundschorf, Grind
scab lesion (crustlike disease lesion) Schorfwunde
scabies/scab/mange (*Sarcoptes scabiei*) Krätze, Milbenkrätze, Räude, Scabies (Krätzmilbe)
scabious/scabby räudig
scabious family/teasel family/ Dipsacaceae Kardengewächse
scabrid/roughened aufgeraut
scabrous/scaly/rough schuppig, rau
scaffold/scaffolding/framework/ stroma/reticulum (web) Rahmen, Gerüst
scaffolding/framework/stroma/ reticulum Gerüst, Netzwerk
scalariform/ladder-shaped leiterförmig
scalariform vessel *bot* Leitertrachee
scale *bot*/*zool* Schuppe
- **antennal scale/scaphocerite** *entom* Antennenschuppe, Scaphocerit
- **bract-scale/secondary bract/ bracteole/bractlet** *bot* Tragschuppe, Deckschuppe, Brakteole, Tragblatt zweiter Ordnung/zweiten Grades
- **chaff/bracts (small dry scales)** *bot* Spreu, Kaff
- **cone scale/cone bract** *bot* Zapfenschuppe
- **coronal scale** Schlundschuppe
- **cosmoid scale** *ichth* Cosmoidschuppe
- **ctenoid scale** *ichth* Ctenoidschuppe, Kammschuppe
- **cycloid scale** *ichth* Cycloidschuppe, Rundschuppe
- **ganoid scale** *ichth* Ganoidschuppe, Schmelzschuppe
- **horn(y) scale** Hornschuppe (der Haut)
- **nectariferous scale** *bot* Honigschuppe
- **ovuliferous scale/seed scale** *bot* Samenschuppe, Fruchtschuppe
- **placoid scale/dermal denticle** *ichth* Placoidschuppe, Zahnschuppe, Hautzahn, Dentikel
- **ramentum/chaffy scale/palea/ pale** *bot* Spreuschuppe
- **scent scale/androconium (*pl* androconia)** Duftfeld, Duftschuppe, Androkonie
- **scute (enlarged scale)** große Schuppe
- **thoracic scale** Thorakalschüppchen
- **ventral scale** Ventralschuppe, Bauchschuppe

schistosomiasis

scale *phys/math* Skala (*pl* Skalen), Maßstab
- **interval scale** Intervallskala
- **laboratory scale/lab scale** Labormaßstab
- **metric scale** metrische Skala
- **nominal scale** Nominalskala
- **ordinal scale** Ordinalskala
- **ranking scale** ranking scale (*used as such: not translated*)
- **rating scale** rating scale (*used as such: not translated*)
- **ratio scale** Verhältnisskala, Ratioskala

scale (weight)/balance (mass) Waage
scale insects/Coccinea Schildläuse
scale-leafed schuppenblättrig
scale-like/scutate schuppenartig
scale-like bracts/scale leaves/ bracteole/bractlet Schuppenblätter
scale-shaped/scutiform schuppenförmig
scale-up/scaling up Maßstabsvergrößerung
scalid (recurved hook) Skalid
scalloped *see* **crenate**
scalpel Skalpell
scalpel blade Skalpellklinge
scaly/scabrous schuppig
scan/screen rastern, abtasten, prüfen
scandent/climbing kletternd, klimmend
scandent plant/climber/(climbing) vine Kletterpflanze
scanner Scanner, Abtaster
scanning Abtastung
scanning calorimetry Raster-Kalorimetrie
scanning electron microscopy (SEM) Rasterelektronenmikroskopie (REM)
scanning tunneling microscopy (STM) Rastertunnelmikroskopie (RTM)
scape/leafless stalk *bot* Blütenschaft (blattlos/bodenbürtig)
scape/scapus (cnidarians) Scapus, Mauerblatt
scape/scapus (feather) *orn* Scapus, Federkiel
scaphocerite/antennal scale Scaphocerit/Antennenschuppe
scaphognathite/bailer/gill bailer Scaphognathit, Atemplatte
scaphoid/navicular/cymbiform navikular, kahnförmig, bootförmig
scaphoid bone Kahnbein, Handwurzelknochen
scapiform schaftförmig
scapula/shoulder blade Schulterblatt

scapular feather *orn* Schulterfeder, Schulterblattfeder
scar/cicatrix/cicatrice Narbe, Wundnarbe, Cicatricula
scarce/rare selten, rar
scarceness/scarcity/rarity Seltenheit, Rarität
scarification (seed treatment) *agr/hort* Skarifizierung
scarification *immun* Skarifikation, Hautritzung
scarify (Boden) auflockern; (Samen) anritzen
scarp Böschung
- **beach scarp** Strandböschung
scat (fecal matter: insects/ wild animal droppings) Exkremente, Losung (Wild)
Scatchard plot Scatchard-Diagramm
scatol/skatole Skatol
scatter/spread/distribute streuen, verstreuen, ausstreuen, verteilen
scatter diagram/scattergram/ scattergraph/scatterplot Streudiagramm
scattering/spreading/distribution Streuung, Verstreuen, Verteilung
scattering angle Streuungswinkel
scavenge (feed on carrion/waste) Aas fressen, Unrat fressen
scavenger *chem* Scavenger, Fänger, Ladungsfänger
scavenger/carrion feeder Aasfresser, Unratfresser
scavenger cell Abraumzelle
scedasticity/heterogeneity of variances *stat* Streuungsverhalten
scent *n* Geruch, Duft; Spürsinn, Witterungssinn
scent *vb* riechen (spühren/fühlen); wittern
scent/stop and test the wind (game) sichern
scent gland/odoriferous gland Duftdrüse, Brunftdrüse, Brunftfeige (Gemse)
scent scale/androconium (*pl* androconia) *ent* Duftfeld, Duftschuppe, Androkonie
scent trail/olfactory trail Duftspur
scent-marking *ethol* Duftmarkierung
scheuchzeria family/arrow-grass family/Scheuchzeriaceae Blumenbinsengewächse, Blasenbinsengewächse
schistosome/blood fluke (*Schistosoma spp.*) Pärchenegel
schistosomiasis/bilharziosis/blood fluke disease (*Schistosoma spp.*) Schistosomiasis, Bilharziose

S Schizaeaceae 298

Schizaeaceae/curly-grass family/ climbing fern family Spaltfarngewächse
schizocarp/schizocarpium Schizokarp, Spaltfrucht
schizocoel/schizocoele/"split coelom" Schizocöl
schizogony/agamogony/merogony Schizogonie
schizomids/Schizomida Zwerggeißelskorpione
Schizophyllum family/ Schizophyllaceae Schizophyllaceae
school/shoal (fish) Schule, Schwarm, Zug
Schwann cell/neurolemma cell/ neurolemmocyte Schwannsche Zelle
Schwann sheath/myelin sheath Schwannsche Scheide, Myelinscheide
SCID (severe combined immune deficiency) schwerer kombinierter Immundefekt
science (sensu lato) Wissenschaft
science/natural sciences Naturwissenschaften
scientific (sensu lato) wissenschaftlich
scientific (pertaining to the natural sciences) naturwissenschaftlich
scientist Wissenschaftler
scientist/natural scientist Naturwissenschaftler
scintillate szintillieren, funkeln, Funken sprühen, glänzen
scintillation counter/scintillometer Szintillationszähler ("Blitz"zähler)
scion/cion Reis (pl Reiser), Edelreis, Pfropfreis
scion grafting hort Reisveredelung, Reiserveredlung, Pfropfen
sciophilous/shade-loving schattenliebend
sciophyll/shade leaf Schattenblatt
sciophyte/skiophyte/skiaphyte/ skiophyte/shade-loving plant/ shade plant Schattenpflanze
scissors lab Schere
- **blunt point scissors/blunt scissors** stumpfe Schere
- **dissecting scissors** Präparierschere
- **iris scissors** Irisschere, Listerschere
- **sharp point scissors** spitze Schere
- **surgical scissors** chirurgische Schere
sclera Sclera, Sklera, Bindegewebshülle
- **sclerotic coat/sclerotica (eye)** Sklera, Lederhaut, harte Augenhaut

sclerenchyma Sclerenchym, Sklerenchym
sclerenchymatous fiber Sklerenchymfaser
sclerification/lignification Sklerotisierung, Verholzung, Lignifizierung
sclerite Sklerit (stark sklerotisierte Platte/Nadel)
sclerocarp Sklerokarp
sclerocyte Sklerozyt, Sclerocyt
scleroderm Scleroderm, Skleroderm, Panzerhaut
sclerophyll/sclerophyllous plant/ hard-leaved plant/hard-leaf Hartlaub, Hartlaubgewächs, Sklerophyll
sclerophyllous forest Hartlaubwald
sclerophyte Sklerophyt
scleroprotein Skleroprotein
sclerosponges/coralline sponges/ Sclerospongiae Sclerospongien
sclerotesta Sklerotesta
sclerotic sklerotisch
sclerotium (pl sclerotia) fung Sklerotium, Dauermyzel
sclerotization/hardening Sklerotisierung
sclerotized/hardened sklerotisiert
sclerotome Sklerotom
scolespore Scolespore
scolex Scolex, Skolex
scolopale Sinnesstift
scolopendromorphs/ Scolopendromorpha Skolopender, Riesenläufer
scolopidium/scolophore/ chordotonal sensilla Scolopidium, Skolopidium, stiftführende Sensille
scopa (bees) Scopa, Schienenbürste
"scope" (sensu: microscope) Mikroskop; o. a. irgendein Beobachtungsinstrument
scopolamine Scopolamin
scorch n (through heat/climate) Versengung, Brandfleck
scorch vb (through heat/climate) versengen
scorpioid cyme/cincinnus (an inflorescence) Wickel
scorpion flies/mecopterans/ Mecoptera Schnabelfliegen
scorpions/Scorpiones Skorpione
scototopia/scotopic vision skotopisches Sehen, Dämmerungssehen
scouring rush/horsetail Schachtelhalm
scout (social insects) Kundschafter, Späher, Pfadfinder
scouting bee/scout bee Spurbiene, Kundschafterin

299   **sea lavender family**   **S**

scrape *vb* schaben
scraper (scrape off food) Kratzer
(Nahrung abkratzen)
scraper/rasp (plectrum)
(>stridulation) Schrillkante
scrapie Traberkrankheit, Scrapie
scraping/scraping mount *micros*
Schabepräparat
scratch *vb* **(chickens)** scharren
screamers & waterfowl (ducks/
geese/swans)/Anseriformes
Entenvögel, Gänsevögel
scree (on mountain slope) (see also
used syn. **with: talus)** Schuttdecke
an einer Schutthalde/Geröllhalde
screen *n* Filter, Schirm, Schutz,
Schutzschirm
  ● **intensifying screens**
  (autoradiography)
  Verstärkerfolien
  (Autoradiographie)
screen off/protect *vb* abschirmen,
schützen
screening Durchmustern,
Durchtesten
screening/screening test *gen/med*
Suchtest, Rasteruntersuchung
screw impeller (bioreactor)
Schneckenrührer
screw-cap vial/screw-cap jar
Schraubgläschen, Schraubgefäß
screw-pine family/Pandanaceae
Schraubenpalmen
scrobiculate/alveolate kleingrubig
Scrophulariaceae/snapdragon
family/foxglove family/figwort
family Braunwurzgewächse,
Rachenblütler
scrotum Skrotum, Hodensack
scrounger Dieb (Nahrung)
scrub/brush/thicket/thick
shrubbery Gestrüpp, Dickicht,
Buschwerk
  ● **sclerophyll scrub/sclerophyllous**
  shrub Hartlaubgebüsch,
  Hartlaubgehölz
sculpins & sea robins (=gurnards)/
Scorpaeniformes Panzerwangen,
Drachenkopffischverwandte,
Drachenkopfartige
sculpture (shells/seeds) Skulptur
scum/film/mat Kahmhaut,
Oberflächenfilm,
Oberflächenhäutchen
(in stehendem Binnengewässer)
scurf Schorf, Blattschorf, Grind
scurfy/scabby/furfuraceous
schorfig, Schorf…
scurvy Skorbut
scutate/scale-like schuppenartig
scute (enlarged scale) große
Schuppe

scutellate/like a small shield
schildchenartig
scutellation Schuppung
scutelliform/shaped like a small
shield schildchenförmig
scutellum (a shield-shaped
structure) Scutellum, Schildchen
(Saugorgan am Keimblatt des
Graskeimlings)
scutiform/scale-shaped
schuppenförmig
scutigeromorphs/Scutigeromorpha
Spinnenasseln
scyphozoans/cup animals/
Scyphozoa Scyphozoen,
Schirmquallen, Scheibenquallen,
Echte Quallen
SDS (sodium dodecyl sulfate)
Natriumdodecylsulfat
sea/lake See (Binnensee)
sea/ocean See, Meer, Ozean
sea anemones/Actiniaria
Seeanemonen
sea butterflies/shelled pteropods/
Thecosomata Seeschmetterlinge,
beschalte Flossenfüßer
(Flügelschnecken)
sea combs/comb jellies/
sea gooseberries/sea walnuts/
ctenophores/Ctenophora
Rippenquallen, Kammquallen,
Ctenophoren
sea cows & manatees & dugongs/
sirenians/Sirenia Seekühe
sea cucumbers/holothurians/
Holothuroidea Seewalzen,
Seegurken, Holothurien
sea daisies/concentricycloids/
concentricycloideans/
Concentricycloidea
Seegänseblümchen,
Concentricycloidea
sea gooseberries/sea combs/comb
jellies/sea walnuts/ctenophores/
Ctenophora Rippenquallen,
Kammquallen, Ctenophoren
sea hares/Aplysiacea/Anaspidea
Seehasen, Breitfußschnecken
sea heath family/alkali-heath
family/Frankeniaceae
Nelkenheidegewächse,
Frankeniengewächse
sea horses & pipefishes and allies/
Syngnathiformes (Syngnathoidei)
Seenadelverwandte,
Seepferdchenverwandte,
Büschelkiemenartige
sea lavender family/
leadwort family/plumbago family/
Plumbaginaceae
Bleiwurzgewächse,
Grasnelkengewächse

S sea level 300

sea level (above/below)/elevation Meeresspiegel, Meereshöhe
sea lilies/crinoids (incl. feather stars)/Crinoidea Seelilien, Crinoiden (inkl. Haarsterne=Federsterne)
  ● **with cirri/Isocrinida** zirrentragende Seelilien
  ● **without cirri/Millericrinida** zirrenlose Seelilien
sea pens/pennatulaceans/ Pennatularia Seefeder
sea purse/mermaid's purse (egg case of skates/sharks) Seemaus, Eikapsel der Knorpelfische
sea scorpions/eurypterids/ Eurypterida Seeskorpione
sea slugs/nudibranchs/ Nudibranchia Nacktkiemer, Meeresnacktschnecken, Nudibranchier
sea spiders/pycnogonids/ pantopods/Pycnogonida/ Pantopoda Asselspinnen, Pycnogoniden, Pantopoden
sea spray/ocean spray Gischt, Spritzwasser
sea squirts/ascidians/Ascideacea Seescheiden, Ascidien
sea urchins/echinoids/Echinoidea Seeigel, Echinoiden
sea wasps/box jellies/ cubomedusas/Cubomedusae/ Cubozoa Würfelquallen
seabed/seafloor/ocean floor Meeresboden, Meeresgrund
seafood Meeresfrüchte
seaboard Küste, Meeresküste
seafloor/seabed/ocean floor Meeresboden, Meeresgrund
seam/border/edge/fringe Saum, Rand
seam/suture/raphe Naht, Fuge, Verwachsungslinie
seamount Tiefseeberg
seashore/seaboard/seacoast Meeresküste, Meeresufer
season(s) *n* Jahreszeit(en)
  ● **spring/springtime** Frühling, Frühjahr
  ● **fall/autumn** Herbst
  ● **summer** Sommer
  ● **winter** Winter
season/store (wood) *vb* lagern, ablagern
seasonal saisonal, jahreszeitlich
  ● **aestival** ästival (früher Sommer)
  ● **autumnal** autumnal (Herbst)
  ● **hibernal** hibernal (Winter)
  ● **prevernal** prävernal (Vorfrühling)
  ● **serotinal** serotinal (Spätsommer)
  ● **vernal** vernal (spätes Frühjahr)

seasonal change Jahreszeitenwechsel
seasonal forest Saisonwald, regengrüner Wald
seasonality Saisonalität
seasoning (wood) Ablagern (Holz)
seastars/starfishes/asteroids/ Asteroidea Seesterne
seawater/saltwater Meerwasser
seawater intrusion/saltwater intrusion Meerwasserintrusion
seaweed Tang, Seetang, Seegras
sebaceous/suety talgig, Talg…
sebaceous cyst Epidermiszyste, Grützbeutel, Atherom
sebaceous duct Talggang
sebaceous follicle Haarbalgdrüse
sebaceous gland Talgdrüse, Haartalgdrüse
sebaceous matter/sebum Talg
secateurs/pruners/pruning shears/ pruning snips *hort* Gartenschere, Trimmschere
second degree relative Verwandter zweiten Grades
second dentition Zahnwechsel
second maxilla 2. Maxille, Maxilla secunda, Unterlippe
second messenger sekundärer Botenstoff, zweiter Bote
second site reversion ausgleichende Reversion
second-order kinetics Kinetik zweiter Ordnung
second-order reaction Reaktion zweiter Ordnung (Reaktionskinetik)
secondary body cavity/coelom/ perigastrium sekundäre Körperhöhle, sekundäre Leibeshöhle, Cölom, Coelom
secondary bract/bracteole/bractlet/ prophyll Vorblatt, Prophyll
secondary consumer Sekundärkonsument
secondary cortex/phelloderm *bot* Korkrinde, Phelloderm
secondary feathers/secondaries (secondary remiges) Armschwingen
secondary growth/secondary thickening *bot* Sekundärwachstum, sekundäres Dickenwachstum
  ● **anomalous s.g.** *bot* anomales sekundäres Dickenwachstum
secondary host Nebenwirt
secondary immune response/ anamnestic response sekundäre Immunantwort, immunologische Sekundärantwort
secondary infection Sekundärinfekt, Sekundärinfektion
secondary meristem Folgemeristem

secondary metabolism
Sekundärstoffwechsel
secondary response
Sekundärantwort
secondary settling tank
Nachklärbecken
secondary sex characteristics
sekundäre Geschlechtsmerkmale
secondary structure
Sekundärstruktur
secondary tectrices Armdecken
secondary wall Sekundärwand
secondary xylem Sekundärxylem
secretagogue *adv/adj* sekretagog,
die Sekretion anregend
secretagogue *n* Sekretagogum,
Sekretogogum
secrete (excrete) sezernieren,
abgeben, ausscheiden (Flüssigkeit)
secretin Secretin, Sekretin
secretion Sekretion, Freisetzung,
Ausscheidung, Sekret
secretor (blood group antigens)
Ausscheider
secretor system Sekretorsystem
secretory sekretorisch
secretory cell Sekretzelle
**secretory component/secretory
piece (antibody)** sekretorische
Komponente (Antikörper)
secretory gland Sekretdrüse
secretory protein Sekretionsprotein,
Sekretprotein,
sekretorisches Protein
secretory tapetum
Sekretionstapetum
secretory tissue Sekretionsgewebe,
Absonderungsgewebe,
Abscheidungsgewebe
secretosome Sekretosom
section/part/moiety
Abschnitt (Teil des Ganzen)
section (cut; *micros* **also: slice)**
Schnitt
  • **cesarean section/cesarean**
  med/vet Kaiserschnitt
  • **cross section** Querschnitt
  • **frozen section** Gefrierschnitt
  • **quick section** Schnellschnitt
  • **sagittal section/median
  longisection** Sagittalschnitt
  (parallel zur Mittelebene)
  • **semithin section**
  Semidünnschnitt
  • **serial sections** *micros/anat*
  Serienschnitte
  • **thickness of section** Schnittdicke
  • **thin section** Dünnschnitt
  • **transverse section/cross section**
  Hirnschnitt, Querschnitt
  • **ultrathin section**
  Ultradünnschnitt

sectionalization/division
Fächerung, Unterteilung
secure *vb* sichern, absichern
security Sicherheit, Absicherung
sedate sedieren, ruhig stellen (ein
Beruhigungsmittel verabreichen)
sedation Sedieren, Sedierung
sedative *adj/adv* sedierend, sedativ,
beruhigend; einschläfernd
sedentary/settled sedentär,
niedergelassen
sedge family/Cyperaceae
Riedgräser, Riedgrasgewächse,
Sauergräser
sedge Segge, Riedgras (Sauergräser)
sediment/deposit *vb* sedimentieren,
ablagern
sediment/pellet *n centrif* Sediment,
Pellet
sedimentation/deposition/deposit
Sedimentation, Ablagerung
  • **filling by sedimentation/silting
  up** Verlandung
sedimentation analysis
Sedimentations-
geschwindigkeitsanalyse
sedimentation coefficient
Sedimentationskoeffizient
**sedum family/stonecrop family/
orpine family/Crassulaceae**
Dickblattgewächse
see/view sehen, anschauen,
erblicken
seed *n* Same
seed/inoculate *vb micb/meteo*
beimpfen
seed/shed seeds *vb* Samen streuen/
ausstreuen
seed/plant (plant seeds) *vb* einsäen
seed bed/seedbed Saatbett
**seed case/seed casing (fruit/
capsule)** Samengehäuse (Frucht/
Samenkapsel)
**seed coat/testa (develops from
integuments)** Samenhülle,
Samenschale
seed company Sämerei
**seed cone/female cone/
megastrobilus/megasporangiate
strobilus** Samenzapfen, weiblicher
Zapfen
seed dormancy Keimruhe,
Samenruhe, Dormanz
seed ferns Samenfarne
seed leaf *see:* cotyledon
seed pan Saatkasten
seed repository Samenbank
seed shrimps/ostracods/Ostracoda
Muschelkrebse, Ostracoden
**seed stalk/ovule stalk/funicle/
funiculus** Nabelstrang, Samenstiel,
Funiculus

S **seed starting** 302

seed starting Anzucht,
Samenanzucht
seed stock/seeds Saatgut
seed tape Saatband
seed-bearing plant/spermatophyte
Samenpflanze, Spermatophyt
seed-cracking beak
Kernbeißerschnabel, Kernbeißer
seedbed Saatbett
seedless kernlos; taub
seedless fruit leerfrüchtig, kenokarp,
kernlose Frucht
seedlessness Kenokarpie,
Kenocarpie, Leerfrüchtigkeit
seedling/sprout Keimling, Sämling,
Setzling
seedpod *see:* pod
seedtime Saatzeit
seeing/vision Sehen
seepage/infiltration Versickerung
segment Segment, Glied
segment/somite Segment, Somit
(Ursegment)
segmentation cavity/blastocoel/
blastocele Furchungshöhle,
primäre Leibeshöhle, Blastocöl,
Blastula-Höhle
segmented neutrophil/filamented
neutrophil/polymorphonuclear
granulocyte
segmentkerniger Neutrophil
segmented worms/annelids/
Annelida Gliederwürmer,
Ringelwürmer, Borstenfüßer,
Anneliden
segregate *gen* aufspalten,
segregieren
segregate/separate out entmischen
segregation *gen* Segregation,
Aufspaltung
segregation line *gen*
Segregationslinie
seizure (sudden attack) plötzlicher
Anfall, Anfall; (convulsion/attack of
epilepsy) epileptischer Anfall
selaginella family/spike-moss
family/small club-moss family/
Selaginellaceae
Moosfarngewächse
selection Selektion, Auslese
 • **artificial s. (selective breeding)**
 künstliche Selektion/Auslese
 • **counterselection** Gegenselektion,
 Gegenauslese
 • **directional s.**
 gerichtete Selektion/Auslese
 • **disruptive s.**
 disruptive Selektion
 • **frequence-dependent s.**
 frequenzabhängige Selektion/
 Auslese
 • **group s.** Gruppenselektion

 • **kin s.** Verwandtenselektion,
 Sippenselektion
 • **natural s./Darwinian s.**
 natürliche Selektion/Auslese
 • **random s.** Zufallsselektion,
 ungerichtete Selektion/Auslese
 • **sexual s.** sexuelle/geschlechtliche
 Selektion/Auslese
 • **stabilizing s./normalizing s.**
 stabilisierende Selektion/Auslese
selection coefficient/coefficient of
selection Selektionswert,
Selektionskoeffizient
selection differential
Selektionsdifferential
selective selektiv, auswählend
selective advantage
Selektionsvorteil
selective breeding/breed selection
Zuchtwahl
selective disadvantage
Selektionsnachteil
selective filter/barrier filter/
stopping filter/selection filter
Sperrfilter
selective medium
Selektivnährmedium,
Elektivnährmedium
selective pressure/selection
pressure Selektionsdruck
selectivity Selektivität
selenium Selen
selenodont/with crescent-shaped
ridges halbmondhöckrig,
selenodont (Zahnhöcker)
selenozone/slit band (gastropod
shell) Schlitzband
self-assembly Selbstassoziierung,
Selbstzusammenbau,
Spontanzusammenbau,
spontaner Zusammenbau
(molekulare Epigenese)
self-consciousness
Selbstbewusstsein
self-dispersal/autochory
Selbstausbreitung
self-fertilization/selfing/autogamy
Selbstbefruchtung, Autogamie
self-fertilize/self (*see* selfing**)**
selbstbefruchten
self-grooming/autogrooming/
autopreening Selbstputzen
self-incompatibility
Selbstinkompatibilität
self-inducting impeller with hollow
impeller shaft selbstansaugender
Rührer mit Hohlwelle
self-ligation Selbst-Ligation
self-limited/self-limiting
selbstbegrenzend
self-marking/automarking *ethol*
Selbstmarkieren, Automarkieren

self-organization Selbstorganisation
self-pollinating/autophilous
selbstbestäubend
self-pollination/autophily
Selbstbestäubung
self-priming Selbst-Priming
self-sterile selbststeril
self-sterility Selbststerilität
self-tolerance Selbsttoleranz,
Eigentoleranz
selfing/self-fertilization/autogamy
(*also sensu:* **self-pollinization**)
Selbstung, Selbstbefruchtung,
Autogamie
selfish selbstsüchtig, egoistisch
selfish DNA egoistische DNA
selfishness Selbstsucht, Eigennutz,
Egoismus
sella turcica (hypophyseal fossa)
Türkensattel
sellaris joint/saddle joint
Sattelgelenk
semelparity (big-bang
reproduction) Semelparitie
(Big-Bang-Fortpflanzung)
semelparous (reproducing only
once) semelpar
semeostome medusas/
Semaeostomeae Fahnenquallen,
Fahnenmundquallen
semi-log graph paper
halblogarithmisches Papier,
Halblogarithmus-Papier
semiarid semiarid, halbtrocken
semicircular canals Bogengänge
semiconservative replication
semikonservative Replikation
semidesert Halbwüste
semidominance Semidominanz
semidominant semidominant
semilog/semilogarithmic
halblogarithmisch,
Halblogarithmus…
semilunar/shape of a halfmoon/
crescent-shaped handmondförmig,
sichelförmig
semilunar bone/lunate bone
Mondbein
semilunar cusp/semilunar flap
Semilunarklappe,
halbmondförmige Klappe
semilunar valve (consisting of three
semilunar cusps/flaps)
Taschenklappe
(aus drei Semilunarklappen)
semilunar zone/navicular zone/
navicular bursa/bursa
podotrochlearis (hoof) Hufrolle,
Fußrolle
seminal Samen…, Sperma…,
Samen betreffend, Sperma
betreffend

seminal discharge/ejaculate
Samenerguss, Ejakulat
seminal discharge/ejaculation
(process) Samenerguss,
Samenausstoß, Ejakulation
seminal duct Samenleiter,
Samengang
seminal fluid Samenflüssigkeit
seminal gland/vesicular gland
Samenblase, Bläschendrüse
seminal receptacle/spermatheca/
sperm chamber/receptaculum
seminis Rezeptakulum seminis,
Samentasche
seminal root Keimwurzel
seminal vesicle/glandula vesiculosa
Samenbläschen, Samenblase,
Bläschendrüse
seminiferous/sperm-forming/
sperm-producing samenbildend
semiparasite/hemiparasite
Halbparasit, Halbschmarotzer,
Hemiparasit
semipermeability
Halbdurchlässigkeit,
Semipermeabilität
semipermeable semipermeabel,
halbdurchlässig
semiplacenta/nondeciduate
placenta/placenta adeciduata
Semiplazenta, Halbplazenta
semiplume/semipluma Semipluma,
Halbdune
semisynthetic halbsynthetisch
semiterrestrial semiterrestrisch
semithin section
Semidünnschnitt
Semper cell Semperzelle
senesce/become old/age altern
senescence/ageing/aging
Seneszenz, Alterung, Altern
senile senil, greis, greisenhaft
senility Senilität, Vergreisung,
Altern, Älterwerden
sensation/perception (feeling)
Empfindung, Gefühl
sense/feel *vb* fühlen, etwas
bemerken
sense/feeling *n* Sinn, Gefühl
sense mutation Sinnmutation
sense of hearing Gehör
(Hörfähigkeit)
sense of taste/gustatory sense
Geschmackssinn
sense organ/sensory organ
Sinnesorgan
sense perception
Sinneswahrnehmung
sense strand (DNA) Sinnstrang
(DNA)
sensibilization/sensitization
Sensibilisierung

S sensibility 304

sensibility/sensitiveness
Empfindbarkeit,
Empfindungsvermögen
sensible sensibel, empfindbar,
empfindlich, reizempfänglich
sensilla Sensille
sensitive sensitiv, empfindlich,
leicht reagierend
sensitive hair/trigger hair Fühlhaar,
Reizhaar
sensitivity Sensitivität,
Empfindlichkeit
sensitize sensibilisieren
sensor/probe/detector Sensor,
Sonde, Messfühler
sensory sensorisch
sensory bristle/sensory chaeta
Sinnesborsten
sensory epithelium Sinnesepithel
sensory hair Sinneshaar
sensory organ/sense organ
Sinnesorgan
**sensory peg/olfactory peg
(sensilla styloconica/basiconica)**
Sinneskegel, Sinnesstäbchen,
Riechkegel
sensory physiology
Sinnesphysiologie
sensory pit Sinnesgrube
sensual sensuell, sinnlich, wollüstig
sentinel/sentry/guard/watch
Wächter
sepal Sepalum (*pl* Sepalen),
Kelchblatt, Blütenkelchblatt,
Blumenhüllblatt
sepal-like bracts Außenkelch,
Hochblatthülle
separate/divide abscheiden,
trennen
separate/fractionate auftrennen,
trennen, fraktionieren
separating gel (running gel)
Trenngel
separation/fractionation
Auftrennung, Trennung,
Fraktionierung
separation method Trennmethode
**separation technique/separation
procedure** Trennverfahren
separator/precipitator Abscheider
separatory funnel *lab*
Scheidetrichter
sepia (defense liquid of cuttlefish)
Sepia (Sekret des Tintenfisches)
sepsis/septicemia/blood poisoning
Sepsis, Septikämie, Blutvergiftung
septal filament Septalfilament
septate/divided/compartmentalized
unterteilt
**septibranch bivalves/septibranchs/
Septibranchia** Siebkiemer,
Verwachsenkiemer

septic septisch, infiziert, faulend,
fäulniserregend
septic tank Faulbehälter
septicidal *bot* **(fruit)** septizid,
wandspaltig, scheidewandspaltig
septicidal capsule septicide
Spaltkapsel, Bruchkapsel
septifragal septifrag, wandbrüchig,
scheidewandbrüchig
septum (*pl* **septa)/partition/
dissepiment/cross-wall/dividing
wall** Septe, Septum, Scheidewand
(*pl* Septen)
sequence *n* Sequenz, Folge, Reihe,
Aufeinanderfolge, Serie
• **amino acid sequence**
Aminosäuresequenz
• **autonomous(ly) replicating
sequence (ARS)** autonom
replizierende Sequenz (ARS)
• **base sequence** Basensequenz
• **centromere DNA sequence
elements (CDE)** CDE-Elemente
(DNA-Sequenzelemente am
Centromer)
• **complimentary base sequence**
komplementäre Basensequenz
• **consensus sequence/canonical
sequence** Consensussequenz,
Konsensussequenz
• **enhancer sequence**
Verstärkersequenz
• **exon/coding sequence/encoding
sequence**
Exon, kodierende Sequenz
• **expressed sequence tag (EST)**
exprimierte sequenzmarkierte
Stelle (EST)
• **flowering sequence** *bot*
Aufblühfolge
• **insertion sequence**
Insertionssequenz
• **intron/intervening sequence/
non-coding sequence** Intron,
intervenierende Sequenz,
dazwischenliegende Sequenz
• **leader sequence** Leadersequenz,
Leitsequenz
• **master sequence** Mastersequenz
• **reaction sequence/reaction
pathway** Reaktionsfolge
• **recombination signal sequences**
Rekombinationssignalsequenzen
• **regulatory sequence**
Regulationssequenz
• **repeat/repetition (of a
sequence)** Wiederholung,
Sequenzwiederholung
• **Shine Dalgarno sequence**
Shine-Dalgarno-Sequenz
• **signal sequence/signal peptide**
Signalsequenz, Signalpeptid

305   setose **S**

- **silencer (sequence)** Silencer, Abschaltsequenz
- **target sequence** Zielsequenz
- **termination sequence/t. codon/ t. factor/stop codon** Terminationssequenz, Stopcodon
- **triplet sequences** Triplettsequenzen
- **untranslated sequence (UTS)** untranslatierte Sequenz (UTS)
sequence *vb* sequenzieren
sequence tagged site (STS) sequenzmarkierte Stelle (STS)
sequencer/sequenator (*esp.* proteins) Sequenzierer, Sequenzierautomat, Sequenzierungsautomat
sequencing Sequenzierung
- **dideoxy s.** Didesoxysequenzierung
- **double strand s.** Doppelstrangsequenzierung
- **genomic s.** genomische Sequenzierung
- **multiplex s.** Multiplex-Sequenzierung
- **plus-minus s.** Plus-Minus-Verfahren
- **transcript s.** Transkript-Sequenzierung
sequential hermaphrodite Sukzedanhermaphrodit
sequential reaction/chain reaction sequentielle Reaktion, Kettenreaktion
sequester/segregate absondern, sequestrieren, abtrennen (z. B. Gewebe, Knochenbruchstücke)
sequestration/segregation Absonderung, Sequestrierung, Abtrennung, Loslösung (z. B. Gewebe, Knochenbruchstücke)
sera (*sg* serum) Seren (*sg* Serum)
seral stage (in ecological succession) Sukzessionsstufe
sere (a successional series) Serie (Sukzessionsfolge)
sericeous/sericate/silky seidenhaarig, seidig
sericin/silk gelatin/silk glue Serizin, Sericin
series Serie (Rangstufe der Klassifizierung)
series elastic component (SEC) serienelastische Komponente
serine Serin
serologic(al) serologisch
serology Serologie
serosa/serous membrane (external membrane: e. g. insect eggs) Serosa, äußere Keimhülle, äußere Eihülle, äußeres Eihüllepithel

serotinous/serotinal/ late in developing (e. g. cone) spät auftretend, spät öffnend, spät aufbrechend (z. B. Zapfen)
serotonin/5-hydroxytryptamine Serotonin, Enteramin, 5-Hydroxytryptamin
serotype/serovar Serotyp, Serovar
serous (pertaining to serum) serös
serous membrane *see* serosa
serpentine mine/heliconome (a leaf mine) *bot* Spiralmine
serrate/serrated/sawed/saw-edged (e. g. leaf margin) sägeförmig gezackt, gesägt
- **finely serrate/serrulate/ serratulate/finely saw-edged** feingesägt, kleingesägt
serrate suture (skull) Sägenaht
serration/serrature (saw-like formation) Auszackung
serrulate/serratulate/finely serrate/ finely saw-edged feingesägt, kleingesägt
serum (*pl* sera or serums) Serum (*pl* Seren)
serum dependence Serumabhängigkeit
serve/cover/leap (copulate by male) decken, begatten, bespringen
service/serving (copulation) Deckakt
sesame family/benne family/ Pedaliaceae Sesamgewächse
sesamoid bone Sesambein, Sesamknöchelchen; Gleichbein
- **distal sesamoid bone (horse)** Strahlbein
sesquiterpene Sesquiterpen
sessile (firmly/permanently attached) sessil, sesshaft, (fest)sitzend, sitzend, (fest)geheftet
sessility Sessilität, Sesshaftigkeit (festsitzend)
session/conference Sitzung, Konferenz
sessoblast Sessoblast, sessiler Statoblast, sitzender Statoblast
seston Seston
set/curdle/coagulate gerinnen, koagulieren
set/layer *n bot* Senker, Absenker
set *vb* **(gel)** fest werden, erstarren
seta Seta, Stiel (Moossporogon)
setaceous/bristle-like borstenartig
setiform/bristle-shaped borstenförmig
setigerous/setiferous/covered with bristles/having setae borstentragend, borstig
setose/bristly/set with bristles borstig

S settle 306

settle/colonize/establish besiedeln, kolonisieren, etablieren
settled/sedentary niedergelassen, sedentär
settlement/colony Siedlung, Kolonie
settlement/establishment Besiedlung, Etablierung
settling tank Absetzbecken, Klärbecken
severe combined immune deficiency (SCID) schwerer kombinierter Immundefekt
sewage/wastewater Abwasser
sewage fields (sewage farm) Rieselfelder
sewage sludge (*esp.*: excess sludge from digester) Faulschlamm (*speziell:* ausgefaulter Klärschlamm)
sewage treatment Klärung
sewage treatment plant Kläranlage, Klärwerk
sewer Kanalisation
sex (*pl* sexes) (male/female/neuter) Geschlecht (männlich/weiblich/neutral)
  ● **heterogametic sex** heterogametisches Geschlecht
  ● **opposite sex** andere Geschlecht
sex *vb* (sexing of chicks) das Geschlecht bestimmen
sex behavior Sexualverhalten
sex cell/gamete Geschlechtszelle, sexuelle Keimzelle, Gamet
sex characteristics Geschlechtsmerkmale
  ● **secondary s.c.** sekundäre Geschlechtsmerkmale
sex chromosome/ heterochromosome Geschlechtschromosom, Heterosom, Gonosom
sex chromosome inactivation Geschlechtschromosominaktivierung
sex determination Geschlechtsbestimmung
sex drive Sexualtrieb, Geschlechtstrieb
sex factor *micb* Sexfaktor, Konjugationsfaktor
sex gland/germ gland/gonad Geschlechtsdrüse, Keimdrüse, Gonade
sex hormone Sexualhormon
sex linkage Geschlechtskopp(e)lung
sex pilus (*pl* sex pili) Sexpilus, Sexualpilus
sex ratio Geschlechterverhältnis
sex reversal Geschlechtsumkehr
sex-induced geschlechtsbedingt
sex-influenced gene geschlechtsbeeinflusstes Gen

sex-limited gene geschlechtsbeschränktes Gen
sex-linked geschlechtsgebunden
sex-linked inheritance geschlechtsgebundene(r) Erbgang/ Vererbung, geschlechtsgekoppelte(r) Erbgang/Vererbung
sexduction Sexduktion
sexhood Geschlechtlichkeit
sexing (of chicks) das Geschlecht bestimmen
sexless/neuter geschlechtslos, ohne Geschlecht, sächlich
sexual sexuell, geschlechtlich
sexual behavior Sexualverhalten, Geschlechtsverhalten
sexual characteristic Sexualmerkmal, Geschlechtsmerkmal
sexual dimorphism Sexualdimorphismus, Geschlechtsdimorphismus
sexual intercourse/coitus Geschlechtsverkehr, Kopulationsakt, Begattungsakt, Koitus
sexual maturity Geschlechtsreife
sexual reproduction sexuelle/ geschlechtliche Fortpflanzung, geschlechtliche Vermehrung
sexual selection sexuelle/ geschlechtliche Selektion/Auslese
sexual swelling (callosity) Brunstschwiele
sexual union/copulation Begattung, Kopulation
sexuality Sexualität, Geschlechtlichkeit
sexually transmitted disease (STD)/ venereal disease (VD) sexuell übertragbare Krankheit, Geschlechtskrankheit, venerische Krankheit
shade *n* Schatten
shade *vb* schattieren
shade leaf/sciophyll Schattenblatt
shade-loving/sciophilous schattenliebend
shade-loving plant/shade plant/ sciophyte Schattenpflanze
shading Beschattung
shadowcasting/shadowing (TEM) Beschattung (Kohle-/Metallbeschattung: Schrägbedampfung)
  ● **rotary shadowing** Rotationsbedampfung
shady schattig
shaft/leafless stem/leafless shoot/ rachis/axis Schaft, Achse
shaft/rachis (feather) Federschaft, Rhachis
shaft/rachis/scape Schaft

307　　　　　　　　　　　　　　　　　**shield** **S**

shake *vb* schütteln
shake *n* **(wood: fissure between growth rings)** Riss
shake culture Schüttelkultur
shake flask Schüttelkolben
shake out ausschütteln
shaker Schüttler
- **circular shaker/rotary shaker** Rundschüttler
- **reciprocating shaker** Reziprokschüttler

shaking dance (bees) Schütteltanz, Schüttelbewegung
shaking water bath Schüttelbad
shallow-rooted plant Oberflächenwurzler
sham attack Scheinangriff
sham feeding *ethol* Scheinfüttern, Scheinfütterung
sham pecking *ethol* Scheinpicken
sham preening/sham grooming *ethol* Scheinputzen
sham rage *ethol* Scheinwut, unechte Wut
shank (leg/tibia or shin) Unterschenkel
shank/tarsus *orn* Tarsus
shape/form/appearance/contour Gestalt
shark tooth comb (gel electrophoresis) Haifischkamm (Gelelektrophorese)
sharks Haie, Haifische
sharks & rays & skates/ Elasmobranchii Plattenkiemer, Haie & Rochen
Sharpey's fiber Sharpeysche Faser
sharpness/focus *micro/photo* Schärfe, Bildschärfe
she-oak family/beefwood family/ Casuarinaceae Streitkolbengewächse
shear/shearing *vb* scheren
shear force Scherkraft
shear gradient Schergefälle, Schergradient
shear rate/rate of shear Scherrate
shear strength/shearing strength/ shear resistance (wood) Scherfestigkeit, Schubfestigkeit
shear stress (shear force per unit area) Scherspannung
shearing scheren
sheath *n* Scheide, Umhüllung; *bot* Blattscheide
sheathe *vb* scheidenförmig umhüllen
sheathed/vaginate scheidenförmig, röhrenförmig umhüllt
shed/barn *n* Stall
shed/drop/abscise abwerfen, abstoßen

shed/slough/sloughing off (skin) abstoßen, ablösen
shedding/abscising/deciduous *adj/adv* abwerfend
shedding/falling off/dropping off/ abscission Abwurf, Abwerfen, Abszission
shedding of leaves/leaf fall Laubfall, Blattfall, Blattabwurf
sheep shearing Schafschur
sheet Blatt (z. B. Papier)
- **beta-sheet/beta-pleated sheet** beta-Faltblatt

sheet erosion *geol* Schichterosion, Schichtfluterosion, Flächenerosion
sheetweb/dome web (with barrier threads) *arach* Baldachinnetz (mit Stolperfäden)
- **horizontal sheetweb** Deckennetz
- **simple sheetweb** Flächennetz

shelf Schelf, Sockel
- **continental shelf** Festlandsockel, Kontinentalsockel, Kontinentalschelf, Schelf

shelf break/shelf edge/shelf margin/continental margin Schelfrand, Schelfkante
shell/bivalve shell Muschel, Muschelschale
shell/carapace (insects/turtles) Schale, Panzer, Carapax
shell/case/casing (mollusks) Gehäuse
shell/test/testa/coat (enclosing cover, e. g. of diatoms) Testa, (harte) Schale, Hülle
shell gape (bivalves) klaffende Schalenöffnung
shell gland/Mehlis' gland (cement gland) Schalendrüse, Mehlissche Drüse
shell membrane Schalenhaut
shell plate/shell valve (chitons) Schalenplatte
shelled/cleidoic beschalt, kleidoisch
shellfish (crustaceans & mollusks) Schalentier
shelterbelt/windbreak Windschutz (Windschutzbäume)
shelterwood Mutterbestand, Schirmstand, Schirmbestand, Plenterwald
shelterwood method (cutting/ felling) Femelhieb, Femelschlag, Plenterschlag (uneven-aged); Schirmschlag (even-aged)
shepherd Schäfer
shield (from radiation) abschirmen (von Strahlung)
shield/clypeus Kopfschild, Clypeus
shield/scute/scutum Schild, Scutum

S shield 308

shield/shell/carapace Schild,
Schale, Carapax
shield/vertebral ossicle (ophiuroids)
Platte
shield budding *bot* Schild-
Okulation (Augenschild/Schildchen)
- **double shield budding**
Doppelschildokulation, Nicolieren
shield grafting/sprig grafting *hort*
seitliches Einspitzen
shield-like/peltate/scale-like/
scutate schildartig, schuppenartig
shield-shaped/peltiform/scutiform
schildförmig, schuppenförmig
shielding (from radiation)
Abschirmung (von Strahlung)
shifting agriculture/shifting
cultivation/swidden agriculture/
swidden cropping
Wanderackerbau
shifting dune Wanderdüne
shikimic acid (shikimate)
Shikimisäure (Shikimat)
shimmering/iridescent schillernd
shimmering body Flimmerkörper
shin (vertebrates: front part of leg
below knee) *n* Schienbein
(Vorderbein zwischen Knie und
Fuß)
shinbone/tibia
Schienbein(knochen), Schiene, Tibia
Shine Dalgarno sequence
Shine-Dalgarno-Sequenz
shingle Schindel
shingleplant family/Marcgraviaceae
Honigbechergewächse
shinleaf family/wintergreen family/
Pyrolaceae Wintergrüngewächse
shiver zittern, schaudern; frösteln,
vor Kälte zittern
shivering Zittern, Schaudern
shoal/school (fish) Schwarm,
Schule, Zug
shoat/shote (young weaned hog/
less than 150 lb/less than 1 year
old) Ferkel
shock freezing Schockgefrieren
shock resistance Stoßfestigkeit
shod *adv/adj* (horses/cattle)
beschlagen
shoe/shoeing *vb* (hoofs) beschlagen
shoestring ferns/Vittariaceae
Vittariaceae
shoot/sprout/sprig Spross, Trieb,
Schoss (kleiner Spross), Schössling
- **adventitious shoot**
Adventivspross, Adventivtrieb,
Zusatztrieb
- **annual shoot/one-year shoot/
annual growth** Jahrestrieb
- **apical shoot/terminal shoot**
Gipfeltrieb, Endtrieb, Terminaltrieb

- **bulbil** Bulbille, Brutspross,
Brutknospe
- **cladode/cladophyll/phylloclade**
Kladodium, Cladodium
(Flachspross eines Langtriebs),
Phyllocladium
- **coppice-shoot** Wassertrieb,
Wasserreis
- **lammas shoot** Johannistrieb
- **lateral shoot/side shoot/
offshoot** Nebentrieb, Seitentrieb
- **long shoot/long axis** Langtrieb
- **main shoot/primary shoot/
leading shoot/main axis/
primary axis** Hauptspross,
Primärspross, Hauptachse
- **offset (short/prostrate lateral
shoot from base of plant)**
kurzer Seitentrieb am Wurzelhals,
kurzer Wurzelhalsschössling
(Wurzelspross/Wurzeltrieb)
- **offshoot/lateral shoot/side
shoot** Nebentrieb, Seitentrieb
- **platyclade** Platycladium,
Flachspross
- **rhizome (horizontal/
underground stem)**
Rhizom, Erdspross,
unterirdischer Ausläufer,
unterirdischer Stolon (geophil)
- **root sucker** wurzelbürtiger
Spross, Wurzelspross,
Wurzeltrieb, Wurzelschössling,
Schosser, Wurzelreis
- **root-collar shoot**
Wurzelhalsschössling
- **runner/sarment (horizontal/
aboveground stem/stolon)**
Kriechspross,
oberirdischer Ausläufer,
oberirdischer Stolon (photophil)
- **short shoot/short axis** Kurztrieb
- **side shoot/lateral shoot/
offshoot** Seitenspross,
Seitentrieb, Nebentrieb
- **slip (softwood/herbaceous
cutting or scion for
propagation/grafting) (e. g.
banana/geraniums)** Ableger,
Steckling, Schnittling; Pfropfreis
- **sobole** Erdspross,
Gehölzausläufer
- **sprout/sprig/small shoot** Schoss,
Schössling (kleiner Spross)
- **spur shoot/fruit-bearing bough
(a short shoot)** Fruchtholz
(Kurztrieb), Lateralorgan;
Infloreszenz-Kurztrieb
- **stolon** Stolon, Stolo, Ausläufer,
Ausläuferspross
- **sucker/sobole (an underground
stolon)** Gehölzausläufer

shrub

- **sucker/tiller (shoot from roots or lower part of stem)** Schössling, Wasserreis, Geiztrieb (an Wurzel oder Baumstumpf), Seitentrieb (am Wurzelhals)
- **sylleptic shoot** sylleptischer Trieb
- **terminal shoot/apical shoot** Terminaltrieb, Endtrieb, Gipfeltrieb
- **water shoot/water sprout/water sucker** Wasserschoss, Wassertrieb, Wasserreis, Geiltrieb, Geiztrieb

shoot apex/shoot tip/vegetative cone Sprossspitze, Sprossscheitel, Sprosspol (Embryo), Vegetationskegel
shoot axis/stem Achsenkörper, Stamm
shoot elongation Sprosszuwachs
shoot tendril Sprossranke
shore/banks/coast Ufer, Küste, Gestade; Strand
- **backshore** Hochstrand, Sturmstrand (trocken)
- **foreshore** Vorstrand, Gezeitenstrand (nass)
- **inshore** an der Küsten, im Küstenbereich
- **nearshore** Küstennähe; küstennah, festlandnah
- **offshore** auf dem Schelf gelegen, (unterhalb der tiefsten Brandungseinwirkung); küstenfern, ablandig

shore/land (vs ocean/water) Land
shore dune Stranddüne
shore jetty Seebuhne, Strandbuhne
shorebirds & gulls & auks/Charadriiformes Watvögel & Möwenvögel & Alken
shoreface oberer Teil der Brandungsplattform, Strandstufe in der Brecherzone
shorefront/beachfront Küstenfront, Strandfront
shoreline/coastline Küste, Strandlinie, Küstenlinie; Uferlinie
- **alluvial coast/shoreline of progradation** Anschwemmungsküste, Anwachsküste
- **emergent shoreline/shoreline of emergence/negative coast/shoreline of elevation** Auftauchküste, Hebungsküste
- **fjord shoreline/fjord coast** Fjordküste
- **ria shoreline/ria coast** Riasküste
- **skerry coast/schären-type shoreline (rocky isle)** Schärenküste
- **submergent shoreline/shoreline of submergence/positive shoreline/shoreline of depression** Untertauchküste, Senkungsküste

shoreline of emergence Auftauchküste (Regressionsküste)
shoreline of submergence Untertauchküste (Transgressionsküste)
short bone/os brevis kurzer Knochen
short circuit Kurzschluss
short period interspersion kurzphasige Einstreuung/Einschub
short shoot/short axis Kurztrieb
short-chain kurzkettig
short-circuit *vb* kurzschließen
short-circuit/short-circuiting *n* Kurzschluss
short-day plant Kurztagspflanze
short-living/short-lived/fugacious/soon disappearing kurzlebig, hinfällig
short-term memory Kurzzeitgedächtnis
shot/injection Spritze, Injektion
shotgun cloning Schrotschussklonierung
shotgun experiment Schrotschussexperiment
shoulder *anat* Schulter, Achsel
shoulder blade/scapula Schulterblatt
shoulder girdle/pectoral girdle Schultergürtel, Brustgürtel
show/display/exhibit/exhibition *n* Schau, Ausstellung
showcase Schaukasten, Vitrine
showy flower auffällige Blüte, prachtvolle Blüte, prächtige Blüte
shredder (large-particle detritivore) Zerkleinerer
shrimps (small)/prawns (large) Garnelen, "Krabben"
shrink *vb* schrumpfen, sich zurückziehen, abnehmen
shrinking/decongestant abschwellend
shrub Strauch
- **arbuscule (tree-like small shrub/dwarf tree)** Bäumchen, baumartiger Strauch, niedriger Baum
- **chamaephyte** Chamaephyt
- **dwarf-shrub/woody chamaephyte** Zwergstrauch, holziger Chamaephyt
- **fruit-bearing shrubs** Beerensträucher
- **half-shrub/suffrutecsent plant** Halbstrauch

S shrub 310

- **nanophanerophyte (shrubs under 2 meters in height)** Nanophanerophyt, Strauch
- **ornamental shrub** Zierstrauch
- **prickly shrub/bramble** stacheliger Strauch
- **sclerophyllous shrub/sclerophyll scrub** Hartlaubgebüsch, Hartlaubgehölz
- **thorny shrub** Dornstrauch, Dornenstrauch

shrub savanna Strauchsavanne
shrubbery/thicket/underbrush (in forest) Buschwerk, Gebüsch
shrubby/bushy/fruticose/frutescent strauchig, buschig
shrubby herb Halbstrauch
shrubland Buschformation
shrublet kleiner Strauch
shunt/diversion/bypass (passage between two natural channels) Nebenschluss, Nebenweg, Ableitung, Bypass
shuttle/shuttling Pendelverkehr, Pendeln, Schleusen (Membran)
shuttle streaming (*Physarium*) Pendelströmung
shuttle vector/bifunctional vector *Shuttle*-Vektor, Schaukelvektor, bifunktionaler Vektor
shy *adj/adv* scheu
shy/take fright/skit *vb* **(horse)** scheuen
shying (horse) Scheuen
shyness Scheu
sialic acid (sialate) Sialinsäure (Sialat)
sib-pair analysis Untersuchung von Geschwistern
sibling/sib (*see also:* siblings) einer von mehreren Geschwistern
sibling species Geschwisterarten
siblings (all offspring having one common parent) Geschwister *pl*
sibship Geschwisterschaft
sick/ill/diseased krank
sickle cell Sichelzelle
sickle dance (bees) Sicheltanz
sickle-shaped/drepanoid/crescent/ falcate/falciform sichelförmig
side body Seitenkörper
side chain *chem* Seitenkette
side cleft grafting/side whip grafting/bottle grafting *hort* seitliche Spaltpfropfung
side crown *bot* Seitenkrone
side effect Nebenwirkung
side grafting *hort* Seitenpfropfung, Seitenveredelung, Veredeln an der Seite
- **spliced side grafting/veneer side grafting/side veneer grafting** Anplatten, seitliches Anplatten

side product Nebenprodukt
side shoot/lateral shoot/sucker Geiz, Geiztrieb, Seitenspross
side-tongue grafting *hort* seitliches Anplatten mit Gegenzunge
side-veneer grafting/veneer side grafting/spliced side grafting *hort* Anplatten, seitliches Anplatten
sidewinding (snakes) Seitenwinden
sieve/sift *vb* sieben
sieve/sifter *n* Sieb
sieve areas *bot* Siebfelder
sieve cell *bot* Siebzelle
sieve element *bot* Siebelement
sieve plate/madreporic plate/ madreporite Siebplatte, Madreporenplatte
sieve plate reactor (bioreactor) Siebbodenkaskadenreaktor, Lochbodenkaskadenreaktor
sieve trachea *arach* Siebtrachee
sieve tube *bot* Siebröhre
sieve tube element *bot* Siebröhrenelement
sieve tube member *bot* Siebröhrenglied
sift *vb* **(e. g. flamingoes)** seihen, sieben
sight/view Sicht
sigillaria family (clubmoss trees)/ Sigillariaceae Siegelbäume
signal forgery Signalfälschung
signal hypothesis Signalhypothese
signal protein Signalprotein, Sensorprotein
signal recognition particle/signal recognition protein (SRP) Signalerkennungspartikel
signal sequence/signal peptide Signalsequenz, Signalpeptid
signal substance Signalstoff
signal thread *arach* Signalfaden
signal transducer Signalwandler
signal transduction Signalübertragung
signal-to-noise ratio Signal-Rausch-Verhältnis
signature protein Erkennungsprotein (ein Membranprotein)
significance level/level of significance (error level) Signifikanzniveau, Irrtumswahrscheinlichkeit
significance test/test of significance *stat* Signifikanztest
silage *n* Silage, Silofutter, Gärfutter
silage *vb* zu Gärfutter silieren, einsäuern
silencer (sequence) *gen* Silencer, Abschaltsequenz
silent gene stummes Gen

single strand **S**

silent infection stumme Infektion,
stille Feiung
silent mutation/samesense
mutation stumme Mutation
silica/silicon dioxide
Siliziumdioxid
silica gel Kieselgel, Silicagel
siliceous Kieselsäure..,
kieselsäurehaltig
siliceous sponges/demosponges/
Silicospongiae Kieselschwämme
(Demospongien)
silicic acid Kieselsäure
silicle Schötchen
silicoflagellates/Silicophyceae
Kieselflagellaten
silicon Silizium, Silicium
silicone (silicoketone) Silikon
silique/siliqua Schote
silk (fibroin/sericin) Seide,
Spinnseide
silk (corn stigma-style) bot
Maisgriffel, Griffelfäden, "Bart"
silk fabric (cobweb/spiderweb)
Spinngewebe
silk gland/spinning gland/
sericterium (caterpillars: labial
gland) Seidendrüse, Spinndrüse,
Sericterium (Labialdrüse: Raupen)
silk gland spigot/tubulus textori
arach Spinndüse
silk gland spool arach Spinnspule
silk thread/silk line Spinnfaden
silk-cotton tree family/cotton-tree
family/kapok-tree family/
Bombacaceae Wollbaumgewächse
silk-oak family/Australian oak
family/protea family/Proteaceae
Proteagewächse,
Silberbaumgewächse
silk-tassel tree family/silktassel-
bush family/Garryaceae
Becherkätzchengewächse
silken seiden, Seiden…
silkworm (larva) Seidenraupe
silky/silk-like/sericate/sericeous
seidig, seidenartig, seidenhaarig
silt Schluff, Silt
silt/warp Schlamm, Flussschlamm
  • silty soil (marsh) leichter Klei/
Kleiboden/Marschboden
silt load/suspension load soil/mar
Schwebfracht
silty loam Schlufflehm,
lehmiger Schluff
Silurian Period/Silurian (geological
time) Silur
silver-vine family/dillenia family/
Dilleniaceae Rosenapfelgewächse
silverline system/argyrome
(ciliates) Silberliniensystem,
Argyrom

silversides & skippers & flying
fishes and others/Atheriniformes
Ährenfischverwandte,
Hornhechtartige
silvics Forstbaumkunde
silviculture Forstkultur, Waldbau
Simaroubaceae/quassia family
Bittereschengewächse,
Bitterholzgewächse
simian Affen (bes.Menschenaffen)
betreffend, affenartig, Affen…
similar-structured gleichgestaltet,
ähnlich gestaltet
simple columnar epithelium
hochprismatisches Epithel, hohes
Zylinderepithel
simple cyme/monochasium (an
inflorescence) eingablige
Trugdolde, Monochasium
simple eye/ocellus (dorsal and
lateral) Einzelauge, Punktauge,
Nebenauge, Ocelle, Ocellus (siehe:
Scheitelauge/Stirnauge)
simple fruit/apocarpous fruit
Einblattfrucht
simple umbel (an inflorescence)
einfache Dolde
simultaneous hermaphrodite/
synchronous hermaphrodite
Simultanhermaphrodit
sinapic acid Sinapinsäure
SINE (short interspersed nuclear
element)
SINE (kurzes eingeschobenes
nukleäres Element)
sing/warble/jug singen, schlagen
(Nachtigall)
single/solitary einzeln, solitär
single animal/solitary animal
Einzeltier
single copy DNA Einzelkopie-DNA,
nichtrepetitive DNA
single copy plasmid
Einzelkopie-Plasmid
single digest einfacher Verdau
single displacement reaction
einfache Verdrängungsreaktion,
Einzel-Verdrängung
single fruit Einzelfrucht
single rowed/uniseriate/uniserial
einreihig
single strand gen Einzelstrang
single strand assimilation
Einzelstrangassimilation
single strand binding protein
einzelstrangbindendes Protein
single strand break
Einzelstrangbruch
single strand conformation
polymorphism (SSCP) gen
Einzelstrang-Konformations-
Polymorphismus (SSCP)

S

single strand exchange 312

single strand exchange *gen*
Einzelstrangaustausch
single sugar/monosaccharide
Einfachzucker, einfacher Zucker,
Monosaccharid
single-cell protein (SCP)
Einzellerprotein
single-celled/unicellular einzellig
**single-chambered/monothalamous/
monothalamic/monothecal**
einkammerig, einkämmrig,
monothalam, monothekal
single-rowed/uniseriate/uniserial
einreihig
single-stranded einsträngig
sink/depression Senke,
Bodensenke
sink (importer of assimilates)
physio Senke, Verbrauchsort
sinoauricular/sinoatrial
sinoaurikulär
**sinoauricular node/sinoatrial node
(SAN)** Sinusknoten,
Sinoatrialknoten, SA-Knoten
**sinuate/sinuous (strongly waved
margin)** *bot*
geschweift, buchtig, gebuchtet
**sinus/cavity/depression/recess/
dilatation/lacuna** Sinus, Höhle,
Vertiefung, Ausweitung, Lakune
sinus gland Sinusdrüse
sinus venosus Sinus venosus
siphon/funnel/infundibulum
Sipho, Trichter, Infundibulum
siphon/tube (tubular cell)
Schlauch; Zellschlauch
siphoneous/siphonaceous/tubular
siphonal, röhrenartig, schlauchartig,
schlauchförmig, tubulär
siphonogamy Siphonogamie,
Pollenschlauchbefruchtung
siphonoglyph Siphonoglyphe
siphonophorans/Siphonophora
Siphonophoren, Staatsquallen
siphonozooid Siphonozooid,
Pumppolyp
siphuncle/siphonet (nautilus)
Siphunkel, Sipho
sire (♂ parent: domestic animals)
Vatertier, männliches Stammtier
(Pferd: Beschäler/Zuchthengst)
siroheme Sirohäm
sister cell Schwesterzelle
sister chromatids
Schwesterchromatiden
  • **sister chromatid exchange**
  Schwesterchromatidenaustausch
  • **unequal exchange of sister
  chromatids (UESCE)**
  ungleicher Austausch von
  Schwesterchromatiden (UESCE)
sister group Schwestergruppe

sister taxa Schwestertaxa,
Adelphotaxa
sister species Schwesterart
site/location/place Fundort, Lage
site assessment *ecol*
Lagebewertung, (*sensu lato*:
Standortbewertung)
site-directed mutagenesis
ortsspezifische Mutagenese
site-specific recombination *gen*
sequenzspezifische Rekombination
sitosterol Sitosterin, Sitosterol
size reduction *photo* Verkleinerung
skates/Rajoidei (Suborder!)
echte Rochen
skates & guitarfishes/Rajiformes
Rochenartige
skeletal musculature
Skelettmuskulatur
skeleton/bones Skelett, Gerippe,
Knochengerüst; Gebein
  • **appendicular skeleton/skeleton
  appendiculare**
  Extremitätenskelett,
  Gliedmaßenskelett
  • **axial skeleton** Achsenskelett,
  Stammskelett, Rumpfskelett
  • **branchial skeleton/skeleton of
  the gills** Branchialskelett,
  Kiemenskelett
  • **cardiac skeleton/skeleton of the
  heart** Herzskelett
  • **cuticular skeleton**
  Kutikularskelett
  • **cytoskeleton** Cytoskelett,
  Zytoskelett
  • **dermal skeleton/
  dermatoskeleton/dermoskeleton**
  Hautskelett, Dermalskelett
  • **endoskeleton/internal skeleton**
  Innenskelett
  • **exoskeleton/dermal skeleton/
  dermatoskeleton/dermoskeleton
  (vertebrates)** Hautskelett,
  Dermalskelett
  • **exoskeleton/external skeleton
  (invertebrates)** Außenskelett
  • **gill arch skeleton** Kiemenskelett
  • **head skeleton/cephalic skeleton**
  Kopfskelett
  • **hydrostatic skeleton**
  hydrostatisches Skelett,
  Hydroskelett
  • **somatic skeleton**
  somatisches Skelett
  • **visceral skeleton/visceroskeleton**
  Viszeralskelett, Visceralskelett,
  Eingeweideskelett
  • **zonoskeleton** Zonoskelett
  (Extremitätengürtel)
skeletonize skelettieren;
skelettbildend, mit Skelett

sluice

skewbald *adj/adv* scheckig (Pferd)
skewbald horse Schecke
skewness *stat* Schiefe
skin Haut
- **cutis (epidermis & dermis/ corium)** Cutis, Kutis (eigentliche Haut)
- **dermis/corium/cutis vera/true skin** Dermis, Corium, Korium, Lederhaut
- **epidermis** Epidermis, Oberhaut
- **mucosa/mucous membrane** Schleimhaut, Schleimhautepithel
- **subcutis/tela subcutanea** Subcutis, Unterhaut, Unterhautbindegewebe

skin/hide/peel Haut, Schale
skin fold/wrinkle Hautfalte
skin graft/skin transplant Hauttransplantat
skin parasite/dermatozoan Hautparasit, Hautschmarotzer, Dermatozoe
skiophilous/umbraticolous schattenliebend
skippers (Hesperiidae: Lepidoptera) Dickkopffalter
- **cheese skippers/cheese maggots (Piophilidae larvas)** Käsefliegenlarve

skull/braincase/cranium Schädel, Hirnkapsel, Kranium, Cranium
skull roof/cranial roof Schädeldecke
skullcap/clavarium Kalotte, Schädeldach, Schädeldecke
slab Platte, Fliese, Tafel; (wood) Holzschwarte
slab gel *biochem/gen* hochkant angeordnetes Plattengel
slack-water zone *mar/limn* Stillwasserzone
slant culture/slope culture *micb* Schrägkultur (Schrägagar)
slash-and-burn Brandrodung
slash-and-burn agriculture Brandrodungsfeldbau
slaughter/butcher *vb* schlachten
slaughter/slaughtering/butchering *n* Schlachtung, Schlachten
slaughter cattle/beef cattle/ slaughter animal Schlachtvieh
slaughterhouse Schlachthof
sleep Schlaf
sleep movement/nyctinasty Schlafbewegung, Nyctinastie
sleep sickness (*Trypanosoma rhodesiense/gambiense*) Schlafkrankheit, Tsetseseuche
sleepiness Schläfrigkeit
sleeping posture Schlafstellung
sleeplessness/insomnia Schlaflosigkeit

sleet/glaze/frozen rain Eisüberzug, überfrorene Nässe, gefrorener Regen; Graupelschauer
slender-stemmed/leptocaulous dünnstämmig, schlankstämmig
slide/microscope slide Objektträger
sliding filament theory Gleitfilamenttheorie
sliding microtome Schlittenmikrotom
sliding stage *micros* Gleittisch
sliding tubule hypothesis Gleittubulushypothese
slime body/slime plug/P-protein body Schleimkörper, Schleimpfropfen, Proteinkörper
slime gland/mucous gland Schleimdrüse
slime molds/Myxomycota Schleimpilze
- **acellular slime molds/plasmodial slime molds/Myxomycetes** echte Schleimpilze (plasmodial)
- **cellular slime molds/ Acrasiomyycetes/ Dictyosteliomycetes (Myxomycota)** Acrasiomyceten, zelluläre Schleimpilze

slime nets/labyrinthulids/ Labyrinthulomycetes Netzschleimpilze
slimy/mucilaginous/glutinous schleimig
slip/sucker/offset/offshoot *bot* Wurzelspross, Wurzeltrieb
slip face (dune) *geol* Rutschfläche
slipped strand mispairing/slippage replication/replication slippage *gen* Fehlpaarung durch Strangverschiebung
slope/incline Hang, Abhang
slope/scarp Böschung, Abhang
sloping terrain/rolling hills hügelig (leicht hügelige Landschaft)
slot blot Schlitzlochplatte
sloths/Pilosa (Xenarthra) Faultiere
slough (off) abstreifen (Haut/Hülle)
slow growing schwachwüchsig, langsamwachsend
slow-twitch fiber langsam-kontrahierende Faser
sludge Faulschlamm, Sapropel
sludge/mud (alluvial) Schlamm, Schlick
sludge/sewage sludge Klärschlamm (>Faulschlamm)
- **activated sludge** Belebtschlamm

sludge gas/sewage gas Faulgas, Klärgas (Methan)
slugs Nacktschnecken
sluice *n* **(membranes)** Schleuse
sluice/channel *vb* schleusen

S slush 314

slush (partly melted/watery snow)
Schneematsch
• nitrogen slush
schmelzender Stickstoff
(zum Einfrieren von Gewebe)
small intestines Dünndarm
small nuclear ribonucleoprotein
(snRNP) kleines nukleäres
Ribonukleoprotein (snRNP)
small nuclear RNA (snRNA) kleine
nukleäre Ribonukleinsäure (snRNA)
small-angle X-ray scattering (SAXS)
Röntgenkleinwinkelstreuung
small-cell lung cancer
kleinzelliger Lungenkrebs
smartweed family/knotweed
family/dock family/buckwheat
family/Polygonaceae
Knöterichgewächse
smear *micros* Abstrich, Ausstrich
smell/scent/odor *n* Geruch, Duft
• pleasant smell/fragrance/scent
angenehmer Duft
• pungent smell
stechender Geruch
• unpleasant smell
unangenehmer Geruch
smell *vb* riechen
Smilacaceae/catbrier family
Stechwindengewächse
smoke *vb* (foods) räuchern
smoker (black/white) *geol*
(schwarzer/weißer) Raucher
smolt Silbersälmling
smooth/even glatt, eben
smooth muscle/plain muscle/
non-striated/unstriped muscle
glatter Muskel (glatte Muskulatur)
smuts/smut fungi/Ustilaginales
(bunt fungi/brand fungi)
Brandpilze, Flugbrandpilze
snag (standing dead tree)
toter Baum,
Baumstumpf (stehender toter Baum)
snag (tree/branch embedded in
lake/stream) Aststumpf, Knorren,
Baumstumpf (in Seen/Flüssen)
snail pollination/malacophily
Schneckenbestäubung
snail shell Schneckenhaus,
Schneckengehäuse
snails/gastropods/Gastropoda
Schnecken, Bauchfüßer,
Gastropoden
snake-like/ophidian schlangenartig,
Schlangen…
snakeflies/Raphidioptera
Kamelhalsfliegen
snakes/serpents/ophidians/
Serpentes/Ophidia (Squamata)
Schlangen
snap schnappen

snap mechanism
Schnappmechanismus,
Klappmechanismus
snap trap Klappfalle, Schlagfalle
snap-back DNA/fold-back DNA
in sich gefaltete DNA,
zurückgebogene DNA
snap-cap bottle/snap-cap vial *lab*
Schnappdeckelglas,
Schnappdeckelgläschen
snapdragon family/foxglove
family/figwort family/
Scrophulariaceae
Braunwurzgewächse, Rachenblütler
snapping reflex Schnappreflex
snare *arach* Fangfaden, Fangnetz
snare trap (*Arthrobotrys*) *fung*
Schlingfalle
snarl (e.g. lion/tiger)/puff fauchen
sneeze *vb* niesen
sniff schnuppern, schnüffeln
sniff at/nuzzle beschnuppern,
beschnüffeln
snip (horse: white spot) Stern
snood (turkey: fleshy protuberance
at base of bill) Stirnzapfen
(an Schnabelbasis des Truthahns)
snoot/snout Schnauze, Maul
snort schnauben
snout/muzzle
Schnauze, Rüssel, Maul
snow drift Schneewehe,
Schneeverwehung
snow line Schneegrenze
snowstorm (*see*: blizzard)
Schneesturm
snRNA (small nuclear RNA) snRNA
(kleine nukleäre Ribonukleinsäure)
snRNP (small nuclear ribonucleic
protein) snRNP (kleines nukleäres
Ribonukleoprotein)
soak/steep/swell (water uptake)
quellen (Wasseraufnahme)
soak up/absorb aufsaugen,
durchtränken, absorbieren
soaked (e.g. soil/ground)
durchtränkt
soaker hose
Tropfberieselungsschlauch
soaking up/absorption
Aufsaugen, Absorption
soapberry family/Sapindaceae
Seifenbaumgewächse
soaring (flight) Segelflug
• dynamic soaring
dynamischer Segelflug
• ridge soaring/slope soaring
Hangsegeln
• static soaring statischer
Segelflug, Gleiten, Gleitflug
• thermal soaring
Thermiksegelflug, Thermiksegeln

315 **soil-moisture tension** S

sobole/root sucker
Wurzelschössling, Erdspross,
Wurzelreis
sobole/sobolifer Wurzelausläufer,
Wurzelspross, Gehölzausläufer
soboliferous wurzelsprossbildend
sociability/gregariousness
Soziabilität, Geselligkeit,
Geselligkeitsgrad
social behavior Sozialverhalten,
soziales Verhalten
social drive Sozialtrieb
social facilitation/mood induction
ethol soziale Verstärkung,
Stimmungsübertragung,
Mach-mit-Verhalten
social fallow Sozialbrache
social grooming (mammals)/social
preening *orn* soziale Körperpflege
social rank soziale Rangstufe,
Rangstellung
sociality Geselligkeit,
Geselligkeitstrieb
society (animal society)
Sozietät, Gesellschaft,
essentielle Vergesellschaftung
(Tiergesellschaft); *entom* Staat
sociobiology Soziobiologie
sociology Soziologie
sod/turf/grass cover (nonforage
grass) Sode, Grasnarbe,
Rasenstück, Rasendecke
sodded mit Rasen bedeckt
sodium Natrium
sodium dodecyl sulfate (SDS)
Natriumdodecylsulfat
soft bast Weichbast
soft corals/alcyonaceans/
Alcyonaria/Alcyonacea
Lederkorallen
soft palate/velum palatinum
weicher Gaumen, Gaumensegel,
Velum
soft wood Weichholz
softener (e. g. foods) Weichmacher
soggy durchnässt, durchweicht
soil/ground/earth Boden, Erde,
Erdboden, Erdreich
- **acidic soil/acid soil**
saurer Boden
- **boulder layer/loose rock layer**
Untergrund (C-Horizont)
- **caliche/lime pan** Caliche,
Kalkanreicherungshorizont
- **crumb structure** Krümelstruktur
- **fermentation layer**
Fermentationsschicht,
Vermoderungshorizont
- **half-bog soil** anmooriger Boden
- **litter layer (forest soil)**
Streuschicht, Streuhorizont, Förna
- **mineral soil** Mineralboden

- **permafrost soil** Permafrostboden,
Dauerfrostboden
- **potting soil** Topferde
- **sandy soil** Sandboden,
sandiger Boden
- **subsoil (zone of accumulation/**
illuviation) Unterboden, unterer
Mineralhorizont (B-Horizont)
- **topsoil (zone of leaching/**
eluviation) Oberboden, Krume,
Bodenkrume, Bodendeckschicht,
oberer Mineralhorizont
- **zone of accumulation/zone of**
illuviation (B-horizon)
Einwaschungshorizont
soil aggregate/ped Gefügekörper
soil compaction Bodenverdichtung
soil components Bodenbestandteile
soil conditioner Bodenverbesserer
soil conditions Bodenbedingungen
soil conservation Bodenschutz
soil erosion Bodenerosion
soil fertility Bodenfruchtbarkeit
soil horizon Bodenhorizont
- **boulder layer/loose rock layer**
(C-horizon) Untergrund
(C-Horizont)
- **fermentation layer (soil)**
Fermentationsschicht,
Vermoderungshorizont
- **litter layer (forest soil)**
Streuschicht, Streuhorizont,
Förna
- **subsoil (zone of accumulation/**
illuviation) Unterboden, unterer
Mineralhorizont (B-Horizont)
- **topsoil (zone of leaching/**
eluviation) Oberboden, Krume,
Bodenkrume, Bodendeckschicht,
oberer Mineralhorizont
- **zone of accumulation/zone of**
illuviation (B-horizon)
Einwaschungshorizont
soil indicator Bodenzeiger
soil organism Bodenorganismus
soil particle Bodenteilchen
soil particle size Korngröße
(Bodenpartikel)
soil profile Bodenprofil
soil science/pedology Bodenkunde,
Pedologie
soil skeleton Bodenskelett
soil structure Bodengefüge
soil surface/ground level
Erdoberfläche
soil texture Bodentextur,
Bodenbeschaffenheit;
Bodenpartikelgröße, Teilchengröße
(Bodenpartikel)
soil type Bodenart
soil-moisture tension/suction
Saugspannung

soilage 316

soilage/green forage/greenstuff
Grünfutter, Grünzeug
**Solanaceae/nightshade family/
potato family**
Nachtschattengewächse
solanine Solanin
solar age Sonnenzeitalter
solar cell/photovoltaic cell
Solarzelle
solar energy Solarenergie,
Sonnenenergie
**solar plexus/celiac plexus/coeliac
plexus** Sonnengeflecht
solar radiation Sonnenstrahlung
solar tracking/heliotropism
Sonnenwendigkeit,
Sonnenorientierung,
Lichtwendigkeit, Heliotropismus
soldier (social insects) Soldat
sole *n* Sohle
sole pad Sohlenballen
solenia (gastrodermal tubes)
Solenia
solenocyte/archinephridium
Solenocyt, Solenozyt
solenogasters/Solenogastres
Furchenfüßer
solenoid Solenoid (helikale
Chromatinstruktur)
solicit *vb* sich bemühen um,
dringend bitten
soliciting behavior
Begattungsaufforderung
solid phase/bonded phase
Festphase
solid phase reactor *biot*
Festphasenreaktor
solifluction Solifluktion
soligenous (bogs) soligen
solitary/single solitär, einzeln
solitary animal/single animal
Einzeltier, Einzelgänger
solitary flower/single flower
Solitärblüte, Einzelblüte
solitary plant Solitärpflanze,
Einzelpflanze
solstice *astr* Sonnenwende
solubility Löslichkeit
solubility product
Löslichkeitsprodukt
solubilization Solubilisierung
soluble löslich
• **insoluble** unlöslich
solute gelöster Stoff
solute potential
Löslichkeitspotential
solution Lösung
solvate *n* solvatisierter Stoff,
gelöster Stoff (Ion/Molekül)
solvate *vb* solvatisieren
solvation Solvatation
solve *math* lösen

solvent Lösungsmittel, Lösemittel
**solvent/mobile solvent/eluent/
eluant (mobile phase)** Laufmittel,
Elutionsmittel, Fließmittel, Eluent
(mobile Phase)
solvent front Lösungsmittelfront;
Laufmittelfront, Fließmittelfront
(DC)
somaclonal variation
somaklonale Variation
somatic somatisch, körperlich,
Körper...
somatic cell/body cell Körperzelle,
Somazelle, somatische Zelle
somatic mutation
somatische Mutation
**somatic nervous system/
voluntary nervous system**
somatisches Nervensystem,
willkürliches Nervensystem
somatic recombination
somatische Rekombination
somatic skeleton
somatisches Skelett
somatocoel/somatocoele
Somatocöl, Somatocoel
**somatoliberin/somatotropin
release-hormone/somatotropin
releasing factor (SRF)/growth
hormone release hormone/factor
(GRH/GRF)** Somatoliberin,
Somatotropin-Freisetzungshormon
somatolysis Somatolyse
**somatomedin/insulin-like growth
factor (IGF) (sulfation factor/
serum sulfation factor)**
Somatomedin
somatopleure Somatopleura,
somatisches Blatt, parietales Blatt,
Hautfaserblatt
**somatostatin/somatotropin release-
inhibiting factor/growth hormone
release-inhibiting hormone (GRIH)**
Somatostatin
**somatotropin (STH)/growth
hormone (GH)** Somatotropin,
somatotropes Hormon,
Wachstumshormon
somite/somatome Somit,
Ursegment
song/singing *orn* Gesang (Vögel)
• **antiphonal singing**
Wechselgesang
• **attracting song** Lockgesang
• **courtship song/mating song**
Balzgesang
• **duetting** Duettgesang,
Paargesang
• **full song** Vollgesang
• **juvenile song/subsong**
Jugendgesang, Dichten
(Jungvögel)

317 spay

- **luring song/soliciting song**
  Lockgesang
- **mating song/courtship song**
  Werbegesang
- **play song** Spielgesang
- **rehearsal song** Studiergesang
- **subsong** Dichten, Jugendgesang
  (Jungvögel)
- **territorial song** Reviergesang

song repertoire *orn*
  Gesangsrepertoire
sonicate beschallen, mit
  Schallwellen behandeln
sonification/sonication
  Sonifikation, Sonikation,
  Beschallung, Schallerzeugung
  (meist bzgl. Ultraschall)
sook (adult blue crab) weibliche
  Blaukrabbe (*see also:* jimmy)
soporific/soporiferous einschläfernd
soralium (*pl* **soralia)** Soral
  (*pl* Sorale)
- **capitate/capitiform sorelium**
  Kopfsoral
- **forniciform sorelium**
  Helmsoral, Gewölbesoral
- **globose soralium** Kugelsoral
- **labriform soralium** Lippensoral
- **maculiform soralium** Flecksoral
- **maniciform soralium**
  Manschettensoral
- **marginal soralium**
  Randsoral, Bortensoral
- **punctiform soralium** Punktsoral
- **rimiform soralium/fissoral
  soralium** Spaltensoral

sorbent Sorbens (*pl* Sorbentien)
sorbic acid (sorbate) Sorbinsäure
  (Sorbat)
sorbitol Sorbit
sore *n med/vet*
  Wunde, wunde Stelle
soredium Soredium
soriferous/bearing sori mit Sori
sorocarp Sorokarp
sorosis/fleshy multiple fruit
  Beerenverband, Beerenfruchtstand
sort/type/kind/variety/cultivar
  Sorte
sorus/"fruit dot" (ferns) Sorus
SOS response SOS-Antwort,
  SOS-Reaktion
**sounder (herd/hoard/party of pigs
  or wild boar)**
  Rotte (Wildschweine)
source Quelle; Produktionsort
source DNA Ausgangs-DNA
source of infection
  Ansteckungsherd, Ansteckungsquelle
source vegetation
  Quellflurvegetation
sourdough Sauerteig

sourgum family/Nyssaceae
  Tupelobaumgewächse
sow *n* (♀ **swine)** Sau
  (Mutterschwein)
sow *vb bot* säen
space web (with barrier threads)
  arach Raumnetz,
  dreidimensionales Netz,
  Fußangelnetz (mit Stolperfäden)
space-filling model Kalottenmodell,
  raumfüllendes Modell
spacer Abstandshalter
spacer (DNA) *gen* Spacer,
  Zwischensequenz
spadix (*pl* **spadices)** Spadix, Kolben,
  Blütenkolben (Infloreszenz)
span *n* (**wings)** Spannweite
**spanworm/measuring worm/
  looper/inchworm (geometer moth
  larva)** Spannerraupe
spar sparren (Scheinhiebe versetzen)
Sparganiaceae/bur-reed family
  Igelkolbengewächse
sparger (in bioreactor) Verteiler,
  Gasverteiler (Düse in Reaktor)
sparteine Spartein
spasm (convulsion) Krampf,
  Verkrampfung (Konvulsion)
spasmodic dance (bees) trippeln
spat *n* Muschel-Laich (bzw. kleine/
  junge Muschel/Auster)
spate/freshet/flood (of a stream)
  limn Überschwemmung,
  Hochwasser (Fluss)
spathaceous/spathal
  scheidenförmig,
  blütenscheidenförmig
spathe *bot* Spatha, Scheide,
  Blütenscheide
spathed/furnished with a spathe
  mit Spatha versehen
spathose (with or like a spathe)
  spatelartig, spatelig
spathulate/spatulate/spoon-shaped
  spatelförmig
spatial isolation räumliche Isolation
spatial summation räumliche
  Summation
spatula *lab* Spatel
spavin (horse) Spat (Entzündung des
  Sprunggelenks)
spawn/mycelium *fung* Pilzmyzel
spawn *n* (**many small eggs of
  aquatic animals: esp. fish/
  mollusks)** Laich
spawn *vb zool* laichen
spawning ground (fish)
  Laichgründe, Laichstätte,
  Laichplatz
**spay (remove ovaries of female
  animal by surgery)**
  die Eierstöcke entfernen

S spear-shaped

spear-shaped/hastate/hastiform
spießförmig
specialist *ecol* Spezialist
specialization Spezialisierung
specialized transduction
spezielle Transduktion
speciation Speziation, Artbildung
• **allopatric/geographic s. ("other country")**
allopatrische/geografische S.
(in getrennten Arealen)
• **parapatric s.** parapatrische S.
• **sympatric s. ("same country")**
sympatrische S. (in gleichen Arealen)
species/kind Spezies, Art
• **accessory species** Begleitart
• **accidental species**
zufällig auftretende Art, Zufallsart
(biotopfremde Art)
• **alien species/immigrant species**
Fremdart, eingewanderte Art,
Zuwanderer
• **biologic(al) species**
biologische Art
• **character species/characteristic species** Charakterart, Leitart
• **chronospecies** Chronospezies
• **coenospecies** Coenospezies
(>Sammelart/Großart)
• **collective species** Kollektivart,
Sammelart (>Superspezies)
• **core species** Kernart
• **cryptic species** verborgene Art
• **diagnostic species** Kennart
• **differential species**
Differentialart, Trennart
• **ecospecies** Ökospezies
• **endangered species** *ecol*
bedrohte Art
• **endemic species/endemic organism/endemic lifeform/ endemic** Endemit
• **evolutionary species**
evolutionäre Art
• **fugitive species/opportunistic species** vagabundierende Art,
opportunistische Art
• **index species/guide species**
Leitart
• **indicator species**
Indikatorart, Zeigerart
• **indifferent species**
indifferente Art
• **indigenous species/ native species/native organism/ native lifeform**
Indigen, einheimische Art
• **keystone species** Schlüsselart
• **macrospecies** Makrospezies,
Großart
• **microspecies** Mikrospezies,
Kleinart

• **monocentric species**
monozentrische Art
• **monotypic species**
monotypische Art
• **morphospecies/morphological species** Morphospezies,
morphologische Art
• **native species** einheimische Art
• **opportunistic species**
opportunistische Art
• **paleospecies (chronospecies)**
Paläospezies (Chronospezies)
• **pioneer species** Pionierart
• **polycentric species**
polyzentrische Art
• **polytypic species**
polytypische Art
• **quasispecies** Quasispezies
• **ring species** Ringart
• **satellite species/marginal species** Satellitenart, Randart
• **sibling species** Geschwisterarten
• **sister species** Schwesterart
• **stem species** Stammart
• **subspecies** Subspezies, Unterart
• **superspecies** Superspezies,
Überart
• **twin species (pair of sibling species)** Zwillingsarten
• **type species** Typus-Art
• **vicarious species**
vikariierende Art, Stellvertreterart
species aggregate/collective group
Artenkreis
species composition
Artenzusammensetzung
species diversity Artenvielfalt,
Artenmannigfaltigkeit
species flock/species swarm
Artenschwarm
species importance value
Artmächtigkeit
species inventory Arteninventar,
Artenbestand
species name Artname
species richness Artenreichtum
species specific *adj/adv*
artspezifisch, arttypisch
species-abundance curve Arten-
Rangkurve, Artenabundanzkurve
species-area curve Arten-Arealkurve
species-specific behavior
artspezifisches/arttypisches
Verhalten
specific gravity (wood density)
spezifisches Gewicht
(Dichte von Holz)
specific name/specific epithet
Artname, Artbezeichnung,
Epitheton
specific spezifisch
• **nonspecific** unspezifisch

319 **spiderwort family** **S**

specificity of action
Wirkungsspezifität
specificity Spezifität
specify spezifizieren
specimen/sample
Exemplar, Muster, Probe
● **preserved specimen** Präparat
● **type specimen** Typexemplar
specimen jar Sammelglas
speckled/patched/spotted/spotty
fleckig
spectacle (snakes) Brille
spectacle eyepiece/high-eyepoint
ocular *micros* Brillenträgerokular
spectrum (*pl* **spectra/spectrums)**
Spektrum (*pl* Spektren)
speech/language Sprache
speech center *neuro* Sprachzentrum
sperm/semen (ejaculate) Sperma,
Samen (Ejakulat)
sperm/sperm cell/spermium/
spermatozoon (male gamete)
Spermium, Samen, Sperma,
Samenzelle, Spermatozoon
(männliche Geschlechtszelle)
sperm chamber/spermatheca/
seminal receptacle/sperm
receptacle/receptaculum seminis
Samentasche, Receptaculum seminis
sperm competition
Spermienkonkurrenz
sperm oil (whale) Walratöl
sperm precedence (*formerly***: sperm**
competition) Spermapräzedenz
(*früher*: Spermienkonkurrenz)
sperm web *arach* Spermanetz
spermaceti/cetaceum Walrat
spermaceti oil/sperm oil Walratöl
spermatheca/sperm chamber/
seminal receptacle/sperm
receptacle/receptaculum seminis
Samentasche, Receptaculum seminis
spermatic cord/funiculus
spermaticus Samenstrang
spermatid/spermatoblast/spermid
Spermatid, Spermid
spermatium *fung* Spermatium
(*pl* Spermatien)
spermatocyte
Spermatocyt, Spermatozyt,
Spermiocyt, Spermiozyt
● **primary s./spermiocyte**
primärer Spermatozyt,
Spermatozyt I. Ordnung
● **secondary s./prespermatid**
sekundärer Spermatozyt,
Spermatozyt II. Ordnung,
Präspermatid
spermatogenesis/spermiogenesis
Spermatogenese, Spermiogenese,
Samenentwicklung
spermatogeny Spermatogenie

spermatogonium/primordial male
germ cell Spermatogonium,
Ursamenzelle
spermatophore/sperm packet
Spermatophore, Samenträger,
Samenpaket
spermatophyte/seed-bearing plant
Spermatophyt, Samenpflanze
spermatozoon (*pl* **spermatozoa)/**
sperm/spermium/sperm cell
Spermatozoon, Samenzelle, Samen,
Sperma, Spermium (männliche
Geschlechtszelle)
spermid *see:* spermatid
spermidine Spermidin
spermine Spermin
spermozeugma Spermiozeugma,
Spermienbündel, Spermiodesmos
sphecophily/wasp pollination
Wespenbestäubung
sphenethmoid bone Gürtelbein
sphenoid bone/os sphenoidale
(skull) Keilbein
sphenoid/wedge-shaped/cuneate/
cuneiform keilförmig
sphenophyllum family/
Sphenophyllaceae
Keilblattgewächse
spherical sphärisch, kugelig
spheroplast Sphäroplast
spherosome Sphärosom
sphincter muscle/musculus
sphincter Sphinkter, Schließmuskel
sphinganine Sphinganin
sphingomyelin Sphingomyelin
sphingophily/hawk moth
pollination Nachtfalterbestäubung
sphingosine Sphingosin
sphragis/mating plug (Lepidoptera)
Sphragis, Begattungssiegel,
Kopulationssiegel
spice Gewürz
spicebush family/strawberry-shrub
family/Calycanthaceae
Gewürzsträucher
spicule/spikelet *bot* Ährchen
spicule *zool* **(e. g. sponges)**
Spiculum, Sklerit, Nadel,
Skelettnadel
spicy (hot) *cul* (stark) gewürzt
(oft *syn* für: hot>scharf)
spiderlike/spidery/arachnoid
spinnenartig
spiderling Jungspinne
spiders/Araneae Spinnen,
Webspinnen
spiderweb/cobweb Spinnennetz,
Spinnwebe
spiderweb-like spinnwebartig,
spinnennetzartig
spiderwort family/Commelinaceae
Commelinengewächse

S spike 320

spike (inflorescence) Ähre
spike/fastigium Spitze
spike/serration/projection Stachel,
Zacke
spike moss (*Selaginella*) Moosfarn
spike-moss family/small club-moss
family/selaginella family/
Selaginellaceae
Moosfarngewächse
spiked/spiky/spikey spitz, stachelig,
stachlig; ährentragend
spikelet/spicule (inflorescence)
Ährchen, Blütenährchen
spiky/spikey/spiny/thorny
stachelig, stachlig
spill *vb* verschütten, vergießen,
umschütten, umwerfen, auslaufen
(Flüssigkeit)
spill *n* Verschütten, Vergießen,
Umschütten, Umwerfen, Auslaufen,
Austreten (Flüssigkeit)
spillage das Übergelaufene, das
Vergossene; Überlaufen, Vergießen
spin *vb* (spiderweb/cocoon)
spinnen, weben
spinal column/vertebral column/
backbone/spine Wirbelsäule,
Rückgrat
spinal cord/neural tube/medullary
canal/nerve cord Rückenmark,
Neuralrohr, Medulla spinalis
spinal ganglion Spinalganglion
spinal nerve Spinalnerv
spindle Spindel
spindle apparatus *cyt*
Spindelapparat
spindle fiber *cyt* Spindelfaser
spindle organ/muscle spindle
Spindelorgan, Muskelspindel
spindle pole body *fung*
Spindelpolkörper
spindle-shaped/fusiform
spindelförmig
spindle-tree family/staff-tree
family/bittersweet family/
Celastraceae
Spindelbaumgewächse,
Baumwürgergewächse
spine/prick/needle Stachel, Nadel,
Dorn
spine/spinal column/vertebral
column/backbone Wirbelsäule,
Rückgrat
spine/thorn *bot* Stachel; Blattdorn,
Nebenblattdorn
spinner (mayfly adult/imago)
Spinner (Imago der Eintagsfliegen);
Blinker (Angeln)
spinneret/sericterium (labial gland)
Spinndrüse, Seidendrüse,
Sericterium (Labialdrüse)
spinneret/spinner *arach* Spinnwarze

spinose/spinous stachelig; dornig
spinous process Spinalfortsatz,
Dornfortsatz
spiny/thorny stachelig, spitz
spiny fishes/Acanthodii
Stachelhaie
spiny-finned stachelflossig
spiny-headed worms/thorny-
headed worms/acanthocephalans/
Acanthocephala Kratzer
spiny-rayed stachelstrahlig
spiracle/ostium/stigma Spiraculum,
Spirakulum, Atemloch, Luftloch,
Tracheenöffnung, Ostium, Stigma
spiral/helix *n* Spirale, Schraube,
Helix
spiral/spiraled/twisted/helical
(spirally twisted) spiralig
(spiralig gewunden)
spiral cecum Spiralcaecum
spiral cleavage Spiralfurchung
spiral coil/gyre Windung, Gyros
spiral flap/spiral valve
(Chondrichthyes) Spiralfalte
spiral grain (wood) Spiraltextur
spiral intestine Spiraldarm
spiral movement/spiral coiling
Windung (Bewegung)
spiral organ/organ of Corti
Cortisches Organ
spiral thread/taenidium (spiral
thickening of intima) Spiralfaden,
Taenidium
spiral valve (frog heart)
Spiralklappe
spiral winding/coiling
Spiralwindung
spiraled/helical/spirally twisted/
spirally coiled/contorted
schraubig, spiralig gewunden,
helical
spirally coiled/strombuliform
spiralig aufgewickelt
spire (inflorescence) Blütenähre
spire (point) zulaufende Spitze,
spitz zulaufender Grashalm
spirilla (*sg* spirillum) Spirillen
spirit/spiritus *chem* (of petroleum/
shale/wood) Spiritus
spirit/distillate Destillat, "Geist"
spirits/distilled alcoholic liquid
Alkohol *sensu lato*
spit/cuspate foreland *n* (small point
of sand/gravel running into
water) *mar* Haken, Sandhaken,
Strandhaken (Sporn aus Sand)
spit *n vulg* Spucke
spit *vb* spucken, speien
spittle Speichel, Schaum (z.B. von
Zikaden: Kuckucksspeichel)
spittlebugs (Auchenorrhyncha)
Schaumzikaden

splanchnocranium
Splanchnocranium, Gesichtsschädel
splanchnopleure Splanchnopleura,
viscerales Blatt, viszerales Blatt,
Darmfaserblatt
splash zone (supralittoral zone)
Spritzwasserzone, Spritzzone,
Gischtwasserzone, Gischtzone
(Supralitoral)
spleen/lien Milz
spleenwort family/Aspleniaceae
Streifenfarngewächse
splenic Milz.., die Milz betreffend
splenic capsule Milzkapsel
splenic fever/anthrax Milzbrand
splenic node/splenic nodule/splenic
corpuscle/splenic follicle/
Malpighian body/Malpighian
corpuscle Milzknötchen,
Malpighi-Körperchen,
Milzkörperchen, Milzfollikel
splenic nodule Malpighi-Körperchen
splenic pulp (red/white) Milzpulpa
(rote/weiße)
splenic trabeculae/trabeculae lienis
Milzbalken, Milztrabekel
spleniform milzförmig, milzartig
splice spleißen
splice acceptor site *gen*
Spleiß-Akzeptorstelle
splice donor site *gen*
Spleiß-Donorstelle
splice grafting/whip grafting *bot*
Kopulation, Kopulieren, Schäften
(Pfropfung); (with stock larger than
scion) Anschäften
splice site *gen* Spleiß-Stelle
• **cryptic splice site**
verborgene Spleißstelle
spliceosome Spleißosom
splicing *gen* Spleißen
• **alternative splicing**
alternatives Spleißen
• **differential splicing**
differentielles Spleißen
splicing junction *gen* Spleiß-
Junktion, Spleiß-Verbindungsstelle
splinkers (sequencing primer
linkers) Splinkers
splint bone/fibula Wadenbein, Fibula
splintwood/sapwood/alburnum
Splintholz
split *chem* aufspalten, zerlegen
split/cleave *vb* spalten
split/cleaved/cracked/...fid *adj/adv*
bot gespalten, spaltig
split gene/interrupted gene
gestückeltes Gen, Mosaikgen
splitting *chem* Aufspaltung,
Zerlegen
spoke/radius Speiche, Radius
(auch: Netzspeiche)

sponges/poriferans/Porifera
Schwämme, Schwammtiere,
Poriferen
• **calcareous sponges/Calcarea**
Kalkschwämme
• **coralline sponges/**
sclerosponges/Sclerospongiae
Sclerospongien
• **glass sponges/Hexactinellida**
Glasschwämme, Hexactinelliden
• **horny sponges/Cornacuspongiae**
Hornschwämme,
Netzfaserschwämme
• **siliceous sponges/demosponges/**
Silicospongiae Kieselschwämme
(Demospongien)
spongiform encephalopathy
spongiforme Enzephalopathie,
Hirnschwammerkrankung,
Hirnschwammkrankheit
spongin (sponge protein) Spongin
spongioblast *neuro/embr*
Spongioblast
spongiocyte Spongocyt, Spongozyt
spongiome Spongiom
spongiose fungus/polypore/pore
fungus Schwammpilz, Porling
spongy bone/cancellous bone
spongiöser Knochen
spongy parenchyma (mesophyll)
Schwammparenchym,
Schwammgewebe
spontaneous generation
hypothesis Urzeugungshypothese
spontaneous mutation
Spontanmutation
spontaneous mutation rate
spontane Mutationsrate
spoon worms/echiuroid worms/
Echiura Igelwürmer,
Stachelschwänze, Echiuriden
spoon-like/cochlear löffelartig,
cochlear
sporadic sporadisch
sporangiocarp Sporangienbehälter
sporangiole Sporangiole
sporangiophore Sporangiophor,
Sporangienträger
sporangium/spore case
Sporangium, Sporenbehälter
spore Spore
spore case/sporangium
Sporenbehälter, Sporangium
spore case/theca Theca, Theka
spore print *fung* Sporenabdruck
spore-former/sporozoans/Sporozoa
Sporentierchen, Sporozoen
sporiferous sporentragend
sporocarp *fung* Sporokarp
sporocyst Sporocyste, Sporozyste
sporocyte Sporozyt,
Sporenmutterzelle

sporogenic

sporogenic/sporogenous sporogen, sporenerzeugend
sporogony/sporogeny/gamogony (in protozoans) Sporogonie
sporophore (spore-bearing structure) Sporophor, Sporenträger
sporophyte Sporophyt
sporosac (hydrozoans) Sporosac, Keimtasche
sporozoite Sporozoit
sport/rogue (deviation usually by somatic mutation) Abart, Spielart, Variation, aus der Art schlagende Pflanze, Missbildung
sportfishing/game fishing Sportfischerei, Sportfischen
spot/blot/stain/stigma Makel, Fleck, Stigma
spot blot/dot blot Rundlochplatte
spot desmosome Plaquedesmosom
spotted/mottled gefleckt, fleckig
spotted fever/typhus/typhus exanthematicus (*Rickettsia* spp.) Fleckfieber, Flecktyphus, Typhus
spout *n* **(whales)** Fontäne
spout/blow *vb* **(water: e.g. whales)** Wasser speien, spritzen, abblasen
spray (young shoot) *bot* junger Zweig, junges Ästchen, kleiner Blütenzweig, Reis
spray/ocean spray *mar* Sprühwasser, Salzwasserspray
spray zone (supralittoral zone) Sprühwasserzone, Sprühzone (Supralitoral)
spread/expand/propagate/disperse/disseminate ausbreiten, verbreiten (*auch:* Krankheiten)
spread/scatter/distribute spreiten, streuen, verstreuen, ausstreuen, verteilen
spread/spreading *n* **(disease)** Ausbreitung
spread-plate method Spatelplattenverfahren
spreading/expansion/propagation/dispersal/dissemination Ausbreitung, Verbreitung, Propagation
spreading/scattering/distribution Spreitung; Streuung, Verstreuen, Verteilung
sprig (shoot/twig/spray) Zweiglein, Schössling
spring/springtime (season) Frühling, Frühjahr
spring/source Quelle
spring fen Quellmoor
spring tide Springtide
springtails/garden fleas/Collembola Springschwänze, Collembolen
springwater Quellwasser

springwood/earlywood Frühlingsholz, Weitholz, Frühholz
sprinkle/spray besprengen, beregnen
sprinkler Sprinkler, Sprenger, Beregnungsanlage, Berieselungsanlage
sprinkler irrigation Spritzbewässerung, Beregnungsbewässerung, künstliche Beregnung
sprout/bud/put forth sprießen, ausschlagen, austreiben, Knospen treiben
sprout/seedling (e.g. bean sprout) Keimling (z. B. Bohnenkeimling)
sprout/sprouting/budding Austrieb, Sprossung, Knospung
spur Sporn
• **floral spur** Blütensporn
spur *immun* **(immunodiffusion)** Sporn
spur shoot/fruit-bearing bough (a short shoot) *bot* Fruchtholz (Kurztrieb), Lateralorgan; Infloreszenz-Kurztrieb
spur vein (horse) Sporvene
spur-and-groove zone/buttress zone (reef) Grat-Rinnen-System
spurge family/Euphorbiaceae Wolfsmilchgewächse
spurge olive family/Cneoraceae Zeilandgewächse, Zwergölbaumgewächse
spurious/false falsch
spurious fruit/pseudocarp/false fruit Scheinfrucht
spurious vein/vena spuria *zool* Scheinader
spurred gesport
sputter *micros* (EM) sputtern, besputtern
sputtering *micros* (EM) Sputtern, Besputtern, Besputterung, Kathodenzerstäubung (auch: Metallbedampfung)
sputtering unit/sputtering appliance *micros* (EM) Besputterungsanlage
squab/chick *orn* Jungvogel, Kücken, Küken
squamata (incl. lizards & amphisbaenians & snakes) Squamata (Eidechsen & Schlangen)
squamate/squamid/scaly (reptiles) schuppig
squamellate/squamelliferous/squamulose mit kleinen Schuppen bedeckt
squamicidal squamicid, squamizid, schuppenspaltig
squamiferous/squamigerous mit Schuppen bedeckt

squamiform/scale-like
schuppenförmig
squamosal *n* **(bone)** Squamosum,
Schuppenbein
squamosal suture/squamous suture
(skull) Schuppennaht
squamous squamös, schuppig,
schuppenförmig,
mit Schuppen bedeckt
squamous epithelium
Plattenepithel, Säulenepithel,
Zylinderepithel
squamous suture/squamosal suture
(skull) Schuppennaht
squamulose/squamulate
squamulös, feinschuppig
squared timber Kantholz
squash/squash mount *micros*
Quetschpräparat
squeal/squeak (pigs/guinea pigs)
quieken, quietschen
squealer (young bird) junger
Vogel
squids/Teuthoidea (Teuthida)
Kalmare
squirrel fishes (primitive
acanthopterygians)/Beryciformes
Schleimköpfe, Schleimkopfartige
SRP (signal recognition particle)
SRP (Signalerkennungspartikel)
SSCP (single strand conformation
polymorphism) SSCP
(Einzelstrang-Konformations-
Polymorphismus)
St. John's wort family/mamey
family/mangosteen family/clusia
family/Clusiaceae/Guttiferae/
Hypericaceae Hartheugewächse,
Johanniskrautgewächse
stab culture *micb* Stichkultur,
Einstichkultur (Stichagar)
stabilimentum (spiderweb) *arach*
Stabilimentum, Stabiliment
stabilization Stabilisierung
stabilizer Stabilisator
stabilizing selection stabilisierende
Selektion/Auslese
stable/stables *n* **(for domesticated**
animals) Stall, Stallung
stable RNA stabile RNA
stack *vb* stapeln
stacked (stack) gestapelt (stapeln)
(z. B. Membranzisternen)
stacked bases *gen* gestapelte Basen
stacked membranes Membranstapel
stacking forces Stapelkräfte
stacking gel Sammelgel
staff-tree family/spindle-tree
family/bittersweet family/
Celastraceae
Spindelbaumgewächse,
Baumwürgergewächse

stag (castrated domesticated
animal) nach der Reife kastriertes
Männchen (Nutztiere)
stag erwachsenes Männchen
(Eber/Schafsbock/Hirsch etc.)
stage/phase
Stadium, Phase (zeitlich)
stage/microscope stage Objekttisch
● **mechanical stage** Kreuztisch
● **plain stage** Standardtisch
● **rotating stage** Drehtisch
● **sliding stage** Gleittisch
stage clip *micros*
Objekttisch-Klammer
stage micrometer *micros*
Objektmikrometer
staggered nicks (e. g. in double
stranded DNA) versetzte
Einschnitte (Einzelstrangbrüche)
(z.B in doppelsträngiger DNA)
stagnant water stehendes Gewässer
stain *n micro* Farbstoff
stain/staining *techn/micros*
(process) Färben, Färbung,
Einfärbung, Kontrastierung
● **counterstain/counterstaining**
Gegenfärbung
● **differential staining/contrast**
staining Differentialfärbung,
Kontrastfärbung
● **Golgi staining method**
Golgi-Anfärbemethode
● **Gram stain** *micb* Gram-Färbung
● **immunogold-silver staining**
Immunogold-Silberfärbung
● **negative staining/negative**
contrasting
Negativkontrastierung
● **periodic acid-Schiff stain**
(PAS stain) Periodsäure,
Schiff-Reagens (PAS-Anfärbung)
● **quick-stain** Schnellfärbung
● **supravital staining**
Supravitalfärbung
● **vital staining** Lebendfärbung,
Vitalfärbung
stain *vb techn/micros* färben,
einfärben, nachfärben, kontrastieren
stain *vb* **(wood)** beizen
stainability *micros* Färbbarkeit
staining Färben, Färbung,
Einfärbung, Kontrastierung
staining dish/staining jar/staining
tray Färbeglas, Färbetrog,
Färbewanne
stake *n* **(for plant support)** Stütze
(zusätzliche Pfahlstütze), Pfahlstütze
für Pflanzen
stake *vb* **(attach an animal to a**
pole) anpflocken
stale/staling (urination of horses/
cattle) harnen, stallen (Vieh)

stalk

stalk/axis/spindle Stiel, Achse, Stengel, Halm, Spindel
- **embryonic stalk/body stalk/ connecting stalk** Bauchstiel
- **optic stalk** Augenbecherstiel
- **pituitary stalk** Hypophysenstiel
stalk/pedicle/pedicel/peduncle *bot* Stiel
stalk/stem *bot* Strunk (Stengel)
stalk cell (pollen of cycads) Stielzelle, Wandzelle, Dislokatorzelle, Dislocatorzelle
stalk game *zool vb* Wild erlegen
stalk prey *zool vb* Opfer erlegen
stalked/petiolate/stipitate/ pedunculate *bot/zool* gestielt
- **not stalked/sessile** ungestielt, sitzend
stalked compound eye (Ephemeroptera) Turbanauge
stalked eye Stielauge
stall *n* Stand, Box (im Pferdestall)
stall *n aer* Sackflug
stall *vb* einstallen, im Stall füttern/mästen
stall *vb aer* absacken, abrutschen (Sackflug)
stallion Hengst (Zuchthengst: stud/studhorse)
stalwart sword fern family/ Oleandraceae Nierenfarngewächse
stamen Staubblatt, "Staubgefäß"
stamina Lebenskraft, Vitalität, Stärke
staminal hair Staubblatthaar
staminate/male staminat, männlich
staminate flower Staubblüte, männliche Blüte
staminode/staminodium (abortive/ sterile stamen) Staminodium (unfruchtbares/steriles Staubblatt)
stampede (e. g. cattle) wilde, panische Flucht (z. B. Rinder)
stand/stock *agr/for* Bestand
- **low-density stand (forest)** lichter Wald
- **small stand/small tree stand/ thicket** Horst
stand of timber Holzbestand
standard (tree stem) Hochstamm
standard deviation/root-mean-square deviation *stat* Standardabweichung
standard error (standard error of the mean = SEM) *stat* Standardfehler (des Mittelwerts), mittlerer Fehler
standard metabolic rate Standardstoffwechselrate
standard petal/banner petal/ vexillum *bot* Fahne (Fabaceen-Blüte)

standardization Standardisierung, Vereinheitlichung
standardize standardisieren, vereinheitlichen
standby (of machine/apparatus/ appliance) Bereitschaft (eines Gerätes)
standing crop Erntebestand, auf dem Halm stehende Ernte
Staphyleaceae/bladdernut family Pimpernussgewächse
staple *agr* Haupterzeugnis
staple crop Hauptanbauprodukt
staple food/basic food/main food source Grundnahrungsmittel, Hauptnahrung, Hauptnahrungsquelle
star activity (of restriction enzymes) Sternaktivität (veränderte Spezifität von Restriktionsenzymen)
star mine/asteronome (a leaf mine) *bot* Sternmine
star-anise family/illicium family/ Illiciaceae Sternanisgewächse
star-shaped/stellate/radial/ actinomorphic sternförmig, radiär, aktinomorph
starch Stärke
starch granule Stärkekorn
starter culture (growth medium) Starterkultur (Anzuchtmedium)
starting (a culture) Anzucht
starting material/basic material/ source material/primary material/ preparation Ausgangsstoff, Ausgangsmaterial, Ansatz, Präparat
startle *vb* erschrecken, überraschen
startle behavior Schreckverhalten
startle display Schreck-Schaustellung
startle reflex Schreckreflex
starvation Hungern
- **death by starvation** Verhungern
starvation phase *micb* Auszehrphase
starve hungern, aushungern
- **die of starvation** verhungern
starwort family/water starwort family/Callitrichaceae Wassersterngewächse
state/condition Zustand
state forest Staatswald
static culture statische Kultur
static soaring *orn* statischer Segelflug
stationary phase/stabilization phase stationäre Phase
statistic/statistic value Kenngröße
- **rank statistics/rank order statistics** Rangmaßzahlen
- **vital statistics** demografische Kennzahlen

325 sterilize

statistical deviation statistische Abweichung
statistical error statistischer Fehler
statistical inference statistische Inferenz
statistics Statistik
• **biostatistics** Biostatistik
• **order statistics** Ordnungsstatistik
statoblast (hibernaculum/winter bud) Statoblast (Dauerknospe/Hibernaculum)
statocyst/apical sense organ (ctenophores) Statocyste, Statozyste, Scheitelorgan, Apikalorgan
statolith Statolith, Schwerestein(chen)
stator-rotor impeller/Rushton-turbine impeller (bioreactor) Stator-Rotor-Rührsystem
stature Statur, Gestalt, Wuchs; Wuchshöhe, Größe
stauromedusas/Stauromedusae Stielquallen, Becherquallen
stay-apparatus, passive (horse) passiver Stehapparat
steady state gleichbleibender Zustand, stationärer Zustand (zeitl.)
steady state/steady-state equilibrium *chem* Fließgleichgewicht, dynamisches Gleichgewicht
stearic acid/octadecanoic acid (stearate/octadecanate) Stearinsäure, Octadecansäure (Stearat/Octadecanat)
steep/soak/swell (water uptake) quellen (Wasseraufnahme)
steer *n* **(male bovine animal castrated early)** Stier
steer *vb* steuern
stelar theory *bot* Stelärtheorie
stele/central cylinder *bot* Stele, Zentralzylinder
stellate/star-shaped sternförmig, Stern…
stellate cell Sternzelle
stellate hair Sternhaar
stellate parenchyma Sternparenchym
stem Stamm
stem/trunk/shaft *bot* Baumstamm, Holzstamm
stem bundle/shoot bundle Sprossbündel
stem cell/initial/primordial cell (precursor cell) Stammzelle, Initiale, Primordialzelle (Vorläuferzelle)
stem culture/stock culture Stammkultur, Impfkultur

stem nematogen (dicyemids) *zool* Stammnematogen
stem reptiles/cotylosaurs/Cotylosauria Stammreptilien
stem rot *bot* Stammfäule
stem species Stammart
stem succulent Stammsukkulente
stem-borne stammbürtig, achsenbürtig (shoot-borne sprossbürtig)
stem-clasping/amplexicaul stengelumfassend, amplexikaul
stem-loop structure Stammschleifenstruktur
stem-tuber Sprossknolle (oberirdisch)
stemlet Stämmchen
stemma (*pl* **stemmata/stemmas**) **(dorsal/lateral ocellus)** Stemma, Punktauge (Einzelauge/Ocelle)
stemma (insect larvas)/lateral eye/lateral ocellus Lateralauge, Lateralocellus, Seitenauge (Punktauge/Einzelauge)
stenoecious/stenecious/stenoecic stenök
stenogastry Stenogastrie
stenohaline stenohalin
stenostomates/stenolaemates/Stenostomata/Stenolaemata ("narrow-throat" bryozoans) Engmäulen
stenotele Stenotele
step/pace/stride (long step) Schritt
step cline Stufen-Kline
steppe Steppe
Sterculiaceae/cacao family/cocoa family Sterkuliengewächse, Kakaogewächse
stere (stack of cordwood: 1 cbm) Ster (Holz) 1 m^3
stereo microscope Stereomikroskop
stereoisomere Stereroisomer
stereoscopic vision/binocular vision stereoskopisches Sehen
stereoselective stereoselektiv
stereospecificity Stereospezifität
steric/sterical/spacial sterisch, räumlich
steric hindrance sterische Hinderung, sterische Behinderung
sterigma Sterigma
sterile/disinfected steril, desinfiziert
sterile/infertile steril, unfruchtbar
sterile bench *lab* sterile Werkbank
sterile filtration *lab* Sterilfiltration
sterility/infertility Sterilität, Unfruchtbarkeit
sterilization/sterilizing Sterilisation, Sterilisierung
sterilize sterilisieren

sternite (insects: ventral sclerite/ part of sternum) Sternit, Bauchstück, Bauchplatte

sternum/breastbone (vertebrates) Sternum, Brustbein

sternum/ventral plate (insects) Sternum, Brustplatte, Brustschild, Bauchteil, Bauchschild

sterol Sterin, Sterol

Stewart's organ Stewart'sches Organ, Gabelblase

stick/cane Stock, Stecken

stick-and-ball model/ball-and-stick model *chem* Stab-Kugel-Model, Kugel-Stab-Model

stick-insects/Phasmida Gespenstheuschrecken & Stabheuschrecken

sticklebacks (and sea horses)/ Gasterosteiformes Stichlingsartige, Stichlingverwandte

sticky/glutinous/viscid klebrig, glutinös

sticky end/cohesive end *gen* klebriges Ende, kohäsives Ende, überhängendes Ende

stiff steif

stiffened versteift

stiffness/pliability Biegsamkeit

stifle (horse) Knie

stifle bone (horse) Kniescheibe

stifle joint (horse) Kniegelenk

stigma/spot Stigma, Fleck, Makel; (eyespot) Augenfleck

stigma (pistil/carpel) *bot* Narbe (Fruchtblattnarbe)

stigma head (clublike swollen stigma) *bot* Narbenkopf

stigmasterol Stigmasterin, Stigmasterol

stillage/distillers' grains Schlempe (moist: Nassschlempe, dry: Trockenschlempe)

stillbirth Totgeburt

stillborn tot geboren

stimulate/excite anregen

stimulation/excitation Stimulierung, Anregung
- **irritation** Reizung

stimulus/incentive/stimulant Anreiz, Ansporn, Stimulans

stimulus/irritation Stimulus, Reiz
- **adequate stimulus** adäquater Reiz
- **conditioned stimulus (CS)** *ethol* bedingter Reiz
- **external stimulus** Außenreiz
- **internal stimulus** Innenreiz
- **key stimulus/sign stimulus (release stimulus)** Schlüsselreiz, Auslösereiz
- **light stimulus** Lichtreiz

- **liminal stimulus/threshold stimulus/minimal stimulus** Schwellenreiz
- **reinforcing stimulus/ reinforcement/amplification** verstärkender Reiz, Verstärkung, Bekräftigung
- **response (to stimulus)** *neuro* Antwort
- **threshold stimulus/liminal s./ minimal s.** Schwellenreiz
- **unconditioned stimulus (US)** *ethol* unbedingter Reiz

stimulus reinforcement Reizverstärkung

stimulus pattern Reizmuster

stimulus threshold Reizschwelle

stimulus transduction Reizumwandlung

sting/"bite" *n* Stich, "Biss"

sting/pierce/prick painfully *vb* stechen

sting *vb* (burning pain) brennen

sting/stinger/piercing stylet Stachel, Stechborsten

sting sheath Stachelscheide

stinger cell/cnidocyte/nematocyte Nesselzelle, Cnidocyt, Cnidozyt, Nematozyt

stinging hair/urticating hair/ urticating trichome Brennhaar

stinging zooid/protective polyp/ dactylozooid Wehrpolyp, Dactylozoid

stingrays/Myliobatiformes Stechrochenartige

stinkhorns/stinkhorn family/ Phallaceae Stinkmorcheln, Rutenpilze

stipe/stalk Blattstiel (Algen, Farne, Palmen), Strunk (Blattstiel), Stengel, kurzer Stiel

stipe (certain algas: *Laminaria*) Algenstiel, Cauloid, Kauloid

stipe *fung* Pilzstiel

stipes *zool* Stipes, Stammstück, Haftglied (Maxille)

stipitate/stalked (petiolate) gestielt

stipular spine *bot* Stipulardorn, Nebenblattdorn

stipule *bot* Stipel, Nebenblatt

stir/agitate rühren

stirps (*pl* stirpes)/lineage Stamm, Linie, Familienzweig; *zool* Überfamilie

stirred cascade reactor (bioreactor) Rührkaskadenreaktor

stirred loop reactor (bioreactor) Rührschlaufenreaktor, Umwurfreaktor

stirred-tank reactor (bioreactor) Rührkesselreaktor

stirrer/impeller/agitator (in bioreactors) Rührer
- **anchor impeller** Ankerrührer
- **crossbeam impeller** Kreuzbalkenrührer
- **disk turbine impeller** Scheibenturbinenrührer
- **flat-blade impeller** Scheibenrührer, Impellerrührer
- **four flat-blade paddle impeller** Kreuzblattrührer
- **gate impeller** Gitterrührer
- **helical ribbon impeller** Wendelrührer
- **hollow stirrer** Hohlrührer
- **marine screw impeller** Schraubenrührer
- **multistage impulse countercurrent impeller** Mehrstufen-Impuls-Gegenstrom (MIG) Rührer
- **off-center impeller** exzentrisch angeordneter Rührer
- **paddle stirrer/paddle impeller** Schaufelrührer, Paddelrührer
- **pitch screw impeller** Schraubenspindelrührer
- **pitched-blade fan impeller/ inclined paddle impeller** Schrägblattrührer
- **profiled axial flow impeller** Axialrührer mit profilierten Blättern
- **rotor-stator impeller/ Rushton-turbine impeller** Rotor-Stator-Rührsystem
- **screw impeller** Schneckenrührer
- **self-inducting impeller with hollow impeller shaft** selbstansaugender Rührer mit Hohlwelle
- **stator-rotor impeller/ Rushton-turbine impeller** Stator-Rotor-Rührsystem
- **two flat-blade paddle impeller** Blattrührer
- **variable pitch screw impeller** Schraubenspindelrührer mit unterschiedlicher Steigung

stirrer/mixer Rührgerät, Mixer
stirrup/stapes (ear) Steigbügel, Stapes
stock/inventory Bestand, Besatz, Inventar
stock/grafting understock *bot/hort* Pfropfgrundlage, Pfropfunterlage
stock/main stem/trunk *bot* Stamm
stock/number/quantity Bestand
stock/prototype/archetype Urform, Urtyp
stock *vb* (a pond with fish) einsetzen (Fische in einen Teich)

stock (mother plant from which cuttings/slips are taken) Mutterpflanze (von der Stecklinge/ Pfropfreißer entnommen werden)
stock horse Zuchtpferd
stock solution *chem/micb* Stammlösung
stocking density/stocking rate (e.g. fish in a pond) Bestandsdichte, Besatzdichte
stoichiometric(al) stöchiometrisch
stolon (aboveground horizontal stem) Stolon, Ausläufer, Ausläuferspross
stolon *alg/fung* rhizomartige Hauptachse (Algen/Zygomyceten)
stolon (stalk-like structure: hydrozoans) Stolo, Stolon, Ausläufer (Hauptachse von Hydrozoen-Kolonien)
stolonial tuber *bot* Ausläuferknolle
stoma/stomatal pore (pl stomata) Spaltöffnung
stomach Magen (*pl* Mägen)
- **cardiac stomach/cardia/gizzard (insects)** Kaumagen, Cardia
- **coronal stomach/coronal sinus** Kranzdarm
- **first stomach/paunch/rumen/ ingluvies/first stomach** Pansen, Rumen
- **fourth stomach/reed/rennet-stomach/abomasum** Labmagen, Abomasus
- **glandular portion of stomach** Drüsenmagen
- **honey stomach/honey crop/ honey sac (bees)** Kropf, Honigmagen, Honigdrüse, Futterdrüse
- **honeycomb stomach/ honeycomb bag/second stomach/reticulum** Netzmagen, Haube, Retikulum
- **primitive stomach/archenteron** *embr* Urdarm, Archenteron
- **pumping stomach** Saugmagen (Vorratsmagen: Kropf der Culiciden)
- **pyloric stomach (echinoderms)** Mitteldarm
- **second stomach/honeycomb stomach/honeycomb bag/ reticulum** Netzmagen, Haube, Retikulum
- **sucking stomach (chelicerates)** Saugmagen
- **third stomach/manyplies/ psalterium/omasum** Blättermagen, Vormagen, Psalter, Omasus

stomach acid

stomach acid/gastric acid
Magensäure
stomach fungi/gastromycetes/
angiocarps/Gasteromycetes/
Gastromycetales Bauchpilze
stomach juice/gastric juice
Magensaft, Magenflüssigkeit
stomochord/buccal tube
Stomochord
stomodaeum/foregut (insects)
Stomodäum, Stomatodäum,
Vorderdarm
stone/drupe/drupaceous fruit *bot*
Steinfrucht
stone/pit/putamen/pyrene *bot*
Stein, Steinkern, Putamen
(Endokarp)
stone canal/hydrophoric canal/
madreporic canal Steinkanal
stone cell/sclereid *bot* Steinzelle,
Sklereide
stone plants lebende Steine
stonecrop family/sedum family/
orpine family/Crassulaceae
Dickblattgewächse
stoneflies/Plecoptera Steinfliegen,
Uferfliegen
stoneworts/stonewort family/
Charophyceae/Charophyta
(Characeae) Armleuchteralgen,
Armleuchtergewächse
stony corals/scleractinians/
Madreporaria/Scleractinia
Steinkorallen, Riffkorallen
stool/feces Stuhl, Fäzes, Kot
stool layering/mound layering *hort*
Ablegervermehrung durch
Anhäufeln (Abrisse nach Anhäufeln)
stool sample Stuhlprobe
stooping/dive-bombing *orn*
herabstoßen (Vögel: im Sturzflug
die Beute ergreifen)
stop codon/termination codon/
translational stop signal *gen*
Stopcodon, Terminationscodon,
Abbruchcodon
stopcock/shutoff cock Absperrhahn
stopper *n* Stopfen, Stöpsel,
Pfropfen, Stopper, Verschlusskappe
stopper *vb* zustöpseln
stopping filter/barrier filter/
selective filter/selection filter
Sperrfilter
storable/durable/lasting haltbar
storage Speicherung
storage chamber Vorratskammer
storage material/reserve material/
food reserve Reservestoff,
Nahrungsreserve
storage parenchyma
Speicherparenchym
storage pest Vorratsschädling

storage protein Speicherprotein,
Reserveprotein
storage root Speicherwurzel
storage tank Lagertank
storage tissue Speichergewebe
storax family/Styracaceae
Storaxgewächse
store *n* **(animal not yet ready for**
slaughter) Masttier,
zur Mast bestimmtes Tier
store *vb* speichern
storied/stratified stockwerkartig,
etagiert, geschichtet
storied cambium/stratified
cambium
Stockwerk-Cambium/Kambium,
etagiertes Cambium/Kambium
"stork's nest"/Storchennest
(stunted treetop/crown; sign of
damage by acid precipitation)
Storchennest
storm *meteo* Sturm
● **snowstorm** Schneesturm
● **thunderstorm** Gewitter, Unwetter
story (space between two floors)
Etage, Stockwerk
stotting/pronking Prellsprung
straight grain (wood) Fasertextur
straight run (bees)
geradliniger Schwänzellauf
strain (caused by stress)
sensu stricto: Belastungsursache
(*siehe*: stress)
strain (e.g. bacterial strain) Stamm
(z.B. Bakterienstamm)
strand/cord Strang (*pl* Stränge)
● **anticoding/antisense strand**
anticodierender Strang,
Nichtsinnstrang, Gegensinnstrang
● **coding strand/sense strand**
(nontranscribed strand)
codierender Strang,
kodierender Strang, Sinnstrang
● **daughter strand** Tochterstrang
● **double strand** Doppelstrang
● **double-stranded DNA**
Doppelstrang-DNA,
doppelsträngige DNA
● **lagging strand** Folgestrang
● **leading strand** Leitstrang
● **minus strand**
Minus-Strang, Negativ-Strang
(nichtcodierender Strang)
● **plus strand** Plus-Strang, Positiv-
Strang (codierender Strang)
● **single strand** Einzelstrang
● **single-strand DNA** Einzelstrang-
DNA, einsträngige DNA
● **template strand/antisense**
strand (transcribed strand)
Matrizenstrang, Nicht-Sinnstrang,
transkribierter Strang, Mutterstrang

329   **string bog** **S**

strand assimilation (DNA)
Strangassimilation
strand break (DNA) Strangbruch
• **double-strand break**
Doppelstrangbruch
• **single-strand break**
Einzelstrangbruch
strand displacement (DNA)
Strangverdrängung
strandline fauna Spülsaumfauna,
Tierwelt des Spülsaums
strange fremd (Gesellschaftstreue)
strangle/throttle würgen, abwürgen,
erdrosseln; drosseln
strangler (tree strangler) Würger
(Baumwürger)
strap muscle bandförmiger Muskel
strap-shaped/ligulate
streifenförmig
Strasburger cell/albuminous cell
Strasburger Zelle, Eiweißzelle
stratification/layering
Stratifizierung, Stratifikation,
Schichtenbildung, Schichtung
stratified/storied geschichtet,
etagiert, stockwerkartig
stratified epithelium
zweischichtiges Epithel,
mehrschichtiges Epithel
stratigraphy Stratigraphie,
Stratigrafie
stratum basale Basal(zell)schicht
**stratum corneum (horny layer of
epidermis)** Hornschicht
**stratum germinativum/germinative
layer** Keimschicht
stratum granulosum Körnerschicht
straw Stroh, Strohhalm
**strawberry fern family/
Hemionitidaceae**
Nacktfarngewächse
**strawberry-shrub family/spicebush
family/Calycanthaceae**
Gewürzsträucher
strawflower Strohblume,
Trockenblume
**stray (domestic animal wandering
at large/lost: e.g. stray dog)**
streunen (z. B. streunender Hund)
streak *vb micb/lab* ausstreichen
streak culture Ausstrichkultur
streak-plate method
Plattenausstrichmethode
stream (big river)
Strom (großer Fluss)
stream/flow *n* Strom (Flüssigkeit);
Strömung
stream/flow *vb* strömen
streambed (riverbed) Strombett
(Flussbett)
streamlet/rivulet Rinnsal,
kleines Bächlein

**strengthening zone/notched zone
(spiderweb)** *arach*
Befestigungszone
streptoneurous nerve pattern
Streptoneurie, Chiastoneurie
streptophyllous gedrehtblättrig
stress Stress, Beanspruchung,
Belastung; *sensu stricto*:
Belastungszustand (*siehe*: strain)
• **oxidative stress** oxidativer Stress
stress *vb* stressen, überlasten,
beanspruchen; betonen
stress fiber Stressfaser
**stress tolerance/maximum stress/
endurance** Belastbarkeit
stressful stressig, aufreibend,
anstrengend
stretch/extend *vb* **(muscle)** dehnen
**stretch/stretching/extension
(muscle)** Dehnung
stretch receptor (muscle)
Dehnungsrezeptor
stretch reflex/myotatic reflex
Dehnungsreflex,
myotatischer Reflex
striate/striated/finely striped
feinstreifig, feingestreift
striate body/corpus striatum
Streifenkörper, Basalkern,
Basalkörper, Corpus striatum
striate veined/striately veined
längsnervig, längsaderig,
streifennervig, streifenadrig
striate venation *bot* **(leaf)**
Längsnervatur, Längsaderung,
Streifennervatur, Streifenaderung
striated feingestreift
striated muscle/striped muscle
gestreifte Muskulatur
striation Streifen, Riefe;
Streifenbildung, Riefenbildung;
Riefung
stride *n* Schritt, Schrittlänge
stride *vb* schreiten
stridulate stridulieren, schrillen
stridulating file Schrill-Leiste,
Schrilleiste (mit Schrill-Rille)
stridulating organ
Stridulationsorgan, Schrillorgan,
Zirporgan
stridulating rasp/scraper (plectrum)
Schrillkante
**strigil/strigilis (antennal comb/
antennal cleaner also file or
scraper)** *entom* Striegel
strigillose *bot*
mit feinen Strichborsten
strigose borstig (mit
kurzgestrichenen Borsten/striegelig)
striker (gas) Anzünder
string bog/aapa mire Strangmoor,
Aapamoor

S stringency 330

stringency Härte, Schärfe, zwingende Kraft; (of reaction conditions) Stringenz (von Reaktionsbedingungen)
stringent conditions stringente Bedingungen, strenge/harte Bedingungen
stringent plasmid stringentes Plasmid
striolate feingestreift, gerieft
strip cropping *agr* Streifenanbau, Streifenkultur
striped gestreift, streifig
• **parallely striped** parallelgestreift
stripped of leaves entlaubt
strobiliform zapfenförmig
strobilization Strobilisation, Strobilation
stroke *n med* Schlaganfall
stroke *n* **(movement)** Schlagbewegung, Schlag, Zug, Stoß
• **downward stroke of wing** Flügelabschlag
• **power stroke/effective stroke (forward stroke)** Kraftschlag, Wirkungsschlag
• **recovery stroke (backstroke)** Erholungsschlag
• **upward stroke of wing** Flügelaufschlag
stroke volume *cardio* Schlagvolumen
stroma Stroma
strombuliform/spirally coiled/ spirally twisted spiralig aufgewickelt, spiralig gewunden
strong ion difference (SID) Starkionendifferenz
Stropharia family/Strophariaceae Träuschlinge, Schuppenpilze
strophiolar plug/operculum *bot* Keimwarze (des Samens)
strophiole *bot* Strophiole, Samenwarze (Auswuchs der Raphe)
STRPs (short tandem repeat polymorphisms) *gen* Polymorphismen von kurzen direkten Wiederholungen
structural analysis Strukturanalyse
structural formula Strukturformel
structural gene Strukturgen
structural protein Strukturprotein, Struktureiweiß
structure Struktur
structure elucidation *chem* Strukturaufklärung
struggle for survival *ethol/evol* Überlebenskampf
strut (rooster) stolzieren
STS (sequence tagged site) *gen* STS (sequenzmarkierte Stelle)
stub Stummel

stubby leg Stummelbein, Stummelfuß
stubby stummelartig (kurz und dick), Stummel...
stubby wings Stummelflügel
stud (group of horses bred and kept by one owner) Gestüt
stud (male animal kept for breeding) Zuchttier
stud/studhorse (see: stallion) Zuchthengst, Schälhengst, Deckhengst, Beschäler
studbook Stammbuch, Zuchtbuch, Zuchtstammbuch, Herdbuch; (horses) Gestütbuch, Stutbuch, Pferdestammbuch
studbull Zuchtbulle, Zuchtstier
studfarm (horses) Gestüt, Pferdezüchterei, Pferdezuchtbetrieb
studhorse Zuchthengst, Schälhengst, Deckhengst, Beschäler
stuffer-DNA Stuffer-DNA
stump/stub/stool/caudex Strunk, Stumpf
stump sprout/sucker/tiller Stumpfaustrieb
stumpage Holz auf dem Stamm; Holzpreis; Schlagrecht, Fällrecht
stunt/dwarf zwergwüchsig, im Wachstum gehemmt
stunted/crippled verkümmert, krüppelig, krüppelhaft, verkrüppelt
stunted forest/miniature forest/ Krummholz Krummholz
stunted growth/stuntedness Krüppelwuchs, Krüppelform
stunted pine Krüppelkiefer
stupefacient/narcotic/narcotizing agent/anesthetic/anesthetic agent *n* Betäubungsmittel, Narkosemittel, Anästhetikum
stupefacient/stupefying/narcotic/ anesthetic *adv/adj* betäubend, narkotisch, anästhetisch
stupefaction/narcosis/anesthesia Betäubung, Narkose, Anästhesie
stupefy/narcotize/anesthetize betäuben, narkotisieren, anästhesieren
sturgeons & sterlets & paddlefishes/Acipenseriformes Störe & Löffelstöre
sty (pigsty/pigpen) Stall (Schweinestall)
style *bot* Griffel, Stylus
stylet/stiletto Stilett
Stylidiaceae/trigger plant family Säulenblumengewächse, Stylidiumgewächse
styliform/prickle-shaped/bristle- shaped stilettförmig, griffelförmig
styloid process Griffelfortsatz

substitute B for A

stylopize stylopisieren, stylepisieren
stylopodium *bot* Stylopodium, Griffelpolster
styptic/hemostatic (astringent) blutstillend (adstringent)
Styracaceae/storax family Storaxgewächse
subalpine subalpin
subalpine zone/subalpine region subalpine Stufe, Gebirgsstufe
subbituminous coal Glanzbraunkohle, subbituminöse Kohle
subbranchial chamber Subbranchialraum, innerer Kiemengang
subcanopy/lower canopy mittlere Kronenregion, mittlere Baumkronenschicht
subcategory Subkategorie, Unterkategorie, Untergruppe
subclimate Subklima
subclimax/preclimax *ecol* Subklimax
subcloning Subklonierung
subconsciousness Unterbewusstsein
subculture/passage (of cell culture) Subkultur, Passage (einer Zellkultur)
subcutis Subcutis, Unterhaut
subdivide untergliedern, unterteilen
subdivided untergliedert
subdivision Untergliederung, Unterteilung
subdual/subduing *ethol* Unterwerfung
subdue unterwerfen
subereous/suberic korkartig, Kork...
suberic acid/octanedioic acid Suberinsäure, Korksäure, Octandisäure
suberification Verkorkung
suberization/suberinization Suberisierung, Suberinanlagerung, Suberinauflagerung
suberize suberisieren, verkorken
suberized layer/lamella Suberinschicht
suberose/suberous/corky verkorkt, korkartig, von korkartiger Beschaffenheit
subesophageal ganglion/ suboesophageal ganglion Subösophagealganglion, Unterschlundganglion
subgenital pit Subgenitaltasche, Trichtergrube
subgerminal cavity Subgerminalhöhle
subgroup Untergruppe
sublethal subletal
sublimate sublimieren

sublimation Sublimation
sublingual gland Unterzungendrüse
sublittoral (continental shelf zone) Sublitoral (Zone des Kontinentalschelfs)
sublittoral zone/subtidal zone Sublitoral (Zone des Kontinentalschelfs)
submental/beneath the chin submental, unter dem Kinn
submerge/submerse untertauchen, unter Wasser sein
submerged/submersed untergetaucht, unter Wasser, submers
submerged culture *micb* Eintauchkultur, Submerskultur
submerged leaf Wasserblatt
submergence Eintauchen, Untertauchen, Versenken
submission/yield Unterwerfung, Demut
submissive gesture/submissive posture Unterwerfungsgebärde, Unterwerfungshaltung, Demutsgebärde, Demutshaltung
suboesophageal ganglion/ subesophageal ganglion Subösophagealganglion, Unterschlundganglion
suborbital gland Unteraugendrüse
suborder Unterordnung
subordinate untergeordnet
subradular organ/sugar gland Zuckerdrüse
subsample *stat* Teilstichprobe
subset selection *stat* Teilmengenauswahl
subsidence Senkung, Absinken, Erdabsenkung
subsidiary cell/accessory cell/ auxiliary cell *bot* Nebenzelle (Spaltöffnung)
subsistence Subsistenz
subsistence economy Subsistenzwirtschaft, Selbstversorgerwirtschaft
subsoil (zone of accumulation/ illuviation) Unterboden, unterer Mineralhorizont (B-Horizont)
subsong *orn* Dichten, Jugendgesang (Jungvögel)
subspeciation Rassenbildung
subspecies Subspezies, Unterart
substage illuminator *micros* Ansteckleuchte
substance P Substanz P
substitute *n* Ersatz
substitute *vb* ersetzen; *chem* substituieren
substitute B for A *A* durch *B* ersetzen, *A* ersetzen durch *B*, *A* durch *B* substituieren

substitute name Ersatzname
substitution Ersatz, Austausch, Substitution
substitution therapy Ersatztherapie
substrate Substrat, Unterlage, Grundlage, Untergrund; Nährboden
- **bisubstrate reaction** Bisubstratreaktion, Zweisubstratreaktion
- **following substrate** Folgesubstrat
- **leading substrate** Leitsubstrat
- **suicide substrate** Selbstmord-Substrat

substrate constant (K_S) Substratkonstante
substrate feeder Substratfresser
substrate inhibition Substrathemmung, Substratüberschusshemmung
substrate recognition Substraterkennung
substrate saturation Substratsättigung
substrate specificity Substratspezifität
substrate-level phosphorylation Substratkettenphosphorylierung
subtend unterliegen (ein Blatt dem anderen)
subtended by unterlegt von
subtending untereinanderliegend
subterranean/underground unterirdisch
subtidal zone/sublittoral zone Sublitoral (Zone des Kontinentalschelfs)
subtractive cloning subtraktive Klonierung
subtractive library subtraktive Genbank, Subtraktionsbank, Subtraktionsbibliothek
subtyping Subtypisierung
subulate/awl-shaped pfriemlich
subumbrella (medusa) Subumbrella, Schirmunterseite
subunit Untereinheit
subunit vaccine Komponentenimpfstoff, Subunitimpfstoff, Subunitvakzine
succession (primary/secondary) *ecol* (primäre/sekundäre) Sukzession (Primärsukzession/Sekundärsukzession)
- **ecological succession** ökologische Sukzession

successional series Sukzessionsserie, Sukzessionsreihe
succinic acid (succinate) Bernsteinsäure (Succinat)
succinylcholine Succinylcholin
succubous sukkub, unterschlächtig

succulence Sukkulenz, Dickfleischigkeit
succulent *adj/adv* sukkulent, dickfleischig
succulent plant/succulent *n* Sukkulente
suck saugen
suck wind (horses) koppen
sucker (attachment organ) Haftscheibe
- **true sucker/acetabulum** *zool* Saugnapf, Acetabulum

sucker/coppice-shoot Wasserreis
sucker/haustellum/proboscis *zool* (adapted for sucking: insects) Saugrüssel, Proboscis
sucker/haustorium (fungi/plants) *bot* Haftscheibe, Saugscheibe, Saugorgan, Haustorium
sucker/sobole (an underground stolon) *n* Gehölzausläufer
sucker/tiller Schössling, Wasserreis (an Wurzel oder Baumstumpf), Seitentrieb (am Wurzelhals)
sucking (insects: haustellate) saugend
sucking lice/Anoplura echte Läuse
sucking pump (*Hymenoptera*: pharynx) Saugpumpe (Pharynx)
sucking stomach (chelicerates) Saugmagen
suckle (nurse/breast-feed) säugen (stillen), Milch saugen, an der Brust saugen
suckling *n* säugendes Jungtier
sucrose/saccharose/table sugar (beet sugar/cane sugar) Saccharose (Rübenzucker/Rohrzucker)
suction disk Saugscheibe, Saugnapf
suction filter/vacuum filter Nutsche, Filternutsche
suction filtration Saugfiltration
suction funnel/suction filter/vacuum filter (Buchner funnel) Filternutsche, Nutsche (Büchner-Trichter)
suction lung Sauglunge
suction pipet/patch pipet Saugpipette
suction reflex Saugreflex
suction root/sucking root *bot* Saugwurzel
suction trap/suctory trap *bot* Schluckfalle, Saugfalle
suctorial saugend, Saug...
suctorial organ/sucker Saugorgan
suet (from abdominal cavity of ruminants) Talg, Nierenfett
suety/sebaceous talgig, Talg...
suffocate ersticken; würgen
suffocation Ersticken, Erstickung

suffrutescent/suffruticose/base
slightly woody halbstrauchig
(am Grunde verholzt)
suffrutescent plant/half-shrub
Halbstrauch
sugar Zucker
● **blood sugar** Blutzucker
● **cane sugar** Rohrzucker
● **double sugar/disaccharide**
Doppelzucker, Disaccharid
● **fruit sugar/fructose**
Fruchtzucker, Fruktose
● **grape sugar/glucose**
Traubenzucker, Glukose, Glucose
● **invert sugar** Invertzucker
● **malt sugar/maltose**
Malzzucker, Maltose
● **milk sugar/lactose**
Milchzucker, Laktose
● **multiple sugar/polysaccharide**
Vielfachzucker, Polysaccharid
● **raw sugar/crude sugar**
(unrefined sugar) Rohzucker
● **reducing sugar**
reduzierender Zucker
● **single sugar/simple sugar/**
monosaccharide
Einfachzucker, einfacher Zucker,
Monosaccharid
● **table sugar/sucrose/saccharose**
(beet sugar/cane sugar)
Saccharose
(Rübenzucker/Rohrzucker)
sugar beet Zuckerrübe
sugar cane Zuckerrohr
sugar gland/subradular organ
Zuckerdrüse
sugar substitute(s)
Zuckeraustauschstoff(e)
sugar-lerp/honeydew Honigtau
suicide Suizid, Selbstmord,
Selbsttötung
suicide inhibition Suizidhemmung
suicide substrate
Selbstmord-Substrat
sulcate/furrowed/grooved/fissured
gefurcht, furchig, gerieft
sulcus/furrow/groove
Furche, Rinne
sulfate Sulfat
sulfur (Br sulphur) Schwefel
sulfur bacteria Schwefelbakterien
sulfur compound/sulfurous
compound Schwefelverbindung,
schwefelhaltige Verbindung
sulfur cycle Schwefelkreislauf
sulfurate/sulfuring *micb* **(vats)**
Schwefeln, Schwefelung
sulfurize *micb* **(vats)** schwefeln
sulfurous/sulfur-containing
schwef(e)lig, schwefelhaltig
sum/total *n* Summe

sum rule Summenregel
sumac family/cashew family/
Anacardiaceae Sumachgewächse
summation (spatial/temporal)
(räumliche/zeitliche) Summation
summer plumage Sommerkleid
summerwood Sommerholz
summit/peak Spitze, Scheitel
sun animalcules/heliozoans
Sonnentierchen, Heliozoen
sun leaf Sonnenblatt, Lichtblatt
sun plant/heliophyte
Lichtpflanze, Heliophyt
sun spiders/false spiders/
windscorpions/solifuges/
solpugids/Solifugae/Solpugida
Walzenspinnen
sun tracking *bot* Solstitialbewegung
sundew family/Droseraceae
Sonnentaugewächse
sunflower family/daisy family/aster
family/composite family/
Compositae/Asteraceae
Köpfchenblütler, Korbblütler
sunscald Sonnenbrand, Rindenbrand
supercoiled *gen*
vertwistet, überspiralisiert,
superspiralisiert, superhelikal
supercoiling *gen* Überspiralisierung
supercool *vb* unterkühlen
supercooling Unterkühlung
superficial oberflächlich
superficial cleavage superfizielle
Furchung, oberflächliche Furchung,
Oberflächenfurchung
supergene family Supergenfamilie
superhelix/supercoil Superhelix
superinfection
Superinfektion, Überinfektion,
zusätzliche Infektion
superior/dominant überlegen,
vorherrschend, dominant
superiority/dominance
Überlegenheit, Dominanz
supernatant *n* Überstand
superorder Überordnung
superovulation Superovulation
superposition eye
Superpositionsauge
● **neural s.e.**
neurales Superpositionsauge
● **optical s.e./clear-zone eye**
optisches Superpositionsauge
supersaturate übersättigen
superspecies Superspezies, Überart
supine supiniert,
auf dem Rücken liegend
supine position Rückenlage
supinate supinieren,
auswärtsdrehen (um Längsachse)
supination Supination,
Auswärtsdrehung (um Längsachse)

supply with blood   334

supply with blood/vascularize
durchbluten, mit Blut versorgen
support *n* Stütze; Unterlage
support *vb* (unter)stützen, erhalten,
(er)tragen
support stand/ring stand/stand *lab*
Stativ, Bunsenstativ
supporting cell/covering cell
Stützzelle, Deckzelle
supporting tissue *bot*
(collenchyma/sclerenchyma)
Stützgewebe, Festigungsgewebe
suppressible unterdrückbar
suppression Suppression,
Unterdrückung
suppressor gene Suppressorgen
suppressor mutation
Suppressormutation
**suppressor T cell/T-suppressor cell
(T_S)/regulator T-cell/regulatory
T-cell** T-Suppressorzelle,
Suppressor T-Zelle
suprabranchial chamber
Suprabranchialraum,
äußerer Kiemengang
supracaudal gland
Schwanzwurzeldrüse
**supraesophageal ganglion/
supraoesophageal ganglion/
"brain"** Supraösophagealganglion,
Oberschlundganglion, "Gehirn"
supralittoral zone/splash zone
Supralitoral, Spritzwasserzone,
Spritzzone
supravital staining
Supravitalfärbung
sural die Wade betreffend,
Waden..., sural
surculose/producing suckers
Ableger treibend
surf/breakers Brandung,
Meeresbrandung
surf zone *mar* Brandungszone
surface *n* Oberfläche
surface *vb* an die Oberfläche
kommen, an der Oberfläche
erscheinen, auftauchen
surface cover/ground cover
Bodenbedeckung
surface culture Oberflächenkultur
surface drift *limn*
Oberflächendrift
surface labeling
Oberflächenmarkierung
surface runoff Oberflächenabfluss
surface tension
Grenzflächenspannung,
Oberflächenspannung
surface water Oberflächenwasser
surface-dwelling bodenbewohnend,
an der Bodenoberfläche lebend
(Erde)

surface-to-volume ratio
Oberflächen-Volumen-Verhältnis
**surfactant/wetter/wetting agent/
spreader** oberflächenaktive
Substanz, Entspannungsmittel
surfactant factor surfactant factor
(not translated> oberflächenaktive
Substanz auf Lungenbläschen)
surplus killing
zusätzliches Erlegen von Beute
(über momentanen Bedarf hinaus)
surra (*Trypanosoma evansi*) Surra
surroyal antler (terminal tine)
Wolfssprost (Geweih)
survey *n* Gutachten, Begutachtung,
Besichtigung; (Land)Vermessung
survey *n stat* Erhebung, Umfrage
survey *vb* prüfen, begutachten,
besichtigen; (Land) vermessen
survey *vb stat* erheben, eine stat.
Erhebung vornehmen
survival Überleben, Überdauerung
survival of the fittest
Überleben des Bestangepassten
survival rate Überlebensrate
survive überleben
survivor Überlebender
survivorship curve Überlebenskurve
susceptible anfällig, empfindlich
susceptibility
Anfälligkeit, Empfindlichkeit
suspend suspendieren, schweben,
schwebend halten; fein verteilen
suspended (floating) suspendiert,
schwebend; fein verteilt;
aufgehängt
suspended animation/anabiosis
latentes Leben, Anabiose
**suspended matter/suspended
material** Schwebstoffe
suspended pupa/pupa suspensa
Stürzpuppe
suspension Suspension,
Aufschwämmung; Schweben
suspension feeder
Suspensionsfresser, Strudler
suspension load/silt load *soil/mar*
Schwebfracht
suspensor Suspensor, Träger,
Embryoträger
suspensor/zygosporophore *fung*
Trägerhyphe
suspicion (of a disease)
Verdacht (auf eine Erkrankung);
Argwohn
suspicious verdächtig, argwöhnig
sustainable development
dauerhaft-umweltgerechte/
nachhaltige Entwicklung
**sutural bone/epactal bone/
wormian bone** Schaltknochen,
Nahtknochen

suture Naht
- **antennal suture** *zool/entom* Antennennaht, Antennalnaht, Fühlerringnaht
- **coronal suture** Kranznaht
- **cranial suture** Schädelnaht
- **dorsal suture/dorsal seam** Dorsalnaht, Rückennaht
- **frontal suture** Stirnnaht
- **labial suture** Labialnaht
- **lambdoid suture** Lambdanaht
- **plane suture** ebenflächige Naht (Schädelknochennaht mit ebenen Flächen)
- **sagittal suture** Pfeilnaht
- **serrate suture** Sägenaht
- **squamosal suture** Schuppennaht
- **ventral suture/ventral seam** *bot* (of carpel) Ventralnaht, Bauchnaht

suture line (ammonite septa) Lobenlinie, Nahtlinie
- **ammonitic suture line** ammonitische Lobenlinie/Nahtlinie
- **ceratitic suture line** ceratitische Lobenlinie/Nahtlinie
- **goniatitic suture line** goniatische Lobenlinie/Nahtlinie

swab Abstrich
- **to take a swab** einen Abstrich machen

swale (tract of low land/usually marshy) *geol* Senke, Mulde, Niederung; Talmulde; Bodensenke; Grundmoränentümpel
swale *mar* Strandpriel
swallow *vb* schlucken
swallowing Schlucken
swamp (wetland dominated by trees/ shrubs > equivalent to European: carr) Sumpf, Flachmoor (Waldmoor)
- **coastal swamp** Küstensumpf
- **freshwater swamp** Süßwassersumpf
- **mangrove swamp** Mangrovensumpf
- **reed swamp** Riedsumpf
- **river swamp** Flusssumpf
- **salt swamp** Salzsumpf
- **tropical swamp** tropischer Sumpf

swamp forest Sumpfwald
- **carboniferous swamp forest** Steinkohlenwälder

swamp meadow Sumpfwiese
swamp woods/swamp forest/ paludal forest Bruchwald, Sumpfwald
swamp-cypress family/redwood family/taxodium family/ Taxodiaceae Sumpfzypressen-gewächse, Taxodiumgewächse
swampland Sumpfland

swampy/boggy sumpfig
swan-necked flask/S-necked flask/ gooseneck flask Schwanenhalskolben
swarm *n* **(e. g. bees/locusts/birds)** Schwarm
swarm *vb* ausschwärmen
swarm cell/zoospore Schwärmer, Zoospore
swarm-forming/schooling schwarmbildend
swarmer/swarm cell/zoospore Schwärmer, Zoospore
swash/uprush (rush of water up the beach from breaking wave) Schwall, Wellenauflauf (*see also:* wave/backwash)
swash (narrow channel within sandbank or between s.b. and shore) kleiner Priel
swash mark Spülmarke, Spülsaum
swash zone Spülzone, Spülstreifen
sweat/perspiration *n* Schweiß
sweat/perspire *vb* schwitzen
sweat gland/sudoriferous gland/ sudoriparous gland Schweißdrüse
sweating/perspiration/hidrosis Schwitzen
sweepback *aer/orn* Pfeilstellung, Flügelpfeilung
sweepstake dispersal Zufallsverbreitung
sweet süß
sweet gale family/bog myrtle family/wax-myrtle family/ Myricaceae Gagelgewächse
sweetener Süßstoff
sweetleaf family/Symplocaceae Rechenblumengewächse
sweetness Süße
swell (massive/crestless wave often continuing after its cause) *n mar* Dünung
swell (swelling/turgescent) *vb* quellen, (an)schwellen, turgeszent
swell off (e. g. swollen tissue) abschwellen
swell up (e. g. infected tissue) anschwellen
swelling Schwellung
swidden agriculture/swidden cropping/shifting agriculture/ shifting cultivation Wanderackerbau
swifts/Apodiformes/ Micropodiformes wood strength Seglervögel, Seglerartige
swimbladder/air bladder Schwimmblase
swimmer's itch (fluke cercaria) Badedermatitis, Cercariendermatitis

swimmeret 336

swimmeret/pleopod (crustaceans)
Schwimmfuß, Schwimmbein,
Bauchfuß, Abdominalbein,
Pleopodium
**swimming bell/nectophore/
nectocalyx** Schwimmglocke,
Nectophore
swimming foot Schwimmfuß
swine/pig/hog Schwein
• **boar (male pig: not castrated)**
Eber (männliches Schwein)
• **gilt** Jungsau, junge Sau
• **piglet/little pig** Ferkel
• **porker** Mastferkel
• **shoat/shote (young weaned
hog/less than 150 lb/less than
1 year old)** Ferkel
• **sow (female pig)**
Sau, Mutterschwein
• **store pig/store/young pig**
Läuferschwein, Läufer
• **wild boar/wild hog/wild pig**
Wildschwein, Schwarzwild
• **>aged male wild boar**
Keiler, Wildeber
• **>wild sow**
Wildschweinsau, Bache
swine fever/hog cholera (viral)
Schweinefieber
**swine influenza/swine flu
(Hemophilus influenzae suis)**
Schweinegrippe
**swine plague (Pasteurella
multocida)** Schweinepest
swinging-bucket rotor centrif
Ausschwingrotor
swirl (liquid in a flask) schwenken
switch n **(tuft of long hairs at end
of tail: bovines/cow)**
Schwanzquaste
switch gene Schaltergen
switch region gen
Switchregion, Schalterregion
**sword fern family/aspidium family/
Aspidiaceae** Schildfarngewächse
**sword-shaped/ensiform/gladiate/
xiphoid** schwertförmig
syconium (a composite fruit: figs)
Syconium
sylleptic shoot bot sylleptischer
Trieb
symbiont Symbiont
symbiosis Symbiose,
symbiotische Lebensgemeinschaft
• **cleaning symbiosis**
Putzsymbiose
symbiotic symbiotisch (sensu lato:
mutualistic gegenseitig)
sympathetic sympathisch
(autonomes Nervensystem)
**sympathetic trunk/Truncus
sympathicus** Grenzstrang

sympatric ("same country") evol
sympatrisch (in gleichen Arealen)
sympatry Sympatrie
sympetalous sympetal,
verwachsenblättrig,
verwachsenblumenblättrig,
verwachsenkronblättrig
symphile Symphile, echter Gast
symphily Symphilie, Gastpflege
symphoriont Aufsiedler,
Symphoriont
symphorism Symphorismus
symphylans/Symphyla Zwergfüßer
symphysis/coalescence Symphyse,
Verwachsung; (Knochen)fuge
• **pubic symphysis/pelvic
symphysis/symphysis pubica/
symphysis pelvina**
Schambeinfuge, Schamfuge
Symplocaceae/sweetleaf family
Rechenblumengewächse
sympodial/determinate sympodial
sympodial branching system
sympodiales Verzweigungssystem
sympodium/pseudaxis Sympodium,
Scheinachse
symport Symport
synandrous synandrisch
synanthropic synanthrop
synanthropic animal Kulturfolger
synapomorphy Synapomorphie
synapse Synapse
synaptic synaptisch
synaptic bulb Synapsenkolben
synaptic cleft/synaptic gap
synaptischer Spalt, Synapsenspalt
synaptic knob/bouton
synaptisches Endknöpfchen
synaptic potential
synaptisches Potential
synaptic vesicle/synaptosome
synaptisches Vesikel,
synaptisches Bläschen, Synaptosom
synapticle (Acrania) Synaptikel
synaptonemal complex
synaptonemaler Komplex
synaptosome/synaptic vesicle
Synaptosom, synaptisches Vesikel,
synaptisches Bläschen
synarthrodial joint/synarthrosis
Synarthrose, Füllgelenk, Fuge, Haft
syncarpous bot synkarp, coenokarp,
coeno-synkarp, verwachsenblättrig
(Fruchtblätter)
syncarpous without septa parakarp
synchondrosis Synchondrose,
Knorpelhaft
synchronizer/Zeitgeber Zeitgeber
synchronous culture micb
Synchronkultur
syncytial syncytial, synzytial
syncytium Syncytium, Synzytium

syndactylism Syndaktylie
syndesmochorial placenta
  syndesmo-choriale Plazenta
syndesmosis Syndesmose, Bandhaft
syndrome/complex of symptoms
  Syndrom, Symptomenkomplex
synecology Synökologie
synergic/synergetic/working together/cooperating
  synergetisch, zusammenwirkend
synergism Synergismus,
  gegenseitige Förderung
synergist Synergist, Mitspieler,
  Förderer
synergistic synergistisch,
  zusammenwirkend
synergy Synergie, Zusammenwirken,
  Zusammenspiel
synflorescence Synfloreszenz
syngamous/syngamic syngam
syngamy/gametogamy
  Syngamie, Gametogamie,
  Gametenverschmelzung
syngeneic/genetically identical
  syngen
syngenesis Syngenese
syngenetic/syngenesious
  syngenetisch
syngenic syngen
syngraft/allograft/homograft
  Homotransplantat, Allotransplantat
syngynous/epigynous epigyn
synnema Synnema
synostosis/synosteosis Synostose,
  Knochenhaft
synovial capsule Gelenkkapsel
synovial cavity/joint cavity
  Gelenkhöhle
synovial fluid Synovialflüssigkeit,
  Gelenkflüssigkeit, Gelenkschmiere
synovial membrane
  Synovialmembran

synovial sac Schleimbeutel
synsarcosis Synsarkose,
  Muskelhaft
syntenic genes syntäne Gene
  (Gene auf *einem* Chromosom)
synteny Syntänie
synthesis Synthese
  ● *de-novo* **synthesis**
    Neusynthese, *de-novo*-Synthese
  ● **semisynthesis** Halbsynthese
synthesize synthetisieren;
  (chemisch) darstellen
synthetic synthetisch
  ● **semisynthetic** halbsynthetisch
synthetic (having same chemical structure as the natural equivalent) naturidentisch
  (synthetisch)
synthetic reactions/synthetic metabolism/anabolism
  Synthesestoffwechsel,
  Anabolismus
synthetic resin Kunstharz
syntype Syntypus
synusia Synusia, Synusie,
  Lebensverein, Verein
synzygy Synzygie
syphilis (*Treponema pallidum*)
  Syphilis, Lues, Schanker
syringe Spritze
syringe filter Spritzenvorsatzfilter,
  Spritzenfilter
syrinx/voice box orn Stimmkopf,
  Syrinx
systematic systematisch
systematic error/bias
  systematischer Fehler, Bias
systematist/taxonomist
  Systematiker, Taxonom
systematics Systematik
systemic systemisch
systems analysis Systemanalyse

T T budding

T budding (shield budding) *hort*
T-Schnitt Okulation (mit T-förmigem
Einschnitt der Rinde)
T cell/T lymphocyte (T = thymic)
T-Zelle
- **cytotoxic T cell/killer T cell/
  T-killer cell (T_K or T_c)**
  cytotoxische T-Zelle
- **effector T cell** T-Effektorzelle
- **helper T cell/T-helper cell (T_H)**
  T-Helferzelle, Helfer T-Zelle
- **pre-T cell/T-cell precursor**
  T-Vorläuferzelle
- **suppressor T cell/T-suppressor
  cell (T_S)/regulator T-cell/
  regulatory T-cell**
  T-Suppressorzelle,
  Suppressor T-Zelle
table reef Plattformriff
**table sugar/sucrose/saccharose
(beet sugar/cane sugar)**
Saccharose
(Rübenzucker/Rohrzucker)
tacca family/Taccaceae
Erdbrotgewächse
tachytelic tachytelisch
tachytely Tachytelie
**tactile disk/Merkel's corpuscle/
Merkel's disk**
Merkelsches Körperchen
tactile hair/vibrissa (*pl* vibrissae)
Vibrissa, Spürhaar, Sinushaar,
Tasthaar (ein Sinneshaar)
tactile organ/touch sense organ
Tastorgan
tactile sense/sense of touch Tastsinn
tactile sensilla Tastkörperchen
tadpole/"polliwog" Kaulquappe
tadpole shrimps/Notostraca
Rückenschaler
taenioglossate radula
taenioglosse Radula, Bandzunge
tag *n* Etikett, Plakette, Anhänger,
Markierung
- **radioactive tag**
  radioaktive Markierung
- **sequence tagged site (STS)** *gen*
  sequenzmarkierte Stelle (STS)
- **skin tag/skin polyp (small
  outgrowth)** Hautlappen,
  Hautzipfel
tag *vb* etikettieren, markieren;
mol/gen markieren
tagged molecule
markiertes Molekül
tagma (fusion of somites) Tagma
(*pl* Tagmata)
tagmatization/tagmosis
Tagmatisierung
taiga (temperate coniferous forest)
Taiga (Nadelwald der gemäßigten
Zone)

tail Schwanz, Rute; Ende;
chem (fat molecule) Schwanz
- **prehensile tail** Greifschwanz
tail fin/caudal fin Schwanzflosse
tail fluke (whale) Schwanzruder,
Schwanzflosse
tail growth Schwanzwachstum,
endständiges Wachstum
tail spine Schwanzstachel
tail wind Rückenwind
tail-to-tail repeats *gen* Schwanz-an-
Schwanz-Wiederholungen
**tail-wagging dance/waggle dance
(bees)** Schwänzeltanz
tailfan Schwanzfächer
take root *vb* anwachsen, bewurzeln,
anwurzeln
take up/take in/ingest aufnehmen,
einnehmen, zu sich nehmen
tallow (extracted from animals)
Talg
**tallowwood family/olax family/
Olacaceae** Olaxgewächse
tally chart Strichliste
talon *orn* *Raubvögel:* Klaue, Kralle,
Fang (meist *pl*: Fänge)
talus (*also used syn with:* scree)
geol Bergschutt, Felsschutt,
Schutthalde (coarse rock debris),
Geröllhalde (rounded/eroded rocks),
Schuttflur
talus/astragalus/ankle bone *zool*
Talus, Sprungbein
talus slope Bergschuttböschung,
Schutthang, Schutthalde/-abhang
(coarse rock debris), Schuttflur,
Geröllhalde/-abhang (rounded/
eroded rocks)
**tamarisk family/tamarix family/
Tamaricaceae** Tamariskengewächse
tame *adj/adv* zahm, gezähmt
tame/domesticate *vb* zähmen,
bändigen; domestizieren
taming/domestication *n* Zähmung,
Bändigung; Domestizierung
tan *adj/adv* lohfarben, gelbbraun
tan *n* (skin) Bräune; (suntan)
Sonnenbräune, Sonnenbräunung
tan *vb* gerben; bräunen
**tanaidaceans/tanaids/Anisopoda/
Tanaidacea** Scherenasseln
tandem duplication/tandem repeat
gen Tandemanordnung,
Tandemwiederholung,
direkte Sequenzwiederholungen
- **short tandem repeat
  polymorphisms (STRPs)**
  Polymorphismen kurzer direkter
  Wiederholungen
tandem integration *gen*
Tandemintegration,
Mehrfachintegration

339   **teat** T

tandem running *ethol* Tandemlauf
tangential section/flatsawn/
  plainsawn (wood)
  Tangentialschnitt, Sehnenschnitt,
  Fladerschnitt (Holz)
tank/water tank (bromeliads)
  Zisterne
tannate (tannic acid)
  Tannat (Gerbsäure)
tannic acid (tannate)
  Gerbsäure (Tannat)
tanniferous gerbsäurehaltig,
  gerbstoffhaltig
tanning agent/tannin
  Gerbstoff, Tannin
tap water Leitungswasser
tape grass family/frog-bit family/
  elodea family/Hydrocharitaceae
  Froschbissgewächse
taper verjüngen, zuspitzen
taper-pointed/acuminate (leaf
  apex) lang zugespitzt
  (konkav zulaufende Blattspitze)
tapering/attenuate verjüngt, spitz
  zulaufend, (keilförmig) zugespitzt
tapetum Tapetum
tapetum lucidum Tapetum lucidum
tapeworms/cestodes/Cestoda
  Bandwürmer, Cestoden
taphonomy Taphonomie,
  Fossilisationslehre
taproot Pfahlwurzel
tardigrades/water bears
  Tardigraden (*sg* Tardigrad *m*),
  Bärtierchen, Bärentierchen
tare *n* (weight of container/
  packaging) Tara (Gewicht des
  Behälters, der Verpackung)
tare *vb* (determine weight of
  container, packaging in order to
  substract from gross weight)
  tarieren, austarieren
  (*Waage: Gewicht des Behälters,
  Verpackung auf Null stellen*)
target *vb* zielen; anvisieren,
  ins Auge fassen, planen
target *n* Ziel
target cell Zielzelle;
  hema Schießscheibenzelle,
  Kokardenzelle, Targetzelle
target organ Zielorgan
target sequence *gen* Zielsequenz
targeted homologous
  recombination *gen*
  Allelen-Austauschtechnik
tarn (small steep-banked mountain
  lake/pool) kleiner Bergsee,
  kleiner Bergweiher (Steilufer)
tarpons/Elopiformes
  Tarpunähnliche
tarsal/tarsus Tarsus, Fußwurzel
tarsal bone Fußwurzelknochen

tarsal gland Tarsaldrüse
tarsus (arthropods) Tarsus, Fuß
tarsus (vertebrates) Fußwurzel
tartaric acid (tartrate) Weinsäure
  (Tartrat)
tassel *bot* (corn: ♂ inflorescence)
  männliche Infloreszenz des Mais
taste *n* Geschmack
taste *vb* schmecken
taste bud/taste corpuscle/taste
  corpuscule Geschmacksknospe,
  Geschmacksbecher,
  Geschmackshügel,
  Geschmackskörperchen
taste cell/gustatory cell
  Schmeckzelle,
  Geschmackssinneszelle
taste hair (microvilli)
  Geschmacksstiftchen (Mikrovilli)
taste pore Geschmackspore,
  Geschmacksporus
taste receptor/gustatory receptor
  Geschmacksrezeptor
TATA box *gen* TATA-Box
taurine Taurin
tautomeric shift tautomere
  Umlagerung
tawny gelb-braun, ocker-braun,
  lohfarben
Taxaceae/yew family
  Eibengewächse
taxidermist *zool* Präparator,
  Tierpräparator (Ausstopfer)
taxidermy (stuffing animals,
  perticularly vertebrates)
  Taxidermie (Ausstopfen von Tieren)
taxis (*pl* taxes) Taxis (*pl* Taxien)
taxodont taxodont
taxon/taxonomic unit Taxon,
  taxonomische Einheit
taxonomy/biological classification
  Taxonomie,
  biologische Klassifizierung
  • numerical taxonomy/
    taxometrics/phenetics
    numerische Taxonomie, Phänetik
  • phylogenetic taxonomy/
    cladistics/phylogenetic
    classification Cladistik, Kladistik
tea family/camellia family/
  Theaceae Teegewächse,
  Kamelliengewächse,
  Teestrauchgewächse
tea herbs Teekräuter
tear duct/lacrimal duct Tränengang,
  Tränenkanal
tear pouch Tränensack
teasel family/scabious family/
  Dipsacaceae Kardengewächse
teat/nipple *zool* Zitze, Brustwarze
  • accessory teat Afterzitze
  • crater teat Stülpzitze

technical lab assistant 340

technical lab assistant/laboratory technician/lab technician Laborant (Laborantin), Laborassistent (Laborassistentin), technischer Assistent (technische Assistentin)
technique/technic Technik
technologic(al) technologisch
technology Technologie
technology assessment Technikfolgenabschätzung
- **OTA (Office of Technology Assessment)** US-Büro für Technikfolgenabschätzung

tectorial bedeckend, abdeckend
tectorial membrane/membrana tectoris Tektorialmembran, Deckmembran
tectrix (pl tectrices)/covert/wing covert/protective feather Decke, Deckfeder
- **primary tectrix** Handdecke
- **secondary tectrix** Armdecke

teeth/dentition Gebiss (see also: tooth)
teethe/cut one's teeth/grow teeth vb zahnen
teething/dentition Zahnen, Zahnung, Zahndurchbruch, Durchbruch der Zähne, Dentition
teg (2-year-old sheep) Schaf im 2. Jahr
tegmentum/protective bud scales bot Tegment, Knospenschuppe, Knospendecke
tegula (tile-shaped structure) Tegula (>Flügelschuppe)
tegular schuppenartig, dachziegelartig
teichoic acid Teichonsäure
teichuronic acid Teichuronsäure
teliospore/teleutospore fung Teliospore, Teleutospore (Winterspore)
telmatophyte (wet meadow plant) Telmatophyt
telocentric chromosome telozentrisches Chromosom
telome Telom
telomere Telomer
telomorphic stage/perfect stage fung Hauptfruchtform
telson (crabs) Telson, Schwanzplatte
temperate/moderate gemäßigt
temperate phage temperenter Phage
temperate zone/temperate region biogeo gemäßigte Zone
temperature gradient gel electrophoresis Temperaturgradienten-Gelelektrophorese

temperature-dependent temperaturabhängig
temperature-sensitive mutation temperatursensitive Mutation
template Matrize
template slippage gen Verrutschen der Matrize
template strand gen Matrizenstrang, Mutterstrang
temple Schläfe, Schläfenregion
temporal bone/os temporale Schläfenbein
temporal fenestrae Schläfenfenster
temporal gland (elephant) Schläfendrüse
temporal lobe/lobus temporalis Schläfenlappen, Temporallappen
temporal summation zeitliche Summation
tenacity/cohesiveness Klebrigkeit, Zähigkeit; Reißfestigkeit, Zugfestigkeit
tenacle/tenaculum Tenaculum
tender/fragile (ecosystem) empfindlich, zerbrechlich
tendinous sehnig, Sehnen...
tendinous cord Sehnenstrang
tendinous sheath/tendon sheath Sehnenscheide
tendon sheath Sehnenscheide
tendril/cirrus/clasper bot Ranke
tendril climber bot Rankenkletterer
tensile strength/breaking strength (wood) Zugfestigkeit, Zerreißfestigkeit, Reißfestigkeit
tension/suction/pull (water conductance) physio Zug, Sog (Wasserleitung)
tension wood Zugholz
tentacle Tentakel, Fanghaar
tentacle sheath (ctenophores) Tentakelscheide
tentacular arm Tentakelarm
tentacular canal (ctenophores) Tentakelgefäß
tentacular plane Tentakelebene
tentaculates (bryozoans & phoronids & brachiopods) Tentaculaten, Kranzfühler, Armfühler, Fühlerkranztiere
tentaculiferans/"tentaculates" (Ctenophora) Tentaculiferen, tentakeltragende Rippenquallen
tentorial pit Tentoriumgrube
tentorial ridge/corpus tentorii Tentoriumbrücke, Corpotentorium
tepals bot Tepalen, gleichartige Blütenhüllblätter
tepidarium/moderately heated greenhouse Tepidarium, Gewächshaus mit mittlerer Temperatur

341

tetralophodont **T**

teratogenic teratogen,
Missbildungen verursachend
teratogeny Teratogenese
(Entstehung von Missbildungen)
teratology Teratologie
(Lehre von Missbildungen)
teratoma Teratom
**terete (somewhat cylindrical with
tapering ends)** annähernd
cylindrisch/zylindrisch (mit
stumpfen Enden); stielrund, walzig
terete *bot* stielrund
tergite/dorsal sclerite *entom* Tergit,
Rückenplatte
**tergum/back/roof/dorsal plate
(consisting of tergites)** Tergum,
Rückenschild, Rückenteil
terminal *n* **(dentrite)** Endigung
terminal/terminate endständig
terminal bud *bot* Terminalknospe,
Endknospe
terminal bulb (nerve cell)
Endknopf
terminal community *ecol*
Schlussgesellschaft
terminal differentiation (e. g. skin)
terminale Differenzierung
(z. B. Haut)
terminal meristem Endmeristem
terminal moraine Frontalmoräne,
Stirnmoräne
terminal redundancy *gen*
terminale Redundanz
terminal shoot/apical shoot
Terminaltrieb, Endtrieb, Gipfeltrieb
**termination codon/terminator
codon** Terminationscodon,
Abbruchcodon, Stoppcodon
• **premature termination codon
(PTC)** vorzeitiges Stoppcodon
terminology Terminologie,
Fachbezeichnungen, Fachsprache
terminus (molecule) Terminus, Ende
(Molekülende)
termites/Isoptera Termiten,
"Weiße Ameisen"
ternate/ternary dreizählig
terpene Terpen
terrace Terrasse
terracing Terrassierung
terrain Terrain, Gelände
terrapins Sumpfschildkröten
terrestrial/land-dwelling
terrestrisch, landlebend
terrestrial alga Landalge, Luftalge
terrestrial ecosystem
Landökosystem
terrestrial plant Landpflanze
terrestrialization *geol/limn*
Verlandung
**terricole/geocole/geodyte/soil
organism** Bodenorganismus

territorial behavior
Territorialverhalten
territoriality Territorialität
territory/range Territorium, Revier,
Wohnbezirk, Gebiet
**tertiary bark/dead outer bark/
rhytidome** Borke, Rhytidom
tertiary structure (polypeptides)
Tertiärstruktur
tertiary swamp forests
Braunkohlenwälder
**Tertiary Period/Tertiary (geological
time)** Tertiärzeit, Tertiär,
Braunkohlenzeit
tessellate(d)/chequered (leaves)
mosaikartig gewürfelt
test *n* Test, Probe, Versuch,
Messung, Prüfung, Analyse,
Nachweis
**test/testa (shell or hard outer
covering)** Testa;
bot (seed coat) Samenschale
test *vb* testen, messen, prüfen,
untersuchen
test medium (for diagnosis)
Testmedium, Prüfmedium
test procedure/testing procedure
Testverfahren
test tube/glass tube Reagenzglas
test tube brush Reagenzglasbürste
test tube holder Reagenzglashalter
test tube rack Reagenzglasständer,
Reagenzglasgestell
test-tube baby Retortenbaby
testaceous beschalt, schalig;
gelb-braun
testcross *gen* Testkreuzung
• **three-point testcross** *gen*
Drei-Faktor-Kreuzung
tester Testpartner
testicle/testis (*pl*** testes)**
Hoden, Samendrüse
testicular testikulär,
den Hoden betreffend, Hoden...
testicular feminization
testikuläre Feminisierung
testis (*pl*** testes)/testicle**
Hoden, Samendrüse
testis-determining factor (TDF)
Testis-Determinationsfaktor
testosterone Testosteron
tetanus/lockjaw (*Clostridium
tetani***)** Tetanus, Wundstarrkrampf
tetanus/spasm Tetanus,
Dauerkontraktion
tetrad Tetrade
tetrad analysis Tetradenanalyse
tetraflagellated viergeißlig
tetrahedral tetraedrisch
**tetralophodont (with four
transverse ridges)** tetralophodont,
mit vier Querjochen

T tetramerous 342

tetramerous/tetrameric/tetrameral
tetramer, vierzählig
tetrandrous *bot* mit 4 Staubblättern
tetraparental tetraparental
tetraploid tetraploid
tetravalent vierwertig
tetrodotoxin Tetrodotoxin
**Texas fever/red-water fever/
hemoglobinuric fever (babesiosis)
(Babesia ssp.)** Texasfieber
texture (*see* grain) Textur, Struktur,
Faser, Fibrillenanordnung (Dichte
der Leitelemente in Jahresring),
Gefüge (Holz)
thallophyte Thallophyt,
Lagerpflanze
thallus (*pl* **thalli/thalluses**)/
thallome Thallus, Lager
thanatosis/feigning death
Thanatose, Totstellen, Totstellung
thaw auftauen
thawing Auftauen
THE (transposable human element)
gen THE (transponierbares
menschliches Element)
**THE-family (transposable human
element) (DNA element)**
THE-Familie (transponierbares
menschliches Element)
(DNA-Element)
**Theaceae/tea family/camellia
family** Teegewächse,
Kamelliengewächse,
Teestrauchgewächse
thebaine Thebain
theca Theka
thecate/armored gepanzert
thecodonts/Thecodontia
Urwurzelzähner
theine/caffeine Thein, Koffein
theliogonum family/Theligonaceae
Hundskohlgewächse
Thelypteridaceae/marsh fern family
Sumpffarngewächse,
Lappenfarngewächse
thelytoky/thelyotoky Thelytokie
theobromine Theobromin
Theophrastaceae/Joe-wood family
Theophrastaceae
theophylline Theophyllin
theoretic/theoretical theoretisch
theory (*not to be confused with:*
hypothesis) Theorie
**theory of evolution/evolutionary
theory** Evolutionstheorie,
Abstammungstheorie,
Deszendenztheorie
thermal conductance
Wärmedurchgangszahl (*C*)
thermal neutral zone
Thermoneutralzone
thermal radiation Wärmestrahlung

thermal soaring Thermiksegelflug,
Thermiksegeln
thermal spring Thermalquelle
thermal stability/thermostability
Thermostabilität, Hitzestabilität,
Hitzebeständigkeit
thermocline Thermokline,
Sprungschicht
thermodynamics Thermodynamik
  • **first law of thermodynamics**
  1.Hauptsatz (der Thermodynamik)
  • **second law of thermodynamics**
  2.Hauptsatz (der Thermodynamik)
thermogenesis Thermogenese
thermometer Thermometer
thermophilic wärmesuchend,
thermophil
thermophobic wärmemeidend,
thermophob
thermostability/thermal stability
Thermostabilität, Hitzestabilität,
Hitzebeständigkeit
therophyte/annual plant
Therophyt, Annuelle,
Einjährige (Pflanze)
thiamine (vitamin B$_1$) Thiamin
thicken eindicken
thickener *techn* Dickungsmittel
thickening (growth)
Dickenwachstum; Verdickung
**thicket/scrub/brush/thick
shrubbery** Gestrüpp, Dickicht
thickness of section *micros*
Schnittdicke
thigh Oberschenkel
thigh bone/femur/os femoris
Oberschenkelknochen, Femur
thigmotaxis (*pl* **thigmotaxes**)
Thigmotaxis (*pl* Thigmotaxien)
thimble *lab* Fingerhut
thin *adj/adv* dünn
thin *vb* ausdünnen
thin out/prune *bot/hort/for*
auslichten, zurückschneiden
thin section/microsection
Dünnschnitt
thin-layer chromatography
Dünnschichtchromatographie
thinning Ausdünnen, Ausdünnung
thiourea Thioharnstoff
third eyelid/nictitating membrane
Blinzelhaut, Nickhaut
**third stomach/manyplies/
psalterium/omasum** Blättermagen,
Vormagen, Psalter, Omasus
third-order reaction (kinetics)
Reaktion dritter Ordnung
(Reaktionskinetik)
thirst *n* Durst; Gier, Verlangen
thirst/crave *vb* dürsten, durstig sein;
verlangen nach
thirsty durstig

343   **thymus** T

thistle-like/thistly distelartig
thoracic cavity
  Thorakalhöhle, Thorakalraum,
  Brusthöhle, Brustraum
thoracic leg Thorakalbein,
  Thorakalfuß, Brustbein, Brustfuß
**thoracic respiration/costal
  breathing**
  Thorakalatmung, Brustatmung
thoracic scale Thorakalschüppchen
thoracic segment/thoracomer
  Thoraxsegment, Thoracomer,
  Rumpfsegment
thoracic vertebra
  Thorakalwirbel, Brustwirbel
thoracopod/thoracic leg
  Thorakalfuß, Thorakopode,
  Thoracopod, Thoraxbein, Rumpfbein
thorax/breast/chest/pectus Thorax,
  Brust, Brustkörper, Brustkasten,
  Oberkörper; (Insekten) Mittelleib
thorn/spine *bot* (*sensu lato*) Dorn
thorn *bot* **(sharp-pointed modified
  branch)** Sprossdorn
thorn woodland Dornwald
thorny/spiny dornig
thorny bush Dornbusch
**thorny corals/black corals/
  antipatharians/Antipatharia**
  Dornkorallen, Dörnchenkorallen
thorny shrub Dornstrauch,
  Dornenstrauch
thorny thicket/thorny brush
  Dorngestrüpp, Dornbuschformation,
  Dornstrauchformation
**thorny-headed worms/spiny-
  headed worms/acanthocephalans/
  Acanthocephala** Kratzer
thoroughbred (horse) Vollblut
thread Faden
  ● **chromatin thread**
    Chromatinfaden
**thread capsule/urticator/cnida/
  nematocyst** Nesselkapsel, Cnide,
  Cnidoblast, Nematocyste
**thread-shaped/filamentous/
  filliform** fadenförmig, filamentös,
  trichal
**threadlike antenna/hairlike
  antenna/filiform antenna**
  Fadenfühler
**threadworms/horsehair worms/
  hairworms/nematomorphans/
  nematomorphs/Nematomorpha**
  Saitenwürmer
**threadworms (causing
  strongyloidiasis) (*Strongyloides
  spp.*)** Zwergfadenwurm
threat Drohung
threat behavior Drohverhalten
threat yawn/threat gape *ethol*
  Drohgähnen

threaten drohen
threatened bedroht
**threatening gesture/threating
  gesture** Drohgebärde, Drohmimik
threatening posture Drohhaltung
**three-dimensional structure/spatial
  structure** Raumstruktur,
  räumliche Struktur
**three-field rotation/three-year
  rotation/three-field system**
  Dreifelderwirtschaft
three-point testcross *gen*
  Drei-Faktor-Kreuzung
threonine Threonin
threshold Schwelle (z. B.
  Reizschwelle/Geschmacksschwelle
  etc.)
threshold current *neuro*
  Schwellenstrom
threshold effect Schwelleneffekt
threshold potential (firing level)
  neuro Schwellenpotential
  (kritisches Membranpotential)
threshold trait Schwellenmerkmal
threshold value Schwellenwert
thrips/Thysanoptera Thripse,
  Fransenflügler, Blasenfüße
thrive/flourish gedeihen, florieren
thriving gedeihend
throat *bot* **(flower)**
  Schlund, Blütenschlund
throat *zool* Kehle, Hals
throatlatch (horse) Kehlgang
thrombin Thrombin
thrombocyte/platelet Thrombozyt,
  Thrombocyt, Plättchen,
  Blutplättchen
throttle *n zool* Kehle, Gurgel,
  Trachee, Luftröhre
throttle *vb* würgen, abwürgen,
  ersticken, erdrosseln, unterdrücken;
  drosseln
throughput Durchsatz,
  Durchsatzmenge
thrust *n phys* Vortrieb, Anschub,
  Schub, Schubkraft
thumb/pollex Daumen, Pollex
thunderstorm *meteo* Gewitter,
  Unwetter
thurl (hip joint in cattle) Hüftgelenk
thylakoid Thylakoid
**Thymelaeaceae/mezereum family/
  daphne family**
  Spatzenzungengewächse,
  Seidelbastgewächse
**thymic corpuscle/Hassall's
  corpuscle** Hassall-Körperchen
thymine Thymin
thymine dimer Thymindimer
thymus (gland) Thymus,
  Thymusdrüse, Bries (Halsthymus/
  Brustthymus)

T thyroid 344

thyroid *see* thyroid gland
thyroid cartilage/cartilago thyreoidea Schildknorpel
thyroid gland/thyreoidea Schilddrüse
thyroliberin/thyreotropin releasing hormone/factor (TRH/TRF) Thyroliberin, Thyreotropin-Freisetzungshormon (TRH/TRF)
thyrotropin/thyroid-stimulating hormone (TSH) Thyreotropin, Tyrotropin, thyreotropes Hormon (TSH)
thyroxine (*also:* thyroxin)/tetraiodothyronine Thyroxin (T_4), Tetraiodthyronin
thyrse/thyrsus (inflorescence) Thyrse, Thyrsus
thysanurans (bristletails & silverfish) Thysanuren (Felsenspringer & Silberfischchen etc.)
tibia/shinbone/shank bone Tibia, Schienbein
tibia (arthropods) Tibia, Schiene
tibiotarsus (birds) Tibiotarsus, Unterschenkelknochen
tick spiders/ricinuleids/Ricinulei/Podogona Kapuzenspinnen
ticks/Ixodides Zecken
tidal channel/tidal flat channel Wattrinne
tidal creek (in tidal flat channel) Priel
tidal current Gezeitenströmung, Gezeitenstrom
tidal flat/tide flat/tideflat/tideland Watt
tidal lift Tidenhub
tidal pool/tidepool Gezeitentümpel, Gezeitenpfütze
tidal rhythm Gezeitenrhythmik, Tidenrhythmik
tidal volume Atemzugvolumen
tidal wave Gezeitenwelle; (seismic wave) Flutwelle
tidal zone/intertidal zone/littoral zone Tidebereich, Gezeitenzone (Eulitoral)
tidbitting/feeding lure *ethol* Futterlocken
tide(s) Tide(n), Gezeiten
• high tide/flood Flut, Tide
• low tide Ebbe
• neap tide Nipptide
• spring tide Springtide
tideflat/tide flat/tidal flat/tideland Watt
tidepool/tidal pool Gezeitentümpel, Gezeitenpfütze
tideway Priel
Tiedemann's body Tiedemannscher Körper, schwammiger Körper

tier Etage, Stockwerk
• arranged in tiers etagenförmig, etagiert, stockwerkartig angelegt
tigelle/tigellum *bot* Keimstengel
tiger sharks/catsharks & sand sharks & requiem sharks & hammerheads and others/Galeomorpha/Carcharhiniformes echte Haie
tight junction (zonula occludens) Tight junction, Kittleiste, Verschlusskontakt, Schlussleiste, Engkontakt
tigrolysis (breakdown of ribosomes) Tigrolyse
Tiliaceae/linden family/lime tree family Lindengewächse
till/cultivate (plowing/sowing/raising crops) bearbeiten, bestellen, bebauen
till/glacial till *n* (unstratified glacial drift) Gletscherschutt, Gletschergeröll, Geschiebemergel (Moräne)
till/turn up the soil/plow *vb* Boden umgraben, pflügen
tillage/cultivation/farming Bodenbestellung, Ackern, Ackerbau
tillage/farmland/arable land Ackerland
tillage farming Pflugbau, wendende Bodenbearbeitung
tiller/shoot/side shoot (esp. flowering shoot of grasses) Schössling, Trieb, Spross, Seitentrieb
tiller/stalk/sprout (from base) Bestockungstrieb
tillering/sprouting (at base) Bestockung, Bestaudung, Seitentriebbildung
tilth/cultivated land Ackerland
tilth/state of aggregation (soil) Anbaufähigkeit/Garezustand/Tiefe des bestellten Bodens
timbal/tympanum (cicadas) *entom* Trommelorgan
timber/lumber (structural) Bauholz, Nutzholz, Brauchholz
• crude timber/crude wood Derbholz
timber industry holzverarbeitende Industrie
timber yield Holzertrag
timberline/tree line Baumgrenze
time horizon *geol* Zeithorizont
tin Zinn
tinamous/Tinamiformes Steißhühner
tinsel flagellum/flimmer flagellum/pleuronematic flagellum Flimmergeißel

tissue/cell association Gewebe
- **areolar tissue**
  lockeres Bindegewebe
- **BALT (bronchus/bronchial/bursa-associated lymphoid tissue)**
  bronchusassoziiertes lymphatisches Gewebe, bronchienassoziiertes lymphatisches Gewebe
- **bone tissue/bony tissue/osseous tissue** Knochengewebe
- **bothryoidal tissue (hirudinids)**
  Bothryoidgewebe
- **cartilaginous tissue**
  Knorpelgewebe
- **cavernous tissue/cavernous bodies/erectile tissue** Corpora cavernosa, Schwellkörper
- **chondroid tissue/pseudocartilage** chondroides Gewebe, Knorpelgewebe (Parenchymknorpel)
- **compact tissue/compact bone** Kompakta
- **conducting tissue/vascular tissue** Leitgewebe
- **conjunctive tissue (monocots)**
  Bindegewebe
- **connective tissue** Bindegewebe
- **dermal tissue/boundary tissue/exodermis** Abschlussgewebe
- **epithelial tissue** Epithelgewebe
- **erectile tissue/cavernous tissue/cavernous bodies** Schwellkörper, Corpora cavernosa
- **excretory tissue/secretory tissue** Ausscheidungsgewebe
- **expansion tissue** mechanisches Gewebe, Expansionsgewebe
- **false tissue/paraplectenchyma/pseudoparenchyma**
  Scheingewebe, Pseudoparenchym
- **fatty tissue/adipose tissue** Fettgewebe
- **fibrous tissue/white fibrous tissue** fibröses Bindegewebe
- **fundamental tissue/ground tissue/parenchyma**
  Grundgewebe, Parenchym
- **GALT (gut-associated lymphatic tissue)** darmassoziiertes lymphatisches Gewebe
- **glandular tissue** Drüsengewebe
- **ground tissue/fundamental tissue/parenchyma**
  Grundgewebe, Parenchym
- **MALT (mucosa-associated lymphoid tissue)**
  schleimhautassoziiertes lymphatisches Gewebe
- **muscular tissue** Muskelgewebe
- **nerve tissue/nervous tissue**
  Nervengewebe

- **nutrient tissue** Nährgewebe
- **osseous tissue/bone tissue/bony tissue** Knochengewebe
- **parenchyma/parenchymatous tissue/ground tissue/fundamental tissue** Parenchym, Grundgewebe
- **permanent tissue/secondary tissue** Dauergewebe
- **resistance tissue (fruit)**
  Widerlagergewebe
- **scar tissue/cicatricial tissue**
  Wundgewebe
- **secretory tissue**
  Sekretionsgewebe, Absonderungsgewebe, Ausscheidungsgewebe
- **storage tissue** Speichergewebe
- **supporting tissue** *bot* (collenchyma/sclerenchyma) Stützgewebe, Festigungsgewebe
- **vascular tissue/conducting tissue** Leitgewebe
- **wound tissue/callus**
  Wundgewebe, Kallus, Callus

tissue culture Gewebekultur
tissue extract Gewebeextrakt
tissue factor Gewebefaktor
tissue fluid Gewebsflüssigkeit
tissue graft/tissue transplant
Gewebetransplantat
titer Titer
titin Titin
titrate titrieren
titration curve Titrationskurve
TLC (thin layer chromatography)
DC (Dünnschichtschromatographie)
TLV (Threshold Limit Value)
Schwellenwert, Grenzwert
toadfishes/Batrachoidiformes
Froschfische
toads Kröten
tobacco mosaic virus
Tabakmosaik-Virus
tobira family/pittosporum family/parchment-bark family/Pittosporaceae
Klebsamengewächse
tocopherol (vitamin E)
Tocopherol, Tokopherol
toe (spur: rotifers) Zeh, Zehe
- **big toe/great toe/hallux**
  großer Zeh, große Zehe, Hallux
- **small toe** kleiner Zeh
- **zygodactyl toe/zygodactylous toe** *orn* Wendezehe

toenail Zehennagel, Fußnagel
togetherness Gemeinsamkeit (Zusammensein)
toggle on/off an-/ausschalten
token stimulus *ethol*
auslösender Reiz

T

tolerance 346

tolerance Toleranz, Verträglichkeit
tolerance limit Toleranzgrenze
tolerogen Tolerogen
toluene Toluol
tom (male animal) männliches Tier;
(male turkey) männlicher
Truthahn
tomcat (male domestic cat) Kater
tomentose filzig
tomentum *bot/fung*
Filz, filzige Behaarung
tomography Tomographie,
Tomografie
  • **computed tomography (CT)**
    Computertomographie,
    Computertomografie
  • **positron emission tomography
    (PET)**
    Positronenemissionstomographie
    (PET)
tone Tonus
  • **muscle tone** Muskelspannung,
    Muskeltonus
tongs *lab* Laborzange
  • **beaker tongs** Becherglaszange
  • **crucible tongs** Tiegelzange
tongue/glossa/lingua Zunge,
Glossa, Lingua
  • **aglossate/without tongue**
    zungenlos, ohne Zunge
  • **ranine (referring to
    undersurface of tongue)**
    Unterzungen…
tongue bar/secondary bar
Zungenbogen, Zwischenbogen
tongue flickering züngeln
tongue flicking
die Zunge herausschnellen lassen
**tongue grafting/whip grafting/
whip-and-tongue grafting** *hort*
Kopulation mit Gegenzunge
(Pfropftechnik)
**tongue-shaped/linguiform/ligular/
oblanceolate** zungenförmig
**tongue-worms/linguatulids/
pentastomids/Pentastomida**
Zungenwürmer, Linguatuliden,
Pentastomitiden
tonic tonisch; stärkend, belebend
tonicity Tonus, Spannungszustand;
Spannkraft
tonofilament Tonofilament
tonoplast Tonoplast
tonsil Tonsille, Mandel
  • **lingual tonsil/tonsilla lingualis**
    Zungenmandel
  • **palatine tonsil/tonsilla palatina**
    Gaumenmandel
  • **pharyngeal tonsil/tonsilla
    pharyngealis** Rachenenmandel
tonus/tonicity/tone Tonus,
Spannungszustand; Spannkraft

tooth (*pl* **teeth**) Zahn (*pl* Zähne)
  • **canine/canine tooth** Eckzahn
  • **cardinal tooth (bivalves)**
    Hauptzahn
  • **carnassial tooth/fang
    (carnivores)** Reißzahn, Fangzahn,
    Fang
  • **cheek tooth** Backenzahn
  • **deciduous tooth/milk tooth/first
    tooth/primary tooth** Milchzahn
  • **egg tooth (reptiles)** Eizahn,
    Eischwiele
  • **eyetooth (canine tooth of upper
    jaw)**
    Augenzahn (oberer Eckzahn)
  • **front tooth/incisor** Vorderzahn,
    Schneidezahn, Beißzahn
  • **grinding tooth** Mahlzahn
  • **hinge teeth (bivalves)**
    Schlosszähne
  • **incisor/front tooth**
    Schneidezahn, Vorderzahn,
    Beißzahn
  • **milk tooth/deciduous tooth/first
    tooth/primary tooth** Milchzahn
  • **molar (multicuspid tooth)/
    grinder** Molar, Backenzahn
  • **permanent tooth**
    bleibender Zahn
  • **poison tooth/venom tooth/fang
    (snakes)** Giftzahn
  • **premolar (bicuspid tooth)**
    Prämolar, vorderer Backenzahn
  • **sabre tooth/saber tooth**
    Säbelzahn
  • **tusk (large teeth)** Stoßzahn,
    Hauer (z. B. Eber)
  • **venom tooth/poison tooth/fang
    (snakes)** Giftzahn
  • **wisdom tooth/third molar**
    Weisheitszahn, dritter Molar
  • **wolf tooth/remnant tooth
    (horse: 1. premolar)** Wolfszahn
tooth bud Schmelzknospe
tooth fungus family/Hydnaceae
Stachelpilze, Stachelinge
tooth germ Zahnanlage, Zahnkeim
tooth replacement Zahnersatz
tooth row/arcade Zahnreihe
**tooth shells/tusk shells/
scaphopods/scaphopodians
(spade-footed mollusks)/
Solenoconchae/Scaphopoda**
Kahnfüßer, Grabfüßer, Scaphopoden
**tooth socket/dental alveolus/
alveolar cavity**
Zahnfach, Zahnalveole
toothed/dentate gezähnt
toothed fungus Stachelpilz
**toothed whales & porpoises &
dolphins/Odontoceti** Zahnwale
toothless/edentate zahnlos

top fermenting (brewing) obergärig
● **bottom fermenting** untergärig
top grafting/top working *hort*
Astpfropfung, Astveredlung
topiary *hort* Formbaum,
Formstrauch (auch Zierschnitt)
topogenic/topogenous topogen
topotype Topotypus, Topotyp,
Topostandard
topsoil (zone of leaching/
eluviation) Oberboden, Krume,
Bodenkrume, Bodendeckschicht,
oberer Mineralhorizont
torchwood family/incense tree
family/frankincense tree family/
Burseraceae Balsambaumgewächse
tormogen cell (socket-forming cell)
(arthropod integument)
tormogene Zelle, Balgzelle
tornado/whirlwind (particularly:
North America) Tornado,
Wirbelsturm (*see also:* cyclone)
tornaria larva Tornaria-Larve
torose (having fleshy swellings)
höckerförmig, knötchenförmig,
wulstartig
torpor Torpor, Starre, Kältestarre,
Winterstarre
torsion Torsion, Drehung
torsion strength/torsional strength
(wood) Drehfestigkeit
torted *neuro* gedreht
(Torsion der Nervenstränge)
tortoise shell/turtle shell Schildpatt
tortoises Landschildkröten
tortuous gewunden, gedreht
torus (*pl* tori) (of wood-cell pit) *bot*
Torus, Tüpfelschließhautverdickung
torus/footpad *zool* Fußballen
torus/protuberance/projection
Wulst, runde Erhebung; Schwellung
torus/receptacle/receptaculum *bot*
Torus, Rezeptakel, Rezeptakulum,
Blütenbasis, Blütenachse,
Blütenboden
total biomass Gesamtbiomasse
total cleavage totale Furchung
total germ count/total cell count
Gesamtkeimzahl
total magnification/overall
magnification
Gesamtvergrößerung
totipalmate swimmers: pelicans
and allies/Pelecaniformes
Ruderfüßer, Ruderfüßler
touch *n* Berührung
touch/boarder *vb* berühren, tasten
touch-me-not family/jewelweed
family/balsam family/
Balsaminaceae
Balsaminengewächse,
Springkrautgewächse

touching/boardering/contiguous
berührend, angrenzend
tough/rigid zäh, hart
toughness/hardness Zähigkeit, Härte
toxic/poisonous toxisch, giftig
toxicity/poisonousness Toxizität,
Giftigkeit
toxiglossate radula (hollow radula
teeth) toxoglosse Radula,
Pfeilzunge (hohl)
toxin/poison Toxin, Gift
toxophore (butterfly larvae)
Toxophorium (Schmetterlingsraupen: Brenn- und Gifthaare)
toxoplasmosis (*Toxoplasma gondii*)
Toxoplasmose
trace/follow up on s.th. *vb*
verfolgen, erforschen, nachgehen,
ausfindig machen
trace/remains/remainder *n*
(meist *pl* remains) Spur, Überrest
(meist *pl* Überreste)
trace/track *n zool* Spur, Fährte
trace *n bot* **(leaf/branch)** Spur
(Blattspur/Astspur)
trace analysis Spurenanalyse
trace element/microelement/
micronutrient
Spurenelement, Mikroelement
trace fossil/ichnofossil
Spurenfossil, Ichnofossil
● **crawling traces/repichnia**
Kriechspuren
● **dwelling structures/domichnia**
Wohnbauten
● **escape traces/fugichnia**
Fluchtspuren
● **feeding burrows/fodinichnia**
Fressbauten
● **grazing traces/pascichnia**
Weidespuren
● **predation traces/praedichnia**
Jagdspuren, Verfolgerspuren
● **resting traces/cubichnia**
Ruhespuren
● **tracks/gradichnia** Schreitfährten
tracer Tracer, Markierungssubstanz,
Markierung, Indikatorsubstanz,
Indikator
tracer enzyme Leitenzym
tracer method Tracer-Methode
trachea/vessel *bot* Trachee, Gefäß
trachea *zool* (*pl* **tracheas/tracheae)**
(windpipe)/breathing tube
Trachee, Luftröhre, Atemröhre
● **sieve trachea** *arach* Siebtrachee
● **tube trachea** Röhrentrachee
tracheal capillary/tracheole
Tracheole
tracheal cartilage/tracheal ring
Trachealknorpel, Knorpelspange der
Luftröhre, Trachealring

T tracheal gill 348

tracheal gill Tracheenkieme
tracheal ring/tracheal cartilage
Trachealring, Trachealknorpel,
Knorpelspange der Luftröhre
tracheal spiracle Tracheenstigma
(Spiraculum)
tracheary elements/xylem Holzteil,
Gefäßteil, Xylem
tracheates/Tracheata Tracheentiere
tracheid Tracheide
tracheole/tracheal capillary
Tracheole
tracheophyte/vascular plant
Tracheophyt, Gefäßpflanze
trachymedusas/trachyline medusas/
Trachymedusae Trachylina
track/pathway/course Pfad, Weg,
Bahn, Route
track/trail/trace/scent Fährte, Spur
tracking dye *electrophor*
Farbmarker
tract Trakt, Kanal (System);
Bahn, Strang
 • **digestive tract/digestive canal/**
  alimentary tract
  Verdauungstrakt, Verdauungskanal
 • **feather tract/pteryla**
  Federflur, Pteryla
 • **gastrointestinal tract**
  Gastrointestinaltrakt,
  Magen-Darm-Trakt
 • **nerve tract (band/bundle/system**
  of nerve fibers) Nervenbahn
 • **olfactory tract/tractus**
  olfactorius Riechbahn
 • **one-way digestive tract**
  durchgängiger Verdauungstrakt
 • **pyramid tract/corticospinal tract**
  Pyramidenbahn
  (verlängertes Rückenmark)
 • **respiratory tract**
  Respirationstrakt, Atemwege,
  Luftwege
 • **urinary tract** Harnwege
 • **urogenital tract** Urogenitaltrakt,
  Urogenitalsystem
  (Harn- & Geschlechtsorgane)
trade winds/trades Passatwinde
tragus (small eminence in ear of
bats) Tragus, Höcker,
Ohrmuschelfortsatz, Ohrklappe
trail Weg, Pfad
trail line *arach* Wegfaden
 • **broad trail line** *arach*
  breiter Schleppfaden
trail pheromone/trail substance
Spurpheromon
trailer segment *gen*
Trailer-Sequenz
trailing flagellum Schleppgeißel
train oil/fish oil (also from whales)
Tran, Fischöl

training/form pruning *hort*
Erziehungsschnitt
trait/characteristic/feature
Merkmal; *ethol* Charakterzug
 • **"have the trait" (to be a carrier/**
  to be heterozygous) *gen* Träger
  sein, heterozygot sein
 • **hereditary trait** Erbmerkmal
trama (of fungal gill)/dissepiment
Trama (Lamellentrama)
tramal plate Tramaplatte
trample burr *bot* Trampelklette
trans-Golgi network
trans-Golgi-Netzwerk
transadenylation Transadenylierung
transamination Transaminierung
transcribe *gen* transkribieren
transcript *gen* Transkript
transcript analysis
Transkriptionsanalyse
transcript sequencing
Transkript-Sequenzierung
transcription *gen* Transkription
 • **3′→5′ (three prime five prime/**
  three prime to five prime)
  3′→5′ (drei Strich-fünf Strich/
  drei Strich nach fünf Strich)
 • **divergent t.**
  divergente Transkription
transcription factor
Transkriptionsfaktor
transcription fusion
Transkriptionsfusion
transcytosis Transcytose
transducer Wandler
transduction *gen* Transduktion
 • **abortive t.** abortive Transduktion
 • **specialized t.**
  spezielle Transduktion
transduction mapping
Transduktionskartierung
transect/cut through *vb*
durchschneiden
transect *n ecol* Transekt
 • **belt transect** Gürteltransekt
 • **line transect** Linientransekt
 • **profile transect/stratum transect**
  Profiltransekt
transection (cutting through)
Durchschnitt
transfection *gen/micb* Transfektion
 • **abortive t.** abortive Transfektion
transfer *n* Transfer, Übertragung
transfer *vb* transferieren, übertragen
transfer cell Transferzelle
transfer host/paratenic host
Transportwirt, paratenischer Wirt,
Sammelwirt, Stapelwirt
transfer loop *lab/micb* Transferöse,
Übertragungsöse
transfer pipet/volumetric pipet
Vollpipette, volumetrische Pipette

transplant

349

transfer rate Transferrate,
Übertragungsgeschwindigkeit
transfer RNA (tRNA) Transfer-RNA
(tRNA)
transformation Transformation,
Umwandlung
transformation into grassland
Versteppung
transformation mapping
Transformationskartierung
transformation series
Transformationsreihe
transformed cell
transformierte Zelle
transforming principle
transformierendes (aktives) Prinzip
transgenic transgen
transgenic animal transgenes Tier
transgenic plant transgene Pflanze
**transhumance/seasonal livestock
movement** Transhumanz
transient vorübergehend, flüchtig,
unbeständig
transient expression vector *gen*
transienter Expressionsvektor
**transillumination/transmitted light
illumination** Durchlicht
transition/developmental transition
Transition, Übergang,
Entwicklungsübergang
transition phase Übergangsphase
transition state (enzyme kinetics)
Übergangszustand
**transition zone/transitional region
(root>shoot)** Übergangszone
(Wurzel>Sproß)
transitional epithelium
Übergangsepithel
transitional fossil Übergangsfossil
transitory bog Übergangsmoor,
Zwischenmoor
**transitory form/transient/
intermediary form**
Übergangsform
translate übersetzen;
gen translatieren
translation Übersetzung;
gen Translation
● **hybrid-arrested translation
(HART)** *gen*
hybridarretierte Translation
● **hybrid-release translation (HRT)**
gen Hybridfreisetzungstranslation
● **reverse translation**
gen reverse Translation
translation fusion
Translationsfusion
translational motion
Translationsbewegung
translator *bot* (part of gynostegium
with caudicles and adhesive
body) Translator

translocation Verlagerung; *gen*
Translokation
● **reciprocal t.** *gen* reziproke
Translokation
translucent/pellucid durchscheinend
transmembrane protein
Transmembranprotein
transmissible/communicable
übertragbar
transmissible/heritable vererbbar
**transmissible disease/
communicable disease**
übertragbare Krankheit
transmission Transmission,
Übertragung (z. B. Krankheit),
Übermittlung (z. B. Reizimpuls),
Weiterreichen; *gen* Vererbung
(Erbmerkmale)
● **horizontal t.**
horizontale Transmission
● **nonpersistent t.** *vir*
nicht-persistente Übertragung
● **vertical t.** vertikale Transmission,
vertikale Übertragung
**transmission electron microscopy
(TEM)** Transmissions-
elektronenmikroskopie,
Durchstrahlungs-
elektronenmikroskopie
transmission genetics
Vererbungslehre
transmission of disease
Krankheitsübertragung
**transmission of signals/impulse
propagation** Erregungsleitung
transmit/pass on übermitteln,
übertragen, weiterleiten,
weiterreichen; *gen* (passing on of
hereditary traits) vererben
transmitter Transmitter, Überträger;
biochem/neuro Überträgerstoff
transmitter of disease
Krankheitsüberträger
transphosphorylation
Transphosphorylierung
transpiration Transpiration;
zool (Haut)Ausdünstung, Schweiß,
Absonderung
transpiration pathway
Transpirationsweg
transpiration pull/tension
Transpirationssog, Transpirationszug
transpiration ratio
Transpirationskoeffizient
transpiration stream
Transpirationsstrom
transpire *vb bot* transpirieren; *zool*
ausdünsten, schwitzen, absondern
transplant/graft *n* Transplantat
transplant/replant *vb bot*
umpflanzen, versetzen, verpflanzen,
pikieren

transplant *vb* *zool* transplantieren, verpflanzen

transplantation Transplantation, Verpflanzung

transport *vb* transportieren

transport/transportation Transport
- **active/uphill transport** aktiver Transport
- **coupled transport/co-transport** gekoppelter Transport
- **facilitated transport** erleichterter Transport

transport protein Transportprotein

transposable element transponierbares Element
- **transposable human element (THE)** transponierbares menschliches Element (THE)

transposition Transposition, Umstellung, Umgruppierung

transposon Transposon

transverse schräg, diagonal, Quer...

transverse canal (ctenophores) Transversalgefäß

transverse process of vertebra/processus transversus Querfortsatz

transverse section/cross section Hirnschnitt, Querschnitt

transverse tubule/T-tubule Transversalkanal, T-Kanal

transversion Transversion

trap *n* Falle

trap *vb* fangen (in einer Falle), einfangen; abfangen (Gase)

trap blossom/trap flower/prison flower Fallenblume, Fallenblüte

trap door *arach* Falltür

trap leaf *bot* Fallenblatt

Trapaceae/water chestnut family Wassernussgewächse

trapezium bone/greater multangular bone großes Vieleckbein

trapezoid bone/lesser multangular bone kleines Vieleckbein

trapper Trapper, Pelztierjäger

traumatic parenchyma Wundparenchym

tray reactor (bioreactor) Gärtassenreaktor

tread/cocktread/germ disk/"eye" Hahnentritt, Fruchthof, Keimscheibe

tread *vb* **(rooster)** treten (begatten: Hahn)

treading/kneading (milk elicitation movement) Milchtritt

treadle/chalaza Hagelschnur, Chalaze

tree Baum
- **arbuscule (tree-like small shrub/dwarf tree)** Bäumchen, baumartiger Strauch, niedriger Baum
- **broadleaf tree (*pl* broadleaves)/hardwood** Laubbaum (*pl* Laubbäume)
- **bronchial tree/arbor bronchialis** *zool/anat* Bronchialbaum
- **coniferous tree/conifer/softwood tree** Nadelbaum
- **crown rosette tree (terminally tufted leaves)** Schopfbaum
- **forest tree** Forstbaum
- **fruit tree/fruit-bearing tree** Obstbaum
- **hardwood tree/hardwood** Laubbaum (speziell: Angiospermen)
- **park tree** Parkbaum
- **phylogenetic tree** *evol* Stammbaum
- **snag (standing dead tree)** toter Baum, Baumstumpf (stehender toter Baum)
- **softwood tree/coniferous tree/conifer** Nadelbaum
- **standard (tree stem)** Hochstamm

tree farm/tree plantation Forst, Wirtschaftswald, Baumplantage

tree fern family/cyathea family/Cyatheaceae Becherfarne, Baumfarne

tree line/timberline Baumgrenze

tree pruning Baumschnitt

tree resin Baumharz

tree savanna Baumsavanne

tree shrews/Scandentia Spitzhörnchen

tree stand/stand of trees//number of trees Baumbestand

tree stratum Baumschicht

tree stump/stub/"stool" Baumstumpf, Stock, Stumpen, Stubbe, Stubben

tree-dwelling/living in trees/arboreal/arboricolous baumbewohnend

treeless baumlos

treelike/arboreal baumartig

treetop/crown Krone, Baumkrone, Gipfel, Baumgipfel, Wipfel, Baumwipfel, Stammkrone

trellis/espalier Spalier

trema/hole/orifice Trema, Loch, Öffnung

tremble/vibrate zittern, vibrieren

tremble dance/trembling dance (bees) Zittertanz

trembling/vibration Zittern, Vibration

tripod

trench/ditch/furrow Graben, Einschnitt, Furche
• **deep-sea trench** Tiefseegraben
triandrous mit drei Staubblättern
triangle (insect wing) Triangulum, Dreieck, Analdreieck
triangular bone/triquestral bone/os triquestrum Dreiecksbein
triangulation number *vir* Triangulationszahl
Triassic Period/Triassic (geological time) Triaszeit, Trias
tribe Tribus (*pl* Triben), Sippe
tributary Zufluss
tricarboxylic acid cycle (TCA cycle)/ citric acid cycle/Krebs cycle Tricarbonsäure-Zyklus, Citrat-Zyklus, Citratcyclus, Zitronensäurezyklus, Krebs-Cyclus
trichina worm (*Trichinella spiralis*) Trichine
trichinosis Trichinose
trichobezoar/pilobezoar/hairball Trichobezoar, Pilobezoar, Haarball
trichobothrium Trichobothrium, Becherhaar
trichocyst/trichite (ciliates) Trichocyste
trichogen cell (seta-forming cell: arthropod integument) trichogene Zelle (haarbildend)
trichogyne Trichogyne, Empfängnishyphe
Tricholoma family/Tricholomataceae Ritterlinge
trichome Trichom; *zool* unechtes Haar; *bot* Pflanzenhaar
• **absorbing trichome (bromeliads)** Saugschuppe, Schuppenhaar
• **multicellular glandular trichome/ colleter** Drüsenzotte, Leimzotte, Colletere
trickle rieseln
trickle irrigation/drip irrigation Tropfbewässerung, Tröpfchenbewässerung, Rieselbewässerung
trickling filter Tropfkörper (Tropfkörperreaktor/ Rieselfilmreaktor)
trickling filter reactor (bioreactor) Tropfkörperreaktor, Rieselfilmreaktor
tricolpate tricolpat
tricuspid *n* Backenzahn
tricuspid(ate) dreizipflig
tricuspid valve (heart) Trikuspidalklappe
tridactyl(ous)/tridigitate tridactyl, dreizehig
tridactyly/tridactylism Tridactylie, Dreizehigkeit

trifid dreispaltig
trifoliate dreiblättrig
trifoliolate dreiblättrig gefiedert
trigger/elicitate (a reaction) auslösen
trigger/elicitor Auslöser, Elicitor
trigger hair/sensitive hair Reizhaar, Fühlhaar
trigger plant family/Stylidiaceae Säulenblumengewächse, Stylidiumgewächse
trigger zone Triggerzone
triggering/elicitation (of a reaction) Auslösung (einer Reaktion)
trigone/triangle Dreieck
triiodothyronine (T₃) Triiodthyronin
trill/warble trillern
trillium family/Trilliaceae Einbeerengewächse, Dreiblattgewächse
trilobites Trilobiten, Dreilapper
trilophosaurs/trilophosaurians (Triassic archosauromorphs)/ Trilophosauria Dreijochzahnechsen
trim/crop *zool/agr* (fur) stutzen, abschneiden (Fell)
trim/prune *bot* trimmen, zuschneiden, beschneiden, zurückschneiden, stutzen, einkürzen
trimerous trimer, dreiteilig
trimester/trimenon (pregnancy) Trimester
trimitic trimitisch
trimming shears Trimmschere
trinocular head *micros* Trinokularaufsatz, Tritubus
trinucleotide expansion Trinucleotid-/Trinukleotidexpansion, -verlängerung
trinucleotide repeat Trinucleotid-Wiederholung, Trinukleotid-Wiederholung
trip-line/barrier thread *arach* Stolperfaden
tripartite dreiteilig, dreigeteilt
tripartite mating *gen* Dreifachpaarung
tripe(s) (esp. ox) Kutteln, Kaldaunen, Gekröse
triphasic triphasisch
tripinnate (three times pinnate) dreifach gefiedert
triple bond *chem* Dreifachbindung
triplet binding assay Triplettbindungsversuch
triplet sequences *gen* Triplettsequenzen
triploid triploid
triploidy Triploidie
triplostemonous triplostemon
tripod *lab* Dreifuß, Dreibein

T

triquestral bone 352

triquestral bone/triangular bone/ os triquestrum Dreiecksbein
triramous/three-branched dreiästig verzweigt
trisomic *gen* trisom
trisomy *gen* Trisomie
tritiate mit Tritium markieren
tritocerebrum (insects) Tritocerebrum, Hinterhirn
triturate zerreiben, zermahlen
triturating mill/gastric mill Magenmühle
trituration Zermahlen, Zerreibung, Pulverisierung
trivalency Dreiwertigkeit
trivalent dreiwertig
trivium (holothurids) Trivium, Kriechsohle
trochanter Trochanter, Schenkelring
trochodendron family/wheelstamen tree family/yama-kuruma family/Trochodendraceae Radbaumgewächse
trochoid(al) joint/pivot joint Walzengelenk
trochophore larva Trochophora-Larve, Wimperkranzlarve
trochus (anterior circlet of cilia) Trochus (vorderer Wimpernkranz)
trogons/Trogoniformes Trogons
Tropaeolaceae/nasturtium family Kapuzinerkressengewächse
trophamnion Trophamnion
trophic egg/nurse egg Nährei
trophic level/feeding level trophische Stufe, Trophiestufe, trophische Ebene, trophisches Niveau, Trophieebene, Trophieniveau
trophogenic trophogen
tropholytic tropholytisch
trophophase (feeding phase) Trophophase (Ernährungsphase)
tropical tropisch
tropical medicine Tropenmedizin
tropics Tropen
tropism Tropismus
troponin Troponin
trot (trotting gait) Trab, Trott (schnelle Gangart)
trough Trog, Traufe; *micros* Trog, Wanne (für/am Mikrotommesser)
trough-shaped muldenförmig
truck farm Gemüsegärtnerei
truck farmer/trucker Gemüsegärtner
true birds/neornithes/Neornithes Neuvögel
true bugs/heteropterans/ Heteroptera (Hemiptera) Wanzen
true flies/Brachycera (Diptera) Fliegen

true-bred/pure-bred (true-breeding/pure-breeding) reinrassig, reinerbig
truffles/Tuberales Trüffeln (Tuberaceae> Speisetrüffel)
trumpet-creeper family/ trumpet-vine family/bignonia family/Bignoniaceae Bignoniengewächse, Trompetenbaumgewächse
truncate *bot* **(leaf)** gestutzt
truncate *math* abgestumpft
truncate synflorescence Rumpfinfloreszenz
truncated gestutzt, verstümmelt, zurechtgeschnitten
truncation *general* Verkürzung, Verstümmelung
truncation *math/stat* Abstumpfung, Schwellenwert
- **constant truncation** konstanter Schwellenwert
- **proportional truncation** proportionaler Schwellenwert
truncation selection *gen* Schwellenwertselektion, Kappungsselektion, Auslesezüchtung
trunk (elephant) Rüssel
trunk/rump Rumpf, Leib, Torso
trunk musculature Rumpfmuskulatur
try *n* Versuch, Probe
try/attempt *vb* versuchen, probieren
trypsine Trypsin
tryptone Trypton
tryptophan Tryptophan
tubal pregnancy Eileiterschwangerschaft, Oviduktträchtigkeit
tube/siphon/ascidium Tube, Schlauch, Röhre
tube/body tube *micros* Tubus
tube/tubing *lab* Schlauch
tube anemones/cerianthids/ Ceriantharia Zylinderrosen
tube brush (test tube brush) *lab* Flaschenbürste
tube cell (pollen) Schlauchzelle, Pollenschlauchzelle
tube cnida/ptychocyst (Ceriantharia) Ptychozyste, Ptychonema (eine Astomocnide)
tube foot/podium (asteroids) *zool* Ambulakralfüßchen, Saugfüßchen
tube trachea/tubular trachea *arach* Röhrentrachee
tube web/funnel web *arach* Röhrennetz, Trichternetz
tube-dweller Röhrenbewohner
tube-dwelling/tubicolous röhrenbewohnend

tubenoses (tube-nosed swimmers: albatrosses & shearwaters & petrels)/Procellariiformes Röhrennasen

tuber *bot/general* Knolle

tuber/tuberosity (an anatomical prominence) *med* Tuber, Höcker, Wulst, Vorsprung; Knoten, Schwellung

tuber/underground stem-tuber *bot* unterirdische Sprossknolle

tubercle/tubercule/tuberculum/ warty protuberance *bot* kleine Knolle, Warze (Höcker/Beule/Wölbung)

tuberculate/tuberculated/vaulted warzig, gewölbt

tuberculous/tubercular höckerig, knotig; tuberkulös

tuberiform knollenförmig

tuberous/tuberose/tuberal/bulbous tuberös, knollig, knollentragend

tubicolous/tube-dwelling röhrenbewohnend

tubiform röhrenförmig

tubiform organ Knollenorgan, tuberöses Organ (Elektrorezeptor)

tubing Schläuche, Schlauchverbindung(en), Rohr, Röhrenmaterial, Rohrstück

tubing clamp Schlauchklemme

tubocurarine Tubocurarin

tubular tubulär, röhrenförmig

tubular bone (long/hollow) Röhrenknochen

tubular flower/disk flower/disk floret Röhrenblüte, Scheibenblüte (Asterales)

tubular heart Röhrenherz, Herzschlauch

tubular loop reactor (bioreactor) Rohrschlaufenreaktor

tubular trachea/tube trachea *arach* Röhrentrachee

tubulin Tubulin

tuft Schopf, Büschel

tuft of cilia/ciliary tuft Wimpernbüschel, Wimpernschopf

tuft of grass/tussock Grasbüschel

tuft of hair Schopf (Haarschopf)

tufted gedrängt

tufted/comose schopfig, büschelig, dichthaarig, haarschopfig

tufted crown Schopfkrone

tularemia (*Pasteurella/Francisella tularensis*) Tularämie, Hasenpest

Tullgren funnel *ecol* Tullgren-Apparat

tumble (bacteria) taumeln

tumbler (mosquito pupa) Schnakenpuppe

tumbleweed Bodenroller, Steppenroller, Steppenhexe

tumor Tumor, Wucherung, Geschwulst

tumor necrosis factor (TNF) Tumornekrosefaktor (TNF)

tungsten Wolfram

tunic/tunica Tunika, Hülle, Hüllschicht, Häutchen (Gewebeschicht)

tunicates Manteltiere, Tunicaten (Urochordaten)

tunnel/gallery/burrow Gang, Grabgang, Fraßgang, Tunnel

tunnel protein Tunnelprotein

tunneling microscopy Tunnelmikroskopie

tunneller (Insekt) Tunnel-Gräber

turbid trüb

turbidimetry Turbidimetrie, Trübungsmessung

turbidostat Turbidostat

turbinate bones/turbinals/ conchae nasalis Nasenmuscheln (schleimhautüberzogene Knorpelplatten)

turbinate spiralig aufgerollt/ gewunden, spiralörmig, wirbelförmig, schneckenförmig

turbine impeller (in bioreactors) Turbinenrührer

turbulent flow turbulente Strömung

turf/sod/grass cover (nonforage grass) Rasendecke

turfgrass Rasengräser

turgescence Turgeszenz, Schwellung

turgescent/swollen turgeszent, geschwollen, angeschwollen, schwellend, prall

turgid/inflated/swollen (slightly swelling with air or water) geschwollen

turgidity Turgidität, Geschwollenheit, Schwellunggrad

turgor/hydrostatic pressure Turgor, hydrostatischer Druck

turgor pressure Turgordruck

turion/turio/detachable winter bud/hibernaculum Turione, Turio, Winterknospe

turn *vb* drehen, wenden

turn *n* Drehung, Umdrehung, Wendung, Wende

● **beta turn/turn (DNA/protein)** beta-Drehung, beta-Schleife

turn anaerobic/become oxygen-deficient/turn over umkippen (Gewässer)

turnera family/Turneraceae Safranmalvengewächse

turnover Umsatz

turnover number

354

turnover number (k_cat) Wechselzahl (katalytische Aktivität)
turnover rate/rate of turnover Umsatzgeschwindigkeit, Umsatzrate
turret *micros* Revolver
turtle shell/tortoise shell Schildkrötenpanzer, Schildpatt
turtles/Chelonia/Testudines Schildkröten
turtles (marine) Seeschildkröten, Meeresschildkröten
tusk (large teeth) Stoßzahn, Hauer (z.B. Eber)
tusk shells/tooth shells/ scaphopods/scaphopodians (spade-footed mollusks)/ Solenoconchae/Scaphopoda Grabfüßer, Kahnfüßer, Scaphopoden
tussock grass Tussockgras, Bültgras
tweezers/forceps Pinzette
twig/limb/branchlet/sprig Zweiglein, dünner Zweig, Ästchen, Schössling, Rute
twig gall Stengelgalle, Zweiggalle
twin(s) Zwilling(e)
 • **dizygous twins/dizygotic twins/ fraternal twins** zweieiige Zwillinge
 • **monozygous twins/monozygotic twins/identical twins** eineiige Zwillinge
twin species (pair of sibling species) Zwillingsarten
twin spots Zwillingsflecken, Zwillingssektoren
twin studies Zwillingsstudien
twine sich emporranken
twiner/winder Schlingpflanze, Windepflanze, Winde
twining/winding windend, rankend
twinning Erzeugung monozygoter Mehrlinge
twist/coil/winding/contortion Wicklung, Windung
twist disease (>trout) (*Myxosoma cerebralis*) Drehkrankheit
twisted/coiled/wound/contorted gewickelt, aufgewickelt, gewunden, gedreht, verdreht
twisted shoot/twisted stem krummschaftig
twisted-winged insects/stylopids/ Strepsiptera Fächerflügler, Kolbenflügler
twisting number (DNA) Umdrehungszahl

twitch *vb* (muscle) zucken
twitching/convulsion Zuckung
twitter/chirp zwitschern; (sing/warble) singen
two flat-blade paddle impeller (in bioreactors) Blattrührer
two-branched/biramous zweiästig, biram
two-branched appendage/ biramous appendage Spaltfuß, Spaltbein (einfach gegabelt)
two-horned/bicornate/bicornuate/ bicornuous zweihörnig
two-parted/bifurcate zweigeteilt
two-row/in two rows/biseriate zweireihig
two-stage impeller (in bioreactors) zweistufiger Rührer
tylosis/thylosis/tylose *bot* Thylle; *med* Schwielenbildung
tylosis formation/tylosis Thyllenbildung
tympanic bone/tympanic Paukenbein, Tympanicum
tympanic canal/scala tympani Paukengang
tympanic cavity/cavum tympani Paukenhöhle, Trommelhöhle
tympanic membrane/eardrum/ tympanum/membrana tympani Tympanalmembran, Trommelfell, Ohrtrommel, Tympanum
tympanic organ Tympanalorgan, Gehörorgan
type Typ, Typus, Standard
type genus Typus-Gattung, Stammgattung
type species Typus-Art, Stammart, den Gattungsnamen festlegende Art
type specimen Typexemplar, Typusexemplar, Typusbeleg, Typbeleg
type-specific antigen typenspezifisches Antigen
Typhaceae/cattail family/ reedmace family Rohrkolbengewächse
typhlosole Typhlosolis
typhoid fever/typhoid/ typhus abdominalis (*Salmonella typhi*) Typhus, Unterleibstyphus
typhus/spotted fever/typhus exanthematicus (*Rickettsia spp.*) Typhus, Fleckfieber, Flecktyphus

unequal

ubichinone Ubichinon
ubiquinone/coenzyme Q
Ubiquinon, Coenzym Q
ubiquist Ubiquist
ubiquitin Ubiquitin
ubiquitous/widespread/existing everywhere ubiqitär, weitverbreitet, überall verbreitet
udder Euter
UESCE (unequal exchange of sister chromatids) ungleicher Austausch von Schwesterchromatiden (UESCE)
uliginose/marshy/boggy/growing in marshes Sumpf..., Moor..., im Sumpf wachsend
Ulmaceae/elm family Ulmengewächse
ulna/elbow bone Ulna, Elle
ultracentrifugation Ultrazentrifugation
ultracentrifuge Ultrazentrifuge
ultrafiltration Ultrafiltration
ultramicrotome Ultramikrotom
ultrasonic *adv/adj* Ultraschall..., den Ultraschall betreffend
ultrasound/ultrasonics Ultraschall
 • **ultrasonography/sonography** Ultraschalldiagnose, Sonographie, Sonografie
ultrastructural ultrastrukturell, Ultrastruktur...
ultrastructure Ultrastruktur, Feinbau
ultrathin section Ultradünnschnitt
ultraviolet spectroscopy/UV spectroscopy UV-Spektroskopie
umbel/sciadium (inflorescence) Dolde, Umbella, Sciadium
 • **compound umbel** zusammengesetzte Dolde
 • **simple umbel** einfache Dolde
umbel-like panicle (a corymb) Doldenrispe, Schirmrispe
umbel-like raceme (a corymb) Doldentraube, Schirmtraube
umbellifer family/parsley family/ carrot family/Apiaceae/ Umbelliferae Umbelliferen, Doldenblütler, Doldengewächse
umbilical/omphalic den Nabel betreffend, Nabel...
umbilical cord Nabelschnur, Nabelstrang
umbilicate/omphaloid/navel-like nabelartig, genabelt, omphaloid
umbilicate foliose lichens Nabelflechten (Blattflechten)
umbilicus/omphamos/navel Nabel
umbo (*pl* **umbones**) Umbo, Wirbel
umbonal cavity (mollusk shell) Wirbelhöhle
umbrella (float of medusa) Schirm

unassigned reading frame (URF) *gen* nicht zugeordnetes Leseraster (URF)
unbiased unvoreingenommen; *stat* unverfälscht, unverzerrt, frei von systematischen Fehlern
unbiasedness Unvoreingenommenheit; *stat* Treffgenauigkeit
unbranched/unramified *bot/zool* unverzweigt
unbranched (chain) *chem* unverzweigt (Kette)
unciform/hamiform/hook-shaped hakig, hakenförmig
unciform bone/hamate bone Hakenbein
uncinate/barbed/hooked hakig, mit Haken versehen
uncoating *vir* Uncoating (*not translated:* Hülle entfernen)
uncompetitive inhibition unkompetitive Hemmung
unconditioned reflex unbedingter Reflex
unconscious unbewusst
uncontaminated unverschmutzt
uncouple entkoppeln
uncoupler Entkoppler
uncoupling agent Entkoppler
undate/undulate gewellt, wellig
undemanding/modest/having low requirements/low demands anspruchslos
undercoat/underfur Unterhaarkleid, Wollhaarkleid
underdominance Unterdominanz
underground/belowground/ subterranean unterirdisch
 • **aboveground/overground/ superterranean** oberirdisch
undergrowth/understory Unterwuchs, Unterholz, Untergehölz
underhair Unterhaar, Wollhaar
underleaf/hypophyll Unterblatt
undernourished (*see:* **malnourished**) unterernährt
undernourishment Unterernährung
undershot *neuro* Nachpotential
underside/undersurface Unterseite
understock (grafting) *hort* Unterlage (Pfropfunterlage)
understory/undergrowth Unterholz, Untergehölz, Unterwuchs
undersurface/underside Unterseite
undulating membrane undulierende Membran
undulation Wellenbewegung
uneatable/inedible ungenießbar, nicht essbar
unequal/different ungleich, nicht identisch, anders

unequal cleavage 356

unequal cleavage inäquale Furchung, ungleichmäßige Furchung
uneven-aged stand/plantation Plenterwald
unfertilized unbefruchtet
unfolding/spreading Entfaltung
unguiculate/clawed (petals) *bot* genagelt
unguis/ungula/nail/claw Unguis, Nagel
ungulate/hoof-like hufartig
ungulate/hoofed/having hoofs mit Hufen, Huf...
ungulates/hoofed mammals/ hoofed animals Huftiere
- **even-toed ungulates/cloven-hoofed animals/artiodactyls/ Artiodactyla** Paarhufer
- **odd-toed ungulates/ perssiodactyls/Perssiodactyla** Unpaarhufer
unguliform/hoof-shaped hufförmig
unguligrade gait Unguligradie, Zehenspitzengang, Hufgang
unicellular/single-celled einzellig
unicellular lifeform Einzeller
unicorn plant family/devil's-claw family/martynia family/ Martyniaceae Gemsbockgewächse
unicornate/unicornuate/ unicornuous/one-horned einhörnig
unidirectional einsinnig, in eine Richtung
- **bidirectional** zweisinnig, in beide Richtungen
unifacial unifazial (ringsum gleiche Oberfläche)
uniform *adj/adv* gleichförmig, einheitlich
uniformitarianism/principle of uniformity Uniformismus
uniformity Gleichförmigkeit, Einheitlichkeit; Übereinstimmung
unilateral unilateral, einseitig
- **bilateral** bilateral, zweiseitig
unilocular unilokulär, einfächerig, einkammerig
- **bilocular** bilokulär, zweifächerig, zweikammerig
uninhabitable unbewohnbar
union/unification/combination Vereinigung, Verbindung
uniparental inheritance uniparentale Vererbung
uniparental mutation uniparentale Mutation
uniparous unipar
unipennate einfach pennat, einfach pinnat, einfach gefiedert
unipolar cell Unipolarzelle
uniport Uniport

uniramous uniram, einästig (ohne Exopodit)
uniseriate/uniserial/single rowed uniseriat, einreihig
unisexual eingeschlechtig (getrenntgeschlechtig)
unisexuality Eingeschlechtigkeit
unison *n* Übereinstimmung, Einklang
unit (measure) Einheit (Maßeinheit)
unit factor unteilbarer Faktor
unit membrane Einheitsmembran
unitary current Einheitsstrom
unite/combine vereinigen
unity Einheit, Einheitlichkeit, Einigkeit
unity *math* Eins, Einheit
univalence Einwertigkeit, Univalenz
univalent/monovalent einwertig, univalent, monovalent; einzeln
universal primer Universalprimer
universal veil/velum universale *fung* Velum universale
univolinism Univolitismus
univoltine univoltin, mit einer Jahresgeneration
unloaded form (e.g. ATP→ADP) entladene Form
- **loaded form (e.g. ADP→ATP)** beladene Form
unnatural unnatürlich
unpalatable ungenießbar, nicht schmackhaft
unsaturated ungesättigt
- **monounsaturated** einfach ungesättigt
- **polyunsaturated** mehrfach ungesättigt
unsaturated fatty acid ungesättigte Fettsäure
unscheduled DNA synthesis außerplanmäßige DNA-Synthese
unstable humus/friable humus/ crustable humus Nährhumus
unstable mutation instabile Mutation
unthriftiness *agr* Kümmerwuchs, mangelndes Gedeihen
unthrifty *agr* kümmernd, nicht gedeihend, schlecht gedeihend, schlecht wachsend; unwirtschaftlich
untranslated region (UTR) *gen* untranslatierte Region
unwinding (of the double helix) Entwinden (der Doppelhelix)
up-mutation Up-Mutation
up-regulation *metabol* Hochregulierung, Heraufregulierung
- **down-regulation** *metabol* Herunterregulierung, Herabregulation, Runterregulierung

357   **urodeum** U

upbringing/rearing/fostering/ nurture/nurturing Aufziehen, Erziehen
upper arm Oberarm
upper critical temperature (UCT) obere kritische Temperatur (OKT)
upper jaw/upper jawbone/maxilla Oberkiefer, Oberkieferknochen, Maxilla
upper leaf surface/adaxial leaf surface Blattoberseite
upper lip/labrum Oberlippe, Labrum
upper montane forest/subalpine conifer forest zone hochmontane Stufe, Nadelwaldstufe
upper surface/above surface/ upperside Oberseite
• **undersurface/underside** Unterseite
Upper Permian Oberperm, Zechstein
Upper Triassic (epoch) Keuper
upright/erect/straight/strict aufrecht
upright gait/orthograde gait (erect) aufrechter Gang, aufrechte Gangart
upright posture/orthograde posture aufrechte Haltung
uproot entwurzeln
upstream stromaufwärts; strangaufwärts, aufwärts (Richtung 5′-Ende eines Polynucleotids)
• **downstream** stromabwärts; strangabwärts, abwärts (Richtung 3′-Ende eines Polynucleotids)
uptake/intake (ingestion) Aufnahme, Einnahme
• **nutrient uptake** Nährstoffaufnahme
• **re-uptake** Wiederaufnahme
upward stroke of wing Flügelaufschlag
upwelling (water) Auftrieb (Auftriebswasser> aufwärtsstrebende Vertikalströmung)
• **downwelling (water)** abwärtsstrebende Vertikalströmung
urban ecology Urbanökologie, Stadtökologie
uracil Uracil
urban forest/community forest Stadtwald, städtischer Wald, Kommunalwald, Gemeindewald
urban landscape Urbanlandschaft
urbanization Urbanisierung, Verstädterung
urceolate/urn-shaped (sensu lato: pitcher-shaped/vase-shaped) urnenförmig (kannen-/vasen-/krugförmig)
urea (ureide) Harnstoff (Ureid)

urea cycle Harnstoffzyklus, Harnstoffcyclus
ureotelic/excreting urea/ urea-excreting ureotelisch, harnstoffausscheidend
ureter/urinary duct Ureter, Harngang, Harnleiter
ureteric bud Ureterknospe
urethra Urethra, Harnröhre
urethral fold Urethralfalte
urethral groove Urethralrinne
urethral plate Urethralplatte
urethral sinus Urethralsinus
URF (unassigned reading frame) gen URF (nicht zugeordnetes Leseraster)
urge vb antreiben, anstacheln, drängen
urge/drive/compulsion/impulse n ethol Trieb, Drang, Impuls
uric acid (urate) Harnsäure (Urat)
uricolytic pathway uricolytischer Weg, urikolytischer Weg
uricotelic/excreting ureic acid uricotelisch, urikotelisch, harnsäureausscheidend
uridine Uridin
uridine triphosphate (UTP) Uridintriphosphat (UTP)
uridylic acid Uridylsäure
urinary bladder Harnblase
urinary duct/ureter Harngang, Harnleiter
urinary tract Harnwege
urinate/micturate urinieren, harnlassen, harnen
urination/micturition Urinieren, Harnlassen, Harnen, Miktion
urine Urin, Harn
• **residual urine** Restharn
urine marking Harnmarkierung
urine sampling Harnprüfen
urine spraying/enurination Harnspritzen
uriniferous tubule Harnkanälchen, Nierenkanälchen (ab Bowman-Kapsel)
urinogenital/urogenital urogenital
urkaryote Urkaryot m
urn-shaped/flask-shaped/urceolate urnenförmig
urn-shaped leaf/pouch leaf/ "flower pot" leaf (Dischidia) Urnenblatt
urocanic acid (urocaninate) Urocaninsäure (Urocaninat), Imidazol-4-acrylsäure
urodeles/salamanders & newts and relatives/Urodela/Caudata Schwanzlurche (Salamander & Molche und Verwandte)
urodeum Urodaeum, Harnraum

U urogastrone 358

urogastrone/epidermal growth factor (EGF) epidermaler Wachstumsfaktor, Epidermiswachstumsfaktor

urogenital/urinogenital urogenital

urogenital cleft Urogenitalspalte

urogenital fold Urogenitalfalte

urogenital groove Urogenitalrinne

urogenital plate Urogenitalplatte

urogenital ridge Urogenitalleiste

uronic acid (urate) Uronsäure (Urat)

urophysis *ichth* Urophyse

uropod Uropod, Uropodium (*pl* Uropodien)

uropygial gland/preen gland/oil gland *orn* Bürzeldrüse

uroseminal duct (archinephric duct conducting both sperm and urine) Harnsamenleiter

urostyl (frogs/toads) Urostyl, Steißbein

ursine Bären betreffend, bärenartig, Bären...

Urticaceae/nettle family Nesselgewächse

urticant/stinging/irritating brennend

urticating hair/urticating trichome/stinging hair *bot* Brennhaar

usnic acid Usninsäure

uteral lining/uteral epithelium/endometrium Gebärmutterwand, Uterusepithel, Endometrium

uterine uterin, die Gebärmutter betreffend, Gebärmutter...

uterine bell Uterusglocke

uterine contraction (during labor) Wehe

uterine gland Uterusdrüse, Milchdrüse

uterine horn/cornu uteri Uterushorn, Gebärmutterzipfel

uterine milk *ichth* Uterusmilch, Uterinmilch

uterine tube/oviduct/Fallopian tube Eileiter

uterine valve Valvula uterina

uterus/womb Uterus, Gebärmutter

utilization *metabol* Verwertung, Verwendung

utilize *metabol* verwerten, verwenden

UTR (untranslated region) *gen* UTR (untranslatierte Region)

utricle/utriculus/small bladder Utriculus, Bläschen; *bot* Fangblase

utriculate/utricular/bladder-like/bladdery/possessing bladders mit kleinen Bläschen versehen, blasenartig

utriform/bladder-shaped blasenförmig

UTS (untranslated sequence) *gen* UTS (untranslatierte Sequenz)

UV radiation/ultraviolet radiation UV-Strahlung, ultraviolette Strahlung

uvula/palatine uvula Gaumenzäpfchen

vaccenic acid/11-octadecenoic acid
Vaccensäure
vaccinate/immunize impfen, immunisieren
vaccination/immunization
Vakzination, Vakzinierung, Schutzimpfung, Immunisierung (Impfung)
vaccine Vakzine, Impfstoff
- **attenuated vaccine** abgeschwächte(r)/attenuierte(r) Impfstoff, Vakzine
- **autogenous vaccine** Autoimpfstoff, Autovakzine
- **caprinized vaccine** durch Ziegen erzeugter Impfstoff
- **combination vaccine/mixed vaccine** Kombinationsimpfstoff, Mischimpfstoff, Mischvakzine
- **conjugate vaccine** Konjugatvakzine, zusammengesetzte Vakzine
- **heterologous vaccine** heterologer Impfstoff, heterologe Vakzine
- **humanized vaccine** durch Menschen erzeugter/passierter Impfstoff
- **inactivated vaccine** inaktivierter Impfstoff, inaktivierte Vakzine
- **killed vaccine** Totimpfstoff, Totvakzine
- **lapinized vaccine** durch Hasen erzeugter Impfstoff
- **live vaccine** Lebendimpfstoff, Lebendvakzine
- **mixed vaccine/combination vaccine** Mischimpfstoff, Mischvakzine, Kombinationsimpfstoff
- **monovalent vaccine** monovalenter Impfstoff, monovalente Vakzine
- **polyvalent vaccine** polyvalenter Impfstoff, polyvalente Vakzine
- **subunit vaccine** Spaltimpfstoff, Spaltvakzine, Komponentenimpfstoff, Subunitimpfstoff, Subunitvakzine
- **toxoid vaccine** Toxoidimpfstoff, Toxoidvakzine
vacuolate/vacuolated vakuolisiert
vacuolation Vakuolisierung
vacuole Vakuole
- **central vacuole** *bot* Zentralvakuole, große Vakuole
- **contractile vacuole** kontraktile Vakuole, pulsierende Vakuole
- **digestive vacuole/secondary lysosome** Verdauungsvakuole
- **food vacuole** Nahrungsvakuole
- **gas vacuole** Gasvakuole

vacuolization/vacuolation
Vakuolisierung, Vakuolenbildung
vacuum Vakuum
vacuum activity *ethol* Leerlaufhandlung
vacuum distillation Vakuumdestillation
vacuum pump *lab* Vakuumpumpe, Absaugpumpe, Unterdruckpumpe
vacuum-metallize aufdampfen, bedampfen
vagile/freely motile vagil, motil, frei beweglich
vagility/motility Vagilität, Motilität, Beweglichkeit, aktive Ausbreitungsfähigkeit
vagina Vagina, Scheide
vaginal pregnancy Vaginalträchtigkeit
vaginal vestibule Scheidenvorhof, Vestibulum vaginae
vaginate/sheated von einer Scheide umgeben
vagrant (moving/shifting unpredictably) wandernd, umherziehend, umherirrend, vagabundierend
valence/valency Valenz, Wertigkeit
valence electron Valenzelektron
valerian family/Valerianaceae Baldriangewächse
valeric acid/pentanoic acid (valeriate, pentanoate) Valeriansäure, Baldriansäure, Pentansäure (Valeriat, Pentanat)
valine Valin
valley Tal
valley flat Flussaue
valvate/chambered fächerig, gefächert, gekammert, klappig
valve/chamber/case Fach, Kammer
valve/locule/loculus/compartment Fach, Lokulament, Loculament, Loculus, Kompartiment
valve (controlling liquid/gas flow/ pressure) Ventil
valve (chitons: individual plates) Schalensegment, Schalenplatte
valve (diatom half-shell) Schalenhälfte (Diatomeen)
valve (single bivalve shell/ diatom shell) Klappe, Schalenklappe, Schalenhälfte (Muscheln/Diatomeen)
valvifer Valvifer
valvula/small valve Valvula
vampire squids/Vampyromorpha Vampirtintenschnecken, Tiefseevampire
vane (feather) Federfahne
vanillic acid Vanillinsäure
vanillin Vanillin

vannal fold

vannal fold/anal fold/vannal fold-line/anal fold-line/plica analis/plica vannalis *entom* **(wing)** Vannalfalte, Analfalte
vannus/anal area *entom* **(wing)** Vannus, Analfeld, Analregion
vapor Dampf; Dunst, Nebel
vapor bath Dampfbad
vapor blasting *micros* Bedampfung, Bedampfen, Aufdampfen
vapor cooling Verdunstungskühlung
vapor pressure Dampfdruck
vaporization/evaporation Verdampfung, Verdunstung
vaporization apparatus *micros* Bedampfungsanlage
vaporize/evaporate verdampfen, verdunsten
variability/diversity Variabilität, Vielfältigkeit, Mannigfaltigkeit; Veränderlichkeit, Wandelbarkeit
• **decay of variability** Variabilitätsrückgang
variable *adj/adv* variabel, veränderlich, wechselnd, unterschiedlich
variable *n stat* Variable, Veränderliche
• **adjustable variable** Stellgröße
variable number of tandem repeats (VNTR) variable Anzahl von Tandemwiederholungen (VNTR)
variable pitch screw impeller (in bioreactors) Schraubenspindelrührer mit unterschiedlicher Steigung
variable region *immun* variable Region
variable residue *math* variabler Rest
variance/mean square deviation *stat* Varianz, mittlere quadratische Abweichung, mittleres Abweichungsquadrat
• **additive genetic variance** additive genetische Varianz
• **dominance variance** *gen* Dominanzvarianz
• **environmental variance** Umweltvarianz, Umweltabweichung
• **genetic variance** genetische Varianz
variance ratio distribution/F-distribution/Fisher distribution *stat* Varianzquotientenverteilung, F-Verteilung, Fisher-Verteilung
variant *adj/adv* abweichend, verschieden, unterschiedlich
variant *n* Variante
variate *n* **(a variable characteristic)** Abweichung (abweichendes Merkmal)
variate/random variable *n* Zufallsvariante
variate *vb* variieren, schwanken
variation Variation, Schwankung
• **genetic v./genotypic v.** genetische Variation
• **phenotypic v.** phänotypische Variation
• **somaclonal v.** somaklonale Variation
varicosity (abnormally swollen) Varikosität
variegated/mottled gescheckt, panaschiert
variegated-leaved buntblättrig
variegation Variegation, Scheckung, Buntblättrigkeit
variety Varietät; Reihe, Mehrzahl; (diversity) Verschiedenheit, Mannigfaltigkeit
variety/sport (usually by somatic mutation) Abart, Spielart
variolation Variolation
varve (one year's sediment deposit) *geol* Warve, Jahresschicht
vary *vb* variieren, wechseln, abweichen, verändern, unterscheiden, unterschiedlich sein
vascular vaskulär, Gefäß..., Gefäße betreffend
vascular bud *bot* Gefäßknospe, Gefäßanlage
vascular bundle/vascular strand *bot* Gefäßbündel, Leitbündel, Leitbündelstrang
• **closed vascular bundle** geschlossenes Leitbündel
• **open vascular bundle** offenes Leitbündel
vascular cylinder *bot* Leitbündelzylinder, Leitbündelring
vascular layer (eye) Gefäßhaut
vascular plant Gefäßpflanze
vascular system Gefäßsystem, Blutgefäßsystem
vascular tissue/conducting tissue Leitgewebe
vascularization Gefäßbildung; *bot* Leitgewebebildung
vascularize/supply with blood durchbluten
vasculum Botanisiertrommel
vasoactive intestinal polypeptide (VIP) vasoaktives intestinales Peptid (VIP)
vasopressin/antidiuretic hormone (ADH) Vasopressin, antidiuretisches Hormon (ADH), Adiuretin
vasotocin Vasotocin
vector Vektor, Überträger
• **bidirectional/dual promoter** bidirektionaler Vektor

361　　　　　　　　　　　　　　　　　　　　　　**venous shunt**

- **bifunctional vector/shuttle
  vector** bifunktionaler Vektor,
  Schaukelvektor
- **containment vector**
  Sicherheitsvektor
- **eviction vector** Apportiervektor
- **multifunctional vector/
  multipurpose vector**
  Vielzweckvektor,
  multifunktioneller Vektor
- **replacement vector**
  Substitutionsvektor
- **transient expression vector**
  transienter Expressionsvektor

vegan *n* orthodoxer Vegetarier
(streng vegetarisch ohne jegliche
tierische Produkte)
vegetable *n* Gemüse; Grünfutter
vegetable *adj/adv* pflanzlich,
Pflanzen...
vegetable dye Pflanzenfarbstoff
vegetable oil Pflanzenöl
vegetable patch Gemüsebeet
vegetal *physiol* vegetativ; *bot*
pflanzlich
vegetal pole/vegetative pole *embr/
cyt* vegetativer Pol
vegetarian *n* Vegetarier
- **lactovegetarian** Laktovegetarier
- **ovovegetarian** Ovovegetarier
vegetation/plant life Vegetation,
Pflanzenwelt
vegetation map Vegetationsplan
vegetation(al) zone/region/belt
Vegetationsstufe
(Höhenstufe *see* altitudinal zone)
vegetative vegetativ
**vegetative cell/somatic cell/body
cell** vegetative Zelle, somatische
Zelle, Körperzelle
vegetative cone/vegetative pole
Vegetationskegel
veil/velum Velum, Schleier; Segel;
fung Pilzhülle
vein *zool* Ader, Nerv
- **cross vein (insect wing)**
  Querader
- **longitudinal vein (insect wing)**
  Längsader
vein/rib *bot* Ader, Nerv, Rippe
vein (blood vessel) Vene (for
individual veins please consult a
medical dictionary)
- **cardinal vein** Cardinalvene,
  Kardinalvene
- **spur vein (horse)** Sporvene
veined/venulous geädert
veliger larva Veligerlarve,
Segellarve
vellozia family/Velloziaceae
Velloziagewächse,
Baumliliengewächse

velocity (vector)/rate
Geschwindigkeit
velum/veil Velum, Schleier; Segel;
fung Pilzhülle; Mooshaube
velutinous/velvet-like/velvety
samtig
velvet Samt; (antlers) Bast
(Geweih)
velvet worms/onychophorans
Stummelfüßer, Onychophoren
velvet-like/velvety/velutinous
samtig
vena cava Hohlvene
venation *bot/zool* Venation,
Aderung, Nervatur, Nervation
- **arched/arciform/arcuate/
  camptodrome venation**
  Bogenaderung, Bogennervatur
- **closed venation** geschlossene
  Aderung, geschlossene Nervatur
- **dichotomous venation**
  Gabeladerung, Fächeraderung
- **open venation** offene Aderung,
  offene Nervatur
- **parallel venation**
  Paralleladerung, Parallelnervatur
- **pinnate venation** Fiederaderung,
  Fiedernervatur
- **reticulate venation/net
  venation/netted venation**
  Netzaderung, Netznervatur
- **striate venation** Längsaderung,
  Längsnervatur, Streifenaderung,
  Streifennervatur
veneer Furnier
**veneer side grafting/side-veneer
grafting/spliced side grafting** *hort*
seitliches Anplatten
**venenation/poisoning/
envenomation** Vergiftung
venereal/genitoinfectious
venerisch,
Geschlechtskrankheiten betreffend
**venereal disease (VD)/sexually
transmitted disease (STD)**
Geschlechtskrankheit,
venerische Krankheit,
sexuell übertragbare Krankheit
venereal transmission
venerische Übertragung
venereology Venerologie
venison Wild, Wildfleisch (Rotwild)
**venom (poison secreted from an
animal)** Gift, Tiergift
**venom tooth/poison tooth/fang
(snakes)** Giftzahn
venomous giftig
venous *adv/adj* venös
**venous angle/Pirogoff's angle/
angulus venosus** Venenwinkel
venous shunt venöse Umgehung,
Kurzschlussdurchblutung

venous sinus/sinus venosus Sinus venosus
vent/anus Anus, After
vent/cloacal opening n Kloakenöffnung
vent vb auftauchen zur Wasseroberfläche zum Luftholen (Otter/Biber etc.)
venter Bauch (Archegonium)
venter cell bot Bauchkanalzelle
ventilation volume Ventilationsvolumen
ventral/front side ventral, bauchseitig, vorderseitig, Bauch...
ventral/ventrally located bauchständig
ventral fin/pelvic fin Bauchflosse
ventral ganglion Ventralganglion, Bauchnervenknoten
ventral gland/renette gland (nematodes) Ventraldrüse
ventral horn neuro Grundplatte (Neuralrohr)
ventral nerve cord ventraler Nervenstrang, Bauchmark, Bauchmarkstrang
ventral pouch Bauchtasche
ventral root/anterior root/motor root neuro Ventralwurzel, motorische Wurzel
ventral scale Ventralschuppe, Bauchschuppe
ventral suture/ventral seam (of carpel) bot Ventralnaht, Bauchnaht
ventrally located bauchständig
ventricle Ventrikel
ventricular cycle cardio Kammerzyklus
ventricular flutter cardio Kammerflattern
ventricular fold/vestibular fold/ false vocal cord/plica ventricularis/plica vestibularis Taschenband
venule (small vein) Venule, Venole
venule/venula (insect wing) entom Äderchen
venulous/veined geädert
verbena family/vervain family/ Verbenaceae Eisenkrautgewächse, Verbenengewächse
vermicular/vermian/wormlike wurmartig
vermiculite Vermiculit
vermiform/worm-shaped wurmförmig
vernacular name/common name Vernakularname, volkstümliche Bezeichnung, volkstümlicher Name
vernalization Vernalisation, Keimstimmung

vernation/ptyxis/prefoliation bot Vernation, Knospenlage der Laubblätter, Blattlage in der Knospe
verruciform/wart-shaped warzenförmig
verrucose/warty/tuberculate warzig, höckerig
vertebra (pl **vertebras/vertebrae**) Wirbel (auch in bezug auf Ophiuroiden)
- **caudal vertebra** Schwanzwirbel, Kaudalwirbel
- **cervical vertebra** Halswirbel, Cervikalwirbel
- **coccygeal vertebra** Steißwirbel, Steißbeinwirbel
- **dorsal vertebrae (thoracic + lumbar)** Dorsalwirbel
- **lumbar vertebra** Lendenwirbel, Lumbarwirbel
- **sacral vertebra** Kreuzbeinwirbel, Sakralwirbel
- **thoracic vertebra** Brustwirbel, Thorakalwirbel

vertebral column/spinal column Wirbelsäule, Rückgrat
vertebral ossicle/shield (ophiurids) Platte
vertebral spine/neural spine/ spinous process of vertebra Dornfortsatz, Processus spinosus
vertebrates Vertebraten, Wirbeltiere
- **terrestrial vertebrates/tetrapods** Landwirbeltiere, Tetrapoden

vertex Vertex, Scheitel
vertical air flow (clean bench with vertical air curtain) vertikale Luftführung (Vertikalflow-Werkbank)
vertical alignment/vertical orientation (of leaves) Profilstellung
vertical flow workstation, hood, unit Fallstrombank
vertical rotor centrif Vertikalrotor
vertical transmission vertikale Transmission, vertikale Übertragung
verticillate wirtelig
vervain family/verbena family/ Verbenaceae Eisenkrautgewächse, Verbenengewächse
very late antigen (VLA) "sehr spätes Antigen" (bildet sich spät in der Entwicklung)
very low density lipoprotein (VLDL) Lipoprotein sehr niedriger Dichte
vesicate/blister/cause blistering vb Blasen treiben/ziehen
vesicating/vesicant blasentreibend, blasenziehend
vesicle Vesikel nt, Bläschen, kleine Blase

363   **virility V**

vesicular/vesiculate/bladderlike
vesikulär, blasenartig, bläschenartig
vesicular gland/seminal gland/
seminal vesicle (♂ accessory
reproductive gland)
Bläschendrüse, Samenblase,
Samenbläschen
vesicular ovarian follicle/Graafian
follicle Graafscher Follikel,
Graaf-Follikel, Tertiärfollikel
vespertine (blooming in the
evening) am Abend blühend
vessel/container Gefäß, Behälter
vessel/trachea Gefäß, Trachee
• **blood vessel** Blutgefäß
• **lymph vessel** Lymphgefäß
• **scalariform vessel** *bot*
Leitertrachee
vessel element/vessel member *bot*
Tracheenglied
vestibular canal/scala vestibuli
Vorhofgang
vestibular gland (♀ vaginal gland)
Vorhofdrüse
• **Bartholin's gland/greater**
vestibular gland/glandula
vestibularis major
Bartholin-Drüse
vestibule Vestibulum, Vorhof,
Vorraum; Präoralhöhle
• **vestibulum labyrinthi (with**
utricle and saccule)
Vestibulum labyrinthi
(mit Utriculus und Sacculus)
vestige/vestigium/remnant/trace
Überbleibsel, Überrest, Spur
vestigial (small and imperfectly
developed)/underdeveloped/
stunted (*sensu lato*: rudimentary)
verkümmert, unterentwickelt
vesture/vesture/body covering
Hülle, Mantel
veterinarian/vet Veterinär, Tierarzt
veterinary clinic/animal hospital
Tierklinik
veterinary medicine
Veterinärmedizin, Tiermedizin,
Tierheilkunde
veterinary practice (for small
mammals) Kleintierpraxis
viability Lebensfähigkeit
viable lebensfähig
vial Gläschen, (Glas)Fläschchen,
Phiole
vibraculum Vibracularie
vibrating dance (bees)/dorsoventral
abdominal vibrating dance
(DVAV) Schütteltanz,
Schüttelbewegung
vibrational motion
Schwingungsbewegung
vibrios *bact* Vibrionen

vicariance *ecol* Vikarianz,
(geographische) Stellvertretung
vicariate vikariieren, für etwas
stehen, stellvertretend sein
vicariism Vikariismus,
Stellvertretertum
vicarious vikariierend,
stellvertretend, mitempfunden,
nachempfunden)
victim/prey Opfer
vigorous kräftig
vigorous growth
kräftiges Wachstum
vigreux column Vigreux-Kolonne
villous/villose/covered with villi
villös, zottig, mit Zotten; *bot*
(having soft long hairs/soft-haired)
weich behaart
villous placenta (hemochorial
placenta)
Zottenplazenta, Topfplazenta
villus (*pl* villi) *cyt/anat* Villus,
Zotte; *bot* Zottenhaar
• **chorionic villi** Chorionzotten
• **intestinal villi** Darmzotten
• **microvillus (*pl* microvilli)**
Mikrovillus (*pl* Mikrovilli);
Stereocilien (Lateralisorgan)
• **vascular villi** Gefäßzotten
vine Rebe, Weinrebe;
rankende Pflanze
vine family/grape family/Vitaceae
Weinrebengewächse
vinegar Essig
• **wood vinegar/pyroligneous acid**
Holzessig
vineyard Weinberg
violet family/Violoaceae
Veilchengewächse
violet gland/supracaudal gland
(fox) Violdrüse
(Schwanzwurzeldrüse des Fuchses)
viral viral
viral burden Virenlast
viral coat Virushülle
viral particle Virusteilchen,
Viruspartikel
viral retroelement
virales Retroelement
viremia Virämie
virgin *n* **(female that has never**
copulated) *zool* Jungfrau
virgin B cell unreife B-Zelle
virgin forest/pristine forest/
primeval forest/jungle Urwald
virgin stand *for/ecol*
Primärbestand
virginity Jungfräulichkeit
virile viril, maskulin, männlich;
(copulative power) potent
virility Virilität, Männlichkeit;
Potenz

virilization/masculinization
Virilisierung, Maskulinisierung, Vermännlichung
virion/viral particle Virion, Viruspartikel, Virusteilchen
virioplasm Virioplasma
viroid Viroid
virology Virologie
viropexis Viropexis
virosis Virose, Viruserkrankung
virostatic *n* Virostatikum, virostatisches Mittel
virtual image *micros* virtuelles Bild
virulence (disease-evoking power/ability of cause disease) Virulenz, Infektionskraft, Ansteckungskraft; *med* Giftigkeit, Bösartigkeit
virulent
virulent, von Viren erzeugt; giftig, bösartig, sehr ansteckend
virus (*pl* viruses) Virus (*pl* Viren)
• **amphotropic virus** amphotropes Virus
• **animal virus** Tiervirus
• **bacterial virus/bacteriophage** Bakteriophage
• **defective virus** defektes Virus
• **ecotropic virus** ecotropes Virus
• **icosahedral virus** ikosaedrisches Virus
• **multicomponent virus** Multikomponentenvirus
• **plant virus** Pflanzenvirus
• **satellite virus** Satellitenvirus
• **xenotropic virus** xenotropes Virus
Viscaceae/christmas mistletoe family Mistelgewächse
viscera/guts/intestines/entrails/bowels Viscera, Viszera, Eingeweide, Gedärme, Splancha
visceral/splanchnic viszeral, zu den Eingeweiden gehörend, Eingeweide...
visceral arch Viszeralbogen; Kiemenbogen
visceral ganglion Visceralganglion, Viszeralganglion
visceral hump/visceral mass (mollusks) Eingeweidesack
visceral musculature viscerale Muskulatur, Eingeweidemuskulatur
visceral pleura//pleura pulmonalis Lungenfell
visceral skeleton Visceralskelett, Viszeralskelett
viscerocranium/splanchnocranium Eingeweideschädel, Visceralcranium, Viscerocranium, Splanchnocranium
viscidium (a sticky disk of orchid gynostemium) Klebscheibe, Klebkörper

viscosity/viscousness Viskosität, Dickflüssigkeit, Zähflüssigkeit
• **coefficient of viscosity** Viskosität, Viskositätskoeffizient
viscous/viscid (glutinous consistency) viskos, viskös, zähflüssig, dickflüssig
visibility Sichtbarkeit; *meteo* Sicht
visible sichtbar
vision/eyesight Gesichtssinn
visual visuell, sichtbar, Sicht.., Seh.., das Sehen betreffend
visual acuity Sehschärfe
visual lobe/optic lobe/lobus opticus Sehlappen
visual pigment Sehpigment
visualize sichtbar machen; sich vorstellen
vital vital; lebensnotwendig, lebenswichtig
vital capacity Vitalkapazität
vital dye/vital stain Vitalfarbstoff
vital functions Vitalfunktionen, lebenswichtige Funktionen
vital power/vital energy Lebenskraft
vital red *micros* Brilliantrot
vital stain/vital dye *micros* Vitalfarbstoff
vital staining *micros* Vitalfärbung, Lebendfärbung
vital statistics demografische Kennzahlen
vitality Vitalität, Lebenskraft; Lebensfähigkeit; Lebensdauer
vitalize vitalisieren, beleben, anregen, kräftigen
vitamin(s) Vitamin(e)
• **ascorbic acid (vitamin C)** Ascorbinsäure
• **biotin (vitamin H)** Biotin
• **carnitine (vitamin B_T)** Carnitin (Vitamin T)
• **carotin/carotene (vitamin A precursor)** Carotin, Caroten, Karotin (Vitamin A Vorläufer)
• **cholecalciferol (vitamin D_3)** Cholecalciferol, Calciol
• **citrin (hesperidin) (vitamin P)** Citrin (Hesperidin)
• **cobalamin (vitamin B_{12})** Cobalamin, Kobalamin
• **ergocalciferol (vitamin D_2)** Ergocalciferol, Ergocalciol
• **folic acid/folacin/pteroyl glutamic acid (vitamin B_2 member)** Folsäure, Pteroylglutaminsäure
• **gadol/3-dehydroretinol (vitamin A_2)** Gadol, 3-Dehydroretinol
• **menadione (vitamin K_3)** Menadion

vulnerable

- menaquinone (vitamin K_2) Menachinon
- pantothenic acid (vitamin B_3) Pantothensäure
- phylloquinone/phytonadione (vitamin K_1) Phyllochinon, Phytomenadion
- pyridoxine/adermine (vitamin B_6) Pyridoxin, Pyridoxol, Adermin
- retinol (vitamin A) Retinol
- riboflavin/lactoflavin (vitamin B_2) Riboflavin, Lactoflavin
- thiamine/aneurin (vitamin B_1) Thiamin, Aneurin
- tocopherol (vitamin E) Tocopherol, Tokopherol

vitamin deficiency Vitaminmangel
vitelline duct Dottergang
vitelline gland/vitellarian gland/ vitellogen/vitellarium/yolk gland Vitellarium, Vitellar, Dotterstock, Dotterdrüse
vitelline layer/vitelline membrane/ membrana vitellina Vitellinmembran, Dotterhaut, Dottermembran, primäre Eihülle
viticulture Weinbau
vitreous body (eye) Glaskörper
vitta/oil tube/oil cavity/resin canal (Apiacean fruit) *bot* Ölstrieme
viviparous/live-bearing vivipar, lebendgebärend
vivipary/viviparity/live-bearing Viviparie, Lebendgebären
VNTR (variable number of tandem repeats) *gen* VNTR (variable Anzahl von Tandemwiederholungen)
vocal cord/vocal fold/true vocal cord/plica vocalis eigentliches Stimmband, Stimmfalte, Stimmlippe
- false vocal cord/ventricular folds/plica vestibularis Taschenfalte, Taschenband, falsche Stimmlippe

vocal ligament/ligamentum vocale Stimmband (*pl* Stimmbänder)
vocal pouch/voise box (birds: larynx) Stimmsack, Stimmbeutel
vocal process Stimmbandfortsatz
vocalization Vokalisation, Lautgebung
vochysia family/Vochysiaceae Ritterspornbaumgewächse

voice Stimme
voice box/larnyx Apparat der Stimmbildung, "Stimmkasten", Kehlkopf, Larynx
voice box/syrinx *orn* Stimmkopf
volar volar, zur Hohlhand gehörend, auf der Hohlhandseite liegend (palmar)
volatile flüchtig
- nonvolatile nicht flüchtig

volcanic lake/maar Maar
Volkmann canal/canal of Volkmann (perforating canal) Volkmannscher Kanal
voltage clamp Spannungsklemme
voltage-sensitive channel/ voltage-gated channel spannungsregulierter/ spannungsgesteuerter Kanal
volumetric volumetrisch
volumetric analysis Maßanalyse
volumetric flask Messkolben
voluntary *med* willkürlich
voluntary musculature willkürliche Muskulatur
voluntomotoricity/voluntary motility Willkürmotorik
volutin granules/metachromatic granules Volutinkörnchen, metachromatische Granula
volva/cup/pouch Volva, Knolle
volva/universal veil *fung* Volva, Velum universale
vomit erbrechen, brechen, sich übergeben
vomiting center Brechzentrum
von Magnus particle/defective interfering particle (DI particle) Von-Magnus-Partikel, DI-Partikel
voracious gefräßig
voracity Gefräßigkeit
vortex/mixer *lab* "Vortex", Mixer, Mixette, Küchenmaschine
vorticose wirbelartig, strudelartig, wirbelig, wirbelbildend, Wirbel...
vorticose veins/vortex veins hintere Ziliarvenen (Auge)
voucher specimen Belegexemplar
vulnerability/vulnerableness Verletzlichkeit
vulnerable verletzlich, verwundbar, verletzbar; anfällig für

W waddle

waddle watscheln
wade waten, schreiten
wading foot *orn* Schreitfuß
waggle/wag wedeln, wackeln
(Schwanz/Kopf)
**waggle dance/tail-wagging dance
(bees)** Schwänzeltanz
waist Taille
**"Waldsterben"/forest
deterioration/forest decline**
Waldsterben
walk *vb* gehen, laufen
(zu Fuß fortbewegen)
walk *n* **(gait of horse)** Schritt
(Gangart des Pferdes)
● **collected walk**
versammelter Schritt
● **extended walk** starker Schritt
● **free walk** freier Schritt
● **medium walk** Mittelschritt
● **working walk** Arbeitsschritt
walking leg/gressorial leg
Laufbein
wall pressure (WP)/turgor pressure
Wanddruck
Wallace's line Wallace-Linie
wallow *n* Suhle
wallow *vb* suhlen
walnut family/Juglandaceae
Walnussgewächse
wanting/lacking/missing fehlend
warble *n* **(swelling under hide)**
Dasselbeule
warble/trill trillern
**Warburg's factor/cytochrome
oxidase** Warburgsches
Atmungsferment, Cytochromoxidase
**warm-blooded/homoiothermic/
homothermic/endothermic**
gleichwarm, warmblütig,
homoiotherm, endotherm
warming Erwärmung
● **global warming**
globale Erwärmung
warmth/heat Wärme, Hitze
**warning behavior/alarm behavior
(aposematic behavior)**
Warnverhalten
**warning coloration/aposematic
coloration** Warnfärbung,
Warntracht, Abschreckfärbung
warp (sediment) angeschwemmtes
Schlicksediment
warp (wood) verziehen, werfen
**warren>rabbit warren
(subterranean living quarters)**
Kaninchenbau; Kaninchengehege,
Kaninchenbrutplatz
wart/tubercle/warty protuberance
Warze (Höcker/Beule/Wölbung)
wart-shaped/verruciform
warzenförmig

warty/verrucose/tuberculate
warzig
wash bottle Spritzflasche
washer *lab* **(thin flat ring for
tightness/preventing leakage)**
Dichtung, Unterlegscheibe,
Dichtungsring
wasp pollination/sphecophily
Wespenbestäubung
wastage/weathering Verwitterung
waste/squander *n* Verschwendung,
Vergeudung
waste/squander *vb* verschwenden,
vergeuden
waste/trash/garbage *n* Abfall, Müll
waste away *vb* verfallen,
dahinsiechen, schwächer werden,
schwinden, Lebenskraft verlieren;
verwelken (Pflanze)
waste heat Abwärme
waste removal Entsorgung
wasteland Ödland
wastewater/sewage Abwasser
wastewater purification plant
Kläranlage (industriell)
watch glass *lab* Uhrglas
water Wasser
● **amniotic fluid/"water"**
Amnionflüssigkeit, Amnionwasser,
Fruchtwasser
● **black water (river)**
Schwarzwasser
● **body of water/water body**
Gewässer
● **brackish water (somewhat salty)**
Brackwasser (leicht salzig)
● **coastal waters** Küstengewässer
● **deionized water**
entionisiertes Wasser
● **distilled water**
destilliertes Wasser
● **drainage water/leachate/
soakage/seepage/gravitational
water** Sickerwasser
● **drinking water** Trinkwasser
● **film water** Haftwasser
● **flowing water (river/stream)**
Fließgewässer (Fluss/Strom)
● **freshwater** Süßwasser
● **gravitational water/seepage
water** Senkwasser, Sickerwasser
● **ground water** Grundwasser
● **hard water** hartes Wasser
● **inland water/inland waterbody**
Binnengewässer
● **meltwater** Schmelzwasser
● **peptone water** Peptonwasser
● **phreatic (pertaining to
groundwater)** phreatisch,
Grundwasser...
● **potable water** trinkbares Wasser,
Trinkwasser

water-conducting element

- **purified water** gereinigtes Wasser, aufgereinigtes Wasser, aufbereitetes Wasser
- **receiving water** Vorfluter
- **retained water** Haftwasser
- **saline water** salziges Wasser
- **saltwater** Salzwasser
- **seawater/saltwater** Meerwasser
- **soft water** weiches Wasser
- **springwater** Quellwasser
- **surface water** Oberflächenwasser
- **tap water** Leitungswasser
- **upwelling (water)** Auftrieb (Auftriebswasser> aufwärtsstrebende Vertikalströmung)
- **wastewater** Abwasser
- **well water** Brunnenwasser
- **white water (Amazon)** Weißwasser

water activity Wasseraktivität, Hydratur, "relative Aktivität"
water balance Wasserbilanz; Wasserhaushalt
water bath Wasserbad
water bears/tardigrades Bärtierchen, Bärentierchen, Tardigraden (*sg* Tardigrad *m*)
water bloom Wasserblüte
water body/body of water Gewässer
water chestnut family/Trapaceae Wassernussgewächse
water clover family/marsilea family/Marsileaceae Kleefarne, Kleefarngewächse
water column Wassersäule
water conductance/conduction/ translocation Wasserleitung, Translokation
water content Wassergehalt
water cycle/hydrologic cycle Wasserkreislauf (der Natur)
water expulsion vesicle/contractile vacuole kontraktile Vakuole, pulsierende Vakuole
water fern family/floating fern family/Parkeriaceae Hornfarngewächse
water fleas/cladocerans/Cladocera Wasserflöhe
water flow Wasserströmung
water free space Water free space (*used as such in German; not translated!*)
water hardness Wasserhärte
water hawthorn family/ cape-pondweed family/ Aponogetonaceae Wasserährengewächse
water hyacinth family/pickerelweed family/Pontederiaceae Hechtkrautgewächse

water level Wasserspiegel
water loss Wasserverlust
water milfoil family/milfoil family/ Haloragaceae Seebeerengewächse, Tausendblattgewächse
water molds/Saprolegniales Wasserschimmel
water molds (oomycetes) wasserlebende Oomyceten
water molds & downy mildews/ oomycetes/Oomycota falsche Mehltaupilze
water movement Wasserbewegung
water nymph family/najas family/ Najadaceae Nixenkrautgewächse
water of crystallization Kristallisationswasser
water of hydration Hydratwasser
water parting/divide Wasserscheide
water pollution Wasserverschmutzung
water potential/suction pressure Wasserpotential, Saugkraft
water pump/filter pump/vacuum filter pump Wasserstrahlpumpe
water purification Wasseraufbereitung
water quality Wasserqualität, Wassergüte
water regime Wasserregime, Wasserhaushalt
water reserve Wasserschutzgebiet
water sample Wasserprobe
water saturation Wassersättigung
water saturation deficit (WSD) Wassersättigungsdefizit
water sprout/water shoot/sucker Geiltrieb, Wassertrieb, Wasserschoss
water starwort family/starwort family/Callitrichaceae Wassersterngewächse
water storage Wasserspeicherung
water stress Wasserstress
water supply Wasserversorgung
water table Grundwasserspiegel
water tank (of certain bromeliads) Zisterne
water tension/water suction Wassersog, Zugspannung (Wasserkohäsion)
water transport Wassertransport
water transport pathway Wassertransportweg
water uptake Wasseraufnahme
water vapor Wasserdampf
water vascular system Wassergefäßsystem, Ambulakralgefäßsystem
water-conducting wasserleitend
water-conducting element/pathway Wasserleitbahn, Wasserleitungsbahn

water-dispersal

water-dispersal/hydrochory Wasserausbreitung, Hydrochorie
water-lily family/Nymphaeaceae Seerosengewächse
water-plantain family/arrowhead family/Alismataceae Froschlöffelgewächse
water-poppy family/Limnocharitaceae Wassermohngewächse
water-repellent/water-resistant wasserabweisend
water-soluble wasserlöslich
water-straining bill Seihschnabel
watercourse/waterway Fließgewässer, Wasserlauf
waterfowl (ducks/geese/swans)/Anseriformes Gänsevögel, Entenvögel
waterfowl *general* Wasservögel
watering place Wasserstelle
waterleaf family/Hydrophyllaceae Wasserblattgewächse
waterlog *vb* mit Wasser vollsaugen
waterlogged mit Wasser vollgesogen
waterlogging/waterlogged (soil) Vernässung, Staunässe (Boden)
waterproof wasserdicht, wasserundurchlässig
watershed/drainage area/drainage district/catchment area/catchment basin Wassereinzugsgebiet, Grundwassereinzugsgebiet
waterway/watercourse Fließgewässer, Wasserlauf
waterwort family/Elatinaceae Tännelgewächse
wattle(s) (fleshy pendant process at head/neck) (e.g. birds/reptiles) Kehllappen (Hautlappen); (Welse) Bartfäden; (Appendices colli: Schaf/Ziege/Schwein) Berlocken, Glöckchen
wave *n* Welle
 • **backrush/backwash** Wellenrücklauf, Wellenrückstrom, Rücksog
 • **contraction wave** *med/physiol* Kontraktionswelle
 • **tidal wave** Gezeitenwelle; (seismic wave) Flutwelle
 • **uprush/swash** Wellenauflauf
wave exposure Wellenexposition
wavelength Wellenlänge
wavy/undulate/repand (slightly undulating) wellig, gewellt
wax Wachs
wax coating Wachsbelag
wax feet *micros* Wachsfüßchen
wax gland/ceruminous gland Wachsdrüse

wax plant Wachsblume
wax-myrtle family/bog myrtle family/sweet gale family/Myricaceae Gagelgewächse
waxy/wax-like/ceraceous wachsartig
weaken *vb* schwächen, abschwächen
weakening *neuro* Abschwächung
weakness Schwäche
wean abstillen, entwöhnen
weaning Abstillen, Entwöhnung
weanling frisch abgestilltes/entwöhntes Jungtier, Läufer
weather Wetter
weathering/wastage Verwitterung
web (skin between digits) Schwimmhaut
web (spiderweb) *arach* Gespinst (Spinnwebe/Spinnennetz), Netz
 • **funnel web/tube web** Trichternetz, Röhrennetz
 • **mesh web** Maschennetz
 • **nursery web (nursery tent)** Brutgespinst, Eigespinst (Schutzgespinst für Jungspinnen)
 • **orb web** Radnetz
 • **purse web** Gespinstschlauch, Röhrennetz, Röhrengespinst
 • **sheetweb, horizontal** Deckennetz
 • **sheetweb, simple** Flächennetz
 • **space web (with barrier threads)** Fußangelnetz (mit Stolperfäden)
 • **sperm web** Spermanetz
 • **spoke/radius** Speiche (Netzspeiche), Radius
 • **tube web** Röhrennetz
webbed schwimmhäutig, mit Schwimmhäuten
webbed foot/swimming foot (e.g. birds) Schwimmfuß
webbing Vernetzung
Weber's line Webersche Linie
Weberian apparatus *ichth* Weberscher Apparat
Weberian ossicle/otolith Webersches Knöchelchen
webspinners/footspinners/embiids/Embioptera Fußspinner, Tarsenspinner, Embien
wedge/peg Keil
wedge grafting/cleft grafting *hort* Spaltpfropfung, Pfropfen in den Spalt
wedge-leaved keilblättrig
wedge-shaped/sphenoid/cuneate/cuneiform keilförmig
weed *n* Krautpflanze; Unkraut (*pl* Unkräuter)
weed *vb* Unkraut jäten
weed control Unkrautbekämpfung, Unkrautvernichtung

369 whip grafting **W**

weigh wiegen
weigh in (after setting tare)
 einwiegen (nach Tara)
weigh out
 abwiegen (eine Teilmenge)
weigh out precisely auswiegen
 (genau wiegen)
weighing table Wägetisch
weight Gewicht
 • gain weight zunehmen
 (Gewichtszunahme)
 • loose weight abnehmen
 (Gewichtsverlust)
weight loss Abnehmen
 (Gewichtsverlust)
weightless schwerelos
weightlessness Schwerelosigkeit
weir Wehr; Fischreuse
weir basket trap *bot* Reusenfalle
well (cell counter/buffer well)
 Vertiefung, Rinne (Pufferrinne/
 Pufferwanne)
well/depression (*electrophor:* gel
 well) Tasche (Geltasche), Vertiefung
well plate *gen/micb* Lochplatte
well water Brunnenwasser
welwitschia/Welwitschiaceae
 Welwitschiagewächse
westerlies/western wind
 Westwinde
 • easterlies/eastern wind
 Ostwinde
western birds/Hesperornithiformes
 Zahntaucher
Western blot/immunoblot
 Western-Blot, Immunoblot
wet nass
wet blotting Nassblotten
wet meadow Nasswiese
wet mount/wet preparation
 Nasspräparat, Frischpräparat,
 Lebendpräparat, Nativpräparat
wet rot Nassfäule
wether Hammel
wetland Feuchtgebiet, Feuchtbiotop
 • billabong (Australian: lagoon/
 backswamp) Lagune,
 Küstensumpf
 • bog (ombrogenic/ombrotrophic
 peatland) Moor (ombrogen/
 oligotroph), Torfmoor; Luch
 • bottomland (river floodplain
 wetland) Tiefland
 (Schwemmland)
 • carr (European fen woodland)
 Bruchmoor, Bruchwald,
 Übergangs-Waldmoor
 • fen/fenland (minerotrophic
 peatland: fed by underground
 water or interior drainage)
 Fehn, Fenn (minerotrophes
 vererdetes Flachmoor/Niedermoor)

 • floodplain/alluvial plain/
 floodland/alluvial land
 Überschwemmungsebene,
 Schwemmland (Flussaue)
 • mangrove(s) Mangrove(n)
 • marsh (dominated by grasses)
 Marsch
 • mire (European: from old Norse
 term)/peatland (peat-forming
 wetlands: bogs & fens) *n* Moor
 • moor/peatland (bogs/fens)
 Moor, Torfmoor; (raised bog)
 Hochmoor; (dry) Bergheide;
 Heidemoor
 • muskeg (Canadian term for
 peatlands) Moor (ombrogen/
 oligotroph), Torfmoor; kanadisches
 Tundramoor
 • peatlands Moor
 • pocosin/"swamp-on-a-hill" (U.S.
 peatland: SE coastal plains)
 amerikanisches Waldmoor
 • swamp Sumpf; (wetland
 dominated by trees/shrubs >
 equivalent to European: carr)
 Waldmoor; Luch
 • wet meadow Nasswiese
wettability Benetzbarkeit
wetting agent/wetter/surfactant/
 spreader oberflächenaktive
 Substanz, Entspannungsmittel
whalebone/baleen Walbein,
 Fischbein
whalebone whales/baleen whales/
 Mysticeti Bartenwale
whaling Walfang
Wharton's duct/submandibular
 duct Wharton-Gang
Wharton's jelly/mucous connective
 tissue Wharton-Sulze
 (Grundgewebe der Nabelschnur)
wheel organ Räderorgan
wheel-stamen tree family/yama-
 kuruma family/trochodendron
 family/Trochodendraceae
 Radbaumgewächse
whelp/cub (fox/wolf/jackal/bear/
 lion) Welpe
whelp/pup/puppy (dog) Welpe,
 junger Hund
whelp *vb* Welpen gebären,
 Welpen werfen
whey Molke
whine *vb* winseln, wimmern;
 quengeln, jammern
whip *n* Peitsche
whip *n* (one year old shoot) *bot*
 Rute
whip *vb* peitschen
whip grafting/splice grafting *hort*
 Kopulation, Kopulieren, Schäften
 (Pfropfung)

whip-and-tongue grafting

whip-and-tongue grafting/whip grafting/tongue grafting Kopulation mit Gegenzunge (Pfropftechnik)
whiplash flagellum/acronematic flagellum Peitschengeißel
whipping peitschend
whipscorpions (incl. vinegarroons)/ Pedipalpi (Uropygi & Amblypygi) Geißelskorpione & Geißelspinnen
whipworm (*Trichuris trichiura*) Peitschenwurm
whirl/swirl/eddy strudeln
whisk fern (*Psilotum*) Gabelblattgewächs
whiskers (cats) Barthaare, Schnurrhaare (Katzen)
whistle vb pfeiffen
white blood cell (WBC)/leukocyte weißes Blutkörperchen, Leukocyt, Leukozyt
white cinnamon family/wild cinnamon family/canella family/ Canellaceae Kaneelgewächse
white frost/hoarfrost (fine/ feathery) fein-flockiger Reif, Raureif
white horse weißer Schimmel
white line/zona alba (hoof) weiße Linie
white mangrove family/Indian almond family/Combretaceae Strandmandelgewächse
white matter weiße Substanz
white ramus/visceral ramus Ramus communicans albus
white rot Weißfäule (Korrosionsfäule)
white rusts/Albuginaceae Weißrost (Pilze)
white-alder family/clethra family/ Clethraceae Scheinellergewächse
whiteflies Mottenschildläuse, "Weiße Fliegen"
whole mount *micros* Totalpräparat
whole mount plastination Ganzkörperplastination
whole-cell patch *neuro* Ganzzell-*Patch*
whole-cell recording *neuro* Ganzzellableitung
whorl *zool* (snail shell) Windung, Umgang
whorl/verticil Quirl, Wirtel
• **false whorl/pseudowhorl** *bot* Scheinquirl, Doppelwickel
whorled (leaf arrangement) quirlständig, wirtelig (Blattstellung)
wide-angle X-ray scattering (WAXS) Röntgenweitwinkelstreuung
widefield *micros* Weitwinkel

widespread/ubiquitous weitverbreitet, ubiquitär
wiggler (mosquito larva) Schnakenlarve
wild/living in the wild (in a state of nature) wildlebend
wild/uncontrolled *adv/adj* wild, ungebändigt
wild/wilderness *n* Wildnis
wild animal reserve/game reserve/ wildlife park Wildreservat, Wildtierpark, Wildpark
wild animal sanctuary Wildschutzgebiet
wild cinnamon family/white cinnamon family/canella family/ Canellaceae Kaneelgewächse
wild pig/wild swine/wild boar/wild hog Wildschwein (*collect.* Schwarzwild); (im ersten Jahr) Frischling
wild sow Bache, Wildschweinsau
wild type Wildform, Wildtyp
wild-type allele Wildallel
wilderness Wildnis
wildflower Wildpflanze, wildwachsende Pflanze
wildfowl Wildgeflügel
wildlife Wild (Tiere in der Natur)
wildlife management Wildmanagement
wildlife park/national park Naturpark, Nationalpark
wildlife sanctuary/wildlife refuge Wildreservat
willow family/Salicaceae Weidengewächse
willowherb family/ evening-primrose family/ Oenotheraceae/Onagraceae Nachtkerzengewächse
wilt *n* Welke, Verwelken
wilt/wither/fade *vb* welken
wilting/flaccid/deficient in turgor welkend
wilting coefficient Welkungskoeffizient
wilting percentage, permanent permanenter Welkungsgrad
wilting point Welkpunkt
wind Wind
• **anabatic wind** Hangaufwind
• **anticyclone (rotating high- pressure wind system)** Antizyklon (Hochdruckgebiet)
• **blizzard** heftiger Schneesturm
• **breeze (sea b./land b.)** Brise (Meeres-/Land-)
• **calm/windlessness** Windstille
• **cyclone (rotating low-pressure wind system or storm)** Zyklone (Tiefdruckgebiet)

wing pad — W

- **cyclone/tropical windstorm**
  Zyklon (trop. Wirbelsturm)
- **downward wind** Abwind
- **dust devil** Staubteufel
- **dust whirl** Staubwirbel,
  Staubhose
- **easterlies (easterly current)**
  Ostwinde
- **gale (51–101 km/h)** Sturmwind
- **geostrophic wind**
  geostrophischer Wind
- **gust** Bö, Windbö
- **head wind** Gegenwind
- **hurricane (>115 km/h)**
  Hurrikan, Orkan
  (mittelamerik. Wirbelsturm)
- **jet stream** Jetstream,
  Strahlströmung
- **katabatic wind** Hangabwind
- **lee/lee side** Lee, Windschatten,
  Windschattenseite (dem Wind
  abgekehrte Seite)
- **luv/windward side** Luv,
  Windseite, Wetterseite
  (in Windrichtung liegende/
  dem Wind zugewandte Seite)
- **offshore wind** Landwind
- **onshore wind** Seewind
- **prevailing wind (direction of
  wind)** vorherrschender Wind
  (Windrichtung)
- **sand whirl** Sandwirbel, Sandhose
- **snowstorm (see: blizzard)**
  Schneesturm
- **squall (sudden violent gusty
  wind)** Sturmbö,
  heftiger Windstoß
- **surface wind** Oberflächenwind
- **tail wind** Rückenwind
- **tornado (North American
  whirlwind)/"twister"** Tornado
  (Nordamerik. Großtrombe/
  Wirbelsturm)
- **trade winds/trades** Passatwinde
- **typhoon** Taifun
  (tropischer Zyklon:
  Philippinen/Chinesisches Meer)
- **upwind/upcurrent** Aufwind
- **waterspout** Wasserhose
  (eine Trombe)
- **westerlies (westerly current)**
  Westwinde
- **whirlwind (violent windstorm)**
  Wirbelsturm, Trombe
- **wind spout/vortex (of a
  tornado)** Windhose
  (eine Trombe)

wind/twist/coil *vb*
winden, sich schlängeln
wind abrasion Windabrasion
wind dispersal/anemochory
Windstreuung, Windausbreitung

wind gap/air gap/wind valley *geol*
Windscharte, Spalt, Pass
**wind intensity/wind force/wind
strength** Windintensität,
Windstärke
wind pollination/anemophily
Windbestäubung, Anemophilie
wind shear/wind abrasion
Windschur
wind speed/wind velocity
Windgeschwindigkeit
wind strength/wind intensity
Windstärke, Windintensität
wind sucking (horse)
Windschnappen, Luftkoppen (Pferd)
wind-pollinated/anemophilous
windblütig, anemophil
**windbreak (breaking of trees by
wind)** Windbruch
windbreak/shelterbelt Windschutz
windchill *n* Windchill
(Windabkühlung)
windchill factor
Windchill-Faktor, Windchill-Index
(eine Abkühlungsgröße)
**windfall/windthrow/blowdown (of
trees)** Windwurf, Sturmwurf
windpipe/trachea/breathing tube
Kehle, Trachee, Trachea, Luftröhre,
Atemröhre
wing Flügel, Fittich, Schwinge
- **forewing/front wing/tegmina**
  Vorderflügel (Oberflügel/
  Deckflügel/Flügeldecke)
- **hindwing**
  Hinterflügel (Unterflügel)
- **hymenopterous wing** Hautflügel
- **spurious wing/bastard wing/
  alula** *orn*
  Daumenfittich, Afterflügel,
  Nebenflügel, Ala spuria,
  Alula (Federngruppe an 1. Finger)
- **stubby wings** Stummelflügel
wing area *see* wing field
wing base Flügelbasis
wing bud Flügelknospe
wing cell Flügelzelle
**wing cover/wing case/elytron/
elytrum (*pl* elytra)** Elytre,
Deckflügel, Flügeldecke
wing field/wing area *entom*
Flügelfeld (Region)
- **anal field/anal area/vannal area/
  vannus** Analfeld, Vannus
- **costal field/costal area**
  Costalfeld, Remigium
- **jugal area/jugal region/jugum/
  neala** Jugalfeld, Jugum, Neala
wing loading *aer*
Flügel-Flächen-Belastung
wing pad (larval developing wing)
Flügelstummel (>Flügelscheide)

wing scale

wing scale Flügelschuppe
wing sheath Flügelscheide
wing spread/wingspan Flügelspannweite
wing stub Flügelstummel
wing surface Flügelfläche
wing tip Flügelspitze
wing venation Flügeladerung, Flügelnervatur
wing-shaped/aliform flügelförmig
wingbeat Flügelschlag
winged/alate geflügelt
winged fruit (samara/key) Flügelfrucht
winged insects/pterygote insects/ Pterygota Fluginsekten, geflügelte Insekten
wingless/lacking wings/exalate/ apterous/apteral/apterygial flügellos, ungeflügelt
winglike/alar/alary flügelartig, schwingenartig
wingspan/wing spread Flügelspannweite
winter bud/hibernaculum/turio/turion Winterknospe, Hibernakel, Turio, Turione
winter fur/winter coat Winterfell
winter hardiness Winterhärte
winter quarters Winterquartier
winter torpor Kältestarre, Winterstarre
Wintera family/winter's bark family/drimys family/ Winteraceae Winterrindengewächse
wintergreen family/shinleaf family/ Pyrolaceae Wintergrüngewächse
wire gauze Drahtnetz
wireworm (elaterid larva) Drahtwurm
wishbone/furcula/fourchette (birds: united clavicles) Gabelbein, Furcula
witch-hazel family/ Hamamelidaceae Zaubernussgewächse
witches' broom *bot* Hexenbesen
wither/wilt/fade (e.g. blossom/flower/plant) welken, verwelken
withers Widerrist
wobble base *gen* Wobble-Base
wobble hypothesis Wobble-Hypothese
wolf teeth/remnant teeth (horse: 1.premolar) Wolfszähne
Wolffian body/deutonephros/ mesonephros Urniere, Mesonephros
Wolffian duct/Leydig's duct/ mesonephric duct Wolffscher Gang, Urnierengang

372

woman (*pl* women) Frau (*pl* Frauen)
womb/uterus Gebärmutter, Uterus
wood/lumber/timber Holz
- brushwood/spray Reisig
- compression wood Druckholz, Rotholz
- cordwood Klafterholz
- crosscut wood/crossgrained timber Hirnholz
- crude wood/crude timber Derbholz
- diffuse porous wood zerstreutporiges Holz
- driftwood Treibholz
- durability Verwitterungsbeständigkeit
- duramen/heartwood Kernholz
- earlywood/springwood Frühholz, Weitholz, Frühlingsholz
- figure/design Maserung, Masertextur, Fladerung, Figur, Zeichnung (Holz)
- firewood/fuelwood Brennholz, Feuerholz
- flatsawn flach-aufgesägt
- grain (form of wood texture) Faser, Faserung, Faserorientierung, Struktur, Fibrillenanordnung (Schnittholz)
- hardwood (tree) Laubbaum (speziell: Angiospermen)
- hardwood (wood of hardwood trees) Hartholz
- heartwood/duramen Kernholz
- ironwood Eisenhölzer
- kind of wood/type of wood Holzart
- latewood Spätholz, Engholz
- manoxylic wood locker gebautes Sekundärholz
- phanerophyte (woody plant; aerial dormant buds) Phanerophyt, Holzgewächs (Bäume/Sträucher)
- plainsawn/flatsawn/tangential section Sehnenschnitt, Fladerschnitt (Holz)
- plywood Sperrholz
- pulpwood Faserholz, Papierholz
- pycnoxylic wood dichtfaseriges Holz
- reaction wood Reaktionsholz
- resinous pinewood Kien, Kienholz
- ring porous wood ringporiges Holz (cyclopor)
- roundwood/log timber Rundholz
- sapwood/alburnum/splintwood Splintholz
- season/store *vb* lagern, ablagern

- **shake (fissure between growth rings)** Riss
- **shelterwood** Mutterbestand, Schirmbestand, Plenterwald
- **soft wood** Weichholz
- **specific gravity (wood density)** spezifisches Gewicht (Dichte von Holz)
- **splintwood/sapwood/alburnum** Splintholz
- **springwood/earlywood** Frühlingsholz, Weitholz, Frühholz
- **stere (stack of cordwood: 1 cbm)** Ster
- **summerwood** Sommerholz
- **tangential section/flatsawn/ plainsawn** Tangentialschnitt, Sehnenschnitt, Fladerschnitt (Holz)
- **tension wood** Zugholz
- **warp** verziehen, werfen
- **xylophilous/thriving in or on wood** in Holz lebend, auf Holz gedeihend

wood alcohol/pyroligneous alcohol/wood spirit/pyroligneous spirit (chiefly: methanol) Holzgeist (zumeist: Methanol/ Methylalkohol)

wood cellulose/xylon Holzzellulose

wood chip Holzschnitzel

wood crate Holzkiste

wood cylinder/wood corpus/wood body Holzkörper

wood felling Holzeinschlag

wood pile Holzhaufen

wood pit Holztüpfel

wood product Holzprodukt

wood pulp Zellstoff

wood ray (xylem ray) Holzstrahl (pith/medullary ray>Markstrahl)

wood rot Holzfäule

wood shavings Holzspäne

wood strength/wood stability Holzfestigkeit, Holzstabilität
- **bending strength** Biegefestigkeit, Tragfähigkeit
- **buckling strength/folding strength/crossbreaking strength** Knickfestigkeit
- **crushing strength/compression resistance (endwise compression)** Druckfestigkeit
- **shear strength/shearing strength** Scherfestigkeit, Schubfestigkeit
- **shock resistance** Stoßfestigkeit
- **tensile strength/breaking strength** Reißfestigkeit, Zerreißfestigkeit, Zugfestigkeit
- **torsion(al) strength** Drehfestigkeit

wood sugar/xylose Holzzucker, Xylose

wood tar Holzteer

wood vinegar/pyroligneous acid Holzessig

wood wool Zellstoffwatte

wood-eating/feeding on wood/ xylophagous holzfressend

wood-sorrel family/Oxalidaceae Sauerkleegewächse

wooden hölzern

woodland Waldsteppe

woodland management/forest management Forstwirtschaft

woodlice/pill bugs/sowbugs/ Isopoda Asseln

woodlot Waldstück

woodpeckers & barbets & toucans and allies/Piciformes Spechtvögel, Spechtartige

woods (see: forest) Wald mittlerer Größe

Woodsiaceae/woodsia family Wimpernfarngewächse, Wimpernfarne

woody/ligneous holzartig, holzig

woody debris Bruchholz

woody plant Gehölz, Holzgewächs

wool Wolle; Haare, Pelz
- **wood wool** Zellstoffwatte

wool fat gland Wollfettdrüse

wooly/lanate wollig

wooly hair Wollhaar

work procedure Arbeitsmethode

work up/process vb aufarbeiten

work up/working up/processing/ down-stream processing n biot Aufarbeitung

worker Arbeiter

worker bee Arbeitsbiene, Arbeiterin

workers' protective clothing Arbeitsschutzkleidung

workplace Arbeitsplatz

workplace protection/safety provisions (for workers) Arbeitsschutz

worldwide/occurring worldwide/ cosmopolitan weltweit verbreitet, kosmopolitisch

worm n Wurm

worm vb entwurmen

worm lizards/amphisbenids/ amphisbenians/Amphisbaenia Wurmschleichen, Doppelschleichen

worm-shaped/vermiform wurmförmig

wormian bone/sutural bone/ epactal bone Schaltknochen, Nahtknochen

wormlike/vermian/vermicular wurmartig

wormy wurmig

Woronin body

Woronin body Woronin-Körper
wort (brewing) Würze
wort/herb/weed Kraut
wound *n* Wunde
wound *vb* verwunden
wound cambium Wundkambium
wound healing Wundheilung
wound hormone Wundhormon
wound overgrowth by bulgy callus
  Überwallung, Wundüberwallung
wound response Wundreaktion
wound tissue/callus Wundgewebe,
  Kallus, Callus
wreath-shaped/coronal
  kranzförmig
wriggle (e.g. nematodes)
  schlängeln, hin und her zucken

wrinkle-leaved runzelblättrig
**wrinkled/rugose/corrugative/
corrugated** runzelig, gerunzelt,
  gewellt, geriffelt
wrist Handwurzel
wrist (joint) Handgelenk
wrist bone/carpal bone
  Handwurzelknochen,
  Handgelenksknochen,
  Carpalia (Ossa carpalia)
writhe *vb* winden, krümmen,
  schlingen, ringeln
writhing number/writhe (*W*) (DNA)
  Windungszahl
wrong host/accidental host
  Fehlwirt, Irrwirt

Xyridaceae

x body (inclusion body)
X-Körper (Einschlusskörper)
X chromosome X-Chromosom
● fragile X chromosome
(syndrome) fragiles
X-Chromosom (Syndrom)
X-chromosome inactivation
X-Chromosom-Inaktivierung
X-linked inheritance
x-chromosomale Vererbung
X-organ X-Organ, Bellonci-Organ
X-ray Röntgenstrahl
X-ray absorption spectroscopy
Röntgenabsorptionsspektroskopie
X-ray crystallography
Röntgenkristallographie,
Röntgenkristallografie
X-ray diffraction Röntgenbeugung
X-ray diffraction method
Röntgenbeugungsmethode
X-ray diffraction pattern
Röntgenbeugungsmuster,
Röntgenbeugungsdiagramm,
Röntgenbeugungsaufnahme,
Röntgendiagramm
X-ray emission spectroscopy
Röntgenemissionsspektroskopie
X-ray microanalysis
Röntgenstrahl-Mikroanalyse
X-ray microscopy
Röntgenmikroskopie
X-ray structural analysis/X-ray
structure analysis
Röntgenstrukturanalyse
xanthan Xanthan
xanthan gum Xanthangummi
xanthene/methylene diphenylene
oxide Xanthen
xanthic acid/xanthonic acid/
xanthogenic acid/
ethoxydithiocarbonic acid
Xanthogensäure
xanthine/2,6-dioxopurine Xanthin
xanthism Xanthismus
xanthophyll Xanthophyll
Xanthorrhoeaceae/grass tree
family/blackboy family
Grasbaumgewächse
xenarthrans/"toothless" mammals/
edentates/Xenarthra/Edentata
Nebengelenktiere, Zahnarme

xenobiosis Xenobiose
xenobiotic adv/adj xenobiotisch
xenobiotic n (pl xenobiotics)
Xenobiotikum (pl Xenobiotika)
xenogamy/cross-fertilization
Xenogamie, Kreuzbefruchtung
xenogeneic/heterologous
(originating from member of
another species
>transplantations) xenogen,
xenogenetisch, heterolog
xenogenic/xenogenous/
exogenous (originating from
outside an organism) xenogen,
durch einen Fremdkörper
hervorgerufen
xenograft (from other species)
Xenotransplantat,
Fremdtransplantat
xenoparasite Xenoparasit
xenospore (immediate
germination) Xenospore
xeric/low moisture content
(dry/arid) gekennzeichnet durch
niedere Feuchtigkeitsmenge
(trocken/arid/wüstenartig)
xeromorphism Xeromorphismus
xerophyte/xeric plant/
xerophilic plant/drought tolerator
Xerophyt, Trockenpflanze,
Trockenheit ertragende Pflanze
xerophytic/drought resistant
trockenresistent
xerosere Xeroserie
xiphoid/ensiform/gladiate/
sword-shaped schwertförmig
xiphoid cartilage/cartilago
xiphoidea Schaufelknorpel
xylem Xylem, Gefäßteil, Holzteil
xylem sap Xylemsaft
xylene/dimethylbenzene Xylol,
Dimethylbenzol
xylitol/xylite Xylit
xylophage n Holzfresser
xylophilous/thriving in or on wood
in Holz lebend,
auf Holz gedeihend
xylose Xylose
xylulose Xylulose
Xyridaceae/yellow-eyed grass
family Xyrisgewächse

Y organ

Y organ (molting gland) Y-Organ, Carapaxdrüse
YAC (yeast artificial chromosome) künstliches Hefechromosom
yam family/Dioscoreaceae Yamswurzelgewächse, Schmerwurzgewächse
yama-kuruma family/wheel-stamen tree family/trochodendron family/ Trochodendraceae Radbaumgewächse
yawn gähnen
yeanling frischgeborenes Lamm, Schäfchen oder Zicklein, Ziegenjunges
**yearling
(short yearling: 9 to 12 months; long yearling: 12 to 18 months)** Jährling, einjähriges Tier (meist Rinder)
yeast Hefe
- **baker's yeast** Bäckerhefe
- **bottom yeast** niedrigvergärende Hefe ("Bruchhefe")
- **brewers' yeast** Bierhefe, Brauhefe
- **dried yeast** Trockenhefe
- **top yeast** hochvergärende Hefe ("Staubhefe")
yeast artificial chromosome (YAC) künstliches Hefechromosom (YAC)
yeast episomal plasmid (YEp) episomales Hefeplasmid (YEp) (Hefevektor)
yeast extract Hefeextrakt
yeast integrative plasmid (YIp) integratives Hefeplasmid (YIp) (Hefevektor)
yellow fever (*Flavivirus*) Gelbfieber
yellow spot/macula lutea (with fovea centralis) gelber Fleck, Macula lutea (mit Fovea centralis)
yellow-eyed grass family/ Xyridaceae Xyrisgewächse
yellow-green algae/Xanthophyta Gelbgrünalgen, Xanthophyten
yelp (dog: shrill bark) kläffen
yelp (dog: shrill cry) jaulen, winseln
yew family/Taxaceae Eibengewächse
yield *n* Ausbeute, Ertrag
yield *n zool* **(honey)** Tracht (Ertrag an Honig)
yield vb *agr/chem* (Ertrag/Ausbeute) ,ergeben, hervorbringen; *agr* tragen, liefern
yield coefficient (Y) Ertragskoeffizient, Ausbeutekoeffizient, ökonomischer Koeffizient

yield reduction Ertragsminderung
yield strength Elastizitätsgrenze, Dehngrenze
yield stress Streckspannung, Fließspannung
yolk/egg yolk/vitellus Dotter, Eidotter, Eigelb
- **centrolecithal (yolk aggregated in center)** zentrolezithal, centrolecithal, Dotter im Zentrum
- **isolecithal (yolk distributed nearly equally)** isolezithal, isolecithal, Dotter gleichmäßig verteilt
- **mesolecithal (with moderate yolk content)** mesolezithal, mesolecithal, mäßig dotterreich
- **oligolecithal (with little yolk)** oligolezithal, oligolecithal, mikrolecithal, dotterarm
- **polylecithal (with large amount of yolk)** polylezithal, polylecithal, makrolecithal, dotterreich
- **telolecithal (yolk in one hemisphere)** telolezithal, telolecithal, Dotter an einem Pol
yolk cell/shell globule (trematodes) Dotterzelle, Vitellophage
yolk duct Vitellodukt, Dottergang
yolk fry/sacfry/alevin (salmon larvae) Dottersackbrut
yolk gland/vitellarian gland/ vitelline gland/vitellarium/ vitellogen Dotterstock, Dotterdrüse, Vitellar, Vitellarium
yolk larva/vitellaria/lecithotroph pericalymma/test-cell larva Hüllglockenlarve, Pericalymma
yolk plug Dotterpfropf
yolk sac Dottersack
yolk-sac placenta/choriovitelline placenta Dottersackplazenta, Dottersackhöhlenplazenta, Omphaloplazenta, omphaloide Plazenta
yolked key/indented key eingerückter Bestimmungsschlüssel
young *n* **(offspring);
see also: Jungtier** Junges (Nachkommen)
young/pup (e.g. whale/seal/rat/ dog)/cub (young carnivore: bear/ fox/lion) Jungtier, Junges (v. a. Säuger)
young forest Jungwald, junger Wald
young plant/juvenile plant Jungpflanze

Z DNA Z-DNA
Z line (Z disk) Z-Linie, Z-Streifen
(Z-Scheibe) (Z = Zwischenscheibe)
Z-scheme Z-Schema (Zickzack-
Schema: Photosynthese)
**Zannichelliaceae/horned pondweed
family** Teichfadengewächse
zeatin Zeatin
zeaxanthin Zeaxanthin
**zebra wood family/connard family/
Connaraceae** Connaragewächse
Zencker's organ
Zenckersches Organ
zero growth Nullwachstum
zero-order kinetics
Kinetik nullter Ordnung
zero-order reaction
Reaktion nullter Ordnung
**zest (peel of orange/lemon etc. for
flavoring)** Zitrusschale zum
Würzen (Orangen/Zitronen etc.)
zeugopodium (amphibians)
Zeugopodium
zinc Zink
zinc finger Zinkfinger
Zingiberaceae/ginger family
Ingwergewächse
zipper Reißverschluss
 • **leucine zipper**
Leucin-Reißverschluss
zipper principle
Reißverschlussprinzip
zippering *gen* Reißverschluss
betätigen, Zippering
(Doppelstrangbildung: kooperativer
Vorgang beim Bilden von
Wasserstoffbrücken)
zoanthids/zoantharians/Zoantharia
Krustenanemonen
zoëa (decapod crustacean larva)
Zoëa
zoecium/zooecium (bryozoans)
Zoecium
zona pellucida/oolemma
Zona pellucida, Oolemma, Eihülle
zonal centrifugation
Zonenzentrifugation
**zonary placenta/annular placenta/
placenta zonaria** Gürtelplazenta
zonation Zonierung
zone electrophoresis
Zonenelektrophorese
zone fossil/zonal fossil/index fossil
Leitfossil
**zone membranelles, adoral (AZM)
(ciliates)** adorales
Membranellenband
**zone of accumulation/zone of
illuviation (B-horizon)**
Einwaschungshorizont
**zone of cell division (region of root
apical meristem)** Wachstumszone

**zone of elongation/region of
expansion (root)** Streckungszone,
Verlängerungszone
zone of equivalence
Äquivalenzzone (Ausfällung
unlöslicher Immunkomplexe)
**zone of leaching/zone of eluviation
(soil: A/E-horizon)**
Auswaschungshorizont
**zone of maturation/root-hair zone
(zone of cell differentiation)**
Wurzelhaarzone
zone of saturation Sättigungszone
**zone sedimentation/zonal
sedimentation**
Zonensedimentation
zonoskeleton Zonoskelett
(Extremitätengürtel)
zonule fibers Zonulafasern
zoo/zoological garden(s) Zoo,
Zoologischer Garten, Tiergarten
zoocecidium Zoocecidium, von
Tieren hervorgerufene Pflanzengalle,
Tiergalle
zoochory/animal-dispersal
Tierausbreitung
**zoocoenosis/zoocenosis/animal
community** Zoozönose,
Tiergemeinschaft
zooecium/zoecium (bryozoans)
Zoecium
zoogamy Zoogamie,
Tierbefruchtung
zoogeography Zoogeographie,
Tiergeographie, Zoogeografie,
Tiergeografie
zoonosis Zoonose
zoophagous zoophag
zoophilous *bot* tierblütig
**zoophily/pollination by animal
vectors** Tierblütigkeit
zoophysiology/animal physiology
Tierphysiologie
zooplankton Zooplankton
zoospore/planospore Zoospore,
Planospore, Schwärmer
zorapterans/Zoraptera Bodenläuse
Zosteraceae/eel-grass family
Seegrasgewächse
zwitterion (*not translated!*)
Zwitterion
zygapophysis ("yoking" process)
Zygapophyse
zygodactyl/zygodactylous *orn*
(e.g. parrots) kletterfüßig
zygodactyl toe/zygodactylous toe
orn Wendezehe
zygomatic arch/arcus zygomaticus
Jochbogen, Backenknochenbogen
**zygomatic bone/malar bone/
cheekbone/os zygomaticum**
Jochbein, Backenknochen

Z zygomorphic

**zygomorphic/zygomorphous/
monosymmetrical/irregular**
zygomorph
**Zygophyllaceae/caltrop family/
creosote bush family**
Jochblattgewächse
zygosity Zygotie
- **autozygosity** Autozygotie
- **dizygosity** Dizygotie,
  Zweieiigkeit
- **hemizygosity** Hemizygotie
- **heterozygosity** Heterozygotie,
  Mischerbigkeit
- **homozygosity** Homozygotie,
  Reinerbigkeit, Reinrassigkeit
- **monozygosity** Monozygotie,
  Eineiigkeit
zygospore Zygospore
**zygospore fungi/bread molds/
zygomycetes (coenocytic fungi)**
Jochpilze, Zygomyceten
zygote Zygote

zygote nucleus/synkaryon
Zygotenkern, Synkaryon
zygotene (during meiotic prophase)
gen Zygotän
zygotic zygotisch
zygotic induction
zygotische Induktion
**zymogen/proenzyme (enzyme
precursor)** Zymogen, Proenzym
(Enzymvorstufe)
zymogenic/zymogenous zymogen
**zymology (science/study of
fermentation)** Zymologie
(Lehre von der Gärung)
zymosis/fermentation Zymose,
Fermentation, Gärung
zymosis *med* **(development of
infectious disease)**
Anfangstadium/Entwicklung einer
Infektion
zymosterol Zymosterin
zymurgy Gärungstechnik

A ersetzen durch B/
A durch B ersetzen/
A durch B substituieren
substitute B for A
A-Bande (Muskel: *anisotrop*)
A band
A-Stelle (Aminoacyl-Stelle) *gen*
A-site (aminoacyl site)
aalartig/anguilliform eel-like,
anguilliform
Aalfische/Aalartige/Anguilliformes
eels
Aalstrich *zool* (dunkler Streifen auf
Rücken: z.B. Pferde)
list (dark stripe on back)
AAM (angeborener auslösender
Mechanismus) *ethol* innate
releasing mechanism (IRM)
Aapamoor (Mischmoor)/
Strangmoor aapa mire, string bog
Aas carrion, decaying carcass
Aasblume carrion flower
Aasfliegenblume/Sapromyiophile
dung-fly flower, sapromyophile
Aasfresser/Unratfresser scavenger,
carrion feeder
Abart/Spielart/Varietät sport,
variety
Abbau *metabol* digestion,
degradative reactions/metabolism,
catabolism; (Zerfall/
Zusammenbruch) breakdown;
(Zersetzung) degradation,
decomposition, breakdown
 ● **biologischer Abbau/**
Biodegradation biodegradation
Abbaubarkeit degradability
abbauen/zersetzen degrade,
decompose, break down
abbauend/katabolisch catabolic
Abbauprodukt degradation product
abbilden *opt* image
abbilden/projezieren project
Abbildung (in einer Fachzeitschrift/
Buch) figure, illustration
Abbildungsmaßstab/
Lateralvergrößerung/
Seitenverhältnis/Seitenmaßstab
lateral magnification
abblättern(d)/abschilfern(d)
exfoliate
Abblätterung/Delamination
(Entodermbildung) delamination
abblühen/verblühen fade
 ● **abgeblüht/verblüht** faded
(withered), deflorate(d)
Abblühen/Verblühen fading,
defloration
Abbruchcodon/Stoppcodon/
Terminationscodon
termination codon,
terminator codon, stop codon

abdampfen evaporate
Abdampfschale evaporating dish
abdecken (Tierkörperbeseitigung)
dispose of animal carcasses
Abdeckerei (Tierkörperbeseitigung)
animal carcass disposal
Abdomen/Hinterleib abdomen
Abdominalbein/Bauchfuß/Propes/
Pes spurius (larval) larval proleg,
false leg
 ● **Schwimmbein/Schwimmfuß/**
Pleopodium swimmeret, pleopod
Abdominalschwangerschaft/
Bauchhöhlenträchtigkeit/
Leibeshöhlenschwangerschaft/
Leibeshöhlenträchtigkeit
abdominal pregnancy
Abdominalsegment/Pleomer
abdominal somite, pleomere
Abdrift/organismische Abdift
organismic drift
Abdruck (Oberflächenabdruck:
EM) *micros* replica; *paleo*
impression (siehe: Abguss)
 ● **Fingerabdruck** fingerprint
 ● **>genetischer Fingerabdruck/**
Fingerprinting fingerprinting,
genetic fingerprinting,
DNA fingerprinting
 ● **Fußabdruck** footprint
Aberration/Abweichung aberration
 ● **Autosomenaberration** *gen*
autosomal aberration
 ● **chromatische Aberration/**
Farbabweichung *opt/micros*
chromatic aberration
 ● **Chromosomenaberration**
chromosome aberration
 ● **Heterosomenaberration** *gen*
sex-chromosome aberration
abfackeln flare, burn off
Abfall waste; trash
abfallend deciduous, falling,
shedding
Abfallfresser/Detritophage
detritivore, detritus-feeder
Abfallstoffe waste materials
 ● **organische Abfallstoffe**
organic debris, organic waste
abferkeln/ferkeln farrow
abfließen run off, drain off,
flow off
Abfluss (Abflussöffnung) outlet;
(Dränung/Drainage) drainage;
Abschwemmung (oberflächlich
abfließend) runoff
abforsten/abholzen/kahlschlagen
(größere Fläche) clearcut, deforest
abfressen
browse (woody shoots/leaves/bark),
graze (herbaceous plants)
abgehärtet hardy

abgeleitet

abgeleitet/fortgeschritten ("höher entwickelt") advanced
abgeleitet *math* derived
abgerundet rounded
abgrasen/grasen graze (herbaceous plants), browse (*esp:* woody shoots/leaves/bark)
Abguss *paleo* mold, cast
Abhang/Hang (Hügel/Berg) hillside, slope, brae (steep bank)
abhängig dependant
• **süchtig** addicted
• **unabhängig** independant
abhärten hardening off
Abhärtung hardening
abholzen fell, clear, clearcut
Abholzung felling, clear cutting, clear felling, deforestation
Abietinsäure abietic acid
abiotisch abiotic
Abklärflasche/Dekantiergefäß decanter
Abkömmling/Deszendent/Nachkomme descendant, offspring, progeny; **Derivat (abgeleitet)** derivative
abkühlen cool
Abkühlung cooling
ablagern/sedimentieren sediment, deposit
Ablagern (Holz) seasoning
Ablagerung/Sedimentation sedimentation, deposition; deposit
Ablaktieren/Ablaktation/Ablaktion approach grafting, inarching
Ablauf/Ausfluss (Austrittsstelle einer Flüssigkeit) outlet; (herausfließende Flüssigkeit) effluent
Ablegen (mehrere Jungpflanzen pro Trieb) *bot/hort* French layering, continuous layering
Ableger *bot/hort* (Absenker) layer, set; (Ausläufer) runner, sucker, offshoot; (Pfropfableger) scion, cutting, sarment
Ableger treibend surculose, producing suckers
Ablegerbildung/Absenkerbildung layering
Ablegervermehrung durch Anhäufeln (Abrisse nach Anhäufeln) stool layering, stooling, mound layering
ableiten *math* derive; *neuro* record
Ableitung *math* derivation; *neuro* recording
ablenken *phys/math* deflect; *psych* distract
Ablenkung *phys/math* deflection; *psych* distraction

ablesen (z. B. Messdaten) read, record
Ablesung (z. B. Messdaten) reading, recording
Ablösungsschicht/Trennschicht/Abszissionsschicht *bot* separation layer, abscission layer
ABM-Papier (Aminobenzyloxymethyl-Papier) ABM paper (aminobenzyloxymethyl paper)
Abmoosen *hort* marcottage using moss (an air layering process)
abnehmen *vb* **(Gewichtsverlust)** loose weight
Abnehmen *n* **(Gewichtsverlust)** weight loss
Abort/Abortus/Abgang/Frühgeburt abortion
• **Abtreibung/Schwangerschaftsabbruch/Abortinduktion** induced abortion
• **Frühabort (Fehlgeburt bis 12. Woche)** early abortion
• **Spätabort (Fehlgeburt nach 12. Woche)** miscarriage
• **Spontanabort/Fehlgeburt** spontaneous abortion, miscarriage
• **verhaltener Abort** missed abortion
abortiv/abgekürzt verlaufend abortiv
abortiv/rudimentär/rückgebildet/verkümmert abortive
abortive Infektion abortive infection
abortive Transduktion abortive transduction
abortive Transfektion abortive transfection
Abortivei/Blasenmole/Molenei/Windei mole
abrichten train; *horse:* break
Abrichtung training; *horse:* breaking-in
Abschaltsequenz/Silencer *gen* silencer (sequence)
abscheiden/absondern exude, secrete, discharge
abscheiden/ausfällen precipitate, deposit
abscheiden/trennen separate
Abscheider separator, precipitator
Abscheidung (Absonderung/Exsudat) exudate, exudation, secretion;
(Ausfällung) precipitate, deposit
abschirmen (von Strahlung) shield (from radiation)
Abschirmung (von Strahlung) shielding (from radiation)
Abschlag (Flug/Flügel) downward stroke

Abschlussgewebe dermal tissue, boundary tissue, exodermis
• **primäres A./Epidermis** epidermis
Abschnitt (Teil des Ganzen) section, part, moiety
abschrecken deter, repel; scare off
Abschreckstoff/Schreckstoff deterrent, repellent
abschwächen/attenuieren (mit herabgesetzter Virulenz) attenuate
Abschwächung *neuro* weakening; (Attenuation) attenuation
abschwellend shrinking, decongestant
Absenken *hort* simple layering
Absenker/Ableger *hort* set, layer
Absenkervermehrung/ Ablegervermehrung *hort* layerage, layering
Absetzbecken/Klärbecken settling tank
absondern/abscheiden (Flüssigkeiten) exude, secrete, discharge
absondern/sequestrieren/ abtrennen (z.B. Gewebe/ Knochenbruchstücke) sequester, segregate
Absonderung/Abscheidung/ Exsudat exudate, exudation, discharge, secretion
Absonderung/Sequestrierung/ Abtrennung/Loslösung (z.B. Gewebe/Knochenbruchstücke) sequestration, segregation
Absonderungsgewebe/ Abscheidungsgewebe secretory tissue
Absorbanz (Extinktion) absorbance, absorbancy (extinction: optical density)
absorbieren absorb
absorbierend/absorptionsfähig absorbent
Absorption absorption
Absorptionsindex absorbance index, absorptivity
Absorptionskoeffizient absorption coefficient
Absorptionsspektrum absorption spectrum
Absorptionsvermögen/ Absorptionsfähigkeit/ Aufnahmefähigkeit absorbency
abstammen von ... descend from ..., originate from ...
Abstammung descent, origin
Abstammungsachse principal axis
abstammungsgeschichtlich/ evolutionär/phyletisch/ phylogenetisch evolutionary, phyletic

Abstammungsgeschichte/ Stammesentwicklung/Evolution/ Phylogenie/Phylogenese evolution, phylogeny
Abstammungslehre evolutionary studies
Abstammungstheorie/ Deszendenztheorie/ Evolutionstheorie theory of evolution, evolutionary theory
absteigend efferent
absterben die off
• **teilweise absterben** dieback
Absterbephase *micb* decline phase, phase of decline, death phase
Absterberate mortality rate
abstillen wean
Abstillen weaning
abstoßen/ablösen (Haut/Hülle/ Rinde) shed, slough, sloughing (off); (Blätter) shed; (Transplantat) reject (graft rejection)
Abstoßung (Transplantat) rejection (graft rejection)
Abstoßungsreaktion rejection reaction
abstreifen (Haut/Hülle) slough (off)
Abstrich *med* swab; *micros* smear
• **einen Abstrich machen** *med* to take a swab
Abstufung/Staffelung/Stufenfolge gradation
Abszission/Abwerfen/Abwurf abscission, falling off, dropping off, shedding
Abszissionsschicht/ Ablösungsschicht/Trennschicht/ Trennungsschicht abscission layer, separation layer
Abteilung/Phylum phylum (division)
abtreiben (eine Fehlgeburt herbeiführen) abort (induce an abortion)
Abtreibung/Abort/Fehlgeburt abortion
Abtreibung/ Schwangerschaftsaabruch (herbeigeführte Fehlgeburt) induced abortion
abundante mRNA abundant mRNA
Abundanz/Individuenzahl/ Individuendichte/ Populationsdichte/ Bevölkerungsdichte abundance, population density; (Artdichte) species density
Abwanderung (Tiere)/ Auswanderung (Mensch)/ Emigration emigration
Abwärme waste heat

abwärts (Richtung 3'-Ende eines Polynucleotids) downstream
• **aufwärts (Richtung 5'-Ende eines Polynucleotids)** upstream
Abwasser wastewater, sewage
• **Rohabwasser** raw sewage
abwechselnd/alternierend alternate
Abwehr/Verteidigung defense
abwehrgeschwächt immunocompromized
Abwehrprotein defense protein
abweichen von ... deviate from ...
Abweichung deviation; (Aberration) aberration
• **Standardabweichung** standard deviation
• **statistische Abweichung** statistical deviation
Abweide-Nahrungskette/Fraß-Nahrungskette grazing food chain
abwerfen shed, drop, abscise; (Baumveredlung) decapitate
abwerfend shedding, abscising, deciduous
abwiegen (eine Teilmenge) weigh out
Abwurf/Abwerfen/Abszission shedding, falling off, dropping off, abscission
Abyssal/Meeresgrund abyssal, abyssal zone, ocean floor
Abzieher/Abduktor/Abductor (Muskel) abductor muscle
Abzug/Dunstabzugshaube hood, fume hood
Abzym abzyme
Acanthaceae/Akanthusgewächse acanthus family
Acanthocephala/Kratzer spiny-headed worms, thorny-headed worms, acanthocephalans
Acanthor-Larve/Hakenlarve acanthor larva
Acarizid/Akarizid acaricide
Aceraceae/Ahorngewächse maple family
Acervulus acervulus
Acetat/Azetat (Essigsäure/Ethansäure) acetate (acetic acid/ethanoic acid)
Acetessigsäure (Acetacetat)/β-Ketobuttersäure acetoacetic acid (acetoacetate), β-ketobutyric acid
Acetylcholin (ACh) acetylcholine
Acetylen acetylene
N-Acetylmuraminsäure N-acetylmuramic acid
Achäne (einblättrige Achäne) achene, akene
• **zweiblättrige Achäne (Asteraceen)** cypsela, bicarpellary achene

Achillessehne Achilles' tendon, tendon of the heel, calcaneal tendon
achlamydeisch achlamydeous
achromatisch/unbunt achromatic
achromatischer Kondensor *micros* achromatic condenser, achromatic substage
achromatisches Objektiv *micros* achromatic objective
Achse axis
• **Abstammungsachse/Hauptachse** principal axis
Achsel/Blattachsel *bot* axil
Achsel/Schulter shoulder
Achsel/Achselhöhle/Achselgrube (Arm) armpit, axilla
Achselbulbille *bot* axillary bulbil
Achselfedern *orn* axillary feathers, axillars
Achselhöhle/Achselgrube armpit, axilla
Achselknospe/Seitenknospe axillary bud, lateral bud
Achselmeristem axillary meristem
achselständig axillary
Achsenbecher hypanthium
achsenbürtig stem-borne
Achsenfaden/Axonema/Axonem central fibril, axial filament, axial rod, axoneme
Achsenkörper/Stamm shoot axis, stem
Achsenskelett/Stammskelett/Rumpfskelett/Skelett des Stammes/Axialskelett axial skeleton
Achsensporn axial spur
Achsenstab (Kinetoplastida) paraxial rod
achsenständig axial, axile
Achterform (der DNA) figure eight (of DNA)
Acidität/Azidität/Säuregrad acidity
Acidose/Azidose acidosis
Acinuszelle acinus cell
Acker/Feld field, land, farmland
Ackerbau (Bebauen des Bodens mit Nutzpflanzen) cropping, plant production, tillage
Ackerbau (auch Viehhaltung) farming
Ackerbaukunde/Ackerbaulehre/Agronomie agronomy (field-crop production & soil management)
Ackerland farmland, tillage, tilth, cultivated land, arable land
Ackerrain/Feldrain field boundary strip, balk
Ackerwirtschaft/Ackerbau *sensu lato* farming
acöl/acoel acelous, acoelous

383   **Adiantaceae** **A**

Aconitsäure (Aconitat)
aconitic acid (aconitate)
Acontium (Anthozoa) acontium
ACR (vorgeschichtliche konservierte
Region) *gen* ACR (ancient
conserved region)
Acrasiomyceten/Acrasiomyycetes/
Dictyosteliomycetes/zelluläre
Schleimpilze (Myxomycota)
cellular slime molds
Acridinfarbstoff acridine dye
Acrosom/Akrosom acrosome
acrostich acrostichal, acrostichoid
Actin/Aktin actin
Actinfilament/Aktinfilament/
Mikrofilament
actin filament, microfilament
Actinidiaceae/
Strahlengriffelgewächse Chinese
gooseberry family, actinidia family
Adamantoblast adamantoblast,
ameloblast, enamel cell
Adams Apfel/Prominentia laryngea
Adam's apple, laryngeal prominence
(largest cartilage of larynx)
Adaptation/Adaption/Anpassung
adaptation
Adaptationsphase/Adaptionsphase/
Anlaufphase/Latenzphase/
Inkubationsphase/lag-Phase
lag phase, latent phase,
incubation phase, establishment phase
adaptiv/anpassungsfähig adaptive
adaptive Landschaft
adaptive landscape, adaptive surface
adäquater Reiz adequate stimulus
additive genetische Varianz
additive genetic variance
Adelphogamie/
Geschwisterbestäubung
adelphogamy
Adelphotaxon/Schwestertaxon
sister taxon
Adenin adenine
Adenohypophyse/
Hypophysenvorderlappen
adenohypophysis,
anterior lobe of pituitary gland
Adenosin adenosine
Adenosindiphosphat (ADP)
adenosine diphosphate (ADP)
Adenosinmonophosphat (AMP)
adenosine monophosphate (AMP)
 • **zyklisches/cyclisches/cyklisches**
  AMP (cyclo-AMP/cAMP)
  cyclic adenosine monophosphate
  (cyclic AMP/cAMP)
Adenosintriphosphat (ATP)
adenosine triphosphate (ATP)
Adenovirus adenovirus
Adenylatcyclase/Adenylylcyclase
adenylate cyclase

Adenylsäure (Adenylat)
adenylic acid (adenylate)
Ader/Blutgefäß (Arterien und
Venen) blood vessel (arteries and
veins) (*siehe auch in veterinär-/*
humanmedizinischen Wörterbüchern)
 • **Halsschlagader/Carotis**
  carotid artery
 • **Körperschlagader, große/Aorta**
  aorta
 • **Pfortader/Vena portae** portal vein
 • **Schlagader/Arterie** artery
Ader/Nerv/Rippe (Insektenflügel)
vein; (Blattader/Blattnerv/
Blattrippe) leaf vein, leaf rib
 • **Antenodalquerader**
  antenodal cross-vein
 • **Humeralquerader**
  humeral cross-vein
 • **Interkalarader/Intercalarader**
  intercalary vein
 • **Jugalader** jugal vein
 • **Längsader** longitudinal vein
 • **Querader** cross-vein
Äderchen (Insektenflügel) venule,
venula
Adergeflecht/Plexus chorioidea
choroid plexus
Aderhaut/Chorioidea/Choroidea
choroid, chorioid
Aderlass/Phlebotomie/Venae sectio
phlebotomy, venesection
Aderung/Nervatur/Nervation/
Venation venation
 • **Bogenaderung/Bogennervatur**
  arched venation, arciform
  venation, arcuate venation
  (camptodrome)
 • **Fächeraderung/Gabeladerung/**
  Gabelnervatur
  dichotomous venation
 • **Fiederaderung** pinnate venation
 • **fingerförmige Aderung/**
  fingerförmige Nervatur
  digitate venation
 • **Längsaderung/Streifennervatur**
  longitudinal venation,
  striate venation
 • **Netzaderung** reticulate venation,
  net venation, netted venation
 • **Paralleladerung** parallel venation
 • **Streifennervatur/Längsaderung**
  striate venation,
  longitudinal venation
Adhärenz/Adhäsion/Anheftung
adherence, adhesion, attachment
Adhärenzfaktor adherence factor
Adhäsin/Adhesin adhesin
Adhäsion adhesion
Adiantaceae/Frauenhaarfarn-
gewächse/Haarfarne adiantum
family, maidenhair fern family

Adipinsäure (Adipat)
adipic acid (adipate)
Adipozyt/Adipocyt/Fettzelle
adipocyte, adipose cell, fat cell
Adjuvans (*pl* **Adjuvantien**)
adjuvant
**Adkrustierung/Akkrustierung
(Kork)** adcrustation, accrustation
**Adlerfarngewächse/
Hypolepidaceae** bracken fern
family, hypolepis family
adorales Membranellenband
adoral zone membranelles
adossiert/rückseitig addorsed,
addossed
Adoxaceae/Moschuskrautgewächse
moschatel family
Adrenalin/Epinephrin adrenaline,
epinephrine
adrenerg adrenergic
adsorbieren adsorb
Adsorption adsorption
Adstringens/adstringierender Stoff
astringent, astringent agent,
astringent substance
adstringent/zusammenziehend
astringent, styptic
Adstringenz astringency
adult/erwachsen adult, grown-up
Adultinsekt/Vollinsekt/Imago
(*pl* **Imagines**) imago (*pl* imagoes/
imagines)
Adultpflanze adult plant
Advektionskälte/Advektionsfrost
advective chill, advective frost
**Adventivspross/Adventivtrieb/
Zusatztrieb** *bot* adventitious shoot
Adventivwurzel adventitious root
Aecidium aecium, aecidium,
cluster cup
Aedeagus/Aedoeagus/Penis *entom*
aedeagus, intromittent organ, penis
aerob (sauerstoffbedürftig)
aerobic
**Aethalium/Aethalie (*pl* Aethalien)/
Sammelfruchtkörper** aethalium
Äffchen monkey(s)
Affen *sensu lato* monkeys and apes
(*siehe auch:* Herrentiere/Primaten)
• **Altweltaffen/Schmalnasenaffen/
Catarrhina** old-world monkeys
(incl. apes)
• **Halbaffen/Prosimii**
lower primates, prosimians
• **Menschenaffen/Pongidae**
great apes, pongids
• **Menschenartige/Hominoidea**
apes, anthropoid apes
• **Neuweltaffen/Breitnasenaffen/
Platyrrhina** new-world monkeys
(South American monkeys and
marmosets)

**affenartig/Affen (bes.
Menschenaffen) betreffend**
simian
Affinade (Zucker) affinated sugar
Affination (Zucker-Filtration)
affination
Affinität affinity
Affinitäts-Blotting affinity blotting
Affinitätskonstante
affinity constant
Affinitätsmarkierung
affinity labeling
Affinitätsreifung *immun*
affinity maturation
Affinitätsverteilung
affinity partitioning
After/Anus anus
Afterfeder/Nebenfeder/Hypopenna
afterfeather, accessory plume,
hypoptile, hypoptilum
Afterfeld/Periprokt periproct
Afterflosse/Analflosse anal fin
**Afterflügel/Nebenflügel/
Daumenfittich/Alula/Ala spuria
(Federngruppe an 1. Finger)** alula,
spurious wing, bastard wing
**Afterfuß (letzter)/Nachschieber/
Postpes/Propodium anale
(Raupen)** anal proleg, anal leg
**Aftergriffel/Cercus
(Schwanzanhang)** cercus,
cercopod (clasping organs)
Afterklaue/Afterzehe dewclaw,
false foot, pseudoclaw
Afterkralle/Arolium (am Prätarsus)
arolium
**Afterlappen/Afterschild/Pygidium
(Telson der Arthropoden)**
pygidium, caudal shield
**Afterraife/Afterfühler/
Schwanzborsten/Cercus** cercus,
cercopod
Afterraupe (Blattwespenlarven)
eruciform larva with more than
5 pairs of abdominal prolegs
(Tenthredinidae)
**Afterröhre/Afterhügel/Analtubus
(Crinoide)** anal cone, anal tube
Afterschaft/Hyporhachis (Feder)
aftershaft, hyporachis, hyporhachis
(median shaft of hypopenna)
**Afterskorpione/Pseudoskorpione/
Pseudoscorpiones/Chelonethi**
pseudoscorpions, false scorpions
Afterzitze accessory teat
Agar agar
• **Blutagar** blood agar
Agardiffusionstest
agar diffusion test
Agarnährboden agar medium
Agarose agarose
Agarplatte agar plate

Agavengewächse/Agavaceae
agava family, century plant family
Agens/Agenz (pl Agenzien) agent
• **interkalierendes Agens**
intercalating agent
• **quervernetzendes Agens**
cross linker, crosslinking agent
Aggregatgefüge (Boden)
aggregate structure
Aggression aggression
Aggressionshemmung/
Angriffshemmung aggressive
inhibition, attack inhibition
Aggressivität/Angriffslust
aggressiveness
Aglycon aglycone
Agranulozyt/agranulärer Leukozyt
agranulocyte, agranular leukocyte
Agrarlandschaft/Ackerlandschaft
agricultural landscape
Agrarökosystem/Agroökosystem
agroecosystem,
agricultural ecosystem
Agroinfektion agroinfection
Agroinokulation agroinoculation
Agronom/diplomierter Landwirt
agronomist
Ahne/Vorfahre ancestor, forebear,
progenitor
Ahnenforschung/
Familienforschung/
Stammbaumforschung/
Genealogie genealogy
Ahnentafel genealogical table
Ahnenreihe ancestral lineage
Ahorngewächse/Aceraceae
maple family
Ährchen spicule, spicula, spikelet
Ährchenachse (Grasblüte) rachilla
Ähre (Infloreszenz) spike, spica;
(Fruchtstand) ear, head (of grain),
spike (infructescence)
Ährenfischverwandte/
Hornhechtartige/Atheriniformes
silversides & skippers & flying
fishes and others
Ährenfüllung/Kornfüllung agr
grain filling (poorly or well-filled)
AIDS (erworbenes
Immunschwächesyndrom/
Immunmangel-Syndrom) acquired
immune deficiency syndrome
(AIDS)
Airliftreaktor/pneumatischer
Reaktor (Mammutpumpenreaktor)
airlift reactor, pneumatic reactor
Aizoaceae/
Mittagsblumengewächse/
Eiskrautgewächse
mesembryanthemum family, fig
marigold family, carpetweed family
Akanthosom acanthosome

Akanthusgewächse/Acanthaceae
acanthus family
Akarizid acaricide
Akinese/Totstellreflex
(reflektorische
Bewegungslosigkeit) akinesis
Akinet/Dauerzelle (unbeweglich)
akinete, resting cell
Akklimatisierung/Akklimatisation
acclimatization (climate/seasons),
acclimation (artificial conditions)
akklimatisieren acclimatize
(climate/seasons), acclimate
(artificial conditions)
Akkommodation accommodation
Akkrustierung/Adkrustierung
(Kork) adcrustation, accrustation
Akkumulierung/Ansammlung
accumulation
akrokarp/gipfelfrüchtig acrocarpic,
acrocarpous
akropetal/basifugal
acropetal, basifugal
Akropetalie
acropetal development
akroplast acroplastic
akroplastes Wachstum
acroplastic growth
Akrosom/Acrosom acrosome
Akrotonie acrotony
akrozentrisches Chromosom
acrocentric chromosome
Aktin/Actin actin
Aktinfilament/Actinfilament/
Mikrofilament actin filament,
microfilament
Aktinkabel actin cable
aktinomorph actinomorphous,
star-shaped, radial
Aktionspotential/Spitzenpotential/
Impuls action potential, impulse,
spike
Aktionsraum ecol home range
Aktivationshormon/
prothoracotropes Hormon
prothoracicotropic hormone (PTTH),
brain hormone
Aktivatorprotein activator protein
aktiver Transport active transport
aktives Zentrum/katalytisches
Zentrum active site, catalytic site
aktivierter Zustand activated state
Aktivierungenergie activation
energy, energy of activation
Aktivitätskurve activity curve
Aktivitätsschub/Burst neuro burst
Aktivkohle activated carbon
akut transformierendes Retrovirus
acute transforming retrovirus
Akutphasenprotein
acute phase protein
Akzeptor/Empfänger acceptor

Akzeptorstamm *biochem*
(Proteinsynthese) acceptor stem
akzessorisch accessory
akzessorische Drüse/Anhangsdrüse
accessory gland
**akzessorisches Chromosom/
zusätzliches Chromosom**
accessory chromosome
akzessorisches Pigment
accessory pigment
Alanin alanine
Alarmsignal/Warnsignal *ethol*
alarm signal(ing)
**Alarmstoff/Schreckstoff/
Alarm-Pheromon** alarm substance, alarm pheromone
**Alarzelle/Blattflügelzelle
(Laubmoose)** alar cell
Albedo/Rückstrahlung albedo, reflective power
Albumin albumin
Älchen (Nematoden)
nematode parasite on plants
Aldehyd/Acetaldehyd aldehyde, acetic aldehyde, acetaldehyde
Aldosteron aldosterone
Aleuronschicht aleurone layer
Alge alga (*pl* algae/algas)
- **Armleuchteralgen/
Armleuchtergewächse/
Charophyceae/Charophyta
(Characeae)** stoneworts, stonewort family
- **Blaualgen/Cyanophyceae/
Cyanobakterien** bluegreen algae, cyanobacteria
- **Braunalgen/Phaeophyceae**
brown algae, phaeophytes
- **Florideen/Florideophyceae
(Rotalgen)** floridean algas, florideans
- **Gelbgrünalgen/Xanthophyten/
Xanthophyta** yellow-green algae
- **Goldalgen/Chrysophyceen/
Chrysophyceae** golden algae, golden-brown algae
- **Grünalgen/Chlorophyceae/
Isokontae** green algae
- **Jochalgen/Conjugaten/
Conjugatae/Acontae/
Zygnematophyceae
(Conjugatophyceae)**
zygnematophycean algas
- **Rotalgen/Rhodophyceae
(Floridaceae)** red algae
- **Zieralgen/Desmidiaceae** desmids
Algenbekämpfungsmittel/Algizid
algicide
Algenblüte algal bloom
Algenfarngewächse/Azollaceae
duckweed fern family, mosquito fern family

**Algenhaftorgan/Algenhaftscheibe/
Rhizoid** holdfast
Algenkunde phycology
**Algenpilze/niedere Pilze/
Phycomycetes** algal fungi, lower fungi
Algenspreite/Phylloid lamina, phyllid
Algenstiel/Kauloid/Cauloid stipe, caulid
Algenteppich/Algenmatte algal mat
Alginsäure (Alginat) alginic acid (alginate)
aliphatisch aliphatic
aliquoter Teil aliquot
Alismataceae/Froschlöffelgewächse
water-plantain family, arrowhead family
alizyklisch alicyclic
Alkali-Blotting alkali blotting
alkalisch/basisch alkaline, basic
Alkaloide alkaloids
Alkalose alkalosis
Alkaptonurie alkaptonuria
Alken/Alcidae (Charadiformes)
auks
Alkohol/Ethanol alcohol, ethanol
**Alkoholreihe/aufsteigende
Äthanolreihe**
graded ethanol series
Allantoin allantoin
Allantoinsäure allantoic acid
Allantois/Harnsack/Harnhaut *embr*
allantois
Allantoisplazenta allantoic placenta
Allel allele
- **multiple Allele** multiple alleles
- **Nullallel** null allele
- **Wildallel** wild-type allele
**Allelausschluss/allele Exclusion/
allele Exklusion** allelic exclusion
Allelen-Austauschtechnik
genetic replacement, gene disruption, gene replacement, gene targeting, gene transplacement, targeted homologous recombination
Allelenfrequenz/Allelenhäufigkeit
allele frequency
Allelopathie allelopathy
**allelspezifisches Oligonucleotid
(ASO)** allele specific oligonucleotide (ASO)
Allensche Regel/Proportionsregel
Allen's law, Allen's rule, proportion rule
allergen *adj/adv* allergenic
Allergen *n* allergen
**Allergie/Überempfindlichkeits-
reaktion** allergy
- **Soforttyp/anaphylaktischer Typ**
immediate-type hypersensitivity reaction

Amarantgewächse

- **Spättyp/verzögerter Typ**
  delayed-type hypersensitivity
  reaction (T_{DTH})
- **allergisch** allergic
- **Alles-oder-Nichts Antwort/Alles-oder-Nichts Reaktion**
  all-or-none response
- **allesfressend/omnivor** omnivorous; polyphag (begrenzte Nahrungsauswahl) polyphagous, polyphagic
- **Allesfresser/Omnivore** omnivore
- **allgemeine Erblichkeit (H^2)**
  broad heritability (H^2)
- **allgemeine Rekombination**
  generalized recombination
- **Allheilmittel/Universalmittel/Wundermittel/Panazee** panacea
- **Alliaceae/Zwiebelgewächse/Lauchgewächse** onion family
- **Allianz/Verband/Assoziationsgruppe** alliance
- **Alloantigen** alloantigen
- **Allogamie/Fremdbefruchtung**
  allogamy, xenogamy, cross-fertilization
- **Allogrooming/Fremdputzen (des Fells bei Säugern)** allogrooming
- **Allometrie** allometry
- **allopatrisch (in getrennten Arealen)**
  allopatric ("other country")
- **allopolyploid/amphidiploid**
  allopolyploid/amphidiploid
- **allosterische Transition/allosterischer Übergang**
  allosteric transition
- **allosterische Wechselwirkung/Interaktion** allosteric interaction
- **Allotransplantat/Homotransplantat**
  allograft (allogeneic graft), homograft, syngraft
- **Allotyp/Allotypus** allotype
- **Aloegewächse/Aloeaceae**
  aloe family
- **Alpha-DNA** alpha DNA
- **Alpha-Komplementation**
  alpha complementation
- **alpin/Hochgebirgs...** alpine
- **Altbestand (Wald)** old-growth (forest), mature forest
- **Alter** age
- **altern** vb age, become old, senesce
- **Altern/Alterung/Seneszenz** ageing, aging, senescence
- **Alternanz/Alternation/Abwechslung** alternation
- **Alternanzregel** alternation rule
- **alternatives Spleißen**
  alternative splicing
- **alternd/alt werdend** ageing, aging, becoming old, senescent

- **alternieren/wechseln/abwechseln zwischen zweien** alternate
- **alternierend/abwechselnd** alternate
- **alternierende Verteilung/Disjunktion (von Chromosomen)**
  alternate disjunction
  (of chromosomes)
- **Altersaufbau/Altersstruktur/Ätilität (Population)** age structure
- **Altersklasse** age class
- **Alterspyramide** age pyramid
  - **umgekehrte A.**
    inverted age pyramid
- **Altersstruktur/Altersaufbau/Ätilität (Population)** age structure
- **Altersstufe** stage of life
- **Altersverteilung** age distribution
- **Alterszusammensetzung** age composition
- **Alterung/Altern/Seneszenz** ageing, aging, senescence
- **Alterungsmutante** ageing mutant, aging mutant
- **Altfische/Chondrostei**
  primitive ray-finned bony fishes
- **Althirn/Paläoencephalon/Paleencephalon** paleoencephalon, paleencephalon
- **Althirnrinde/Archicortex/Palaeocortex** paleocortex
- **Altlauf/Altwasser** backwater, dead channel, slew
- **Altlungenschnecken/Archaeopulmonata**
  archeopulmonates
- **Altruismus/Selbstlosigkeit/Selbstaufopferung/Uneigennützigkeit/Gemeinnutz**
  altruism, self-sacrifice
- **Altschnecken/Schildkiemer/Archaeogastropoda/Diotocardia**
  limpets and allies, archeogastropods
- **Altweibersommer (Spinnengewebe)** gossamer
  (film of cobwebs floating in air)
- **Altweltaffen/Schmalnasenaffen/Catarrhina** old-world monkeys
  (incl. apes)
- **Alu-Familie** *Alu* family
- **Alula (Insekten: Flügellappen)**
  alula (insects wing: small lobe)
- **Alula/Daumenfittich/Afterflügel/Nebenflügel/Ala spuria (Federngruppe an 1. Finger)** alula, spurious wing, bastard wing
- **Aluminium** aluminum
- **alveolär** alveolar
- **amakrine Zelle** amacrine cell
- **Amarantgewächse/Fuchsschwanzgewächse/Amaranthaceae** amaranth family, cockscomb family, pigweed family

Amaryllisgewächse

**Amaryllisgewächse/
Narzissengewächse/
Amaryllidaceae**
daffodil family, amaryllis family
Amboss/Incus (Ohr) anvil (bone), incus
Ambrosiazelle (Pilzzelle)
ambrosia cell
Ambulakralfurche
ambulacral groove
Ambulakralfüßchen/Saugfüßchen
ambulacral foot, tube foot, podium
Ambulakralplatte ambulacral plate
**Ambulakralring/Ringkanal/
Radiärkanal** ring canal,
radial canal
**Ambulakralsystem/
Wassergefäßsystem** ambulacral system, water-vascular system
Ameisen/Formicidae ants
- **Dinergat/Soldat** dinergate, soldier
- **Ergat/Arbeiterin** ergate, worker
- **Gamergat** gamergate
  (fertilized, ovipositing worker)
- **Makraner (große ♂ Ameise)**
  macraner (large-size ♂ ant)
- **Makroergat** macrergate
- **Makrogyne (große
  Ameisenkönigin)** macrogyne
  (large ♀ ant/queen)
- **Mikraner (kleine ♂ Ameise)**
  micraner (dwarf ♂ ant)
- **Mikrergat/Zwergarbeiterin**
  micrergate, microergate,
  dwarf worker
- **Mikrogyne (kleine ♀ Ameise)**
  microgyne (dwarf ♀ ant)
- **Pterergat (Arbeiterin mit
  Stummelflügeln)** pterergate
  (with rudiments of wings)
Ameisenausbreitung
myrmecochory, ant-dispersal
**Ameisenbären/Vermilingua
(Xenarthra)** ant eaters
Ameisenhaufen anthill, mound
Ameisenjungfer antlion (adult)
**Ameisenlöwe (Larve der
Ameisenjungfer)**
doodlebug (antlion larva)
Ameisensäure (Format)
formic acid (formate)
Ameisenspinnen
ant-mimicking spiders
Amensalismus amensalism
Ames-Test Ames test
Amid amide
Amidierung amidation
amiktisch amictic
Amin amine
Aminierung amination
Aminoacyl-Stelle (A-Stelle)
aminoacyl site (A-site)

Aminoacyl-tRNA-Synthetase
aminoacyl-tRNA synthetase
Aminoacylierung aminoacylation
γ-Aminobuttersäure (GABA)
gamma-aminobutyric acid (GABA)
Aminosäure amino acid
Aminozucker amino sugar
Amme caretaker, nurse (e. g. asexual individual in social insects)
Ammenaufzucht foster raising
**Ammenbiene (Arbeitsbiene/
Arbeiterin)** nurse bee (worker bee)
**Ammenveredelung/Anhängen/
Vorspann geben** *hort* inarching
Ammoniak ammonia
ammoniotelisch/ammonotelisch
ammoniotelic, ammonotelic
Ammoniten/Ammonoidea
ammonites
ammonitische Lobenlinien
ammonitic suture lines
Ammonshorn/Pes hippocampi
Ammon's horn,
anterior hippocampus
Amnion/innere Keimhülle amnion, "bag of waters"
- **Faltamnion/Pleuramnion**
  pleuramnion
- **Spaltamnion/Schizamnion**
  schizamnion
Amnionfalte amniotic fold
Amnionflüssigkeit/Fruchtwasser
amniotic fluid
Amnionhöhle amniotic cavity
**amniotischer Strang/
Simonart'-Band** amniotic band
**Amniozentese/Amnionpunktion/
Fruchtwasserpunktion**
amniocentesis
**Amöben/Wechseltierchen/
Wurzeltierchen/Rhizopoden/
Amoebozoa** amebas, amoebas
- **beschalte Amöben/
  Thekamöben/Testacea**
  testate amebas
- **nackte Amöben/Gymnamoebia**
  naked amebas
**Amöbenruhr/Amöbiasis
(*Entamoeba histolytica*)**
amebic dysentery, amebiasis
amöboid ameboid
Amöbozyt/Amöbocyt amebocyte
amorph amorphous
AMP (Adenosinmonophosphat)
AMP (adenosine monophosphate)
- **Cyclo-AMP/Zyklo-AMP/
  zyklisches AMP (cAMP)**
  cyclic AMP (cAMP)
Amphiarthrose amphiarthrosis
Amphibien amphibians
amphibisch amphibian, amphibious
amphibol amphibolic

amphiboler Stoffwechselweg
amphibolic pathway,
central metabolic pathway
amphicöl/amphicoel amphicelous
amphicribrales Bündel amphicribral
(vascular) bundle
Amphid (Nematoden) amphid
amphidiploid/allopolyploid
amphidiploid, allopolyploid
amphidrom amphidromous
**Amphiesmalbläschen/
Amphiesmalvesikel**
amphiesmal vesicle
**Amphigastrium/Bauchblatt
("Unterblatt")** amphigastrium,
ventral leaf, underleaf
**Amphikarpie (verschiedenförmige
Früchte)** *bot* amphicarpy
amphiphil amphiphilic
Amphitokie amphitoky,
amphitokous parthenogenesis
amphitrich amphitrichous
amphivasales Bündel
amphivasal bundle
amphoter/amphoterisch amphoteric
amphotrop/amphotropisch *vir*
amphotropic
**Amplifikation/Vervielfältigung/
Vermehrung** amplification
amplifizieren amplify
amplifiziertes Gen amplified gene
Amplimer amplimer
Ampullardrüse ampullary gland
(of deferent duct)
Ampulle ampoule, ampulla; Blase
(*Utricularia*) ampulla, bladder;
Schlund/Geißelsäckchen (*Euglena*)
reservoir
Ampullenorgan *ichth*
(Elektrorezeptor) ampullary organ
Amylopektin amylopectin
Anabiose/latentes Leben anabiosis,
suspended animation
anabol anabolic (synthetic reactions)
Anacardiaceae/Sumachgewächse
sumac family, cashew family
anadrom anadromous
anaerob anaerobic
**Anaerobiose/Anerobiose/
Anoxibiose** anerobiosis,
anoxybiosis
Anagenese anagenesis
Analbeutel/Sinus paranalis
anal sac, paranal sinus
Analdreieck/Dreieck/Triangulum
triangle
Analdrüse (Hai/Säuger) anal gland,
rectal gland
Analfächer *entom* anal fan
Analfalte/Plica analis/Plica vannalis
anal fold, vannal fold, anal/vannal
fold-line

389

Androkonie

Analfeld/Vannus *entom* anal field,
anal area, vannal area, vannus
Analfurche/Gesäßspalte/Rima ani
anal cleft
Analgrube/Analgrübchen *embr*
anal pit
Analhügel *arach* anal tubercle,
anal papilla
analog/funktionsgleich analogous
Analogie analogy
analogisieren analogize
Analogon (*pl* **Analoga)** analog,
analogue
Analplatte *embr* anal plate
Analrand anal margin
Analschild/Pygidium (Käfer)
caudal shield, pygidium
Analschleife (Libellen) anal loop
Analysator analyzer
Analyse analysis (*pl* analyses)
analysenrein/zur Analyse *lab*
reagent grade
Analysenwaage analytical balance
analysieren analyze
analytisch analytic(al)
Analzelle *entom* anal cell
Anamorphose anamorphosis
Ananasgalle pineapple gall
**Ananasgewächse/Bromelien/
Bromeliaceae** pineapple family,
bromeliads, bromelia family
Anaphylaxe anaphylaxis
**anaplerotische Reaktion/
Auffüllungsreaktion**
anaplerotic reaction
Anapophyse anapophysis
**Anatomie (Morphologie der
inneren Gestalt)** anatomy
anatomisch anatomic(al)
anatrop anatropous
Anbau *agr* cultivation, cropping
Anbaueignung cultivability
anbauen cultivate, till, crop, grow
anbaufähiger Boden arable land,
tillable land
Anbaufähigkeit/Garezustand tilth
Anbaumethode/Anbauverfahren
cropping method/technique/
procedure
Anbaurotation/Fruchtwechsel
crop rotation
Ancestrula ancestrula
Andockprotein/Docking-Protein
docking protein
androgen androgenic
Androgen androgen
Androgenese androgenesis
Androgynophor androgynophore,
gynandrophore
Androkonie/Duftschuppe/Duftfeld
androconium (*pl* androconia),
scent scale(s)

Androsteron androsterone
Anellus anellus
Anemophilie/Windbestäubung anemophily, wind pollination
Anerobiose/Anaerobiose/Anoxibiose anerobiosis, anoxybiosis
aneuploid aneuploid
Aneuploidie aneuploidy
anfällig sein susceptible
Anfälligkeit (Empfindlichkeit) susceptibility; (Veranlagung/Disposition) disposition
Anfangsgeschwindigkeit (v_0: Enzymkinetik) initial velocity (vector), initial rate
anfärbbar dyeable, stainable
Anfärbbarkeit dyeability, stainability
anfärben dye, stain
Anfärbung dyeing, staining
angeboren/ererbt/kongenital/konnatal innate, inborn, congenital, connate
angeborener auslösender Mechanismus (AAM) ethol innate releasing mechanism (IRM)
angeborener Fehler/Erbleiden inborn error
Angebot offer, offering
- **Futterangebot/Nahrungsmittelangebot** feed supply, nutrient supply
Angehöriger (einer Gruppe/Tierart) member; (Verwandter) relative
Angel ichth fishing pole
Angelfaden/Wurffaden (Spinnfaden) casting line, "fishing line"
Angelhaken hook
Angelköder bait
angeln/fischen fish
Angelstück/Cardo (pl Cardines) cardo (basal segment of maxilla)
angepasst/adaptiert adapted
angepasst/beeinflusst ethol conditioned
angeregt excited
angeregter Zustand/erregter Zustand excited state
angewachsen/verwachsen (der Länge nach) adnate
angewandt applied
angewandte Botanik applied botany
angewandte Zoologie applied zoology
angiokarp angiocarpic, angiocarpous
Anglerfische/Armflosser/Lophiiformes anglerfishes
angrenzend/anliegend/anstoßend contiguous, adjoining, bordering
angrenzend/benachbart adjacent

Angriff/Attacke attack
- **Gegenangriff/Gegenattacke** counterattack
Angriffs-Mimikry/Peckhammsche Mimikry aggressive mimicry, Peckhammian mimicry
Angriffshemmung/Aggressionshemmung attack inhibition, aggressive inhibition
Angriffswinkel angle of attack
anhaltende Infektion/persistente Infektion persisting infection
Anhang appendage
Anhängen/Vorspann geben/Ammenveredelung inarching
Anhangsdrüse/akzessorische Drüse accessory gland
Anhängsel/Anhangsgebilde appendage, appendix
anhäufeln (Pflanzen) to ridge up
Anhäufung/Akkumulation accumulation
Anheftung/Adhärenz/Adhäsion adherence, adhesion
Anheftung/Befestigung attachment, affixment
Anheftungsorgan (Bryozoen) adhesive sac, metasomal sac, internal sac
animaler Pol animal pole
animalisch/tierisch adj (z.B. tierisches Fett) animal (e.g. animal fat)
Anionenaustauscher anion exchanger
Anisaldehyd anisic aldehyde, anisaldehyde
Anisogamie anisogamy
Anisophyllie/Heterophyllie/Verschiedenblättrigkeit/Ungleichblättrigkeit anisophylly, heterophylly
Ankerwurzel anchorage root, adhesion root
Ankylose ankylosis
Anlage/Keim/Ansatz/Primordium "anlage", precursor, preformation, early form, primordium
Anlage/öffentliche Grünanlage/Park public gardens, public park
Anlagerung/Apposition adsorption, apposition
Anlandung ecol aggradation
Anlaufphase/Latenzphase/Inkubationsphase/Verzögerungsphase/Adaptationsphase/lag-Phase lag phase, latent phase, incubation phase, establishment phase
anliegend (entlang anderem Gegenstand) accumbent (along/against other body)

anlocken/locken lure, attract
Anlockung luring, attraction
Anmoor early bog, half-bog
anmooriger Boden half-bog soil
annähern/näherkommen/sich annähern/erreichen (z. B. einen Wert) approach (e.g. a value)
Annattogewächse/Bixaceae annatto family, bixa family
Annealing/Doppelstrangbildung/ Reannealing/Renaturierung/ Reassoziation (DNA) reassociation, annealing, reannealing, renaturation, reassociation
Anneliden/Ringelwürmer/ Gliederwürmer annelids, segmented worms
● **Gürtelwürmer/Clitellaten** clitellates (oligochetes & hirudineans)
● **Vielborster/Borstenwürmer/ Polychaeten** bristle worms, polychaetes, polychetes, polychete worms
● **Wenigborster/Oligochaeten/ Oligochaeten** oligochetes
Annonaceae/ Schuppenapfelgewächse custard apple family, cherimoya family
Annuelle/Einjährige/Therophyt annual (plant), therophyte
annulierte Lamellen annulate lamellae/lamellas
Annulus/Anulus annulus
Annulus inferus/Ring/Kragen (Rest des Velum partiale) inferior annulus, ring
Annulus superus/Manschette/ Armilla superior annulus, manchette, armilla
Anogenitalkontrolle anogenital control
Anogenitalmassage anogenital licking
anomal/unregelmäßig/irregulär anomalous, irregular
Anomalie/Unregelmäßigkeit anomaly, irregularity
anonyme DNA anonymous DNA
Anordnung (Position) arrangement
anorganisch inorganic
anorganische Chemie/"Anorganik" inorganic chemistry
Anoxie anoxia
anoxisch anoxic
anpassen/akklimatisieren adapt, adjust to, acclimate, acclimatize
Anpassung/Adaptation adaptation, acclimation, acclimatization
anpassungsfähig/adaptiv adaptive
Anpassungsfähigkeit/Adaptabilität adaptability

Anpassungsgipfel adaptive peak
Anpassungswert/Selektionswert adaptive value, selective value
anpflanzen plant, cultivate, grow
Anpflanzung/Pflanzung/Plantage plantation
anpflocken stake
Anplatten *hort* side grafting
● **seitliches Anplatten** side-veneer grafting, veneer side grafting, spliced side grafting
● **seitliches Anplatten mit Gegenzunge** side-tongue grafting
● **seitliches Anplatten mit langer Gegenzunge** flap grafting
anregen stimulate, excite
Anregung stimulation, excitation
anreichern enrich; fortify
Anreicherung enrichment
Anreicherung durch Filter filter enrichment
Anreicherungskultur enrichment culture
Ansatz (Versuchsansatz/ Versuchsaufbau) arrangement, set-up
Ansatz/Charge batch
Ansatz/Methode approach, method
Ansatz/Präparat starting material, preparation
Ansatz/Versuch attempt
Ansatzstelle/Anheftungsstelle attachment site
Ansatzstück *lab* attachment, extension (piece)
ansäuern acidify
Anschäften *hort* splice grafting, whip grafting (with stock larger than scion)
ansetzen (z. B. eine Lösung) start, prepare, mix, make, set up
anspruchslos undemanding, modest, having low requirements/demands
anspruchsvoll demanding, having high requirements/demands
anstecken/infizieren infect
ansteckend/ansteckungsfähig/ infektiös contagious, infectious
ansteckende Krankheit/infektiöse Krankheit contagious disease, infectious disease
Ansteckleuchte *micros* substage illuminator
Ansteckung/Infektion contagion, infection
Ansteckungsfähigkeit/ Infektionsvermögen infectivity
Ansteckungsherd/ Ansteckungsquelle source of infection
Ansteckungskraft/Virulenz virulence

Anstellwinkel

Anstellwinkel angle of attack
Antagonismus antagonism
Anteil/Hälfte/Teil moiety
Antennalorgan antennal organ
Antenne/Fühler antenna, feeler
(Antennentypen *siehe unter:* Fühler)
- **Geißelantenne/Ringelantenne/ amyocerate Antenne**
amyocerate antenna
- **Gliederantenne/myocerate Antenne** myocerate antenna

Antennendrüse/ Antennennephridium/ grüne Drüse antennal gland, antennary gland, green gland
Antennengrube/Fühlergrube antennal furrow/pit
Antennenkomplex (von Chlorophyllmolekülen) antenna complex (of chlorophyll molecules)
Antennennaht/Antennalnaht/ Fühlerringnaht antennal suture
Antennenpigment antenna pigment
Antennenschuppe/Scaphocerit antennal scale, scaphocerite
Antennensegment/Fühlersegment/ Antennomer antennal segment, antennomer
Antennenträger/Antennifer antennifer, socket of antenna
Antennula (1.Antenne: Crustaceen) antennule
Antenodalquerader antenodal cross vein
Anthere anther
Antheridium antheridium
Antheridiumzelle/generative Zelle (Cycadeenpollen) antheridial/ generative cell
Anthese/Blütezeit/Floreszenz anthesis, flowering period, florescence
Anthocarp/Anthokarp anthocarp
Anthocladium anthoclade
Anthraknose/Brennfleckenkrankheit anthracnose
Anthranilsäure anthranilic acid, 2-aminobenzoic acid
Anthrazen anthracene
anthropogen anthropogenic
Anthropologie anthropology
Anti-Müller-Hormon (AMH)
Mullerian inhibiting hormone (MIH)
Antibiose/Widersachertum antibiosis
Antibiotikum (*pl* **Antibiotika)** antibiotic
- **Resistenz gegen Antibiotika**
antibiotic resistance

anticodierender Strang/Nicht-Sinnstrang anticoding strand, antisense strand

Anticodon anticodon
antidiuretisches Hormon (ADH)/ Adiuretin/Vasopressin antidiuretic hormone (ADH), vasopressin
antidrom antidromic
antigen *adv/adj* antigenic
antigen antigenic
Antigen *n* antigen
- **"sehr spätes Antigen" (bildet sich spät in der Entwicklung)**
very late antigen (VLA)
- **Differenzierungsantigen**
differentiation antigen
- **gruppenspezifisches Antigen**
group-specific antigen (gag)
- **kreuzreagierendes Antigen**
cross-reacting antigen
- **prozessiertes Antigen/ weiterverarbeitetes Antigen**
processed antigen
- **typenspezifisches Antigen**
type-specific antigen

Antigen-Processing/ Antigenweiterverarbeitung
antigen processing
Antigenbindungsstelle/ Antigenbindestelle/Paratop
antigen combining site, antigen binding site, paratope
Antigendeterminante/Epitop
antigenic determinant, epitope
Antigendrift antigen drift; antigenic drift
antigene Determinante/Epitop
antigenic determinant, epitope
antigene Variation
antigenic variation
Antigenität antigenicity
Antigenpräsentation
antigen presentation
antigenpräsentierende Zelle
antigen-presenting cell (APC)
- **professionelle a.Zelle**
professional antigen-presenting cell
Antigenrezeptor antigen receptor
Antigenshift antigen(ic) shift
Antigenvarianz antigenic variance
Antigenvariation antigen variation
Antigenweiterverarbeitung/ Antigen-Processing
antigen processing
Antikörper antibody
- **Autoantikörper** autoantibody
- **bispezifischer A.** bispecific antibody, hybrid antibody
- **katalytischer A.**
catalytic antibody
- **monoklonaler A. (mAb)**
monoclonal antibody (MAb)
- **polyklonaler A.**
polyclonal antibody
Antimetabolit antimetabolite

Appetenz

Antimon antimony
**antiphlogistisch/
  entzündungshemmend**
  antiphlogistic, anti-inflammatory
Antipode antipode, antipodal cell
Antischaummittel/Entschäumer
  antifoam
**Antisense-Oligonucleotid/Anti-
  Sinn-Oligonucleotid/Gegensinn-
  Oligonucleotid** antisense
  oligonucleotide
**Antisense-RNA/Anti-Sinn-RNA/
  Gegensinn-RNA** antisense RNA
**Antisense-Technik/Anti-Sinn-
  Technik/Gegensinntechnik** *gen*
  antisense technique
Antiserum antiserum
**Antiterminationsprotein/
  Antiterminator** *gen*
  antitermination protein
Antizipation anticipation, antedating
Antrieb/Trieb drive; Voranbringen
  (Fortbewegung) propulsion
Antriebskraft/Triebkraft propulsive
  force
Antriebspotential driving potential
Antwort (auf Reiz) response
antworten answer; respond
anwachsen/anwurzeln/bewurzeln
  take root
Anzeige (Gerät) display
anzeigen display, show
Anzeiger/Indikator indicator
Anzeigerpflanze/Indikatorpflanze
  indicator plant
**Anzieher/Adduktor/Adductor
  (Muskel)** adductor muscle
Anzucht (*einer Kultur*) starting
  • **Samenanzucht** seed starting
anzüchten (*einer Kultur*) establish,
  start (a culture)
Anzuchtmedium starter medium
  (growth medium)
Anzüchtung (*einer Kultur*)
  establishing growth, starting growth
anzünden ignite, strike, start a fire
Anzünder (Gas) striker
äolisch *geol* aeolian
Äon *m (pl* **Äonen)/Weltalter
  (größte geochronologische
  Einheit)** eon
  • **Archaikum (Altpräkambrium/
    frühes Präcambrium)** Archean
    Eon, Archeozoic Eon, Archeozoic
    (early Precambrian)
  • **Phanerozoikum** Phanerozoic
    Eon, Phanerozoic
  • **Proterozoikum
    (Jungpräkambrium/
    spätes Präcambrium/Eozoikum)**
    Proterozoic Eon, Proterozoic
    (late Precambrian)

Aorta/große Körperschlagader
  aorta
Aortenbogen aortic arch
Aortenklappe aortic valve
Aortenkörper/Zuckerkandl'-Organ
  aortic body, Zuckerkandl's body
**AP-Stelle (purin- oder pyrimidinlose
  Stelle)** *gen* AP site (apurinic or
  apyrimidinic site)
**Apertur (Blende)/Öffnung/
  Mündung** aperture, opening,
  orifice
**Aperturblende/Kondensorblende
  (Irisblende)** condensor diaphragm
  (iris diaphragm)
apetal apetalous
Apfelfrucht/Pomum pome,
  pomaceous fruit, core-fruit
Äpfelsäure (Malat)
  malic acid (malate)
Apfelschimmel (Pferd)
  dapple-gray horse
aphotisch aphotic
Aphrodisiakum (*pl* Aphrodisiaka)
  aphrodisiac
Aphyllophorales gymnocarps
  (coral fungi & pore fungi and allies)
**Apiaceae/Umbelliferae/
  Umbelliferen/Doldenblütler/
  Doldengewächse** carrot family,
  parsley family, umbellifer family
Apikalkomplex (Apicomplexa)
  apical complex
**Aplacophoren/Wurmmollusken/
  Wurmmolluscen/Aplacophora**
  aplacophorans
Apnoe apnea
apochlamydeisch apochlamydeous
**Apocynaceae/Hundsgiftgewächse/
  Immergrüngewächse**
  periwinkle family,
  dogbane family
Apoenzym apoenzyme
apokarp/chorikarp apocarpous
apokrine Drüse apocrine gland
Apomorphie apomorphism
apomorph apomorphic
**Aponogetonaceae/
  Wasserährengewächse**
  cape-pondweed family,
  water hawthorn family
apopetal apopetalous
Apoptose/programmierter Zelltod
  apoptosis, programmed cell death
Apopyle apopyle
**Appendicularien/Appendicularia/
  Larvacea/geschwänzte Schwimm-
  Manteltiere** appendicularians
Appetenz appetence, appetency
  (fixed/strong desire)
  • **bedingte Appetenz**
    conditioned appetence

Appetenzverhalten

Appetenzverhalten
appetitive behavior
Apportiervektor
eviction vector
Appositionsauge
apposition eye
Appressorium/Haftscheibe/ Haftorgan *allg* appressorium, holdfast
apyren (Spermien) apyrene
Aquakultur aquaculture
Aquarium aquarium, fishtank
Äquatorialfurchung
equatorial cleavage
Äquatorialteilung
equatorial division
Äquidistanz equidistance
Aquifoliaceae/ Stechpalmengewächse
holly family
Äquität/Äquitabilität *ecol* evenness, equitability
Äquivalenzzone (Ausfällung unlöslicher Immunkomplexe)
zone of equivalence
Ära (*pl* **Ären)/Zeitalter (erdgeschichtliches Zeitalter) (***siehe auch:* **Äon/Epoche/Periode)**
era (*pl* eras), age, geological era, geological age
- **Eophytikum** Eophytic Era
- **Erdaltertum/Paläozoikum**
 Paleozoic, Paleozoic Era
- **Erdmittelalter/Mesozoikum**
 Mesozoic, Mesozoic Era
- **Erdneuzeit/Neozoikum/ Känozoikum/Kaenozoikum**
 Neozoic, Neozoic Era, Cenozoic, Cenozoic Era
 (Cainozoic Era/Caenozoic Era)
- **Känozoikum/Kaenozoikum/ Erdneuzeit/Neozoikum**
 Cenozoic, Cenozoic Era, Neozoic Era
 (Cainozoic Era/Caenozoic Era)
- **Mesophytikum** Mesophytic Era
- **Mesozoikum/Erdmittelalter**
 Mesozoic, Mesozoic Era
- **Neozoikum/Erdneuzeit/ Känozoikum/Kaenozoikum**
 Neozoic, Neozoic Era, Cenozoic, Cenozoic Era
 (Cainozoic Era/Caenozoic Era)
- **Paläophytikum/Florenaltertum**
 Paleophytic Era
- **Paläozoikum/Erdaltertum**
 Paleozoic, Paleozoic Era
- **Präkambrium/Präcambrium**
 Precambrian, Precambrian Era
Araceae/Aronstabgewächse
Arum family, calla family, aroid family

Arachidonsäure arachidonic acid, icosatetraenoic acid
Arachinsäure/Arachidinsäure/ Eicosansäure arachic acid, arachidic acid, icosanic acid
Araliaceae/Efeugewächse
ivy family, ginseng family
Aräometer areometer (hydrometer)
Ärathem (geochronologisch) *geol* era
Araukariengewächse/ Schmucktannengewächse/ Araucariaceae araucaria family, monkey-puzzle tree family
Arbeiter worker
Arbeiterin/Ergat (Ameisen) worker, ergate
- **Dinergat/Soldat** dinergate, soldier
- **Gamergat** gamergate (fertilized, ovipositing worker)
- **Makroergat** macrergate
- **Mikrergat/Zwergarbeiterin**
 micrergate, microergate, dwarf worker
- **Pterergat (Arbeiterin mit Stummelflügeln)** pterergate (with rudiments of wings)
Arbeitsbiene/Arbeiterin worker bee
Arbeitsmethode work procedure
Arbeitsplatz workplace
Arbeitsplatzhygiene
occupational hygiene
Arbeitsplatzkonzentration, zulässige
permissible workplace exposure
Arbeitsplatzsicherheit *lab*
occupational safety, workplace safety
Arbeitsschutz workplace protection, safety provisions (for workers)
Arbeitsschutzkleidung
workers' protective clothing
Arbeitsstoffwechsel/ Leistungsstoffwechsel
active metabolism
Arbeitsumsatz/Leistungsumsatz
active metabolic rate
Arboretum arboretum
arborikol arboricole, arboricolous
Archaikum (erdgeschichtliche Zeit)
Archean Eon, Archean, Archeozoic Eon (early Precambrian)
Archäzyt/Archäocyt/Archaeozyt/ Archaeocyt archaeocyte, archeocyte
Archegonium archegonium
Archenmuscheln/Arcidae arks
Archespor archespore, sporoblast
Archicöl/Archicoel (Blastocöl/ Bastocoel) archicoel, archicele (blastocoel/blastocele)

Areal/Verbreitungsgebiet area of distribution, geographic range

Arealkarte/Verbreitungskarte range chart, range map, distribution map

Arealkunde/Chorologie/ Verbreitungslehre chorology, biogeography

Arecaceae/Palmae/Palmen palm family

Areole areole

Arginin arginine

Argyrom/Silberliniensystem (Ciliaten) argyrome, silverline system

Arillus aril

Arista arista

Aristolochiaceae/ Osterluzeigewächse birthwort family, Dutchman's-pipe family

arithmetisches Mittel *stat* arithmetic mean

arithmetisches Wachstum arithmetic growth

Arktis arctic

arktisch/polar arctic, polar

Arm arm; (Crinoiden) arm, pinnule

armartig/Arm.../brachial brachial, arm-like

Armbeuge/Ellenbeuge bend of the arm, crook of the arm, inside of the elbow, front of elbow

Armdecken *orn* secondary tectrices

Armflosser/Anglerfische/ Lophiiformes anglerfishes

Armfüßer/"Lampenmuscheln"/ Brachiopoden lampshells, brachiopods

Armgerüst/Brachidium (Brachiopoden) brachidium

Armilla/Manschette/Annulus superus armilla, manchette, annulus superus

Armleuchteralgen/ Armleuchtergewächse/ Charophyceae/Charophyta (Characeae) stoneworts, stonewort family

Armleuchterbäume/Didieraceae didierea family

Armpalisade (Coniferen) arm-palisade

ARMS (System der amplifizierungsresistenten Mutation) ARMS (amplification refractory mutation system)

Armschwingen secondaries, secondary feathers (secondary remiges)

Armskelettplatte/Brachiale brach, brachial

Armstachel (Ophiuroidea) arm spine

armtragend/mit Armen brachiate, brachiferous, having arms

Armwirbler/Süßwasserbryozoen/ Lophopoda/Phylactolaemata phylactolaemates, "covered throat" bryozoans

Aroma (Wohlgeruch) aroma, fragrance, (pleasant) odor; (Wohlgeschmack) flavor, taste (pleasant)

Aromastoff flavoring, aromatic substance

aromatisch aromatic

Aronstabgewächse/Araceae arum family, calla family, aroid family

arretieren (Chromosomen in Metaphase) arrest

Arretierung *tech/mech* (z. B. am Mikroskop) stop

arrhenotok arrhenotokous

Arrhenotokie/arrhenotoke Parthenogenese arrhenotoky, arrhenotokous parthenogenesis

ARS (autonom replizierende Sequenz) ARS (autonomously replicating sequence)

Arsen arsenic

Art/Spezies kind, species
- **biologische Art/Biospezies** biological species, biospecies
- **Charakterart/Leitart** character species
- **Chronospezies** chronospecies
- **Differentialart/Trennart** differential species
- **einheimische Art** native species
- **Elementarart/Kleinart/ Mikrospezies/Jordanon** microspecies, jordanon
- **eurytope Art** eurytopic species
- **evolutionäre Art** evolutionary species
- **Folgeart** successional species
- **Fremdart/eingewanderte Art/ Zuwanderer** alien species, immigrant species
- **Geschwisterarten/Zwillingsarten** sibling species
- **Großart/Makrospezies** macrospecies
- **indifferente Art** indifferent species
- **Indikatorart/Zeigerart** indicator species
- **Kennart** diagnostic species
- **Kernart** core species
- **Kleinart/Elementarart/ Mikrospezies/Jordanon** microspecies, jordanon
- **Kollektivart/Kollektivspezies/ Sammelart (Superspezies)** aggregated species, agg. species, collective species (superspecies)

A Art

- **Leitart** index species, guide species
- **monotypische Art** monotypic species
- **monozentrische Art** monocentric species
- **Morphospezies/morphologische Art** morphospecies
- **Ökospezies** ecospecies
- **Paläospezies (Chronospezies)** paleospecies (chronospecies)
- **Pionierart** pioneer species
- **polytypische Art** polytypic species
- **polyzentrische Art** polycentric species
- **Quasispezies** quasispecies
- **Ringart** ring species
- **Sammelart/Großart/Coenospezies** coenospecies
- **Sammelart/Kollektivart/Kollektivspezies (agg.)** aggregated species, collective species
- **Sammelart/Überart/Superspezies** superspecies
- **Satellitenart/Randart** satellite species, marginal species
- **Schlüsselart** keystone species
- **Schwesterart** sister species
- **Stammart** stem species
- **Stellvertreterart/vikariierende Art** vicarious species
- **Superspezies/Überart/Sammelart** superspecies, aggregated species
- **taxonomische Art** taxonomic species, taxospecies
- **Typus-Art** type species
- **Überart/Superspezies** superspecies
- **Unterart/Subspezies** subspecies
- **vagabundierende Art** fugitive species
- **Zeigerart/Indikatorart/Bioindikator** indicator species, bioindicator
- **Zönospezies/Coenospezies** coenospecies, cenospecies
- **Zwillingsarten/Geschwisterarten** sibling species

Art und Weise/Modus/Modalwert mode

Art-Arealkurve species-area curve

Artbildung/Speziation species formation, speciation
- **allopatrische A.** allopatric speciation
- **parapatrische A.** parapatric speciation
- **sympatrische A.** sympatric speciation

Artefakt artifact, artefact

arteigen adv/adj characteristic of a species

Artenbestand/Arteninventar species inventory

Artendichte/Artdichte/Artenabundanz species density, species abundance

Artenfehlbetrag species deficiency

Artengefälle species gradient

Artengruppe collective group, species-group, species aggregate

Arteninventar/Artenbestand species inventory

Artenkreis/Überart/Superspezies superspecies

Artenmächtigkeit siehe Artmächtigkeit

Artenreichtum species richness

Artenschutz protection of species, conservation of species

Artenschutzabkommen treaty on the protection of species
- **Washingtoner Artenschutzübereinkommen** Convention on International Trade in Endangered Species (CITES), Washington 1975

Artenschutzverordnung legal regulations on species protection

Artenschwarm species flock, species swarm

Artenschwund species impoverishment, species loss, species decline (steady decrease in number of species)

Artenvielfalt/Artenmannigfaltigkeit species diversity

Artenvorkommen (number of) occurring species, occurrence of species

Arterhaltung preservation of species, species preservation; reproductive survival, reproductive continuance

Arterie artery (die verschiedenen Arterienbezeichnungen finden Sie in einem Wörterbuch der Human- bzw. Veterinärmedizin)

Arterien../arteriell arterial

Arterkennung species recognition

artfremd (Eiweiss) foreign

artfremd ethol untypical/uncommon/unusual of the species

Artgenosse conspecific (member of the same species), peer

Artgrenze/Kreuzungsbarriere species barrier

Arthrobranchie arthrobranch, arthrobranchia, joint gill

Arthropoden/Gliederfüßer/Arthropoda arthropods

Arthrospore/Arthrokonidium/ Oidium/Oidie arthrospore, arthroconidium, oidium
Artmächtigkeit cover abundance index, species importance value
Artname/Artbezeichnung species name (*see also:* specific epithet)
artspezifisch/arttypisch species specific
arttypisches Verhalten/ artspezifisches Verhalten species-specific behavior
Arznei/Arzneimittel/Medizin medicine, medication, drug
Arzneibuch pharmacopeia
Arzneikunde/Arzneilehre/ Pharmazie pharmacy
Arzneipflanze/Heilpflanze medicinal plant
Asche ash
Aschelminthen/Nemathelminthen/ Schlauchwürmer/Rundwürmer *sensu lato* aschelminths, nemathelminths
Asclepiadaceae/ Schwalbenwurzgewächse/ Seidenpflanzengewächse milkweed family
Ascorbinsäure (Ascorbat) ascorbic acid (ascorbate)
asexuell/ungeschlechtlich asexual
Askus ascus
Askushaken(zelle) crozier
ASO (allelspezifisches Oligonucleotid) *gen* ASO (allele specific oligonucleotide)
Asparagaceae/Spargelgewächse asparagus family
Asparagin asparagine, aspartamic acid
Asparaginsäure (Aspartat) asparagic acid, aspartic acid (aspartate)
Aspidiaceae/Dryopteridaceae/ Schildfarngewächse/ Wurmfarngewächse aspidium family, sword fern family
Aspleniaceae/Streifenfarngewächse spleenwort family
Asseln/Isopoda pill bugs, woodlice, sowbugs
Asselspinnen/Pycnogonida/ Pantopoda pycnogonids, pantopods, sea spiders
Assemblierung/Zusammenbau assembly
Assimilat assimilate
Assimilation assimilation, anabolism
Assimilationsparenchym/ Chlorenchym chlorenchyma
Assimilationswurzel assimilative root

assimilatorisch assimilatory
Assimilatstrom assimilate stream
assimilieren assimilate
assortative Paarung/bewusste Paarung assortative mating
Assoziationsfeld *neuro* association area
Assoziationskoeffizient *stat* coefficient of association
Assoziationslernen/ Erfahrungslernen associative learning, learning by experience
Ast branch, limb
● **absteigender Ast (Henlesche Schleife)** *nephro* descending limb
● **aufsteigender Ast (Henlesche Schleife)** *nephro* ascending limb
Ästchen twig, branchlet, sprig
Aster/Polstrahl (meist *pl:* **Asteren/ Polstrahlen)** *cyt* aster (in mitosis)
Asteraceae/Compositae/ Korbblütler/Köpfchenblütler sunflower family, daisy family, aster family, composite family
Ästhet (pl **Ästheten)** esthete, aesthete (photosensitive structure in chitons)
Ästhetask/Riechschlauch esthetasc, aesthetasc
astich astichous
Astigmatismus astigmatism
Ästivation/Knospendeckung estivation, aestivation
Astpfropfung/Astveredlung *hort* top grafting, top working
Astring/Astwulst branch collar
Astrozyt astrocyte
Astschere *hort* loppers
Aststumpf/Knorren snag
Ästuar/Ästuarium (pl **Ästuarien)/ trichterartige Flussmündung** estuary
Äsung (äsen) grazing, browsing
Atavismus atavism
Atem breath
Atemgifte/Fumigantien respiratory toxin, fumigants
Atemloch/Luftloch spiracle
Atemminutenvolumen (AMV) minute respiratory volume
Atemöffnung *bot* air pore
Atemöffnung *zool* breathing aperture
Atemplatte/Scaphognathide/ Scaphognathit bailer, gill bailer, scaphognathite
Atemschutzgerät breathing apparatus
Atemwurzel pneumatophore, air root, airial root, aerating root
Atemzentrum respiratory center
Atemzugvolumen tidal volume

Äthanol

Äthanol/Ethanol/Äthylalkohol/ Ethylalkohol/"Alkohol" ethanol, ethyl alcohol, alcohol
Äther/Ether ether
ätherisches Öl ethereal oil, essential oil
Äthylen/Ethylen ethylene
Athyriaceae/Frauenfarngewächse lady fern family
Ätilität/Altersstruktur/Altersaufbau (Population) age structure
Ätiologie etiology
atmen breathe, respire
- **ausatmen** breathe out, exhale
- **einatmen** breathe in, inhale

Atmosphäre atmosphere
Atmung breathing, respiration
- **aerobe Atmung** aerobic respiration
- **anaerobe Atmung** anaerobic respiration
- **Ausatmung/Ausatmen/ Expiration/Exhalation** expiration, exhalation
- **Bauchatmung/Zwerchfellatmung** abdominal breathing, diaphragmatic respiration
- **Brustatmung/Thorakalatmung** thoracic respiration, costal breathing
- **Einatmung/Einatmen/ Inspiration/Inhalation** inspiration, inhalation
- **Hautatmung** cutaneous respiration/breathing, integumentary respiration
- **Lichtatmung/Photoatmung/ Photorespiration** photorespiration
- **Plastronatmung** plastron breathing
- **Zellatmung** cellular respiration

Atmungsepithel/ Respirationsepithel/ respiratorisches Epithel respiratory epithelium
Atmungsgift respiratory poison
Atmungskette/ Elektronentransportkette/ Elektronenkaskade (Endoxydation) respiratory chain, electron transport chain
Atmungsquotient/respiratorischer Quotient respiratory quotient
atok atoke
Atoll/Atollriff/Lagunenriff atoll
Atomabsorptionsspektroskopie (AAS) atomic absorption spectroscopy (AAS)
atomar/Atom... atomic
atomar verseucht radioactively contaminated
Atomgewicht atomic weight
Atomkraft/Atomenergie nuclear/atomic power, nuclear/atomic energy
Atommüll nuclear waste
Atomzahl atomic number
ATP (Adenosintriphosphat) ATP (adenosine triphosphate)
atriales natriuretisches Peptid (ANP)/Atriopeptin atrial natriuretic peptide (ANP), atrial natriuretic factor (ANF), atriopeptin
Atrioventrikularknoten/ Aschoff-Tawara-Knoten (Herz: sek. Autonomiezentrum) atrioventricular node, A-T node
atrop atropous, orthotropous
Atropin atropine
Attached X-Chromosomen/ verbundene X-Chromosomen/ verklebte X-Chromosomen attached X-chromosomes
Attenuation/Abschwächung attenuation
attenuieren/abschwächen (die Virulenz vermindern/mit herabgesetzter Virulenz) attenuate
Attenuierung attenuation
Attraktans (pl Attraktantien)/ Lockmittel/Lockstoff attractant
ätzen vb med cauterize
ätzen vb metall/techn/micros etch (*siehe:* Gefrierätzen)
ätzen/korrodieren vb chem eat into, corrode
Ätzen/Ätzung/Ätzverfahren med cauterization
Ätzen/Ätzung/Ätzverfahren metall/ techn/micros etching (*siehe:* Gefrierätzen)
Ätzen/Ätzung/Korrosion corrosion
ätzend/korrosiv chem caustic, corrosive, mordant
Ätzmittel chem caustic agent
Ätzmittel metall/techn/micros etchant
Aue riverine floodplain
Auenwald/Auwald floodplain forest
Auenwiese/Auwiese/ Überschwemmungswiese riverine floodplain meadow, bottomland meadow
aufarbeiten lab/biot work up, process
Aufarbeitung lab/biot work up, working up, processing, down-stream processing
Aufbau (Struktur) construction, structure, body plan, anatomy
Aufbau metabol **Synthesestoffwechsel** anabolism, synthetic reactions/metabolism

auftauen

aufbauend/anabol *metabol* anabolic
Aufblühfolge flowering sequence
aufdampfen/bedampfen *micros* vacuum-metallize
Auferstehungspflanze/ Wiederauferstehungspflanze resurrection plant
Aufforstung/Wiederaufforstung/ Wiederbewaldung afforestation, reforestation, reafforestation
Auffrischimpfung (Auffrischinjektion) booster vaccination (booster shot)
Auffüllreaktion/ Auffüllungsreaktion/ anaplerotische Reaktion fill-in reaction, filling in reaction, anaplerotic reaction
aufgeblasen inflated
aufgeraut/aufgerauht roughened, scabrid
aufgerollt (schneckenförmig) circinate, coiled, volute
aufgewickelt coiled, twisted, wound
Aufheller/Aufhellungsmittel (optischer Aufheller) brightener, clearant, clearing agent (optical brightener)
Aufklärung (Strukturen/ Zusammenhänge) elucidation
Aufklärungshof/Lysehof/Hof/ Plaque plaque
Auflagehumus (*allg*: ungenauer Begriff) organic layer
• **saurer Auflagehumus/Rohhumus** raw humus, mor humus
Auflagerung (Zellwandwachstum durch) apposition, accretion
auflauern/aus dem Hinterhalt angreifen ambush
Auflicht/Auflichtbeleuchtung epiillumination, incident illumination
auflösen *chem* dissolve
auflösen *opt* resolve
Auflösung *chem* dissolution
• **optische Auflösung** optical resolution
Auflösungsgrenze *opt* limit of resolution
Auflösungsvermögen *opt* resolving power
Aufnahme/Annahme acceptance; acquisition
Aufnahme/Aufschreiben/ Registration recording, registration
Aufnahme/Bild picture, image
• **mikroskopische A./ mikroskopisches Bild** *photograph* micrograph, microscopic picture/image

Aufnahme/Einnahme uptake/intake; ingestion
Aufnahmezeit *vir* acquisition time
aufnehmen/aufschreiben/ registrieren record, register
aufnehmen/einnehmen/zu sich nehmen take up, take in; ingest
aufplatzen/aufspringen break open, dehisce
• **zum Platzen bringen** burst
Aufplatzen/Aufspringen/Dehiszenz breaking open, dehiscence
aufrecht erect, strict, upright, straight
aufrecht (Samenanlage) *bot* atropous, orthotropous
aufrechter Gang (Gangart) upright gait
aufsaugen/absorbieren soak up, absorb
Aufsaugen/Absorption soaking up, absorption
Aufschlag (Flug/Flügel) upward stroke
aufschließen *chem* dissolve, disintegrate, break up
Aufschluss *chem* dissolution, disintegration
Aufschluss *geo* outcrop
Aufschluss/Zellaufschluss (Öffnen der Zellmembran) cell lysis
Aufschluss/Zellfraktionierung cell fractionation
Aufschluss/Zellhomogenisierung cell homogenization
aufschmelzen/schmelzen melt
Aufsiedler/Symphoriont epizoon, symphoriont
aufsitzend/sesshaft sessile
Aufsitzerpflanze/Epiphyt/ Luftpflanze epiphyte, air plant, aerial plant, aerophyte
aufspalten/segregieren *gen* segregate
aufspalten/spalten/öffnen *chem* crack, break down, open
aufspalten/verteilen distribute
aufspalten/zerlegen *chem* split
Aufspaltung/Öffnen *chem* cracking, opening
Aufspaltung/Segregation *gen* segregation
• **Nicht-Mendelsches Aufspaltungsverhältnis** non-Mendelian ratio
Aufspaltung/Zerlegen *chem* splitting
aufsperren (Schnabel) gape, gaping
Aufspießen *ethol* impaling (shrikes)
Aufspringen/Dehiszenz dehiscence
aufsteigend afferent, rising
auftauen *vb* thaw

Auftauen

Auftauen *n* thawing
auftragen/"plotten" *math/geom* plot
auftragen/applizieren *chromat* apply
Auftragung/Applikation *chromat* application
auftrennen/trennen/fraktionieren separate, fractionate
Auftrennung/Trennung/Fraktionierung separation, fractionation
Auftrieb *mar* (vertikale Aufwärtsströmung) upwelling
Auftrieb (im Wasser: Fische) buoyancy
Auftrieb (in Luft: Vögel) lift
aufwachsen grow up
aufwärts (Richtung 5′-Ende eines Polynucleotids) upstream
  • **abwärts (Richtung 3′-Ende eines Polynucleotids)** downstream
aufwinden coil up
Aufwinden coiling
Aufwuchs/Bewuchs/Periphyton *limn* periphyton, attached algae
Aufwuchs/Nachkommenschaft descendants, descendents
Aufwuchs/Waldanpflanzung/Schonung *for* protected young forest plantation
aufziehen/erziehen bring up, rear, foster, nurture
Aufziehen/Erziehen/Aufzucht upbringing, rearing, fostering, nurture, nurturing
Aufzucht rearing, fostering, nursing
Aufzuchtbeet *hort* nursery bed
Augapfel/Bulbus oculi eyeball
Auge eye; *bot* (Holz) knot; (Knospe) *bot* eye, node, bud
  • **Appositionsauge** apposition eye
  • **Becherauge/Pigmentbecherauge** pigment cup eye, inverted eye
  • **Blasenauge** retinal cup eye, everted eye
  • **Einzelauge/Punktauge/Ocelle/Ocellus (siehe: Nebenauge/Scheitelauge/Stirnauge)** simple eye, ocellus (dorsal and lateral)
  • **Einzelauge/Punktauge/Stemma (Seitenauge/Lateralocellus bei Insekten-Larven)** stemma (*pl* stemmata/stemmas), lateral ocellus, lateral eye
  • **Facettenauge/Komplexauge/Netzauge/Seitenauge** compound eye, facet eye
  • **Hauptauge** main eye
  • **Lateralauge/Seitenauge (Ocellus)** lateral eye, lateral ocellus
  • >**Stemma (Punktauge/Einzelauge bei Insekten-Larven)** stemma (*pl* stemmata/stemmas), lateral eye/ocellus
  • **Linsenauge** lens eye, lenticular eye
  • **Mittelauge/Medianauge (Stirnocelle)** median eye, midline eye (a dorsal ocellus)
  • **Naupliusauge** naupliar eye (median eye)
  • **Nebenauge/Punktauge/Ocelle** simple eye, ocellus
  • **Netzauge/Facettenauge/Komplexauge/Seitenauge** facet eye, compound eye
  • **Parietalauge (*Sphenodon*)** parietal eye
  • **Pinealauge (bei Neunaugen: *Petromyxon*)** pineal eye, epiphyseal eye (median eye)
  • **Punktauge/Nebenauge/Ocelle** simple eye, ocellus
  • **Seitenauge/Lateralauge/Lateralocelle/Stemma (Ocellen)** lateral eye/ocellus, stemma
  • **Stielauge** stalked eye
  • **Superpositionsauge** superposition eye
  • >**neurales S.** neural superposition eye
  • >**optisches S.** optical superposition eye, clear-zone eye
  • **Turbanauge (*Ephemeroptera*)** stalked compound eye
Augenbecher *embr* eyecup, optic cup
Augenbecherstiel *embr* optic stalk
Augenbewegung eye movement
Augenblase/Augenbläschen/Vesicula ophthalmica *embr* optic vesicle
Augenbraue eyebrow
Augenbutter/Augenschmalz/Sebum palpebrale (Meibom-Drüsensekret) palpebral sebum ("sleep")
Augendeckel/Augenlid/Lid eyelid
Augenfacette eye facet
Augenflagellaten/Euglenophyta euglenoids, euglenids
Augenfleck/Stigma eyespot, stigma
Augengruß *ethol* eyebrow flash, eyebrow raise
Augenhaut, harte/Lederhaut/Sklera sclera, sclerotic coat, sclerotica
Augenhöhle/Orbita orbit, eyepit, eye socket
Augenkammer eye chamber
Augenlid/Lid/Augendeckel/Palpebra eyelid

Augenlinse/Okularlinse eye lens, ocular lens

Augensteckling eye cutting, single eye cutting, bud cutting, leaf bud cutting

Augenstiel eyestalk

Augenveredlung/Äugeln/ Okulieren/Okulation *hort* bud grafting, budding

Augenwimper/Wimper eyelash

Augenzahn (oberer Eckzahn) eyetooth (canine tooth of upper jaw)

Augenzwinkern wink of the eye

Augspross (Geweih) brow tine

Aurikel/Atrium/Herzvorhof/ Herzvorkammer auricle, atrium

Aurikel/Pollenschieber (Bienen) auricle, pollen press, pollen packer

ausatmen *vb* expire, exhale, breathe out

Ausatmen/Ausatmung/Expiration/ Exhalation expiration, exhalation

Ausbeute/Ertrag yield

ausbeuten (Rohstoffe) exploit

ausbleichen/bleichen bleach

ausbleichen *vb* (*passiv*, z. B. **Fluoreszenzfarbstoffe**) fade (*siehe* bleichen)

Ausbleichen *n* (*passiv*, z. B. **Fluoreszenzfarbstoffe**) fading (*siehe* bleichen)

ausbreiten (z. B. Krankheit/Flügel) spread

Ausbreitung (z. B. Krankheit/Flügel) spread, spreading

Ausbreitung/Propagation spreading, expansion; propagation, dispersal, dissemination

Ausbreitungseinheit/ Propagationseinheit/ Fortpflanzungseinheit/Diaspore dispersal unit, propagule, diaspore

Ausbreitungsgebiet area of expansion

ausbrüten (Eier/Junge) hatch, brood

Ausbuchtung outpocketing, protrusion

Ausdauer/Dauerhaftigkeit endurance, persistence, hardiness, perseverance

ausdauernd (wiederstandsfähig) hardy, persistent, enduring

ausdauernd/perennierend (mehrjährig) perennial

Ausdehnung/Verlängerung extension

Ausdrucksverhalten expressive behavior

ausdünnen *vb* thin

Ausdünnen/Ausdünnung thinning

Auseinandersetzung/Konflikt conflict

ausfällen/fällen precipitate

Ausfällung/Ausfällen/Fällung/ Fällen precipitation

Ausfluss/Abfluss *techn* discharge, outflow, efflux, draining off

Ausfluss *med* discharge, secretion, flux

Ausformung/Aushaltung/ Holzaushaltung (Holzstamm-Zuschnitt) bucking

ausführen/wegführen/ableiten (Flüssigkeit) discharge, drain, lead out, lead/carry away

ausführend/wegführend/ableitend (Flüssigkeit) efferent

Ausführgang/Ausführkanal duct, passageway

Ausgangs-DNA source DNA

Ausgangsgestein/Grundgestein/ Muttergestein bedrock, rock base, parent rock

Ausgangspopulation initial population

Ausgangsprodukt primary product, initial product

Ausgangsstoff/Ausgangsmaterial starting material, basic material, source material, primary material

Ausgangsstoff/ Reaktionsteilnehmer/Reaktand reactant

Ausgangsverteilung *stat* initial distribution

ausgeizen *hort* removing side shoots, removing suckers

ausgerandet emarginate

Ausgesetztsein/Gefährdung exposure

ausgestorben extinct, died out

ausgewachsen full-grown

ausgleichende Reversion *gen* second site reversion

ausgraben *geol* excavate, dig (out)

Ausgrabung *geol* excavation, dig

aushärten/vulkanisieren *chem* **(Polymere)** cure, vulcanize

aushungern starve

auskreuzen/herauskreuzen *gen* cross out

Auskreuzen/Herauskreuzen *gen* outcrossing

auslaufend/astlos in die Spitze auslaufend (Baum) excurrent

Ausläufer/Ausläuferspross/Stolon/ Stolo *allg* stolon

Ausläufer/Erdspross (unterirdischer Stolon/geophil) rhizome

Ausläufer/Kriechspross (oberirdischer Stolon/photophil) runner, sarment

Ausläufer bildend stoloniferous, sarmentose

Ausläuferknolle stolonial tuber
Ausläufersspross/Stolon/Stolo stolon
auslaugen (Boden) leach
Auslaugung (Boden) leaching
Auslese/Selektion selection
- **disruptive Auslese** disruptive selection, diversifying selection
- **frequenzabhängige Auslese** frequence-dependent selection
- **gerichtete Auslese** directional selection
- **geschlechtliche/sexuelle Auslese** sexual selection
- **künstliche Auslese** artificial selection
- **natürliche Auslese** natural selection
- **stabilisierende Auslese** stabilizing selection

Auslesezüchtung/ Schwellenwertselektion/ Kappungsselektion truncation selection
auslichten/zurückschneiden thin out, prune
Auslösemechanismus (AM) *ethol* releasing mechanism (RM)
- **angeborener A. (AAM)** innate releasing mechanism (IRM)
- **erworbener A. (EAM)** acquired releasing mechanism (ARM)

auslösen (Reaktion) trigger, elicitate
Auslöser *ethol* releaser
Auslöser/Elicitor (Reaktion) trigger, elicitor (e.g. stimulating phytoalexin production)
Auslösung (Reaktion) triggering, elicitation
ausmerzen eliminate, eradicate
ausnehmen/ausweiden (Fische) gut, eviscerate, disembowel
Ausreißer *allg* runaway, fugitive
Ausreißer *stat* outlier
ausrotten/ausmerzen eradicate, eliminate, extirpate
Ausrottung/Ausmerzung *med* **(z. B. Schädlinge)** eradication, elimination, extirpation
Aussaat/Saat/Saatgut (das Ausgesäte) seed(s)
Aussaat/Säen/Aussäen sowing, seed sowing
Aussackung/Divertikel diverticulum
aussäen/säen sow
Aussalzchromatographie salting-out chromatography
aussalzen *vb* salt out
Aussalzen *n* salting out
ausscheiden *allg* secrete
ausscheiden (Exkrete/Exkremente) egest, excrete

Ausscheider (Blutgruppenantigene) secretor
Ausscheidung *allg* secretion
Ausscheidung (Exkrete) excretion
Ausscheidung/Exkretion egestion, excretion
Ausscheidungen/Exkrete/ Exkremente excreta, excretions
Ausscheidungsgewebe excretory/secretory tissue
Ausscheidungsorgan/ Exkretionsorgan excretory organ
ausschlagen/sprießen (Bäume/ Blätter) leaf (leafing), bud (budding), sprout (sprouting); (Wurzel/Spross) sprout
Ausschlagswald coppice forest, sprout forest
Ausschlagvermögen *bot* budding potential/rate
Ausschleuderorganelle/Extrusom extrusive organelle, extrusome
ausschlüpfen hatch
Ausschluss/Exklusion exclusion
Ausschluss der Vaterschaft paternity ecxlusion
Ausschlussprinzip exclusion principle
Ausschnittszeichnung cutaway drawing
ausschütteln shake out
Ausschüttelung shaking out
Ausschüttung (z. B. Hormone/ Neurotransmitter) release
- **Blattausschüttung** (rapid) leaf flushing, bud bursting

ausschwärmen swarm, swarming
Ausschwingrotor *centrif* swinging-bucket rotor
Außengruppe (Kladistik) outgroup
Außenhaut/Exodermis exodermis
Außenkelch/Hochblatthülle epicalyx, sepal-like bracts
Außenkelch/Nebenkelch calycle, calyculus
Außenlade/Galea galea, outer lobe of maxilla
Außenparasit/Exoparasit/ Ektoparasit (Hautparasit) exoparasite, ectoparasite, epizoon
Außenreiz external stimulus
Außenschale/Ectocochlea external shell, ectocochlea
Außenschicht outer layer, exterior layer
Außenschicht/Exine (Pollen/Spore) exine
Außenskelett/Hautpanzer/ Exoskelett dermal skeleton, dermatoskeleton, exoskeleton
äußerlich/von außen/extern external, extrinsic

außerplanmäßige DNA-Synthese unscheduled DNA synthesis
außerzellulär/extrazellulär extracellular
aussetzen (einem Schadstoff/ einer Strahlung aussetzen) expose to (hazardous chemical/ radiation)
Ausspritzungsgang/ Samenausführgang/Samengang/ Ductus ejaculatorius ejaculatory duct
ausstellen exibit, show, display
Ausstellung exhibition, show, display
aussterben become extinct, die out
Aussterben extinction, dying out
ausstreichen *micb* (z. B. Kultur) streak, smear
ausstreuen disseminate, disperse, spread, release
Ausstreuung dissemination, dispersal, spreading, releasing
Ausstrich *micb* smear
Ausstrichkultur/Abstrichkultur *micb* streak culture, smear culture
Ausstrom efflux
Ausströmöffnung (Egestionsöffnung) excurrent/ exhalant aperture (egestive aperture)
ausstülpbar (z. B. Pharynx der Turbellarien) protrusive
ausstülpen evert, evaginate, protrude, turn inside out
Ausstülpung evagination, protrusion; (z. B. Darmdivertikel) outpocketing
austarieren (Waage: Gewicht des Behälters/Verpackung auf Null stellen) tare (determine weight of container/packaging as to substract from gross weight > set reading to zero)
Austausch exchange
Austausch (zwischen Chromosomen)/reziproke Translokation/interchromosomale Umordnung interchange, interchromosomal rearrangement
Austauschreaktion exchange reaction
Austrag *ecol* output
Australheidegewächse/ Epacridaceae epacris family
austreiben sprout, put forth; (Blätter) producing leaves, coming into leaf
Austrieb sprout, sprouting, budding
Austrittspupille/Augenpunkt *micros* exit pupil, eyepoint
austrocknen/entwässern desiccate, dry up, dry out

Austrocknung/Entwässerung desiccation
Austrocknungsschaden/Trocknis desiccation damage
Austrocknungstoleranz desiccation tolerance
Austrocknungsvermeidung desiccation avoidance
Auswaschung (feste Bodenbestandteile in Suspension) eluviation
Auswaschung (gelöste Bodenmineralien) leaching
ausweiden/ausnehmen (z. B. Fisch) gut, eviscerate, disembowel
auswerten (z. B. Ergebnisse) evaluate (e. g. results)
Auswertung (z. B. von Ergebnissen) evaluation (e. g. of results)
auswiegen (genau wiegen) weigh out precisely
Auswuchs *allg* outgrowth, protrusion; (Höcker/Beule) protuberance; (bei *Lycophyta*) enation
• **halbkugelförmiger Auswuchs (an bestimmten Bäumen)** lignotuber, burl, woody outgrowth
Auszackung serration
Auszehrphase starvation phase
Auszug/Extrakt extract
Autapomorphie (evolutive Neuheit) autapomorphy
Autoalloploidie autoalloploidy
Autoantigen autoantigen, self-antigen
Autoantikörper autoantibody
Autogamie autogamy, orthogamy, self-fertilization
autogene Kontrolle autogeneous control
autoimmun autoimmune
Autoimmunität autoimmunity
Autoimmunkrankheit autoimmune disease
Autokatalyse autocatalysis
Autoklav autoclave
autoklavieren autoclave
Autökologie autecology
autokrin/autocrin autocrine
autolog autologous
Autolyse autolysis
Automarkieren/Selbstmarkieren *ethol* automarking, self-marking
Automatiezentrum (Sinusknoten) automaticity center (sinus node)
automiktisch automictic
autonom replizierende Sequenz (ARS) *gen* autonomous(ly) replicating sequence (ARS)
autonomes Kontrollelement autonomous control element

Autoparasit

Autoparasit autoparasite
Autophagie autophagy
Autophagosom autophagosome, autosome
Autophilie autophily, self-pollination
autopolyploid autopolyploid
Autoradiographie autoradiography, radioautography
Autosom autosome
autosomal autosomal
autosomal-dominant autosomal-dominant
autosomal-rezessiv autosomal-recessive
Autotomie/Selbstverstümmelung autotomy
autotomieren/selbst verstümmeln autotomize
Autotransplantat autograft (autologous graft)
autotroph autotroph, autotrophic
Autotrophie autotrophy
autözisch autoecious
Autozygotie autozygosity
Auwald/Auenwald floodplain forest
Auwiese/Auenwiese (Überschwemmungswiese) riverine floodplain meadow
Auxine auxins
auxotroph auxotrophic
Auxotrophie auxotrophy
Avicularie/"Vogelköpfchen" avicularium (*pl* avicularia)
Avidität avidity
Axialfilament axial filament
Axialkorn/Axosom axosome
Axialorgan (Axialdrüse/braune Drüse) (Echinoderma) axial organ, axial gland
Axialskelett/Achsenskelett axial skeleton
Axialzelle/Zylinderzelle (Mesozoa) axial cell
Axis/Epistropheus/ zweiter Halswirbel axis, second cervical vertebra
Axocöl/Axocoel/Protocöl/Protocoel protocoel
Axolemm/Axolemma/ Mauthnersche Scheide axolemma
Axonema/Axonem/Achsenfaden axoneme, axial rod, axial complex
Axonhügel/Axonkegel axon hillock
Axopodium axopodium
Axosom/Axialkorn axosome
Azelainsäure azelaic acid
azentrisches Chromosom acentric chromosome
azeotrop azeotropic
azeotropes Gemisch azeotropic mixture
azid/acid/sauer acid
Azidität/Acidität/Säuregrad acidity
Azidose/Acidose acidosis
Azollaceae/Algenfarngewächse duckweed fern family, mosquito fern family
AZT (Azidothymidin) AZT (azidothymidine)

Balkenzunge

B-Form/B-Konformation (der DNA)
B form (of DNA)
B-Lymphocyt/B-Zelle B lymphocyte,
B cell
B-Zelle/B-Lymphozyt/B-Lymphocyt
B cell, B lymphocyte (B = Bursa)
● **unreife B-Zelle** virgin B cell
BAC (künstliches
Bakterienchromosom) BAC
(bacterial artificial chromosome)
Bach brook, creek
Bache/Wild(schwein)sau wild sow
Backe/Wange/Gena cheek, gena
Backen/Hitzebehandlung baking,
heat treatment
Backendrüse cheek pouch (gland)
Backenknochen/Os zygomaticum
cheekbone, zygomatic bone,
malar bone
Backenknochenbogen
zygomatic arch
Backenregion/Jochbeingegend *orn*
malar region
Backentasche (z. B. Hamster)
cheek pouch
Backenzahn/Molar cheek tooth,
molar, grinder
● **Prämolar/vorderer Backenzahn**
premolar (bicuspid tooth)
● **Weisheitszahn/dritter Molar**
wisdom tooth, third molar
Bäckerhefe baker's yeast
Backhefe/Bäckerhefe baker's yeast
Badedermatitis/Cercariendermatitis
swimmer's itch
Bahn/Tractus (Nervenbahn) *neuro/*
anat tract
bahnen (bahnend) *neuro*
facilitate (facilitating)
Bahnung/Facilitation facilitation
Bahnungsneuron
facilitator neuron
Bakteriämie bacteremia
Bakterie/Bakterium (*pl* **Bakterien)**
bacterium (*pl* bacteria)
● **Bazillen/Bacillen (Stäbchen)**
bacilli (*sg* bacillus) (rods)
● **denitrifizierende Bakterien/**
Denitrifikanten
denitrifying bacteria
● **Fäulnisbakterien**
putrefactive bacteria
● **Knallgasbakterien/**
Wasserstoffbakterien
hydrogen bacteria (aerobic
hydrogen-oxidizing bacteria)
● **Knöllchenbakterien**
nodule bacteria
● **Kokken/Coccen (kuglig)** cocci
(*sg* coccus) (spherical forms)
● **Leuchtbakterien**
luminescent bacteria

● **Myxobakterien/**
Schleimbakterien myxobacteria
● **Rickettsien (Stäbchen- oder**
Kugelbakterien)
rickettsias, rickettsiae (*sg* rickettsia)
(rod-shaped to coccoid)
● **Schwefelbakterien**
sulfur bacteria
● **Spirillen (schraubig gewunden)**
spirilla (*sg* spirrilum)
(spiraled forms)
● **stickstofffixierende/**
stickstofffixierende Bakterien
nitrogen-fixing bacteria
● **Vibrionen (meist gekrümmt)**
vibrios (mostly comma-shaped)
● **wärmesuchende Bakterien/**
thermophile Bakterien
thermophilic bacteria
bakteriell bacterial
bakterielle Infektion
bacterial infection
Bakterienflora bacterial flora
bakterienfressend bacterivorous,
bactivorous
Bakterienknöllchen bacterial nodule
Bakterienkultur bacterial culture
Bakterienrasen bacterial lawn
Bakteriologie bacteriology
bakteriologisch bacteriologic,
bacteriological
Bakteriophage/Phage
bacteriophage, phage,
bacterial virus
Bakteriose bacteriosis
bakterizid/keimtötend
bacteriocidal, bactericidal
balanciert letal balanced lethal
balancierte Translokation
balanced translocation
balancierter Polymorphismus
balanced polymorphism
Balanophoraceae/
Kolbenträgergewächse/
Kolbenschmarotzer
balanophora family
Balbiani-Ringe Balbiani rings
Baldachinnetz (mit Stolperfäden)
arach sheetweb/dome web
(with barrier threads)
Baldriangewächse/Valerianaceae
valerian family
Balg/Balgfrucht/Follikel (eine
Fruchtform) follicle
Balgzelle/tormogene Zelle
(Arthropodenintegument)
tormogen cell (socket-forming cell)
Balken beam
Balken/Corpus callosum callosum,
corpus callosum
Balkenzunge/docoglosse Radula
docoglossate radula

B Ballaststoffe

Ballaststoffe (dietätisch)
dietary fiber
Ballen (Heu/Stroh etc.) bale
Ballen (Finger/Fuß/Zehen/Sohlen)
pad (finger/foot/toe/paw)
Ballon (für Flüssigkeiten) *chem/lab*
bottle with faucet
(carboy with spigot)
**Ballung (lokale Häufung/
Kumulation/Aggregation)**
concentration, accumulation,
aggregation
Balsam balsam (a plant exudate)
Balsambaumgewächse/Burseraceae
torchwood family, incense tree
family, frankincense tree family
**Balsaminengewächse/
Springkrautgewächse/
Balsaminaceae**
balsam family, jewelweed family,
touch-me-not family
Balz courtship, mating behavior;
display
Balzarena lek (communal mating
ground)
Balzgesang courtship song,
mating song
Bananengewächse/Musaceae
banana family
Band/Ligament ligament
Bande *electrophor/chromat* band
• **Hauptbande** main band
• **Satellitenbande** satellite band
**Bänderungsmuster/Bandenmuster
(von Chromosomen)**
banding pattern
Bänderungstechnik
banding technique
bandförmig band-shaped, fascial,
fasciate
Bandhaft/Syndesmose syndesmosis
bändigen (wilde Tiere) break, tame
**Bandscheibe/
Zwischenwirbelscheibe/Discus
intervertebralis** intervertebral disk
Bandwürmer/Cestoden tapeworms,
cestodes
Bandzunge/taenioglosse Radula
taenioglossate radula
Bank/Bibliothek/Klonbank library,
bank (clone bank)
Bannwald (in Austria) protected
forest for stabilizing slopes etc.
**Bannwald/Naturwaldreservat
(in S/W Germany)** protected forest
(no commercial usage)
**Barbadoskirschengewächse/
Malpighiengewächse/
Malpighiaceae** malpighia family,
Barbados cherry family
Barberfalle *ecol* Barber trap,
pitfall trap

**Bärentierchen/Bärtierchen/
Tardigrade** water bears, tardigrades
**Bärlappbäume/Lepidophyten/
Lepidodendrales** lepidophyte trees,
club-moss trees
Bärlappgewächse/Lycopodiaceae
club mosses, lycopods,
clubmoss family
barophil barophilic, barophilous
barophiler Organismus barophile
Barrierefunktion barrier function
Barriereriff/Wallriff barrier reef
Barrkörperchen Barr body
**Barschfische/Barschartige/
Perciformes**
perch & perchlike fishes
Barschlachse/Percopsiformes
pirate perch and freshwater relatives
Bart beard
"Bart"/Maisgriffel/Griffelfäden *bot*
silk (corn stigma-style)
Barteln barbels, barbs, beard
Barten baleen plates
(cornified tissue sheets)
Bartenwale/Mysticeti
whalebone (baleen) whales
Bartfäden/Barthaare *ichth* barbs,
barbels, whiskers; wattles
Barthaare (Katzen) whiskers
**Bartholin-Drüse/Gl. vestibularis
major** Bartholin's gland,
greater vestibular gland
**Bärtierchen/Bärentierchen/
Tardigraden (*sg* Tardigrad *m*)**
water bears, tardigrades
Bartwürmer/Bartträger
beard worms, beard bearers,
pogonophorans
Basalfalte/Plica basalis basal fold
**Basalganglion/Stammganglion/
Corpus striatum** basal ganglion,
cerebral nucleus
**Basalkern/Basalkörper/
Streifenkörper/Corpus striatum**
striate body, corpus striatum
**Basalkörper/Kinetosom/
Blepharoplast** basal body, flagellar
basal body/corpuscle/granule,
kinetosome, mastigosome,
blepharoplast, blepharoblast
Basalmembran/Basallamina
basement membrane, basal lamina
Basalplatte basal plate
Basalschicht basal layer, basal zone
**Basalumsatz/
basale Stoffwechselrate/
Grundstoffwechselrate/
Grundumsatz** basal metabolic rate,
base metabolic rate (BMR)
Basalzelle basal cell
Basalzellschicht/Stratum basale
basal layer

Basapophyse basapophysis
Base base
• **stickstoffhaltige Base/"Base"
(Purine/Pyrimidine)**
nitrogenous base
Baseität/Basizität basicity
**Basellaceae/
Schlingmeldengewächse**
Madeira vine family, basella family
Basenanalogon (pl Basenanaloga)
base analogue, base analog
Basenaustausch base substitution
Basendefizit base deficit
Basenfehlpaarung/Fehlpaarung
mismatch
basenhold/basophil basophile,
basophilic, basophilous
basenmeidend basifuge, basifugous
(calcifuge/calcifugous)
Basenpaar base pair
Basenpaarung base pairing
Basenpaarungsregeln
base-pairing rules
Basenstapelung base-stacking
Basensubstitution
base substitution
Basenüberschuss base excess
Basenzusammensetzung
base composition, base ratio
Basidie (pl Basidien) basidium
(pl basidia)
basifix (Anthere) basifixed
Basilarmembran basilar membrane
basipetal basipetal
Basipetalie basipetal development
basiplast basiplastic
basiplastes Wachstum
basiplastic growth
Basipodit basipodite
basisch/alkalisch basic, alkaline
Basisnährboden basal medium
Basitonie basitony
Basizität/Baseität basicity
Bast (Geweih) velvet
Bast/sekundäres Phloem bast,
secondary phloem, secondary bark
Bastard/Hybride bastard, hybrid
Bastardisierung/Hybridisierung
bastardization, hybridization
Bastardsterilität/Hybridensterilität
hybrid sterility
Bastardwüchsigkeit/Heterosis
hybrid vigor, heterosis
Bastfaser bast fiber
Baststrahl bast ray
Bastteil/Siebteil/Phloem phloem
Batessche Mimikry
Batesian mimicry
Bathyal/Bathyalzone mar
(Kontinentalrand/-abhang)
bathyal zone (upper: continental
slope; lower: continental rise)

**Bathypelagial/mittlerer
Tiefseebereich/mittlere
Tiefseezone** bathypelagic zone
Batisgewächse/Bataceae
batis family, saltwort family
Bau/Bauplan construction,
body plan
Bau/Höhle/Erdhöhle/Loch burrow,
hole (rabbit), earth, cave
Bau/Lager (höhlenartig)
lodge (beaver), couch (otter)
Bau/Lager/Rastplatz lair (resting/
living place: game/wild animal),
den (bear/lion; often a hollow or
cavern)
Bauch/Abdomen belly, venter,
abdomen
Bauch (Archegonium) bot venter
Bauch.../bauchseitig/ventral
ventral
Bauchatmung/Zwerchfellatmung
abdominal breathing,
diaphragmatic respiration
**Bauchblatt/Amphigastrium
("Unterblatt")** ventral leaf,
underleaf, amphigastrium
Bauchfell/Peritoneum peritoneum
Bauchflosse ventral fin, pelvic fin
**Bauchfuß/Propes/Pes spurius
(larvales Abdominalbein)** proleg,
ventral proleg
**Bauchfüßer/Schnecken/
Gastropoden/Gastropoda** snails,
gastropods
**Bauchhaarlinge/Bauchhärlinge/
Flaschentierchen/Gastrotrichen**
gastrotrichs
Bauchhöhle abdominal cavity,
peritoneal cavity
Bauchkanalzelle venter cell
**Bauchmark/Bauchmarkstrang/
ventraler Nervenstrang**
ventral nerve cord
**Bauchnaht/Ventralnaht (des
Fruchtblattes)** ventral suture/seam
**Bauchnervenknoten/
Ventralganglion** ventral ganglion
Bauchnetz/Netz/Omentum
omentum
**Bauchpanzer/Bauchplatte/
Brustschild/Plastron
(Schildkröten/Vögel)** plastron,
plastrum
Bauchpilze/Boviste/"Stäublinge"
puffballs,
globose puffball-type fungi
**Bauchpilze/Gasteromycetes/
Gastromycetales** stomach fungi,
gastromycetes, angiocarps
**Bauchplatte/Sternit (Insekten:
ventraler Sklerit)** sternite,
ventral sclerite

B Bauchrippen 408

Bauchrippen/Gastralrippe/Gastralia (Reptilien) abdominal ribs, gastralia
Bauchschild/Bauchteil/Brustplatte/ Sternum (Insekten) sternum, ventral plate
Bauchschuppe/Ventralschuppe ventral scale
Bauchseite belly, undersurface of animal's body
bauchseitig/Bauch.../ventral ventral
Bauchspeichel/Pankreassaft pancreatic juice
Bauchspeicheldrüse/Pankreas pancreas
bauchständig ventrally located
Bauchstück/Sternit sternite
Bauchtasche ventral pouch
Bauchteil/Bauchschild/Brustplatte/ Sternum (Insekten) sternum, ventral plate
Bauholz (structural) timber, lumber
Baum tree
Baum.../baumartig treelike, arboreal
- **Formbaum/Formstrauch (auch Zierschnitt)** topiary
- **Forstbaum** forest tree
- **Heister (junger Laubbaum aus Baumschule)** sapling
- **Hochstamm (Wuchsform eines Baumes)** standard tree, standard
- **Laubbaum** broadleaf, broadleaf tree, hardwood, hardwood tree (*pl* broadleaves/hardwoods)
- **Nadelbaum/Konifere/Conifere** coniferous tree, conifer, (softwood tree)
- **Obstbaum** fruit tree, fruit-bearing tree
- **Parkbaum** park tree
- **Schnurbaum/ Schnurspalierbaum/Kordon** cordon
- **Schopfbaum** tree with terminally tufted leaves
- **Zierbaum** ornamental tree
- **Zwergbaum** dwarf tree
baumartig arboroid, dendroid, dentritic
baumartig verwachsen/verzweigt/ sich ausbreitend arborescent
Baumbestand tree stand, stand/ number of trees
- **alter B.** old-growth forest
- **dichter B.** close-set stand, dense stand
baumbewohnend tree-dwelling, living in trees, arboreal, arboricolous
Bäumchen/Bäumlein sapling

Bäumchenschnecken/ Dendronotacea dendronotacean snails, dendronotaceans
Baumfarn tree fern
Baumfarngewächse (Cyatheaceae & Dicksoniaceae) tree fern family (cyathea family & dicksonia family)
Baumgarten orchard
Baumgrenze timberline, tree line
Baumhain orchard, grove
Baumharz tree resin
Baumkrebs canker
Baumkrone treetop, crown
Baumkronenbereich canopy
Baumkunde/Gehölzkunde/ Dendrologie (*sensu stricto:* Taxonomie der Holzgewächse) dendrology
Baumliliengewächse/ Velloziagewächse/Velloziaceae vellozia family
baumlos treeless
Baumplantage tree farm, plantation
Baumsavanne tree savanna
Baumschicht tree stratum
Baumschnitt tree pruning
Baumschule tree nursery
Baumschwamm, konsolenförmiger/ Baumpilz bracket fungus, shelf fungus, tree fungus
Baumstamm stem, trunk, bole
- **gefällter Baumstamm** log
Baumstumpf/Stubbe/Stock/ Stumpen tree stump, stub, "stool"
Baumstumpf/toter stehender Baum (in Sümpfen) snag
Baumwipfel treetop
Baumwipfelzone (Wald) canopy
Baumwürgergewächse/ Spindelbaumgewächse/ Celastraceae spindle-tree family, staff-tree family, bittersweet family
Baumzucht (speziell Ziersträucher/ Zierbäume) arboriculture
Baumzüchter arborist, arboriculturist
Bauplan body plan, construction, structure; blueprint
Baustein/Bauelement building block
bazillär/Bazillen.../bazillenförmig/ stäbchenförmig bacillary
beblättert in leaf, leaved, bearing leaves, foliar
Beblätterung/Belaubung foliation
bebrüten/brüten/inkubieren brood, breed, incubate
Bebrütung/Bebrüten/Inkubation incubation
Becherauge/Pigmentbecherauge pigment cup eye, inverted eye
Becherchen/Helotiaceae helotium family

Becherfarne/Becherfarngewächse/ Cyatheaceae cup fern family, tree fern family
becherförmig cup-shaped, cyathiform
Becherfrüchtler/Buchengewächse/ Fagaceae beech family
Becherglas/Zylinderglas beaker
Becherglaszange *lab* beaker tongs
Becherhaar/Trichobothrium trichobothrium
Becherkätzchengewächse/ Garryaceae silk-tassel tree family, silktassel-bush family
Becherkeim/Becherlarve/Gastrula gastrula
Becherquallen/Stielquallen/ Stauromedusae stauromedusas
Becherzelle/Schleimzelle (mucus-producing) Goblet cell
Becken *geol* basin
Becken/Pelvis pelvis
Beckenboden pelvic floor
Beckengürtel pelvic girdle, hip girdle
Beckenhöhle pelvic cavity
Beckenknochen (Darmbein & Sitzbein) pelvic bone, pelvis (ilium & ischium)
Bedampfung/Bedampfen/ Aufdampfen *micros* vapor blasting
Bedampfungsanlage *micros* vaporization apparatus
Bedecktsamer/Decksamer/ Angiosperme angiosperm, anthophyte
bedingt conditional
bedrohen *ethol* threaten
bedrohen/gefährden *ecol* endanger
bedroht *ecol* endangered (threatened by extinction)
bedroht *ethol* threatened (e.g. by an attack)
bedrohte Art *ecol* endangered species
Bedrohung threat; endangerment
Beere berry
Beerenobst berries
Beerensträucher fruit-bearing shrubs
Beerenverband/Beerenfruchtstand sorosis, fleshy multiple fruit
Beerenzapfen/Zapfenbeere fleshy cone, "berry"
Beet bed, patch
 • **Blumenbeet** flowerbed
 • **Frühbeet/Anzuchtkasten (unbeheizt)** cold frame
 • **Frühbeet/Mistbeet/Treibbeet (beheizt)** forcing bed, hotbed
 • **Gemüsebeet** (vegetable) patch

Befall (Schädlingsbefall) infestation (with pests/parasites)
 • **Wiederbefall** reinfestation
befallen (Schädlingsbefall) infest (pests/parasites)
Befallsrate degree/level/rate of infestation
Befestigungszone (Radnetz) *arach* strengthening zone, notched zone
befeuchten moisten, humidify, dampen
Befeuchtung moistening, humidification, dampening
befranst fimbriate(d), fringed
Befriedungsgebärde appeasement gesture
Befriedungsritual appeasement ritual
befruchten fertilize, fecundate
befruchtet fertilized
 • **unbefruchtet** unfertilized
Befruchtung/Fekundation fecundation
Befruchtung/Fertilisation *sensu stricto:* fertilization (pollination *see* Bestäubung)
Befruchtungshügel fertilization cone
Befruchtungsmembran fertilization membrane
Befund findings, result
begasen fumigate
Begasung fumigation
begatten/paaren/kopulieren mate, copulate
Begattung/Kopulation sexual union, copulation
Begattungsakt/Kopulationsakt/ Geschlechtsverkehr/Koitus sexual intercourse, copulation, coitus
Begattungsaufforderung soliciting behavior
Begattungsfaden (männl. Spinne) mating line
Begattungsfuß/Genitalfuß/ Gonopodium gonopodium, gonopod
Begattungsorgan (männliches)/ Penis copulatory organ, intromittent organ, penis
Begattungstasche/Bursa copulatrix genital pouch, bursa copulatrix
Begattungstasche/Sphragis sphragis
Begattungsvorspiel precopulatory rite, precopulatory behavior
Begehung/Besichtigung (z. B. Geländebegehung) inspection (on-site inspection)
begeißelt/flagellat flagellated
Begeißelung flagellation

B Begleitprotein 410

**Begleitprotein/Chaperonin/
molekulares Chaperon** chaperone
protein, chaperone,
molecular chaperone
Begleitzelle *bot* companion cell
Begoniaceae/Schiefblattgewächse
begonia family
begrannt/Grannen tragend awned,
aristate
**begrenzender Faktor/
limitierender Faktor/Grenzfaktor**
limiting factor
Begrüßungsverhalten
greeting behavior
**Begutachtungsverfahren
(wissenschaftl. Manuskripte)**
peer review
behaart (haarig) pilose, piliferous,
piligerous, bearing hairs (hairy)
● **unbehaart/haarlos** hairless,
glabrous
● **weich behaart** villose, villous,
soft-haired
Behaarung hair, hairiness, pilosity;
(Haarkleid/Indument) indumentum,
hair-covering
● **Flaumbehaarung/Feinbehaarung**
pubescence
Behausung/Wohnquartier
dwelling
**Behennussgewächse/
Bennussgewächse/
Moringagewächse/
Pferderettichgewächse/
Moringaceae** horseradish tree
family
Behenöl ben oil, benne oil
Behensäure/Docosansäure
behenic acid, docosanoic acid
behuft/mit Hufen/Huf... hoofed,
hooved, ungulate
Beieierstock/Paroophoron
paroophoron
Beiknospe/akzessorische Knospe
accessory bud
Beiknospengruppe eye cluster,
bud cluster
Beimischung admixture
beimpfen/inokulieren inoculate
Beimpfung/Inokulation inoculation
Bein (Laufextremitäten) leg
Bein/Keule (Geflügel) leg,
drumstick
Bein/Knochen bone
**Beiname/zusätzliche Bezeichnung/
Zusatzbezeichnung/Epitheton
(Artname/Artbezeichnung)**
epithet (specific epithet)
Beintastler/Protura proturans
**Beischilddrüse/Nebenschilddrüse/
Epithelkörperchen**
parathyroid gland

Beischlaf cohabitation
beißen bite
beißend (z. B. mit Zähnen/Kiefer)
biting; (Geruch/Geschmack) sharp,
pungent, acrid
beißend-kauend (Mundwerkzeuge)
biting-chewing, mandibulate
(mouthparts)
**Beiwurzel/Nebenwurzel/
Adventivwurzel** supplementary
root, adventitious root
beizen (Holz) stain
beizen (Saatgut) dress (coat/treat
with fungicides/pesticides)
Beizenfärbungsmittel/Beize
mordant
Beizmittel (zur Saatgutbehandlung)
fungicide treatment,
pesticide treatment (of seeds)
Bekräftigung reinforcement
beladene Form (z. B. ATP)
loaded form
Belastbarkeit stress tolerance,
maximum stress, endurance
● **Grenze der ökologischen
Belastbarkeit/
Kapazitätsgrenze/
Umweltkapazität**
carrying capacity
belasten (belastet/verschmutzt)
contaminate(d)
Belastung (Verschmutzung)
contamination
Belastung/Traglast/Last load,
burden; (Gewicht) weight
**Belastungsfähigkeit/
Grenze der ökologischen
Belastbarkeit/Kapazitätsgrenze**
carrying capacity
Belastungsursache strain
Belastungszustand stress
Belaubung/Beblätterung foliation
Belaubung/Blattwerk foliage
beleben (belebt) animate(d)
● **unbelebt** inaminate(d), lifeless,
nonliving
● **wiederbeleben (wiederbelebt)**
reaminate(d)
Belebtschlamm/Rücklaufschlamm
activated sludge
**Belebtschlammbecken/
Belebungsbecken (Kläranlage)**
aeration tank
Belegexemplar voucher specimen
**Belegknochen/Deckknochen/
Hautknochen** dermal bone
Belegzelle (HCl-Produktion) *zool*
parietal cell, oxyntic cell
Belegzelle (Holzparenchym) *bot*
contact cell
Belemniten/Belemnitida belemnites
beleuchten illuminate

Beschleunigungsspannung B

Beleuchtung illumination
- **Auflicht/Auflichtbeleuchtung** epiillumination, incident illumination
- **Durchlicht/ Durchlichtbeleuchtung** transillumination, transmitted light illumination
- **Köhlersche Beleuchtung** Koehler illumination
- **künstliche Beleuchtung** artificial light(ing)

Beleuchtungsstärke illuminance
Belichtung (z. B. Film/Pflanzen) exposure (to light)
bellen (Hunde) bark
Belohnung (Lockmittel der Blüte) reward
Beltsche Körperchen/Futterkörper (Acacia) Beltian bodies, Belts' bodies, food bodies
belüften aerate
Belüftungsbecken (Belebungs- becken) aeration tank, aerator
benachbart/angrenzend adjacent
Benadelung needle arrangement
Benennung/Bezeichnung/ Namensgebung naming, designation, nomenclature
Benetzbarkeit wettability
benetzen wet
benigne/gutartig benign
Benignität/Gutartigkeit benignity, benign nature
Benthal/Bodenzone (von Gewässern) benthic zone
benthisch benthic
Benthos (Organismen des Benthal) benthos, benthos community
Benzoesäure (Benzoat) benzoic acid (benzoate)
Benzofuran/Cumaron benzofuran, coumarone
Benzol benzene
Berberitzengewächse/ Sauerdorngewächse/ Berberidaceae barberry family
beregnen (künstlich) sprinkle, spray
Beregnung/Bewässerung irrigation
- **B. von oben** overhead irrigation
- **künstliche B.** sprinkling irrigation

Beregnungsanlage/ Berieselungsanlage/Sprinkler sprinkler, sprinkler irrigation system
Bereicherungszone enrichment zone, paracladial zone
Bereinigung math/stat adjustment
Bereitschaftspotential readiness potential
Berghang mountainside, mountain slope

Bergheide upland moor, moorland, montane heathland
Bergkamm/Bergrücken/Berggrat mountain crest, mountain ridge
Bergkette mountain chain, mountain range
Bergregenwald montane rain forest
Bergschlucht ravine
Bergstufe/Bergwaldstufe/montane Stufe montane zone, montane region
Bergwald allg mountain forest
Bergwald (immergrüne Coniferenstufe) montane forest
Bergwiese/alpine Matte alpine meadow
berieseln sprinkle, spray; irrigate
Berieselung sprinkle irrigation
Berindung/Kortikation cortication
Beringung (z. B. Vögel) banding (*Br* ringing)
Berlese-Apparat Berlese funnel
Berlocken/Glöckchen/Appendices colli (Schaf/Ziege/Schwein) wattles
Bernstein amber
Bernsteinsäure (Succinat) succinic acid (succinate)
Beruhigungsgeste/ Beschwichtigungsgeste reassuring gesture
berühren touch, boarder
berührend touching, boardering, contiguous
besamen/inseminieren inseminate
Besamung/Insemination/ Inseminierung/Samenübertragung insemination
Besatzdichte stocking density
Beschaffenheit/Konsistenz consistency
Beschäler/Schälhengst/Zuchthengst stud, studhorse
beschallen/mit Schallwellen behandeln sonicate
Beschallung/Sonifikation/ Sonikation sonication
Beschälseuche (*Trypanosoma equiperdum*) dourine
beschalt/kleidoisch shelled, cleidoic
beschaltes Ei/kleidoisches Ei cleidoic egg, shelled egg, "land egg"
Beschattung allg shading
Beschattung (Schrägbedampfung bei TEM) shadowcasting
- **Metallbeschattung** metallizing

beschicken micb charge, feed
Beschleunigungsphase/ Anfahrphase acceleration phase
Beschleunigungsspannung (EM) accelerating voltage

B beschneiden

beschneiden cut, prune, pruning;
(Präputium/Clitoris) circumcise
Beschneiden cutting, pruning;
(Präputium/Clitoris) circumcision,
circumcising
beschnuppern/beschnüffeln
smell at, sniff at, nuzzle
**beschränkt/begrenzt/bestimmt
(Wachstum)** determinate, restricted
• **unbeschränkt/unbegrenzt/
unbestimmt (Wachstum)**
indeterminate, indefinite,
unrestricted
**Beschwichtigungsgeste/
Beruhigungsgeste**
reassuring gesture
besiedeln/etablieren settle,
establish
besiedeln/kolonisieren colonize
Besiedlung/Etablierung settlement,
establishment
**Besiedlung/Kolonisation/
Kolonisierung** *micb* colonization
**bespitzt/kleinspitzig/
mit aufgesetzter Spitze**
blunt with a point
besprengen sprinkle
bespringen (begatten) mount, cover
besputtern *vb micros* **(EM)** sputter
Besputtern/Kathodenzerstäubung
n micros **(EM)** sputtering
Besputterungsanlage *micros* **(EM)**
sputtering unit/appliance
Bestand population; stand, standing
crop, stock, number, quantity
beständig/resistent resistant
Beständigkeit/Resistenz resistance
Bestandsaufnahme (to make an)
inventory
Bestandsdichte/Populationsdichte
population density
Bestäuber pollinator
Bestäubung *sensu stricto:*
pollination (fertilization *see*
Befruchtung)
• **Bienenbestäubung**
bee pollination, melittophily
• **Fledermausbestäubung/
Fledermausblütigkeit/
Chiropterophilie** bat pollination,
chiropterophily
• **Insektenbestäubung/
Insektenblütigkeit/Entomophilie**
insect pollination, entomophily
• **Käferbestäubung**
beetle pollination, cantharophily
• **Mottenbestäubung**
moth pollination, phalaenophily
• **Schmetterlingsbestäubung**
butterfly pollination, psychophily
• **Schneckenbestäubung**
snail pollination, malacophily

• **Tierbestäubung/Tierblütigkeit/
Zoophilie** pollination by animal
vectors, zoophily
• **Vogelbestäubung/
Vogelblütigkeit/Ornithophilie**
bird pollination, ornithophily
• **Wasserbestäubung/
Wasserblütigkeit/Hydrophilie**
pollination by water, hydrophily
• **Wespenbestäubung** wasp
pollination, sphecophily
• **Windbestäubung/
Windblütigkeit/Anemophilie**
wind pollination, anemophily
**Bestäubungstropfen/
Befruchtungströpfchen**
pollination drop, pollination droplet
**Bestaudung/Bestockung/
Seitentriebbildung** tillering,
sprouting (at base)
bestehend/existierend existing,
extant
bestellen *agr* (Feld) cultivate
Bestie/wildes Tier beast
bestimmen (Pflanzen/Tiere)
identify
bestimmen *chem* determine,
identify
Bestimmung (Pflanzen/Tiere)
identification
**Bestimmung/Determinierung/
Determination** determination
Bestimmungsbuch manual
Bestimmungsschlüssel key
bestocken/bestauden reforest,
reafforest
Bestockung/Bestandsdichte
density of a stand
**Bestockung/Seitentriebbildung/
Bestaudung** tillering,
sprouting (at base)
Bestockung/Wiederaufforstung
reforestation, reafforestation
Bestockungstrieb tiller,
stalk/sprout (from base)
bestrahlen irradiate
Bestrahlung irradiation
**Bestrahlungsintensität/
Bestrahlungsdichte** irradiance,
fluence rate, radiation intensity,
radiant-flux density
**beta-Drehung/beta-Schleife (DNA/
Proteine)** beta-turn, β turn
beta-Faltblatt beta-sheet,
beta-pleated sheet
beta-Fass β-barrel
beta-Mäander β-meander
beta-Rohr/beta-Fass beta-barrel
**beta-Schleife/β-Schleife/
Haarnadelschleife/Winkelschleife/
Umkehrschleife** beta turn, β bend,
hairpin loop, reverse turn

Bewegung

Betain betaine, lycine, oxyneurine, trimethylglycine
betäuben/narkotisieren/ anästhesieren stupefy, narcotize, anesthetize
betäubend/narkotisch/anästhetisch stupefacient, stupefying, narcotic, anesthetic
Betäubung/Narkose/Anästhesie stupefaction, narcosis, anesthesia
Betäubungsmittel/Narkosemittel/ Anästhetikum stupefacient, narcotic, narcotizing agent, anesthetic, anesthetic agent
Betriebsstoffwechsel maintenance metabolism
Bettelverhalten begging behavior
Betulaceae/Birkengewächse birch family
Beuge flexure, bend
- **Armbeuge/Ellenbeuge** bend of the arm, crook of the arm, inside of the elbow, front of elbow
- **Brückenbeuge** *embr* pontine flexure
- **Hirnbeuge** *embr* cerebral flexure
- **Leistenbeuge/Leistengegend/ Inguinalgegend/Leiste/ Hüftbeuge/Regio inguinalis** groin, inguinal zone
- **Nackenbeuge** *embr* cervical flexure
- **Scheitelbeuge** *embr* cephalic flexure, cranial flexure

beugen (Muskel/Arm/Bein) flex, bend, curve back
Beuger/Flexor (Muskel) flexor
Beugung (Muskel) flexion
Beulenkrankheit (*Myxobolus pfeifferi*) boil disease
Beulenpest/Bubonenpest (*Yersinia pestis*) bubonic plague
Beute/Jagdbeute/Beutetier prey
Beutefesselfäden (Hyptiotes) *arach* swathing band (orb weavers)
Beutel/Brutbeutel/Marsupium pouch, marsupium
Beutel/Sack/Tasche/Bursa pouch, sac, sac-like cavity, pocket, bursa
Beutelknochen/Os marsupialis marsupial bone
Beutelsäuger/Metatheria/Didelphia pouched mammals, metatherians
Beuteltiere/Marsupialia marsupials, pouched mammals
Bevölkerungsdichte/ Populationsdichte/Abundanz population density
Bevölkerungsgröße/ Populationsgröße population size
Bevölkerungspyramide population pyramid
Bevölkerungsschwankung fluctuation of population
Bevölkerungswachstum/ Populationszuwachs population growth
Bevölkerungszusammenbruch/ Populationszusammenbruch population crash
bewachen guard
bewahren/erhalten/preservieren preserve, keep, maintain
Bewahrung/Erhaltung/ Preservierung preservation
bewaldet forested, wooded, arboreous
Bewässerung irrigation
- **Beregung von oben** overhead irrigation
- **Beregnungsbewässerung** sprinkler irrigation
- **Bewässerung durch Überflutung** flood irrigation
- **Grabenbewässerung/ Furchenbewässerung** furrow irrigation
- **Rieselbewässerung** trickle irrigation
- **Schwallbewässerung** surge irrigation
- **Spritzbewässerung** sprinkler irrigation
- **Tropfbewässerung/ Tröpfchenbewässerung** drip irrigation

Bewässerungsgraben irrigation ditch
Bewässerungskultur irrigated crop
beweglich/mobil/vagil (Ortsveränderung des Gesamtorganismus) mobile, vagile, wandering
beweglich/motil/bewegungsfähig (Bewegung eines Körperteils) motile
Beweglichkeit/Mobilität/Vagilität (Ortsveränderung des Gesamtorganismus) mobility, vagility
Beweglichkeit/Motilität/ Bewegungsvermögen (Bewegung eines Körperteils) motility
Bewegung/Fortbewegung/ Lokomotion movement, motion, locomotion
- **Rotationsbewegung** rotational motion
- **saltatorische Bewegung** saltatory movement
- **Schwingungsbewegung** vibrational motion
- **Translationsbewegung** translational motion

Beweidung pasturing
Bewertung/Erfassung assessment
**bewimpert/zilientragend/
cilientragend** ciliated, bearing
cilia, cilium-bearing, ciliferous
Bewimperung ciliation
bewohnbar inhabitable
● **unbewohnbar** uninhabitable
bewohnen inhabit, lodge, occupy,
dwell, reside
Bewohner inhabitant, dweller
Bewuchs growth, cover, stand
● **unterer B. (Waldschicht)**
undergrowth
bewurzeln root
Bewurzelung radication, rootage,
rooting
bewusst *psych* conscious
● **unbewusst** unconscious,
unknowing(ly)
bewusstlos/ohnmächtig sein
unconscious (have fainted/have
passed out)
**bewusste Paarung/assortive
Paarung** assortive mating
Bewusstheit awareness
Bewusstsein consciousness
● **Bewusstlosigkeit**
unconsciousness
**Bezahnung/Zahnsystem
(*siehe auch:* Gebiss oder Zahn)**
dentition
**Bezeichnung/Benennung/Name/
Namensgebung (Nomenklatur)**
name, term, designation,
nomenclature
● **zusätzliche Bezeichnung/
Zusatzbezeichnung/Beiname/
Epitheton (Artbezeichnung/
Artname)** epithet (specific
epithet)
Bezeichnungssystem/Nomenklatur
nomenclature
Bezoar (Magenkugel) bezoar
(stomach ball/hair ball)
**Bi-Bi-Reaktion (zwei Substrate/
zwei Produkte)** Bi Bi reaction
Bibergeil/Castoreum castor,
castoreum
Bibliothek/Bank/Klonbank library,
bank (clone bank)
● **Expressionsbibliothek**
expression library
● **subgenomische Bibliothek/
subgenomische Genbank**
subgenomic library
● **subtraktive Genbank/
Subtraktionsbank/
Subtraktionsbibliothek**
subtractive library
Bidder'sches Organ (Kröten)
Bidder's organ

bidirektionale Replikation
bidirectional replication
bidirektionaler Vektor dual
promoter vector, bidirectional vector
Biegefestigkeit/Tragfähigkeit (Holz)
bending strength
Biegesteifigkeit (Holz) bending
resistance
biegsam flexible, pliable
Biegsamkeit flexibility, pliability;
stiffness
Biene/Imme bee
● **Ammenbiene** nurse bee,
nursery bee
● **Arbeitsbiene/Arbeiterin**
worker bee
● **Drohne/Drohn** drone
● **Nachläuferin** follower bee
● **Sammelbiene/Sammlerin/
Flugbiene/Trachtbiene** forager,
field bee, flying bee
● **Spurbiene/Kundschafter/
Pfadfinder** scout
● **Stockbiene** house bee
● **Wächterbiene/Wehrbiene**
guard bee
● **Weisel/Bienenkönigin** queen bee
Bienenbrot/Cerago beebread, cerago
Bienenhaus/Bienenstand apiary,
bee yard
Bienenkönigin/Weisel queen bee
Bienenkorb beehive, hive
Bienensprache bee language
**Bienenstock/Bienenkorb
(künstliche Behausung)**
beehive (artificial nest)
Bienentanz bee dance
● **Dauertanz** persistent dance
● **Drängeln** jostling dance,
jostling run
● **Nachttanz** night dance
● **Rucktanz** jerk dance
● **Rumpellauf** bumping run
● **Rundtanz** round dance
● **Rütteltanz** shaking dance
● **Schütteltanz/Schüttelbewegung**
vibrating dance, vibratory dance,
dorsoventral abdominal vibrating
dance (DVAV)
● **Schwänzeltanz** waggle dance,
wagging dance, tail-wagging
dance, figure-eight dance
● **Schwirrlauf** buzzing run,
breaking dance
● **Sicheltanz** sickle dance
(bowed figure 8)
● **Sterzeln** fanning with lifted
abdomen (exposing Nasanov
organ)
● **Trippeln** spasmodic dance
● **Zittertanz** tremble dance,
trembling dance, quiver dance

415 binsenartig **B**

Bienenvolk bee colony
Bienenwachs beeswax
Bienenzucht/Imkerei beekeeping,
apiculture
Bienenzuchtbetrieb/Imkerei apiary
Bienenzüchter/Imker beekeeper,
apiarist
**Bienenzüchterei
(Lehre/Studium der Bienenzucht)**
apiology
**Biestmilch/Vormilch/
Kolostralmilch/Colostrum**
foremilk, colostrum
**bifazial/zweiseitig/dorsiventral/
zygomorph** bifacial, dorsiventral,
zygomorph
**bifunktionaler Vektor/
Schaukelvektor**
bifunctional vector, shuttle vector
**Big-Bang-Fortpflanzung/
Big-Bang-Reproduktion/
Semelparitie**
big-bang reproduction, semelparity
**Bignoniengewächse/
Trompetenbaumgewächse/
Bignoniaceae** bignonia family,
trumpet-creeper family,
trumpet-vine family
**Bilanz (Energiebilanz/
Stoffwechselbilanz)** balance
**Bilateralfurchung/
bilaterale Furchung**
bilateral cleavage
Bilateralsymmetrie bilateral
symmetry
Bild picture, image
• **elektronenmikroskopisches Bild/
elektronenmikroskopische
Aufnahme** electron micrograph
• **Endbild** *micros* final image
• **mikroskopisches Bild/
mikroskopische Aufnahme**
microscopic image,
microscopic picture, micrograph
• **reelles Bild** *micros* real image
• **virtuelles Bild** *micros*
virtual image
**bilden (entwickeln)
(z. B. Gase/Dämpfe)**
generate (develop)
Bildpunkt *opt* image point
Bildungsgewebe/Meristem
meristem
**Bildungsplasma/Eiplasma/
Ooplasma** ooplasm
**Bilharziose/Schistosomiasis
(*Schistosoma spp.*)**
bilharziosis, schistosomiasis,
blood fluke disease
bimodale Verteilung
bimodal distribution,
two-mode distribution

**binäre/binominale Nomenklatur
(zweigliedrige Bezeichnung)**
binary/binomial nomenclature
Bindegewebe connective tissue
Bindegewebshülle/Faszie fascia
(ensheating band of connective
tissue)
Bindegewebshülle/Sklera/Sclera
sclera
Bindeglied (Brückentier)
connecting link
binden *chem* bond, link
Bindung *chem* bond, linkage
• **Atombindung** atomic bond
• **chemische Bindung** chemical
bond
• **Disulfidbindung (Disulfidbrücke)**
disulfide bond, disulfide bridge
• **Doppelbindung** double bond
• **Dreifachbindung** triple bond
• **energiereiche Bindung** high
energy bond
• **glykosidische Bindung**
glycosidic bond/linkage
• **heteropolare Bindung**
heteropolar bond
• **homopolare Bindung** homopolar
bond, nonpolar bond
• **hydrophile Bindung** hydrophilic
bond
• **hydrophobe Bindung**
hydrophobic bond
• **Ionenbindung** ionic bond
• **Kohlenstoffbindung** carbon
bond
• **konjugierte Bindung** conjugated
bond
• **kooperative Bindung**
cooperative binding
• **kovalente Bindung** covalent
bond
• **Mehrfachbindung** multiple bond
• **Peptidbindung** peptide bond,
peptide linkage
Bindungsenergie binding energy,
bond energy
Bindungskurve binding curve
Bindungswinkel bond angle
Binneneber/Spitzeber cryptorchid
pig
Binnengewässer inland waterbody
**Binnenklima/Kontinentalklima/
Landklima** continental climate
**Binnenmeer (Salzwasser)/
Binnensee (Süßwasser)** inland sea
Binokular binoculars
Binomialverteilung binomial
distribution
binomische Formel binomial
formula
binsenartig rushy, rushlike,
juncaceaous

B binsenförmig

binsenförmig rush-shaped, junciform
bioanorganisch bioinorganic
Bioäquivalenz bioequivalence
Biochemie biochemistry
biochemischer Sauerstoffbedarf/ biologischer Sauerstoffbedarf (BSB) biochemical oxygen demand, biological oxygen demand (BOD)
Biochorion/Choriotop biochore
Biocönose/Biozönose/Biozön biocenosis
Biodegradation/biologischer Abbau biodegradation
Bioenergetik bioenergetics
Bioethik bioethics
biogen biogenic
biogenetische Regel/biogenetisches Grundgesetz biogenetic law/principle, Haeckel's law
Biogeographie biogeography
Biogeozönose biogeocoenosis
Bioindikator/Indikatorart/Zeigerart/ Indikatororganismus bioindicator, indicator species
Biolistik biolistics, microprojectile bombardment
Biologe/Biologin biologist, bioscientist, life scientist
Biologie/Biowissenschaften biology, bioscience, life sciences
biologisch/biotisch biologic(al), biotic
biologisch abbaubar biodegradable
biologische Abbaubarkeit biodegradability
biologische Art biological species
biologische Kriegsführung biological warfare, biowarfare
biologische Membran biomembrane
biologische Schädlingsbekämpfung biological pest control
biologische Sicherheit(smaßnahmen) biological containment
biologische Uhr biological clock
biologische Verfahrenstechnik/ Biotechnik/Bioingenieurwesen bioengineering
biologischer Abbau/Biodegradation biodegradation
biologischer Kampfstoff biological warfare agent
biologischer Sauerstoffbedarf/ biochemischer Sauerstoffbedarf (BSB) biological oxygen demand, biochemical oxygen demand (BOD)
biologischer Test bioassay, biological assay
biologisches Gleichgewicht biological equilibrium
Biolumineszenz bioluminescence

Biom/Bioformation biome (biogeographical region/formation)
Biomasse biomass
Biomathematik biomathematics
Biomechanik biomechanics
Biomedizin biomedicine
Biometrie biometry, biometrics
Biomolekül biomolecule
Bionik bionics
Biophysik biophysics
Bioreaktor (Reaktortypen *siehe* **Reaktor)** bioreactor
Biorhythmik biorhythmicity
Biorhythmus biorhythm
Biosphäre biosphere
Biostatik biostatics
Biostatistik biostatistics
Biosynthese biosynthesis
Biosynthesereaktion biosynthetic reaction (anabolic reaction)
biosynthetisch biosynthetic(al)
biosythetisieren biosynthesize
Biotechnik/biologische Verfahrenstechnik/ Bioingenieurwesen bioengineering
Biotechnologie biotechnology
Biotin (Vitamin H) biotin (vitamin H)
Biotin-Markierung/Biotinylierung biotin labelling, biotinylation
biotisch/biologisch biotic, biological
● **abiotisch** abiotic
● **präbiotisch** prebiotic
Biotop/Lebensraum biotope, life zone
biotopfremd/bodenfremd allochthonous
Biotopprägung habitat imprinting
Biotransformation/Biokonversion biotransformation, bioconversion
Bioverfügbarkeit bioavailability
Biowissenschaft bioscience (meist *pl* biosciences), life science (meist *pl* life sciences)
Biozid biocide
Biozön/Biozönose/Biocönose/ Lebensgemeinschaft/ Organismengemeinschaft biocenosis, biotic community
biparental biparental
biped/bipedisch/zweibeinig/ zweifüßig bipedal
Bipedie/Bipedität/Zweibeinigkeit/ Zweifüßigkeit bipedalism, bipedality
Bipolarzelle bipolar cell
Birbeck-Granula *immun* Birbeck's granules
Birkengewächse/Betulaceae birch family
birnenförmig pear-shaped, pyriform

417

Blatt B

birnenförmiges Organ (Bryozoen)
pyriform organ, piriform organ
bispezifischer Antikörper hybrid
antibody, bispecific antibody
Biss bite
Bisubstratreaktion/
Zweisubstratreaktion bisubstrate
reaction
● **Bi-Bi-Reaktion (zwei Substrate/**
zwei Produkte) Bi Bi reaction
● **doppelte Verdrängungsreaktion/**
Doppel-Verdrängung (Pingpong-
Reaktion) double displacement
reaction (ping-pong reaction)
● **einfache Verdrängungsreaktion/**
Einzel-Verdrängung
single displacement reaction
● **zufällige Verdrängungsreaktion/**
nicht-determinierte
Verdrängungsreaktion
random displacement reaction
● **geordnete**
Verdrängungsreaktion
ordered displacement reaction
bitter bitter
Bittereschengewächse/
Bitterholzgewächse/
Simaroubaceae quassia family
Bitterkeit bitterness
Bitterkleegewächse/
Fieberkleegewächse/
Menyanthaceae bogbean family
Bitterstoffe bitters
bivalent bivalent
Biwak (Wanderameisen) bivouac
Bixaceae/Annattogewächse
annatto family, bixa family
bizistronisch/bicistronisch
bicistronic
blähen bloat
Blähschlamm bulking sludge
Blähungen/Flatulenz bloating, gas
Bläschen/Vesikel vesicle
bläschenartig bubblelike, bullate
Bläschendrüse/Samenblase/
Samenbläschen/Glandula
vesiculosa (♂ akzessorische Drüse)
vesicular gland, seminal gland,
seminal vesicle
bläschenförmig bubble-shaped,
bulliform
Blase bladder
Blase/Ampulle (*Utricularia*) bladder,
ampulla
Blasen-Linker-PCR *gen*
bubble linker PCR
blasenartig/blasenförmig
bladderlike, bladdery, utriculate,
utricular
● **vesikulär** bladderlike, vesicular
Blasenauge retinal cup eye,
everted eye

Blasenbinsengewächse/
Blumenbinsengewächse/
Scheuchzeriaceae arrow-grass
family, scheuchzeria family
Blasenfüße/Fransenflügler/Thripse/
Thysanoptera thrips
Blasenhaar bladder hair
Blasenkeim/Keimblase/Blastula
blastula
Blasenmole/Mole (entartete Frucht)
med mole
Blasensäulen-Reaktor
bubble column reactor
Blasensprung rupture of fetal
membranes/amniotic membrane
blasentreibend/blasenziehend
vesicating, vesicant
Blasenwurm/Finne/Cysticercus
(Bandwurmlarve) bladderworm,
cysticercus
Blasenzelle bladder cell
blasig bullous, with blisters,
vesiculate
Blasloch/Spritzloch/Spiraculum
(Wale) blowhole, vent, spiracle
Blastocöl blastocoel, blastocoele
Blastocyste/Blastozyste/
Keimbläschen blastocyst
Blastoderm blastoderm
Blastoderm/Keimscheibe
blastoderm
Blastoporus/Protostom/Urmund
blastopore, protostoma
Blastostyle (Fruchtpolyp)
blastostyle (reduced gonozooid)
Blastozooid blastozooid
Blastozyste/Blastocyste/
Keimbläschen blastocyst
Blastula-Höhle/Blastocöl/primäre
Leibeshöhle blastocoel, blastocele
Blatt (*pl* Blätter) *bot* leaf (*pl* leaves)
Blatt/Keimblatt/Keimschicht *embr*
germ layer
● **Amphigastrium/Bauchblatt**
("Unterblatt") amphigastrium,
ventral leaf, underleaf
● **Blumenhüllblatt/**
Blütenkelchblatt/Kelchblatt
sepals
● **Blütenkelchblatt/Kelchblatt/**
Blumenhüllblatt/Sepale sepals
● **Deckblatt/Tragblatt (Blüte)**
bract, subtending bract
● **Fahnenblatt/Fähnchenblatt**
flag leaf
● **Fallenblatt** trap leaf
● **Farnblatt (eingerolltes junges)**
crozier, fiddlehead
● **Fiederblatt (ganzes!)**
compound leaf, divided leaf
● **Fiederblättchen (ersten Grades)**
pinna

B Blattabwurf

- **Fiederblättchen (zweiten Grades)** pinnule, pinnula
- **Fiederchen/Pinnula** pinnule
- **Folgeblatt/Laubblatt** foliage leaf; metaphyll
- **Fruchtblatt/Karpel/Carpell** carpel
- **gefenstertes Blatt** fenestrated leaf
- **herunterhängendes Blatt** drooping leaf
- **Hochblatt** *allg* hypsophyll
- **Hochblatt/Braktee** floral bract
- **Honigblatt** nectariferous leaf
- **Involukralblatt/-schuppe** phyllary, involucral bract
- **Kannenblatt/Schlauchblatt** pitcher leaf, ascidiate leaf
- **Karpel/Fruchtblatt** carpel
- **Keimblatt/Kotyledone/ Cotyledone** cotyledon, seminal leaf
- **Keimblattscheide/Keimscheide/ Koleoptile/Coleoptile** coleoptile, plumule sheath
- **Kelchblatt/Blütenkelchblatt/ Blumenhüllblatt/Sepale/Sepalum** sepal
- **Kotyledone/Cotyledone/ Keimblatt** cotyledon, seminal leaf
- **Laubblatt/Folgeblatt** foliage leaf
- **Liesche/Lieschenblatt (Hüllblatt an Maiskolben)** (corn) husk
- **Mantelblatt/Nischenblatt** nest leaf
- **Nebenblatt/Stipel** stipule
- **Nektarblatt/Honigblatt** nectar leaf, honey leaf
- **Niederblatt** cataphyll (a bud scale/scale leaf/bulb scale or cotyledon)
- **Nischenblatt/Mantelblatt** nest leaf
- **Oberblatt (Spreite & Stiel)** leaf blade and leaf stalk
- **Phyllodium/Blattstielblatt** phyllode
- **Prophyll/Vorblatt** prophyll, first leaf
- **Samenblatt/Makrosporophyll** macrosporophyll
- **Schattenblatt** shade leaf, sciophyll
- **Schildblatt/peltates Blatt** peltate leaf
- **Schlauchblatt (Genlisea)** lobster pot
- **Schlauchblatt/Kannenblatt (Nepenthes)** siphonaceous leaf, ascidiform leaf, ascidium, pitcher-leaf

- **schneckenhausförmig eingerolltes junges Farnblatt** fiddlehead, crozier
- **Schwimmblatt** floating leaf
- **Sonnenblatt/Lichtblatt** sun leaf
- **Spreublatt/Spreuschuppe** ramentum, chaffy scale, palea, palet, pale
- **Staubblatt** stamen
- **Tragblatt/Deckblatt** bract, subtending bract
- **Tragblatt zweiter Ordnung/ Tragschuppe/Deckschuppe/ Brakteole** bract-scale, bracteole, bractlet, secondary bract
- **Trichterblatt** funnel leaf
- **Unterblatt** underleaf, hypophyll
- **Unterblatt/Bauchblatt/ Amphigastrium** underleaf, ventral leaf, amphigastrium
- **Urnenblatt (Dischidia)** urn-shaped leaf, pouch leaf, "flower pot" leaf
- **Vorblatt/Bracteola** secondary bract, bracteole, bractlet
- **Vorblatt/Prophyll** first leaf, prophyll
- **Wasserblatt** submerged leaf
- **zusammengesetztes Blatt** compound leaf
- **Zwischenblatt** metaxyphyll

Blattabwurf abscission

Blattachse leaf axis

Blattachse eines gefiederten Blattes/Rhachis/Blattspindel rachis (midrib of compound leaf)

Blattachsel leaf axil

Blattader/Blattnerv/Blattrippe leaf vein, leaf rib

Blattaderung/Blattnervatur/ Blattnervation/Blattvenation leaf venation

- **bogenförmige B./Bogennervatur** arched/arciform/arcuate venation
- **fiederförmige B./Fiedernervatur/ Fiederaderung** striate venation
- **fingerförmige B. (fingernervig/handnervig)** digitate venation
- **gabelige B./Gabeladerung** dichotomous venation
- **geschlossene B.** closed venation
- **netzförmige B./Netznervatur** reticulate venation, net venation
- **offene B.** open venation
- **parallele B./Parallelnervatur** parallel venation
- **streifenförmige B./ Längsnervatur/Streifenaderung** striate venation

Blattanlage/Blattprimordium leaf primordium

Blattanordnung/Blattstellung
leaf arrangement
blattartig/blattförmig leaf-like,
phylloid, phylloidal, foliaceous,
foliose
Blattausschüttung/
Laubausschüttung leaf flushing
Blattaustrieb production of leaves,
coming into leaf
Blattbasis leaf base
Blattbein/Phyllopodium (*pl***
Phyllopodien)** phyllopod
Blattbeine/Blattfüße/Kiemenbeine/
Buchkiemen (dichtstehende
Kiemenlamellen: Xiphosuriden)
book gills (gill book)
Blattbildung foliation
Blattbündel leaf bundle
blattbürtig leaf-borne
Blättchen leaflet
 • **Fiederblättchen** pinna (leaflet of
pinnate leaf)
Blattdorn spine
Blattdüngung foliar feeding
Blattentfaltung leafing,
unfolding of leaves
Blattentstehung leaf origin
Blattentwicklung foliation, leaf
development/ontogeny
Blätterdach (Wald) (forest) canopy
Blättermagen/Vormagen/Omasus/
Psalter (Wiederkäuer) third
stomach, omasum, manyplies,
psalterium
Blätterpilz/Lamellenpilz gill fungus,
gill mushroom
Blattfall/Laubfall leaf abscission,
shedding of leaves; (frühzeitiger)
leaf drop
Blattflächenindex (BFI)
leaf area index (LAI)
Blattflächenverhältnis (BFV)
leaf area ratio (LAR)
Blattflechten/Laubflechten
foliose lichens
Blattfolge am Spross phyllotaxy,
phyllotaxis, leaf sequence (relation
of leaves on stem)
Blattform leaf shape *(die*
verschiedenen Blattformen im
Englisch-Deutsch Teil unter:
leaf shape)
blattfressend/blätterfressend
leaf-eating, folivorous
Blattfüße/Blattbeine/Kiemenbeine/
Buchkiemen (dichtstehende
Kiemenlamellen: Xiphosuriden)
book gills (gill book)
Blattfußkrebse/Kiemenfüßer/
Phyllopoda/Branchiopoda
phyllopods
Blattgalle leaf gall

Blattgemüse leaf vegetable, leafy
vegetable; (gekochtes B.) potherbs
Blattgrün foliage green
Blattgrund leaf base
Blatthäutchen/Ligula (Gräser) ligule
Blatthöcker (frühe Blattanlage)
leaf buttress
Blattkieme/Lamellibranchie/
Eulamellibranchie lamellar gill,
sheet gill, lamellibranch,
eulamellibranch
Blattkiemer/Lamellenkiemer/
Eulamellibranchia
eulamellibranch bivalves
Blattkissen/Blattpolster/
Gelenkpolster/Pulvinus
leaf cushion, leaf pulvinus
Blattknospe/Laubknospe
foliage bud
Blattknospenlage/Vernation
vernation, ptyxis, prefoliation
Blattkräuselkrankheit lead curl,
leaf roll, crinkle
Blattlage in Knospe/Vernation
vernation, ptyxis, prefoliation
Blattläuse/Aphidina aphids
blattlos leafless, aphyllous
Blattlosigkeit/Aphyllie aphylly,
absence of leaves
Blattlücke leaf gap, foliar gap
Blattnarbe leaf scar
Blattnerv/Blattader/Blattrippe
leaf vein, leaf rib
Blattnervatur leaf venation
(*siehe* Blattaderung)
Blattoberfläche leaf surface
Blattoberseite upper/superior/
adaxial leaf surface
Blattöhrchen/Auricula (Gräser)
auricle
Blattorgan/Phyllom phyllome
Blattpolster/Blattkissen/
Gelenkpolster/Pulvinus
leaf pulvinus, leaf cushion
Blattrand leaf margin, leaf edge
Blattranke leaf tendril
blättrig/plättchenartig geschichtet
laminar, laminiform, laminous
Blattrippe leaf rib, leaf vein
Blattroller (Insekt) leaf-roller,
leaf-tier
Blattrosette rosette of leaves,
whorl of leaves
Blattscheide leaf sheath
Blattschneider leaf cutter
Blattschopf comal tuft
Blattspindel/Rhachis rachis
Blattspitze leaf tip, leaf apex
Blattspreite leaf blade, leaf lamina
(verschiedene Formen der
Blattspreite im Englisch-Deutsch
Teil unter: leaf shape)

B Blattspreitengrund 420

**Blattspreitengrund/
Blattspreitenbasis**
base of leaf blade
Blattspreitenrand margin/edge of
leaf blade, leaf blade margin/edge
Blattspreitenspitze leaf apex
Blattspur leaf trace, foliar trace
Blattspurstrang/Blattspurbündel
leaf trace bundle
Blattsteckling leaf cutting
**Blattstellung/Blattanordnung/
Beblätterung/Phyllotaxis**
phyllotaxis, phyllotaxy,
leaf arrangement/position
• **gedrängte B.**
crowded leaf arrangement
• **gegenständige B.**
opposite leaf arrangement
• **kreuzgegenständige B.
(dekussiert)**
decussate leaf arrangement
• **schraubige B.**
spiral leaf arrangement
• **wechselständige B.**
alternate leaf arrangement
• **wirtelige/quirlige B.**
whorled leaf arrangement
• **zerstreute B. (dispers)**
scattered leaf arrangement
• **zweizeilige B. (distich)**
distichous, distichate, two-ranked/
two-rowed leaf arrangement
Blattstiel *allg* leaf stalk, petiole
Blattstiel (Algen/Farne/Palmen)
stipe
**Blattstiel (an gefiedertem Blatt)/
Rhachis** rachis
**Blattstielkissen/Blattkissen/
Blattpolster/Gelenkpolster/
Pulvinus** leaf pulvinus,
leaf cushion
Blattstreu leaf litter
Blattunterseite lower/abaxial leaf
surface
Blattwedel frond
blattwerfend deciduous,
leaf-dropping
Blattwerk/Belaubung/Beblätterung
foliage
**Blaualgen/Cyanophyceae/
Cyanobakterien** bluegreen algae,
cyanobacteria
blaugrün glaucous
**Blaukorallen/Coenothecalia/
Helioporida** blue corals
bläulich-grün glaucous
Blausäure/Zyanwasserstoff
hydrogen cyanide, hydrocyanic acid,
prussic acid
Blechnaceae/Rippenfarngewächse
blechnum family, deer fern family
Blei lead

bleich/blass pale
Bleiche *chem* bleach
Bleiche/bleiche Farbe/Blässe
paleness
bleichen/ausbleichen (*activ*: **weiss
machen/aufhellen**) bleach
Bleicitrat (EM) lead citrate
**Bleiwurzgewächse/
Grasnelkengewächse/
Plumbaginaceae**
sea lavender family, leadwort family,
plumbago family
Blende/Diaphragma *micros*
diaphragm
Blende/Öffnung/Apertur *micros*
aperture
Blendenöffnung *micros*
diaphragm aperture
**Blepharoplast/Kinetosom/
Basalkörper** blepharoplast,
blepharoblast, mastigosome,
kinetosome, flagellar basal granule/
corpuscle/body
Blesse (z.B. Pferd)
blaze (white stripe)
Blickfeld/Sehfeld/Gesichtsfeld
field of view, scope of view,
field of vision, range of vision,
visual field
**Blinddarm/Darmdivertikulum/
Caecum** blind-ended diverticulum,
cecum
Blinddarmkot/Vitaminkot (Hasen)
reingested soft/greenish pellets
blinder Fleck blind spot
(optic disk)
Blindheit blindness
Blindsack/Divertikulum
diverticulum
Blindwert blank
**Blindwühlen/Gymnophiona/
Caecilia/Apoda**
gymnophionas, caecilians,
wormlike amphibians (legless)
Blinzelhaut/Nickhaut
nictitating membrane, third eyelid
Blinzknorpel *orn*
cartilage of third eyelid
Blockhalter *micros* block holder
Blockierungsreagenz
blocking reagent
Blockverfahren block synthesis
blöken bleat
**blotten (klecksen/Flecken machen/
beflecken)** blot
Blotten/Blotting blotting,
blot transfer
• **Affinitäts-Blotting**
affinity blotting
• **Alkali-Blotting** alkali blotting
• **Diffusionsblotting**
capillary blotting

Blüte B

- **genomisches Blotting**
  genomic blotting
- **Liganden-Blotting** ligand blotting
- **Nassblotten** wet blotting
- **Trockenblotten** dry blotting
**Blotting-Elektrophorese/
Direkttransfer-Elektrophorese**
  direct blotting electrophoresis,
  direct transfer electrophoresis
blühen flower, bloom
Blühinduktion/Evocation evocation
Blühphase bloom stage,
  blooming stage
Blühreife (*Klebs*) (ripeness to
  flower) ripeness to respond
Blümchen/Blütchen/kleine Blüte
  floret, tiny flower
Blümchen/Pflänzchen plantlet
Blume/Blüte/Anthium flower,
  blossom
- **Aasblume** carrion flower
- **Einzelblüte** solitary flower,
  single flower
- **Fallenblume/Fallenblüte** trap
  blossom, trap flower, prison flower,
  pitcher plant
- **Gallenblüte (Feigen)**
  gall flower (figs)
- **Klemmfallenblume/
  Klemmfallenblüte**
  pinch-trap flower
- **Merianthium/Teilblume**
  merianthium, partial flower
- **Napfblume** bowl-shaped flower
- **Röhrenblüte (verwachsene
  Kronblätter)** corolla tube,
  tubular corolla
- **Röhrenblüte/Scheibenblüte
  (Asterales)** disk flower,
  disk floret, tubular flower
- **Staubblüte/männliche Blüte**
  staminate flower, male flower
- **Stempelblüte/weibliche Blüte**
  carpellate/pistillate flower,
  female flower
- **Strahlenblüte (Zungenblüte)**
  ray floret, ligulate flower
- **Täuschblume** deceptive flower
- **Teilblume/Merianthium**
  partial flower, merianthium
- **Trichterblüte** funnel-shaped
  corolla, funnel-shaped flower
- **unvollständige Blüte**
  incomplete flower
- **vollständige Blüte**
  complete flower
- **Zungenblüte (Strahlblüte)**
  ray flower, ray floret
- **zweigeschlechtige Blüte/
  Zwitterblüte** bisexual flower,
  hermaphroditic flower, perfect
  flower

Blume/Pflanze flower, plant
- **Kesselfallenblume/
  Gleitfallenblume**
  pitfall trap, slippery-trap flower,
  slippery-slide flytrap
- **Schnittblume** cut flower
- **Strohblume/Trockenblume**
  strawflower
Blumenbeet flowerbed
**Blumenbinsengewächse/
Blasenbinsengewächse/
Scheuchzeriaceae**
  arrow-grass family,
  scheuchzeria family
Blumenhändler florist
**Blumenhüllblatt/Blütenkelchblatt/
Kelchblatt** sepals
Blumenkrone/Krone/Corolla
  corolla
Blumennesselgewächse/Loasaceae
  loasa family
Blumenrohrgewächse/Cannaceae
  canna family,
  Queensland arrowroot family
Blumenschau horticultural show/
  exhibition
**Blumentiere/Blumenpolypen/
Anthozoen** flower animals,
  anthozoans
Blumentopf flower pot
Blumenzucht floriculture
Blumenzüchter/Blumengärtner
  floriculturist
Blumenzwiebel bulb
Blut blood
Blut-Ersatz blood substitute
Blut-Hirn-Schranke
  blood-brain barrier
Blutagar blood agar
"Blutarmut"/Anämie anemia
- **bösartige Blutarmut/
  perniziöse Anämie**
  pernicious anemia
Blutausstrich *micros* blood smear
Blutbank blood bank
Blutbild/Blutstatus/Hämatogramm
  blood count, hematogram
**Blutbildung/Blutzellbildung/
Hämatopoese/Haematopoese**
  hematopoiesis
Blutdruck blood pressure
Blüte/Blume flower, blossom
- **Aasblume** carrion flower
- **eingeschlechtige Blüte**
  unisexual flower, imperfect flower
- **Einzelblüte** solitary flower,
  single flower
- **Fallenblüte** trap blossom,
  trap flower, prison flower
- **Gallenblüte (Feigen)** gall flower
- **Klemmfallenblüte**
  pinch-trap flower

B Blüte 422

- **Merianthium/Teilblume** merianthium, partial flower
- **Napfblume** bowl-shaped flower
- **radiäre Blüte/strahlenförmige Blüte** radial flower, actinomorphic flower, regular flower
- **Röhrenblüte (verwachsene Kronblätter)** corolla tube, tubular corolla
- **Röhrenblüte/Scheibenblüte (Asterales)** disk flower, disk floret, tubular flower
- **Staubblüte/männliche Blüte** staminate flower, male flower
- **Stempelblüte/weibliche Blüte** carpellate flower, pistillate flower, female flower
- **Strahlenblüte (Zungenblüte)** ray floret, ligulate flower
- **Täuschblume** deceptive flower
- **Teilblume/Merianthium** partial flower, merianthium
- **Trichterblüte** funnel-shaped corolla, funnel-shaped flower
- **unvollständige Blüte** incomplete flower
- **vollständige Blüte** complete flower
- **Zungenblüte (Strahlblüte)** ray flower, ray floret
- **zweigeschlechtige Blüte/ Zwitterblüte** bisexual flower, hermaphroditic flower, perfect flower
- **zwittrige Blüte/ zweigeschlechtige Blüte/ Zwitterblüte** bisexual flower, hermaphroditic flower, perfect flower

Blutegel/Hirudineen/Hirudinea leeches, hirudineans
bluten *vb* bleed
Bluten *n* bleeding; (Pflanzenwunde) bleeding
Blütenachse/Torus/Blütenboden receptacle, torus
- **vergrößerte B./ scheibenförmige B.** hypanthium
Blütenährchen spikelet, spicule
Blütenbasis receptacle
Blütenbau flower structure
Blütenbecher/Cupula/Kupula flower cup, floral cup, cupule, cupula
Blütenbiologie floral biology
Blütenblätter floral leaves
Blütenboden/Blütenachse/Torus receptacle, torus
- **vergrößerter B./ scheibenförmiger B.** hypanthium
Blütendiagramm flower diagram, floral diagram

Blütenduft flower scent, flower perfume
Blütenentfaltung anthesis
Blütenfall flower abscission
Blütenhüllblätter, gleichartige tepals
Blütenhüllblattkreis (differenzierter B.) perianth; (einheitlicher B.) perigon
Blütenhüllkreis/Blütenhülle (differenzierte B.) perianth; (einheitliche B.) perigon
Blütenkelch (aus Sepalen) calyx
Blütenkelchblatt/Kelchblatt/ Blumenhüllblatt/Sepale sepals
Blütenknäuel glomerule, flower cluster
Blütenknospe flower bud, floral bud
Blütenkolben/Spadix spadix
Blütenköpfchen capitulum, flower head
Blütenkronblätter petals
Blütenkrone corolla
blütenlos ananthous, flowerless
blütenlose Pflanze cryptogam
Blütenmal floral guide
Blütenökologie pollination ecology, anthecology
Blütenorgan flower organ
Blütenpflanze flowering plant, angiosperm, anthophyte
Blütenrispe/Rispe panicle
Blütenröhre/Röhrenblüte (Kronblätter) corolla tube, tubular corolla; (Blütenboden) hypanthium, floral tube
Blütenschaft peduncle, flower stalk; (blattlos) scape, leafless stalk
Blütenscheide/Spatha spathe
blütenscheidenartig/-förmig spathaceous, spathal
Blütenschlund flower funnel
Blütenschopf flower tuft
Blütenstand inflorescence
Blütenstandsstiel peduncle
Blütenstaub pollen
Blütenstengel flower stalk
Blütenstiel flower stalk, peduncle
Blütenstiel (einzelner Blüte in Blütenstand) pedicel
Blütenstiel (einzelner Grasblüte) rachilla
Blütentange/Podostemaceae riverweed family
Blütenzapfen cone, flower cone
Blütenzweig flowering branch; (kleiner Blütenzweig) spray
Bluter bleeder, hemophiliac
Bluterguss/Hämatom bruise, hematoma
Bluterkrankheit/Hämophilie bleeder's disease, hemophilia

Blütezeit/Anthese/Floreszenz flowering period, anthesis, florescence

Blutfaktor blood factor

Blutfaserstoff/Fibrin fibrin

Blutgefäß/Ader blood vessel

Blutgerinnsel/Blutkoagulum blood clot

Blutgerinnung blood clotting

Blutgerinnungsfaktoren blood clotting factors

Blutgerinnungskaskade blood clotting cascade

Blutgruppe blood group

Blutgruppenbestimmung blood-typing

Blutgruppenunverträglichkeit blood group incompatibility

Blutinsel blood island

Blutkörperchen blood cell, blood corpuscle, blood corpuscule
- **basophiler Granulocyt/ Granulozyt** basophil granulocyt
- **eosinophiler Granulocyt/ Granulozyt** eosinophil(ic) granulocyte
- **Granulocyt/Granulozyt (polymorphkerniger Leukozyt)** granulocyte (polymorphonuclear)
- **Leukocyt/Leukozyt/weißes Blutkörperchen** leukocyte, white blood cell (WBC)
- **Lymphozyt/Lymphocyt** lymphocyte
- **Monozyt/Monocyt** monocyte
- **neutrophiler Granulocyt/ Granulozyt** neutrophil(ic) granulocyte
- **Retikulocyt/Reticulozyt/ Proerythrozyt** reticulocyte, proerythrocyte (immature RBC)
- **rotes Blutkörperchen/ Erythrocyt/Erythrozyt** red blood cell (RBC), erythrocyte
- **segmentkerniger Granulocyt/ Granulozyt** segmented granulocyte, filamented neutrophil
- **stabkerniger Granulocyt/ Granulozyt** band granulocyte, stab cell, band cell, rod neutrophil
- **Thrombozyt/Thrombocyt/ Blutplättchen** thrombocyte, blood platelet
- **weißes Blutkörperchen/ Leukocyt/Leukozyt** white blood cell (WBC), leukocyte

Blutkörperchenzählung/ Blutzellzahlbestimmung blood count

Blutkreislauf circulation, bloodstream
- **kleiner Blutkreislauf/ Lungenkreislauf** pulmonary circulation

Blutkreislaufsystem/ Zirkulationssystem circulatory system

Blutkultur blood culture

Blutmahlzeit blood meal

Blutmehl blood meal

Blutplasma blood plasma

Blutplättchen/Thrombozyt/ Thrombocyt blood platelet, thrombocyte

Blutplättchen-Wachstumsfaktor/ Plättchenwachstumsfaktor/ Plättchenfaktor platelet-derived growth factor (PDGF)

blutsaugend/sich von Blut ernährend blood-sucking, sanguivorous, hematophagous

Blutsenkung blood sedimentation

Blutstäubchen/Hämokonia blood dust, hemoconia, hemokonia

blutstillend (adstringent) styptic, hemostatic (astringent)

Blutstrom/Blutkreislauf bloodstream

Blutsverwandschaft/ Konsanguinität consanguinity

blutsverwandt/konsanguin consanguineous

Bluttest blood test

Blutung/Hämorrhagie bleeding, hemorrhage (esp. profuse bleeding)
- **Monatsblutung/Menstruation** menstruation

Blutvergiftung/Sepsis blood poisoning

Blutversorgung blood supply

Blutweiderichgewächse/ Weiderichgewächse/Lythraceae loosestrife family

Blutzellbildung/Blutbildung/ Hämatopoese/Haematopoese hematopoiesis

Blutzelle/Hämatozyt/Hämatocyt hematocyte
- **Blutkörperchen** blood cell, blood corpuscle, blood corpuscule

Blutzellzahlbestimmung/ Blutkörperchenzählung blood count

blutzersetzend/hämorrhagisch hemorrhagic

Blutzucker blood sugar

Blutzuckerspiegel (erhöhter/ erniedrigter) blood sugar level (elevated/reduced)

Blutzufuhr/Blutversorgung blood supply

B Bö

Bö/Gust gust
- **böiger Wind** gusty wind
- **Sturmbö/heftiger Sturmwind** squall

Bock (Schafbock)/Widder ram
Bock (Ziegenbock/Rehbock)/ Männchen buck
Boden (Meeresboden/ Gewässeruntergrund) bottom
Boden/Erdboden soil, ground, earth
Bodenart soil type
Bodenbearbeitung, wendende tillage farming
Bodenbedeckung surface cover, ground cover
Bodenbedingungen soil conditions
Bodenbeschaffenheit soil consistency
Bodenbestandteile soil components
Bodenbestellung/Ackern/Ackerbau farming, tillage, cultivation
bodenbewohnend (Erde: Bodenoberfläche) surface-dwelling; (Ozean) benthic
Bodendecker ground cover, herbaceous soil cover; *agr* cover crop
Bodenerosion soil erosion
Bodenfeuchte soil moisture, soil humidity
Bodenfracht *ecol/mar* bed load
bodenfremd allochthonous
Bodenfrost ground frost
Bodenfruchtbarkeit soil fertility
Bodengefüge soil structure
Bodenhaltung *agr* (Geflügel) free-running (house-confined/ *not:* free-ranging)
Bodenhorizont soil horizon
Bodenkrume/Oberboden topsoil
Bodenkunde/Pedologie soil science, pedology
Bodenläuse/Zoraptera zorapterans
Bodenorganismus soil organism, geodyte, geocole, terricole
Bodenpartikelgrößen soil texture
Bodenpicken *zool/orn* ground pecking
Bodenplatte *neuro* floor plate, subplate
Bodenprofil soil profile
Bodensanierung soil decontamination
Bodenschicht ground stratum, ground layer
Bodenschutz soil conservation
Bodenskelett soil skeleton (inert quartz fraction)
bodenständig/autochthon autochthonous
bodenstet restricted to certain soil type

Bodenteilchen soil particle
Bodentextur soil texture
bodenvag indifferent to soil type
Bodenverbesserer soil conditioner
Bodenverbesserung soil conditioning, soil amelioration
Bodenverdichtung soil compaction
Bodenversalzung soil salinization
Bodenversiegelung surface sealing
Bodenzeiger soil indicator
Bodenzone/Meeresbodenzone benthic zone
Bogengänge semicircular canals
Bogennervatur/Bogenaderung arched venation, arciform venation, arcuate venation (camptodrome)
Bohle plank
Bohrkern/Kern *geol/paleo* drill core, core
Bollinger Körper (viraler Einschlusskörper) Bollinger body, Bollinger's granule (inclusion body)
Bolzenflug/Bogenflug bounding flight
Bombacaceae/Wollbaumgewächse cotton-tree family, silk-cotton tree family, kapok-tree family
Bonitierung *agr* **(Boden)** classification of soil, valuation
Bonitur *stat* notation, scoring
Bor boron
Bor(r)etschgewächse/ Rau(h)blattgewächse/ Boraginaceae borage family
Borke/Rhytidom tertiary bark, dead outer bark, rhytidome
Borste bristle, seta, chaeta
Borste (Sinnesborste) sensory bristle, sensory seta, sensory chaeta
Borsten.../mit Borsten versehen/ borstig bristly, bristle-bearing, setose, setaceous, chaetigerous, chaetiferous, chaetiphorous
borstenartig bristle-like, setaceous, chaetaceous
Borstenbildungszelle/Chaetoblast (Anneliden) chaetoblast
borstenförmig bristle-shaped, setiform, chaetiform
"Borstenfüßer"/Ringelwürmer/ Gliederwürmer/Anneliden segmented worms, annelids
Borstenkiefer/Pfeilwürmer/ Chaetognathen arrow worms, chaetognathans
Borstenporlinge/ Hymenochaetaceae hymenochaete family
Borstenschwänze/Thysanuren (Felsenspringer/Fischchen) bristletails, thysanurans

Braunwurzgewächse

Borstenwürmer/Chaetopoden/ Chaetopoda (Polychaeten & Oligochaeten) bristle worms, chaetopods (annelids with chaetae: polychetes and oligochetes)
Borstenwürmer/Vielborster/ Polychaeten bristle worms, polychetes, polychete worms
Borstgras matgrass
borstig (kurzborstig) hispid
borstig (mit kurzgestrichenen Borsten/striegelig) strigose
borstig (rauhaarig) hirsute
borstig/mit Borsten versehen/ Borsten... bristly, bristle-bearing, setose, setaceous, chaetigerous, chaetiferous, chaetiphorous
Bortensoral/Randsoral (Flechten) marginal soralium
bösartig/maligne malignant
Bösartigkeit/Malignität malignancy
Böschung (künstliche) embankment
Böschung (Uferböschung/ Flußböschung) bank, riverbank, embankment
Böschung/steiler Abhang/ Steilabbruch slope, scarp, escarpment
Botanik botany
Botaniker botanist
Botanischer Garten botanical garden, botanic garden
Botanisiertrommel vasculum
Boten-RNA/mRNA/Messenger-RNA messenger RNA, mRNA
Botenstoff messenger
 • **sekundärer Botenstoff/zweiter Bote** second messenger
Bothrosom/Sagenosom/ Sagenogenetosom bothrosome, sagenogen, sagenogenetosome
Bothryoidgewebe (Hirudineen) bothryoidal tissue
bovine spongiforme Enzephalopathie (BSE) bovine spongiform encephalopathy (BSE)
Boviste/Stäublinge/Lycoperdales puffballs
Bowman-Kapsel/Bowmansche Kapsel Bowman's capsule, glomerular capsule
brach liegen lie fallow
Brache/Brachland/Brachfeld (unbebauter Acker) *agr* fallow
Brachialmuskulatur brachiomeric musculature
Brachiation/Hangeln/ Schwingklettern brachiation
Brachidium/Armgerüst (Brachiopoden) brachidium
brachliegen lie fallow

Brachsenkrautgewächse/Isoetaceae quillwort family
Brachsenregion *limn* bream zone
Brackwasser brackish water (somewhat salty)
Bradytelie bradytely
bradytelisch bradytelic
Brakteole/Tragschuppe/ Deckschuppe/Tragblatt zweiter Ordnung bracteole, bract-scale, bractlet, secondary bract
Branchialbogen/Kiemenbogen branchial arch, gill arch, visceral arch, gill bar
Branchialmuskulatur branchiomeric musculature
Branchiostagalmembran/ Kiemenhaut branchiostegal membrane, branchiostegous membrane, gill membrane
Branchiostegalstrahl/ Kiemenhautstrahl (Radius branchiostegus) *ichth* branchiostegal ray
Brandfläche burned area, area devastated by fire
Brandpilze/Flugbrandpilze/ Ustilaginales smuts, smut fungi (bunt funi/brand fungi)
Brandrodung slash-and-burn
Brandrodungsfeldbau slash-and-burn agriculture
Brandschutz fire protection, fire prevention
Brandung surf, breakers
Brandungslängsströmung/ Brandungslängsstrom longshore current
Brandungsrückströmung rip current
Brandungszone surf zone
Branntwein brandy
Brassicaceae/Kreuzblütler/ Kreuzblütlergewächse/Cruciferae cabbage family, mustard family
Brauchholz lumber, timber
Braunalgen/Phaeophyceae brown algae, phaeophytes
brauner Körper (Bryozoen) brown body
Braunfäule/Destruktionsfäule brown rot
Braunkohle brown coal, lignite
Braunkohlenwälder tertiary swamp forests
Braunmoor fen
Brauntang kelp (brown seaweed: Laminariales)
Braunwurzgewächse/ Rachenblütler/Scrophulariaceae figwort family, foxglove family, snapdragon family

Brautflug nuptial flight
brechen (bei Übelkeit) vomit
Brechnussgewächse/
Strychnosgewächse/Loganiaceae
logania family
Brechung, optische/Refraktion
optical refraction
Brechungsindex/
Brechungskoeffizient/Brechzahl
refractive index, index of refraction
Brechungsvermögen refractivity
Brechungswinkel refracting angle
Brechzentrum vomiting center
Breiapfelgewächse/
Sapotegewächse/Sapotaceae
sapodilla family
Breitbandantibiotikum
broad-spectrum antibiotic
Breitengrad degree of latitude;
parallel
Breitfußschnecken/Seehasen/
Aplysiacea/Anaspidea sea hares
brennbar combustible, flammable
Brennbarkeit combustibility,
flammability
Brennebene focal plane
Brennerei distillery
Brennereihefe distiller's yeast
Brennfleckenkrankheit/
Anthraknose anthracnose
Brennhaar stinging hair,
urticating hair, trichome
Brennholz firewood, fuelwood
Brennpunkt focal point, focus
Brennweite focal length
Brennwert caloric value
Brenztraubensäure (Pyruvat)
pyruvic acid (pyruvate)
Brett board, plank
Brettwurzel buttress root,
plank buttress
Bries/Thymus/Thymusdrüse (Hals-/
Brustthymus) thymus (gland)
Brille (z. B. Schlangen) spectacle
Brillenträgerokular *micros* spectacle
eyepiece, high-eyepoint ocular
Brilliantrot *micros* vital red
Brise *meteo* breeze
• **Landbrise** land breeze
• **Meeresbrise** sea breeze
Bromatium/Gongylidie/"Kohlrabi"
bromatium, gongylidium
(*pl* gongylidia)
Bromcyan-aktiviertes Papier (CBA-
Papier) cyanogen bromide activated
paper (CBA-paper)
Bromcyanspaltung
cyanogen bromide cleavage
Bromelain bromelain
Bromelien/Ananasgewächse/
Bromeliaceae bromeliads,
bromelia family, pineapple family

Bronchiole/Bronchiolus/Bronchulus
bronchiole
Bronchus/Ast der Luftröhre
(*pl* **Bronchien**) bronchus
(*pl* bronchi)
Broschüre/Informationsschrift
brochure, pamphlet
Bruch-Fusion/Bruch und
Wiedervereinigung *gen* breakage-
fusion, breakage and reunion
Bruch-Fusions-Brücke *gen*
breakage-fusion bridge
Bruchfrucht/Gliederhülse/
Gliederfrucht/Klusenfrucht
loment, lomentum,
lomentaceous fruit, jointed fruit
Bruchholz woody debris
Bruchkapsel *bot* septicidal capsule
Bruchstelle *gen* breakpoint
Bruchwald/Bruchwaldmoor/
Bruchmoor/Sumpfwald/Waldmoor
carr (fen woodland),
swamp woods/forest,
wooded swamp, paludal forest
Bruchwaldtorf/Fen fen
Brücke/Varolsbrücke/Pons varoli
pons varolii
Brücke-Muskel/Brückescher Muskel
Brucke's muscle (meriodonal fibers
of ciliary muscle)
Brückenbeuge *embr* pontine flexure
Brückenechsen/Rhynchocephalia
(*Sphenodon*) rhynchocephalians
Brückenfaden (Spinnennetz) *arach*
bridge line
brüllen (Raubtiere) roar, bellow
brummen (Bär) growl
Brunft (Hirsch) rut, courting (deer)
Brunftdrüse/Brunftfeige (Gemse)
scent gland
Brunftzeit/Paarungssaison (Hirsch)
rutting time/season,
courting/mating season
Brunnenkrebse/Bathynellacea
bathynellaceans
Brunnenwasser well water
Brunnersche Drüse/Duodenaldrüse
Brunner's gland, duodenal gland
Brunst rut (male), heat (female)
Brunst/Östrus estrus
• **Diöstrus** diestrus
• **Nachbrunst/Metöstrus**
metestrus
• **Vorbrunst/Proöstrus** proestrus
brunsten/brunften rut (male),
be in heat (female), court
brünstig/in der Brunst/
geschlechtlich erregt
rutting (male), in heat (female),
sexually aroused
Brunstschwiele sexual swelling
(callosity)

BSE **B**

Brunstzeit rutting season, season of heat
Brunstzyklus/Östruszyklus estrous cycle, estrus cycle, estral cycle
Brust breast (pectus); thorax, chest
• **weibliche Brust/Busen** breast, mamma, bosom
Brustatmung/Thorakalatmung thoracic respiration, costal breathing
Brustbein/Brustfuß/Thorakalfuß/ Thorakalbein thoracic leg
Brustbein/Clavicula (Geflügel) collarbone, clavicle (wishbone *see* Gabelbein)
Brustbein/Sternum breastbone, sternum
Brustbeinkamm/Carina breastbone keel/ridge, carina
Brustdrüse/Milchdrüse mammary gland
Brustfell/Pleura parietalis pleura
Brustflosse pectoral fin
Brustfuß (an Pereion) pereiopod, walking leg (attached to pereion)
Brustgang thoracic duct
Brustgürtel pectoral girdle
Brusthöhendurchmesser (BHD) diameter at breast height (dbh)
Brusthöhle/Brustraum/ Thorakalraum thoracic cavity
Brustkasten/Brustkorb (Thorax) rib basket, rib cage (chest/thorax)
Brustplatte/Bruststück/Sternum (Insekten) ventral plate, sternum
Brustraum/Thorakalraum/ Brusthöhle thoracic cavity
Brustschild/Bauchplatte/ Bauchpanzer (Schildkröten/Vögel) plastron, plastrum
brustständig (Flossen) located in the shoulder region
Bruststück (Schlachtvieh) brisket
Brustwarze nipple
• **Warzenhof/Areola mammae** areola
Brustwirbel/Thorakalwirbel thoracic vertebra
Brut *allg* brood, hatch
Brut/Gelege/Eigelege/Nest mit Eiern *orn* clutch
• **Fischbrut** fry
Brutablösung nest relief
Brutbecher/Brutkörbchen gemma cup
Brutbeutel/Marsupium brood pouch, marsupium
Brutdauer/Inkubationszeit breeding period, incubation period
brüten brood, breed, incubate
Brutfleck *orn* brood spot, brood patch

Brutfürsorge/Brutpflege brood provisioning, brood care, brooding, parental care
Brutgebiet breeding grounds, breading area
Brutgespinst/Eigespinst (Schutzgespinst für Jungspinnen) nursery web (nursery tent)
Brutkapsel brood capsule
Brutkleid nuptial dress, nuptial plumage, breeding plumage, courtship plumage
Brutknolle (*Gladiolus*) cormel, cormlet
Brutknospe/Brutspross/Bulbille brood bud, bulbil
Brutkörper (unterirdischer Zwiebelbrutkörper) bulblet
Brutkörper/Brutkörperchen brood body, gemma (*pl* gammae)
Brutparasit/Brutschmarotzer brood parasite
Brutparasitismus/ Brutschmarotzertum brood parasitism
Brutpflänzchen (z. B. Kalanchoe) adventitious plantlet, foliar plantlet
Brutpflege/Brutfürsorge brood provisioning, brood care (parental care), brood protection
Brutpflegesystem, kooperatives communal breeding system, cooperative breeding
Brutplatte/Oostegite (Krebse) oostegite
Brutraum (*Daphnia*) brood chamber
Brutschmarotzer/Brutparasit brood parasite
Brutschrank incubator
Brutspross/Brutknospe/Bulbille bulbil
Brutstätte breeding place, breeding ground, spawning ground (fish)
Bruttoprimärproduktion gross primary production (GPP)
Bruttoproduktion/ Gesamtproduktion gross production
Bruttoproduktivität gross productivity
Brutzeit (Dauer des Ausbrütens bis zum Schlüpfen) hatching time, breeding period
Brutzeit/Brützeit (Jahreszeit) breeding season
Brutzwiebel/Zeh (Reservestoffe in Blattorganen) offset bulb, bulblet, bulbil
BSE (bovine spongiforme Enzephalopathie) BSE (bovine spongiform encephalopathy)

BTA (biologisch-technischer Assistent) biology lab technician, biological lab assistant

Bubonenpest/Beulenpest (*Yersinia pestis*) bubonic plague

Buccalapparat/Oralapparat/ Mundapparat (Ciliaten) oral apparatus, ingestion apparatus, mouth

Buccalhöhle (Ciliaten) buccal cavity

Buchengewächse/Becherfrüchtler/ Fagaceae beech family

Bücherläuse/Psocoptera/ Copeognatha book lice, psocids

Buchkiemen/Kiemenbeine/ Blattbeine/Blattfüße (dichtstehende Kiemenlamellen: Xiphosuriden) book gills (gill book)

Buchlunge/Fächerlunge/ Fächertrachee book lung

Buchmagen/Blättermagen/ Vormagen/Omasus/Psalter (Wiederkäuer) third stomach, omasum, manyplies, psalterium

Buchsbaumgewächse/Buxaceae box family

Bucht *mar* bay, bight; *geol* basin; *anat/bot* sinus
 • **kleine Bucht (am Meer mit kleiner Mündung)** cove

Buchtenfarngewächse/ Hypolepidaceae (inkl. Adlerfarngewächse) hypolepis family (incl. bracken ferns)

buchtig/gebuchtet sinuate

Buckel/Erhebung hump, bulge, knoll, mound

Buddlejaceae/ Sommerfliedergewächse buddleja family

Bufonin bufonin

Bufotenin bufotenine

Bufotoxin bufotoxin

Bug (Schlachtvieh: Bugstück/ Schulterstück) shoulder, bladebone; chuck (beef)

Buggelenk point of shoulder

Bulbille (oberirdische Brutknospe) bulbil
 • **Brutknöllchen/Achsenbulbille** axillary stem tuber
 • **Brutzwiebel/Zeh (Reservestoffe in Blattorganen)** offset bulb, bulblet
 • **Wurzelbulbille (*Ficaria verna*)** root bulbil

Bulbourethraldrüse/Cowpersche Drüse/Cowper-Drüse bulbourethral gland, Cowper's gland

Bulbus olfactorius/Riechhügel/ Riechkolben olfactory bulb, olfactory dome (sensory dome)

Bulle (adultes männliches Tier: Rind/Elefant/Wal/Seelöwe) bull

Bult/Bülte hummock, hillock, tussock

Bültgras/Tussockgras tussock gras

Bündel/Faszikel/Faszikulus bundle, fascicle (bundle/tuft of fibers)

Bündel/Leitbündel bundle, vascular bundle

Bündelrohr (Farne) hollow vascular cylinder with internal pith (ferns)

Bündelscheide bundle sheath
 • **erweiterte Bündelscheide** bundle-sheath extension

buntblättrig variegated-leaved

Buntblättrigkeit/Scheckung/ Variegation variegation

Buntsandstein (Epoche) Lower Triassic

Bürette buret, burette

Burg (Wohnquartier) *zool* burrow, lodge
 • **Bieberburg** beaver lodge

Burgess-Schiefer Burgess shale

Burmanniagewächse/ Burmanniaceae burmannia family

Bursa copulatrix/Begattungstasche bursa copulatrix, genital pouch

Bursa Fabricii bursa of Fabricius

Bursalorgan Lang's vesicle

Burseraceae/Balsambaumgewächse torchwood family, incense tree family, frankincense tree family

Bürstensaum/Stäbchensaum/ Mikrovillisaum/Rhabdorium brush border

Bürstenzunge/hystrichoglosse Radula hystrichoglossate radula

Bursterneuron burster neuron

Bürzel rump, tail, uropygium

Bürzeldrüse uropygial gland, preen gland, oil gland

Busch bush

Büschel bunch, cluster, tuft

Büschel/Faszikel/Faszikulus (Infloreszenz) cyme with very short pedicles, fascicle

Büschelkiemenartige/ Seenadelverwandte/ Seepferdchenverwandte/ Syngnathiformes sea horses & pipefishes and allies

Büschelwurzelsystem (Gräser) fibrous root system

Buschfeuer brush fire

Buschformation shrubland

buschig bushy, shrubby, fruticose

Buschland brush, scrubland

Buschwald maquis
Buschwerk scrub, shrubbery
Butomaceae/
 Wasserlieschgewächse/
 Schwanenblumengewächse
 flowering rush family

Buttersäure/Butansäure (Butyrat)
 butyric acid, butanoic acid
 (butyrate)
Buxaceae/Buchsbaumgewächse
 box family

C-Banding

C-Banding/Centromer-Banding
C-banding (centromere-banding)
C-Form/C-Konformation (der DNA)
C form (of DNA)
c-onc (zelluläres Onkogen) c-*onc*
(cellular oncogene)
C_0t-Analyse/C_0t-Wert
(*sprich* kott; Produkt aus
DNA-Gesamtkonzentration zur
Zeit 0 und Hybridisierungszeit t)
C_0t-analysis/value (*pronounce* cot;
product of DNA concentration at
time 0 and hybridization time t)
CAAT-Box (Teil der
Nucleotidsequenz im
eukaryotischen Promotor) *gen*
CAAT box (component of
nucleotide sequence of eukaryotic
promoter)
Cactaceae/Kaktusgewächse/
Kakteen cactus family
Cadaverin cadaverine
Cadherin cadherin
Caecotrophie/Coecotrophie/
Coecophagie (Kaninchen/
Meerschweinchen) refection
Caesalpinogewächse/
Johannisbrotgewächse/
Caesalpiniaceae
caesalpinia family
Cala-Azar/Kala-Azar/schwarzes
Fieber/viszerale Leishmaniasis
Cala-Azar, kala azar
Calamiten/Schachtelhalmbäume/
Calamitaceae calamites, calamite
family, giant horsetail family
calcicol/kalzikol/kalkhold calcicole
Calciferol/Ergocalciferol
(Vitamin D_2) calciferol,
ergocalciferol
calcifug/kalzifug/kalkfliehend/
kalkmeidend calcifuge
Calciol/Cholecalciferol (Vitamin D_3)
cholecalciferol
Caldarium/Warmhaus caldarium,
heated greenhouse, hot-house
Callitrichaceae/
Wassersterngewächse water
starwort family, starwort family
Callus-Kultur/Kallus-Kultur
callus culture
Calmodulin calmodulin
Calvin-Zyklus/Calvin-Cyclus
Calvin cycle
Calycanthaceae/Gewürzsträucher
spicebush family,
strawberry-shrub family
Calyceraceae/Kelchhorngewächse
calycera family
Calyx (kelchförmiger Körper der
Crinoiden)
calyx, crown

Cambium/Kambium cambium
● **etagiertes Cambium/Stockwerk-**
Cambium storied cambium,
stratified cambium
● **Faszikularcambium**
fascicular cambium
● **Fusiformcambium**
fusiform cambium
● **Korkcambium/Phellogen**
cork cambium, phellogen
● **nichtetagiertes Cambium**
nonstoried cambium,
nonstratified cambium
● **Wundcambium** wound cambium
● **Zwischenbündelcambium**
interfascicular cambium
Cambrium/Kambrium
(erdgeschichtliche Periode)
Cambrian, Cambrian Period
Campanulaceae/
Glockenblumengewächse
bellflower family, bluebell family
campylotrop/kampylotrop
(Samenanlage) campylotropous,
bent
Canellaceae/Kaneelgewächse
wild cinnamon family,
white cinnamon family, canella family
Cannabaceae/Hanfgewächse
hemp family
Cannaceae/Blumenrohrgewächse
canna family,
Queensland arrowroot family
CAP-Stelle (Anheftungspunkt des
Katabolitaktivatorproteins)
CAP site (attachment point for the
catabolite activator protein)
Cap-Struktur (modifiziertes 5'-Ende
im eukaryotischen mRNA-
Molekül) cap structure (modified
5' end of eukaryotic mRNA)
Capparidaceae/Capparaceae/
Kaperngewächse caper family
Caprifoliaceae/Geißblattgewächse
honeysuckle family
Caprinsäure/Decansäure (Caprinat/
Decanat) capric acid, decanoic acid
(caprate/decanoate)
Capronsäure/Hexansäure
(Capronat/Hexanat) caproic acid,
capronic acid, hexanoic acid
(caproate/hexanoate)
Caprylsäure/Octansäure (Caprylat/
Octanat) caprylic acid, octanoic
acid (caprylate/octanoate)
Capsid/Kapsid capsid (viral shell)
Capsomer/Kapsomer capsomere
(virion: morphological unit)
Captacula/Fangfädenbüschel
(Scaphopoda) captacula
Carapax (Insekten/Rückenpanzer
der Schildkröte u. a.) carapace

431 chaotrope Substanz C

Carapaxdrüse/Y-Organ molting
gland, Y organ
**Carbonsäuren/Karbonsäuren
(Carbonate/Karbonate)** carboxylic
acids (carbonates)
Carboxysom carboxysome,
polyhedral body
carcinoembryonales Antigen (CEA)
carcinoembryonic antigen
Cardia/Cardiaregion cardiac region
Caricaceae/Melonenbaumgewächse
papaya family
Carinaten *orn* carinate birds
Carnitin (Vitamin T) carnitine
(vitamin B_T)
Carotidenkörper/Glomus caroticum
carotid body
**Carotin/Caroten/Karotin (Vitamin A
Vorläufer)** carotin, carotene
(vitamin A precursor)
Carpell/Karpel/Fruchtblatt carpel
Carpopodit carpopodite
Carrageen/Carrageenan
carrageenan, carrageenin
(*Irish moss* extract)
Caruncula/Karunkula caruncle
Caryophyllaceae/Nelkengewächse
pink family, carnation family
Casein casein
Cäsiumchloridgradient
cesium chloride gradient
Casparischer Streifen
Casparian strip
**Casuarinaceae/
Streitkolbengewächse**
she-oak family, beefwood family
Catenan/Concatenat catenane,
concatenate
Catenation/Ringbildung catenation
Caudex/Strunk/Wurzelstock
caudex, rootstock, stem base
**Caudex (Stamm von Palmen und
Baumfarnen)** caudex, trunk of tree
(palms and treeferns)
Caudicula/Kaudikula/Stielchen
caudicle (stalk of pollinium)
**Cauloid/Kauloid/Stämmchen
(Algen/Moose)**
caulid, stemlet, stipe
CBA-Papier CBA-paper (cyanogen
bromide activated paper)
**cccDNA (DNA aus kovalent
geschlossenen Ringen)** cccDNA
(covalently closed circles DNA)
CD (Differenzierungscluster) *gen*
CD (cluster of differentiation)
**CDE-Elemente (DNA-
Sequenzelemente am Centromer)**
gen CDE (centromere DNA
sequence elements)
cDNA (komplementäre DNA) cDNA
(complementary DNA)

cDNA-Bibliothek cDNA library
**CDR (komplementaritäts-
bestimmende Region)**
CDR (complementarity-determining
region)
**Celastraceae/
Spindelbaumgewächse/
Baumwürgergewächse**
spindle-tree family, staff-tree family,
bittersweet family
**Centimorgan (Einheit für
genetische Rekombination)**
centimorgan (unit of genetic
recombination)
Centriol/Zentriol centriole
centroacinäre Zelle
centro-acinar cell
Centromer/Zentromer centromere
Centromer-Banding/C-Banding
centromere banding, C-banding
Centroplast/Zentralkorn
centroplast, central granule,
axoplast
Cephalisation/Kopfbildung
cephalization, head development
Cephalotaceae/Krugblattgewächse
Australian pitcher-plant family
**Cephalotaxaceae/
Kopfeibengewächse** cephalotaxus
family, plum yew family
Cerago/Bienenbrot cerago,
beebread
ceratitische Lobenlinien
ceratitic suture lines
**Ceratophyllaceae/
Hornblattgewächse**
hornwort family
Cercarie/Zerkarie/Schwanzlarve
cercaria
Cercidiphyllaceae/Katsuragewächse
katsura-tree family
**Cercus/Aftergriffel
(Schwanzanhang)** cercus,
cercopod
Cerebralganglion cerebral ganglion
Cerebralkommissur
cerebral commissure
Cerebrosid cerebroside
Cerotinsäure/Hexacosansäure
cerotic acid, hexacosanoic acid
**CGH (vergleichende
Genomhybridisierung)** CGH
(comparative genome hybridization)
Chaetotaxie chaetotaxy
**Chagas Krankheit (*Trypanosoma
cruzi*)** Chagas disease
**Chamaephyt (Halb- und
Zwergsträucher)** chamaephyte
Chaostheorie chaos theory
chaotrope Reihe chaotropic series
chaotrope Substanz
chaotropic agent

C Chaperon 432

**Chaperon/molekulares Chaperon/
Begleitprotein** chaperone protein,
chaperone, molecular chaperone
Chaperonin chaperonin
Charakterart/Leitart
character species
Charakterzug/Eigenschaft/Merkmal
trait, character
chasmogam chasmogamous
Chelat/Komplex chelate
Chelatbildner/Komplexbildner
chelating agent, chelator
Chelatbildung/Komplexbildung
chelation, chelate formation
**Chelicere/Chelizere/Kieferfühler/
Scherenkiefer/Klaue/Fresszange**
chelicera, fang, cheliceral fang
**Chelicerengrundglied/
Chelicerenbasalsegment/Paturon**
paturon, basal segment of chelicera
Chelifore chelifore
Chemiosmose chemiosmosis
**chemiosmotische Hypothese/
Theorie**
chemiosmotic hypothesis/theory
chemische Bindung chemical bond
chemische Komplexität
chemical complexity
chemische Kriegsführung
chemical warfare
chemischer Kampfstoff
chemical warfare agent
**chemischer Sauerstoffbedarf
(CSB)** chemical oxygen demand
(COD)
**Chemisorption/chemische
Adsorption** chemisorption
Chemoaffinitäts-Hypothese
chemoaffinity hypothesis
chemoheterotroph
chemoheterotroph(ic)
Chemoheterotrophie
chemoheterotrophy
Chemokline (chem. Sprungschicht)
limn chemocline
chemolithotroph/chemoautotroph
chemolithotroph(ic),
chemoautotroph(ic)
**Chemolithotrophie/
Chemoautotrophie**
chemolithotrophy,
chemoautotrophy
Chemomorphose chemomorphosis
chemoorganotroph
chemoorganotroph(ic)
Chemoorganotrophie
chemoorganotrophy
Chemostat chemostat
Chemosynthese chemosynthesis
Chemotaxis (*pl* Chemotaxien)
chemotaxis (*pl* chemotaxes)
Chemotherapie chemotherapy

**Chenopodiaceae/
Gänsefußgewächse**
goosefoot family
Chi-Form Chi form
Chi-Quadrat-Test chi-square test
**Chiasma (*pl* Chiasmata)/
Überkreuzung** chiasma
Chiastoneurie/Streptoneurie
streptoneurous nerve pattern
**Chimäre (Pfropfhybride/
Zellhybride)** chimera
● **Meriklinalchimäre**
mericlinal chimera
● **Periklinalchimäre**
periclinal chimera
● **Pfropfchimäre/Pfropfbastard**
graft chimera
● **Sektorialchimäre**
sectorial chimera
**Chimären/Seedrachen/Seekatzen/
Holocephali** chimaeras, ratfishes,
rabbit fishes
Chinasäure chinic acid, kinic acid,
quinic acid (quinate)
Chinin chinine, quinine
Chinolin chinoline, quinoline
Chinolsäure chinolic acid
Chinon chinone
**Chipveredelung/Chipveredlung/
Span-Okulation** *hort*
chip budding
chiral chiral
Chiralität chirality
**Chiropatagium (Flughaut der
Fledermäuse)** chiropatagium
Chitin chitin
chitinös chitinous
Chitinschale chitinous shell
Chlamydospore/Gemme
chlamydospore
Chlor chlorine
Chloragogzelle (Oligochaeten)
chloragogen cell, chloragogue cell
**Chlorenchym/
Assimilationsparenchym**
chlorenchyma
Chloridzelle/Ionocyt *ichth*
chloride cell
chlorieren chlorinate
Chlorierung chlorination
Chlorogensäure chlorogenic acid
Chlorophyll chlorophyll
Chloroplast chloroplast
Chlorosom (Chlorobium-Vesikel)
chlorosome
**Choanozyt/Kragengeißelzelle/
Kragenzelle** collar cell, choanocyte
Cholecalciferol/Calciol (Vitamin D$_3$)
cholecalciferol
**Cholecystokinin-Pankreozymin
(CCK-PZ)** cholecystokinin-
pancreozymin (CCK-PZ)

Chromatographie

Cholesterin/Cholesterol cholesterol
cholinerg cholinergic
Cholsäure (Cholat) cholic acid (cholate)
Chorda dorsalis/Rückensaite/ Notocorda notochord
Chordalplatte chordal plate, notochordal plate
Chordascheide notochordal sheath
Chordatiere/Rückgrattiere/ Chordaten chordates
chordotonal chordotonal
Chordotonalorgan/Saitenorgan chordotonal organ
chorikarp/apokarp apocarpous
Chorioallantoisplazenta/ Zottenplazenta chorioallantoic placenta
Chorion/Eischale (Insektenei) chorion, eggshell (insect egg)
Chorion frondosum/Zottenhaut (mittlere Eihaut) chorion
Chorion laeve/Zottenglatze chorion laeve (nonvillous chorion)
Choriongonadotropin (hCG) human chorionic gonadotropin (hCG)
Chorionplazenta chorionic placenta
Chorionzotten chorionic villi
Chorionzotten-Biopsie chorion villi biopsy, chorionic villus biopsy, chorionic villus sampling
choripetal choripetalous
Chorisminsäure (Chorismat) chorismic acid (chorismate)
Chrom chromium
chromaffin chromaffin, chromaffine, chromaffinic
Chromatide chromatid
- **UESCE (ungleicher Austausch von Schwesterchromatiden)** UESCE (unequal exchange of sister chromatids)

Chromatidenkonversion chromatid conversion
Chromatin chromatin
Chromatinfaden chromatin thread
Chromatogramm chromatogram
Chromatograph chromatograph
Chromatographie chromatography
- **Affinitätschromatographie** affinity chromatography
- **Ausschlusschromatographie/ Größenausschluss- chromatographie** size exclusion chromatography (SEC)
- **Aussalzchromatographie** salting-out chromatography
- **Dünnschichtchromatographie (DC)** thin-layer chromatography (TLC)
- **enantioselektive Chromatographie** chiral chromatography
- **Festphasenchromatographie** bonded-phase chromatography
- **Flüssigkeitschromatographie** liquid chromatography (LC)
- **Gaschromatographie** gas chromatography
- **Gas-Flüssig-Chromatographie** gas-liquid chromatography
- **Gelpermeationschromato- graphie/Molekularsieb- chromatographie** gel permeation chromatography, molecular sieving chromatography
- **Größenausschlusschromato- graphie/Ausschluss- chromatographie** size exclusion chromatography (SEC)
- **Hochdruckflüssigkeitschromato- graphie/Hochleistungs- chromatographie** high-pressure liquid chromatography, high performance liquid chromatography (HPLC)
- **Immunaffinitätschromato- graphie** immunoaffinity chromatography
- **Ionenaustauschchromatographie** ion-exchange chromatography
- **Kapillarchromatographie** capillary chromatography
- **Molekularsiebchromatographie/ Gelpermeationschromato- graphie/Gelfiltration** molecular sieving chromatography, gel permeation chromatography, gel filtration
- **Papierchromatographie** paper chromatography
- **präparative Chromatographie** preparative chromatography
- **Säulenchromatographie** column chromatography
- **überkritische Fluidchromatographie/ superkritische Fluid- Chromatographie/ Chromatographie mit überkritischen Phasen** supercritical fluid chromatography (SFC)
- **Umkehrphasenchromatographie** reversed phase chromatography, reverse-phase chromatography
- **Verteilungschromatographie** partition chromatography
- **Zirkularchromatographie/ Rundfilterchromatographie** circular chromatography, circular paper chromatography

C Chromomer 434

Chromomer chromomere
Chromoplast chromoplast
Chromosom chromosome
- **akrozentrisches Chromosom**
  acrocentric chromosome
- **akzessorisches/zusätzliches Chromosom**
  accessory chromosome
- **azentrisches Chromosom**
  acentric chromosome
- **dizentrisches Chromosom**
  dicentric chromosome
- **ektopische Paarung**
  ectopic pairing
- **Fluoreszenzmarkierung ganzer Chromosomen**
  chromosome painting
- **Harlekin-Chromosomen**
  harlequin chromosomes
- **homologes Chromosom**
  homologous chromosome
- **isodizentrisches Chromosom**
  isodicentric chromosome
- **Lampenbürstenchromosom**
  lampbrush chromosome
- **Metaphasenchromosom**
  metaphase chromosome
- **metazentrisches Chromosom**
  metacentric chromosome
- **Minichromosom** artificial
  chromosome, minichromosome
- **polytäne Chromosomen**
  polytene chromosomes
- **Riesenchromosom**
  giant chromosome
- **Ringchromosom**
  ring chromosome
- **Satellitenchromosom**
  satellite chromosome
- **submetazentrisches Chromosom**
  submetacentric chromosome
- **telozentrisches Chromosom**
  telocentric chromosome
- **Urchromosom**
  ancestral chromosome
Chromosom mit mehreren Replikationsgabeln
multiforked chromosome
chromosomale Aberration/ Chromosomenaberration
chromosomal aberration, chromosome aberration
Chromosomen-vermittelter Gentransfer chromosome-mediated gene transfer
Chromosomenaberration/ chromosomale Aberration
chromosomal aberration, chromosome aberration
Chromosomenbestand/ Chromosomensatz chromosome complement, chromosome set

Chromosomenfehlverteilung/ Non-Disjunction nondisjunction
Chromosomenhopsen/-springen/ -wandern chromosome hopping, jumping, walking
Chromosomeninstabilität
chromosome instability
Chromosomenpaar/Bivalent
bivalent
Chromosomenpuff
chromosome puff
Chromosomensatz/ Chromosomenbestand
chromosome set, chromosome complement
Chromosomentheorie (der Vererbung) chromosome theory (of inheritance)
Chromozentrum chromocenter
chronisch chronic, chronical
Chronospezies chronospecies
chronotrop chronotropic
Chrysalis (Puppe der holometaboler Insekten) chrysalis (*pl* chrysalids/chrysalides/ chrysalises)
Chrysobalanaceae/ Goldpflaumengewächse cocoa-plum family, coco-plum family
Chylus/Darmlymphe chyle
Chylusblindsack chylific ventricle/ cecum
Chymosin/Labferment/Rennin
chymosin, lab ferment, rennin
Chymotrypsin chymotrypsine
Chymus/Speisebrei/Magenbrei
chyme
Ciliarkörper/Ziliarkörper/ Corpus ciliare ciliary body
Cilie/Zilie/Wimper/Flimmerhärchen
cilium
Ciliengrube (*Gnathostomulida*)
cilary pit
cilientragend/zilientragend/ bewimpert bearing cilia, cilium-bearing, ciliated, ciliferous
Cinnamonsäure/Zimtsäure (Cinnamat) cinnamic acid
Circulardichroismus/ Zirkulardichroismus
circular dichroism
Circumapicalband (Rotatorien)
circumapical band
Circumnutation circumnutation
Cirre cirrus
Cirrusbeutel cirrus pouch
***cis*-aktiver Lokus** *gen*
cis-acting locus
CISS (chromosomale *in-situ* Suppressionshybridisierung)
chromosomal *in situ* suppression hybridization

Cistron *gen* cistron
Cistrosengewächse/
Zistrosengewächse/
Sonnenröschengewächse/
Cistaceae rockrose family
Citrat-Zyklus/Citratcyclus/
Zitronensäurezyklus/
Tricarbonsäure-Zyklus/
Krebs-Cyclus citric acid cycle,
tricarboxylic acid cycle
(TCA cycle), Krebs cycle
Citronensäure/Zitronensäure
(Citrat) citric acid (citrate)
Citrullin/Zitrullin citrulline
Cladodium/Kladodium (Flachspross
eines Langtriebs) cladode,
cladophyll
Cladogenese/Kladogenese
cladogenesis
Cladogramm/Kladogramm
cladogram
Clearance/Klärung clearance
Cleptobiose/Kleptobiose
cleptobiosis
Clethraceae/Scheinellergewächse
clethra family, white-alder family,
pepperbush family
Clitoris/Klitoris/Kitzler clitoris
Clusiaceae/Guttiferae/
Hypericaceae/Hartheugewächse/
Johanniskrautgewächse
St. John's wort family,
mamey family, mangosteen family,
clusia family
Clusteranalyse cluster analysis
Cneoraceae/
Zwergölbaumgewächse/
Zeilandgewächse
spurge olive family
Cnide/Nesselkapsel/Nematocyste
cnida, thread capsule, urticator,
nematocyst
- **Durchschlagskapsel/Penetrant**
penetrant
- **Klebkapsel/Haftkapsel/Glutinant**
glutinant
- **Ptychonema/Ptychozyste (eine**
Astomocnide: *Ciergantharia*)
ptychocyst, tube cnida
- **Wickelkapsel/Volvent** volvent
Cnidoblast/Cnidocyt/Nesselzelle
cnidoblast, nematoblast, nematocyte,
stinging cell
coated pit/Stachelsaumgrübchen
coated pit
coated vesicle/Stachelsaumvesikel/
Stachelsaumbläschen/Korbvesikel
coated vesicle
Cochlospermaceae/
Nierensamengewächse
cochlospermum family, buttercup
tree family (silk-cotton tree family)

Code code
- **Ein-Buchstaben-Code**
one-letter-code
codieren/kodieren code, encode
codierender Strang/
kodierender Strang/Sinnstrang
coding strand, sense strand
Codierungskapazität/
Kodierungskapazität
coding capacity
codominant/kodominant
codominant
Codominanz/Kodominanz
codominance
Codon *gen* codon
- **Abbruchcodon/Stoppcodon/**
Terminationscodon
stop codon, termination codon,
terminator codon
- **Initiationscodon/Startcodon**
initiation codon
- **Nichtsinncodon/Nonsense-**
Codon nonsense codon
- **PTC (vorzeitiges Stoppcodon)**
PTC (premature termination
codon)
- **Satzzeichencodon**
punctuation codon
Codon-Nutzung codon usage
Codon-Präferenz codon preference
Coecotrophie/Caecotrophie/
Coecophagie (Kaninchen/
Meerschweinchen)
refection
Coenenchym coenenchyme
coeno-parakarp paracarpous
Coenobium/Cönobium/Zönobium
(*pl* Coenobien) coenobium
(*pl* coenobia), cell family
coenocytisch/coenozytisch
coenocytic, cenocytic (aseptate)
coenokarp/coeno-synkarp/synkarp
syncarpous
Coenosark coenosarc
Coenozyt/Coenocyt coenocyte,
cenocyte
coenozytisch/coenocytisch
coenocytic, cenocytic (aseptate)
Coenzym/Koenzym coenzyme
Coevolution/Koevolution
coevolution
Cofaktor cofactor
Coinzidenzfaktor/Koinzidenzfaktor
coefficient of coincidence
Cokonversion coconversion
Colchicaceae/Krokusgewächse/
Zeitlosengewächse
crocus family
Colchicin/Kolchizin colchicine
Coleoptile/Koleoptile/Keimscheide/
Keimblattscheide coleoptile,
plumule sheath

Coleorhiza 436

**Coleorhiza/Koleorhiza/
Wurzelscheide** coleorhiza,
root sheath, radicle sheath
Colinearität/Kolinearität
colinearity
Collenchym/Kollenchym
collenchyma
- **Kantencollenchym/
Eckencollenchym**
angular collenchyma
- **Lückencollenchym**
lacunar collenchyma
- **Plattencollenchym**
lamellar collenchyma,
tangential collenchyma
Collenzyt/Collencyt collencyte
**Colletere/Kolletere (Leimzotte/
Drüsenzotte)** colleter,
multicellular glandular trichome
(sticky/viscous secretions)
Colloblast/Kolloblast/Klebzelle
colloblast, lasso cell, adhesive cell
Cölom coelom, celom
Cölomflüssigkeit coelomic fluid,
celomic fluid
**Colon/Kolon (Vertebraten:
Grimmdarm)** colon
Colulus colulus
Columella/Gewebesäule columella
**Combretaceae/
Strandmandelgewächse**
Indian almond family,
white mangrove family
**Commelinengewächse/
Commelinaceae**
spiderwort family
**Compositae/Asteraceae/
Köpfchenblütler/Korbblütler**
sunflower family, daisy family,
aster family, composite family
**compound heterozygot/
zusammengesetzt-heterozygot**
compound heterozygote
Computertomographie
computed tomography (CT)
Concatemer concatemer
Concatenat/Catenan concatenate,
catenane
**Concentricycloidea/
Seegänseblümchen**
sea daisies, concentricycloids,
concentricycloideans
congene Stämme (Mäuse)
congenic strains (mice)
Conidie/Konidie/Knospenspore
conidium
Conidienträger/Konidienträger
conidiophore
Coniferylalkohol coniferyl alcohol,
coniferol
Connaraceae/Connaragewächse
connard family, zebra wood family

Conodonten/Conodontophorida
conodonts ("fascinating little
whatzits")
**Consensussequenz/
Konsensussequenz**
consensus sequence
Convolvulaceae/Windengewächse
bindweed family, morning glory
family, convolvulus family
Coomassie-Blau Coomassie Blue
Coprodaeum/Kotdarm coprodeum
**coprophag/koprophag/
kotfressend/dungfressend**
coprophagous, coprophagic,
dung-feeding
**Coprophage/Koprophage/
Kotfresser/Dungfresser**
coprophagist, dung feeder
**Coprophagie/Koprophagie/
Dungfressen/Kotfressen**
coprophagy, dung-feeding
**coprophil/koprophil/
mistbewohnend/dungbewohnend**
coprophilic, coprophilous
(thriving on dung or fecal matter)
Core/Kern/Mark core
Core-Enzym core enzyme
Core-Octamer core octamer
Corepressor/Korepressor
corepressor
Cori-Zyklus/Cori-Cyclus Cori cycle
**Coriariaceae/Gerbersträucher/
Gerberstrauchgewächse**
coriaria family
Cormidium (Siphonophoren)
cormidium
**Cornaceae/Hartriegelgewächse/
Hornstrauchgewächse**
dogwood family
Corona ciliata corona ciliata,
ciliary loop
Corpora allata corpora allata
Corpora cardiaca corpora cardiaca
Corpora cavernosa/Schwellkörper
cavernous bodies, erectile tissue
Corpora pedunculata/Pilzkörper
corpora pedunculata, pedunculate
bodies, mushroom bodies
Cortexschicht cortical layer,
cortical zone
**Corticoliberin/Corticotropin-
freisetzendes Hormon/
corticotropes Releasing-Hormon
(CRH)** corticoliberin, corticotropin-
releasing hormone (CRH),
corticotropin-releasing factor (CRF)
**Corticotropin/Kortikotropin/
adrenocorticotropes Hormon/
adrenokortikotropes Hormon
(ACTH)** corticotropin,
adrenocorticotropic hormone
(ACTH)

Cymodoceaceae

Cortischer Kanal cochlear duct
Cortisches Organ organ of Corti, spiral organ
Cortisol/Hydrocortison cortisol, hydrocortisone
Cortison/Kortison cortisone
Corylaceae/Haselnussgewächse hazel family
cos-**Stelle** *gen cos* site
Cosmid cosmid
Cosmin cosmine
Cosmoidschuppe cosmoid scale
Costa (große Längsader)/Kosta/ Rippe *entom* costa, rib
Costa/Rippe (Blattrippe) *bot* costa, rib, vein
Costalfalte costal fold
Costalfeld/Remigium costal field
Cotransduktion/Kotransduktion cotransduction
Cotransfektion/Kotransfektion cotransfection
Cotransformation/ Kotransformation cotransformation
cotranslational cotranslational
Coulter-Zellzählgerät Coulter counter
Cousin(e)/Kousin(e) (ersten/ zweiten Grades) (first/second) cousin
Cowpersche Drüse/Cowper-Drüse/Bulbourethraldrüse Cowper's gland, bulbourethral gland
Coxaldrüse coxal gland
Coxopodit coxopodite
Crassulaceae/Dickblattgewächse stonecrop family, sedum family, orpine family
Cremaster cremaster
Crinophagie crinophagy
Crista (*pl* **Cristae) (mitochondrial)** mitochondrial crista (*pl* cristae/cristas)
CRM+ (positiv für kreuzreagierendes Material) *gen* CRM+ (positive for cross-reacting material)
Cro-Repressor Cro repressor
Crossing over/ Überkreuzungsaustausch (homologer Chromatidenabschnitte)/ Überkreuzungsstelle crossing over, crossover
• **Mehrfachaustausch** compound crossing over
• **ungleiches Crossing over** unequal crossing over
Crotonsäure/Transbutensäure crotonic acid, α-butenic acid

Cruciferae/Kreuzblütler/ Kreuzblütlergewächse/ Brassicaceae cabbage family, mustard family
Cryptogrammaceae/ Rollfarngewächse parsley fern family, rock-brake fern family
Cryptophyt/Kryptophyt/Geophyt/ Erdpflanze/Staudengewächs cryptophyte, geophyte, geocryptophyte
Ctenoidschuppe/Kammschuppe ctenoid scale
Cucurbitaceae/Kürbisgewächse gourd family, pumpkin family, cucumber family
Cumaceen/Cumacea cumaceans
Cunoniaceae lightwood family
Cupressaceae/Zypressengewächse cypress family
Cupula/Kupula/Blütenbecher/ Fruchtbecher cupula, cupule
Curare curare
Cuscutaceae/Seidengewächse dodder family
Cuticula/Kutikula cuticle, cuticula
Cutikularisierung/Cutin-Auflagerung/Cutin-Anlagerung cuticularization
Cutinisierung/Cutin-Einlagerung cutinization
Cutis/Haut/eigentliche Haut cutis, skin
Cuvier'sche Schläuche Cuvierian tubules, tubules of Cuvier
Cyanastrumgewächse/ Cyanastraceae cyanastrum family
Cyanelle cyanelle
Cyatheaceae/Becherfarne/ Baumfarne cyathea family, tree fern family
Cyathium/Zyathium cyathium
Cycadaceae/Palmfarngewächse cycads, cycad family, cycas family
Cyclanthaceae/Scheinpalmen panama-hat family, jipijapa family, cyclanthus family
cyclisches AMP/zyklisches AMP/ Cyclo-AMP/Zyklo-AMP/cAMP (Adenosinmonophosphat) cyclic AMP, cAMP (adenosine monophosphate)
Cycloidschuppe/Rundschuppe cycloid scale
Cydippe-Larve cydippid larva
Cyme/Cymus/Zyma/Zyme/Zymus/ cymöser Blütenstand/zymöser Blütenstand cyme, cymose inflorescence
Cymodoceaceae/Tanggrasgewächse manatee-grass family

cymös/cymos/zymös/trugdoldig/ sympodial verzweigt cymose, sympodially branched
Cyperaceae/Riedgräser/ Riedgrasgewächse/Sauergräser sedge family
Cyphelle (*pl* **Cyphellen**) **(Flechten)** cyphella
• **Pseudocyphelle** pseudocyphella
Cypris-Larve cypris larva
Cyrillaceae/Lederholzgewächse cyrilla family, leatherwood family, titi family
Cystacanthus/Hakencyste (Acanthocephala) cystacanth (acanthocephalans)
Cyste/Zyste cyst
Cysteamin cysteamine
Cystein cysteine
Cysteinsäure cysteic acid
Cystenwand/Zystenwand cyst wall
Cystid cystid, *sensu lato:* zooecium
Cystidie cystidium
Cystin cystine
Cystokarp/Hüllfrucht cystocarp, cystocarpium
Cystozygote/Zygotenfrucht/ Zygokarp (Oospore) cystozygote (oospore)
Cytidin/Zytidin cytidine
• **Desoxycytidin** deoxycytidine
Cytidintriphosphat cytidine triphosphate
Cytochemie/Zytochemie/ Zellchemie cytochemistry
Cytochrom cytochrome
Cytogenetik cytogenetics
• **molekulare Cytogenetik** molecular cytogenetics
Cytohet cytohet
Cytokeratin/Zytokeratin cytokeratin
Cytokin/Zytokin cytokine (biological response mediator)
Cytokinese cytokinesis
Cytologie/Zytologie/Zellenlehre/ Zellbiologie cytology, cell biology
cytolytisch/zytolytisch cytolytic
cytopathisch/zytopathisch/ zellschädigend (zytotoxisch) cytopathic (cytotoxic)
Cytopharynx/Zytopharynx/ Zellschlund cytopharynx, gullet
Cytoplasma/Zytoplasma cytoplasm
cytoplasmatisch/zytoplasmatisch cytoplasmic
cytoplasmatische Vererbung/ zytoplasmatische Vererbung cytoplasmic inheritance
cytoplasmatischer Plaque/ zytoplasmatischer Plaque cytoplasmic plaque
Cytoproct/Zytoproct/Cytopyge/ Zytopyge/Zellafter cytoproct, cytopyge, cell-anus
Cytosin cytosine
Cytoskelett/Zytoskelett cytoskeleton
Cytosol/Zytosol cytosol
Cytosom cytosome, microbody
Cytostatikum/Zytostatikum (meist *pl* **Zytostatika/Cytostatika)** cytostatic agent, cytostatic
cytotoxisch/zytotoxisch cytotoxic

439 Darwinsche Fitness **D**

Dachpilze/Pluteaceae
Pluteus family
**dachziegelartig/dachig/
schindelartig überlappend**
imbricate, overlapping
Dactylopodit dactylopodite
Damm *allg* dam
- **Deich (am Meer)** dike
- **Erddamm/Erdwall** mound
- **Flussdamm** river embankment,
levee
Damm/Perineum perineum
**Damm.../den Damm betreffend/
Perineal.../perineal** perineal
Dammdrüse/Perinealdrüse
perineal gland
Dämmerung/Zwielicht
crepuscule, twilight
- **Abenddämmerung** dusk
- **Morgendämmerung** dawn
**dämmerungsaktiv/im Zwielicht
erscheinend** crepuscular
**Dämmerungssehen/skotopisches
Sehen** scotopic vision
Dämmerzone/dysphotische Zone
limn dysphotic zone
Dammschwiele (Schimpansen)
perineal swelling
Dampf vapor
- **Wasserdampf** water vapor, steam
Dampfraum-Gaschromatographie
head-space gas chromatography
**Damwild/Damhirsch (Gattung
Dama)** fallow deer
Dansylierung dansylation
Darm (*pl* **Därme**) intestine
(*pl* intestines)
- **Afterdarm/Enddarm/Hinterdarm
(Colon & Rectum)/Proctodaeum**
hindgut, proctodeum
- **Blinddarm/Darmdivertikulum/
Caecum** blind-ended
diverticulum, cecum
- **Colon/Kolon (Vertebraten:
Grimmdarm)** colon
- **Coprodaeum/Kotdarm**
coprodeum
- **Dickdarm** large intestines
- **Dünndarm** small intestines
- **Enddarm/Afterdarm/Hinterdarm
(Colon & Rectum)/Proctodaeum**
hindgut, proctodeum
- **Grimmdarm/Colon/Kolon** colon
- **Hinterdarm/Enddarm/Afterdarm/
Proctodaeum** hindgut,
proctodeum
- **Hüftdarm/Ileum** ileum
- **Kiemendarm/Pharynx** branchial
gut, branchial basket, pharynx
- **Kopfdarm** head gut, foregut
- **Kotdarm/Kotraum/Coprodaeum**
coprodeum

- **Kranzdarm/Jejunum
(Wiederkäuer)** coronal stomach,
coronal sinus
- **Leerdarm/Jejunum** jejunum
- **Mastdarm/Rectum** rectum
- **Mitteldarm** *sensu stricto*
mid intestine, midgut
- **Mitteldarm/Mesenteron/
"Magen"/Ventriculus (Insekten)**
midgut, mesenteron, ventricle,
ventriculus, "stomach"
- **Mitteldarm (Echinodermen)**
pyloric stomach
- **Nahrungsdarm** midgut, intestine
- **Rumpfdarm** midgut
- **Schwanzdarm** tail gut,
postanal gut
- **Spiraldarm** spiral intestine
- **Urdarm/Gastrocöl/Archenteron**
primitive gut, gastrocoel,
archenteron
- **Urodaeum** urodeum
- **Vorderdarm/Stomodaeum**
foregut, stomodeum
- **Vorderdarm/Vormagen/Kropf/
Ingluvies (Insekten/Vögel)** crop
- **Wurmfortsatz des Blinddarms/
Appendix/Appendix vermiformis**
appendix, vermiform appendix
- **Zwölffingerdarm/Duodenum**
duodenum
**darmassoziiertes lymphatisches
Gewebe**
gut-associated lymphatic tissue
(GALT)
Darmbein/Ilium/Os ilium
flank bone, ilium
**Darmblindsack/Darmdivertikulum/
Caecum** blind-ended diverticulum,
cecum
**Darmdivertikel (Mollusken/
Echinodermen)** digestive gland
**Darmentleerung/Stuhlgang/
Defäkation** defecation, egestion
**Darmfaserblatt/viscerales Blatt/
viszerales Blatt/Splanchnopleura**
splanchnopleure
Darmflora intestinal flora
Darmkanal *allg* gut
Darmkanal (Echinodermen)
digestive gland duct
Darmlymphe/Chylus chyle
Darmzotten/Villi intestinales
intestinal villi
Darre/Darrofen kiln, kiln oven
(for drying grain/lumber/tobacco)
darren kiln-dry
darstellen/synthetisieren *chem*
synthesize
Darstellung/Synthese *chem* synthesis
Darwinsche Fitness
Darwinian fitness

Dasselbeule

Dasselbeule (*Hypoderma bovis* et al.) cattle grub & botfly infestation
Datenanalyse data analysis
- **explorative D.** explorative data analysis
- **konfirmatorische D.** confirmatory data analysis

Datiscaceae/Scheinhanfgewächse datisca family, Durango root family
Dauerei/Latenzei resting egg, dormant egg (winter egg)
Dauerfrostboden/Permafrostboden permafrost soil
Dauergebiss/bleibende Zähne/ zweite Zähne/Dentes permanentes permanent dentition, permanent teeth
Dauergesellschaft *ecol* permanent community
Dauergewebe permanent tissue, secondary tissue
Dauerknospe/Hibernaculum hibernaculum, winter bud
Dauerkontraktion/Tetanus tetanus
Dauerkultur *agr* permanent crop, maintenance crop, permaculture
Dauerlarve dauer larva (*not translated*: temporarily dormant larva)
Dauerpräparat *micros* permanent mount/slide
Dauerspore/Hypnospore persistent spore, dormant spore, resting spore, hypnospore
Dauerweide *agr* permanent pasture
Dauerzelle permanent cell; (*unbeweglich:* Akinet) resting cell, akinete
Daumen/Pollex thumb, pollex
Daumenfeder alular quill
Daumenfittich/Afterflügel/ Nebenflügel/Alula/Ala spuria (Federn am 1. Finger) alula, spurious wing, bastard wing
Daune/Dune/Dunenfeder/ Flaumfeder/dunenartige Feder down, down feather, plumule
Davalliaceae/ Hasenfußfarngewächse davallia family, bear's-foot fern family
DC (Dünnschichtchromatographie) TLC (thin layer chromatography)
Deckakt serving
Deckblatt/Tragblatt (Blüte) bract, subtending bract
Decke (z.B. Körperdecke) cover
Decke/Deckfeder/Tectrix *orn* covert, wing covert, protective feather, deck feather, tectrix (*pl* tectrices)
- **kleine Decke (Flügeldecke/ Flügelfeder)** lesser covert, minor covert
- **mittlere Decke** median covert
- **Randdecke** marginal covert

Decke/Haube/Tegmentum (Gehirn) tegmentum
Deckel/Operkulum/Operculum (z.B. Kiemendeckel) lid, opercle, operculum
Deckel.../gedeckelt/mit Deckel versehen operculate, opercular, operculiferous, bearing a lid
deckelförmig/deckelartig lid-like, operculiform
Deckelkapsel *bot* lid capsule, circumscissile capsule, pyxis, pyxidium
Deckeltopfbäume/Lecythidaceae brazil-nut family, lecythis family
decken/begatten/bespringen cover, serve, leap
Deckenmoor blanket bog, blanket mire, climbing bog
Deckennetz *arach* horizontal sheetweb; (Baldachinspinnen: Zeltdachnetz) hammock web
Deckfeder/Decke/Tectrix *orn* covert, wing covert, protective feather, deck feather, tectrix (*pl* tectrices)
Deckflügel/Vorderflügel forewing, front wing, tegmina
Deckglas *micros* coverslip, coverglass
Deckhaar (Leithaare + Grannenhaare) guard hair
Deckhengst/Zuchthengst/ Schälhengst/Beschäler studhorse, stud
Deckknochen/Hautknochen/ Belegknochen dermal bone
Deckmembran/Tektorialmembran/ Membrana tectoris tectorial membrane
Decksamer/Bedecktsamer/ Angiosperme angiosperm, anthophyte
Deckschuppe *bot* bract-scale, subtending bract, secondary bract
Deckspelze lemma, lower palea, outer palea
Deckstück (Siphonophora) bract
Deckungsgrad coverage percentage, coverage level
Deckungswert cover value
Deckzelle/Stützzelle cover cell, covering cell, supporting cell
Dedifferenzierung/ Entdifferenzierung dedifferentiation
Defäkation/Darmentleerung/ Stuhlgang/Klärung/Koten defecation, egestion
Defäkationsdrang defecation urge
Defäkationsreflex defecation reflex

Defäkationsstörung defecation disturbance
Defäkationstraining bowel training
Defektgen defective gene
Defektmutante defective mutant
Defensivmedizin defensive medicine
Deflation/Ausblasung *geol* (*siehe auch:* **Korrasion**) deflation
Deflationskessel/Windmulde blowout, deflation basin
Deformation/Verformung/Formänderung deformation
Degeneration degeneracy
degenerieren/entarten degenerate
Dehnbarkeit expansivity
dehnen (Muskel) stretch, extend
Dehnung (Muskel) stretch, stretching, extension
Dehnungsreflex/myotatischer Reflex stretch reflex, myotatic reflex
Dehnungsrezeptor (Muskel) stretch receptor
Dehydratation/Entwässerung dehydration
dehydratieren/entwässern dehydrate
dehydrieren dehydrogenate
Dehydrierung dehydrogenation
Deich (Fluss) bank, embankment, levee, dike
Deich (Meer) dike
Dekontamination/Dekontaminierung/Reinigung/Entseuchung decontamination
dekontaminieren/reinigen/entseuchen decontaminate
Dekussation/Wirtelung decussation
dekussiert/gekreuzt/kreuzgegenständig decussate, crossed
Deletion (Mutation unter Verlust von Basenpaaren) deletion
Deletionsanalyse *gen* deletion analysis
Deletionskartierung *gen* deletion mapping
Deletionsmutation deletion mutation
Dem deme
Demographie demography (study of populations: growth rates/age structure)
Demökologie population ecology
denaturieren denature
denaturierendes Gel denaturing gel
Denaturierung denaturation, denaturing
Dendrit/Markfortsatz dendrite
Dendritenscheidezelle dendritic sheath

Dendritenspine dendritic spine
Dendritenzelle/dendritische Zelle dendritic cell
Dendrogramm (phylogenetische Beziehungen) dendrogram
Dendrologe dendrologist
Dendrologie (Lehre von Bäumen u. Holzgewächsen) dendrology (study of trees)
Dennstaedtiaceae (Schüsselfarngewächse) dennstaedtia family (cup fern family)
dephosphorylieren dephosphorylate
Dephosphorylierung dephosphorylation
Depolarisation depolarization
depolarisieren depolarize
Depositfresser deposit feeder
Depurinisierung *gen* depurination
derb coarse, sturdy, rough, robust, tough, hard
Derbholz crude timber, crude wood
Derivat derivative
Derivatisation derivatization
derivatisieren derivatize
dermal dermal, dermic, dermatic
Dermaldrüse/Hautdrüse dermal gland
Dermatom dermatome
Dermis/Korium/Corium/Lederhaut dermis, corium, cutis vera, true skin
dermo-epidermale Junktionszone dermo-epidermal junction zone
Dermomyotom dermomyotome
Dermoptera/Pelzflatterer/Riesengleitflieger dermopterans, colugos, flying lemurs
Desamidierung deamidation, deamidization, desamidization
Desaminierung deamination, desamination
Desinfektionsmittel disinfectant
desinfizieren disinfect
Desinfizierung/Desinfektion disinfection
Desmonem desmoneme
Desmosom/Macula adhaerens desmosome, bridge corpuscule, bridge corpuscle, macula adherens
- **Gürteldesmosom/Banddesmosom** belt desmosome
- **Halbdesmosom** hemidesmosome
- **Plaquedesmosom** spot desmosome

Desoxycytidin deoxycytidine
Desoxyribonucleinsäure/Desoxyribonukleinsäure (DNS/DNA) deoxyribonucleic acid (DNA)
Destillat distillate
Destillation distillation
destillieren distil, distill, still

Destilliergerät distilling apparatus, still
Destillierkolben/Retorte distilling flask, retort
Destruent/Zersetzer/Reduzent decomposer
Destruktionsfäule/Braunfäule brown rot
Deszendenztheorie/ Abstammungstheorie/ Evolutionstheorie descent theory, theory of evolution, evolutionary theory
Detergens/Reinigungsmittel detergent
Determination/Determinierung/ Bestimmung determination
Determinationskarte/ Schicksalskarte *gen* fate map
detritivor detritivorous
Detritivorie detritivory
Detritus detritus
Detritusernährer/Detritusfresser/ Detritivor detritus-feeder, detritivore
Detritusnahrungskette detritus food chain, detrital food chain
Deuter deuter cell, pointer cell, eurycyst
Deuteromyceten/Deuteromycetes/ unvollständige Pilze/Fungi imperfecti imperfect fungi, deuteromycetes
Deuterotokie deuterotoky
Devon (erdgeschichtliche Periode) Devonian, Devonian Period
DFG (Deutsche Forschungsgemeinschaft) "German Research Society" (German National Science Foundation)
DI-Partikel/Von-Magnus-Partikel defective interfering particle (DI particle), von Magnus particle
diadrom diadromous
Diagnose diagnosis
- **Differentialdiagnose** differential diagnosis
- **pränatale Diagnose** antenatal diagnosis, prenatal diagnosis
- **präsymptomatische Diagnose** presymptomatic diagnosis

diagnostisch diagnostic
Diagonalgang/Kreuzgang diagonal gait
Diagramm (*auch* **Kurve**) *math/graph* diagram, plot, graph
- **Blütendiagramm** *bot* flower diagram, floral diagram
- **Histogramm/Streifendiagramm** histogram, strip diagram
- **Kreisdiagramm** pie chart
- **Lineweaver-Burk-Diagramm** Lineweaver-Burk plot, double-reciprocal plot
- **Phasendiagramm** phase diagram
- **Punktdiagramm** dot diagram
- **Ramachandran-Diagramm** Ramachandran plot
- **Röntgenbeugungsdiagramm/ Röntgenbeugungsmuster/ Röntgenbeugungsaufnahme/ Röntgendiagramm** X-ray diffraction pattern
- **Scatchard-Diagramm** Scatchard plot
- **Spindeldiagramm** spindle diagram
- **Stabdiagramm** bar diagram, bar graph
- **Strahlendiagramm** *opt* ray diagram
- **Streudiagramm** scatter diagram (scattergram/scattergraph/ scatterplot)
- **Strichdiagramm** line diagram

Diakinese diakinesis
Dialyse dialysis
dialysieren dialyze
Diapause diapause
Diapensiagewächse/Diapensiaceae diapensia family
Diapophyse (Rippenfortsatz) diapophysis (transverse process)
Diarrhö diarrhea
Diarthrose/Gelenk/echtes Gelenk/ Articulatio diarthrosis, diarthrodial joint, synovial joint
Diaspore/Ausbreitungseinheit/ Disseminule diaspore, disseminule
Diät diet
- **ausgewogene Diät** balanced diet

Diät.../diät/die Diät betreffend dietary
Diätetik dietetics
diätetisch dietetic
Diatomeen/Kieselalgen/ Bacillariophyceae diatoms
Diatomeenerde/Kieselerde diatomaceous earth
Diatropismus diatropism
Dichasium/zweigablige Trugdolde dichasium, dichasial cyme
Dichlordiphenyldichlorethylen (DDE) dichlorodiphenyltrichloroethylene (DDE)
Dichlordiphenyltrichlorethan (DDT) dichlorodiphenyltrichloroethane (DDT)
Dichogamie dichogamy
dichotom/gabelig verzweigt dichotomous, forked

443　　　　　　　　　　　　　　　　　　**diphasisch** **D**

Dichotomie/Gabelung/Gabelteilung
dichotomy, (repeated) forking,
bifurcation
dicht (Masse/Vol) dense
Dichte (Masse/Vol) density
dichteabhängig density dependent
Dichtegradient density gradient
Dichtegradientenzentrifugation
density gradient centrifugation
Dichten/Jugendgesang (Jungvögel)
subsong
dichteunabhängig
density independent
dichtfaseriges Holz
pycnoxylic wood
Dickblattgewächse/Crassulaceae
stonecrop family, sedum family,
orpine family
Dickdarm large intestines
Dickenwachstum thickening,
growth
● **primäres D.** primary growth,
primary thickening
● **sekundäres D.** secondary growth,
secondary thickening
dickfleischig succulent
Dickfleischigkeit succulence
dickflüssig/zähflüssig/viskos/viskös
viscous, viscid
Dickicht brush, thicket,
thick shrubbery
Dicksoniaceae/Baumfarngewächse
dicksonia family
Dickung *for* young plantation,
young forest stand
Dickungsmittel thickener
Dictyosom/Diktyosom dictyosome,
Golgi body
Dictyotän dictyotene
Didesoxynucleotid/
Didesoxynukleotid
dideoxynucleotide
Didesoxysequenzierung
dideoxy sequencing
Didieraceae/Armleuchterbäume
didierea family
Dieb (Nahrung) scrounger
diedrische Symmetrie
dihedral symmetry
Dielektrizitätskonstante
dielectric constant
Differential-Interferenz (Nomarski)
differential interference
Differentialart/Trennart
differential species
Differentialdiagnose
differential diagnosis
Differentialfärbung/
Kontrastfärbung differential
staining, contrast staining
Differentialgleichung
differential equation

differentielle Genexpression
differential gene expression
differentieller Display (Form der
RT-PCR) *gen* differential display
(Form of RT-PCR)
differentielles Spleißen *gen*
differential splicing
differenzieren differentiate
Differenzierung differentiation
Differenzierungscluster (CD)
cluster of differentiation (CD)
diffundieren diffuse
Diffusionsblotting
capillary blotting
Diffusionskoeffizient
diffusion coefficient
Diffusionstest/Agardiffusionstest
agar diffusion test
Digitigradie/Zehengang
digitigrade gait
Digitoxin digitoxin
Digoxin digoxin
Dihybridkreuzung dihybrid cross
Dikaryophase/Paarkernphase
dikaryotic phase
Dikotyle/Dikotyledone dicotyledon,
dicot
Dilatation/Ausweitung expansion,
dilation, dilatation
Dilleniaceae/Rosenapfelgewächse
silver-vine family, dillenia family
Dimegetismus/sexueller
Größenunterschied dimegaly
Dimer dimer
● **Cyclobutyldimer**
cyclobutyl dimer
● **Thymindimer** thymine dimer
dimerisieren dimerize
Dimerisierung dimerization
dimiktisch *limn* dimictic
dimitisch dimitic
Dimorphismus dimorphism
Dinergat/Soldat dinergate, soldier
Dinoflagellaten/Pyrrhophyceae
dinoflagellates
Dinosaurier dinosaurs
Dioptrie (*Einheit***)** diopter (D)
dioptrisch dioptric
Dioscoreaceae/
Yamswurzelgewächse/
Schmerwurzgewächse yam family
Diöstrus/Dioestrus diestrus
Diözie/Zweihäusigkeit/
Getrenntgeschlechtigkeit
(speziell postnatal:
Gonochorismus) dioecy, dioecism
(gonochory)
diözisch/zweihäusig/
getrenntgeschlechtig (speziell
postnatal: gonochor) dioecious,
diecious (gonochoric/gonochoristic)
diphasisch diphasic

**diphycerk/diphyzerk/protocerk/
protozerk** diphycercal
**diphyodont (einmaliger
Zahnwechsel)** diphyodont
diploid diploid
Diplosom diplosome
diplostemon *gen* diplostemonous
Diplotän diplotene
Dipol dipole
Dipolmoment dipole moment
Dipsacaceae/Kardengewächse
teasel family, scabious family
**Dipterocarpaceae/
Zweiflügelfruchtgewächse/
Flügelnussgewächse**
meranti family,
dipterocarpus family
direkte Genetik direct genetics
direkte Sequenzwiederholungen
gen direct repeats
**Direkttransfer-Elektrophorese/
Blottingelektrophorese**
direct transfer electrophoresis,
direct blotting electrophoresis
**disjunkt/zerstückelt/voneinander
isoliert** disjunct, disjunctive
Disjunktion/Isolierung disjunction,
discontinuity, isolation
**Disjunktion/Verteilung/Trennung
(der Tochterchromosomen)**
disjunction
• **alternierende Disjunktion**
alternate disjunction
**Diskalzelle/Discalzelle/
Discoidalzelle** discal cell,
discoidal cell
Disklimax (Störungsklimax)
disclimax
diskoidal discoidal, disk-like,
disc-like
diskontinuierliche Replikation
discontinuous replication
**diskontinuierliches Gen/
gestückeltes Gen/Mosaikgen**
discontinuous gene, split gene,
mosaic gene
Diskontinuität discontinuity
**Dislokation/Verlagerung (von
Chromosomenabschnitten)**
dislocation
**Dislokatorzelle/Dislocatorzelle/
Wandzelle/Stielzelle
(Cycadeenpollen)**
stalk cell
dislozieren/verlagern dislocate
disom disomic
Disomie disomy
dispergieren disperse
Dispergierung/Dispersion
dispersion
disperse Replikation *gen* dispersive
replication

Dispersion/Verteilung *ecol*
dispersion
• **Überdispersion/Hyperdispersion
(gehäufte Verteilung)**
overdispersion, overdispersed
distribution, hyperdispersion
(contagious distribution)
• **Unterdispersion/Hypodispersion
(regelmäßige Verteilung)**
underdispersion, underdispersed
distribution, hypodispersion
(uniform distribution)
Dispersion *bot* **(wechselständige/
zerstreute Blattstellung)** alternate
leaf arrangement
**Disposition/Veranlagung/
Anfälligkeit** disposition
Dissimilation dissimilation,
catabolism
• **anaerobe Dissimilation/
anaerobe Gärung**
anaerobic fermentation
dissimilatorisch dissimilatory
Dissoziationsgeschwindigkeit
dissociation rate
Dissoziationskonstante (K_i)
dissociation constant
dissoziieren dissociate
**dissymmetrisch/asymmetrisch/
unsymmetrisch** dissymmetrical,
asymmetrical
distelartig thistle-like, thistly
Distichie distichy
Distylie distyly,
dimorphic heterostyly
Disulfidbindung/Disulfidbrücke
disulfide bond, disulfide bridge,
disulfhydryl bridge
disymmetrisch/bilateral
disymmetrical, bilateral, biradial,
bilaterally symmetrical,
radially symmetrical
**dithalam/zweikammerig/
zweikämmrig/dithekal**
dithalamous, dithalamic,
with two chambers, dithecal
Dityp ditype
• **nicht-parentaler Dityp (NPD)**
non-parental ditype (NPD)
• **parentaler Dityp (PD)**
parental ditype (PD)
**Diurese/Harnfluss/
Harnausscheidung** diuresis
divergente Transkription
divergent transcription
Divergenz divergence, divergency
divergieren diverge
Divertikel/Aussackung/Blindsack
diverticulum
Diversität diversity
dizentrisches Chromosom
dicentric chromosome

Dominanz D

dizygot/zweieiig dizygotic, dizygous
Dizygotie/Zweieiigkeit dizygosity
DNA/DNS (Desoxyribonucleinsäure/ Desoxyribonukleinsäure) DNA (deoxyribonucleic acid)
- **3′→5′ (drei Strich-fünf Strich/ drei Strich nach fünf Strich)** 3′→5′ (three prime five prime/ three prime to five prime)
- **A-Form/A-Konformation** A form
- **Achterform** figure eight
- **Alpha-DNA** alpha-DNA
- **anonyme DNA** anonymous DNA
- **B-Form/B-Konformation** B form
- **C-Form/C-Konformation** C form
- **cccDNA (DNA aus kovalent geschlossenen Ringen)** cccDNA (covalently closed circles DNA)
- **cDNA (komplementäre DNA)** cDNA (complementary DNA)
- **egoistische DNA** selfish DNA
- **Einzelkopie-DNA/nichtrepetitive DNA** single copy DNA
- **extragene DNA** extragenic DNA
- **Fremd-DNA** foreign DNA
- **in sich gefaltete DNA/ zurückgebogene DNA** fold-back DNA, snap-back DNA
- **kreuzförmige DNA** cruciform DNA
- **Linker-DNA** linker DNA
- **Minisatelliten-DNA** minisatellite DNA
- **native DNA** native DNA
- **oc-DNA (offene ringförmige DNA)** oc-DNA (open circle DNA)
- **passagere DNA/Passagier-DNA** passenger DNA
- **promiskuitive DNA** promiscuitive DNA
- **rekombinierte DNA-Technologie (Methoden mit Hilfe rekombinierter DNA)/ rekombinante DNA-Technologie** recombinant DNA technology
- **rekombiniertes DNA-Molekül/ rekombinantes DNA-Molekül** recombinant DNA molecule
- **repetitive DNA** repetitive DNA
- **Satelliten-DNA** satellite DNA
- **Stuffer-DNA** stuffer-DNA
- **unnütze DNA/überflüssige DNA/ wertlose DNA** junk DNA
- **vorgeschichtliche DNA** ancient DNA
- **Z-DNA/Z-Konformation** Z DNA
DNA-abhängige-DNA Polymerase DNA-dependent DNA polymerase
DNA-Bank/DNA-Bibliothek DNA bank, DNA library

DNA-Biegung/DNA-Verbiegung DNA bending
DNA-Fingerprinting/genetischer Fingerabdruck DNA profiling, DNA fingerprinting
DNA-Fußabdruck/DNA-Footprint DNA footprint
DNA-getriebene Hybridisierung DNA-driven hybridization
DNA-Polymerase DNA polymerase
DNA-Reparatur DNA repair
- **Excisionsreparatur/ Exzisionsreparatur** excision repair
- **Fehlpaarungsreparatur** mismatch (DNA) repair
- **lichtunabhängige DNA-Reparatur** dark repair, light independent DNA repair
- **Lichtreparatur** light repair
DNA-Sequenzierung DNA sequencing
DNA-Sequenzierungsautomat DNA sequencer
DNA-Synthese DNA synthesis
- **außerplanmäßige D.** unscheduled DNA synthesis
DNA-Tumorvirus DNA tumor virus
DNA-Welt DNA world
Docking-Protein/Andockprotein docking protein
Doggenhaiartige/ Heterodontiformes horn sharks
Dolde/Umbella/Sciadium (Infloreszenz) umbel, sciadium
- **einfache Dolde** simple umbel
- **zusammengesetzte Dolde** compound umbel
Doldengewächse/Doldenblütler/ Umbelliferen/Apiaceae/ Umbelliferae carrot family, parsley family, umbellifer family
Doldentraube corymb
doldig umbel-like, sciadioid
Doliolaria/Tönnchenlarve (Vitellaria) doliolaria, vitellaria larva
Doliporus dolipore
Domäne (Tertiärstruktur) domain
Domestikation domestication
domestisch domestic
domestizieren domesticate
dominant dominant
Dominanz dominance
- **Überdominanz** overdominance
- **Unterdominanz** underdominance
- **unvollständige Dominanz/ Semidominanz/Partialdominanz** incomplete dominance
- **variable Dominanz** shifting dominance
- **verzögerte Dominanz** delayed dominance

Dominanzvarianz
dominance variance
dominieren dominate
Donnerkeil/Rostrum (Belemniten Schalenspitze)
thunderbolt (fossil belemnite shell), rostrum, guide
Donor/Spender donor
Donorzelle donor cell
DOP-PCR (PCR mit degeneriertem Oligonucleotidprimer) *gen* DOP-PCR (degenerate oligonucleotide primer PCR)
DOP-Vernebelung (Dioctylphthalat-Vernebelung)
DOP smoke (dioctyl phthalate smoke)
Dopamin dopamine
Doppelachäne (Schizokarp der Umbelliferen) cremocarp
Doppelähre double spike
Doppelbindung double bond
Doppelblindversuch double blind assay, double-blind study
doppelbrechend birefringent, double-refracting
doppelbrechend/anisotrop (Muskel) anisotropic
Doppelbrechung birefringence, double refraction
doppelchromosomig/bivalent bivalent
Doppeldiffusion/Doppelimmundiffusion double diffusion, double immunodiffusion (Ouchterlony technique)
Doppeldolde double umbel
Doppelfüßer/Diplopoden/Tausendfüßler/Myriapoden/Myriapoda millipedes, diplopods, myriapodians
Doppelhelix double helix
Doppelinfektion double infection
doppelklappig/zweiklappig bivalve
doppelköpfiges Zwischenprodukt/janusköpfiges Zwischenprodukt
double-headed intermediate
Doppelkreuzung double cross
Doppelmembran double membrane
Doppelschaler/Krallenschwänze/Diplostraca/Onychura clam shrimps & water fleas
Doppelschicht double layer, bilayer
Doppelschildokulation/Nicolieren
double shield budding
Doppelschleichen/Wurmschleichen/Amphisbaenia worm lizards, amphisb(a)enids, amphisbenians
Doppelschwänze/Diplura japygids, diplurans
Doppelspreitigkeit/Diplophyllie
diplophyly

Doppelstrang *gen* double strand
Doppelstrangbildung/Annealing/Reannealing/Reassoziation (DNA)
annealing, reannealing, reassociation, renaturation
Doppelstrangbruch
double strand break
Doppelstrangsequenzierung
double-strand sequencing
doppelte Befruchtung
double fertilization
doppelte Rekombination
double recombination
Doppeltraube (Infloreszenz)
double raceme
Doppelverdau *gen/biochem*
double digest
Doppelwendel-Dimer (superspiralisierte Helices)
coiled coil (superspiraled helices/helixes)
Doppelzucker/Disaccharid
double sugar, disaccharide
Dormanz (endogen bedingte Ruheperiode) dormancy
Dorn (Blattdorn/Nebenblattdorn)
spine; **(Sprossdorn)** thorn (sharp-pointed modified branch)
Dornbusch thorny bush
Dornbuschformation/Dornstrauchformation
thorny thicket, thorny brush
Dornfortsatz/Spinalfortsatz/Processus spinalis vertebrae
vertebral spine, neural spine, spinous process of vertebra
• **oberer Dornfortsatz/oberer Spinalfortsatz//Neurapophyse**
neurapophysis
• **unterer Dornfortsatz/unterer Spinalfortsatz/Hämapophyse**
haemapophysis, hemapophysis
Dorngestrüpp thorny thicket
dornig thorny, spiny
Dornkorallen/Dörnchenkorallen/Antipatharia thorny corals, black corals, antipatharians
Dornrückenaale/Notacanthiformes
notacanthiforms
Dornstrauch/Dornenstrauch/Dornbusch thorn shrub, thorny thicket, thorn brush
Dornwald thorn woodland
Dorsalnaht/Rückennaht
dorsal suture, dorsal seam
Dorsalwurzel *neuro* dorsal root
Dorschfische/Gadiformes
codfishes & haddock & hakes
dorsifix (Anthere) dorsifixed
dorsiventral dorsiventral, bifacial
dosieren dose (give a dose), measure out

dressieren

**Dosierung/Dosieren (im Verhältnis/
anteilig)** proportioning
Dosis dose, dosage
- **letale Dosis/Letaldosis/tödliche
Dosis** lethal dose
- **mittlere letale Dosis (LD$_{50}$)**
median lethal dose (LD$_{50}$)
- **Überdosis** overdose
Dosis-Wirkungskurve
dose-response curve
Dosiseffekt dosage effect
Dotter *nt*/**Eidotter/Eigelb** yolk,
vitellus
- **Dotter gleichmäßig verteilt/
isolezithal/isolecithal**
isolecithal (yolk distributed nearly
equally)
- **Dotter im Zentrum/
zentrolezithal/centrolecithal**
centrolecithal (yolk aggregated in
center)
- **Dotter an einem Pol/telolezithal/
telolecithal** telolecithal (yolk in
one hemisphere)
- **dotterarm/oligolezithal/
oligolecithal/mikrolecithal**
oligolecithal (with little yolk)
- **dotterreich/polylezithal/
polylecithal/makrolecithal**
polylecithal (with large amount of
yolk)
- **mäßig dotterreich/mesolezithal/
mesolecithal** mesolecithal (with
moderate yolk content)
Dottergang vitelline duct
**Dotterhaut/Dottermembran/
primäre Eihülle/Membrana
vitellina** vitelline layer,
vitelline membrane
Dotterpfropf yolk plug
Dottersack yolk sac
Dottersackbrut (Lachs) yolk fry,
sacfry, alevin
**Dottersackplazenta/
Dottersackhöhlenplazenta/
Omphaloplazenta/omphaloide
Plazenta** yolk-sac placenta,
choriovitelline placenta
**Dotterstock/Dotterdrüse/Vitellar/
Vitellarium** yolk gland,
vitellarian gland, vitelline gland,
vitellarium, vitellogen
Dotterzelle/Vitellophage yolk cell;
shell globule (trematodes)
Down-Mutation down mutation
**Drachenbaumgewächse/
Dracaenaceae**
dragon-blood tree family
**Drachenkopfartige/
Drachenkopffischverwandte/
Panzerwangen/Scorpaeniformes**
sculpins & sea robins

**Drachenwurm/Medinawurm/
Guineawurm (*Dracunculus
medinensis*)** guinea worm,
medina worm
Drahtnetz *chem*/*lab* wire gauze
Drahtwurm (Elateriden-Larve)
wireworm (clickbeetle larva)
Drang/Trieb/Impuls urge, drive,
compulsion, impulse
Drängeln (Bienen) jostling run
Dränung/Drainage drainage
drehen/verdrehen contort
Dreher/Drehmuskel/Rotator
rotator muscle
Drehfestigkeit (Holz) torsion(al)
strength
**Drehkrankheit (*Myxosoma
cerebralis*)** twist disease (trout)
Drehsinn/Rotationssinn
rotational sense, sense of rotation
Drehtisch *micros* rotating stage
Drehung/Torsion torsion
Drei-Faktor-Kreuzung
three-point testcross
dreiästig verzweigt three-branched,
triramous
**Dreiblattgewächse/
Einbeerengewächse/Trilliaceae**
trillium family
dreiblättrig trifoliate
dreiblättrig gefiedert trifoliolate
Dreieck/Triangulum (Insektenflügel)
triangle
Dreiecksknochen/Os triquetrum
triangular bone, triquestral bone
Dreifachbindung triple bond
Dreifachpaarung *gen*
tripartite mating
Dreifelderwirtschaft three-field
rotation, three-year rotation,
three-field system
Dreifuß/Dreibein tripod
Dreijochzahnechsen/Trilophosauria
trilophosaurs, trilophosaurians
(Triassic archosauromorphs)
Dreilapper/Trilobiten trilobites
dreischneidig (Initiale) *bot*
with three cutting faces
dreispaltig trifid
dreiteilig tripartite
dreiwertig trivalent
Dreiwertigkeit trivalency
Dreizackgewächse/Juncaginaceae
arrowgrass family
dreizählig ternate, ternary
dreizehig/tridactyl tridactyl
(having three digits)
Dreizehigkeit/Tridactylie
tridactyly
dreschen thresh
dressieren/abrichten
condition, train

Dressur

Dressur (animal) conditioning, training
Drillfurche/Saatfurche/Saatrille drill, drill furrow
Drillreihe drill row
Drillsaat (Reihensaat) drill, drilling, drill sowing (row seeding)
Droge drug
- **Pflanzendroge** herbal drug
- **unter Drogen stehen** to be drugged

drogenabhängig/drogensüchtig addicted to drugs
Drogenabhängigkeit/Sucht drug addiction
Drogenkunde/Pharmakognosie/ pharmazeutische Biologie pharmacognosy
Drogenmissbrauch drug abuse
Drogenpflanze/Arzneipflanze medicinal plant
Drohen threat, threatening
- **Aggressivdrohen** offensive threat
- **Defensivdrohen** defensive threat

drohen threaten
Drohgähnen threat yawn, threat gape
Drohgebärde/Droh-Mimik threatening gesture
Drohhaltung threatening posture
Drohmimik threating gesture
Drohne/Drohn drone
Drohung threat
Drohverhalten threat behavior
Droseraceae/Sonnentaugewächse sundew family
Drosselvene/Vena jugularis jugular vein
Druckfestigkeit (Holz) crushing strength, compression resistance (endwise compression)
Druckholz/Rotholz compression wood
Druckstromtheorie/ Druckstromhypothese pressure-flow theory/hypothesis
Druse druse, granule
Drüse gland
- **akzessorische Drüse/ Anhangsdrüse** accessory gland
- **Ampullardrüse** ampullary gland (of deferent duct)
- **Analdrüse (Hai)** anal gland, rectal gland (shark)
- **Analdrüse/Pygidialdrüse** anal gland, pygidial gland
- **Anhangsdrüse/akzessorische Drüse** accessory gland
- **Antennendrüse/ Antennennephridium/grüne Drüse** antennal gland, antennary gland, green gland
- **apokrine Drüse** apocrine gland
- **Backendrüse** cheek gland
- **Bartholin-Drüse/Glandula vestibularis major** Bartholin's gland, greater vestibular gland
- **Beischilddrüse/ Nebenschilddrüse/ Epithelkörperchen** parathyroid gland, parathyroidea
- **Bläschendrüse/Samenblase/ Samenbläschen (♂ akzessorische Drüse)** vesicular gland, seminal gland, seminal vesicle
- **Brunftdrüse/Brunftfeige/ Duftdrüse (Gemse)** odoriferous gland, scent gland
- **Brunnersche Drüse/ Duodenaldrüse** Brunner's gland, duodenal gland
- **Brustdrüse/Milchdrüse** mammary gland
- **Bulbourethraldrüse/Cowpersche Drüse/Cowper-Drüse** bulbourethral gland, Cowper's gland
- **Bürzeldrüse** uropygial gland, preen gland, oil gland
- **Carapaxdrüse/Y-Organ** molting gland, Y organ
- **Cowpersche Drüse/Cowper-Drüse/Bulbourethraldrüse** Cowper's gland, bulbourethral gland
- **Coxaldrüse** coxal gland
- **Dermaldrüse/Hautdrüse** dermal gland
- **Dotterstock/Dotterdrüse/ Vitellar/Vitellarium** yolk gland, vitellarian gland, vitelline gland, vitellarium, vitellogen
- **Dufour-Drüse** Dufour's gland
- **Duftdrüse/Brunftdrüse/ Brunftfeige (Gemse)** odoriferous gland, scent gland
- **Duodenaldrüse/Brunnersche Drüse** duodenal gland, Brunner's gland
- **Eischalendrüse/Schalendrüse/ Nidamentaldrüse** nidamental gland
- **Eiweißdrüse** albumen gland
- **Enddarmdrüse/Proktodaemdrüse** protodeal gland
- **endokrine Drüse** endocrine gland
- **Follikulärorgan/Follikulärdrüse (Schenkeldrüse: Eidechsen)** follicular gland (femoral gland)
- **Frontaldrüse/Stirndrüse/ Kopfdrüse** frontal gland, cephalic gland
- **Fundusdrüse** fundus gland

Drüse

- **Fußdrüse/Klebdrüse/Kittdrüse** pedal gland, adhesive gland, cement gland (rotifers)
- **Futterdrüse/Futtersaftdrüse/ Honigdrüse/Hypopharynxdrüse** hypopharyngeal gland
- **Gasdrüse** gas gland
- **Geschlechtsdrüse/Keimdrüse/ Gonade** sex gland, germ gland, gonad
- **Giftdrüse** poison gland
- **grüne Drüse/Antennendrüse/ Antennennephridium** green gland, antennal gland, antennary gland
- **Haftdrüse/Klebdrüse/Kittdrüse/ Zementdrüse** adhesive gland, colleterial gland, cement gland (insects)
- **Hardersche Drüse (Auge)** Harderian gland, Harder's gland
- **hedonische Drüse** hedonic gland (amphibians)
- **Hirnanhangdrüse/Hypophyse** pituitary gland, hypophysis
- **interdigitale Drüse/ Interdigitaldrüse/ Zwischenzehendrüse** interdigital gland
- **Kalkdrüse** calciferous gland
- **Keimdrüse/Geschlechtsdrüse/ Gonade** sex gland, germ gland, gonad
- **Kittdrüse/Klebdrüse/ Zementdrüse (Lepidoptera: Glandula sebacea)** colleterial gland, adhesive gland, cement gland (insects)
- **Klauendrüse** claw gland
- **Kopfdrüse/Stirndrüse/ Frontaldrüse** cephalic gland, frontal gland
- **Körnerdrüse** granular gland, poison gland (amphibians)
- **Kornsekretdrüse** prostatic gland (annelids: spermiducal gland)
- **Labialdrüse** labial gland
- **Leistendrüse** inguinal gland
- **Lymphdrüse** lymphatic gland
- **Magendrüse** gastric gland
- **Mandibulardrüse/ Unterkieferdrüse** mandibular gland
- **Mehlissche Drüse/Schalendrüse** Mehlis' gland, shell gland (cement gland)
- **Meibom-Drüse/Glandula tarsalis** Meibomian gland, tarsal gland
- **merokrine Drüse** merocrine gland
- **Metatarsaldrüse** metatarsal gland
- **Milchdrüse/Brustdrüse** mammary gland
- **Mitteldarmdrüse/"Leber"** midgut gland, digestive gland, "liver"
- **Mitteldarmdrüse/Darmdivertikel** digestive gland, "liver" (mollusks, echinoderms)
- **Mitteldarmdrüse/ Hepatopankreas** midgut gland, hepatopancreas
- **Moschusdrüse (Duftdrüse)** musk gland (scent gland)
- **Nassanov Drüse/ Nassanoffsche Drüse** Nassanov gland, Nassanov's gland
- **Nebenniere** adrenal gland
- **Nebenschilddrüse/ Beischilddrüse/ Epithelkörperchen** parathyroid gland, parathyroidea
- **Nickhautdrüse** gland of third eyelid
- **Nidamentaldrüse/ Eischalendrüse/Schalendrüse** nidamental gland
- **Ohrdrüse/Parotoiddrüse/ Parotisdrüse/ Duvernoysche Drüse** parotoid gland (amphibians)
- **Ohrenschmalzdrüse** ceruminous gland, wax gland
- **Ohrspeicheldrüse/Parotis** parotid gland, parotis, parotid (mammals: salivary gland)
- **Öldrüse/Schmierdrüse** oil gland
- **Parotis/Ohrspeicheldrüse** parotid gland, parotis, parotid (mammals: salivary gland)
- **Parotoiddrüse/Parotisdrüse/ Ohrdrüse/Duvernoysche Drüse** parotoid gland (amphibians)
- **Perinealdrüse/Dammdrüse** perineal gland
- **Pharynxdrüse** pharyngeal gland
- **Pinealorgan/Epiphyse/ Zirbeldrüse** pineal gland, pineal body, conarium, epiphysis
- **Präputialdrüse** preputial gland, castor gland (beaver)
- **Prothoraxdrüse** prothoracic gland (an ecdysial/molting gland)
- **Purpurdrüse** purple gland
- **Pygidialdrüse/Analdrüse** pygidial gland, anal gland
- **Rectaldrüse/Rektaldrüse (see: Klebdrüse/Kittdrüse/ Zementdrüse)** rectal gland
- **Salzdrüse** salt gland
- **Schalendrüse/Maxillendrüse/ Maxillennephridium** maxillary gland
- **Schalendrüse/Mehlissche Drüse** shell gland, Mehlis' gland (cement gland)

D Drüsengewebe 450

- **Schalendrüse/Nidamentaldrüse** shell gland, nidamental gland
- **Schenkeldrüse (Follikulärorgan: Eidechsen)** femoral gland (follicular gland: lizards)
- **Schilddrüse** thyroid gland, thyreoidea
- **Schläfendrüse (Elefant)** temporal gland
- **Schleimdrüse** slime gland, mucous gland
- **Schlunddrüse (Gastropoden)** esophageal gland
- **Schmierdrüse/Öldrüse** oil gland
- **Schwanzdrüse (Rektaldrüse)** caudal gland (rectal gland)
- **Schwanzwurzeldrüse** supracaudal gland
- **Schweißdrüse** sweat gland, sudoriferous gland, sudoriparous gland
- **Seidendrüse/Spinndrüse/ Sericterium (Labialdrüse: Raupen)** silk gland, spinning gland, sericterium (caterpillars: labial gland)
- **Sekretdrüse** secretory gland
- **Sinusdrüse** sinus gland
- **Speicheldrüse** salivary gland
- **Spüldrüse/von Ebnersche Drüse/ von Ebner' Drüse** gustatory gland
- **Stinkdrüse (*siehe:* Wehrdrüse)** repugnatorial gland
- **Stirndrüse/Frontaldrüse/ Kopfdrüse** frontal gland, cephalic gland
- **Talgdrüse/Haartalgdrüse** sebaceous gland
- **Tarsaldrüse** tarsal gland
- **Thymus/Thymusdrüse/Bries (Halsthymus/Brustthymus)** thymus (gland)
- **Tintendrüse/Tintensack/ Tintenbeutel** ink gland, ink sac
- **Tränendrüse** lacrimal gland
- **Unteraugendrüse** suborbital gland
- **Unterkieferdrüse/ Mandibulardrüse** mandibular gland
- **Unterzungendrüse** sublingual gland
- **Urethraldrüse** urethral gland
- **Ventraldrüse** ventral gland, renette gland (nematodes)
- **Verdauungsdrüse** digestive gland
- **Violdrüse (Schwanzwurzeldrüse des Fuchses)** violet gland, supracaudal gland
- **Voraugendrüse/Antorbitaldrüse** preorbital gland, antorbital gland

- **Vorhofdrüse (Scheidenvorhof)** vestibular gland
- **Wachsdrüse** wax gland, ceruminous gland
- **Wehrdrüse (*Peripatus*: Schleimdrüse)** defensive gland (slime gland)
- **Wollfettdrüse** wool fat gland
- **Y-Organ/Carapaxdrüse** Y organ, molting gland
- **Zeis-Drüse** gland of Zeis, sebaceous ciliary gland
- **Zementdrüse/Kittdrüse/ Klebdrüse** cement gland, adhesive gland
- **Zirkumanaldrüse** circumanal gland
- **Zuckerdrüse** sugar gland, subradular organ
- **Zungendrüse** lingual gland
- **Zwischenzehendrüse/ interdigitale Drüse/ Interdigitaldrüse** interdigital gland
- **Zwitterdrüse** hermaphroditic gland, hermaphroditic gonad, ovotestis

Drüsengewebe glandular tissue
Drüsengürtel/Clitellum clitellum
Drüsenhaar *bot* glandular hair, glandular trichome
drüsenlos eglandulous, eglandular
Drüsenmagen *orn* glandular portion of stomach
Drüsensekret glandular secretion
Drüsenzelle gland cell
Drüsenzotte/Leimzotte/Colletere/ Kolletere *bot* colleter, multicellular glandular trichome
drüsig glandular
Dryopteridaceae/ Wurmfarngewächse (Aspidiaceae) male fern family, dryopteris family, aspidium family
Dschungel jungle
Duettgesang/Paargesang *orn* duetting
Dufour Drüse Dufour's gland
Duft/Geruch smell, odor, scent
- **angenehmer Duft/Geruch** fragrance, scent, pleasant smell
- **unangenehmer Duft/Geruch** unpleasant smell
Duftdrüse scent gland, odoriferous gland
duftend (angenehm) fragrant
Duftfeld/Duftschuppe/Androkonie scent scale, androconium (*pl* androconia)
Duftmarkierung scent-marking
Duftpinsel/Haarpinsel (Schmetterlinge) hair pencil, tibial tuft

Duftschuppe/Androkonie
(♂ **Lepidoptera**) scent scale, androconium (*pl* androconia)
Duftspur scent trail, olfactory trail
Duftstoffe
scents, odiferous substances
Dune/Daune/Dunenfeder/
Flaumfeder/dunenartige Feder
down, down feather, plumule
Düne dune
• **Binnendüne/Inlandsdüne/**
Innendüne/Festlandsdüne/
Kontinentaldüne inland dune
• **Braundüne** brown dune
• **Deflationsdüne/Haldendüne**
blowout dune
• **Kuppeldüne/Haufendüne**
dome dune
• **Kupstendüne/Kupste**
shrub-coppice dune, nebkha
• **Küstendüne** coastal dune
• **Längsdüne/Longitudinaldüne/**
Seif longitudinal dune, seif dune
• **Paraboldüne/Parabeldüne**
parabolic dune
• **Primärdüne** primary dune
• **Querdüne/Tranversaldüne**
transverse dune
• **Seif/Längsdüne/**
Longitudinaldüne seif dune,
longitudinal dune
• **Sekundärdüne (Weißdüne/**
Gelbdüne) secondary dune
(white dune/yellow dune)
• **Sicheldüne/Bogendüne/Barchan**
crescentic dune,
crescent-shaped dune, brachan
• **Sterndüne/Pyramidendüne**
star dune
• **Stranddüne** shore dune
• **Strichdüne/Silk-Düne** lineal dune
• **Tertiärdüne (Graudüne)**
tertiary dune (gray dune)
• **Tranversaldüne/Querdüne**
transverse dune
• **Vordüne** foredune
• **Wanderdüne** shifting dune,
mobile dune, migratory dune
• **Zungendüne** linguoid dune
Dung (tierische Exkremente) dung,
manure
Dung/Mist/tierische Exkremente/
Tierkot dung, manure
düngen fertilize, manure
Dünger/Düngemittel
fertilizer, plant food, manure
Dungfressen/Kotfressen/
Koprophagie/Coprophagie
coprophagy
dungfressend/kotfressend/
koprophag/coprophag
coprophagous, coprophagic

Dungfresser/Kotfresser/
Koprophage/Coprophage
coprophagist
Düngung fertilization
• **Überdüngung** overfertilization,
excessive fertilization
Dunkelfeld *micros* dark field
Dunkelkammer *micros/photo*
darkroom
Dunkelkeimer germinating in
darkness, dark germinator
Dunkelkeimung dark germination
Dunkelreaktion dark reaction
Dünndarm small intestines
Dünnschnitt thin section,
microsection
• **Semidünnschnitt**
semithin section
• **Ultradünnschnitt** ultrathin section
dünnstämmig slender-stemmed,
leptocaulous
Dunstabzugshaube/Abzug
fume hood, hood
dunstig misty
Duodenaldrüse/Brunnersche Drüse
duodenal gland, Brunner's gland
Duodenum/Zwölffingerdarm
duodenum
durchbluten supply with blood,
vascularize
Durchblutung circulation,
blood supply, blood circulation
Durchbrenner (überlebender Träger
einer Letalmutation) *gen*
break through
durchfließen percolate, flow through
Durchfluss percolation, flowing
through, flux
Durchflussgeschwindigkeit
flow rate
Durchflussrate/Verdünnungsrate
biot dilution rate
Durchflussreaktor (Bioreaktor)
flow reactor
Durchflusszytometrie/
Durchflusscytometrie
flow cytometry
durchlässig/permeabel pervious,
permeable
• **halbdurchlässig/semipermeabel**
semipermeable
• **undurchlässig/impermeabel**
impervious, impermeable
durchlässige Mutante leaky mutant
Durchlässigkeit/Permeabilität
perviousness, permeability
• **Halbdurchlässigkeit/**
Semipermeabilität
semipermeability
• **Undurchlässigkeit/**
Impermeabilität imperviousness,
impermeability

D Durchlasszelle

Durchlasszelle passage cell
Durchlesen *gen* read-through
Durchlicht/Durchlichtbeleuchtung transillumination, transmitted light illumination
durchlüften aerate
Durchlüftung aeration
Durchlüftungsgewebe/Aerenchym aerenchyma
Durchmustern/Durchtesten screening
durchnässt/durchweicht soggy
Durchsatz/Durchsatzmenge throughput
durchscheinend translucent, pellucid
Durchschlagskapsel/Penetrant (Nematocyste) penetrant
durchschneiden transect, cut through
Durchschnitt (Mittelmaß) average, mean
Durchschnitt (schneiden) transection
Durchschnittsertrag average yield
durchsickern *vb* percolate
Durchsickern *n* percolation
durchtränken (durchtränkt) soak (soaked)
durchwachsen(blättrig) perfoliate
Durchzügler migratory animal; *orn* bird of passage, migratory bird
Dürre drought

dürreertragend/dürreüberdauernd drought-enduring
Dürrehärte/Dürrefestigkeit/ Dürrebeständigkeit drought hardiness, drought tolerance
dürremeidend drought-avoiding
dürreresistent/dürrefest drought-resistant
Dürreresistenz drought resistance
dürretolerant/dürreduldend drought-tolerant
Dürrevermeidung drought avoidance
Durst thirst
durstig thirsty
Düsenumlaufreaktor/Strahl- Schlaufenreaktor jet loop reactor
Düsenumlaufreaktor/Umlaufdüsen- Reaktor nozzle loop reactor, circulating nozzle reactor
Dy/Torfschlamm/Torfmudde dy, gel mud
Dyade dyad
Dynein dynein
Dynorphin dynorphin
Dysmorphie dysmorphy
dysphotische Zone/Dämmerzone *limn* dysphotic zone
Dysplasie dysplasia
Dyspnoe dyspnea
dystroph/schlecht ernährt/ mangelhaft ernährt dystrophic, wrongly nourished, inadequately nourished

453 **Eichel** E

Ebbe low tide, ebb tide, ebb
eben plane, level
Ebene/ebene Fläche *math/geom*
plane (flat/level surface)
- **Brennebene** focal plane
- **Sagittalebene**
(parallel zur Mittellinie)
median longitudinal plane
- **Schlundebene** pharyngeal plane
- **Schnittebene/Schnittfläche**
cutting face, cutting plane
- **Trophieebene/Trophieniveau/**
trophisches Niveau trophic level
Ebene/Flachland *geol/ecol* plain
(level area)
- **Flussebene** river plain,
fluvial plain
- **Hochebene/Hochfläche** plateau,
elevated plain, tableland
- **Küstenebene** coastal plain
- **Tiefebene** lowland,
low-lying plain
- **Tiefseeebene** abyssal plain
ebenerdig at ground level
Ebenholzgewächse/Ebenaceae
ebony family
Ebenstrauß/Corymbus (inkl.
Schirmrispe und Schirmtraube)
corymb
Eber (männliches Schwein) boar
(male pig: not castrated)
Ecdysialflüssigkeit/Exuvial-
flüssigkeit/Häutungsflüssigkeit
molting fluid, exuvial fluid
Ecdyson ecdysone
Echinozyt/Stechapfelform
echinocyte, crenocyte, burr cell
Echiuriden/Igelwürmer/
Stachelschwänze/Echiura
echiuroid worms, spoon worms
Echolotpeilung/Echoortung
echolocation
Echsen/Eidechsen/Lacertilia lizards
Echsenbecken-Dinosaurier/
Saurischia lizard-hipped dinosaurs,
reptile-like dinosaurs,
saurischian reptiles
Eckenkollenchym/
Kantenkollenchym (auch:
Collenchym) angular collenchyma
Eckstrebe/Pars inflexa (Huf) bar
Eckstrebenwinkel/Sohlenwinkel
(Huf) seat of corn
Eckzahn canine
ecotrop *vir* ecotropic
edaphisch edaphic
Edelfäule noble rot
Edelgas inert gas, rare gas
Edelreis (*pl* **Edelreiser)/Pfropfreis/**
Pfröpfling/Reis *hort* scion (cion),
graft, (slip>softwood or herbaceous
cutting for grafting)

Edentata/Zahnarme/Xenarthra/
Nebengelenktiere edentates,
"toothless" mammals, xenarthrans
Ediacara-Fauna Ediacaran fauna
Edman-Abbau/Edmanscher Abbau
Edman degradation
Efeugewächse/Araliaceae
ivy family, ginseng family
Egel/Hirudineen/Hirudinea
(Anneliden) leeches, hirudineans
Egel/Saugwürmer/Trematoden
(Plathelminthen)
flukes, trematodes
Egerlinge/Agaricaceae
Agaricus family
Egge *agr* harrow
eggen harrow
Egoismus/Eigennutz selfishness
egoistische DNA selfish DNA
Ei (Fortpflanzungseinheit: Embryo/
Nährstoffe/Schale)
egg (reproductive body: embryo/
nutrients/hard shell)
Ei/Eizelle (weibliche
Geschlechtszelle) egg, egg cell,
ovum (female gamete)
- **beschaltes Ei/kleidoisches Ei**
(Reptilien/Vögel) cleidoic egg,
shelled egg, "land egg"
- **Dotter an einem Pol/telolezithal/**
telolecithal telolecithal
(yolk in one hemisphere)
- **Dotter gleichmäßig verteilt/**
isolezithal/isolecithal isolecithal
(yolk distributed nearly equally)
- **Dotter im Zentrum/**
zentrolezithal/centrolecithal
centrolecithal
(yolk aggregated in center)
- **dotterarm/oligolezithal/**
oligolecithal/mikrolecithal
oligolecithal (with little yolk)
- **dotterreich/polylezithal/**
polylecithal/makrolecithal
polylecithal
(with large amount of yolk)
- **mäßig dotterreich/mesolezithal/**
mesolecithal mesolecithal
(with moderate yolk content)
Eiablage/Oviposition
egg-laying, egg deposition,
deposit of eggs, oviposition
Eibefruchtung/Oogamie
oogamy
Eibengewächse/Taxaceae
yew family
Eibildung/Oogenese
oogenesis
Eichel *bot* acorn
Eichel (Prosoma der
Enteropneusten) acorn, proboscis
Eichel/Glans penis glans penis

E Eichelwürmer

Eichelwürmer/Enteropneusten
acorn worms, enteropneusts
eichen/kalibrieren gauge, calibrate
Eidechsen/Lacertilia (Squamata)
lizards
**Eidonomie (Morphologie der
äußeren Gestalt)** external
morphology (descriptive)
Eier (Crustaceen: Hummer) roe
Eier legen/ablegen lay eggs,
deposit eggs
**Eierfruchtbaumgewächse/
Hernandiaceae** Hernandia family
eierlegend/ovipar egg-laying,
oviparous
Eierleim egg glue
Eiernährboden egg culture medium
Eierstock/Ovar/Ovarium ovary
Eiertasche/Eierbeutel egg case
eiförmig ovate, egg-shaped
Eigelb/Dotter/Eidotter
yolk, egg yolk
Eigelege/Gelege clutch, egg clutch
(nest of eggs)
Eigelenk/Ellipsoidgelenk
ellipsoidal joint, condyloid joint
Eigenname proper name
**Eigenrand/Excipulum proprium
(Flechten)** proper margin,
exciple without algae
Eigenschaft/Merkmal
characteristic, character;
(Charakterzug) character
• **erworbene E.** acquired character
Eigentoleranz/Selbsttoleranz
self-tolerance
**Eigespinst/Brutgespinst
(Schutzgespinst)** nursery web
(nursery tent)
Eignung/Fitness fitness, suitability
**Eihaut/Eihülle/"Eimembran"/
Oolemma** egg membrane
• **äußere E./Chorion** chorion
(external extraembryonic
membrane)
• **innere E./inneres Eihüllepithel/
innere Keimhülle/Amnion**
amnion
**Eihülle/Eihaut/"Eimembran"/
Oolemma** egg membrane
• **äußere E./äußeres Eihüllepithel/
äußere Keimhülle/Serosa** serosa,
serous membrane (external
membrane, e. g., of insect eggs)
• **innere E./inneres Eihüllepithel/
innere Keimhülle/Amnion**
amnion
• **primäre E./Dotterhaut/
Dottermembran/
Vitellinmembran/
Membrana vitellina**
vitelline layer/membrane

Eikapsel/Oothek egg capsule,
ovicapsule, ootheca
Eikapsel/Seemaus (Knorpelfische)
sea purse, mermaid's purse
Eiklar/natives Eiweiss
native egg white
Eikokon/Eipaket/Eisack *arach*
egg sac, "cocoon"
Eikultur (Hühnerei)
chicken embryo culture
Eilarve/Junglarve/Primärlarve
primary larva
Eileiter/Ovidukt oviduct;
Fallopian tube, uterine tube
(oviduct of mammals)
Eileiterenge/Isthmus
isthmus of oviduct
**Eileiteröffnung/Muttertrompete/
Ostium tubae** ostium (of oviduct)
**Eileiterschwangerschaft/
Oviduktträchtigkeit**
tubal pregnancy
**Eileitertrichter/Flimmertrichter/
Wimperntrichter/Infundibulum
(mit Ostium tubae)** fimbriated
funnel of oviduct, infundibulum
Ein Enzym-ein Gen-Theorie
one-enzyme-one-gene theory
**Ein Gen-ein Protein-Theorie/
Ein Gen-ein Polypeptid-Theorie**
one-gene-one-protein theory;
one gene-one polypeptide theory
Ein-Buchstaben-Code *gen*
one-letter-code
einästig/uniram (ohne Exopodit)
uniramous
einatmen *vb* breathe in, inhale
Einatmen *n* inhalation
**Einatmung/Einatmen/Inspiration/
Inhalation** inspiration, inhalation
einbalsamieren enbalm
**Einbeerengewächse/
Dreiblattgewächse/Trilliaceae**
Trillium family
**Einbettautomat/
Einbettungsautomat** *micros*
embedding machine,
embedding center
einbetten *micros* embed
Einbettung *micros* embedding
Einbettungsmittel/Einschlussmittel
mountant, mounting medium
Einblattfrucht simple fruit,
apocarpous fruit, unicarpellary fruit,
monocarpellate fruit
Einbuchtung indentation
einbürgern introduce, establish,
naturalize, acclimate
Einbürgerung establishment,
settlement, naturalization,
acclimatization
eindringen intrude

Einschlusskörperchen

Eindringling intruder
Einehe/Monogamie monogamy
eineiig/monozygot monozygous, monozygotic
Eineiigkeit/Monozygotie monozygosity
Einemsen *orn* anting
einengen/konzentrieren reduce, concentrate
einfachblumenblättrig/ monochlamydeisch/ haplochlamydeisch monochlamydeous, haplochlamydeous
einfachbrechend/isotrop isotropic
Einfachkreuzung single cross
Einfachzucker/einfacher Zucker/ Monosaccharid single sugar, monosaccharide
einfrieren freeze
einführen introduce; import
eingebuchtet (Blattspitze) retuse
eingedrückt indented
eingeführt introduced, allochthonous; imported
eingekerbt/gekerbt/kerbig nicked, notched
eingerollt rolled up, coiled, coiled up;
(eingewickeltes Farnblatt) circinate
- **nach hinten eingerollt/ zurückgerollt** revolute, rolled backward
- **seitlich eingewickelt/ übereinandergerollt (Blattränder)** convolute, rolled up
eingeschlechtig unisexual
eingeschnitten (gleichmäßig) incised, cut; (ungleichmäßig) lacerate, torn
eingeschoben/intercalar/interkalar intercalary (inserted between others)
eingewachsen ingrown
Eingeweide/Gedärme/Viscera/ Viszera/Splancha viscera, guts, intestines, entrails, bowels
Eingeweidefische/Ophidiiformes brotulas & cusk-eels & pearlfishes
Eingeweidemuskeln visceral muscles
Eingeweidemuskulatur/viscerale Muskulatur visceral musculature
Eingeweidesack (Mollusken) visceral hump, visceral mass
Eingeweideschädel/ Visceralcranium/Viscerocranium/ Splanchnocranium viscerocranium, splanchnocranium
Eingewöhnung acclimation, acclimatization
Eingewöhnungsphase establishment phase
einhäusig monecious, monoecious
einheimisch indigenous, native, endemic
Einheit (Maßeinheit) unit (measure)
einheitlich uniform
Einheitsmembran unit membrane
einhörnig one-horned, unicornate, unicornuate, unicornuous
einjährig annual
Einjährige/Annuelle/Therophyt annual (plant), therophyte
einjähriges Tier/Jährling yearling
einkammerig/einkämmrig/ monothalam/monothekal single-chambered, monothalamous, monothalamic, monothecal
Einkapselung encapsulation
einkeimblättrig monocotyledonous
Einkeimblättrige/Monokotyledone/ Monokotyle monocotyledon, monocot
Einkerbung/Kerbung indentation, crenation, indenture, notching
Einlagerung inclusion, intercalation
Einlagerung/Intussuszeption (Zellwandwachstum durch) intussusception
einlagige Schicht/monomolekulare Schicht monolayer, monomolecular layer
einmieten *agr* ensile; store in a pit/silo (frost-protection)
einnehmen/etwas zu sich nehmen ingest
einnisten/implantieren implant
Einnistung/Implantation/Nidation implantation, nidation
einordnen/einstufen/klassifizieren rank, classify
einpflanzen (Organe) implant
Einpflanzung (Organe) implantation
- **Implantat** implant
Einplatter/Monoplacophora monoplacophorans
einreihig/uniseriat single rowed, uniseriate, uniserial
einreiten gait
- **eingeritten** gaited
Einrollung (seitlich eingewickelt/ zusammengerollt; z.B. Blätter) convolution
Einrollung/Involution involution
Einsalzen/Einsalzung *chem* salting in
einscheidig monodelphic
Einschichtzellkultur monolayer cell culture
Einschluss inclusion
Einschlusskörperchen inclusion body
- **Bollinger-Körper** Bollinger body, Bollinger's granule

einschneidig

- **Guarnierischer Einschlusskörper** Guarnieri body
- **Kerneinschlusskörper** nuclear inclusion body
- **Negrisches Körperchen/Negri Körper** Negri body
- **X-Körper** x body
- **Zelleinschluss (Inklusion)** cell inclusion, cellular inclusion

einschneidig (Scheitelzelle) with one cutting face
Einschnitt incision, cut; indentation
Einschnürung constriction
einseitig/unilateral unilateral
einsetzen (z. B. Fische in einen Teich) stock
einsinnig unidirectional
Einspitzen, seitliches *hort* shield grafting, sprig grafting
einspritzen/injizieren inject
Einspritzung/Injektion injection
Einstichkultur/Stichkultur (Stichagar) stab culture
einsträngig *gen* single-stranded
Einstreuung interspersion
- **kurzphasige Einstreuung** short period interspersion
- **langphasige Einstreuung** long period interspersion

Einstrom influx
Einströmöffnung/ Ingestionsöffnung incurrent/ inhalant/ingestive aperture
Einstufung/Kategorisierung categorization
Einstülpung/Einfaltung/Embolie/ Invagination emboly, invagination
Eintagsfliegen/Ephemeroptera mayflies
Eintauchkultur submerged culture
Einteilung/Gruppeneinteilung/ Klassifizierung classification
eintopfen pot
Eintrag *ecol* input
Eintrittspforte route of entry
einverleiben/verschlingen engulf
Einverständniserklärung nach ausführlicher Aufklärung informed consent
einwandern/zuwandern/ immigrieren immigrate
einwandern *embr* ingress
Einwanderung/Zuwanderung/ Immigration immigration
Einwanderung *embr* ingression
Einwaschung illuviation
Einweghandschuhe disposable gloves
Einwegspritze disposable syringe
einwertig/univalent/monovalent *chem* univalent, monovalent
Einwertigkeit/Univalenz *chem* univalence

einwiegen (nach Tara) weigh in (after setting tare)
einwirtig/homoxen monoxenous, monoxenic
Einzapfung/Gomphose gomphosis
einzehig one-toed
Einzelauge/Punktauge/Ocelle/ Ocellus (*siehe:* Nebenauge/ Scheitelauge/Stirnauge) simple eye, ocellus (dorsal and lateral)
Einzelauge/Punktauge/Stemma (Seitenauge/Lateralocellus bei Insekten-Larven) stemma (*pl* stemmata/stemmas), lateral ocellus, lateral eye
Einzelblüte solitary flower, single flower
Einzelfrucht simple fruit
Einzelgänger solitary animal, loner, lone wolf
Einzelkopie-DNA/nichtrepetitive DNA single copy DNA
Einzelkopie-Gen single copy gene
Einzelkopie-Plasmid single copy plasmid
Einzeller unicellular lifeform
"Einzeller"/Urtierchen/Urtiere/ Protozoen protozoans
Einzellerprotein single-cell protein (SCP)
einzellig single-celled, unicellular
einzeln/solitär single, solitary
Einzelstammentnahme *for* single stem removal
Einzelstrang *gen* single strand
Einzelstrang-Konformations-Polymorphismus (SSCP) single strand conformation polymorphism (SSCP)
Einzelstrangassimilation single strand assimilation
Einzelstrangaustausch single strand exchange
einzelstrangbindendes Protein single strand binding protein
Einzelstrangbruch single strand break
- **versetzte Einschnitte/versetzte Einzelstrangbrüche (z.B in doppelsträngiger DNA)** staggered nicks (e. g. in double stranded DNA)

Einzeltier single animal, solitary animal
einziehbar/zurückziehbar retractile
Einzugsgebiet/ Wassereinzugsgebiet catchment basin, drainage basin
Eipilze/Oomyzeten/Oomyceten/ Oomycota oomycetes (water molds & downy mildews)

ektopische Schwangerschaft **E**

Eiplasma/Bildungsplasma/
  Ooplasma ooplasm
Eiplatte/Eischiffchen (Gastropoden/
  Culiciden) egg raft
Eiröhre/Eischlauch/Ovariole/
  Ovariolschlauch (Insekten)
  egg tube, ovarian tube, ovariole
Eirollbewegung egg-rolling
Eis ice
Eisack ovisac, brood pouch, egg
  case; (*Spinnen:* Eipaket/Eikokon)
  egg sac, "cocoon"
Eisäckchen (Copepoden) egg sac
Eisbad ice-bath
Eisbein (Hüftbein & Kreuzbein beim
  Schalenwild) *hunt* hipbone &
  sacrum (hoofed game)
Eisbein *cul* cured and cooked pork
  knuckle/forehock
Eischale eggshell; (*Insektenei:*
  Chorion) eggshell, chorion
Eischalendrüse/Schalendrüse/
  Nidamentaldrüse (Gastropoden)
  nidamental gland
Eischiffchen/Eiplatte (Gastropoden/
  Culiciden) egg raft
Eischlauch/Eiröhre/Ovariole/
  Ovariolschlauch (Insekten)
  egg tube, ovariole
Eischnur egg-string
Eisen iron
Eisen-Schwefel-Protein
  iron-sulfur protein
Eisenbakterien iron bacteria
Eisenhölzer ironwood
Eisenkrautgewächse/
  Verbenengewächse/Verbenaceae
  verbena family, vervain family
eisenregulierender Faktor
  iron-regulating factor (IRF)
Eisessig glacial acetic acid
Eiskernaktivität *micb*
  ice nucleating activity
Eiskrautgewächse/
  Mittagsblumengewächse/
  Aizoaceae mesembryanthemum
  family, fig marigold family,
  carpetweed family
Eisprung/Follikelsprung/Ovulation
  ovulation
Eisspross (Geweih) bez tine
Eisüberzug/überfrorene Nässe/
  gefrorener Regen sleet, glaze,
  frozen rain
Eiszeit *allg* glacial period
  • Nacheiszeit postglacial period
  • Voreiszeit preglacial period
  • Zwischeneiszeit/Interglazial
    interglacial (period)
Eiszeit/Glazialzeit/Pleistozän/
  Diluvium Ice Age, Glacial Epoch,
  Pleistocene Epoch, Diluvial

Eiszeitrefugium glacial refugium,
  Pleistocene refuge
Eiszeitrelikt glacial relict
EITB (enzymgekoppelter
  Immunoelektrotransfer)
  EITB (enzyme-linked immuno-
  electrotransfer blot)
Eiter pus
eitragend/eiführend/oviger
  ovigerous
Eiträger/Oviger (Brutbein
  bestimmter Arachniden/
  Pantopoden)
  oviger (egg-carrying leg)
Eiweiß (Ei) egg white, egg albumen
  • aus Eiweiß bestehend/Eiweiß.../
    proteinartig/proteinhaltig/
    Protein... proteinaceous
  • denaturiertes Eiweiß
    denatured egg white
  • natives Eiweiß/Eiklar
    native egg white
Eiweiß/Protein protein
Eiweißdrüse albumen gland
eiweißlos exalbuminous
Eizahn/Eischwiele (Reptilien)
  egg tooth
Eizahn/Oviruptor (Insekten)
  egg burster, hatching spine
Eizelle/Ei (weibliche
  Geschlechtszelle) egg cell, egg,
  ovum (female gamete)
Eizelle/Ovozyt/Oozyt/Ovocyt/Oocyt
  (vor und während Meiose)
  egg cell, ovocyte, oocyte (before and
  during meiosis)
Ejectisom ejectisome, ejectosome,
  ejectile body, ejectile organelle
Ekdyse/Ecdysis/Häutung/
  Federverlust/Haarverlust ecdysis,
  molt, molting
ekkrin eccrine
Eklektor *ecol* eclector,
  emergence trap
Eklipse *vir* eclipse period, eclipse
Ektokarp ectocarp, epicarp, exocarp
Ektoparasit/Exoparasit/
  Außenparasit (*siehe*: Hautparasit)
  ectoparasite, exoparasite, epizoon
Ektopie (an unüblicher Stelle/
  auf unübliche Weise) ectopy
ektopisch (an unüblicher Stelle/auf
  unübliche Weise) ectopic
ektopisch/verlagert (an unüblicher
  Stelle liegend) ectopic
ektopische Paarung (unspezifische
  Paarung von Chromosomen)
  ectopic pairing (of chromosomes)
ektopische Schwangerschaft/
  Extrauteringravidität (EUG)
  ectopic pregnancy,
  extrauterine pregnancy (EUP)

E Elaeagnaceae

Elaeagnaceae/Ölweidengewächse
oleaster family
Elaeocarpusgewächse/
Elaeocarpaceae makomako family
Elaiosom elaiosome
Elaphoglossaceae/
Zungenfarngewächse
elephant's-ear fern family
Elastin elastin
elastisch elastic
elastische Faser elastic fiber
Elastizität elasticity
Elastizitätsgrenze/Dehngrenze
yield strength
Elatere/Schleuderzelle elater
Elatinaceae/Tännelgewächse
waterwort family
electrophiler Angriff
electrophilic attack
Elefantenvögel/
Madagaskarstrauße/
Aepyornithiformes elephant birds
elektoneutral electroneutral
(electrically silent)
Elektroencephalogramm (EEG)
electroencephalogram
elektrogen electrogenic
Elektroimmunodiffusion
electroimmunodiffusion,
counter immunoelectrophoresis
Elektrokardiogramm (EKG)
electrocardiogram
Elektrolyt electrolyte
elektromotorische Kraft (EMK)
electromotive force (emf/E.M.F.)
Elektron electron
Elektronen-Energieverlust-
Spektroskopie electron energy loss
spectroscopy (EELS)
Elektronenakzeptor
electron acceptor
Elektronendonor/
Elektronenspender electron donor
Elektronenmikroskopie
electron microscopy (EM)
- **Rasterelektronenmikroskopie**
(REM) scanning electron
microscopy (SEM)
- **Transmissionselektronen-**
mikroskopie/
Durchstrahlungselektronen-
mikroskopie transmission
electron microscopy (TEM)
- **Höchstspannungselektronen-**
mikroskopie high voltage
electron microscopy (HVEM)
- **Immun-Elektronenmikroskopie**
immunoelectron microscopy
(IEM)
Elektronenraffer/
Elektronenempfänger
electron acceptor

Elektronenspender/
Elektronendonor electron donor
Elektronenspinresonanz (ESR)
electron spin resonance (ESR),
electron paramagnetic resonance
(EPR)
Elektronentransport
electron transport
- **nichtzyklischer/nichtcyclischer/**
linearer Elektronentransport
noncyclic electron transport
- **zyklischer/cyclischer**
Elektronentransport
cyclic electron transport
Elektronentransportkette
electron-transport chain
Elektronenüberträger
electron carrier
Elektronenübertragung
electron transfer
elektronisch electronic
Elektronspinresonanz
electron spin resonance (ESR)
Elektrophorese electrophoresis
- **Direkttransfer-Elektrophorese/**
Blotting-Elektrophorese
direct transfer electrophoresis,
direct blotting electrophoresis
- **Diskelektrophorese/**
diskontinuierliche
Elektrophorese
disk electrophoresis
- **freie Elektrophorese**
free electrophoresis
(carrier-free electrophoresis)
- **Gegenstromelektrophorese/**
Überwanderungselektrophorese
countercurrent electrophoresis
- **Gelelektrophorese**
gel electrophoresis
- **Kapillarelektrophorese**
capillary electrophoresis
- **Papierelektrophorese**
paper electrophoresis
- **Puls-Feld-Gelelektrophorese**
pulsed field gel electrophoresis
(PFGE)
- **Tasche/Vertiefung**
(Elektrophorese-Gel) well,
depression (at top of gel)
- **Trägerelektrophorese**
carrier electrophoresis
- **Überwanderungselektro-**
phorese/
Gegenstromelektrophorese
countercurrent electrophoresis
- **Wechselfeld-Gelelektrophorese**
alternating field gel
electrophoresis
- **Zonenelektrophorese**
zone electrophoresis
elektrophoretisch electrophoretic

empfängnisverhütendes Mittel

elektrophoretische Mobilität
electrophoretic mobility
Elektroplaque (*pl* **Elektroplaques/**
slang: **Elektroplaxe)** electroplaque
Elektroporation electroporation
Elektroretinogramm (ERG)
electroretinogram
elektrotonisches Potential
electrotonic potential
Element (Vogelgesang) *orn*
element, phrase element
Elementarkörperchen
elementary body
Elfenbein ivory
ELISA (enzymgekoppelter
Immunadsorptionstest/
enzymgekoppelter
Immunnachweis) ELISA (enzyme-
linked immunosorbent assay)
Ellagsäure ellagic acid, gallogen
Elle/Ulna elbow bone, ulna
Ellenbeuge/Armbeuge
bend of the arm, crook of the arm,
inside of the elbow
Ellenbogen/Cubitus elbow, cubitus
Ellenbogenhöcker/
Ellenbogenspitze/
Ellenbogenfortsatz/Olekranon
point of the elbow, olecranon
elliptisch elliptic, elliptical
Elongationsfaktor elongation factor
elterliche Fürsorge parental care
Eltern parents
Elternaufwand *ethol* parental
investment
Elternhocker/Tragling parent-clinger
Elternmittelwert midparent value
Elternschaft parenthood
Elternteil parent
Eluat eluate
eluieren eluate
eluotrope Reihe
(Lösungsmittelreihe)
eluotropic series
Elutionskraft eluting strength
(eluent strength)
Elutionsmittel/Eluens (Laufmittel)
eluent, eluant
Elytre/Deckflügel/Flügeldecke
elytron, elytrum (*pl* elytra),
wing sheath
Embden-Meyerhof-Weg
Embden-Meyerhof pathway,
Embden-Meyerhof-Parnas pathway
(EMP pathway), hexosediphosphate
pathway, glycolysis
Embolie (Obstruktion der Blutbahn)
embolism
Embolie/Invagination/Einfaltung/
Einstülpung emboly, invagination
Embolium (Insektenflügel)
embolium

Embolus embolus
Embryo embryo
Embryoblast/Embryonalknoten
embryoblast, inner cell mass
Embryologie embryology
● **vergleichende E.**
comparative embryology
embryonal embryonal, embryonic
Embryonalentwicklung/
Embryogenese/Embryogenie/
Keimesentwicklung embryonal/
embryonic development,
embryogenesis, embryogeny
Embryonalhülle/Keimhülle/
extraembryonale Membran
extraembryonic membrane
Embryonalschale/
Embryonalgewinde/Larvenschale/
Primärschale/Protoconch
(Mollusken: Gastropoden)
embryonic shell, nuclear whorls,
protoconch
Embryonalschale/Larvenschale/
Primärschale/Prodissoconch
(Mollusken: Muscheln)
embryonic shell, prodissoconch
Embryosack/Keimsack
(Gametophyt) *bot* embryo sac
Embryosack/Ovicelle/Ooecie
(Bryozoen) ovicell
Embryosackmutterzelle/
Makrosporenmutterzelle/
Megasporenmutterzelle *bot*
megaspore mother cell, macrospore
mother cell, megasporocyte
Embryoträger/Suspensor suspensor
Embryotransfer embryo transfer
Emergenz emergence
Emission/Ausstoß/Ausstrahlung
emission
Emissionskoeffizient
emissivity coefficient
(absorptivity coefficient)
Empetraceae/
Krähenbeerengewächse
crowberry family
Empfänger/Rezeptor receptor
Empfänger/Rezipient
(z. B. Transplantate)
recipient (also: host)
Empfängerzelle recipient cell
empfänglich receptive
Empfängnis (Befruchtung einer
Eizelle) conception, fertilization
Empfängnishügel fertilization cone
Empfängnishyphe/Trichogyne *fung*
trichogyne
empfängnisverhütend/kontrazeptiv
contraceptive
empfängnisverhütendes Mittel/
Verhütungsmittel/Kontrazeptivum
contraceptive

E Empfängnisverhütung 460

Empfängnisverhütung/
Kontrazeption contraception
empfindbar perceptible, sensible
Empfindbarkeit sensibility,
sensitiveness
empfinden/fühlen/spüren
feel, sense, perceive
empfindlich (reizempfänglich)
irritable, sensible
empfindlich (sensitiv/leicht
reagierend) sensitive
empfindlich/zerbrechlich (Pflanze/
Ökosystem) tender, fragile
Empfindlichkeit *photo/micros*
sensitivity
Empfindlichkeit/Anfälligkeit
susceptibility
Empfindlichkeit/Gekränktsein
sensitiveness, touchiness
Empfindung sensation, perception
empfohlener täglicher Bedarf
recommended daily allowance
(RDA)
empirisch empiric(al)
empirische Formel
empirical formula
Emulgator emulsifier,
emulsifying agent
emulgieren emulsify
Emulsion emulsion
Enantiomer enantiomere
Endbild *micros* final image
Enddarm/Hinterdarm (Colon &
Rectum)/Proctodaeum hindgut,
proctodeum
Enddarmdrüse/Proktodaemdrüse
protodeal gland
Ende/Terminus (Molekülende)
terminus
Endemie endemic
endemisch endemic
Endemismus endemism, endemicity
Endemit endemic species, endemic
organism/lifeform, endemic
endergon/energieverbrauchend
endergonic
Endfüßchen (Astrozyt) end-foot
Endgruppenanalyse
end-group analysis
Endgruppenbestimmung end-group
analysis, terminal residue analysis
Endhandlung *ethol* end act
(consummatory behavior)
Endhirn/Großhirn/Telencephalon
endbrain, cerebrum, telencephalon
Endigung (Dendriten) terminal
Endit endite
Endknopf/Endknöpfchen/Bouton
neuro terminal bulb, bouton
Endknospe/Terminalknospe
terminal bud
Endmarkierung end labelling

Endmeristem terminal meristem
Endmoräne end moraine
Endocuticula endocuticle
Endocytose endocytosis
• **rezeptorvermittelte/**
rezeptorgekoppelte E.
receptor-mediated endocytosis
Endoderm endodermis
endodermal endodermal
endogen endogenic, endogenous
Endokarp endocarp
endokrin endocrine
endokrine Drüse endocrine gland
Endolymphe endolymph
Endomitose endomitosis
Endoparasit/Endosit/Innenparasit
endoparasite, endosite
Endophyt endophyte
endophytisch endophytic
endoplasmatisches Retikulum (ER)
(glattes/raues ER) endoplasmic
reticulum (ER) (smooth/rough ER)
Endopodit (Innenast) *zool* endopod,
endopodite (inner branch)
Endopolyploidie endopolyploidy
endorheisch (interne
Entwässerung) *limn* endorheic,
endoreic (internal drainage)
Endorphin endorphin
Endosom/Endozytosevesikel
endosome, endocytic vesicle
Endosperm (Nährgewebe:
Embryosack) endosperm
Endostyl/Hypobranchialrinne
endostyle, hypobranchial furrow,
hypobranchial groove
Endosymbiont endosymbiont
Endosymbiontentheorie
endosymbiont theory
Endothel endothelium
endotherm endothermic
Endotoxin endotoxin
Endozytose/Endocytose
endocytosis
Endozytosevesikel/Endosom
endocytic vesicle, endosome
Endplatte, motorische motor
endplate, myoneural junction
Endplattenpotential
endplate potential (epp)
Endplattenstrom endplate current
Endprodukthemmung/
Rückkopplungshemmung
end-product inhibition, feedback
inhibition
Endpunktverdünnungsmethode
(Virustitration) end-point dilution
technique
endständig terminal, terminate
endständiges Wachstum/
Schwanzwachstum tail growth
Endwirt final host

entomophag

461

Endysis/Federneubildung/
Fellneubildung endysis
Energetik energetics
Energie energy
Energiebarriere energy barrier
Energiebedarf energy requirement
Energiebilanz energy balance,
energy budget
Energieerhaltungssatz
law of conservation of energy
Energiefluss
energy flux, energy flow
Energieladung energy charge
Energieprofil energy profile
Energiequelle energy source
energiereich energy-rich
energiereiche Bindung
high energy bond
energiereiche Verbindung
high energy compound
Energiestoffwechsel
energy metabolism
Energieübergang/Energietransfer
energy transfer
Energieverlust-Spektroskopie
electron energy loss spectroscopy
(EELS)
Engelhaie/Engelhaiartige/
Squatiniformes monkfishes,
angel sharks
Engerling (*sensu lato*: im Boden
lebende Larve der
Blatthornkäfer) grub
(scarabaeiform larva)
Engerling (*sensu stricto*: Larve des
Maikäfers) grub (of the
cockchafer), *Br* "rookworm"
Enghalsflasche narrow-neck bottle,
narrow-mouth flask
Engholz/Spätholz summerwood,
latewood
Engkontakt/Verschlusskontakt/
Schlussleiste/Kittleiste/Tight
junction (Zonula occludens)
tight junction
Engmünder/Stenostomata/
Stenolaemata (Bryozoen)
stenostomates, stenolaemates,
"narrow throat" bryozoans
Engpass/Flaschenhals bottleneck
Enhancer/Verstärker(sequenz) *gen*
enhancer (sequence)
Enkephalin enkephalin
entarten/degenerieren degenerate
Entartung degeneration,
degeneracy
Entblättern/Entblätterung
defoliation
entblättert defoliated, denuded
Entdifferenzierung/
Dedifferenzierung
dedifferentiation

Entelechie entelechy
Entenvögel/Gänsevögel/
Anseriformes screamers and
waterfowl (ducks/geese/swans)
Enterich/Erpel drake
● wilder E. mallard
enterisch enteral, enteric
Enterocöl enterocoel,
"intestine coelom"
Enterotoxin enterotoxin
Entfaltung unfolding; spreading
Entfeuchter demister
entflammbar/brennbar/
entzündlich flammable,
inflammable
● flammbeständig/flammwidrig
flame-resistant
● nicht entflammbar/nicht
brennbar nonflammable,
incombustible
● schwer entflammbar flameproof,
flame-retardant
Entflammbarkeit/Brennbarkeit/
Entzündbarkeit flammability
entgasen degas, outgas
Entgasen/Entgasung degassing,
gassing-out
entgiften detoxify
Entgiftung detoxification
Entgiftungszentrale/
Entgiftungsklinik poison control
center, poison control clinic
Enthalpie enthalpy
Enthemmung/Disinhibition
disinhibition
Entjungferung/Defloration
defloration
Entkalkung/Dekalzifizierung
decalcification
entkernt (Frucht) pitted
entkernt (Zelle) enucleate
entkoppeln decouple, uncouple,
release
Entkoppler uncoupler,
uncoupling agent
Entkopplung decoupling,
uncoupling, release
entladene Form (z.B. ATP → ADP)
unloaded form
Entladung *neuro* discharge
entlauben strip of leaves, denude
Entlaubung/Entblätterung
defoliation, denudation,
stripping of leaves
entmischen segregate, separate out
Entökie/Schutzeinmietung/
Einmietung entoecism
Entomologie/Insektenkunde
entomology, study of insects
entomophag/insektivor/
insektenfressend entomophagous,
insectivorous

Entparaffinierungsmittel *micros* decerating agent (for removing paraffin)
Entpuppung eclosion
entrinden/schälen (Rinde) decorticate, debark, bark, ross
Entrindung decortication
Entropie entropy
entsalzen desalinate
Entsalzung desalination
Entschäumer/Antischaummittel antifoam, antifoaming agent
Entseuchung/Dekontamination/ Dekontaminierung/Reinigung decontamination
entsorgen dispose of, remove
Entsorgung waste removal
entspannen *physiol* relax
entspannt/relaxiert (Konformation) relaxed
Entspannung *physiol* relaxation
Entspannungsmittel/ oberflächenaktive Substanz surfactant
entspitzen/pinzieren *hort* pinch off, tip
entsprechend corresponding
Entvölkerung depopulation
Entwaldung deforestation
entwässern/dehydratisieren dehydrate
entwässern/drainieren drain
Entwässerung/Dehydratation dehydration
Entwässerung/Drainage drainage, draining
Entwässerungsgraben drainage ditch
entwickeln/entstehen develop, emerge, unfold
Entwicklung development
- **dauerhaft-umweltgerechte/ nachhaltige Entwicklung** sustainable development
- **Embryonalentwicklung/ Embryogenese/Embryogenie/ Keimesentwicklung** embryonal/ embryonic development, embryogenesis, embryogeny
- **Mosaikentwicklung** mosaic development
- **regulative Entwicklung** regulative development
- **Rückentwicklung** retrogressive development, retrogressive evolution
- **Stammesentwicklung/ Stammesgeschichte/ Abstammungsgeschichte/ Phylogenie/Phylogenese/ Evolution** phylogeny, phylogenesis, evolution

Entwicklungs-Zyklus life cycle, life history
Entwicklungsbiologie developmental biology
Entwicklungsfeld developmental field
Entwicklungsgang course of development
Entwicklungsgenetik developmental genetics
entwicklungsgeschichtlich/ ontogenetisch ontogenetic
Entwicklungsgeschichte (des Einzelorganismus)/Ontogenese/ Ontogenie ontogeny, development
Entwicklungshilfe developmental aid
Entwicklungsländer developing countries, less developed countries (LDCs)
Entwicklungsmechanik causal morphology
Entwicklungsschwankung developmental noise
Entwicklungsstadium (*pl* **Entwicklungsstadien)/ Entwicklungsphase** developmental stage, developmental phase
Entwicklungsstufe developmental level
Entwicklungszyklus *biol* developmental cycle
Entwinden (der Doppelhelix) *gen* unwinding
entwöhnen (vom Säugen) wean
Entwöhnung weaning
entwurmen worm
entzündbar ignitable
entzünden/entflammen/anbrennen *chem* inflame, ignite
entzündet *med* inflamed
entzündlich/entflammbar/brennbar *chem* flammable, inflammable
entzündlich *med* inflammed, inflammatory
Entzündung *chem*/*med* inflammation
Enziangewächse/Gentianaceae gentian family
Enzym/Ferment enzyme
- **Apoenzym** apoenzyme
- **Coenzym/Koenzym** coenzyme
- **Holoenzym** holoenzyme
- **Isozym/Isoenzym** isozyme, isoenzyme
- **Kernenzym (RNA-Polymerase)** core enzyme
- **Leitenzym** tracer enzyme
- **Multienzymkomplex/ Multienzymsystem/Enzymkette** multienzyme complex, multienzyme system

Epithel

- **Proenzym/Zymogen** proenzyme, zymogen
- **progressiv arbeitendes Enzym** processive enzyme
- **Reparaturenzym** repair enzyme
- **Restriktionsenzym** restriction enzyme
- **Schlüsselenzym** key enzyme
- **Verdauungsenzym** digestive enzyme

Enzym-Substrat-Komplex/Enzym-Substrat-Zwischenverbindung enzyme-substrate complex

Enzymaktivierung enzyme activation, activation of enzyme

Enzymaktivität (*katal*) enzyme activity (*katal*)

enzymatisch enzymatic

enzymatische Reaktionskette enzymatic pathway

enzymatischer Abbau enzymatic degradation

enzymgekoppelter Immunadsorptionstest/ enzymgekoppelter Immunnachweis (ELISA) enzyme-linked immunosorbent assay (ELISA)

enzymgekoppelter Immunoelektrotransfer enzyme-linked immunotransfer blot (EITB)

Enzymhemmung enzymatic inhibition, repression of enzyme, inhibition of enzyme

Enzymimmunoassay/ Enzymimmuntest (EMIT-Test) enzyme immunoassay, enzyme immunoassay (EIA)

Enzymkaskade enzyme cascade

Enzymkatalyse enzymatic catalysis

Enzymkinetik enzyme kinetics

Enzymkopplung enzymatic coupling

Enzymreaktion enzymatic reaction

Enzymspezifität enzymatic specificity, enzyme specificity

enzystieren/zystieren encyst

Enzystierung/Encystierung encystment

Eophytikum Eophytic Era

Eozän (erdgeschichtliche Epoche) Eocene, Eocene Epoch

Epacridaceae/ Australheidegewächse epacris family

epaxionisch (Rumpfmuskulatur) epaxial

Epeirologie/Festlandsökologie/ terrestrische Ökologie terrestrial ecology

Ependymzelle ependymal cell

Ephedraceae/Meerträubelgewächse joint-pine family, mormon tea family, ephedra family

Ephemere ephemere

Ephippium ephippium

Epibolie (Umwachsung) epiboly

Epicotyl/Epikotyl epicotyl

Epicuticula/Grenzlamelle epicuticle

epideiktisches Verhalten epideictic behavior

Epidemie epidemic

Epidemiologie epidemiology

epidemiologisch epidemiologic(al)

epidermal/Haut../die Haut betreffend epidermal, cutaneous

Epidermis epidermis

Epidermiswachstumsfaktor/ epidermaler Wachstumsfaktor epidermal growth factor (EGF)

Epifauna epifauna

epigäisch epigeous, epigean, epigeal

epigäische Keimung epigean/ epigeal germination

epigenetisch epigenetic

epigenetische Faktoren epigenetic factors

Epigyne epigynum

Epikard epicardium

Epikotyl/Epicotyl epicotyl

epimeletisches Verhalten epimeletic behavior

epimerisieren epimerize

Epimerisierung epimerization

Epinephrin/Adrenalin epinephrine, adrenaline

epipetal epipetalous

epiphyll (auf Blättern wachsend) epiphyllous

Epiphyse epiphysis

Epiphyt/Aufsitzerpflanze/ Luftpflanze epiphyte, air plant, aerial plant, aerophyte

Epipodit epipod, epipodite

episepal episepalous

Episom (integrationsfähiges Plasmid) episome

episomales Hefeplasmid (YEp) (Hefevektor) yeast episomal plasmid (YEp)

Epistase (Unterdrückung des Phänotyps eines nichtallelen Gens) epistasis

Epithel (*pl* Epithelien) epithelium
- **Atmungsepithel** respiratory epithelium
- **Drüsenepithel** glandular epithelium
- **Flimmerepithel/Wimperepithel/ Geißelepithel** ciliated epithelium
- **hochprismatisches Epithel (hohes Zylinderepithel)** simple columnar epithelium

Epithelgewebe

- **hohes-mehrschichtiges Epithel** pseudostratified columnar epithelium
- **Keimepithel** germinal epithelium
- **Pflasterepithel/kubisches Epithel** cuboidal epithelium
- **Plattenepithel** squamous epithelium
- **Riechepithel** olfactory epithelium
- **Säulenepithel/Zylinderepithel** squamous epithelium
- **Schleimhaut/Schleimhautepithel** mucous membrane
- **Sinnesepithel** sensory epithelium
- **Übergangsepithel** transitional epithelium
- **zweischichtiges/mehrschichtiges Epithel** stratified epithelium
- **Zylinderepithel/Säulenepithel** columnar epithelium

Epithelgewebe epithelial tissue
Epithelkörperchen/ Nebenschilddrüse/Beischilddrüse/ Parathyreoidea parathyroid gland
Epithelstäbchen/Rhabdit (Turbellaria) rhabdite
Epitheton/zusätzliche Bezeichnung/ Zusatzbezeichnung/Beiname epithet
- **Artname/Artbezeichnung (zweiter, kleingeschriebener Teil des Artnames)** specific epithet

Epithezium epithecium (*pl* epithecia)
epitok epitoke
Epitop/Antigendeterminante epitope, antigenic determinant
- **Konformationsepitop** conformational/discontinuous epitope
- **kontinuierliches/lineares Epitop** continuous/linear epitope

Epizootie epizootic disease
Epoche (frühe/späte) (*siehe auch:* Äon/Periode/Erdzeitalter) epoch (lower/upper *or* early/late)
- **Buntsandstein** Lower Triassic
- **Eiszeit/Glazialzeit/Pleistozän/ Diluvium** Ice Age, Glacial Epoch, Pleistocene Epoch, Diluvial
- **Eozän** Eocene, Eocene Epoch
- **Holozän/Jetztzeit/Alluvium** Holocene, Recent, Holocene Epoch, Recent Epoch
- **Jetztzeit/Holozän/Alluvium** Recent, Holocene, Recent Epoch, Holocene Epoch
- **Keuper** Upper Triassic
- **Miozän** Miocene, Miocene Epoch
- **Muschelkalk** Middle Triassic
- **Oligozän** Oligocene, Oligocene Epoch
- **Paläozän** Paleocene, Paleocene Epoch
- **Pleistozän/Diluvium/Glazialzeit/ Eiszeit** Pleistocene Epoch, Glacial Epoch, Diluvial, Ice Age
- **Pliozän** Pliocene, Pliocene Epoch

Epökie/Aufsiedlung epoecism
Epoophoron epoophoron
Equisetaceae/ Schachtelhalmgewächse horsetail family
Erbanalyse genetic analysis
erben/ererben inherit
Erbfaktor/Gen gene
Erbgang/Vererbungsmodus mode of inheritance
- **geschlechtsgebundener Erbgang** sex-linked inheritance
- **monogener Erbgang** monogenic inheritance
- **multifaktorieller/polygener Erbgang** multifactorial/polygenic inheritance

Erbgut/Genom hereditary material, genome
Erbhygiene/Eugenik eugenics, eugenetics
Erbinformation hereditary information, genetic information
Erbkoordination fixed action pattern
Erbkrankheit/erbliche Erkrankung hereditary disease, genetic disease, inherited disease, heritable disorder, genetic defect, genetic disorder
- **monogene/polygene E.** monogenic/polygenic disease

Erblast/genetische Last/genetische Bürde/genetische Belastung genetic load, genetic burden, genetic bond
Erbleiden/angeborener Fehler inborn error
erblich/hereditär hereditary, heritable
Erblichkeit im engeren Sinne (h₂) heritability in the narrow sense (h_2)
- **allgemeine Erblichkeit (H₂)** broad heritability (H_2)

Erblichkeitsgrad/Heritabilität heritability
Erbmerkmal hereditary trait
Erbschaden/genetischer Schaden genetic hazard
Erbsenbein pisiform bone, pisiform
Erbträger/Erbsubstanz hereditary material
Erdaltertum/Paläozoikum (erdgeschichtliches Zeitalter) Paleozoic, Paleozoic Era
Erdatmosphäre global atmosphere
Erdausläufer rhizome; rootstock

Erdbeschleunigung acceleration of gravity

Erdboden/Erdreich/Erde soil, ground, earth

Erdbrotgewächse/Taccaceae tacca family

Erde/Erdboden/Erdreich soil, ground, earth

Erde/Welt Earth, world

Erdeessen/Geophagie geophagy, geophagism

erdeessend/geophag geophagous

Erdferkel/Röhrchenzähner/ Tubulidentata aardvark (of Africa)

Erdgas natural gas

Erdgeschichte earth history, history of the Earth, geologic history

erdgeschichtlich/geologisch geological

Erdhöhle/Erdloch/Bau hole, burrow

Erdhügel mound

Erdkunde/Geographie geography

erdkundlich/geographisch geographical

Erdläufer/Geophilomorpha geophilomorphs

Erdmiete *agr* pit, pit silo, trench silo (excavated silo; frost protection for silage/feed)

Erdmittelalter/Mesozoikum (erdgeschichtliches Zeitalter) Mesozoic, Mesozoic Era

Erdneuzeit/Neozoikum/ Känozoikum/Kaenozoikum (erdgeschichtliches Zeitalter) Neozoic Era, Cenozoic, Cenozoic Era (Cainozoic Era/Caenozoic Era)

Erdnussartige (Pilze)/ Hymenogastrales (Hymenogastraceae) gilled puffballs

Erdoberfläche soil surface, ground level

Erdöl petroleum, crude oil

Erdölschiefer oil shale

Erdrauchgewächse/Fumariaceae bleeding heart family, fumitory family

Erdreich/Erdboden/Erde soil, ground, earth

Erdspross/Rhizom rhizome, creeping underground stem, rootstock

Erdsterne/Geastraceae earth balls

Erdwall mound

Erdzeitalter (*siehe auch:* Periode/ Epoche) geological era
- **Archaikum**
  Archean Era, Archeozoic Era (early Precambrian)
- **Eophytikum** Eophytic Era
- **Erdaltertum/Paläozoikum**
  Paleozoic, Paleozoic Era

- **Erdmittelalter/Mesozoikum**
  Mesozoic, Mesozoic Era
- **Erdneuzeit/Neozoikum/ Känozoikum/Kaenozoikum**
  Neozoic, Neozoic Era, Cenozoic, Cenozoic Era (Cainozoic Era/ Caenozoic Era)
- **Känozoikum/Kaenozoikum/ Erdneuzeit/Neozoikum**
  Cenozoic, Cenozoic Era, Neozoic Era (Cainozoic Era/Caenozoic Era)
- **Mesophytikum** Mesophytic Era
- **Mesozoikum/Erdmittelalter**
  Mesozoic, Mesozoic Era
- **Neozoikum/Erdneuzeit/ Känozoikum/Kaenozoikum**
  Neozoic, Neozoic Era, Cenozoic, Cenozoic Era (Cainozoic Era/ Caenozoic Era)
- **Paläozoikum/Erdaltertum**
  Paleozoic, Paleozoic Era
- **Paläophytikum/Florenaltertum**
  Paleophytic Era
- **Präkambrium/Präcambrium**
  Precambrian, Precambrian Era
- **Proterozoikum** Proterozoic, Proterozoic Era (late Precambrian)

Erdzungen/Geoglossaceae earth tongues

Erektion/Erigieren/Aufrichtung erection

Eremolepidaceae catkin-mistletoe family

erfassen/bewerten assess

Erfassung/Bewertung assessment

Erfolgsorgan effector organ

ergastische Substanz ergastic substance

Ergastoplasma/endoplasmatisches Retikulum endoplasmic reticulum

Ergotamin ergotamine

Erhaltungsenergie maintenance energy

Erhaltungskoeffizient maintenance coefficient (m)

Erhaltungsschnitt *hort* maintenance pruning

erheben *math/stat* survey

Erhebung (Hügel) elevation, hill, mound

Erhebung *allg/med* prominence, elevation, torus, bump, swelling

Erhebung *math/stat* survey

erhitzen heat

erholen recover

Erholung recovery

Erholungsphase (Muskel) relaxation period

Erholungsschlag *aer/orn* recovery stroke

Erholungswald amenity forest, recreational forest

Ericaceae/Heidekrautgewächse
heath family
erigieren erect
Erigieren/Erektion erection
Eriocaulongewächse/Eriocaulaceae
pipewort family
Erkältung (viraler Infekt) cold
**Erkältung/Kälteschaden
(Schädigung durch Unterkühlung)**
chilling injury
erkennen recognize
Erkennung recognition
**Erkennungsprotein
(Membranprotein)**
signature protein
Erkennungssequenz *gen*
recognition site
**Erkennungssequenz-
Affinitätschromatographie**
recognition site affinity
chromatography
erkranken fall ill, get sick, sicken,
contract a disease
Erkrankung illness, sickness,
disease, disorder (Störung)
Erkundungsverhalten
exploratory behavior
erleichterter Transport
facilitated transport
Erlenmeyer Kolben
Erlenmeyer flask
ermüden fatigue; tire,
become tired
Ermüdung fatigue, tiring
ernähren/nähren/füttern nurture,
feed
**ernähren/sich von etwas ernähren/
leben von** (Mensch) eat something,
live on; (Tiere) feed on something
**Ernährung/Füttern (z.B. eines
Tieres)** feeding, nourishing
Ernährung/Nahrung food, diet,
nourishment, nutrition
Ernährungswissenschaft/Diätetik
nutrition (nutrition science/nutrition
studies), dietetics
Erneuerungsknospe *bot* renewal bud
Ernte harvest
● **reiche Ernte** heavy crop
**Erntebestand/stehende Ernte/auf
dem Halm** standing crop
Ernteertrag crop, crop yield, harvest
**Erntemilbenlarve/
Herbstmilbenlarve (parasitäre
rote Milben)** chigger, "red bug",
harvest mite
ernten harvest (a crop), pick (fruits)
Erosion erosion
● **Bodenerosion** soil erosion
● **Grabenerosion/rinnenartige
Erosion (Schluchterosion)**
gully erosion

● **Regenerosion** pluvial erosion
● **Schichterosion/
Schichtfluterosion/
Flächenerosion** sheet erosion
erregbar excitable, irritable,
sensitive
Erregbarkeit excitability, irritability,
sensitivity
erregen excite, irritate
erregend/exzitatorisch excitatory
Erreger (Fluoreszenzmikroskopie)
exciter
● **Krankheitserreger**
disease-causing agent, pathogen
**Erregerfilter
(Fluoreszenzmikroskopie)**
exciter filter
**erregter Zustand/angeregter
Zustand** *chem/med/physiol*
excited state
Erregung/Aufregung arousal,
excitement
Erregung/Impuls impulse
Erregung/Irritation excitation,
irritation
Erregungsleitung transmission of
signals, impulse propagation
**erreichen/sich annähern/
näherkommen/annähern
(z.B. einen Wert)**
approach (e.g. a value)
Ersatz substitute
Ersatzname substitute name
Ersatztherapie
substitution therapy
**Ersatztrieb/Stresstrieb/
Proventivtrieb/Proventivspross**
bot/hort proventitious shoot,
latent shoot
Erscheinungsbild/Erscheinungsform
appearance
erschlaffen (z.B. Muskel) relax
Erschlaffung relaxation
Erschöpfungshybridisierung
exhaustion hybridization
Erstarkungswachstum
establishment growth (*Zimmermann/
Tomlinson*),
corroborative growth (*Troll*)
erstarren freeze
Erstarrungsgestein/Eruptivgestein
igneous rock
Erstbesiedlung/primäre Sukzession
ecol primary settlement,
primary succession
erstgebärend primiparous
**Erstmünder/Urmundtiere/
Urmünder/Protostomia**
protostomes
Ertrag/Ausbeute yield
Ertragsklasse/Ertragsniveau/Bonität
yield level, quality class

Eutelie

**Ertragskoeffizient/
Ausbeutekoeffizient/
ökonomischer Koeffizient**
yield coefficient (Y)
Ertragsminderung yield reduction
Ertragssteigerung yield increase
Erucasäure/Δ^{13}-Docosensäure
erucic acid, (Z)-13-docosenoic acid
erwärmen warm (warm up)
  • **erhitzen** heat (heat up)
Erwärmung warming
Erwartungspotential contingent
negative variation (CNV)
Erweiterer/Dilator (Muskel)
dilator muscle
**Erweiterung/
Erweiterungswachstum/
Dilatationswachstum** expansion,
dilation, dilatation
erwerben acquire
erworbene Eigenschaft
acquired character
**Erythrocyt/Erythrozyt/rotes
Blutkörperchen** erythrocyte,
red blood cell (RBC)
Erythropoetin erythropoietin,
erythropoiesis-stimulating factor
(ESF)
**Erythroxylaceae/
Kokastrauchgewächse**
coca family
**Erythrozyt/Erythrocyt/rotes
Blutkörperchen** erythrocyte,
red blood cell (RBC)
Erythrozytenreifung/Erythropoese
erythropoiesis
**Erythrozytenschatten/Schatten
(leeres/ausgelaugtes rotes
Blutkörperchen)** erythrocyte ghost
erzeugen produce, make
Erzeuger/Produzent producer
Erzeugung monozygoter Mehrlinge
twinning
Erziehungsschnitt *hort* training,
form pruning, shape pruning
Erzlaugung, mikrobielle
microbial metal-ore leaching,
microbial leaching of metal ores
Eserin/Physostigmin eserine,
physostigmine
ESR (Elektronenspinresonanz)
ESR (electron spin resonance)
essbar edible, eatable
  • **nicht essbar** inedible, uneatable
Essbarkeit edibility, edibleness
essen eat
Essen food; (Mahlzeit) meal
essentiell essential
essentielle Aminosäure
essential amino acids
Essenz *chem/pharm* essence
  • **Fruchtessenz** fruit essence

Essig vinegar
Essigsäure/Ethansäure (Acetat)
acetic acid, ethanoic acid (acetate)
  • **"aktivierte Essigsäure"/
  Acetyl-CoA**
  acetyl CoA, acetyl coenzyme A
Essigsäureanhydrid
acetic anhydride, ethanoic anhydride,
acetic acid anhydride
**EST (exprimierte sequenzmarkierte
Stelle)** *gen*
EST (expressed sequence tag)
Estron/Östron estrone
etablierte Zellinie
established cell line
Etage/Stockwerk story
etagenförmig arranged in tiers
Etagenmeristem tiered meristem
etagiert/stockwerkartig/geschichtet
storied, stratified
Ethogramm/Verhaltensinventar
ethogram, behavioral inventory
Ethökologie/Verhaltensökologie
behavioral ecology
Etikett tag
  • **Namensetikett** name tag
etikettieren/markieren tag
Etiolement/Vergeilung etiolation
Etioplast etioplast
Euchromatin euchromatin
**Eucommiaceae/
Guttaperchagewächse**
eucommia family
Eucyt/Euzyt/Euzyte eukaryotic cell
Eugenik/Erbhygiene/Rassenhygiene
eugenics, eugenetics
Eugenik/Eugenetik eugenics,
eugenetics
Euglenophyta/Augenflagellaten
euglenoids, euglenids
**Eukaryont/Eukaryot (Eucaryont/
Eucaryot)** eukaryote (eucaryote)
**eukaryontisch/eukaryotisch
(eucaryontisch/eucaryotisch)**
eukaryotic (eucaryotic)
Eulen/Strigiformes owls
Eulitoral eulittoral, eulittoral zone
**Euphorbiaceae/
Wolfsmilchgewächse**
spurge family
euphotische Zone euphotic zone
Eupnoe eupnea
eupyren (Spermien) eupyrene
euryhalin euryhaline
euryök/euryözisch euryoecious,
euryecious, euryoecic
Eurytele eurytele
**Eustachische Röhre/Eustach'-Röhre/
Ohrtrompete** eustachian tube,
auditory tube
Eutelie/Zellkonstanz
eutely, cell constancy

Euter udder
Euthyneurie/ sekundäre Orthoneurie
euthyneural nerve pattern
eutroph (nährstoffreich) eutrophic
eutrophieren eutrophicate
Eutrophierung eutrophication
Evaporimeter/Verdunstungsmesser
evaporimeter, evaporation gauge, evaporation meter
evers eversed
Evertebraten/Invertebraten/ Wirbellose invertebrates
Evolution/Phylogenie/Phylogenese/ Stammesgeschichte/ Stammesentwicklung/ Abstammungsgeschichte
evolution, phylogeny, phylogenesis
- **iterative Evolution**
  iterative evolution
- **Koevolution** coevolution
- **konvergente Evolution**
  convergent evolution
- **netzartige Evolution**
  reticulate evolution
- **phyletische Evolution**
  phyletic evolution
- **Quantenevolution**
  quantum evolution
- **Rückentwicklung**
  retrogressive development, retrogressive evolution

evolutionär/ abstammungsgeschichtlich/ phylogenetisch/phyletisch
evolutionary, phylogenetic, phyletic
evolutionär stabile Strategie
evolutionarily stable strategy (ESS)
Evolutionstheorie/ Abstammungstheorie/ Deszendenztheorie
theory of evolution, evolutionary theory
Evolutionsumkehr
counterevolution
evoziertes Potential
evoked potential
Excipulum (Flechten) exciple
- **Eigenrand/Excipulum proprium (Flechten)** proper margin, exciple without algae
- **Ektalexcipulum** ectal exciple
- **Entalexcipulum/inneres Excipulum** medullary exciple
- **Lagerrand** thalline exciple, excipulum thallium

Excision/Exzision/Herausschneiden
excision
Excisionsreparatur/ Exzisionsreparatur *gen*
excision repair

Exclusion/Exklusion/Ausschluss
exclusion
- **allele Exclusion/Allelausschluss** *gen* allelic exclusion

Exemplar/Muster/Probe specimen, sample
exergon/energiefreisetzend
exergonic
Exine (Pollen) *bot* exine
"Existenzkampf"
struggle for survival
existierend/bestehend existing, extant
Exit (Arthropoden) exite
Exklusion/Ausschluss exclusion
Exkremente excretions
Exkret/Exkretion excretion
Exkretionskanal excretory canal
Exkretionssystem/ Ausscheidungssystem
excretory system
Exkretzelle excretory cell
Exkursion excursion, field trip
exogen exogenic, exogenous
Exokonjugant exoconjugant
Exonklonierung exon cloning
Exonmischung/Exonshuffling (Hin- und Herschieben von Exons)
exon shuffling
Exopodit (Außenast) exopod, exopodite (outer branch)
exorheisch (Entwässerung in den Ozean) *limn* exorheic, exoreic
exotherm exothermic
Exozytose exocytosis
Expiration/Ausatmen expiration
Explantat explant
explodieren explode
Explodierfrucht explosive fruit
Explodierkapsel/Explosionskapsel (Springkapsel) explosive capsule
Explosion explosion
explosiv explosive
Explosivstoffe explosives
exponentielle Wachstumsphase/ exponentielle Entwicklungsphase
exponential growth phase
Expression expression
- **Überexpression** *gen*
  overexpression, high level expression

Expressionsbibliothek *gen*
expression library
Expressionskassette *gen*
expression cassette, expression cartridge
Expressionsklonierung *gen*
expression cloning
Expressionsvektor *gen*
expression vector
Expressivität expressivity
exprimieren express

exprimierte sequenzmarkierte Stelle (EST) *gen*
expressed sequence tag (EST)
Exsikkator desiccator
Exsudat/Absonderung/ Abscheidung exudate, exudation, secretion
Extensivierung extensification
Extensivwirtschaft extensive agriculture, extensive farming
Extensor/Strecker extensor
Extinktionskoeffizient extinction coefficient, absorptivity
extrachromosomales Gen extrachromosomal gene
extragene DNA extragenic DNA
extrahieren/herauslösen extract
Extrakt/Auszug extract
- **Fleischextrakt** *micb* meat extract
- **Hefeextrakt** yeast extract
- **Rohextrakt** crude extract
- **Zellextrakt** cell extract
- **zellfreier Extrakt** cell-free extract
Extraktion extraction
extranukleäres Gen extranuclear gene
extrapolieren (hochrechnen) extrapolate
extrazellulär/außerzellulär extracellular
Extremität extremity, limb
Extremitätenmuskel limb muscle, appendicular muscle
Extremitätenmuskulatur appendicular musculature
Extremitätenskelett appendicular skeleton
extrinsisch extrinsic, extrinsical
extrors extrorse
Extrusom/Ausschleuderorganelle extrusome, extrusive organelle
- **Discobolocyst** discobolocyst
- **Ejectisome** ejectisome/ejectosome/ejectile body
- **Haptocyste** haptocyst
- **Kinetocyste** kinetocyst
- **Mucocyste** mucocyst
- **Nematocyste** nematocyst
- **Rhabdocyste** rhabdocyst
- **Schleimsack** muciferous body
- **Spindeltrichocyste** spindle trichocyst
- **Toxicyste** toxicyst
Exuvialflüssigkeit/ Häutungsflüssigkeit/ Ecdysialflüssigkeit
molting fluid, exuvial fluid
Exuvie exuvia
(cast-off skin/shell etc.)
Exzisionsreparatur/ Excisionsreparatur *gen*
excision repair
exzitatorisch/erregend excitatory
exzitatorisches postsynaptisches Potential
excitatory postsynaptic potential (EPSP)

F⁺-Zelle F⁺ cell
F-duktion F-duction
F-Faktor (Fertilitäts-Faktor)
F factor (fertility factor)
F-Plasmid F plasmid
F-Verteilung/Fisher-Verteilung/
Varianzquotientenverteilung *stat*
F-distribution, Fisher distribution,
variance ratio distribution
Fab-**Fragment** *Fab* (antigen-binding
fragment of an Ig)
Fabaceae/Papilionaceae/
Hülsenfruchtgewächse/
Hülsenfrüchtler/
Schmetterlingsblütler
(Leguminosae) pea family, bean
family, legume family, pulse family
Facettenauge/Komplexauge/
Netzauge/Seitenauge
compound eye, facet eye
Fach (Kapsel) *bot* valve
Fach/Lokulament/Loculament/
Loculus/Kompartiment (von Ovar/
Anthere/Sporangium) locule,
loculus, compartment
Fachbezeichnungen/Terminologie
terminology
Fächel (Infloreszenz) rhipidium
(fan-shaped cyme)
fächeln *vb* (Bienen) fan
Fächeln *n* (Bienen) fanning
Fächer fan
Fächeraderung/Gabeladerung/
Gabelnervatur
dichotomous venation
Fächerflügler/Kolbenflügler/
Strepsiptera
twisted-winged insects, stylopids
fächerförmig fan-shaped, flabellate
Fächerfühler (Antenne) *entom*
fan-shaped antenna,
flabelliform/flabellate antenna
fächerig/gefächert/gekammert
valvate, chambered
Fächerlunge/Fächertrachee/
Buchlunge book lung
Fächerpalme fan palm
Fächertrachee *siehe* Fächerlunge
Fächerung/Kompartimentierung/
Unterteilung
compartmenta(liza)tion,
sectionalization, division
Fächerzunge/rhipidoglosse Radula
rhipidoglossate radula
fachspaltig/lokulizid/loculicid
loculicidal
Fachsprache/Fachterminologie
terminology
FACS (fluoreszenzaktivierte
Zelltrennung/Zellsortierung)
FACS (fluorescence-activated cell
sorting)

FAD/FADH₂ (Flavin-adenin-
dinucleotid) FAD/FADH₂
(flavin adenine dinucleotide)
Faden filament, thread
• **Chromatinfaden**
chromatin thread
Fadendune/Fadenfeder/Filopluma
orn filoplume
Fadenflechten/Haarflechten
filamentous lichens
Fadenflug/"Luftschiffen"/
"Ballooning" (Spinnen:
Altweibersommer) *arach*
ballooning
fadenförmig/fadenartig/
haarförmig/trichal filiform,
filamentous, threadlike, hairlike
Fadenfühler (Antenne) *entom*
threadlike/hairlike antenna,
filiform antenna
Fadenkieme/Filibranchie
filibranch gill
Fadenkiemer/Filibranchia
filibranch bivalves
Fadenschnecken/Aeolidiacea/
Eolidiacea aeolidacean snails,
aeolidaceans
Fadensegelfische/Aulopiformes
aulopiform fishes
Fadenthallus filamentous thallus
Fadenwürmer/Rundwürmer/
Nematoden roundworms,
nematodes
Fagaceae/Buchengewächse/
Becherfrüchtler beech family
Fähe (♀ Tier v. a. des Raubwildes der
Niederjagd) bitch (♀ dogs and
other carnivores)
Fahne (Fabaceen-Blütenblatt)
standard, banner (petal)
• **Federfahne** *orn* vane
Fahnenblatt/Fähnchenblatt
flag leaf
Fahnenquallen/
Fahnenmundquallen/
Semaeostomea
semeostome medusas
Fährte/Spur track, trail, trace; scent
Fäkalien (Kot & Harn) fecal matter
(incl. urin) (*siehe:* Fäzes/Kot)
Faktor factor
• **begrenzender Faktor** *ecol*
limiting factor
• **dichteabhängiger Faktor** *ecol*
density-dependent factor
• **dichteunabhängiger Faktor** *ecol*
density-independent factor
• **unteilbarer Faktor** *gen*
unit factor
• **Umweltfaktoren**
environmental factors
fakultativ facultative, optional

fakultatives Heterochromatin
facultative heterochromatin
Falbe (Pferd) dun
Falkner falconer, hawker
Falknerei falconery
Fall *med* case
Fallaub *siehe* Falllaub
Fallaubwald *siehe* Falllaubwald
Falle trap
fallen fall
fällen *for* fell
fällen/ausfällen/präzipitieren *chem*
precipitate
Fällen felling
• **Baumfällen** felling trees, logging
Fällen/Ausfällen/Ausfällung/
Präzipitation *chem* precipitation
Fallenblatt *bot* trap leaf
Fallenblume/Fallenblüte
trap blossom, trap flower,
prison flower, pitcher plant
Falllaub fallen leaves
Falllaubwald deciduous forest
Fallschirm (Spannhaut) parachute
Fallstrombank *lab* vertical flow
workstation/hood/unit
Falltür *arach* trap door
Fällung/Ausfällung/Präzipitat
precipitate
Fällung/Ausfällung/Präzipitation
precipitation
• **fraktionierte Fällung**
fractional precipitation
Fallzahl *stat* sample size
falsch false, spurious
Faltamnion/Pleuramnion
pleuramnion
β-Faltblatt β-sheet, pleated sheet
Falte fold, plication, wrinkle
Faltenfilter folded filter
faltig folded, pleated, plicate(d)
Familie family
• **mit mehreren befallenen**
Mitgliedern multiplex family
Familienmerkmal family trait
Fang (Fangzähne) fangs
Fang (gefangene/erbeutete Fische)
catch
Fang (Jagdbeute) bag, kill, catch
Fang (Raubvögel: meist *pl* **Fänge)/**
Krallen/Klauen talons, claws
Fangarm/Tentakelarm
catch tentacle, tentacular arm
Fangblase/Utriculus *bot*
bladder trap, utricle, utricle
Fangfaden *arach* catching thread,
trapline
Fanghaar (*Drosera*) tentacle
Fangmaske (Libellenlarven)
prehensile mask (retractible
prehensile labium)
Fangnetz *arach* snare

Fangpolyp/Fresspolyp/Autozooid/
Zooid (*Octocorallia*)
autozooid
Fangschlauch *arach* purse web
Fangschrecken &
Gottesanbeterinnen/Mantodea/
Mantoptera mantids
Fangschreckenkrebse/Maulfüßer/
Hoplocarida/Stomatopoda
mantis shrimps
Fangwolle *arach* cribellate silk,
catching wool
Farbanpassung color-matching
Färbbarkeit *micros* stainability
Färbeglas/Färbetrog/Färbewanne
micros staining dish, staining jar,
staining tray
färben/einfärben
dye, add color, add pigment;
(kontrastieren) *techn/micros* stain
Färben/Färbung/Einfärbung/
Kontrastierung *techn/micros*
stain, staining
Farbensehen color vision
Farbmarker *electrophor*
tracking dye
Farbstoff/Pigment dye, colorant,
pigment; *micros* stain;
(in Nahrungsmitteln) colors,
coloring
• **künstliche Farbstoffe**
artificial colors,
artificial coloring
• **natürliche Farbstoffe**
natural colors, natural coloring
• **Supravitalfarbstoff**
supravital dye, supravital stain
• **Vitalfarbstoff/Lebendfarbstoff**
vital dye, vital stain
Färbung (durch Farbstoffzugabe)
micros staining
• **Lebendfärbung/Vitalfärbung**
vital staining
• **Supravitalfärbung**
supravital staining
Färbung/Farbton/Pigmentation
color, shade, tint, tone, pigmentation
• **unauffällige(r) F.**
(z. B. Süßwasserfische)
obliterative shading
Farbzelle/Pigmentzelle/
Chromatophore pigment cell,
chromatophore
Farn fern
Farnbaum/Baumfarn tree fern
Farnblatt (eingerolltes junges)
crozier, fiddlehead
Farnblattentwicklung aus
aufgerollter Knospenlage
circinate vernation
Färse (junge Kuh: noch nicht
gekalbt) heifer

Faser fiber
- **Ballaststoffe (dietätisch)** dietary fiber
- **Bastfaser** *bot* bast fiber
- **elastische Faser** elastic fiber
- **Holzfaser (Libriformfaser)** wood fiber (libriform fiber)
- **Intrafusalfaser** intrafusal fiber
- **Kernhaufenfaser** *cyt* nuclear bag fiber
- **Kernkettenfaser** *cyt* nuclear chain fiber
- **Kletterfaser** *neuro* climbing fiber
- **langsam-kontrahierende Faser** slow-twitch fiber
- **Libriformfaser/Holzfaser** libriform fiber
- **Moosfaser** mossy fiber
- **Müller-Faser** *ophthal* Müllerian fiber, fiber of Müller
- **Muskelfaser** muscle fiber
- **Nervenfaser** nerve fiber
- **Parallelfaser** parallel fiber
- **phasische Faser** phasic fiber
- **Purkinje-Faser** Purkinje fiber, conduction myofiber
- **Riesenfaser/Mauthnersche Zelle/ Mauthner-Zelle** *ichth* giant fiber, Mauthner's cell
- **schnell-kontrahierende Faser** fast-twitch fiber
- **Sclerenchymfaser** *bot* sclerenchyma fiber, sclerenchymatous fiber
- **Sharpeysche Faser** Sharpey's fiber
- **Spindelfaser** *cyt* spindle fiber
- **Stressfaser** stress fiber
- **Zonulafasern** zonule fibers
- **Zugfaser** mantle fiber

Faser/Faserung (Schnittholz) grain
Faserholz pulpwood
faserig/fasrig fibrous, stringy
Faserknochen/Geflechtknochen fibrous bone
Faserknorpel fibrous cartilage, fibrocartilage
Faserpflanze fiber plant, fiber crop
Faserproteine/fibrilläre Proteine fibrous proteins
Faserschicht (der Anthere) fibrous layer
Fasertextur (Holz) straight grain
Fasertracheide fiber tracheid
Faserung (Schnittholz) grain
Faserwurzel fibrous root
Faserzelle (Placozoa) fiber cell
fasrig/faserig fibrous, stringy
Fass/Fass-Struktur (Proteinstruktur) barrel
fasten *vb* fast

Fasten *n* fasting
Fasziation/Verbänderung fasciation
Faszikularkambium fascicular cambium
fauchen hiss (e.g. snake); snarl (e.g. lion/tiger), puff
faul/modernd foul, rotten, decaying, decomposing
Faulbehälter septic tank
Faulbrut (Bienen) foulbrood
Fäule rot, mold, mildew, blight
faulen rot, decay, decompose, disintegrate; (im Faulturm der Kläranlage) digest
Faulgas/Klärgas (Methan) sludge gas, sewage gas
Fäulnis decay, rot, putrefaction
Fäulnisbakterien putrefactive bacteria
Fäulnisbewohner saprobe, saprobiont
Fäulnisernährer/Fäulnisfresser/ Saprovore/Saprophage saprophage, saprotroph, saprobiont
fäulniserregend/saprogen saprogenic
Fäulnispflanze/Faulpflanze/ Saprophyt saprophyte
Faulschlamm/Sapropel sludge, sapropel
- **Halbfaulschlamm/Grauschlamm/ Gyttia/Gyttja** gyttja, necron mud

Faulschlamm (*speziell:* ausgefaulter Klärschlamm) sewage sludge (*esp.*: excess sludge from digester)
Faultiere/Pilosa (Xenarthra) sloths
Faulturm digester, digestor, sludge digester
Fauna/Tierbestimmungsbuch fauna, faunal work (manual: with key)
Fauna/Tierwelt (einer bestimmten Region) fauna, animal life (within a restricted area)
Faunenelement faunal element
Faunenkomplex faunal complex
Faunenprovinz faunal province
Faunenregion/tiergeographische Region faunal region, zoogeographical region
Faunenreich faunal realm
Faunenschnitt faunal break
Faunistik faunistics
faunistisch faunal
Faust fist
Faustgang (Handknöchel) fist-walking
Fäzes/Kot feces (Stuhl > human feces)
Fazies facies
Fc Fragment *Fc* (crystallizable fragment of an Ig)

feinbehaart/flaumig **F**

FCKW (Fluorchlorkohlenwasserstoffe) CFCs (chlorofluorocarbons/chlorofluorinated hydrocarbons)
Fechser/Ausläufer *bot*/*hort* runner
Feder feather
- **Armschwingen/Unterarmschwungfedern** secondaries, secondary feathers (secondary remiges)
- **Deckfeder/Decke/Tectrix** covert, wing covert, protective feather, tectrix (*pl* tectrices)
- **dunenartige Feder/Dune/Dunenfeder/Flaumfeder** down feather, plumule
- **Fadenfeder/Fadendune/Filopluma** filoplume
- **Handschwingen/Hautschwingen/Handschwungfedern** primaries, primary feathers (primary remiges)
- **Konturfeder/Umrissfeder** contour feather
- **Ohrenfeder** auricular (ear covert)
- **Schwungfeder/Remex (*pl* Remiges)** remex (*pl* remiges)
- **Steuerfeder/Rectrix (*pl* Rectrices)** rectrix (*pl* rectrices)
Federast/Ramus barb (main branch of feather)
Federbalg/Federpapille feather papilla
Federfahne vane
Federflur/Pteryla pteryla, feather tract
federförmig/fiedrig/gefiedert pinnate
federig/fedrig feathery, plumose
Federkiel/Scapus scape, scapus (feather)
Federkleid/Gefieder/Ptilosis plumage, ptilosis
- **Brutkleid/Hochzeitskleid** breeding plumage, nuptial plumage, courtship plumage
- **Jugendkleid** juvenile plumage, juvenal plumage
- **Prachtkleid** display plumage, conspicuous plumage
- **Schlichtkleid** basic plumage; eclipse plumage, inconspicuous plumage (♂ ducks)
- **Tarnkleid/Tarntracht** camouflage plumage, cryptic plumage
Federling feather parasite (bird louse/body louse: *Mallophaga*)
Federlinge & Haarlinge/Mallophaga biting lice, chewing lice
federlos featherless, apterial
Federmistelgewächse/Misodendraceae feathery mistletoe family

Federrain/federlose Stelle/Apterium featherless space, apterium (*pl* apteria)
Federschaft/Rhachis shaft, rhachis (feather)
Federschale pen tray
Federseele (Pulpa) pulp cavity of quill
Federspule/Calamus quill, calamus (feather)
Federstrahl/Radius (Bogenstrahl/Hakenstrahl) barbule (notched/hooked barbule)
Federvieh/Geflügel fowl (Hausgeflügel>poultry)
Federzunge/ptenoglosse Radula ptenoglossate radula
fedrig/federig feathery, plumose
Feeder-Zelle feeder cell
Feenlämpchen (Spinnenkokon) Japanese lantern
Fegehaar (an Pappus) brush hair (hair-like capillary bristle/pappus hair)
Fegen (Geweih) antler rubbing
Fehlbildung malformation
fehlend lacking, missing, wanting
Fehler error, mistake; defect
- **statistischer Fehler** statistical error
- **systematischer Fehler/Bias** systematic error, bias
- **zufälliger Fehler/Zufallsfehler** random error
fehlernährt malnourished
Fehlernährung malnutrition
Fehlgeburt/Abort/Abtreibung abortion
Fehlgeburt/Spontanabort miscarriage
Fehlingsche Lösung Fehling's solution
Fehlpaarung *gen* mispairing, mismatch
- **Basenfehlpaarung** mismatch/mispairing of bases
- **Chromosomenfehlpaarung** mispairing of chromosomes
- **Fehlpaarung durch Strangverschiebung** *gen* slipped strand mispairing, slippage replication, replication slippage
Fehlpaarungsreparatur mismatch repair
Fehlsinnmutation/Missense-Mutation missense mutation
Fehlwirt/Irrwirt wrong host, accidental host
Feinbau/Feinstruktur fine structure
Feinbau/Ultrastruktur ultrastructure
feinbehaart/flaumig pilose, downy, pubescent

Feind enemy
- **natürlicher Fressfeind** natural enemy

feingesägt finely notched, serrulate

Feinjustierschraube/Feintrieb *micros* fine adjustment knob

Feinjustierung/Feineinstellung *micros* fine adjustment, fine focus adjustment

feinkerbig/feingekerbt crenulate, finely notched

Feinstruktur/Feinbau fine structure

Feinwaage precision balance

Feiung/stille Feiung/stumme Infektion silent infection

Fekundität fecundity

Feld field

Feldbau plant production, cropping

Feldbiene fielder bee

Feldblende *opt/micros* field diaphragm

Feldfrucht crop, produce

Feldführer field guide

Feldkapazität (Boden) field capacity, field moisture capacity, capillary capacity

Feldlinse *micros* field lens

Feldrain/Ackerrain balk, field boundary strip

Feldversuch/Freilanduntersuchung/ Freilandversuch field study, field investigation, field trial

Fell fur, coat; hide (esp. large/heavy skins: cowhide)

Fellsträuben/Haarsträuben piloerection

Fels (Gestein) rock; (Klippe) cliff

Felsenbein/Os perioticum/ Perioticum periotic bone, periotic

Felsenbein/Pars petrosa (des Schläfenbeins) petrous bone, petrosal bone (of temporal bone)

felsig rocky

Felspflanze petrophyte, rock plant

Felsrasen/Felssteppe (Hochland) fellfield

Felstümpel/Felsentümpel rockpool

Femelschlag/Femelhieb/ Plenterschlag *for* uneven shelterwood method, femel coupe

Femelwald/Plenterwald shelterwood: uneven-aged stand, uneven-aged plantation (with selective logging)

Femur/Oberschenkelknochen/Os femoris femur, femoral bone, thighbone

Femur/Schenkel (Arthropoden) femur

Fenn/Fen/Fehn/Feen/Vehn (Moorland/Sumpf) fen

Fenster window
- **ovales Fenster/Fenestra vestibuli** oval window
- **rundes Fenster/Fenestra cochlea** round window

fensterspaltig/foraminizid/ foraminicid foraminicidal

Fenstertüpfel *bot* fenistiform pit

Ferkel piglet, little pig; (Mastferkel) porker

ferkeln (Wurf kleiner Schweine hervorbringen) farrow (bring forth young pig litter)

Ferment/Enzym enzyme

Fermentationsschicht/ Vermoderungshorizont (Boden) fermentation layer, F-layer

Fermenter/Gärtank (siehe auch: Reaktor) fermenter

fermentieren/gären ferment

Ferntransport *bot/physio* long-distance transport

Ferse heel, calcaneus

Fersenbein/Hypotarsus (Vögel) heel, calcaneum, calcaneus, hypotarsus

Fersenbein/Os calcis/Kalkaneus/ Calcaneus heelbone, calcaneal bone, calcaneum, calcaneus

Fersenbeinhöcker/Tuber calcanei calcaneal tuber, calcaneal tubercle, tuberosity of calcaneus

Fertigplatte *chromat* precoated plate

Fertilität/Fruchtbarkeit fertility

Fertilitätsfaktor (F-Faktor) fertility factor (F factor)

Ferulasäure ferulic acid

Fessel (Pferd) pastern

Fesselbein/Fesselknochen/ proximale Phalanx pastern bone, long pastern bone, first phalanx

Fesselbeingelenk pastern joint

Fesselgelenk (Pferd)/Fesselkopf/ Köte fetlock

Festbettreaktor (Bioreaktor) fixed bed reactor, solid bed reactor

festgewachsen/festsitzend/ aufsitzend/festgeheftet/sessil firmly attached (permanently), sessile

Festigungsgewebe supporting tissue (collenchyma/sclerenchyma)

Festland mainland

Festlandsockel/Kontinentalsockel/ Kontinentalschelf/Schelf continental shelf

Festlandsökologie/terrestrische Ökologie/Epeirologie terrestrial ecology

Festphase solid phase, bonded phase

475   **fiederspaltig** **F**

festsitzend/festgewachsen/
festgeheftet/aufsitzend/sessil
firmly attached (permanently),
sessile
Festwinkelrotor *centrif*
fixed-angle rotor
Fet/Fötus fetus
fetal/fötal fetal
fetales Kälberserum
fetal calf serum (FCS)
Fett fat
• **braunes Fett** brown fat
Fett.../fettartig/fetthaltig fatty,
adipose
Fettflosse adipose fin
Fettgewebe fatty tissue,
adipose tissue
Fettkörper/Corpus adiposum
fat body
fettlöslich fat-soluble
Fettröpfchen *siehe* Fetttröpfchen
Fettsäure fatty acid
• **einfach ungesättigte Fettsäure**
monounsaturated fatty acid
• **gesättigte Fettsäure**
saturated fatty acid
• **mehrfach ungesättigte Fettsäure**
polyunsaturated fatty acid
• **ungesättigte Fettsäure**
unsaturated fatty acid
Fettspeicher/Fettreserve fat storage,
fat reserve
Fettsucht adiposity
Fetttröpfchen/Fett-Tröpfchen
fat droplet
Fettwiese rich meadow,
rich pasture
Fettzelle/Adipozyt/Adipocyt
fat cell, adipocyte, adipose cell
feucht humid, damp, moist
Feuchtbiotop humid biotope,
wetland
Feuchte moistness
Feuchte-Orgel/Feuchtigkeitsorgel
ecol humidity-gradient apparatus
Feuchtgebiet wetland
Feuchtigkeit humidity, dampness,
moisture
• **Luftfeuchtigkeit (absolute/**
relative) (absolute/realtive) air
humidity
feuerbeständig fire-resistant
feuerfest fireproof, flameproof
feuerhemmend/flammenhemmend
fire-retardant, flame-retardant
Feuerkorallen/Milleporina
milleporine hydrocorals,
stinging corals, fire corals
Feuerlöscher fire extinguisher
Feuerlöschmittel
fire-extinguishing agent
feuern *neuro* fire, firing

Feuerwalzen/Pyrosomida
pyrosomes
Feuerwehr fire brigade,
fire department
Feuerwehrmann firefighter, fireman
fibrillär fibrillar
Fibrille fibril
Fibrin (Blutfaserstoff) fibrin
Fibrinogen fibrinogen
Fibrinolysin/Plasmin plasmin,
fibrinolysin
Fibroblast fibroblast
Fibroblastenkultur fibroblast culture
Fibroin fibroin
Fichtenspargelgewächse/
Monotropaceae Indian pipe family
Ficksche Diffusionsgleichung
Fick diffusion equation
Fidelität/Standorttreue fidelity
Fieber fever
• **schwarzes Fieber/Kala-Azar/**
Cala-Azar/viszerale
Leishmaniasis
Cala-Azar, kala azar
Fieberkleegewächse/
Bitterkleegewächse/
Menyanthaceae bogbean family
Fieder/Pinna pinna
Fieder/Blattfieder/Fiederblättchen/
Teilblatt/Blättchen leaflet, foliole,
pinna
fiederaderig/fiedernervig
pinnately veined, pinnately nerved,
penninerved
Fiederaderung pinnate venation
Fiederblatt (ganzes!)
compound leaf, divided leaf
Flederblattachse/Rhachis/
Blattspindel rachis
Fiederblättchen (ersten Grades)
pinna
Fiederblättchen (zweiten Grades)
pinnule, pinnula
Fiederchen/Pinnula pinnule
Fiederfuß pinnate appendage/leg
fiederig/fiedrig/fiederblättrig
pinnate, pinnated, foliolate
Fiederkieme/Kammkieme/Ctenidie/
Ctenidium gill plume, gill comb,
ctenidium
Fiederkiemer/Kammkiemer/
Protobranchiata (Bivalvia)
protobranch bivalves
fiederlappig pinnately lobed,
pinnatilobate
fiedernervig/fiederadrig
pinnately veined, penninerved
Fiederpalme pinnately-leaved palm
fiederschnittig pinnately incised,
pinnatisect
fiederspaltig pinnately split,
pinnately cleft, pinnatifid

Fiedersträuben

Fiedersträuben *orn* feather ruffling
fiederteilig pinnately parted, pinnately partite, pinnatipartite
Fiederung pinnation
fiedrig/fiederig/gefiedert/pinnat/ pennat (fiederblättrig) pinnate, pennate (foliolate)
Figur (Holz) figure
Filament (z. B. des Staubblattes) filament
- **intermediäres Filament** intermediate filament
- **Mikrofilament/Aktinfilament/ Actinfilament** microfilament, actin filament

Filarien filarial worms
Filialgeneration/Tochtergeneration (erste/zweite) (F1/F2) (first/ second) filial generation (F1/F2)
Filopodium filopodium
Filter filter
- **Anreicherung durch Filter** filter enrichment
- **Erregerfilter (Fluoreszenzmikroskopie)** exciter filter
- **Faltenfilter** folded filter
- **HOSCH-Filter (Hochleistungsschwebstoffilter)** HEPA-filter (high efficiency particulate air filter)
- **Membranfilter** membrane filter
- **Polarisationsfilter/"Pol-Filter"/ Polarisator** polarizing filter, polarizer
- **Rauschfilter** noise filter
- **Rundfilter** *chem/lab* round filter, filter paper disk
- **Sperrfilter** *micros* selective filter, barrier filter, stopping filter, selection filter
- **Spritzenvorsatzfilter/ Spritzenfilter** syringe filter

Filteranreicherung *lab* filter enrichment
Filterblättchenmethode filter disk method
Filtermagen/Pylorus (Crustaceen) pyloric stomach, pylorus (posterior region of gizzard)
Filtern (Nahrungsfiltern) filter-feeding
Filternetzwerk filter network, filtering network
Filternutsche/Nutsche (Büchner-Trichter) suction funnel, suction filter, vacuum filter (Buchner funnel)
Filterpumpe *lab* filter pump
Filterträger *micros* filter holder
Filtrat filtrate
Filtration filtration

filtrieren/passieren filter, pass through
Filtrierer/Filterer filter feeder
Filtrierflasche/Filtrierkolben/ Saugflasche filter flask, vacuum flask
Filtrierrate/Filtrationsrate filtering rate
Filtrierung/Filtrieren filtering
filzig felty, felt-like, tomentose
Finalismus finalism
Finger/Digitus finger, digit
Fingerabdruck fingerprint
Fingerbeere/Fingerballen/Torulus tactilis (Unterseite der Fingerspitze) soft volar portion of fingertip (finger pulp/digital pulp)
Fingerflughaut/Dactylopatagium dactylopatagium
fingerförmig/handförmig fingerlike, fingershaped, digitiform
fingerförmige Aderung/ fingerförmige Nervatur digitate venation
Fingerfruchtgewächse/ Lardizabalaceae akebia family, lardizabala family
Fingerglied/Zehenglied/Phalanx phalanx (*pl* phalanges)
Fingerhut *chem* thimble
Fingerkuppe/Fingerspitze fingertip
Fingernagel fingernail
Fingerprinting/genetischer Fingerabdruck fingerprinting, genetic fingerprinting, DNA fingerprinting
Fingerspitze/Fingerkuppe fingertip
Finne/Blasenwurm/Cysticercus (Bandwurmlarve) bladderworm, cysticercus
finnig (finniges Fleisch) measly (containing larval tapeworms)
Firn/Gletschereis firn, névé
Firnregion firn region/zone
First crest
Fisch (kulinarisch) fish
Fischbein/Walbein baleen, whalebone
Fische/Pisces fishes
Fischer-Projektion/Fischer-Formel/ Fischer-Projektionsformel Fischer projection, Fischer formula, Fischer projection formula
Fischerei fishing; (Gewerbe) fishery, fishing industry
fischfressend fish-eating, piscivorous
Fischfresser fish-eater, piscivore
Fischgründe fishing grounds
Fischkunde/Ichthyologie ichthyology

Fischlaich/Fischeier (Rogen)
 fish eggs (roe)
Fischläuse/Karpfenläuse/
 Kiemenschwänze/Branchiura/
 Argulida fish lice
Fischsaurier/Ichthyosauria
 fish-reptiles, ichthyosaurs
 (ocean-living reptiles)
Fischschwarm school of fish
Fischsterben fish kill
Fischteich fishpond
Fischvögel/Ichthyornithiformes
 fish birds
Fischzucht fish culture, pisciculture
FISH (*in situ* Hybridisierung mit
 Fluoreszenzfarbstoffen) FISH
 (fluorescence activated *in situ*
 hybridization)
Fissiparie fissiparity
Fitness fitness
 • **Darwinsche Fitness**
 Darwinian fitness
 • **frequenzabhängige Fitness**
 frequency-dependent fitness
Fittich/Flügel/Schwinge *orn* wing
fixieren (befestigen/fest machen)
 affix, attach
fixieren (mit Fixativ härten) fix
Fixiermittel/Fixativ fixative
Fixierung/Fixieren fixation
flach-aufgesägt (Holzstamm)
 flatsawn
Flachbrustvögel/Ratiten ratite birds
 (flightless birds)
Flächenbelastung
 (Flügelflächenbelastung)
 wing loading
Flächennetz *arach* simple sheetweb
Flächenquelle non-point source
flächenständige Plazentation/
 laminale Plazentation laminary/
 lamellate placentation
Flachkrebse/Flohkrebse/Amphipoda
 beach hoppers, sand hoppers and
 relatives
Flachland lowland, plain,
 flat country
Flachmeerzone/neritische Region
 neritic zone, neritic province
Flachmoor/Niedermoor/
 Wiesenmoor/Braunmoor/Fen
 fen (minerotrophic mire)
Flachspross/Phyllocladium cladode,
 cladophyll, phylloclade;
 (Platycladium) platyclade
Flacourtiagewächse/Flacourtiaceae
 flacourtia family,
 Indian plum family
Fladerschnitt (Holz) tangential
 section, flatsawn, plainsawn
Fladerung/Maserung (Holz) figure,
 design

Flagellariaceae/Peitschenklimmer
 flagellaria family
Flagellomer flagellomer
Flagellum/Flagelle/Geißel flagellum
Flamingos/Phoenicopteriformes
 flamingoes and allies
flammbeständig flame-resistant
Flamme (*siehe auch:* Feuer...)
 flame
flammenhemmend/feuerhemmend
 flame-retardant
Flammenionisationsdetektor
 flame ionization detector (FID)
Flammenzelle *zool* **(exkretorisch)**
 flame cell, (terminal) flame bulb
Flammpunkt flash point
Flammschutzmittel flame retardant,
 flame retarder
flammsicher/flammfest (schwer
 entflammbar) flameproof
Flanke flank, side (of horse)
Flankenkiemer/Notaspidea
 notaspideans
Flankenmeristem flank meristem,
 peripheral meristem
flankierende Region *gen*
 flanking region
Flarke (Moor) flark
Flaschenbürste tube brush (test tube
 brush), bottle brush (beaker/jar/
 cylinder brush)
Flaschenhals/Engpass *stat*
 bottleneck
Flaschentierchen/Bauchhaarlinge/
 Bauchhärlinge/Gastrotrichen
 gastrotrichs
Flatterflug/Schlagflug
 flapping flight
Flatterhaut/Flughaut/Gleithaut/
 Spannhaut/Patagium patagium
flattern (mit den Flügeln schlagen)
 flutter, flap (the wings)
Flattern (Blätter) leaf flutter
flatternde Blätter fluttering leaves
Flaum down
Flaumbehaarung pubescence
Flaumfeder/Dune/Daune/
 Dunenfeder/dunenartige Feder
 orn down feather, down, plumule
flaumig/feinstflaumig downy,
 pubescent
Flavinmononukleotid (FMN)
 flavine mononucleotide (FMN)
Flavonoid flavonoid
Flechte(n) lichen(s)
 • **Gallertflechten** gelatinous lichens
 • **Haarflechten/Fadenflechten**
 hairlike lichens,
 filamentous lichens
 • **Krustenflechte** crustose lichen
 • **Laubflechte/Blattflechte**
 foliose lichen

Flechtensäure

- **Nabelflechten (Blattflechten)** umbilicate foliose lichens
- **Strauchflechte** fruticose lichen, shrub-like lichen

Flechtensäure lichen acid
Flechtgewebe/Plectenchym plectenchyma
Fleck/Stigma spot, stigma
- **blinder Fleck** blind spot (optic disk)
- **gelber Fleck/Macula lutea (mit Fovea centralis)** yellow spot, macula lutea (with fovea centralis)

Fleckenkrankheit (*Nosema bombycis*) pebrine
Fleckenriff patch reef, bank reef
Fleckfieber/Flecktyphus/Typhus (*Rickettsia spp.*) spotted fever, typhus
fleckig speckled, patched, spotted, spotty
Flecksoral (Flechten) maculiform soralium
Fledermausausbreitung bat-dispersal, chiropterochory
Fledermausbestäubung/ Fledermausblütigkeit/ Chiropterophilie bat-pollination, chiropterophily
Fledermausblume bat-pollinated flower, chiropterophile
fledermausblütig/chiropterophil bat-pollinated, chiropterophilous
Fledermausblütigkeit/ Fledermausbestäubung/ Chiropterophilie bat-pollination, chiropterophily
Fledermäuse/Flattertiere/Chiroptera bats, chiropterans
flehmen flehmen (not translated!), lip-curling
Fleisch flesh, meat
Fleischbeschau meat inspection
Fleischbrühe/Kochfleischbouillon cooked-meat broth
Fleischextrakt *micb* meat extract
Fleischflosser/Sarcopterygii/ Choanichthyes fleshy-finned fishes, sarcopterygians
fleischfressend/karnivor/carnivor flesh-eating, meat-eating, carnivorous
Fleischfresser/Karnivor/Carnivor flesh eater, meat eater, carnivore
fleischig fleshy
Fleischwasser/Fleischbrühe/ Fleischsuppe *micb* meat infusion (meat digest/tryptic digest)
flensen (abhäuten/Walspeck abziehen) flense
Flexor/Beuger flexor

fliegen *vb* fly
Fliegen *n* flight
Fliegen/Brachycera (Diptera) true flies
Fliegenblume/Myiophile fly-pollinated flower, myiophile
Fliegenschimmel (Pferd) flea-bitten gray horse, flea-bitten white horse
fliehen flee
Fließbettreaktor fluid bed reactor
fließen flow
Fließfähigkeit/Fluidität fluidity
Fließgeschwindigkeit flow rate
Fließgewässer/fließendes Gewässer/lotisches Gewässer (Fluss/Strom) flowing water, running water, lotic water, waterway, watercourse (river/stream)
Fließgleichgewicht/dynamisches Gleichgewicht steady state, steady-state equilibrium
Fließmittel *chromat* solvent (mobile phase)
Fließmittelfront *chromat* solvent front
Fließrichtung direction of flow
Flimmerepithel/Wimperepithel/ Geißelepithel ciliated epithelium
Flimmergeißel flimmer flagellum, tinsel flagellum, pleuronematic flagellum
Flimmerhaar/Kinozilie/Kinozilium (Haarzelle) cilium (*pl* cilia)
Flimmerhärchen flimmer, tinsel
Flimmerkörper shimmering body
Flimmertrichter/Wimperntrichter/ Eileitertrichter/Infundibulum (mit Ostium tubae) fimbriated funnel of oviduct, infundibulum
Flip-Flop-Mechanismus (Membranlipide/Genexpression) flip-flop mechanism (membrane lipids/gene expression)
Flitterzelle/Iridocyt/Iridozyt/ Leucophor/Guanophor iridocyte, iridophore, leucophore, guanophore
flocken/ausflocken flock
flockig floccose
Flockulation flocculation
Flockung flocking
Flöhe/Siphonaptera/Aphaniptera/ Suctoria fleas
Flohkrebse/Flachkrebse/Amphipoda beach hoppers, sand hoppers and relatives
Flora flora
Floreneinheit floristic unit
Florenelement floristic element
Florengebiet floristic region
Florengefälle/Gesellschaftsgefälle/ Zönokline plant community gradient, coenocline

479 **Flügeladerung** **F**

Florenreich floral realm,
floral kingdom
Floreszenz/Blütezeit/Anthese
florescence, flowering period,
anthesis
Florideen (Rotalgen) floridean
algas, florideans
Florideenstärke floridean starch
florieren/gedeihen flourish, thrive
Florist (Blumenzüchter bzw.
Blumenverkäufer) florist (person
raising and/or selling flowers/plants)
Floristik/Florenkunde floristics
Floristik/Zierpflanzenbau/
Blumenzucht floriculture
flößen (Holz) raft, float (wood)
Flosse fin
• **Afterflosse/Analflosse** anal fin
• **Bauchflosse**
ventral fin, pelvic fin
• **Brustflosse** pectoral fin
• **Fettflosse** adipose fin
• **paddelartige Flosse (z.B.**
Brustflosse der Delphine/Wale)
flipper
• **Quastenflosse** lobe fin
• **Rückenflosse** dorsal fin
• **Ruderflosse/Ruder**
(Schwanzflosse der Wale) rudder
• **Schwanzflosse**
tail fin, caudal fin
• >**Schwanzruder/Fluke (Wale)**
tail fluke
• **Schwimmflosse (groß/fleischig)/**
Paddel flipper, fluke
• **Strahlenflosse** ray fin
Flösselhechte/
Flösselhechtverwandte/
Polypteriformes bichirs &
reedfishes and allies
Flößen/Treiben (Holztransport zu
Wasser) rafting (of timber/logs)
Flossenfüßer/Flügelschnecken/
Pteropoda pteropods
Flossenfüßer/Robben/Pinnipedia
marine carnivores (seals, sealions,
walruses)
Flossenkammer (*Branchiostoma*)
fin box
flossenlos without fins, apterygial
Flossensaum continuous fin,
elongated fin
Flossenstrahl (aus Hautknochen)/
Dermotrichium dermotrichium
Flossenträger/Radius/Pterygophor
pterygiophore
Flottoblast (Bryozoen) floatoblast
Flotzmaul (Drüsenmaul der
Boviden) muzzle (glandular muzzle
of bovids)
Flucht flight, escape
• **wilde Flucht** rout, stampede

Fluchtdistanz escape distance
flüchtig volatile
• **schwerflüchtig/nicht flüchtig**
nonvolatile
Fluchtreaktion flight reaction,
escape reaction
Flug flight
• **Bolzenflug** bounding flight
(intermittent flight)
• **Brautflug** nuptial flight
• **dynamischer Segelflug**
dynamic soaring
• **Fadenflug/"Luftschiffen"/**
"Ballooning" (Spinnen:
Altweibersommer) ballooning
• **Formationsflug** formation flight
• **Geradeausflug**
straight-in approach
• **Gleitflug** glide, gliding (flight)
• **Hangsegeln** slope soaring,
ridge soaring
• **Kompensationsflug** *entom*
compensation flight
• **Kraftflug** powered flight,
propulsive flight
• **Kreisen** circling (flight)
• **Langstreckenflug**
long distance flight
• **Orientierungsflug (junge**
Bienen) orientation flight
• **Schlagflug (Flatterflug)**
flapping flight
• **Schwirrflug/Schwebeflug**
(Kolibris: Rüttelflug)
hovering flight, hovering
• **Segelflug (*siehe auch dort*)**
soaring (flight)
• **statischer Segelflug**
static soaring
• **Steigflug** climb
• **Sturzflug** dive
• **Thermiksegelflug/Thermiksegeln**
thermal soaring
• **Wanderflug** migratory flight
• **Wellenflug** undulating flight
(intermittent flight)
• **Zielflug (Heimkehrvermögen/**
Heimfindevermögen) homing
Flugbahn flight path
Flugbalz aerial courtship
Flügel/Fittich/Schwinge wing;
bot Ala (Fabaceen-Blüte) wing, ala
• **Afterflügel/Nebenflügel/**
Daumenfittich/Alula/Ala spuria
(Federngruppe an 1. Finger)
alula, spurious wing, bastard wing
Flügel-Flächen-Belastung
wing loading
Flügelabschlag downward stroke of
wing
Flügeladerung/Flügelnervatur
wing venation

Flügelanlage 480

Flügelanlage/Flügelknospe
wing bud
flügelartig/schwingenartig
winglike, alar, alary
Flügelaufschlag
upward stroke of wing
Flügelbasis wing base
**Flügelbein/Pterygoid/
Os pterygoides**
pterygoid bone/process
Flügeldecke/Deckflügel/Elytre
entom elytrum, elytron, wing case,
wing sheath, wing cover
Flügeldreieck/Triangulum triangle,
triangulum
**Flügelfarngewächse/Schwertfarne/
Pteridaceae** pteris family
Flügelfeld (Region) *entom*
wing field, wing area
• **Analfeld/Vannus** anal field,
anal area, vannal area, vannus
• **Costalfeld/Remigium**
costal field, costal area
• **Jugalfeld/Jugum/Neala**
jugal field, jugal area,
jugal region, neala
Flügelfläche wing surface
flügelförmig wing-shaped, aliform
Flügelfrucht winged fruit
**Flügelhäkchen/Frenalhäkchen/
Hamulus** hamulus (*pl* hamuli)
Flügelkiemer/Pterobranchia
pterobranchs
Flügelknospe wing bud
flügellos/ungeflügelt wingless,
lacking wings, exalate, apterous,
apteral, apterygial
Flügelmal/Pterostigma pterostigma
Flügelnervatur/Flügeladerung
wing venation
Flügelnuss (Fruchtform) samara,
key
**Flügelnussgewächse/
Dipterocarpaceae** meranti family,
dipterocarpus family
Flügelpfeilung/Pfeilstellung *aer*
sweepback
Flügelplatte (Neuralrohr) dorsal
horn
Flügelprofil airfoil section,
aerofoil section
Flügelrandmal/Pterostigma (Makel)
pterostigma, stigma
**Flügelscheide (sich entwickelnder
Flügel)** wing pad, wing sheath
Flügelschlag wingbeat
**Flügelschnecken/Flossenfüßer/
Pteropoda** pteropods
**Flügelschüppchen/Jugum/
Antitegula** jugum
Flügelschuppe/Tegula wing scale,
tegula

Flügelspannweite wing spread,
wingspan
Flügelspitze wing tip
Flügelstreckung aspect ratio
**Flügelstummel (Larve: sich
entwickelnder Flügel)** wing pad
Flügelviereck/Quadrangulum
quadrangle
Flügelzelle *entom* wing cell
Flugfeder remex (*pl* remiges)
flügge (flugfähig) fully fledged,
full-fledged (able to fly)
flügge werden fledge
**Flughaut/Flatterhaut/Spannhaut/
Gleithaut/Patagium** patagium
• **Fingerflughaut/Dactylopatagium**
dactylopatagium
• **Schwanzflughaut/Uropatagium**
uropatagium
• **Vorderflughaut/Propatagium**
propatagium
**Fluginsekten/geflügelte Insekten/
Pterygota** winged insects,
pterygote insects
Flugloch entrance (to hive/nest/
shelter), entrance hole
Flugsand wind-borne sand
Flugsaurier/Pterosauria pterosaurs
(extinct flying reptiles/winged
reptiles)
flugunfähig unable to fly, flightless
Flugunfähigkeit unableness to fly,
flightlessness
Flugweite/Flugentfernung
flight distance
Fluidität/Fließfähigkeit fluidity
Fluktuation fluctuation
• **gerichtete F.** steady drift
• **ungerichtete F.** random drift
Fluktuationsanalyse/Rauschanalyse
fluctuation analysis, noise analysis
Fluktuationstest fluctuation test
Fluor fluorine
Fluoreszenz fluorescence
**Fluoreszenz-*in-situ*-Hybridisation
(FISH)** fluorescence-*in-situ*-
hybridization (FISH)
**fluoreszenzaktivierter Zellsorter/
Zellsortierer**
fluorescence-activated cell sorter
**fluoreszenzaktivierte
Zellsortierung/Zelltrennung**
fluorescence-activated cell sorting
(FACS)
**Fluoreszenzerholung nach
Lichtbleichung** fluorescence
photobleaching recovery,
fluorescence recovery after
photobleaching (FRAP)
Fluoreszenzlöschung
fluorescence quenching
fluoreszieren fluoresce

481 **Forstkundler** **F**

fluoreszierend fluorescent
Fluoridierung fluoridation
fluorieren fluorinate
Flur (Feld) field, plain, open fields; meadowland, pasture
Flurbereinigung reallocation of arable land, consolidation of arable land
Flurenmuster/Pterylographie pterylosis
Fluss *geol* river
Fluss (Licht/Energie) flux
● **diffuser Fluss** diffuse flux
Fluss (Volumen pro Zeit pro Querschnitt) flux
Flussaue riverine meadow, valley flat
Flussbett riverbed
Flussblindheit/Onchocercose (*Onchocerca volvulus*) river blindness
Flussebene fluvial plain
Flusseinzugsgebiet catchment area
flüssig fluid, liquid
Flüssigkeit fluid, liquid
Flüssigkeitschromatographie liquid chromatography (LC)
Flüssigmosaikmodell fluid-mosaic model
Flusskrebse crayfishes, crawdads
Flussmarsch estuarine marsh
Flussmündung river mouth; (Flussdelta/Ästuar) estuary
Flussniederung/Flusstal river plain, river valley
Flussrate fluence
Flusstal/Flussniederung river valley, river plain
Flussufer riverbank
Flut/Tide high tide, flood
fMet (N-Formylmethionin) fMet (N-formyl methionine)
FMN (Flavin**m**ono**n**ucleotid) FMN (**f**lavin **m**ono**n**ucleotide)
Fohlen/Füllen foal
● **männliches Fohlen** colt (under 4 years)
● **weibliches Fohlen** filly (under 4 years)
Föhn *meteo* foehn, föhn
Föhre pine
Föhrengewächse/Kieferngewächse/ Tannenfamilie/Pinaceae pine family, fir family
fokusbildende Einheit focus-forming unit (ffu)
Fokusbildung focus formation
Fokuskarte *gen* focus map
fokussieren focus, focussing
Folgeblatt/Laubblatt foliage leaf; metaphyll
Folgemeristem secondary meristem
Folgestrang *gen* lagging strand

folgsam/gelehrig docile
Folgsamkeit/Gefügigkeit/ Gelehrigkeit docility
Follikel follicle
● **Graafscher Follikel/Graaf- Follikel/Tertiärfollikel** Graafian follicle, vesicular ovarian follicle
● **Haarfollikel/Haarbalg** hair follicle
● **Lymphfollikel/Lymphknötchen** lymph follicle, lymph nodule
● **Milzfollikel/Milzknötchen/ Milzkörperchen/Malpighi- Körperchen** splenic follicle, splenic corpuscle, splenic nodule, splenic node
● **Primärfollikel** primary follicle
● **Sekundärfollikel** secondary follicle
● **Tertiärfollikel/Graafscher Follikel/Graaf-Follikel** Graafian follicle, vesicular ovarian follicle
Follikulärorgan/Follikulärdrüse (Schenkeldrüse: Eidechsen) follicular gland (femoral gland)
Follitropin/follikelstimulierendes Hormon (FSH) follicle-stimulating hormone (FSH)
Folsäure (Folat)/ Pteroylglutaminsäure folic acid (folate), pteroylglutamic acid
Fontäne (Wale) spout
Fontanelle fontanel, fontanelle
Forensik/forensische Medizin/ Gerichtsmedizin/Rechtsmedizin forensics, forensic medicine
forensisch/gerichtsmedizinisch forensic
formale Genetik formal genetics
Formänderung/Verformung/ Deformation deformation
Formation formation
Formationsflug formation flight
Formbaum/Formstrauch (auch Zierschnitt) topiary
formkonstante Verhaltenselemente fixed action pattern
Formylmethionin formyl methionine
Forst/Kulturwald/Wirtschaftswald cultivated forest, tree plantation
Forstbaum forest tree
Forstbaumkunde silvics
Förster/Forstaufseher forester, forest ranger
Forstkultur (Pflanzung) forest plantation
Forstkultur/Waldbau silviculture
Forstkunde/Forstwissenschaft silviculture, forest science, science of forestry
Forstkundler/Forstwissenschaftler forest scientist, forestry scientist

Forstverwaltung forest administration, forest service
Forstwart forest warden, forest ranger, ranger
Forstwirtschaft woodland management, forest management, forest economy, forestry
Forstwissenschaft/Forstkunde silviculture, forest science, science of forestry
Fortbewegung/Bewegung/ Lokomotion movement, motion, locomotion
fortleiten/weiterleiten (Nervenimpuls) propagate
Fortleitung/Weiterleitung (Nervenimpuls) propagation
fortpflanzen/vermehren/ reproduzieren propagate, reproduce
Fortpflanzung/Vermehrung/ Reproduktion propagation, reproduction
• **geschlechtliche/sexuelle F.** sexual reproduction
• **ungeschlechtliche/vegetative F.** asexual/vegetative reproduction
fortpflanzungsfähig/fruchtbar/fertil fertile
Fortpflanzungsfähigkeit/ Fruchtbarkeit/Fertilität fertility
Fortpflanzungsorgane (Gesamtheit) reproductive system
Fortpflanzungsrate reproductive rate
Fortpflanzungsverhalten reproductive behavior
Fortpflanzungszelle reproductive cell
Fortpflanzungszyklus reproductive cycle
Fortsatz (Nerven) process
fortwachsend/weiterwachsend accrescent
Fossil (*pl* **Fossilien)** fossil
• **lebendes Fossil** living fossil
• **Leitfossil/Faziesfossil** index fossil, zone fossil, zonal fossil
• **Spurenfossil/Ichnofossil** trace fossil, ichnofossil
• **Übergangsfossil** transitional fossil
fossile Überreste fossil remains
fossile(r) Brennstoffe fossil fuel(s)
fossilienführend/Fossilien enthaltend (Erdschichten) fossiliferous
Fossilisationslehre/Taphonomie taphonomy
fossilisieren/versteinern fossilize
fossilisiert/versteinert fossilized
Fossilisierung/Versteinerung fossilization

fötal/fetal fetal
Fötus/Fet fetus
Fouquieriaceae/Ocotillogewächse ocotillo family
Fracht (Flüssigkeit/Abwasser) load, freight
fragiles X-Chromosom (Syndrom) fragile X chromosome (syndrome)
Fraktion fraction
fraktionieren fractionate
Fraktioniersäule *lab* fractionating column
Fraktionierung fractionation
Fraktionssammler *lab* fraction collector
Frankeniaceae/ Frankeniengewächse/ Nelkenheidegewächse sea heath family, alkali-heath family
fransenartig fimbriate
Fransenflügler/Blasenfüße/Thripse/ Thysanoptera thrips
fransenspaltig/fimbricid/frimbrizid fimbricidal
fransig frayed, fringed, fimbriate(d)
fräsen (Holz) mill, shape
Fraß/Fressen feed (*anthrop* often *sensu:* junk food)
Fraß (Insektenfäkalien in Bohrgängen von Holz) insect frass (feces)
Fraßgang burrow
Fraßhemmer/fraßverhinderndes Mittel antifeeding agent, antifeeding compound, feeding deterrent
Fraßnahrungskette/Abweide- Nahrungskette grazing food chain
Fraßschaden feeding damage, browsing damage (injury)
Frauenfarngewächse/Athyriaceae lady fern family
Frauenhaarfarngewächse/ Haarfarne/Adiantaceae adiantum family, maidenhair fern family
frei/mit freien Blütenorganen free-flowering
frei schwebend free-floating, pendulous
Freiblättler/Wulstlinge/ Amanitaceae Amanita family
freiblättrig (Fruchtblatt) apocarpous
freie Wildbahn in the wild, free-ranging
freie Zone (Radnetz) *arach* free zone
Freiheitsgrad *stat* degree of freedom (df)
freikronblättrig/freiblumenblättrig dialypetalous, choripetalous, with free petals
Freiland range, field, outdoors

Froschlurche

Freilanduntersuchung/ Freilandversuch/vor-Ort-Untersuchung/Feldversuch field study, field investigation, field trial
freilaufend (Geflügel etc.) running free, free-running, free-ranging
freilaufender Rhythmus free-running rhythm
freilebend free-living
freisetzen *vb* release
Freisetzung *n* release
Freisetzung/Sekretion secretion
Freisetzungsexperiment deliberate release experiment, environmental release experiment
Freisetzungsfaktor release factor
Freisetzungshormon/ Freisetzungsfaktor/freisetzendes Hormon/freisetzender Faktor releasing hormone, release hormone, releasing factor, release factor
- **Gonadoliberin/Gonadotropin-Freisetzungshormon** gonadoliberin, gonadotropin releasing hormone/factor (GnRH/ GnRF)
- **Corticoliberin/Kortikoliberin/ Corticotropin-Freisetzungshormon/ corticotropes Releasing-Hormon** corticoliberin, corticotropin-releasing hormone, corticotropin-releasing factor (CRH/CRF)
- **Prolaktoliberin/ Prolaktin-Freisetzungshormon** prolactoliberin, prolactin releasing hormone, prolactin releasing factor (PRH/PRF)
- **Somatoliberin/Somatotropin-Freisetzungshormon** somatoliberin, somatotropin release-hormone, somatotropin releasing factor (SRF), growth hormone release hormone/factor (GRH/GRF)
- **Thyroliberin/Thyreotropin-Freisetzungshormon (TRH/TRF)** thyroliberin, thyreotropin releasing hormone/factor (TRH/TRF)
Freiwasserzone/Pelagial/pelagische Zone pelagial zone
freiwillig/aus freier Entscheidung voluntary, at free will
freiwilliger Induktor (Transkription) *gen* gratuitous inducer
fremd (Gesellschaftstreue) strange
Fremd-DNA foreign DNA
Fremdaufzucht cross-fostering
Fremdbefruchtung/ Kreuzbefruchtung/Allogamie cross-fertilization, allogamy, xenogamy

Fremdbestäubung/ Kreuzbestäubung cross-pollination
Fremdgen heterologous gene, foreign gene
Fremdpaarung disassortive mating
Fremdputzen (z.B. Fell bei Säugern) allogrooming
- **Gefiederkrauelen** *orn* allopreening
Fremdtransplantat/ Xenotransplantat xenograft
Frequenz/Häufigkeit frequency
frequenzabhängige Fitness frequency-dependent fitness
Fressbauten/Fodinichnia *paleo* feeding burrows
fressen feed (on something), ingest (etwas zu sich nehmen)
Fressfeind/Räuber/Raubfeind/ Raubtier/Jäger/Prädator predator, predatory animal (enemy)
Fressgewohnheiten feeding habits
Fresspolyp/Nährtier/Gasterozoid/ Gastrozooid/Autozooid/ Trophozoid feeding/nutritive polyp, gastrozooid, trophozooid
Fresssucht/Esssucht/Gefräßigkeit/ Hyperphagie hyperphagia
Freundsches Adjuvans Freund's adjuvant
Frischgewicht (*sensu stricto*: Frischmasse) fresh weight (*sensu stricto*: fresh mass)
Frischling young wild pig/swine/hog (1. year)
Fritte *lab* frit
Front *meteo* front
- **Kaltfront** cold front
- **Okklusion** occlusion, occluded front
- **stationäre Front** stationary front
- **Warmfront** warm front
Frontaldrüse/Kopfdrüse/Stirndrüse cephalic gland, frontal gland
Frontalmembran (Bryozoen) frontal membrane
Frontalmoräne/Stirnmoräne terminal moraine
froschartig ranine, froglike; raniform
Froschbissgewächse/ Hydrocharitaceae frog-bit family, tape-grass family, elodea family
Froschfische/Batrachoidiformes toadfishes
Froschlöffelgewächse/Alismataceae water-plantain family, arrowhead family
Froschlurche (Frösche und Kröten)/ Salientia/Anura frogs and toads, anurans

F Frost 484

Frost frost, rime frost, white frost
frostbeständig/frostresistent
frost-resistant, frost hardy
Frostbrand frost blight, nip,
winter burn
frösteln/vor Kälte zittern shiver
frostempfindlich frost-tender,
susceptible to frost
Frosthärte/Frostbeständigkeit
frost hardiness
Frosthärtung frost hardening
Frostkeimer germinating after
freezing, frost germinator
Frostloch frost pocket
frostresistent/frostbeständig
frost-resistant
Frostriss frost crack, trunk splitting
due to frost
Frostschaden/Frostschädigung frost
damage, frost injury, freezing injury
Frostschütte (Blattabwurf) *bot/for*
leaf cast (abscission of leaves) due
to frost
Frostschutzberegnung
frost-protective irrigation
Frostschutzmittel cryoprotectant
frostsicher frost-proof
Frosttrocknis frost drought damage,
frost desiccation damage,
winter desiccation damage
Frostverträglichkeit frost tolerance
Frucht fruit
- **Achäne (einblättrige)** achene,
akene
- **Achäne (zweiblättrige)**
(Asteraceen) cypsela,
bicarpellary achene
- **Apfelfrucht/Pomum**
pome, core-fruit
- **Balg/Balgfrucht/Follikel** follicle
- **Beere** berry
- **Beerenverband/**
Beerenfruchtstand
sorosis, fleshy multiple fruit
- **Bruchfrucht/Gliederhülse/**
Gliederfrucht/Klausenfrucht
loment, lomentum,
lomentaceous fruit, jointed fruit
- **Bruchkapsel** septicidal capsule
- **Deckelkapsel** lid capsule,
circumscissile capsule, pyxis,
pyxidium
- **Einblattfrucht** simple fruit,
apocarpous fruit, unicarpellary
fruit, monocarpellate fruit
- **Einzelfrucht** simple fruit
- **Explodierfrucht** explosive fruit
- **Explodierkapsel/**
Explosionskapsel (Springkapsel)
explosive capsule
- **Flügelfrucht** winged fruit
- **Flügelnuss** samara, key

- **Gliederfrucht/Gliederhülse/**
Klausenfrucht/Bruchfrucht
loment, lomentum, jointed fruit
- **Gliederschote** lomentose siliqua
- **Gurkenfrucht/Kürbisfrucht/**
Panzerbeere pepo, gourd
- **Hackfrucht** root crop
- **Halmfrucht (Getreide)** cereal
- **Haselnussfrucht** filbert
- **Hesperidium/Citrusfrucht/**
Zitrusfrucht (eine Panzerbeere)
hesperidium
- **Hüllfrucht/Cystokarp** cystocarp,
cystocarpium
- **Hülse** legume, pod
- **Kapsel** capsule (*siehe auch dort*)
- **Karyopse/Caryopse/**
"Kernfrucht"/Kornfrucht
caryopsis, grain
- **Katapultfrucht/Katapultkapsel**
catapult fruit, catapult capsule
- **Klause** cell, mericarpic nutlet
(one-seeded segment/fruitlet of
loment)
- **Klausenfrucht/Gliederfrucht/**
Gliederhülse/Bruchfrucht
loment, lomentum, jointed fruit
- **Klettenfrucht/Klettenfrucht**
bur, burr, burry fruit
- **Kornfrucht/Caryopse/Karyopse/**
"Kernfrucht" (Grasfrucht)
caryopsis, grain
- **Kürbisfrucht/Gurkenfrucht (eine**
Panzerbeere) pepo, gourd
- **Lochkapsel/Löcherkapsel/**
Porenkapsel/porizide Kapsel
poricidal capsule, porose capsule
- **Merikarp (***pl* **Merkarpien)/**
Teilfrucht mericarp
- **Nuss** nut
- **Nüsschen** nutlet, nucule
- **Öffnungsfrucht/Streufrucht/**
Springfrucht dehiscent fruit
- **Panzerbeere (Hesperidium und**
Kürbisfrucht, *siehe dort*)
berry with hard rind
(hesperidium and pepo/gourd)
- **Saftfrucht** fleshy fruit
- **Sammelfrucht** aggregate fruit,
composite fruit
- **Scheinfrucht** pseudocarp,
pseudofruit, false fruit,
spurious fruit
- **Schizokarp/Spaltfrucht**
schizocarp, schizocarpium
- **Schlauchfrucht/Utriculus** utricle,
utriculus
- **Schleuderfrucht/ballistische**
Frucht ballistic fruit, ballist
- **Schleuderkapsel** ballistic capsule
- **Schließfrucht** indehiscent fruit
- **Schötchen** silicle

Fruchtwasseruntersuchung

- **Schote** silique
- **Spaltfrucht/Schizokarp** schizocarp, schizocarpium
- **Spaltkapsel** longitudinally dehiscent capsule
- **>dorsicide/dorsizide** dorsicidal capsule
- **>loculicide/lokulizide/ fachspaltige Kapsel** loculicidal capsule
- **>septicide/septizide/ wandspaltige Kapsel** septicidal capsule
- **Springfrucht/Streufrucht/ Öffnungsfrucht** dehiscent fruit
- **Steinfrucht** stone, drupe, drupaceous fruit
- **Steinfruchtverband (Feige)** multiple drupe (fig: syconium)
- **Streufrucht/Springfrucht/ Öffnungsfrucht** dehiscent fruit
- **Syconium/Sykonium (Steinfruchtverband/ Feigenfrucht)** syconium, syconus (a composite fruit: multiple drupe)
- **Teilfrucht/Karpidium/Karpid (ein ganzes Karpell)** fruitlet
- **Teilfrucht/Merikarp (Teil eines Karpells)** mericarp
- **Trockenfrucht** dry fruit
- **Zerfallfrucht** fissile fruit
- **Zitrusfrucht/Citrusfrucht/ Hesperidium (eine Panzerbeere)** hesperidium

fruchtbar/fertil fertile, fecund
- **unfruchtbar/steril** infertile, sterile

fruchtbar machen/befruchten fertilize, fecundate

Fruchtbarkeit/Fertilität fertility; (Fekundität) fecundity
- **Unfruchtbarkeit/Sterilität** infertility, sterility

Fruchtbecher/Cupula cupula, cupule

Fruchtbildung fructification

Fruchtblase/Fruchtwassersack/ Fruchtsack *zool/embr* amniotic sac

Fruchtblatt/Karpel/Carpell carpel

Früchtchen/Karpidium/Karpid fruitlet

fruchten *vb* set fruit

Fruchten *n* fruitage

fruchtend/fruchttragend fruiting, bearing fruit, fructiferous

Fruchtfach locule

Fruchtfall fruit abscission, fruit drop

Fruchtfleisch fruit pulp

Fruchtfolge/Fruchtwechsel/ Anbaurotation crop rotation

Fruchtform *fung* a developmental stage in fungi

- **Hauptfruchtform** perfect stage, telomorphic stage (sexual stage)
- **Nebenfruchtform** imperfect stage, anamorphic stage (asexual stage)

fruchtfressend/frugivor/fruktivor/ karpophag fruit-eating, feeding on fruit, frugivorous, carpophagous

Fruchtfresser/Frugivor/Fruktivor frugivore, fructivore

Fruchtgeschmack fruity taste

Fruchthalter/Fruchtträger/ Karpophor carpophore, receptacle

Fruchthaut/Fruchtschicht/ Hymenium (Pilze) hymenium

Fruchtholz/Tragholz (Kurztrieb) spur shoot, fruit-bearing bough (short shoot)

Fruchtknoten/Ovar/Ovarium ovary

Fruchtknotenhöhle ovary cavity

Fruchtknotenhülle (Gräser) perigynium

Fruchtknotenwand ovary wall

Fruchtkörper/Karposoma fruiting body, fruitbody, fructification, carposoma; konsolenförmiger F. (Pilz) conk (fruiting body of bracket fungus)

Fruchtkuchen *bot/hort* cluster base, knob, bourse

Fruchtlager/Fruchtschicht/ Hymenium hymenium

Fruchtmark/Obstpulpe (fruit) pulp

Fruchtmus fruit pulp

Fruchtrute (Beerensträucher) floricane (second-year cane)

Fruchtsack perigynium

Fruchtschale fruit skin, peel

Fruchtschicht/Fruchthaut/ Hymenium hymenium

Fruchtstand/Fruchtverband (Zönokarpium) multiple fruit, infructescence
- **Beerenfruchtstand** sorosis, fleshy multiple fruit

Fruchtstiel fruit stalk

Fruchtträger/Fruchthalter/ Karpophor carpophore, receptacle

Fruchtverband/Fruchtstand multiple fruit, infructescence

Fruchtwand/Fruchtknotenwand/ Perikarp fruit wall, ovary wall, pericarp

Fruchtwasser/Amnionwasser/ Amnionflüssigkeit amniotic fluid

Fruchtwasserpunktion/ Amniozentese/Amnionpunktion amniocentesis

Fruchtwasseruntersuchung analysis of amniotic fluid (for prenatal diagnosis)

F Fruchtwechsel 486

Fruchtwechsel/Fruchtfolge/ Anbaurotation crop rotation
Fruchtzucker/Fruktose fruit sugar, fructose
Frühbeet/Anzuchtkasten (unbeheizt) cold frame
Frühbeet/Mistbeet/Treibbeet (beheizt) forcing bed, hotbed
frühblühend (vor der Beblätterung) precocious (flowering before leaf formation)
Frühblüher early bloomer
frühes Gen early gene
Frühgeburt premature birth
Frühholz/Weitholz/Frühlingsholz earlywood, springwood
Frühjahrszirkulation *limn* spring overturn
Frühling/Frühjahr spring, springtime
Frühlingsblumen spring flowers
Frühprotein *micb/vir* early protein
Frühsommer-Meningoenzephalitis (FSME) tick-borne encephalitis (TBE), Central European encephalitis (CEE), Russian spring-summer encephalitis (RSSE)
frühsommerlich estival, aestival
Fruktifikation/Fruchtkörper fructification, fruit body, fruiting body
Fruktose/Fructose (Fruchtzucker) fructose (fruit sugar)
Fuchsschwanzgewächse/ Amarantgewächse/ Amaranthaceae cockscomb family, pigweed family, amaranth family
Fuge/Haft/Füllgelenk/Synarthrose synarthrodial joint, synarthrosis
Fuge/Naht/Verwachsungslinie seam, suture, raphe
fühlen feel, sense
Fühler/Antenne antenna
- **blätterförmig/lamellenartig** lamellate antenna
- **borstenartig** bristlelike antenna, setaceous antenna
- **fächrig/gefächerter Fühler/ Fächerfühler** fan-shaped antenna, flabelliform/flabellate antenna
- **fadenförmig/Fadenfühler** threadlike/hairlike antenna, filiform antenna
- **gekämmt** comblike antenna, pectinate antenna
- **gekeult/keulenförmig** clubbed antenna, clavate antenna
- **gekniet** elbowed antenna, geniculate antenna
- **gesägt** sawlike antenna, serrate antenna
- **grannenartig** aristate antenna

- **keulenförmig/gekeult** clubbed antenna, clavate antenna
- **kolbenförmig** capitate antenna
- **lamellenartig/blätterförmig** lamellate antenna
- **mit kurzen Fühlern** brachycerous, with short antennae
- **rosenkranzförmig/ Perlschnurfühler** moniliform antenna
- **stilettförmig** stylate antenna
- **Wendeglied/Pedicellus** (antennal) pedicel
Fühler besitzend antennate
Fühler/Sensor (techn: z.B. Temperaturfühler) sensor, detector
Fühlerfurche/Antennalfurche/ Fühlerrinne antennal sulcus
Fühlergeißel/Flagellum flagellum
Fühlergelenk articular pivot of antenna
Fühlergrube/Antennengrube/Fossa antennalis antennal furrow
Fühlerkeule antennal club
Fühlerkranztiere/Kranzfühler/ Armfühler/Tentaculaten tentaculates (bryozoans/phoronids/ brachiopods)
Fühleröffnung/Antennalforamen antennal aperture
Fühlerpfanne antennal socket
Fühlerringnaht/Antennennaht/ Antennalnaht antennal suture
Fühlerschaft/Scapus (antennal) scape
Fühlersegment/Antennensegment/ Antennalsegment/Antennomer antennal segment, antennomer
Fühlhaar/Reizhaar sensitive hair, trigger hair
Führer (Broschüre/ Informationsschrift) guide, pamphlet, brochure
Führer (Führungsperson) guide, tour guide
Führung (z.B. Zoo/Gelände) guided tour
Fukose/Fucose/6-Desoxygalaktose fucose, 6-deoxygalactose
Füllen/Fohlen foal
- **männliches Füllen/männliches Fohlen** colt (under 4 years)
- **weibliches Füllen/weibliches Fohlen/Stutenfohlen** filly (under 4 years)
Füllstoff filler
Fumariaceae/Erdrauchgewächse bleeding heart family, fumitory family
Fumarsäure (Fumarat) fumaric acid (fumarate)

furunkulös/Furunkel...

Fundort/Lage site, location
Fundusdrüse fundus gland
fünfblättrig gefiedert
quinquefoliolate
fünffingerig/fünffingrig/
fünfstrahlig/pentadaktyl
(Fünfzahl von Fingern/Zehen)
limb with five digits, pentadactyl
(five-toed)
Fünffingerigkeit/Pentadaktylie
pentadactylism
fünfstrahlig/fünfteilig/pentamer
pentamerous
fünfteilig quinquepartite
fünfwertig pentavalent
Fünfzahl von Fingern und Zehen/
Pentadaktylie pentadactyly
fünfzählig pentameric
Fungi imperfecti/
unvollständige Pilze/
Deuteromyceten/
Deuteromycetes imperfect fungi,
deuteromycetes
Fungizid fungicide
Funiculus/Nabelstrang funicle,
funiculus, seed stalk, ovule stalk
Funktion function
● **Verteilungsfunktion**
distribution function
● **Wahrscheinlichkeitsfunktion**
likelihood function
funktionelle Gruppe
functional group
Funktionseinheit/Modul module
Funktionsgewinnmutation
gain of function mutation
funktionsgleich/analog analogous
Funktionskreis functional system,
behavior system
Funktionsverlustmutation
loss of function mutation
Furan furan
Furche/Rinne/Sulcus groove, furrow,
sulcus
● **große Furche/große Rinne/tiefe**
Rinne (DNA-Struktur)
major groove (DNA structure)
● **kleine Furche/kleine Rinne/**
flache Rinne (DNA-Struktur)
minor groove (DNA structure)
furchen cleave; groove, striate,
furrow, fissure
Furchenberieselung corrugation
irrigation
Furchenbewässerung/
Grabenbewässerung
furrow irrigation
Furchenfüßer/Solenogastres/
Neomeniomorpha
solenogasters
furchig furrowed, grooved, fissured,
sulcate

Furchung/Furchungsteilung/
Eifurchung cleavage; segmentation
● **äquale Furchung/gleichmäßige**
Furchung equal cleavage
● **äquatoriale Furchung/**
Äquatorialfurchung
equatorial cleavage
● **bilaterale Furchung/**
Bilateralfurchung
bilateral cleavage
● **determinative Furchung/**
determinierte Furchung
(nichtregulative)
determinate cleavage
● **diskoidale/discoidale/**
scheibenförmige Furchung
discoidal cleavage
● **holoblastische/vollständige/**
totale Furchung
holoblastic/complete cleavage
● **inäquale Furchung/**
ungleichmäßige Furchung
unequal cleavage
● **Meridionalfurchung**
meridional cleavage
● **meroblastische/unvollständige/**
partielle Furchung meroblastic/
incomplete cleavage
● **nichtdeterminative Furchung**
(regulative)
indeterminate cleavage
● **Oberflächenfurchung/**
oberflächliche Furchung/
superfizielle Furchung
superficial cleavage
● **Radialfurchung** radial cleavage
● **regulative Furchung**
(nichtdeterminativ)
regulative cleavage
● **Spiralfurchung** spiral cleavage
● **superfizielle Furchung/**
oberflächliche Furchung/
Oberflächenfurchung
superficial cleavage
● **totale Furchung** total cleavage
● **unregelmäßige Furchung**
irregular cleavage
Furchungshöhle/primäre
Leibeshöhle/Blastocöl/Blastocoel/
Blastula-Höhle segmentation
cavity, blastocoel, blastocele
Furchungsteilung/Blastogenese
blastogenesis
Furchungszelle/Blastomere
blastomere
Furnier veneer
Fürsorge care, provisioning
fürsorglich caring, providing
Furunkel med/vet
furuncle, boil
furunkulös/Furunkel...
furuncular

F Fusiformkambium 488

Fusiformkambium/
Fusiformcambium
fusiform cambium
Fusion fusion
Fusionsgen fusion gene
Fusionsprodukt (Fusion zweier
Replikons) cointegrate structure
(fusion of two replicons)
Fusionsprotein fusion protein
Fuß foot; (Haustorium) foot,
haustorium; (Stativfuß) *micros* base,
foot (supporting stand);
(Arthropoden: Tarsus) tarsus
- **Afterfuß (letzter)/Nachschieber/**
Postpes/Propodium anale
(Raupen) anal proleg, anal leg
- **Bauchfuß/Propes/Pes spurius**
(larvales Abdominalbein) proleg,
ventral proleg
- **Begattungsfuß/Genitalfuß/**
Gonopodium gonopodium,
gonopod
- **Brustfuß/Brustbein/Thorakalfuß/**
Thorakalbein thoracic leg
- **Brustfuß (an Pereion)** pereiopod,
walking leg (attached to pereion)
- **Fiederfuß** pinnate appendage,
pinnate leg
- **Genitalfuß/Begattungsfuß/**
Gonopodium gonopod,
gonopodium
- **Greiffuß** grasping foot
- **Greiffuß/Fang (Raubvogel)**
raptorial claw
- **Kieferfuß/Maxilliped/**
Maxillarfuß/Pes maxilliaris
maxilliped, maxillipede,
gnathopodite, jaw-foot, foot-jaw
- **Klammerfuß/Pes semicoronatus**
(Bauchfüße der Larven:
Großschmetterlinge) proleg with
crochets on planta in a row
- **Kletterfuß** climbing foot
- **Kopffuß/Cephalopodium**
(Mollusken) head-foot
- **Kranzfuß/Pes coronatus**
(Bauchfüße der Larven:
Kleinschmetterlinge) proleg with
crochets on planta in a circle
- **Kriechfuß (Mollusken)**
creeping foot
- **kurzfüßig/mit kurzem Fuß**
brachypodous, with short legs
- **Maxillarfuß/Maxillipes/**
Kieferfuß/Pes maxilliaris
maxilliped, maxillipede,
gnathopodite, jaw-foot, foot-jaw
- **Mittelfuß/Metatarsus**
metatarsal
- **Rankenfuß (Thoracopod der**
Cirripedier) cirrus, feeding leg
- **Scherenfuß** cheliped

- **Schreitfuß** wading foot
- **Schwimmfuß (z. B. Vögel)**
webbed foot, swimming foot
- **Schwimmfuß/Schwimmbein/**
Bauchfuß/Abdominalbein/
Pleopodium (Crustaceen)
swimmeret, pleopod
- **Spaltfuß/Spaltbein**
branched appendage/leg
- **>einfach gegabelter Spaltfuß**
two-branched appendage,
biramous appendage
- **Stummelfuß/Stummelbein**
stubby leg
- **Thorakalfuß/Thorakopode/**
Thoracopod/Thoraxbein/
Rumpfbein thoracopod,
thoracic leg
Fußabdruck footprint
Fußabdruckmethode footprinting
Fußangelnetz (mit Stolperfäden)
arach space web (with barrier
threads)
Füßchen (Ambulakralfüßchen)
podium, tube-foot
Füßchenzelle/Podozyt/Podocyt
podocyte
Fußdrüse/Klebdrüse (Rotatorien)
pedal gland, adhesive gland,
cement gland
fußförmig pedate
Fußgalle/Kreuzgalle/Sporn (Pferde)
ergot
Fußgelenk ankle, ankle joint
Fußgelenksknochen/
Fußwurzelknochen/Tarsal
ankle bone, tarsal
Fußknöchel/Malleolus malleolus
fußlos/beinlos/apod apod
Fußretraktor pedal retractor muscle
Fußrolle/Hufrolle/Bursa
podotrochlearis navicular zone,
semilunar zone, navicular bursa
Fußscheibe (z. B. Anthozoen)
basal disk, pedal disk
Fußsohle/Planta
(auch: Kriechsohle/
Kriechfußsohle) foot sole,
pedal sole, planta
Fußspinner/Tarsenspinner/Embien/
Embioptera webspinners,
embiids
Fußwurzel/Tarsus tarsal, tarsus
Fußwurzelknochen/
Fußgelenksknochen/Tarsal
ankle bone, tarsal
Futter feed
Futterbetteln *ethol* food begging
Futterdrüse/Honigdrüse/
Honigmagen/Kropf (Biene)
honey crop
Futterhorten *ethol* food hoarding

Fütterung

Futterlocken *ethol* tidbitting,
 feeding lure
füttern *vb* feed
Füttern *n* feeding (allofeeding)
Futternapf feeding dish,
 feeding bowl
Futterpflanze fodder,
 forage (plant)

Futterraufe feeding rack
Futterstelle/Futterplatz
 feeding grounds, feeding place,
 feeding area
Futtertrog feeding trough, manger
Fütterung feeding
 (zoo: feeding time)

G

G-Banding

G-Banding/Giemsa-Banding
G banding
G1-Phase (von "gap = Lücke")
G1 phase
G2-Phase G2 phase
**Gabeladerung/Gabelnervatur/
Fächeraderung**
dichotomous venation
Gabelbein/Furcula wishbone,
united clavicles of birds, furcula,
fourchette
Gabelblase/Stewart'sches Organ
Stewart's organ
**Gabelfarngewächse/
Gabelblattgewächse/Psilotaceae**
psilotum family, whisk ferns
gabeln (gegabelt) fork (forked/
V-shaped)
Gabelschwanz forked tail,
V-shaped tail
Gabelung (Forstbaum)
forking of trunk (at midhight)
Gabelung/Gabelteilung/Dichotomie
forking, bifurcation, dichotomy
Gagelgewächse/Myricaceae bog
myrtle family, wax-myrtle family,
sweet gale family, bayberry family
gähnen yawn
Gaia-Hypothese Gaia hypothesis
Galaktosämie galactosemia
Galaktosamin galactosamine
Galaktose galactose
Galakturonsäure galacturonic acid
**Galerie/unterirdischer Gang/
Stollen/Laufgang** gallery
Galeriewald gallery forest,
fringing forest
Gall.../gallenbewohnend
gallicolous
Gallapfel gall apple
Galle/Gallflüssigkeit bile
Galle/Pflanzengalle/Cecidium
gall, cecidium
- **Ananasgalle** pineapple gall
- **Bechergalle** button gall
- **Beutelgalle** pouch gall
- **Blattgalle** leaf gall
- **Blattrandgalle** fold gall
- **Blattstielgalle** petiolar gall
- **Eichenrose (Andricus
  fecundator)** artichoke gall
- **Eichenschwammgalle (Biorhiza
  pallida)** oak apple
- **Filzgalle** filz gall
- **Kegelgalle** cone gall
- **Knoppergalle** knopper gall
- **Linsengalle, große** button gall,
  spangle gall (oak)
- **Markgalle** medullar gall,
  mark gall
- **Pflanzengalle/Cecidium** gall,
  cecidium

- **Rollgalle** roll gall
- **Schlafapfel/Bedeguar (Diplolepis
  rosae)** pincushion gall, bedeguar
- **Schwammkugelgalle (Andricus
  kollari)** marble gall
- **Stengelgalle/Zweiggalle**
  twig gall
- **Tiergalle** zoocecidium
- **Umwallungsgalle** covering gall
- **von Pilzen hervorgerufene
  Galle/Phytocecidium**
  phytocecidium
- **von Tieren hervorgerufene
  Galle/Zoocecidium** zoocecidium
- **Weidenrose** camellia gall
- **Wurzelgalle** root gall
gallenbewohnend/Gall.../gallicol
gallicolous
Gallenblase gall bladder
Gallenblüte (Feigen) gall flower
Gallengang bile duct
**Gallenkunde/Lehre von den Gallen/
Cecidologie** cecidology
Gallensalze bile salts
Gallerreger (sg & pl) galler(s),
gallmaker(s)
gallertartig/gelartig/gelatinös
gelatinous, gel-like
Gallerte/Gelatine jelly, gelatin, gel
Gallertflechten gelatinous lichens
Gallertgeißel/Pseudocilie
pseudocilium (pl pseudocilia)
Gallertkuppe/Cupula cupule,
cupula
Gallertpilze/Zitterpilze/Tremellales
jelly fungi
gallerzeugende Tiere/Cecidozoen
cecidozoa
Gallussäure gallic acid
Galopp (Sprunglauf) gallop
(fast three-beat gait)
- **Kreuzgalopp** disunited canter
- **leichter Galopp/Kanter** canter,
  Canterbury gallop, slow gallop
galoppieren gallop
Gamet/Keimzelle/Geschlechtszelle
gamete, sex cell
**Gametangienträger/
Gametangienstand** gametophore
Gametocyst/Gametozyst
gametocyst
Gametocyt/Gametozyt gametocyte
Gametogamie/Syngamie
gametogamy, syngamy
Gametogonie/Gamogonie
gametogony, gamogony
Gametophor gametophore
Gametophyt gametophyte
Gammakörper gamma particle
Ganasche (Pferd) lower jaw
**Gang (unterirdischer)/Stollen/
Laufgang/Galerie** gallery

Gaskammer G

Gang/Gangart gait, pace
- **aufrechter Gang/aufrechte Gangart** upright gait (erect), orthograde gait
- **Diagonalgang/Kreuzgang** diagonal gait
- **Faustgang (Handknöchel)** fist-walking
- **Galopp (Sprunglauf)** gallop (fast three-beat gait), run
- **Kanter/leichter-mittelschneller/ alternierender Galopp** canter, Canterbury gallop, lope, slow gallop
- **Knöchelgang (Fußknöchel)** knuckle-walking
- **Kreuzgalopp** disunited canter
- **Pasos** paso
- **Passgang** pace
- **Schritt** walk
- **>Arbeitsschritt** working walk
- **>freier Schritt** free walk
- **>Mittelschritt** medium walk
- **>starker Schritt** extended walk
- **>versammelter Schritt** collected walk
- **Sohlengang/Plantigradie** plantigrade gait
- **Spreizgang** sprawling gait
- **Tölt (Pferde)** running walk
- **Trab** trot
- **Zehengang/Digitigradie** digitigrade gait
- **Zehenspitzengang/Hufgang/ Unguligradie** unguligrade gait

Gang/Grabgang/Fraßgang/Tunnel tunnel, gallery, burrow
Ganghöhe (*DNA-Helix:* Anzahl Basenpaare pro Windung) pitch (DNA: helix periodicity)
Ganglion/Nervenknoten ganglion (*see also dictionaries of medicine and veterinary science*)
- **Basalganglion/Stammganglion/ Corpus striatum** basal ganglion, cerebral nucleus
- **Bauchnervenknoten/ Ventralganglion** ventral ganglion
- **Cerebralganglion** cerebral ganglion
- **Oberschlundganglion/ Supraösophagealganglion/ "Gehirn"** supraesophageal ganglion, "brain"
- **Pedalganglion** pedal ganglion
- **Spinalganglion** spinal ganglion
- **Unterschlundganglion/ Subösophagealganglion** subesophageal ganglion
- **Ventralganglion/ Bauchnervenknoten** ventral ganglion

- **Viszeralganglion** visceral ganglion
- **Zerebralganglion** cerebral ganglion

ganglionär ganglionic
Gangliosid ganglioside
Gangunterschied *opt* path difference
Ganoidschuppe/Schmelzschuppe ganoid scale
Ganoin ganoine
Gänsefußgewächse/ Chenopodiaceae goosefoot family
Gänsehaut gooseflesh, goose pimples, goose bumps
Gänserich/Ganter gander
Gänsevögel/Entenvögel/ Anseriformes waterfowl (ducks/ geese/swans)
Ganter/Gänserich gander
ganzrandig (Blatt) entire, simple
Ganzzellableitung *neuro* whole-cell recording
Gare (Boden) mellowness
gären/fermentieren ferment
- **obergärig** top fermenting
- **untergärig** bottom fermenting

Gärmittel/Gärstoff/Treibmittel leavening
Garnelen/"Krabben" shrimps (small), prawns (large)
Gärröhrchen/Einhorn-Kölbchen fermentation tube
Garryaceae/ Becherkätzchengewächse silk-tassel tree family, silktassel-bush family
Gärtassenreaktor tray reactor
Garten garden
Gartenbau horticulture, gardening
Gartenbauausstellung horticultural show, horticultural exhibit
Gartenlaube arbor, bowery
Gartenpflanze garden plant
Gartenschau/Blumenschau horticultural show, flower show
Gartenschere pruners, pruning shears, *Brit:* secateurs
Gärtner gardener, horticulturist
Gärtnerei garden/gardening market, horticulture shop
Gärtnereibedarf gardening supplies
Gärung/Fermentation fermentation
- **obergärig** top fermenting
- **untergärig** bottom fermenting

Gasaustausch gas exchange, gaseous interchange, exchange of gases
Gasbehälter/Pneumatophor float, air sac, pneumatophore
Gasdrüse gas gland
Gaskammer (Nautilus) gas chamber

Gaskieme (aquatische Insekten) gaseous plastron
Gaskonstante gas constant
Gasmaske gas mask
Gast guest
- **echter Gast/Symphil** symphile
Gaster (Hymenoptera: geschwollener Teil des Abdomens) gaster
Gastpflege/Symphilie symphily
Gastraea-Theorie (Häckel) gastrea theory
Gastralfilament gastric filament
Gastraltasche/Darmsack gastric pouch
Gastricsin (Pepsin C) gastricsin (pepsin C)
Gastrodermis-Kanal/Solenie gastrodermal tube, solenia
gastrointestinal-inhibitorisches Peptid/gastrisches Inhibitor-Peptid (GIP)/glucoseabhängiges Insulin-releasing-Peptid gastric inhibitory peptide (GIP), glucose-dependent insulin-release peptide
Gastrointestinaltrakt/Magen-Darm-Trakt gastrointestinal tract
Gastrolith/Magenstein/ Magensteinchen/Hummerstein gastrolith
Gastrovaskularsystem gastrovascular system
Gastrozooid/Gasterozoid/ Autozooid/Trophozoid/Nährtier/ Fresspolyp gastrozooid, trophozooid, feeding/nutritive polyp
Gatter (Zaun: Weide) fence
Gattung genus (*pl* genera)
Gattungsname genus name, generic name
Gaumen palate, roof of the mouth (vertebrates), roof of the pharynx (insects)
- **harter Gaumen/Palatum durum** hard palate
- **weicher Gaumen/Palatum molle/ Gaumensegel/Velum/Velum palatinum** soft palate, velum palatinum
Gaumenbein/Palatinum/ Os palatinum palatine bone
Gaumenbogen palatal arch
Gaumenleiste/Ruga palatina ridge of palate
Gaumenmandel/Tonsilla palatina palatine tonsil
Gaumensegel/weicher Gaumen/ Velum/Velum palatinum soft palate, palatal velum, velum, velum palatinum
Gaumenspalte palatine cleft

Gaumenzäpfchen uvula, palatine uvula
Gauß-Kurve/Gauß'sche Kurve *stat* Gaussian curve
Gauß-Verteilung/Normalverteilung/ Gauß'sche Normalverteilung *stat* Gaussian distribution (Gaussian curve/normal probability curve)
Gaze gauze
GC (Gaschromatographie) GC (gas chromatography)
GC-Box *gen* GC box
geädert veined, venulous
Geäse *hunt* **(Maul des wiederkäuenden Schalenwildes)** muzzle of ruminant hoofed game
Geäst *bot/hort/for* branches, boughs, branchwork
Gebälk (eine Holzkonstruktion aus Balken) rafters; framework, timberwork, timber construction
gebändert/breit gestreift banded, fasciate
Gebärde/Geste/Haltung gesture, posture
gebären/niederkommen/Junge bekommen give birth, bear young, bear offspring
Gebären/Niederkunft giving birth, delivery, parturition
Gebärmutter/Uterus uterus
Gebärmutter.../die Gebärmutter betreffend/uterin uterine
Gebärmutterhals/Zervix/Cervix cervix
Gebärmuttermund/Muttermund mouth/orifice of the uterus, orificium uteri
Gebärmutterwand/Uterusepithel/ Endometrium uteral lining, uteral epithelium, endometrium
gebärtet bearded, barbate (having hair tufts)
Gebein/Knochengerüst bones, skeleton
Gebeine/sterbliche Hülle corpse
Gebiet/Territorium territory
- **Verbreitungsgebiet** geographic range, area of distribution
Gebietsassoziation regional association
Gebirge mountains
gebirgig mountainous
Gebirgsbach mountain stream
Gebirgskamm mountain crest, mountain ridge
Gebirgskette mountain chain, mountain range
Gebirgspflanze/Bergpflanze/ Oreophyt/Orophyt orophyte (subalpine plant)

493 gefächert **G**

Gebirgsstufe/subalpine Stufe
subalpine zone, subalpine region
Gebirgswald mountain forest,
montane forest
Gebiss dentition, teeth
- **akrodont/auf der Kieferkante
stehend (Teleostei/Echsen)**
acrodont, attached to outer surface
of bone/summit of jaws (teleosts/
lizards)
- **brachyodont/niedrigkronig**
brachydont, brachyodont,
with low crowns
- **bunodont/rundhöckrig/
stumpfhöckrig** bunodont,
with low crowns and cusps
- **Dauergebiss/bleibende Zähne/
Dentes permanentes** permanent
dentition, permanent teeth
- **diphyodont (einmaliger
Zahnwechsel)** diphyodont
(with two sets of teeth)
- **gleichartig bezahnt/homodont**
homodont, isodont
- **halbmondhöckrig/selenodont**
crescentic, with crescent-shaped
ridges, selenodont
- **heterodont/ungleichzähnig**
heterodont, anisodont
- **hochkronig/hypsodont/
hypselodont** with high crowns,
hypsodont, hypselodont
- **homodont/gleichartig bezahnt**
homodont, isodont
- **hypsodont/hypselodont/
hochkronig** hypsodont,
hypselodont (high crowns/short
roots)
- **lophodont/mit Querjochen**
lophodont,
with transverse ridges
- **Milchgebiss/Milchzähne/Dentes
decidui** milk dentition,
deciduous dentition
- **monophyodont (einfaches
Gebiss/ohne Zahnwechsel)**
monophyodont (only one set of
teeth)
- **niedrigkronig/brachyodont**
with low crowns, brachydont,
brachyodont
- **pleurodont/an der
Kieferinnenseite** pleurodont,
attached to inside surface of jaws
- **plicodont (mit gefalteten
Zahnhöckern: Elefanten)**
plicodont
- **polyphyodont (mehrfacher
Zahnwechsel)** polyphyodont
- **rundhöckrig/stumpfhöckrig/
bunodont** with low crowns and
cusps, bunodont

- **selenodont/halbmondhöckrig
(Zahnhöcker)**
selenodont, crescentic,
with crescent-shaped ridges
- **stumpfhöckrig/rundhöckrig/
bunodont**
with low crowns and cusps,
bunodont
- **tetralophodont/mit vier
Querjochen** tetralophodont,
with four transverse ridges
- **thekodont/in Zahnfächern
verankert**
thecodont, teeth in sockets
- **triconodont/dreihöckrig (in einer
Reihe)** triconodont (three crown
prominences in a row)
- **ungleichzähnig/heterodont**
heterodont, anisodont
**Gebrech/Maul/Rüssel
(Schwarzwild)** *hunt* snout
gebuchtet/buchtig sinuate
gebündelt bundled, fasciculate
Geburt birth
- **Fehlgeburt/Abort/Abtreibung**
abortion
- **Fehlgeburt/Spontanabort**
miscarriage
- **Frühgeburt** premature birth
- **Mehrlingsgeburt**
multiple birth
- **Totgeburt** stillbirth
Geburtenkontrolle birth control
**Geburtenrate/Geburtenzahl/
Geburtsrate/Natalität**
birthrate, natality
- **Bruttogeburtenrate**
crude birthrate
Geburtsfehler birth defect
Geburtsgewicht birth weight
Geburtshilfe obstetrics
**Geburtsrate/Geburtenrate/
Geburtenzahl/Natalität**
birthrate, natality
Geburtsvorgang birthing process
Gebüsch bushes, shrubbery, thicket,
underbrush (in forest)
Gedächtnis memory
Gedächtniszelle memory cell
**Gedärme/Eingeweide/Innereien/
Viscera/Splancha** intestines,
entrails, innards, guts, viscera
(Mensch: bowels, intestines, guts)
gedeihen/florieren thrive, flourish
Gedeihstörung failure to thrive
gedrängt (Blätter) crowded, tufted
**gedreht (Torsion der
Nervenstränge)** torted
gedreht/verdreht/gewunden
twisted, contorted
gefächert/fächerig/gekammert
valvate, chambered

G Gefahr

Gefahr/Risiko
danger, hazard, risk, chance
- **biologische Gefahr/
biologisches Risiko** biohazard

Gefahr am Arbeitsplatz
occupational hazard

gefährden endanger

Gefährdung endangerment

Gefahrenbereich/Gefahrenzone
danger area, danger zone

Gefahrencode/Gefahrenkennziffer
hazard code

Gefahrenquelle
hazard, source of danger
- **biologische G.** biohazard

**Gefahrenstufe/Gefahrenklasse/
Risikostufe** hazard class

Gefahrensymbol hazard icon

Gefahrenzone danger zone

Gefahrgut dangerous goods,
hazardous materials

Gefahrgutbestimmungen
hazardous materials regulations

gefährlich/riskant dangerous,
hazardous, risky

Gefahrstoff dangerous substance,
hazardous material
- **biologischer G.** biohazard

Gefälle/Gradient *chem* gradient

gefaltet folded, pleated, plicate

Gefänge (Geweih) antlers

gefangen captive; captured, caught;
trapped

Gefäß vessel; (Behälter) container;
(Trachee) trachea
- **Blutgefäß (Arterie/Vene)**
blood vessel (artery/vein)
- **Herzkranzgefäße**
coronary blood vessels
- **Leitertrachee** *bot*
scalariform vessel
- **Lymphgefäß** lymph vessel

**Gefäßbündel/Leitbündel/
Leitbündelstrang** *bot* vascular
bundle, vascular strand

Gefäßhaut (Auge) vascular layer

Gefäßhaut (Hirn: *Pia mater*)
pia mater

Gefäßpflanze vascular plant

Gefäßteil/Holzteil/Xylem xylem

gefenstert fenestrated

gefenstertes Blatt
fenestrated leaf

Gefieder/Federkleid/Ptilosis
plumage, ptilosis
- **Brutkleid/Hochzeitskleid**
breeding plumage, nuptial
plumage, courtship plumage
- **Jugendkleid** juvenile plumage,
juvenal plumage
- **Prachtkleid** display plumage,
conspicuous plumage

- **Schlichtkleid** basic plumage;
eclipse plumage, inconspicuous
plumage (♂ ducks)
- **Tarnkleid/Tarntracht** camouflage
plumage, cryptic plumage

Gefiederputzen/Gefiederkraulen
allg (gegenseitig)
allopreening; (selbst) autopreening

Gefiedersträuben plumage ruffling,
feather ruffling

gefiedert/pennat/pinnat pennate,
pinnate
- **einfach gefiedert** unipennate,
unipinnate
- **paarig gefiedert** equally pennate/
pinnate, even-pinnate, paripinnate
- **unpaarig gefiedert** odd-pinnate,
unequally pinnate, imparipinnate
(with single terminal leaflet or
tendril)
- **zweifach gefiedert/doppelt
gefiedert** bipennate, bipinnate

gefiedert *zool/orn* feathered,
plumed, with plumage

Gefiederwechsel/Mausern *orn*
molt, molting

gefingert fingered, digitate

Geflechtknochen/Faserknochen
fibrous bone

gefleckt spotted, mottled

Geflügel fowl
- **Hausgeflügel** poultry

geflügelt winged, alate
- **ungeflügelt** unwinged, exalate,
apterous, wingless

gefranst fimbriate(d), fringed

gefräßig voracious

Gefräßigkeit voracity

gefrierätzen freeze-etch

Gefrierätzung freeze-etching
- **Tiefenätzung** deep etching

Gefrierbruch *micros*
freeze-fracture, freeze-fracturing,
cryofracture

gefrieren freeze

Gefrierfach freezer compartment

**Gefrierkonservierung/
Kryokonservierung** freeze
preservation, cryopreservation

Gefrierlagerung freeze storage

Gefriermikrotom freezing
microtome, cryomicrotome

Gefrierpunkt freezing point

Gefrierschnitt *micros*
frozen section

Gefrierschutz cryoprotection

Gefrierschutzmittel cryoprotectant

gefriertrocknen/lyophilisieren
freeze-dry, lyophilize

Gefriertrocknung/Lyophilisierung
freeze-drying, lyophilization

Gefriertruhe freezer

495 Geißelskorpione G

Gefüge (Holz) texture
Gefühl feeling, sensation
gefurcht/gerieft *allg* furrowed, grooved, fissured, sulcate
gefurcht/gyrencephal *neuro* gyrencephalous, gyrencephalic (convoluted surface)
● **ungefurcht/lissencephal** *neuro* lissencephalous (no/few convolutions)
gegabelt forked, furcate (V-shaped)
● **einfach gegabelt/dichotom** bifurcate, dichotomous
Gegenangriff counterattack
Gegenauslese/Gegenselektion counterselection
gegenfärben *micros* counterstain
Gegenfärbung *micros* counterstain, counterstaining
Gegengift antidote, antitoxin, antivenin (tierische Gifte)
Gegenkraft/Rückwirkungskraft reactive force
Gegenschattierung countershading
gegenseitig mutual, mutualistic
Gegenseitigkeit/Mutualismus mutualism
Gegenselektion/Gegenauslese counterselection
Gegensinn-Oligonucleotid/ Antisense-Oligonucleotid/ Anti-Sinn-Oligonucleotid antisense oligonucleotide
Gegensinn-RNA/Antisense-RNA/ Anti-Sinn-RNA antisense RNA
Gegensinntechnik/Antisense-Technik/Anti-Sinn-Technik *gen* antisense technique
gegenständig/gegenüberliegend opposite, opposing
● **gekreuzt-gegenständig/ kreuzgegenständig/dekussiert** decussate, crossed
Gegenstrom countercurrent
Gegenstromextraktion countercurrent extraction
Gegenstromverteilung countercurrent distribution
Gegenwind head wind
gegliedert/unterteilt divided
gegliedert/mit Gelenk articulate
Gehäuse shell, case, casing
Gehege (Tiergehege/Wildgehege) enclosure (game preserve/game reserve)
gehen/laufen (zu Fuß fortbewegen) walk
Geheul howling
Gehirn brain (*siehe auch:* Hirn)
Gehirn-Rückenmark-Flüssigkeit/ Liquor cerebrospinalis cerebrospinal fluid (CSF)

Gehirnhaut/Hirnhaut/Meninx (*pl* Meninga) cerebral membrane, cerebral meninx (*pl* meninga)
Gehirnschädel/Hirnschädel/ Neurocranium neurocranium
Gehölz woody plant
Gehölzausläufer sucker, stolon, sobole
Gehölzkunde/Baumkunde/ Dendrologie dendrology
Gehölzschnitt pruning of woody plants
Gehör hearing; (Hörfähigkeit) sense of hearing
Gehörgang auditory canal, auditory meatus
Gehörknöchelchen auditory ossicle/ ossiculum
Gehörn/Hörner horns (Geweih > antlers)
gehörnt horned
Gehörsinn hearing, sense of hearing
Gehörstein/Hörsteinchen/Otolith "ear bone", "ear stone", otolith
geifern (Speichelfluss) *zool* drivel, drool, dribble, slobber (saliva flow from mouth/jaw)
geigenförmig fiddle-shaped, panduriform
Geigenrochen/Rhinobatoidei guitarfishes
Geiger-Zähler Geiger counter
Geiltrieb/Wasserschoss *bot/hort* water sprout, water shoot
Geiß doe (goat doe: nanny, nanny goat)
Geißblattgewächse/Caprifoliaceae honeysuckle family
Geißel/Flagelle flagellum (*pl* flagella/flagellums)
● **Flimmergeißel/pleuronematische Geißel** tinsel flagellum, flimmer flagellum, pleuronematic flagellum
● **Peitschengeißel/akronematische Geißel** whiplash flagellum, acronematic flagellum
● **Schleppgeißel** trailing flagellum
● **Schubgeißel** pushing flagellum
● **Zuggeißel** pulling flagellum
Geißelantenne/Ringelantenne/ amyocerate Antenne amyocerate antenna (muscles only in base segment)
Geißelhärchen mastigonema
Geißelsäckchen/Ampulle/Schlund (Euglena) flagellar pocket, reservoir, anterior pocket
Geißelskorpione & Geißelspinnen/ Pedipalpi (Uropygi & Amblypygi) whipscorpions and tailless whipscorpions (incl. vinegarroons)

G Geißeltierchen 496

Geißeltierchen/Geißelträger/
Flagellaten/Mastigophora
flagellates, mastigophorans
Geißfußpfropfung/
Geißfußveredelung
(Triangulation) *hort*
inlay grafting
Geitonogamie/Nachbarbestäubung
geitonogamy
(*sensu stricto*: geitonophily)
Geiz/Geiztrieb *hort* side shoot,
lateral shoot, sucker
Geizen/Ausgeizen
removal of side shoots/suckers
gekammert/gefächert/fächerig
chambered, valvate
gekämmt/kammartig/kammförmig
comblike, rakelike, ctenoid,
pectinate, pectiniform
gekerbt/kerbig notched, nicked,
crenate
gekielt keeled, having a keel,
carinate
geklärtes Lysat cleared lysate
gekoppelt (koppeln) coupled
(couple), linked (link)
gekoppelte Reaktion coupled
reaction
gekreuzt-gegenständig/
kreuzgegenständig/dekussiert
decussate, crossed
Gekröse/Bauchfellfalte/
Mesenterium mesentery
Gekröse/Kutteln (Rind: Kaldaunen)
tripes
gekrümmt/campylotrop/
kampylotrop (Samenanlage)
campylotropous, bent
Gel gel
 • denaturierendes Gel
 denaturing gel
 • hochkant angeordnetes
 Plattengel slab gel
 • horizontal angeordnetes
 Plattengel
 flat bed gel, horizontal gel
 • natives Gel native gel
 • Sammelgel stacking gel
 • Trenngel
 running gel, separating gel
Gel-Sol-Übergang gel-sol-transition
Gelände land, tract of land,
area, country, ground;
(Terrain) terrain
 • hügeliges Gelände hilly terrain
 • offenes Gelände open country,
 open terrain
Geländeaufnahme
topographic survey
Geländekartierung
topographic mapping
Geländeübung field exercise

gelappt/lappig lobed, lobate
gelartig/gallertartig/gelatinös
gelatinous, gel-like
Gelatine gelatin, gelatine
gelber Fleck/Macula lutea (mit
Fovea centralis) yellow spot,
macula lutea (with fovea centralis)
Gelbfieber (*Flavivirus*) yellow fever
Gelbgrünalgen/Xanthophyten/
Xanthophyta yellow-green algae
Gelbkörper corpus luteum
Gelbreife (Getreide) yellow ripeness
Gelee jelly
Gelege/Eigelege
clutch (nest of eggs)
Gelegenheitsparasit/fakultativer
Parasit facultative parasite
gelehrig/folgsam docile, obedient
Gelehrigkeit/Folgsamkeit docility,
obedience
Geleitzelle *bot* companion cell
Gelelektrophorese
gel electrophoresis
 • Feldinversions-
 Gelelektrophorese field
 inversion gel electrophoresis
 (FIGE)
 • Gradienten-Gelelektrophorese
 gradient gel electrophoresis
 • Pulsfeld-Gelelektrophorese
 pulsed field gel electrophoresis
 (PFGE)
 • Temperaturgradienten-
 Gelelektrophorese temperature
 gradient gel electrophoresis
 • Wechselfeld-Gelelektrophorese
 alternating field gel
 electrophoresis
Gelenk/Verbindung/Angelpunkt
joint, hinge, articulation
 • Buggelenk point of shoulder
 • Eigelenk/Ellipsoidgelenk
 ellipsoidal joint, condyloid joint
 • Ellenbogengelenk/Articulatio
 cubiti
 cubital joint, cubital articulation
 • Fesselbeingelenk pastern joint
 • Fesselgelenk (Pferd)/Fesselkopf/
 Köte fetlock
 • Fühlergelenk
 articular pivot of antenna
 • Fußgelenk ankle
 • Gleitgelenk/ebenes Gelenk
 gliding joint, plane joint
 (arthrodia)
 • Handgelenk wrist, wrist joint
 • Hufgelenk coffin joint
 • Hüftgelenk hip joint, coxal joint,
 femoral articulation; hip joint,
 coxa (Arthropoden)
 • Kniegelenk knee joint; stifle
 (Pferd/Hund)

Gen

- **Krongelenk/Fesselbeingelenk**
  pastern joint
- **Kugelgelenk/Nussgelenk/**
  Enarthrose/Articulatio cotylica
  ball-and-socket joint, spheroid
  joint, spheroidal joint, enarthrodial
  articulation, enarthrosis
- **mit Gelenk/gegliedert** articulate
- **Mittelfußgelenk (Vögel)** hock
- **Sattelgelenk** saddle joint,
  sellaris joint
- **Scharniergelenk/Ginglymus**
  hinge joint, ginglymus joint
- **Sprunggelenk** ankle joint;
  hock (horse)
- **Wackelgelenk/Amphiarthrosis**
  (straffes Gelenk)
  amphiarthrodial joint
- **Walzengelenk/Articulatio**
  bicondylaris condylar joint
- **Zapfengelenk/Radgelenk/**
  Drehgelenk/Articulatio
  trochoidea trochoid joint,
  pivot joint, rotary joint
Gelenk.../Glieder.../mit Gelenk/
  gegliedert jointed, hinged, articular
Gelenk (Immunglobulinmolekül/
  Kollagen Typ VII-Molekül) hinge
  (immunglobulin molecule/collagen
  type VII molecule)
Gelenkflüssigkeit/Gelenkschmiere/
  Synovialflüssigkeit synovial fluid
Gelenkfortsatz articular process
Gelenkhöcker/Epikondyle (siehe
  auch: Gelenkkopf) epicondyle
Gelenkhöhle joint cavity
gelenkig/gelenkartig verbunden
  articulate, jointed
Gelenkkapsel joint capsule,
  articular capsule
Gelenkknorpel articular cartilage
Gelenkkopf/Gelenkhöcker/
  Capitulum/Kondyle/Condylus
  rounded articular prominence/
  eminence/extremity, capitulum,
  condyle
Gelenkpfanne/Gelenkgrube/
  Acetabulum articular socket,
  acetabulum
Gelenkschmiere/Gelenkflüssigkeit/
  Synovialflüssigkeit synovial fluid
Gelenkspalt joint cavity
Gelenkverbindung/Gelenk
  articulation
Gelenkzelle/motorische Zelle/
  Motorzelle (im Schwellkörper
  des Blattes) bulliform cell,
  motor cell
Gelenkzwischenscheibe/Diskus
  joint disk, articular disk
Gelenkzwischenscheibe/Meniskus
  joint meniscus, articular meniscus

Gelfiltration/
  Molekularsiebchromatographie/
  Gelpermeations-Chromatographie
  gel filtration,
  molecular sieving chromatography,
  gel permeation chromatography
gelieren vb gel
Gelieren n gelation
Geliermittel gelling agent
Gelierpunkt gelling point
gelöst (lösen) dissolved
gelöster Stoff solute
Gelpräzipitationstest/
  Immunodiffusionstest
  immunodiffusion (siehe dort)
Gelretardationsexperiment
  mobility shift experiment
Gelretentionsanalyse gel retention
  analysis, band shift assay
Gelretentionstest gel retention
  assay, electrophoretic mobility shift
  assay (EMSA)
gemähnt/mit einer Mähne maned
gemäßigt temperate, moderate
gemäßigte Zone temperate zone,
  temperate region
gemein/gewöhnlich/einfach/normal
  common, usual, normal
Gemeinsamkeit (Zusammensein)
  togetherness
Gemeinschaft community,
  association
Gemeinschwämme/Demospongiae
  demosponges
Gemenge/Mischung mixture
Gemischtgeschlechtigkeit/
  Einhäusigkeit/Monözie monecy,
  monoecy, monecism, monoecism
Gemsbockgewächse/Martyniaceae
  unicorn plant family,
  devil's-claw family, martynia family
Gemüse vegetable
Gemüseanbau olericulture
Gemüsebeet (vegetable) patch
Gen gene
- **amplifiziertes Gen**
  amplified gene
- **Einzelkopie-Gen**
  single copy gene
- **extrachromosomales Gen**
  extrachromosomal gene
- **extranukleäres/extranucleäres**
  Gen extranuclear gene
- **Fremdgen** heterologous gene,
  foreign gene
- **frühes Gen** early gene
- **gewebespezifisches Gen**
  tissue-specific gene
- **Haushaltsgen/Haushaltungsgen/**
  konstitutives Gen
  housekeeping gene
- **Hitzeschockgen** heat shock gene

Gen-Farming

- **homöotisches Gen** homeotic gene
- **ineinandergesetzte Gene/ ineinandergeschachtelte Gene** nested genes
- **Kandidatengen** candidate gene
- **Luxusgen** luxury gene
- **Meistergen** master gene
- **Minigen** minigene
- **Modifikationsgen** modifier gene
- **Mosaikgen/gestückeltes Gen/ diskontinuierliches Gen** mosaic gene, split gene, discontinuous gene
- **Regulationsgen** regulatory gene
- **Reportergen** reporter gene
- **Resistenzgen** resistance gene
- **Schaltergen** switch gene
- **spätes Gen** late gene
- **springendes Gen** jumping gene
- **Strukturgen** structural gene
- **stummes Gen** silent gene
- **Suppressorgen** suppressor gene
- **syngene Gene (Gene auf** *einem* **Chromosom)** syngenic genes
- **überlappende Gene** overlapping genes
- **unvollständig gekoppelte Gene** incompletely linked genes
- **Vererbungslinienbestimmung** gene tracking
- **zellspezifisches Gen** cell-specific gene

Gen-Farming gene farming
Gen-Knockout (Unterbrechung von Genen durch homologe Rekombination) gene knockout
Gen-Targeting/Allelen-Austauschtechnik gene targeting, gene disruption, gene replacement
Gen-Verstärkung/Genamplifikation gene amplification
Genaktivierung gene activation
Genamplifikation gene amplification
Genaustausch gene exchange
Genbank/DNA-Bibliothek bank, clone bank, DNA-library
Gendiagnostik/Bestimmung des Genotyps genetic diagnostics, genotyping
Gendosis gene dosage
Gendosiseffekt gene dosage effect
Gendrift/genetische Drift genetic drift
Genealogie/Stammbaumforschung/ Ahnenforschung/ Familienforschung genealogy
Genegoismus gene egoism
Generalist generalist

Generation generation
- **Filialgeneration/ Tochtergeneration** filial generation

Generationsdauer generation period
Generationswechsel alternation of generations
- **antithetischer/heterophasischer Generationswechsel (Heterogenese)** antithetic theory, interpolation theory of alternation of generations
- **homologer Generationswechsel** homologous theory, transformation theory of alternation of generations

Generationszeit (Verdopplungszeit) generation time (doubling time)
Generatorpotential *neuro* generator potential
Genetik/Vererbungslehre genetics (study of inheritance)
- **direkte Genetik** direct genetics
- **Entwicklungsgenetik** developmental genetics
- **Eugenik/"Erbhygiene"** eugenics, eugenetics
- **formale Genetik** formal genetics
- **Humangenetik** human genetics
- **Immungenetik** immunogenetics
- **klinische Genetik** clinical genetics
- **Molekulargenetik** molecular genetics
- **Ökogenetik** ecogenetics
- **Pflanzengenetik** plant genetics, phytogenetics
- **Phänogenetik** phenogenetics
- **Pharmakogenetik** pharmacogenetics
- **Populationsgenetik** population genetics
- **reverse Genetik** reverse genetics
- **Tiergenetik** animal genetics, zoogenetics
- **Verhaltensgenetik** behavior genetics

genetische Anfälligkeit genetic susceptibility
genetische Belastung/genetische Last/Erblast genetic bond
genetische Beratung genetic counsel(l)ing
genetische Bürde/genetische Belastung/genetische Last/Erblast genetic load
genetische Dissektion genetic dissection
genetische Fixierung genetic fixation
genetische Immunisierung genetic immunization

Geobotanik

genetische Kolonisierung
genetic colonization
genetische Prädisposition
genetic predisposition
genetische Varianz genetic variation
genetischer Abstand
genetic distance
genetischer Code genetic code
genetischer Fingerabdruck/DNA-Fingerprinting DNA profiling, DNA fingerprinting
genetischer Hintergrund/ genotypischer Hintergrund
genetic background
genetischer Marker genetic marker
genetischer Suchtest
genetic screening
genetisches Risiko genetic risk
Genexpression gene expression
• **differentielle G.** differential gene expression
Genexpressionskontrolle/Kontrolle der Genexpression
control of gene expression
Genfamilie gene family
Genfluss/Genwanderung gene flow
Genfrequenz/Genhäufigkeit
gene frequency
Gengruppe/Gencluster gene cluster
Genhäufigkeit/Genfrequenz
gene frequency
Genick/Nacken nape (back of the neck), nucha; poll (crest/apex of skull: Hinterkopf)
genießbar/essbar comestible, eatable, edible
• **ungenießbar/unessbar**
uneatable, inedible
genießbar/schmackhaft palatable
• **ungenießbar/nicht schmackhaft**
unpalatable
Genitalanlage genital primordium
Genitalfalte genital fold
Genitalfuß/Begattungsfuß/ Gonopodium gonopod, gonopodium
Genitalhöcker/Geschlechtshöcker/ Tuberculum genitale
genital tubercle
Genitalien genitals, genitalia, genital organs, sexual organs
Genitalleiste/Keimdrüsenleiste
genital ridge
Genitalplatte genital plate
Genitalporus/Genitalöffnung/ Geschlechtsöffnung/Gonopore
genital opening, genital aperture, gonopore
Genitalpräsentieren *ethol*
genital display
Genitaltaster gonopalpon
Genkarte gene map, genetic map

Genkartierung gene mapping, genetic mapping
Genklonierung gene cloning
Genkomplex gene complex
Genkonversion/Konversion/ Umwandlung/Übergang
(gene) conversion
Genkopplung gene linkage
Genkopplungskarte
gene linkage map, linkage map
Genlocus gene locus
Genmanipulation gene manipulation
• **Gentechnik/Gentechnologie**
genetic engineering, gene technology
Genom genome
• **Kerngenom** nuclear genome
• **Menschliches Genomprojekt (HUGO)**
Human Genome Project (HUGO)
Genomanalyse genome analysis
genomische Bibliothek/genomische Genbank genomic library
genomische Prägung
genomic imprinting
genomische Sequenzierung
genomic sequencing
genomisches Blotting
genomic blotting
Genotyp/Genotypus genotype
Genpool gene pool
Genprodukt gene product
Genrückgewinnung gene eviction, gene rescue
Gensuperfamilie gene superfamily
Gentechnik/Gentechnologie/ Genmanipulation genetic engineering, gene technology
gentechnisch verändert
genetically engineered
gentechnisch veränderter Organismus (GVO) genetically engineered organism (GVO)
Gentechnologie/Gentechnik/ Genmanipulation gene technology, *sensu lato*: genetic engineering (>Gentechnik)
Gentherapie gene therapy, gene surgery
• **Keimbahngentherapie**
germ line gene therapy
• **somatische Gentherapie**
somatic gene therapy
Gentianaceae/Enziangewächse
gentian family
Gentisinsäure gentisic acid
Genträger gene carrier
Gentransfer/Genübertragung
gene transfer
Geobotanik/Pflanzengeographie
geobotany, plant geography, plant biogeography, phytogeography

G Geocline 500

Geocline
geocline, geographical cline
geogen geogenous
Geographie/Geografie/Erdkunde
geography
- **Pflanzengeographie/Geobotanik**
plant geography, plant
biogeography, phytogeography,
geobotany
- **Tiergeographie/Zoogeographie**
zoogeography
geöhrt auriculate, eared, ear-like
Geologie (Erdgeschichte)
geology (Earth science)
Geonastie geonasty
Geoökologie geo-ecology
geophag geophagous
Geophagie/Erdeessen geophagy,
geophagism
geophil geophilous (living in/on
soil)
**Geophyt/Erdpflanze/Cryptophyt/
Kryptophyt/Staudengewächs**
geophyte, geocryptophyte,
cryptophyte *sensu lato*
geotaktisch geotactic
Geotaxis (*pl* **Geotaxien)**
geotaxis (*pl* geotaxes)
gepanzert armored, thecate
gepökelt/eingesalzen corned
(>corned beef: gepökeltes
Rindfleisch)
gepunktet punctuated
Geradflügler/Orthoptera
orthopterans
geradläufig (Samenanlage) *bot*
atropous, orthotropous, orthotropic
**Geraniengewächse/
Storchschnabelgewächse/
Geraniaceae** geranium family,
cranesbill family
Geraniumsäure geranic acid
Geranylacetat geranyl acetate
Geräusch sound, noise
gerben tan
Gerben tanning
**Gerberstrauchgewächse/
Gerbersträucher/Coriariaceae**
coriaria family
Gerbsäure (Tannat) tannic acid
(tannate)
gerbsäurehaltig/gerbstoffhaltig
tanniferous
Gerbstoff tanning agent, tannin
gerichtete Fluktuation steady drift
gerichtete Mutagenese
directed mutagenesis
**Gerichtsmedizin/Rechtsmedizin/
Forensik/forensische Medizin**
forensics, forensic medicine
gerinnen/koagulieren set; curdle,
coagulate; (Milch) curdle; (Blut) clot

Gerinnsel (z. B. Blut) clot (e. g. blood
clot)
Gerinnung clotting
Gerinnungsfaktor clotting factor
- **Blutgerinnungsfaktor**
blood clotting factor
Gerippe/Skelett/Knochengerüst
skeleton
Geröll *geol*
rock debris (rounded by erosion),
loose stones (of various size)
Geröllhalde *geol* scree, talus (slope)
Geruch *allg* smell, scent, odor
- **angenehmer Geruch/Duft**
pleasant smell, fragrance, scent,
odor
- **stechender Geruch** pungency
- **unangenehmer Geruch**
unpleasant smell
Geruchsfährte/Geruchsspur
odor trail
Geruchssinn/olfaktorischer Sinn
olfactory sense
Geruchsstoff (angenehmer G.)
fragrance, perfume (stronger scent);
(unangenehmer/abweisender G.)
repugnant substance
Gerüst scaffolding, framework,
stroma, reticulum
Gerüst (Netz) *arach* scaffolding
Gerüsteiweiß/Stützeiweiß
structural protein, fibrous protein
Gerüstnetz (Theridiiden) *arach*
scaffold web
**Gerüstregion (von
Immunglobulinen)** framework
region (of immunoglobulins)
Gerüstschnitt (Baumschnitt)
frameworking
gesägt serrate, serrated, sawed,
saw-edged
- **fein gesägt** serrulate,
finely serrate, finely notched
Gesamtbiomasse total biomass
Gesamteignung inclusive fitness
Gesamtkeimzahl *micb*
total germ count, total cell count
Gesamtpopulation total population
Gesamtvergrößerung *micros* total
magnification, overall magnification
Gesang *orn* **(Vogelgesang)** singing,
song (*siehe auch:* Lied)
- **Balzgesang** courtship song,
mating song
- **Dichten/Jugendgesang
(Jungvögel)** subsong
- **Duettgesang/Paargesang**
duetting
- **Element** element, phrase element
- **Lied** song
- **Lockgesang** attracting song,
luring song, soliciting song

Geschmacksorgan

- **Motiv** motive, theme
- **Paargesang/Duettgesang** duetting
- **Phrase/Tour** phrase
- **Reviergesang** territorial song
- **Schlag (Gesang der Nachtigall)** caroling, elaborate song with many different verses
- **Spielgesang** play song
- **Strophe** verse (part of song)
- **Studiergesang** rehearsal song
- **Vollgesang** full song
- **Wechselgesang** antiphonal singing
- **Werbegesang** mating song, courtship song

Gesangsrepertoire song repertoire

Gesäß buttocks, posterior, behind, rump

Gesäßbein/Sitzbein/ Sitzknochen/Os ischii ischium

Gesäßfalte gluteal fold

Gesäßschwiele/Sitzschwiele/ Analkallosität ischial callosity, sitting pad, anal callosity

Gesäßweisen buttocks display

gesättigt (sättigen) saturated (saturate)

- **ungesättigt** unsaturated

gescheckt variegated

Geschein (Rispe des Weinstocks) cluster, flower cluster of vine (a panicle)

geschichtet (schichten) laminated (laminate)

Geschlecht (männlich/weiblich/neutral) sex (male/female/neuter), gender

- **heterogametisches Geschlecht** heterogametic sex
- **homogametisches Geschlecht** homogametic sex

Geschlechterverhältnis sex ratio

geschlechtlich sexual

- **ungeschlechtlich/asexuell** asexual

Geschlechtlichkeit sexhood

Geschlechtsarm/ Geschlechtstentakel/Hectocotylus hectocotylus, hectocotylized arm, heterocotylus

Geschlechtsbestimmung sex determination

Geschlechtschromosom/Heterosom/ Gonosom sex chromosome, heterochromosome

Geschlechtschromosominaktivierung sex chromosome inactivation

Geschlechtsdimorphismus/ Sexualdimorphismus sexual dimorphism

Geschlechtsdrüse/Keimdrüse/ Gonade sex gland, gonad

geschlechtsgebunden sex-linked

geschlechtsgebundener Erbgang sex-linked inheritance

Geschlechtshöcker/Genitalhöcker/ Tuberculum genitale genital tubercle

Geschlechtshormone sex hormones

Geschlechtskopplung *gen* sex linkage

Geschlechtskrankheit/ venerische Krankheit/ sexuell übertragbare Krankheit venereal disease (VD), sexually transmitted disease (STD)

Geschlechtsmerkmal sexual characteristic

Geschlechtsöffnung/ Begattungsöffnung/Genitalporus/ Genitalöffnung/Gonopore genital opening/pore/aperture, gonopore

Geschlechtspartner mate, mating partner

Geschlechtspolyp/Gonozoid/ Gonozooid (Fruchtpolyp) reproductive polyp, gonozooid

Geschlechtsreife sexual maturity

Geschlechtstentakel/ Geschlechtsarm/Hectocotylus hectocotylus, hectocotylized arm, heterocotylus

Geschlechtstrennung/ Getrenntgeschlechtigkeit/Diözie (speziell postnatal: Gonochorismus) dioecism (speziell postnatal: gonochorism)

Geschlechtsumkehr sex reversal

Geschlechtsverkehr/ Kopulationsakt/Begattungsakt/ Koitus sexual intercourse, copulation, coitus

Geschlechtszelle/Keimzelle/Gamet sex cell, gamete

geschlitzt/zerschlitzt (gleichmäßig) incised, evenly notched/cut; (ungleichmäßig) lacerate, torn

geschlossene Knospe (mit Knospenschuppen) protected bud

geschlossener Promotorkomplex closed promoter complex

geschlossenes Leseraster closed reading frame

Geschmack taste

Geschmacksknospe/ Geschmacksbecher/ Geschmackshügel taste bud

Geschmackskörperchen taste corpuscle, taste corpuscule

Geschmacksorgan gustatory organ

G Geschmackspapille

Geschmackspapille/Zungenpapille
gustatory papilla, lingual papilla
- **Blattpapille/blättrige Papille**
foliate papilla
- **Pilzpapille** fungiform papilla
- **linsenförmige Papille**
lentiform papilla
- **fadenförmige Papille** threadlike
papilla, filiform papilla
- **Wallpapille** vallate papilla
Geschmackspore taste pore
Geschmacksrezeptor taste receptor,
gustatory receptor
Geschmackssinn sense of taste,
gustatory sense/sensation
**Geschmackssinneszelle/Schmeck-
zelle** taste cell, gustatory cell
Geschmacksstiftchen (Mikrovilli)
taste hairs (microvilli)
Geschmackstoff(e) flavor,
flavoring
- **künstliche G.** artificial flavor,
artificial flavoring
- **natürliche G.** natural flavor,
natural flavoring
Geschöpf/Kreatur/Wesen creature,
being
geschützt protected (protect)
geschwanzt (Blattspitze) caudate,
tail-pointed (leaf apex)
geschweift/leicht gewellt repand
Geschwindigkeit
speed; velocity (vector); rate
**geschwindigkeitsbegrenzende(r)
Schritt/Reaktion** rate-limiting step/
reaction
**geschwindigkeitsbestimmende(r)
Schritt/Reaktion**
rate-determining step/reaction
**Geschwindigkeitskonstante
(Enzymkinetik)** rate constant
Geschwister *pl* siblings
- **Halbgeschwister** half-sibs
- **Untersuchung von Geschwistern**
sib-pair analysis
Geschwisterarten (Zwillingsarten)
sibling species
Geschwisterschaft sibship
geschwollen (schwellen) turgid,
swollen (swell)
Geschwollenheit/Turgidität
turgidity
gesellig sociable, gregarious
**Geselligkeit/Geselligkeitsgrad/
Soziabilität** sociability,
gregariousness
Geselligkeitstrieb/Geselligkeit
sociality
**Gesellschaft (z. B. Tierges./
Pflanzenges.)** community
Gesellschaft/Beisammensein
company

**Gesellschaft/Vergesellschaftung
(Gruppe)** society
Gesellschaftstreue fidelity to a
particular community
Gesetz der großen Zahlen *stat*
law of large numbers
**Gesetz der konstanten
Proportionen
(Mischungsverhältnisse)**
law of combining ratios
**Gesetz von der Erhaltung der
Masse**
law of the conservation of mass
Gesicht face
Gesichtsausdruck/Physiognomie
facial expression, physiognomy
Gesichtsfeld/Sehfeld/Blickfeld
field of vision, field of view, scope
of view, range of vision, visual field
Gesichtsfeldblende/Okularblende
micros ocular diaphragm, eyepiece
diaphragm, eyepiece field stop
Gesichtsschädel/Splanchnocranium
splanchnocranium
Gesichtssinn vision, eyesight
Gesichtszüge facial features
Gesneriengewächse/Gesneriaceae
gesneria family, gesneriad family,
gloxinia family,
African violet family
gespalten split, cracked
**Gespenstheuschrecken &
Stabheuschrecken/Phasmida**
stick-insects
Gespinst (Raupenkokon) cocoon
Gespinst (Spinnwebe) web
gespornt spurred
gesprenkelt mottled
**Gestagen/Progestin/
Corpus-luteum-Hormon/
"Schwangerschaftshormon"**
gestagen, progestin
Gestalt shape, form, appearance,
contour
**gestapelt (stapeln) (z. B.
Membranzisternen)** stacked
(stack)
gestapelte Basen *gen* stacked bases
gestaucht/zusammengezogen
compressed, contracted
Gestein rock
- **Ausgangsgestein/Grundgestein/
Muttergestein**
bedrock, rock base, parent rock
- **Effusivgesteine/Ergussgesteine/
Extrusivgesteine/
Ausbruchsgesteine**
extrusive rocks
- **Eindampfungsgesteine/
Evaporite** evaporites
- **Erstarrungsgestein/
Eruptivgestein** igneous rock

503 · Gewässerzonierung G

- **Ganggesteine** gangue rock
- **Intrusivgesteine** intrusive rocks
- **Muttergestein/Ausgangsgestein/ Grundgestein** parent rock, bedrock, rock base
- **Sedimentgestein/Absetzgestein/ Schichtgestein** sedimentary rock
- **Umwandlungsgestein** metamorphic rock
- **Urgestein** primary rock, primitive rock; basement complex

gestielt stalked, petiolate, stipitate
- **ungestielt/sitzend** not stalked, sessile

Gestik/Geste/Gebärde gesture
gestreift striped
- **breit gestreift** broadly striped, fasciate
- **fein gestreift** finely striped, striated

Gestrüpp thicket, scrub, brush
gestückeltes Gen/ diskontinuierliches Gen/ Mosaikgen split gene, discontinuous gene, mosaic gene
Gestüt (Zuchttiere: alle Pferde eines Gestüts) stud (group of horses bred and kept by one owner)
Gestüt/Pferdezüchterei/ Pferdezuchtbetrieb studfarm
gestutzt/verstümmelt/zurecht-geschnitten truncated; (Baum/Ast) pruned, trimmed; (Blatt) truncate
geteilt divided, parted, partite (divided into parts)
- **ungeteilt** undivided, not divided

Getier animals (*collect.*,i. e., various kinds of animals)
Getreide cereals, grain
Getreideflocken cereal
Getreidemehl *grob:* meal, *fein:* flour
Getreideschrot whole meal
getrenntblumenblättrig/ freiblumenblättrig/ freikronblättrig dialypetalous, choripetalous
getrenntgeschlechtig/zweihäusig/ diözisch (speziell postnatal: gonochor) dioecious, diecious (postnatal: gonochoric)
Getrenntgeschlechtigkeit/ Zweihäusigkeit/ Geschlechtstrennung/Diözie (speziell postnatal: Gonochorie) dioecy, dioecism (postnatal: gonochory)
getüpfelt pitted
Gewächs plant, growth, wort
Gewächshaus/Treibhaus greenhouse, hothouse, forcing house
Gewächshauseffekt/Treibhauseffekt greenhouse effect

Gewässer body of water, waterbody
- **Binnengewässer** inland waterbody
- **Fließgewässer/fließendes Gewässer/lotisches Gewässer (Fluss/Strom)** flowing water, running water, lotic water, water-way, watercourse (river/stream)
- **Stehgewässer/Stillgewässer/ stehendes Gewässer/ lenitisches Gewässer** stagnant water, standing water, lenitic water, lentic water

Gewässergüte/Wassergüte water quality
Gewässerufer shore, banks
Gewässerzonierung (lacustrine/ riverine/marine) zonation
- **Abyssal/Meeresgrund** abyssal, abyssal zone, ocean floor
- **aphotic zone** aphotic zone
- **Bathyal (Meeresboden)** bathyal zone (upper: continental slope; lower: continental rise)
- **Bathypelagial/mittlerer Tiefseebereich/mittlere Tiefseezone** bathypelagic zone
- **Benthal/Bodenzone** benthic zone
- **Eulimnion** eulimnion (upper warmer water)
- **Hadal/Tiefseegrabenzone (Hänge)** hadal zone
- **Hypolimnion** hypolimnion (cold bottom water zone)
- **Litoral/Litoralzone/ Litoralbereich/Uferzone (*mar* Gezeitenzone/Küstenzone)** littoral, littoral zone
- **Metalimnion** metalimnion (zone of steep temperature gradient)
- **neritische Region** neritic zone
- **ozeanische Region/Hochsee** oceanic zone/region, pelagic zone
- **Pelagial/pelagische Zone/ Freiwasserzone** pelagic zone (*limn* also: limnetic zone)
- **photische Zone** photic zone, euphotic zone
- **Profundal/profundale Zone (aphotische Zone)** profundal zone (aphotic zone)
- **Spritzwasserzone/Spritzzone/ Gischtwasserzone/Gischtzone (Supralitoral)** splash zone (supralittoral zone; *mar* supratidal zone/surf zone)
- **Sublitoral (Zone des Kontinentalschelfs)** sublittoral (continental shelf zone)
- **Uferzone/Uferregion/Litoral/ Litoralzone/Litoralbereich (*mar* Gezeitenzone/Küstenzone)** littoral, littoral zone

G Gewebe 504

Gewebe (Zellassoziation) tissue; (z. B. Spinngewebe) fabric, mesh, network
- **Abschlussgewebe** boundary tissue, dermal tissue, exodermis
- **>primäres A./Epidermis** epidermis
- **Absonderungsgewebe/ Abscheidungsgewebe** secretory tissue
- **Ausscheidungsgewebe** excretory tissue
- **Bindegewebe** connective tissue
- **Bothryoidgewebe (Hirudineen)** bothryoidal tissue
- **chondroides Gewebe/ Parenchymknorpel** chondroid tissue, pseudocartilage
- **darmassoziiertes lymphatisches Gewebe** gut-associated lymphatic tissue (GALT)
- **Dauergewebe** permanent tissue, secondary tissue
- **Drüsengewebe** glandular tissue
- **Epithelgewebe** epithelial tissue
- **Festigungsgewebe** supporting tissue (collenchyma/sclerenchyma)
- **Fettgewebe** fatty tissue, adipose tissue
- **fibröses Bindegewebe** fibrous tissue, white fibrous tissue
- **Grundgewebe/Parenchym** ground tissue, fundamental tissue, parenchymatous tissue, parenchyma
- **Knochengewebe** bone tissue, bony tissue, osseous tissue
- **Knorpelgewebe** cartilaginous tissue
- **>chondroides Gewebe/ Parenchymknorpel** chondroid tissue, pseudocartilage
- **Kompakta** compact tissue, compact bone
- **Leitgewebe** conducting tissue, vascular tissue
- **mechanisches Gewebe/ Expansionsgewebe** expansion tissue
- **Muskelgewebe** muscular tissue
- **Nährgewebe** *allg* nutritive tissue, nutrient tissue
- **Nährgewebe (Embryosack)** endosperm
- **Nährgewebe (nucellar)** perisperm
- **Scheingewebe/ Pseudoparenchym** false tissue, paraplectenchyma, pseudoparenchyma
- **Schwammgewebe** spongy tissue

- **Schwellkörper/ Corpora cavernosa** erectile tissue, cavernous tissue, cavernous bodies
- **Sekretionsgewebe** secretory tissue
- **Speichergewebe** storage tissue
- **Stützgewebe** supporting tissue
- **Widerlagergewebe (Frucht)** resistance tissue
- **Wundgewebe** *zool/med* scar tissue, cicatricial tissue
- **Wundgewebe/Wundcallus/ Wundholz** wound tissue, callus

Gewebeabstoßung tissue rejection
Gewebefaktor tissue factor
Gewebekultur tissue culture
Gewebekulturflasche/ Zellkulturflasche tissue culture flask
Gewebelehre/Histologie histology
gewebespezifisches Gen tissue-specific gene
Gewebetiere/Mitteltiere/ "Vielzeller"/Metazoa metazoans
Gewebeunverträglichkeit/ Histoinkompatibilität histoincompatibility
Gewebeverträglichkeit/ Histokompatibilität histocompatibility
Gewebswanderzelle/Gewebs- Makrophage/Histiozyt (*eigentlich:* Makrophage) histiocyte (*actually:* macrophage)
Geweih antlers
- **Augspross** brow tine
- **Eisspross** bez tine
- **Mittelspross** royal antler
- **Schaufel** palm, palmated antler
- **Stange** beam, main beam
- **Wolfsspross** surroyal antler

Geweihbasis/Hornbasis burr (base of antler)
Geweihrose pedicel
Geweihzacke/Geweihspross prong, spike
gewellt undate, undulate
- **leicht gewellt/geschweift** repand

Gewicht weight
- **Atomgewicht** atomic weight
- **Bruttogewicht** gross weight
- **Frischgewicht (*sensu stricto:* Frischmasse)** fresh weight (*sensu stricto:* fresh mass)
- **Lebendgewicht** live weight
- **Molekulargewicht/ relative Molekülmasse (M_r)** molecular weight, relative molecular mass (M_r)
- **Nettogewicht** net weight

505   **Glasstab** G

- **spezifisches Gewicht (Holz)**
  specific gravity
- **Tara (Gewicht des Behälters/der Verpackung)** tare (weight of container/packaging)
- **Trockengewicht (*sensu stricto*: Trockenmasse)** dry weight (*sensu stricto*: dry mass)

gewimpert/bewimpert ciliate(d)
Gewinde/Spirale spiral, coil
Gewinde (Schneckenschale) spire
Gewitter thunderstorm
gewöhnen/anpassen habituate, get used to, adapt
Gewöhnung/Anpassung habituation, habit-formation, adaptation
Gewöhnungslernen/Habituation habituation
Gewölbe/Fornix (Hirngewölbe) fornix
gewölbt tuberculate, vaulted
Gewölle (Raubvögel) pellets
gewunden twisted, coiled, wound
Gewürz spice
Gewürzsträucher/Calycanthaceae spicebush family, strawberry-shrub family
gezackt/gesägt serrate
gezähnelt denticulate
gezähnt toothed, dentate
Gezeiten/Tiden tides
- **Ebbe** low tide
- **Flut** high tide
- **Nipptide** neap tide
- **Springtide** spring tide
- **Tidenhub** tidal lift

Gezeitenstromrinne tidal channel
Gezeitentümpel tidal pool, tide pool
Gezeitenwechsel/Gezeitenzyklus tide cycle, tidal cycle
Gezeitenzone/Tidebereich/Eulitoral tidal zone, intertidal zone, littoral zone, eulittoral zone
gezielte Konstruktion von Proteinen protein engineering
GFC (Gas-Flüssig-Chromatographie) GLC (gas-liquid chromatography)
Ghost (leere Zellhülle) ghost
Gibberelline gibberellins
Gibberellinsäure gibberellic acid
gießen pour, irrigate, water the plants
Gift/Toxin poison, toxin
- **Tiergift** venom

Giftdrüse poison gland
giftig (Tiere) venomous
giftig/toxisch poisonous, toxic
Giftigkeit/Toxizität poisonousness, toxicity

Giftinformationszentrale poison information center
Giftklaue poison claw, forcipule, prehensor (Chilopoda); poison fang, venomous fang (unguis) (Arachnida)
Giftpflanze poisonous plant
Giftstoffe poisonous materials, poisonous substances
Giftzahn (Schlangen) poison tooth, venom tooth, fang
Gilde guild
Ginkgogewächse/Ginkgoaceae ginkgo family
Gipfel crown, treetop, apex, tip
gipfelfrüchtig/akrokarp acrocarpous, acrocarpic
Gipfelknospe apical bud, terminal bud
Gipfeltrieb/Terminaltrieb/Endtrieb *bot* apical shoot, terminal shoot
Gischt (vom Wind getrieben/ aufsprühend) spray, sea spray (spoondrift)
Gischt/Schaum der Wellen (aufschäumend) foam, froth, spume
Gischtwasserzone/Gischtzone/ Spritzwasserzone/Spritzzone splash zone
Gitter/Netz/Gitternetz/ Probenträger(netz) (für Elektronenmikroskop) *micros* grid
Gitterstichprobenverfahren *stat* lattice sampling, grid sampling
Gittertheorie/Netzwerktheorie *immun* network theory
Gladius/Rückenfeder (pergamentartige Schulpe) pen
glänzend glossy
Glanzfische/Glanzfischartige/ Gotteslachsverwandte/ Lampriformes moonfishes
Glanzkohle/Anthrazit hard coal, anthracite
Glanzkugel (Placozoa) refractile body
Glanzstreifen/Kittlinie (Muskel) intercalated disk
Gläschen/Glasfläschchen/Phiole vial
Glashomogenisator ("Potter"; Dounce) glass homogenizer (Potter-Elvehjem homogenizer; Dounce homogenizer)
Glaskörper/Corpus vitreum vitreous body
Glasschwämme/Hexactinelliden glass sponges
Glasstab *lab* glass rod

G Glasstößel 506

Glasstößel/Glaspistill (Homogenisator) glass pestle
glatt smooth, even
Glatteis glaze
glattes Ende/bündiges Ende *gen*
blunt end, flush end
• **Ligation glatter Enden**
blunt end ligation
Glattferser/Litopterna litopterns
Glazialgeschiebe glacial drift
gleich/identisch (völlig gleich/ein u. dasselbe) equal, same, identical
• **ungleich/nicht identisch/anders**
unequal, different, nonidentical
gleichartig/sehr ähnlich
very similar
gleichartig/verwandt/kongenial
congenial
Gleichartigkeit resemblance
Gleichbein/Sesambein/ Sesamknöchelchen (Pferd)
proximal sesamoid bone
gleichbleibender Zustand/ stationärer Zustand steady state
gleichen *math* equate
• **sich gleichen/gleichartig sein**
resemble
gleichförmig uniform
Gleichförmigkeit uniformity
gleichgestaltet similar-structured
Gleichgewicht balance, equilibrium
• **Fließgleichgewicht/dynamisches Gleichgewicht** steady state,
steady-state equilibrium
• **Ionengleichgewicht**
ion equilibrium, ionic steady state
• **natürliches Gleichgewicht (Naturhaushalt)** natural balance
• **ökologisches Gleichgewicht**
ecological balance,
ecological equilibrium
• **Säure-Basen-Gleichgewicht**
acid-base balance
• **Ungleichgewicht** imbalance,
disequilibrium
Gleichgewichtsdialyse
equilibrium dialysis
Gleichgewichtskonstante
equilibrium constant
Gleichgewichtsorgan (statisches/ dynamisches) equilibrium organ
(static/dynamic)
Gleichgewichtspotential
equilibrium potential
Gleichgewichtszentrifugation
equilibrium centrifugation,
equilibrium centrifuging
Gleichgewichtszustand
equilibrium state
Gleichneriaceae gleichneria family
gleichrichten rectify
Gleichrichter rectifier

Gleichrichtung rectification
• **anomale G.**
anomalous rectification
• **verzögerte G.**
delayed rectification
Gleichung *math* equation
Gleichung xten Grades
equation of the xth order
gleichwarm/warmblütig/ homoiotherm/endotherm
warm-blooded, homoiothermic,
homeothermic, endothermic
gleichzählig/isomer isomerous
Gleitbahn *aer/orn* glide path
Gleitfallenblume/Kesselfallenblume
slippery-trap flower,
slip-slide flytrap
Gleitfilamentmodel (Muskel)
sliding-filament model
Gleitfilamenttheorie
sliding-filament theory
Gleitflug glide, gliding (flight)
Gleitgelenk/ebenes Gelenk
gliding joint, plane joint (arthrodia)
Gleithaut/Flughaut/Spannhaut/ Flatterhaut/Patagium patagium
Gleittubulushypothese
sliding-tubule hypothesis
Gleitwinkel *aer* glide angle,
gliding angle
Gletscher glacier
Gletschermoränenschutt/ Gletschergeröll/Glazialschutt
glacial till, glacial detritus
Gletschersee glacial lake
Gleybildung/Vergleyung (Boden)
gleization
Gliazelle glial cell
Glied/Segment segment
• **männliches Glied/Penis** penis
Glied.../Glieder../Gelenk../ gegliedert articulate
Gliederantenne/myocerate Antenne
myocerate antenna (muscles in each
antennal segment)
Gliederborste articular bristle
Gliederfrucht/Gliederhülse/ Klausenfrucht/Bruchfrucht
loment, lomentum, jointed fruit
Gliederfüßer/Arthropoden
arthropods
Gliederkette/Proglottidenkette/ Strobilus (Bandwürmer)
chain of proglottids, strobilus
gliedern/einteilen divide
• **untergliedern/unterteilen**
subdivide
gliedern/klassifizieren classify
Gliederschote (Fruchtform)
lomentose siliqua
gliederspaltig/segmenticid/ segmentizid segmenticidal

Gonade G

507

Gliedertiere/Articulaten articulates, articulated animals
Gliederung/Einteilung division
• **Untergliederung/Unterteilung** subdivision
Gliederung/Klassifikation classification
Gliedmaße/Extremität (*pl* **Gliedmaßen/Extremitäten**) limb, extremity, appendage (articulated)
Globalstrahlung global radiation
globuläres Protein/Sphäroprotein globular protein
Globulariaceae/ Kugelblumengewächse globe daisy family, globularia family
Globulin globulin
Glochidium (Larve) glochidium
Glocke (Medusen) bell
Glockenblumengewächse/ Campanulaceae bellflower family, bluebell family
glockenförmig bell-shaped, campanular, campaniform
Glockenkern entocodon
Glockenkurve (Gauß'sche Kurve) bell-shaped curve (Gaussian curve)
Glockenwindengewächse/ Nolanaceae nolana family
glomeruläre Filtrationsrate glomerular filtration rate
Glomerulus/Gefäßknäuel glomerulus, network of blood capillaries
Glomerulusfiltration glomerular filtration
glomuläre Filtrationsrate glomerular filtration rate (GFR)
Glucarsäure glucaric acid, saccharic acid
Glucocorticoid glucocorticoid
Gluconeogenese gluconeogenesis
Gluconsäure (Gluconat) gluconic acid (gluconate), dextronic acid
Glucuronsäure (Glukuronat) glucuronic acid (glucuronate)
Glukosamin/Glucosamin glucosamine
Glukose/Glucose (Traubenzucker) glucose (grape sugar)
Glukosurie/Glycosurie glucosuria, glycosuria
Glutamin glutamine
Glutaminsäure (Glutamat)/ 2-Aminoglutarsäure glutamic acid (glutamate), 2-aminoglutaric acid
Glutarsäure glutaric acid
Glutathion glutathione
Glycin/Glyzin/Glykokoll glycine, glycocoll

Glycyrrhetinsäure glycyrrhetinic acid
Glykämie glycemia
Glykogen glycogen
Glykokalyx glycocalyx (cell coat)
Glykokoll/Glycin/Glyzin glycocoll, glycine
Glykolaldehyd/Hydroxyacetaldehyd glycol aldehyde, glycolal, hydroxyaldehyde
Glykolsäure (Glykolat) glycolic acid (glycolate)
Glykosaminoglykan glycosaminoglycan, mucopolysaccharide
glykosidische Bindung glycosidic bond, glycosidic linkage
Glykosurie/Glukosurie glycosuria, glucosuria
Glyoxalatzyklus glyoxylate cycle
Glyoxalsäure (Glyoxalat) glyoxalic acid (glyoxalate)
Glyoxylsäure (Glyoxylat) glyoxylic acid (glyoxylate)
Glyoxysom glyoxysome
Glyphosat glyphosate
Glyzerin/Glycerin/Propantriol glycerol
Glyzerinaldehyd/Glycerinaldehyd glyceraldehyde, dihydroxypropanal
Glyzin/Glycin/Glykokoll glycine, glycocoll
GM-CSF (Granulocyten-Makrophagen-stimulierender Faktor) GM-CSF (granulocyte-macrophage stimulating factor)
Gnathochilarium gnathochilarium
Gnathopod gnathopod
Gnathos gnathos
Gnathosoma/Capitulum gnathosoma, capitulum
Gnetumgewächse/ Gnemonbaumgewächse/ Gnetaceae joint-fir family
Goldalgen/Chrysophyceen/ Chrysophyceae golden algae, golden-brown algae
Goldmarkierung gold-labelling
Goldpflaumengewächse/ Chrysobalanaceae cocoa-plum family, coco-plum family
Golgi-Anfärbemethode Golgi staining method
Golgi-Apparat Golgi apparatus, Golgi complex
Golgi-Vesikel Golgi vesicle
Gonade/Keimdrüse/ Geschlechtsdrüse gonad, sex gland
• **Eierstock/Ovar/Ovarium** ovary
• **Hoden (sg/pl)** testicle, testis (*pl* testes)

Gonadenhöhle

Gonadenhöhle/Gonocoel
perigonadial cavity, gonocoel
Gonadotropin gonadotropin
**Gonadotropin-Releasing Hormon/
Gonadoliberin** gonadotropin
releasing hormone/factor (GnRH/
GnRF), gonadoliberin
Gongylidie/Bromatium/"Kohlrabi"
gongylidium (*pl* gongylidia),
bromatium
goniatische Lobenlinien
goniatitic suture lines
gonochor gonochoric,
gonochoristic
Gonochorismus gonochory,
gonochorism
Gonozyt/Gonocyt gonocyte
Gonys gonys
Gössel/Gänseküken gosling
**Gottesanbeterinnen &
Fangschrecken/Mantodea/
Mantoptera** mantids
**Gotteslachsverwandte/
Glanzfischartige/Glanzfische/
Lampriformes** moonfishes
**Graafscher Follikel/Graaf-Follikel/
Tertiärfollikel** Graafian follicle,
vesicular ovarian follicle
Grab grave
Grab.../grabend digging, fossorial;
burrowing
Grabbein fossorial leg
graben dig
• **einen Gang graben/eine Höhle
graben** burrow
Graben ditch
**Grabenbewässerung/
Furchenbewässerung**
furrow irrigation
**Grabfüßer/Kahnfüßer/
Solenoconchae/Scaphopoden**
tusk shells, tooth shells, scaphopods,
scaphopodians (spade-footed
mollusks)
Grabstätte/Friedhof graveyard
Gradienten-Hypothese
gradient hypothesis
Gradualismus gradualism
**graduiert/mit einer Gradeinteilung
versehen** graduated
Gram-Färbung Gram stain,
Gram's method
**Gramineae/Poaceae/Süßgräser/
Gräser** grass family
Grammäquivalent gram equivalent
gramnegativ gram-negative
grampositiv gram-positive
Granatapfelgewächse/Punicaceae
pomegranate family
Grandrysches Körperchen
Grandry's corpuscle
Granne *bot* awn

Grannen tragend *bot* aristate,
awned
Grannenhaar *zool* short guard hair
granulär granular
Granulaviren granulosis viruses
**Granulocyt/Granulozyt
(polymorphkerniger Leukozyt)**
granulocyte (polymorphonuclear)
• **basophiler G.**
basophil(ic) granulocyte
• **eosinophiler G.**
eosinophil(ic) granulocyte
• **neutrophiler G.**
neutrophil(ic) granulocyte
• **segmentkerniger G.** segmented
granulocyte, filamented neutrophil
• **stabkerniger G.**
band granulocyte, stab cell,
band cell, rod neutrophil
**Granulocyten-Makrophagen-
stimulierender Faktor (GM-CSF)**
granulocyte-macrophage stimulating
factor (GM-CSF)
Granum (*pl* Grana)
granum (*pl* grana)
Graptolithen/Graptolithina
graptolites
Gras grass, lawn
grasartig graminoid, graminaceous,
grassy
**Grasbaumgewächse/
Xanthorrhoeaceae**
grass tree family, blackboy family
grasblättrig graminifoliose
Grasbüschel tuft of grass, tussock
grasbüschelartig/rasig/rasenartig
cespitose, caespitose, caespitulose
(growing densely in tufts)
grasen/abgrasen/abfressen/weiden
graze (herbs), browse (twigs/leaves
of shrubs)
grasendes Tier grazer (herbs),
browser (twigs/leaves of shrubs)
**Gräser/Süßgräser/Gramineae/
Poaceae** grasses, grass family
Grashalm (Blattspreite)
blade of grass
Grashalm (Stengel) culm, haulm,
halm, spire
Grashalmspitze spire
Grasheidenstufe grass heath
(a tussock community)
Grasland grassland
Grasnarbe/Rasenstück/Sode
sod, turf
**Grasnelkengewächse/
Bleiwurzgewächse/
Plumbaginaceae** sea lavender
family, leadwort family, plumbago
family
Grasrispe (Infloreszenz)
juba, loose panicle of grasses

Grünanlage

Grasstengel
culm, haulm, halm, spire
Grat ridge
• **Berggrat** mountain ridge
Grat-Rinnen-System (Riff)
spur-and-groove zone, buttress zone
Gräte/Fischgräte bone
Grätenblattgewächse/Ochnaceae
ochna family
graue Substanz *neuro* gray matter
graugrün/blaugrün gray-green, glaucous
Graupel/Graupelschauer/ Schneeregen
graupel, sleet, soft hail
Graupen hulled barley, barley groats
Gravidität/Trächtigkeit/ Schwangerschaft gravidity, pregnancy
• **Bauchhöhlengravidität/ Bauchhöhlenträchtigkeit/ Leibeshöhlenträchtigkeit/ Leibeshöhlenschwangerschaft/ Abdominalschwangerschaft**
abdominal pregnancy
• **Eileiterschwangerschaft/ Oviduktträchtigkeit**
tubal pregnancy
• **Extrauteringravidität (EUG)/ ektopische Schwangerschaft**
extrauterine pregnancy (EUP)
Greif.../zum Greifen geeignet/ zupackend/ergreifend
grasping, prehensile, able to grasp, raptorial
Greiffuß/Fang (Raubvogel)
raptorial claw
Greifhaken (Chaetognathen)
grasping spines
Greifhand/Greiffuß (Säuger/Vögel)
prehensile hand/foot, grasping hand/foot
Greiforgan prehensile organ
Greifschwanz prehensile tail
Greifvogel/Raubvogel bird of prey, predatory bird, raptorial bird
Greifvögel/Falconiformes diurnal birds of prey (falcons and others)
Greifzange/Chelizere
grasping claws, chelicera
Greifzange/Haltezange/Klasper
grasping claws, clasper(s), clasps
Grenzdifferenz *stat*
least significant difference, critical difference
Grenzfaktor/begrenzender Faktor/ limitierender Faktor *ecol*
limiting factor
Grenzfläche interface
Grenzflächenspannung
surface tension
Grenzfrequenz corner frequency

Grenzkonzentration
limiting concentration
Grenzplasmolyse *bot*
incipient plasmolysis
Grenzschicht boundary layer
Grenzstrang/ Truncus sympathicus
sympathetic trunk
Grenzwert/Schwellenwert limit, liminal value
Griff (klammernd) clutch; (zupackend/festhaltend) grip, grasp
Griffel/Stylus *bot* style
Griffelbein/Griffelbeinknochen (Nebenmittelfußknochen)
splint bone (small metacarpal)
Griffelfäden (Mais)
silk (corn stigma-style)
Griffelfortsatz styloid process
Griffelpolster/Stylopodium
stylopodium
Griffelsäule/Gynostemium
gynostemium
Grimmdarm/Colon/Kolon colon
Grind/Schorf scab
grobfaserig coarse-grained
Grobjustierschraube/Grobtrieb *micros* coarse adjustment knob
Grobjustierung/Grobeinstellung (Grobtrieb) *micros*
coarse adjustment, coarse focus adjustment
Großart/Makrospezies macrospecies
großblättrig megaphyllous
große Furche/große Rinne/ tiefe Rinne (DNA-Struktur)
major groove (DNA structure)
Großhirn/Endhirn/Telencephalon
cerebrum, endbrain, telencephalon
Großhirnrinde cerebral cortex
• **Hörrinde** auditory cortex
• **Neocortex/Neokortex** neocortex
• **Sehrinde** optic cortex
Großhirnsichel/Falx cerebri
falx cerebri (sickle-shaped fold in dura mater)
Grossulariaceae/ Stachelbeergewächse
gooseberry family, currant family
grubbern *agr* grubbing
Grübchen/kleine Grube
fovea, small pit
Grube pit, crypt
Grubenorgan (Schlangen) pit organ
grubig pitted, foveate
• **kleingrubig** foveolate, having small depressions
Grün (floristisch) green, greenery
Grünalgen/Chlorophyceae/ Isokontae green algae
Grünanlage (öffentliche)
public park, public gardens

Grund (*pl* **Gründe**)/**Erdboden**
soil; (Boden) bottom, bed, floor; (Gebiet/Revier) ground; (Senkung) depression
- **Blattgrund** leaf base
- **Blattspreitengrund/ Blattspreitenbasis**
base of leaf blade
- **Fischgrund/Fischgründe**
fishing ground(s)
- **Jagdgrund/Jagdgründe**
hunting ground(s), hunting range, hunting territory
- **Meeresgrund** bottom of the sea, seabed, ocean floor
- **Moorgrund** quagmire
- **Schlickgrund** mud bottom
- **Talgrund/Talsohle** valley floor
- **Wiesengrund**
lowlying meadow in a valley

Grundbaustein basic building block
Grundbewohner bottom dweller
Grundeis *limn* anchor ice
gründeln (Wasservögel) dabble
Gründereffekt founder effect
Gründermaus founder mouse
Gründerpolyp/Primärpolyp
founder polyp, primary polyp
Gründerprinzip founder principle
Grundgestein/Muttergestein/ Ausgangsgestein bedrock, rock base, parent rock
Grundgewebe/Parenchym
ground tissue, fundamental tissue, parenchyma
Grundkörper *chem*
parent compound, parent molecule (backbone)
Grundlage base, foundation
- **Pfropfgrundlage** stock

Grundlagenforschung
basic research
Grundmeristem ground meristem
Grundmoräne/Untermoräne
ground moraine, basal moraine
Grundnahrungsmittel staple food, basic food
Grundplatte (Neuralrohr)
ventral horn
Grundstoff/Rohstoff base material, starting material, raw material
Grundstoffwechsel/ Ruhestoffwechsel
basal metabolism
Grundstoffwechselrate/ Grundumsatz/basale Stoffwechselrate/Basalumsatz
basal metabolic rate, base metabolic rate (BMR)
Grundsubstanz/Grundgerüst/Matrix
base material, ground substance, matrix
Grundumsatz/Basalumsatz/ Grundstoffwechselrate/basale Stoffwechselrate base metabolic rate, basal metabolic rate
Gründünger green manure
Grundwasser groundwater
Grundwassereinzugsgebiet
catchment area/basin, watershed, drainage area/district
Grundwasserspiegel/ Grundwasseroberfläche
water table
Grundzustand ground state
grüne Drüse/Antennendrüse/ Antennennephridium green gland, antennal gland, antennary gland
grüne Revolution green revolution
Grünfutter/Grünzeug soilage, green forage, greenstuff
Grünland grassland
Grünpflanze/Blattpflanze (floristisch) foliage plant, leafy plant
grunzen *vb* grunt
Grunzen *n* grunt
Gruppe group, assemblage; *gen* cluster
Gruppe (derselben Organisationsstufe) grade
Gruppe von hoher Beweglichkeit (HMG-Box) *gen*
high mobility group (HMG-box)
Gruppenbalz communal courtship
Gruppenmächtigkeit
group importance value
Gruppenselektion group selection
Gruppenübertragung
group transfer
Gruppenwert group value
Gruppierung assemblage
Grütze/Grieß/Grießmehl groats, grits
- **Maisgrütze/Maisgrieß** corn grits, hominy grits

GT-AC-Regel *gen* GT-AC rule
Guajazulen guaiazulene
Guanidin guanidine
Guanin guanine
Guano guano
Guanophore/Iridocyt guanophore, iridocyte
Guanosin guanosine
Guanosintriphosphat (GTP)
guanosine triphosphate
Guanylsäure (Guanylat)
guanylic acid (guanylate)
Guar-Gummi/Guarmehl guar gum, guar flour
Guar-Samen-Mehl guar meal, guar seed meal
Guarnierischer Einschlusskörper
Guarnieri body

Guide-RNA guide RNA
Gularplatte/Schlundplatte/ Kehlplatte (Fische/prognathe Insekten) gular plate, gula
Gülle/Flüssigmist manure, liquid manure (liquid: total excretions diluted with water)
Gulonsäure (Gulonat) gulonic acid (gulonate)
Gummi arabicum/Arabisches Gummi/Acacia Gummi gum arabic
Gummiharz resinous gum
Gurkenfrucht/Kürbisfrucht/ Panzerbeere pepo, gourd
Gürtel/Gurt/Cingulum girdle, cingulum
Gürtelbein/Sphenethmoid/ Os en ceinture sphenethmoid bone
Gürteldesmosom/Banddesmosom belt desmosome
Gürteln/Ringelung (Baumrinde) girdling, ringing
Gürtelpuppe/Pupa cingulata girdled pupa
Gürteltiere (Cingulata/Loricata: Xenarthra) armadillos
Gürteltransekt belt transect
Gürtelwürmer/Clitellaten clitellata
Gussplattenmethode/ Plattengussverfahren *micb* pour-plate method
Gutachten/Expertise expertise
gutartig/benigne benign
 ● **bösartig/maligne** malignant

Gutartigkeit/Benignität benignity, benign nature
 ● **Bösartigkeit/Malignität** malignancy, malignant nature
Guttaperchagewächse/ Eucommiaceae eucommia family
Guttation/Tropfenabscheidung/ Exsudation guttation, droplet secretion, exudation
GVO (gentechnisch veränderter Organismus) GVO (genetically engineered organism)
Gynandromorph/Gynander gynandromorph, gyander, sex mosaic
Gynandromorphismus gynandromorphism
Gynoandrophor gynoandrophore, androgynophore
Gynogenese gynogenesis
Gynophor gynophore
Gynostegium (Asclepiadaceen) gynostegium
Gynostemium/Säule/Säulchen/ Griffelsäule (Orchideen) gynostemium, column
gyrencephal/gefurcht (Gehirn) gyrencephalous, gyrencephalic (convoluted surface)
gyrocon (Cephalopoden: Gehäuse) gyroconic
Gyrus dentatus dentate gyrus
Gyttia/Gyttja/Grauschlamm/ Halbfaulschlamm gyttja, necron mud

H

H-Zelle

512

H-Zelle (Exkretionszelle:
Nematoden) renette cell
H-Zone (Muskel) H zone
Haar (Trichom) hair (trichome)
- **Barthaare/Schnurrhaare (Katzen)**
  whiskers
- **Becherhaar/Trichobothrium**
  trichobothrium
- **Blasenhaar** bladder hair
- **Brennhaar** stinging hair,
  urticating hair, trichome
- **Deckhaar (Leithaare u.**
  Grannenhaare) guard hair
- **Drüsenhaar** glandular hair,
  glandular trichome
- **Fanghaar (*Drosera*)** tentacle
- **Fegehaar (an Pappus)** brush hair
  (hair-like capillary bristle/pappus
  hair)
- **Flimmerhaar/Kinozilie/**
  Kinozilium (Haarzelle) cilium
  (*pl* cilia)
- **Fühlhaar/Reizhaar** sensitive hair,
  trigger hair
- **Grannenhaar/Deckhaar**
  short guard hair
- **Hörhaar/Becherhaar/**
  Trichobothrium
  acoustical hair, trichobothrium,
  vibratory sensory hair
- **Kötenhaare/Fesselhaare/**
  Kötenbehang fetlock hair,
  feather
- **Leithaar**
  long and smooth guard hair
- **Oberflächenhäutchen**
  cuticle of the hair
- **Pflanzenhaar** trichome
- **Reizhaar/Fühlhaar** trigger hair,
  sensitive hair
- **Safthaar/Paraphyse** paraphysis,
  paranema
- **Schnurrhaare** whiskers
- **Schuppenhaar/Saugschuppe**
  (Bromelien) squamiform hair,
  peltate trichome,
  absorbing trichome
- **Sinneshaar** *allg* sensory hair
- **>Riechhärchen** olfactory hair
- **>Spürhaar/Sinushaar/Tasthaar/**
  Vibrissa tactile hair, vibrissa
- **Spürhaar/Sinushaar/Tasthaar/**
  Vibrissa (ein Sinneshaar)
  tactile hair, vibrissa
- **Sternhaar** stellate hair
- **Tasthaar/Sinushaar/Spürhaar/**
  Vibrissa tactile hair, vibrissa
- **unechtes Haar/Trichom** trichome
- **Unterhaar** underhair
- **Wimper/Augenwimper** eyelash
- **Wollhaar** wooly hair
- **Wurzelhaar/-härchen** root hair

haarartig piliform, trichoid
Haarbalg/Haarfollikel hair follicle
Haarbedeckung der Säugetiere
pelage, furcoat (hairy covering of
mammals)
Haarbulbus/Haarzwiebel hair bulb
Haarfarne/Frauenhaarfarn-
gewächse/Adiantaceae
adiantum family,
maidenhair fern family
Haarflechten/Fadenflechten
hairlike lichens, filamentous lichens
Haarflügler/Köcherfliegen/
Trichoptera caddis flies
Haarfollikel/Haarbalg hair follicle
Haargefäß/Kapillare capillary
haarig (*siehe*: behaart) hairy
Haarkleid hair-covering,
indumentum; coat of hair, furcoat,
pelage
Haarkranz/Krone (am Huf) coronet
Haarlinge & Federlinge/
Mallophaga
biting lice, chewing lice
haarlos/unbehaart/kahl hairless,
glabrous, bald
Haarmark hair medulla
Haarmuskel/Musculus arrector pili
hair erector muscle,
arrector pili muscle
Haarnadelschleife/
Haarnadelstruktur/Winkelschleife/
Umkehrschleife/-Schleife *gen*
hairpin loop, hairpin, reverse turn,
beta turn,bend
Haarpinsel/Duftpinsel
(Schmetterlinge) hair pencil,
tibial tuft
Haarrinde hair cortex
Haarschaft hair shaft
Haarschleierpilze/Schleierlinge/
Cortinariaceae
cortinarius family
Haarschopf/Haarbüschel/Haarkranz
bot (an Samen) coma;
zool hair-tuft
Haarsensille hair sensilla
Haarsterne/Federsterne/
Comatuliden/Comatulida
feather stars, comatulids
Haarsträuben/Fellsträuben
piloerection
Haartalgdrüse sebaceous gland
Haarwurzel hair root
Haarwurzelscheide hair root sheath
Haarzelle hair cell
Haarzwiebel/Haarbulbus hair bulb
Hachse/Bein leg;
(Beine der Schlachttiere) shank(s);
(Sprunggelenk der Schlachttiere)
hock
Hackfrucht root crop

513   **halbmondförmige Klappe** **H**

Hackkultur/Hackbau hoe culture,
hoe cultivation, hoe agriculture
Hackordnung *ethol* peck order,
pecking order
**Hadal/hadische Zone/
Tiefseegrabenzone (Hänge)**
hadal zone
**Haematopoese/Hämatopoese/
Blutbildung/Blutzellbildung**
hematopoiesis
**Haemodorumgewächse/
Haemodoraceae** bloodwort family,
redroot family, kangaroo paw family,
haemodorum family
**Haff (Küstensee durch Nehrung
abgetrennt vom Meer)**
haff (*North German term*), lagoon
Haftborste/Frenulum (Lepidoptera)
frenulum
Haftdrüse adhesive gland
**Hafte/echte Netzflügler/
Planipennia/Neuroptera**
neuropterans
(dobson flies/ant lions)
**Haftkapsel/Klebkapsel/Glutinant
(Nematocyste)** glutinant
**Haftläppchen/Afterkralle/Arolium
(am Prätarsus)** arolium
**Haftlappen/Pulvillus/Lobulus
lateralis** adhesive pad, pulvillus
Haftorgan *allg* attachment organ
Haftorgan (Haie/Rochen)
clasper (sharks/rays)
Haftplatte/Haftscheibe/Saugnapf
adhesive disk, suction disk
**Haftscheibe/Haftorgan/
Appressorium** *allg* holdfast,
appressorium; (Kletterorgan)
adhesive disk
**Haftscheibe/Saugscheibe/
Saugorgan/Haustorium
(parasitäre Pilze)** sucker,
haustorium
Haftwasser film water,
retained water
Haftwurzel holdfast root,
clinging root
Hagebutte (rose)hip
Hagel *meteo* hail
Hagelschnur/Chalaze chalaza,
treadle, tread
**Hageman-Faktor (Blutgerinnungs-
Faktor XII)** Hageman factor (blood
clotting factor XII)
Hahn *zool* male chicken, cock,
rooster
Hähnchen cockerel (under 1 year)
**Hahnenfußgewächse/
Ranunculaceae**
buttercup family, crowfoot family
Hahnenkamm/Crista galli (Schädel)
cock's comb, crista galli

**Hahnentritt/Fruchthof/Keimscheibe
(Keimfleck im Ei)**
cocktread, germ disk, germinal disk,
blastodisc, cicatricle, "eye"
**Haie, echte/Galeomorpha/
Carcharhiniformes** tiger sharks/
catsharks & sand sharks & requiem
sharks & hammerheads and others
Haifische sharks
Haifischkamm (Gelelektrophorese)
shark tooth comb
Hain/Gehölz/Waldung grove
Häkchen/Hamulus (an Feder)
hooklet, barbicel, hamulus
(*pl* hamuli)
Haken hook
● **Sandhaken/Strandhaken** *mar*
spit, cuspate foreland
hakenartig/hakig hooked, hamate,
hamulose
Hakenbein/Os hamatum
hamate bone, unciform bone
hakenförmig hook-shaped,
unciform, hamiform
Hakenlarve/Acanthor-Larve
acanthor larva
● **Sechshakenlarve/Oncosphaera-
Larve (Cestoda)** hexacanth larva,
hooked larva, oncosphere
Hakenrüßler/Kinorhyncha
kinorhynchs
**Hakenwürmer (*Ancylostoma/
Necator spp.*)** hookworms
hakig hooked, hook-like, uncinate,
hamate
Halbaffen/Prosimii prosimians,
lower primates
Halbchromatidenkonversion
half-chromatid conversion
Halbdecke/Hemielytre
hemielytron, hemelytron
Halbdune/Semipluma
semiplume, semipluma
halbdurchlässig/semipermeabel
semipermeable
**Halbdurchlässigkeit/
Semipermeabilität**
semipermeability
**Halbflügler/Hemiptera/Rhynchota/
Schnabelkerfe (Heteroptera &
Homoptera)** bugs, hemipterans
Halbgeschwister half-sibs
Halbinsel peninsula
Halblebenszeit (Enzyme) half-life
Halbmond, grauer gray crescent
halbmondförmig crescentic,
crescent-shaped, semilunar
**halbmondförmige Klappe/
Semilunarklappe (Herz: als Teile
der Taschenklappe)** semilunar
cusp, semilunar flap (parts of
semilunar valve)

halbmondhöckrig 514

halbmondhöckrig/selenodont (Zahnhöcker) crescentic, with crescent-shaped ridges, selenodont
Halbparasit/Halbschmarotzer/ Hemiparasit semiparasite, hemiparasite
Halbplazenta/Semiplazenta/ Placenta adeciduata semiplacenta, nondeciduate placenta
Halbsättigungskonstante/ Michaeliskonstante (K_M) Michaelis constant, Michaelis-Menten constant
Halbstrauch half-shrub, semishrub, shrubby herb, suffrutecsent plant
halbstrauchig (am Grunde verholzt) suffruticose, suffrutescent, base somewhat woody
halbsynthetisch semisynthetic
halbtrocken semiarid
Halbtrockenrasen semiarid grassland
Halbwertsbreite *math/stat* full width at half-maximun (fwhm), half intensity width
Halbwertszeit half-life
Halbwüste semidesert
Hälfte/Anteil/Teil moiety
Hallersches Organ (Zecken) Haller's organ
halluzinogen hallucinogenic
Halluzinogen hallucinogen
Halm *bot* blade, stalk
• **Grashalm** culm, haulm, halm, spire; blade of grass
• **Strohhalm** straw
Halmfrucht (Getreide) cereal
halmtragend culmiferous
Halophyt ("Salzpflanze") halophyte
Haloragaceae/Seebeerengewächse/ Tausendblattgewächse water milfoil family, milfoil family
Hals neck
Hals/Kehle throat
Hals/Tubusträger *micros* neck
Halskanalzelle *bot* neck canal cell
Halsschild cervical sclerite; (Collum: in Diplopoda) collum
Halsschlagader carotid artery
Halswirbel/Cervikalwirbel cervical vertebra
haltbar storable, durable, lasting
Haltbarkeit storability, durability, shelf-life
Haltere/Schwingkölbchen haltere, balancer
Haltezange (Insektenmännchen)/ Harpagon/Harpe male clasper, harpagone, harpe

Haltung/Stellung/Lage posture, stance
• **Drohhaltung** threatening posture
• **aufrechte Haltung** upright posture, orthograde posture
Häm heme
Hämadsorptionshemmtest (HADH) hemadsorption inhibition test (HAI test)
Hämagglutinationshemmtest (HHT) hemagglutination inhibition test (HI test)
Hämalbogen/Haemalbogen hemal arch
Hämalkanal/Haemalkanal hemal canal, hemal duct
Hämalsystem/Haemalsystem hemal system
Hamamelidaceae/ Zaubernussgewächse witch-hazel family
Hämatokrit hematocrit
Hämatozyt/Hämatocyt/Hämocyt/ Hämozyt/Blutzelle hematocyte, hemocyte
Hammel wether
Hammer/Malleus (Ohr) hammer, malleus
Hämoglobin hemoglobin
Hämolymphe (Blutplasma einiger Invertebraten) hemolymph
hamstern *vb* hoard
Hamstern *n* hoarding
Handdecken *orn* primary tectrices
handförmig hand-shaped, palmate
Handgelenk wrist joint
Handgelenksknochen/ Handwurzelknochen/Carpalia/ Ossa carpalia wrist bone, carpal bone
handnervig palmately veined, palmate
Handschuhkasten/ Handschuhschutzkammer glove box
Handschwingen/Hautschwingen/ Handschwungfedern primaries, primary feathers (primary remiges)
Handwurzel/Karpus/Carpus carpus
Handwurzelknochen/ Handgelenksknochen/Carpalia/ Ossa carpalia wrist bone, carpal bone
Hanfgewächse/Cannabaceae hemp family
Hang slope, incline
• **Berghang** mountain slope, hillslope
Hangaufwind anabatic wind
hangeln brachiate
Hangeln/Schwingklettern/ Brachiation brachiation

Harzsäure

hängend pendulous
Hanglage hillside location, slope location
Hangler/Schwingkletterer brachiator
Hangmoor slope fen
Hangsegeln slope soaring, ridge soaring
Hanke (Hüfte des Pferdes) hip
hapaxanth/monokarpisch hapaxanthic, hapaxanthous, hapanthous, monocarp, monocarpic
haploid haploid
Haploidisierung haploidization
Haploinsuffizienz haploinsufficiency
haplostemon *bot* haplostemonous
Haplotyp haplotype
Haplotypanalyse/Bestimmung des Haplotyps haplotyping
Haptocyste (ein Extrusom) haptocyst
Haptor (Haftorgan: Trematoden) haptor
Hardersche Drüse (Auge) Harderian gland, Harder's gland
Hardy-Weinberg-Gesetz Hardy-Weinberg law
Hardy-Weinberg-Gleichgewicht Hardy-Weinberg equilibrium
Harem harem
Harlekin-Chromosomen harlequin chromosomes
Harmonikabewegung/ Regenwurmbewegung (Schlangen) concertina movement
Harn/Urin urine
- **Primärharn/Glomerulusfiltrat** glomerular ultrafiltrate
- **Sekundärharn** secondary urine
Harnblase urinary bladder
harnen/urinieren/miktuieren urinate, micturate
Harnen/Harnlassen/Urinieren/ Miktion urination, micturition
Harnfluss/Harnausscheidung/ Diurese diuresis
Harngang/Harnleiter ureter, urinary duct
Harnhaut/Harnsack/Allantois allantois
Harnkanälchen/Nierenkanälchen uriniferous tubule
Harnleiter urinary duct
- **früher primärer Harnleiter/ Pronephros-Gang** pronephric duct
- **sekundärer Harnleiter/Ureter** ureter
- **später primärer Harnleiter/ Wolffscher Gang** mesonephric duct, Wolffian duct

Harnleiterklappe ureteral valve
Harnmarkierung urine marking
Harnprüfen urine sampling
Harnraum/Urodaeum urodeum
Harnröhre/Urethra urethra
Harnröhrendrüse/Urethraldrüse/ Littre'-Drüse urethral gland
Harnsack/Harnhaut/Allantois allantois
Harnsamenleiter uroseminal duct (archinephric duct conducting both sperm and urine)
Harnsäure (Urat) uric acid (urate)
Harnspritzen urine spraying, enurination
Harnstoff (Ureid) urea (ureide)
Harnstoffzyklus/Harnstoffcyclus urea cycle
Harpagon/Harpe (Valven bei Insekten) harpagone, harpe (male claspers)
Harsch/Harschschnee wind-slab, crusted snow
Hartbast hard bast
Hartboviste/Sclerodermataceae earth balls
Härte hardness, toughness
härten harden
härten/aushärten *vb polym* cure
Härten/Aushärten *n polym* curing
Härter/Aushärtungskatalysator *polym* curing agent
Härtezeit/Abbindezeit *polym* curing period
Hartheugewächse/ Johanniskrautgewächse/ Hypericaceae/Clusiaceae/ Guttiferae St. John's wort family, mamey family, mangosteen family, clusia family
Hartholz hard wood
Hartig'sches Netz *fung* Hartig net
Hartlaub/Hartlaubgewächs/ Sklerophyll hard-leaf, hard-leaved plant, sclerophyllous plant, sclerophyll
Hartlaubgebüsch/Hartlaubgehölz scrub, sclerophyll shrub
Hartlaubgewächs/Sklerophyll hard-leaf, hard-leaved plant, sclerophyllous plant, sclerophyll
Hartlaubwald sclerophyllous forest
Hartriegelgewächse/ Hornstrauchgewächse/Cornaceae dogwood family
Harz resin
harzabsondernd resiniferous
Harzgalle resin gall
Harzgang/Harzkanal resin duct, resin canal
harzig resinous
Harzsäure resin acids

Haselnussfrucht

Haselnussfrucht filbert
Haselnussgewächse/Corylaceae hazel family
Hasen/Lagomorpha rabbits, lagomorphs
Hasenfußfarngewächse/ Davalliaceae davallia family, bear's-foot fern family
Hasenpest/Tularämie (*Pasteurella/ Francisella tularensis*) tularemia
Hasenscharte/Lippenspalte hare lip, cleft lip
Hassall-Körperchen Hassall's corpuscle, thymic corpuscle
Hassen/Hassverhalten mobbing behavior
Hatscheksche Grube Hatschek's pit, Hatschek's groove
Hatz/Hetzjagd chase, hunt, hunting
Haube/Decke/Tegmentum (Gehirn) tegmentum
Haube/Kalyptra *bot* calyptra
Haube/Netzmagen/Retikulum *zool* honeycomb stomach, honeycomb bag, reticulum, second stomach
Haubennetz *arach* dome web
Hauer (z.B. Eber) tusk, fang (large teeth)
Hauerzahnsaurier/Anomodontia anomodonts
häufig frequent, abundant
Häufigkeit/Frequenz frequency (of occurrence), abundance
• **relative H.** *stat* frequency ratio
Häufigkeitshistogramm frequency histogram
Häufigkeitsverteilung *stat* frequency distribution (FD)
Häufungsgrad/Häufigkeitsgrad kurtosis
Hauptachse main axis, principal axis
• **rhizomartige H. (Algen/ Zygomyceten)** stolon
Hauptachse/Stolo (Hydrozoen-Kolonien) *zool* stolon
Hauptanbauprodukt staple crop
Hauptassoziation chief association
Hauptauge main eye
Hauptbande *chromat/electrophor* main band
Haupterzeugnis staple
Hauptfruchtform/Teleomorph *fung* perfect stage, telomorphic stage
Haupthistokompatibilitätskomplex major histocompatibility complex (MHC)
Hauptsatz (1.Hauptsatz/ 2.Hauptsatz der Thermodynamik) first/second law of thermodynamics
Hauptspross/Primärspross/ Hauptachse leading/main/primary shoot, main/primary axis

Hauptwirt primary host, main host, definitive host
Hauptwurzel/Primärwurzel main root, primary root
Hauptwurzelanlage radicula
Hauptzelle (Magen) chief cell
Haushalt household
• **Naturhaushalt (natürliches Gleichgewicht)** natural balance
• **Stoffwechsel/Metabolismus** metabolism
• **Wasserhaushalt/Wasserregime** water regime
Haushaltsgen/Haushaltungsgen/ konstitutives Gen housekeeping gene
Haustier/domestiziertes Tier (für landwirtschaftliche Nutzwecke) domesticated animal
• **zahmes Haustier (Liebhaberei)** pet
Haustorium/Fuß (auch bei Muschellarve) haustorium, foot
Haustorium/Saugorgan haustorium, sucker
Haut skin; hide, peel; integument
• **Kutis/Cutis (eigentliche Haut; Epidermis & Dermis)** skin, cutis
• **Lederhaut/Korium/Corium/ Dermis** cutis vera, true skin, corium, dermis
• **Oberhaut/Epidermis** epidermis
• **Schleimhaut/Schleimhautepithel** mucous membrane, mucosa
• **Unterhaut/ Unterhautbindegewebe/ Subcutis/Tela subcutanea** subcutis
Haut.../dermal dermal, dermic, dermatic
Haut.../die Haut betreffend epidermal, cutaneous
Hautatmung cutaneous respiration/ breathing, integumentary respiration
Hautdrüse/Dermaldrüse dermal gland
häuten molt, shed skin
Hautfalte flap of skin, skin fold, wrinkle
Hautfarngewächse/ Schleierfarngewächse/ Hymenophyllaceae filmy fern family
Hautfaserblatt/parietales Blatt/ somatisches Blatt/Somatopleura somatopleure
Hautflügel hymenopterous wing
Hautflügler/Hymenoptera hymenopterans
Hautknochen/Deckknochen/ Belegknochen dermal bone
Hautlappen/Fleischlappen lappet

Hautmuskelschlauch
epitheliomuscular tube
Hautmuskelzelle epitheliomuscular
cell, epitheliomuscle cell
Hautmuskulatur
dermal musculature
Hautpapille/Dermispapille
dermal papilla
Hautparasit/Hautschmarotzer/
Dermatozoe skin parasite,
dermatozoan
Hautpilz/Dermatophyt
dermatophyte
Hautpilze/Hymenomycetes
exposed hymenium fungi
Hautplatte dermal plate
Hautschwingen/Handschwingen
primaries, primary feathers (primary
remiges)
Hautskelett/Dermalskelett dermal
skeleton, integumentary skeleton,
dermatoskeleton (exoskeleton)
Häutung/Ecdysis/Ekdyse molt,
molting, ecdysis, shedding skin
Häutungsflüssigkeit/
Ecdysialflüssigkeit/
Exuvialflüssigkeit molting fluid,
exuvial fluid
Hautzahn/Zahnschuppe/
Placoidschuppe/Dentikel
dermal denticle, placoid scale
Haversscher Kanal Haversian canal
(central canal)
Haverssches System/Osteon
Haversian system, osteon
Haworth-Projektion/Haworth-
Formel Haworth projection,
Haworth formula
Hebelmechanismus
leverage mechanism
Heber/Levator (Muskel)
levator, lifter
(muscle: raising an organ or part)
hecheln (z. B. Hund) pant, panting
Hechtkrautgewächse/
Pontederiaceae pickerel-weed
family, water hyacinth family
Hecke hedge
Heckenpflanze hedge plant
Heckenschere hedge clippers,
hedge trimmers
hedonische Drüse (Amphibien)
hedonic gland
Hefe yeast
• **Backhefe/Bäckerhefe**
baker's yeast
• **Bierhefe/Brauhefe** brewers' yeast
• **Brennereihefe** distiller's yeast
• **hochvergärende Hefe**
("Staubhefe") top yeast
• **Mineralhefe**
mineral accumulating yeast

• **niedrigvergärende Hefe**
("Bruchhefe") bottom yeast
• **Trockenhefe** dried yeast
Hefechromosom, künstliches (YAC)
yeast artificial chromosome (YAC)
Hefeextrakt yeast extract
Hefeplasmid yeast plasmid
• **episomales H. (YEp)** yeast
episomal plasmid (YEp)
Hege (Wild) *hunt* preservation,
care and protection
hegen (schützen/bewahren/
pflegen) preserve/maintain
(Wild/Forst), tend/nurse (Garten/
Pflanzen)
Hegewald/Hegeschlag (geschonter
Wald) protected forest
Hegezeit/Schonzeit (Jagd/Wild)
hunt close (closed) season
Heide heath
Heidegras heath sedge
Heidekraut heather (*Calluna vulg.*);
generell: heath
Heidekrautgewächse/Ericaceae
heath family
Heideland heath, heathland,
moorland
Heidemoor heath moor
Heidewald heath forest
heilen cure, heal
Heilpflanze/Arzneipflanze
medicinal plant
Heilung cure, healing
heimisch local, endemic
Heimkehrvermögen/
Heimfindevermögen/Zielflug
homing instinct
Heister (junger Laubbaum aus
Baumschule) sapling
heizen heat
Heizplatte hot plate
Heizschlange *lab* heating coil
Helfervirus helper virus
Helferzelle helper cell
heliophil heliophilic, heliophilous
Heliophyt/Sonnenpflanze/
Starklichtpflanze heliophyte
Heliotropismus/Lichtwendigkeit/
Sonnenwendigkeit heliotropism,
solar tracking
Helix/Spirale (*pl* Helices) helix
(*pl* helices or helixes), spiral
Helix-Loop-Helix (Strukturmotiv)
helix-loop-helix
Helix-Turn-Helix (Strukturmotiv)
helix-turn-helix
Hellfeld *micros* bright field
Hellkeimer/Lichtkeimer (Samen)
light-induced germination of seed,
photodormant seed
Hellkeimung light-induced
germination (photodormancy)

helmartig

helmartig (Blütenblatt) hood-like, cucullate
Helminthologie helminthology
Helokrene/Sickerquelle/ Sumpfquelle helocrene
Helotismus helotism
Hemerocallidaceae/ Tagliliengewächse daylily family
hemiangiokarp hemiangiocarpic, hemiangiocarpous
Hemibranchie hemibranch
Hemichordaten/Kragentiere/ Branchiotremata hemichordates
hemihomocerk/hemihomozerk hemihomocercal
Hemikryptophyt (Überdauerungsknospen an Erdoberfläche) hemicryptophyte
hemimetabole Entwicklung hemimetabolic/hemimetabolous development
Hemionitidaceae/ Nacktfarngewächse strawberry fern family
Hemipenis hemipenis
Hemisphäre (Halbkugel) hemisphere
- **Erdhalbkugel/Erdhälfte/ Erdhemisphäre** hemisphere, global hemisphere
- **Hirnhemisphäre/Großhirnhälfte** cerebral hemisphere
- **Nordhalbkugel/Nordhemisphäre (Erde)** Northern Hemisphere
- **Osthemisphäre (Erde)** Eastern Hemisphere
- **Südhalbkugel/Südhemisphäre (Erde)** Southern Hemisphere
- **Westhemisphäre (Erde)** Western Hemisphere

hemizygot hemizygous
Hemizygotie hemizygosity
hemizyklisch hemicyclic
hemmen inhibit
hemmend/inhibierend/inhibitorisch inhibitory
Hemmkonzentration inhibitory concentration
- **minimale Hemmkonzentration (MHK)** minimal inhibitory concentration, minimum inhibitory concentration (MIC)

Hemmstoff inhibitor
Hemmung/Inhibition inhibition
- **irreversible Hemmung** irreversible inhibition
- **kompetitive Hemmung/ Konkurrenzhemmung** competitive inhibition
- **nichtkompetitive Hemmung** noncompetitive inhibition
- **reversible Hemmung** reversible inhibition
- **Suizidhemmung** suicide inhibition
- **unkompetitive Hemmung** uncompetitive inhibition

Hemmungsneuron inhibitory neuron
Hemmzone inhibition zone
Henderson-Hasselbalch-Gleichung/ Henderson-Hasselbalchsche Gleichung Henderson-Hasselbalch equation
Hengst stallion (Zuchthengst: stud, studhorse)
Henle-Schleife/Henlesche Schleife loop of Henle, loop of the nephron, nephronic loop
Henne hen
- **Legehenne** layer (hen)

Heparin heparin
Hepatopankreas/Mitteldarmdrüse midgut gland
Heptamer heptamer
herabhängend/herablaufend decurrent
- **schlaff herabhängend** drooping

Herabregulation down regulation
herabstoßen *orn* **(im Sturzflug die Beute ergreifen)** stooping, dive-bombing
Heraufregulation up regulation
herauskommen/hervorkommen/ auftauchen emerge
Herauskreuzen/Auskreuzen outcrossing
herausragen emerge; (hervorstehen) protrude, stand out
herausschneiden/exzidieren excise
Herausschneiden/Excision/Exzision excision
Herbar herbarium
Herbivor/Pflanzenfresser/ pflanzenfressendes Tier herbivore
Herbizid/ Unkrautvernichtungsmittel/ Unkrautbekämpfungsmittel herbicide, weed killer
Herbst fall, autumn
Herbstfärbung autumn/fall coloration
Herbstlaub autumn/fall foliage
herbstlich fall, autumnal
Herbstsches Körperchen *orn* Herbst corpuscle
Herde herd, flock
Herden (Hüteverhalten) herding (guarding behavior)
Herden.../in Herden lebend/ gesellig (Herdentiere/Insekten) gregarious
Herdentrieb/Herdeninstinkt herd instinct, herding instinct

heterocerk

Heringsfische/Heringsverwandte/ Clupeiformes
herrings and relatives
Heritabilität/Erblichkeit(sgrad)
heritability
Herkunft/Abstammung origin, descent, provenance (Provenienz)
Hermaphrodit/Zwitter
hermaphrodite
- **protandrischer/proterandrischer Hermaphrodit**
protandric hermaphrodite
- **Pseudohermaphrodit/ Scheinzwitter**
pseudohermaphrodite
- **Simultanhermaphrodit**
simultaneous/synchronous hermaphrodite
- **Suczedanhermaphrodit**
sequential hermaphrodite
hermaphroditisch/zwittrig
hermaphroditic
Hermaphroditismus/Zwittertum
hermaphrodism
hermatypisch/riffbildend
hermatypic, reef-building
Hernie hernia; (z.B. Kohlhernie) club-root
Herpetologie (Amphibien- und Kriechtierkunde) herpetology
Herrentiere/Primaten primates
herunterhängendes Blatt
drooping leaf
Herunterstufung/Hinunterstufung
downward classification
HERV-Familie (menschliche endogene Retroviren-Familie) (DNA-Element) HERV-family (human endogeneous retrovirus family) (DNA element)
hervorkommen/herauskommen/ auftauchen emerge
hervorstehen/herausragen protrude
Herz heart
- **Kiemenherz (Cephalopoda)**
gill heart, branchial heart
- **Lateralherz** lateral heart
- **Nebenherz/auxilläres Herz**
auxillary heart
- **Röhrenherz/Herzschlauch**
tubular heart
- **Stirnherz**
frontal heart, frontal sac
Herzaktivität cardiac activity
Herzausstoß cardiac output
Herzbeutel/Perikard/Pericard
pericardium
Herzblattgewächse/Parnassiaceae
grass of Parnassus family
Herzfäule heart rot
herzförmig cordate, cordiform, heart-shaped

Herzfrequenz heart rate
Herzgallerte cardiac jelly
Herzgewichtsregel/Reihenregel/ Hessesche Regel evol
heart-weight rule, Hesse's rule
Herzigel/Herzseeigel/Spatangoida
heart urchins
Herzklappe heart valve, cardiac valve, coronary valve
- **Aortenklappe** aortic valve
- **Mitralklappe/Bikuspidalklappe/ Zweisegelklappe/zweizipflige Segelklappe**
mitral valve, bicuspid valve (with two cusps/flaps)
- **Pulmonalklappe**
pulmonary valve, pulmonic valve
- **Segelklappe/ Atrioventrikularklappe**
atrioventricular valve
- **Taschenklappe** semilunar valve (consisting of three semilunar cusps/flaps)
- **Trikuspidalklappe/ Dreisegelklappe/dreizipflige Segelklappe** tricuspid valve (with three cusps/flaps)
Herzklopfen beating/throbbing of the heart
Herzknochen/Os cordis
cardiac bone, heart ossicle
Herzkranzfurche coronary groove, coronary sulcus
Herzkranzgefäße (Arterien/Venen)
coronary blood vessels (arteries/ veins)
Herzminutenvolumen (HMV)
cardiac output per minute
Herzmuskel cardiac muscle
Herzohren/Auriculae cordis
auricles of the heart
Herzschlag (einfache Kontraktion des Herzens) heart beat
Herzschlag/Herzinfarkt
myocardial infarction
Herzschlauch/Röhrenherz
tubular heart
Herzskelett cardiac skeleton, skeleton of the heart
Herzstillstand cardiac arrest
Herzvorhof/Herzvorkammer/ Aurikel/Atrium auricle, atrium
Herzwurm (*Dirofilaria spp.*)
heartworm
Hesperidium/Citrusfrucht/ Zitrusfrucht (eine Panzerbeere)
hesperidium
Heterobasidie/Phragmobasidie
fung heterobasidium
(*pl* heterobasidia)
heterocerk/heterozerk
heterocercal

heterochlamydeisch
heterochlamydeous
Heterochromatin heterochromatin
• **fakultatives H.**
facultative heterochromatin
• **konstitutives H.**
constitutive heterochromatin
heterochron heterochronous
Heterochronie heterochrony, heterochronism
heterocöl/heterocoel heterocelous
Heteroduplex heteroduplex
Heteroduplex-Kartierung
heteroduplex mapping
heterogam heterogamous
heterogametisches Geschlecht
heterogametic sex
Heterogamie heterogamy
heterogen/ungleichartig/ verschiedenartig/andersartig
heterogeneous (consisting of dissimilar parts)
heterogen/unterschiedlicher Herkunft heterogenous (of different origin)
heterogene Kern-RNA
heterogeneous nuclear RNA (hnRNA)
Heterogenese heterogenesis, heterogeny
heterogenetisch/genetisch unterschiedlichen Ursprungs
heterogenetic
Heterogenie/unterschiedlicher Herkunft heterogeny
Heterogenität/Ungleichartigkeit/ Verschiedenartigkeit/ Andersartigkeit heterogeneity
Heterogenote f heterogenote
Heterogonie/ zyklische Parthenogenese
heterogony, heterogamy
heterolog heterologous
heterologe Sonde
heterologous probe
heteromorph/anders gestaltet/ verschiedengestaltig
heteromorphous
Heterophyllie/Anisophyllie/ Verschiedenblättrigkeit/ Ungleichblättrigkeit heterophylly, anisophylly
Heteropolymer heteropolymer
Heterosis/Bastardwüchsigkeit
heterosis, hybrid vigor
heterospor heterosporous
heterostyl/verschiedengriffelig
heterostylous
Heterostylie/ Verschiedengriffeligkeit
heterostyly
Heterothermie heterothermy

heterothermisch heterothermic
heterotroph heterotroph, heterotrophic
Heterotrophie heterotrophy
heterotypisch heterotypic
Heterözie/Heteröcie heteroecy, heteroecism
heterözisch heteroecious, heterecious, heteroxenous
heterozygot/mischerbig
heterozygous
• **zusammengesetzt-heterozygot**
compound heterozygous
Heterozygotenvorteil heterzygote advantage
Heterozygotie/Mischerbigkeit
heterozygosity
• **Verlust der Heterozygotie/ Heterozygotieverlust**
loss of heterozygosity (LoH)
heterozyklisch heterocyclic
hetzen inciting, chasing
Heu hay
Heuaufguss hay infusion
heulen (Heuleraffe/Koyote) howl
heulen/schreien (Eulen) hoot
Heulen n howling
Heuler (Robbenjunges) young seal, seal pup (*literally*: wailing seal pup)
heuristisch heuristic
Hexacorallia hexacorallians, hexacorals
Hexamer/Hexon vir hexamer, hexon
Hexenbesen witches' broom
Hexenei fung immature, closed fructification (fruit body) of *Phallales* (stinkhorn)
Hexenring fung fairy ring
Hexosemonophosphatweg/ Pentosephosphatweg/ Phosphogluconatweg hexose monophosphate shunt (HMS), pentose phosphate pathway, pentose shunt, phosphogluconate oxidative pathway
Hfr-Zelle (hohe Rekombinationshäufigkeit)
Hfr cell (*from:* high frequency of recombination)
Hibernakel hibernaculum, winter bud
Hierarchie hierarchy
hierarchisch hierarchical
Hilfsstoff/Adjuvans auxiliary drug, adjuvant
Hill-Auftragung Hill plot
Hill-Gleichung Hill equation
Hill-Koeffizient/ Kooperativitätskoeffizient
Hill coefficient, Hill constant
Hill-Reaktion Hill reaction
Hilum hilum

Hirnbläschen

**Himmelsleitergewächse/
Sperrkrautgewächse/
Polemoniaceae** phlox family
**Hin-und Herschieben von Exons/
Exonmischung/Exonshuffling** *gen*
exon shuffling
**Hinfallhaut/Siebhaut/Dezidua/
Decidua** decidua
hinführend/zuführend/zuleitend
afferent
**Hinteraugenschild/Postoculare
(Schlangen)** postocular
Hinterbeine hindlegs, posterior legs
Hinterbrust/Metathorax metathorax
Hinterdarm/Enddarm/Proctodaeum
hindgut, proctodeum
Hinterextremität hindlimb
Hinterflügel (Unterflügel) hindwing
**Hinterfühlerorgan/
Postantennalorgan**
postantennal organ
Hintergrund background
Hintergrundsfärbung (Tarnung)
background camouflage, mimesis
Hinterhand (Pferd) hindquarter,
haunch
Hinterhaupt/Occiput occiput
(dorsal/posterior part of head)
Hinterhauptbein/Os occipitale
occipital bone
**Hinterhauptslappen/
Okzipitallappen/
Lobus occipitalis**
occipital lobe
Hinterhirn/Tritocerebrum (Insekten)
tritocerebrum
**Hinterkiemenschnecken/
Hinterkiemer/Opisthobranchia**
opisthobranch snails, opisthobranchs
Hinterkopf back of the head;
poll (crest/top/apex/back of head:
esp horses/cattle)
**Hinterkörper/Hinterleib/
Abdomen/Opisthosoma**
abdomen; opisthosoma
Hinterleibsstiel/"Taille"/Petiolus
podeon, podeum, petiole
**Hinterschenkel/Hose
(Unterschenkel des Pferdes)**
gaskin
Hinterteil/Hinterleib/Hinterviertel
hindquarter, haunch
• **Kruppe (Pferd)** croup
hinweglesen über (ein Stoppcodon)
gen read through (a stop codon)
**Hippocastanaceae/
Rosskastaniengewächse**
horse chestnut family,
buckeye family
**Hippuridaceae/
Tannenwedelgewächse**
marestail family, mare's-tail family

**Hirn/Gehirn (Encephalon/
Enzephalon)** brain (encephalon)
• **Althirn/Paläoencephalon/
Paleencephalon**
paleoencephalon, paleencephalon
• **Endhirn/Großhirn/Telencephalon**
endbrain, cerebrum, telencephalon
• **Großhirn/Endhirn/Telencephalon**
cerebrum, endbrain, telencephalon
• **Hinterhirn/Metencephalon
(Pons + Cerebellum)**
afterbrain, metencephalon
• **Hinterhirn/Tritocerebrum
(Insekten)** tritocerebrum
• **Kleinhirn/Hinterhirn/Cerebellum**
cerebellum, epencephalon
• **Markhirn/Myelencephalon**
marrow brain, medullary brain,
myelencephalon
• **Mittelhirn/Deutocerebrum
(Insekten)** deutocerebrum
• **Mittelhirn/Mesencephalon
(Vertebraten)** midbrain,
mesencephalon
• **Nachhirn/Metencephalon
(Pons + Cerebellum)**
afterbrain, metencephalon
• **Neuhirn/Neoencephalon/
Neencephalon** neoencephalon,
neencephalon
• **Neukleinhirn/Neocerebellum**
neocerebellum (lateral lobes of
cerebellum)
• **Rautenhirn/Rhombencephalon**
hindbrain, rhombencephalon
• **Riechhirn/Rhinencephalon**
"nose brain", olfactory brain,
rhinencephalon
• **Urhirn/Archencephalon**
primitive brain, archencephalon
• **verlängertes Rückenmark/
Medulla oblongata**
medulla, medulla oblongata
• **Vorderhirn/Prosencephalon
(Vertebraten)** forebrain,
prosencephalon (telencephalon +
diencephalon)
• **Vorderhirn/Protocerebrum
(Insekten)** protocerebrum
• **Zwischenhirn** diencephalon,
interbrain, betweenbrain
Hirn-Herz-Infusionsagar
brain-heart infusion agar
Hirnanhangdrüse/Hypophyse
pituitary gland, pituitary,
hypophysis
Hirnbeuge *embr* cerebral flexure
• **Brückenbeuge** pontine flexure
• **Nackenbeuge** cervical flexure
• **Scheitelbeuge** cephalic flexure,
cranial flexure
Hirnbläschen *embr* cerebral vesicle

H Hirnflüssigkeit

Hirnflüssigkeit/Gehirn-Rückenmarks-Flüssigkeit/Liquor cerebrospinalis
cerebrospinal fluid (CSF)
Hirngewölbe/Fornix
fornix of cerebrum
Hirnhaut/Gehirnhaut/Meninx (pl Meninga) cerebral meninx (pl meninga/meninges),
cerebral membrane
• **harte Hirnhaut/Pachymeninx (Dura mater)** pachymeninx
• **weiche Hirnhaut/Leptomeninx (Arachnoidea & Pia mater)**
leptomeninx, pia-arachnoid membrane
Hirnholz cross-grained timber, crosscut wood
Hirnkapsel/Schädel/Cranium
braincase, skull, head capsule
Hirnrinde/Großhirnrinde
cerebral cortex
Hirnschädel/Gehirnschädel/Neurocranium neurocranium
Hirnschenkel/Hirnstiel/Crura cerebri
cerebral peduncle
Hirnschnitt/Querschnitt (Holz)
cross section, transverse section
Hirnstamm/Truncus cerebri
brain stem
Hirsche/Cervidae cervids
• **erwachsenes/adultes Hirschmännchen ("Hirsch")**
stag (red deer male: hart)
• **Hirschkuh**
doe (red deer female: hind)
His-Bündel/Hissches Bündel/Fasciculus atrioventricularis/tert. Autonomiezentrum (Herz)
bundle of His,
atrioventricular bundle
Histamin histamine
Histidin histidine
Histiozyt/Gewebswanderzelle/Gewebs-Makrophage histiocyte (actually: macrophage)
Histogramm/Streifendiagramm stat
histogram, strip diagram
Histoinkompatibilität/Gewebeunverträglichkeit
histoincompatibility
Histokompatibilität/Gewebeverträglichkeit
histocompatibility
Histokompatibilitätsantigen
histocompatibility antigen
• **Haupthistokompatibilitätsantigene**
major histocompatibility antigens
• **Nebenhistokompatibilitätsantigene**
minor histocompatibility antigens

Histokompatibilitätskomplex
histocompatibility complex
• **Haupthistokompatibilitätskomplex** major histocompatibility complex (MHC)
Histon histone
histonartiges Protein
histone-like protein
Hitze heat
Hitzebehandlung/Backen
heat treatment, baking
hitzebeständig heat-resistant, heat-stable
Hitzeerschöpfung heat exhaustion
Hitzekrämpfe heat cramps
hitzemeidend/thermophob
thermophobic
Hitzeschlag heatstroke
Hitzeschock heat shock
Hitzeschockgen heat shock gene
Hitzeschockprotein
heat shock protein
Hitzeschockreaktion heat shock reaction, heat shock response
hitzeverträglich heat-tolerant
HLA-Komplex (menschlicher Leukozytenantigen-Komplex)
HLA complex (human leucocyte antigen complex)
HMG-Box (Gruppe von hoher Beweglichkeit) gen
HMG-box (high mobility group)
Hochblatt hypsophyll;
(Braktee) floral bract
Hochblatthülle/Außenkelch
sepal-like bracts
Hochdruck/Bluthochdruck
hypertension
Hochdruckflüssigkeitschromatographie/Hochleistungschromatographie
high-pressure liquid chromatography, high performance liquid chromatography (HPLC)
Hochfläche/Hochebene plateau, elevated plane, tableland
Hochgebirge alpine mountains, alpine mountain chain
Hochgebirgsmoor alpine mire/bog
Hochgebirgsregion alpine region
Hochgebirgssee alpine lake
Hochgebirgsstufe alpine zone
hochkant angeordnetes Plattengel
slab gel
hochkronig/hypsodont/hypselodont (Zähne)
with high crowns, hypsodont
Hochland highland
hochmolekular high-molecular
Hochmoor (ombrotroph)
raised bog, raised mire,
(upland/high) moor, peat bog

Holozän

Hochmoortorf highmoor peat, sphagnum peat, moss peat
Hochmoorwald (upland) bog forest
Hochsee/offenes Meer/ Hochseebereich/ozeanische Region open sea, pelagic zone, oceanic zone/province
Hochstamm (Wuchsform eines Baumes) standard tree, standard
Hochstaude tall/montane perennial herb
Hochstaudenflur tall/montane herbaceous vegetation zone
Höchsterträge maximum yield
Hochwald high forest
hochwürgen/wiederaufstoßen/ regurgitieren regurgitate
Hochwürgen/Wiederaufstoßen/ Regurgitation regurgitation
Hochzeitsflug mating flight, nuptial flight
Hochzeitskleid *orn* nuptial dress, nuptial plumage
"Hochzeitslaube" (schwarze Witwe) *arach* mating bower
Höcker/Wölbung/Tuberkel (Erhebung) knob, tuber, tubercle (protuberance); (Vogelschnabel: Buckel/Wölbung) gibbosity
- **Blatthöcker/frühe Blattanlage** *bot* leaf buttress
- **Fetthöcker (Kamele/Rinder)** hump
- **Geschlechtshöcker/ Genitalhöcker/Tuberculum genitale** genital tubercle
höckerig/buckelig *allg* humped; (Vogelschnabel) gibbose, gibbous
Hoden (sg/pl) testicle, testis (*pl* testes)
Hoden.../den Hoden betreffend testicular
Hodensack/Skrotum scrotum
Hof/Lysehof/Aufklärungshof/ Plaque plaque
Hofmeistersche Reihe/lyotrope Reihe Hofmeister series, lyotropic series
Hoftüpfel *bot* bordered pit
Hogness-Box *gen* Hogness box
Höhe height
Höhe über dem Meeresspiegel altitude, above sea level
Höhenlage altitude, elevation, higher location
Höhenstufe/Vegetationsstufe altitudinal zone/region/belt, vegetation(al) zone/region/belt
Höhentrieb/Haupttrieb *bot/hort* leader
höhere Pflanzen higher plants
hohl hollow

Höhle *allg* cave, crypt, cavity
Höhle/Kammer/Ventrikel (kleine Körperhöhle) cavity, chamber, ventricle
höhlenbewohnend/kavernikol cave-dwelling, cavernicolous (troglophilic)
Höhlenbewohner cave dweller, cavernicole (troglophile)
Hohlknochen/pneumatischer Knochen/Os pneumaticum hollow bone, pneumatic bone
Hohlraum/Höhlung/Lumen cavity, lumen; (Blattparenchym) airspace
Hohlspiegel concave mirror
Hohlstachler/Coelacanthiformes coelacanths
Hohltiere/Nesseltiere/ Coelenteraten/Cnidaria cnidarians, coelenterates
Höhlung crypt, cavity, cave
Hohlvene/Vena cava vena cava
Hohlwelle (Rührer) hollow impeller shaft
Holandrie *gen/zool* holandry
holandrisch *gen/zool* holandric
hold/preferentiell (Boden/ Gesellschaftstreue) preferential, favorably associated
Holismus holism
holistisch holistic
Holliday-Struktur *gen* Holliday structure, Holliday junction
Holobasidie/Homobasidie holobasidium, homobasidium (*pl* -basidia)
Holobranchie holobranch
Holocoen/Ökosystem holocoen, ecosystem
Holoenzym holoenzyme
hologyn hologynic
Hologynie hologyny
holokrin holocrine
holometabol holometabolous, holometabolic
holometabole Entwicklung holometabolic development
Holometabolie holometabolism
holomiktisch holomictic
Holonephros holonephros
Holoparasit/Vollschmarotzer/ Vollparasit holoparasite, obligate parasite
Holoplankton holoplankton
Holotypus/Holotyp/Holostandard holotype (type specimen)
Holozän/Jetztzeit/Alluvium (erdgeschichtliche Epoche) Holocene, Recent, Holocene Epoch, Recent Epoch

H Holz

Holz wood
- **abholzen** fell, clear, clearcut
- **Abholzung** felling, clear cutting, clear felling, deforestation
- **Ablagern** seasoning
- **Auge** knot
- **Bauholz** (structural) timber, lumber
- **beizen** stain
- **Belegzelle (Holzparenchym)** contact cell
- **Brauchholz** lumber, timber
- **Brennholz** firewood, fuelwood
- **Bruchholz** woody debris
- **Derbholz** crude timber, crude wood
- **dichtfaseriges Holz** pycnoxylic wood
- **Druckholz/Rotholz** compression wood
- **Engholz/Spätholz** summerwood, latewood
- **Faser/Faserung (Schnittholz)** grain
- **Faserholz** pulpwood
- **Faserung (Schnittholz)** grain
- **Figur** figure
- **flach-aufgesägt (Holzstamm)** flatsawn
- **Fladerschnitt** tangential section, flatsawn, plainsawn
- **Fladerung/Maserung** figure, design
- **fräsen** mill, shape
- **Fruchtholz/Tragholz (Kurztrieb)** spur shoot, fruit-bearing bough (short shoot)
- **Frühholz/Weitholz/Frühlingsholz** earlywood, springwood
- **Gefüge** texture
- **Gewicht, spezifisches** specific gravity
- **Hartholz** hard wood
- **Hirnholz** cross-grained timber, crosscut wood
- **Hirnschnitt/Querschnitt** cross section, transverse section
- **in Holz lebend/ auf Holz gedeihend** xylophilous, thriving in/on wood
- **Kalamitätennutzung (Holzernte)** salvage logging, salvage felling
- **Kantholz** cant, squared timber, square-edged lumber, squared log
- **Kernholz** heartwood, duramen
- **Kien/Kienholz** resinous pinewood
- **Knorren (am Baum)/Holzmaser/ Maser/Maserknolle** gnarl, burl, burr
- **Krummholz** stunted, miniature forest; Krummholz
- **lagern** season, store

- **Liane (verholzte Kletterpflanze)** liana, woody climber
- **Maserknolle/Kropf** burl, burr, gnarl, woody outgrowth, wood knot (with wavy grain)
- **Maserknolle, ebenerdige (durch Feuer/Trockenheit)** lignotuber
- **Maserung/Fladerung** *allg* figure, design
- **Maserung (Faserorientierung)** grain
- **morsch** decayed, rotten; brittle; frail, fragile
- **Nutzholz** timber, lumber
- **Oberholz/Oberstand/ Schirmbestand** overstory
- **Papierholz** pulpwood
- **Querschnitt/Hirnschnitt** cross section, transverse section
- **Radialschnitt/Spiegelschnitt** radial section, quartersawn
- **Reaktionsholz** reaction wood
- **Riss (Holz: zwischen Jahresringen)** shake
- **Rundholz** roundwood, log timber
- **Sehnenschnitt** tangential section
- **Sekundärholz, lockeres** manoxylic wood
- **Sommerholz** summer wood
- **Span** (*pl* Späne)**/Holzspäne** (wood) chips, shavings
- **Spätholz/Herbstholz/Engholz** latewood
- **Sperrholz** plywood
- **Spiegelschnitt/Radialschnitt** radial section, quartersawn
- **Splintholz** sapwood, splintwood, alburnum
- **Stammholz** log, lumber
- **Ster** stere (cordwood: 1 cbm)
- **Struktur/Textur/Faser/ Fibrillenanordnung** grain
- **Treibholz** driftwood
- **Unterholz/Untergehölz** understory
- **verholzt/lignifiziert** lignified
- **Verholzung/Lignifizierung** lignification, sclerification
- **verziehen** warp
- **vierteilig-aufgesägt (Holzstamm)** quartersawn
- **Vogelaugenholz** bird's eye (wood texture)
- **Weichholz** soft wood
- **Weitholz/Frühholz/Frühlingsholz** earlywood, springwood
- **werfen/verziehen** warp
- **Wundholz/Wundgewebe/ Wundcallus** wound tissue, callus
- **Zeichnung/Fladerung** figure
- **zerstreutporig** diffuse porous
- **Zugholz** tension wood

Holzapfel crab apple
Holzart kind/type of wood
holzartig woody
Holzbalken beam
Holzbestand stand of timber
Holzbewohner lignophile, xylophile
Holzeinschlag wood felling; felling quantity
hölzern wooden
Holzertrag timber yield
Holzessig wood vinegar, pyroligneous acid
Holzfällen logging, lumbering, felling of trees
Holzfäller lumberjack, woodcutter, woodchopper
Holzfaser (Libriformfaser) wood fiber (libriform fiber)
Holzfäule wood rot
Holzfestigkeit/Holzstabilität wood strength, wood stability
- **Biegefestigkeit/Tragfähigkeit** bending strength
- **Drehfestigkeit** torsion(al) strength
- **Druckfestigkeit** crushing strength, compression resistance (endwise compression)
- **Knickfestigkeit** buckling strength, folding strength, crossbreaking strength
- **Reißfestigkeit/Zerreißfestigkeit/ Zugfestigkeit** tensile strength, breaking strength
- **Scherfestigkeit/Schubfestigkeit** shear strength, shearing strength
- **Stoßfestigkeit** shock resistance
- **Zerreißfestigkeit/Reißfestigkeit/ Zugfestigkeit** tensile strength, breaking strength
holzfressend wood-eating, feeding on wood, xylophagous
Holzfresser lignivore, xylophage
Holzgeist wood spirit, wood alcohol, pyroligneous spirit, pyroligneous alcohol (chiefly: methanol)
Holzgewächs (Phanerophyt) woody plant (phanerophyte)
Holzhaufen wood pile
holzig/faserig woody, ligneous, fibrous
Holzkiste wood crate
Holzkohle charcoal
Holzkörper wood cylinder, wood corpus, wood body
Holzprodukt wood product
Holzqualität wood/lumber/timber quality
Holzschnitzel wood chips
Holzschwarte slab
Holzspäne wood shavings
Holzstamm (gefällt) log, lumber
Holzstrahl wood ray

Homöobox-Gen

Holzteer wood tar
Holzteil/Gefäßteil/Xylem tracheary elements, xylem
holzverarbeitende Industrie timber industry
Holzwirtschaft lumber industry, timber industry
holzzersetzend decomposing wood, xylophilous
homocerk/homozerk homocercal
homocöl/homocoel homocoelous
Homoduplex *gen* homoduplex
homogametisches Geschlecht homogametic sex
homogam homogamous
Homogamie homogamy
homogen/einheitlich/gleichartig homogeneous (having same kind of constituents)
homogen/gleicher Herkunft homogenous (of same origin)
Homogenie/gleicher Herkunft homogeny (of same origin)
Homogenisation homogenization
Homogenisator homogenizer
homogenisieren homogenize
Homogenisierung homogenization
Homogenität/Einheitlichkeit/ Gleichartigkeit homogeneity (with same kind of constituents)
Homogenote *f* homogenote
Homogentisinsäure homogentisic acid
homoiochlamydeisch/mit gleichartigen Hüllblättern homoiochlamydeous, homochlamydeous
homoiosmotisch homoiosmotic, homeosmotic
homoiotherm/gleichwarm/ endotherm/warmblütig homoiothermic, homeothermic, endothermic, warm-blooded
Homoiothermie/Warmblütigkeit homoiothermy, homeothermy, homoiothermism, warm-bloodedness
homolog/ursprungsgleich homologous
homologe Chromosomen homologous chromosomes
homologe Rekombination homologous recombination
Homologie homology
homologisieren homologize
homonom homonomous
homonym *adv/adj* homonymous, homonymic
Homonym *n* homonym
Homonymie homonymy
Homöobox homeobox
Homöobox-Gen/*Hox*-Gen homeobox gene, *Hox* gene

Homöodomäne

Homöodomäne homeodomain
Homöostase/Homöostasie homeostasis
homöotische Mutation homeotic mutation
homöotisches Gen homeotic gene
Homopolymer homopolymer
Homoserin homoserine
homospor homosporous
Homotransplantat/Allotransplantat homograft, syngraft, allograft (syngeneic graft)
Homotyp homotype
homotypisch homotypic
homozygot/reinerbig/reinrassig homozygous
Homozygotie/Reinerbigkeit/ Reinrassigkeit homozygosity
Honig honey
- **Scheibenhonig** comb honey

Honigbechergewächse/ Marcgraviaceae shingleplant family
Honigblatt bot nectariferous leaf
Honigdrüse/Nektarium bot nectar gland, nectary
Honigmagen/Kropf (Biene) honey crop, honey stomach, honey sac
Honigmal bot honey guide
Honigschuppe bot nectariferous scale
Honigstrauchgewächse/ Melianthaceae honeybush family
Honigtau honeydew (*Australien:* sugar-lerp)
Honigwabe honeycomb
hoppeln (z. B. Hase) hop
Hörbarkeit audibility
hören (vernehmen) hear
- **zuhören** listen

Hörgrenze hearing limit, auditory limit, limit of audibility
Hörhaar/Becherhaar/ Trichobothrium acoustical hair, trichobothrium, vibratory sensory hair
horizontal angeordnetes Plattengel horizontal gel, flat bed gel
horizontale Transmission horizontal transmission
Hörkölbchen/Rhopalium rhopalium
Hormocyste (Flechten) hormocyst
Hormocystangium hormocystangium
Hormon hormone *(siehe auch unter individuellen Begriffen)*
- **Adiuretin/antidiuretisches Hormon (ADH)/Vasopressin** antidiuretic hormone (ADH), vasopressin
- **Adrenalin/Epinephrin** adrenaline, epinephrine
- **Aktivierungshormon/ prothorakotropes Hormon** prothoracicotropic hormone (PTTH), brain hormone (BH)
- **Aldosteron** aldosterone
- **Androsteron** androsterone
- **Anti-Müller-Hormon (AMH)** Mullerian inhibiting hormone (MIH)
- **antidiuretisches Hormon (ADH)/ Adiuretin/Vasopressin** antidiuretic hormone (ADH), vasopressin
- **Calcitonin** calcitonin
- **Corticoliberin/Corticotropin- freisetzendes Hormon/ corticotropes Releasing-Hormon (CRH)** corticoliberin, corticotropin-releasing hormone (CRH), corticotropin-releasing factor (CRF)
- **Corticotropin/Kortikotropin/ adrenocorticotropes Hormon/ adrenokortikotropes Hormon (ACTH)** corticotropin, adrenocorticotropic hormone (ACTH)
- **Cortisol/Hydrocortison** cortisol, hydrocortisone
- **Cortison/Kortison** cortisone
- **Endorphin/Endomorphin** endorphin
- **Follitropin/follikelstimulierendes Hormon (FSH)** follicle-stimulating hormone (FSH)
- **Freisetzungshormon/ Freisetzungsfaktor/freisetzendes Hormon/freisetzender Faktor** releasing hormone, release hormone, releasing factor, release factor
- **Gastrin** gastrin
- **Geschlechtshormone** sex hormones
- **Gestagen/Progestin/Corpus- luteum-Hormon/ "Schwangerschaftshormon"** gestagen, progestin
- **Glukagon/Glucagon** glucagon
- **Glukokortikoide** glucocorticoids
- **Gonadotropin** gonadotropin
- **Gonadotropin-Releasing Hormon/Gonadoliberin** gonadotropin releasing hormone/ factor (GnRH/GnRF), gonadoliberin
- **häutungshemmendes Hormon** molt-inhibiting hormone (MIH)
- **Insulin** insulin
- **Juvenilhormon** juvenile hormone (JH)

- **Kortikotropin/Corticotropin/ adrenokortikotropes Hormon/ adrenocorticotropes Hormon (ACTH)** corticotropin, adrenocorticotropic hormone (ACTH)
- **Kortisol/Cortisol/Hydrocortison** cortisol, hydrocortisone
- **Kortison/Cortison** cortisone
- **Lutropin/Luteotropin/ Luteinisierendes Hormon (LH)/ Zwischenzellstimulierendes Hormon** luteinizing hormone (LH), interstitial-cell stimulating hormone (ICSH)
- **Melanoliberin/Melanotropin-Freisetzungshormon** melanoliberin, melanotropin releasing hormone, melanotropin releasing factor (MRH/MRF)
- **Melanotropin/ Melanozytenstimulierendes Hormon (MSH)** melanocyte-stimulating hormone (MSH)
- **Melatonin** melatonin
- **Norepinephrin/Noradrenalin** norepinephrine, noradrenaline
- **Östradiol** estradiol, progynon
- **Östrogen** estrogen
- **Östron/Estron** estrone
- **Oxytocin/Oxytozin** oxytocin
- **Parathormon/Parathyrin/ Nebenschilddrüsenhormon (PTH)** parathyrin, parathormone, parathyroid hormone (PTH)
- **Progesteron** progesterone
- **Progestin** progestin
- **Prolaktin/Prolactin (PRL)/ Mammatropin/Mammotropes Hormon/Lactotropes Hormon/ Luteotropes Hormon (LTH)** prolactin (PRL), luteotropic hormone (LTH)
- **Prolaktoliberin/Prolaktin-Freisetzungshormon** prolactoliberin, prolactin releasing hormone, prolactin releasing factor (PRH/PRF)
- **Prostaglandin(e)** prostaglandin(s)
- **prothorakotropes Hormon/ Aktivierungshormon** prothoracicotropic hormone (PTTH), brain hormone (BH)
- **Relaxin** relaxin
- **Sekretin/Secretin** secretin
- **Sexualhormon** sex hormone
- **Somatoliberin** somatoliberin, somatotropin release-hormone, somatotropin releasing factor (SRF), growth hormone release hormone/factor (GRH/GRF)

- **Somatomedin** somatomedin, insulin-like growth factor (IGF) (sulfation factor/serum sulfation factor)
- **Somatostatin** somatostatin, somatotropin release-inhibiting factor, growth hormone release-inhibiting hormone (GRIH)
- **Somatotropin/somatotropes Hormon/Wachstumshormon** somatotropin (STH), growth hormone (GH)
- **Testis-Determinationsfaktor (TDF)** testis-determining factor
- **Testosteron** testosterone
- **Thyr(e)otropin/Tyrotropin/ thyreotropes Hormon/ thyreoideastimulierendes Hormon (TSH)** thyrotropin, thyroid-stimulating hormone (TSH)
- **Thyroliberin/Thyreotropin-Freisetzungshormon (TRH/TRF)** thyroliberin, thyreotropin releasing hormone/factor (TRH/TRF)
- **Thyroxin (T$_4$)** thyroxine (*also:* thyroxin), tetraiodothyronine
- **Triiodthyronin (T$_3$)** triiodothyronine
- **vasoaktives intestinales Peptid (VIP)** vasoactive intestinal polypeptide
- **Vasopressin/antidiuretisches Hormon (ADH)/Adiuretin** vasopressin, antidiuretic hormone (ADH)
- **Vasotocin** vasotocin
- **Wachstumshormon/ Somatotropin/somatotropes Hormon** growth hormone (GH), somatotropin
- **>menschliches W. (Somatotropin/somatotropes Hormon)** human growth hormone (hGH), human somatotropin

hormonal/hormonell hormonal
Horn horn
Horn.../aus Horn horny
Hornballen/Hufballen/Torus ungulae pad of the hoof, digital pad, bulb
Hornblattgewächse/ Ceratophyllaceae hornwort family
Hörner/Gehörn horns
Hörnerv auditory nerve
Hornfarngewächse/Parkeriaceae water fern family, floating-fern family
Hornhaut (verhornte Haut) callus (hyperkeratosis)
Hornhaut/Cornea (Auge) cornea

Hornhechtartige

**Hornhechtartige/
Ährenfischverwandte/
Atheriniformes**
silversides & skippers &
flying fishes and others
**Hornkorallen/Rindenkorallen/
Gorgonaria** horny corals,
gorgonians, gorgonian corals
Hornmoose (Anthocerotae)
hornworts
Hornröhrchen (Huf) horn tubule
Hornscheide cornified sheath,
horn sheath
Hornschicht/Stratum corneum
stratum corneum, horny layer
(of epidermis)
**Hornschuh/Hornkapsel/Hufkapsel
(Huf)** horny hoof, horny capsule,
hoof capsule
**Hornschwämme/
Netzfaserschwämme/
Cornacuspongiae** horny sponges
Hornsohle/Solea cornes (Huf)
horny sole
**Hornstrauchgewächse/
Hartriegelgewächse/
Cornaceae**
dogwood family
Horotelie horotely
horotelisch horotelic
Hörschwelle hearing threshold,
auditory threshold
Horst *hort/for* small (tree) stand,
thicket
Horst/Raubvogelnest *zool* nest
(esp. of predatory birds)
- **Adlerhorst** eagle nest, eyrie, aerie
Hörsteinchen/Gehörstein/Otolith
"ear bone", "ear stone", otolith
horsten (Raubvögel) nest
**Hortensiengewächse/
Hygrangeaceae**
hydrangea family
Hörvermögen/Gehör audition
Hörzentrum auditory center
**HOSCH-Filter
(Hochleistungsschwebstofffilter)**
HEPA-filter (high efficiency
particulate air filter)
Höschen (Biene) pollen in
corbiculum of hindlegs
Hose *orn* leg feathers
**Hose/Hinterschenkel
(Unterschenkel des Pferdes)**
gaskin (lower thigh)
**Hot-Spot/sensible Position
(Stelle in einem Gen mit hoher
Mutabilität)**
hot spot
Hude/Viehweide pasture
hudern *orn* take (chicks) under its
wing, gathering under wings

Huf/Ungula hoof (*pl* hooves/hoofs)
- **Ballen/Torus ungulae**
bulb (of heel), pad, digital pad
- **Eckstrebe/Pars inflexa** bar
- **Eckstrebenwinkel/Sohlenwinkel**
angle of sole, seat of corn*
(*corn = hardening/thickening of
epidermis)
- **Fesselbein/Fesselknochen/
proximale Phalanx** pastern bone,
long pastern bone, first phalanx
- **Hornschuhwand/Hornwand/
Hufwand** horny wall
- **Kronbein/mittlere Phalanx**
coronary bone, small/short pastern
bone, second phalanx
- **Krone/Corona** coronet
- **Seitenwand/Seitenteil (der
Hufwand)/Pars lateralis** quarter
- **Strahlbein/distale Sesambein/
Os sesamoideum distale**
navicular bone, distal sesamoid bone
- **Trachte** (horny) heel (buttress of
heel/angle of heel/angle of wall)
- **Trachtenwand/Trachtenteil**
wall of the heel, heel wall
- **Zehenwand/Zehenteil/
Rückenteil** wall of the toe,
toe wall
Huf.../mit Hufen/hufartig hoofed,
hoof-like, ungulate
Hufballen/Torus ungulae
pad of the hoof, digital pad, bulb
**Hufbein/Os ungulare/Phalanx
distalis** coffin bone, distal phalanx
Hufbeinknorpel/Hufknorpel
coffin bone cartilage
Hufeisen horseshoe
**Hufeisenwürmer/Phoroniden/
Phoronidea** phoronids
hufförmig hoof-shaped, unguliform
**Hufgang/Zehenspitzengang/
Unguligradie** unguligrade gait
Hufgelenk coffin joint
Hufkapsel/Hornschuh hoof capsule,
horny capsule
**Hufkissen/Pulvinus digitalis
(Strahlkissen = Strahlpolster +
Ballenkissen = Kronkissen =
Ballenpolster)**
digital cushion, plantar cushion
Hufkrone (Haarkranz am Huf)
coronet
- **Kronfurche/Sulcus coronalis**
coronary groove
- **Kronlederhaut**
coronary dermis (corium)
Huflederhaut hoof dermis (corium)
Hufplatte/Hornplatte hoof plate
Hufrehe thrush
Hufrolle/Fußrolle/Podotrochlea
navicular zone, semilunar zone

Hufrollenschleimbeutel/Bursa podotrochlearis navicular bursa
Hufsaum/Limbus limbus
Hufsohle hoof sole, horny sole
Hufstrahl/Cuneus ungulae frog
- **Hahnenkamm/Spina cunei** frog-stay, spine of the frog
- **Hornstrahl** horny frog
- **Strahlfurche, mittlere/Sulcus cunealis centralis** cleft of frog, central groove, central sulcus
- **Strahlfurche, seitliche/Sulcus paracunealis** paracuneal groove, collateral groove, commissure
- **Strahlkissen/Pulvinus cunealis** cuneal cushion
- **Strahllederhaut/Corium cunei** dermis of frog
- **Strahlschenkel/Crus cunei** crus of frog
- **Strahlspitze/Apex cunei** point of frog

Hüftband (Vertebraten) cotyloid ligament
Hüftbein/Hüftknochen/Os coxae hipbone, coxal bone, innominate bone
Hüftbeuge/Leistenbeuge/ Leistengegend/Leiste/ Inguinalgegend/Hüftbeuge/Regio inguinalis groin, inguinal zone
Hüftdarm/Ileum ileum
Hüftgelenk (Vertebraten) hip joint, coxal joint, femoral articulation
Hüftgelenk/Coxa (Arthropoden) hip joint, coxa
Hüftgelenkpfanne/Hüftpfanne/ Acetabulum (Vertebraten) cotyloid cavity
Hüfthöcker/Tuber coxae coxal tuber, point of hip
Huftier hoofed animal/mammal, ungulate
Hüftknochen/Hüftbein/Os coxae (Beckenhälfte: Darmbein & Sitzbein & Schambein) hipbone, innominate bone (lateral half of pelvis)
Hüftmünder/Merostomata merostomes, merostomates
Hüftpfanne/Hüftgelenkpfanne/ Acetabulum (Vertebraten) cotyloid cavity
Hufwand/Hufwall/Paries corneus hoof wall, wall of hoof
Hügel hill
- **kleiner H.** mound, knoll, hummock (rounded knoll)
Hügelbeet ridge (ridge bed)
Hügelbeetkultur ridging
hügelig (leicht hügelige Landschaft) hilly (sloping terrain/rolling hills)

Humus

Hügelland hill country, hilly terrain, rolling countryside
Hügelstufe/Hügellandstufe/kolline Stufe/Vorgebirge foothills, foothill zone
Huhn chicken, hen (Henne)
Hühnchen young chicken, young hen, pullet
- **Brathühnchen** broiler (chicken)
hühnerartig gallinaceous
Hühnerstall chicken coop
Hühnervögel/Galliformes gallinaceous birds, fowl-like birds
Hüllblätter/Blumenhüllblätter sepals, calyx
Hüllblattkreis/Hüllkelch/Involukrum (Infloreszenz) involucre
Hülle/Häutchen/Tunika (Gewebeschicht) tunic
Hülle/Involukrum *bot* envelope, hull, involucre
Hülle/Mantel body covering, vesture, vestiture
- **Bakterienhülle** bacterial envelope
- **Virenhülle** viral envelope
Hülle (z. B. Wasser) envelope, jacket
Hüllfrucht/Cystokarp *fung* cystocarp, cystocarpium
Hüllglockenlarve/Pericalymma (Yoldia) pericalymma larva (lecithotroph), test-cell larva (trochophore of *Yoldia*)
Hüllkelch/Hüllblattkreis/Involukrum (Compositen) involucre
Hüllprotein coat protein
Hüllspelze *bot* glume
Hüllzelle (Dicyemida) *zool* jacket cell
Hülse *bot* (Fruchtform) legume, pod
Hülsenfrüchtler *allg* legume, leguminous plant
Hülsenfrüchtler/ Hülsenfruchtgewächse/ Schmetterlingsblütler/Fabaceae/ Papilionaceae (Leguminosae) pea family, bean family, legume family, pulse family
Humanbiologie human biology
Humangenetik/Anthropogenetik human genetics
Humanökologie human ecology
Humeralflügel *orn* humeral feathers, humerals, tertiaries, tertial feathers
Humeralqueader *entom* humeral cross-vein
humifizieren humify
Humifizierung/Humifikation/ Humusbildung humification
Huminsäure humic acid
Huminstoffe humic substances
Humus humus

Humusabbau/Humusdegradation humus degradation
Humusauflage humus layer
Hund dog
• **Hündin** female dog
• **Rüde** male dog
Hundemeute kennel, pack of dogs
Hundepension/Hundeheim (*siehe auch:* Tierheim) dog kennel
Hundertfüßer/Chilopoden centipedes, chilopodians
Hundespulwurm (*Toxocara canis*) canine ascarid
Hundezwinger (staatl. Tierheim für verwaiste Tiere) dog pound
Hundsgiftgewächse/ Immergrüngewächse/ Apocynaceae periwinkle family, dogbane family
Hundskohlgewächse/Theligonaceae theliogonum family
Hunger hunger
hungern *vb micb* starve
Hungern *n micb* starvation
hungrig hungry
Huperziaceae/ Teufelsklauengewächse fir clubmoss family
hüpfen hop, jump, skip, leap
husten *vb* cough
Husten *n* cough
Hut hat, cap
• **Pilzhut** cap, pileus
hutförmig/konsolenförmig/pileat cap-shaped, pileate, pileiform
Hyacinthaceae/ Hyazinthengewächse hyacinth family
Hyalinzelle/Hyalocyt hyaline cell
Hyaluronsäure hyaluronic acid
hybrid/durch Kreuzung erzeugt hybrid, crossbred
Hybrid-DNA hybrid DNA, chimeric DNA
Hybrid-Freisetzungstranslation hybrid-release translation (HRT)
hybridarretierte Translation hybrid-arrested translation (HART)
Hybride hybrid, crossbreed
hybridisieren hybridize
Hybridisierung/Bastardisierung hybridization, bastardization
• **CISS (chromosomale in-situ Suppressionshybridisierung)** chromosomal in situ suppression hybridization
• **DNA-getriebene Hybridisierung** DNA-driven hybridization
• **Erschöpfungshybridisierung** exhaustion hybridization
• *in situ* **Hybridisierung** *in situ* hybridization
• **Kreuzhybridisierung** cross hybridization
• **RNA-getriebene Hybridisierung** RNA-driven hybridization
• **Sandwich-Hybridisierung** sandwich hybridization
• **Sättigungshybridisierung** saturation hybridization
• **vergleichende Genomhybridisierung (CGH)** comparative genome hybridization (CGH)
Hybridisierungszone/ Bastardisierungszone hybid zone
Hybridom hybridoma
Hybridschwarm/Bastardschwarm (Bastardpopulation) hybrid swarm
Hybridsterilität/Bastardsterilität hybrid sterility
Hybridzelle hybrid cell
Hydathode/Wasserspalte hydathode, water pore, water stoma
Hydatide hydatid
Hydnoraceae/ Lederblumengewächse hydnora family
Hydranth (Cnidaria) hydranth
Hydrat hydrate
Hydratation/Hydratisierung/ Solvation (Wassereinlagerung/ Wasseranlagerung) hydration, solvation
Hydrathülle/Wasserhülle/ Hydratationsschale hydration shell
Hydratwasser water of hydration
hydrieren/hydrogenieren hydrogenate
Hydrierung (Wasserstoffanlagerung) hydrogenation
hydrisch hydric
Hydrocharitaceae/ Froschbissgewächse frog-bit family, tape-grass family, elodea family
Hydrocoel hydrocoel, hydrocoele
Hydrocotylaceae/ Wassernabelgewächse pennywort family
Hydrokultur hydroponics (soil-less culture/solution culture)
Hydrologie hydrology
Hydrolyse/Wasserspaltung hydrolysis
hydrolytisch/wasserspaltend hydrolytic
hydrophil (wasseranziehend/ wasserlöslich) hydrophilic (water-attracting/water-soluble)
Hydrophilie (Wasserlöslichkeit) hydrophilicity (water-attraction/ water-solubility)

531 Hypophysentasche H

hydrophob (wasserabweisend/ wasserabstoßend/wasserunlöslich) hydrophobic (water-repelling/water-insoluble)
hydrophobe Bindung hydrophobic bond
Hydrophobie (Wasserabweisung/ Wasserunlöslichkeit) hydrophobicity (water-insolubility)
Hydrophyllaceae/ Wasserblattgewächse waterleaf family
Hydrophyt/Wasserpflanze hydrophyte, aquatic plant
Hydroskelett/hydrostatisches Skelett hydrostatic skeleton
Hydrosphäre/Wasserhülle hydrosphere
Hydrostachyaceae/ Wasserröhrengewächse hydrostachys family
hydrostatischer Druck hydrostatic pressure
Hydrotaxis hydrotaxis
hydrothermaler Schlot hydrothermal vent
Hydroxyapatit hydroxyapatite
Hydroxylierung hydroxylation
Hydroxyprolin hydroxyproline
Hydrozoen/Hydroidea hydrozoans, hydra-like animals, hydroids
Hygiene hygiene
hygienisch hygienic
Hygrangeaceae/ Hortensiengewächse hydrangea family
Hygrophyt (an feuchten Standorten) hygrophyte
hygroskopisch hygroscopic
Hymenophyllaceae/ Hautfarngewächse/ Schleierfarngewächse filmy fern family
Hyoidbogen/Zungenbeinbogen (Gesamtheit der Teile) hyoid arch; (nur Knorpelspange) hyoid bar (skeleton only)
Hyoideum/Zungenbein/ Os hyoideum hyoid bone, lingual bone
Hyolithen/Hyolithida hyolithids
Hypanthium hypanthium
hypaxionisch (Rumpfmuskulatur) hypaxial
Hyperämie hyperemia
Hyperchromasie hyperchromasia, hyperchromia, hyperchromatism
Hyperchromie hyperchromicity, hyperchromism
Hyperchromizität hyperchromicity, hyperchromic effect, hyperchromic shift

Hyperglykämie hyperglycemia
hyperglykämisch hyperglycemic
Hypericaceae/Clusiaceae/ Guttiferae/ Johanniskrautgewächse/ Hartheugewächse St. John's wort family, mamey family, mangosteen family, clusia family
Hyperkalzämie hypercalcemia
Hyperkapnie hypercapnia
Hypermorphose hypermorphosis
Hypernatriämie hypernatremia
Hyperparasit hyperparasite
Hyperphagie/Esssucht/Fresssucht/ Gefräßigkeit hyperphagia
hyperploid hyperploid
Hyperploidie hyperploidy
Hyperpnoe hyperpnea
Hyperpolarisierung hyperpolarization
Hypersensibilität/Allergie hypersensitivity, allergy
Hypertonie hypertonicity, hypertonia
hyperton(isch) hypertonic
hypertroph hypertrophic
Hypertrophie hypertrophy
hypervariable Region (Ig) *immun* hypervariable region
Hyphe hypha (*pl* hyphas/hyphae)
Hypnospore/Dauerspore hypnospore, persistent spore, dormant spore, resting spore
Hypnozygote/Zygospore zygospore
Hypoblast epiblast
Hypobranchialrinne/Endostyl hypobranchial furrow/groove, endostyle
hypocerk/hypozerk hypocercal
Hypodermis (Epidermis einiger Wirbelloser) epidermis
hypogäisch hypogeous, hypogean, hypogeal
hypogäische Keimung hypogean/ hypogeal germination
Hypoglykämie hypoglycemia
hypoglykämisch hypoglycemic
hypognath hypognathous
Hypokotyl hypocotyl
Hypokotylknolle (unterirdische) corm (swollen shoot base)
Hypolepidaceae (inkl. Adlerfarn) hypolepis family (incl. bracken fern)
Hypolimnion hypolimnion
Hypophyse/Hirnanhangdrüse hypophysis, pituitary, pituitary gland
Hypophysenhinterlappen/ Neurohypophyse neurohypophysis, posterior lobe of pituitary gland
Hypophysentasche/Rathkesche Tasche hypophyseal pouch/sac, Rathke's pouch

Hypophysenvorderlappen/Adenohypophyse adenohypophysis, anterior lobe of pituitary gland
hypoploid hypoploid
Hypoploidie hypoploidy
hypopneustisch hypopneustic
hyporheisch hyporheic
Hypostasie hypostasis
Hypostracum/Perlmutterschicht hypostracum, nacreous layer
hypothermisch hypothermic
Hypothese hypothesis
hypothetisch hypothetic, hypothetical

Hypotonie hypotonicity, hypotonia
hypoton(isch) hypotonic
hypotroph hypotrophic
Hypotrophie hypotrophy
Hypoxie/Sauerstoffmangel hypoxia
hypoxisch hypoxic
hypsodont/hypselodont hypsodont, hypselodont (high crowns/short roots)
Hypurale *ichth* hypural (fused hemal spines)
Hysterese hysteresis
Hysterotelie *entom* hysterotely
Hysterothecium (Pilze/Flechten) hysterothecium

533   **Immunelektrophorese**   I

I-Bande (Muskel: *isotrop*) I band
Ichnofossil/Spurenfossil ichnofossil,
  trace fossil
Ichnologie/Spurenkunde
  ichnology
identisch identical
identisch aufgrund gemeinsamer
  Abstammung
  identity by desecent (IBD)
identisch aufgrund von Zufällen
  identity by state (IBS)
Idioblast idioblast
Idiophase (Produktionsphase)
  idiophase
Idioplasma/Keimplasma idioplasm,
  germ plasm, gonoplasm
Idiotop idiotope
igelborstig echinate
Igelkolbengewächse/Sparganiaceae
  bur-reed family
Igelwürmer/Stachelschwänze/
  Echiuriden/Echiura spoon worms,
  echiuroid worms
ikosaedrisch *vir* icosahedral
illegitime Rekombination *gen*
  illegitimate recombination
Illiciaceae/Sternanisgewächse
  star-anise family, illicium family
Imaginalanlage imaginal anlage
Imaginalring imaginal ring
Imaginalscheibe imaginal disk,
  imaginal bud
Imago (*pl* Imagines)/Vollinsekt/
  Adultinsekt
  imago (*pl* imagoes/imagines)
imbibieren/hydratieren imbibe,
  hydrate
Imbibition/Hydratation imbibition,
  hydration
Imidazol imidazole
Iminosäure imino acid
Imker/Bienenzüchter beekeeper,
  apiarist
Imkerei/Bienenzucht beekeeping,
  apiculture
Imkerei/Bienenzuchtbetrieb apiary
Imme/Biene bee
immergrün evergreen
Immergrüngewächse/
  Hundsgiftgewächse/Apocynaceae
  periwinkle family, dogbane family
Immigration immigration
Immission/Einwirkung immission,
  injection, admission, introduction
Immission (Belastung durch
  Luftschadstoffe)
  exposure level of air pollutants
immobil/fixiert/bewegungslos
  immobile, fixed, motionless
Immobilisation immobilization
immobilisieren immobilize (to make
  immobile)

Immobilität/Bewegungslosigkeit
  immobility, motionlessness
immortalisierte Zelle
  immortalized cell
immun immune
Immun-Elektronenmikroskopie
  (IEM)
  immunoelectron microscopy (IEM)
Immunadhärenz immune adherence
Immunadsorptionstest,
  enzymgekoppelter (ELISA)
  enzyme-linked immunosorbent assay
  (ELISA)
Immunaffinitätschromatographie
  immunoaffinity chromatography
Immunantwort immune response
  ● **sekundäre I./Sekundärantwort**
  secondary immune response,
  anamnestic response
  ● **zellvermittelte I.** cell-mediated
  immune response
Immundefekt immune deficiency
  ● **erworbenes**
  Immunschwächesyndrom
  acquired immune deficiency
  syndrome (AIDS)
  ● **schwerer kombinierter**
  Immundefekt severe combined
  immune deficiency (SCID)
Immundiffusion immunodiffusion
  ● **doppelte radiale**
  Immundiffusion (Ouchterlony-
  Methode) double radial
  immunodiffusion (DRI)
  (Ouchterlony technique)
  ● **Doppelimmundiffusion** double
  diffusion, double immunodiffusion
  ● **einfache Immundiffusion/**
  lineare Immundiffusion
  (Oudin-Methode) single
  immunodiffusion (Oudin test)
  ● **einfache radiale Immundiffusion**
  (Mancini-Methode) single radial
  immunodiffusion (SRI)
  (Mancini technique)
  ● **Identität** identity
  ● **radiale Immundiffusion**
  radial immunodiffusion (RID)
  ● **Teilidentität/partielle**
  Übereinstimmung partial identity
  ● **Verschiedenheit (Nicht-Identität)**
  nonidentity
Immunelektrophorese
  immunoelectrophoresis
  ● **Tandem-**
  Kreuzimmunelektrophorese
  charge-shift
  immunoelectrophoresis
  ● **Kreuzimmunelektrophorese**
  crossed immunoelectrophoresis,
  two-dimensional
  immunoelectrophoresis

Immunerkennung

- **Linienimmunelektrophorese**
  immunoelectrophoresis
- **Raketenimmunelektrophorese**
  rocket immunoelectrophoresis
- **Überwanderungs-immunelektrophorese/ Überwanderungselektrophorese**
  countercurrent
  immunoelectrophoresis,
  counterelectrophoresis

Immunerkennung
immune recognition

Immunfluoreszenz
immunofluorescence

Immunfluoreszenzchromatographie
immunofluorescence
chromatography

Immunfluoreszenzmikroskopie
immunofluorescence microscopy

Immungenetik immunogenetics

Immunglobulin immunoglobulin

Immunglobulinfaltung
immunoglobulin fold

immunisieren/impfen immunize,
vaccinate

Immunisierung/Impfung
immunization, vaccination

Immunisierungsstärke/ Immunogenität immunogenicity

Immunität immunity
- **begleitende I./Prämunität**
  concomitant immunity,
  premunition
- **erworbene I. (aktive/passive)**
  acquired immunity,
  adaptive immunity (active/passive)
- **Kreuzimmunität/ übergreifender Schutz**
  cross protection
- **künstliche I.** artificial immunity
- **natürliche I.** natural immunity
- **passive I.** passive immunity
- **zelluläre I.** cellular immunity

Immunitätsregion immunity
region

Immunkompetenz
immunocompetence,
immunologic competence

Immunkomplex immune complex

Immunkrankheit/Immunopathie
immunopathy

Immunoassay immunoassay

Immunoblot/Western-Blot
immunoblot, Western blot

immunogen adv/adj immunogenic

Immunogen n immunogen

Immunogenität/ Immunisierungsstärke
immunogenicity

Immunogold-Silberfärbung (IGSS)
immunogold-silver staining (IGSS)

Immunologie immunology

immunologisch immunologic(al)

immunologische Überwachung/ Immunüberwachung
immunosurveillance,
immunological surveillance

immunologisches Gedächtnis
immunological memory

immunoradiometrischer Assay
immunoradiometric assay (IRMA)

Immunpräzipitation
immunoprecipitation

Immunprophylaxe
immunoprophylaxis

Immunreaktion immune reaction

Immunschwäche immune
deficiency, immunodeficiency

Immunschwächesyndrom/ Immunmangel-Syndrom
immune deficiency syndrome
- **erworbenes Immunschwächesyndrom**
  acquired immune deficiency
  syndrome (AIDS)
- **schwerer kombinierter Immundefekt** severe combined
  immune deficiency (SCID)

Immunscreening immunoscreening

Immunsuppression
immunosuppression,
immune suppression

Immuntoleranz immune tolerance,
immunological tolerance

Immunüberwachung/ immunologische Überwachung
immunosurveillance,
immunologic(al) surveillance

impermeabel/undurchlässig
impermeable, impervious

Impermeabilität/Undurchlässigkeit
impermeability, imperviousness

Impfdraht inoculating wire

impfen med inoculate, vaccinate

impfen micb inoculate, seed

Impfen/Impfung/Vakzination (Immunisierung) inoculation,
vaccination

Impfnadel inoculating needle

Impföse inoculating loop

Impfstoff/Inokulum/Inokulat/ Vakzine
inoculum, vaccine
- **abgeschwächte(r)/attenuierte(r) Impfstoff/Vakzine**
  attenuated vaccine
- **Autoimpfstoff/Autovakzine**
  autogenous vaccine
- **heterologer Impfstoff/ heterologe Vakzine**
  heterologous vaccine
- **inaktivierter Impfstoff/ inaktivierte Vakzine**
  inactivated vaccine

535   infizieren   **I**

- **Kombinationsimpfstoff/ Mischimpfstoff/Mischvakzine** combination vaccine, mixed vaccine
- **Komponentenimpfstoff/ Spaltimpfstoff/Spaltvakzine/ Subunitimpfstoff/ Subunitvakzine** subunit vaccine
- **Konjugatimpfstoff/ zusammengesetzte Vakzine** conjugate vaccine
- **Lebendimpfstoff/Lebendvakzine** live vaccine
- **Mischimpfstoff/Mischvakzine/ Kombinationsimpfstoff** mixed vaccine, combination vaccine
- **polyvalenter Impfstoff/ polyvalente Vakzine** polyvalent vaccine
- **Spaltimpfstoff/Spaltvakzine/ Subunitimpfstoff/ Subunitvakzine** split-protein vaccine, SP vaccine, subunit vaccine
- **Totimpfstoff/Totvakzine** killed vaccine
- **Toxoidimpfstoff/Toxoidvakzine** toxoid vaccine
- **zusammengesetzter Impfstoff/ Konjugatvakzine** conjugate vaccine

Impfung/Immunisierung vaccination, immunization

Impfung/Inokulation/Vakzination (Immunisierung) inoculation, vaccination (immunization)

imponieren/Eindruck machen impress, be impressive

Imponierverhalten/ Imponiergehabe/ Imponiergebaren display behavior

in Blüte in bloom, in blossom

in sich gefaltete DNA/ zurückgebogene DNA fold-back DNA, snap-back DNA

in situ-Hybridisierung *in situ* hybridization
- **FISH (*in situ* Hybridisierung mit Fluoreszenzfarbstoffen)** FISH (fluorescence activated *in situ* hybridization)

in vitro-Mutagenese *in vitro* mutagenesis

in vitro-Verpackung *in vitro* packaging

in Windrichtung leeward

inaktiv inactive

indifferente Art indifferent species

Indigen indigenous species, native species/organism/lifeform

Indikan/Indoxylsulfat indican, indoxyl sulfate

Indikatorpflanze/Anzeigerpflanze indicator plant

Indikatororganismus/Indikatorart/ Bioindikator bioindicator

indirekte Endmarkierung *gen* indirect end-labeling

indirekte Sequenzwiederholungen *gen* indirect repeats

individuell individual(ly)

Individuum individual

Indolessigsäure indolyl acetic acid, indoleacetic acid (IAA)

Induktion induction

Induktor inducer
- **freiwilliger Induktor** gratuitous inducer

Indusium/Schleierchen indusium

Industriemelanismus industrial melanism

induzierbar inducible

induzieren induce

induzierte Anpassung/ induzierte Passform induced fit

induzierte Passform induced fit

ineinandergesetzte Gene/ ineinandergeschachtelte Gene nested genes

Infektion/Ansteckung infection
- **Agroinfektion** agroinfection
- **abortive Infektion** abortive infection
- **anhaltende/ persistierende Infektion** persisting Infektion
- **Doppelinfektion** double infection
- **latente Infektion** latent infection
- **lytische Infektion** lytic infection
- **produktive Infektion** productive infection
- **stumme Infektion/stille Feiung** silent infection
- **Superinfektion/Überinfektion** superinfection
- **unvollständige Infektion** incomplete infection

Infektionsdosis infectious dose ($ID_{50} = 50\%$ infectious dose)

Infektionskrankheit infectious disease

Infektionsmultiplizität multiplicity of infection

Infektionsvermögen/ Ansteckungsfähigkeit infectivity

infektiös/ansteckend infectious

infektiöser Abfall infectious waste

Inferenz inference
- **statistische Inferenz** statistical inference

infizieren/anstecken infect

I

Infloreszenz 536

Infloreszenz inflorescence,
flower cluster
- **Ähre** spike, spica
- **Büschel/Faszikel/Faszikulus**
cyme with very short pedicles,
fascicle
- **Cyme/Cymus/Zyma/Zyme/
Zymus/cymöser Blütenstand**
cyme, cymose inflorescence
- **Ebenstrauß/Corymbus (inkl.
Schirmrispe und Schirmtraube)**
corymb
- **Fächel** rhipidium (fan-shaped
cyme)
- **geschlossene Infloreszenz**
determinate inflorescence
- **Knäuel**
cyme with sessile flowers
- **Korb/Körbchen/Köpfchen/
Capitulum/Cephalium** capitulum,
cephalium, flower head
- **offene Infloreszenz**
indeterminate inflorescence
- **Rispe/Blütenrispe** panicle
- **Rumpfinfloreszenz**
truncate synflorescence
- **Scheindolde/Trugdolde/Cyme/
Zymus** cyme
- **Scheindolde/Pseudosciadioid**
contracted cymoid, cymose umbel,
pseudosciadioid
- **Schirmtraube
(ein Ebenstrauß/Corymbus)**
umbel-like raceme
- **Schirmrispe (ein Ebenstrauß/
Corymbus)**
umbel-like panicle
- **Schraubel (cymöse Infloreszenz)**
bostryx (helicoid cyme)
- **Sichel/Drepanium** drepanium
(a helicoid cyme)
- **Spirre/Trichterrispe** anthela
- **Teilinfloreszenz/Teilblütenstand**
partial inflorescence
- **Traube/Botrys** raceme, botrys
- **Trichterrispe/Spirre** anthela
- **Trugdolde/Scheindolde/Cymus/
Zymus/Cyme** cyme
- **>eingablige Trugdolde**
simple cyme, monochasium
- **>zweigablige Trugdolde**
compound cyme, dichasial cyme,
dichasium
- **Wickel (cymöse Infloreszenz)**
cincinnus (scorpioid cyme)
- **Zyme/Zyma/Zymus/Cymus/
Cyme/cymöser Blütenstand**
cyme, cymose inflorescence
Infloreszenz-Kurztrieb spur shoot
Influent (*pl* Influenten) *ecol*
influent
Infralitoral infralitoral

**Infrarot-Spektroskopie/
IR-Spektroskopie**
infrared spectroscopy
infusiform infusiform
Infusorigen infusorigen
**Ingerartige/Inger/Schleimaale/
Myxiniformes (bzw. Myxinida)**
hagfishes
**Inguinaltasche/Sinus inguinalis
(Schaf)**
inguinal sinus, inguinal pouch
Ingwergewächse/Zingiberaceae
ginger family
inhibitorisch/hemmend inhibitory
**inhibitorisches postsynaptisches
Potential** inhibitory postsynaptic
potential (IPSP)
**Initiale/Stammzelle
(Primordialzelle/Primane)**
initial, stem cell (primordial cell)
- **dreischneidige**
initial with three cutting faces
- **zweischneidige**
initial with two cutting faces
**Initialsegment (myelinisierte
Fasern)** initial segment
Initiationsfaktor initiation factor
Initiationskomplex
initiation complex
Inititationscodon/Startcodon *gen*
initiation codon
**Injektion/Spritze
(eine I./S. geben/bekommen)**
injection, shot
injizieren/spritzen inject, shoot
Injunktion injunction
Inkohlung *paleo/geol* carbonization,
coalification
inkompatibel incompatible
Inkompatibilität incompatibility
Inkompatibilitätsgruppe
incompatibility group
Inkrustierung incrustation,
encrustation
Inkubation (Bebrütung/Bebrüten)
incubation
Inkubationszeit incubation period
inkubieren/brood/breed incubate,
brüten, bebrüten
Innenhaut/Endodermis
endodermis
Innenlade/Lacinia
lacinia, inner lobe of maxilla
Innenohr inner ear
Innenparasit/Endoparasit
endoparasite
Innenschicht inner layer, interior
layer; (Pollen/Spore: Intine) intine
Innenskelett/Endoskelett
internal skeleton, endoskeleton
innerartlich intraspecific
innere Zellmasse inner cell mass

Innereien/Eingeweide
entrails, innards, viscera, guts
(fish viscera etc.)
**Innereien/Eingeweide (von
Schlachttieren: Schweine/Rinder)**
pluck
**Innereien (essbare Gedärme des
Schweins)** chitterlings, chitlins
(pork intestines)
**Innereien (essbare Organe des
Geflügels)**
giblets (edible viscera of fowl)
innerlich/von innen/intern internal,
intrinsic
Innervation/Innervierung innervation
innervieren innervate
Inokulation/Einimpfung/Impfung
inoculation
inokulieren/einimpfen/impfen
inoculate
Inosin inosine
Inosinmonophosphat (IMP)
inosine monophosphate,
inosinic acid
Inosintriphosphat (ITP)
inosine triphosphate
Inosit/Inositol inositol
inotrop inotropic
Inquilinismus/Einmietung/Synökie
inquilinism
Insekt (*pl* Insekten) insect
● **geflügelte Insekten/
Fluginsekten/Pterygota**
winged insects, pterygote insects
● **ungeflügelte Insekten/
Apterygota** wingless insects
**Insektenbekämpfungsmittel/
Insektizid** insecticide
**Insektenbestäubung/
Insektenblütigkeit/Entomophilie**
insect pollination, entomophily
Insektenblume/Entomophile insect-
pollinated flower, entomophile
Insektenfalle insect-trap
insektenfressend/insektivor
insectivorous
Insektenfresser/Insectivoren
insectivores
Insektenkunde/Entomologie
entomology
Insektenplage insect pest
**Insektenvernichtungsmittel/
Insektizid** insecticide
Insel *biogeo/evol* island
Insel/Inselfeld *neuro* insula
Inselbiogeographie
island biogeography
Inselchen/kleine Insel islet
Inselökologie island ecology
**Inselorgan/Langerhanssche Insel/
Pankreasinsel** islet organ,
islet of Langerhans, pancreatic islet

Inseltheorie
theory of island biogeography,
MacArthur-Wilson theory
inserieren (inseriert) insert
(inserted)
Insertion *gen* insertion
Insertionsaktivierung
insertional activation
Insertionsinaktivierung
insertional inactivation
Insertionsmutation
insertion mutation
Insertionssequenz *gen*
insertion sequence
Insertionsvektor insertion vector
**Inside-out Vesikel (Vesikel mit der
Innenseite nach außen)**
inside-out vesicle
Inspiration/Einatmen inspiration
inspirieren/einatmen inspire
instabil unstable (instable)
instabile Mutation unstable mutation
Instinkt instinct
● **Heimkehrvermögen/
Heimfindevermögen/Zielflug**
homing instinct
● **Herdeninstinkt/Herdentrieb**
herd instinct, herding instinct
● **Sexualtrieb/Geschlechtstrieb**
sexual instinct, life instinct, eros
● **Todestrieb** death instinct,
aggressive instinct
● **Verschränkung**
interlocking (instinct)
Instinkt-Dressur-Verschränkung
instinct-training-interlocking
instinktiv instinctive, by instinct
Instinktverhalten/Triebverhalten
instinctive behavior,
instinct behavior
Instinktverschränkung
instinct interlocking
**integrale Proteine (intrinsische
Proteine)** integral proteins
(intrinsic proteins)
integrales Membranprotein
integral membrane protein
**integratives Hefeplasmid (YIp)
(Hefevektor)**
yeast integrative plasmid (YIp)
**integrierte Schädlingsbekämpfung/
integrierter Pflanzenschutz**
integrated pest management (IPM)
Integrin integrin
**Integument/Decke/Hülle (z. B.
Körperdecke/Haut)** integument,
covering (e. g. body covering/skin)
Interaktion interaction
● **allosterische Interaktion**
allosteric interaction
Interaktionsvarianz
interaction variance

interdisziplinäre Forschung

interdisziplinäre Forschung
  interdisciplinary research
Interferenz-Mikroskopie
  interference microscopy
Interferenzassay interference assay
Interferon interferon
intergene Region intercistronic
  region, intergenic region
interkalar/eingeschoben
  intercalary (inserted between others)
Interkalarader/Intercalarader *entom*
  intercalary vein
interkalares Meristem
  intercalary meristem
Interkalation intercalation
interkalierendes Agens
  intercalation agent,
  intercalating agent
intermediäres Filament/
Intermediärfilament
  intermediate filament
Intermembranraum
  intermembrane space
Internationale Maßeinheit/
SI Einheit International Unit (IU),
  SI unit (*fr:* Système Internationale)
internationales
Maßeinheitensystem/
SI Einheitensystem international
  unit system, SI unit system
  (*fr:* Système Internationale)
Internodium/Zwischenknoten *bot*
  internode
Interphase interphase
interpolieren interpolate
Interradie/Bivium (Holothurien)
  bivium
interspezifisch/zwischenartlich
  interspecific
Interstitialfauna/Sandlückenfauna
(Meiofauna)
  interstitial fauna (meiofauna)
Interstitialflüssigkeit/
interstitielle Flüssigkeit
  interstitial fluid (ISF), tissue fluid
Interstitialraum/
(Gewebs)Zwischenraum/
Interstitium interstitial space,
  interstice (*pl* interstices)
Interstitialzelle/Zwischenzelle
  interstitial cell
interstitiell interstitial
interstitielle Region
  interstitial region
Intervall interval
Intervallskala *stat* interval scale
intervenierende Sequenz/
dazwischenliegende Sequenz/
Intron *gen*
  intervening sequence, intron
Interzellulare/Zwischenzellraum
  intercellular space

interzellulär intercellular
interzelluläre Verbindung/
interzelluläre Junktion
  intercellular junction
Intine *bot* intine
intraallele Komplementation
  intraallelic complementation
intrachromosomale Umordnung
  intrachange, intrachromosomal
  recombination
Intrafusalfaser intrafusal fiber
intragene Komplementation
  intragenic complementation
Intrakörper
(intrazellulärer Antikörper)
  intrabody (intracellular antibody)
Intramembran-Partikel
  intramembrane particle,
  membrane intercalated particle
intraspezifisch/innerartlich
  intraspecific
intrazellulär intracellular
Intrinsic-Faktor/
hämopoetischer Faktor
  intrinsic factor, hemopoietic factor
intrinsisch intrinsic, intrinsical
Introgression introgression
Intron/intervenierende Sequenz/
dazwischenliegende Sequenz *gen*
  intron, intervening sequence
intrors introrse
Introvert introvert
Intussuszeption intussusception
Invagination/Einstülpung/
Einfaltung/Embolie invagination,
  emboly
Invasivität invasiveness
Inventar inventory
invers inverted
inverse Polymerasekettenreaktion
gen
  inverse polymerase chain reaction,
  inverse PCR
Inversion inversion
  • **parazentrische Inversion** *gen*
    paracentric inversion
Inversionsmutation
  inversion mutation
Invertebraten/Evertebraten/
Wirbellose invertebrates
invertierte Sequenzwiederholung/
gegenläufige -/umgekehrte
Sequenzwiederholung *gen*
  inverted repeat, inverted repetition
Invertzucker invert sugar
Involukralblatt/Involukralschuppe
  phyllary, involucral bract
Involukrum/Hülle (siehe: Hüllkelch)
  involucre
involutiv (Blatt-/Knospenlage: nach
oben eingerollte Spreitenflügel)
  involute, rolled inward

Inzest incest
Inzucht/Reinzucht inbreeding, endogamy
Inzucht betreiben inbreed
Inzuchtlinie inbred line
Inzuchtstamm inbred strain
Ionenaustaucher ion exchanger
Ionenaustauscherharz ion-exchange resin
Ionenbindung ionic bond
Ionengleichgewicht ion equilibrium, ionic steady state
Ionenkanal (Membrankanal) ion channel (membrane channel)
Ionenkopplung ionic coupling
Ionenleitfähigkeit ionic conductivity
Ionenpaar ion pair
Ionenpore ion pore
Ionenprodukt ion product
Ionenpumpe ion pump
Ionenradius ionic radius
Ionenschleuse gated ion channel
Ionenstärke ionic strength
Ionenstrom ionic current
Ionentransport ion transport
Ionisation ionization
ionisch ionic
ionisieren ionize
ionisierende Strahlen/ ionisierende Strahlung ionizing radiation
Ionophor ionophore
Ionophorese/Iontophorese ionophoresis
Iridaceae/Schwertliliengewächse iris family
Iridocyt/Iridozyt/Flitterzelle/ Leucophor/Guanophor iridocyte, iridophore, leucophore, guanophore
Irisblende *micros* iris diaphragm
IRMA (immunoradiometrischer Assay) immunoradiometric assay (IRMA)
IRP (inselspezifische PCR) *gen* IRP (island rescue PCR)
irreversibel irreversible
isabellfarben (gelb-olivbraun) isabelline
Ischämie ischemia
Ischiopodit ischiopodite
Isidie isidium
Isoakzeptoren isoacceptors
isoelektrische Fokussierung/ Isoelektrofokussierung isoelectric focus(s)ing

isoelektrischer Punkt isoelectric point
Isoetaceae/Brachsenkrautgewächse quillwort family
isogam isogamous
Isogamie isogamy
Isolationsmechanismus *ecol* isolating mechanism
Isolationsmedium *micb* isolation medium
Isoleucin isoleucine
isolezithal/isolecithal (mit gleichmäßig verteiltem Dotter) isolecithal
isolieren/abtrennen isolate, separate
isomer *adv/adj* isomeric
Isomer *n* isomer
Isomeratzucker/Isomerose high fructose corn syrup
Isomerie isomerism, isomery
Isomerisation isomerization
isomerisieren isomerize
Isophän *nt* isophene
isopyknische Zentrifugation isopycnic centrifugation
isosmotisch/isoosmotisch isosmotic, iso-osmotic
isospor isosporous
Isosystem isosystem
Isotachophorese isotachophoresis
Isotherm isotherm
Isotonie isotonicity
isotonisch isotonic
Isotop isotope
Isotopenversuch isotope assay
Isotypus/Isotyp/Isostandard isotype
Isotypwechsel/Klassenwechsel isotype switching
Isovaleriansäure isovaleric acid
Isozönose isocoenosis (*pl* isocoenoses)
Isozym/Isoenzym isozyme, isoenzyme
Istwert actual value, effective value
iterative Evolution iterative evolution
Iteroparitie iteroparity
ITR (umgekehrte terminale Repetitionen) *gen* ITR (inverted terminal repetitions/ inverted terminal repeats)
IVS (intervenierende Sequenz/ dazwischenliegende Sequenz)/ Intron IVS (intervening sequence), intron

J Jackobslachs 540

Jackobslachs/Bartolomäuslachs grilse

Jacobsonsches Organ/ vomeronasales Organ Jacobson's organ, vomeronasal organ

Jagd (Raub) predation

Jagd/Jägerei hunt, hunting
- **Hetzjagd/Hatz** chase
- **Treibjagd** drive

Jagdbeute/verfolgtes Wild quarry, prey

Jagdgeflügel game birds (legally hunted)

Jagdgründe hunting range, hunting grounds, hunting territory

Jagdspiel mock-hunting

jagen hunt, prey

Jäger (Mensch) hunter

Jäger (Räuber) predator

Jägerei/Jagd hunt, hunting

Jahresrhythmus circannual rhythm

Jahresring annual ring, growth ring

Jahrestrieb annual shoot, one-year shoot, annual growth

Jahreswachstum annual growth

Jahreszeit season

Jahreszeitenwechsel seasonal change

jahreszeitlich/saisonal seasonal

Jahreszuwachs annual growth

Jährling/einjähriges Tier (meist Rinder) yearling
(short yearling: 9 to 12 months; long yearling: 12 to 18 months)

janusköpfiges Zwischenprodukt/ doppelköpfiges Zwischenprodukt double-headed intermediate

Jasmonsäure jasmonic acid

jäten weed

Jauche liquid manure (urine)

Jetztzeit/Holozän (erdgeschichtliche Epoche) Recent, Holocene, Recent Epoch, Holocene Epoch

Jochalgen/Conjugaten/Conjugatae/ Acontae/Zygnematophyceae (Conjugatophyceae) zygnematophycean algas

Jochbein/Jugale jugal (bone)

Jochblattgewächse/Zygophyllaceae caltrop family, creosote bush family

Jochbogen/Arcus zygomaticus zygomatic arch

Jochpilze/Zygomyceten zygospore fungi, bread molds, zygomycetes (coenocytic fungi)

Jod iodine

Jodessigsäure iodoacetic acid

jodieren (mit Jod/Jodsalzen versehen) iodize

Jodierung (mit Jod reagieren/ substituieren) iodination; (mit Jod/ Jodsalzen versehen) iodization

Jodzahl iodine number, iodine value

Johannisbrotgewächse/ Caesalpinogewächse/ Caesalpiniaceae caesalpinia family

Johannisbrotkernmehl/ Karobgummi locust bean gum, carob gum

Johanniskrautgewächse/ Hartheugewächse/Hypericaceae/ Clusiaceae/Guttiferae St. John's wort family, mamey family, mangosteen family, clusia family

Johannistrieb lammas shoot

Johnston's organ Johnstonsches Organ

Jordansches Organ/Chaetosoma/ Chaetosoma Jordan's organ, chaetosoma, chaetosoma

Jugalader jugal vein

Jugalfalte jugal fold

Jugalfeld/Jugum/Neala jugal field, jugal area, jugal region, neala

Jugalzelle jugal cell

Jugend (Jugendzeit/Jugendphase/ Jugendstadium) adolescence, juvenile stage, juvenile phase

Jugend/Jugendlichkeit juvenility

Jugendform juvenile form

Jugendgesang/Dichten (Jungvögel) juvenile song, subsong

Jugendstadium/Jugendphase/ Jugendzeit adolescence, juvenile stage, juvenile phase

Juglandaceae/Walnussgewächse walnut family

Jullienesches Organ organ of Jullien

Juncaceae/Simsengewächse rush family

Juncaginaceae/Dreizackgewächse arrowgrass family

junge Sprösslinge abfressendes Tier browser

Junge werfen bear young, litter

Junges (Nachkommen) young (offspring); see also: **Jungtier**

Jungfer virgin

Jungfernflug (Bienenkönigin) maiden flight (queen bee)

Jungfernfrüchtigkeit/ Parthenokarpie parthenocarpy

Jungfernhäutchen/Hymen hymen

Jungfernzeugung/Parthenogenese parthenogenesis

Jungfrau virgin

jungfräulich virginal, virgin

Jungfräulichkeit virginity

Junggeselle bachelor

Junglarve/Eilarve/Primärlarve primary larva

Jungpflanze young/juvenile plant

Jungsau gilt

Jungspinne spiderling

Juvenilhormon

Jungtier/Junges (v. a. Säuger)
young, pup (e. g. whale/seal/rat/dog),
cub (young carnivore: bear/fox/lion)
• **säugendes Jungtier** suckling
Jungvogel/Kücken/Küken squab,
chick
Jungwald/junger Wald
young forest
Junktion/Verbindung
junction (meeting point)
• **interzelluläre Junktion/
interzelluläre Verbindung**
intercellular junction
Junktionszone junction zone
• **dermo-epidermale J.**
dermo-epidermal junction

**Jura/Jurazeit (erdgeschichtliche
Periode)** Jurassic, Jurassic Period
**justieren/fokussieren
(Scharfeinstellung des
Mikroskops: fein/grob)** adjust,
focus (*fine/coarse*)
**Justierschraube/Justierknopf/
Triebknopf** *micros* adjustment
knob, focus adjustment knob
**Justierung/Fokussierung
(Scharfeinstellung des
Mikroskops: fein/grob)**
adjustment, focus adjustment,
focus (*fine/coarse*)
Juvenilhormon
juvenile hormone (JH)

K-Selektion

K-Selektion K selection
K-Stratege K strategist, K-selected species
Kabeltheorie cable theory
Kadaver/Tierleiche cadaver, carcass, corpse
Käfer/Coleoptera beetles
Käferblume/Coleopterophile/Cantharophile beetle-pollinated flower, coleopterophile, cantharophile
Käferschnecken/Placophora placophorans (incl. chitons)
Kaffeesäure caffeic acid
Käfig cage
• **kleiner Tierkäfig/kleiner Verschlag** hutch, pen, coop
kahl bare, barren; bald, glabrous
Kahlfraß (durch Schädlinge) complete defoliation (by pests)
Kahlhechte/Amiiformes (Schlammfisch) modern bowfin
Kahlschlag *for* clear-cut, clearing, clearance
kahlschlagen *for* clear-cutting, clear-felling, land clearing
Kahmhaut/Oberflächenhäutchen (auf Teich) scum, film (pond scum)
Kahnbein/Fußwurzelknochen/Os naviculare navicular bone
Kahnbein/Handwurzelknochen/Os scaphoideum scaphoid bone
kahnförmig/bootförmig/navikular navicular, scaphoid, cymbiform, resembling/having the shape of a boat
Kahnfüßer/Grabfüßer/Solenoconchae/Scaphopoden tooth shells, tusk shells, scaphopods, scaphopodians (spade-footed mollusks)
Kai wharf, quay
Kairomon kairomone
Kakaogewächse/Sterkuliengewächse/Sterculiaceae cacao family, cocoa family
Kakteen/Kaktusgewächse/Cactaceae cactus family
Kala-Azar/Cala-Azar/schwarzes Fieber/viszerale Leishmaniasis (*Leishmania donovani*) Cala-Azar, kala azar
Kalamitätennutzung (Holzernte) *for* salvage logging, salvage felling
Kalb/Jungtier calf
kalben *vb* calf
Kalben *n* calving
kalibrieren calibrate
Kalibrierung calibration
Kalium potassium
Kalk lime
Kalkablagerung lime(stone) deposit
Kalkalge calcareous alga
Kalkanreicherungshorizont/Caliche *geol* caliche, lime pan
Kalkdrüse calciferous gland
Kalkeinlagerung/Verkalkung/Calcifikation calcification
kalken lime, calcify
Kalkflieher *bot* calcifuge, basifuge
kalkig/kalkartig/kalkhaltig limy, limey, calcareous
Kalkkörper calcareous corpuscle, calcareous body
kalkliebend/kalziphil/kalzikol/kalkhold calciphile, calcicole
kalkmeidend/kalkfliehend/kalziphob/kalzifug calciphobe, calcifuge, basifuge
Kalkplättchen/Kalkkörperchen/Kokkolit/Coccolit coccolith
Kalkschale calcareous shell
Kalkschwämme/Calcarea calcareous sponges
Kalkstein limestone
Kalkung liming
Kallikrein kallikrein
Kallus/Callus callus
• **Wundkallus/Wundcallus/Wundgewebe/Wundholz** *bot* wound tissue, callus
Kallus-Kultur/Callus-Kultur callus culture
Kalmare/Teuthoidea (*bzw.* Teuthida) squids
Kalmen(gürtel) *meteo* doldrums
Kalorie calorie
Kalorimeter calorimeter
Kalorimetrie calorimetry
Kalotte/Schädelkappe/Clavarium skullcap
Kalottenmodel *chem* space-filling model
kälteempfindlich/kältesensitiv cold-sensitive
Kältepflanze/Kryophyt cryophyt, plant preferring low temperatures
Kälteresistenz cold resistance
Kälteschaden/Kälteschädigung chilling damage/injury
Kälteschock cold shock
Kälteschütte *bot/for* abscission of leaves due to chilling
Kältestarre/Winterstarre winter torpor
Kältetoleranz cold hardiness
Kältewüste cold desert
Kalthaus/Frigidarium (kühles Gewächshaus) cold house
kalzifug/calcifug/kalkmeidend calcifuge
kalzikol/calcicol/kalkhold calcicole
Kalzium/Calcium calcium

543 — Kapazitätsgrenze **K**

Kambium/Cambium cambium
(*siehe unter*: Cambium)
Kambrium/Cambrium
(erdgeschichtliche Periode)
Cambrian, Cambrian Period
Kamelhalsfliegen/Raphidioptera
snakeflies
Kamerunbeule (*Loa loa*)
African eyeworm disease, loa
Kamm comb, pecten
Kamm/Crista (Hahnenkamm) comb,
crest, ridge
• **Pollenkamm** pollen rake, pecten
kammartig/gekämmt comblike,
rakelike, ctenoid, ctenose, pectinate
Kammer/Fach *bot/zool* chamber,
valve, case
Kammer *electrophor* chamber
Kammerflattern *cardio*
ventricular flutter
Kammerflüssigkeit (Nautilus)
cameral fluid
Kammerpore, zuführende prosopyle
Kammerung/Fächerung/
Unterteilung
compartmentation,
compartmentalization,
septation, division
Kammerwasser/Humor aquaeus
(Auge) aqueous humor
Kammerzyklus *cardio*
ventricular cycle
kammförmig comb-shaped,
cteniform, pectiniform
Kammkieme/Fiederkieme/Ctenidie/
Ctenidium gill plume, gill comb,
ctenidium
Kammkiemer/Fiederkiemer/
Protobranchiata (Bivalvia)
protobranch bivalves
Kammlage (Berg/Gebirge)
along crest, ridge zone
Kammlinie (Berg/Gebirge)
(mountain) crest, ridge
Kammmuskel/Pektineus pectineus
Kammmünder/Ctenostomata
(Bryozoen) ctenostomates
Kammmuscheln/Pectinidae pen shells
Kammquallen/Rippenquallen/
Ctenophoren
sea gooseberries, sea combs, comb
jellies, sea walnuts, ctenophores
Kammschuppe/Ctenoidschuppe
ctenoid scale
Kammünder *siehe* Kammmünder
Kammuscheln *siehe* Kammmuscheln
Kammzelle comb cell
Kampfspiel play-flight(ing)
Kampfverhalten fighting behavior
kampylotrop/campylotrop
(Samenanlage)
campylotropous, bent

Kanal *neuro* **(Membrankanal)**
channel (membrane channel); (zum
Weiterleiten von Flüssigkeiten)
canal, duct, tube
• **Ionenkanal** ion channel
• **ligandenregulierter/**
ligandengesteuerter Kanal
ligand-gated channel
• **mechanisch gesteuerter Kanal**
mechanically gated channel
• **Ruhemembrankanal/Leckkanal**
resting channel, leakage channel
• **spannungsregulierter/**
spannungsgesteuerter Kanal
voltage-sensitive channel,
voltage-gated channel
Kanalisation (Abwasser) sewer
kanalisiertes Merkmal
canalized character
Kanalprotein/Tunnelprotein *neuro*
channel protein
Kanalstrom *neuro* channel current
Kanaltor *neuro* channel gate
Kaneelgewächse/Canellaceae
wild cinnamon family,
white cinnamon family,
canella family
kannenartig/krugartig/sackartig/
schlauchartig ascidiate
Kannenblatt/Schlauchblatt
pitcher leaf, ascidiate leaf
kannenförmig/krugförmig/
schlauchförmig ascidiform
Kannenpflanze pitcher plant
Kannenpflanzengewächse/
Nepenthaceae East Indian pitcher
plant family, nepenthes family
Kannibalismus cannibalism
Kanonenbein/Sprungbein
(Mittelfußknochen der Huftiere)
cannon bone
Känozoikum/Kaenozoikum/
Erdneuzeit/Neozoikum
(erdgeschichtliches Zeitalter)
Cenozoic, Cenozoic Era, Neozoic
Era (Cainozoic Era/Caenozoic Era)
Kantenkollenchym/
Eckenkollenchym (*auch*:
Collenchym) angular collenchyma
Kanter (kurzer/leichter Galopp)
canter (slow gallop)
Kantholz cant, squared timber,
square-edged lumber, squared log
kantig angular
Kanüle cannula
Kapaun (kastrierter Hahn) capon
Kapazität capacity
• **elektrische K.** capacitance (C)
Kapazitätsgrenze/Grenze der
ökologischen Belastbarkeit/
Tragfähigkeit (Ökosystem)
carrying capacity

Kapazitätskontrollsystem 544

**Kapazitätskontrollsystem,
 limitiertes** limited capacity control
 system (LCCS)
kapazitiver Strom capacitative
 current
**Kaperngewächse/Capparidaceae/
 Capparaceae** caper family
Kapillare/Haargefäß capillary
- **Blutkapillare** blood capillary
Kapillarelektrophorese capillary
 electrophoresis
Kapillarpipette capillary pipet,
 capillary pipette
kapnophil/kohlendioxidliebend
 capnophilic
kappen/köpfen (Baum) pollard,
 pollarding, beheading of tree,
 decapitation of tree
Kappenzelle (Scolopidium) cap cell
**Kappungsselektion/
 Auslesezüchtung/
 Schwellenwertselektion**
 truncation selection
kapsal/capsal/kokkal/coccal nonmotile unicellular
Kapsel *bot* capsule
- **Deckelkapsel** lid capsule, pyxis, pyxidium
- **dorsizide Spaltkapsel** dorsicidal capsule
- **fachspaltige Kapsel/lokulizide Spaltkapsel** loculicidal capsule
- **Katapultkapsel/Katapultfrucht** catapult capsule, catapult fruit
- **Lochkapsel/Porenkapsel** poricidal capsule
- **wandspaltige Kapsel/ septizide Spaltkapsel** septicidal capsule
Kapsid/Capsid *vir* **(Virenhülle)** capsid (viral shell)
Kapsomer/Capsomer *vir*
 (morphologische Untereinheit des
 Virions) capsomere (*virion:* morphological unit)
**Kapuzenspinnen/Ricinulei/
 Podogona**
 ricinuleids, "tick spiders"
**Kapuzinerkressengewächse/
 Tropaeolaceae** nasturtium family
**Karbon/Steinkohlenzeit
 (erdgeschichtliche Periode)**
 Carboniferous, Carboniferous
 Period, "Coal Age"
Karbonisation carbonization
Kardengewächse/Dipsacaceae
 teasel family, scabious family
karnivor/carnivor/fleischfressend
 carnivorous, flesh-eating,
 meat-eating
Karnivor/Carnivor/Fleischfresser
 carnivore, flesh-eater, meat-eater

**Karobgummi/
 Johannisbrotkernmehl** carob gum,
 locust bean gum
Karotinoide/Carotinoide carotinoids
Karpel/Fruchtblatt carpel
Karpellodium *bot* carpellode
**Karpfenfische/Karpfenartige/
 Cypriniformes** carps & characins
 & minnows & suckers & loaches
**Karpfenläuse/Fischläuse/
 Kiemenschwänze/Branchiura/
 Argulida** fish lice
Karpogon/Carpogon (Algen)
 carpogonium
**Karpophor/Carpophor/Fruchthalter/
 Fruchtträger** *bot* carpophore,
 receptacle
Karpose carposis
Karposoma/Fruchtkörper *fung*
 carposoma, fruiting body, fruitbody
**Karpospore/Carpospore/
 Carpogonidie (Algen)** carpospore
Karstsee karst lake
Karte/Landkarte map
- **biologische Karte** biological map
- **Determinationskarte/
 Schicksalkarte** *gen* fate map
- **Fokuskarte** focus map
- **genetische Karte** genetic map
- **pysikalische Karte** physical map
Karteneinheit map unit
kartieren map, plot
Kartierung mapping, plotting
- **Deletionskartierung** *gen*
 deletion mapping
- **Geländekartierung**
 terrain mapping
- **Genkartierung** gene mapping, genetic mapping
- **Konjugationskartierung** *gen* conjugation mapping
- **Positionskartierung** *gen* positional mapping
- **Schicksalskartierung** *gen* fate mapping
- **Transduktionskartierung** *gen* transduction mapping
- **Transformationskartierung** *gen* transformation mapping
Kartierungsfunktion mapping function
Karton (feste Pappe) cardboard, paperboard, fiberboard
Karunkula/Caruncula *bot* caruncle
**Karyogamie/Kernvereinigung/
 Kernverschmelzung** karyogamy, nuclear fusion
Karyogramm/Karyotyp karyogram, karyotype
**Karyopse/Caryopse/"Kernfrucht"/
 Kornfrucht** caryopsis, grain
Karyotyp karyotype

Karyotypanalyse/Bestimung des Karyotyps karyotyping
karzinogen/carcinogen/ krebserzeugend carcinogenic
Karzinogen *n* carcinogen
Karzinom carcinoma
Käscher/Kescher (Fangnetz für Fische) landing net, aquatic net (collecting net for fish)
Käschernetz/Keschernetz *arach* (Dinopis) retiarius web
Käsefliegenlarve (Piophilidae) cheese-skipper
Kaskade/Kascade cascade
Kaskadensystem (Enzyme) cascade system
Kassette cartridge, cassette
Kassetten-Mutagenese cassette mutagenesis
Kaste caste
kastrieren castrate, geld, neuter
kastriertes Tier (z.B. Pferd) gelding
Kastrierung/Kastration castration
Kasuare & Emus/Casuariiformes cassowaries & emus
katabol/catabol catabolic (degradative reactions)
Katabolit-Repression/katabolische Repression catabolite repression
Katabolitaktivatorprotein catabolite activator protein (CAP)
Katabolitrepression (Hemmung) catabolite repression
katadrom catadromous
Katalepsie catalepsy
Katalepsis catalepsis
Katalysator catalyst
Katalyse catalysis
katalysieren catalyze
katalytisch catalytic, catalytical
katalytische Einheit/Einheit der Enzymaktivität (*katal*) catalytical unit, unit of enzyme activity (*katal*)
katalytischer Antikörper catalytic antibody
Katapultfrucht/Katapultkapsel catapult fruit, catapult capsule
Katastrophentheorie catastrophism
Katecholamin catecholamine
Kater tomcat (male domestic cat)
katharob *limn* katharobic
Katharobiont/Katharobie katharobiont, katharobe, katharobic organism
Kationenaustauscher cation exchanger
Katsuragewächse/Cercidiphyllaceae katsura-tree family
Kätzchen *bot* **(Infloreszenz)** catkin, ament, amentum
Kätzchen/junge Katze/ Katzenjunges kitten

Kätzchen/Katzenjunges kitten
Katzenschrei-Syndrom cri-du-chat syndrome
Kaudikula/Caudicula/Stielchen caudicle (stalk of pollinium)
kauen/zerkauen chew, masticate
Kauen/Zerkauen chewing, mastication
kauend chewing, masticatory
kauend-beißend (Mundwerkzeuge) chewing-biting (mouthparts)
Kaufläche masticatory surface
Kaugummi chewing gum
Kaulade (Crustaceen) gnathobase, blade
Kaulade (Insekten: aus Galea und Lacinia) (maxilliary/maxilliped) plate
Kauleszenz/Cauleszenz *bot* caulescence
kauliflor/stammblütig cauliflorous
Kauliflorie/Stammblütigkeit cauliflory
Kauloid/Cauloid/Stämmchen (Algen/Moose) caulid, stemlet, stipe
Kaulquappe tadpole, "polliwog"
Kaumagen/Cardia cardiac stomach, cardia
Kaumagen/Pharynx/Mastax (Rotatorien) pharynx, mastax
Kaumagen/Proventriculus (Insekten/Crustaceen) gizzard, proventriculus
Kaumittel (Gummiharz) masticatory, gum
Kaumuskel masticatory muscle, muscle of mastication
kausal causal, causative
kausaler Zusammenhang correlation, connection, causal interrelationship
Kausalität causality, causation
Kautschuk caoutchouc, rubber, india rubber
Kavitation (von Leitelementen) *bot* cavitation (with rupture of water column)
Kegel cone; (Sensille: Insektenantenne) peg
kegelförmig/konisch cone-shaped, conical
Kegelgalle cone gall
Kegelspirale (Gastropoda) helicone
Kehldeckel/Epiglottis epiglottis
Kehldeckelknorpel/Schließknorpel epiglottic cartilage
Kehle/Hals throat
Kehle/Luftröhre windpipe, trachea
Kehle/Speiseröhre esophagus, oesophagus; gullet
Kehlgang (Pferd) throatlatch

Kehlhautsack *orn* pouch
Kehlkopf/Larynx (siehe: Adamsapfel) larynx
Kehllappen (Vögel/Reptilien: Hautlappen) dewlap, wattle
Kehlplatte/Schlundplatte/ Gularplatte (prognathous insects) gula, gular plate
Kehlritze/Aditus laryngis laryngeal aditus
Kehlsack (Pelikan/Frosch) gular pouch
kehlständig (Bauchflossen) located near the "chin"
Kehrwert/reziproker Wert reciprocal
Keil wedge, peg
Keilbein/Os cuneiforme (Fuß) cuneiform bone
Keilbein/Os sphenoidale (Schädel) sphenoid bone
Keilbeinflügelknochen alisphenoid bone
Keilblattgewächse/ Sphenophyllales/ Sphenophyllaceae sphenophyllum family
keilblättrig wedge-leaved
Keiler/Wildeber male wild boar (aged)
keilförmig cuneate, cuneiform, sphenoid, wedge-shaped
keilförmig zugespitzt attenuate, tapering
Keim (Mikroorganismus) germ
Keim/Keimling/Embryo germ, embryo
Keimbahn germ line
Keimbahngentherapie germ line gene therapy
Keimbahnhypothese/-theorie germline hypothesis/theory
Keimbahnmosaik germline mosaic, germinal mosaic, gonadal mosaic, gonosomal mosaic
Keimbahnmutation germ-line mutation
Keimbläschen/Blastozyste/ Blastocyste germinal vesicle, blastocyst; (großer Oocytenkern) germinal vesicle
Keimblase/Blasenkeim/Blastula blastula
Keimblatt/Blatt/Keimschicht *embr* germ layer
• **primäres K.** (äußeres Keimblatt/ Ectoderm) ectoderm; (inneres Keimblatt/Entoderm/Endoderm) entoderm, endoderm
• **sekundäres K./Mesoderm** mesoderm
Keimblatt/Kotyledone/Cotyledone *bot* cotyledon, seminal leaf

Keimblattscheide/Keimscheide/ Koleoptile/Coleoptile *bot* coleoptile, plumule sheath
Keimdrüse/Geschlechtsdrüse/ Gonade sex gland, germ gland, gonad
Keimdrüsenleiste/Genitalleiste *embr* genital ridge
keimen germinate, sprout
Keimepithel germinal epithelium
Keimesbewegung/Blastokinese blastokinesis
Keimfähigkeit germinability
Keimfleck/Macula germinativa germ spot
keimfrei/steril germ-free, sterile
Keimhülle, äußere/äußere Eihülle/ äußeres Eihüllepithel/Serosa serosa, serous membrane
Keimhülle, innere/innere Eihülle/ Amnion amnion
Keimknospe/Plumula/ Stammknospe/Sprossknospe/ terminale Embryoknospe plumule, terminal embryonic bud
Keimling/Embryo embryo
Keimling/Keimpflanze *bot* sprout, seedling
Keimmund micropyle
Keimplasma/Idioplasma germ plasm, idioplasm, gonoplasm
Keimpore (Pollen) germination/ germinating aperture
Keimruhe seed dormancy
Keimsack/Embryosack embryo sac
Keimscheibe/Embryonalschild/ Blastodiskus/Diskus/Discus germinal disk, germ disk, blastodisc
Keimscheide/Keimblattscheide/ Koleoptile/Coleoptile coleoptile, plumule sheath
Keimschicht/Stratum germinativum Malpighian layer, germinal layer, germinative layer, stratum germinativum;
(*Echinococcus*-Blase) germinal layer
Keimschlauch *fung* germ-tube
Keimstelle (Pollen) aperture
Keimstimmung/Vernalisation vernalization
Keimstock/Germarium ovary, germarium
Keimstrang *embr* germinal cord
Keimstreifen (Insektenei) germ band
Keimstreifen/Primitivstreifen (Gastrulation) germinal streak, primitive streak
Keimtasche/Sporosac (Hydrozoen) sporosac
keimtötend/bakterizid bacteriocidal, bactericidal

Kernnährelemente

Keimung germination
- **Dunkelkeimung** dark germination
- **epigäische Keimung** epigean/epigeal germination
- **Hellkeimung** light-induced germination (photodormancy)
- **hypogäische Keimung** hypogean/hypogeal germination

Keimwarze (des Samens) strophiolar plug, operculum
Keimwurzel/Radicula embryonic root, radicle
Keimzahl (Anzahl von Mikroorganismen) cell count, germ count; (Samenkeimung) germination percentage
Keimzelle *allg* germ cell (any reproductive cell, i.e. spores/zygote/gametes); (Gamet) sex cell, gamete
Keimzentrum germinal center
Kelch calyx
- **Nebenkelch/Außenkelch** calycle, calyculus

Kelchblatt/Blütenkelchblatt/Blumenhüllblatt/Sepale/Sepalum sepal
kelchförmig cup-shaped, calyciform
Kelchhorngewächse/Calyceraceae calycera family
Kelchwürmer/Nicktiere/Kamptozooen/Entoprocta kamptozoans, entoprocta
Kellerschwämme/Warzenschwämme/Coniophoraceae dry rot family
Kelter fruit/juice press
keltern press (fruit/grapes)
Kennart diagnostic species
Kenngröße/Parameter parameter; *math* dimensionless group/quantity/number
Kennwert characteristic value
Kennzahl basic number, characteristic number; (Chiffre) key, cipher
Kennzahl/Kennziffer *stat* index number, indicator
Kennzahl/statistische Maßzahl statistic, statistic value
kenokarp/leerfrüchtig seedless fruit
Kenokarpie/Kenocarpie/Leerfrüchtigkeit seedlessness
Keratinfilament keratin filament
keratinisieren (verhornen) keratinize (cornify)
Keratinisierung (Verhornung) keratinization (cornification)
Keratinozyt keratinocyte
Kerbe indentation, notch
Kerbe/Schlitz/Bruchstelle/Einzelstrangbruch *gen* nick

kerbig/gekerbt notched, nicked, crenate
Kerbtiere/Kerfe/Insekten insects
Kermesbeerengewächse/Phytolaccaceae pokeweed family
Kern/Zentrum (Mark/Core) core, center
- **Bohrkern** *geol* drill core
- **Fruchtkern/Obstkern** *bot* kernel, seed; pip (einer vielsamigen Frucht)
- **Viruskern (zentrale Virionstruktur)** core
- **Zellkern** nucleus, karyon

Kernäquivalent/Nukleoid/Karyoid/"Bakterienkern" nucleoid, nuclear body
Kernart core species
kernassoziiertes Organell nucleus-associated organelle
Kernbeißerschnabel seed-cracking beak
Kerndimorphismus nuclear dimorphism
Kerndualismus nuclear dualism
Kernenzym (RNA-Polymerase) core enzyme
Kernfaserschicht/Kernlamina nuclear lamina
Kernfäule heart rot
Kerngehäuse (Frucht) (fruit) core
Kerngenom nuclear genome
Kerngerüst nucleoskeleton
Kerngrundsubstanz/Kernmatrix nuclear matrix
Kernhaufenfaser nuclear bag fiber
Kernholz heartwood, duramen
Kernhülle nuclear envelope
Kernkappe nuclear cap
Kernkettenfaser nuclear chain fiber
Kernkeulen/Clavicipitacae ergot fungi, ergot family
Kernkörperchen/Nukleolus nucleolus
Kernlamina/Kernfaserschicht nuclear lamina
kernlos *bot* seedless
kernlos *cyt* anucleate
kernlose Zelle enucleate cell, anucleate cell
kernmagnetische Resonanz/Kernspinresonanz nuclear magnetic resonance (NMR)
kernmagnetische Resonanzspektroskopie/Kernspinresonanz-Spektroskopie nuclear magnetic resonance spectroscopy, NMR spectroscopy
Kernmatrix/Kerngrundsubstanz nuclear matrix
Kernmembran nuclear membrane
Kernnährelemente macronutrients

Kernobst

Kernobst pomaceous fruit
Kernpartikel core particle
Kernphase nuclear phase
Kernphasenwechsel
  alternation of nuclear phase
Kernplasma/Karyoplasma/Nucleoplasma karyoplasm, nucleoplasm
Kernpolyederviren
  nuclear polyhedrosis viruses (NPV)
Kernpore nuclear pore
Kernspinresonanz/kernmagnetische Resonanz
  nuclear magnetic resonance (NMR)
Kernspinresonanz-Spektroskopie/kernmagnetische Resonanzspektroskopie nuclear magnetic resonance spectroscopy, NMR spectroscopy
Kernspintomographie (KST)/Magnetresonanztomographie (MRT)
  magnetic resonance imaging (MRI), nuclear magnetic resonance imaging
Kernteilung/Mitose
  nuclear division, mitosis
Kerntransplantation nuclear transfer, nuclear transplantation
Kernverschmelzung *cyt*
  fusion of nuclei, caryogamy
Kescher/Käscher (Fangnetz für Fische) landing net, aquatic net (collecting net for fish)
Keschernetz/Käschernetz *arach*
  (Dinopis) retiarius web
Kessel/Brutkammer (geschlossene Bruthöhle der Ameisenkönigin)
  claustral cell
Kesselfallenblume/Gleitfallenblume
  pitfall trap, slippery-trap flower, slippery-slide flytrap
Kesselhieb *for*
  patch clear-cutting
Ketoaldehyd ketoaldehyde, aldehyde ketone
Keton ketone
Ketonkörper ketone body (acetone body)
Ketonurie ketonuria, acetonuria
Ketosäure keto acid
Kette (verzweigte/unverzweigte)
  chain (branched/unbranched)
  • **leichte Kette (L-Kette)**
    light chain (L chain)
  • **schwere Kette (H-Kette)**
    heavy chain (H chain)
Kettenabbruchverfahren *gen*
  chain-terminating technique
Kettenform *chem*
  chain form, open-chain form
Kettenformel
  chain formula, open-chain formula
Kettenlänge chain length
Kettenreaktion chain reaction
Keule/Oberschenkel (Schlachtvieh)
  haunch, hindquarters
Keule/Schlegel/Bein (Geflügel)
  drumstick, leg
keulenartig club-like, clubbed, clavate
Keulenpilz club fungus
Keuper (Epoche) Upper Triassic
Kiefer *zool* jaw
Kiefer/Trophi (Rotatorien)
  (pharyngeal) jaws, trophi
Kieferbogen/Mandibularbogen
  (Gesamtheit der Teile) mandibular arch; (nur Knorpelspange) mandibular bar (skeleton only)
Kieferfühler/Scherenkiefer/Klaue/Fresszange/Greifzange/Chelicere
  chelicera, fang, cheliceral fang
Kieferfuß/Maxilliped/Maxillarfuß/Pes maxilliaris maxilliped, maxillipede, gnathopodite, jaw-foot, foot-jaw
Kieferfüßer/Maxillopoda
  maxillopods
Kieferknochen jawbone
Kieferlose/Agnathen/Agnatha
  jawless fishes, agnathans
Kiefermäuler/Kiefermündchen/Gnathostomuliden
  gnathostomulids
Kiefermünder/Gnathostomata
  jawed vertebrates, jaw-mouthed animals, gnathostomatans
Kieferngewächse/Föhrengewächse/Tannenfamilie/Pinaceae
  pine family, fir family
Kieferschluss/Okklusion occlusion
Kiefertaster/Maxillartaster/Maxillarpalpus/Palpus maxillaris
  maxillary palp
Kiel keel, carina;
  bot (Schiffchen) keel
Kieme gill
  • **Arthrobranchie** arthrobranch, arthrobranchia, joint gill
  • **Außenkieme/äußere Kieme/Ektobranchie** external gill
  • **Blattkieme/Lamellibranchie/Eulamellibranchie** lamellar gill, sheet gill, lamellibranch, eulamellibranch
  • **Buchkiemen/Kiemenbeine/Blattbeine/Blattfüße (dichtstehende Kiemenlamellen: Xiphosuriden)**
    book gills (gill book)
  • **Dendrobranchie (Crustaceen)**
    dendrobranchiate gill
  • **Fadenkieme/Filibranchie**
    filibranch gill

Kindel K

- **Fiederkieme/Kammkieme/ Ctenidie/Ctenidium** gill plume, gill comb, ctenidium
- **Gaskieme (aquat. Insekten)** gaseous plastron, air-bubble gill
- **Hemibranchie** hemibranch
- **Holobranchie** holobranch
- **Innenkieme/innere Kieme/ Entobranchie** internal gill
- **Kammkieme/Fiederkieme/ Ctenidie/Ctenidium** gill plume, gill comb, ctenidium
- **kompressible Gaskieme** compressible gill
- **Phyllobranchie (Crustaceen)** phyllobrachiate gill
- **physikalische Kieme** physical gill
- **Pleurobranchie** pleurobranch, "side gill"
- **Podobranchie** podobranch, "foot gill"
- **Scheinblattkieme/ Pseudolamellibranchie** pseudolamellar gill, pseudolamellibranch
- **Tracheenkieme** tracheal gill
- **Trichobranchie (Crustaceen)** trichobranchiate gill

Kiemenbalken (Cephalochordaten) gill rod (dorsal lamina/languets)

Kiemenbeine/Blattbeine/Blattfüße/ Buchkiemen (dichtstehende Kiemenlamellen: Xiphosuriden) book gills (gill book)

Kiemenblatt/Kiemenblättchen/ Kiemenlamelle/Hemibranchie (Fische/Muscheln) gill lamella

Kiemenbogen/Branchialbogen/ Viszeralbogen (Gesamtheit der Teile) gill arch, branchial arch, visceral arch; (nur Knorpelspange) gill bar, branchial bar, visceral bar (skeleton only)

Kiemenbürste/Flabellum gill cleaner

Kiemendarm/Pharynx branchial gut, branchial basket, pharynx

Kiemendeckel/Operculum/ Operkulum gill cover, operculum

Kiemendorn gill raker (bristle-like process on gill arch)

Kiemenfaden/Kiemenfilament gill filament

Kiemenfüßer/Blattfußkrebse/ Phyllopoda/Branchiopoda phyllopods, branchiopods

Kiemengang branchial chamber
- **äußerer K./Suprabranchialraum** suprabranchial chamber
- **innerer K./Subbranchialraum** subbranchial chamber

Kiemenhautstrahl/ Branchiostegalstrahl/ Radius branchiostegus branchiostegal ray

Kiemenherz (Cephalopoda) gill heart, branchial heart

Kiemenhöhle/Kiemenkammer gill cavity, gill chamber

Kiemenkorb gill basket

Kiemenöffnung gill opening, gill aperture

Kiemenpumpe branchial pump

Kiemenreuse gill rakers (sifting/ straining apparatus formed by the total of all gill rakers)

Kiemensack/Kiementasche gill pouch, branchial sac, pharyngeal pouch

Kiemenschwänze/Karpfenläuse/ Fischläuse/Branchiura/Argulida fish lice

Kiemenskelett/Branchialskelett branchial skeleton, skeleton of the gills

Kiemenspalte/Viszeralspalte gill slit, pharyngeal slit, gill cleft, branchial cleft, pharyngeal cleft

Kiemenstrahl gill ray

Kiementasche/Kiemensack gill pouch, branchial sac, pharyngeal pouch

kiementragend branchiferous

Kien/Kienholz resinous pinewood

Kienapfel pinecone, "pine"

kienig/harzreich resinous, resiny

Kienspan chip of pinewood, pinewood chip

Kies gravel

Kieselalgen/Diatomeen/ Bacillariophyceae diatomes

Kieselerde diatomaceous earth

Kieselflagellaten/Silicophyceae silicoflagellates

Kieselgel/Silicagel silica gel

Kieselgur kieselguhr (loose/porous diatomite; diatomaceous/infusorial earth)

Kieselsäure silicic acid

kieselsäurehaltig siliceous

Kieselschwämme/Silicospongiae (Demospongien) siliceous sponges, demosponges

Kieselstein pebble

Kiesgrube gravel pit

Killer-Zelle/Killerzelle killer cell, K cell

Kilosequenzierung kilosequencing

Kindbettfieber/Wochenbettfieber/ Puerperalfieber (bakteriell) childbed fever, puerperal fever

Kindel/Kindl (Bromelien/Bananen) sucker

Kindelbildung

Kindelbildung (Kartoffel: Knollenmissbildung) formation of miniature stolons due to water stress
Kindermord infanticide
Kindersterblichkeit childhood mortality
Kindheit childhood
Kindspech/Mekonium/Meconium meconium
kinematische Viskosität kinematic viscosity
Kinese kinesis
Kinet kinetium, kinety
Kinetik (nullter/erster/zweiter Ordnung) (zero-/first-/second-order...) kinetics
- **Reaktionskinetik** reaction kinetics
- **Reassoziationskinetik** reassociation kinetics

Kinetin kinetin, zeatin
kinetische Komplexität kinetic complexity
Kinetochor kinetochore
Kinetosom/Basalkörper kinetosome, basal body
Kinorhyncha/Hakenrüßler kinorhynchs
Kippmoment (Vogelflug) pitch(ing) moment
Kistenbretter (Holz) crate planks/boards
Kitt/Kittsubstanz adhesive, cement
Kittdrüse/Klebdrüse/Zementdrüse/Fußdrüse adhesive gland, cement gland, pedal gland
Kittleiste/Verschlusskontakt/Schlussleiste/Engkontakt/Tight junction (Zonula occludens) tight junction
Kitz (Rehkitz) fawn; (Zicklein) kid, young goat
Kitzler/Klitoris/Clitoris clitoris
Kiwis/Apterygiformes kiwis
Kladistik/Cladistik cladistics, phylogenetic analysis
Kladodium/Cladodium (Flachspross eines Langtriebs) cladode, cladophyll; *also*: phylloclade
Kladogenese cladogenesis
Kladogramm cladogram
klaffen/offen stehen gape
klaffende Schalenöffnung shell gape
Klaffmuskel diductor muscle
Klammer *lab* clamp, clip
- **Objekttisch-Klammer** *micros* stage clip

Klammerfuß/Pes semicoronatus (Bauchfüße der Larven: Großschmetterlinge) proleg with crochets on planta in a row
Klammergriff grasp
Klammerreflex clasp reflex
Klappe (auch: Schalenklappe = Muschelschalenhälfte) valve
- **Aortenklappe** aortic valve
- **halbmondförmige Klappe/Semilunarklappe (Herz: Teile der Taschenklappe)** semilunar cusp, semilunar flap (parts of semilunar valve)
- **Harnleiterklappe** ureteral valve
- **Mitralklappe/Bikuspidalklappe (siehe auch: Herzklappe)** mitral valve, bicuspid valve
- **Pulmonalklappe** *cardio* pulmonary valve
- **Spiralklappe (Froschherz)** spiral valve
- **Taschenklappe** *cardio* semilunar valve (consisting of three semilunar cusps/flaps)
- **Trikuspidalklappe** *cardio* tricuspid valve (with three cusps/flaps)

Klappfalle/Schlagfalle *bot* snap trap
klappig valvate
Klappmechanismus/Schnappmechanismus snap mechanism
Kläranlage (kommunal) sewage treatment plant; (industriell) waste-water purification plant
Klärbecken/Absetzbecken settling tank
klären (z. B. absetzen/entfernen von Schwebstoffen aus einer Flüssigkeit) clear, clarify, purify
klären/filtrieren filtrate
klarer Plaque clear plaque
Klärfaktor *physio* clearance, clearing factor
Klärgas/Faulgas (Methan) sludge gas
Klärgrube cesspool, cesspit
Klärschlamm (>Faulschlamm) sludge, sewage sludge
Klärung/Filtrierung/Filtration filtration
Klärung (z. B. absetzen/entfernen von Schwebstoffen aus einer Flüssigkeit) clarification, purification
Klärung/Abwasseraufbereitung sewage treatment
Klärwerk/Kläranlage (Abwasser) sewage treatment plant
Klasse class
Klassenhäufigkeit/Besetzungszahl/absolute Häufigkeit *stat* class frequency, cell frequency

Klima

Klassenmerkmal class trait
Klassenwechsel/Klassensprung *immun* class switch, class-switching (isotype switching)
Klassierung *stat* grouping of classes
klassifizieren classify
Klassifizierung/Klassifikation classifying, classification
• **biologische K./Taxonomie** taxonomy
klassische Konditionierung *ethol* classical conditioning
Klaue/Unguis claw
Klauenbein/Os ungulare ungual bone, unguicular bone (distal phalanx)
Klauendrüse claw gland
Klauenglied/Krallenglied/ Krallensegment/Krallensockel/ Prätarsus pretarsus
Klause (Fruchtform) cell, mericarpic nutlet (one-seeded segment/fruitlet of loment)
Klausenfrucht/Gliederfrucht/ Gliederhülse/Bruchfrucht loment, lomentum, jointed fruit
Klebdrüse/Kittdrüse/Zementdrüse adhesive gland, cement gland
• **Fußdrüse** pedal gland
• **Schwanzdrüse/Rektaldrüse (Nematoden)** rectal gland
Klebfaden (Spinnennetz) *arach* adhesive thread, viscid/sticky line
Klebfalle *bot* adhesive trap, flypaper trap
Klebkapsel/Haftkapsel/Glutinant (Nematocyst) glutinant
klebrig/glutinös sticky, glutinous, viscid
klebriges Ende/kohäsives Ende/ überhängendes Ende *gen* sticky end, cohesive end, protruding end, protruding extension
Klebsamengewächse/ Pittosporaceae pittosporum family, tobira family, parchment-bark family
Klebscheibe/Klebkörper (Orchideen) viscidium (a sticky disk)
Klebzelle/Kolloblast/Colloblast/ Collocyt (Ctenophora) adhesive cell, colloblast, lasso cell
Kleeblatt cloverleaf
Kleefarngewächse/Kleefarne/ Marsileaceae marsilea family, water clover family
Klei/Kleiboden/Marschboden heavy marshland soil
kleidoisches Ei/beschaltes Ei cleidoic egg, shelled egg, "land egg"
Kleie bran
Kleinart/Mikrospezies microspecies

kleinborstig/kleindornig echinulate, with small bristles/prickles
kleine Furche/kleine Rinne/flache Rinne (DNA Struktur) minor groove (DNA structure)
kleine nukleäre Ribonukleinsäure (snRNA) small nuclear RNA (snRNA)
kleines nukleäres Ribonukleoprotein (snRNP) small nuclear ribonucleoprotein (snRNP)
kleingesägt/feingesägt serrulate, finely serrate, finely notched
kleingrubig scrobiculate, alveolate
Kleinhirn/Hinterhirn/Cerebellum cerebellum, epencephalon
Kleinkärpflinge/ Cyprinodontiformes killifishes
Kleintiere/Kleinsäuger small mammals
Kleintierpraxis (Tierarzt) veterinary practice for small mammals
Kleintierzüchterverein *literally:* small mammal breeders club
kleinzelliger Lungenkrebs small-cell lung cancer
kleistogam cleistogamous
Kleistogamie cleistogamy
Kleistothecium cleistothecium, cleistocarp
Klemme clamp; clip
• **Arterienklemme** artery forceps, artery clamp
• **Krokodilklemme** *lab* alligator clip
• **Schlauchklemme/Quetschhahn** *lab* tubing clamp, pinch clamp
• **Spannungsklemme** voltage clamp
Klemmfalle pinch trap
Klemmfallenblume/ Klemmfallenblüte pinch-trap flower
Klemmkörper (Asclepiadaceen) adhesive body, clamp, corpuscle, corpusculum
Kleptobiose/Cleptobiose cleptobiosis
Kleptoparasit kleptoparasite, cleptoparasite
Klette/Klettenfrucht/Klettenfrucht bur, burry fruit
Kletterfaser *neuro* climbing fiber
Kletterfuß climbing foot
kletterfüßig *orn* **(z. B. Papageien)** zygodactyl, zygodactylous
kletternd/klimmend climbing, scandent
Kletterpflanze climber, scandent plant, (climbing) vine; (holzig) liana
Klicklaut (Zahnwale) clicking sound
Klima climate

Klimaanpassung

Klimaanpassung acclimation, acclimatization
Klimafaktoren climatic factors
Klimagürtel climatic belt
Klimakterium/Wechseljahre (Menopause) climacteric (menopause)
Klimaveränderung/Klimaänderung climatic change
Klimax/Höhepunkt climax; (Orgasmus) orgasm
Klimaxformation climax formation
Klimaxgesellschaft climax community
Klimaxvegetation climax vegetation
klimmen/klettern climb
klimmend/kletternd climbing, scandent
Kline/Klin/Cline/Merkmalsgefälle/ Merkmalsgradient cline, phenotypic gradient, character gradient
klinisch getested/geprüft clinically tested
klinische Genetik clinical genetics
klinisches Merkmal clinical feature
klinisches Symptom clinical symptom
Klitoris/Clitoris/Kitzler clitoris
Kloake cloaca
Kloakentiere/Monotremata (Prototheria) monotremes (prototherians)
Klon clone
klonale Selektionstheorie clonal selection theory
Klonausgangspflanze/ Klonmutterpflanze/Ortet ortet (original single ancestor of a clone)
Klonbank/Bibliothek clone bank, bank, library
klonieren clone
klonierte Zellinie cloned cell line
Klonierung cloning
- **Exonklonierung** exon cloning
- **Expressionsklonierung** expression cloning
- **Positionsklonierung** positional cloning
- **Subklonierung** subcloning
- **subtraktive Klonierung** subtractive cloning
Klonierungsstelle cloning site
- **multiple K./ Vielzweckklonierungsstelle** multiple cloning site (MCS)
Klonierungsvektor cloning vector
Klonindividuum/Klonmitglied/ Einzelpflanze eines Klons/Ramet (>Zweig/Steckling eines Ortets) ramet (individual member of clone)

Klonselektionstheorie/ klonale Selektionstheorie clonal selection theory
klopfen (Hasen) thump, thumping
Klopphengst/Spitzhengst (Kryptorchide) ridgeling, ridgling (cryptorchid)
knabbern (Hasen) nibble
Knabenkrautgewächse/Orchideen/ Orchidaceae orchids, orchid family, orchis family
Knallgasbakterien/ Wasserstoffbakterien hydrogen bacteria (aerobic hydrogen-oxidizing bacteria)
Knäuel (Infloreszenz) cyme with sessile flowers
Knäuelkonformation/ Schleifenkonformation coil conformation, loop conformation
Knickfestigkeit (Holz) buckling strength, folding strength, crossbreaking strength
Knie knee; *bot* knee, knee-root
Knie/Patella (Arthropoden) genu, patella
Knie.../knieartig genicular
knieförmig gebogen geniculate, bent like a knee
Kniegelenk knee joint; (bei Pferd/ Hund) stifle
Kniehöcker *neuro* geniculate body
Kniekehle back of the knee, hollow behind knee, bend of the knee, poples, popliteal fossa, popliteal space
Kniescheibe/Patella kneecap, knee bone, patella; (beim Pferd) stifle bone
Knöchel (Fußknöchel) ankle; (Handknöchel) knuckle
Knöchelchen small bone, ossicle
Knöchelgang (Fußknöchel) knuckle-walking
Knochen bone
- **Amboss/Incus (Ohr)** anvil (bone), incus
- **Backenknochen/Os zygomaticum** cheekbone, zygomatic bone, malar bone
- **Beckenknochen (Darmbein & Sitzbein)** pelvic bone, pelvis (ilium & ischium)
- **Belegknochen/Deckknochen/ Hautknochen** dermal bone
- **Beutelknochen/Os marsupialis** marsupial bone
- **Brustbein/Clavicula (Geflügel)** collarbone, clavicle (wishbone see Gabelbein)
- **Brustbein/Sternum** breastbone, sternum

Knochen

- **Darmbein/Ilium/Os ilium** flank bone, ilium
- **Dreiecksknochen/Os triquetrum** triangular bone, pyramidal bone, triquestral bone
- **Erbsenbein** pisiform bone, pisiform
- **Faserknochen/Geflechtknochen** fibrous bone
- **Felsenbein/Os perioticum/ Perioticum** periotic bone, periotic
- **Felsenbein/Pars petrosa (des Schläfenbeins)** petrous bone, petrosal bone (of temporal bone)
- **Fersenbein/Hypotarsus (Vögel)** heel, calcaneum, calcaneus, hypotarsus
- **Fersenbein/Os calcis/Kalkaneus/ Calcaneus** heelbone, calcaneal bone, calcaneum, calcaneus
- **Fesselknochen/Fesselbein/ proximale Phalanx** pastern bone, long pastern bone, first phalanx
- **Fischbein/Walbein** baleen, whalebone
- **Flügelbein/Pterygoid/ Os pterygoides** pterygoid bone/process
- **Fußwurzelknochen/ Fußgelenksknochen/Tarsal** ankle bone, tarsal (bones)
- **Gabelbein/Furcula** wishbone, united clavicles of birds, furcula, fourchette
- **Gaumenbein/Palatinum/ Os palatinum** palatine bone
- **Geflechtknochen/Faserknochen** fibrous bone
- **Gesäßbein/Sitzbein/ Sitzknochen/Os ischii** ischium
- **Gleichbein/Sesambein/ Sesamknöchelchen (Pferd)** proximal sesamoid bone
- **Griffelbein/Griffelbeinknochen (Nebenmittelfußknochen)** splint bone (small metacarpal)
- **Gürtelbein/Sphenethmoid/Os en ceinture** sphenethmoid bone
- **Hakenbein/Os hamatum** hamate bone, unciform bone
- **Hammer/Malleus (Ohr)** hammer, malleus
- **Handgelenksknochen/ Handwurzelknochen/ Carpalia/Ossa carpalia** wrist bone, carpal bone
- **Handwurzelknochen/Ossa carpi** carpal bones
- **Hautknochen/Deckknochen/ Belegknochen** dermal bone
- **Herzknochen/Os cordis** cardiac bone, heart ossicle
- **Hinterhauptbein/Os occipitale** occipital bone
- **Hohlknochen/pneumatischer Knochen/Os pneumaticum** hollow bone, pneumatic bone
- **Hufbein/Os ungulare/Phalanx distalis** coffin bone, distal phalanx
- **Hüftknochen/Hüftbein/Os coxae (Beckenhälfte: Darmbein & Sitzbein & Schambein)** hipbone, coxal bone, innominate bone (lateral half of pelvis)
- **Jochbein/Jugale** jugal (bone)
- **Kahnbein/Fußwurzelknochen/ Os naviculare** navicular bone
- **Kahnbein/Handwurzelknochen/ Os scaphoideum** scaphoid bone
- **Kanonenbein/Sprungbein (Mittelfußknochen der Huftiere)** cannon bone
- **Keilbein/Os cuneiforme (Fuß)** cuneiform bone
- **Keilbein/Os sphenoidale (Schädel)** sphenoid bone
- **Kieferknochen** jawbone
- **Klauenbein/Os ungulare** ungual bone, unguicular bone (distal phalanx)
- **kompakter Knochen** compact bone, dense bone
- **Kopfbein/Kapitatum/ Os capitatum** capitate bone, capitate, capitatum
- **Kreuzbein/Sakrum/Os sacrum** sacrum
- **Kronbein/mittlere Phalanx (Pferd)** coronary bone, small/short pastern bone, second phalanx
- **kurzer Knochen/Os brevis** short bone
- **Lamellenknochen/lamellärer Knochen** laminar bone
- **langer Knochen/Röhrenknochen/ Os longum** long bone (hollow/ tubular bone)
- **Mittelfußknochen/ Os metatarsalis** metatarsal bone
- **Mittelhandknochen/ Os metacarpalis** metacarpal bone
- **Mondbein/Os lunatum** lunate bone, semilunar bone
- **Nasenbein/Os nasale** nasal bone
- **Oberarmknochen/Oberarmbein/ Humerus** arm bone, humerus
- **Oberkieferbein/Os maxillare** maxillary bone
- **Oberschenkelknochen/ Os femoris/Femur** thighbone, femur, femoral bone
- **Paukenbein/Os tympanicum** tympanic bone, tympanic

Knochenbildung

- **Penisknochen/Os penis** penis bone, baculum
- **Pflugscharbein/Vomer** ploughshare bone, vomer
- **platter Knochen/Os planum** flat bone
- **pneumatischer Knochen/ Hohlknochen/Os pneumaticum** pneumatic bone, hollow bone
- **Quadratbein/Quadratum** quadrate bone
- **Rabenbein/Rabenschnabelbein/ Coracoid** coracoid
- **Rippe/Costa** rib
- **Röhrbein/Kanonenbein/ Sprungbein (Pferd: Hauptmittelfußknochen)** cannon bone
- **Röhrenknochen/langer Knochen/ Os longum** long bone (hollow/ tubular bone)
- **Rollbein/Sprungbein/Os tali/ Talus** ankle bone, talus, astragalus
- **Rüsselbein/Os rostrale** rostral bone
- **Schaltknochen/Nahtknochen** sutural bone, epactal bone, wormian bone
- **Schambein/Os pubis** pubic bone
- **Scheitelbein/Os parietale** parietal bone
- **Schienbein/Schiene/Tibia** shinbone, tibia
- **Schläfenbein/Os temporale** temporal bone
- **Schlüsselbein/Clavicula** collarbone, clavicle
- **Schulterblatt/Scapula** shoulder blade
- **Schuppenbein/Squamosum** squamosa
- **Sesambein/Sesamknöchelchen/ Os sesamoideum** sesamoid bone
- **Siebbein/Os ethmoidale** ethmoid bone
- **Sitzbein/Gesäßbein/ Sitzknochen/Os ischii** ischium
- **Sparrknochen/Chevron (ventraler Wirbelbogen)** chevron (bone), hemal arch
- **Speiche/Radius** radius
- **spongiöser Knochen** spongy bone, cancellous bone
- **Sprungbein/Kanonenbein (Mittelfußknochen der Huftiere)** cannon bone
- **Sprungbein/Talus** ankle bone, talus, astragalus
- **Steigbügel/Stapes** stirrup
- **Steiß/Steißbein/Os coccygis** coccyx
- **Steißbein/Urostyl (frogs/toads)** urostyl
- **Sternum/Brustbein** sternum, breastbone
- **Stirnbein/Os frontale** frontal bone
- **Strahlbein/distales Sesambein/ Os sesamoideum distale** navicular bone, distal sesamoid bone
- **Tränenbein/Os lacrimale** lacrimal bone
- **Unterkieferknochen/Unterkiefer/ Mandibel** lower jawbone, lower jaw, submaxilla, submaxillary bone, mandible
- **Unterschenkelknochen/ Tibiotarsus (Vögel)** tibiotarsus
- **Vieleckbein, großes/Os trapezium** trapezium bone, greater multangular bone
- **Vieleckbein, kleines/Os trapezoideum** trapezoid bone, lesser multangular bone
- **Wadenbein/Fibula** splint bone, fibula
- **Würfelbein/Os cuboideum** cuboid bone
- **Zungenbein/Hyoideum/ Os hyoideum** hyoid bone, lingual bone
- **Zwischenkieferknochen/ Os incisivum** incisive bone
- **Zwischenscheitelbein/ Os interparietale** interparietal bone

Knochenbildung/ Knochenentstehung/Osteogenese osteogenesis
Knochenbildung/Ossifikation ossification
Knochengewebe bone tissue, osseous tissue
Knochenhaft/Synostose synostosis, synosteosis
Knochenhaut/Beinhaut/Periost periosteum
Knochenhechte/Lepisosteiformes gars
Knochenkamm/Crista bony ridge, bone crest
Knochenmark bone marrow
Knochenmarkzelle/Myelozyt/ Myelocyt bone marrow cell, myelocyte
Knochenmehl bone meal
Knochenzüngler/ Knochenzünglerartige/ Osteoglossiformes osteoglossiforms
knöchern/Knochen... bony
Knockout-Mutation knockout mutation

Knotengeflecht

Knöllchen nodule
Knöllchenbakterien nodule bacteria (nitrogen-fixing bacteria)
Knolle tuber, bulb
- **kleine Knolle** tubercle, tuberculum
- **unterirdische Sprossknolle** tuber, underground stem-tuber
- **Zwiebelknolle** bulb

knollenförmig bulb-shaped
Knollenorgan/tuberöses Organ (Elektrorezeptor) tubiform organ
knollig tuberous, bulbous, bulbose
Knorpel cartilage
- **Blinzknorpel** *orn* cartilage of third eyelid
- **elastischer Knorpel** elastic cartilage
- **Faserknorpel** fibrous cartilage, fibrocartilage
- **Gelenksknorpel** articular cartilage
- **Hufbeinknorpel/Hufknorpel** coffin bone cartilage
- **hyaliner Knorpel** hyaline cartilage
- **Kehldeckelknorpel/Schließknorpel** epiglottic cartilage
- **Meckelscher Knorpel (Mandibulare)** Meckel's cartilage, mandibular cartilage
- **Ringknorpel/Cartilago cricoidea** annular cartilage, cricoid cartilage
- **Rippenknorpel** costal cartilage
- **Schaufelknorpel/Cartilago xiphoidea** xiphoid cartilage
- **Schildknorpel/Cartilago thyreoidea** thyroid cartilage
- **Schließknorpel/Resilium** resilium (flexible horny hinge)
- **Stellknorpeln/Cartilagines arytaenoideae** arytenoid cartilages
- **verkalkter Knorpel** calcified cartilage

Knorpelfische/Chondrichthyes cartilaginous fishes, chondrichthians
Knorpelgewebe/chondroides Gewebe (Parenchymknorpel) chondroid tissue
Knorpelhaft/Synchondrose synchondrosis
Knorpelhaut/Perichondrium perichondrium
knorpelig cartilaginous
Knorpelschädel/Chondrokranium cartilaginous neurocranium, chondrocranium
Knorpelzelle/Chondrozyt cartilage cell, chondrocyte

Knorren (an Baum)/Holzmaser/Maser/Maserknolle gnarl, burl, burr
knorrig gnarled
Knospe *bot* bud
- **Achselknospe/Seitenknospe** axillary bud, lateral bud
- **Adventivknospe** adventitious bud
- **Beiknospe/akzessorische Knospe** accessory bud
- **Blattknospe/Laubknospe** foliage bud
- **Blütenknospe** flower bud, floral bud
- **Endknospe/Terminalknospe** terminal bud
- **Erneuerungsknospe** renewal bud
- **Ersatzknospe/Proventivknospe** latent bud
- **geschlossene Knospe (mit K.schuppen)** protected bud
- **Gipfelknospe** apical bud
- **Keimknospe/Stammknospe/Sprossknospe/terminale Embryoknospe/Plumula** terminal embryonic bud, plumule
- **nackte Knospe/offene Knospe (ohne K.schuppen)** naked bud
- **ruhende Knospe/schlafende Knospe** resting bud, quiescent bud, dormant bud
- **Winterknospe/Überwinterungsknospe/Hibernakel** winter bud, perennating bud, hibernaculum, turio, turion

Knospe/Frustel/Frustula (Polypen) (asexual) bud, frustule
knospen/knospend bud, budding
Knospenanlage bud primordium
knospend budding
Knospendeckung/Ästivation/Aestivation estivation, aestivation
Knospenhülle bud (envelope) bracts
Knospenlage/Vernation vernation, ptyxis, prefoliation
- **aufgerollte K. (Farnblattentwicklung)** circinate vernation

Knospenlücke bud gap
Knospenruhe bud dormancy
Knospenschuppe/Knospendecke/Tegment (protective) bud scale, tegmentum
Knospenspore/Conidie conidium
Knospenstrahler/Blastoida (Echinoderma) blastoids
Knospung budding
Knospung/Frustulation (Polypen) (asexual) budding, frustulation
Knoten/Nodium node
Knotengeflecht nodal plexus

Knöterichgewächse

Knöterichgewächse/Polygonaceae
buckwheat family, dock family, knotweed family, smartweed family
knotig knotty, nodose; nodular, papular
Knüppeldamm corduroy road/walkway (made of logs)
knurren growl, snarl (wütend)
Koazervat coacervate
Kobalt/Cobalt cobalt
Koch's Postulat/Koch'sches Postulat Koch's postulate
Kochblutagar/Schokoladenagar chocolate agar
kochen cook, boil
Köcher (*Trichoptera*) case (*in some species:* tube)
Köcherfliegen/Haarflügler/Trichoptera caddis flies
Kochfleischbouillon/Fleischbrühe cooked-meat broth
Kochsalz (NaCl) table salt
Kochsalzlösung saline
• **physiologische K.** saline, physiological saline solution
Köder bait
kodieren/codieren encode, code
kodierender Strang/codierender Strang/Sinnstrang *gen* coding strand, sense strand
Kodierungskapazität/Codierungskapazität *gen* coding capacity
kodominant/codominant codominant
Kodominanz/Codominanz codominance
Kodon/Codon *gen* codon
Koevolution coevolution
Koexistenz coexistance
koexistieren coexist
Koffein/Thein caffeine, theine
Kohäsion cohesion
Kohäsionstheorie cohesion theory (cohesion-tension theory)
kohäsiv cohesive
Kohäsivität cohesiveness
Kohl cabbage, cole
Kohle coal
• **Anthrazit/Kohlenblende** anthracite, hard coal
• **Glanzbraunkohle/subbituminöse Kohle** subbituminous coal
• **Steinkohle/bituminöse Kohle** bituminous coal
• **Weichbraunkohle & Mattbraunkohle/Lignit** lignite
Kohlendioxid carbon dioxide
kohlendioxidliebend/kapnophil capnophilic
Kohlenhydrat carbohydrate

Kohlensäure (Karbonat/Carbonat) carbonic acid (carbonate)
Kohlenstoff carbon
Kohlenstoffbindung carbon bond
Kohlenstoffquelle carbon source
Kohlenstoffverbindung carbon compound
Kohlenwasserstoff hydrocarbon
• **Fluorchlorkohlenwasserstoffe (FCKW)** chlorofluorocarbons, chlorofluorinated hydrocarbons (CFCs)
• **Fluorkohlenwasserstoff** fluorinated hydrocarbon
• **chlorierter Kohlenwasserstoff** chlorinated hydrocarbon
Köhlersche Beleuchtung *micros* Koehler illumination
"Kohlrabi"/Bromatium/Gongylidie bromatium, gongylidium (*pl* gongylidia)
Kohnsche Pore pore of Kohn
Kohorte cohort
Koinzidenz coincidence
Koinzidenzfaktor/Coinzidenzfaktor coefficient of coincidence
Koitus/Kopulationsakt/Begattungsakt/Geschlechtsverkehr coitus, coition, copulation, sexual intercourse
Kojisäure kojic acid
Kokain cocaine
Kokastrauchgewächse/Erythroxylaceae coca family
kokkal/coccal coccal
kokkoid coccoid
Kokkolit/Coccolit/Kalkplättchen/Kalkkörperchen coccolith
Kokkus/Kugelbakterium (*pl* **Kokken**) coccus (*pl* cocci)
Kokon cocoon
Kokon (Spinnen: Eikokon) cocoon, egg sac
• **aufgehängter Kokon** pendant egg sac
• **Feenlämpchen** Japanese lantern
Kolben (Mais) ear, cob
Kolben/Blütenkolben (Inflorezenz) spadix
Kolben *lab/chem* flask
• **Destillierkolben** distilling flask, retort
• **Erlenmeyer Kolben** Erlenmeyer flask
• **Filtrierkolben/Filtrierflasche/Saugflasche** filter flask, vacuum flask
• **Messkolben** volumetric flask
• **Rundkolben/Siedegefäß** boiling flask with round bottom
• **Schüttelkolben** shake flask

kompetent

- **Schwanenhalskolben**
  swan-necked flask,
  S-necked flask, gooseneck flask
- **Stehkolben/Siedegefäß** Florence
  boiling flask, Florence flask
  (boiling flask with flat bottom)

Kolbenflügler/Fächerflügler/
 Strepsiptera
 twisted-winged insects, stylopids
Kolbenträgergewächse/
 Kolbenschmarotzer/
 Balanophoraceae
 balanophora family
Kolchizin/Colchicin colchicine
Koleoptile/Coleoptile *bot*
 coleoptile, plumule sheath
Koleorhiza/Coleorhiza/
 Wurzelscheide coleorhiza,
 root sheath, radicle sheath
Kolibris/Trochiliformes
 hummingbirds
kolinear/colinear colinear
Kolinearität/Colinearität
 colinearity
Kolk/Moorauge/Blänke pothole,
 deep pool
kollabieren (Lunge) collapse, deflate
Kollagen collagen
Kollaps collapse
kollateral collateral
Kollektorblende/Leuchtfeldblende
 field diaphragm
Kollektorlinse collector lens,
 collecting lens
Kollenchym/Collenchym
 collenchyma
- **Kantenkollenchym/**
  Eckenkollenchym
  angular collenchyma
- **Lückenkollenchym**
  lacunar collenchyma
- **Plattenkollenchym**
  lamellar collenchyma,
  tangential collenchyma

Kolletere/Colletere (Leimzotte/
 Drüsenzotte) colleter,
 multicellular glandular trichome
 (sticky/viscous secretions)
kollidieren collide
kolligative Eigenschaft
 (Teilchenzahl) colligative property
Köllikersche Grube (Branchiostoma)
 Kölliker's pit
Kollimationsblende/Spaltblende
 micros collimating slit
Kollimator collimator
kolline Stufe/Hügellandstufe/
 Vorgebirge foothills, foothill zone
Kollision (Enzymkinetik) collision
Kolobom coloboma
Kolon/Colon (Vertebraten:
 Grimmdarm) colon

kolonial/koloniebildend colonial,
 colony-forming
Kolonie colony
Koloniebank *gen* colony bank
koloniebildend/kolonial
 colony-forming, colonial
koloniebildende Einheit (KBE)/
 plaquebildende Einheit (PBE)
 (im Knochenmark gebildete
 Vorläuferzelle/Stammzelle)
 colony-forming unit (CFU),
 plaque-forming unit (PFU)
Kolonne/Turm (Bioreaktor) column
Kolonscheibe/aufsteigender Kolon/
 Colon ascendens (Wiederkäuer)
 ascending colon
Kolophonium colophony, rosin
Kolossalfaser/Riesenaxon/
 Riesenfaser giant axon
Kolossalzelle/Riesenzelle giant cell
Kolostralmilch/Biestmilch/Vormilch/
 Colostrum foremilk, colostrum
Kombinationsimpfstoff/
 Mischimpfstoff/Mischvakzine
 combination vaccine, mixed vaccine
Kombinationsregel/
 Unabhängigkeitsregel (Mendel)
 law/principle of random
 (independent) assortment
kommalos (DNA-Code)
 comma-less (DNA-code)
Kommandofunktion
 command function
Kommandoneuron
 command neuron
Kommensale/Mitesser commensal
Kommensalismus commensalism
Kommentkampf/Turnierkampf (z. B.
 bei Schlangen) ritualized fight
Kommissur commissure
Kommunikationskontakt/Macula
 communicans/Nexus/Gap junction
 (Zellkontakte) gap junction
Kompartiment compartment
kompartimentieren
 compartmentalize
Kompartimentierung
 compartmentalization,
 compartmentation
Kompasspflanze/Medianpflanze
 compass plant, heliotropic plant
kompatibel/verträglich compatible
Kompatibilität/Verträglichkeit
 compatibility
Kompensationsflug *entom*
 compensation flight
Kompensationspunkt
 compensation point
Kompensationstiefe/
 Kompensationsebene *mar*
 compensation depth
kompetent (Zelle/Kultur) competent

Kompetenz

Kompetenz (zur Blühinduktion) competence
Kompetition/Konkurrenz/Wettbewerb competition
Kompetitionshybridisierung gen competition hybridization
kompetitiv competitive
komplementär complementary
komplementäre Basenpaarung gen complementary base pairing
komplementäre DNA (cDNA) complementary DNA (cDNA)
komplementaritätsbestimmende Region (CDR) gen complementary determining region (CDR)
Komplementation complementation
- **alpha-Komplementation** gen alpha complementation
- **intraallele Komplementation** gen intraallelic complementation

Komplementationsgruppen complementation groups
Komplementbindung complement fixation
Komplementbindungsreaktion (KBR) complement binding reaction, complement fixation reaction
Komplettmedium complete medium, rich medium
Komplexbildner/Chelatbildner chelating agent, chelator
Komplexbildung/Chelatbildung chelation, chelate formation
komplexieren chelate
Komplexität complexity
- **chemische K.** chemical complexity
- **kinetische K.** kinetic complexity

Komponentenimpfstoff/Subunitimpfstoff/Subunitvakzine subunit vaccine
Kompost compost
Konchyliologie conchology
Kondensation condensation
Kondensationspunkt condensing point
Kondensationsreaktion/Dehydrierungsreaktion condensation reaction, dehydration reaction
Kondensator opt condenser; elektr capacitor
kondensieren condense
Kondensorblende/Aperturblende condenser diaphragm (iris diaphragm)
Kondensortrieb micros condenser adjustment knob, substage adjustment knob
konditional letale Mutation/bedingt letale Mutation conditional-lethal mutant

konditionieren ethol/med/chromat condition
konditioniertes Medium conditioned medium
Konditionierung ethol/med/chromat conditioning
- **klassische K./Pawlowsche K.** ethol classical conditioning, Pavlovian conditioning
- **operante K./operative K./instrumentelle K.** ethol operant conditioning, instrumental conditioning (trial-and-error learning)

Konfidenzgrenze/Vertrauensgrenze/Mutungsgrenze stat confidence limit
Konfidenzintervall/Vertrauensintervall/Vertrauensbereich stat confidence interval
Konfidenzniveau/Konfidenzwahrscheinlichkeit stat confidence level
konfluent cyt confluent
Konformation conformation
- **Knäuelkonformation/Schleifenkonformation** gen coil conformation, loop conformation
- **relaxiert/entspannt** relaxed (conformation)
- **Repulsionskonformation** gen repulsion conformation
- **Ringform** ring form, ring conformation
- **Schleifenkonformation/Knäuelkonformation** gen loop conformation, coil conformation
- **Sesselform (Cycloalkane)** chem chair conformation
- **Wannenform (Cycloalkane)** chem boat conformation

Konformationsepitop conformational/discontinuous epitope
Konformationspolymorphismus conformation polymorphism
kongenial/verwandt/gleichartig congenial
kongenital/angeboren/ererbt congenital
Konidie/Conidie conidium
Konidienträger/Conidienträger conidiophore
Konifere/Conifere/Nadelbaum conifer, coniferous tree
Königin queen
Königin-Futtersaft/Gelée Royale royal jelly, bee milk
Königin-Substanz queen substance

Königsfarngewächse/ Rispenfarngewächse/ Osmundaceae flowering fern family, cinnamon fern family, royal fern family
Konjugation conjugation
Konjugationsfortsatz/ Konjugationsrohr/Pilus (Bakterien) pilus
Konjugationskartierung conjugation mapping
konjugatives Plasmid conjugative plasmid, self-transmissible plasmid, transferable plasmid
konjugierte Bindung *chem* conjugated bond
Konjunktivalsack (Auge) conjunctival sac
Konkatamer concatemer
Konkauleszenz concaulescence
Konkrementvakuole (Placozoa) concrement vacuole
Konkurrent competitor
Konkurrenz/Kompetition/ Wettbewerb competition
Konkurrenz-Ausschluss-Prinzip/ Konkurrenz-Exklusions-Prinzip principle of competitive exclusion, exclusion principle (Gause's rule/ principle)
Konkurrenzhemmung/kompetitive Hemmung competitive inhibition
Konkurrenzvermeidung evasion of competition
konnatal/angeboren connate
Konnektiv/Mittelband connective
konsanguin/blutsverwandt consanguineous
konsanguine Ehe/Ehe unter Blutsverwandten/Verwandtenehe consanguineous marriage
Konsanguinität consanguinity
Konsensussequenz/ Consensussequenz *gen* consensus sequence
Konservatorium conservatory
konservieren/präservieren/ haltbar machen/erhalten conserve, preserve
Konservierungsstoff preservative
Konsistenz/Beschaffenheit consistency
konsistieren/beschaffen sein consist
Konsole (Fruchtkörper von Baumpilzen, z.B. *Fomes***)** bracket, conk (shelf-like sporophyte)
konspezifisch (von der gleichen Art) conspecific
Konspezifität conspecificity
konstante Region (*lg***)** constant region

konstanter Schwellenwert constant truncation
Konstanz constancy
Konstitution constitution
konstitutive Mutante constitutive mutant
konstitutives Gen/ Haushaltungsgen/Haushaltsgen housekeeping gene
konstitutives Heterochromatin constitutive heterochromatin
Konsum consumption
Konsument/Verbraucher consumer
Konsumhandlung consummatory act
Konsumverhalten consummatory behavior
Kontagionsindex/Infektionsindex contagion index
Kontagiosität contagiousness
Kontakthemmung/ Kontaktinhibition contact inhibition
Kontaktinsektizid contact insecticide
Kontaktparenchym boundary parenchyma
Kontaktpestizid contact pesticide
Kontaktpunktanalyse missing contact analysis
Kontaktverhalten huddling
Kontamination/Verunreinigung contamination
kontaminieren/verunreinigen contaminate
Kontinentalböschung continental slope
Kontinentalfuß continental rise
Kontinentalklima/Binnenklima/ Landklima continental climate
Kontinentallage continental location
Kontinentalrand continental fringe, continental edge
Kontinentalsockel/Festlandsockel/ Kontinentalschelf continental shelf
Kontinentalverschiebung/ Kontinentaldrift continental drift
Kontingenzkoeffizient coefficient of contingency
kontrahieren/zusammenziehen contract
kontraktile Vakuole/pulsierende Vakuole contractile vacuole, water expulsion vesicle (WEV)
Kontraktion contraction
Kontraktionsphase contraction period
Kontraktur contracture
Kontrast contrast
Kontrastbetonung *evol* character displacement

Kontrastfärbung/
  Differentialfärbung contrast staining, differential staining
kontrastieren contrast
kontrastieren/färben/einfärben *techn/micros* stain
Kontrastierung/Färben/Färbung/
  Einfärbung *techn/micros* stain, staining
Kontrazeption/
  Empfängnisverhütung contraception
kontrazeptiv/empfängnisverhütend contraceptive
Kontrazeptivum/
  empfängnisverhütendes Mittel/
  Verhütungsmittel contraceptive
Kontrolle control
  • **autogene Kontrolle** autogeneous control
  • **Genexpressionskontrolle/**
    Kontrolle der Genexpression control of gene expression
Kontrollelement, autonomes *gen* autonomous control element
Konturfeder/Umrissfeder *orn* contour feather
konventionelles Pseudogen conventional pseudogene
konvergent convergent
Konvergenz convergence
konvergieren converge
Konversion/Umwandlung/
  Übergang conversion
  • **Chromatidenkonversion** chromatid conversion
  • **Genkonversion** gene conversion
Konzentrationsgefälle/
  Konzentrationsgradient concentration gradient
konzentrieren concentrate
Konzeptakel (*Fucus*) conceptacle
Koog young/juvenile marsh
kooperative Bindung cooperative binding
Kooperation/Zusammenarbeit cooperation, colaboration
Kooperativität cooperativity
Kooperativitätskoeffizient/
  Hill-Koeffizient Hill coefficient, Hill constant
kooperieren/zusammenarbeiten cooperate, colaborate
Koordination coordination
koordinieren coordinate
Kopf/Cephalon/Caput head
Kopf (Fettmolekül) head
Kopf-an-Kopf-Wiederholungen *gen* head-to-head repeats
Kopf-Rumpf-Länge head-body length

Kopfbein/Kapitatum/Os capitatum capitate bone, capitate, capitatum
Kopfbrust(stück)/Cephalothorax cephalothorax
Köpfchen/Korb/Körbchen/
  Capitulum/Cephalium
  (Infloreszenz) capitulum, cephalium, flower head
Köpfchenblütler/Korbblütler/
  Asteraceae/Compositae sunflower family, daisy family, aster family, composite family
Kopfdrüse/Stirndrüse/Frontaldrüse cephalic gland, frontal gland
Kopfeibengewächse/
  Cephalotaxaceae cephalotaxus family, plum yew family
köpfen/kappen (Baum) pollarding, beheading, decapitation (of tree)
Kopffortsatz/Chordafortsatz head process, notochordal process
Kopffuß/Cephalopodium
  (Mollusken) head-foot
Kopffüßer/Cephalopoden cephalopods
kopfig capitate
Kopfkappe (Nautilus/
  Chaetognathen) hood
Kopflappen/Acron/Prostomium acron, prostomium
Kopfschild (Trilobiten) cephalon
Kopfschild/Clypeus shield, clypeus
Kopfschildschnecken/
  Kopfschildträger/Kephalaspidea/
  Cephalaspidea bubble shells
Kopfschlagader/Halsschlagader carotid, carotid artery
Kopfskelett head skeleton, cephalic skeleton
Kopfsoral (Flechten) capitate/capitiform sorelium
Kopfwachstum/kopfseitiges
  Wachstum head growth
Kopienzahl copy number
Koppel/Koppelweide *agr* enclosed pasture, fenced pasture
Koppel (Gruppe: Wale/Delphine/
  Seehunde) pod (whales/dolphins/seals)
koppeln/aneinander festmachen/
  verbinden couple
koppen (Pferde) suck wind
Kopplung *gen* linkage
  • **partielle Kopplung** partial linkage
Kopplungsanalyse *gen* linkage analysis
Kopplungsgleichgewicht linkage equilibrium
Kopplungsgruppe *gen* linkage group
  • **partielle K.** partial linkage group
Kopplungskarte *gen* linkage map

561   **Körperflüssigkeit** **K**

Kopplungspotential
coupling potential
Kopplungsungleichgewicht
linkage disequilibrium
koprophag/coprophag/
dungfressend/kotfressend
coprophagous, coprophagic
Koprophage/Coprophage/
Dungfresser/Kotfresser
coprophagist
Koprophagie/Coprophagie/
Dungfressen coprophagy
koprophil/coprophil coprophilic,
coprophilous
Kopulation/Begattung/Paarung
copulation, sexual union, mating
Kopulation/Kopulieren/Schäften
(Pfropfung) *hort* splice grafting,
whip grafting
Kopulation mit Gegenzunge
(Pfropfverfahren) *hort*
whip grafting, tongue grafting,
whip-and-tongue grafting
Kopulationsakt/Koitus/
Begattungsakt/
Geschlechtsverkehr
copulation, coitus,
sexual intercourse
kopulieren/begatten/paaren
copulate, mate
Kopulieren/Kopulation/Schäften
(Pfropfung) *hort*
splice grafting, whip grafting
Kopulieren mit Gegenzunge *hort*
whip and tongue grafting
Korallen corals
• **Blaukorallen/Coenothecalia/**
Helioporida blue corals
• **Dornkorallen/Dörnchenkorallen/**
schwarze Edelkorallen/
Antipatharia thorny corals,
black corals, antipatharians
• **Feuerkorallen/Milleporina**
milleporine hydrocorals,
stinging corals, fire corals
• **Hornkorallen/Rindenkorallen/**
Gorgonaria horny corals,
gorgonians, gorgonian corals
• **Lederkorallen/Alcyonaria/**
Alcyonacea soft corals,
alcyonaceans
• **Riffkorallen/Steinkorallen/**
Madreporaria/Scleractinia
stony corals,
madreporarian corals,
scleractinians
• **Rindenkorallen/Hornkorallen/**
Gorgonaria gorgonians,
gorgonian corals, horny corals
• **Stylasterida**
stylasterine hydrocorals
Korallenpfeiler coral/reef pinnacle

Korallenpilze/Keulenpilze/
Clavariaceae coral fungus family
Korallenriff coral reef
Korazidium/Coracidium
(Schwimmlarve: Cestoda)
coracidium
Korb/Körbchen/Köpfchen/
Capitulum/Cephalium
(Infloreszenz) capitulum,
cephalium, flower head
Korbblütler/Köpfchenblütler/
Asteraceae/Compositae
sunflower family, daisy family,
aster family, composite family
Körbchen/Pollenkörbchen/
Corbiculum (Bienen) pollen
basket, corbiculum
Korbvesikel/Stachelsaumbläschen/
Stachelsaumvesikel
coated vesicle
Korbzelle *neuro* basket cell
Kordon/Schnurbaum *hort* cordon
Koremie/Koremium/Coremium
coremium
Korepressor/Corepressor
corepressor
Kork/Phellem cork, phellem,
secondary bark
korkartig corky, suberose, suberous
Korkholzgewächse/Leitneriaceae
corkwood family
Korkkambium cork cambium,
phellogen
Korkrinde/Phelloderm
secondary cortex, phelloderm
Korksäure/Suberinsäure/
Octandisäure suberic acid,
octanedioic acid
Kormophyt/Achsenpflanze/
Sprosspflanze (Gefässpflanze)
cormophyte (vascular plant)
Kormus cormus
Korn kernel, corn, grain
Korn (Getreide) grain
Körnerdrüse/Giftdrüse (Amphibien)
granular gland, poison gland
Körnerfresser/Granivor *orn*
granivorous animal (bird)
Körnerschicht granule cell layer
Körnerzelle (Cerebellum)
granule cell
Kornfrucht/Caryopse/Karyopse/
"Kernfrucht" (Grasfrucht)
caryopsis, grain
Korngröße (Bodenpartikel)
soil particle size
Kornsekretdrüse prostatic gland
Körnung grain
Körper body, soma
Körperdecke/Integument body
covering, integument (skin)
Körperflüssigkeit body fluid

Körperhöhle/Leibeshöhle
body cavity
- **primäre K./Blastocöl** blastocoel, blastocele
- **sekundäre K./Cölom/Coelom** secondary body cavity, coelom, perigastrium

Köperoberfläche body surface; (als spez. Maß) body surface area

Körperpflege preening
- **soziale Körperpflege** social grooming (mammals), social preening (birds)

Körpertemperatur body temperature
Körperumriss body contours
Körperzelle/Somazelle/somatische Zelle body cell, somatic cell
Korrasion (*sensu lato* mechanische Erosion durch Wind/Wasser/ Schnee) *geol* corrasion
- **Windkorrasion/Sandschliff/ Windschliff** sand blasting, wind carving

Korrekturlesen proofreading
Korrelationskoeffizient *stat* correlation coefficient
- **Maßkorrelationskoeffizient/ Produkt-Moment-Korrelationskoeffizient** product-moment correlation coefficient
- **Rangkorrelationskoeffizient** rank correlation coefficient
- **Teilkorrelationskoeffizient** partial correlation coefficient

Korrosionsfäule/Weißfäule *fung* white rot
Korsetttierchen/Panzertierchen/ Loriciferen corset bearers, loriciferans
Kortikotropin/Corticotropin/ adrenokortikotropes Hormon/ adrenocorticotropes Hormon (ACTH) corticotropin, adrenocorticotropic hormone (ACTH)
Kortison/Cortison cortisone
Kosmopolit cosmopolitan, cosmopolite
kosmopolitisch/weltweit verbreitet cosmopolitan, occurring worldwide
Kost/Essen/Speise/Nahrung/Diät diet, food, feed, nutrition
Kosta/Costa/Rippe costa, rib
kostal costal
Kosten-Nutzen-Analyse cost-benefit analysis
Kot/Fäkalien feces
Kotdarm/Kotraum/Coprodaeum coprodeum
Köte/Fesselkopf/Fesselgelenk des Pferdes fetlock (metatarso-phalangeal articulation)
Kötenbehang/Kötenhaare/ Fesselhaare fetlock hair, feather
Kotransduktion/Cotransduktion cotransduction
Kotransfektion/Cotransfektion cotransfection
Kotwurst (*Arenicola*) casting (lugworm)
Kotyledone/Cotyledone/Keimblatt cotyledon, seminal leaf
Kousin(e)/Cousin(e) (ersten/zweiten Grades) (first/second) cousin
Kovarianz covariance
Kovarianzanalyse covariance analysis
Kraftflug *orn* powered flight, propulsive flight
kräftiges Wachstum vigorous growth
Kraftmikroskopie force microscopy
Kraftschlag/Wirkungsschlag *aer/orn* power stroke, effective stroke
Kragen/Ring/Annulus inferus (Rest des Velum partiale) *fung* ring, inferior annulus
Kragen (z. B. Mesosoma der Enteropneusten) collar
Kragengeißelzelle/Kragenzelle/ Choanozyt collar cell, choanocyte
Kragenlappen/Mantellappen (Brachiopoden) mantle lobe
Kragentiere/Hemichordaten/ Branchiotremata/Hemichordata hemichordates
krähen crow
Krähenbeerengewächse/ Empetraceae crowberry family
Kraken/Octobrachia/Octopoda octopuses, octopods
Kralle/Klaue *orn* claw, talon
Kralle/Nagel/Klaue/Unguis claw, nail
Kralle/Unguiculus small claw, nail, unguiculus
Krallenschwänze/Doppelschaler/ Diplostraca/Onychura clam shrimps & water fleas
Krallensegment/Krallenglied/ Krallensockel/Klauenglied/ Prätarsus (Insekten) pretarsus
Krallensohle/Subunguis sole of claw
Krallenwall wall of claw
Kranichvögel/Kranichverwandte/ Gruiformes cranes & rails and allies
krank sick, ill, diseased
krankhaft/pathologisch pathological
krankhafte Veränderung/Störung/ Läsion lesion

Kreislaufschock

Krankheit disease, illness
- **ansteckende Krankheit/ infektiöse Krankheit** contagious disease, infectious disease
- **Erbkrankheit** inheritable disease
- **erbliche Erkrankung/ Erbkrankheit** hereditary disease, genetic disease, inherited disease, heritable disorder
- **monogene Krankheit** monogenic disease
- **polygene Krankheit** polygenic disease
- **übertragbare Krankheit** transmissible disease, communicable disease
- **Zivilisationskrankheiten** diseases of civilization ("affluent peoples' diseases")

krankheitserregend/pathogen disease-causing, pathogenic
Krankheitserreger disease-causing agent, pathogen
Krankheitsüberträger transmitter of disease
Kranzdarm/Jejunum (Wiederkäuer) coronal stomach, coronal sinus
kranzförmig coronal, wreath-shaped
Kranzfühler/Fühlerkranztiere/ Tentaculaten tentaculates (bryozoans/phoronids/brachiopods)
Kranzfuß/Pes coronatus (Bauchfüße der Larven: Kleinschmetterlinge) proleg with crochets on planta in a circle
Kranzgefäße/Herzkranzgefäße coronary blood vessels (arteries/ veins)
Kranznaht (Schädeldach) coronal suture
Kranzquallen/Tiefseequallen/ Coronata coronate medusas
Krappgewächse/ Labkrautgewächse/ Rötegewächse/Rubiaceae madder family, bedstraw family
Kratersee crater lake
Krätze/Räude/Scabies/Milbenkrätze (Krätzmilbe: *Sarcoptes scabiei*) scabies; scab (domestic animals), mange
Kratzer (Nahrung abkratzend) scraper
Kratzer/Acanthocephala spiny-headed worms, thorny-headed worms
Krause (Federn/Haar) *zool* ruff
Krause-Körperchen/Krause'sches Körperchen Krause's end bulb, bouton
Kräuselkamm/Calamistrum calamistrum
Kräuselkrankheit (Blatt) leaf curl, crinkle
kräuseln (gekräuselt) (z. B. Blatt) ruffle (ruffled) (leaf with strongly wavy margin)
Kraut (*siehe*: Krautpflanze) herb (annual and biennial); wort, weed
Kräuter (Küchenkräuter) herbs
Kräuterbuch herbal
Kräutergarten herb garden
krautig herbaceous
krautige Pflanze herb, herbaceous plant; (nicht Gräser) forb
Krautpflanze herb, herbaceous plant; (Unkraut) weed; (nicht Gräser) forb (nongraminoid herbaceous plant)
Krautschicht herbaceous plant layer
Kreatin creatine
Krebs cancer (malignant neoplasm)
Krebs-Cyclus/Citrat-Zyklus/ Citratcyclus/Zitronensäurezyklus/ Tricarbonsäure-Zyklus Krebs cycle, citric acid cycle, tricarboxylic acid cycle (TCA cycle)
krebsartig cancerous
Krebse/Echte Krabben/Brachyura crabs
krebserregend/karzinogen/ carcinogen carcinogenic
krebserzeugend/onkogen/oncogen oncogenic, oncogenous
Krebsschere/Chela crab pincers, chela
Krebstiere/"Krebse"/Crustaceen crustaceans
Kreide/Kreidezeit (erdgeschichtliche Periode) Cretaceous, Cretaceous Period
Kreis circle
- **Blütenhüllkreis** perianth
- **Staubblattkreis** androecium

Kreisdiagramm pie chart
Kreisen (Fischschwärme) milling
Kreisen *orn* circling (flight)
kreisförmig/kreisrund/rund/ zirkular/zirkulär orbicular, circular, round
- **fast rund** orbiculate, nearly round

Kreislauf cycle
- **Blutkreislauf** circulation
- **Lebenskreislauf/Lebenszyklus** life cycle
- **Lungenkreislauf/ kleiner Blutkreislauf** pulmonary circulation
- **Nahrungskreislauf/Stoffkreislauf** *ecol* nutrient cycle
- **Wasserkreislauf** water cycle, hydrologic cycle

Kreislaufschock/Kreislaufkollaps circulatory shock

Kreislaufsystem

Kreislaufsystem/Zirkulationssystem circulatory system
Kreismünder/Rundmünder/Cyclostomata cyclostomes
Kreiswirbler/Stelmatopoda/Gymnolaemata gymnolaemates, "naked throat" bryozoans
kreißen be in labor
Kremplinge/Paxillaceae Paxillus family
Krenal/Quellzone crenal
Krenon (Lebensgemeinschaft der Quellzone) crenon
Kreuzband *anat* cruciate ligament
Kreuzbefruchtung/Fremdbefruchtung/Allogamie cross-fertilization, allogamy, xenogamy
Kreuzbein/Sakrum/Os sacrum sacrum
Kreuzbeinwirbel/Sakralwirbel sacral vertebra
Kreuzbestäubung/Fremdbestäubung cross-pollination
Kreuzblumengewächse/Kreuzblümchengewächse/Polygalaceae milkwort family
Kreuzblütler/Kreuzblütlergewächse/Cruciferae/Brassicaceae cabbage family, mustard family
Kreuzdorngewächse/Rhamnaceae buckthorn family, coffeeberry family
kreuzen/züchten cross, crossbreed, breed, interbreed
kreuzförmige DNA cruciform DNA
kreuzförmige Struktur cruciform structure
Kreuzgalopp disunited canter
kreuzgegenständig/gekreuztgegenständig/dekussiert decussate
Kreuzhybridisierung cross hybridization
Kreuzimmunität (übergreifender Schutz) cross protection
Kreuzlähme (*Trypanosoma equinum*) Mal de Calderas
Kreuzprobe *immun* cross-matching
kreuzreagierendes Antigen cross-reacting antigen
Kreuzreaktion cross-reaction
- **CRM+ (positiv für kreuzreagierendes Material)** CRM+ (positive for cross-reacting material)
kreuzreaktiv cross-reactive
Kreuzreaktivität cross-reactivity
Kreuzstrom-Filtration cross flow filtration

Kreuztisch *micros* mechanical stage
Kreuzung/Züchtung crossing, cross, crossbre(e)d, breed, crossbreeding, interbreeding; (Kreuzungsprodukt) cross, breed
- **aus der Kreuzung entfernt oder nicht verwandter Individuen gezüchtet** outbred
- **Dihybridkreuzung** dihybrid cross
- **Doppelkreuzung** double cross
- **Drei-Faktor-Kreuzung** three-point testcross
- **Einfachkreuzung** single cross
- **Herauskreuzen/Auskreuzen** outcrossing
- **Monohybridkreuzung** monohybrid cross
- **nicht verwandte Individuen kreuzen** outbreed
- **Testkreuzung** testcross
- **Überbrückungskreuzung** bridging cross
Kreuzungsbarriere/Artgrenze species barrier
Kreuzungstyp/Paarungstyp mating type
Kreuzwirbel/Sakralwirbel sacral vertebra
kriechen crawl
kriechend (am Boden entlang/an Nodien bewurzelnd) *bot* creeping, crawling, repent
Kriechfrustel (Polypenknospe) creeping frustule (asexual polyp bud)
Kriechfuß (Mollusken) creeping foot
Kriechpflanze creeper, trailing plant
Kriechsohle/Trivium (z.B. Holothuriden) foot sole, trivium
Kriechspross/oberirdischer Ausläufer (photophil) runner, sarment
Kriechspuren/Repichnia *paleo* crawling traces
Kriechtiere/Reptilien reptiles
Krill/Leuchtkrebse/Euphausiacea krill (and allies)
Kristallisation cristallization
Kristallisationskern/Kristallisationskeim cristallization nucleus
kristallisieren crystallize
Kristallkegel/Kristallkörper/Linsenzylinder/Conus crystalline cone
Kristallographie crystallography
Kristallstiel (Muscheln) crystalline style
Kristallstruktur crystal structure, crystalline structure
Kristallzelle/Cristallogenzelle/Sempersche Zelle crystal cell

Kritisch-Punkt-Trocknung
critical point drying (CPD)
kritischer Punkt critical point
Krokodile/Panzerechsen/Crocodilia
crocodiles
Krokodilklemme *lab* alligator clip
Krokusgewächse/
Zeitlosengewächse/Colchicaceae
crocus family
Kronbein/mittlere Phalanx (Pferd)
coronary bone, small/short pastern
bone, second phalanx
Kronblätter petals, corolla
Krone crown
• **Baumkrone/Stammkrone**
treetop, crown
• **Blütenkrone/Blumenkrone**
corolla
• **Haarkranz (am Huf)** coronet
Kronendach *for* (forest) canopy
Kronenregion/Kronenschicht
canopy
• **mittlere K. (Baumkrone)**
subcanopy, lower canopy
• **obere K. (Baumkrone)** canopy,
crown layer, upper canopy;
overstory
Kronenveredlung *hort*
crown grafting
Krongelenk/Fesselbeingelenk
pastern joint
Kronlederhaut (Pferd)
coronary dermis
Kronsaum/Kronband (Pferd)
coronary band, coronary ring,
coronary cushion
Kropf/Vormagen crop
Kropfmilch/Kropfsekret (Tauben)
crop milk, pigeon milk (milky
secretion from crop lining)
Kröten toads
Krötenschlangensterne/
Phrynophiurida phrynophiurids
Krugblattgewächse/Cephalotaceae
Australian pitcher-plant family
krugförmig pitcher-shaped,
urceolate
Krugpflanzengewächse/
Schlauchpflanzengewächse/
Sarraceniaceae
pitcher-plant family
Krume/Bodenkrume/Oberboden
topsoil
Krümelstruktur (Boden)
crumb structure
Krummholz *for* stunted, miniature
forest; Krummholz
krummläufig (Samenanlage)
campylotropous
krummschaftig twisted shoot,
contorted stem
Krümmung contortion, bending

Krümmungsbewegung
campylokinesis
Kruppe (Hinterteil: Pferd) croup
Krüppelfüße/Stummelfüßchen/
Crepidotaceae Crepidotus family,
crep fungus family
krüppelig/krüppelhaft stunted
Krüppelkiefer stunted pine
Krüppelwuchs/Krüppelform
stunted growth, stuntedness
Krustenanemonen/Zoantharia
zoanthids
krustenbildend encrusting
Krustenflechten crustose lichens
Krustenpilz crustose fungus
krustig crusty, crustose, crustaceous
Kryophyt/Kältepflanze cryophyt,
plant preferring low temperatures
Kryostat cryostat
Kryostatschnitt *micros*
cryostat section
Kryoultramikrotomie
cryoultramicrotomy
Kryptobiose cryptobiosis
Kryptogame cryptogam
Kryptomonaden/Cryptophyceae
cryptomonads
Kryptophyt/Cryptophyt/Geophyt/
Erdpflanze/Staudengewächs
cryptophyte, geophyte,
geocryptophyte *sensu lato*
Küchenkräuter herbs, culinary herbs
Kuckucksspeichel (Schaumzikaden)
spittle ("cuckoo spit" of spittlebugs)
Kuckucksvögel/Cuculiformes
cuckoos and turacos and allies
Kugel-Stab-Model/Stab-Kugel-
Model *chem* ball-and-stick model,
stick-and-ball model
Kugelbakterium/Kokkus
(*pl* **Kokken)** coccus (*pl* cocci)
Kugelbettreaktor (Bioreaktor)
bead-bed reactor
Kugelblumengewächse/
Globulariaceae globe daisy family,
globularia family
Kugelfischverwandte/
Tetraodontiformes/Plectognathi
plectognath fishes
Kugelgelenk ball-and-socket joint,
spheroid joint
kugelig/sphärisch spherical
Kuhfladen (*vulg* **"Kuhplatscher")**
cow dropping, cow pat, cow dung
kühlen (gefrieren) cool (freeze)
• **in den Kühlschrank stellen**
refrigerate
Kühler *lab* condenser
• **Liebigkühler** Liebig condenser
• **Rückflusskühler** reflux condenser
Kühlfach/Gefrierfach
freezer compartment

Kühlfinger *lab* cold finger (finger-type condenser)
Kühlhaus cold store
Kühlmantel *lab* condenser jacket
Kühlraum/Gefrierraum cold-storage room, cold store, "freezer"
Kühlschlange *lab* cooling coil, condensing coil
Kühlschrank refrigerator, fridge
Kühltruhe/Gefriertruhe/ Gefrierschrank freezer
Küken chick
kulinarisch culinary
kultivierbar cultivatible, arable
kultivierbares Land arable land
kultivieren *agr* cultivate; *micb* culture, culturing
Kultur culture
- **Anreicherungskultur** enrichment culture
- **Ausstrichkultur** streak culture
- **Blutkultur** blood culture
- **diskontinuierliche Kultur/Batch-Kultur/Satzkultur** batch culture
- **Eikultur** chicken embryo culture
- **Einstichkultur/Stichkultur (Stichagar)** stab culture
- **Eintauchkultur** submerged culture
- **Erhaltungskultur** maintenance culture
- **Gewebekultur** tissue culture
- **kontinuierliche Kultur** continuous culture, maintenance culture
- **Mischkultur** mixed culture
- **Oberflächenkultur** surface culture
- **Perfusionskultur** perfusion culture
- **Reinkultur** pure culture, axenic culture
- **Rollerflaschenkultur** roller tube culture
- **Satzkultur/Batch-Kultur/ diskontinuierliche Kultur** batch culture
- **Schrägkultur (Schrägagar)** slant culture, slope culture
- **Schüttelkultur** shake culture
- **Stammkultur** stem culture, stock culture
- **statische Kultur** static culture
- **Stichkultur/Einstichkultur (Stichagar)** stab culture
- **Submerskultur** submerged culture
- **Synchronkultur** synchronous culture
- **Verdünnungs-Schüttelkultur** dilution shake culture
- **Zellkultur** cell culture

Kulturfolger synanthropic species, anthropophilic species (plant or animal)
Kulturform domestic variety, cultivated variety, cultivar
Kulturlandschaft cultural landscape
Kulturmedium/Medium/ Nährmedium medium, culture medium
- **Anreicherungsmedium** enrichment medium
- **Differenzierungsmedium** differential medium
- **Elektivmedium/Selektivmedium** selective medium
- **Komplettmedium/Vollmedium** complete medium
- **komplexes Medium** complex medium
- **Mangelmedium** deficiency medium
- **Minimalmedium** minimal medium
- **Selektivmedium** selective medium
- **synthetisches Medium (chem. definiertes Medium)** defined medium
- **Vollmedium** complete medium

Kulturpflanze crop plant, cultivated plant
Kulturschale culture dish
Kulturwald/Forst cultivated forest, tree plantation
Kumazeen/Cumacea cumaceans
Kümmerwuchs/Nanismus dwarfishness, nanism, microsomia
Kundschafter/Späher/Pfadfinder (z. B. soziale Insekten; Spurbiene) scout
Kunstharz synthetic resin
künstliche Befruchtung artificial insemination
künstliches Bakterienchromosom (BAC) bacterial artificial chromosome (BAC)
künstliches Hefechromosom (YAC) yeast artificial chromosome (YAC)
Kupfer copper
Kupfernetz *micros* copper grid
Kupffer-Zelle/Kupffer-Sternzelle Kupffer cell, stellate reticuloendothelial cell
Kuppel (Ctenophoren) dome
Kupula/Cupula/Blütenbecher/ Fruchtbecher cupula, cupule
Kürbisfrucht/Gurkenfrucht (eine Panzerbeere) pepo, gourd
Kürbisgewächse/Cucurbitaceae gourd family, pumpkin family, cucumber family
kurzborstig hispid

Kybernetik

kurzer Knochen/Os brevis
short bone
kurzflüglig brachypterous,
with short wings
kurzfüßig/mit kurzem Fuß
brachypodous, with short legs
kurzgehörnt short-horned,
brachycerous
kurzkettig short-chain
kurzlebig/hinfällig short-living,
short-lived, ephemeral, fugacious,
soon disappearing; (früh abfallend/
früh verblühend) fugacious,
falling off unusually early
● **langlebig** long-lived, long-living
kurzlebige(s) Pflanze/Tier
ephemeral
kurzphasige Einstreuung
short period interspersion
● **langphasige Einstreuung**
long period interspersion
Kurzschluss short-circuit(ing)
**kurzschwänzig (Krabben: Abdomen
und den Thorax geklappt)**
brachyural, brachyurous
Kurzstreckentransport
short-distance transport
Kurztagpflanze short-day plant
Kurztrieb short shoot, short axis
Kurzweg-Destillation
flash distillation
Kurzzeitgedächtnis
short-term memory
Küste (*siehe auch:* Küstenlinie)
coast, seaboard, shore
● **Anschwemmungsküste/
Anwachsküste** alluvial coast,
shoreline of progradation
● **Fjordküste** fjord(ed) coast,
fjord shoreline
● **Flachküste** low coast
● **Kliffküste** cliffed coast

● **Riasküste** ria coast, ria shoreline
● **Schärenküste** skerry coast,
schären-type shoreline
● **Steilküste** steep coast,
steep shore
**küstenbewohnend/uferbewohnend
(Meeresküste)** littoral
Küstendüne coastal dune
Küstenebene coastal plain
Küstengewässer coastal waters
Küstenklima/Meeresklima
maritime/coastal/oceanic climate
Küstenlinie coastline, shoreline,
waterline
● **Auftauchküste/Hebungsküste**
shoreline of emergence,
shoreline of elevation
● **Untertauchküste/Senkungsküste**
shoreline of submergence,
shoreline of depression
Küstenriff/Strandriff/Saumriff
fringing reef
Küstensaum/Ufersaum
littoral fringe
Küstenstreifen/Küstenstrich
coastline, coastal strip
Küstensumpf coastal swamp/marsh
Küstenvegetation
maritime/coastal vegetation
Küstenwüste coastal desert
Küstenzone/Uferzone coastal zone,
littoral zone
kutan/Haut... cutaneous
Kutikula/Cuticula cuticle, cuticula
Kutikularskelett cuticular skeleton
**Kutis/Cutis (eigentliche Haut;
Epidermis & Dermis)** skin, cutis
Kutteln/Gekröse (Rind: Kaldaunen)
tripes
Küvette (für Spektrometer) cuvette,
spectrophotometer tube
Kybernetik cybernetics

Labferment

Labferment/Rennin/Chymosin rennet, lab ferment, rennin, chymosin
Labialdrüse labial gland
Labialnaht labial suture
Labialpalpus/Labialtaster/ Lippentaster/Palpus labialis labial palp, labipalp, labial feeler, palp, palpus
labidognath labidognathous
Labium/Schamlippe labium
Labium/Unterlippe (Vertebraten) labium, lower lip
• **2. Maxille (Insekten)** labium, second maxilla
Labkrautgewächse/Rötegewächse/ Krappgewächse/Rubiaceae bedstraw family, madder family
Labmagen/Abomasus abomasum, fourth stomach, reed, rennet-stomach
Labor laboratory, lab
Laborant(in) lab worker
Laborassistent(in)/technische(r) Assistent(in) technical lab assistant, laboratory technician, lab technician
Laborbedarf labware, laboratory supplies, lab supplies
Laborbefund laboratory findings
Laboreinrichtung/Laborausstattung laboratory facilities, lab facilities
Laborgerät laboratory equipment, lab equipment
Laborkittel laboratory coat, labcoat
Labormaßstab laboratory scale, lab scale
Laborschürze laboratory apron
Laborsicherheit laboratory safety
Laborsicherheitsstufe physical containment (level)
Labortisch/Labor-Werkbank laboratory table, lab table, laboratory bench, lab bench
Laborwaage laboratory balance
Laborzange tongs
Labrum/Oberlippe labrum, upper lip
Labyrinthplazenta labyrithine placenta, hemoendothelial placenta
lac-**Operon** *lac* operon
Lache/Pfütze puddle
Lachse & Lachsverwandte/ Salmoniformes salmon & trout
Lackglanz glossiness
Lactose-Repressor (*lac***-Repressor)** lactose repressor
Laden (der Palpen) filling
Ladung/elektrische Ladung charge
Ladungstrennung charge separation

lag-Phase/Adaptationsphase/ Anlaufphase/Latenzphase/ Inkubationsphase lag phase, incubation phase, latent phase, establishment phase
Lage (in Bezug)/Position position
Lage (Ort) location
Lager (Bau) den (lions), lair (game)
Lager/Thallus *bot* thallus
lagern (Holz) season, store
Lagerpflanze/Thallophyt thallophyte
Lagerrand/Excipulum thallinum (Flechten) thalline exciple
Lagertank storage tank
Lagg (Randsumpf von Hochmooren) lagg (drainage channel within a bog)
Lagune lagoon
lahm lame
Lähme lameness
lahmen *vb* lame
lähmen/paralysieren paralyze
Lähmung/Paralyse paralysis, paralyzation
Laich spawn (many small eggs of aquatic animals: *esp.* fish/ mollusks)
• **Fischlaich/Fischeier (Rogen)** fish eggs (roe)
• **Froschlaich** frog eggs
• **Muschellaich** spat (bivalves)
laichen spawn
Laichkrautgewächse/ Potamogetonaceae pondweed family
Laichplatz/Laichstätte/Laichgründe spawning ground
Laichschnur/Laichkette egg string
Laichwanderung *ichth* run
Laktat (Milchsäure) lactate (lactic acid)
Laktatgärung/Milchsäuregärung lactic acid fermentation, lactic fermentation
Laktation lactation
Laktose/Lactose (Milchzucker) lactose (milk sugar)
Lakune/Spalt/Hohlraum lacuna, space, cavity
Lakunensystem lacunar system
Lambdanaht/Sutura lambdoidea (Schädeldach) lambdoid suture
Lamblienruhr/Giardiasis (*Giardia lamblia***)** giardiasis
Lamelle/Lamina lamella, lamina
• **Pilzlamelle** gill
Lamellenkiemer/Blattkiemer/ Eulamellibranchia eulamellibranch bivalves
Lamellenknochen/lamellärer Knochen laminar bone

Längenwachstum

Lamellenkörperchen/ Endkörperchen/Pacinisches Körperchen lamellated corpuscle, Pacinian body, Pacinian corpuscle
Lamellenpilz/Blätterpilz gill fungus, gill mushroom
Lamellenpilze/Agaricales agarics
Lamellentrama gill trama, dissepiment
Lamiaceae/Labiatae/ Lippenblütengewächse/ Lippenblütler deadnettle family, mint family
Lamina/Lamelle (Platte/Spreite/Blatt) lamina (thin layer), lamella (blade)
- **Basallamina/Basalmembran** basement membrane, basal lamina
- **Blattspreite** leaf blade, leaf lamina *(verschiedene Formen der Blattspreite im Englisch-Deutsch Teil unter: leaf shape)*
- **Kernlamina/Kernfaserschicht** nuclear lamina
- **Zahnleiste** dental lamina
laminale Plazentation/ flächenständige Plazentation *bot* laminar/laminate/lamellate placentation
laminare Strömung/ Schichtströmung laminar flow
Lamm/Schäfchen lamb, little sheep
- **frischgeborenes Lamm** yeanling
- **Lamm gebären** lambing
Lampenbürstenchromosom lampbrush chromosome
Lampenmuscheln/Armfüßer/ Brachiopoden lampshells, brachiopods
Land (*pl*** Länder)** land, soil, ground; country (*pl* countries); (political) nation
- **Entwicklungsländer** developing countries, peripheral countries, less-developed countries (LDCs)
- **Industrieländer** developed countries, industrialized nations, core countries, more-developed countries (MDCs)
- **Schwellenländer** semi-peripheral countries
Landbauprodukt/ landwirtschaftliches Produkt crop; (leicht verkäufliches L.) cash crop
Landbevölkerung rural population
landeinwärts inland
Ländereien lands, land property/ properties (extensive), domain
Landerzeugnis/Naturerzeugnis produce, crop

Landgewinnung land reclamation
Landleben land life, life on land, terrestrial life; *anthrop* country life, rural life
landlebend/terrestrisch terrestrial, land-dwelling
ländlich rural
ländlicher Raum rural environment
Landlungenschnecken/ Stylommatophora (Pulmonata) land snails
Landökosystem terrestrial ecosystem
Landpflanze terrestrial plant
Landraubtiere/Fissipedia terrestrial carnivores
Landschaft landscape, countryside; (Gelände/Terrain) terrain
- **Hügellandschaft** hill country, hilly terrain, rolling countryside
- **Kulturlandschaft** cultural landscape
- **Moorlandschaft** moorland
- **Naturlandschaft** natural landscape, natural environment, natural setting
- **Sumpflandschaft/Sumpfland** swampland, moorland
Landschaftsbau landscape designing
Landschaftsgestaltung landscaping, landscape architecture
Landschaftsökologie landscape ecology
Landschaftspflege landscape management
Landschaftspfleger environmental warden, landscape manager
Landschaftsplaner/ Landschaftsarchitekt landscape architect
Landschaftsplanung/ Landschaftsarchitektur landscaping, landscape planning, landscape architecture
Landschaftsschutzgebiet wilderness reserve, wilderness sanctuary
Landschildkröten tortoises
Landtiere land animals, terrestrial animals
Landwind offshore wind
Landwirbeltiere/Tetrapoden terrestrial vertebrates, tetrapods
Landwirt farmer; (diplomierter L./ Agronom) agronomist
Landwirtschaft agriculture, farming
lange terminale Sequenzwiederholung (LTR) *gen* long terminal repeat (LTR)
Längengrad degree of longitude
Längenwachstum longitudinal growth

langer Knochen 570

- **Streckungswachstum (Zuwachs)** elongational growth, extension growth
- **langer Knochen/Röhrenknochen/ Os longum** long bone (hollow/tubular bone)
- **Langerhanssche Insel/ Pankreasinsel/Inselorgan** islet of Langerhans, pancreatic islet
- **Langerhanssche Zelle/Langerhans-Zelle** *immun* Langerhans cell
- **langes eingeschobenes nukleäres Element (LINE)** *gen* long interspersed nuclear element (LINE)
- **Langhans-Riesenzelle** Langhans giant cell
- **Langhans-Zelle** Langhans cell
- **Langhölzer** logs
- **langkettig** long-chain
- **langlebig** long-lived, long-living
  - **kurzlebig/hinfällig** short-living, short-lived, ephemeral, fugacious, soon disappearing; (früh abfallend/ früh verblühend) fugacious, falling off unusually early
- **Langlebigkeit** longevity
- **länglich** oblong
- **langphasige Einstreuung** long period interspersion
- **Längsader** *entom* longitudinal vein
- **längsaderig/längsnervig/ streifennervig** *bot* striate veined
- **Längsaderung/Streifennervatur** *bot* longitudinal venation, striate venation
- **langsamwachsend** slow-growing
- **Längskonstante** length constant
- **Längsmuskel** longitudinal muscle
- **Längsschnitt** longisection, longitudinal section, long section
- **Längsteilung** *cyt* longitudinal division, fission
- **Langstreckentransport** *bot/physio* long-distance transport
- **Langtagspflanze** long-day plant
- **Langtrieb** *bot* long shoot, long axis
- **Langzeitgedächtnis** long-term memory
- **Langzeitpotenzierung** long-term potentiation (LTP)
- **Lanosterin/Lanosterol** lanosterol
- **Lanthionin** lanthionine
- **Lanzenseeigel/Cidaroida** cidaroids
- **Lanzettfischchen/Cephalochordaten (bzw. Amphioxiformes)** lancelet, cephalochordates
- **lanzettförmig/lanzettlich** lanceolate
- **Lappen** lobe
  - **Afterlappen/Afterschild/ Pygidium (Telson der Arthropoden)** pygidium, caudal shield
  - **Haftlappen/Pulvillus/Lobulus lateralis** adhesive pad, pulvillus
  - **Hautlappen/Fleischlappen** lappet
  - **Hinterhauptslappen/ Okzipitallappen/Lobus occipitalis** *neuro* occipital lobe
  - **Hypophysenhinterlappen/ Neurohypophyse** neurohypophysis, posterior lobe of pituitary gland
  - **Hypophysenvorderlappen/ Adenohypophyse** adenohypophysis, anterior lobe of pituitary gland
  - **Kehllappen (Vögel/Reptilien: Hautlappen)** dewlap, wattle
  - **Kopflappen/Acron/Prostomium** acron, prostomium
  - **Kragenlappen/Mantellappen (Brachiopoden)** mantle lobe
  - **Narbenlappen** stigmatic lobe
  - **Pollappen (Furchung)** polar lobe
  - **Randlappen (Scyphozoa)** lappet(s), flap(s)
  - **Scheitellappen/ Lobus parietalis** *neuro* parietal lobe
  - **Schläfenlappen/Lobus temporalis** *neuro* temporal lobe
  - **Sehlappen/Lobus opticus** visual lobe, optic lobe
  - **Stirnlappen/Frontallobus** *neuro* frontal lobe
- **Lappenfarne/Sumpffarngewächse/ Thelypteridaceae** marsh fern family
- **Lappenmünder/Lippenmünder/ Cheilostomata (Bryozoen)** cheilostomates
- **Lappentaucher/Podicipediformes** grebes
- **lappig/gelappt** lobed, lobate
- **Lardizabalaceae/ Fingerfruchtgewächse** akebia family, lardizabala family
- **Larve** larva (*pl* larvas/larvae)
  - **Acanthor-Larve/Hakenlarve** acanthor larva
  - **Afterraupe (Blattwespenlarven)** eruciform larva with more than 5 pairs of abdominal prolegs (Tenthredinidae)
  - **Ameisenlöwe (Larve der Ameisenjungfer)** doodlebug (antlion larva)
  - **Auricularia-Larve** auricularia
  - **Becherkeim/Becherlarve/ Gastrula** gastrula
  - **Blasenwurm/Finne/Cysticercus (Bandwurmlarve)** bladderworm, cysticercus

Latenzzeit — L

- **Cercarie/Zerkarie/Schwanzlarve**
  cercaria
- **Coracidium-Larve/Korazidium
  (Schwimmlarve: Cestoda)**
  coracidium
- **Cydippe-Larve** cydippid larva
- **Cypris-Larve** cypris larva
- **Dauerlarve** dauer larva (*not
  translated!*: temporarily dormant
  larva)
- **Doliolaria-Larve/Tönnchenlarve
  (Vitellaria-Larve)**
  doliolaria larva (vitellaria larva)
- **Dotterlarve/Vitellaria-Larve**
  vitellaria larva
- **Drahtwurm (Elateriden-/
  Schnellkäfer-Larve)** wireworm
  (clickbeetle larva)
- **Eilarve/Junglarve/Primärlarve**
  primary larva
- **Engerling (im Boden lebende
  Larve der Blatthornkäfer)**
  grub (scarabaeiform larva)
- **Erntemilbenlarve (parasitäre
  rote Milben)** chigger, "red bug",
  harvest mite
- **Finne/Blasenwurm/Cysticercus
  (Bandwurmlarve)** bladderworm,
  cysticercus
- **Glochidium** glochidium
- **Hakenlarve/Acanthor-Larve**
  acanthor larva
- **>Sechshakenlarve/
  Oncosphaera-Larve (Cestoda)**
  hexacanth larva, hooked larva,
  oncosphere
- **Hüllglockenlarve/Pericalymma
  (*Yoldia*)** pericalymma larva
  (lecithotroph), test-cell larva
  (trochophore of *Yoldia*)
- **Junglarve/Eilarve/Primärlarve**
  primary larva
- **Käsefliegenlarve (Piophilidae)**
  cheese-skipper
- **Korazidium/Coracidium-Larve
  (Schwimmlarve: Cestaoda)**
  coracidium
- **Larva coarctata/Scheinpuppe/
  Pseudocrysalis** coarctate larva,
  coarctate pupa, pseudocrysalis
- **Lasidium** lasidium
- **Made (apode Larve)**
  maggot (apodal larva)
- **Metacercarie** metacercaria,
  adolescaria
- **Mirazidium/Miracidium (*pl*
  Miracidien) (Digenea-Larve)**
  miracidium (fluke larva)
- **Mitraria-Larve (eine
  Metatrochophora)** mitraria
- **Müllersche Larve** Müller's larva
- **Mysis-Larve** mysis larva

- **Naupliuslarve/Nauplius**
  naupliar larva, nauplius
- **Onkosphäre/Oncosphaere
  (unbewimperte Cestodenlarve)**
  oncosphere
- **Pilidium-Larve (Nemertini)**
  pilidium larva
- **Primärlarve/Junglarve/Eilarve**
  primary larva
- **Procercoid/Prozerkoid (Cestoda-
  Postlarve)** procercoid
- **Rattenschwanzlarve** rat-tailed
  larva, rat-tailed maggot
- **Redie (Trematoden)** redia
- **Rotiger/Pseudotrochophora**
  rotiger, pseudotrochophore
- **Schwanzlarve/Zerkarie/Cercarie**
  cercaria
- **Sechshakenlarve/Oncosphaera-
  Larve** hexacanth larva,
  hooked larva, oncosphere
- **Segellarve/Veliger/Veligerlarve**
  veliger larva
- **Tönnchenlarve/Doliolaria-Larve
  (Vitellaria-Larve)**
  doliolaria larva (vitellaria larva)
- **Tornaria-Larve** tornaria larva
- **Veliger-Larve/Segellarve**
  veliger larva
- **Wimperkranzlarve/Trochophora**
  trochophore larva
- **Wimperlarve** ciliated larva
- **Zerkarie/Cercarie/Schwanzlarve**
  cercaria
- **Zoëa (Decapoden-Larve)**
  zoëa (decapod crustacean larva)

larvenförmig larviform
**Larvenschale/Embryonalschale/
Embryonalgewinde/Protoconch**
protoconch (nuclear whorls)
larvipar larviparous
Larviparie larvipary
Lasidium (Larve) lasidium
**Läsion/Schädigung/Verletzung/
Störung** lesion
**Last (Ausmaß eines
Parasitenbefalls)** burden
Lasttier/Tragtier beast of burden,
pack animal
**latent/verborgen/unsichtbar/
versteckt** latent
Latenz latency
Latenzei/Dauerei resting egg,
dormant egg (winter egg)
**Latenzphase/Adaptationsphase/
Anlaufphase/Inkubationsphase/
lag-Phase**
latent phase, incubation phase,
establishment phase, lag phase
Latenzzeit (Inkubationszeit)
latency period, latent period
(incubation period)

L lateral 572

lateral/seitlich lateral
Lateralauge/Seitenauge (Ocellus)
lateral eye, lateral ocellus
- Stemma (Punktauge/Einzelauge
bei Insekten-Larven)
stemma (*pl* stemmata/stemmas),
lateral eye/ocellus
Lateralherz lateral heart
Lateralorgan spur shoot
Lateralpore lateral pore
Lateralvergrößerung/
Seitenverhältnis/Seitenmaßstab/
Abbildungsmaßstab *micros*
lateral magnification
Laterit (Boden) laterite
Lateritisierung/Laterisation/
Lateritbildung (Boden)
laterization, latosolization
Laterne des Aristoteles
Aristotle's lantern
- Aurikel auricle
- Epiphyse epiphysis
- Interpyramidalmuskel/
interpyramidaler Muskel
interpyramid muscle,
comminator muscle
- Kompass compass
- Pyramide pyramid
- Rotula rotule
- Zahnführung tooth guide
Laternenfische/Myctophiformes
lanternfishes & blackchins
Latte (aus Holz) lath, plank
Laub foliage, leaves, leafage
laubartig foliage-like, leaflike,
foliaceous
Laubausbruch foliage eruption
Laubbaum broadleaf, broadleaf tree,
hardwood, hardwood tree
(*pl* broadleaves/hardwoods);
deciduous tree
Laubblatt/Folgeblatt foliage leaf
Laubdach canopy of leaves
Laube arbor, bowery
Laubfall/Blattfall
shedding of leaves, leaf fall
Laubflechten/Blattflechten
foliose lichens
laubförmig/blättrig foliose, leaflike
Laubgehölze broadleaves,
hardwoods
Laubmoose mosses
Laubschicht leaf litter layer
Laubstreu leaf litter
Laubwald deciduous forest,
broadleaf forest
Laubwechsel change of foliage
Laubwerfen deciduousness,
dropping of leaves, leaf-dropping
laubwerfend
deciduous, dropping of leaves
Laubwerk foliage

Lauchgewächse/Zwiebelgewächse/
Alliaceae
onion family
Lauf (Huftier) hoof
Lauf (Vögel) leg, foot
Lauf.../zum laufen geeignet
gressorial, adapted for walking
Laufbein walking leg, gressorial leg
laufen/gehen (zu Fuß fortbewegen)
walk
Laufen/Gehen (zu Fuß
fortbewegen) walk, walking
laufen/rennen run
Laufen/Rennen run, running
Läufer/Läuferschwein young pig,
store pig, store
läufig/brünstig (z.B. Hündin)
in heat
Laufmittel/Elutionsmittel/
Fließmittel/Eluent (mobile Phase)
solvent, mobile solvent, eluent,
eluant (mobile phase)
Laufmittelfront solvent front
Laufsäugling follower
Laufschlag (Gang) leg beat
Laufvögel/Strauße/Straußenvögel/
Struthioniformes ostriches
Lauge *chem* lye
Lauge (Bodenauslaugung) leachate
Lauraceae/Lorbeergewächse
laurel family
Laurer ambush predator
Laurer-Kanal/Laurerscher Kanal
Laurer's canal (vestigial copulatory
canal)
Laurinsäure/Dodecansäure
(Laurat/Dodecanat)
lauric acid, decylacetic acid,
dodecanoic acid (laurate/
dodecanate)
Lauscher (Hase: Ohren) ears
Läuse (*sg* Laus) lice (*sg* louse)
- echte Läuse/Anoplura
sucking lice
lausen/entlausen delouse
Lausen/Entlausung delousing
Lauskerfe/Läuslinge/Phthiraptera
phthirapterans
Laut/Ton sound, noise
Lautäußerung utterance of sound
Lautbildung/Artikulation
articulation
Lautbildung/Stimmbildung/
Phonation phonation
Lautgebung vocalization
Lautstärke volume, loudness
Lävan levan
Lävulinsäure levulinic acid
LCR (Lokus-Kontrollregion) *gen*
LCR (locus control region)
LD$_{50}$ (mittlere letale Dosis)
LD$_{50}$ (median lethal dose)

573   **Lehm**   **L**

LDL (Lipoproteinfraktion niedriger Dichte) LDL (low density lipoprotein)
Leader-Sequenz *gen* leader segment
leben *vb* live
Leben *n* life
lebend alive, living; biological, biotic
lebendes Fossil living fossil
Lebendfärbung/Vitalfärbung vital staining
lebendgebärend/vivipar live-bearing, viviparous
Lebendgeburt/Lebendgebären/ Viviparie live-birth, live-bearing, vivipary, viviparity
Lebendfang live catch
Lebendgewicht live weight
lebendig alive
Lebendimpfstoff/Lebendvakzine live vaccine
Lebendkeimzahl live germ count
Lebendkultur live culture, living culture
Lebensdauer life span
Lebenserwartung life expectancy, expected life, expected lifetime
lebensfähig viable
● **lebensunfähig** nonviable
Lebensfähigkeit viability
Lebensgemeinschaft/ Organismengemeinschaft/Biozön/ Biozönose/Biocönose life community, biotic community, biocenosis; (Pflanzen) guild
Lebensgröße life size
Lebenskreislauf/Lebenszyklus life cycle
Lebensmittel foodstuff, nutrients
Lebensmittelchemie food chemistry
Lebensmittelkonservierungsstoff food preservative
Lebensmittelüberwachung/ Lebensmittelkontrolle food inspection
Lebensmittelzusatzstoff food additive
Lebensqualität quality of life
Lebensraum/Habitat habitat
Lebensraum/Lebenszone/Biotop life zone, biotope
Lebensspanne life span
lebensunfähig not viable, nonviable
Lebensvielfalt/biologische Vielfalt biodiversity, biological diversity, biological variability
Lebensvorgänge life processes
Lebensweise lifestyle, mode/way of life, habits
lebenswichtig/lebensnotwendig/ vital essential for life, vital
Lebenszeit lifetime

Lebenszone/Lebensraum/Biotop life zone, biotope
Lebenszyklus/Lebenskreislauf life cycle, "life history"
Leber liver
Lebermoos liverwort
Lebersack/Lebersäckchen/ Leberblindsack (*siehe:*** Mitteldarmdrüse)** hepatic sacculation
leberschädigend/hepatotoxisch hepatotoxic
Lebertran cod-liver oil
Lebewesen/Organismus lifeform, organism
leblos/tot lifeless, inanimate, dead
Lecithin lecithin
lecken lick
Leckstrom *neuro* leak current
Lectotypus/Lectotyp/Lectostandard lectotype
Lecythidaceae/Deckeltopfbäume brazil-nut family, lecythis family
Lederblumengewächse/ Hydnoraceae hydnora family
Lederhaut/harte Augenhaut/Sklera sclera
Lederhaut/Korium/Corium/Dermis cutis vera, true skin, corium, dermis
Lederholzgewächse/Cyrillaceae leatherwood family, cyrilla family
Lederkorallen/Alcyonaria/ Alcyonacea soft corals, alcyonaceans
ledrig/lederartig coriaceous, leathery
Leerdarm/Jejunum jejunum
leerfrüchtig/kenokarp/samenlos seedless (fruit)
Leerfrüchtigkeit/Kenokarpie/ Samenlosigkeit seedlessness, phenospermy (abortive seed condition)
Leerlauf-Zyklus/Leerlaufzyklus *biochem* futile cycle
Leerlaufhandlung *ethol* vacuum activity
Leerlaufreaktion *ethol* idling reaction
Lefze (v. a. Hund) flews
Legeapparat/Legeorgan/Ovipositor (Insekten) egg-laying apparatus/ organ, egg depositor, ovipositor
Legebohrer ovijector (piercing/ boring)
Legehenne layer (hen)
Legerohr/Legeröhre oviposition tube
Legesäbel sword-shaped ovipositor
Legescheide ovipositor sheath
Legestachel/Spicula spicule
Lehm loam

Lehmboden loamy soil
Leibeshöhle/Körperhöhle body cavity
• **primäre L./Furchungshöhle/ Blastocöl/Blastula-Höhle** blastocoel, blastocele
• **sekundäre L./Cölom/Coelom** secondary body cavity, coelom, perigastrium
Leibeshöhlenträchtigkeit/ Bauchhöhlenträchtigkeit/ Leibeshöhlenschwangerschaft/ Abdominalschwangerschaft abdominal pregnancy
Leiche/Kadaver (Tierleiche) corpse, carcass, cadaver
Leichengeruch cadaverous smell
Leichenschau inspection of corpse, postmortem examination
Leichenstarre/Totenstarre rigor mortis
Leichnam body, dead body, corpse
leichte Kette (L-Kette) light chain (L-chain)
Leid suffering
leiden suffer
Leiden/anhaltende Krankheit *med* condition
leierförmig/lyraförmig lyre-shaped, lyrate, lyriform
Leimbola *arach* bola
Leimtropfen (an Spinnfaden) *arach* viscid ball
Leimzotte/Colletere/Kolletere (Drüsenzotte) *bot* colleter, multicellular glandular trichome (sticky/viscous secretions)
Leinblattgewächse/Santalaceae sandalwood family
Leingewächse/Linaceae flax family
Leishmaniose leishmaniasis
• **kutane Leishmaniose/ Hautleishmaniose/Orientbeule (*Leishmania* spp.)** cutaneous leishmaniasis, oriental sore
• **viszerale Leishmaniose/Kala-Azar/Cala-Azar/schwarzes Fieber (*Leishmania donovani*)** visceral leishmaniasis, kala azar, Cala-Azar
Leiste/Leistenbeuge/ Leistengegend/Hüftbeuge/Regio inguinalis groin, inguinal zone
Leisten.../inguinal inguinal
Leistendrüse inguinal gland
Leistengegend/Leistenbeuge/ Leiste/Hüftbeuge/Regio inguinalis inguinal region, inguinal zone, groin
Leistenkanal/Canalis inguinalis inguinal canal
Leistenpilze/Cantharellales chanterelles

Leistung achievement, performance; *phys* power
Leistungsstoffwechsel/ Arbeitsstoffwechsel active metabolism
Leistungszahl performance value, performance coefficient
Leitart index species, guide species
Leitband/Gubernaculum testis gubernaculum testis
Leitbündel/Gefäßbündel/ Leitbündelstrang/Faszikel *bot* vascular bundle, vascular strand, fascicle
• **amphicribrales Bündel** amphicribral bundle
• **amphivasales Bündel** amphivasal bundle
• **geschlossenes Leitbündel** closed bundle
• **offenes Leitbündel** open bundle
Leitbündelring/Leitzylinder/ Leitbündelzylinder vascular cylinder
Leitbündelscheide bundle sheath
Leitbündelzylinder/Leitzylinder/ Leitbündelring vascular cylinder
leiten (Elektrizität/Flüssigkeiten) conduct, transport, translocate, lead
Leitenzym tracer enzyme
Leiter *electr* conductor
leiterförmig ladder-shaped, scalariform
Leiternetz *arach* ladder web
Leitertrachee *bot* scalariform vessel
leitfähig conductive
Leitfähigkeit conductivity; *neuro* conductance (G)
Leitfossil/Faziesfossil index fossil
Leitgewebe *bot* conducting tissue, vascular tissue
Leithaar *zool* long and smooth guard hair
Leithorizont *paleo* key bed, marker bed
Leitneriaceae/Korkholzgewächse corkwood family
Leitpflanze/Charakterart character species
Leitstrang *gen* leading strand
Leittier leader
Leittrieb/Haupttrieb/Hauptspross *bot/hort* leader
Leitung conduction, conductance, transport, translocation
Leitungswasser tap water
Leitzylinder/Leitbündelzylinder/ Leitbündelring *bot* vascular cylinder
Lektin lectin
Lemnaceae/Wasserlinsengewächse duckweed family

Lichtquelle

Lemnisk (*pl* **Lemnisken**) lemniscus (*pl* lemnisci)
Lende loin
Lenden.../lumbar lumbar
Lendenwirbel/Lumbalwirbel lumbar vertebra
Lennoagewächse/Lennoaceae lennoa family
Lentibulariaceae/ Wasserschlauchgewächse bladderwort family, butterwort family
lentisch (in stehendem Gewässer lebend) lentic
lenitisches Gewässer/stehendes Gewässer/Stehgewässer/ Stillgewässer lenitic water, lentic water, stagnant water, standing water
Lentizelle/Korkpore lenticel
Lepidodendraceae/Schuppenbäume lepidodendron family (clubmoss trees)
Lepidophyten/Lepidodendrales/ Bärlappbäume lepidophyte trees, club-moss trees
Leptospirose/Weil-Krankheit/ Weilsche Krankheit (*Leptospira interrogans*) leptospirosis, Weil's disease, swamp fever, infectious anemia
Leptotän leptotene
lernen learn
Lernen learning
- **instrumentelles Lernen** operant conditioning
Lernfähigkeit/Lernvermögen ability to learn, ability of learning
Lernverhalten/erlerntes Verhalten learned behavior
Leseraster/Leserahmen *gen* reading frame
- **geschlossenes Leseraster** closed reading frame
- **nicht zugeordnetes Leseraster (URF)** unassigned reading frame (URF)
- **offenes Leseraster (ORF)** open reading frame (ORF)
- **unbekanntes Leseraster** unidentified reading frame
Leserasterverschiebung(smutation) frameshift (mutation)
Lesetreue *gen* reading fidelity
Lestobiose lestobiosis
letal/tödlich lethal, deadly
- **balanciert letal** balanced lethal
- **bedingt letal/konditional letal** conditional lethal
letale Dosis lethal dose
letale Mutation lethal mutation
Letalfaktor/letaler Faktor lethal factor

Letalität lethality
Letalmutante lethal mutant
Leuchtbakterien luminescent bacteria
Leuchte *micros* illuminator
Leuchtfeldblende/Kollektorblende *micros* field diaphragm
Leuchtkraft luminosity
Leuchtkrebse/Krill/Euphausiacea krill and allies, euphausiaceans
Leuchtorgan/Photophore luminous organ, light-emitting organ, photophore
Leucin leucine
Leucin-Reißverschluss leucine zipper
Leukämie/"Weißblütigkeit" leukemia
Leukocyt/Leukozyt/ weißes Blutkörperchen leucocyte, white blood cell (WBC)
Leukocytose leukocytosis
Leukopenie leucopenia
Leukoplast leucoplast
Leukozyt/Leukocyt leukocyte
- **polymorphonuklearer L./ Granulocyt/Granulozyt** polymorphonuclear leukocyte, granulocyte
Leydigsche Zwischenzelle Leydig cell
Liane (verholzte Kletterpflanze) liana, woody climber
Libellen/Odonata dragonflies (anisopterans) and damselflies (zygopterans)
Libriformfaser/Holzfaser libriform fiber
Lichenin (Flechtenstärke/ Moosstärke) lichenin
lichenisiert lichenized
Lichenisierung lichenization
Lichtatmung/Photorespiration photorespiration
lichtbeständig photostable
Lichtbeständigkeit photostability
Lichtbleichung photobleaching
Lichtbrechung optical refraction
Lichtdurchlässigkeit light permeability
Lichtempfindbarkeit light sensitivty
lichtempfindlich (leicht reagierend) light-sensitive
lichter Wald low-density stand
Lichtmikroskop light microscope (compound microscope)
Lichtorgel *ecol* light-gradient apparatus
Lichtpflanze/Heliophyt sun plant, heliophyte
Lichtpunkt point of light
Lichtquelle light source

Lichtreaktion 576

Lichtreaktion *bot/physio* light reaction
Lichtreiz light stimulus
Lichtreparatur *gen* light repair
Lichtrückenreflex/Licht-Rücken-Orientierung dorsal light response, dorsal light orientation
Lichtsammelkomplex light-harvesting complex (LHC)
Lichtstärke/Lichtintensität light intensity
Lichtstrahl/Lichtbündel beam of light
Lichtstreuung light scattering
lichtunabhängige DNA-Reparatur dark repair, light-independent DNA repair
Lichtung/Schneise clearing, glade, aisle
Lichtwahrnehmung photoperception
Lichtwendigkeit/Sonnenwendigkeit/Heliotropismus heliotropism
Lid/Augenlid/Augendeckel/Palpebra eyelid
Lidschlag/Wimpernschlag (Auge) bat of an eye (lid)
Lidspalte/Rima palpebrarum palpebral fissure
Lieberkühnsche Krypte crypt of Lieberkühn, intestinal gland
Lieberkühnsches Organell (hymenostome Ciliaten) Lieberkühn's organelle, watchglass organelle
Liebespfeil (Gastropoda/Stylommatophora) dart, love dart
Liebeswerbung courtship
Lied (Vogelgesang) song
liegend/niederliegend prostrate, procumbent, trailing, lying
Liesche/Lieschenblatt (Hüllblatt an Maiskolben) (corn) husk
Ligament/Band ligament
Ligamentsack ligament sac
Ligand ligand
Liganden-Blotting ligand blotting
Ligation/Verknüpfung ligation
 • **Selbst-Ligation** *gen* self ligation
Ligation glatter Enden *gen* blunt end ligation
ligationsvermittelte Polymerasekettenreaktion ligation-mediated PCR
Lignifizierung lignification
Lignin lignin
Lignocerinsäure/Tetracosansäure lignoceric acid, tetracosanoic acid
Liliengewächse/Liliaceae lily family
limitierender Faktor/begrenzender Faktor/Grenzfaktor *ecol* limiting factor
limitiertes Kapazitätskontrollsystem limited capacity control system (LCCS)
Limnanthaceae/Sumpfblumengewächse false mermaid family, meadowfoam family
limnisch/im Süßwasser lebend limnetic, limnal, limnic
Limnocharitaceae/Wassermohngewächse water-poppy family
Limnokinetik limnokinetics
Limnokrene/Tümpelquelle limnocrene
Limnologie (Binnengewässerkunde) limnology
Limnosaprobität limnosaprobity
Limonen limonene
Linaceae/Leingewächse flax family
Lindengewächse/Tiliaceae lime tree family, linden family
LINE (langes eingeschobenes nukleäres Element) LINE (long interspersed nuclear element)
linealisch/linear lineal, linear
Lineweaver-Burk-Diagramm Lineweaver-Burk plot, double-reciprocal plot
Linienstichprobenverfahren *stat/ecol* line transect method
Linientransekt *ecol* line transect
Linker-DNA linker DNA
 • **Polylinker** polylinker
linksdrehend/linkswindend/sinistrorse sinistrorse
linksgängig left-handed
linkshändig left-handed, sinistral
Linolensäure linolenic acid
Linolsäure linolic acid, linoleic acid
Linse (*also:* lense)
Linsenauge lens eye, lenticular eye
Linsenbläschen (Auge) lens vesicle
linsenförmig lentil-shaped, lentiform, lenticular
Linsenkern/Nucleus lentiformis *neuro* lenticular nucleus
Linsenpapier *micros* lens tissue
Linsenplakode lens placode
Linsenzylinder/Kristallkegel/Kristallkörper/Conus crystalline cone
Lipid lipid
Lipiddoppelschicht (biol. Membran) lipid bilayer
Lipofektion lipofection
Liponsäure/Dithiooctansäure/Thioctsäure/Thioctansäure (Liponat) lipoic acid (lipoate), thioctic acid
lipophil lipophilic

Lipoprotein hoher Dichte
high density lipoprotein (HDL)
Lipoprotein mittlerer Dichte
intermediate density lipoprotein
(IDL)
Lipoprotein niedriger Dichte
low density lipoprotein (LDL)
Lipoprotein sehr niedriger Dichte
very low density lipoprotein
(VLDL)
Liposom liposome
Lipoteichonsäure lipoteichoic acid
Lippe/Labellum lip, labellum;
bot mesopetalum
Lippenblütler/
Lippenblütengewächse/
Lamiaceae/Labiatae
deadnettle family, mint family
Lippenmünder/Lappenmünder/
Cheilostomata (Bryozoen)
cheilostomates
Lippensoral (Flechten)
labriform soralium
Lippenspalte/Hasenscharte
cleft lip, hare lip
Lippentaster/Labialtaster/
Labialpalpus/Palpus labialis
labial palp, labipalp, labial feeler,
palp, palpus
lissencephal (ungefurcht-glattes
Gehirn) lissencephalous
(no/few convolutions)
Lithothelma/Gesteinstümpel
lithothelma, rockpool
lithotroph lithotroph(ic)
Lithotrophie lithotrophy
Litocholsäure litocholic acid
Litoral/Litoralzone/Litoralbereich/
Uferzone (*mar* Gezeitenzone)
littoral, littoral zone
Loasaceae/Blumennesselgewächse
loasa family
Lobeliengewächse/Lobeliaceae
lobelia family
Lobenlinien/Nahtlinien
(Ammoniten) suture lines
● **ammonitische L.**
ammonitic suture lines
● **ceratitische L.**
ceratitic suture lines
● **goniatische L.**
goniatitic suture lines
Lobopodium lobopodium
Lobus opticus optic lobe
Lochbodenkaskadenreaktor/
Siebbodenkaskadenreaktor
sieve plate reactor
löcherig/perforiert perforated
Lochkapsel/Löcherkapsel/
Porenkapsel/porizide Kapsel
poricidal capsule
Lochplatte *gen/micb* well plate

lochspaltig/porizid/poricid
poricidal
Lochträger/Foraminiferen
foraminiferans, forams
locken/anlocken attract, lure
locker gebautes Sekundärholz
manoxylic wood
Lockgesang *orn* attracting song
Lockmittel/Lockstoff/Attraktans
attractant
Lockruf *orn* attracting call
Loculament/Lokulament locule,
loculus
Lod-Wert lod score ("logarithm of
the odds ratio")
Löffel (Hasenohren) rabbit ears
löffelartig/cochlear spoon-like,
cochlear
Loganiaceae/Strychnosgewächse/
Brechnussgewächse
logania family
logarithmische Phase
logarithmic phase (log-phase)
Lognormalverteilung/
logarithmische Normalverteilung
lognormal distribution, logarithmic
normal distribution
LoH (Verlust der Heterozygotie/
Heterozygotieverlust)
loss of heterozygosity (LoH)
Lokomotion/Bewegung
(Ortsveränderung) locomotion
Lokomotorik locomotory
lokulizid/rückenspaltig loculicidal,
dorsally dehiscent
Lokus/Ort locus
● *cis*-**aktiver Lokus**
cis-acting locus
Lokus aus mehreren eng
gekoppelten Genen
compound locus
Lokus-Kontrollregion (LCR) *gen*
locus control region (LCR)
Lomasom lomasome
London-Dispersionskräfte
London dispersion forces
lophotrich lophotrichous
Loranthaceae/Mistelgewächse/
Riemenblumengewächse
mistletoe family
(showy mistletoe family)
Lorbeergewächse/Lauraceae
laurel family
Lorchelpilze/Helvellaceae
saddle fungi & false morels
Lorenzinische Ampulle/
Ampullenrezeptor Lorenzini flask
Lorica (Gehäuse einiger
Chrysophyceen) lorica
Loriciferen/Korsettierchen/
Panzertierchen corset bearers,
loriciferans

löschen

löschen extinguish, put out
Löschgerät/Feuerlöscher
fire extinguisher
Löschmittel/Feuerlöschmittel
fire-extinguishing agent
lösen *chem* (in einem Lösungsmittel) dissolve
lösen *math* solve
löslich soluble
• **unlöslich** insoluble
Löslichkeit solubility
• **Unlöslichkeit** insolubility
Löslichkeitspotential
solute potential
Löslichkeitsprodukt
solubility product
Löss loess
Losung (animal) droppings, dung
Lösung solution
Lösungsmittel solvent
Lösungsmittelfront solvent front
lotisch (in fließendem Gewässer lebend) lotic
lotisches Gewässer/fließendes Gewässer/Fließgewässer (Fluss/Strom) lotic water, flowing water, running water, waterway, watercourse (river/stream)
lotsen/lenken/führen piloting
Lotte (unverholzter Langtrieb des Weinstocks) lateral shoot, summer shoot (unlignified long-shoot of vine)
Lotusblumengewächse/Nelumbonaceae lotus lily family, Indian lotus family
LTR (lange terminale Sequenzwiederholung) *gen* LTR (long terminal repeat)
Luch bog, swamp
Lücke gap
Lückencollenchym
lacunar collenchyma
Lues/Schanker/Syphilis (*Treponema pallidum*) syphilis
Luftablegerverfahren (Vermehrung) *hort* air layering, Chinese layering/layerage, marcottage
Luftabsenker *bot* adventitious root
Luftalge terrestrial alga
luftatmend air-breathing
luftdicht airtight, airproof
Luftembolie air embolism (due to cavitation)
Luftfeuchtigkeit air humidity
Luftkammer (im Ei) *zool* air space
Luftkammer (*Marchantia*)
air chamber
Luftkapazität air capacity
Luftkapillare air capillary
Luftknolle/Pseudobulbe
pseudo-bulb

Luftröhre/Atemröhre/Kehle/Trachee/Trachea windpipe, trachea; breathing tube
Luftröhre/Pharynx pharynx
Luftsack (Pollen) *bot* air pocket/bag/sac, vesiculum
Luftsack/Saccus aerophorus *orn*
air sac
Luftschadstoff air pollutant
"Luftschiffen"/"Ballooning"/Fadenflug (Spinnen: Altweibersommer) *arach*
ballooning
Luftverschmutzung air pollution
Luftvorhang (Vertikalflow-Biobench) *lab* air curtain
Luftwurzel aerial root, air root
lumineszent luminescent
Lumineszenz luminescence
• **Biolumineszenz** bioluminescence
Lunarperiodik/Lunarperiodizität/Mondperiodik lunar periodicity
Lunge lung; bellows
• **Buchlunge/Fächerlunge/Fächertrachee** book lung
• **Sauglunge** suction lung
• **Wasserlunge (Holothurien)** respiratory tree
Lungen.../die Lunge betreffend
pulmonary
Lungenarterie pulmonary artery
Lungenbläschen/Lungenalveole/Alveole
pulmonary alveolus, alveola
Lungenentzündung pneumonia
Lungenfell/Pleura pulmonalis
pulmonary pleura, visceral pleura
Lungenfische/Dipnoi lungfishes
Lungenhöhle (Pulmonata)
pulmonary cavity, pulmonary sac
Lungenkreislauf/kleiner Blutkreislauf
pulmonary circulation
Lungenpest (*Yersinia pestis*)
pneumonic plague
Lungenpfeife/Parabronchus (*pl* Parabronchien)
parabronchus (*pl* parabronchi)
Lungenschnecken/Pulmonaten/Pulmonata pulmonate snails (freshwater & land snails and slugs)
Lungenvene pulmonary vein
Lunte (Fuchsschwanz) brush, tail
Lupe/Vergrößerungsglas
lens, magnifying glass
Lurche/Amphibien amphibians
Lutropin/Luteotropin/luteinisierendes Hormon (LH)/zwischenzellstimulierendes Hormon luteinizing hormone (LH), interstitial-cell stimulating hormone (ICSH)

Luxusgen luxury gene
Lycopodiaceae/Bärlappgewächse
  clubmoss family
lymphatisch lymphatic
Lymphdrüse lymphatic gland
Lymphe lymph
Lymphgefäß lymph vessel,
  lymphatic vessel
Lymphknötchen/Lymphfollikel
  lymph nodule, lymph follicle
Lymphknoten lymph node
Lymphokin (lymphozytäres
  Zytokin/Cytokin) lymphokine
Lymphozyt/Lymphocyt lymphocyte
Lymphsystem/Lymphgefäßsystem
  lymphatic system
Lyonisierung *gen* lyonization
Lyophilisierung/Gefriertrocknung
  lyophilization, freeze-drying
Lysat lysate
Lyse lysis
Lysehof/Aufklärungshof/Hof/
  Plaque lytic plaque, plaque

Lysergsäure lysergic acid
lysieren lyse
lysigen lysigenic, lysigenous
Lysin lysine
lysogen (temperent)
  lysogenic (temperate)
lysogene Konversion
  lysogenic conversion
Lysogenie lysogeny
Lysosom lysosome
  • **sekundäres Lysosom**
    secondary lysosome,
    phagolysosome
Lysozym lysozyme
Lythraceae/
  Weiderichgewächse/
  Blutweiderichgewächse
  loosestrife family
lytisch lytic
lytische Infektion lytic infection
lytischer Hof/Lysehof lytic plaque
lytischer Phage lytic phage
lytischer Zyklus lytic cycle

M

M-Linie 58C

M-Linie/M-Streifen (M-Scheibe:
Mesophragma) M line (M disk)
M-Phase (Mitosephase des
Zellzyklus) M phase
(mitotic phase of cell cycle)
β-Mäander β meander
Macaedium/Mazaedium *fung*
macaedium, mazaedium
Macchia/Macchie/Maquis maquis
Made (apode Larve)
maggot (apodal larva)
Madenkrankheit/Myiasis
(Dipterenlarven) myiasis
Madenwurm (*Enterobius*
***vermicularis*)** pinworm
Madreporenköpfchen
madreporian body
Madreporenplatte/Siebplatte
madreporic plate, madreporite,
sieve plate
Magen (*pl* Mägen) stomach
- **Blättermagen/Vormagen/**
Omasus/Psalter (Wiederkäuer)
third stomach, omasum, manyplies,
psalterium
- **Drüsenmagen** *orn*
glandular portion of stomach
- **Filtermagen/Pylorus (Krebse)**
pyloric stomach, pylorus (posterior
region of gizzard)
- **Haube/Netzmagen/Retikulum**
honeycomb stomach, honeycomb
bag, reticulum, second stomach
- **Honigmagen/Honigdrüse/**
Futterdrüse/Kropf (Biene)
honey crop
- **Kaumagen/Cardia**
cardiac stomach, cardia
- **Kaumagen/Pharynx/Mastax**
(Rotatorien) pharynx, mastax
- **Kaumagen/Proventriculus**
(Insekten/Crustaceen) gizzard,
proventriculus
- **Labmagen/Abomasus**
fourth stomach, abomasum, reed,
rennet-stomach
- **Muskelmagen** *orn* gizzard
- **Netzmagen/Haube/Retikulum**
honeycomb stomach, honeycomb
bag, reticulum, second stomach
- **Pansen/Rumen** paunch, rumen,
first stomach, ingluvies
- **Saugmagen** sucking stomach
- **Saugmagen (Vorratsmagen:**
Culiciden) pumping stomach
- **Schleudermagen/Pansenvorhof/**
Atrium ruminis atrium ruminis
- **Vormagen/Blättermagen/Psalter/**
Omasus (Wiederkäuer) third
stomach, omasum, psalterium
- **Vormagen/Vorderdarm/Kropf/**
Ingluvies (Insekten/Vögel) crop

Magen-Darm-Trakt/
Gastrointestinaltrakt
gastrointestinal tract
Magenblindsack/Magendivertikel
gastric/digestive cecum,
gastric/digestive diverticulum
Magenbrei/Speisebrei/Chymus
chyme
Magendivertikel/Magenblindsack
gastric/digestive cecum,
gastric/digestive diverticulum
Magendrüse (Rotatorien)
gastric gland
Magenflüssigkeit/Magensaft
stomach juice, gastric juice
Magengrübchen/Foveola gastrica
gastric pit
Magenmühle gastric mill,
triturating mill
Magenmund/Mageneingang/
Kardia/Cardia
cardia
Magenrinne/Sulcus ventriculi
reticular groove
Magensaft/Magenflüssigkeit
stomach juice, gastric juice
Magensäure stomach acid
Magenschleimhaut/Tunica mucosa
gastric mucosa, mucous tunic
(mucosal layer of stomach)
Magenstein/Magensteinchen/
Hummerstein/Gastrolith
gastrolith
Magenstiel/Mundrohr/Manubrium
hypostome, oral cone, peduncle;
gullet, pharynx; manubrium
Magerwiese poor grassland,
rough pasture/meadow
Magnesium magnesium
Magnetosom magnetosome
Magnetresonanztomographie
(MRT)/
Kernspintomographie (KST)
magnetic resonance imaging (MRI),
nuclear magnetic resonance imaging
Magnetrührer *lab* magnetic stirrer
Magnoliengewächse/Magnoliaceae
magnolia family
Mahd cut grass, hay, mowing
Mahlzahn grinding tooth
Mahlzeit meal
Mähne mane
Mähwiese hay meadow,
mowed meadow
Maiapfelgewächse/
Fußblattgewächse/
Podophyllaceae
may apple family
Maiglöckchengewächse/
Convallariaceae
lily-of-the-valley family
Maische mash

Männchen

Maiskolben (gesamter Fruchtstand)
  ear (of corn);
  (Fruchtstandachse) cob (of corn)
Maisquellwasser cornsteep liquor
Maisstengel cornstalk
MAK-Wert (maximale Arbeitsplatz-Konzentration)
  maximum permissible workplace concentration,
  maximum permissible exposure
Makel/Fleck spot, blot, stain, stigma
Makel am Flügelrand/ Flügelrandmal/Pterostigma *entom*
  pterostigma, stigma
makrandrisch macrandrous
Makrander (große ♂ Pflanze)
  macrander (large ♂ plant)
Makraner (große ♂ Ameise)
  macraner (large-size ♂ ant)
Makrelenhaiverwandte/ Lamniformes
  mackerel sharks and relatives
Makroergat (großer Ameisensoldat)
  macrergate (large worker ant)
Makrofauna macrofauna
Makrogyne (große Königin)
  macrogyne (large ♀ ant/queen)
Makrokonsument macroconsumer
Makromer macromere
Makromolekül macromolecule
Makronukleus macronucleus
Makrophage macrophage
Makrophyt macrophyte
makroskopisch macroscopic
Makrospore/Megaspore
  macrospore, megaspore
Makrozyt/Makrocyt macrocyte
Malakologie/Weichtierkunde
  malacology, study of mollusks
Malaria/Sumpffieber/Wechselfieber
  malaria
Maleinsäure (Maleat)
  maleic acid (maleate)
maligne/bösartig malignant
  • **benigne/gutartig** benign
Malignität/Bösartigkeit malignancy
Malonsäure (Malonat)
  malonic acid (malonate)
Malpighi-Gefäß/Malpighisches Gefäß/Malpighi-Schlauch/ Malpighischer Schlauch
  Malpighian tubule
Malpighi-Körperchen/ Milzknötchen/Milzkörperchen/ Milzfollikel Malpighian body,
  Malpighian corpuscle,
  splenic nodule
Malpighi-Körperchen/ Nierenkörperchen Malpighian body, Malpighian corpuscle,
  renal corpuscle

Malpighiengewächse/ Barbadoskirschengewächse/ Malpighiaceae malpighia family,
  Barbados cherry family
Maltose (Malzzucker)
  maltose (malt sugar)
Malvengewächse/Malvaceae
  mallow family
Malz malt
Malzzucker/Maltose malt sugar, maltose
Mamillarkörper/Corpus mamillare
  neuro mamillary body
Mamille/Brustwarze/Zitze mamilla,
  mammilla, nipple (multiple ducts),
  teat (single duct)
Manca-Stadium
  manca (prejuvenile peracarids)
Mandel/Tonsille tonsil
  • **Gaumenmandel/Tonsilla palatina**
    palatine tonsil
  • **Rachenmandel/Tonsilla pharyngealis** pharyngeal tonsil
  • **Zungenmandel/Tonsilla lingualis**
    lingual tonsil
Mandelkern/Mandelkörper/ Mandelkernkomplex/ Nucleus amygdalae/ Corpus amygdaloideum
  amygdaloid nucleus,
  amygdaloid nuclear complex
Mandelsäure/Phenylglykolsäure
  mandelic acid, phenylglycolic acid,
  amygdalic acid
Mandibel mandible
Mandibularbogen/Kieferbogen
  mandibular arch
Mangan manganese
Mangel/Defizienz deficiency
Mangelerscheinung/ Defizienzerscheinung/ Mangelsymptom
  deficiency symptom
Mangelmedium
  deficiency medium
mangelnd/Mangel../defizient
  deficient, lacking
Mangrove mangrove
Mangrovenformation/ Mangrovenwald/Gezeitenwald/ Mangrove *biogeo* mangal
Mangrovengewächse/ Rhizophoraceae mangrove family,
  red mangrove family
Mangrovensumpf mangrove swamp
Mannbarkeit/Geschlechtsreife/ Fähigkeit zur Fruktifikation
  sexual maturity
Männchen male;
  (Eber/Hirsch etc.) stag
  • **nach der Reife kastriertes Männchen (Nutztiere)** stag

Mannigfaltigkeit

Mannigfaltigkeit/Vielfalt/ Variabilität diversity, variability
Mannit mannitol
männlich/männlichen Geschlechts male; *bot* (Blüte) male, staminate
Mannuronsäure mannuronic acid
Manschette/Armilla/Annulus superus *fung* manchette, armilla, superior annulus
Manschettensoral (Flechten) maniciform soralium
Mantel/Pallium (Mollusken) mantle, pallium
Mantel/Tunica mantle, tunic
Mantelblatt/Nischenblatt *bot* nest leaf
Mantelgürtel/Gürtel/Perinotum (Käferschnecken) mantle girdle
Mantelhöhle mantle cavity, pallial cavity
Mantellappen/Kragenlappen (Brachiopoden) mantle lobe
Mantellinie/Palliallinie pallial line
Manteltiere/Tunicaten (Urochordaten) tunicates
Mantelzelle (Moose) jacket cell
Maquis/Macchia/Macchie maquis
Marantaceae/Pfeilwurzelgewächse arrowroot family, prayer plant family
Marattiaceae marattia family
Marcgraviaceae/ Honigbechergewächse shingleplant family
marginal/randständig marginal
Marginalfalte/Mantelfalte (Mantelrand) mantle fold
Mark medulla, pith, core
Mark.../medullär/markhaltig/ markig medullar, medullary, pithy
Marker/Markersubstanz (genetischer/radioaktiver) marker (genetic/radioactive)
Markfortsatz/Dendrit dendrite
Markhirn/Myelencephalon marrow brain, medullary brain, myelencephalon
Markhöhle marrow cavity, medullary cavity
markieren/etikettieren tag; *chem* label; (kennzeichnen) mark, brand, earmark
markiertes Molekül tagged molecule
Markierung label(l)ing
• **Immunmarkierung** immunolabeling
• **radioaktive M.** radiolabeling
Markottage *hort* marcotage
Markscheide/Myelinscheide *neuro* medullary sheath, myelin sheath

Markstrahl *bot* pith ray, medullary ray
Markstrahlinitiale ray initial
Markstrahlparenchym ray parenchyma
Markstrang *neuro* nerve cord
Marsch marsh (dominated by grasses)
• **Brackmarsch** brackish marsh
• **Flussmarsch** riverine marsh; (an der Flussmündung/im Flussdelta) estuarine marsh
• **Flussmündungsmarsch** river-mouth marsh
• **Gezeitenmarsch/Tidenmarsch** tidal marsh
• **Hochmarsch** high marsh
• **Koog** young marsh, juvenile marsh
• **Küstenmarsch/Seemarsch** coastal marsh
• **Salzmarsch (Salzwiese)** salt marsh (salt meadow)
• **Süßwassermarsch** freshwater marsh
• **Tiefmarsch** shallow marsh, low marsh
• **Torfmarsch** peat marsh
Marschland marsh, marshland, fen
Marsileaceae/Kleefarne/ Kleefarngewächse marsilea family, water clover family
Martyniaceae/Gemsbockgewächse unicorn plant family, devil's-claw family, martynia family
Maschennetz *arach* mesh web
maschig meshy
Maserknolle/Kropf (Holz) burl, burr, gnarl, woody outgrowth, wood knot (with wavy grain)
• **ebenerdige Maserknolle (durch Feuer/Trockenheit)** lignotuber
Maserung/Fladerung *allg* figure, design; (Faserorientierung) grain
Maß measure
Maßanalyse volumetric analysis
Masse mass
• **Biomasse** biomass
• **"Frischmasse" (Frischgewicht)** "fresh mass" (fresh weight)
• **Molekülmasse ("Molekulargewicht")** molecular mass ("molecular weight")
• **Molmasse/molare Masse ("Molgewicht")** molar mass ("molar weight")
• **relative Molekülmasse/ Molekulargewicht (M_r)** relative molecular mass, molecular weight (M_r)
• **Trockenmasse/Trockensubstanz** dry mass, dry matter

583 | **Medikament** **M**

Massenerhaltungssatz law of conservation of matter
Massenspektroskopie (MS) mass spectroscopy
Massensterben mass extinction
Massenströmung (Wasser) mass flow, bulk flow
Massenübergang/Massentransfer/ Stoffübergang mass transfer
Massenvermehrung mass reproduction, mass spread, outbreak
Massenwechsel *ecol* population changes
Massenwirkungsgesetz law of mass action
Massenwirkungskonstante mass action constant
Maßkorrelationskoeffizient/ Produkt-Moment-Korrelationskoeffizient product-moment correlation coefficient
Maßstab scale
Maßstabsvergrößerung scale-up, scaling up
Maßstabzahl *micros* initial magnification
Mast (Viehmast/Tiermast) mast, fattening; stuffing
Mastdarm/Rectum rectum
mästen *allg* fatten; (Geflügel) cram, stuff
Mastersequenz *gen* master sequence
Mastjahr *bot/for* mast year
Mastferkel porker
Mastzelle mast cell
matern/mütterlich maternal; motherly
maternale Vererbung maternal inheritance
maternaler Effekt/ maternale Prädetermination maternal effect
matriarchalisch matriarchal
Matrix matrix; (*Chloroplast*: Stroma) stroma
Matrize *biochem* template
Matrizenstrang/Mutterstrang *gen* template strand
matrokline Vererbung matroclinous inheritance
Matte/Mattenstufe alpine grassland
Matte/Teppich (z.B. Algen) mat, layer (e.g. algal mat)
mauerartig/mauerförmig muriform
Mauerblatt/Scapus (Cnidaria) scape
Maul/Schlund/Mundöffnung jaw
Maul/Schnauze muzzle, snout
Maul-und-Klauenseuche (*Aphthovirus*) foot-and-mouth disease, aphthous fever

Maulbeergewächse/Moraceae mulberry family, fig family
Maulbeerkeim/Morula morula
Maulbrüten mouthbreeding, oral gestation, buccal incubation
Maulbrüter mouthbreeder
Maulesel (Pferdehengst x Eselstute) hinny
Maulfüßer/Fangschreckenkrebse/ Stomatopoda mantis shrimps
Maultier (Pferdestute x Eselhengst) mule
Mauser molt, molting season, deplumation
mausern molt, shed feathers
Mauserzeit molting time/period/season
Mausvögel/Coliiformes mousebirds, colies
Mauthnersche Zellen/Riesenfasern Mauthner's cells
Maxillarfuß/Maxillipes/Kieferfuß/ Pes maxilliaris maxilliped, maxillipede, gnathopodite, jaw-foot, foot-jaw
Maxillartaster/Kiefertaster/ Maxillarpalpus/Palpus maxillaris maxillary palp
Maxille/Kiefer (Insekten) maxilla, (*pl.* maxillas/maxillae)
Maxille/Oberkiefer (Wirbeltiere) maxilla, upper jawbone
Maxilliped/Maxillarfuß/Kieferfuß/ Pes maxilliaris maxilliped, maxillipede, gnathopodite, jaw-foot, foot-jaw
Maximalgeschwindigkeit (V_{max}: Enzymkinetik/Wachstum) maximum rate
Mayacaceae/Moosblümchen mayaca family, bogmoss family
Mazaedium/Macaedium *fung* mazaedium, macaedium
Mazeration maceration
mazerieren macerate
mechanisches Gewebe/ Expansionsgewebe expansion tissue
Meckelscher Knorpel (Mandibulare) Meckel's cartilage, mandibular cartilage
Meckelsches Divertikel Meckel's diverticulum
meckern (Ziege) bleak
Medianwert/Zentralwert *stat* median value
Medianzelle/Media media
Medikament/Medizin/Droge medicine, drug
  ● **zielgerichtete "Konstruktion" neuer Medikamente am Computer** drug design

Medium

Medium/Kulturmedium/Nährmedium medium, culture medium, nutrient medium
- **Anreicherungsmedium** enrichment medium
- **Basisnährmedium** basal medium
- **Differenzierungsmedium** differential medium
- **Eiermedium/Eiernährmedium** egg medium
- **Elektivmedium/Selektivmedium** selective medium
- **Erhaltungsmedium** maintenance medium
- **Komplettmedium** complete medium, rich medium
- **komplexes Medium** complex medium
- **konditioniertes Medium** conditioned medium
- **Mangelmedium** deficiency medium
- **Minimalmedium** minimal medium
- **Selektivmedium/Elektivmedium** selective medium
- **synthetisches Medium (chemisch definiertes Medium)** defined medium
- **Testmedium/Prüfmedium (zur Diagnose)** test medium
- **Vollmedium/Komplettmedium** rich medium, complete medium

Medizin medicine
- **Biomedizin** biomedicine
- **Defensivmedizin** defensive medicine
- **Forensik/forensische Medizin/Gerichtsmedizin/Rechtsmedizin** forensics, forensic medicine
- **Präventivmedizin** preventive medicine
- **Umweltmedizin** environmental medicine
- **Veterinärmedizin/Tiermedizin/Tierheilkunde** veterinary medicine, veterinary science
- **vorhersagende Medizin** predictive medicine

Medizin/Medikament/Droge medicine, drug

medizinische Untersuchung medical examination, medical exam, physical examination, physical

Medullarplatte/Neuralplatte/Markplatte neural plate

Medullarrohr/Neuralrohr/Markrohr/Tubus medullaris *embr* medullary tube, neural tube

Medullarwulst/Neuralwulst/Neuralfalte/Markfalte neural fold

Meduse/"Qualle" medusa

Meer sea, ocean
- **offenes Meer/Hochsee** open sea, pelagic zone

Meeresbecken oceanic basin, ocean basin

meeresbewohnend/marin marine

Meeresbiologie marine biology

Meeresboden/Meeresgrund seafloor, ocean floor
- **den Meeresboden bewohnend** benthic, benthonic

Meeresbodenbereich/Benthal benthic zone

Meeresbodenorganismen/Benthos benthos

Meeresbrandung surf, breakers

Meeresbusen bay, gulf

Meeresgrund/Meeresboden ocean floor, seafloor

Meereshöhe sea level, elevation

Meeresklima/ozeanisches Klima maritime/oceanic/marine climate

Meereskunde/Ozeanographie marine sciences, oceanography

Meeresküste/Meeresufer seashore, seaboard, seacoast

Meeresküstenlage oceanic location, coastal location

Meeresleuchten marine phosphorescence

Meeresnacktschnecken sea slugs

Meeresspiegel sea level

Meeresstrand (ocean) beach

Meeresströmung ocean current

Meerestier marine animal

Meerrettichperoxidase horse radish peroxidase

Meerträubelgewächse/Ephedraceae joint-pine family, mormon tea family, ephedra family

Meerwasser seawater, saltwater

Meerwasserintrusion seawater intrusion, saltwater intrusion

Megaspore/Makrospore megaspore, macrospore

Megasporenmutterzelle/Makrosporenmutterzelle megaspore/macrospore mother cell

Mehl fluor
- **Blutmehl** blood meal
- **Getreidemehl** *grob:* meal, *fein:* flour
- **Grießmehl/Grütze/Grieß** groats, grits
- **Guarmehl/Guar-Gummi** guar gum, guar flour
- **Guar-Samen-Mehl** guar meal, guar seed meal
- **Johannisbrotkernmehl/Karobgummi** locust bean gum, carob gum
- **Knochenmehl** bone meal

Membran

- **Sägemehl** sawdust
- **ungebleichtes Mehl** unbleached fluor
- **Vollkornmehl** whole-grain fluor

Mehlbleichung fluor bleaching
mehlig mealy, farinaceous
Mehlische Drüse/Schalendrüse Mehlis' gland, shell gland
Mehltaupilze mildews
- **echte M./Erysiphales** powdery mildews
- **falsche M./Peronosporaceae** downy mildews
- **schwarze M./Meliolales** black mildews

Mehlwurm mealworm
Mehrfachaustausch *gen* compound crossing over
Mehrfachbindung *chem* multiple bond
Mehrfaktortheorie/Polygentheorie *gen* multiple-factor hypothesis
mehrgestaltig/polymorph/ pleomorph polymorphic, pleomorphic
Mehrgestaltigkeit/Polymorphismus/ Pleomorphismus polymorphism, pleomorphism
mehrjährig/ausdauernd perennial
mehrjährig wachsend bis zur Blüte (*Agave*) pluriennal
mehrkammerig/vielkammerig/ vielkämmrig/polythalam/ polythekal polythalamous, polythalamic, with many chambers, polythecal
Mehrlinge
progeny of a multiple birth
- **Erzeugung monozygoter Mehrlinge** twinning

Mehrlingsgeburt multiple birth
mehrreihig/vielreihig/multiseriat multiseriate, multiple rowed, in several rows
mehrstufig multistage
mehrwirtig/heteroxen polyxenous, polyxenic
mehrzellig/vielzellig multicellular
Meibom Drüse Meibomian gland
Meidereaktion *ethol* avoidance reaction, avoiding reaction
- **elektrische Meidereaktion** *phyio* jamming avoidance reaction

Meideverhalten avoidance behavior
Meiofauna meiofauna
Meiose/Reifeteilung/ Reduktionsteilung meiosis, reduction division
Meissner-Körperchen/Meissner- Tastkörperchen Meissner's corpuscle, corpuscle of touch
Meistergen master gene

Melanoliberin/Melanotropin- Freisetzungshormon melanoliberin, melanotropin releasing hormone, melanotropin releasing factor (MRH/MRF)
Melanophage melanophage
Melanotropin/ melanozytenstimulierendes Hormon (MSH) melanocyte-stimulating hormone (MSH)
Melanozyt/Melanocyt melanocyte
Melastomataceae/ Schwarzmundgewächse meadow-beauty family, melastome family
Melatonin melatonin
Meliaceae/Zedrachgewächse mahogany family
Melianthaceae/ Honigstrauchgewächse honeybush family
melken milk
Melone (Wale) melon
Melonenbaumgewächse/Caricaceae papaya family
Membran membrane
- **Außenmembran** outer membrane
- **Basalmembran/Basallamina** basement membrane, basal lamina
- **Basilarmembran** basilar membrane
- **Befruchtungsmembran** fertilization membrane
- **Doppelmembran** double membrane
- **Dottermembran/ Vitellinmembran/Dotterhaut/ primäre Eihülle** vitelline membrane, vitelline layer, membrana vitellina
- **"Eimembran"/Eihaut/Eihülle** egg membrane
- **Elementarmembran/ Doppelmembran** unit membrane, double membrane
- **extraembryonale Membranen/ Embryonalhülle/Keimhülle** extraembryonic membranes
- **Frontalmembran (Bryozoen)** frontal membrane
- **Hirnhaut/Meninx** cerebral membrane, meninx
- **Kernmembran** nuclear membrane
- **peritrophische Membran** peritrophic membrane
- **Plasmamembran/Zellmembran/ Ektoplast/Plasmalemma** plasma membrane, (outer) cell membrane, unit membrane, ectoplast, plasmalemma
- **Schleimhaut/Schleimhautepithel** mucous membrane, mucosa

Membran-Angriffskomplex

- **Tympanalmembran/Trommelfell/ Ohrtrommel/Tympanum**
 tympanic membrane, eardrum, tympanum
- **undulierende Membran**
 undulating membrane
- **Vitellinmembran/Dotterhaut/ Dottermembran/primäre Eihülle**
 vitelline membrane, vitelline layer, membrana vitellina
- **Zellmembran/Plasmamembran/ Ektoplast/Plasmalemma** (outer) cell membrane, plasma membrane, unit membrane, ectoplast, plasmalemma

Membran-Angriffskomplex *immun*
 membrane attack complex
Membran-Ghost (künstlich hergestellte leere Membran)
 membrane ghost
Membrandurchfluss membrane flux
Membranelle membranelle
Membranellenband, adorales (Ciliaten)
 adoral zone membranelles
Membranfilter *lab* membrane filter
Membranfluss membrane flow
Membranfusion membrane fusion
membrangebunden
 membrane-bound
Membrankanal membrane channel
- **Ionenkanal** ion channel
- **ligandenregulierter/ ligandengesteuerter Kanal**
 ligand-gated channel
- **mechanisch gesteuerter Kanal**
 mechanically gated channel
- **Ruhemembrankanal/Leckkanal**
 resting channel, leakage channel
- **spannungsregulierter/ spannungsgesteuerter Kanal**
 voltage-sensitive channel, voltage-gated channel

Membrankapazität
 membrane capacitance
Membranlängskonstante (Raumkonstante) membrane length constant (space constant)
Membranleitfähigkeit
 membrane conductance
membranös membraneous
Membranpotential
 membrane potential
Membranreaktor (Bioreaktor)
 membrane reactor
Membranstapel stacked membranes
Membrantransport
 membrane transport
membranumgeben
 membrane coated
Memnospore memnospore (remains at place of origin)

Menachinon (Vitamin K$_2$)
 menaquinone
Menadion (Vitamin K$_3$)
 menadione
Menarche (erste Menstruation)
 menarche (first menstruation)
mendeln mendelize
mendelnd/nach den Mendelschen Gesetzen vererbt *adj/adv*
 mendelian
Mendelsche Vererbung
 Mendelian inheritance
Mendelsche Vererbung beim Menschen
 Mendelian Inheritance in Man (MIM)
Mendelsches Gesetz Mendel's law
Meniskus (Gelenkmeniskus)
 meniscus, disk
Menispermaceae/ Mondsamengewächse
 moonseed family
Menopause menopause (cessation of ovulation/menstruation)
Mensch human
menschenartig/menschenähnlich
 manlike (*besser:* humanlike)
Menschheit/Menschengeschlecht (Gesamtheit der Menschen)
 humanity, mankind (*besser:* humankind/humans)
menschlich/den Menschen betreffend human
menschlich (wie ein guter Mensch handelnd/hilfsbereit/selbstlos)
 humane
menschlicher Leukozytenantigen- Komplex (HLA-Komplex)
 human leucocyte antigen complex (HLA complex)
menschliches Genomprojekt (HUGO)
 Human Genome Project (HUGO)
Menstruation/Blutung/ Monatsblutung/Periode/Regel
 menstruation, period
Menstruationszyklus
 menstrual cycle
menstruieren menstruate
Menyanthaceae/ Bitterkleegewächse/ Fieberkleegewächse
 bogbean family
MER (mittlere Wiederholungs- häufigkeit) (>DNA-Element)
 MER (medium reiteration frequency) (>DNA element)
Mergel *geol* marl
Merianthium/Teilblume
 merianthium, partial flower
Meridionalfurchung meridional cleavage
Merikarp/Teilfrucht mericarp

Messkolben

Meriklinalchimäre
mericlinal chimera
Meristem/Bildungsgewebe
meristem
- **Achselmeristem**
axillary meristem
- **Endmeristem** terminal meristem
- **Etagenmeristem** tiered meristem
- **Flankenmeristem** flank meristem, peripheral meristem
- **Folgemeristem**
secondary meristem
- **Grundmeristem** ground meristem
- **interkalares Meristem/ Restmeristem**
intercalary meristem
- **laterales Meristem**
lateral meristem
- **offenes Meristem** open meristem, indetermiante meristem
- **Plattenmeristem** plate meristem
- **Randmeristem**
marginal meristem
- **Rippenmeristem** file meristem, rib meristem
- **Spitzenmeristem/ Scheitelmeristem/ Apicalmeristem/ Vegetationspunkt**
apical meristem, growing point
Meristemmantel, primärer
primary thickening meristem
Merkelsches Körperchen Merkel's corpuscle, Merkel's disk, tactile disk
Merkmal trait, characteristic, feature
- **abgeleitetes Merkmal**
derived characteristic
- **erworbenes Merkmal**
acquired characteristic
- **Familienmerkmal** family trait
- **kanalisiertes Merkmal**
canalized character
Merkmal/Charakterzug/Eigenschaft
trait, character
Merkmalsdivergenz
character divergence
Merkmalsgefälle/ Merkmalsgradient/Cline/Kline/ Klin cline,
phenotypic/character gradient
Merkmalsphylogenetik
character phylogeny
Merkmalsunterschied
character difference
Merkmalsverschiebung
character displacement
Merocyt/Merozyt merocyte
Merogamie merogamy
Merogenese/Segmentierung
merogenesis, segmentation
Merognathit merognathite
Merogonie merogony

merokrine Drüse merocrine gland
meromiktisch meromictic
Meromyosin meromyosin
Meroplankton meroplankton
Meropodit meropodite
Merospermie merospermy
Mertenssche Mimikry
Mertensian mimicry
Merzvieh cull
Merzvieh aussondern cull, culling
Mesenchym (embryonales Bindegewebe) mesenchyme
mesenchymatisch mesenchymal
Mesenterium mesentery
Mesocöl/Mesocoel mesocoel
Mesafauna mesafauna
Mesoglöa/Stützschicht mesogloea, mesoglea
Mesohyl (Schwämme) mesohyl
Mesokarp (Frucht) mesocarp
Mesomerie mesomerism
Mesonephros/Urniere/Wolffscher Körper mesonephros, middle kidney, midkidney
mesophil (20–45°C) mesophil, mesophilic
Mesophile mesophile
Mesophyll (Schwamm- & Palisadenparenchym) mesophyll
Mesophyt mesophyte
Mesophytikum (erdgeschichtliches Zeitalter) Mesophytic Era
Mesosaprobien mesosaprobes
Mesosom mesosome
Mesothel mesothelium
Mesothorakalschild (dorsal)/ Mesonotum mesonotum
mesotroph (mittlerer Nährstoffgehalt) mesotrophic
Mesozoikum/Erdmittelalter (erdgeschichtliches Zeitalter)
Mesozoic, Mesozoic Era
Messbecher *lab* measuring cup
Messbereich range of measurement
messen/ablesen read, record
messen/abmessen measure
messen/prüfen test
Messenger-RNA/Boten-RNA/mRNA
messenger RNA, mRNA
Messfehler error in measurement, measuring mistake
Messfühler/Sensor/Sonde *lab*
sensor, probe
Messgenauigkeit
accuracy of measurement
Messgerät
meter, measuring apparatus
Messglied *math* **(Größe)**
measuring unit, measuring device
Messgröße quantity to be measured
Messkolben volumetric flask

Messpipette graduated pipette, measuring pipet
Messtechnik metrology
messtechnisch metrological
Messung measurement, test, testing, reading, recording
Messverfahren measuring procedure
Messzylinder graduated cylinder
Metabiose metabiosis
metabolisches Spektrum/ Stoffwechselspektrum metabolic scope, index of metabolic expansibility
Metabolismus/Stoffwechsel metabolism
Metabolismusrate/ Stoffwechselrate/ Energieumsatzrate metabolic rate
Metabolit/Stoffwechselprodukt metabolite
Metacercarie metacercaria, adolescaria
metachromatische Granula (*pl***)** metachromatic granules, volutin granules
Metallothionein metallothionein
Metamer/echtes Segment metamere, segment
Metamerie/Segmentierung metamerism, segmentation
metamorph/metamorphisch/sich verändernd metamorphic
Metamorphose/Verwandlung metamosphosis, transformation
metamorphosieren/verwandeln/die Gestalt ändern metamorphose, metamorphize, transform
Metanephros/Nachniere/definitive Niere metanephros, hind kidney, definitive kidney
Metaphasenchromosom metaphase chromosome
Metaphyt (pflanzlicher Vielzeller) metaphyte
Metasaprobität metasaprobity
Metastase/Tochtergeschwulst metastasis
Metatarsaldrüse metatarsal gland
Metathorakalschild (dorsal)/ Metanotum metanotum
metazentrisches Chromosom metacentric chromosome
Methan methane
methanbildend/methanogen methanogenic
Methanbildner methanogenic organism, methanogen
methanophil methanophile
Methionin methionine
• **fMet (N-Formylmethionin)** fMet (N-formyl methionine)

Methode der kleinsten Quadrate *stat* least squares method
Methroxat methroxate
methylieren methylate
Methylierung/Methylieren methylation
Metöstrus/Metoestrus metestrus
metrische Skala metric scale
Meute (Hunde) kennel, pack (of dogs)
Mevalonsäure (Mevalonat) mevalonic acid (mevalonate)
Micelle micelle
Micellierung micellation
Michaelis-Menten-Gleichung Michaelis-Menten equation
Michaeliskonstante/ Halbsättigungskonstante (K_M) Michaelis constant, Michaelis-Menten constant
Miene/Gesichtsausdruck facial expression
Mienenspiel play of facial features, changing facial expressions
Miesmuscheln/Mytiloidea mussels
Miete (Grube) *agr* pit, silo (for feed storage)
Migration/Wanderung migration
Mikraner (kleine ♂ Ameise) micraner (dwarf ♂ ant)
Mikrergat (kleine Ameisenarbeiterin) micrergate
Mikrobe/Mikroorganismus microbe, microorganism
mikrobiell microbial
Mikrobody/Mikrokörperchen microbody
Mikrofauna/Kleintierwelt microfauna
Mikrofilament/Aktinfilament/ Actinfilament microfilament, actin filament
Mikroflora microflora
Mikrogliazelle microglial cell
Mikrogyne (kleine ♀ Ameise) microgyne (dwarf ♀ ant)
Mikrohabitat microhabitat
Mikroinjektion microinjection
Mikroklima microclimate
Mikrokosmos microcosm
Mikromanipulation micromanipulation
Mikromanipulator micromanipulator
Mikromer micromere
Mikrometerschraube *micros* micrometer screw, fine-adjustment, fine-adjustment knob
Mikronema (*pl* **Mikronemen)** micronema (*pl* **micronemas**)
Mikronukleus micronucleus

Milbenforschung

**Mikroorganismus
(pl Mikrorganismen)/Mikrobe**
microorganism, microbe
Mikrophyten microphytes
Mikropipette micropipet
Mikropipettenspitze
micropipet tip
Mikropräparat
prepared microscope slide
Mikropyle micropyle
**Mikropylenwulst/
Mikropylenwarze/Karunkula/
Caruncula** caruncle
Mikrosatellit microsatellite
Mikroskop microscope
• **Kursmikroskop**
course microscope
• **Polarisationsmikroskop**
polarizing microscope
• **Präpariermikroskop**
dissecting microscope
• **Stereomikroskop**
stereo microscope
• **zusammengesetztes Mikroskop**
compound microscope
Mikroskopie microscopy
• **Dunkelfeld-Mikroskopie**
darkfield microscopy
• **Hellfeld-Mikroskopie**
brightfield microscopy
• **Hochspannungselektronen-
mikroskopie** high voltage
electron microscopy (HVEM)
• **Immun-Elektronenmikroskopie**
immunoelectron microscopy (IEM)
• **Interferenzmikroskopie**
interference microscopy
• **konfokale Laser-Scanning
Mikroskopie** confocal laser
scanning microscopy
• **Kraftmikroskopie**
force microscopy
• **Lichtmikroskopie**
light microscopy (LM)
(compound microscope)
• **Phasenkontrastmikroskopie**
phase contrast microscopy
• **Polarisationsmikroskopie**
polarizing microscopy
• **Rasterelektronenmikroskopie
(REM)** scanning electron
microscopy (SEM)
• **Rastertunnelmikroskopie (RTM)**
scanning tunneling microscopy
(STM)
• **Rasterkraftmikroskopie**
atomic force microscopy (AFM)
• **Transmissionselektronen-
mikroskopie/
Durchstrahlungselektronenmikroskopie** transmission
electron microscopy (TEM)

mikroskopieren vb
examine under a microscope,
use a microscope
Mikroskopieren n
examination under a microscope,
usage of a microscope
Mikroskopierleuchte
microscope illuminator
Mikroskopierverfahren
microscopic procedure
Mikroskopierzubehör
microscopy accessories
mikroskopisch microscopic,
microscopical
**mikroskopische Aufnahme/
mikroskopisches Bild**
micrograph,
microscopic image
mikroskopisches Präparat
microscopical preparation/mount
Mikroskopzubehör
microscope accessories
Mikrosphäre microsphere
Mikrospore microspore
Mikrotom microtome
• **Gefriermikrotom**
freezing microtome,
cryomicrotome
• **Rotationsmikrotom**
rotary microtome
• **Schlittenmikrotom**
sliding microtome
• **Ultramikrotom** ultramicrotome
**Mikrotom-Präparatehalter/
Objekthalter (Spannkopf)**
microtome chuck
Mikrotomie microtomy
• **Kryoultramikrotomie**
cryoultramicrotomy
Mikrotommesser
microtome blade
Mikrotrabekulargeflecht
microtrabecular network
mikrotubuliassoziertes Protein
microtubule-associated protein
(MAP)
Mikrotubulus microtubule
**Mikrotubulus-
Organisationszentrum**
microtubule organizing center
(MTOC)
Mikroverfahren microprocedure
Mikrovillus (pl Mikrovilli)
microvillus (pl microvilli)
Mikrozelle microcell
miktisch mictic
Milben & Zecken/Acari/Acarina
mites & ticks
Milbenbekämpfungsmittel/Akarizid
acaricide
Milbenforschung/Acarologie
acarology

M Milch

Milch milk
- **geronnene Milch** curd
- **Kropfmilch/Kropfsekret (Tauben)** crop milk, pigeon milk (milky secretion from crop lining)
- **Vormilch/Biestmilch/ Kolostralmilch/Colostrum** foremilk, colostrum
- **Uterusmilch/Uterinmilch** *ichth/ entom* uterine milk

Milchbrustgang/Ductus thoracicus (Lymphbahn) ductus thoracicus
Milchdrüse *allg* milk gland
Milchdrüse/Brustdrüse mammary gland
Milcheinschuss lactation (often surge-like/actual onset of lactation after colostrum)
Milchfischverwandte/Sandfische/ Gonorhynchiformes milkfishes and relatives
milchführend lactiferous
Milchgang milk duct, lactiferous duct
Milchgebiss deciduous dentition, lacteal dentition, primary dentition
Milchkuh dairy cow
Milchleiste milk line, mammary ridge
Milchprodukt dairy product
Milchreife/Grünreife (Getreide) milk ripeness, milk stage
Milchröhre/Milchsaftröhre *bot* latex tube, lactifer, laticifer
- **gegliederte Milchröhre** articulated lactifer/laticifer

Milchsaft/Latex *bot* latex
Milchsäure (Laktat) lactic acid (lactate)
Milchsäureamid/Laktamid/Lactamid lactamide
Milchsäuregärung/Laktatgärung lactic acid fermentation, lactic fermentation
- **heterofermentative Milchsäuregärung** heterolactic fermentation
- **homofermentative Milchsäuregärung** homolactic fermentation

Milchtritt treading, kneading (milk elicitation movement)
Milchvieh dairy cattle
Milchwirtschaft dairy (dairy husbandry)
Milchzähne milk teeth, deciduous teeth, first teeth, primary teeth
Milchzisterne/Milchsinus milk cistern, lactiferous sinus
Milchzucker/Laktose milk sugar, lactose
Milieutheorie *ethol* learning theory

Milz spleen
Milzbalken/Trabeculae lienis splenic trabeculae
Milzkapsel splenic capsule
Milzknötchen/Milzkörperchen/ Malpighi-Körperchen/Milzfollikel splenic corpuscle, splenic nodule, splenic node, splenic follicle
Milzpulpa (rote/weiße) splenic pulp (red/white)
Milzstrang splenic cord
Mimese/äußere,schützende Ähnlichkeit (Hintergrundsfärbung) mimesis
Mimik mimic
Mimikry/schützende Nachahmung/ Schutztracht/Angleichung mimicry
- **Angriffs-Mimikry/Peckhamsche Mimikry** aggressive mimicry, Peckhamian mimicry
- **Automimikry** automimicry
- **Batessche Mimikry** Batesian mimicry
- **Mertenssche Mimikry** Mertensian mimicry
- **Müllersche Mimikry** Muellerian mimicry
- **Peckhamsche Mimikry/Angriffs-Mimikry** Peckhamian mimicry, aggressive mimicry
- **Verhaltensmimikry/ Ethomimikry** behavioral mimicry, ethomimicry
- **Verteidigungs-Mimikry** protective mimicry

Mimosengewächse/Mimosaceae mimosa family
Mine (Fraßgang) mine
- **Blattmine/Fraßgang** mine
- **Platzmine** blotch mine
- **Spiralmine/Heliconom/ Heliconomium** serpentine mine, heliconome
- **Sternmine/Asteronom** star mine, asteronome

Mineralboden mineral soil
Mineraldünger mineral fertilizer, inorganic fertilizer
Mineralokortikoid/ Mineralocorticoid mineralocorticoid
Mineralöl mineral oil
Mineralquelle mineral spring
Mineralstoffe/Mineralien minerals
Mineralwasser mineral water
minerotroph minerotrophic
Miniaturenplattenpotential (MEPP) miniature endplate potential
Minichromosom minichromosome, artificial chromosome
Minigen minigene

minimale Hemmkonzentration (MHK) minimal inhibitory concentration, minimum inhibitory concentration (MIC)
Minimalmedium minimal medium
Miniprep/Minipräparation miniprep, minipreparation
Minisatelliten-DNA minisatellite DNA
Minus-Strang/Negativ-Strang (nichtcodierender Strang) *gen* minus strand
Miozän (erdgeschichtliche Epoche) Miocene, Miocene Epoch
Mirazidium/Miracidium (*pl* Miracidien) (Digenea-Larve) miracidium (fluke larva)
Mischantiserum mixed antiserum
mischbar miscible
• **unvermischbar** immiscible
mischerbig/heterozygot heterozygous
Mischerbigkeit/Heterozygotie heterozygosity
mischfunktionelle Oxidase mixed-function oxidase
Mischkultur *agr* mixed crop, mixed stand; *micb* mixed culture
Mischvererbung blending inheritance
Mischwald mixed forest
Mischzylinder *lab* volumetric flask
Misodendraceae/ Federmistelgewächse feathery mistletoe family
Missbildungen verursachend/ teratogen teratogenic
Missense-Mutation/ Fehlsinnmutation missense mutation
misshandeln/quälen maltreat, mistreat, abuse, torment, being cruel
Misshandlung/Quälerei maltreatment, mistreatment, abusement, cruelty
Mist/Dung manure, dung; droppings (Tierkot)
Mistbeet/Frühbeet forcing bed, hotbed
Mistelgewächse/ Riemenblumengewächse/ Loranthaceae mistletoe family (showy mistletoe family)
Mistelgewächse/Viscaceae christmas mistletoe family
Mistpilze/Bolbitiaceae bolbitius family
mitbewohnend (Muschel/Schnecke etc.) inquiline
mitochondriale Vererbung mitochondrial inheritance
Mitochondrium/Mitochondrion (*pl* Mitochondrien) mitochondrion (*pl* mitochondria)

Mitose/Kernteilung mitosis, nuclear division, duplication division
• **Endomitose** endomitosis
Mitosezyklus mitotic cycle
mitotisch mitotic
mitotische Rekombination mitotic recombination
Mitralklappe/Bikuspidalklappe (*siehe auch:* Herzklappe) mitral valve, bicuspid valve
Mitralzelle mitral cell
Mittagsblumengewächse/ Eiskrautgewächse/Aizoaceae mesembryanthemum family, fig marigold family, carpetweed family
Mittel/Durchschnittswert (*siehe auch:* Mittelwert) mean, average
Mittelauge/Medianauge (Stirnocelle) median eye, midline eye (a dorsal ocellus)
Mittelband/Konnektiv (Staubblatt) connective
Mittelbrust/Mesothorax mesothorax
Mitteldarm *sensu stricto* mid intestine
Mitteldarm/Mesenteron/"Magen"/ Ventriculus (Insekten) midgut, mesenteron, ventricle, ventriculus, "stomach"
Mitteldarm (Echinodermen) pyloric stomach
Mitteldarmdivertikel/ Mitteldarmventrikel/ Mitteldarmdrüse (Blindsack) midgut diverticulum/cecum, midgut gland
Mitteldarmdrüse/ Mitteldarmdivertikel/ Mitteldarmventrikel (Blindsack) midgut gland, midgut diverticulum/cecum
Mitteldarmdrüse (Mollusken/ Echinodermen) digestive gland, "liver"
Mitteldarmdrüse/Hepatopankreas (Decapoden) hepatopancreas
Mittelfuß/Metatarsus metatarsal
Mittelfußgelenk (Vögel) hock
Mittelfußknochen metatarsal bone
Mittelgebirge low mountain range, highlands
Mittelhand/Metacarpus metacarpal
Mittelhandknochen metacarpal bone
Mittelhirn/Deutocerebrum (Insekten) deutocerebrum
Mittelhirn/Mesencephalon (Vertebraten) midbrain, mesencephalon
Mittelhirndach/Tectum opticum optic tectum, optic lobe

Mittellamelle 592

Mittellamelle (Zellwand) *bot*
  middle lamella
Mittelleib/Thorax (Insekten) thorax
Mittelleittrieb *bot* central leader
Mittelmehl *agr* middlings, shorts
Mittelmoräne medial moraine
Mittelohr middle ear, midear
Mittelrippe/Costa midrib, midvein,
  costa
**Mittelrippe eines Fiederblattes/
  Fiederblattachse/Rhachis/
  Blattspindel** rachis
Mittelschicht median layer,
  median zone
**Mittelschnecken/Mesogastropoda/
  Taenioglossa (Kammkiemer/
  Monotocardia)** mesogastropods:
  periwinkles & cowries
mittelständig *bot* perigynous
**Mitteltiere/Gewebetiere/
  "Vielzeller"/Metazoa** metazoans
Mittelwald middle-aged forest
**Mittelwert/Mittel/arithmetisches
  Mittel/Durchschnittswert** *stat*
  mean value, mean, arithmetic mean,
  average
  • **bereinigter Mittelwert/
    korrigierter Mittelwert**
    adjusted mean
  • **Elternmittelwert** midparent value
  • **harmonisches Mittel**
    harmonic mean
  • **Quadratmittel** quadratic mean
  • **Regression zum Mittelwert**
    regression to the mean
**Mixer/Mixette/Küchenmaschine
  (Vortex)** mixer, blender (vortex)
Mixis mixis
mixoploid mixoploid
Mixoploidie mixoploidy
mixotrope Reihe mixotropic series
mixotroph mixotrophic, mesotrophic
Mizelle micelle
Moas/Dinornithiformes moas
Modalwert/Modus/Art und Weise
  mode
Modalwert *stat* modal value
Modellbau model building
Moder (Schimmel) mould, mildew
moderig/faulend/verfaulend
  rotting, decaying, putrefying,
  decomposing; (Geruch) mouldy,
  putrid, musty
**modern/vermodern/faulen/
  verfaulen** rot, decay, putrefy,
  decompose
Modifikationsgen modifier gene
Modul/Funktionseinheit module
Modus/Art und Weise/Modalwert
  mode
Mohngewächse/Papaveraceae
  poppy family

**molare Masse/Molmasse
  ("Molgewicht")** molar mass
  ("molar weight")
**Molchschwanzgewächse/
  Saururaceae** lizard's tail family
Mole breakwater, jetty, mole
Molekül molecule
Molekularbiologie
  molecular biology
**molekulare Cytogenetik/
  molekulare Zytogenetik**
  molecular cytogenetics
Molekulargenetik
  molecular genetics
**Molekulargewicht/relative
  Molekülmasse (M_r)** molecular
  weight, relative molecular mass (M_r)
Molekularsieb/Molekülsieb
  molecular sieve
Molekülion molecular ion
**Molekülmasse
  ("Molekulargewicht")**
  molecular mass
  ("molecular weight")
Molke whey
Molkerei dairy
Molkereiprodukt dairy product
**Molmasse/molare Masse
  ("Molgewicht")** molar mass
  ("molar weight")
Molvolumen molar volume
Molybdän molybdenum
monadal/monadoid/monadial
  motile unicellular
monask monascous
**Monatsblutung/Menstruation/
  Periode** menstruation, period
Mondbein/Os lunatum lunate bone,
  semilunar bone
Mondrhythmus/Lunarrhythmus
  lunar rhythm, circamonthly rhythm
**Mondsamengewächse/
  Menispermaceae** moonseed family
Mondzyklus/Lunarzyklus (28 Tage)
  lunar cycle
Monimiengewächse/Monimiaceae
  monimia family, boldo family
**Monochasium/eingablige
  Trugdolde** monochasium,
  simple cyme, monochasial cyme
**monochlamydeisch/
  haplochlamydeisch/
  einfachblumenblättrig/
  mit einfacher Blütenhülle**
  monochlamydeous,
  haplochlamydeous
monocistronisch/monozistronisch
  monocistronic
monocolpat monocolpate
Monocyt/Monozyt monocyte,
  mononuclear leucocyte
monogam monogamous

Monogamie/Einehe monogamy
monogen monogenic
monogene (Erb)Krankheiten
  monogenic diseases
monogonont monogonont
monogyn/einweibig monogynous
Monohybridkreuzung
  monohybrid cross
monokarpisch/hapaxanth
  monocarp, monocarpic, hapaxanthic,
  hapaxanthous, hapanthous
monoklonal monoclonal
monoklonaler Antikörper
  monoclonal antibody
Monokotyle/Monokotyledone
  monocotyledon, monocot
Monokultur monoculture
monolektisch monolectic
monomiktisch monomictic
monomitisch monomitic
monomorph monomorphic,
  monomorphous
Monomorphismus monomorphism
mononukleär/mononucleär
  mononuclear
monophag/monotroph
  monophagous, monotrophic,
  univorous
monophasisch monophasic
monophyletisch monophyletic
monophyodont monophyodont
monopodial monopodial,
  indeterminate
Monopodium monopodium
monospezifisch monospecific
Monospezifität monospecificity
monöstrisch monoestrous,
  monestrous
monosymmetrisch/zygomorph
  monosymmetrical, zygomorphic
monothalam/einkammerig/
  einkämmrig/monothekal
  monothalamous, monothalamic,
  single-chambered, monothecal
monothetisch monothetic
monotok monotokous
monotrich monotrichous
Monotropaceae/
  Fichtenspargelgewächse
  Indian pipe family
Monözie/Einhäusigkeit/
  Gemischtgeschlechtigkeit
  monecy, monoecy, monecism,
  monoecism
monözisch/einhäusig/
  gemischtgeschlechtig
  monoecious, monecious
monozygot/eineiig monozygotic,
  monozygous
Monozygotie/Eineiigkeit
  monozygosity
Monsunwald monsoon forest

Moor moor(land), peatland, bog
  (ombrotroph), fen (minerotroph),
  mire (European: from old Norse
  term); muskeg (Canadian bog/fen)
● **Aapamoor/Strangmoor**
  aapa mire, string bog
● **Anmoor** early bog, half-bog
● **Braunmoor/Flachmoor/**
  Niedermoor (minerotroph)
  fen (minerotrophic/alkaline mire)
● **Bruchwaldmoor/Bruchmoor/**
  Bruchwald/Sumpfwald/
  Waldmoor carr (fen woodland),
  swamp woods/forest, wooded
  swamp, paludal forest
● **Deckenmoor** blanket mire,
  blanket bog
● **Fenn/Fen/Fehn/Feen/Vehn**
  (Moorland/Sumpf) fen
● **Flachmoor/Niedermoor/**
  Wiesenmoor/Braunmoor/Fen
  (minerotroph) fen
  (minerotrophic/alkaline mire),
  fenland
● **Flarke** flark
● **Hangmoor** slope fen
● **Heidemoor** heath, heath moor
● **Hochgebirgsmoor**
  alpine mire/bog
● **Hochmoor** raised bog, raised
  mire, (upland/high) moor, peat bog
● **Lagg (Randsumpf von Hoch-**
  mooren) lagg (drainage channel/
  water trough within bog/fen)
● **Mudde/organogener Schlamm**
  peat clay, organic silt
● **Niedermoor/Flachmoor/**
  Wiesenmoor/Braunmoor/Fen
  (minerotroph) fen
  (minerotrophic/alkaline mire),
  fenland
● **Niederungsmoor/Talmoor**
  valley bog
● **Palsenmoor/Torfhügelmoor**
  palsa bog
● **Quellmoor** spring fen
● **Randgehänge** rand/slope
  community of raised bog
● **Rülle** bog drainage rill
● **Schlenke** bog hollow, bog ditch,
  bog rivulet (in raised bog)
● **Schwingmoor/Schwingrasen**
  quaking bog, quagmire,
  floating mat
● **Sphagnum-Moor** peat bog
● **Strangmoor (Aapamoor)** string
  bog, patterned mire (aapa mire)
● **Sumpfmoor** swamp-marsh
● **Talmoor/Niederungsmoor**
  valley bog, head water bog
● **terrainbedeckendes Moor**
  blanket bog, blanket mire

Moorauge

- **topogene Moore** topogenic bogs
- **Torfhügelmoor/Palsenmoor** palsa bog
- **Torfmoor** bog, peat bog, peat moor, muskeg
- **Tundramoor** tundra bog, tundra muskeg (Canada)
- **Übergangs-Waldmoor** carr
- **Übergangsmoor/Zwischenmoor (ombrominerotroph)** transitory bog, transition bog (poor fen/weakly soligenous bog)

Moorauge/Kolk/Blänke pothole, deep pool
Moorgrund quagmire
Moorlandschaft moorland
Moorpflanze/Sumpfpflanze helophyte, bog plant, marsh plant
Moortorf/Hochmoortorf highmoor peat, sphagnum peat, moss peat
Moos moss
Moosblümchen/Mayacaceae mayaca family, bogmoss family
Moosblüte moss flower
Moosdecke moss mat, moss cover
Moosfarngewächse/Selaginellaceae selaginella family, spike-moss family, small club-moss family
Moosfaser mossy fiber
Mooshaube moss cap, haircap, calyptra
Mooskunde/Bryologie bryology
Moospolster/Mooskissen/ Moosrasen moss cushion, moss carpet
Moosschicht moss layer
Moosstiel/Kauloid/Cauloid stemlet
Moosteppich moss carpet
Moostierchen/Bryozoen/ Ectoprocta/Polyzoa moss animals, bryozoans
- **Armwirbler/ Süßwasserbryozoen/ Lophopoda/Phylactolaemata** phylactolaemates, "covered throat" bryozoans, freshwater bryozoans
- **Engmünder/Stenostomata/ Stenolaemata** stenostomates, stenolaemates, "narrow throat" bryozoans
- **Kammmünder/Ctenostomata** ctenostomates
- **Kreiswirbler/Stelmatopoda/ Gymnolaemata** gymnolaemates, "naked throat" bryozoans
- **Lippenmünder/Lappenmünder/ Cheilostomata** cheilostomates

594

Moraceae/Maulbeergewächse mulberry family, fig family
Moräne/Gletschermoräne/ Gletscherschutt/Gletschergeröll moraine, till, glacial till
- **Endmoräne** end moraine
- **Frontalmoräne/Stirnmoräne** terminal moraine
- **Grundmoräne/Untermoräne** ground moraine, basal moraine
- **Mittelmoräne** medial moraine
- **Seitenmoräne** lateral moraine

Morast (sumpfiges Land/ schlammiger Boden) quagmire, swampy/muddy ground; (Schlamm) mud; mire (wet spongy earth of bog or marsh)
Morbidität (Häufigkeit der Erkrankungen) morbidity
Morcheln/Morchellaceae morels, morel family
Morcheltrüffeln/Gautieriales plated puffballs
Morgagnische Tasche/Recessus laryngis piriform recess
Moringagewächse/ Bennussgewächse/ Behennussgewächse/ Pferderettichgewächse/ Moringaceae horseradish tree family
Morphe morph, shape, form
Morphogenese morphogenesis
morphogenetisch morphogenetic
Morphologie morphology
morphologisch morphologic, morphological
Morphometrie morphometrics
Morphopoese morphopoesis
Morphospezies/morphologische Art morphospecies
morsch (Holz) decayed, rotten; brittle; frail, fragile
Mörser mortar
Mortalität/Sterblichkeit/Sterberate mortality
Morula/Maulbeerkeim morula
Mosaik mosaic
- **Keimbahnmosaik** germline mosaic, germinal mosaic, gonadal/gonadic mosaic, gonosomal mosaic

Mosaikdoppelschichtmodel mosaic bilayer model
Mosaikei mosaic egg
Mosaikentwicklung mosaic development
Mosaikgen/diskontinuierliches Gen/gestückeltes Gen mosaic gene, split gene, discontinuous gene
- **Vorkommen eines Gens im Mosaik/als Mosaik** mosaicism

Moschusbeutel musk bag
Moschusdrüse (Duftdrüse)
musk gland (scent gland)
Moschuskrautgewächse/Adoxaceae
moschatel family
Motiv (Vogelgesang) motive, theme
Motoneuron motoneuron,
motor neuron
motorisch motoric, motor ...
motorische Einheit motor unit
Motorzelle/motorische Zelle/
Gelenkzelle
(im Schwellkörper des Blattes)
motor cell, bulliform cell
Motten/Heterocera moths
Mottenbestäubung
moth pollination, phalaenophily
Mottenblume
moth-pollinated flower
Mottenschildläuse/"Weiße Fliegen"
whiteflies
Möwenvögel & Watvögel & Alken/
Charadriiformes
gulls & shorebirds & auks
mRNA/Boten-RNA mRNA,
messenger RNA
● **abundante mRNA**
abundant mRNA
MS (Massenspektroskopie)
MS (mass spectroscopy)
MSQ-Schätzung (Methode der
kleinsten Quadrate)
LSE (least squares estimation)
MTA (medizinisch-technische(r)
AssistentIn) medical technician,
medical assistant
(*auch:* Sprechstundenhilfe >
doctor's assistant)
MTLA (medizinisch-technische(r)
LaborassistentIn)
medical lab technician,
medical lab assistant
Mucin mucin
Mücken & Schnaken/Nematocera
(Diptera) mosquitoes
Mucoviszidose/Mukoviszidose/
zystische Fibrose
mucoviscidosis, cystic fibrosis
Mud/Mudde (Schlamm/Morast/
Schlick) mud
Mudde/Organopelit (limnische
Sedimente) lacustrine sediments
● **Dy/Torfschlamm/Torfmudde**
dy, gel mud
● **Faulschlamm/Sapropel** sludge,
sapropel
● **Gyttia/Gyttja/Grauschlamm/**
Halbfaulschlamm gyttja
(sedimentary peat), necron mud
● **Torfmudde**
(organogener Schlamm)
peat clay (organic silt/mud)

Muffe *lab* clamp holder
muhen (Rinder) moo
Mukoviszidose/Mucoviszidose/
zystische Fibrose
mucoviscidosis, cystic fibrosis
Mulch mulch
mulchen mulch
Mulchung/Mulchen mulching
Mulde depression, basin
muldenförmig trough-shaped
Muldensee kettle lake
Mull (fast neutraler Auflagehumus/
milder Dauerhumus)
mull humus, mull
Müll waste, trash, garbage
Mülldeponie/Müllplatz/
Müllabladeplatz/Müllkippe
waste disposal site, waste dump;
(Müllgrube: geordnet) landfill,
sanitary landfill
Müller-Faser *ophthal*
Müllerian fiber, fiber of Müller,
Muellerian fiber
Müller-Stützzelle *ophthal*
Müller cell
Müllersche Larve Müller's larva
Müllersche Mimikry
Müllerian mimicry,
Muellerian mimicry
Müllersche Muskel Müllerian
muscle, Muellerian muscle,
Mueller's muscle
Müllerscher Gang/
Ductus paramesonephricus
Mueller's duct, Müller's duct,
Müller's canal, Müllerian duct,
paramesonephric duct
Müllverbrennungsanlage
waste incineration plant,
incinerator
Müllvermeidung waste avoidance
Müllverwertungsanlage
(waste) recycling plant
Mulm/Fäule rot, decaying matter,
mold
Multienzymkomplex/
Multienzymsystem/Enzymkette
multienzyme complex,
multienzyme system
multifaktorieller Erbgang/
polygener Erbgang
multifactorial inheritance,
polygenic inheritance
Multigenfamilie multigene family
Multikomponentenvirus
multicomponent virus
multipar multiparous
multiple Allele multiple alleles
Multiplex-Sequenzierung *gen*
multiplex sequencing
Multiplizität der Infektion
multiplicity of infection (m.o.i.)

multiseriat

multiseriat/vielreihig/mehrreihig
multiseriate, multiple rowed
multivesikulärer Körper
multivesicular body
multivoltin/polyvoltin
multivoltine
**multizistronisch/multicistronisch/
polyzistronisch/polycistronisch**
multicistronic, polycistronic
**Mumienpuppe/bedeckte Puppe/
Pupa obtecta** obtect pupa
Mund/Öffnung mouth, opening,
orifice
**Mund-Kiemenhöhle/
Orobranchialhöhle**
orobranchial cavity
Mundarm (Polypen/Echinodermen)
oral arm
Mundarmgefäß arm canal,
brachial canal
Mundarmscheibe arm disk
Mundbucht/Stomodaeum foregut,
stomodaeum, stomodeum
Mundbucht/Vestibulum
oral vestibule
**Munddarm/Mundbucht/
Stomodaeum** foregut,
stomodaeum, stomodeum
Mundfeld/Buccalfeld/Peristom
mouth, buccal field, peristome
**Mundgliedmaße (*pl*
Mundgliedmaßen)** mouthpart,
oral appendage
Mundhaken *sg/pl* mouth hook(s)
Mundhöhle mouth cavity,
oral cavity, buccal cavity
Mundlappenanhang (Muscheln)
palp proboscis, palp appendage
**mundlos/ohne Mund/ohne
Öffnung/astom** mouthless,
astomous, astomatous
Mundöffnung opening of the mouth
Mundrohr/Magenstiel/Manubrium
gullet, pharynx; hypostome,
oral cone; manubrium
Mundsaugnapf buccal sucker
(prohaptor in flukes)
Mundscheibe/Oralscheibe/Peristom
oral disk, peristome, peristomium
Mundspalte/Rima oris
opening of the mouth
Mundwerkzeuge mouthparts
Muraminsäure muramic acid
Murein murein
Musaceae/Bananengewächse
banana family
Muscarin muscarine
**muscarinischer Rezeptor/
muskarinischer Rezeptor**
muscarinic receptor
Muschel (populär für Schale)
shell (*siehe auch:* Muscheln)

Muschelkalk (Epoche)
Middle Triassic
**Muschelkrebse/Ostracoden/
Ostracoda** seed shrimps, ostracods
Muschellaich spat
**Muscheln/Bivalvia/Pelecypoda/
Lamellibranchiata** bivalves,
pelecypods, "hatchet-footed
animals" (clams: sedimentary,
mussels: freely exposed)
Muschelschalenhälfte valve
Muschelschaler/Conchostraca
clam shrimps
**Muskatnussgewächse/
Myristicaceae** nutmeg family
Muskel muscle *(weitere
Muskelbezeichnungen finden Sie in
einem Wörterbuch der Human- bzw.
Veterinärmedizin)*
- **Abzieher/Abduktor/Abductor**
  abductor muscle
- **Anzieher/Schließmuskel/
  Adduktor/Adductor**
  adductor muscle
- **Dreher/Drehmuskel/Rotator**
  rotator muscle
- **Eingeweidemuskeln**
  visceral muscles
- **Erweiterer/Dilator** dilator muscle
- **Extremitätenmuskel**
  limb muscle, appendicular muscle
- **Fußretraktor**
  pedal retractor muscle
- **Haarmuskel/Musculus arrector
  pili** hair erector muscle,
  arrector pili muscle
- **Heber/Levator** levator, lifter
  (muscle: raising an organ or part)
- **Herzmuskel** cardiac muscle
- **Interpyramidalmuskel/
  interpyramidaler Muskel
  (Laterne des Aristoteles)**
  interpyramid muscle,
  comminator muscle
- **Kaumuskel** masticatory muscle,
  muscle of mastication
- **Klaffmuskel** diductor muscle
- **Längsmuskel** longitudinal muscle
- **Niederleger/Senker/Depressor**
  depressor muscle
- **Retraktormuskel/Retraktor/
  Rückzieher** retractor muscle
- **Ringmuskel** ring muscle,
  circular muscle
- **Rückzieher/Rückwärtszieher/
  Rückziehmuskel/Retraktor/
  Retraktormuskel** retractor muscle
- **Schließmuskel/Anzieher/
  Adduktor/Adductor**
  adductor muscle
- **Schließmuskel/Sphinkter**
  sphincter muscle

Mutation

- Senker/Niederleger/Depressor
  depressor muscle
- Vorzieher/Protractor
  protractor muscle

Muskelansatz muscle insertion
Muskelbauch/Venter musculi
  muscle belly
Muskelbinde/Muskelfaszie
  muscle fascia
Muskelbündel/Muskelfaserbündel
  muscle bundle, muscle fascicle
Muskelfaser muscle fiber
Muskelfibrille/Myofibrille
  myofibril
Muskelhaft/Synsarkose
  synsarcosis
Muskelkontraktion
  muscular contraction
Muskelkriechsohle (Gastropoden)
  muscular foot sole
Muskelleiste (Reptilienherz)
  muscular ridge, horizontal septum
Muskelmagen *orn* gizzard
Muskelspannung/Muskeltonus
  muscle tone
Muskeltonus muscle tone
Muskelursprung muscle origin
Muskelzucken muscle twitching
muskulär/die Muskeln betreffend
  muscular
Muskulatur musculature, muscles
- **Branchialmuskulatur**
  branchiomeric musculature/muscle
- **Eingeweidemuskulatur/ viscerale Muskulatur/ viszerale Muskulatur**
  visceral musculature/muscle
- **gestreifte Muskulatur**
  striated muscle, striped muscle
- **glatte Muskulatur**
  smooth muscle, plain muscle,
  non-striated muscle,
  unstriped muscle
- **Hautmuskulatur**
  dermal musculature/muscle
- **Rumpfmuskulatur**
  trunk musculature/muscle
- **schräggestreifte Muskulatur**
  obliquely striated musculature/
  muscle
- **unwillkürliche Muskulatur**
  involuntary musculature/muscle
- **viscerale Muskulatur/ viszerale Muskulatur/ Eingeweidemuskulatur**
  visceral musculature/muscle
- **willkürliche Muskulatur**
  voluntary musculature/muscle

muskulös very muscular,
  with big muscles (strong)
Mustang mustang
  (halbwildes Präriepferd)

Muster/Vorlage/Modell pattern,
  sample, model; specimen;
  (Musterung/Zeichnung) pattern,
  design
- **Verhaltensmuster**
  behavioral pattern

Musterbildung *gen*
  pattern formation
Mustererkennung *neuro*
  pattern recognition
mutagen/mutationsauslösend
  mutagenic
Mutagen/mutagene Substanz
  mutagen
Mutagenese mutagenesis
- *in vitro*-**Mutagenese**
  in vitro mutagenesis
- **oligonucleotidgesteuerte/ oligonukleotidgesteuerte Mutagenese** oligonucleotide-directed mutagenesis
- **sequenzspezifische Mutagenese**
  site-specific mutagenesis
- **ortsspezifische Mutagenese**
  site-directed mutagenesis

Mutagenität mutagenicity
Mutante mutant
- **durchlässige Mutante**
  leaky mutant
- **konstitutive Mutante**
  constitutive mutant
- **Letalmutante** lethal mutant
- **Petite-Mutante** petite mutant

Mutarotation mutarotation
Mutation
- **ARMS (System der amplifizierungsresistenten Mutation)** ARMS (amplification refractory mutation system)
- **Deletionsmutation**
  deletion mutation
- **Down-Mutation** down mutation
- **durchlässige Mutation**
  leaky mutation
- **Funktionsgewinnmutation**
  gain of function mutation
- **Funktionsverlustmutation**
  loss of function mutation
- **Insertionsmutation**
  insertion mutation
- **instabile Mutation**
  unstable mutation
- **Inversionsmutation**
  inversion mutation
- **Keimbahnmutation**
  germ-line mutation
- **Knockout-Mutation**
  knockout mutation
- **Leserasterverschiebung-(smutation)** frameshift mutation
- **letale Mutation/Letalmutation**
  lethal mutation

M Mutation   598

- **Missense-Mutation/ Fehlsinnmutation**
  missense mutation
- **Neumutation** new mutation
- **Nonsense-Mutation/ Nichtsinnmutation**
  nonsense mutation
- **pleiotrope Mutation**
  pleiotropic mutation
- **polare Mutation** polar mutation
- **Prä-Mutation** pre-mutation
- **Punktmutation** point mutation
- **Rückmutation** back mutation, reverse mutation
- **somatische Mutation**
  somatic mutation
- **Spontanmutation**
  spontaneous mutation
- **stumme Mutation**
  silent mutation, samesense mutation
- **Suppressormutation**
  suppressor mutation
- **temperatursensitive Mutation**
  temperature-sensitive mutation
- **uniparentale Mutation**
  uniparental mutation
- **Up-Mutation** up-mutation
- **Vorwärtsmutation**
  forward mutation

Mutationsbelastung/Mutationslast
mutational load
Mutationsrate mutation rate
Mutierbarkeit/ Mutationsfähigkeit/Mutabilität
mutability
mutieren mutate
Mutter mother
Mutterbestand original stand
Muttergestein/Ausgangsgestein/ Grundgestein
bedrock, rock base, parent material/rock
Mutterkuchen/Plazenta placenta
Muttermund/Gebärmuttermund
mouth/orifice of the uterus, orificium uteri
Mutterpflanze mother plant, female plant
Mutterschaft maternity, motherhood
Mutterstrang/Matrizenstrang
template strand
Muttersubstanz parent substance
Muttertier mother animal (female parent), dam
Muttertrompete/Eileiteröffnung/ Ostium tubae
ostium (of oviduct)
Mutterzelle mother cell
Mutualismus/Gegenseitigkeit
mutualism
myelinisiert/markhaltig myelinated

Myelinisierung *neuro* myelination, myelinization
myelinlos/nichtmyelinisiert/ marklos/markfrei
nonmyelinated
Myelinscheide/Markscheide *neuro*
myelin sheath, medullary sheath
Myelom myeloma
Myelozyt/Myelocyt/ Knochenmarkzelle
myelocyte, bone marrow cell
Mykobiont/Mycobiont/Pilzpartner
mycobiont
Mykologe/Kenner bzw. Erforscher der Pilze mycologist
Mykologie/Pilzkunde mycology
mykophag/myzetophag/fungivor/ pilzfressend mycophagous, mycetophagous, fungivorous, fungus-eating, feeding on fungus
Mykophagie/Mycetophagie
mycophagy
Mykoplasma (*pl* **Mykoplasmen**)
mycoplasma (*pl* myoplasmas)
Mykorrhiza/"Pilzwurzel"
mycorrhiza
Mykose mycosis
Mykotoxin mycotoxin
Myofibrille/Muskelfibrille
myofibril
myogen myogenic
Myomer/Myotom
myomere, myotome
Myonem myoneme
myotatischer Reflex/ Dehnungsreflex
myotatic reflex, stretch reflex
Myotubulus (*pl* **Myotubuli**)
myotubule (*pl* myotubules)
Myricaceae/Gagelgewächse
bog myrtle family, wax-myrtle family, sweet gale family, bayberry family
Myristicaceae/ Muskatnussgewächse
nutmeg family
Myristinsäure/Tetradecansäure (Myristat) myristic acid, tetradecanoic acid (myristate/ tetradecanate)
Myrmekochorie (Ameisenausbreitung) *bot*
myrmecochory, ant-dispersal
myrmekophil myrmecophilous
Myrmekophyt/Ameisenpflanze
myrmecophyte, myrmecoxenous plant
Myrsinaceae myrsine family
Mysis-Larve mysis larva
Myxobakterien/Schleimbakterien
myxobacteria
Myxomatose myxomatosis

Myzel/Pilzgeflecht (*pl* Myzelien)
mycelium (*pl* mycelia)
- **Dauermyzel/Mycelium perenne**
  persistent mycelium
- **Luftmyzel** aerial mycelium
- **Paarkernmyzel/Sekundärmyzel**
  dikaryotic mycelium,
  secondary mycelium
- **Pilzbrut/Mycelium fecundum**
  spawn
- **Primärmyzel/Einkernmyzel**
  primary mycelium

- **Raquettemyzel/Keulenmyzel
  (Raquettehyphen/
  Keulenhyphen)**
  raquet mycelium (raquet hyphae/
  raquet hyphas)
- **Sekundärmyzel/Paarkernmyzel**
  secondary mycelium

Myzelstrang mycelial cord
**myzetophag/fungivor/
pilzfressend**
mycetophagous, fungivorous,
feeding on fungi

N Nabe

Nabe (Netznabe) *arach* hub, nub
Nabel/Hilum *bot* hilum,
funiculus scar
Nabel *zool* navel, umbilicus,
omphamos
nabelartig/omphaloid navel-like,
umbilicate, omphaloid
Nabelflechten (Blattflechten)
umbilicate foliose lichens
Nabelschnur/Nabelstrang
umbilical cord
Nabelstrang/Funiculus *bot* seed
stalk, ovule stalk, funicle, funiculus
Nach-Hyperpolarisation *neuro*
after-hyperpolarization
**nachahmen/nachmachen/mimen
("nachäffen")** imitate, mimic
**nachahmend/mimetisch/fremde
Formen nachbildend** imitating,
mimetic
Nachahmung (Mimikry) imitation
(mimikry)
Nachauflaufbehandlung *agr*
post-emergence treatment
Nachbalz postcopulatory behavior
Nachblüte/Postfloration
postfloration
Nachbrunst/Metöstrus metestrus
Nacheiszeit postglacial period
nachfeuern *neuro* afterdischarge
Nachfeuerung/Nachentladung
neuro afterdischarge
Nachfolgeverhalten following
behavior
Nachgeburt (Plazenta etc.)
afterbirth
**nachhaltige Entwicklung/
dauerhaft-umweltgerechte
Entwicklung**
sustainable development
Nachhaltigkeit sustained yield
**Nachhirn/Metencephalon
(Pons + Cerebellum)** afterbrain,
metencephalon
Nachklärbecken
secondary settling tank
Nachkomme progeny, descendant,
offspring
Nachkommen/Nachkommenschaft
progeny
**Nachniere/definitive Niere/
Metanephros** hind kidney,
definitive kidney, metanephros
Nachpotential *neuro* undershot
Nachreifen *bot/hort/agr*
after-ripening
**Nachschieber/letzter Afterfuß/
Postpes/Propodium anale
(Raupen)** anal proleg, anal leg
nachtaktiv night-active, nocturnal
● **tagaktiv** day-active, diurnal
Nachteil disadvantage

**Nachtkerzengewächse/Onagraceae/
Oenotheraceae** willowherb family,
evening-primrose family
Nachtpflanze/Nachtblüher
nocturnal plant
**Nachtschattengewächse/
Solanaceae**
nightshade family, potato family
Nachtschwalben/Caprimulgiformes
nightjars, goatsuckers, oilbirds
Nachttier nocturnal animal
nachwachsen regenerate, regrow,
grow back, reestablish
Nachweis detection, proof
Nachweis verkürzter Proteine (PTT)
protein truncation test (PTT)
nachweisen detect, prove
Nachweisgrenze detection limit
Nachweismethode
detection method
Nachwuchs offspring, young
Nacken/Genick
nape, back of the neck, nucha
Nackenband/Ligamentum nuchae
nuchal ligament
Nackenbeuge *embr* cervical flexure
Nackenkamm nuchal crest
nackt naked, nude
**nackte Knospe/offene Knospe
(ohne Knospenschuppen)**
naked bud
**Nacktfarngewächse/
Hemionitidaceae**
strawberry fern family
Nacktheit nakedness, nudeness,
nudity
**Nacktkiemer/
Meeresnacktschnecken/
Nudibranchier**
sea slugs, nudibranchs
Nacktmaus nude mouse
Nacktsamer/Gymnosperme
naked-seed plant, gymnosperm
nacktsamig gymnosperm
Nacktschnecken slugs
Nacréschicht/Perlmuttschicht
nacreous layer
Nacréwand Nacré wall,
nacreous wall
NAD/NADH (Nikotinamid-**a**denin-
dinucleotid) NAD/NADH
(**n**icotinamide **a**denine **d**inucleotide)
Nadel needle
Nadel/Kanüle/Hohlnadel (Spritze)
hypodermic needle
Nadel/Spiculum/Sklerit spicule
Nadelbaum/Konifere/Conifere
coniferous tree, conifer, softwood
tree
nadelförmig needle-shaped, acicular
Nadelschicht (Boden) *for* needle
litter layer

Nadelstreu needle litter
Nadelwald coniferous forest
Nadelwaldstufe/hochmontane Stufe upper montane/subalpine conifer forest zone
NADP/NADPH (Nikotinamid-**a**denin-**d**inucleotid-**p**hosphat) NADP/ NADPH (**n**icotinamide **a**denine **d**inucleotide **p**hosphate)
Nagana/Naganaseuche (*Trypanosoma spp.*) nagana (disease)
Nagel nail
Nagel (des Kronblattes)/Unguis *bot* claw, unguis
nagen/an etwas nagen gnaw
nagend gnawing
Nager gnawer
Nagetiere/Rodentia rodents, gnawing mammals (except rabbits)
näherkommen/annähern/sich annähern/erreichen *math/stat* approach (e. g. a value)
Näherung *math* approximation
Nähragar nutrient agar
Nährboden/Nährmedium/ Kulturmedium/Medium/Substrat (*siehe auch:* Medium/ Kulturmedium) nutrient medium (solid and liquid), culture medium, substrate
Nährbouillon/Nährbrühe nutrient broth
Nährei trophic egg, nurse egg
Nährgewebe *allg* nutritive tissue, nutrient tissue; *bot* (nucellar) perisperm; *bot* (Embryosack) endosperm
nahrhaft/nährend/nutritiv nutritious, nutritive
Nährhumus unstable humus, friable humus, crustable humus
Nährlösung nutrient solution, culture solution
Nährmedium/Kulturmedium/ Medium nutrient medium, culture medium
- **Anreicherungsmedium** enrichment medium
- **Basisnährmedium** basal medium
- **Differenzierungsmedium** differential medium
- **Eiermedium/Eiernährmedium** egg medium
- **Elektivmedium/Selektivmedium** selective medium
- **Erhaltungsmedium** maintenance medium
- **komplexes Medium** complex medium
- **konditioniertes Medium** conditioned medium
- **Mangelmedium** deficiency medium
- **Minimalmedium** minimal medium
- **Selektivmedium/Elektivmedium** selective medium
- **synthetisches Medium (chemisch definiertes Medium)** defined medium
- **Testmedium/Prüfmedium (zur Diagnose)** test medium
- **Vollmedium/Komplettmedium** rich medium, complete medium
Nährmuskelzelle (Cnidaria) nutritive-muscular cell
Nährsalz nutrient salt
Nährstoff nutrient
nährstoffarm nutrient-deficient, oligotroph(ic)
nährstoffarm und humusreich/ dystroph dystroph(ic)
Nährstoffarmut nutrient deficiency
Nährstoffaufnahme nutrient uptake
Nährstoffbedarf nutrient demand, nutrient requirement
Nährstoffhaushalt nutrient budget
Nährstoffkreislauf/Stoffkreislauf (*siehe auch dort*) nutrient cycle
Nährstoffmangel nutritional deficit
Nährstoffprotein nutrient protein
nährstoffreich/eutroph nutrient-rich, eutroph, eutrophic
Nährstoffverhältnis nutritive ratio, nutrient ratio
Nährtier/Fresspolyp/Gasterozoid/ Gastrozooid/Trophozoid nutritive/feeding polyp, gasterozooid, trophozooid
Nahrung/Ernährung nutrition
Nahrung/Essen/Fressen food, feed
Nahrung/Nährstoff nutrient
Nahrungsaufnahme ingestion, food intake
Nahrungsbedarf (*pl* Nahrungsbedürfnisse) nutritional requirements
Nahrungsbedürfnisse nutritional requirements
Nahrungsdarm midgut, intestine
Nahrungsdotter/Deutoplasma deutoplasm
Nahrungsgefüge/Nahrungsnetz *ecol* food web
Nahrungskette *ecol* food chain
- **Detritusnahrungskette** detritus food chain, detrital food chain
- **Fraßnahrungskette/ Weidenahrungskette/ Weidekette** grazing food chain
Nahrungskonkurrenz food competition

**Nahrungskreislauf/
Nährstoffkreislauf** nutrient cycle
Nahrungsmangel nutrient
deficiency, food shortage
Nahrungsmenge food quantity
Nahrungsmittelkonservierung
food preservation
Nahrungsmittelvergiftung
food poisoning
Nahrungsnetz/Nahrungsgefüge
ecol food web
Nahrungspflanze food crop,
forage plant, food plant
Nahrungspflanzenanbau food crop
production
Nahrungspyramide *ecol*
biotic pyramid
Nahrungsquelle food source,
nutrient source
Nahrungssuche search for food,
forage, foraging
Nahrungsvakuole food vacuole,
gastriole
Nahrungswahl nutrient selection
Nährwert food value,
nutritive value
Nährwert-Tabelle nutrient table,
food composition table
Nährwurzel *bot* feeder root
Naht/Fuge/Verwachsungslinie
seam, suture, raphe
**Nahtlinien/Lobenlinien
(Ammoniten)** suture lines
• **ammonitische N./L.**
ammonitic suture lines
• **ceratitische N./L.**
ceratitic suture lines
• **goniatische N./L.**
goniatitic suture lines
Najadaceae/Nixenkrautgewächse
najas family, water nymph family
Name name, term
• **Ersatzname** substitute name
• **volkstümlicher Name/
Vernakularname** common name,
vernacular name
Namensbezeichnung name, term,
designation, nomenclature
Namensetikett/Namensschildchen
name tag
**Namensgebung/Benennung/
Bezeichnung (Nomenklatur)**
naming, designation (nomenclature)
Nandus/Rheiformes rheas
Nanismus nanism, dwarfishness,
dwarfism
nannandrisch nanandrous
Nannandrium/Zwergmännchen
nanander (male dwarf plant)
Nanophanerophyt (Sträucher)
nanophanerophyte
(shrubs under 2 m in height)

Nanoplankton/Nannoplankton
nanoplankton, nannoplankton
Napfblume bowl-shaped flower
Naphthalin naphthalene
Narbe/Wundnarbe/Cicatricula scar,
cicatrix, cicatrice
• **Fruchtblattnarbe** stigma
Narbenfäden (Mais) silk
Narbenkopf *bot* stigma head
(clublike swollen stigma)
Narbenlappen *bot* stigmatic lobe
Narkomedusen/Narcomedusae
narcomedusas
**Narzissengewächse/
Amaryllisgewächse/
Amaryllidaceae** daffodil family,
amaryllis family
Nase nose
Nasenbein/Os nasale nasal bone
Nasengaumengang
nasopalatine duct
Nasenhöhle nasal cavity/chamber
Nasenkapsel nasal capsule
Nasenloch/Nasenöffnung
nasal opening, nasal aperture;
(Vertebraten) nare,
naris (mostly plural: nares),
nostril of vertebrates
Nasenmuscheln/Conchae nasalis
turbinals, turbinate bones
Nasenöffnung nasal opening,
nasal aperture; (Vertebraten) nare,
naris (mostly plural: nares),
nostril of vertebrates
• **innere N./Choane**
internal nostril, choana
Nasenrachengang nasopharyngeal
duct
Nasenrücken bridge of the nose
Nasenschleimhaut olfactory
epithelium, nasal mucosa
**Nasensoldat/Nasutus-Soldat
(Nasutitermiten)** nasute
**Nassanov Drüse/
Nassanoffsche Drüse**
Nassanov gland, Nassanov's gland
Nassblotten wet blotting
Nassfäule wet rot
**Nasspräparat (Frischpräparat/
Lebendpräparat/Nativpräparat)**
wet mount
Nasswiese damp meadow,
wet meadow, wetland
Nastie nastic movement
nastisch nastic
**Natalität/Geburtenrate/
Geburtenzahl/Geburtenziffer**
natality, birthrate
Nationalpark national park
nativ (nicht-denaturiert)
native (not denatured)
Natrium sodium

Natriumdodecylsulfat
sodium dodecyl sulfate (SDS)
Natternzungengewächse/
Rautenfarngewächse/
Ophioglossaceae
adder's tongue family,
grape fern family
Naturdenkmal natural monument
naturfern/künstlich/synthetisch
man-made, artificial, synthetic
Naturforscher research scientist,
natural scientist
Naturführer nature guide
Naturgeschichte natural history
Naturgesetz natural law
Naturgewalten forces of nature,
natural forces, natural powers
Naturhaushalt (natürliches
Gleichgewicht) natural balance
naturidentisch (synthetisch)
synthetic (having same chemical
structure as the natural equivalent)
Naturkatastrophe natural
catastrophe, natural disaster
Naturkunde/Biologie life science,
biology
Naturkundemuseum
natural history museum
Naturlandschaft natural landscape,
natural environment, natural setting
Naturlehrpfad
nature trail, nature walk
natürlich natural
• **unnatürlich** unnatural
naturnah near-natural
Naturpark wildlife park
Naturreservat nature reserve
Naturschutz environmental
protection, nature protection,
conservation/preservation
Naturschutzbewegung
nature conservation movement
Naturschützer conservationist
Naturschutzgebiet nature/wildlife
reserve, wildlife sanctuary,
protected area; national park
Naturschutzverein/
Naturschutzbund
nature protection group,
nature protection league,
environmental group
Naturstoff natural product
Naturstoffchemie natural product
chemistry
Naturwiese native meadow
Naturwissenschaften
natural sciences, science
Naturwissenschaftler
natural scientist, scientist
naturwissenschaftlich scientific
Naupliusauge
naupliar eye (median eye)

Naupliuslarve/Nauplius naupliar
larva, nauplius
Nautilusverwandte/Nautiloidea
nautilus (*pl* nautili)
Nebel fog
• **leichter Nebel** mist
nebelig foggy
• **leicht nebelig** misty
Nebelwald cloud/fog forest, humid/
perhumid forest, montane rainforest
Nebelwüste fog desert
Nebenauge/Punktauge/Ocelle
simple eye, ocellus
Nebenblatt/Stipel stipule
Nebenblattdorn/Stipulardorn
stipular spine
nebenblattlos/ohne Stipeln
exstipulate, astipulate, estipulate
Nebenbuhler rival
Nebeneierstock/Nebenovar/
Epoophoron pampiniform body,
Rosenmüller's body, proovarium,
parovarium, epoophoron
Nebenfahne/Hypovexillum *orn*
aftervane
Nebenfeder/Afterfeder/Hypopenna
orn afterfeather, accessory plume,
hypoptile, hypoptilum
Nebenfittich *siehe* Nebenflügel
Nebenflügel/Afterflügel/
Daumenfittich/Alula/Ala spuria
orn alula, spurious wing,
bastard wing
Nebenfruchtform *fung*
imperfect stage, asexual stage,
anamorphic stage
Nebengelenktiere/Xenarthra/
Zahnarme/Edentata xenarthrans,
"toothless" mammals, edentates
Nebenherz/auxilläres Herz
auxillary heart
Nebenhistokompatibilitätsantigen
minor histocompatibility antigen
Nebenhistokompatibilitätskomplex
minor histocompatibility complex
Nebenhoden/Epididymis
epididymis
Nebenhöhle (Schädel) frontal sinus
Nebenkelch/Außenkelch *bot*
calycle, calyculus
Nebenkrone/Parakorolle *bot*
paracorolla
Nebenniere adrenal gland
Nebenovar/Nebeneierstock/
Epoophoron pampiniform body,
Rosenmüller's body, proovarium,
epoophoron
Nebenprodukt by-product,
residual product, side product
Nebenschilddrüse/Beischilddrüse/
Epithelkörperchen/Parathyreoidea
parathyroid gland

Nebenschilddrüsenhormon/
Parathyrin/Parathormon (PTH)
parathyroid hormone, parathyrin,
parathormone (PTH)
Nebentrieb/Seitentrieb *bot*
offshoot, lateral shoot
Nebenwirkung(en) side effect(s)
Nebenwirt secondary host
Nebenwurzel/Beiwurzel/
Adventivwurzel supplementary
root, adventitious root
Nebenwurzel/Seitenwurzel
lateral root, secondary root
Nebenzelle (Spaltöffnung)
subsidiary/accessory/auxiliary cell
Nebenzunge/Paraglossa
paraglossa
Nebulin nebulin
Nectophor/Nektophor
(Schwimmglocke) nectophore
(swimming bell)
Negativ-Strang/Minus-Strang
(nichtcodierender Strang) *gen*
minus strand (noncoding strand)
Negativkontrastierung *micros*
negative staining,
negative contrasting
Negrisches Körperchen/Negri
Körper Negri body
Nehrung (einem *Haff* vorgelagerter
Landstreifen)
baymouth bar, bay bar, bay barrier
Neigung/Neigungswinkel
inclination
Nekrophyt necrophyte
Nekrose necrosis
nekrotisch necrotic
nekrotroph necrotroph, necrotrophic
Nektar nectar
Nektarblatt/Honigblatt nectar leaf, honey leaf
Nektarium/Nektardrüse/Honigdrüse
nectar gland, nectary
• extraflorales Nektarium
extrafloral nectary
Nekton (Organismen mit starker
Eigenbewegung) nekton
(organisms with high mobility)
Nektophor/Nectophor
(Schwimmglocke) nectophore
(swimming bell)
Nelkengewächse/Caryophyllaceae
pink family, carnation family
Nelkenheidegewächse/
Frankeniengewächse/
Frankeniaceae sea heath family,
alkali-heath family
Nemathelminthen/Aschelminthen/
Schlauchwürmer/Rundwürmer
sensu lato
nemathelminths,
aschelminths

Nematocyste/Nesselkapsel/Cnide
nematocyst, thread capsule,
urticator, cnida
Nematoden/Fadenwürmer/
Nematoda nematodes,
roundworms
Nematodenbekämpfungsmittel/
Nematizid nematicide
Nematogen (*Dicyemida*) nematogen
Nematophore nematophore,
nematocalyx
Neogen/Jung-Tertiär
(erdgeschichtliche Periode)
Neogene
Neophyt/Neubürger/
Neuankömmling neophyte
Neotenie neoteny
Neotypus/Neotyp/Neostandard
neotype
Neozoikum/Erdneuzeit/
Känozoikum/Kaenozoikum
(erdgeschichtliches Zeitalter)
Neozoic Era, Cenozoic,
Cenozoic Era
(Cainozoic Era/Caenozoic Era)
Nepenthaceae/
Kannenpflanzengewächse
East Indian pitcher plant family,
nepenthes family
Nephelometrie/Streulichtsmessung
nephelometry
Nephrocyt/Nephrozyt nephrocyte
Nephron/Nierenelement/
"Elementarapparat" nephron
(functional unit of kidney)
• gewundenes Kanälchen
(distal/proximal)
convoluted tubule
(distal/proximal)
• Henlesche Schleife loop of
Henle, loop of the nephron
• Sammelrohr/Ductus papillaris
collecting duct/tubule,
papillary duct
• Überleitungsstück narrow
descending limb of loop of Henle
• Verbindungsstück
junctional section
Nephrostom nephrostome
Nephrotom/Nierenplatte
nephrotome, renal plate
Neptungrasgewächse/
Neptunsgräser/Posidoniaceae
posidonia family
neritische Region/Flachmeerzone
("küstennahe" Zone) *mar*
neritic zone, neritic province
Nernst-Gleichung/Nernstsche
Gleichung Nernst equation
Nerv/Ader/Rippe (Blattnerv/
Blattader/Blattrippe) *bot*
vein, rib

Nerv *neuro* nerve *(weitere Nervenbezeichnungen finden Sie in einem Wörterbuch der Human- bzw. Veterinärmedizin)*
- **Hörnerv/Akustikus/ Vestibulokochlearis** auditory nerve, acoustic nerve, otic nerve (vestibulocochlear nerve)
- **Radiärnerv/Nervus radialis** radial nerve (musculospiral nerve)
- **Riechnerv/Nervus olfactorius** olfactory nerve
- **Sehnerv/Nervus opticus** optic nerve
- **Spinalnerv/Nervus spinalis** spinal nerve

Nervatur/Nervation/Aderung/ Venation venation *(siehe Blattaderung)*

Nervenbahn/Nervenstrang nerve cord

Nervenbündel nerve bundle

Nervenendigung nerve ending

Nervenfaser nerve fiber

Nervenknoten/Ganglion ganglion *(siehe auch unter: Ganglion)*

Nervenleitung impulse propagation

Nervennetz/Nervengeflecht nerve net *(invertebrates)*; neuronal network

Nervenstrang/Nervenbahn nerve cord

Nervensystem nerve system, nervous system
- **autonomes/vegetatives/ viscerales/unwillkürliches Nervensystem** autonomic nervous system (ANS), vegetative/ visceral/involuntary nerve system
- **peripheres Nervensystem** peripheral nervous system (PNS)
- **somatisches/willkürliches/ animales/animalisches Nervensystem** somatic nervous system (SNS), voluntary nervous system
- **Strickleiternervensystem** double-chain nerve system, ladder-type nerve system
- **Zentralnervensystem (ZNS)** central nervous system (CNS)

Nervenwachstumsfaktor nerve growth factor (NGF)

Nervenwurzel nerve root

Nervonsäure/Δ^{15}-Tetracosensäure nervonic acid, (Z)-15-tetracosenoic acid, selacholeic acid

Nesselbatterie/Cnidophore battery of nematocysts, cnidophore

Nesselgewächse/Urticaceae nettle family

Nesselkapsel/Cnide/Nematocyste thread capsule, urticator, cnida, nematocyst
- **Durchschlagskapsel/Penetrant** penetrant
- **Klebkapsel/Haftkapsel/Glutinant** glutinant
- **Ptychonema/Ptychozyste (eine Astomocnide: Ceriantharia)** ptychocyst, tube cnida
- **Wickelkapsel/Volvent** volvent

Nesselsack/Acrorhagus acrorhagus (tubercle with stinging cells)

Nesseltiere/Hohltiere/Cnidarien/ Coelenteraten cnidarians, coelenterates

Nesselzelle/Cnidozyt/Nematozyt/ Cnidoblast stinging cell, cnidocyte, nematocyte, cnidoblast

Nestdune/Neossoptile/Neoptile natal down, neossoptile, neoptile (a down feather)

Nestflüchter precocial animal, nidifugous (bird)

Nestgeruch nest odor

Nesthocker altricial animal, nidicolous (bird)

Nestling nestling

Nestparasitismus nest parasitism

Nestpilze/Nestlinge/Vogelnestpilze/ Teuerlinge/Nidulariaceae bird's-nest fungi, bird's-nest family

Nettoprimärproduktion net primary production (NPP)

Netz *arach* net, web
- **Baldachinnetz (mit Stolperfäden)** sheetweb, dome web (with barrier threads)
- **Deckennetz (Ageleniden)** horizontal sheetweb
- **Deckennetz (Linyphiiden)** hammock web
- **dreidimensionales Schutzgewebe** barrier web
- **Fangnetz (innerhalb des Radnetzes)** capturing zone
- **Fangschlauch** purse web
- **Flächennetz** simple sheetweb
- **Fußangelnetz mit Stolperfäden** tripping web with barrier threads
- **Gerüstnetz (Theridiiden)** scaffold web
- **Haubennetz (Theridiiden)** dome web (theridiid web)
- **"Hochzeitslaube"** mating bower (black widow)
- **Käschernetz (*Dinopis*)** retiarius web
- **klebriges Netz** viscid web
- **Leiternetz** ladder web
- **Maschennetz** mesh web
- **Radnetz** orb web

N Netzaderung

- **Rahmennetz** frame web
- **Raumnetz/dreidimensionales Netz** space web
- **rituelle Fesselfäden** bridal veil (crab spiders)
- **Röhrennetz/Trichternetz** tube web, funnel web
- **Schlagfallennetz (Hyptiotes)** sprung web
- **Spermanetz** sperm web
- **Spinnennetz** spider web
- **Wurfnetz (Netzwerferspinnen)** casting web
- **Zeltdachnetz** tent web (polygonal sheet web)

Netzaderung *bot* reticulate venation, net venation, netted venation

netzadrig reticulately veined

netzartig/netzförmig/retikulär net-like, reticulate, reticular

Netzauge/Facettenauge/ Komplexauge/Seitenauge facet eye, compound eye

Netzflügler, echte/Hafte/ Planipennia/Neuroptera neuropterans (dobson flies/antlions)

netzförmig net-like, reticulate

Netzhaut/Retina retina

Netzmagen/Haube/Retikulum honeycomb stomach, honeycomb bag, reticulum, second stomach

Netznabe *arach* hub, nub

Netznervatur/Netzaderung *bot* reticulate venation, net venation, netted venation

netznervig/netzrippig reticulately veined

Netzschleimpilze/ Labyrinthulomycetes/ Labyrinthulomycota slime nets, labyrinthulids

Netzspeiche/Radius *arach* spoke, radius, radial thread

Netzwarte *arach* retreat

Netzwerfen *arach* web-casting, web-throwing

Netzwerk network

- **neuromotorisches Netzwerk** neuromotor network
- **neuronales Netzwerk/ neuronales Netz** neural network
- **trans-Golgi-Netzwerk** trans-Golgi network

Netzwerk-Theorie *immun* network theory

Netzwerktheorie/Gittertheorie *immun* network theory

Neubesiedlung ecesis (pioneer stage of dispersal to a new habitat)

Neubürger/Neuankömmling/ Neophyt *bot* neophyte

neugeboren/Neugeborene betreffend/neonatal neonatal

Neugeborene neonate

Neugier/Neugierverhalten curiosity, inquisitiveness

neugierig curious, inquisitive

Neuhirn/Neoencephalon/ Neencephalon neoencephalon, neencephalon

Neuhirnrinde/Neocortex/ Neopallium neocortex, neopallium

Neukleinhirn/Neocerebellum neocerebellum (lateral lobes of cerebellum)

Neumundtiere/Neumünder/ Zweitmünder/Deuterostomia deuterostomes

Neumutation new mutation

Neunaugen/Neunaugenartige/ Petromyzontida lampreys

Neuordnung/Neusortierung *gen* reassortment

neural neural, neuric

Neuralbogen/oberer Wirbelbogen/ Basidorsale neural arch

Neuralfalte/Markfalte/Neuralwulst/ Medullarwulst neural fold

Neuralkranium/Neurokranium/ Hirnschädel neurocranium, cerebral cranium

Neuralleiste/Ganglienleiste neural crest

Neuralplatte/Medullarplatte/ Markplatte neural plate

Neuralrinne/Medullarrinne neural groove

Neuralrohr/Medullarrohr/Markrohr/ Tubus medullaris *embr* neural tube, medullary tube

Neuralwulst/Neuralfalte/Markfalte/ Medullarwulst neural fold

Neuraminsäure neuraminic acid

Neurit neurite

Neurobiologie neurobiology

Neurofilament neurofilament

Neurogliazelle neuroglial cell

Neurohämalorgan neurohemal organ

Neurohypophyse/ Hypophysenhinterlappen neurohypophysis, posterior lobe of pituitary gland

Neurokranium/Neuralkranium/ Hirnschädel neurocranium, cerebral cranium

Neurolemm/Neurilemm neurolemma

Neurologie neurology

Neuromer neuromere

neuromotorisches Netzwerk neuromotor network

Niederkunft

Neuron/Nervenzelle neuron, neurone, nerve cell
- **Bipolarzelle** bipolar cell
- **Bursterneuron** burster neuron
- **Hemmungsneuron** inhibitory neuron
- **Kommandoneuron** command neuron
- **Motoneuron** motoneuron, motor neuron
- **Multipolarzelle** multipolar cell
- **Pionierneuron** pioneer neuron
- **Pseudounipolarzelle** pseudounipolar cell
- **Relaisneuron** relay neuron
- **Unipolarzelle** unipolar cell
- **Wegweiserneuron** guidepost neuron
- **Zwischenneuron/Interneuron** interneuron

neuronal/neuronisch neuronal
neuronaler Schaltkreis neuronal circuit
neuronales Netz/Netzwerk neural network
neuronisch/neuronal neuronal
Neuropeptid neuropeptide
Neuropil neuropil
neurosekretorisch neurosecretory
neurotoxisch neurotoxic
Neurotransmitter neurotransmitter
Neuschnecken/Schmalzüngler/ Neogastropoda/Stenoglossa neogastropods: whelks & cone shells
Neuston neuston (organisms of surface water/surface film)
Neusynthese/*de-novo*-Synthese *de-novo* synthesis
Neutralisationstest (NT) neutralization test (NT)
neutrophil *adv/adj* neutrophil(ic)
Neutrophil *nt*/ **neutrophiler Granulozyt** neutrophil, neutrophilic granulocyte
- **segmentkerniger Neutrophil** segmented neutrophil, filamented neutrophil, polymorphonuclear granulocyte
- **stabkerniger Neutrophil** rod neutrophil, band neutrophil, stab neutrophil, stab cell

Neuvögel/Neornithes true birds, neornithes
Neuweltaffen/Breitnasenaffen/ Platyrrhina new-world monkeys (South American monkeys and marmosets)
Newton'sche Flüssigkeit Newtonian fluid
Nexin nexin
nicht-homologe Rekombination nonhomologous recombination

nicht-konjugatives Plasmid nonconjugative plasmid
nicht-Mendelsches Aufspaltungsverhältnis non-Mendelian ratio
nicht-Newton'sche Flüssigkeit non-Newtonian fluid
nicht-parentaler Dityp (NPD) nonparental ditype (NPD)
nicht-permissiv (Zelle/Wirt) nonpermissive (cell/host)
nicht-persistente Übertragung *vir* nonpersistent transmission
nicht-repetitive DNA/Einzelkopie-DNA single copy DNA
Nicht-Sinnstrang/anticodierender Strang *gen* antisense strand, noncoding strand
nicht-überlappend *gen* non-overlapping
nicht-zufallsgemäße Verteilung nonrandom disjunction
nichtessentiell nonessential
Nichtmatrizenstrang (nichtcodierender Strang) *gen* nontemplate strand (noncoding strand)
Nichtsättigungskinetik nonsaturation kinetics
Nichtsinncodon/Nonsense-Codon *gen* nonsense codon
Nichtsinnmuation/Nonsense-Mutation nonsense mutation
Nichtstrukturprotein nonstructural protein
Nick-Translation nick translation
Nickel nickel
nicken (nickend) nod (nodding)
Nickhaut/Blinzelhaut nictitating membrane, third eyelid
Nickhautdrüse/Glandula lacrimalis accessoria gland of third eyelid
Nicktiere/Kelchwürmer/ Kamptozoen/Entoprocta kamptozoans, entoprocta
Nicolieren/Doppelschildokulation *hort* double shield budding
Nicotin/Nikotin nicotine
Nidamentaldrüse (Schalendrüse) nidamental gland (shell gland)
Niederblatt cataphyll (a bud scale/scale leaf/bulb scale or cotyledon)
niedere Pflanzen lower plants, primitive plants
niedergedrückt/niederliegend mit aufrecht wachsender Sprossspitze decumbent, lodged (cereals)
niedergelassen/sedentär settled, sedentary
Niederkunft/Gebären parturition, delivery

N Niederkunft 608

Niederkunft/Kindesgeburt childbirth
niederliegend prostrate, procumbent, trailing, lying
niedermolekular low-molecular
Niedermoor/Flachmoor/Braunmoor/ Fen fen
Niederschlag *meteo* precipitation
Niederschlag/Sediment/Präzipitat *chem* deposit, sediment, precipitate
Niederschlagsmenge amount of precipitation
Niederung/Tiefland lowland
Niederungsmoor fen, fenland, valley bog
Niederwald (durch Rückschnitt) coppice
niedrigkronig/brachyodont (Zähne) with low crowns, brachydont, brachyodont
Niere kidney
- **einwarzige Niere/unipyramidale Niere** unilobular kidney, monopyramidal kidney, unipyramidal kidney
- **Holonephros/Archinephros** holonephros, archinephros
- **mehrwarzige Niere/ zusammengesetzte Niere/ gelappte Niere/multipyramidale Niere** multilobular kidney, multipyramidal kidney, polypyramidal kidney
- **Nachniere/definitive Niere/ Metanephros** hind kidney, definitive kidney, metanephros
- **Nebenniere** adrenal gland
- **Rumpfniere/Opisthonephros** opisthonephros
- **Urniere/Wolffscher Körper/ Mesonephros** middle kidney, midkidney, mesonephros
- **Vorniere/Pronephros** fore-kidney, primitive kidney, primordial kidney, head kidney, pronephros
- **zusammengesetzte Niere/ mehrwarzige Niere/gelappte Niere/multipyramidale Niere** multilobular kidney, multipyramidal kidney, polypyramidal kidney
Nieren.../die Niere betreffend renal
Nierenbecken/Pelvis renalis renal pelvis, pelvis of the kidney
Nierendurchblutung renal blood flow (RBF)
Nierenfarngewächse/Oleandraceae stalwart sword fern family
Nierenfett/Talg perirenal fat, perinephric fat; suet (from abdominal cavity of ruminants)

Nierenfettkapsel/Capsula adiposa adipose capsule of kidney, fatty capsule of kidney
nierenförmig kidney-shaped, reniform
Nierengang, primärer/ Urnierengang/Wolffscher Gang Wolffian duct, mesonephric duct
Nierenkanälchen/Nierentubulus/ Tubulus renalis renal tubule
- **distaler (gewundener) Tubulus/ Schaltstück** distal convoluted tubule
- **gewundenes Kanälchen** convoluted tubule
- **Henle-Schleife/Henlesche Schleife** loop of Henle, loop of the nephron, nephronic loop
- **proximaler (gewundener) Tubulus/Hauptstück** proximal convoluted tubule
- **Sammelrohr/Ductus papillaris** collecting duct/tubule, papillary duct
- **Überleitungsstück** narrow descending limb of loop of Henle
- **Verbindungsstück** junctional section
Nierenkapsel renal capsule
Nierenkelch/Calix renalis renal calix (*pl* calices), infundibula of kidney
Nierenkörperchen/Malpighi- Körperchen/Malpighisches Körperchen renal corpuscle
Nierenlappen renal lobule
Nierenleiste nephric ridge, nephrogenic ridge
Nierenmark/Medulla renis renal medulla
Nierenöffnung/Nephridialöffnung/ Nephridialporus nephridiopore
Nierenpapille renal papilla
Nierenpforte/Nierenstiel/ Nierenhilus/Hilus renalis renal hilus
Nierenplasmadurchströmung renal plasma flow (RPF)
Nierenpyramide renal pyramid
Nierenrinde renal cortex
Nierensamengewächse/ Cochlospermaceae cochlospermum family, buttercup tree family (silk-cotton tree family)
Nierenstiel/Nierenpforte/ Nierenhilus/Hilus renalis renal hilus
niesen sneeze
Nikotin/Nicotin nicotine
nikotinischer/nicotinischer Rezeptor nicotinic receptor

Nikotinsäure/Nicotinsäure (Nikotinat) nicotinic acid (nicotinate), niacin
Nikotinsäureamid/Nicotinsäureamid nicotinamide
Nilhechte/Mormyriformes mormyrids
NIOSH (National Institute for Occupational Safety and Health) U.S. Institut für Sicherheit und Gesundheit am Arbeitsplatz
Nipptide neap tide
Nische (Wirkungsfeld) *ecol* niche, ecological niche
Nischenblatt/Mantelblatt *bot* nest leaf
Nischengröße *ecol* niche size
Nischenbreite *ecol* niche breadth, niche width
Nischenüberlappung *ecol* niche overlap
Nischenverschiebung *ecol* niche shift
Nisse nit
Nissl-Schollen/Tigroidschollen (raues ER) Nissl granules (rough ER with ribosomes)
nisten build a nest
nistend (eingebettet in einer Aushöhlung) nesting, nestling, nidulant
Nistkasten nesting box
Nistkolonie nesting colony, crèche
Nistplatz nesting site
Nitrat nitrate
Nitrifikanten nitrifier, nitrifying bacteria
Nitrifikation/Nitrifizierung nitrification
nivale Stufe nival zone
Nixenkrautgewächse/Najadaceae najas family, water nymph family
Nodium/Knoten node
Nolanaceae/ Glockenwindengewächse nolana family
Nomenklatur/Fachausdruck/Name nomenclature, designation, name; (Gesamtheit der Fachausdrücke) nomenclature (system of terms)
● **binäre Nomenklatur/ binominale Bezeichnung/ zweigliedrige Benennung** binary/binomial nomenclature
Nominalskala *stat* nominal scale
Non-Disjunktion/ Chromosomenfehlverteilung *gen* nondisjunction
● **meiotische Non-Disjunction/ Chromosomenfehlverteilung in der Meiose** meiotic nondisjunction

Nonsense-Codon/Nichtsinn-Codon *gen* nonsense codon
Nonsense-Mutation/ Nichtsinnmutation nonsense mutation
Noosphäre noosphere
nördlich *biogeo* northern, boreal
Norepinephrin/Noradrenalin norepinephrine, noradrenaline
Normalverteilung *stat* normal distribution
Nosemaseuche (*Nosema apis*) nosema disease, nosemosis
Nosokomialinfektion/nosokomiale Infektion/Krankenhausinfektion nosocomial infection, hospital-acquired infection
Notfall emergency
Notfalleinsatz emergency response
Nothosaurier/Nothosauria nothosaurs
Notopodium notopodium
Notum (dorsale Thorakalplatte/ Tergum der Thorakalsegmente) notum (*pl* nota) = tergum of thoracic segment
Nozizeption/Nociception/ Schmerzwahrnehmung nociception
nozizeptiv/nociceptiv/Schmerz empfindend/schmerzempfindlich nociceptive
Nuchalorgan nuchal organ
Nüchternheit *med/physio* emptiness (of stomach); soberness
nüchtern (ohne Nahrung) with an empty stomach
Nucleinsäure/Nukleinsäure nucleic acid
Nucleinsäurehybridisierung/ Nukleinsäurehybridisierung nucleic acid hybridization
Nucleoid/Nukleoid/Kernäquivalent/ Karyoid/"Bakterienkern" nucleoid, nuclear body
Nucleokapsid/Nukleokapsid nucleocapsid
Nucleolus/Nukleolus nucleolus
nucleophiler Angriff *chem* nucleophilic attack
Nucleoplasma/Nukleoplasma nucleoplasm
Nucleosid/Nukleosid nucleoside
Nucleosom/Nukleosom nucleosome
Nucleotid/Nukleotid nucleotide
Nucleotidpaaraustausch nucleotide-pair substitution
Nucleus/Nukleus/Zellkern nucleus, karyon
nukleär/nucleär nuclear
Nukleinsäure/Nucleinsäure nucleic acid

Nukleinsäurehybridisierung/ Nucleinsäurehybridisierung nucleic acid hybridization
Nukleoid/Nucleoid/Kernäquivalent/ Karyoid/"Bakterienkern" nucleoid, nuclear body
Nukleokapsid/Nucleokapsid nucleocapsid
Nukleolus/Nucleolus nucleolus
Nukleolus-Organisator/Nucleolus-Organisator nucleolar organizer, nucleolus organizer (NOR)
nukleophiler Angriff nucleophilic attack
Nukleoplasma/Nucleoplasma nucleoplasm
Nukleosid/Nucleosid nucleoside
Nukleosom/Nucleosom nucleosome
Nukleotid/Nucleotid nucleotide
Nukleotidpaaraustausch nucleotide-pair substitution
Nukleus/Nucleus/Zellkern nucleus, karyon
Null-Zelle null cell
Nullallel null allele
Nullhypothese null hypothesis
nullipar nulliparous
nullisom nullisomic
nullizygot nullizygous
numerische Taxonomie/Phänetik numerical taxonomy, phenetics, taxometrics
Nuss nut
nussartig nutlike (shape), nutty (flavor)
Nüsschen nutlet, nucule
Nussmuscheln/Nuculacea nut clams
Nüster/Nasenloch nostril
Nutation nutation
Nutsche/Filternutsche *lab* suction filter, vacuum filter
nutzen utilize, use; (anwenden) apply
Nutzen benefit, use; (Vorteil) advantage; (Anwendung) application
nützen benefit
Nutzfläche, landwirtschaftliche cultivated land
Nutzholz timber, lumber
Nutzinsekt beneficial insect, beneficient insect
nützlich beneficial, useful
• **schädlich** harmful, causing damage
Nützling/Nutzart beneficial species, beneficient species
• **Schädling/Ungeziefer** pest
nutznießen profit
Nutznießer profiteer
Nutznießung/Probiose profiting, probiosis
Nutzpflanze economic plant, useful plant, crop plant
Nutztier domestic animal
Nutzung utilization, use
Nutzvieh livestock; (Rinder) domestic cattle
Nuzellus/Nucellus/"Knospenkern" nucellus
Nyctaginaceae/ Wunderblumengewächse four-o'clock family
Nymphaeaceae/Seerosengewächse water-lily family
Nymphe nymph
nymphipar nymphiparous
Nymphiparie nymphipary
Nyssaceae/Tupelobaumgewächse sourgum family

611   **Obst** **O**

obdiplostemon *bot*
obdiplostemonous
Oberarm upper arm
  ● **Unterarm** forearm
Oberarmknochen/Oberarmbein/
Humerus arm bone, humerus
Oberblatt (Blattspreite & Blattstiel)
leaf blade and leaf stalk
  ● **Unterblatt** underleaf, hypophyll;
  (Bauchblatt/Amphigastrium)
  underleaf, ventral leaf,
  amphigastrium
Oberboden
(Auswaschungshorizont/
A-Horizont) topsoil (zone of
leaching/eluviation)
  ● **Unterboden**
  (Einwaschungshorizont/
  B-Horizont) subsoil (zone of
  accumulation/illuviation)
Oberfläche surface
  ● **Blattoberfläche** leaf surface
Oberflächen-Volumen-Verhältnis
surface-to-volume ratio
Oberflächenabfluss
surface runoff
oberflächenaktive Substanz/
Entspannungsmittel surfactant
Oberflächenfilm/
Oberflächenhäutchen (in
stehendem Binnengewässer)
scum, film, mat
Oberflächenkultur *micb*
surface culture
Oberflächenmarkierung
surface labeling
Oberflächenspannung
surface tension
Oberflächenwasser surface water
Oberflächenwellen surface waves
Oberflächenwurzler
shallow-rooted plant
oberflächlich on the surface,
superficial
obergärig (Fermentation: Bier)
top fermenting
Oberhaut/Epidermis epidermis
Oberholz/Oberstand/
Schirmbestand overstory
oberirdisch aboveground,
overground, superterranean
Oberkiefer/Oberkieferknochen/
Maxilla upper jaw, upper jawbone,
maxilla
  ● **Unterkiefer/Mandibel/**
  Unterkieferknochen lower jaw,
  lower jawbone, submaxilla,
  submaxillary bone, mandible
Oberkieferbein/Os maxillare
maxillary bone
Oberkörper/Brustkorb/Brustkasten/
Thorax chest, thorax

Oberlippe/Labrum upper lip, labrum
Oberschenkel thigh
Oberschenkel/Keule (Schlachtvieh)
haunch, hindquarters
  ● **Unterschenkel** shank
Oberschenkelknochen/Os femoris/
Femur
thighbone, femur, femoral bone
  ● **Unterschenkelknochen/**
  Tibiotarsus (Vögel) tibiotarsus
oberschlächtig/incub overshot,
incubous
  ● **unterschlächtig/succub**
  undershot, succubous
Oberschlundganglion/
Supraösophagealganglion/
"Gehirn" supraesophageal
ganglion, "brain"
  ● **Unterschlundganglion/**
  Subösophagealganglion
  subesophageal ganglion
Oberseite upperside, upper surface
  ● **Unterseite** underside,
  undersurface
oberständig hypogynous
  ● **unterständig** *bot* epigynous
Objektiv *micros* objective
Objektivrevolver/Revolver *micros*
nosepiece, nosepiece turret
  ● **Dreifachrevolver** triple nosepiece
  ● **Fünffachrevolver**
  quintuple nosepiece
  ● **Vierfachrevolver**
  quadruple nosepiece
  ● **Zweifachrevolver**
  double nosepiece
Objektmikrometer *micros*
stage micrometer
Objekttisch *micros* stage,
microscope stage
Objekttisch-Klammer *micros*
stage clip
Objektträger (microscope) slide
Objektträger mit Vertiefung
microscope depression slide,
concavity slide, cavity slide
obligatorisch/obligat obligatory,
obligate
Obst fruit
  ● **Beerenobst** berries
  ● **Fallobst** windfall (fruit dropped
  from trees; e.g. apple/pear/plum/
  cherry..)
  ● **Kernobst** pomaceous fruit
  ● **Spalierobst** espalier fruit
  ● **Steinobst** stone fruit
  (drupaceous fruit) (with pit)
  ● **Strauchbeerenobst** bush fruit
  (*Ribes*: currents, gooseberries, etc.)
  ● **Streuobst** fruit from irregularly
  planted trees within otherwise
  cultivated farmland

Obstbau fruit growing
Obstbaukunde pomology
Obstbaum fruit tree,
  fruit-bearing tree
Obstplantage fruit orchard
Obststein pit, stone
Obturator (Gewebewucherung)
  obturator (outgrowth)
oc-DNA (offene ringförmige DNA)
  oc-DNA (open circle DNA)
Occlusor occlusor
Ocelle/Ocellus ocellus
Ocellenstiel ocellar pedicel
Ocellenzentrum ocellar center
Ochnaceae/Grätenblattgewächse
  ochna family
Ochrea/Tute *bot* ochrea, ocrea,
  mantle
Ochse ox (*pl* oxen)
ocker ochre
Ocotillogewächse/Fouquieriaceae
  ocotillo family
Octocorallia octocorallians,
  octocorals
Ödland *ecol/biogeo* wasteland,
  barren
Oenocyt oenocyte
**Oenotheraceae/Onagraceae/
  Nachtkerzengewächse**
  willowherb family,
  evening-primrose family
offene ringförmige DNA (oc-DNA)
  open circle DNA (oc-DNA)
offener Promotorkomplex *gen*
  open promoter complex
offenes Leseraster *gen*
  open reading frame (ORF)
Offenzeit/Öffnungszeit *neuro*
  open time
öffnend opening, dehiscent
Öffnung/Mund/Mündung opening,
  aperture, orifice, mouth, perforation,
  entrance
Öffnungsdauer (Membrankanal)
  life-time
**Öffnungsfrucht/Streufrucht/
  Springfrucht** dehiscent fruit
Öffnungswinkel *micros*
  angular aperture
Öffnungszeit/Offenzeit *neuro*
  open time
Ohr ear
Ohrenfedern auriculars (ear coverts)
**Ohrenknöchelchen/
  Gehörknöchelchen** auditory
  ossicle, ear ossicle, ossiculum
Ohrenöffnung ear opening,
  auditory meatus
Ohrenschmalz/Cerumen earwax,
  cerumen
Ohrenschmalzdrüse
  ceruminous gland

Ohrgrübchen *embr* otic pit,
  otic depression
Ohrkapsel otic capsule
Ohrläppchen earlobe
Ohrlappenpilze/Auriculariales
  Old man's ears and allies
Ohrmuschel/äußeres Ohr/Pinna
  ear conch, auricle, external ear,
  outer ear, pinna
Ohrplakode otic placode
Ohrspeicheldrüse/Parotis
  parotid gland
Ohrtrommel/Trommelfell eardrum,
  tympanic membrane
**Ohrtrompete/Eustachische Röhre/
  Eustach'-Röhre** auditory tube,
  eustachian tube
Ohrwürmer/Dermaptera
  earwigs
Oidie/Oidium oidium
Okazaki-Fragment *gen*
  Okazaki fragment
Öko-Audit/Umweltaudit
  environmental audit
Ökobilanz life cycle assessment,
  life cycle analysis (LCA)
Ökogenetik ecogenetics
Ökogramm ecogram
Ökokline ecocline (gradient of
  vegetation and biotopes)
Ökologie ecology
 ● **Autökologie** autecology
 ● **Demökologie**
  population ecology
 ● **Ethökologie/Verhaltensökologie**
  behavioral ecology
 ● **Festlandsökologie/terrestrische
  Ökologie/Epeirologie**
  terrestrial ecology
 ● **Geoökologie** geo-ecology
 ● **Humanökologie** human ecology
 ● **Landschaftsökologie**
  landscape ecology
 ● **Palökologie** paleoecology
 ● **Pflanzenökologie/
  Vegetationsökologie/
  Vegetationskunde** plant ecology,
  phytoecology
 ● **Stadtökologie/Urbanökologie**
  urban ecology
 ● **Standortlehre** habitat ecology
 ● **Synökologie** synecology
 ● **Systemökologie** systems ecology
 ● **Vegetationsökologie/
  Pflanzenökologie/
  Vegetationskunde**
  plant ecology, phytoecology
 ● **Verhaltensökologie/Ethökologie**
  behavioral ecology
 ● **vorausschauende Ökologie/
  voraussagende Ökologie**
  predictive ecology

oligomiktisch

ökologisch ecological
ökologische Amplitude
  ecological amplitude,
  range of tolerance
ökologische Effizienz/
ökologischer Wirkungsgrad
  ecological efficiency
ökologische Genetik/Ökogenetik
  ecological genetics
ökologische Nische ecological niche
ökologische Potenz
  ecological potency
ökologische Toleranz/
ökologische Verträglichkeit/
Reaktionsbreite
  ecological tolerance, tolerance range
ökologische Valenz
  ecological valency
ökologisches Gleichgewicht
  ecological balance,
  ecological equilibrium
Ökophän nt ecophene
Ökophänotypie ecophenotypy
Ökophysiologie/
ökologische Physiologie
  ecophysiology,
  ecological physiology
Ökospezies ecospecies
Ökosphäre ecosphere
Ökosystem ecosystem
 • Süßwasserökosystem
   freshwater ecosystem
 • Landökosystem
   terrestrial ecosystem
 • marines Ökosystem
   marine ecosystem
Ökoton/Übergangsgesellschaft
  ecotone
Ökotop ecotope
Ökotyp ecotype
Oktade octad
Okular micros ocular, eyepiece
 • Binokular binoculars
 • Brillenträgerokular spectacle
   eyepiece, high-eyepoint ocular
 • Trinokularaufsatz/Tritubus
   trinocular head
 • Zeigerokular pointer eyepiece
Okularblende/Gesichtsfeldblende
des Okulars ocular diaphragm,
  eyepiece diaphragm,
  eyepiece field stop
Okularlinse/Augenlinse ocular lens
Okularmikrometer
  ocular micrometer
Okulation/Okulieren/
Augenveredlung/Äugeln hort
  bud grafting, budding
 • Chipveredelung/Chipveredlung/
   Span-Okulation chip budding
 • Doppelschildokulation/
   Nicolieren double shield budding

 • Platten-Okulation patch budding,
   plate budding
 • Ring-Okulation/Ringveredlung
   ring budding, annular budding
   (flute budding)
 • Schild-Okulation
   (Augenschild/Schildchen)
   shield budding
 • T-Schnitt Okulation
   (mit T-förmigem Einschnitt der
   Rinde)
   T budding (shield budding)
Öl oil
 • ätherisches Öl essential oil,
   ethereal oil
 • Baumwollsaatöl cotton oil
 • Behenöl ben oil, benne oil
 • Erdnussöl peanut oil
 • Erdöl crude oil, petroleum
 • Fuselöl fusel oil
 • Jungfernöl virgin oil (olive)
 • Kokosöl coconut oil
 • Kürbiskernöl pumpkinseed oil
 • Lebertran cod-liver oil
 • Leinöl linseed oil
 • Maisöl corn oil
 • Mineralöl mineral oil
 • Olivenkernöl olive kernel oil
 • Olivenöl olive oil
 • Palmöl palm oil
 • Pflanzenöl vegetable oil
 • Rizinusöl castor oil, ricinus oil
 • Safloröl safflower oil
 • Schmieröl lubricating oil
 • Senföl mustard oil
 • Sesamöl sesame oil
 • Sojaöl soybean oil
 • Sonnenblumenöl
   sunflower seed oil
 • Speise-Rapsöl/Rüböl
   canola oil
   (rapeseed/colza oil)
 • Walratöl sperm oil (whale)
Olaxgewächse/Olacaceae
  olax family, tallowwood family
Ölbad lab oil bath
Ölbaumgewächse/Oleaceae
  olive family
Ölbehälter bot oil cavity
Öldrüse/Schmierdrüse zool/orn
  oil gland
Oleaceae/Ölbaumgewächse
  olive family
Oleandraceae/Nierenfarngewächse
  stalwart sword fern family
Oleosom bot oleosome
ölig oily
Oligodendrozyt/Oligodendrocyt
  oligodendrocyte
oligomer adj/adv oligomerous
Oligomer n oligomer
oligomiktisch oligomictic

Oligonucleotid 614

Oligonucleotid/Oligonukleotid
oligonucleotide
- **Antisense-Oligonucleotid/ Anti-Sinn-Oligonucleotid/ Gegensinn-Oligonucleotid**
antisense oligonucleotide
- **ASO (allelspezifisches Oligonucleotid)** ASO (allele specific oligonucleotide)

oligonucleotidgesteuerte Mutagenese oligonucleotide-directed mutagenesis
oligophag oligophagous
Oligosaccharid oligosaccharide
oligotroph/nährstoffarm oligotrophic, nutrient-deficient
Oligozän (erdgeschichtliche Epoche) Oligocene, Oligocene Epoch
Olivenkern *neuro* olivary nucleus
Ölkatastrophe *ecol* oil spill
Ölkörper (Samen) elaiosome
Ölpest/Ölverschmutzung oil pollution
Ölquelle *geol* oil well
Ölsaat *bot* oilseed; (ölliefernde Pflanzen) oil crops, oil seed crops
Ölsäure/Δ⁹-Octadecensäure (Oleat) oleic acid, (Z)-9-octadecenoic acid (oleate)
Ölschiefer/Brandschiefer *geol* oil shale
Ölstrieme/Vitta (Apiaceenfrüchte) vitta, oil tube, oil cavity, resin canal
Ölteppich *geol* oil slick
Ölverschmutzung/Ölpest oil pollution
Ölvorkommen/ölführende Schicht *geol* oil reservoir
Ölweidengewächse/Elaeagnaceae oleaster family
Omega-Schleife/Ω-Schleife (Proteine) Omega loop, Ω loop
omnivor/pantophag/allesfressend omnivorous, pantophagous
Omnivor omnivore
Omphaloplazenta/omphaloide Plazenta/Dottersackplazenta/ Dottersackhöhlenplazenta yolk-sac placenta, choriovitelline placenta
Onagraceae/Oenotheraceae/ Nachtkerzengewächse willowherb family, evening-primrose family
onkogen/oncogen/krebserzeugend oncogenic, oncogenous
Onkogen oncogene, onc gene
Onkogenität oncogenicity
Onkologie oncology
Onkosphäre/Oncosphaere (unbewimperte Cestodenlarve) oncosphere

onkotischer Druck/ kolloidosmotischer Druck oncotic pressure
Ontogenese/Ontogenie/ Entwicklungsgeschichte des Einzelorganismus ontogenesis, ontogeny, development
Oocyste oocyst
Oocyt oocyte
Ooecium/Ovicelle (Bryozoa/ Ectoprocta) ooecium, ovicell
Oogamie/Eibefruchtung oogamy
Oogon/Oogonium oogonium
Ookinet ookinete
Oolemma/Eihülle/Eihaut egg membrane
Oomycota/falsche Mehltaupilze water molds & downy mildews, oomycetes
Ooplasma/Bildungsplasma/ Eiplasma ooplasm
Oothek ootheca
Ootyp ootype
Oozooid/"Amme" (Ascidien) oozooid
Operator operator
Operon operon
Opfer victim, prey
Opfer erlegen stalk prey (Wild erlegen: stalk game)
opfern sacrifice
Ophioglossaceae/ Natternzungengewächse/ Rautenfarngewächse adder's-tongue family, grape fern family
Opiat opiate
opisthocöl/opisthocoel opisthocelous
Opisthonephros/Rumpfniere opisthonephros
opistognath opisthognathous
opponierbar/entgegenstellbar opposable
Opportunist opportunist, opportunistic species
opportunistisch opportunistic
Opsin opsin, scotopsin
Opsonierung/Opsonisation/ Opsonisierung opsonization
Opsonin opsonin
Optik optics
optische Dichte/Absorption optical density, absorbance
optische Spezifität optical specificity
Oralapparat/Buccalapparat (Ciliaten) oral apparatus
Oralscheibe/Mundscheibe/Peristom oral disk, peristome, peristomium
Orchideen/Knabenkrautgewächse/ Orchidaceae orchids, orchid family, orchis family

Ordinalskala *stat* ordinal scale
Ordnung order
- **Gleichung *x*ter Ordnung**
  equation of the *x*th order
Ordnungsstatistik order statistics
Ordovizium (erdgeschichtliche Periode)
  Ordovician, Ordovician Period
Oreophyt/Gebirgspflanze orophyte
  (subalpine plant)
ORF (offenes Leseraster) *gen*
  ORF (open reading frame)
Organ organ
Organbildung/Organentwicklung/ Organogenese organogenesis
Organell *nt*/**Organelle** *f* organelle
Organisationsstufe organizational
  level, grade of organization
Organisationstyp/ Organisationsform
  organizational form
Organisator (dorsale Blastoporenlippe) organizer
organisch organic
organische Chemie/"Organik"
  organic chemistry
organische Substanz organic matter
organisches Material organic matter
organismisch organismal
Organismus organism, lifeform
Organopelit/organischer Sclamm
  limn siehe Mudde
**Orientbeule/Hautleishmaniose/ kutane Leishmaniose
  (*Leishmania spp.*)** oriental sore,
  cutaneous leishmaniasis
Orientierung/ Orientierungsverhalten
  orientation, orientational behavior
Orientierungsbewegung/Taxie/Taxis
  orientational movement,
  taxy, taxis (*pl* taxes)
Ornithin ornithine
Ornithin-Harnstoff-Zyklus
  ornithine-urea cycle
Ornithologe/Vogelkundler
  ornithologist, birds specialist
Ornithologie/Vogelkunde
  ornithology, study of birds
ornithologisch/vogelkundlich
  ornithological
Orobanchaceae/ Sommerwurzgewächse
  broomrape family
Orotsäure orotic acid
Orsellinsäure orsellic acid,
  orsellinic acid
orten locate
Ortet/Klonausgangspflanze/ Klonmutterpflanze ortet (original
  single ancestor of a clone)
orthodrom *neuro* orthodromic

Orthogenese orthogenesis
orthognath orthognathous
ortholog *gen* ortholog(ous)
Orthologie orthology
orthostich orthostichous
orthotrop orthotropous, orthotropic,
  atropous
Orthotropismus orthotropism
ortsspezifische Mutagenese
  site-directed mutagenesis
Ortstein/Eisenstein ironpan, ortstein
  (a hardpan)
Ortstreue/Philopatrie philopatry,
  homing
Osmaetherium osmeterium,
  osmaterium
osmiophil (OsO$_4$-bindend)
  osmiophilic
Osmiumsäure osmic acid
Osmiumtetroxid osmium tetraoxide
Osmokonformer osmoconformer
Osmolalität osmolality
Osmolarität/osmotische Konzentration osmolarity,
  osmotic concentration
osmophil osmophilic
Osmoregulation osmoregulation
Osmoregulierer osmoregulator
Osmose osmosis
osmotisch osmotic
osmotischer Druck osmotic pressure
osmotischer Schock osmotic shock
osmotisches Potential
  osmotic potential
osmotroph osmotrophic
Osmundaceae/ Königsfarngewächse/ Rispenfarngewächse flowering
  fern family, cinnamon fern family
Osphradium (*pl* Osphradien)
  osphradium
Ossifikation/Verknöcherung/ Knochenbildung ossification
Osteoblast/knochenbildende Zelle
  osteoblast, bone-forming cell
Osteocyt/Osteozyt osteocyte
Osteoklast osteoclast
Osterluzeigewächse/ Aristolochiaceae birthwort family,
  Dutchman's-pipe family
Östradiol estradiol, progynon
östrisch/östral/Brunst... estrous,
  estral
Östrogen estrogen
Östron/Estron estrone
Östrus (Brunst) estrus
- **Diöstrus/Dioestrus** diestrus
- **Metöstrus/Metoestrus/ Nachbrunst** metestrus
- **Proöstrus/Vorbrunst** proestrus
Östruszyklus estrous cycle,
  estrus cycle, estral cycle

Ostwind(e)

Ostwind(e)
easterly wind (easterlies)
● **Westwind(e)**
westerly wind (westerlies)
otisch otic
**Ovar/Ovarium/Eierstock/
Fruchtknoten** ovary
Ovarialballen ovarian ball
**Ovariole/Ovariolschlauch/
Eischlauch/Eiröhre (Insekten)**
ovariole, egg tube
**Ovariolenträchtigkeit/
Ovarialträchtigkeit/
Ovarialschwangerschaft**
ovarian pregnancy
**Ovicelle/Ooecium (Bryozoa/
Ectoprocta)** ooecium, ovicell
**Oviduktträchtigkeit/
Eileiterschwangerschaft**
tubal pregnancy
oviger/eitragend/eiführend
ovigerous
ovipar/eierlegend oviparous,
egg-laying
Oxalbernsteinsäure (Oxalsuccinat)
oxalosuccinic acid
(oxalosuccinate)
Oxalessigsäure (Oxalacetat)
oxaloacetic acid (oxaloacetate)

Oxalidaceae/Sauerkleegewächse
wood-sorrel family
Oxalsäure (Oxalat)
oxalic acid (oxalate)
Oxidation oxidation
Oxidationsmittel
oxidizing agent, oxidant
oxidativ oxidative
oxidative Phosphorylierung
oxidative phosphorylation,
carrier-level phosphorylation
oxidieren oxidize
oxidierend oxidizing
Oxoglutarsäure (Oxoglutarat)
oxoglutaric acid (oxoglutarate)
Ozean ocean
ozeanisch oceanic
ozeanische Region
oceanic zone/region, pelagic zone
ozeanisches Klima/Meeresklima
oceanic climate, marine climate
**Ozeanographie/Ozeanografie/
Ozeanologie** oceanography,
oceanology
Ozon ozone
Ozonabbau ozone depletion
Ozonloch ozone hole
Ozonschicht/Ozonosphäre
ozone layer

P-Stelle/Peptidyl-Stelle
P-site, peptidyl-site
Paar pair, couple
paaren/begatten/kopulieren
pair, mate, copulate
Paargesang/Duettgesang duetting
Paarhufer/Artiodactyla even-toed
ungulates, cloven-hoofed animals,
artiodactyls
Paarkernphase/Dikaryophase
dikaryotic phase
Paarung pairing, mating
• **assortive Paarung/bewusste
Paarung** assortive mating
• **Basenpaarung** *gen* base pairing
• **Dreifachpaarung** *gen*
tripartite mating
• **Fehlpaarung/
Chromosomenfehlpaarung** *gen*
mispairing of chromosomes
• **Fremdpaarung**
disassortive mating
• **unterbrochene Paarung**
interrupted mating
• **zufällige Paarung/Panmixie/
Zufallspaarung**
random mating
Paarungsbevorzugung
mating preference
Paarungsrad (Libellen) mating
wheel (dragonflies: mating in wheel
position)
Paarungsruf mating call
Paarungsschranke mating barrier
Paarungstyp/Kreuzungstyp
mating type
Paarungsverhalten
mating behavior
Paarungszeit/Paarungssaison
pairing/mating season
paarweise liegende Zisternen
paired cisternae
paarzehig cloven-hoofed
Pachytän pachytene
**Pacini Körperchen/Pacinisches
Körperchen/Lamellenkörperchen/
Endkörperchen**
Pacinian body, Pacinian corpuscle,
lamellated corpuscle
Paddelechsenartige/Sauropterygia
sauropteryians
Pädogamie pedogamy
Pädogenese pedogenesis
Pädomorphose pedomorphosis
Paeoniaceae/Pfingstrosengewächse
peony family
Paläobotanik paleobotany
Paläoendemit/Reliktendemit
paleoendemic
**Paläogen/Alt-Tertiär
(erdgeschichtliche Periode)**
Paleogene

Paläontologie paleontology
Paläoökologie/Palökologie
paleoecology
**Paläophytikum/Florenaltertum
(erdgeschichtliches Zeitalter)**
Paleophytic Era
Paläospezies paleospecies
**Paläozän (erdgeschichtliche
Epoche)**
Paleocene, Paleocene Epoch
**Paläozoikum/Erdaltertum
(erdgeschichtliches Zeitalter)**
Paleozoic, Paleozoic Era
**Palindrom/umgekehrte Repetition/
umgekehrte Wiederholung/
invertierte Sequenzwiederholung**
gen palindrome, inverted repeat
Palingenese palingenesis
Palisadenparenchym
palisade parenchyma
**Palisadenwurm (*Strongylus
equinus* et al.)** palisade worm
Palliallinie/Mantellinie pallial line
Pallialraum pallial sinus
Pallium/Mantel pallium, mantle
Palmen/Palmae/Arecaceae
palm family
Palmfarn/Cycadee palmfern, cycad
Palmfarngewächse/Cycadaceae
cycads, cycad family, cycas family
**Palmitinsäure/Hexadecansäure
(Palmat/Hexadecanat)** palmitic
acid, hexadecanoic acid (palmate/
hexadecanate)
**Palmitoleinsäure/
Δ^9-Hexadecensäure** palmitoleic
acid, (*Z*)-9-hexadecenoic acid
Palmwedel palm frond
Palökologie/Paläoökologie
paleoecology
Palpe/Taster/Tastfühler palp, palpus
• **Kiefertaster/Maxillartaster/
Maxillarpalpus/Palpus maxillaris**
maxilliary palp
• **Lippentaster/Labialpalpus/
Palpus labialis** labial palp,
labipalp, labial feeler, palp, palpus
• **Pedipalpe** pedipalp
Palpigraden/Palpigradi
microwhipscorpions, palpigrades
Palsenmoor/Torfhügelmoor
palsa bog
**Palynologie/Pollenkunde/
Pollenanalyse** palynology,
pollen analysis; study of spores
panaschiert variegated, mottled
**Pandanaceae/
Schraubenbaumgewächse/
Schraubenpalmen**
screw-pine family
Pandemie pandemic
pandemisch pandemic

P Pankreas 618

Pankreas/Bauchspeicheldrüse
pancreas
**Pankreasinsel/Inselorgan/
Langerhanssche Insel** pancreatic
islet, islet organ, islet of Langerhans
pankreatisches Polypeptid (PP)
pancreatic polypeptide (PP)
Panmixie/zufällige Paarung
random mating
Pansen/Rumen rumen, paunch,
first stomach, ingluvies
Pansenpfeiler ruminal pillar
**Pansenvorhof/Schleudermagen/
Atrium ruminis** atrium ruminis
Panspermie *evol* panspermia,
panspermatism
Pantoinsäure pantoic acid
pantophag pantophagous,
pantophagic
Pantothensäure pantothenic acid
Panzer armor, test, theca; lorica,
case
**Panzer/Schale/Carapax
(Schildkröte)** shell, carapace
● **Bauchpanzer/Bauchplatte/
Brustschild/Plastron
(Schildkröten/Vögel)** plastron,
plastrum
● **Hautpanzer/Außenskelett/
Exoskelett** dermal skeleton,
dermatoskeleton, exoskeleton
● **Rückenpanzer/Rückenschild/
Carapax (Schildkröte)** dorsal
shield, carapace
**Panzerbeere (Hesperidium und
Kürbisfrucht, siehe dort)** berry
with hard rind
(hesperidium and pepo/gourd)
Panzerechsen/Krokodile/Crocodilia
crocodiles
**Panzerfische (Placodermen &
Ostracodermen)**
placoderms & ostracoderms
Panzergeißler/Dinoflagellaten
dinoflagellates
Panzerhaut/Skleroderm scleroderm
Panzerplatte/Lorica lorica (a girdle-
like skeleton), case
**Panzertierchen/Korsetttierchen/
Loriciferen** corset bearers,
loriciferans
**Panzerwangen/
Drachenkopffischverwandte/
Scorpaeniformes**
sculpins & sea robins
**PAP-Färbung/
Papanicolaou-Färbung**
PAP stain, Papanicolaou's stain
Papageien/Psittaciformes
parrots & parakeets
Papaveraceae/Mohngewächse
poppy family

Papierholz pulpwood
Papille papilla
● **Blattpapille/blättrige Papille
(Zunge)** foliate papilla
● **fadenförmige Papille (Zunge)**
threadlike papilla,
filiform papilla
● **linsenförmige Papille (Zunge)**
lentiform papilla
● **Pilzpapille (Zunge)**
fungiform papilla
● **Wallpapille (Zunge)**
vallate papilla
**Pappus/Haarkelch/Federkelch
(Haarkranz des Blütenkelchs)**
pappus (tuft of calyx appendages)
Papula (Asteroiden) dermal papula
Parabiose parabiosis
Paradidymis/Giraldessches Organ
paradidymis
Paraflagellarkörper paraflagellar
body, flagellar swelling
parakarp *bot* syncarpous without
septa
parakrin/paracrin paracrine
Paralleladerung/Parallelnervatur
bot parallel venation
paralleladrig parallely veined
Parallelfaser parallel fiber
parallelgestreift parallely striped
Parallelismus/parallele Evolution
ecol parallelism, parallel evolution
parallelnervig parallely veined
paralog *gen* paralog(ous)
Paralogie paralogy
Paramer paramere
Parameter parameter
paranemisch *gen* paranemic,
anorthospiral
paranemische Verbindung
paranemic joint
Parapatrie/Kontakt-Allopatrie
parapatry
parapatrisch parapatric
paraphyletisch paraphyletic
Paraphyse/Saftfaden paraphysis
Parapinealorgan paraparietal organ,
parietal organ
Parapodium parapod, side-foot
Pararetrovirus pararetrovirus
Parasit/Schmarotzer parasite
● **Außenparasit/Exoparasit/
Ektoparasit/Episit (Hautparasit)**
exoparasite, ectoparasite, episite,
epizoon
● **Brutparasit/Brutschmarotzer**
brood parasite
● **Gelegenheitsparasit/fakultativer
Parasit** facultative parasite
● **Halbparasit/Halbschmarotzer/
Hemiparasit** semiparasite,
hemiparasite

partikuläre Vererbung

- **Hautparasit/Hautschmarotzer/ Dermatozoe** skin parasite, dermatozoan
- **Hyperparasit** hyperparasite
- **Innenparasit/Endoparasit/ Endosit** endoparasite, endosite
- **Kleptoparasit** kleptoparasite, cleptoparasite
- **Phytoparasit/Pflanzenparasit (Schmarotzer in/auf Pflanzen)** plant parasite (thriving in/on plants)
- **tierischer Parasit/parasitierendes Tier (Zooparasit)** zooparasite, animal parasite (a parasitic animal) (*see:* zoophagous parasite)
- **Vollparasit/Vollschmarotzer/ obligater Parasit** obligate parasite
- **Zooparasit (Schmarotzer in/auf Tieren)** zoophagous parasite (thriving in/on animals)

Parasitämie parasitemia
parasitär/parasitisch/schmarotzend parasitic
parasitieren/schmarotzen parasitize
Parasitismus/Schmarotzertum parasitism
Parasitoide parasitoid
Parasitose parasitosis
parasomaler Sack *zool* parasomal sac
parasympathisch (autonomes Nervensystem) parasympathetic
Parathion (E 605) parathion
Parathyrin/Parathormon/ Nebenschilddrüsenhormon (PTH) parathyrin, parathormone, parathyroid hormone (PTH)
Paratop/Antigenbindestelle/ Antigenbindungsstelle paratope, antigen combining site, antigen binding site
Paratypus/Parastandard paratype
parazentrische Inversion paracentric inversion
Pärchenegel (*Schistosoma spp.*) schistosome, blood fluke
Parenchym/Grundgewebe parenchyma, ground tissue, fundamental tissue
- **Assimilationsparenchym/ Chlorenchym** chlorenchyma
- **Kontaktparenchym** boundary parenchyma
- **Markstrahlparenchym** ray parenchyma
- **Palisadenparenchym** palisade parenchyma
- **Pseudoparenchym** pseudoparenchyma, paraplectenchyma

- **Rindenparenchym** cortical parenchyma
- **Schwammparenchym** spongy parenchyma
- **Speicherparenchym** storage parenchyma
- **Sternparenchym** stellate parenchyma
- **Wundparenchym** traumatic parenchyma

parenchymatisch parenchymatous
Parenchymknorpel/chondroides Gewebe chondroid tissue
parentaler Dityp (PD) parental ditype (PD)
parenteral parenteral
parenterale Ernährung parenteral nutrition, parenteral feeding/ alimentation/food uptake
Parenthosom/Parenthesom/ Porenkappe parenthosome, parenthesome, septal pore cap
Parietalauge (*Sphenodon*) parietal eye
Parietalorgan/Parapinealorgan parietal organ of epiphysis, parapineal organ
Parietalplazentation/ wandständige Plazentation *bot* parietal placentation
Parkbaum park tree
Parkeriaceae/Hornfarngewächse water fern family, floating fern family
Parklandschaft/Parkwald parkland
Parnassiaceae/Herzblattgewächse grass of Parnassus family
Parökie/Beisiedlung paroecism
Paroophoron/Beieierstock paroophoron, parovarium
Parotis/Parotisdrüse/ Ohrspeicheldrüse/Glandula parotis (Säuger) parotis, parotid, parotid gland
Parotoiddrüse/Parotoide/ Duvernoysche Drüse/Ohrdrüse (Amphibien) parotoid gland
Paroxysmus paroxysm
Parthenogenese/Jungfernzeugung parthenogenesis
parthenokarp/jungfernfrüchtig parthenocarpic
Parthenokarpie/ Jungfernfrüchtigkeit parthenocarpy
Partialdruck partial pressure
Partialverdau partial digest
partielle Kopplung partial linkage
partielle Kopplungsgruppe *gen* partial linkage group
partikuläre Vererbung particulate inheritance

P Partner

Partner mate, partner, companion;
(Geschlechtspartner) mate
Partnerfüttern mate feeding
Partnerschaft relationship,
companionship;
(Geschlechtsbeziehung) mating
relationship
**Partnerschaftssystem/
Paarungssystem** *ethol*
mating system
**PAS-Anfärbung (Periodsäure/Schiff-
Reagens)** PAS stain
(periodic acid-Schiff stain)
Passage/Subkultivierung passage,
subculture
passagere DNA/Passagier-DNA
passenger DNA
Passatwinde trade winds, trades
Passgang/Passschritt amble, pace
Passgänger pacer, side-wheeler
**Passionsblumengewächse/
Passifloraceae**
passionfruit family
Pasteur-Effekt Pasteur effect
pasteurisieren pasteurize
Pasteurisierung/Pasteurisieren
pasteurizing, pasteurization
Pasteurpipette Pasteur pipet
Patch-Clamp Verfahren patch clamp
technique (patch = Flicken)
Patching/Verklumpung patching
Patella/Kniescheibe
patella, knee bone, genu
pathogen/krankheitserregend
pathogenic (causing or capable of
causing disease)
Pathogenität pathogenicity
**Pathologie/Lehre von den
Krankheiten** pathology
pathologisch/krankhaft
pathological
(altered or caused by disease)
patriarchalisch patriarchal
Paukenbein/Os tympanicum
tympanic bone
Paukengang/Scala tympani
tympanic canal
Paukenhöhle/Cavum tympani
tympanic cavity, middle ear cavity,
tympanum
PCR (Polymerasekettenreaktion)
PCR (polymerase chain reaction)
- **Blasen-Linker-PCR**
bubble linker PCR
- **differentieller Display (Form der
RT-PCR)** differential display
(form of RT-PCR)
- **DOP-PCR
(PCR mit degeneriertem
Oligonucleotidprimer)** DOP-PCR
(degenerate oligonucleotide primer
PCR)

- **inverse PCR**
inverse PCR
- **IRP (inselspezifische PCR)**
IRP (island rescue PCR)
- **ligationsvermittelte PCR**
ligation-mediated PCR
- **RACE-PCR/(schnelle
Vervielfältigung von cDNA-
Enden)-PCR** RACE-PCR, (rapid
amplification of cDNA ends)-PCR
- **RT-PCR (PCR mit reverser
Transcriptase)** RT-PCR
(reverse transcriptase-PCR)
Peckhamsche Mimikry
Peckhamian mimicry
Pedaliaceae/Sesamgewächse
sesame family, benne family
Pedipalpe pedipalp
Pedipalpenlade palpal endite
(with scapula)
Pedizellarie/Pedicellarie
pedicellaria
- **gezähnte Beißzange/
ophiocephale Pedicellarie/
globifere Pedicellarie**
ophiocephalous pedicellaria
- **Giftzange/
gemmiforme Pedicellarie/
globifere Pedicellarie**
poison pedicallaria,
toxic pedicallaria,
gemmiform pedicellaria,
globiferous pedicellaria,
glandular pedicellaria
- **Klappzange/
tridactyle Pedicellarie/
tridentate Pedicellarie**
tridentate pedicellaria
- **Putzzange/
trifoliate Pedicellarie/
triphyllate Pedicellarie**
trifoliate pedicellaria,
triphyllous pedicellaria
Pedizellus/Pedicellus pedicellus,
pedicel
peitschen (peitschend)
whip (whipping)
Peitschengeißel whiplash flagellum,
acronematic flagellum
Peitschenklimmer/Flagellariaceae
flagellaria family
Peitschenwurm (*Trichuris trichiura*)
whipworm
Pektin pectin
Pektinelle pectinella
Pektinsäure (Pektat) pectic acid
(pectate)
**Pelagial/pelagische Zone/
Freiwasserzone** pelagic zone
pelagisch/pelagial (offenes Wasser)
pelagic, pelagial
(open-water/open-sea)

Periode

Pelagos (Organismen des Pelagial)
pelagic organisms,
pelagic community
Peleusball (Pipettierball) *lab*
safety pipet filler, safety pipet ball
Pellicula pellicle, pellicula
Pelz fur
Pelzfarngewächse/Sinopteridaceae
sinopterids
**Pelzflatterer/Riesengleitflieger/
Dermoptera** flying lemurs,
colugos, dermopterans
Pendelströmung (*Physarium*)
shuttle streaming
Pendelverkehr/Pendeln (Membran)
shuttle, shuttling
Penetranz penetrance
• **unvollständige P.**
 incomplete penetrance
• **vollständige P.**
 complete penetrance
penetrieren/eindringen
(sexuell) penetrate; *ethol/med* invade
Penetrieren/Penetration/Eindringen
(sexuell) penetration;
ethol/med invasion
Penicillansäure penicillanic acid
Penis/Aedeagus/Aedoeagus *entom*
penis, aedeagus, intromittent organ
**Penis/Phallus/männliches Glied/
männliches Begattungsorgan/
Rute** penis (*pl* penes), phallus,
copulatory organ, intromittent
organ; pizzle (esp. bull)
**Penisblase/Praeputialsack
(Insekten)** preputial sac
Penisknochen/Os penis penis bone,
baculum
**Pentadaktylie/Fünfzahl von Fingern
und Zehen** pentadactyly
Pentamer/Penton *vir* pentamer,
penton
**Pentosephosphatweg/
Hexosemonophosphatweg/
Phosphogluconatweg** pentose
phosphate pathway, pentose shunt,
phosphogluconate oxidative
pathway, hexose monophosphate
shunt (HMS)
Pentosurie pentosuria
**Peperomiaceae/Zwergpfeffer-
gewächse** peperomia family
Peplomer peplomer
Pepsin (Pepsin A) pepsin (pepsin A)
Peptid peptide
Peptidbindung peptide bond,
peptide linkage
Peptidkette peptide chain
Peptidoglykan/Mukopeptid
peptidoglycan, mucopeptide
Peptidyl-Stelle (P-Stelle)
peptidyl-site (P-site)

Peptidyltransferase
peptidyl transferase
Pepton peptone
peptonisieren peptonize
Peptonwasser peptone water
Perameisensäure performic acid
**Pereion/Peraeon/Pereon (Brust/
Thorax bei Crustaceen)** pereion,
pereon
**Pereiopode/Peraeopode
(Schreitbein des Pereon)**
pereiopod, pereopod (walking leg of
pereon)
perennierend perennial
Perforationsplatte perforation plate
perforieren (perforiert/löcherig)
perforate(d)
perforierte Endwand (Xylem)
perforation plate
Perfusionskultur perfusion culture
pergamentartig (z. B. Flügel)
parchmentlike
Perianth perianth
Peribranchialraum
peribranchial cavity, atrial cavity
Periderm periderm, outer bark
Peridie peridium
Perigon perigon
Perikambium/Perizykel
pericambium, pericycle
**Perikardhöhle/Perikardialraum/
Perikardialsack/Perikardialbeutel/
Perikardialsinus** pericardial cavity,
pericardial chamber, pericardial sac,
pericardial sinus
Perikardialseptum/Diaphragma
pericardial septum, diaphragm
**Perikaryon/Zellkörper/Soma
(Nervenzellen)** perikaryon,
cell body, soma
Periklinalchimäre periclinal chimera
Perilymphe perilymph
**perinukleärer Raum/perinukleärer
Spaltraum/perinukleäre Zisterne/
Cisterna karyothecae** perinuclear
space, perinuclear cistern
**Periode (erdgeschichtliche Zeit)
(*siehe auch: Zeitalter/Epoche*)**
period
• **Cambrium/Kambrium** Cambrian,
 Cambrian Period
• **Devon** Devonian,
 Devonian Period
• **Jura/Jurazeit** Jurassic,
 Jurassic Period
• **Kambrium/Cambrium** Cambrian,
 Cambrian Period
• **Karbon/Steinkohlenzeit**
 Carboniferous,
 Carboniferous Period, 'Coal Age'
• **Kreide/Kreidezeit** Cretaceous,
 Cretaceous Period

P Periodensystem 622

- **Neogen/Jung-Tertiär** Neogene
- **Ordovizium** Ordovician, Ordovician Period
- **Paläogen/Alt-Tertiär** Paleogene
- **Perm** Permian, Permian Period
- **Quartär** Quaternary, Quaternary Period
- **Silur** Silurian, Silurian Period
- **Tertiär/Tertiärzeit/ Braunkohlenzeit** Tertiary, Tertiary Period
- **Trias** Triassic, Triassic Period

Periodensystem (der Elemente) periodic table (of the elements)
periodisch periodic(al)
Periodizität periodicity
Periodsäure/Schiff-Reagens (PAS-Anfärbung) periodic acid-Schiff stain (PAS stain)
Periost/Knochenhaut periosteum
Periostracum/Schalenhäutchen periostracum
peripher peripheral
periphere (extrinsische) Proteine peripheral (extrinsic) proteins
Periphyse periphysis
Periphyton/Aufwuchs/Bewuchs *limn* periphyton, attached algae
periplasmatischer Raum periplasmic space
Periplasmodialtapetum plasmodial tapetum
Periplast periplast
Periprokt periproct
Peristaltik peristalsis
peristaltisch peristaltic
Peristom (Protozoen: Zellmundhöhlung) peristome, peristomium, buccal cavity
Peritoneum/Bauchfell peritoneum
peritrich peritrichous
peritrophische Membran peritrophic membrane
Perizykel/Perikambium pericycle, pericambium
Perizyt/Pericyt/Rougetsche Zelle pericyte, pericapillary cell (type of macrophage)
Perjodsäure *siehe* Periodsäure
Perle pearl
Perlit/Perlstein perlite
Perlmutt/Perlmutter nacre, mother-of-pearl
perlmuttartig glänzend/ perlmutterartig glänzend nacreous
Perlmutterschicht/Nacréschicht nacreous layer
perlmuttfarben/perlmutterfarben nacrine, mother-of-pearl colored

perlschnurartig/moniliat arranged like a chain of beads, moniliform
Perlschnurfühler *entom* moniliform antenna
Perlschnurstruktur (von Chromatin) beads-on-a-string structure
Perm (erdgeschichtliche Periode) Permian Period
Permafrost permafrost
permeabel/durchlässig permeable, pervious
- **impermeabel/undurchlässig** impermeable, impervious
- **semipermeabel/halbdurchlässig** semipermeable

Permeabilität/Durchlässigkeit permeability
Permeant(en) *zool* permeant(s)
permissive Zelle (permissiver Wirt) permissive cell
- **nichtpermissive Zelle** nonpermissive cell

Permissivität permissivity, permissive conditions
permuttfarben/perlmutterfarben nacrine, mother-of-pearl colored
Peroxisom peroxisome
persistent persistent
persistente Infektion/anhaltende Infektion persisting infection
Persistenz/Beharrlichkeit/Ausdauer persistence
persistieren/verharren/ausdauern persist
Perthophyt perthophyte
perthotroph/perthophytisch perthotrophic, perthophytic
Perubalsam Peruvian balsam, balsam of Peru
Perzeption/Wahrnehmung perception
perzipieren/sinnlich wahrnehmen perceive
Pest plague
- **Beulenpest/Bubonenpest (*Yersinia pestis*)** bubonic plague
- **Lungenpest (*Yersinia pestis*)** pneumonic plague

Pestizid/ Schädlingsbekämpfungsmittel/ Biozid pesticide, biocide
- **Algenbekämpfungsmittel/ Algizid** algicide
- **Insektenbekämpfungsmittel/ Insektizid** insecticide
- **Kontaktpestizid** contact pesticide
- **Milbenbekämpfungsmittel/ Akarizid** acaricide
- **Nematodenbekämpfungsmittel/ Nematizid** nematicide
- **Schneckenbekämpfungsmittel/ Molluskizid** molluscicide

Pestizidanreicherung pesticide accumulation
Pestizidresistenz pesticide resistance
Pestizidrückstand pesticide residue
PET (Positronenemissionstomographie) PET (positron emission tomography)
Petalodium petaloid
Petasma petasma
Petersfischartige/Petersfische und Eberfische/Zeiformes dories (John Dory) and others
Petite-Mutante petite mutant
Petrischale Petri dish
Petroläther petroleum ether
Peyerscher Plaque Peyer's patch
Pfad/Weg path, pathway, way, route
Pfahlstütze (für Pflanzen) stake
Pfahlwurzel taproot
Pfanne geol pan
Pfeffergewächse/Piperaceae pepper family
pfeiffen whistle
Pfeil (Mollusken) dart
pfeilförmig arrowhead-shaped, sagittate, sagittiform
Pfeilgift arrow poison
Pfeilnaht/Sutura sagittalis (Schädelnaht) sagittal suture
Pfeilsack (Mollusken) dart sac
Pfeilschwanzkrebse/Xiphosura horseshoe crabs
Pfeilstellung/Flügelpfeilung aer/orn sweepback
Pfeilwürmer/Borstenkiefer/Chaetognathen arrow worms, chaetognathans
Pfeilwurzelgewächse/Marantaceae arrowroot family, prayer plant family
Pfeilzunge (hohl)/toxoglosse Radula toxiglossate radula (hollow radula teeth)
Pferch pen
Pferd horse
• **Apfelschimmel** dapple gray horse
• **Falbe** dun
• **Fliegenschimmel** flea-bitten gray
• **Fohlen/Füllen, männliches** colt (male horse/pony under 4 years)
• **Fohlen/Füllen, weibliches** filly (female horse/pony under 4 years)
• **Hengst** stallion (Zuchthengst: stud/studhorse)
• **Klopphengst/Spitzhengst (Kryptorchide)** ridgeling, ridgling (cryptorchid)
• **Maulesel (Eselstute x Pferdehengst)** hinny (♀ ass/donkey x ♂ horse stallion)
• **Maultier (Pferdestute x Eselhengst)** mule (♀ horse x ♂ ass/donkey)
• **Mustang (verwildertes Präriepferd)** mustang (naturalized horse of western plains)
• **Muttertier** dam (mother animal)
• **Rappe** black horse
• **Schecke** skewbald horse
• **Schimmel** gray horse
• **>Apfelschimmel** dapple gray horse
• **>Fliegenschimmel** flea-bitten gray
• **>Rotschimmel** (red/strawberry) roan
• **>weißer Schimmel** white horse
• **Spitzhengst/Klopphengst (Kryptorchide)** ridgeling, ridgling (cryptorchid)
• **Vatertier/männliches Stammtier (Beschäler/Zuchthengst)** sire (male parent)
• **Vollblut** thoroughbred
• **Wallach (Pferd)** gelding
• **Zuchthengst/Schälhengst/Beschäler** stud, studhorse (see: stallion)
• **Zuchtpferd** stock horse
Pferderettichgewächse/Bennussgewächse/Behennussgewächse/Moringagewächse/Moringaceae horseradish tree family
Pfifferlinge/Cantharellaceae chanterelle family, chantarelle family
Pfingstrosengewächse/Paeoniaceae peony family
Pflänzchen plantlet
Pflanze plant, "flower", wort
pflanzen plant
Pflanzenabfälle plant waste
Pflanzendecke plant cover, vegetational cover, vegetation
Pflanzendroge herbal drug
Pflanzenfarbstoff plant pigment
pflanzenfressend plant-eating, phytophagous, herbivorous
Pflanzenfresser herbivore
Pflanzengalle/Cecidium gall
Pflanzengeographie/Geobotanik plant geography, plant biogeography, phytogeography, geobotany
Pflanzengesellschaft/Pflanzengemeinschaft (allgemein/abstrakt) phytocoenon, community type, nodum, abstract plant community
Pflanzengesellschaft/Pflanzengemeinschaft (spezifische) phytocoenose, concrete plant community
Pflanzenhaar trichome

P Pflanzeninhaltsstoff 624

Pflanzeninhaltsstoff plant chemical, phytochemical
Pflanzenkonsument plant consumer
Pflanzenkörper plant body
Pflanzenkrankheit plant disease
Pflanzenkultur crop
pflanzenlos devoid of plants
Pflanzenmaterial plant specimens
• **sich zersetzendes Pflanzenmaterial** plant debris
Pflanzenöl (diätetisch) vegetable oil
Pflanzenpresse plant press
Pflanzenreich plant kingdom
Pflanzensaft sap, xylem/phloem fluid
Pflanzensauger/Homoptera homopterans (cicadas & aphids & scale insects)
pflanzenschädlich/phytotoxisch phytotoxic
Pflanzenschädling plant pest
Pflanzenschau plant show
Pflanzenschauhaus greenhouse (open to the public)
Pflanzenschutz plant protection
Pflanzenschutzmittel plant-protective agent, pesticide
Pflanzensoziologie plant sociology, phytosociology
Pflanzensystematik plant systematics, plant classification
Pflanzenvielfalt plant diversity
Pflanzenvirus plant virus
Pflanzholz/Setzholz *hort* (Werkzeug zum Einpflanzen) dibber, dibble
Pflanzung planting, plantation
Pflasterepithel/kubisches Epithel cuboidal epithelium
Pflasterzahnsaurier/Placodontia placodonts, placodontians (mollusk-eating euryapsids)
Pflege/Erziehung care, fostering, nurture
Pflegeeltern foster parents
Pflegekinder foster children
Pflegemutter nursing mother, foster mother
pflegen (Pferd: striegeln) grooming (horse: currycomb)
pflegen/versorgen care for, provide for
Pflegetier/Phorozoid phorozooid
Pflock peg
Pflug plow, plough
Pflugbau tillage farming
pflügen plowing, till, tilling
Pflugscharbein/Vomer ploughshare bone, vomer
Pfortader/Vena portae portal vein
Pförtner/Pylorus pylorus
Pfosten post
• **Zaunpfosten** fence post

Pfote/Tatze paw
pfriemlich awl-shaped, subulate
pfropfen *vb* graft
Pfropfen *n* grafting
Pfropfgrundlage/Pfropfunterlage stock, understock
Pfropfkopf stub, grafting stub
Pfropfreis/Edelreis/Pfröpfling/Reis scion (cion), graft (slip>softwood or herbaceous cutting for grafting)
Pfropfstelle graft union
Pfropfströmung plug flow
Pfropfung *hort* grafting
• **Ablaktieren/Ablaktion/ Ablaktation** approach grafting
• **Ammenveredelung/Anhängen/ Vorspann geben** inarching
• **Anschäften** splice grafting, whip grafting (with stock larger than scion)
• **Astpfropfung/Astveredlung** top grafting, top working
• **Augenveredlung/Okulieren/ Okulation** bud grafting, budding
• **Geißfußpfropfung/ Geißfußveredelung (Triangulation)** inlay grafting
• **in den Spalt pfropfen** cleft grafting, wedge grafting
• **Kopulieren/Kopulation/Schäften** splice grafting, whip grafting
• **Kopulieren mit Gegenzunge** whip and tongue grafting
• **Kronenveredlung** crown grafting
• **Rindenpfropfung/Pfropfen hinter die Rinde** rind grafting, bark grafting
• **Sattelschäften** saddle grafting
• **Schäften/Kopulation/Kopulieren (Pfropfung)** splice grafting, whip grafting
• **Seitenpfropfung/ Seitenveredelung/Veredeln an die Seite** side grafting
• **seitliches Anplatten** side grafting; veneer side grafting, side-veneer grafting, spliced side grafting
• **seitliches Anplatten mit Gegenzunge** side-tongue grafting
• **seitliches Anplatten mit langer Gegenzunge** flap grafting
• **seitliches Einspitzen** shield grafting, sprig grafting
• **seitliche Spaltpfropfung** side cleft grafting, side whip grafting, bottle grafting
• **Spaltpfropfung/ Pfropfen in den Spalt** wedge grafting, cleft grafting
• **Tischveredelung** bench grafting

phosphorhaltig

- **Überbrückung/
  Wundüberbrückung**
  bridge grafting, repair grafting
- **Veredeln auf den Kopf**
  apical grafting
- **Wurzelpfropfung/
  Wurzelveredlung** root grafting
- **Zwischenveredlung/
  Zwischenpfropfung** double-
  working (grafting with interstock)
Pfropfunterlage/Pfropfgrundlage
  stock
**Pfropfwachs
  (Wundverschlussmittel)**
  grafting wax (a grafting sealant)
Pfütze puddle
Phage/Bakteriophage phage,
  bacteriophage, bacterial virus
- **lytischer Phage** lytic phage
- **temperenter Phage**
  temperate phage
- **virulenter Phage** virulent phage,
  lysogenizing phage
Phagemid phagemid
Phagosom/Heterophagosom
  phagosome, heterophagosome
  (*siehe auch*: Autophagosom)
phagotroph phagotrophic
Phagozyt/Phagocyt phagocyte
phagozytieren phagocytize
Phagozytose phagocytosis
**Phallus/Penis/männliches Glied/
  männliches Begattungsorgan**
  phallus, penis, copulatory organ,
  intromittent organ
Phallusdrohen phallic threat
Phän *nt* phene
- **Isophän** isophene
- **Ökophän** ecophene
**Phanerophyt/Holzgewächs (Bäume/
  Sträucher; hochliegende
  Erneuerungsknospen)**
  phanerophyte (*siehe*:
  Nanophanerophyt)
Phänetik/numerische Taxonomie
  phenetics, numerical taxonomy,
  taxometrics
Phänogenese phenogenesis
Phänogenetik phenogenetics
**Phänogramm/
  Ähnlichkeitsdendrogramm**
  phenogram
Phänokopie phenocopy
Phänologie phenology
Phänotyp/Phaenotypus phenotype
- **quantitativer P.** quantitative
  phenotype
Phäomelanin phaeomelanin
Phäophytin pheophytin
pharat (coarctate Puppe) pharate
  (cloaked adult/coarctate pupa)
Pharmakogenetik pharmacogenetics

Pharmakognosie pharmacognosy
Pharmakologie pharmacology
pharmazeutisch pharmaceutical
**Pharmazie/Arzneilehre/
  Arzneikunde** pharmacy
Pharynxdrüse pharyngeal gland
Phasendiagramm phase diagram
Phasengrenze phase boundary
Phasenkontrast phase contrast
Phasenkontrastmikroskop
  phase contrast microscope
Phasenring phase ring,
  phase annulus
Phasenübergang phase transition
Phasenübergangstemperatur
  phase transition temperature
Phasenveränderung phase variation
Phasmid/Schwanzpapillendrüse
  phasmid
Phellem/Kork phellem, cork,
  secondary bark
Phellogen/Korkcambium phellogen,
  cork cambium
Phenanthren phenanthrene
Phenol phenol
Phenylalanin phenylalanine
Phenylketonurie phenylketonuria
Pheromon pheromone
Phloem/Siebteil/Bastteil phloem
- **externes Phloem**
  external phloem
- **internes Phloem/inneres
  Phloem/Innenphloem**
  internal phloem
- **interxylares Phloem**
  interxylary phloem
- **intraxylares Phloem**
  intraxylary phloem
Phloembeladung/-entladung
  phloem loading/unloading
Phloemsaft phloem sap
Phorbolester phorbol ester
Phoresie phoresis, phoresy, phoresia
Phormiumgewächse/Phormiaceae
  flax lily family
Phorozooid/Tragtier phorozooid
Phosphat phosphate
Phosphatidsäure phosphatidic acid
Phosphatidylcholin
  phosphatidylcholine
Phosphodiesterbindung
  phosphodiester bond
**Phosphogluconatweg/
  Pentosephosphatweg/
  Hexosemonophosphatweg**
  phosphogluconate oxidative
  pathway, pentose phosphate pathway,
  pentose shunt, hexose
  monophosphate shunt (HMS)
Phosphor *n* phosphorus
**phosphorhaltig/phosphorig/
  Phosphor...** *adj/adv* phosphorous

P

Phosphorsäure 626

Phosphorsäure phosphoric acid
Phosphorylierung phosphorylation
● **nichtzyklische/nichtcyclische/ lineare P.**
noncyclic phosphorylation
● **oxidative P.**
oxidative phosphorylation, carrier-level phosphorylation
● **Substratkettenphosphorylierung**
substrate-level phosphorylation
● **zyklische/cyclische P.**
cyclic phosphorylation
photoallergen photoallergenic
Photoatmung/Lichtatmung/ Photorespiration photorespiration
photoautotroph photoautotrophic
photoheterotroph
photoheterotrophic
photolithotroph/photoautotroph
photolithotrophic, photoautotrophic
Photonenstromdichte
photosynthetic photon flux (PPF)
photoorganotroph
photoorganotrophic(al)
Photoperiodismus photoperiodism
Photoperiodizität photoperiodicity
Photophore/Leuchtorgan
photophore, luminous organ, light-emitting organ
Photoreaktivierung
photoreactivation
Photorespiration/Photoatmung/ Lichtatmung photorespiration
Photosensibilisierung
photosensibilization
Photosynthese photosynthesis
Photosynthese-Einheit
photosynthetic unit
Photosyntheseprodukt
photosynthetic product, photosynthate
Photosynthesequotient/ Assimilationsquotient
photosynthetic quotient
Photosynthesereaktionszentrum/ Reaktionszentrum reaction center
photosynthetisch photosynthetic
photosynthetisch aktive Strahlung
photosynthetically active radiation (PAR)
photosynthetisieren
photosynthesize
phototroph/photosynthetisch
phototroph(ic), photosynthetic
Phototrophie phototrophy
phototropisch phototropic
Phototropismus phototropism
Phragmokon/Phragmoconus
phragmocone
Phrase/Tour (Vogelgesang) phrase
Phratrie phratry
Phreatophyt phreatophyte

Phrenologie (nach Gall) phrenology
Phrymagewächse/Phrymaceae
lopseed family
Phthalsäure phthalic acid
phyletisch phyletic
phyletische Evolution
phyletic evolution
Phyllochinon (Vitamin K1)
phylloquinone, phytonadione
Phyllocladium (Flachspross eines Kurztriebs) phylloclade
Phyllodie/Verlaubung phyllody
(transformation of floral organ into leaflike structure)
Phyllodium/Blattstielblatt phyllode
Phylloid (*pl*** Phyllidien) (blattartiges Organ)** phylloid; (Algensreite/ Moosblättchen) phyllid, leaflet, blade, lamina
Phyllom phyllome
Phyllopodium (*pl*** Phyllopodien)/ Blattbein** phyllopod
Phylogenese/Phylogenie/ Stammesgeschichte/ Stammesentwicklung/ Abstammungsgeschichte/ Evolution phylogenesis, phylogeny, evolution
phylogenetisch/phyletisch/ stammesgeschichtlich/evolutionär
phylogenetic, phyletic, evolutionary
physikalische Karte physical map
physikalische Sicherheit(smaßnahmen) *lab*
physical containment
Physiologe physiologist
Physiologie physiology
physiologisch physiological
Physogastrie physogastry
Phytansäure phytanic acid
Phytinsäure phytic acid
Phytoalexin phytoalexin
Phytobezoar (Magenkugel)
phytobezoar (stomach ball)
Phytocecidium/Pflanzengalle (von Pilzen hervorgerufen) plant gall
Phytol phytol
Phytolaccaceae/ Kermesbeerengewächse
pokeweed family
phytophag/herbivor/ pflanzenfressend phytophagous, herbivorous, plant-eating, feeding on plants
Phytoplankton phytoplankton
Phytosterin phytosterol
Phytothelme phytothelma (*adj* phytotelmic)
PIC (Informationsgehalt eines Polymorphismus) PIC (polymorphism information content)
picken peck, pick

627   **plagiotrop** **P**

piepen/piepsen (Maus) peep,
 squeak; (Vögel) chirp, cheep
Pier pier, quay
Pigmentierung pigmentation
Pigmentschicht (Auge)
 pigment layer
Pigmentzelle/Farbzelle/
 Chromatophore pigment cell,
 chromatophore
pikieren *hort* transplant
Pikrinsäure picric acid
pileat/hutförmig/konsolenförmig
 pileate, pileiform, cap-shaped
Pilus/Konjugationsrohr/
 Konjugationsfortsatz
 (auf Bakterienoberfläche)
 pilus (on bacterial surface)
 ● **Sexpilus** sex pilus
Pilz fungus, mushroom
Pilzbefall fungal infestation
Pilzbekämpfungsmittel/Fungizid
 fungicide
Pilzfaden/Hyphe hypha
Pilzfleisch/Fleisch flesh
pilzfressend/fungivor/mykophag/
 myzetophag fungus-eating,
 feeding on fungus, fungivorous,
 mycophagous, mycetophagous
Pilzgarten fungus garden
Pilzgeflecht/Myzel
 (*Myzeltypen unter: Myzel*)
 mycelium
Pilzhülle/Velum veil, velum
Pilzhut cap, pileus
Pilzkörper *pl*/**Corpora pedunculata**
 mushroom bodies,
 pedunculate bodies,
 corpora pedunculata
Pilzkunde/Mykologie mycology
Pilzvergiftung mushroom poisoning,
 mycetism
Pimelinsäure pimelic acid
Pimpernussgewächse/
 Staphyleaceae bladdernut family
Pinaceae/Kieferngewächse/
 Föhrengewächse/Tannenfamilie
 pine family, fir family
Pinakozyt/Pinacocyt pinacocyte
Pinealauge (bei Neunauge:
 ***Petromyxon*)** pineal eye,
 epiphyseal eye (median eye)
Pinealorgan/Epiphyse/Zirbeldrüse
 pineal body, pineal gland, conarium,
 epiphysis
Pingpong-Reaktion/doppelte
 Verdrängungsreaktion *biochem*
 ping-pong reaction,
 double-displacement reaction
Pinguine/Sphenisciformes penguins
Pinozytose pinocytosis
Pinselfüßer/Pselaphognatha
 pselaphognaths

Pinzette tweezers, forceps
 ● **Arterienklemme** artery forceps,
 artery clamp
 ● **Gewebepinzette** tissue forceps
 ● **Knorpelpinzette** cartilage forceps
 ● **Mikropinzette/Splitterpinzette/**
 Uhrmacherpinzette
 microdissection forceps,
 watchmaker forceps
 ● **Sezierpinzette/anatomische**
 Pinzette dissection tweezers/
 forceps
pinzieren/entspitzen pinch off, tip
Pionierart pioneer species
Pionierneuron pioneer neuron
Pionierorganismus pioneer organism
Pionierpflanze pioneer plant
Pionierstadium pioneer stage
Piperaceae/Pfeffergewächse
 pepper family
Piperazin piperazine
Piperidin piperidine
Piperin piperine
Pipette pipet, pipette
 ● **Kapillarpipette** capillary pipet
 ● **Messpipette** graduated pipet,
 measuring pipet
 ● **Pasteurpipette** Pasteur pipet
 ● **Saugpipette** suction pipet (patch
 pipet)
 ● **Tropfpipette/Tropfglas** dropper
 ● **Vollpipette/**
 volumetrische Pipette
 transfer pipet, volumetric pipet
Pipettenflasche dropping bottle,
 dropper vial
Pipettierball/Pipettierbällchen
 pipet bulb, rubber bulb
 ● **Peleusball** safety pipet filler,
 safety pipet ball
pipettieren pipet
Pipettierhilfe pipet helper
Pipettierhütchen/Pipettenhütchen/
 Gummihütchen pipeting nipple,
 rubber nipple
Pirsch stalk (hunt/pursue prey
 stealthily)
pirschen stalk
Pistill (*zu Mörser*) pestle
Pistill/Stempel *bot* pistil
pistillat *bot* pistillate, carpellate
Pittosporaceae/
 Klebsamengewächse
 pittosporum family, tobira family,
 parchment-bark family
Placebo/Plazebo/Scheinarznei
 placebo
Placoidschuppe/Zahnschuppe/
 Hautzahn/Dentikel placoid scale,
 dermal denticle
plagiotrop plagiotropic,
 plagiotropous, obliquely inclined

Plagiotropismus

Plagiotropismus plagiotropism
Plakode placode
- **Linsenplakode** lens placode
- **Ohrplakode** otic placode

Plan-Hohlspiegel/Plankonkav
plano-concave mirror
Planarien planarians
Planation planation
Planke plank
Plankter plankter,
planktonic organism
Plankton (passiv schwebend)
plankton (passive drifters)
- **Femtoplankton** femtoplankton
- **Mikroplankton** microplankton
- **Nanoplankton/Nannoplankton**
nanoplankton, nannoplankton
- **Phytoplankton/pflanzliches Plankton** phytoplankton
- **Pikoplankton** picoplankton
- **Potamoplankton/Flussplankton**
potamoplankton
- **Ultraplankton** ultraplankton
- **Zooplankton/tierisches Plankton**
Zooplankton

Planktonfresser planktotroph
planktonisch planktonic
Planktonseiher plankton strainer (a food-strainer)
Planogamet planogamete
Planozygote planozygote
Planspiegel plane mirror, plano-mirror
planspiral/flach-scheibenförmig gewunden planispiral
Plantage plantation, orchard, grove
Plantaginaceae/Wegerichgewächse
plantain family
Plantigradie/Sohlengang
plantigrade gait
Plaque plaque (siehe: Zahnbelag; siehe: Lysehof/Aufklärungshof)
- **klarer Plaque** clear plaque

Plaque-bildende Einheit (PBE)/ Kolonie-bildende Einheit (KBE)
plaque-forming unit (PFU)
Plaque-Test plaque assay
Plaquedesmosom spot desmosome
Plasmaabschöpfung plasma skimming
Plasmalogen plasmalogen
Plasmamembran/Zellmembran/ Ektoplast/Plasmalemma
plasma membrane, (outer) cell membrane, unit membrane, ectoplast, plasmalemma
Plasmaströmung/Dinese
plasma streaming,
cytoplasmic streaming, cyclosis
plasmatisch plasmatic
Plasmazelle plasma cell
Plasmensäure plasmenic acid

Plasmid plasmid
- **Einzelkopie-Plasmid**
single copy plasmid
- **konjugatives Plasmid**
conjugative plasmid,
self-transmissible plasmid,
transferable plasmid
- **kryptisches Plasmid**
cryptic plasmid
- **mobilisierbares Plasmid**
mobilizable plasmid
- **nicht-konjugatives Plasmid**
non-conjugative plasmid
- **relaxiertes Plasmid/schwach kontrolliertes Plasmid**
relaxed plasmid
- **stringentes Plasmid**
stringent plasmid

Plasmid mit breitem Wirtsbereich
broad host range plasmid
Plasmidamplifikation
plasmid amplification
Plasmidinkompatibilität
plasmid incompatibility
Plasmidinstabilität
plasmid instability
Plasmidkurierung (Entfernung eines Plasmid aus einer Wirtszelle) plasmid curing
Plasmidmobilisierung
plasmid mobilization
Plasmidpromiskuität
plasmid promiscuity
Plasmin/Fibrinolysin plasmin, fibrinolysin
Plasmodesmos/Plasmodesma (pl Plasmodesmen/ Plasmodesmata)
plasmodesm, plasmodesma
(pl plasmodesmas/plasmodesmata)
Plasmodiokarp plasmodiocarp
Plasmolyse plasmolysis
- **Grenzplasmolyse**
incipient plasmolysis

Plastide plastid
Plastilin plasticine
Plastination plastination
- **Ganzkörperplastination**
whole mount plastination

Plastizität plasticity
Plastom plastome
Plastronatmung plastron breathing
Platanengewächse/Platanaceae
plane family, plane tree family,
sycamore family
Plättchenwachstumsfaktor/ Blutplättchen-Wachstumsfaktor/ Plättchenfaktor platelet-derived growth factor (PDGF)
Platte (Kronblatt) blade, lamina
Platte (Ophiuroiden) shield, vertebral ossicle

629 Plenterschlag **P**

Platten-Okulation *hort*
patch budding
Platten-Test plate assay
Plattenausstrichmethode
streak-plate method
Plattencollenchym/
Plattenkollenchym
lamellar collenchyma,
tangential collenchyma
Plattenepithel squamous epithelium
Plattengussverfahren/
Gussplattenmethode
pour-plate method
Plattenhäuter/Placodermen/
Placodermi placoderms
Plattenkiemer/Haie & Rochen/
Elasmobranchii
sharks & rays & skates
Plattenmeristem plate meristem
Plattenverfahren *micb* **(Kultur)**
plate assay, plating
Plattenzählverfahren *micb*
plate count
platter Knochen/Os planum
flat bone
Plattfische/Pleuronectiformes
flatfishes
Plattformriff table reef
Plattierung/Plattieren *micb*
plating (plating out)
● **Replikaplattierung**
replica-plating
Plattierungseffizienz efficiency of
plating
Plattwürmer/Plathelminthen/
Plathelminthes flatworms,
platyhelminths (Platyhelminthes)
Platycladium/Flachspross
platyclade
Platzmine *bot* blotch mine
Plazenta/Mutterkuchen *zool*
placenta; *bot* (Samenleiste) placenta
● **Allantoisplazenta**
allantoic placenta
● **Chorioallantoisplazenta**
chorioallantoic placenta
● **Chorionplazenta**
chorionic placenta
● **Dottersackplazenta/**
Dottersackhöhlenplazenta/
Omphaloplazenta/
omphaloide Plazenta
yolk-sac placenta,
choriovitelline placenta
● **endothelio-choriale Plazenta**
endotheliochorial placenta
● **epithelio-choriale Plazenta**
epitheliochorial placenta
● **Gürtelplazenta/Placenta zonaria**
zonary placenta
● **haemo-choriale Plazenta**
hemochorial placenta

● **Halbplazenta/Semiplazenta/**
Placenta adeciduata
semiplacenta,
nondeciduate placenta
● **Labyrinthplazenta**
hemoendothelial placenta
● **Placenta bidiscoidalis**
bidiscoidal placenta
● **Placenta cotyledonaria/Placenta**
multiplex cotyledonary placenta
● **Placenta diffusa** diffuse placenta
● **Placenta discoidalis**
discoidal placenta
● **Semiplazenta/Halbplazenta/**
Placenta adeciduata
semiplacenta,
nondeciduate placenta
● **syndesmo-choriale Plazenta**
syndesmochorial placenta
● **Vollplazenta/Placenta vera/**
Placenta deciduata deciduate
placenta
● **Zottenplazenta/Topfplazenta**
villous placenta (hemochorial
placenta)
Plazentalaktogen human placento
lactogen (HPL), human chorionic
somatomammotropin (HCS)
Plazentatiere/Placentalia/Eutheria
placentals, eutherians
Plazentation placentation
● **grundständige/basale/basiläre P.**
basal placentation
● **laminale/flächenständige P.**
laminary/lamellate placentation
● **randständige P.**
marginal placentation
● **wandständige P./**
Parietalplazentation
parietal placentation
● **zentralwinkelständige P.**
axile placentation
● **Zentralplazentation**
free central placentation
Plazentom placentome
Plectenchym/Plektenchym/
Flechtgewebe plectenchyma
Pleiochasium/vielgablige
Trugdolde (Infloreszenz)
pleiochasium
pleiotrop pleiotrop, pleiotropic
Pleiotropie pleiotropy
Pleistozän/Diluvium/Glazialzeit/
Eiszeit (erdgeschichtliche
Teilepoche) Pleistocene Epoch,
Glacial Epoch, Diluvial, Ice Age
plektonemische Windung
plectonemic winding
Plenterschlag/Plenterung/
Plenterbetrieb/Femelschlag/
Femelbetrieb *for*
uneven shelterwood method

Plenterwald 630

Plenterwald/Femelwald
shelterwood: selectively cut/uneven-aged stand, uneven-aged forest/plantation
Pleomorphismus/Polymorphismus/
Mehrgestaltigkeit pleomorphism, polymorphism
Pleon (Abdomen der Crustaceen)
pleon
Plerocercoid plerocercoid
Plesiomorphie plesiomorphism
plesiomorph plesiomorphic
Plesiosaurier/Plesiosauria
plesiosaurs
Pleurapophyse pleurapophysis
Pleuridium faucet gland (of bucket orchid)
Pleurobranchie pleurobranch, pleurobranchia
pleurokarp/seitenfrüchtig
pleurocarpic, pleurocarpous
Pleustal pleustal
Pleuston pleuston (free-floating organisms)
Pliozän (erdgeschichtliche Epoche)
Pliocene, Pliocene Epoch
Ploidie ploidy
Plotter/Kurvenzeichner *lab* plotter
Plumbaginaceae/
Bleiwurzgewächse/
Grasnelkengewächse
plumbago family, sea lavender family, leadwort family
Plumula/Keimknospe/
Stammknospe/Sprossknospe/
terminale Embryoknospe
plumule, terminal embryonic bud
plurilokulär/mehrkammerig
plurilocular, multilocular
Plus-Minus-Verfahren *gen*
plus-minus sequencing
Plus-Strang/Positiv-Strang
(codierender Strang) *gen*
plus strand (coding strand)
Pneumatode/Atemöffnung *bot*
pneumathode
Pneumatophore/Atemwurzel
pneumatophore, aerating root
Podaxales desert inky cap fungi
Podetium podetium
Podobranchie podobranch, podobranchia, foot-gill
Podocarpaceae/
Steineibengewächse
podocarpus family
Podophyllaceae/Fußblattgewächse/
Maiapfelgewächse
may apple family
Podostemaceae/Blütentange
riverweed family
Podozyt/Podocyt/Füßchenzelle
podocyte

Podsol (Boden) podsol, podzol
poikilotherm/wechselwarm/
ektotherm poikilothermal, poikilothermous, cold-blooded, ectothermal, heterothermal
Poissonsche Verteilung/Poisson
Verteilung Poisson distribution
pökeln (Fleisch) cure; (sauer einlegen: Gurken/Hering etc.) pickle
Pökeln (Fleisch) curing;
(in Salzlake oder Essig einlegen: Gurken/Hering etc.) pickling
Pol pole
• **animaler Pol** animal pole
• **vegetativer Pol** vegetal pole, vegetative pole
polar polar
• **unpolar** apolar
polare Mutation polar mutation
polares Wachstum polar growth
Polarimeter polarimeter
Polarisationsfilter/"Pol-Filter"/
Polarisator polarizing filter, polarizer
Polarisationsmikroskop polarizing microscope
polarisiertes Licht polarized light
• **linear p.L.** plane-polarized light
Polarzelle/Polzelle (*Dicyemida***)**
polar cell, calotte cell
Polarzellen-Kappe/Kalotte
(*Dicyemida***)** polar cap, calotte
Polemoniaceae/
Sperrkrautgewächse/
Himmelsleitergewächse
phlox family
Polfaden (Mikrotubulus) polar fiber (microtubule)
Polfelder (Coelenteraten)
polar plates
Polgranula polar granules
Polische Blase polian vesicle
Polkappe polar cap (within polar sac)
Polkapsel polar capsule
Polkern polar nucleus
Polkörper/Richtungskörper
polar body
pollakanth pollakanthic
Pollappen (Furchung) polar lobe
Pollen (Furchung) pollen
Pollenanalyse (Pollenkunde/
Palynologie)
pollen analysis (palynology)
Pollenbürstchen (Bienen)
pollen brush
Pollenfach/Pollensack *bot*
pollen sac (saccus/locule/loculus)
Pollenkamm (Bienen) pollen comb, pecten
Pollenkammer *bot* (im oberen Bereich der Samenanlage)
pollen chamber

Polyp

Pollenkitt *bot* pollenkitt, pollen coat
Pollenkörbchen/Corbiculum (Bienen) pollen basket, corbiculum
Pollenkorn pollen grain
Pollenkunde/Palynologie palynology
Pollensack/Pollenfach *bot* pollen sac (saccus/locule/loculus)
Pollensackgruppe/Theka pollen case, theca
Pollenschlauch pollen tube
Pollenschlauchbefruchtung/ Siphonogamie "pollen tube fertilization", siphonogamy
Pollenschlauchzelle pollen tube cell
Pollenübertragung pollen transfer
Pollinarium pollinarium
Pollinium pollinium
Polplasma polar plasm, pole plasm
Polplatte (Ctenophoren) polar field
Polplatte *cyt* pole plate
Polring (Apicomplexa) polar ring
Polster *bot* mat
Polsterarterie/Sperrarterie artery with intimal cushions
polsterförmig/kissenförmig pulvinate, cushion-shaped
Polsterpflanze cushion plant
Polstervegetation mat-like vegetation
Poly(A)-Schwanz *gen* poly(A) tail
Polyacrylamid polyacrylamide
Polyadenylierung *gen* polyadenylation
Polyandrie/Vielmännerei polyandry
polyaxenisch polyaxenic
Polychaeten/Vielborster/ Borstenwürmer bristle worms, polychaetes, polychetes, polychete worms
polycistronisch/polyzistronisch polycistronic
polycistronische mRNA polycistronic mRNA
polyedrische Symmetrie polyhedral symmetry
polyenergid polyenergid
Polygalaceae/ Kreuzblümchengewächse/ Kreuzblumengewächse milkwort family
polygam/in Vielehe lebend polygamous
Polygamie/Vielehe polygamy
polygen polygenic
polygene (Erb-)Krankheiten polygenic diseases
Polygentherorie/Mehrfaktortheorie multiple-factor hypothesis
Polygonaceae/Knöterichgewächse buckwheat family, dock family, knotweed family, smartweed family

polygyn/vielweibig polygynous
Polygynie/Vielweiberei polygyny
polyklonaler Antikörper polyclonal antibody
polylektisch polylectic
Polymerasekettenreaktion polymerase chain reaction (PCR)
polymiktisch polymictic
Polymorphismus/Pleomorphismus/ Mehrgestaltigkeit polymorphism, pleomorphism
  • **balancierter Polymorphismus** *gen* balanced polymorphism
  • **Konformationspolymorphismus** *gen* conformation polymorphism
  • **PIC (Informationsgehalt eines Polymorphismus)** PIC (polymorphism information content)
  • **Restriktionsfragmentlängen- polymorphismus** restriction fragment length polymorphism (RFLP)
  • **SSCP (Einzelstrang- Konformations-Polymorphismus)** *gen* SSCP (single strand conformation polymorphism)
  • **STRPs (Polymorphismen von kurzen direkten Wiederholungen)** STRPs (short tandem repeat polymorphisms)
polymorphonuklearer Leukozyt/ Granulocyt/Granulozyt polymorphonuclear leukocyte, granulocyte
Polynucleotid/Polynukleotid polynucleotide
polynukleär/polynucleär polynuclear
polyöstrisch polyestrous
Polyp polyp, hydroid
  • **Fresspolyp/Fangpolyp/ Autozooid (*Octocorallia*)** feeding polyp, autozooid
  • **Fresspolyp/Nährtier/ Gasterozoid/Gastrozooid/ Trophozoid** feeding polyp, nutritive polyp, gastrozooid, trophozooid
  • **Fruchtpolyp/Blastostyle (>Gonozooid)** blastostyle (reduced gonozooid)
  • **Geschlechtspolyp/Gonozoid/ Gonozooid (Fruchtpolyp)** reproductive polyp, gonozooid
  • **Pumppolyp/Siphonozooid** siphonozooid
  • **Scyphopolyp** scyphozoan polyp
  • **Wehrpolyp/Dactylozoid/ Dactylozooid** stinging zooid, protective polyp, defensive polyp, dactylozooid

Polypenstadium polypoid stage
polyphag polyphagous
Polyphänie/Pleiotropie
 polypheny, pleiotropy, pleiotropism
Polyphänismus polyphenism
polyphyletisch polyphyletic
polyphyodont polyphyodont
Polypid polypide
polyploid polyploid
Polyploidie polyploidy
Polypnoe polypnea
**Polypodiaceae/
 Tüpfelfarngewächse**
 polypody family, fern family
Polyprotein polyprotein
Polysom/Polyribosom polysome, polyribosome
polystemon polystemonous
polytäne Chromosomen
 polytene chromosomes
**polythalam/vielkammerig/
 mehrkammerig/vielkämmrig/
 polythekal**
 polythalamous, polythalamic,
 with many chambers, polythecal
polythetisch polythetic
polytok polytokous
polytypische Art polytypic species
**polyvoltin/plurivoltin/
 mit mehreren Jahresgenerationen**
 polyvoltine, multivoltine
Polzelle/Polarzelle (*Dicyemida*)
 polar cell, calotte cell
**Pontederiaceae/
 Hechtkrautgewächse**
 pickerel-weed family,
 water hyacinth family
**"Pool" (Gesamtheit einer
 Stoffwechselsubstanz)** pool
 (whole quantity of a particular
 substance: body substance/
 metabolite etc)
**poolen/vereinigen/
 zusammenbringen**
 pool, combine, accumulate
**Population/Bevölkerung/
 Fortpflanzungsgemeinschaft**
 population, reproductive group
Population/Bevölkerung population
**Populationsdichte/
 Bevölkerungsdichte**
 population density
**Populationsdruck/
 Bevölkerungsdruck**
 population pressure
Populationsdynamik
 population dynamics
Populationsgenetik
 population genetics
**Populationsgröße/
 Bevölkerungsgröße**
 population size

**Populationskontrolle/
 Bevölkerungskontrolle**
 population control
**Populationskurve/
 Bevölkerungskurve**
 population curve
**Populationsschwankung/
 Populationsfluktuation/
 Bevölkerungsschwankung**
 fluctuation of population
**Populationszusammenbruch/
 Bevölkerungszusammenbruch**
 population crash
**Populationszuwachs/
 Populationswachstum/
 Bevölkerungswachstum**
 population growth
Pore pore
**Porenkappe/Parenthosom/
 Parenthesom** parenthosome,
 parenthesome, septal pore cap
**Porenkapsel/porizide Kapsel/
 Lochkapsel/Löcherkapsel**
 poricidal capsule
Porenplatte (Sensille) plate
porig/porös/durchlässig porous
Porin porin
Porling pore mushroom, pore fungus, polypore
**Porlinge/Echte Porlinge/Poriales/
 Polyporaceae** bracket fungi,
 polypore family
porös/porig/durchlässig porous
Porosität/Durchlässigkeit porosity
Portulakgewächse/Portulacaceae
 purslane family
**Posidoniaceae/
 Neptungrasgewächse/
 Neptunsgräser** posidonia family
Positionskartierung *gen*
 positional mapping
Positionsklonierung
 positional cloning
**Positiv-Strang/Plus-Strang
 (codierender Strang)** *gen*
 plus strand (coding strand)
**Positronenemissionstomographie
 (PET)** positron emission tomography (PET)
Postabdomen/Metasoma
 postabdomen, metasoma
**Postantennalorgan/
 Hinterfühlerorgan**
 postantennal organ
 • **Tömösvarysches Organ**
 organ of Tömösvary
posttetanische Potenzierung
 posttetanic potentiation (PTP)
posttranslational posttranslational
Potamal (Zone des Tieflandflusses)
 potamal
potamodrom potamodromous

**Potamogetonaceae/
Laichkrautgewächse**
pondweed family
**Potamon (Lebensgemeinschaft des
Potamal)** potamon
Potamoplankton/Flussplankton
potamoplankton
Potential potential
- **Aktionspotential**
  action potential
- **Bereitschaftspotential**
  readiness potential
- **elektrotonisches Potential**
  electrotonic potential
- **Endplattenpotential**
  end plate potential (epp)
- **Erwartungspotential** contingent
  negative variation (CNV)
- **evoziertes Potential**
  evoked potential
- **exzitatorisches postsynaptisches
  Potential** excitatory postsynaptic
  potential (EPSP)
- **Generatorpotential**
  generator potential
- **Gleichgewichtspotential**
  equilibrium potential
- **graduiertes Potential**
  graded potential
- **inhibitorisches postsynaptisches
  Potential** inhibitory postsynaptic
  potential (IPSP)
- **Kopplungspotential**
  coupling potential
- **lokales Potential**
  localized potential
- **Löslichkeitspotential**
  solute potential
- **Membranpotential**
  membrane potential
- **osmotisches Potential**
  osmotic potential
- **Ruhepotential** resting potential
- **Schrittmacherpotential**
  pacemaker potential
- **Schwellenpotential**
  threshold potential
- **Summenpotential**
  gross potential
- **Umkehrpotential**
  reversal potential
Potentialdifferenz/Spannung
potential difference, voltage
potentiell potential
"Potter" (Glashomogenisator)
Potter-Elvehjem homogenizer
(glass homogenizer)
Prä-mRNA/Vorläufer-mRNA
pre-mRNA (precursor mRNA)
Prä-Mutation pre-mutation
Prä-Proinsulin/Präproinsulin
pre-proinsulin, preproinsulin

Prä-rRNA/Vorläufer-rRNA pre-rRNA
(precursor rRNA)
Prä-Startkomplex *gen*
prepriming complex
Prä-tRNA/Vorläufer-tRNA pre-tRNA
(precursor tRNA)
präbiotisch prebiotic, prebiotical
präbiotische Suppe prebiotic soup
präbiotische Synthese
prebiotic synthesis
Prachtkleid *orn* display plumage,
conspicuous plumage (nuptial/
breeding plumage)
Prädetermination, maternale *gen*
maternal effect
Prädilektionsstelle *gen*
predilection site
Prädisposition/Veranlagung
predisposition
Prägung *ethol/gen* imprinting
- **genomische Prägung**
  genomic imprinting
**Präimmunität/Prämunität/
Prämunition/
begleitende Immunität**
premunition, concomitant immunity
**Präimplantationstest
(Untersuchung vor Einnistung
des Eis)** preimplantation testing
Präinitiationskomplex *gen*
preinitiation complex
**Präkambrium/Präcambrium
(erdgeschichtliches Zeitalter)**
Precambrian, Precambrian Era
Präkursor/Vorläufer precursor
prall/schwellend/turgeszent
turgescent
**Prallblech/Prallplatte/Ablenkplatte
(Strombrecher z. B. an Rührer
von Bioreaktoren)** baffle plate
Prämaxille premaxilla
Prämolaren/vordere Backenzähne
premolars, bicuspid teeth
**Prämunität/Präimmunität/
Prämunition/begleitende
Immunität** premunition,
concomitant immunity
**pränatale Diagnose/
Pränataldiagnose** antenatal
diagnosis, prenatal diagnosis
pränatale Diagnostik
prenatal diagnostics
Pranke (Tatze großer Raubtiere)
paw (of big carnivores)
Präoralhöhle/Vestibulum vestibule
Präparat preparation (*Lebewesen:*
preserved specimen)
- **Dauerpräparat** *micros*
  permanent mount
- **mikroskopisches Präparat**
  microscopical preparation,
  microscopic mount

P Präparat 634

- **Nasspräparat (Frischpräparat/ Lebendpräparat/Nativpräparat)** wet mount
- **Quetschpräparat** *micros* squash (mount)
- **Schabepräparat** *micros* scraping (mount)
- **Totalpräparat** whole mount

Präparation *anat* dissection
präparativ preparative
Präparator/Tierpräparator taxidermist
Präparierbesteck dissecting instruments (dissecting set)
präparieren *allg* prepare; *anat* dissect; *micros* mount
Präpariernadel dissecting needle, probe
Präparierschale dissecting dish, dissecting pan
Präpatenz prepatent period
Präproinsulin/Prä-Proinsulin preproinsulin, pre-proinsulin
Präputialdrüse/Kastordrüse (Bieber) preputial gland, castor gland
Präputialsack preputial sac
Prärie prairie
präsymptomatische Diagnose presymptomatic diagnosis
präsymptomatische Diagnostik presymptomatic diagnostics
Prävalenz prevalence, prevalency
Prävention prevention
Präventivmedizin/vorbeugende Medizin preventive medicine
Präzipitat/Fällung precipitate
Präzipitation/Fällung precipitation
präzipitieren/fällen/ausfällen precipitate
präzis/genau precise, exact
Präzision/Genauigkeit precision, exactness
Präzisionsgriff precision grip
Pregnenolon pregnenolone
Prellsprung stotting, pronking (horse/donkey)
Prenylierung prenylation
Prephensäure (Prephenat) prephenic acid
Pressspan (Holz) pressboard
Priapswürmer/Priapuliden priapulans
Pribnow-Box *gen* Pribnow box
Priel (Gezeitenstromrinne bei Ebbe im Watt) swash (narrow channel between sandbank and shore), tideway, tidal gully, tidal creek
Primane *bot* primordial cell (*siehe* Initiale)
Primärantwort primary response
Primärblätter/Erstlingsblätter primary foliage leaves, first foliage leaves

Primärharn/Glomerulusfiltrat glomerular ultrafiltrate
Primärkonsument primary consumer
Primärkultur primary culture
Primärlarve/Junglarve/Eilarve primary larva
Primärmetabolit/ Primärstoffwechselprodukt primary metabolite
Primärproduktion primary production
- **Bruttoprimärproduktion** gross primary production (GPP)
- **Nettoprimärproduktion** net primary production (NPP)

Primärstoffwechsel primary metabolism
Primärstoffwechselprodukt/ Primärmetabolit primary metabolite
Primärstruktur (Proteine) primary structure
Primärtranskript *gen* primary transcript
Primärwachstum primary growth
Primärwand primary wall
Primärwurzel/Hauptwurzel primary root, main root
Primärxylem primary xylem
Primelgewächse/ Schlüsselblumengewächse/ Primulaceae primrose family
Primer *gen* primer
- **Universalprimer** universal primer
- **verschachtelte Primer** nested primer

Primer-Extension/ Primer-Verlängerung *gen* primer extension
Primer-Extensionsanalyse (Verfahren zur Bestimmung des 5'-Endes einer mRNA) primer extension analysis
Primitivgrube *embr* primitive pit
Primitivknoten/Urmundknoten/ Hensenscher Knoten/ Hensen'-Knoten *embr* primitive node, Hensen's node, primitive knot, Hensen's knot
Primitivplatte *embr* primitive plate
Primitivrinne *embr* primitive groove
Primitivstreifen/Keimstreifen (Gastrulation) primitive streak, germinal streak
Primordium/Anlage primordium, anlage
Primosom primosome
Prion prion
Prioritätsregel priority rule
Prisma prism
Prismenschicht prismatic layer
Proanura protofrogs, proanurans

Proband/Propositus propositus
Probe (Teilmenge eines zu untersuchenden Stoffes) sample
Probe/Probensubstanz/ Untersuchungsmaterial assay material, test material, examination material
Probe/Versuch/Untersuchung/Test/ Prüfung assay, test, trial, examination, exam, investigation
Probennahme/Probeentnahme taking a sample, sample-taking
Probenvorbereitung sample preparation
probieren/versuchen try, attempt
Probiose/Nutznießung probiosis
Procercoid/Prozerkoid (Cestoda-Postlarve) procercoid
procöl/procoel *adj* procelous
Procöl/Procoel *n* procoel, procoele
Prodissoconch/Prodissoconcha (Muscheln: Larvenschale) prodissoconch (premetamorphic shell)
Produkt product
Produkthemmung product inhibition
produktive Infektion productive infection
Produktivität productivity
Produktregel product rule
Produzent/Erzeuger/Hersteller producer
produzieren/erzeugen/herstellen produce, manufacture, make
Proenzym/Zymogen proenzyme, zymogen
professionelle antigenpräsentierende Zelle professional antigen presenting cell
Profilstellung *bot* vertical alignment of leaves, vertical orientation of leaves
Proflavin proflavin
Profundal/profundale Zone (aphotische Zone) profundal zone (aphotic zone)
Progenese progenesis (precocious reproduction)
Progenot *m* progenote
Progesteron progesterone
Progestin progestin
Proglottide/"Segment" (Bandwürmer) proglottis, proglottid, tape "segment"
prognath/vorkiefrig prognathous, prognathic
Prognathie/Vorkiefrigkeit prognathy, prognathism
Prognose prognosis
Progymnospermen progymnosperms
Proinsulin proinsulin

Projektionsfeld *neuro* projection field, projection area
Prokaryont/Prokaryot *m* (Procaryont/Procaryot) prokaryote (procaryote)
prokaryontisch/prokaryotisch (procaryontisch/procaryotisch) prokaryotic (procaryotic)
prokurv/procurv procurved
Prolaktin/Prolactin (PRL)/ Mammatropin/Mammotropes Hormon/Lactotropes Hormon/ Luteotropes Hormon (LTH) prolactin (PRL), luteotropic hormone (LTH)
Prolaktoliberin/Prolaktin-Freisetzungshormon prolactoliberin, prolactin releasing hormone, prolactin releasing factor (PRH/PRF)
Prolamellarkörper prolamellar body
Prolepsis/Vorzeitigkeit prolepsis, early development
Proliferation proliferation
Proliferationszone/Sprossungszone (Cestoda) proliferative zone, budding zone
proliferieren proliferate
Prolin proline
Promiskuität promiscuity
promiskuitiv/in Promiskuität lebend promiscuous
promiskuitive DNA promiscuous DNA
Promotor (starker/schwacher) *gen* promoter (strong/weak)
Promotorkomplex (offener/geschlossener) *gen* (open/closed) promoter complex
Pronation pronation
Pronationsstellung pronated position
Pronephros/Vorniere pronephros, fore-kidney, primitive kidney, primordial kidney, head kidney
Pronotum/Prothorakalschild/ Halsschild (dorsal) pronotum
Proöstrus proestrus
Propagationseinheit/ Fortpflanzungseinheit/ Ausbreitungseinheit/Diaspore propagule, diaspore, disseminule, dispersal unit
propagative Übertragung *vir* propagative transmission
propagieren propagate
Prophage prophage
Prophase prophase
prophylaktisch prophylactic
Prophylaxe prophylaxis
Prophyll/Vorblatt prophyll, first leaf

P Propionaldehyd 636

Propionaldehyd propionic aldehyde, propionaldehyde
Propionsäure (Propionat) propionic acid (propionate)
Proplastide proplastid
Propodit propodite
proportionaler Schwellenwert proportional truncation
Proportionsregel/Allen'sche Regel proportion rule, Allen's law, Allen's rule
Propositus/Proband propositus
Proprioceptor proprioceptor, proprioreceptor
Propupa/Präpupa/Vorpuppe/ Semipupa propupa, prepupa
Prosenchym *bot/fung* prosenchyma
prosenchymatisch prosenchymatous
Prostaglandin prostaglandin
Prostansäure prostanoic acid
Prostata/Prostatadrüse/ Vorsteherdrüse prostate, prostate gland
prosthekat/prostekat prosthecate
prosthetische Gruppe prosthetic group
Prostomium/Kopflappen prostomium
protandrisch protandric, protandrous
Proteagewächse/ Silberbaumgewächse/Proteaceae protea family, Australian oak family, silk-oak family
Proteasom proteasome
Protein/Eiweiß protein
- **Abwehrprotein** defense protein
- **Akutphasenprotein** acute phase protein
- **Einzellerprotein** single-cell protein (SCP)
- **fibrilläre Proteine/Faserproteine** fibrous proteins
- **Gerüstprotein/Stützprotein** structural protein, fibrous protein
- **gezielte Konstruktion von Proteinen** protein engineering
- **globuläre Proteine** globular proteins
- **integrale (intrinsische) Proteine** integral (intrinsic) proteins
- **kontraktiles Protein/ motiles Protein** contractile protein, motile protein
- **Nachweis verkürzter Proteine/ Test auf verkürzte Proteine (PTT)** protein truncation test (PTT)
- **Nährstoffprotein** nutrient protein
- **Nichtstrukturprotein** nonstructural protein
- **periphere (extrinsische) Proteine** peripheral (extrinsic) proteins

- **Regulatorprotein/regulatives Protein/regulatorisches Protein** regulative protein, regulatory protein
- **Reserveprotein/Speicherprotein** storage protein
- **Schlepperprotein/Trägerprotein** carrier protein
- **Schutzprotein** protective protein
- **Sekretionsprotein/Sekretprotein/ sekretorisches Protein** secretory protein
- **Signalprotein/Sensorprotein** signal protein
- **Skleroprotein** scleroprotein
- **Speicherprotein/Reserveprotein** storage protein
- **Sphäroprotein/globuläres Protein** globular protein
- **Steuerung von Proteinen** protein targeting
- **Strukturprotein** structural protein
- **Trägerprotein/Schlepperprotein** carrier protein
- **Transportprotein** transport protein
- **vesikelassoziierte Proteine** vesicle-associated proteins

Protein-Tagging/Proteinmarkierung protein tagging
proteinartig/proteinhaltig/ Protein.../aus Eiweiß bestehend/ Eiweiß... proteinaceous
Proteinfaltung protein folding
Proteinkörper protein body
Proteinkörper (in Siebröhren) *bot* P-protein body, phloem protein body, slime body/plug (in sieve tube cells)
Proteinmarkierung/Protein-Tagging protein tagging
Proteinoid proteinoid
Proteinsynthese protein synthesis
Proteinurie proteinuria
Proteoglycan proteoglycan
Proteolyse proteolysis
proteolytisch proteolytic
Proterozoikum (erdgeschichtliches Zeitalter) Proterozoic Eon (late Precambrian)
Prothallium/Vorkeim (Farne) prothallus
Prothorakalschild/Halsschild/ Pronotum (dorsal) pronotum
prothorakotropes Hormon/ Aktivationshormon prothoracicotropic hormone (PTTH), brain hormone
Prothoraxdrüse prothoracic gland (an ecdysial/molting gland)
Prothrombin prothrombin, thrombinogen

Protisten/Protista protists
Protobranchie protobranch
Protocöl/Protocoel/Axocöl/Axocoel
protocoel
Protocyt/Protocyte prokaryotic cell
Protofilament protofilament
Protokormus-ähnlicher Körper
protocorm-like body
Protomer protomer
Protonengradient proton gradient
protonenmotorische Kraft
proton motive force
Protonenpumpe proton pump
Protonensonde proton microprobe
Protoonkogen proto-oncogene
Protoplast protoplast
Protopodit (Sympodit) protopod,
protopodite (basal part)
Protozelle protocell
Protozoen/Protozoa protozoans
Proventivknospe/Ersatzknospe
latent bud
Proventivspross/Proventivtrieb/
Ersatztrieb/Stresstrieb
proventitious shoot, latent shoot
Provirus provirus
proximal/ursprungsnah proximal
Prozentsatz/prozentualer Anteil
percentage
Prozerkoid/Procercoid
(Cestoda-Postlarve) procercoid
Prozess-Kontrolle process control
prozessieren/weiterverarbeiten
process
prozessiertes Antigen/
weiterverarbeitetes Antigen
processed antigen
prozessiertes Pseudogen/
weiterverarbeitetes Pseudogen
processed pseudogene
Prozessierung/Verarbeitung
processing
Prozessivität processivity
prüfen/untersuchen/testen/
probieren/analysieren investigate,
examine, test, try, assay, analyze
Prüfung/Untersuchung/Test/Probe/
Analyse investigation, examination
(exam), test, trial, assay, analysis
prusten/schnauben snort
Psammon psammon (interstitial
flora/fauna)
psammophil/sandliebend
psammophilous,
living in sandy habitats
Pseudobulbe/Luftknolle pseudobulb
Pseudodominanz pseudodominance
Pseudogamie pseudogamy,
pseudomixis
Pseudogen pseudogene
• **konventionelles P.**
conventional pseudogene

• **prozessiertes/**
weiterverarbeitetes Pseudogen
processed pseudogene
Pseudoparenchym/Scheingewebe
pseudoparenchyma,
paraplectenchyma, false tissue
pseudopregnant/scheinschwanger
pseudopregnant
Pseudoskorpione/Afterskorpione/
Pseudoscorpiones
pseudoscorpions, false scorpions
pseudostigmatisches Organ
(Orbatiden)
pseudostigmatic organ
Psilotaceae/Gabelfarngewächse
psilotum family
Psychrometer (ein
Luftfeuchtigkeitsmessgerät)
psychrometer,
wet-and-dry-bulb hygrometer
psychrophil psychrophilic
(thriving at low temperatures)
Psychrophyt (kälteangepasste
Pflanze) psychrophyt
psychrotroph psychrotrophic
PTC (vorzeitiges Stoppcodon) *gen*
PTC (premature termination codon)
Pteridaceae/Flügelfarngewächse/
Schwertfarne pteris family
PTT (Nachweis verkürzter Proteine)
PTT (protein truncation test)
Ptychonema/Ptychozyste (eine
Astomocnide: Ceriantharia)
ptychocyst, tube cnida
Pubertät puberty
pubertieren go through puberty
Puderdune/Pulvipluma *orn*
powder-down feather, pulviplume
Puffer buffer
Pufferkapazität buffering capacity
puffern buffer
Pufferung buffering
Pufferzone buffer zone
Pulk *zool* group, bunch, assemblage
Pulmonalklappe (Herz)
pulmonary valve, pulmonic valve
Pulpa/Pulpe pulp
Pulpahöhle pulp cavity
Pulpe pulp
Puls pulse
Puls-Feld-Gelelektrophorese/
Wechselfeld-Gelelektrophorese
pulsed field gel electrophoresis
(PFGE)
pulsieren pulsate, throb, beat
Pulsmarkierung pulse labeling,
pulse chase
Pulsschlag pulsation, pulse beat,
throb
Pulsstrom/Pulsströmung
pulsatile flow
Pulszahl pulse rate

pulvinat (Flechten) leprose (scurfy/scaly)
Pumppolyp/Siphonozooid siphonozooid
Punicaceae/Granatapfelgewächse pomegranate family
Punktauge/Stemma (Einzelauge/Ocelle) (*siehe:* **Nebenauge/Scheitelauge/Stirnauge**) stemma (dorsal and lateral ocelli)
Punktdiagramm dot diagram
punktieren puncture, tap
Punktion puncture (needle biopsy)
Punktmutation point mutation
Punktquelle point source
Punktualismus punctualism, punctuated equilibrium theory
Punnett-Schema *gen* Punnett square
Pupa adectica adecticous pupa
Pupa cingulata/Gürtelpuppe girdled pupa
Pupa dectica decticous pupa
Pupa exarata/gemeißelte Puppe exarate pupa (free appendages)
Pupa libra/freie Puppe free pupa
Pupa suspensa/Stürzpuppe suspended pupa
Puparium puparium; pupal instar
Pupille pupil
Pupillenerweiterung pupil dilatation
pupipar pupiparous
Pupiparie pupipary
Puppe pupa
  • **bedeckte Puppe/Mumienpuppe/Pupa obtecta** obtect pupa
  • **Chrysalis (holometaboler Insekten)** chrysalis (*pl* chrysalids)
  • **freie Puppe/Pupa libra** free pupa
  • **gemeißelte Puppe/Pupa exarata** exarate pupa
  • **Gürtelpuppe/Pupa cingulata** girdled pupa
  • **Mumienpuppe/bedeckte Puppe/Pupa obtecta** obtect pupa
  • **Pupa adectica** adecticous pupa
  • **Pupa dectica** decticous pupa
  • **Pupa exarata/gemeißelte Puppe** exarate pupa (free appendages)
  • **Scheinpuppe/Larva coarctata/Pseudocrysalis** coarctate larva, coarctate pupa, pseudocrysalis
  • **Stürzpuppe/Pupa suspensa** suspended pupa
  • **Tönnchenpuppe/Pupa coarctata** coarctate pupa
  • **Vorpuppe/Propupa/Präpupa/Semipupa** propupa, prepupa
Puppenhülle/Kokon cocoon
Purin purine

purin- oder pyrimidinlose Stelle (AP-Stelle) apurinic or apyrimidinic site (AP site)
Purkinje-Faser Purkinje fiber, conduction myofiber
Purkinje-Zelle Purkinje cell
Purpurdrüse purple gland
Purpurmembran purple membrane
Pusule (Dinoflagellaten) pusule
Putamen putamen
Putrescin/Putreszin putrescine
putzen clean; cleanse; *zool* groom, preen
  • **Selbstputzen/am eigenen Körper putzen** *zool/ethol* autopreening, autogrooming, self-grooming
  • **Fremdputzen/an fremdem Körper putzen** *zool/ethol* allopreening
Putzsymbiose cleaning symbiosis
Putzverhalten preening behavior
Pygidialdrüse/Analdrüse pygidial gland, anal gland
Pygidium/Afterlappen/Afterschild/Analschild (Arthropoden) pygidium (caudal shield)
Pygostyl/Schwanzstiel (Pflugscharbein/Vomer der Vögel) pygostyle (ploughshare bone/vomer of birds)
Pyknidie/Pycnide/Pyknidium/Pyknosporenlager pycnium, pycnidium (*pl* pycnidia)
Pyknose (Kernverdichtung/Karyoplasmaagglutination) pyknosis, pycnosis
Pyknospore pycnospore, pycnidiospore
Pylorus-Anhang pyloric cecum
Pyramidenbahn (verlängertes Rückenmark) pyramid tract, corticospinal tract
Pyramidenkrone *bot* pyramid-shaped treetop/crown, excurrent treetop, conical treetop
Pyramidenzelle pyramidal cell
Pyran pyran
Pyrethrin pyrethrin
Pyrethrinsäure pyrethric acid
Pyridoxin/Pyridoxol/Adermin/Vitamin B$_6$ pyridoxin, adermine, vitamin B$_6$
Pyrimidin pyrimidine
Pyrolaceae/Wintergrüngewächse wintergreen family, shinleaf family
Pyrophyt (stark feuerresistente Pflanze/durch Brände gefördert) pyrophyte
Pyrrhophyten/Feueralgen pyrrhophytes
Pyrrol pyrrole
Pyrrolidin pyrrolidine

Quaddel welt (weal)
Quadratbein/Quadratum
 quadrate bone
Quadratmethode *ecol*
 quadrat method, quadrat sampling
quaken (Ente) quack; (Frosch)
 croak
quälen/misshandeln
 torment, being cruel, maltreat,
 mistreat, abuse
Quälerei/Misshandlung
 tormenting, cruelty, maltreatment,
 mistreatment, abuse
"Qualle"/Meduse medusa
Quallen jellyfishes
 • **Becherquallen/Stielquallen/**
 Stauromedusae stauromedusas
 • **Fahnenquallen/**
 Fahnenmundquallen/
 Semaeostomea
 semeostome medusas
 • **Kammquallen/Rippenquallen/**
 Ctenophoren sea gooseberries,
 sea combs, comb jellies,
 sea walnuts, ctenophores
 • **Kranzquallen/Tiefseequallen/**
 Coronata coronate medusas
 • **Rippenquallen/Kammquallen/**
 Ctenophoren sea gooseberries,
 sea combs, comb jellies,
 sea walnuts, ctenophores
 • **Scheibenquallen/Schirmquallen/**
 Scyphozoen (echte Quallen)
 cup animals, scyphozoans
 • **Staatsquallen/Siphonophora**
 siphonophorans
 • **tentakeltragende**
 Rippenquallen/
 Tentaculiferen (Ctenophora)
 tentaculiferans, "tentaculates"
 • **Tiefseequallen/Kranzquallen/**
 Coronatae coronate medusas
 • **Würfelquallen/Cubozoa**
 box jellies, sea wasps,
 cubomedusas
 • **Wurzelmundquallen/**
 Rhizostomeae
 rhizostome medusas
Quantenevolution
 quantum evolution
quantifizieren *med/chem* quantify,
 quantitate
Quantifizierung *med/chem*
 quantification, quantitation
Quantil/Fraktil *stat* quantile,
 fractile
Quantität quantity
quantitativer Phänotyp
 quantitative phenotye
Quarantäne quarantine
Quartär (erdgeschichtliche Periode)
 Quaternary, Quaternary Period

Quartärstruktur (Proteine)
 quarternary structure
Quartil/Viertelswert *stat* quartile
Quasi-Äquivalenz-Theorie
 quasi-equivalence theory
Quasispezies quasispecies
Quaste (Schwanz) tassel, tuft
Quastenflosse lobe fin
Quastenflosser/Crossopterygii
 lobe-finned fishes, crossopterygians
Quecksilber mercury
Quelle spring, source
 • **heiße Quelle** hot spring
 • **Mineralquelle** mineral spring
 • **Schmelzwasserquelle**
 meltwater spring
 • **Sickerquelle/Sumpfquelle/**
 Helokrene helocrene
 • **Sturzquelle/Rheokrene**
 flowing spring, rheocrene
 • **Thermalquelle** thermal source,
 thermal spring
 • **Tümpelquelle/Limnokrene**
 limnocrene
Quelle/Produktionsort source
quellen (Wasseraufnahme)
 soak, steep
 • **anschwellen** swell
 • **hervorquellen** emanate
Quellflur/Quellflurvegetation
 source vegetation
Quellgebiet/Quellbereich (der
 Flüsse) headwaters
Quellmoor/Quellsumpf spring fen
Quellschüttung spring flow (flow
 rate of a source)
Quellwasser springwater
Querader *entom* cross vein
Querbrücke (Myosinfilament)
 cross bridge
Querfaserung (Holz) crossgrain
Querfortsatz/Processus transversus
 transverse process of vertebra
Querjoch (*pl* Querjoche)
 transverse ridge (teeth)
 • **mit Querjochen/lophodont**
 with transverse ridges, lophodont
 • **mit vier Querjochen/**
 tetralophodont
 with four transverse ridges,
 tetralophodont
Querschnitt/Hirnschnitt cross
 section, transverse section
Querstrombank *lab*
 laminar flow workstation,
 laminar flow hood,
 laminar flow unit
Querstromfiltration
 cross-flow filtration
quervernetzendes Agens
 cross linker, crosslinking agent
quervernetzt cross-linked

Quervernetzung cross-linking
Querwand crosswall
Quetschpräparat *micros* squash (mount)
quieken/quietschen (Schwein/ Meerschweinchen) squeal, squeak
Quieszenz (exogen bedingte Ruheperiode) quiescence
Quirl/Wirtel whorl, verticil
quirlständig/wirtelig (Blattstellung) whorled, verticillate
Quotient/Verhältnis ratio, relation

R-Banding/Revers-Banding *gen*
R-banding
R-Faktor/Resistenz-Faktor R factor,
resistance factor
r-**Selektion** *r* selection
(*rapid* development)
r-**Strategie** *r*-strategy
(*actually not a "strategy"*)
Rabatte border, bordered flowerbed
Rabenbein/Rabenschnabelbein/
Coracoid coracoid
RACE-PCR/(schnelle
Vervielfältigung von cDNA-
Enden)-PCR RACE-PCR/(rapid
amplification of cDNA ends)-PCR
racemös/razemös/racemos/traubig
(monopodial verzweigt)
racemose, botryose
(monopodially branched)
Rachenblütler/
Braunwurzgewächse/
Scrophulariaceae figwort family,
foxglove family, snapdragon family
Rachenmandel/Rachentonsille/
Tonsilla pharyngealis/Tonsilla
pharyngica pharyngeal tonsil
rachenspaltig/oricid oricidal
rachitisch rachitic, rickety
Rackenvögel/Coraciiformes
kingfishers & bee-eaters & hoopoes
& rollers & hornbills
Radbaumgewächse/
Trochodendraceae trochodendron
family, wheel-stamen tree family,
yama-kuruma family
Räderorgan/Krone (Rotatorien)
ciliated crown, ciliated organ,
corona
Räderorgan wheel organ
Rädertiere/Rotatorien rotifers
Radialgliazelle radial glial cell
Radialschild (Ophiuroidea)
radial shield
Radialschnitt/Spiegelschnitt (Holz)
radial section, quartersawn
Radialzelle/Radius radial cell,
radius
radiär/radiärsymmetrisch/zyklisch/
strahlenförmig/aktinomorph
radial, cyclic, radially symmetrical,
regular, actinomorphic
Radiärfurchung radial cleavage
Radiärkanal radial canal
Radiärnerv radial nerve
Radiärsymmetrie radial symmetry
• **fünfstrahlige Radiärsymmetrie/**
Pentamerie pentamery
Radiation, adaptive *evol*
adaptive radiation
Radikal radical
• **freies Radikal** free radical
Radikalfänger radical scavenger

Radio-Allergo-Sorbent Test
radioallergosorbent test (RAST)
radioactiv (Atomzerfall)
radioactive (nuclear disintegartion)
radioaktive Markierung
radiolabelling
radioaktiver Marker
radioactive marker
Radioaktivität radioactivity
Radioimmunassay/
Radioimmunoassay
radioimmunoassay
Radioimmunelektrophorese
radioimmunoelectrophoresis
Radiokarbonmethode/
Radiokohlenstoffmethode
radiocarbon method
Radiolarien/Strahlentierchen/
Radiolaria radiolarians
Radiolarienschlamm
radiolarian ooze
Radionuklid/Radionuclid
radionuclide
Radnetz *arach* orb web
Radula/Reibplatte/"Zunge" radula
• **docoglosse Radula/Balkenzunge**
docoglossate radula
• **hystrichoglosse Radula/**
Bürstenzunge
hystrichoglossate radula
• **ptenoglosse Radula/Federzunge**
ptenoglossate radula
• **rhachiglosse Radula/**
stenoglosse Radula/
Schmalzunge
rachiglossate radula
• **rhipidoglosse Radula/**
Fächerzunge
rhipidoglossate radula
• **stenoglosse Radula/**
rhachiglosse Radula/
Schmalzunge
rachiglossate radula
• **taenioglosse Radula/Bandzunge**
taenioglossate radula
• **toxoglosse Radula/Pfeilzunge**
(hohl) toxiglossate radula (hollow
radula teeth)
Radulapolster/Odontophor
radula support, odontophore
raffen/horten hoarding
Rafflesiaceae/
Schmarotzerblumengewächse
rafflesia family
Rahmen/Gerüst *arach* scaffold,
scaffolding
Rahmenfaden *arach* frame line
Rahmennetz *arach* frame web
Rain field boundary,
margin of a field, balk
Ramachandran-Diagramm
Ramachandran plot

R

Ramet 642

Ramet/Klonindividuum/
Klonmitglied/
Einzelpflanze eines Klons
(>Zweig/Steckling eines Ortet)
ramet (individual member of clone)
rammeln/kopulieren rut, mate,
copulate
Rammler (Schafbock/Widder)
ram; (männlicher Hase) buck
Rand edge, margin
Randart/Satellitenart
satellite species, marginal species
Randdecken *orn* marginal tectrices,
marginal *coverts*
Randeffekt *ecol* edge effect
Randgehänge (Moor) rand/slope
community of raised bog
Randkörper/Rhopalium/
"Hörkölbchen" (Randsinnesorgan)
rhopalium (tentaculocyst)
Randlappen (Scyphozoa) lappet(s),
flap(s)
randomisieren *stat* randomize
Randomisierung *stat* randomization
Randpopulation
marginal population
Randsoral/Bortensoral (Flechten)
marginal soralium
randständige Plazentation *bot*
marginal placentation
Randverteilung *stat*
marginal distribution
randwellig (Blatt) repand
Rang rank
Rangkorrelationskoeffizient *stat*
rank correlation coefficient
Rangmaßzahlen *stat* rank statistics,
rank order statistics
Rangordnung/Rangfolge/
Stufenfolge/Hierarchie
order of rank, ranking, hierarchy
Rangstufe/Kategorie category
• **soziale Rangstufe** social rank
Ranke tendril, cirrus, clasper,
capreolus (>Sprossranke)
Ranken.../mit Ranken versehen
capreolate
ranken (rankend) twine, climb,
creep (twining/climbing/creeping)
Rankenfuß (Thoracopod der
Cirripedier) cirrus, feeding leg
Rankenfüß(l)er/Cirripeden/
Cirripedier/Cirripedia barnacles,
cirripedes
Rankengewächs twiner, creeper,
climber
Rankenkletterer tendril climber
Rankenpflanze/rankende Pflanze
tendril climber; vine, cane, sarment
Ranunculaceae/
Hahnenfußgewächse
buttercup family, crowfoot family

Ranvierscher Schnürring
Ranvier's node, node of Ranvier,
neurofibral node
Ranz/Brunft heat
Ranzenkrebse/Peracarida
peracarids
Raphe/Samennaht/Samenwulst
raphe
Rappe black horse
Raquettehyphen/Keulenhyphen
(Raquettemyzel/Keulenmyzel)
raquet hyphae/hyphas, raquet
mycelium
Rasen *micb/bact* lawn
Rasendecke grass cover, sod, turf
(nonforage grass)
Rasengräser turfgrass
Rasenkultur lawn culture
Rasierklinge *lab* razor blade
rasig/rasenartig/grasbüschelartig
cespitose, caespitose, caespitulose
(growing densely in tufts)
Rasse race
Rassendiskriminierung/
Rassendiskrimination
racial discrimination
Rassenhygiene (*Nazi term for
***Aryan eugenics*)/Erbhygiene**
race hygiene, racial hygiene
Rassenkreis/polytypische Art
polytypic species
rassig/reinrassig thoroughbred
rassisch/Rassen... racial
Rassismus racism
Rassist racist
rassistisch racist
Raster grid, screen, raster
Raster-Kalorimetrie scanning
calorimetry
Rasterelektronenmikroskop (REM)
scanning electron microscope
(SEM)
Rasterkarte *ecol* grid map
Rasterkartierung *ecol/biogeo*
frame raster mapping, grid mapping
Rasterkraftmikroskopie
atomic force microscopy (AFM)
Rastermethode grid method
Rastermutation frameshift mutation
rastern scan, screen
Rasterstichprobenerhebung *ecol*
grid sampling
Rasterschub-Mutation
frameshift mutation
Rastertunnelmikroskopie
scanning tunneling microscopy
Rasteruntersuchung/
Reihenuntersuchung *med*
screening
Rasterverschiebung *gen* frameshift
Rasterverschiebungsmutation
frameshift mutation

Rathkesche Tasche/
Hypophysentasche Rathke's pouch, hypophyseal pouch/sac
Ratiten/Flachbrustvögel ratite birds (flightless birds)
Rattenschwanzlarve rat-tailed larva, rat-tailed maggot
rau/schuppig rough, scabrous
Raub predation
Raub.../räuberisch predatory, raptorial (greifend)
Räuber/Raubfeind/Raubtier/
Fressfeind/Jäger/Prädator predator, predatory animal
Räuber-Beute-Verhältnis predator-prey relationship
räuberisch/Raub... predatory, raptorial (greifend)
Räubertum predation
Raubgastgesellschaft/Synechthrie hostile commensalism, synechthry, synechthry
Raubinsekt/räuberisches Insekt predatory insect
Raublattgewächse/
Bor(r)etschgewächse/
Boraginaceae borage family
raublättrig rough-leaved, trachyphyllous
Raubtier/Carnivor/Karnivor carnivore
Raubtier/Räuber/Raubfeind/
Fressfeind/Jäger/Prädator predator, predatory animal
Raubtiere/Carnivora carnivores
Raubvogel/Greifvogel bird of prey, predatory bird, raptorial bird, raptor
Rauchabzug/Abzug *lab* fume hood
Raucher, schwarzer/weißer
(Tiefsee) black/white smoker
Rauchgase flue gases
Räude/Krätze (Milbenkrätze) scabies, scab (domestic animals), mange (mites)
räudig scabious, scabby
Raufutter roughage
rauh *siehe* rau
Raum (Länge-Breite-Höhe) room, compartment; space
 • **Brustraum/Thorakalraum/**
 Brusthöhle thoracic cavity
 • **Brutraum (***Daphnia***)**
 brood chamber
 • **Harnraum/Urodaeum** urodeum
 • **Hohlraum/Höhlung/Lumen** cavity, lumen; (Blattparenchym) airspace
 • **Intermembranraum** intermembrane space
 • **Interzellulare/Zwischenzellraum** intercellular space

Rautenhirn

 • **Kühlraum/Gefrierraum** cold-storage room, cold store, "freezer"
 • **Pallialraum** pallial sinus
 • **Peribranchialraum** peribranchial cavity, atrial cavity
 • **perinukleärer Raum/**
 perinukleärer Spaltraum/
 perinukleäre Zisterne/Cisterna
 karyothecae perinuclear space, perinuclear cistern
 • **periplasmatischer Raum** periplasmic space
 • **Reinraum** clean room (*auch:* Reinstraum)
 • **Sicherheitsraum/Sicherheitslabor**
 (S1-S4) biohazard containment (laboratory) (classified into biosafety containment classes)
 • **Totraum** deadspace
 • **Zwischenzellraum/Interzellulare** intercellular space
Raum/Gebiet/Gegend/Region/Zone area, region, zone, territory
 • **ländlicher Raum** rural environment
 • **Lebensraum/Lebenszone/Biotop** life zone, biotope
Raum/Platz place
Raum/Weite/Ausdehnung space; expanse
Raumkonkurrenz spatial competition
räumlich spatial, of space; (dreidimensional) three-dimensional
räumliche Orientierung spatial orientation
räumliche Wahrnehmung spatial perception
Raumstruktur/räumliche Struktur three-dimensional structure, spatial structure
Raupe caterpillar
Raupenbewegung/
Integumentbewegung
(Schlangen) caterpillar movement, rectilinear movement
Raureif (fest aufgefroren) rime
Raureif/Reif/Raufrost (fein/flockig) hoarfrost, white frost
Rauschanalyse/Fluktuationsanalyse noise analysis, fluctuation analysis
Rauschen *neuro* noise
Rauschfilter noise filter
Rautenfarngewächse/
Natternzungengewächse/
Ophioglossaceae adder's tongue family, grape fern family
rautenförmig/rhombisch rhomboid
Rautengewächse/Rutaceae rue family
Rautenhirn/Rhombencephalon hindbrain, rhombencephalon

R razemös

**razemös/racemös/racemos/traubig
(monopodial verzweigt)**
racemose, botryose (monopodially
branched)
Reagenz/Reagens (*pl* Reagenzien)
reagent
Reagenzglas test tube, glass tube
**Reagenzglasbefruchtung/
In-vitro-Fertilisation**
in-vitro fertilization (IVF)
Reagenzglasbürste test tube brush
Reagenzglashalter
test tube holder
**Reagenzglasständer/
Reagenzglasgestell** test tube rack
reagieren react
**Reaktand/Reaktionsteilnehmer/
Ausgangsstoff** reactant
**Reaktion (nullter/erster/zweiter..
Ordnung) (Reaktionskinetik)**
(zero-order/first-order/second-
order..) reaction
**Reaktion *ethol* (bedingte/
unbedingte R.)** response
(conditioned/unconditioned r.)
Reaktionsfolge reaction sequence,
reaction pathway
**Reaktionsgeschwindigkeit/
Reaktionsrate** reaction rate
Reaktionsholz reaction wood
Reaktionskette reaction pathway
Reaktionskinetik reaction kinetics
Reaktionsnorm norm of reaction
**Reaktionszentrum/
Photosynthesereaktionszentrum**
reaction center
Reaktionszwischenprodukt
reaction intermediate
Reaktor/Bioreaktor *biot* reactor,
bioreactor
- **Airliftreaktor/
pneumatischer Reaktor**
airlift reactor, pneumatic reactor
- **Blasensäulen-Reaktor**
bubble column reactor
- **Druckumlaufreaktor**
pressure cycle reactor
- **Durchflussreaktor** flow reactor
- **Düsenumlaufreaktor/
Umlaufdüsen-Reaktor**
nozzle loop reactor,
circulating nozzle reactor
- **Fedbatch-Reaktor/
Fed-Batch-Reaktor/Zulaufreaktor**
fedbatch reactor, fed-batch reactor
- **Festbettreaktor**
fixed bed reactor, solid bed reactor
- **Festphasenreaktor**
solid phase reactor
- **Filmreaktor** film reactor
- **Fließbettreaktor**
moving bed reactor

- **Füllkörperreaktor/
Packbettreaktor**
packed bed reactor
- **Gärtassenreaktor** tray reactor
- **Kugelbettreaktor**
bead-bed reactor
- **Lochbodenkaskadenreaktor/
Siebbodenkaskadenreaktor**
sieve plate reactor
- **Mammutpumpenreaktor/
Airliftreaktor** airlift reactor
- **Mammutschlaufenreaktor**
airlift loop reactor
- **Membranreaktor**
membrane reactor
- **Packbettreaktor/
Füllkörperreaktor**
packed bed reactor
- **Pfropfenströmungsreaktor/
Kolbenströmungsreaktor**
plug-flow reactor
- **Rohrschlaufenreaktor**
tubular loop reactor
- **Rührkammerreaktor**
fermentation chamber reactor,
compartment reactor, cascade
reactor, stirred tray reactor
- **Rührkaskadenreaktor**
stirred cascade reactor
- **Rührkesselreaktor**
stirred-tank reactor
- **Rührschlaufenreaktor/
Umwurfreaktor**
stirred loop reactor
- **Säulenreaktor/Turmreaktor**
column reactor
- **Schlaufenradreaktor**
paddle wheel reactor
- **Schlaufenreaktor/Umlaufreaktor**
loop reactor
- **Siebbodenkaskadenreaktor/
Lochbodenkaskadenreaktor**
sieve plate reactor
- **Strahlreaktor** jet reactor
- **Strahlschlaufenreaktor/Strahl-
Schlaufenreaktor** jet loop reactor
- **Tauchflächenreaktor**
immersing surface reactor
- **Tauchkanalreaktor**
immersed slot reactor
- **Tauchstrahlreaktor**
plunging jet reactor, deep jet
reactor, immersing jet reactor
- **Tropfkörperreaktor/
Rieselfilmreaktor**
trickling filter reactor
- **Turmreaktor/Säulenreaktor**
column reactor
- **Umlaufdüsen-Reaktor/
Düsenumlaufreaktor**
nozzle loop reactor,
circulating nozzle reactor

- **Umlaufreaktor/Umwälzreaktor/ Schlaufenreaktor** loop reactor, circulating reactor, recycle reactor
- **Umwurfreaktor/ Rührschlaufenreaktor** stirred loop reactor
- **Wirbelschichtreaktor/ Wirbelbettreaktor** fluidized bed reactor
- **Zulaufreaktor/Fedbatch-Reaktor/ Fed-Batch-Reaktor** fedbatch reactor, fed-batch reactor

Reannealing/Annealing/ Doppelstrangbildung/ Reassoziation/Renaturierung *gen* reannealing, annealing, reassociation, renaturation (of DNA)

Rearrangement/Umordnung/ Neuordnung rearrangement (DNA/genes/genome)

Reassoziation/Reannealing/ Annealing/Doppelstrangbildung/ Renaturierung reassociation, annealing, reannealing, renaturation (of DNA)

Reassoziationskinetik reassociation kinetics

Rebe vine

Receptaculum seminis/Samentasche seminal receptacle, spermatheca

Rechen (Kläranlage) grate, bar screen

Rechenblumengewächse/ Symplocaceae sweetleaf family

Rechengebissechsen/ Mesosaurier/Mesosauria mesosaurs

rechtsgängig right-handed

rechtshändig right-handed, dextral

Rechtsmedizin/Gerichtsmedizin/ Forensik/forensische Medizin forensics, forensic medicine

rechtswindend/rechtsdrehend/ dextrors dextrorse

Rectaldrüse/Rektaldrüse rectal gland

Redie redia

Redigieren von RNA RNA editing

Redoxpotential redox potential

Redoxreaktion oxidation-reduction reaction

Reduktion reduction

Reduktionsmittel reducing agent

Reduktionsteilung/Reifeteilung/ Meiose reduction division, meiosis

Redundanz redundancy

Reduzenten *ecol* reducers

reduzieren reduce

reelles Bild *micros* real image

Referenzstamm *micb* reference strain

Reflex reflex
- **bedingter Reflex** conditioned reflex (CR)
- **Dehnungsreflex/myotatischer Reflex** stretch reflex, myotatic reflex
- **Klammerreflex** clasp reflex
- **myotatischer Reflex/ Dehnungsreflex** myotatic reflex, stretch reflex
- **Saugreflex** suction reflex
- **Schnappreflex** snapping reflex
- **Schreckreflex** startle reflex
- **Totstellreflex/Sichtotstellen/ Katalepsie (Akinese)** shamming dead reflex, catalepsis, catalepsy (akinesis)
- **unbedingter Reflex** unconditioned reflex (UCR)
- **Zuckreflex** jerk

Reflexbogen reflex arc

Refraktärzeit/Refraktärphase/ Refraktärstadium refractory period, refractory stage

Refraktion/Brechung refraction

Refraktometer refractometer

Refugium refuge

Regel/Menstruation menstruation

Regelglied control element, control unit

Regelgröße controlled variable, controlled condition

Regelkreis feedback system, feedback control system

regelmäßig regular
- **unregelmäßig** irregular

regeln/kontrollieren regulate, control

Regelstrecke control system of a process

Regenbogenhaut/Iris iris

regenerieren regenerate

Regenerierung/Regeneration regeneration

Regenfälle rain showers

Regenmesser pluviometer, rain gauge

Regenschatten rain shadow

Regenschattenwüste rain-shadow desert

Regenwald rain forest

Regenwasser rainwater

Regenwurmbewegung/ Harmonikabewegung (Schlangen) concertina movement

Regenzeit/Pluvialzeit rainy season

Regression *stat* regression

Regression zum Mittelwert regression to the mean

Regressionsanalyse *stat* regression analysis

Regressionskoeffizient *stat*
regression coefficient,
coefficient of regression
**regressiv/zurückbildend/
zurückentwickelnd** regressive
Regulationsei regulative egg
Regulationsgen regulatory gene
Regulationsmechanismen
regulatory mechanisms
**Regulatorprotein/regulatives
Protein/regulatorisches Protein**
regulative protein, regulatory protein
Reh/Rehwild (*Capreolus spp.*)
roe deer
- **Ricke/Rehgeiß** (weibl. Reh nach 1. Wurf) doe (adult female)
Rehbock roebuck (adult male)
Rehgeiß/Ricke (weibl. Reh nach 1. Wurf) doe (adult female)
Rehkitz fawn
Rehydratation/Rehydratisierung
rehydration
Reibplatte/"Zunge"/Radula (siehe auch dort) radula
Reichweite (Strahlung) range
reif mature, ripe
- **unreif** unripe, immature
Reif/Raureif rime, hoarfrost,
white frost
Reife maturity, ripeness
- **Unreife** immaturity, immatureness
reifen *vb* mature, ripen
Reifen *n* maturing, ripening
**Reifeteilung/Reduktionsteilung/
Meiose** reduction division, meiosis
Reifung maturation
Reifungs-Förderfaktor
maturation promoting factor
Reihe row; series
- **Alkoholreihe/
aufsteigende Äthanolreihe**
graded ethanol series
- **chaotrope Reihe**
chaotropic series
- **Drillreihe** drill row
- **eluotrope Reihe
(Lösungsmittelreihe)**
eluotropic series
- **Hofmeistersche Reihe/
lyotrope Reihe**
Hofmeister series, lyotropic series
- **mixotrope Reihe**
mixotropic series
- **Transformationsreihe**
transformation series
- **Versuchsreihe**
experimental series
**Reihenregel/Hessesche Regel/
Herzgewichtsregel** *evol*
Hesse's rule, heart-weight rule
rein/sauber
clean; (ohne Zusatz) pure

Reinbestand pure stand
reinerbig/reinrassig/homozygot
homozygous, true-bred, pure-bred
reinerbig sein breed true, breed pure
**reinerbige Linie/reine Linie/
reinerbiger Stamm/reiner Stamm**
pure breeding line,
pure breeding strain
Reinig-Linie Reinig's line
**Reinigung/Dekontamination/
Dekontaminierung/Entseuchung**
decontamination
Reinigungsmittel/Detergens
detergent
Reinigungsverfahren purification
procedure, purification technique
Reinkultur pure culture,
axenic culture
reinrassig *bot* true-bred, pure-bred;
zool thoroughbred (e. g. horses)
Reinraum/Reinstraum clean room
reinst *lab/chem* highly pure
(superpure/ultrapure)
Reis (*pl* Reiser) (Zweiglein/junger
Zweig) young shoot, twig, spray;
(Pfropfreis/Edelreis) scion (cion),
graft
Reischlinge/Fistulinaceae
beef-steak fungi
Reisig spray, brushwood
reißen (Hengst) geld, castrate
reißen (Wild) attack and rend
reißen (z. B. Wassersäule) break,
cavitate
**Reißfestigkeit/Zerreißfestigkeit/
Zugfestigkeit (Holz)**
tensile strength, breaking strength
**Reißnersche Membran/Reißner-
Membran/Membrana vestibularis**
Reissner's membrane
Reißverschluss zipper
- **Leucin-Reißverschluss**
leucine zipper
**Reißverschluss betätigen/Zippering
(Doppelstrangbildung:
kooperativer Vorgang beim
Bilden von Wasserstoffbrücken)**
zippering
Reißverschlussprinzip
zipper principle
**Reißzahn/Fangzahn/Fang
(Raubtiere)** fang, carnassial tooth
**Reisveredelung/Reiserveredelung/
Pfropfen** *hort* scion grafting
reiten *vb* ride
Reiten riding, equitation
Reiz/Stimulus irritation, stimulus
- **adäquater Reiz**
adequate stimulus
- **Außenreiz** external stimulus
- **bedingter Reiz** *ethol*
conditioned stimulus

Replikation

- **Lichtreiz** light stimulus
- **Schlüsselreiz/Auslösereiz** key stimulus, sign stimulus (release stimulus)
- **unbedingter Reiz** *ethol* unconditioned stimulus

reizbar irritable
Reizbarkeit irritability
reizempfänglich irritable, excitable, sensitive
reizen/anregen/stimulieren excite, stimulate
reizen/irritieren *med/physio/chem* irritate
Reizhaar/Fühlhaar trigger hair, sensitive hair
Reizschwelle stimulus threshold
Reizumwandlung stimulus transduction
Reizung/Stimulation irritation, stimulation
Rekapitulations-Theorie recapitulation theory, principle of recapitulation
Rekauleszenz recaulescence
Rekombinante (Zelle) recombinant (cell)
Rekombination recombination
- **allgemeine R.** general recombination
- **doppelte R.** double recombination
- **homologe R.** homologous recombination
- **illegitime R.** illegitimate recombination
- **intrachromosomale Umordnung** intrachange, intrachromosomal recombination
- **mitotische R.** mitotic recombination
- **nichthomologe R.** non-homologous recombination
- **sequenzspezifische R.** site specific recombination

Rekombinationsfrequenz recombination frequency
Rekombinationsknoten recombination nodule
Rekombinationssignalsequenzen recombination signal sequences
rekombinieren recombine
rekombiniert/rekombinant recombinant
rekombinierte DNA-Technologie (Methoden mit Hilfe rekombinierter DNA)/ rekombinante DNA-Technologie recombinant DNA technology
rekombiniertes DNA-Molekül/ rekombinantes DNA-Molekül recombinant DNA molecule

rekombiniertes Protein/ rekombinantes Protein recombinant protein
rekonstituieren reconstitute
Rekonstitution reconstitution
Rektaldrüse/Rectaldrüse rectal gland
rekultivieren recultivate, replant
rekurv/recurv recurved, bent backwards
Relaisneuron/Projektionsneuron/ Hauptneuron relay neuron
Relaiszelle relay cell
relative Häufigkeit relative frequency
Relaxation relaxation
relaxiert/entspannt relaxed (conformation)
relaxiertes Plasmid/ schwach kontrolliertes Plasmid relaxed plasmid
Relaxin relaxin
Relief relief
Relikt relict
Remigium *entom* remigium
renaturieren renature
Renaturierung renaturation, renaturing
Renaturierung/Annealing/ Reannealing/Reassoziation/ Doppelstrangbildung *gen* annealing, reannealing, reassociation (of DNA)
Renin renin (angiotensinogen>angiotensin)
rennen/laufen run
rennend/Renn... running, cursorial
Rennin/Labferment/Chymosin rennin, lab ferment, chymosin
Rensch'sche Haarregel Rensch's rule
Reparaturenzym repair enzyme
Reparaturmechanismus *gen* repair mechanism
Repellens (*pl* Repellentien) repellent
Replikaplattierung replica-plating
Replikation replication
- **bidirektionale R.** bidirectional replication
- **diskontinuierliche R.** discontinuous replication
- **disperse R.** dispersive replication
- **Rollender-Ring-Replikation** rolling circle replication
- **saltatorische R.** saltatory replication
- **semidiskontinuierliche R.** semidiscontinuous replication
- **semikonservative R.** semiconservative replication
- **Überreplikation** overreplication

Replikationsblase *gen* replication bubble, replication eye
Replikationsgabel *gen* replication fork
Replikationskomplex/Replisom *gen* replisome
Replikationsursprung/ Replikationsstartpunkt *gen* replication origin, origin or replication (ori)
replikative Form *gen* replicative form
Replikon/Replikationseinheit *gen* replicon, unit of replication
Replisom/Replikationskomplex *gen* replisome
Reportergen reporter gene
reprimieren/unterdrücken/hemmen *gen/med/tech* repress, control, suppress, subdue
Reprimierung/Unterdrückung/ Hemmung repression, control, suppression
Reproduzierbarkeit reproducibility
reproduzieren reproduce
Reptilien/Kriechtiere/Reptilia reptiles
Reptilienkunde & Amphibienkunde/ Herpetologie herpetology
Repulsionskonformation *gen* repulsion conformation
Resedagewächse/ Resedengewächse/Waugewächse/ Resedaceae mignonette family
Reservat reserve
- **Naturreservat** nature reserve
- **Naturwaldreservat/Bannwald (in S/W Germany)** protected forest (no commercial usage)
- **Wildreservat/Wildtierpark/ Wildpark** wildlife reserve, wildlife park, wild animal reserve, game reserve

Reservestoff reserve material, storage material, food reserve
Reservevolumen reserve volume
Reservoir-Wirt reservoir host
Residualkörper residual body
Residualvolumen residual volume
resistent resistant
Resistenz resistance
Resistenz gegen Antibiotika antibiotic resistance
Resistenz-Faktor (R-Faktor) resistance factor (R factor)
Resistenzgen resistance gene
resorbieren resorb
Resorption resorption
Respirationsepithel/respiratorisches Epithel/Atmungsepithel respiratory epithelium

respiratorischer Quotient/ Atmungsquotient respiratory quotient
Ressource/Rohstoffquelle resource
Ressourcennutzung resource utilization
Ressourcenschonung resource conservation
Rest (z. B. Aminosäuren-Seitenkette) rest, residue
- **unveränderter Rest/invarianter Rest** *math* invariant residue
- **variabler Rest** *math* variable residue

Restiogewächse/Restionaceae restio family
restituieren/wiederherstellen restitute
Restitution/Wiederherstellung restitution
Restmeristem intercalary meristem
Restriktionsendonuclease restriction endonuclease
Restriktionsenzym restriction enzyme
Restriktionsfragmentlängen-polymorphismus restriction fragment length polymorphism (RFLP)
Restriktionsschnittstelle *gen* restriction site
Resupination resupination (inversion)
Reten retene
Retentionszeit/Verweildauer/ Aufenthaltszeit retention time
Retikulopodium/Reticulopodium reticulopodium, reticulopod
Retikulozyt/Reticulocyt/ Proerythrozyt reticulocyte, proerythrocyte (immature red blood cell)
Retinal retinal, retinene
Retinol (Vitamin A) retinol
Retinsäure retinic acid
Retinulazelle retinular cell
Retorte retort
Retraktormuskel/Retraktor/ Rückzieher retractor muscle
Retrocerebralorgan/ Retrocerebralkomplex retrocerebral organ
Retrocerebralsack retrocerebral sac
Retroelement, virales viral retroelement
Retrogen retrogene
Retrotransposon retrotransposon
retroviral retroviral
Retrovirus retrovirus
- **akut transformierendes Retrovirus** acute transforming retrovirus

Rhizophoraceae

Reusenfalle *bot* weir basket trap
Reusengeißelzelle/Cyrtocyte (siehe:
Flammenzelle) fenestrated flame
cell (protonephridia)
Revers-Banding/R-Banding
R-banding
reverse Genetik reverse genetics
reverse Transkriptase/Revertase/
Umkehrtranskriptase
reverse transcriptase
reverse Transkription
reverse transcription
reverse Translation
reverse translation
reversibel/umkehrbar reversible
Reversibilität/Umkehrbarkeit
reversibility
reversible Hemmung
reversible inhibition
Reversion/Umkehrung reversion
• **ausgleichende Reversion** *gen*
second site reversion
Reversosmose/Umkehrosmose
reverse osmosis
Reversphase/Umkehrphase
reverse phase
Revertase/Umkehrtranskriptase/
reverse Transkriptase
reverse transcriptase
Revier/Wohnbezirk/Gebiet/
Territorium territory, range
Reviermarkierung *ethol* marking of
territory, territorial marking
Revolver/Objektivrevolver *micros*
nosepiece, nosepiece turret
Reynold'sche Zahl/Reynolds-Zahl/
Reynoldsche Zahl
Reynolds number
rezent/gegenwärtig/heute lebend
recent, contemporary, extant
Rezeptakel/Rezeptakulum
receptacle, receptaculum
Rezeptakulum seminis/
Samentasche seminal receptacle,
spermatheca, sperm chamber
Rezeptor/Empfänger receptor
• **adrenerger Rezeptor**
adrenergic receptor
• **Barorezeptor** baroreceptor
• **cholinerger Rezeptor**
cholinergic receptor
• **Antigenrezeptor** antigen receptor
• **Chemorezeptor** chemoreceptor
• **Dehnungsrezeptor (Muskel)**
stretch receptor
• **Geschmacksrezeptor**
taste receptor, gustatory receptor
• **Mechanorezeptor**
mechanoreceptor
• **muscarinischer Rezeptor/**
muskarinischer Rezeptor
muscarinic receptor

• **nikotinerger Rezeptor**
nicotinic receptor
• **Osmorezeptor** osmoreceptor
• **phasischer Rezeptor**
phasic receptor
• **postsynaptischer Rezeptor**
postsynaptic receptor
• **Photorezeptor** photoreceptor
• **Thermorezeptor** thermoreceptor
Rezeptor-Ausdünnungsregulation
receptor-down regulation
rezeptorvermittelte Endozytose/
rezeptorgekoppelte Endozytose
receptor-mediated endocytosis
rezessiv recessive
reziprok reciprocal
reziproke Translokation
reciprocal translocation
Reziprokschüttler
reciprocating shaker
RFLP (Restriktionsfragmentlängen-
polymorphismus) RFLP
(restriction fragment length
polymorphism)
Rhabdit/Epithelstäbchen
(Turbellaria) rhabdite
Rhabdom rhabdome
Rhabdomer rhabdomere
Rhachis/Blattspindel/
Fiederblattachse (Mittelrippe
eines Fiederblattes) rachis
• **kleine sekundäre Rhachis**
rachilla
Rhagon (Schwämme) rhagon
Rhamnaceae/Kreuzdorngewächse
buckthorn family,
coffeeberry family
Rheokrene/Sturzquelle rheocrene,
flowing spring
rheophil (in der Strömung lebend)
rheophilous, rheophilic (preferring
running water)
Rheophyt (Pflanze der
Fließgewässer) rheophyte
Rheotaxis rheotaxis
Rhinarium/Riechplatte rhinarium
Rhithral rhithral
Rhithron rhithron
Rhizine rhizine, rhizina
Rhizodermis/Wurzelepidermis
rhizodermis, epiblem(a)
Rhizoid (Algen/Moose) holdfast
(algas); rhizoid, rootlet (mosses)
Rhizom/Erdspross/Wurzelstock
rhizome, creeping underground stem
rhizomartige Hauptachse (Algen/
Zygomyceten) stolon
Rhizomknolle rhizomatous tuber,
rhizome tuber
Rhizophoraceae/
Mangrovengewächse mangrove
family, red mangrove family

Rhizosphäre

Rhizosphäre rhizosphere
Rhodopsin/Sehpurpur rhodopsin, rose-purple
Rhombogen (Mesozoa) rhombogen
Rhopalium/Randkörper/ "Hörkölbchen" (Randsinnesorgan) rhopalium (tentaculocyst)
Rhopalonema rhopaloneme
Rhoptrie (*pl* **Rhoptrien**) rhoptry (*pl* rhoptries)
Rhythmik rhythm, rhythmics
  • **Gezeitenrhythmik/Tidenrhythmik** tidal rhythm
  • **Tagesrhythmik/circadiane Rhythmik** circadian rhythm
Rhythmus rhythm
  • **Nachtrhythmus (Gegensatz zu: Tagrhythmus)** nocturnal rhythm
  • **Tag-Nacht-Rhythmus/ Tag-Nacht-Periodizität (24-Stunden-Takt)** diel periodicity, diel pattern
  • **Tagesrhythmus (Gegensatz zu: Nachtrhythmus)** diurnal rhythm
Rhythmus-Anpassung (circadiane) entrainment (rhythm adjustment)
Riboflavin/Lactoflavin (Vitamin B$_2$) riboflavin, lactoflavin (vitamin B$_2$)
Ribonucleinsäure/Ribonukleinsäure (RNA/RNS) ribonucleic acid (RNA)
Ribonucleoprotein/ Ribonukleoprotein ribonuclear protein
Ribosom ribosome
  • **Polyribosom/Polysom** polyribosome, polysome
ribosomale RNA (rRNA) ribosomal RNA (rRNA)
Ribosomenbindungsstelle ribosome binding site
Ribosonde/RNA-Sonde riboprobe
Ribozym ribozyme
Richtigkeit/Genauigkeit *stat* correctness, exactness, accuracy
Richtungskörper/Polkörper polar body
Richtungsorientierung directional orientation
Ricke (♀ Reh nach 1. Wurf) doe (adult female)
Riechbahn/Tractus olfactorius *neuro* olfactory tract
riechen smell
Riechepithel olfactory epithelium
Riechgrube olfactory pit (a sensory pit)
Riechhirn/Rhinencephalon "nose brain", rhinencephalon
Riechhügel/Riechkolben/Bulbus olfactorius olfactory dome (sensory dome), olfactory bulb
Riechnerv olfactory nerve
Riechorgan olfactory organ
Riechplatte/Porenplatte (Sensilla placodea) olfactory plate (sensory plate); (Rhinarium) rhinarium
Riechschleimhaut olfactory mucosa
Riechschwelle olfactory threshold
Ried reed
Riedgras/Segge (Sauergräser) sedge
Riedgrasgewächse/Riedgräser/ Sauergräser/Cyperaceae sedge family
Riedsumpf reed swamp
Riemenblumengewächse/ Mistelgewächse/Loranthaceae mistletoe family (showy mistletoe family)
Rieselfelder (Abwasser-Kläranlage) sewage fields, sewage farm
Rieselfilm falling liquid film
Rieselfilmreaktor/ Tropfkörperreaktor trickling filter reactor
rieseln trickle
Riesenaxon giant axon
Riesenchromosom giant chromosome
Riesenfaser/Mauthnersche Zelle/ Mauthner-Zelle *ichth* giant fiber, Mauthner's cell
Riesenläufer/Skolopender/ Scolopendromorpha scolopendromorphs
Riesenzelle giant cell
Riff reef
  • **Atollriff/Atoll/Lagunenriff** atoll
  • **Barriereriff/Wallriff** barrier reef
  • **Fleckenriff** patch reef, bank reef
  • **Plattformriff** table reef
  • **Rückriff** rear reef
  • **Saumriff/Strandriff/Küstenriff** fringing reef
  • **Vorriff** fore reef
riffbildend/hermatypisch reef-building, hermatypic
  • **nicht riffbildend/ahermatypisch** not reef-building, non-hermatypic
Riffdach reef flat
Riffhang reef slope
Riffkante reef edge
Riffkorallen/Steinkorallen/ Madreporaria/Scleractinia stony corals, madreporarian corals, scleractinians
Riffkrone reef crest
Right side-out Vesikel (Vesikel mit der richtigen Seite nach außen) right side-out vesicle
Rinde bark, cortex; (Haut/Schale) skin, peel; shell

651   **Rispe** R

Rindenbildung cortication
Rindenbrand/Sonnenbrand
sunscald
Rindenkorallen/Hornkorallen/
Gorgonaria gorgonians,
gorgonian corals, horny corals
Rindenparenchym cortical
parenchyma
Rindenpfropfung/
Pfropfen hinter die Rinde *hort*
rind grafting, bark grafting
Rindenpilze/Corticiaceae
crust fungus family
Rindenschichtpilze/Stereaceae
parchment fungus family
Rinderwahnsinn mad cow disease
(bovine spongiform encephalopathy
= BSE)
Rinderwirtschaft cattle ranching
Ring/Kragen/Annulus inferus
(Rest des Velum partiale) *fung*
ring, inferior annulus
Ring-Okulation/Ringveredlung *hort*
ring budding, annular budding
(flute budding)
Ringart ring species
ringartig ringlike, annular
Ringbildung/Catenation catenation
Ringblende *micros* disk diaphragm
(annular aperture)
Ringchromosom ring chromosome
Ringelantenne/Geißelantenne/
amyocerate Antenne *entom*
amyocerate antenna
Ringelborke/Ringborke ringbark
Ringelung/Gürteln (Baumrinde)
ringing, girdling
Ringelwürmer/Gliederwürmer/
Borstenfüßer/Anneliden
segmented worms, annelids
Ringerlösung/Ringer-Lösung
Ringer's solution
Ringform *chem* ring form,
ring conformation
Ringformel ring formula
ringförmig/zyklisch annular, cyclic
Ringfurche/Coronalfurche
coronal groove
Ringkanal/Radiärkanal/
Ambulakralring
ring canal, radial canal
Ringknorpel/Cartilago cricoidea
annular cartilage, cricoid cartilage
Ringmuskel ring muscle,
circular muscle
ringporig (cyclopor) (Holz)
ring porous
Ringschluss *chem* ring formation,
cyclization
Ringschluss/Zirkularisierung
circularization
Ringspaltung *chem* ring cleavage

Rinne/Furche *anat/morph/gen*
groove, furrow
● **große Rinne/tiefe Rinne/große**
Furche (DNA-Struktur)
major groove (DNA structure)
● **kleine Rinne/flache Rinne/kleine**
Furche (DNA-Struktur)
minor groove (DNA structure)
Rinnsal/kleines Bächlein rill,
rivulet, streamlet
Rippe/Costa rib, costa; *bot* vein
● **Bauchrippe/Gastralrippe/**
Gastralia (Reptilien)
abdominal rib, gastralia
● **echte Rippe/Costa vera** true rib
● **falsche Rippe/unechte Rippe/**
Costa spuria false rib
● **frei endende Rippen/**
Costae fluitantes floating ribs
● **gerippt/mit Rippen** ribbed, costate
● **Halsrippe/Costa cervicalis**
cervical rib
● **Mittelrippe** *bot* midrib; *(eines*
Fiederblattes/Rhachis/Blattspindel/
Fiederblattachse) rachis
● **Sakralrippe/Kreuzbeinrippe**
sacral rib
● **Thorakalrippe/Costa thoracalis**
thoracic rib
Rippel *geol* ripple
Rippelmarke *geol* ripple mark
Rippenbogen/Arcus costalis
costal arch
Rippenfarngewächse/Blechnaceae
blechnum family, deer fern family
Rippenfell/Pleura parietalis
(thorakal: Pleura costalis) parietal
pleura (thoracic: costal pleura)
Rippenfortsatz/Diapophyse
diapophysis (transverse process of
neural arch for rib attachment)
Rippenfurche costal groove,
costal sulcus
Rippengefäß/Meridionalkanal
(Ctenophoren) meridional canal,
gastrovascular canal
Rippenhals/Collum costae
neck of rib, rib collar
Rippenknorpel costal cartilage
Rippenmeristem rib meristem,
file meristem
Rippenquallen/Kammquallen/
Ctenophoren sea gooseberries,
sea combs, comb jellies,
sea walnuts, ctenophores
Risiko (*pl* Risiken)/Gefahr risk,
danger
Rispe/Blütenrispe (Infloreszenz)
panicle
● **Geschein (Rispe des Weinstocks)**
cluster, flower cluster of vine
(a panicle)

Rispenfarngewächse 652

- **Grasrispe**
  juba, loose panicle of grasses
- **Schirmrispe (ein Ebenstrauß/ Corymbus)** umbel-like panicle
- **Trichterrispe/Spirre** anthela

Rispenfarngewächse/ Königsfarngewächse/ Osmundaceae flowering fern family, cinnamon fern family, royal fern family

rispig/paniculat paniculate, panicular

Riss/Fissur/Furche/Einschnitt fissure; (Spalte) crevice; (Holz: zwischen Jahresringen) shake

Riss/Riss (Beute des Raubwildes) rendered prey

Ritterlinge/Tricholomataceae tricholoma family

Rittersportbäume/Vochysiaceae vochysia family

rituelle Fesselfäden *arach* bridal veil (crab spiders)

Ritus rite

Rivale rival

rivalisieren rival, be rivals, compete

Rivalität rivalry

RNA/RNS (Ribonucleinsäure/ Ribonukleinsäure) RNA (ribonucleic acid)

- **3′ → 5′ (drei Strich-fünf Strich/ drei Strich nach fünf Strich)** 3′ → 5′ (three prime five prime/ three prime to five prime)
- **Antisense-RNA/Anti-Sinn-RNA/ Gegensinn-RNA** antisense RNA
- **Messenger-RNA/Boten-RNA/ mRNA** messenger RNA (mRNA)
- **Prä-mRNA/Vorläufer-mRNA** pre-mRNA
- **Prä-rRNA/Vorläufer-rRNA** pre-rRNA (precursor rRNA)
- **Prä-tRNA/Vorläufer-tRNA** pre-tRNA (precursor tRNA)
- **Redigieren von RNA** RNA editing
- **ribosomale RNA (rRNA)** ribosomal RNA (rRNA)
- **snRNA/kleine nucleäre-RNA** snRNA, small nuclear RNA
- **stabile RNA** stable RNA
- **tRNA/Transfer-RNA** tRNA, transfer RNA

RNA-getriebene Hybridisierung RNA-driven hybridization

RNA-Polymerase RNA polymerase

RNA-Priming RNA priming

RNA-Processing/ RNA-Weiterverarbeitung RNA processing

RNA-Sonde/Ribosonde riboprobe

RNA-Transkript RNA transcript

RNA-Welt RNA-world

RNase (Ribonuclease/Ribonuklease) RNase (ribonuclease)

Rochen/Batoidea (Überordnung) rays & skates

- **echte Rochen/Rajoidei (Unterordnung)** skates

Rochenartige/Rajiformes guitarfishes & skates

Rodung felling, clearing

- **Brandrodung** clearing by fire (intentional forest fires)

Rogen (Fischeier innerhalb der Eierstöcke) (*siehe auch*: Fischlaich) roe (esp. fish-eggs within ovarian membrane)

Rohabwasser raw sewage

Rohdichte green density

Rohextrakt crude extract

Rohhumus/saurer Auflagehumus/ Trockentorf mor (humus)

Rohr/Röhre pipe, tube

- **Schilfrohr** cane

Röhrbein/Kanonenbein/Sprungbein (Pferd: Hauptmittelfußknochen) cannon bone

Röhrchenzähner/Erdferkel/ Tubulidentata aardvark (of Africa)

röhren (Hirsch) bell

röhrenbewohnend tube-dwelling, tubicolous

Röhrenbewohner tube-dweller

Röhrenblüte (verwachsene Kronblätter) corolla tube, tubular corolla

Röhrenblüte/Scheibenblüte (Asterales) disk flower, disk floret, tubular flower

röhrenförmig/schlauchförmig siphoneous, siphonaceous, tubular

Röhrenherz/Herzschlauch tubular heart

Röhrenknochen/langer Knochen long bone (hollow/tubular bone)

Röhrennasen/Procellariiformes tubenoses, tube-nosed swimmers: albatrosses & shearwaters & petrels

Röhrennetz/Trichternetz *arach* funnel web

Röhrenpilze/Röhrlinge/Boletaceae boletes, bolete mushroom family, boletus mushroom family

Röhrentrachee *arach* tube trachea, tubular trachea

Röhricht reed bank, reeds

Rohrkolben cat's-tail, reedmace

Rohrkolbengewächse/Typhaceae reedmace family, cattail family

Röhrling/Porling pore mushroom, pore fungus, boletus mushroom

Rohrstock cane

Rohzucker/Rübenzucker/ Saccharose/Sukrose/Sucrose cane sugar, beet sugar, table sugar, sucrose
Rohschlamm raw sludge
Rohstoff raw material, resource
• **erneuerbare Rohstoffe** renewable resources
• **nachwachsende Rohstoffe** regenerating resources, replenishable resources
• **natürliche Rohstoffe** natural resources
• **nichterneuerbare Rohstoffe** nonrenewable resources
Rohstoffquelle/Ressource resource
Rohzucker raw sugar, crude sugar (unrefined sugar)
Rollender-Ring-Replikation *gen* rolling-circle replication
Rollerflaschenkultur roller tube culture
Rollfarngewächse/ Cryptogrammaceae parsley fern family, rock-brake fern family
Röntgenabsorptionsspektroskopie X-ray absorption spectroscopy
Röntgenbeugung X-ray diffraction
Röntgenbeugungsdiagramm/ Röntgenbeugungsaufnahme/ Röntgendiagramm X-ray diffraction pattern
Röntgenbeugungsmethode X-ray diffraction method
Röntgenbeugungsmuster X-ray diffraction pattern
Röntgenemissionsspektroskopie X-ray emission spectroscopy
Röntgenkleinwinkelstreuung small-angle X-ray scattering (SAXS)
Röntgenkristallographie X-ray crystallography
Röntgenmikroskopie X-ray microscopy
Röntgenstrahl X-ray
Röntgenstrahl-Mikroanalyse X-ray microanalysis
Röntgenstrukturanalyse X-ray structural analysis, X-ray structure analysis
Röntgenweitwinkelstreuung wide-angle X-ray scattering (WAXS)
Rosaceae/Rosengewächse rose family
Rosenapfelgewächse/Dilleniaceae silver-vine family, dillenia family
Rosengewächse/Rosaceae rose family
Rosette rosette, whorl

Rosettenpflanze rosette plant
Rosettenplatte (Bryozoen) rosette plate
Rossbreiten horse latitudes
rossig/brünstig (Stute) in heat
Rossigkeit (Stute) heat
Rosskastaniengewächse/ Hippocastanaceae horse chestnut family, buckeye family
Rostellum/Klebkörper (Gynostemium) rostellum, adhesive body
rösten/rötten (Flachsrösten) retting
Rostpilze/Uredinales rusts, rust fungi
rostrot ferruginous
Rotalgen/Rhodophyceae (Floridaceae) red algae
Rotationsbewegung rotational motion
Rotationsmikrotom rotary microtome
Rotationssinn/Drehsinn rotational sense, sense of rotation
Rotationsverdampfer rotary evaporator
Rote Liste Red Data Book
rote-Königin-Hypothese *evol* Red Queen's hypothesis
Rötegewächse/Labkrautgewächse/ Krappgewächse/Rubiaceae bedstraw family, madder family
Rötelpilze/Rotblättler/ Rhodophyllaceae/Entolomataceae entoloma family
Rotenon rotenone
roter Körper (Schwimmblase) *ichth* red body
Rothirsch/Rotwild/Edelhirsch/ Edelwild (*Cervus elaphus*) red deer
Rotiger/Pseudotrochophora (Larve) rotiger, pseudotrochophore
Rotor rotor
• **Ausschwingrotor** *centrif* swinging-bucket rotor
• **Festwinkelrotor** *centrif* fixed-angle rotor
• **Vertikalrotor** *centrif* vertical rotor
• **Winkelrotor** *centrif* angle rotor, angle head rotor
Rotte (Wildschweine) sounder (herd/hoard/party of pigs or wild boar)
rötten/rösten (Flachsrösten) retting
Rotwild/Rothirsch/Edelwild/ Edelhirsch (*Cervus elaphus*) red deer

rRNA 654

rRNA/ribosomale RNA rRNA, ribosomal RNA
RT-PCR (PCR mit reverser Transcriptase) RT-PCR (reverse transcriptase-PCR)
RTLV-Familie (Familie retrovirusartiger Elemente) (DNA-Element) RTLV-family (reverse transcriptase-like virus family) (DNA-element)
Rübe (*Beta*) beet
Rübe/Speicherwurzel fleshy taproot, storage root
rübenartig turnip-like, napaceous
rübenförmig turnip-shaped, napiform
Rübenzucker/Rohrzucker/Sukrose/ Sucrose beet sugar, cane sugar, table sugar, sucrose
Rubiaceae/Labkrautgewächse/ Rötegewächse/Krappgewächse madder family, bedstraw family
rückbilden degenerate, regress
Rückbildung degeneration, regression
Rückdrehung/Detorsion (Gastropoda: Nervensystem) detorsion
Rücken back; (Schlachttiere) saddle, chine
Rückenanhänge/Cerata (Nudibranchia) cerata
Rückenfeder/Gladius (pergamentartige Schulpe) pen, gladius
Rückenflosse dorsal fin
Rückenmark/Medulla spinalis spinal cord, spinal medulla, medullary canal, nerve cord
• **verlängertes R./ Medulla oblongata** medulla oblongata, medulla
Rückennaht/Dorsalnaht (Mittelrippe des Fruchtblattes) dorsal suture, dorsal seam
Rückenpanzer/Rückenschild/ Carapax (Schildkröte) dorsal shield, carapace
Rückenplatte/Rückenschild/Tergit (Insekten: dorsaler Sklerit) tergite, dorsal sclerite
Rückensaite/Chorda dorsalis/ Notocorda notochord
Rückenschaler/Notostraca tadpole shrimps
Rückenschild/Rückenpanzer/ Carapax (Schildkröte) dorsal shield, carapace
Rückenschild/Rückenteil/Tergum (Insekten: Rückenteil der Körpersegmente) tergum (dorsal plate: esp. abdominal segments)

rückenspaltig/lokulizid *bot* loculicidal
Rückenstück/Tergit (dorsale Sklerite) tergite
Rückenteil/Rückenschild/Tergum (Insekten: Rückenteil der Körpersegmente) tergum (dorsal plate: esp. abdominal segments)
Rückentwicklung retrogressive development, retrogressive evolution
Rückenwind *aer/orn* tail wind
Rückfallfieber (*Borrelia recurrentis*) relapsing fever
Rückflusskühler reflux condenser
rückgebildet/abortiv/rudimentär/ verkümmert abortive
Rückgrat/Wirbelsäule spinal column, vertebral column, backbone
Rückkopplung feedback
• **negative Rückkopplung/ Rückkopplungshemmung/ Endprodukthemmung** feedback inhibition, end-product inhibition
Rückkopplungshemmung/ Endprodukthemmung feedback inhibition, end-product inhibition
Rückkopplungsschleife feedback loop
Rückkreuzung backcrossing, backcross
Rücklaufschlamm/Belebtschlamm activated sludge
Rückmutation back-mutation, reverse mutation
Rückriff rear reef
Rückschlag/Atavismus (ursprüngliches Merkmal) atavism, throwback
Rückschnitt (bis auf den Stumpf für Neuaustrieb) coppice, coppicing
Rückschnitt (Gehölzrückschnitt) pruning, pruning back
rückseitig/dorsal dorsal
Rückstand *chem* residue
Rückstrahlvermögen/Albedo albedo
rückwärts/nach unten gerichtet/ gebogen *bot* retrorse
Rückzieher/Rückwärtszieher/ Rückziehmuskel/Retraktor/ Retraktormuskel retractor muscle
Rückzug/Versteck/Schlupfwinkel *arach* retreat
Rüde (Hund) male dog
Rudel/Meute/Koppel pride (lions), pack (dogs/wolves), party (wild boar), pod (whales/dolphins/seals)
Ruder/Ruderflosse (Schwanzflosse der Wale) rudder
ruderal/auf Schutt wachsend ruderal
Ruderalpflanze ruderal plant

Rumpfbein

Ruderfüßer/Ruderfüßler/ Pelecaniformes totipalmate swimmers: pelicans and allies
Ruderfußkrebse/Ruderfüßer/ Copepoda copepods
Ruderplatte/Ruderplättchen/ Schwimmplatte/"Kamm"/Ctene/ Wimperplättchen (Ctenophoren) ctene, swimming plate, ciliary comb
Ruderschnecken/Gymnosomata/ nackte Flossenfüßer (Flügelschnecken) naked pteropods
Rudiment rudiment (*sensu lato*: vestige)
rudimentär rudimentary (*sensu lato*: vestigial)
rudimentär/abortiv/rückgebildet/ verkümmert abortive
Rudist rudistid, rudistan
Ruf *orn* (Lautäußerung) call, call note
rufen call
Ruffini'sches Körperchen Ruffini's endings, Ruffini's organ, corpuscles of Ruffini
Ruhekern *cyt* resting nucleus
Ruhemembrankanal/Leckkanal resting channel, leakage channel
ruhen rest, lie dormant
ruhend resting, quiescent, dormant
ruhende Knospe/ schlafende Knospe resting bud, dormant bud, quiescent bud
ruhendes Zentrum quiescent center, quiescent zone
Ruhephase/Ruheperiode resting period, quiescent period, dormancy period
Ruhepotential resting potential
Ruhespuren/Cubichnia *paleo* resting traces
Ruhestadium resting stage; dormant stage; quiescent stage
Ruhestellung resting posture
Ruhestoffwechsel/ Grundstoffwechsel basal metabolism
Ruhezustand inactive state, dormant state, dormancy
- **endogen bedingter Ruhezustand/Dormanz** dormancy
- **exogen bedingter Ruhezustand/ Quieszenz** quiescence

Ruhr dysentery
Rührbehälter/Rührkessel agitator vessel
rühren stir, agitate
Rührer/Rührwerk *biot* stirrer, impeller, agitator
- **Ankerrührer** anchor impeller
- **Axialrührer mit profilierten Blättern** profiled axial flow impeller
- **Blattrührer** two flat-blade paddle impeller
- **exzentrisch angeordneter Rührer** off-center impeller
- **Gitterrührer** gate impeller
- **Hohlrührer** hollow stirrer
- **Kreuzbalkenrührer** crossbeam impeller
- **Kreuzblattrührer** four flat-blade paddle impeller
- **Mehrstufen-Impuls-Gegenstrom (MIG) Rührer** multistage impulse countercurrent impeller
- **Propellerrührer** propeller impeller
- **Rotor-Stator-Rührsystem** rotor-stator impeller, Rushton-turbine impeller
- **Schaufelrührer/Paddelrührer** paddle stirrer, paddle impeller
- **Scheibenrührer/Impellerrührer** flat-blade impeller
- **Scheibenturbinenrührer** disk turbine impeller
- **Schneckenrührer** screw impeller
- **Schrägblattrührer** pitched-blade fan impeller, pitched-blade paddle impeller, inclined paddle impeller
- **Schraubenrührer** marine screw impeller
- **Schraubenspindelrührer mit unterschiedlicher Steigung** variable pitch screw impeller
- **Schraubenspindelrührer** pitch screw impeller
- **selbstansaugender Rührer mit Hohlwelle** self-inducting impeller with hollow impeller shaft
- **Stator-Rotor-Rührsystem** stator-rotor impeller, Rushton-turbine impeller
- **Turbinenrührer** turbine impeller
- **Wendelrührer** helical ribbon impeller
- **zweistufiger Rührer** two-stage impeller

Rührerwelle impeller shaft
Rührgerät/Mixer stirrer, mixer
Rührkessel/Rührbehälter agitator vessel
Rührwerk impeller
Rülle (im Moor) bog drainage rill
Rumen/Pansen rumen, paunch, first stomach, ingluvies
Rumpf/Leib/Torso trunk, rump
Rumpfbein/Thoraxbein/ Thorakalfuß/Thoracopod (Malacostraca) thoracopod, thoracic leg

Rumpfinfloreszenz

Rumpfinfloreszenz
truncate synflorescence
Rumpfniere/Opisthonephros
opisthonephros
Rumpfsegment/Thoraxsegment/ Thoracomer thoracic segment, thoracomer
Rumpfskelett/Stammskelett/ Achsenskelett/Axialskelett
axial skeleton
Rumposom (Flagellaten)
rumposome
rundblättrig rotundifolious
Rundfraß (Wildverbiss) ringing
rundhöckrig/stumpfhöckrig/ bunodont (Zähne)
with low crowns and cusps, bunodont
Rundholz roundwood, log timber
Rundkolben/Siedegefäß
boiling flask with round bottom
rundlich/abgerundet roundish, rounded, rotund
Rundlochplatte dot blot, spot blot
Rundmäuler/Kreismünder/ Cyclostomata cyclostomes
Rundschuppe/Cycloidschuppe
cycloid scale
Rundschüttler circular shaker, rotary shaker
Rundtanz (Bienen) round dance
Rundwürmer (*sensu lato*)/ Schlauchwürmer/ Nemathelminthen/Aschelminthen
nemathelminths, aschelminths
Rundwürmer (*sensu stricto*)/ Fadenwürmer/Nematoden
roundworms, nematodes
Runterstufung/Herunterstufung
downward classification
runzelig/gerunzelt/gewellt/geriffelt
wrinkled, rugose; corrugative, corrugated
Runzelkorallen/Rugosa
rugose corals
rupfen (Federn) pluck
Ruppiaceae/Saldengewächse
ditch-grass family
Rüssel (Elefant) trunk
Rüssel/Proboscis *entom* proboscis
● **mit kurzem Rüssel** *entom*
brachystomatous,
with a short proboscis
Rüssel/Schnabel/Rostrum (Wanzen)
beak, rostrum
Rüssel/Schnauze (Schwein) snout
Rüsselbein/Os rostrale rostral bone
Rüsselscheibe/Planum rostrale (Schwein) rostral plate
Rüsselscheide/Rhynchocoel
proboscis receptacle, rhnychocoel
Rüsselscheidenretraktor
proboscis receptacle retractor
Rüsselspringer/Macroscelidea
African elephant shrews
Rüsseltiere/Proboscidea
elephants and relatives
Rutaceae/Rautengewächse
rue family
Rute/Gerte (langer/dünner Zweig)
rod, switch; whip
Rute/männliches Glied/Penis
penis; pizzle (esp. bull)
Rute/Schwanz tail
● **Beerenrute** cane
rutenförmig rod-shaped

S-Phase (Synthesephase im Zellzyklus) S phase (synthesis phase during cell cycle)
Saat/Saatgut/Aussaat (das Ausgesäte) seed(s)
Saat/Säen/Aussäen/Aussaat sowing, seed sowing, seeding
Saatband seed tape
Saatbeet seedbed
Saatgut seed stock, seeds
Saatgutbeizmittel dressing agent (pesticides/fungicides)
Saatkasten seed pan
Saatzeit seedtime
Säbelzahn sabre tooth, saber tooth
Saccharimeter saccharimeter
Saccharose/Sucrose (Rübenzucker/ Rohrzucker) sucrose (beet sugar/ cane sugar)
sackförmig/taschenförmig pouched, saccate
Sackschnecken/Schlauchschnecken/ Schlundsackschnecken/ Sacoglossa/Saccoglossa sacoglossans
säen/aussäen sow
Säen/Aussäen/Aussaat sowing, seed sowing
Safranmalvengewächse/ Turneraceae turnera family
Saft juice
• **Pflanzensaft** sap, xylem/phloem fluid
Saftfrucht fleshy fruit
Safthaar/Paraphyse paraphysis, paranema
saftig juicy
Saftmal/Honigmal nectar guide, honey guide
Saftwaage *hort* pruning a tree's branches to an equally horizontal level
Sägehaie/Pristiophoriformes sawsharks
Sägemehl sawdust
Sägenaht/Sutura serrata (Schädelnaht) serrate suture
Sägerochen/Sägefische/Pristiformes sawfishes
Sägewerk sawmill, timber mill
sagittal/in Pfeilrichtung/ in Pfeilebene sagittal, median longitudinal
Sagittalebene (parallel zur Mittellinie) median longitudinal plane
Sagittalkamm/Scheitelkamm sagittal crest
Sagittalschnitt sagittal section, median longisection
saisonal/jahreszeitlich seasonal
Saisonalität seasonality

Samen

Saisonwald seasonal forest
Saitenwürmer/Nematomorpha horsehair worms, hairworms, gordian worms, threadworms, nematomorphans, nematomorphs
Salamander salamanders
Saldengewächse/Ruppiaceae ditch-grass family
Salicaceae/Weidengewächse willow family
Salicylsäure (Salicylat) salicic acid (salicylate)
Salinität/Salzgehalt salinity, saltiness
Salmler/Characiformes characins: tetras & piranhas
Sälmling/Lächsling (junge Lachsbrut) parr (stage between fry and smolt)
• **Silbersälmling** smolt
Salmonidenregion *limn* salmonid zone
Salpen/Thaliaceen/Thaliacea salps, thaliceans
• **eigentliche Salpen/Salpida** salps
saltatorisch saltatory, saltatorial (adapted for/used in jumping)
saltatorische Erregungsleitung saltatory conduction
saltatorische Replikation saltatory replication
Salvadoraceae/Senfbaumgewächse mustard-tree family
Salve *neuro* burst
Salviniaceae/ Schwimmfarngewächse salvinia family
Salzbrücke (Ionenpaar) salt bridge (ion pair)
Salzdrüse salt gland
salzen salt
• **versalzen (Essen)** oversalt
Salzgehalt/Salzigkeit salinity, saltiness
salzig salty, saline
Salzigkeit saltiness
Salzmarsch saltmarsh
Salzpfanne *geol* saltpan, salina
Salzpflanze halophyte
Salzsee salt lake
Salzsteppe salt flat
Salzsumpf/Salzmarsch saltmarsh
Salzwasser saltwater
Salzwiese salt meadow
Same *bot* seed
Samen/Sperma (Ejakulat) sperm, semen (ejaculate)
Samen/Spermium/Samenzelle/ Spermatozoon (männliche Geschlechtszelle) sperm, spermium, sperm cell, spermatozoon (male gamete)

S

Samen...   658

**Samen.../Sperma.../
Samen betreffend/
Sperma betreffend** seminal
Samenanlage *bot* ovule
Samenausbreitung *bot*
dissemination, seed dispersal
**Samenausführgang/Samengang/
Ausspritzungsgang/
Ductus ejaculatorius**
ejaculatory duct
Samenbank seed repository
samenbildend seminiferous
**Samenbläschen/Samenblase/
Bläschendrüse/Glandula
vesiculosa** seminal vesicle
Samenblatt/Makrosporophyll
macrosporophyll
Samendrüse testicle, testis
**Samenerguss/Samenausstoß/
Ejakulation** seminal discharge,
ejaculation; (Ejakulat) seminal
discharge, ejaculate
Samenfarne seed ferns
Samenflüssigkeit *zool* seminal fluid
**Samengang/Samenausführgang/
Ductus ejaculatorius**
ejaculatory duct
Samengehäuse *bot* seed casing,
fruit
Samenhülle *bot* seed coat
(develops from integuments)
Samenkapsel *bot* seed case, capsule
Samenkeimung seed germination
Samenleiste/Plazenta *bot* placenta
**Samenleiter, primärer
(Wolffscher Gang)**
seminal duct, Wolffian duct
Samenmantel *bot* aril
Samennabel/Hilum *bot* hilum,
funiculus scar
Samennaht/Raphe *bot* raphe
Samenpflanze/Spermatophyt
seed-bearing plant, spermatophyt
Samenruhe (Dormanz/Quieszenz)
bot seed dormancy (dormancy/
quiescence)
Samenschale *bot* seed coat, testa
Samenschuppe/Fruchtschuppe *bot*
ovuliferous scale, seed scale
Samenstiel *bot* funicle, seed stalk
Samenstrang/Funiculus spermaticus
spermatic cord
Samentasche/Receptaculum seminis
sperm chamber, spermatheca,
seminal receptacle, sperm receptacle
**Samenträger/Samenpaket/
Spermatophore** spermatophore,
sperm packet
Samenverbreitung seed dispersal
Samenwarze *bot* aril
**Samenwarze/Karunkula (Auswuchs
an der Mikropyle)** *bot* caruncle

**Samenwarze/Strophiole (Auswuchs
der Raphe)** *bot* strophiole
Samenwulst/Raphe *bot* raphe
Samenzapfen *bot* seed cone,
female cone
**Samenzelle/Sperma (männliche
Geschlechtszelle)** sperm cell,
sperm (male gamete)
Sämerei/Samen *bot/hort* seeds
Sämerei/Samenproduzent *hort*
seed company
Sämling *bot* seedling
Sammelart/Großart/Coenospezies
coenospecies
**Sammelart/Kollektivart/
Kollektivspezies (agg.)** aggregated
species, collective species (agg.)
Sammelart/Überart/Superspezies
superspecies
Sammelbecken *geol* catchment area
Sammelbegriff/Sammelname
generic name
Sammelbiene/Sammlerin forager,
field bee
Sammelfrucht aggregate fruit
Sammelfruchtkörper/Aethalium
fung aethalium
Sammelgel *electrophor* stacking gel
Sammelglas *lab* specimen jar
Sammellinse *micros* collecting lens,
focusing lens
• **parallel-richtende Sammellinse**
collimating lens
sammeln
collect, put/come/bring together
**sammeln/versammeln/
zusammenscharen/zu Scharen
zusammenkommen** *orn* flock
**Sammelwirt/Stapelwirt/
paratenischer Wirt/Transportwirt**
paratenic host, transfer host
Sammler gatherer, collector
Sammlung/Kollektion collection
samtig velutinous, velvet-like,
velvety
Sand sand
• **im Sand lebend/
den Sand bewohnend**
arenicolous, sand-dwelling
• **Treibsand** quicksand
• **Wüstensand** desert sand
**sandartig/in sandigem Boden
lebend** arenaceous
Sandbank sandbank
• **längliche Sandbank/Sandbarre**
sandbar
Sandboden sandy soil
**Sanddollars/Schildseeigel/
Clypeasteroida** (true) sand dollars
Sanddüne sand dune
Sandelholzgewächse/Santalaceae
sandalwood family

659 **Säugerkunde** **S**

Sander *geol* outwash, outwash plain
Sandfang (Kläranlage) grit chamber
Sandfische/Milchfischverwandte/
Gonorhynchiformes
milkfishes and relatives
sandliebend/psammophil
psammophilous,
living in sandy habitats
Sandlückenfauna/Interstitialfauna
interstitial fauna
Sandverwehung sand drift
Sandwich-Hybridisierung *gen*
sandwich hybridization
Santalaceae/Sandelholzgewächse/
Leinblattgewächse
sandalwood family
Sapindaceae/Seifenbaumgewächse
soapberry family
Sapotaceae/Sapotegewächse/
Breiapfelgewächse
sapodilla family
Saprobie saprobity
Saprobien (Organismen) saprobes,
saprobionts
Saprobiensystem saprobity system
Saprobiont saprobiont
saprobiotisch/saprophag
saprobio(n)tic, saprophagous
saprogen/fäulniserregend
saprogenic
saprophag/saprotroph
saprophagous, saprotrophic
Saprophage/Fäulnisernährer/
Fäulnisfresser saprophage,
saprotroph, saprobiont
Saprophagie saprophagy
saprophil/saprob saprophilic,
saprobic
Saprophyt saprophyte
saprotroph/saprophag saprotrophic,
saprophagous
Saprozoen saprozoic lifeforms
Sarcolemm/Sarkolemm sarcolemma
Sarcosin sarcosine
Sarcosom/Riesenmitochondrion
sarcosome
sarkoplasmatisches Retikulum (SR)
sarcoplasmatic reticulum
Sarkotesta sarcotesta
sarkotubuläres System
sarcotubular system
Sarraceniaceae/
Schlauchpflanzengewächse/
Krugpflanzengewächse
pitcher-plant family
Satelliten-DNA satellite DNA
(sat-DNA)
Satellitenart/Randart
satellite species, marginal species
Satellitenchromosom
satellite chromosom
Satellitenvirus satellite virus

satt/gesättigt full,
having eaten enough, saturated
Sattelgelenk saddle joint,
sellaris joint
Sattelschäften *hort* saddle grafting
sättigen (gesättigt)
saturate (saturated)
Sättigung saturation
Sättigungsbereich/Sättigungszone
range of saturation,
zone of saturation
Sättigungshybridisierung
saturation hybridization
Sättigungskinetik
saturation kinetics
Sättigungsverlust/Sättigungsdefizit
saturation deficit
Satzkultur/diskontinuierliche
Kultur/Batch-Kultur batch culture
Satzverfahren batch process
Satzzeichencodon *gen*
punctuation codon
Sau (Mutterschwein)
sow (female swine)
sauer/azid acid, acidic
Sauerdorngewächse/
Berberitzengewächse/
Berberidaceae barberry family
Sauergräser/Riedgrasgewächse/
Riedgräser/Cyperaceae
sedge family
Sauergräser/Seggen/Riedgräser
sedges
Sauerkleegewächse/Oxalidaceae
wood-sorrel family
säuerlich acidic
Sauerstoff oxygen
Sauerstoffbedarf oxygen demand
● **biologischer S. (BSB)**
biological oxygen demand (BOD)
● **chemischer S. (CSB)**
chemical oxygen demand (COD)
sauerstoffbedürftig/aerob aerobic
Sauerstoffpartialdruck
oxygen partial pressure
Sauerstoffschuld/Sauerstoffverlust/
Sauerstoffdefizit oxygen debt
Sauerstofftransferrate
oxygen transfer rate (OTR)
Sauerstoffverlust/Sauerstoffschuld/
Sauerstoffdefizit oxygen debt
Säuerung acidification
saugen *zool* suck
saugen/aufsaugen absorb, take up,
soak up
säugen/stillen nurse, suckle,
breast-feed
Säugen/Stillen nursing, suckling,
breast-feeding
saugend sucking (insects: haustellate)
Säugerkunde/Säugetierkunde/
Mammalogie mammalogy

S säugetierähnliche Reptilien 660

säugetierähnliche Reptilien/
Therapsida mammallike reptiles
(advanced synapsids)
Säugetiere/Säuger/Mammalia
mammals
Saugfalle/Schluckfalle suction trap,
suctory trap
Saugfiltration suction filtration
Saugfischverwandte/Schildfische/
Spinnenfischartige/
Gobiescociformes clingfishes
Saugflasche/Filtrierflasche *lab*
filter flask, vacuum flask
Sauggrube/Bothrium bothrium
Saugkraft/Wasserpotential
water potential
Säugling infant
Säuglingsalter/frühe Kindheit
infancy
Säuglingssterblichkeit
infant mortality
Sauglunge suction lung
Saugmagen (Cheliceraten)
sucking stomach
Saugmagen (Vorratsmagen: Kropf
der Culiciden) pumping stomach
Saugnapf/Acetabulum
(true) sucker, acetabulum
Saugnapf/Saugscheibe suction disk
Saugorgan suctorial organ, sucker
Saugorgan/Haustorium sucker,
haustorium
Saugorgan/Schildchen/Scutellum
(Keimblatt des Graskeimlings) *bot*
scutellum
Saugpipette suction pipette
(patch pipette)
Saugpumpe (Hymenoptera:
Pharynx) sucking pump (pharynx)
Saugreflex suction reflex
Saugrüssel/Proboscis
sucker, haustellum, proboscis
(adapted for sucking)
Saugschuppe/Schuppenhaar
(Bromelien) absorbing trichome,
squamiform hair, peltate trichome
Saugspannung soil-moisture
tension; suction, suction force
Saugstellung (Säuger)
nursing position
Saugwürmer/Egel/Trematoden
flukes, trematodes
Saugwurzel suction root, seeker
Säule pillar, column; (des
Mikroskops) pillar
Säule/Säulchen/Griffelsäule/
Gynostemium (Orchideen)
column, gynostemium
Säulenblumengewächse/
Stylidiaceae trigger plant family
Säulenchromatographie
column chromatography

Säulenepithel/Zylinderepithel
squamous epithelium
Säulenreaktor/Turmreaktor
column reactor
säulenspaltig/columnicid/
columnizid *bot* columnicidal
Saum/Rand seam, border, edge,
fringe
Saumgesellschaft *biogeo* fringe
community, gallery community
Saumriff fringing reef
Saumschlag (Waldschlag) *for*
aisle clearing, strip felling
saure Niederschläge
acid precipitation
Säure acid
● **Abietinsäure** abietic acid
● **Acetessigsäure (Acetacetat)/**
γ-Ketobuttersäure
acetoacetic acid (acetoacetate),
β-ketobutyric acid
● **Aconitsäure (Aconitat)**
aconitic acid (aconitate)
● **Adenylsäure (Adenylat)**
adenylic acid (adenylate)
● **Adipinsäure (Adipat)** adipic acid
(adipate)
● **"aktivierte Essigsäure"/**
Acetyl-CoA
acetyl CoA, acetyl coenzyme A
● **Alginsäure (Alginat)** alginic acid
(alginate)
● **Allantoinsäure** allantoic acid
● **Ameisensäure (Format)**
formic acid (formate)
● **γ-Aminobuttersäure**
gamma-aminobutyric acid
● **Aminosäure** amino acid
● **Anthranilsäure** anthranilic acid,
2-aminobenzoic acid
● **Äpfelsäure (Malat)** malic acid
(malate)
● **Arachidonsäure** arachidonic acid,
icosatetraenoic acid
● **Arachinsäure/Arachidinsäure/**
Eicosansäure arachic acid,
arachidic acid, icosanic acid
● **Ascorbinsäure (Ascorbat)**
ascorbic acid (ascorbate)
● **Asparaginsäure (Aspartat)**
asparagic acid, aspartic acid
(aspartate)
● **Azelainsäure/Nonandisäure**
azelaic acid, nonanedioic acid
● **Behensäure/Docosansäure**
behenic acid, docosanoic acid
● **Benzoesäure (Benzoat)**
benzoic acid (benzoate)
● **Bernsteinsäure (Succinat)**
succinic acid (succinate)
● **Brenztraubensäure (Pyruvat)**
pyruvic acid (pyruvate)

Säure S

- **Buttersäure/Butansäure (Butyrat)** butyric acid, butanoic acid (butyrate)
- **Caprinsäure/Decansäure (Caprinat/Decanat)** capric acid, decanoic acid (caprate/decanoate)
- **Capronsäure/Hexansäure (Capronat/Hexanat)** caproic acid, capronic acid, hexanoic acid (caproate/hexanoate)
- **Caprylsäure/Octansäure (Caprylat/Octanat)** caprylic acid, octanoic acid (caprylate/octanoate)
- **Carbonsäuren/Karbonsäuren (Carbonate/Karbonate)** carboxylic acids (carbonates)
- **Cerotinsäure/Hexacosansäure** cerotic acid, hexacosanoic acid
- **Chinasäure** chinic acid, kinic acid, quinic acid (quinate)
- **Chinolsäure** chinolic acid
- **Chlorogensäure** chlorogenic acid
- **Cholsäure (Cholat)** cholic acid (cholate)
- **Chorisminsäure (Chorismat)** chorismic acid (chorismate)
- **Cinnamonsäure/Zimtsäure (Cinnamat)** cinnamic acid
- **Citronensäure/Zitronensäure (Citrat/Zitrat)** citric acid (citrate)
- **Crotonsäure/Transbutensäure** crotonic acid, α-butenic acid
- **Cysteinsäure** cysteic acid
- **einwertige/einprotonige Säure** monoprotic acid
- **Eisessig** glacial acetic acid
- **Ellagsäure** ellagic acid, gallogen
- **Erucasäure/Δ¹³-Docosensäure** erucic acid, (Z)-13-docosenoic acid
- **Essigsäure/Ethansäure (Acetat)** acetic acid, ethanoic acid (acetate)
- **Ferulasäure** ferulic acid
- **Fettsäure** (*siehe auch dort*) fatty acid
- **Flechtensäure** lichen acid
- **Folsäure (Folat)/ Pteroylglutaminsäure** folic acid (folate), pteroylglutamic acid
- **Fumarsäure (Fumarat)** fumaric acid (fumarate)
- **Galakturonsäure** galacturonic acid
- **Gallussäure (Gallat)** gallic acid (gallate)
- **Gentisinsäure** gentisic acid
- **Geraniumsäure** geranic acid
- **Gerbsäure (Tannat)** tannic acid (tannate)
- **Gibberellinsäure** gibberellic acid
- **Glucarsäure/Zuckersäure** glucaric acid, saccharic acid

- **Gluconsäure (Gluconat)** gluconic acid (gluconate)
- **Glucuronsäure (Glukuronat)** glucuronic acid (glucuronate)
- **Glutaminsäure (Glutamat)/ 2-Aminoglutarsäure** glutamic acid (glutamate), 2-aminoglutaric acid
- **Glutarsäure (Glutarat)** glutaric acid (glutarate)
- **Glycyrrhetinsäure** glycyrrhetinic acid
- **Glykolsäure (Glykolat)** glycolic acid (glycolate)
- **Glyoxalsäure (Glyoxalat)** glyoxalic acid (glyoxalate)
- **Glyoxylsäure (Glyoxylat)** glyoxylic acid (glyoxylate)
- **Guanylsäure (Guanylat)** guanylic acid (guanylate)
- **Gulonsäure (Gulonat)** gulonic acid (gulonate)
- **Harnsäure (Urat)** uric acid (urate)
- **Homogentisinsäure** homogentisic acid
- **Huminsäure** humic acid
- **Hyaluronsäure** hyaluronic acid
- **Ibotensäure** ibotenic acid
- **Iminosäure** imino acid
- **Indolessigsäure** indolyl acetic acid, indoleacetic acid (IAA)
- **Isovaleriansäure** isovaleric acid
- **Jasmonsäure** jasmonic acid
- **Kaffeesäure** caffeic acid
- **Ketosäure** keto acid
- **Kohlensäure (Karbonat/ Carbonat)** carbonic acid (carbonate)
- **Kojisäure** kojic acid
- **Laktat (Milchsäure)** lactate (lactic acid)
- **Laurinsäure/Dodecansäure (Laurat/Dodecanat)** lauric acid, decylacetic acid, dodecanoic acid (laurate/dodecanate)
- **Lävulinsäure** levulinic acid
- **Lignocerinsäure/Tetracosansäure** lignoceric acid, tetracosanoic acid
- **Linolensäure** linolenic acid
- **Linolsäure** linolic acid, linoleic acid
- **Liponsäure/Thioctsäure (Liponat)** lipoic acid (lipoate), thioctic acid
- **Lipoteichonsäure** lipoteichoic acid
- **Litocholsäure** litocholic acid
- **Lysergsäure** lysergic acid
- **Magensäure** stomach acid, gastric acid
- **Maleinsäure (Maleat)** maleic acid (maleate)

S Säure 662

- **Malonsäure (Malonat)**
  malonic acid (malonate)
- **Mandelsäure/Phenylglykolsäure**
  mandelic acid, phenylglycolic
  acid, amygdalic acid
- **Mannuronsäure** mannuronic acid
- **Mevalonsäure (Mevalonat)**
  mevalonic acid (mevalonate)
- **Milchsäure (Laktat)** lactic acid
  (lactate)
- **Muraminsäure** muramic acid
- **Myristinsäure/Tetradecansäure
  (Myristat)** myristic acid,
  tetradecanoic acid (myristate/
  tetradecanate)
- **N-Acetylmuraminsäure**
  N-acetylmuramic acid
- **Nervonsäure/Δ^{15}-Tetracosensäure**
  nervonic acid, (Z)-15-tetracosenoic
  acid, selacholeic acid
- **Neuraminsäure** neuraminic acid
- **Nikotinsäure (Nikotinat)**
  nicotinic acid (nicotinate), niacin
- **Ölsäure/Δ^{9}-Octadecensäure
  (Oleat)** oleic acid,
  (Z)-9-octadecenoic acid (oleate)
- **Orotsäure** orotic acid
- **Orsellinsäure** orsellic acid,
  orsellinic acid
- **Osmiumsäure** osmic acid
- **Oxalbernsteinsäure
  (Oxalsuccinat)** oxalosuccinic acid
  (oxalosuccinate)
- **Oxalsäure (Oxalat)** oxalic acid
  (oxalate)
- **Oxoglutarsäure (Oxoglutarat)**
  oxoglutaric acid (oxoglutarate)
- **Palmitinsäure/Hexadecansäure
  (Palmat/Hexadecanat)**
  palmitic acid, hexadecanoic acid
  (palmate/hexadecanate)
- **Palmitoleinsäure/
  Δ^{9}-Hexadecensäure** palmitoleic
  acid, (Z)-9-hexadecenoic acid
- **Pantoinsäure** pantoic acid
- **Pantothensäure (Pantothenat)**
  pantothenic acid (pantothenate)
- **Pektinsäure (Pektat)** pectic acid
  (pectate)
- **Penicillansäure** penicillanic acid
- **Perameisensäure** performic acid
- **Phosphatidsäure**
  phosphatidic acid
- **Phosphorsäure (Phosphat)**
  phosphoric acid (phosphate)
- **Phthalsäure** phthalic acid
- **Phytansäure** phytanic acid
- **Phytinsäure** phytic acid
- **Pikrinsäure (Pikrat)**
  picric acid (picrate)
- **Pimelinsäure** pimelic acid
- **Plasmensäure** plasmenic acid

- **Prephensäure (Prephenat)**
  prephenic acid (prephenate)
- **Propionsäure (Propionat)**
  propionic acid (propionate)
- **Prostansäure** prostanoic acid
- **Pyrethrinsäure** pyrethric acid
- **Retinsäure** retinic acid
- **Salicylsäure (Salicylat)**
  salicic acid (salicylate)
- **Schleimsäure/Mucinsäure**
  mucic acid
- **Shikimisäure (Shikimat)**
  shikimic acid (shikimate)
- **Sialinsäure (Sialat)** sialic acid
  (sialate)
- **Sinapinsäure** sinapic acid
- **Sorbinsäure (Sorbat)** sorbic acid
  (sorbate)
- **Stearinsäure/Octadecansäure
  (Stearat/Octadecanat)**
  stearic acid, octadecanoic acid
  (stearate/octadecanate)
- **Suberinsäure/Korksäure/
  Octandisäure** suberic acid,
  octanedioic acid
- **Teichonsäure** teichoic acid
- **Teichuronsäure** teichuronic acid
- **Uridylsäure** uridylic acid
- **Urocaninsäure (Urocaninat)/
  Imidazol-4-acrylsäure**
  urocanic acid (urocaninate)
- **Uronsäure (Urat)** uronic acid
  (urate)
- **Usninsäure** usnic acid
- **Valeriansäure/Pentansäure
  (Valeriat/Pentanat)** valeric acid,
  pentanoic acid (valeriate/
  pentanoate)
- **Vanillinsäure** vanillic acid
- **Weinsäure (Tartrat)** tartaric acid
  (tartrate)
- **Zimtsäure/Cinnamonsäure
  (Cinnamat)** cinnamic acid
- **Zitronensäure/Citronensäure
  (Zitrat/Citrat)** citric acid (citrate)
- **Zuckersäure/Aldarsäure
  (Glucarsäure)** saccharic acid,
  aldaric acid (glucaric acid)
- **zweiwertige/zweiprotonige
  Säure** diprotic acid

Säure-Basen-Gleichgewicht
acid-base balance
Säureamid acid amide
säurebildend/säurehaltig acidic
Säurebildung acidification
Säureester acid ester
säurefest acid-fast
Säurefestigkeit acid-fastness
Säuregrad/Säuregehalt/Azidität
acidity
- **ansäuern** acidify
saurer Boden acid soil, acidic soil

663   **Schalenlose** **S**

saurer Regen acid rain
Saururaceae/
Molchschwanzgewächse
lizard's tail family
Savanne savanna
- **Baumsavanne** tree savanna
- **Strauchsavanne** shrub savanna
**Savisches Bläschen (*Torpedo:* an
elektr. Organ)** *ichth* Savi vesicle
Saxifragaceae/Steinbrechgewächse
saxifrage family
Scaphognathit/Atemplatte
scaphognathite, bailer, baler
Scatchard-Diagramm
Scatchard plot
schaben scrape
Schaben/Blattodea cockroaches
Schabepräparat *micros*
scraping (mount)
Schachtelhalm horsetail,
scouring rush
Schachtelhalmgewächse/
Equisetaceae horsetail family
schächten kill/slaughter according to
Jewish rites
Schädel/Hirnkapsel/Cranium skull,
braincase, cranium
- **Bindegewebsschädel/**
Desmokranium *embr*
desmocranium
(precursor of chondrocranium)
- **Gesichtsschädel/Viscerokranium**
facial skeleton, visceral cranium,
viscerocranium
- **Hautknochenschädel/**
Dermatokranium
dermatocranium
- **Hirnschädel/Neurokranium/**
Neuralkranium cerebral cranium,
neurocranium
- **Kiemenschädel/Kiemenskelett/**
Branchiokranium branchial
cranium, branchiocranium
- **Knochenschädel/Osteokranium**
osteocranium
- **Knorpelschädel/Chondrokranium**
cartilaginous cranium,
chondrocranium
Schädelbasis skull base,
base of skull, cranial base,
(interne S.) cranial floor
Schädeldecke/Schädeldach/
Clavarium
skull roof, cranial roof, skullcap
Schädelgrube cranial fossa
Schädellose/Acrania acranians
Schädelnaht/Sutura cranial suture
Schaden damage
Schädigung damage
Schädigungskurve *ecol*
damage response curve
Schadinsekt pest insect

schädlich harmful, causing damage
- **unschädlich**
harmless, not harmful; inactive
Schädling(e)/Ungeziefer pest(s)
Schädlingsbefall pest infestation
Schädlingsbekämpfung/
Schädlingskontrolle pest control
- **biologische**
Schädlingsbekämpfung
biological pest control
- **integrierte**
Schädlingsbekämpfung/
integrierter Pflanzenschutz
integrated pest management (IPM)
Schädlingsbekämpfungsmittel/
Pestizid/Biozid pesticide, biocide
Schädlingsbekämpfungsmittel-
resistenz/Pestizidresistenz
pesticide resistance
Schadorganismus harmful organism,
harmful lifeform
Schaf sheep
Schafbock/Widder/Rammler ram
Schäfer shepherd
Schafhaut/Amnion amnion
Schaft shaft, leafless stem, leafless
shoot, rachis, trunk; (dünner) cane
- **Blütenschaft**
peduncle, flower stalk
Schäften/Kopulation/Kopulieren
(Pfropfung)
splice grafting, whip grafting
Schale *allg*
shell, testa; husk, coat, cover
- **Diatomeenschale** frustule
- **harte Schale/Testa (z. B.**
Mollusken) shell, test, testa
- **Haut** skin, peel
- **Panzer/Carapax** shell, carapace
- **Schneckenschale/Muschelschale**
shell
- **Schutzschicht/Hülle** husk, coat,
cover
schälen/entrinden decorticate,
debark, bark, ross
Schalendrüse/Maxillendrüse/
Maxillennephridium
maxillary gland
Schalendrüse/Mehlissche Drüse
shell gland, Mehlis' gland,
cement gland
Schalendrüse/Nidamentaldrüse
shell gland, nidamental gland
Schalenhälfte (Diatomeen/
Muscheln) valve
Schalenhaut (Ei) shell membrane
Schalenhäutchen/Periostracum
periostracum
Schalenhäuter/Ostracodermata
ostracoderms
Schalenlose/Kiemenfüße/Anostraca
fairy shrimps, anostracans

Schalenplatte (Käferschnecken)
shell plate, shell valve
Schalensschließmuskel (Muscheln)
adductor muscle
Schalensegment/Schalenplatte (Chiton) valve
Schalentier (Crustaceen & Mollusken) shellfish
Schalenweichtiere/Conchifera conchiferans
Schalenwild/Schaltier hoofed game
Schalenzone (Muschelschalen)/ Litoriprofundal *limn* littoriprofundal zone
Schälhengst/Beschäler/Zuchthengst stud, studhorse
Schall/Geräusch sound
Schall/Widerhall resonance, echo, reverberation
Schallblase (Frösche) resonance pouch, vocal sac, vocal pouch
Schallwellen sound waves
Schaltergen switch gene
Schalterregion/Switchregion *gen* switch region
Schaltknochen/Nahtknochen sutural bone, epactal bone, wormian bone
Schaltkreis/Schaltsystem *neuro* circuit (neural circuit)
- **divergenter S.** divergent circuit
- **konvergenter S.** convergent circuit
- **zurückwirkender S.** reverberating circuit

Scham/Schamgegend/ Schambeinregion pubic region, pubic zone
Schambein/Os pubis pubic bone
Schambeinfuge/Schamfuge/ "Symphyse"/Symphysis pubica/ Symphysis pelvina pubic symphysis, pelvic symphysis
Schambogen pubic arch
Schamhügel/Venushügel/ Mons pubis pubic prominence
Schamlippe/Labium vulvae labium (folds at margin of vulva)
Schampräsentieren *ethol* vulva presentation
Schamspalte/Rima pudendis urogenital cleft, pudendal fissure
Schamweisen *ethol* pubic presentation
Schanker/Lues/Syphilis (*Treponema pallidum*) syphilis
Schar (Vogelschar) flock, flight (birds)
scharf *micro/photo* in focus, sharp
- **unscharf** *micro/photo* not in focus, out of focus, blurred

Schärfe *micro/photo* sharpness, focus
- **Sehschärfe** visual acuity
- **Unschärfe** *micro/photo* blurredness, blur, obscurity, unsharpness

Scharfeinstellung focussing
Schärfentiefe/Tiefenschärfe depth of focus, depth of field
Scharfstellung/Akkommodation (*opt:* **Auge**) accommodation
Scharnier/Schloss/Schlossleiste (Muscheln) hinge
Scharniergelenk hinge joint, ginglymus joint
scharren (Hühner) scratch
scharren (Pferde) pawing, paw the ground (scraping the ground)
scharrend (Geflügel) rasorial
Schatten *allg* shade; (eines bestimmten Gegenstandes) shadow
Schatten/Erythrozytenschatten (ausgelaugtes rotes Blutkörperchen/leere Erythrozytenmembran) erythrocyte ghost
Schattenblatt shade leaf, sciophyll
schattenliebend shade-loving, sciophilous, umbraticolous
Schattenpflanze shade-loving plant, shade plant, sciophyte, sciaphyte, skiophyte, skiaphyte
schattieren shade
schattig shady
schätzen/annehmen estimate, assume
Schätzfehler *stat* error of estimation
Schätzung/Annahme estimate, estimation, assumption
Schätzverfahren *stat* method of estimation
Schätzwert estimate
Schau show
Schaufel (Geweih) palm, palmated antler
Schaufelknorpel/Cartilago xiphoidea xiphoid cartilage
Schaufler (mit schaufelförmigem Geweih) stag/elk with palmated antlers
Schaukasten/Vitrine showcase
Schaukelvektor/ bifunktionaler Vektor shuttle vector, bifunctional vector
Schaum foam
Schaum/Speichel (z. B. von Zikaden: Kuckucksspeichel) spittle ("cuckoo spit")
Schaumhemmer *chem/lab* anti-foaming agent
Schaumzikaden (Auchenorrhyncha) spittlebugs

Schaustellung (protzig) display
Schecke (Pferd) skewbald horse
scheckig (Pferd) skewbald
Scheckung/Variegation variegation
Scheibenblüte/Röhrenblüte disk
flower, disk floret, tubular flower
scheibenförmig disk-shaped
**Scheibenquallen/Schirmquallen/
Scyphozoen (echte Quallen)**
cup animals, scyphozoans
Scheide/Umhüllung sheath
• **Blattscheide** sheath
• **Blütenscheide/Spatha** spathe
Scheide/Vagina vagina
**scheidenförmig/
blütenscheidenförmig** *bot*
spathaceous, spathal
scheidenförmig/röhrenförmig
sheathed, vaginate
**Scheidenvorhof/Vestibulum
vaginae** vaginal vestibule
Scheidetrichter *lab* separatory funnel
Scheidewand/Septe/Septum
dividing wall, cross-wall, partition,
dissepiment, septum
**scheidewandbrüchig/wandbrüchig/
septifrag** *bot* septifragal
scheidewandspaltig/septizid *bot*
septicidal
Scheinachse/Sympodium
sympodium
Scheinader/Vena spuria
spurious vein
Scheinangriff *ethol* sham attack
**Scheinblattkieme/
Pseudolamellibranchie**
pseudolamellar gill,
pseudolamellibranch
Scheinblüte/Pseudanthium
pseudanthium
**Scheindolde/Pseudosciadioid
(Infloreszenz)** contracted cymoid,
cymose umbel, pseudosciadioid
**Scheindolde/Trugdolde/Cyme/
Zymus (Infloreszenz)** cyme
Scheinellergewächse/Clethraceae
clethra family, white-alder family
Scheinfrucht
pseudocarp, pseudofruit,
false fruit, spurious fruit
Scheinfüßchen/Pseudopodium
pseudopod
Scheinfüttern/Scheinfütterung
ethol sham feeding
Scheingewebe/Pseudoparenchym
false tissue, paraplectenchyma,
pseudoparenchyma
Scheinhanfgewächse/Datiscaceae
datisca family, durango root family
Scheinpalmen/Cyclanthaceae
panama-hat family, jipijapa family,
cyclanthus family

Scheinpicken *ethol* sham pecking
**Scheinpuppe/Larva coarctata/
Pseudocrysalis** coarctate larva,
pseudocrysalis
Scheinputzen pseudogrooming,
sham grooming
**Scheinquirl/Scheinwirtel/
Doppelwickel (Infloreszenz)** false
whorl, pseudowhorl, verticillaster
scheinschwanger/pseudopregnant
pseudopregnant
Scheinstamm/Blattstamm (*Musa*)
false stem, pseudostem, leafy stem
Scheinwut/unechte Wut *ethol*
sham rage
Scheitel/Vertex crest, vertex (crown/
top of the head)
Scheitelauge/Stirnauge (Ocelle)
dorsal ocellus
Scheitelbein/Os parietale
parietal bone
Scheitelbeuge/Mittelhirnbeuge
embr cephalic flexure,
cranial flexure
Scheitelfurche/Scheitelgrube
apical furrow
Scheitelhöcker/Tuber parietale
parietal tuber, parietal tuberosity
Scheitelhöhe top height, total
height, crown height
**Scheitelkamm/Sagittalkamm/Crista
sagittalis** sagittal crest
Scheitellappen/Lobus parietalis
parietal lobe
Scheitelmeristem apical meristem
**Scheitelorgan/Apikalorgan/
Statocyste (Ctenophoren)**
apical scnsc organ, statocyst
Scheitelpunkt apex, peak (highest
among other high points), vertex,
summit
Scheitelwert/Höchstwert/Maximum
peak value, maximum (value)
Scheitelzelle apical cell
• **dreischneidige S.**
apical cell with three cutting faces
• **einschneidige S.**
apical cell with one cutting face
• **zweischneidige S.** apical cell
with two cutting faces
Schelf shelf
• **Kontinentalsockel/
Festlandsockel/Kontinentalschelf**
continental shelf
Schelfrand/Schelfkante
shelf braek, shelf edge, shelf margin,
continental margin (edge of shelf)
Schenkel/Femur femur
• **Oberschenkel** thigh
• **Unterschenkel** shank
Schenkel *chem/biochem/immun*
arm

Schenkeldrüse

Schenkeldrüse (Follikulärorgan: Eidechsen) femoral gland (follicular gland)
Schenkelring/Trochanter trochanter
Schenkelwanderung (DNA) *gen* branch migration
Schere *lab* scissors
- **chirurgische Schere** surgical scissors
- **Irisschere/Listerschere** iris scissors
- **Präparierschere** dissecting scissors
- **spitze Schere** sharp point scissors
- **stumpfe Schere** blunt point scissors

Schere/Klaue/Zange/Chela (Crustaceen) crab pincers, forceps, chela
scheren shearing
Scheren/Stutzen/Beschneiden clipping
Scheren.../mit Scheren versehen/ scherentragend cheliferous
scherenartig/zangenartig chelate, cheliform, pincerlike, clawlike
Scherenasseln/Anisopoda/ Tanaidacea tanaidaceans, tanaids
Scherenfinger/Chelicerenklaue (Unguis) cheliceral fang
scherenförmig forficulate, forficiform
Scherenfuß cheliped
Scherfestigkeit/Schubfestigkeit (Holz) shear strength, shearing strength
Schergefälle/Schergradient shear gradient
Scherkraft shear force; shear stress (shear force per unit area)
Scherrate shear rate, rate of shear
Scherspannung shear stress (shear force per unit area)
scheu shy
Scheu shyness
Scheuchzeriaceae/ Blumenbinsengewächse/ Blasenbinsengewächse arrow-grass family, scheuchzeria family
scheuen *vb* (Pferd) shy, take fright, skit
Scheuen *n* (Pferd) shying
Schicht layer, story, stratum, sheet
Schichtenbildung stratification (act/process of stratifying)
Schichtung stratification (state of being stratified), layering
Schicksalskarte/ Determinationskarte *gen* fate map
Schicksalskartierung *gen* fate mapping

schief oblique
Schiefblattgewächse/Begoniaceae begonia family
Schienbein/Schiene/Tibia shinbone, tibia
Schiene/Tibia (Insekten) tibia
Schienenbürste/Scopa (Bienen) scopa
schießen (früh in Blüte) bolting, shooting
Schiffchen/Kiel (Fabaceen-Blüte) keel
Schild/Carapax shield, shell, carapace
Schild/Scutum shield, scutum, scute
Schild-Okulation (Augenschild/ Schildchen) *hort* shield budding
schildartig/schuppenartig shield-like, peltate, scale-like, scutate
Schildblatt/peltates Blatt peltate leaf
Schildchen/Scutellum *bot* (Saugorgan an Graskeimling) scutellum (a shield-shaped structure)
Schildchen/Scutellum/ Mesoscutellum (Wanzen) scutellum, mesoscutellum
schildchenartig scutellate, like a small shield
schildchenförmig scutelliform, shaped like a small shield
Schilddrüse/Thyreoidea thyroid gland
Schildfarngewächse/Aspidiaceae sword fern family, aspidium family
Schildfische/Saugfischverwandte/ Spinnenfischartige/ Gobiescociformes clingfishes
schildförmig/schuppenförmig shield-shaped, peltiform, peltate, scutiform
Schildfüßer/Caudofoveata caudofoveates
Schildkiemer/Altschnecken/ Archaeogastropoda/Diotocardia limpets and allies, archeogastropods
Schildknorpel/Cartilago thyreoidea thyroid cartilage
Schildkröten/Chelonia/Testudines turtles
- **Landschildkröten** tortoises

Schildläuse/Coccinea scale insects
Schildpatt tortoise/turtle shell
Schildseeigel/Sanddollars/ Clypeasteroida (true) sand dollars
Schilf/Schilfrohr/Schilfgras/ Schilfröhricht reed
schillernd shimmering; iridescent
Schimmel/Moder mould, mildew
Schimmel (Pferd) gray horse
- **Apfelschimmel** dapple gray horse
- **Fliegenschimmel** flea-bitten gray

Schlauchblatt

- **Rotschimmel**
  (red/strawberry) roan
- **weißer Schimmel** white horse
Schimmelpilz mould
Schindel shingle
schindelig shingled, imbricate,
  overlapping
Schirm (Meduse) umbrella (float)
Schirmbestand/Mutterbestand *for*
  shelterwood
**Schirmoberseite/Exumbrella
(Meduse)** exumbrella
**Schirmquallen/Scheibenquallen/
Scyphozoen (echte Quallen)**
  cup animals, scyphozoans
**Schirmrispe (ein Ebenstrauß/
Corymbus) (Infloreszenz)**
  umbel-like panicle
**Schirmschlag/Schirmhieb
(>Waldschlag)** *for* shelterwood
  method, selective logging/cutting
  (even-aged stand, even-aged forest/
  plantation)
Schirmstand *for* shelterwood
**Schirmtraube (ein Ebenstrauß/
Corymbus) (Infloreszenz)**
  umbel-like raceme
**Schirmunterseite/Subumbrella
(Meduse)** subumbrella
Schizaeaceae/Spaltfarngewächse
  curly-grass family,
  climbing fern family
Schizocöl schizocoel, schizocoele,
  "split coelom"
schizogen schizogenic
Schizogenie schizogeny
Schizogonie schizogony, agamogony,
  merogony
Schizokarp/Spaltfrucht schizocarp,
  schizocarpium
Schizophyllaceae
  schizophyllum family
schlachten slaughter, butcher
Schlachter/Fleischer/Metzger
  butcher
Schlachthof slaughterhouse
Schlachtung/Schlachten slaughter,
  slaughtering, butchering
Schlachtvieh slaughter cattle, beef
  cattle, slaughter animal
Schlaf sleep
Schlafbewegung/Nyctinastie
  sleep movement, nyctinasty
Schläfe temple
schlafen sleep
Schläfenbein/Os temporale
  temporal bone
Schläfendrüse (Elefant)
  temporal gland
Schläfenfenster temporal fenestra
Schläfenlappen/Lobus temporalis
  temporal lobe

schlaff (welk) limp
**Schlafkrankheit/Tsetseseuche
(*Trypanosoma rhodesiense/
gambiense*)** sleep sickness
Schlafstätte/Ruheplatz
  resting place, resting site
Schlafstellung sleeping posture
Schlag (Gesang der Nachtigall)
  caroling, elaborate song with many
  different verses
Schlag (Tierrasse) breed, stock
Schlag/Waldschlag clearing
Schlagader/Arterie artery
- **Halsschlagader/Carotis**
  carotid artery
- **Körperschlagader, große/Aorta**
  aorta
Schlaganfall *med* stroke
schlagen/hauen beat, hit, strike
schlagen (Nachtigall) *orn* sing,
  warble, jug
Schlagfallennetz (Hyptiotes) *arach*
  sprung web
Schlagflug *orn* flapping flight
Schlagvolumen *cardio*
  stroke volume
Schlamm mud
- **Flussschlamm** silt, warp
Schlammfliegen/Megaloptera
  megalopterans: dobsonflies/fishflies/
  alderflies (neuropterans)
**Schlängelbewegung/Schlängeln/
seitliche bzw. horizontale
Wellenbewegung (Schlangen)**
  lateral undulation,
  lateral undulatory movement
**schlängeln/hin und her zucken
(Nematoden)** wriggle
- **sich schlängeln** wind
**Schlangen/Serpentes/Ophidia
(Squamata)** snakes, serpents,
  ophidians
schlangenartig/Schlangen...
  snake-like, ophidian
Schlangenkühler coil condenser,
  coiled-tube condenser, spiral
  condenser
Schlangenkunde/Ophiologie
  ophiology
Schlangensterne/Ophiuroiden
  brittle stars/serpent stars;
  basket stars (*Gorgonocephalus,
  Astrophyton*)
schlankstämmig *bot*
  slender-stemmed, leptocaulous
Schlauch tube, siphon, ascidium;
  lab tube, tubing
**schlauchartig/röhrenartig/siphonal/
tubulär** siphoneous, siphonaceous,
  tubular
Schlauchblatt (*Genlisea*)
  lobster pot

S   Schlauchblatt

Schlauchblatt/Kannenblatt (Nepenthes)
siphonaceous leaf, ascidiform leaf, ascidium, pitcher-leaf

Schlauchfrucht/Utriculus utricle, utriculus

Schlauchklemme/Quetschhahn *lab* tubing clamp, pinch clamp

Schlauchpflanzengewächse/ Sarraceniaceae pitcher-plant family

Schlauchpilze/Ascomyceten/ Ascomycetes sac fungi, cup fungi, ascomycetes, "spore shooter"

Schlauchschnecken/Sackschnecken/ Schlundsackschnecken/ Sacoglossa/Saccoglossa
sacoglossans

Schlauchwürmer/Rundwürmer/ Aschelminthen/Nemathelminthen (Pseudocölomaten) aschelminths, nemathelminths, pseudocoelomates

Schlauchzelle tube cell

Schlaufe *gen/biochem* loop

Schlaufenradreaktor
paddle wheel reactor

Schlaufenreaktor/Umlaufreaktor
loop reactor, circulating reactor, recycle reactor

schlecken lick

Schlegel/Keule/Bein (Geflügel)
drumstick, leg

Schleier/Cortina (Rest des Velum partiale/universale am Hutrand) *fung* cortina

Schleier/Hülle/Pilzhülle/Velum *fung* veil, velum

Schleierchen/Indusium indusium

Schleierfarngewächse/ Hautfarngewächse/ Hymenophyllaceae
filmy fern family

Schleierlinge/Haarschleierpilze/ Cortinariaceae
cortinarius family, cortinarias

Schleierzone *fung* cortinal zone

Schleifenkonformation/ Knäuelkonformation loop conformation, coil conformation

Schleim mucus, slime, ooze; mucilage (speziell pflanzlich)

Schleimaale/Inger/Ingerartige/ Myxiniformes (bzw. Myxinida)
hagfishes

Schleimbeutel *zool* synovial sac

Schleimdrüse *bot* mucilage gland; *zool* slime gland, mucous gland

schleimführender Kanal
mucilaginous canal

Schleimhaut/Schleimhautepithel
mucous membrane, mucosa

schleimig slimy, mucilaginous, glutinous

Schleimköpfe/Schleimkopfartige/ Beryciformes squirrel fishes (primitive acanthopterygians)

Schleimkörper/Schleimpfropfen/ Proteinkörper slime body, slime plug, P-protein body

Schleimpilze/Myxomycota
slime molds

- **echte S./Myxomycetes (plasmodial)** acellular slime molds, plasmodial slime molds
- **haploide S./Urschleimpilze/ Protosteliomycetes** protostelids
- **zelluläre S./Acrasiomyceten/ Acrasiomyycetes/ Dictyosteliomycetes (Myxomycota)**
cellular slime molds

Schleimsäure/Mucinsäure
mucic acid

Schleimzelle mucilage cell

Schlempe (Trockenschlempe/ Nassschlempe) *agr* stillage (dry or wet), distillers' grains

Schlenke (im Hochmoor)
bog hollow, bog ditch, bog rivulet (in raised bog)

Schlepperprotein/Trägerprotein
carrier protein

Schleppfaden/Schleppleine/ Zugleine *arach* dragline
- **breiter Schleppfaden**
broad trail-line

Schleppgeißel trailing flagellum

Schleppleine/Schleppfaden/ Zugleine *arach* dragline

Schleppnetz dragnet, trawlnet

Schleppnetzfischerei trawling

Schleuderausbreitung
ballistic dispersal

Schleuderfrucht/ballistische Frucht
ballistic fruit, ballist

Schleuderkapsel ballistic capsule

Schleudermagen/Pansenvorhof/ Atrium ruminis atrium ruminis

Schleudermechanismus (Samenausbreitung)
ballistic dispersal mechanism

Schleudervorrichtung (Frucht)
ejection device, ballistic device

Schleuderzelle/Elatere elater

Schleuse (Membran) sluice

schleusen sluice, channel

Schleusenmechanismus
gating mechanism

Schlichtkleid *orn* basic plumage, inconspicuous plumage (male ducks: eclipse plumage)

Schlick (alluvial) mud, silt, sludge

Schlickgrund (See/Fluss) mud bottom

Schlicksediment (angeschwemmt)
warp

Schliefer/Hyracoidea conies
Schließfrucht indehiscent fruit
Schließknorpel/Resilium resilium
(flexible horny hinge)
Schließmechanismus
closing mechanism
**Schließmuskel/Anzieher/Adduktor/
Adductor** adductor muscle
Schließmuskel/Sphinkter
sphincter muscle
Schließzelle *bot* guard cell
Schlinger gorger (animal which
gulps down entire prey)
Schlingfalle (*Arthrobotrys*)
snare trap
**Schlingmeldengewächse/
Basellaceae** Madeira vine family,
basella family
Schlingpflanze/Windepflanze *allg*
winder, twiner; (verholzte) liana
Schlittenmikrotom
sliding microtome
Schlitzband (Schneckenschale)
slit band, selenozone
Schlitzlochplatte slot blot
Schloss (Verschluss) lock
**Schloss/Schlossleiste/Scharnier
(Muscheln)** hinge
Schloss-Schlüssel-Prinzip
lock-and-key principle
**Schlossligament/Schlossband
(Muscheln)** hinge ligament
Schlosszähne (Muscheln)
hinge teeth
Schlot *mar/geol* vent
● **hydrothermaler Schlot**
hydrothermal vent
**Schlucht/Bergschlucht/Klamm/
Hohlweg** canyon, gorge; ravine
schlucken *vb* swallow
Schlucken *n* swallowing
Schluckfalle/Saugfalle suction trap,
suctory trap
Schluff/Silt *geol* silt
Schlund/Blütenschlund *bot* throat
Schlund/Kehle/Rachen throat, jaw
Schlund/Pharynx pharynx, gullet
**Schlundbogen/Zungenbeinbogen/
Hyoidbogen** *embr* hyoid arch
Schlunddrüse (Gastropoden)
esophageal gland
Schlundebene pharyngeal plane
**Schlundplatte/Kehlplatte/
Gularplatte (prognathous insects)**
gula, gular plate
Schlundring (Nervenring)
buccal nerve ring
Schlundrohr/Manubrium manubrium
**Schlundsackschnecken/
Sackschnecken/
Schlauchschnecken/Sacoglossa/
Saccoglossa** sacoglossans

Schlundschuppe *bot* coronal scale
Schlundtasche pharyngeal pouch
schlüpfen *vb* (aus dem Ei) hatch,
emerge
Schlüpfen *n* hatching, emerging
Schlüpfrate/Schlüpfzahlen
hatching rate, emergence
Schlupfwinkel/Retraite *arach*
retreat
Schlüssel (Bestimmungsschlüssel)
key
**Schlüssel-Schloss-Prinzip/
Schloss-Schlüssel-Prinzip**
lock-and-key principle
Schlüsselart keystone species
Schlüsselbein collarbone, clavicle
**Schlüsselblumengewächse/
Primelgewächse/Primulaceae**
primrose family
Schlüsselenzym/Leitenzym
key enzyme
Schlüsselinnovation, evolutionäre
key evolutionary innovation (KEI)
Schlüsselräuber *ethol*
keystone predator
Schlüsselreiz/Auslösereiz
key stimulus, sign stimulus
(release stimulus)
Schlüsselsubstanz key substance
Schlussgesellschaft terminal
community
**Schlussleiste/Verschlusskontakt/
Engkontakt/Kittleiste/Tight
junction (Zonula occludens)**
tight junction
**Schmalz/Schweineschmalz/
Schweinefett** lard
**Schmalzunge/rhachiglosse Radula/
stenoglosse Radula**
rachiglossate radula
**Schmalzüngler/Neuschnecken/
Neogastropoda/Stenoglossa**
neogastropods:
whelks & cone shells & allies
schmarotzen/parasitieren parasitize
schmarotzend/parasitisch/parasitär
parasitic
**Schmarotzer/Parasit (*siehe auch:
Parasit*)** parasite
**Schmarotzerblumengewächse/
Rafflesiaceae** rafflesia family
**Schmarotzertum/Parasitismus
(*siehe auch: Parasitismus*)**
parasitism
schmecken taste
Schmelz (Zahnschmelz) enamel
schmelzen/aufschmelzen *chem/gen*
melt
Schmelzknospe (Zahn) tooth bud
Schmelzkurve *chem* melting curve
Schmelzorgan enamel organ
Schmelzpunkt *chem* melting point

S   Schmelzschuppe   670

Schmelzschuppe/Ganoidschuppe
ichth ganoid scale
Schmelztemperatur
melting temperature
Schmelztiegel *lab* crucible
Schmelzwasser meltwater
Schmerwurzgewächse/
Yamswurzelgewächse/
Dioscoreaceae yam family
Schmerz pain
schmerzen hurt, be painful
Schmerzgefühl pain sensation
schmerzhaft painful
Schmetterlingsblume
butterfly-pollinated flower
schmetterlingsblütig butterfly-like,
papilionaceous
Schmetterlingsblütler/
Hülsenfrüchtler/
Hülsenfruchtgewächse/Fabaceae/
Papilionaceae (*Leguminosae*)
pea family, bean family,
legume family, pulse family
Schmierlinge/Gomphidiaceae
gomphidius family
schmollen pouting
Schmucktannengewächse/
Araukariengewächse/
Araucariaceae araucaria family,
monkey-puzzle tree family
Schnabel *orn* bill (general term),
beak (strong/short/broad bill)
• **Hakenschnabel** decurved bill
• **Kernbeißerschnabel**
seed-cracking beak
• **Seihschnabel** water-straining
beak, filter-feeding bill
• **Stocherschnabel/**
Sondenschnabel probing bill
Schnabel (Sepia)/Kiefer jaw, beak
Schnabel/Rüssel/Rostrum (Wanzen)
beak, rostrum
Schnabelaufsatz (Nashornvögel)
casque
Schnabelfirst/Culmen culmen
Schnabelfliegen/Mecoptera
scorpion flies, mecopterans
Schnabelkerfe/Rhynchota/
Halbflügler/Hemiptera
(Heteroptera & Homoptera) bugs,
hemipterans
Schnabelköpfe/Rynchocephalia/
Sphenodonta rynchocephalians,
sphenodontids
schnäbeln *orn* bill
Schnabelspalt (beak/bill) gape
Schnakenpuppe tumbler (mosquito
pupa)
Schnalle (Basidiomyceten) clamp
Schnallenverbindung
(Basidiomyceten)
clamp connection

Schnappdeckel/Schnappverschluss
lab snap cap
Schnappdeckelglas/
Schnappdeckelgläschen *lab*
snap-cap bottle, snap-cap vial
schnappen snap
• **nach Luft schnappen** gape
Schnappreflex snapping reflex
schnattern (Gänse) cackle;
(Affen) chatter, jabber
schnauben/prusten snort
Schnauze/Maul snout, muzzle
• **schnabelähnliche Schnauze** bill
Schnecken/Bauchfüßer/
Gastropoden
snails, gastropods
Schnecke/Ohrenschnecke/Cochlea
anat cochlea
Schneckenbekämpfungsmittel/
Molluskizid molluscicide
Schneckengang/Scala media
cochlear duct
Schneckenhaus/Schneckengehäuse
snail shell
schneckenhausförmig eingerollt/
schneckenhausartig gewunden/
cochlear coiled like a snail's shell,
cochleate, cochleiform
schneckenhausförmig eingerolltes
junges Farnblatt
fiddlehead, crozier
Schnee snow
• **Neuschnee** new-fallen snow,
fresh-fallen snow
Schneegrenze snow line
Schneematsch slush
Schneeschmelze snowmelt
Schneesturm snowstorm;
(heftiger S.) blizzard
Schneewehe/Schneeverwehung
snow drift (a bank of drifted snow)
Schneidezahn incisor
Schneise clearing, aisle
Schnellfärbung *micros* quick-stain
Schnellgefrieren rapid freezing
Schnitt section
• **Dünnschnitt** thin section
• **Gefrierschnitt** frozen section
• **Hirnschnitt/Querschnitt**
transverse section, cross section
• **Kaiserschnitt** *med/vet*
cesarean section, cesarean
• **Querschnitt** cross section
• **Sagittalschnitt** sagittal section,
median longisection
• **Schnellschnitt** quick section
• **Semidünnschnitt** semithin
section
• **Serienschnitte** *micros/anat*
serial sections
• **Ultradünnschnitt**
ultrathin section

Schnittblume cut flower
Schnittdicke thickness of section, section thickness
Schnittfläche/Schnittebene cutting face, cutting plane
schnittig/geschnitten/ eingeschnitten cut, incised
Schnittstelle *gen* cleavage site
schnuppern/schnüffeln sniff
Schnurbaum/ Schnurspalierbaum/Kordon *hort* cordon
schnüren (Fuchs) move/run in straight line
schnurren (Katze) purr
Schnurrhaare whiskers
Schnurwürmer/Nemertini/ Nemertea/Rhynchocoela nemertines, nemerteans, proboscis worms, rhynchocoelans, ribbon worms(broad/flat), bootlace worms(long)
Schockgefrieren shock freezing
Schokoladenagar/Kochblutagar chocolate agar
Scholle/Erdscholle lump of soil
schollig lumpy
Schonung (junger/geschützter Wald) young, protected forest plantation
Schonwald/Schutzwald protected forest (only for limited specified use)
Schopf/Büschel *zool/bot* tuft
  • **Haarschopf** tuft of hair; (Pferd) forelock
Schopfbaum tree with terminally tufted leaves
schopfig/dichthaarig/haarschopfig tufted, comose
Schopfkrone *bot* tufted crown
Schopfrosettenpflanze/ Schopfpflanze crown rosette plant, caulescent perennial herb, giant rosette plant, giant leaf-rosette plant
Schorf *bot* (Blatt) scurf
Schorf (Wundschorf)/Grind *zool/med* scab
schorfig/Schorf... scurfy, scabby, furfuraceous
Schorfwunde scab lesion (crustlike disease lesion)
Schorre/Abrasionsplatte/ Brandungsplatte/ Abrasionsterrasse/ Abrasionsplattform/ Brandungsplattform *mar* platform, shore platform, abrasion platform, wave-cut shelf
Schoss/Schössling/Spross/ junger Trieb *bot* shoot

Schössling/Schoss (kleiner Spross) shoot, sprout, sprig
Schössling/Wasserreis (an Wurzel oder Baumstumpf) sucker, tiller; (speziell: Zuckerrohr/Banane; Stauden) ratoon
Schösslinge treiben (v. a. Stauden) ratoon
Schötchen (Frucht) silicle
Schote (Frucht) silique
Schotter *geol* gravel
Schotterbank *mar* gravel bar
schräg oblique
Schrägkultur (Schrägagar) *micb* slant culture
Schranke barrier
  • **physiologische Schranke** physiologic barrier
schränken/kreuzweise übereinanderlegen/verschränken (>Beine: kreuzen) fold (arms) (>legs: cross)
Schraube/Spirale/Helix spiral, helix
Schraubel (cymöse Infloreszenz) bostryx (helicoid cyme)
schraubenartig gewunden/ schneckenhausartig gewunden/ cochlear irregularly helical (like a snail shell), cochleate
Schraubengläschen *lab* screw-cap vial, screw-cap jar
Schraubenpalmen/ Schraubenbaumgewächse/ Pandanaceae screw-pine family
schraubig/spiralig/helical spiraled, helical, spirally twisted, contorted
Schreck-Schaustellung *ethol* startle display
Schreckfärbung/Schrecktracht *ethol* fright coloration
Schreckreflex startle reflex
Schreckstoff/Abschreckstoff deterrent, repellent
Schreckstoff/Alarmstoff/ Alarm-Pheromon alarm substance, alarm pheromone
Schreckverhalten startle behavior
Schreitbein des Pereon/Pereiopode/ Peraeopode pereiopod, pereopod
schreiten wade
Schreitfährten/Gradichnia (Spurenfossile) tracks
Schreitfuß wading foot
Schreitvögel/Stelzvögel/ Ciconiiformes herons & storks & ibises and allies
Schrill-Leiste/Schrilleiste/Pars stridens (mit Schrill-Rille) *entom* file, stridulating file
schrillen/stridulieren/zirpen stridulate, chirp

Schrillkante 672

Schrillkante/Plectrum *entom*
scraper, rasp
**Schrillorgan/Stridulationsorgan/
Zirporgan/Organum stridens**
entom stridulating organ
Schritt step, pace; stride (long step)
Schritt/Gangart *allg (siehe auch
unter:* **Gangart)** gait
• **(als Gangart des Pferdes)** walk
Schrittlänge stride
Schrittmacher pacemaker *(siehe:*
Sinusknoten)
Schrittmacherpotential
pacemaker potential
schrotsägeförmig runcinate,
hook-backed, retroserrate
Schrotschussexperiment
shotgun experiment
Schrotschussklonierung
shotgun cloning
Schub *aer* thrust
**Schubfestigkeit/Scherfestigkeit
(Holz)** shear strength,
shearing strength
Schubgeißel pushing flagellum
Schubkraft/Vortriebkraft thrust,
forward thrust
Schule/Schwarm/Zug school, shoal
(fish), pod (whales/dolphins/seals),
covey (quails), flight/flock (birds)
Schulp/Schulpe cuttlebone
• **pergamentartiger Schulp/
Rückenfeder/Gladius**
pen, gladius
Schulter/Achsel shoulder
Schulterblatt *zool* shoulder blade,
scapula
Schulterfeder/Schulterblattfeder
orn scapular feather
(meist pl: scapulars)
Schultergürtel shoulder girdle,
pectoral girdle
Schüppchen *bot/allg* squamella
(small scale or bract)
Schüppchen (Grasblüte) lodicule,
paleola, glumellule
Schuppe scale
• **Antennenschuppe/Scaphocerit**
entom antennal scale,
scaphocerite
• **Bauchschuppe/Ventralschuppe**
bot ventral scale
• **Cosmoidschuppe** *ichth*
cosmoid scale
• **Ctenoidschuppe/Kammschuppe**
ichth ctenoid scale
• **Cycloidschuppe/Rundschuppe**
ichth cycloid scale
• **Deckschuppe** *bot* bract-scale,
subtending bract, secondary bract
• **Ganoidschuppe/Schmelzschuppe**
ichth ganoid scale

• **große Schuppe**
scute (enlarged scale)
• **Honigschuppe** nectariferous scale
• **Involukralschuppe/
Involukralblatt** *bot*
phyllary, involucral bract
• **Kammschuppe/Ctenoidschuppe**
ichth ctenoid scale
• **Knospenschuppe/
Knospendecke/Tegment**
(protective) bud scale, tegmentum
• **mit kleinen Schuppen bedeckt**
squamellate, squamelliferous,
squamulose
• **mit Schuppen bedeckt**
squamiferous, squamigerous,
squamose
• **Placoidschuppe/Zahnschuppe/
Hautzahn/Dentikel** placoid scale,
dermal denticle
• **Rundschuppe/Cycloidschuppe**
ichth cycloid scale
• **Samenschuppe/Fruchtschuppe**
bot ovuliferous scale, seed scale
• **Saugschuppe/Schuppenhaar
(Bromelien)** absorbing trichome
• **Schlundschuppe** *bot* coronal scale
• **Schmelzschuppe/Ganoidschuppe**
ichth ganoid scale
• **Spreuschuppe/Spreublatt** *bot*
ramentum, chaffy scale, palea,
pale, palet
• **Tragschuppe/Deckschuppe/
Tragblatt zweiter Ordnung/
Brakteole** bract-scale, bracteole,
bractlet, secondary bract
• **Ventralschuppe/Bauchschuppe**
bot ventral scale
• **Zahnschuppe/Placoidschuppe**
ichth dermal denticle,
placoid scale
• **Zapfenschuppe** *bot* cone scale,
cone bract
**Schuppen/Kopfschuppen/
Haarschuppen/Hautschuppen**
dandruff
**Schuppenapfelgewächse/
Annonaceae** custard apple family,
cherimoya family
schuppenartig scale-like, scutate
Schuppenbäume/Lepidodendraceae
lepidodendron family
(clubmoss trees)
**Schuppenbein/Squamosum
(Schädelknochen)** squamosa
Schuppenblätter scale-like bracts,
scale leaves, bracteole, bractlet
schuppenblättrig scale-leafed
Schuppenborke scale bark
**Schuppenflügler/Lepidoptera
(Schmetterlinge u. Motten)**
lepidopterans (butterflies and moths)

Schwangerschaft

schuppenförmig scale-shaped, scale-like, squamiform, scutiform
Schuppenhaar/Saugschuppe (Bromelien) squamiform hair, peltate trichome, absorbing trichome
Schuppenkriechtiere/Lepidosauria lepidosaurs
Schuppennaht/Sutura squamosa (Schädel) squamosal suture
Schuppenpilze/Träuschlinge/ Strophariaceae stropharia family
schuppenspaltig/squamicid squamicidal
Schuppentiere/Pholidota pangolins, scaly anteaters
schuppig scaly, squamid, squamate; scabrous (see: schorfig)
 ● **feinschuppig** squamulose, squamulate
Schuppung scaling, scutellation
Schüsselfarngewächse (Dennstaedtiaceae) cup fern family (dennstaedtia family)
Schutt geol rubble, debris, detritus
 ● **Gesteinsschutt** rock debris, rubble
Schütte/Blattschütte/Nadelschütte (Kiefernadeln) leaf cast, needle cast (caused by frost/dryness/fungal disease)
 ● **Blattausschüttung/ Laubausschüttung** leaf flushing
Schüttelbad lab shaking water bath
Schüttelklette bot shake burr, rattle burr
Schüttelkolben lab shake flask
Schüttelkultur micb shake culture
Schütteltanz/Schüttelbewegung (Bienen) shaking dance
Schutthalde/Schuttflur talus, scree
Schüttler shaker
 ● **Reziprokschüttler** reciprocating shaker
 ● **Rundschüttler** circular shaker, orbital shaker, rotary shaker
Schuttpflanze/Ruderalpflanze ruderal plant
Schutzanpassung ethol protective adaptation, protective resemblance
Schutzbrille lab goggles, safety goggles, safety spectacles
Schutzfärbung/Tarnfärbung ethol cryptic coloration, concealing coloration
Schutzgebiet ecol reserve
Schutzgespinst, dreidimensionales arach barrier web
Schutzhandschuhe protective gloves
Schutzhaube protective hood
Schutzimpfung protective immunization, vaccination

Schutzkleidung protective clothing
Schutzmaßnahme lab protective/precautionary measure
Schutztracht/Mimikry mimicry
Schutzversuch/Schutzexperiment protection assay, protection experiment
Schutzvorrichtung lab protective device
Schutzwald (Schonwald/Hegewald) protected forest (only limited specified use)
schwachwüchsig/gering wachsend (comparatively) slow growing
Schwalbenschwanzverbindung micros dovetail connection
Schwalbenwurzgewächse/ Seidenpflanzengewächse/ Asclepiadaceae milkweed family
Schwall mar swash
Schwallbewässerung agr surge irrigation
Schwämme/Schwammtiere/ Poriferen sponges, poriferans
Schwammgewebe spongy tissue
Schwammparenchym spongy parenchyma
Schwammpilz/Porling spongiose fungus, polypore, pore fungus
Schwanenblumengewächse/ Wasserlieschgewächse/ Butomaceae flowering rush family
Schwanenhalskolben lab swan-necked flask, S-necked flask, gooseneck flask
schwanger/trächtig pregnant, gravid, gestational
Schwangerschaft/Trächtigkeit/ Gestation pregnancy, gravidity, gestation
 ● **Abdominalschwangerschaft/ Bauchhöhlenträchtigkeit/ Leibeshöhlenschwangerschaft/ Leibeshöhlenträchtigkeit** abdominal pregnancy
 ● **Eileiterschwangerschaft/ Oviduktträchtigkeit** tubal pregnancy
 ● **ektopische Schwangerschaft/ Extrauteringravidität (EUG)** ectopic pregnancy, extrauterine pregnancy (EUP)
 ● **Leibeshöhlenschwangerschaft/ Leibeshöhlenträchtigkeit/ Bauchhöhlenträchtigkeit/ Abdominalschwangerschaft** abdominal pregnancy
 ● **Ovarialschwangerschaft/ Ovariolenträchtigkeit/ Ovarialträchtigkeit** ovarian pregnancy

Schwangerschaftsperiode
gestational period
Schwangerschaftswoche
week of gestation,
week of pregnancy
schwanken/fluktuieren fluctuate
schwanken/variieren variate
Schwankung/Fluktuation fluctuation
Schwankung/Variation variation
Schwannsche Scheide/
Myelinscheide
Schwann sheath, myelin sheath
Schwannsche Zelle Schwann cell,
neurilemma cell
Schwanz tail; (Fettmolekül) tail
Schwanz stutzen/anglisieren
(Pferde) dock, docking
Schwanz-an-Schwanz-
Wiederholungen *gen*
tail-to-tail repeats
Schwanzanhang/Cercus cercus,
cercopod
Schwanzansatz base of tail
Schwanzdarm tail gut, postanal gut
Schwanzdrüse (Rektaldrüse)
caudal gland (rectal gland)
Schwänzeltanz (Bienen)
tail-wagging dance, waggle dance
Schwanzfächer tailfan
• **Hypurale** *ichth* hypural fan
Schwanzfeder quill feather
Schwanzflosse tail fin, caudal fin
• **Schwanzruder/Fluke (Wale)**
tail fluke
Schwanzflughaut/Uropatagium
uropatagium
Schwanzfortsatz coccyx
Schwanzgabel caudal furca
Schwanzlarve (Ascidien)
tadpole larva
Schwanzlarve/Zerkarie/Cercarie
cercaria
Schwanzlurche/Urodela/Caudata
urodeles (salamanders & newts)
Schwanzplatte/Telson (Krebse)
telson
Schwanzquaste (Rinder) switch
Schwanzruder/Schwanzflosse
(Wale) tail fluke
Schwanzschild/Pygidium
(Trilobiten) caudal plate, pygidium
Schwanzstachel tail spine
Schwanzstiel/Pygostyl pygostyle
Schwanzstummel (Pferde) dock
Schwanzstumpf dock
Schwanzwachstum/
endständiges Wachstum
tail growth
Schwanzwirbel caudal vertebra
(*see:* coccygial vertebra)
Schwanzwurzeldrüse
supracaudal gland

Schwanzzange pincer
Schwarm (z. B. Bienen) swarm
Schwarm/Schule/Zug (fish) shoal,
school; (whales/dolphins/seals) pod;
(quails) covey; (birds) flight/flock
schwarmbildend swarm-forming,
schooling
schwärmen/ausschwärmen
swarm, swarm off, swarm out
Schwärmer (Zoospore)
swarmer (zoospore/swarm cell)
Schwarte (dicke/zähe Haut) thick
skin, hide; rind (esp. pork rind)
• **Holzschwarte** slab
• **Speckschwarte** (bacon) rind
schwarzer Körper black body
Schwarzmundgewächse/
Melastomataceae meadow beauty
family, melastome family
Schwarztorf black peat
Schwarzwasserfieber
blackwater fever
Schwarzwasserfluss
blackwater river
Schwarzwild/Wildschweine
wild boar
Schwebedichte/Schwimmdichte
buoyant density
schweben (schwebend) float
(floating), suspend (suspended); *orn*
(in der Luft schweben/stehen) hover
Schweben/Schwebeflug *orn* **(in der**
Luft schweben/stehen) hovering
Schwebeorgan float
Schwebfracht *ecol/mar* silt load,
suspension load
Schwebstoffe suspended matter
Schwefel sulfur
Schwefelbakterien sulfur bacteria
schwefelhaltig sulfurous,
sulfur-containing
Schwefelkreislauf sulfur cycle
schwefeln (z. B. Fässer) sulfurize
(e. g. vats)
Schwefeln/Schwefelung (Fässer)
sulfuring
Schwefelverbindung/
schwefelhaltige Verbindung
sulfur compound
schweflig sulfurous
Schweif/Schwanz tail
Schweifkern/Nucleus caudatus
caudate nucleus
Schweifrübe (Pferd:
Schwanzansatz) root of tail, dock
Schwein swine, pig, hog
• **Bache/Wildschweinsau**
wild sow
• **Eber (männliches Schwein)** boar
(male pig: not castrated)
• **Ferkel** piglet, little pig;
(Mastferkel) porker

Schwimmflosse

- **Frischling** wild pig/swine/boar/ hog in its 1. year
- **Jungsau** gilt
- **Keiler/Wildeber** male wild boar (aged)
- **Läuferschwein/Läufer** young pig, store pig, store
- **Mutterschwein/Sau** sow (female pig)
- **Spanferkel (junges/noch gesäugtes Schwein)** sucking pig/piglet/porkling
- **Wildschweinsau/Bache** wild sow
- **Wildschwein/Schwarzwild** wild pig, wild hog, boar

Schweinefieber *vir* swine fever, hog cholera

Schweinegrippe (*Hemophilus influenzae suis*) swine influenza, swine flu

Schweinepest (*Pasteurella multocida*) swine plague

Schweiß sweat, perspiration

Schweißdrüse sweat gland, sudoriferous gland, sudoriparous gland

Schwelle (z.B. Reizschwelle/ Geschmacksschwelle etc.) threshold

schwellen/anschwellen/turgeszent swell, swelling, turgescent

schwellend/prall/turgeszent turgescent

Schwelleneffekt threshold effect

Schwellenländer semi-peripheral countries

Schwellenmerkmal threshold trait

Schwellenpotential (kritisches Membranpotential) threshold potential (firing level)

Schwellenstrom threshold current

Schwellenwert threshold value
- **konstanter S.** *gen* constant truncation
- **proportionaler S.** *gen* proportional truncation

Schwellenwertselektion/ Kappungsselektion/ Auslesezüchtung truncation selection

Schwellkörper/Corpora cavernosa erectile tissue, cavernous tissue, cavernous bodies

Schwellkörper/Lodicula *bot* (Grasblüte) lodicule, paleola, glumellule

Schwellung swelling; (Turgeszenz) turgescence

Schwellunggrad turgidity

Schwemmkegel/Schwemmfächer alluvial fan

Schwemmland floodland, flood plain, alluvial plain

schwenken (Flüssigkeit in Kolben) swirl

schwere Kette (H-Kette) heavy chain (H chain)

Schwerefeld gravitational field

Schwerelosigkeit weightlessness

Schweresinn gravitational sense

Schwerestein(chen)/Statolith statolith

schwerflüchtig nonvolatile
- **flüchtig** volatile

Schwerkraft gravity, gravitational force

Schwermetallbelastung heavy metal contamination

schwertförmig sword-shaped, ensiform, gladiate, xiphoid

Schwertliliengewächse/Iridaceae iris family

Schwesterart sister species

Schwesterchromatiden sister chromatids

Schwesterchromatidenaustausch sister chromatid exchange
- **UESCE (ungleicher Austausch von Schwesterchromatiden)** UESCE (unequal exchange of sister chromatids)

Schwestergruppe (Kladistik) sister group

Schwestertaxon/Adelphotaxon sister taxon

Schwesterzelle sister cell

Schwiele/Kallosität callosity

Schwimm../zum Schwimmen geeignet swim.., swimming .., natatorial

Schwimm-Manteltiere, geschwänzte/Appendicularien/ Appendicularia/Larvacea appendicularians

Schwimmbein (Insekten) natatorial leg, swimming leg

Schwimmbein/Schwimmfuß/ Bauchfuß/Abdominalbein/ Pleopodium (Crustaceen) swimmeret, pleopod

Schwimmblase (Algen) air bladder, float, pneumatophore

Schwimmblase (Fische) swimbladder, air bladder

Schwimmblatt floating leaf

Schwimmdichte/Schwebedichte buoyant density

Schwimmfarngewächse/ Schwimmfarne/Salviniaceae salvinia family

Schwimmflosse (groß/fleischig)/ Paddel flipper, fluke

S Schwimmfuß

Schwimmfuß (z. B. Vögel)
webbed foot, swimming foot
Schwimmfuß/Schwimmbein/
Bauchfuß/Abdominalbein/
Pleopodium (Crustaceen)
swimmeret, pleopod
Schwimmglocke/Nectophor
swimming bell, nectophore
Schwimmhaut web
schwimmhäutig/
mit Schwimmhäuten webbed
Schwimmplatte/Ruderplatte/
Ruderplättchen/"Kamm"/
Wimperplättchen (Ctenophoren)
ctene, swimming plate,
ciliary comb
Schwinge/Feder (*siehe:*
Armschwingen/Handschwingen)
orn feather
Schwinge/Fittich/Flügel *orn* wing
Schwingkletterer/Hangler
brachiator
Schwingklettern/Hangeln/
Brachiation brachiation
Schwingkölbchen/Haltere balancer,
haltere
Schwingphase swing phase,
suspension phase
Schwingrasen (Moor) quaking bog,
quagmire
Schwingungsbewegung
vibrational motion
Schwirrflug/Schwebeflug (Kolibris:
Rüttelflug) hovering flight
Schwirrflügler/Apodiformes (Segler
& Kolibris) swifts & hummingbirds
Schwirrgeräusch (Bienen beim
Schwänzeltanz) buzzing sound
Schwirrlauf (Bienen) buzzing run
schwitzen *vb* sweat, perspire
Schwitzen *n* sweating, perspiration,
hidrosis
Schwundschicht (Anthere)
disappearing layer
Schwungfeder/Remex
flight feather, remex (*pl* remiges)
Sclerenchym/Sklerenchym
sclerenchyma
Sclerenchym-Steinzelle/Sclereide/
Sklereide stone cell, sclereid,
sclereide, sclerid
sclerenchymatisch
sclerenchymatous
Sclerenchymfaser sclerenchyma
fiber, sclerenchymatous fiber
Sclerospongien/Sclerospongiae
sclerosponges, coralline sponges
Scolespore scolespore
Scolopidium/stiftführende Sensille
scolopidium, scolophore,
chondrotonal sensilla
Scopolamin scopolamine

Scrophulariaceae/
Braunwurzgewächse/
Rachenblütler
figwort family, foxglove family,
snapdragon family
Scutellum/Schildchen (Saugorgan
des Graskeimlings) scutellum
Scyphistoma (Semaeostomea)
scyphistoma (scyphozoan polyp)
Scyphopolyp scyphozoan polyp
Scyphozoen/Schirmquallen/
Scheibenquallen (echte Quallen)
scyphozoans, cup animals
Sechshakenlarve/Oncosphaera-
Larve (Cestoda) hexacanth larva,
hooked larva, oncosphere
sedentär/niedergelassen sedentary
Sediment sediment; *centrif* **(Pellet)**
pellet
Sedimentationsgeschwindigkeits-
analyse *biochem*
sedimentation analysis
Sedimentationskoeffizient
sedimentation coefficient
Sedimentfresser deposit feeders
Sedimentgestein/Absatzgestein/
Schichtgestein sedimentary rock
See (Binnensee) sea, inland sea,
lake; (Ozean) sea, ocean
• **Alkalisee** alkaline lake
• **Braunwassersee/dystropher See**
dystrophic lake
• **Hochsee/offenes Meer**
open sea, pelagic zone
• **Pluvialsee** pluvial lake
• **Salzsee** salt lake
• **See mit Abfluss** drainage lake
• **Sickersee** seepage lake
• **Stausee** reservoir, atrificial lake
• **Süßwassersee** freshwater lake
• **Tiefsee** deep sea
• **Tropensee** tropical lake
Seeanemonen/Actiniaria
sea anemones
Seeäpfel/Cystoida (Echinoderma)
cystoids
Seebeerengewächse/
Tausendblattgewächse/
Haloragaceae
water milfoil family, milfoil family
Seedrachen/Seekatzen/Chimären/
Holocephali
chimaeras, ratfishes, rabbit fishes
Seefeder/Pennatularia
sea pens, pennatulaceans
Seegänseblümchen/
Concentricycloidea *zool*
sea daisies, concentricycloids,
concentricycloideans
Seegras/Seetang/Tang seaweed
Seegrasgewächse/Zosteraceae
eel-grass family

Seidelbastgewächse S

**Seehasen/Breitfußschnecken/
Aplysiacea/Anaspidea** sea hares
Seeigel/Echinoiden/Echinoidea
sea urchins, echinoids
Seekühe/Sirenia sea cows &
manatees & dugongs, sirenians
**Seelilien/Crinoiden (inkl.
Haarsterne=Federsterne)/
Crinoidea** sea lilies, crinoids
(incl. feather stars)
● **zirrenlose Seelilien/Millericrinida**
sea lilies without cirri
● **zirrentragende Seelilien/
Isocrinida** sea lilies with cirri
**Seemaus (im Plankton suspendierte
Eikapsel einiger Knorpelfische)**
sea purse
**Seemotten/Pegasiformes
(Pegasidae)** seamoths
**Seenadelverwandte/
Seepferdchenverwandte/
Büschelkiemenartige/
Syngnathiformes (bzw.
Syngnathoidei)**
sea horses & pipefishes and allies
Seenkunde/Limnologie limnology
Seepflanze lacustrine plant
Seerosengewächse/Nymphaeaceae
water-lily family
Seescheiden/Ascidien/Ascideacea
sea squirts, ascidians
Seeschildkröten turtles
**Seeschmetterlinge/Thecosomata/
beschalte Flossenfüßer
(Flügelschnecken)**
sea butterflies, shelled pteropods
Seeskorpione/Eurypterida
sea scorpions, eurypterids
Seesterne/Asteroidea seastars,
starfishes
Seetaucher/Gaviiformes divers, loons
Seeufer lakeshore,
shore/banks of a lake
**Seewalzen/Seegurken/Holothurien/
Holothuroidea** sea cucumbers,
holothurians
Seewind onshore wind
Segel (z. B. Velellina) sail
Segelflug *orn* soaring (flight)
**Segelklappe/Atrioventrikularklappe
(*siehe auch:* Herzklappe/
Taschenklappe)**
atrioventricular valve
● **Dreisegelklappe/dreizipflige
Segelklappe/Trikuspidalklappe**
tricuspid valve (with three cusps/
flaps)
● **Zweisegelklappe/zweizipflige
Segelklappe/Mitralklappe/
Bikuspidalklappe** mitral valve,
bicuspid valve (with two cusps/
flaps)

Segellarve/Veliger/Veligerlarve
veliger larva
segeln sail
**Seglervögel/Seglerartige/
Apodiformes/Micropodiformes**
swifts (hummingbirds *see* Kolibris)
Segment/Somit (Ursegment)
segment, somite
segmentieren segment
Segmentierung segmentation
segmentkerniger Neutrophil
segmented neutrophil,
filamented neutrophil,
polymorphonuclear granulocyte
Segregation/Aufspaltung
segregation
Segregationslinie segregation line
segregieren/aufspalten segregate
sehen/anschauen/erblicken *vb*
see, view
Sehen *n* seeing, vision
Sehfeld/Blickfeld/Gesichtsfeld field
of view, scope of view, field of
vision, range of vision, visual field
Sehfeldblende/Gesichtsfeldblende
field stop (a field diaphragm)
● **Gesichtsfeldblende des Okulars/
Okularblende** ocular diaphragm,
eyepiece diaphragm,
eyepiece field stop
Sehkeil/Ommatidium ommatidium,
facet, stemma
Sehkraft/Sehvermögen eyesight
Sehlappen/Lobus opticus
visual lobe, optic lobe
Sehne tendon
Sehnenscheide tendon sheath
Sehnenschnitt (Holz)
tangential section
Sehnerv/Nervus opticus optic nerve
**Sehnervkreuzung/
Sehnervenkreuzung/Chiasma
opticum** optic chiasma
Sehpigment visual pigment
Sehpurpur/Rhodopsin rose-purple,
rhodopsin
Sehschärfe visual acuity
Sehvermögen
vision, sight; eyesight;
(Sehstärke) strength of vision
Sehweite range of vision, visual
distance
Sehzentrum visual center
Seiche *limn* seiche (standing wave
oscillation)
seicht/flach/niedrig (Wasser)
limn/mar shallow, low
Seide silk (fibroin/sericin)
**Seidelbastgewächse/
Spatzenzungengewächse/
Thymelaeaceae** daphne family,
mezereum family

S seiden 678

seiden/Seiden... silken
seidenartig/seidenhaarig/seidig
silky, sericeous, sericate
Seidendrüse/Spinndrüse/
Sericterium silk gland, sericterium;
(Labialdrüse: Raupen) labial gland
Seidengewächse/Cuscutaceae
dodder family
seidenhaarig sericeous, sericate,
silky
Seidenpflanzengewächse/
Schwalbenwurzgewächse/
Asclepiadaceae milkweed family
Seidenraupe silkworm
Seife *geol* placer
Seifenbaumgewächse/Sapindaceae
soapberry family
seihen (z.B. Flamingo) sift
Seihschnabel water-straining bill
Seite/Flanke side, flank, latus
Seitenachse lateral axis,
lateral branch
Seitenast *bot* lateral branch,
offshoot
Seitenauge/Lateralauge/
Lateralocelle/Stemma (Ocellen)
lateral eye/ocellus, stemma
seitenfrüchtig/pleurokarp
pleurocarpic, pleurocarpous
Seitenkette *chem* side chain
Seitenknospe/Achselknospe
lateral bud, axillary bud
Seitenkörper (Flagellaten)
side body
Seitenkrone *bot* side crown
Seitenlinienorgan/Lateralisorgan
lateralis organ
Seitenliniensystem
lateral line system, lateralis system,
acoustico-lateralis system
Seitenmoräne lateral moraine
Seitenpfropfung/Seitenveredelung/
Veredeln an die Seite *hort* side
grafting
Seitenplatte (des Mesoderms) *embr*
lateral plate
Seitenplatte/Seitenstück/Pleurit
(Insekten: lateraler Sklerit)
pleurite, lateral sclerite
Seitenspross/Seitentrieb/
Nebentrieb
side shoot, lateral shoot
Seitenteil/Pleuron
pleuron, lateral plate
Seitentrieb/Nebentrieb
lateral shoot, side shoot, offshoot;
(am Wurzelhals) sucker
• **kurzer S. (am Wurzelhals)**
offset
seitenwendig/lateral lateral
Seitenwinden (Schlangen)
sidewinding

Seitenwurzel/Nebenwurzel
lateral root
seitlich/lateral lateral
seitliche Spaltpfropfung *hort*
side cleft grafting,
side whip grafting, bottle grafting
seitliches Anplatten *hort* veneer
side grafting, side-veneer grafting,
spliced side grafting
seitliches Anplatten mit
Gegenzunge *hort*
side-tongue grafting
seitliches Anplatten mit langer
Gegenzunge *hort* flap grafting
seitliches Einspitzen *hort*
shield grafting, sprig grafting
Sekret secretion
sekretagog/die Sekretion anregend
secretagogue
Sekretagogum/Sekretogogum
secretagogue
Sekretdrüse secretory gland
Sekretion secretion
Sekretionsgewebe secretory tissue
Sekretionsprotein/Sekretprotein/
sekretorisches Protein
secretory protein
Sekretionstapetum *bot*
secretory tapetum
sekretorisch secretory
sekretorische Komponente
(Antikörper) secretory component/
piece (antibody)
Sekretorsystem secretor system
Sekretosom secretosome
Sekretzelle secretory cell
Sektorialchimäre sectorial chimera
Sekundärantwort
secondary response
• **immunologische S.**
secondary immune response,
anamnestic response
Sekundärbiotop (Ersatzbiotop)
secondary biotope/habitat
Sekundärholz, lockeres
manoxylic wood
Sekundärinfekt/Sekundärinfektion
secondary infection
Sekundärkonsument
secondary consumer
Sekundärstoffwechsel
secondary metabolism
Sekundärstruktur (Proteine)
secondary structure
Sekundärwachstum
secondary growth
Sekundärwand secondary wall
Sekundärxylem secondary xylem
Selaginellaceae/Moosfarngewächse
selaginella family, small club moss
family, spike-moss family
Selbst-Ligation *gen* self-ligation

Selbst-Priming *gen* self-priming
Selbstassoziierung/
Selbstzusammenbau/spontaner
Zusammenbau (molekulare
Epigenese) self-assembly
Selbstausbreitung self-dispersal,
autochory
Selbstbefruchtung/Selbstung/
Autogamie self-fertilization,
selfing, autogamy
Selbstbestäubung self-pollination,
autophily
selbstentzündlich *chem*
autoignitable
Selbstentzündung *chem*
autoignition
Selbstheilung autotherapy
Selbstinkompatibilität
self-incompatibility
Selbstmord-Substrat
suicide substrate
Selbstorganisation self-organization
Selbstputzen *ethol* self-grooming,
autogrooming, autopreening
Selbstreinigung self-cleansing
Selbststerilität self-sterility
(self-incompatibility)
Selbsttoleranz/Eigentoleranz
self-tolerance
Selbstung/Selbstbefruchtung/
Autogamie selfing, self-
fertilization, autogamy
Selbstversorgerwirtschaft
subsistence economy
Selbstzusammenbau/
Spontanzusammenbau/
Selbstassoziierung/spontaner
Zusammenbau (molekulare
Epigenese) self-assembly
selektieren/auslesen select
Selektion/Auslese selection
• **disruptive S.** disruptive selection,
diversifying selection
• **frequenzabhängige S.**
frequency-dependent selection
• **Gegenselektion/Gegenauslese**
counterselection
• **gerichtete S.** directional selection
• **geschlechtliche/sexuelle S.**
sexual selection
• **Gruppenselektion** group
selection
• **künstliche S.** artificial selection
• **natürliche S.** natural selection
• **stabilisierende S.**
stabilizing selection
• **Verwandtenselektion**
kin selection
• **Verwandtschaftsselektion**
kinship selection
• **Zuchtwahl** selective breeding,
breed selection

Selektionsdifferential
selection differential
Selektionsdruck selective pressure,
selection pressure
Selektionsnachteil selective
disadvantage
Selektionsvorteil
selective advantage
Selektionswert/
Selektionskoeffizient
selection coefficient,
coefficient of selection
selektiv selective
Selektivität selectivity
Selen selenium
selten/rar scarce, rare
seltene Art rare species
Seltenheit/Rarität scarcity, rarity
Semaphoront semaphoront
semelpar/unipar semelparous,
uniparous (reproducing only once)
Semelparitie (Big-Bang-
Reproduktion) semelparity
(big-bang reproduction)
semidominant semidominant
Semidominanz semidominance
Semidünnschnitt semithin section
semikonservative Replikation
semiconservative replication
Semilunarklappe/
halbmondförmige Klappe
(Herz: als Teile der Taschenklappe)
semilunar cusp,
semilunar flap
(parts of semilunar valve)
semiterrestrisch semiterrestrial
Semperzelle/Sempersche Zelle/
Kristallzelle/Cristallogenzelle
Semper cell, crystal cell
Senfbaumgewächse/Salvadoraceae
mustard-tree family
Senföl mustard oil
Senke *geol* sink; depression, valley
Senke/Verbrauchsort (von
Assimilaten) sink (importer of
assimilates)
Senker/Absenker *bot* set, layer
Senker/Depressor *zool* **(Muskel)**
depressor muscle
Senker/Haustorium holdfast,
haustorium
Senkung/Absinken/Erdabsenkung
subsidence
Senkwasser/Sickerwasser/
Gravitationswasser
seepage water, soakage,
gravitational water
Sensille sensilla
• **stiftführende Sensille/**
Skolopidium/Scolopidium
chordotonal sensilla, scolopidium,
scolophore

S   Sensitivität   680

Sensitivität/Empfindlichkeit
sensitivity
sensorisch sensory
Sepale/Sepalum (*pl* Sepalen)
sepal
Sepia (Sekret des Tintenfisches)
sepia
Sepsis/Septikämie/Blutvergiftung
sepsis, septicemia,
blood poisoning
Septalfilament septal filament
Septe/Septum/Scheidewand
(*pl* Septen) septum (*pl* septa),
partition, dissepiment, cross-wall,
dividing wall
septifrag/scheidewandbrüchig/
wandbrüchig septifragal
sequentielle Reaktion/
Kettenreaktion
sequential reaction, chain reaction
Sequenz *gen* sequence
• **Abschaltsequenz**
silencer (sequence)
• **Aminosäurensequenz**
amino acid sequence
• **Basensequenz** base sequence
• **Insertionssequenz**
insertion sequence
• **intervenierende Sequenz/**
dazwischenliegende Sequenz/
Intron intervening sequence,
non-coding sequence, intron
• **kodierende Sequenz/Exon**
coding sequence,
encoding sequence, exon
• **komplementäre Basensequenz**
complementary base sequence
• **Leadersequenz/Leitsequenz**
leader segment/sequence
• **Mastersequenz** master sequence
• **Rekombinationssignalsequenzen**
recombination signal sequences
• **Shine-Dalgarno-Sequenz**
Shine Dalgarno sequence
• **Signalsequenz/Signalpeptid**
signal sequence, signal peptide
• **Terminationssequenz/Stopcodon**
termination sequence/codon/factor,
stop codon
• **Trailer-Sequenz** trailer segment
• **Triplettsequenzen**
triplet sequences
• **untranslatierte Sequenz (UTS)**
untranslated sequence (UTS)
• **Verstärker(sequenz)/Enhancer**
enhancer (sequence)
• **Zielsequenz** target sequence
• **Zwischensequenz/Spacer**
spacer
Sequenzierer/Sequenzierautomat
sequencer, sequenator
(*esp.* proteins)

Sequenzierung sequencing
• **Didesoxysequenzierung**
dideoxy sequencing
• **Doppelstrangsequenzierung**
double strand sequencing
• **Multiplex-Sequenzierung**
multiplex sequencing
• **Plus-Minus-Verfahren**
plus-minus sequencing
• **Transkript-Sequenzierung**
transcript sequencing
Sequenzierungsautomat sequencer
sequenzmarkierte Stelle (STS) *gen*
sequence tagged site (STS)
sequenzspezifische Mutagenese
site-specific mutagenesis
sequenzspezifische Rekombination
site-specific recombination
Sequenzwiederholung
repeat, repetition
• **direkte S.** direct repeat(s)
• **invertierte S./Palindrom**
inverted repeat, palindrome
• **indirekte S.** indirect repeats
• **Kopf-an-Kopf-Wiederholungen**
head-to-head repeats
• **lange terminale S.**
long terminal repeats (LTR)
• **Palindrom/umgekehrte**
Repetition/umgekehrte
Wiederholung palindrome,
inverted repeat
• **Schwanz-an-Schwanz**
Wiederholungen
tail-to-tail repeats
• **STRPs (Polymorphismen von**
kurzen direkten
Wiederholungen) STRPs (short
tandem repeat polymorphisms)
• **Tandemwiederholung/**
Tandemanordnung
tandem repeat, tandem duplication
• **umgekehrte terminale**
Repetitionen
inverted terminal repetitions,
inverted terminal repeats (ITR)
• **verstreut liegende S.** dispersed
repeats, interspersed repeats
serale Stadien/Pionierstadien
seral stages, pioneer stages
Sericin/Serizin sericin, silk gelatin,
silk glue
Serie (Rangstufe/Klassifizierung)
series
Serie (Sukzessionsfolge)
sere (a successional series)
serienelastische Komponente
series elastic component (SEC)
Serienschnitte serial sections
Serin serine
Serizin/Sericin sericin, silk gelatin,
silk glue

Sicherheitswerkbank

serologisch serologic(al)
serös serous
Serosa/äußere Keimhülle/äußere Eihülle/äußeres Eihüllepithel serosa, serous membrane
Serotonin/Enteramin/ 5-Hydroxytryptamin serotonin, 5-hydroxytryptamine
Serotyp serotype, serovar
Serum (pl Seren) serum (pl sera or serums)
Serumabhängigkeit serum dependence
Sesambein/Sesamknöchelchen sesamoid bone
Sesamgewächse/Pedaliaceae sesame family, benne family
Sesquiterpen sesquiterpene
Sesselform (Cycloalkane) chem chair conformation
sesshaft/sessil sessile
Sesshaftigkeit/Sessilität sessility
sessil/sesshaft/festsitzend/sitzend/ festgeheftet sessile (firmly and permantently attached)
Sessilität/Sesshaftigkeit sessility
Sessoblast (sessiler/sitzender Statoblast) sessoblast
Seston seston
Seta seta
Setzhase/Häsin doe
Setzholz/Pflanzholz hort/agr dibber, dibble
Setzling bot seedling
Setzling/junge Fischbrut> Dottersackbrut (v. a. Lachs) alevin, sacfry, yolk fry
Seuche/Epidemie epidemic
Sexduktion sexduction
Sexfaktor/Konjugationsfaktor micb sex factor
Sexpilus (pl Sexpili) sex pilus (pl sex pili)
Sexualdimorphismus/ Geschlechtsdimorphismus sexual dimorphism
Sexualhormon sex hormone
Sexualität sexuality
Sexualtrieb/Geschlechtstrieb sex drive
Sexualverhalten/ Geschlechtsverhalten sexual behavior
sexuell sexual
sexuell übertragbare Krankheit/ Geschlechtskrankheit/venerische Krankheit sexually transmitted disease (STD), venereal disease (VD)
sexuelle Fortpflanzung sexual reproduction
sexueller Größenunterschied/ Dimegetismus dimegaly

sezernieren/abgeben (Flüssigkeit) secrete (excrete)
Sezierbesteck dissection equipment (dissecting set)
Seziernadel dissecting needle (teasing needle); (Stecknadel) dissecting pin
Sezierpinzette dissecting forceps
Sezierschere dissecting scissors
sezieren dissect
Sezierung dissection
Sharpeysche Faser Sharpey's fiber
Shikimisäure (Shikimat) shikimic acid (shikimate)
Shine-Dalgarno-Sequenz Shine Dalgarno sequence
Sialinsäure (Sialat) sialic acid (sialate)
Sich-Totstellen ethol feigning death
Sichel/Drepanium (Infloreszenz) drepanium (a helicoid cyme)
sichelförmig sickle-shaped, drepanoid, crescent, falcate, falciform
Sicheltanz (Bienen) sickle dance
Sichelzelle sickle cell
sicher techn safe; (personal protection) secure
Sicherheit techn safety; (personal protection) security
Sicherheitbestimmungen safety regulations
Sicherheitsdatenblatt safety data sheet
Sicherheitsfaden arach safety line, dragline, securing thread
Sicherheitsmaßnahmen security measures, safety measures, containment
  ● **biologische S.** biological containment
  ● **physikalische S.** physical containment
Sicherheitsmaßnahme/ Sicherheitsmaßregel safety measure
Sicherheitsraum/Sicherheitsbereich/ Sicherheitslabor (S1-S4) biohazard containment (laboratory) (classified into biosafety containment classes)
Sicherheitsrichtlinien safety guidelines
Sicherheitsstufe/Risikostufe risk class, security level
Sicherheitsvektor containment vector
Sicherheitsvorkehrung/ Sicherheitsvorbeugemaßnahme safety precaution
Sicherheitsvorrichtung safety device
Sicherheitswerkbank clean bench

Sicherheitswirt

Sicherheitswirt containment host
sichern (Wild) stop and test the wind, scent
sichern/absichern secure
Sicht sight, view
sichtbar visible
• **unsichtbar** invisible
Sickerquelle/Sumpfquelle/Helokrene helocrene
Sickerwasser/Senkwasser/Gravitationswasser (Boden) soakage, seepage water, gravitational water
Sieb sieve, sifter
• **Molekularsieb/Molekülsieb** molecular sieve
Siebbein/Os ethmoidale ethmoid bone
Siebbeinplatte/Lamina cribrosa ossis ethmoidalis cribriform plate (cribriform lamina of ethmoid bone)
Siebbodenkaskadenreaktor/Lochbodenkaskadenreaktor sieve plate reactor
Siebelement/Siebröhrenelement *bot* sieve tube element
sieben sieve, sift
siebenteilig heptamerous
Siebfeld *bot* sieve area
Siebhaut/Hinfallhaut/Dezidua/Decidua decidua
Siebkiemer/Verwachsenkiemer/Septibranchia septibranch bivalves, septibranchs
Siebplatte *bot* sieve plate
Siebplatte/Madreporenplatte *zool* sieve plate, madreporic plate, madreporite
Siebröhre *bot* sieve tube
Siebröhrenelement/Siebelement *bot* sieve tube element
Siebröhrenglied *bot* sieve tube member
Siebröhrenmutterzelle/Siebzellenmutterzelle/Phloemmutterzelle sieve tube mother cell
Siebteil/Bastteil/Phloem phloem
Siebtrachee *arach* sieve trachea
Siebzelle *bot* sieve cell
Siedegefäß *lab* boiling flask
sieden/kochen boil
Siedepunkt boiling point
Siedestein/Siedesteinchen *lab* boiling stone, boiling chip
Siedeverzug *chem* defervescence, delay in boiling
Siegelbäume/Sigillariaceae sigillaria family (clubmoss trees)
Signal-Rausch-Verhältnis signal-to-noise ratio

Signalerkennungspartikel (SRP) signal recognition particle (SRP), signal recognition protein (SRP)
Signalfaden *arach* signal thread, signal line
Signalfälschung signal forgery
Signalhypothese signal hypothesis
signalisieren signal
Signalprotein/Sensorprotein signal protein
Signalsequenz/Signalpeptid signal sequence, signal peptide
Signalstoff signal substance
Signalübertragung signal transduction
Signalwandler signal transducer
Signifikanzniveau/Irrtumswahrscheinlichkeit significance level, level of significance (error level)
Signifikanztest *stat* significance test, test of significance
Silberbaumgewächse/Proteagewächse/Proteaceae protea family, Australian oak family, silk-oak family
Silberliniensystem/Argyrom (Ciliaten) silverline system, argyrome
Silbersälmling smolt
Silencer/Abschaltsequenz *gen* silencer (sequence)
Silikon silicone (silicoketone)
Silizium/Silicium silicon
Siliziumdioxid silica, silicon dioxide
Silur (erdgeschichtliche Periode) Silurian, Silurian Period
Simaroubaceae/Bittereschengewächse quassia family
Simse rush
Simsengewächse/Juncaceae rush family
Sinapinalkohol sinapic alcohol
Sinapinsäure sinapic acid
SINE (kurzes eingeschobenes nukleäres Element) *gen* SINE (short interspersed nuclear element)
Sinnesborsten sensory bristle
Sinnesgrube sensory pit
Sinneshaar sensory hair
• **Riechhärchen** olfactory hair
• **Spürhaar/Sinushaar/Tasthaar/Vibrissa** tactile hair, vibrissa
Sinneshügel/Neuromaste neuromast
Sinneskegel/Sinnesstäbchen/Riechkegel/Sensilla styloconica/Sensilla basiconica sensory peg, olfactory peg
Sinneskuppel *entom* campaniform sensillum (*pl* -a)

Skolopidium

Sinnesorgan sensory organ, sense organ
Sinnesphysiologie sensory physiology
Sinnespolster/Macula macula
Sinnesstift scolopale
Sinnstrang (DNA) sense strand (DNA)
sinoaurikulär sinoauricular, sinoatrial
Sinopteridaceae/Pelzfarngewächse sinopterids
Sinus/Höhle/Vertiefung/ Ausweitung/Lakune sinus, cavity, depression, recess, dilatation, lacuna
Sinusdrüse sinus gland
Sinusknoten/Sinoatrialknoten/ SA-Knoten (Herz: prim. Autonomiezentrum) sinoauricular node, sinoatrial node (SAN), primary sinus node
Sipho/Trichter/Infundibulum siphon, funnel, infundibulum
siphonal/röhrenartig siphoneous, siphonaceous, tubular
Siphonoglyphe siphonoglyph
Siphunkel/Siphunculus/Sipho (Nautilus) siphuncle, siphonet
Sippe/Geschlecht/Verwandtschaft/ Familie kin
Sippe/Großfamilie/Klan clan
Sippe/Tribus tribe
Sirohäm siroheme
Sitosterin/Sitosterol sitosterol
Sitzbein/Gesäßbein/Sitzknochen/ Os ischii ischium
Sitzbeinhöcker/Tuber ischiadicum ischial tuber, ischial tuberosity
sitzend/festsitzend/sesshaft/ festgeheftet sessile, sedentary
Sitzschwiele/Gesäßschwiele/ Analkallosität sitting pad, ischial callosity, anal callosity
Skala (pl Skalen) scale
Skalid scalid (recurved hook)
Skalpell scalpel
Skalpellklinge scalpel blade
Skatol scatol, skatole
Skelett skeleton
- **Achsenskelett/Axialskelett/ Stammskelett/Rumpfskelett** axial skeleton
- **Außenskelett/Exoskelett/ Hautpanzer** exoskeleton, external skeleton
- **Dermalskelett/Hautskelett** dermal skeleton, dermatoskeleton, dermoskeleton, integumentary skeleton
- **Eingeweideskelett/ Viszeralskelett/Visceralskelett** visceral skeleton, visceroskeleton
- **Extremitätenskelett/ Gliedmaßenskelett** appendicular skeleton, skeleton appendiculare
- **Hautskelett/Dermalskelett** dermal skeleton, dermatoskeleton, dermoskeleton, integumentary skeleton
- **Herzskelett** cardiac skeleton, skeleton of the heart
- **Hydroskelett/hydrostatisches Skelett** hydrostatic skeleton
- **Innenskelett/Endoskelett** endoskeleton, internal skeleton
- **Kiemenskelett/Branchialskelett** branchial skeleton, skeleton of the gills (gill arch skeleton)
- **Kopfskelett** head skeleton, cephalic skeleton
- **Kutikularskelett** cuticular skeleton
- **Rumpfskelett/Stammskelett/ Achsenskelett/Axialskelett** axial skeleton
- **somatisches Skelett** somatic skeleton
- **Viszeralskelett/Visceralskelett/ Eingeweideskelett** visceral skeleton, visceroskeleton
- **Zonoskelett (Extremitätengürtel)** zonoskeleton
- **Zytoskelett/Cytoskelett** cytoskeleton
Skelettmuskulatur skeletal musculature
Skelettnadel (Schwämme) spicule
Skene-Gänge Skene's tubules, Skene glands, paraurethral glands
Sklavenhaltung/Sklavenhalterei/ Dulosis (Ameisen) dulosis
Sklerenchym/Sclerenchym (*siehe dort*) sclerenchyma
sklerifiziert sclerified
Sklerifizierung sclerification
Sklerit (stark sklerotisierte Platte/ Nadel) sclerite
Sklerokarp sclerocarp
Sklerophyt sclerophyte
Skleroprotein scleroprotein
Sklerotesta sclerotesta
sklerotisch sclerotic
sklerotisiert sclerotized, hardened
Sklerotisierung sclerotization, hardening
Sklerotium/Dauermyzel sclerotium, hypothallus
Sklerotom sclerotome
Sklerozyt/Sclerocyt sclerocyte
Skolex scolex
Skolopidium/Scolopidium/ stiftführende Sensille scolopidium, scolophore, chordotonal sensilla

Skorbut scurvy
Skorpione/Scorpiones scorpions
Skrotalnaht/Skrotalraphe
scrotal raphe
Skrotalwulst scrotal swelling
Skrotum/Hodensack scrotum
Skulptur (Schalen-/
Samentopographie) sculpture
Smilacaceae/Stechwindengewächse
catbrier family
snRNP (kleines nukleäres
Ribonukleoprotein) snRNP (small
nuclear ribonucleic protein)
Sodbrennen heartburn,
acid indigestion
Sog/Zug (Wasserleitung) tension,
suction, pull
Sohlenballen sole pad
Sohlengang/Plantigradie
plantigrade gait
Sohlengänger (Plantigrade)
plantigrade
Solanaceae/
Nachtschattengewächse
nightshade family, potato family
Solanin solanine
Solarenergie/Sonnenenergie
solar energy
Solarzelle solar cell,
photovoltaic cell
Soldat (Insekten) soldier
Solenia solenia (gastrodermal tubes)
Solenoid (helikale
Chromatinstruktur) solenoid
Solenozyt solenocyte,
archinephridium
Solifluktion solifluction
soligen (Moore) soligenous
solitär/einzeln solitary, single
Solitärpflanze solitary plant
Sollwert nominal value, rated value,
desired value
Solstitialbewegung *bot*
sun tracking
Solubilisierung solubilization
Solvatation solvation
solvatisieren solvate
solvatisierter Stoff (Ion/Molekül)
solvate
somaklonale Variation
somaclonal variation
somatische Mutation
somatic mutation
somatische Rekombination
somatic recombination
somatische Zelle/Körperzelle
somatic cell, body cell
Somatoliberin somatoliberin,
somatotropin release-hormone,
somatotropin releasing factor (SRF),
growth hormone release hormone/
factor (GRH/GRF)

Somatolyse somatolysis
Somatomedin somatomedin, insulin-
like growth factor (IGF) (sulfation
factor/serum sulfation factor)
Somatopleura/somatisches Blatt/
Hautfaserblatt/parietales Blatt
somatopleure
Somatostatin somatostatin,
somatotropin release-inhibiting
factor, growth hormone release-
inhibiting hormone (GRIH)
Somatotropin/somatotropes
Hormon/Wachstumshormon
somatotropin (STH),
growth hormone (GH)
Somazelle/Körperzelle somatic cell,
body cell
Somit/Ursegment somite, somatome
Sommerfliedergewächse/
Buddlejaceae
buddleja family
sommergrün deciduous
Sommerholz summer wood
Sommerkleid *orn* summer plumage
sommerlich summer .., aestival
(appearing in summer)
Sommerschlaf *zool* summer sleep
Sommerschlaf/Trockenschlaf/
Ästivation *bot* estivation,
aestivation
Sommerwurzgewächse/
Orobanchaceae
broomrape family
Sonde (Mikrosonde) probe,
microprobe
● **mit Hilfe einer heterologen**
Sonde heterologous probing
● **Protonensonde**
proton microprobe
● **Ribosonde/RNA-Sonde**
riboprobe
Sondermüll hazardous waste
Sonifikation/Beschallung/
Ultraschallbehandlung
sonification, sonication
Sonnenblatt/Lichtblatt sun leaf
Sonnenbrand/Rindenbrand
sunscald
Sonneneinstrahlung insolation
Sonnenenergie/Solarenergie
solar energy
Sonnengeflecht/Plexus solaris/
Plexus coeliacus *neuro*
solar plexus, celiac plexus
Sonnenorientierung solar tracking
Sonnenröschengewächse/
Cistrosengewächse/
Zistrosengewächse/Cistaceae
rockrose family
Sonnenstich sunstroke
(heatstroke: Hitzschlag)
Sonnenstrahlung solar radiation

Sonnentaugewächse/Droseraceae
sundew family
Sonnentierchen/Heliozoen
sun animalcules, heliozoans
Sonnenwende *astr* solstice
Sonnenwendigkeit/
Lichtwendigkeit/Heliotropismus
heliotropism, solar tracking
Sonnenzeitalter solar age
Sonogramm sonogram
Sonographie/Ultraschalldiagnose
sonography, ultrasound,
ultrasonography
Soral (*pl* **Sorale**) **(Flechten)**
soralium (*pl* soralia)
• **Bortensoral/Randsoral**
marginal soralium
• **Flecksoral** maculiform soralium
• **Helmsoral/Gewölbesoral**
forniciform sorelium
• **Kopfsoral**
capitate/capitiform sorelium
• **Kugelsoral** globose soralium
• **Lippensoral** labriform soralium
• **Manschettensoral**
maniciform soralium
• **Punktsoral** punctiform soralium
• **Randsoral/Bortensoral**
marginal soralium
• **Spaltensoral** rimiform soralium,
fissoral soralium
Sorbens (*pl* **Sorbentien**) sorbent
Sorbinsäure (Sorbat) sorbic acid
(sorbate)
Sorbit sorbitol
Soredium soredium
Sorokarp sorocarp
Sorte sort, type, kind, variety,
cultivar
Sortenreinheit purity of variety,
variety purity
Sorus sorus, "fruit dot"
SOS-Antwort/SOS-Reaktion
SOS response
Soziabilität/Geselligkeitsgrad
sociability, gregariousness
Sozialbrache social fallow
Sozialisation sozialization
sozialisieren socialize
Sozialtrieb social drive
Sozialverhalten social behavior
Soziobiologie sociobiology
Soziologie sociology
Spacer/Zwischensequenz *gen* spacer
Spadix spadix (*pl* spadices)
Späher/Pfadfinder/Kundschaftler
(z. B. soziale Insekten) scout
Spalier *hort* espalier, trellis
Spalierobst espalier fruit
Spalt/Spalte cleft, crack, slit, crevice
Spaltamnion/Schizamnion
schizamnion

Spaltbarkeit cleavage
Spaltbein/Schizopodium
biramous appendage (schizopodal)
Spalte crevice, crack
spalten cleave, break, open, crack,
split, break down
Spaltensoral (Flechten) fissoral
soralium, rimiform soralium
Spaltfarngewächse/Schizaeaceae
curly-grass family,
climbing fern family
Spaltfrucht/Schizokarp schizocarp,
schizocarpium
Spaltfuß/Spaltbein
branched appendage/leg
• **einfach gegabelter Spaltfuß**
two-branched appendage,
biramous appendage
Spaltfüßer/Mysidacea
opossum shrimps
Spaltfusion cleavage fusion
spaltig/gespalten (Blattrand)
split, *suffix*: -fid
(pinnatifid > fiederspaltig)
Spaltkapsel longitudinally dehiscent
capsule
• **dorsicide/dorsizide S.**
dorsicidal capsule
• **septicide/septizide S.**
septicidal capsule
Spaltöffnung *bot* stoma,
stomatal pore
Spaltpfropfung/Pfropfen in den
Spalt *hort* wedge grafting,
cleft grafting
• **seitliche Spaltpfropfung**
side cleft grafting,
side whip grafting, bottle grafting
Spaltung cleavage, breakage,
opening, cracking, splitting,
breakdown; (Furchung) cleavage
Spaltungsregel (Mendel)
law/principle of segregation
Span-Okulation/Chipveredelung/
Chipveredlung chip budding
spannen (Fortbewegung bei
Spannerraupen) looping
Spannerraupe (Geometridae)
looper, measuring worm, inchworm,
spanworm
Spannfaden *arach* mooring thread,
guyline (e. g. spiral guyline)
Spannhaut/Flughaut/Gleithaut/
Flatterhaut/Patagium patagium
Spannkraft *physiol* tonicity
Spannung/Potentialdifferenz
potential difference, voltage
Spannungsklemme voltage clamp
spannungsregulierter Kanal
voltage-sensitive channel,
voltage-gated channel
Spannweite (Flügel) (wing) span

S Spannweite 686

Spannweite *stat* range
Spanplatte (Holz) chipboard
Sparganiaceae/Igelkolbengewächse bur-reed family
Spargelgewächse/Asparagaceae asparagus family
sparren (Scheinhiebe versetzen) spar
Sparrknochen/Chevron chevron
Spartein sparteine
spät auftretend/öffnend/ aufbrechend (z. B. Zapfen) serotinous, late in developing (e.g. cone)
Spatel *lab* spatula
spatelartig/spatelig spathose
spatelförmig spathulate, spatulate
Spatelplattenverfahren *micb* spread-plate method
spätes Gen late gene
Spatha/Blütenscheide spathe
Spätholz/Herbstholz/Engholz latewood
Spatzenzungengewächse/ Seidelbastgewächse/ Thymelaeaceae daphne family, mezereum family
Spechtvögel/Spechtartige/ Piciformes woodpeckers and barbets and toucans and allies
Speck (Schweinespeck) bacon
Speck/Tran/Blubber (Walspeck: Unterhautfettschicht) blubber
Speckschwarte (Schwein) pork rind
Speiche/Radius (*auch:* Netzspeiche) *arach* spoke, radius
Speichel saliva
Speicheldrüse salivary gland
Speicheltasche/Salivarium salivarium
Speichergewebe storage tissue
speichern/anreichern/akkumulieren store, save, accumulate
Speicherparenchym storage parenchyma
Speicherprotein storage protein
Speicherung storage
Speicherwurzel storage root
speien spit
Speisebrei/Magenbrei/Chymus chyme
Speisepilz culinary/edible mushroom
Speiseröhre/Ösophagus esophagus, oesophagus
Speisetrüffel/Tuberaceae truffles
Spelze husk, glume (small bract)
● **Deckspelze** lemma, lower palea, outer palea
● **Hüllspelze** glume
● **Vorspelze** palea, palet, pale, glumella, inner glume

Spender/Donor donor
sperlingartig passeriform
Sperlingsvögel/Passeriformes passerines, passeriforms (perching birds)
Sperma/Samen sperm
Spermanetz *arach* sperm web
Spermatogenese spermatogenesis
Spermatophore/Samenträger/ Samenpaket spermatophore, sperm packet
Spermatophyt/Samenpflanze spermatophyte, seed-bearing plant
Spermidin spermidine
Spermienbündel/Spermiozeugme/ Spermiodesmos spermozeugma
Spermienkonkurrenz sperm competition (*now:* sperm precedence)
Spermin spermine
Spermium/Spermatozoon/Samen/ Samenzelle sperm, spermium, spermatozoon
Sperrarterie/Polsterarterie artery with intimal cushions
sperren (Schnabel) gape, gaping
Sperrfilter *micros* selective filter, barrier filter, stopping filter, selection filter
Sperrholz plywood
Sperrkrautgewächse/ Himmelsleitergewächse/ Polemoniaceae phlox family
Spezialisierung specialization
Spezialist *ecol* specialist
spezielle Transduktion specialized transduction
spezifisch specific
● **unspezifisch** nonspecific
spezifische Wärme specific heat
spezifisches Gewicht (Dichte von Holz) specific gravity, wood density
Spezifität specificity
spezifizieren specify
Sphagnum-Moor peat bog
Sphäroplast spheroplast
Sphäroprotein/globuläres Protein globular protein
Sphärosom spherosome
Sphenophyllales/ Sphenophyllaceae/ Keilblattgewächse sphenophyllum family
Sphinganin sphinganine
Sphingosin sphingosine
Spiculum (Kopulationshaken: Nematoden) penial/copulatory spicule
Spiegel (Reh) rump patch

687 Spiralwindung S

Spiegelschnitt/Radialschnitt (Holz)
radial section, quartersawn
Spielgesang play song
Spieltheorie game theory
Spielverhalten play, play behavior
Spießer/Spießbock/Heldbock
buck of first year (roebuck/elk/stag:
with unbranched antlers)
spießförmig spear-shaped, hastate,
hastiform
Spinalfortsatz spinous process
• **oberer Dornfortsatz/
Neurapophyse** neurapophysis
Spinalganglion spinal ganglion
(dorsal root ganglion/posterior root
ganglion)
Spinalnerv spinal nerve
Spindel stalk, axis, spindle;
(Columella: Schneckenschale)
columella
• **Blattspindel/Rhachis/
Fiederblattachse (Mittelrippe
eines Fiederblattes)** rachis
Spindelapparat spindle apparatus
**Spindelbaumgewächse/
Baumwürgergewächse/
Celastraceae** spindle-tree family,
staff-tree family, bittersweet family
Spindeldiagramm spindle diagram
Spindelfaser spindle fiber
spindelförmig spindle-shaped,
fusiform
Spindelorgan/Muskelspindel
spindle organ, muscle spindle
Spindelpolkörper *fung*
spindle pole body
Spinnborste spinning bristle
**Spinndrüse/Seidendrüse/
Sericterium (Labialdrüse)**
spinneret, sericterium (labial gland)
Spinndrüse silk gland,
spinning gland
Spinndüse/Tubulus textori *arach*
(silk gland) spigot
spinnen (Netz/Kokon) spin
Spinnen/Webspinnen/Araneae
spiders
spinnenartig spiderlike, spidery,
arachnoid
spinnenartiges Tier arachnid,
arachnoid
Spinnenasseln/Scutigeromorpha
scutigeromorphs
**Spinnenfischartige/
Saugfischverwandte/Schildfische/
Gobiescociformes** clingfishes
Spinnenforscher/Spinnenkundler
araneologist, arachnologist *sensu lato*
Spinnenkunde araneology,
arachnology *sensu lato*
**Spinnenläufer/Notostigmophora/
Scutigeromorpha** scutigeromorphs

Spinnennetz spiderweb, cobweb
(Netztypen *siehe* Netz)
Spinnentiere/Arachniden arachnids
Spinnfaden silk thread, silk line
• **Angelfaden/Wurffaden**
casting line, "fishing line"
• **Beutefesselfäden (Hyptiotes)**
swathing band (orb weavers)
• **Fangfaden** catching thread,
trapline
• **Fangwolle** cribellate silk,
catching wool
• **rituelle Fesselfäden** bridal veil
(crab spiders)
• **Sicherheitsfaden** dragline,
securing thread
• **Stolperfaden** trip-line,
tripping line, barrier thread
• **Tropfenfaden** viscid line
• **Wurffaden/Angelfaden**
casting line
• **Zickzackfaden** pendulum line
Spinngewebe cobweb, spiderweb;
silk fabric
Spinngriffel/Cercus (Chilopoda)
spinneret, cecus
Spinnplatte/Spinnsieb/Cribellum
cribellum
Spinnseide silk
Spinnsieb/Spinnplatte/Cribellum
cribellum
Spinnspule (silk gland) spool
Spinnwarze spinner, spinneret
spinnwebartig/spinnennetzartig
spider-web like
Spinnwebe (Netz *oder* **Faden)**
spiderweb, cobweb *or*
silk thread/line
**Spinnwebenhaut/Arachnoidea
(mittlere Hirnhaut)** arachnoid
Spiraculum/Stigma spiracle, stigma
Spiralcaecum spiral cecum
Spiraldarm spiral intestine
Spirale/Helix spiral, helix
Spiralfaden/Taenidium
spiral thread, taenidium
(spiral thickening of intima)
Spiralfalte (Chondrichthyes)
spiral flap/valve
Spiralfurchung spiral cleavage
spiralig spiral, spiraled, twisted,
helical
spiralig aufgewickelt
spirally coiled, strombuliform
Spiralklappe (Froschherz)
spiral valve
**Spiralmine/Heliconomium
(Blattmine)** *bot* serpentine mine,
heliconome
Spiraltextur (Holz) spiral grain
Spiralwindung spiral winding,
coiling

Spirillen (sg Spirille) spirilla (sg spirillum)
Spiritus spirit
Spirometrie respirometry
Spirre/Trichterrispe/Anthela (Infloreszenz) anthela
spitz acute, sharp, pointed, sharp-pointed
spitz zulaufen (spitz zulaufend) taper (tapering/tapered), attenuate
Spitze point, tip, spike, fastigium
Spitze (zulaufende) spire
Spitze/Gipfel/Scheitelpunkt/ Höhepunkt apex, summit, peak
• **Blattspitze** leaf tip
Spitzeber/Binneneber cryptorchid pig
Spitzenmeristem apical meristem
Spitzenwachstum apical growth
Spitzhengst/Klopphengst (Kryptorchide) ridgeling, ridgling (cryptorchid)
Spitzhörnchen/Scandentia tree shrews
spitzig (mit steifer/harter Blattspitze) pungent
Splanchnopleura/Darmfaserblatt/ viscerales Blatt/viszerales Blatt splanchnopleure
Spleiß-Akzeptorstelle *gen* splice acceptor site
Spleiß-Donorstelle splice donor site
Spleiß-Junktion/ Spleiß-Verbindungsstelle splicing junction
Spleiß-Stelle splice site
• **verborgene S.** cryptic splice site
spleißen *gen* splice
Spleißen *gen* splicing
• **alternatives Spleißen** alternative splicing
• **differentielles Spleißen** differential splicing
Spleißosom spliceosome
Splinkers *gen* splinkers (sequencing primer linkers)
Splintholz sapwood, splintwood, alburnum
Spongiom spongiome
Spongozyt/Spongocyt spongocyte
Spontanabort/Fehlgeburt miscarriage
spontane Mutationsrate spontaneous mutation rate
Spontanmutation spontaneous mutation
Spontanzusammenbau/ Selbstzusammenbau self-assembly
Sporangienbehälter sporangiocarp
Sporangienträger sporangiophore
Sporangiole sporangiole
Sporangium sporangium

Spore spore
• **Arthrospore/Arthrokonidium/ Oidium/Oidie** arthrospore, arthroconidium, oidium
• **Chlamydospore/Gemme** chlamydospore (thick-walled resting spore)
• **Dauerspore/Hypnospore** persistent spore, dormant spore, resting spore, hypnospore
• **Diaspore/Ausbreitungseinheit/ Disseminule** diaspore, disseminule
• **Hypnospore/Dauerspore** hypnospore, persistent spore, dormant spore, resting spore
• **Karpospore** carpospore
• **Knospenspore/Conidie** conidium
• **Megaspore/Makrospore** megaspore, macrospore
• **Memnospore** memnospore (remains at place of origin)
• **Mikrospore** microspore
• **Pyknospore** pycnospore, pycnidiospore
• **Scolespore** scolespore
• **Teleutospore** teleutospore, teliospore
• **Zoospore/Schwärmer** zoospore, swarm cell
• **Zygospore/Hypnozygote** zygospore
Sporenbehälter/Sporangium spore case, sporangium
Sporenornament spore sculpture
Sporentierchen/Sporozoen spore-former, sporozoans
Sporenträger spore-bearing structure, sporophore
Sporn (Immunodiffusion) spur
Sporn *bot* **(z. B. an Blüte)** (floral) spur
Sporn/Calcar *zool* **(Knochen-/ Knorpelspange)** spur, calcar
Sporn/Fußgalle *zool* **(Pferd: Horn am hinteren Fesselgelenk)** ergot
Sporogonie sporogony, gamogony (in protozoans)
Sporokarp sporocarp
Sporophyt sporophyte
Sportfisch game fish
Sportfischerei sportfishing
Spottverhalten mocking behavior
Sprache language, speech
• **Körpersprache** body language
• **Symbolsprache** symbol language
• **Tanzsprache (Bienen)** dance language
• **Tiersprache** animal language
sprechen speak; talk
Spreite blade, lamina; phyllid; frond
spreiten spread

Spross

Spreitenrand/Blattspreitenrand
 leaf blade margin/edge
Spreitenspitze/Blattspreitenspitze
 leaf tip
spreitig/spreitenförmig laminar,
 laminiform, laminous
Spreitung spreading
Sprenger/Sprinkler (Bewässerung)
 sprinkler
Spreu/Kaff *bot* chaff (small dry
 scales/bracts)
spreuartig/voller Spreu chaffy,
 paleaceous
Spreuschuppe/Spreublatt
 ramentum, chaffy scale, palea, palet,
 pale
sprießen sprout, grow, bud
Springbein/Sprungbein jumping
 leg, saltatory leg, saltatorial leg
springen jump, spring, bound, leap
springendes Gen jumping gene
Springfrucht/Streufrucht/
Öffnungsfrucht dehiscent fruit
Springkrautgewächse/
Balsaminengewächse/
Balsaminaceae
 balsam family, jewelweed family,
 touch-me-not family
Springschwänze/Collembolen
 springtails, garden fleas
Springtide spring tide
Sprinkler/Sprenger (Bewässerung)
 sprinkler
Spritzbewässerung
 sprinkler irrigation
Spritze syringe
 • **Kanüle/Hohlnadel** needle
Spritze/Injektion shot, injection
Spritzenvorsatzfilter/Spritzenfilter
 lab syringe filter
Spritzflasche *lab* wash bottle
Spritzloch/Blasloch/Spiraculum
 (Wale) blow hole, vent, spiracle
Spritzwasser (Gischt *siehe auch*
 *dort***)** splash water
Spritzwasserzone/Spritzzone/
Gischtwasserzone/Gischtzone
(Supralitoral) splash zone
 (supralittoral zone)
Spritzwürmer/Sternwürmer/
Sipunculiden peanut worms,
 sipunculoids, sipunculans
Sprödblätterpilze/Sprödblättler/
Russulaceae Russula family
Spross/Trieb (junger Trieb) *bot*
 shoot, sprout
 • **Adventivspross/Adventivtrieb/**
  Zusatztrieb adventitious shoot
 • **Ausläufer (oberirdisch)/**
  Kriechspross/oberirdischer
  Stolon (photophil) runner,
  sarment

• **Ausläufer (unterirdisch)/**
 Erdspross/Rhizom/unterirdischer
 Stolon (geophil) rhizome
• **Ausläufer/Ausläuferspross/**
 Stolon/Stolo *allg* stolon
• **Brutspross/Brutknospe/Bulbille**
 bulbil
• **Erdspross/Rhizom/unterirdischer**
 Ausläufer/unterirdischer Stolon
 (geophil) rhizome
• **Flachspross/Phyllocladium**
 cladode, cladophyll, phylloclade
• **Flachspross/Platycladium**
 platyclade
• **Geiltrieb/Wasserschoss**
 water sprout, water shoot
• **Geiztrieb** sucker, side shoot,
 lateral shoot
• **Gipfeltrieb/Terminaltrieb/**
 Endtrieb apical shoot,
 terminal shoot
• **Hauptspross/Primärspross/**
 Hauptachse leading/main/primary
 shoot, main/primary axis
• **Infloreszenz-Kurztrieb** spur shoot
• **Jahrestrieb** annual shoot,
 one-year shoot, annual growth
• **Johannistrieb** lammas shoot
• **Kladodium/Cladodium**
 (Flachspross eines Langtriebs)
 cladode, cladophyll; *also*:
 phylloclade
• **Kriechspross/oberirdischer**
 Ausläufer/oberirdischer Stolon
 (photophil) runner, sarment
• **Kurztrieb** short shoot, short axis
• **Langtrieb** long shoot, long axis
• **Nebentrieb/Seitentrieb**
 lateral shoot, side shoot, offshoot
• **Schössling/Schoss (kleiner**
 Spross) (small) shoot, sprout,
 sprig
• **Seitenspross/Seitentrieb/**
 Nebentrieb side shoot,
 lateral shoot, offshoot
• **sylleptischer Trieb**
 sylleptic shoot
• **Terminaltrieb/Endtrieb/Gipfel-**
 trieb terminal shoot, apical shoot
• **Wassertrieb/Wasserreis**
 watershoot, water sprout,
 water sucker, coppice-shoot
• **wurzelbürtiger Spross/**
 Wurzelspross/Wurzeltrieb
 root sucker, offshoot, offset, slip;
 Erdspross/Gehölzausläufer> sobole
• **Wurzelhalsschössling** root-collar
 shoot, offshoot (kurz> offset)
• **Wurzelspross/Wurzelschössling/**
 Wurzelreis/Erdspross
 (Gehölzausläufer) root sucker,
 sobole; offset (short)

Sprossbündel shoot bundle (stem bundle)
sprossbürtige Wurzel shoot-borne root
Sprossdorn thorn (sharp-pointed modified branch)
Sprossknolle/Stengelknolle (oberirdisch) storage stem, stem-tuber; (unterirdisch) tuber, underground tuber, underground storage stem; (mit gedrungen-aufrechter Achse: *Gladiolus*) corm
Sprössling *bot* sapling
Sprosspflanze/Kormophyt cormophyte
Sprosspol (Embryo) *bot* shoot apex
Sprossranke (shoot) tendril, capreolus
Sprossscheitel/Sprossvegetationspunkt (Apicalmeristem des Sprosses) shoot apex, apical meristem of shoot
Sprossspitze shoot apex, shoot tip
Sprossung/Knospung sprouting, budding; (Hefe) budding
Sprossungsnarbe (Hefe) bud scar
Sprosszuwachs shoot elongation
Sprühgerät/Zerstäuber atomizer
Sprühregen mist, drizzle
Sprühwasser *mar* spray, ocean spray, sea spray
Sprühwasserzone spray zone
Sprungbein/Kanonenbein (Mittelfußknochen der Huftiere) cannon bone
Sprungbein/Springbein saltatory leg, saltatorial leg, jumping leg
Sprungbein/Talus talus, astragalus
Sprunggabel (Furca) furca; (Furcula) furcula
Sprunggabelhalter/Retinaculum furcal retinaculum
Sprunggelenk ankle joint; hock (horse)
Sprungschicht/Thermokline thermocline
Spüldrüse/von Ebnersche Drüse/von Ebner' Drüse gustatory gland
Spule spool, coil
• **Federspule** quill
• **Spinnspule** spool (silk gland spool)
Spülsaum *mar* drift line (on shore), swash mark, intertidal fringe
Spulwürmer (*Ascaris spp*) ascarid worms
Spur (Blattspur/Astspur) *bot* trace
Spur/Fährte track; scent
Spur/Überrest (meist *pl* Überreste) trace, remainder (meist *pl* remains)

Spurbiene/Kundschafterin scouting bee, scout bee
Spurenanalyse trace analysis
Spurenelement/Mikroelement trace element, microelement, micronutrient
Spurenfossil/Ichnofossil trace fossil, ichnofossil
• **Fluchtspuren/Fugichnia** escape traces
• **Fressbauten/Fodinichnia** feeding burrows
• **Jagdspuren/Verfolgerspuren/Praedichnia** predation traces
• **Kriechspuren/Repichnia** crawling traces
• **Ruhespuren/Cubichnia** resting traces
• **Schreitfährten/Gradichnia** tracks
• **Weidespuren/Pascichnia** grazing traces
• **Wohnbauten/Domichnia** dwelling structures
Spurenkunde/Ichnologie ichnology
Spürhaar/Sinushaar/Tasthaar/Vibrissa (ein Sinneshaar) tactile hair, vibrissa
Spurnaht/Sulcus groove, sulcus
Spurpheromon trail pheromone, trail substance
Spürsinn/Witterungssinn scent
sputtern/besputtern sputter
Sputtern/Besputtern/Besputterung (Metallbedampfung) sputtering
Squamata (Eidechsen & Schlangen) squamata (incl. lizards & amphisbaenians & snakes)
SRP (Signalerkennungspartikel) SRP (signal recognition particle)
SSCP (Einzelstrang-Konformations-Polymorphismus) *gen* SSCP (single strand conformation polymorphism)
Staat/Tierstaat animal colony
staatsbildend *entom* colony-forming
Staatsquallen/Siphonophora siphonophorans
Staatswald state forest
Stab-Kugel-Model/Kugel-Stab-Model *chem* stick-and-ball model, ball-and-stick model
Stäbchen/Stäbchenbakterien/Bazillen rods, bacilli
Stäbchen/Stäbchenzelle rod, rod cell
Stabdiagramm bar graph, bar diagram,
stabil stable
• **instabil/nicht stabil** unstable (instable)
stabile RNA stable RNA

691   **stammesgeschichtlich**   **S**

Stabiliment/Stabilimentum (Netz)
arach stabilimentum
● **zickzackförmiges S.**
  hackled band, zig-zag silk
Stabilisator stabilizer
stabilisieren stabilize
Stabilisierung stabilization
stabkerniger Neutrophil
  rod neutrophil, band neutrophil,
  stab neutrophil, stab cell
Staborgan/Staborganell (*Peranema*)
  rod organ, ingestion rod
  (cytopharyngeal basket)
Stachel (Epidermisauswuchs) prick,
  prickle
Stachel (Igel/Stachelschwein) quill
Stachel *sensu lato* spine, spike
Stachel/Stechborsten sting, stinger,
  piercing stylet
**Stachelbeergewächse/
  Grossulariaceae** gooseberry family,
  currant family
**Stachelfische/"Dornhaie"/
  Acanthodier/Acanthodii**
  spiny fishes, acanthodians
Stachelhaie/Squaliformes
  bramble sharks & dogfishes sharks
  and allies
**Stachelhäuter/Echinodermen/
  Echinodermata** echinoderms
stachelig/stachlig *sensu lato* spiky,
  spikey, spiny, thorny;
  (Epidermisauswüchse) prickly
**Stachelpilze/Stachelinge/
  Hydnaceae** tooth fungus family,
  toothed fungi
Stachelsaum bristle-like coat
  (cell surface: clathrin)
**Stachelsaumbläschen/
  Stachelsaumvesikel/Korbvesikel**
  coated vesicle
Stachelsaumgrübchen coated pit
Stachelscheide sting sheath
**Stachelschwänze/Igelwürmer/
  Echiuriden/Echiura**
  spoon worms, echiuroid worms
**stachelspitz (mit von Costa
  abgesetzter Spitze)** mucronate
  (hard-sharp pointed)
stachelspitzig cuspidate
**Stachelweichtiere/Aculifera/
  Amphineura**
  amphineurans
**Stachelzellschicht/Stratum
  spinosum epidermidis**
  spinous layer of epidermis
Stadium (*pl* Stadien) stage
Stadtökologie/Urbanökologie
  urban ecology
**Stadtwald/städtischer Wald/
  Kommunalwald//Gemeindewald**
  urban forest, community forest

Stall stable, sty (pigs), shed (cows),
  barn
Stallung(en) stables
staminat/männlich staminate, male
Staminodium staminode,
  staminodium
Stamm/Achsenkörper (*pl* Stämme)
  stem, shoot axis
● **Baumstamm/Holzstamm** stem,
  trunk, bole, shaft
● **>gefällter Baumstamm**
  log, lumber
● **Beerenrute** cane
● **Caudex/Stamm von Palmen und
  Baumfarnen** caudex, trunk of tree
  (palms and treeferns)
● **Hochstamm (Wuchsform eines
  Baumes)** standard tree, standard
● **Scheinstamm/Blattstamm (*Musa*)**
  false stem, pseudostem, leafy stem
Stamm *micb* strain
● **Bakterienstamm** bacterial strain
● **Inzuchtstamm** inbred strain
● **Referenzstamm** reference strain
Stamm (z.B. bei Siphonophora)
  stem
Stamm *syst/tax* (z.B. Tierstamm)
  phylum (*pl* phyla/phylums)
Stammablauf (Wasser an Bäumen)
  stem flow
Stammart stem species
● **Typus-Art** type species
Stammbaum family tree,
  genealogical diagram, dendrogram;
  gen pedigree
**Stammbaumforschung/
  Ahnenforschung/
  Familienforschung/Genealogie**
  genealogy
**stammbildend/stengeltreibend/
  cauleszent** caulescent
  (with stem above ground)
stammblütig/kauliflor/cauliflor
  cauliflorous
**Stammblütigkeit/Kauliflorie/
  Cauliflorie** cauliflory
**Stammbuch/Zuchtbuch/Herdbuch
  (*siehe:* Stutbuch)** studbook
Stammbündel/Stammleitbündel
  axial bundle, cauline bundle
stammbürtig stem-borne,
  arising from the stem, cauline
Stämmchen stemlet
**Stammesgeschichte/
  Stammesentwicklung/
  Abstammungsgeschichte/
  Phylogenie/Phylogenese/
  Evolution**
  phylogeny, phylogenesis, evolution
**stammesgeschichtlich/phyletisch/
  phylogenetisch/evolutionär**
  phylogenetic, phyletic, evolutionary

Stammfäule stem rot
Stammform/Urform primitive form, basic form, parent form
Stammfuß/Stammanlauf/Wurzelanlauf root butt, buttress (supportive ridge at base of tree trunk)
Stammholz log, lumber
Stammkrone crown
Stammkultur/Impfkultur stem culture, stock culture
stammlos acaulescent
Stammlösung stock solution
Stammnematogen stem nematogen
Stammreptilien/Cotylosauria stem reptiles, cotylosaurs
Stammschicht/Stammregion stem region, zone of tree trunks
Stammschleifenstruktur stem-loop structure
Stammskelett/Rumpfskelett/Achsenskelett/Axialskelett axial skeleton
Stammstück/Haftglied/Stipes (Maxille) stipes
Stammsukkulente stem succulent
Stammzelle (Vorläuferzelle) stem cell (precursor cell); (Initiale) initial
Standard/Typus type
Standardabweichung *stat* standard deviation, root-mean-square deviation
Standardbedingung standard condition
Standardfehler/mittlerer Fehler *stat* standard error (standard error of the means)
Standardisierung/Vereinheitlichung standardization
Standardtisch *micros* plain stage
Ständerpilz mushroom
Ständerpilze/Basidiomycetes club fungi
Standort habitat, place of growth *sensu stricto*; site, location (*see:* Fundort)
Standortansprüche habitat requirements
Standorttreue habitat fidelity
Standortbewertung habitat assessment (*sensu lato*: site assessment)
Standortlehre habitat ecology
Standvogel nonmigratory bird, resident
Stange pole
Stange (Geweih) main beam
Stängel *siehe* Stengel
Stapel stack
 • **Membranstapel** stacked membranes

Stapelkräfte stacking forces
stapeln stack
Staphyleaceae/Pimpernussgewächse bladdernut family
Stärke starch
Stärkekorn starch granule
Starkionendifferenz strong ion difference (SID)
Starklichtpflanze/Heliophyt heliophyte
Starre torpor
Startcodon/Initiationscodon *gen* initiation codon
Starterkultur (Anzuchtmedium) starter culture (growth medium)
stationäre Phase stationary phase, stabilization phase
stationärer Zustand/gleichbleibender Zustand steady state
Statistik statistics
statistische Abweichung statistical deviation
statistische Verteilung statistical distribution
statistischer Fehler statistical error
Stativ/Bunsenstativ *lab* support stand, ring stand, stand
Stativring *lab* ring (for support stand/ring stand)
Statoblast (Dauerknospe/Hibernaculum) statoblast (hibernaculum/winter bud)
Statozyste statocyst
Staubbeutel *bot* anther
Staubblatt *bot* stamen
Staubblattkreis androecium
Staubblüte/männliche Blüte staminate flower, male flower
Staubfaden *bot* filament
Staubgefäß *bot* stamen
Staubläuse/Psocoptera psocids
Stäublinge/Boviste/Lycoperdales puffballs
Staubzelle/Körnchenzelle/Rußzelle (Alveolarmakrophage) dust cell (large alveolar macrophage)
stauchen compress
Stauchung compression
Staude hardy/perennial herbaceous plant (*see:* Staudengewächs/Geophyt)
Staudengewächs/Geophyt/Erdpflanze/Kryptophyt/Cryptophyt geophyte, geocryptophyte, cryptophyte *sensu lato*
Staudruck-Ventilation *ichth* ram ventilation
Staunässe (Boden) waterlogging, waterlogged soil

693   **stenohalin** S

Stearinsäure/Octadecansäure (Stearat/Octadecanat) stearic acid, octadecanoic acid (stearate/octadecanate)

Stechapfelform/Echinozyt (Erythrozyt) burr cell, echinocyte, crenocyte

Stechborsten/Stachel sting, stinger, piercing stylet

stechen sting, pierce, puncture

stechend/beizend/ätzend (Geruch) pungent

stechend-saugend (Mundwerkzeuge) piercing-sucking, stylate-haustellate (mouthparts)

Stechpalmengewächse/ Aquifoliaceae holly family

Stechrochenartige/Myliobatiformes stingrays

Stechrüssel *entom* beak, "stinger"

Stechsauger *entom* piercing-sucking mouthparts

Stechwindengewächse/Smilacaceae catbrier family

Steckling *hort* cutting (slip>herbaceous or softwood)

Steckling mit Astring (Stammsteckling) heel cutting

Stecklingsvermehrung *hort* cuttage, propagation by cuttings

Stehapparat, passiver (Pferd) passive stay-apparatus

Stehgewässer/Stillgewässer/ stehendes Gewässer/ lenitisches Gewässer stagnant water, standing water, still water, lenitic water, lentic water

Stehkolben/Siedegefäß *lab* Florence boiling flask, Florence flask (boiling flask with flat bottom)

steifhaarig hispid

Steigbügel/Stapes (Ohr) stirrup, stapes

steigen (Flug/Gelände) climb

Steilufer (am Fluss) river bluff

Stein/Steinkern/Putamen (Endokarp) stone, pit, putamen, pyrene

Steinbrechgewächse/Saxifragaceae saxifrage family

Steinbruch *geol* quarry

Steineibengewächse/ Podocarpaceae podocarpus family

Steinfliegen/Uferfliegen/Plecoptera stoneflies

Steinfrucht stone, drupe, drupaceous fruit

Steinkanal stone canal, hydrophoric canal, madreporic canal

Steinkern (Putamen)/Stein stone, pit, putamen, pyrene

Steinkohle bituminous coal, soft coal (*siehe unter:* Kohle)

Steinkohlenwälder carboniferous swamp forests

Steinkorallen/Riffkorallen/ Madreporaria/Scleractinia stony corals, madreporarian corals, scleractinians

Steinläufer/Lithobiomorpha lithobiomorphs

Steinläuse/Petrophagaeae petrophagous lice (*see also: Pschyrembel 258th edn.*)

Steinobst stone fruit, drupaceous fruit

Steinschale pit casing

Steinzelle (in Blättern/Saftfrüchten) grit cell

Steinzelle (isodiametrisch/ palisadenförmig)/Sclereide/ Sklereide (Sclerenchym) stone cell, sclereid, sclereide, sclerid

Steiß/Steißbein/Os coccygis coccyx

Steißbein/Urostyl (frogs/toads) urostyl

Steißbeinwirbel/Steißwirbel coccygeal vertebra

Steißhühner/Tinamiformes tinamous

Stelärtheorie stelar theory

Stele stele, central cylinder

Stellenäquivalenz ecological equivalence

Stellglied controlling element, adjuster, actuator

Stellgröße adjustable variable

Stellknorpel/Áryknorpel/Cartilago arytaenoidea arytenoid cartilage

Stelzvögel/Schreitvögel/ Ciconiiformes herons & storks & ibises and allies

Stemmphase stance phase

Stempel/Pistill *bot* pistil

Stempel-Methode *micb* replica plating

Stempelblüte/weibliche Blüte carpellate/pistillate flower, female flower

Stengel/Stängel stalk; *bot* stipe

Stengelgemüse leaf stalk vegetable

Stengelknolle/Sprossknolle stem-tuber, storage stem
- **unterirdische S.** tuber, underground tuber, underground storage stem; (mit gedrungen-aufrechter Achse: *Gladiolus*) corm

Stengelmark pulp

stengelumfassend/amplexikaul stem-clasping, amplexicaul

Stenogastrie stenogastry

stenohalin stenohaline

stenök stenoecious, stenecious, stenoecic
stenophag stenophagous
Stenotele stenotele
stenotherm stenothermic, stenothermous, stenothermal
Steppenroller *bot* tumbleweed
Ster (Holz) stere (stack of cordwood: 1 cbm)
sterben *vb* die
Sterben *n* dying
Sterbetafel life table
sterblich mortal
- **unsterblich** immortal

Sterblichkeit/Sterberate/Mortalität mortality, death rate
- **Bruttosterberate** crude death rate
- **Unsterblichkeit/Immortalität** immortality

Stereocilien (Lateralisorgan) microvilli
Stereoisomer stereoisomer
stereoskopisches Sehen stereoscopic vision, binocular vision
steril/desinfiziert sterile, disinfected
steril/unfruchtbar sterile, infertile
- **fertil/fruchtbar** fertile

sterile Werkbank sterile bench
Sterilfiltration sterile filtration
Sterilisation/Sterilisierung sterilization, sterilizing
sterilisieren/unfruchtbar machen sterilize
Sterilität/Unfruchtbarkeit sterility, infertility
- **Fertilität/Fruchtbarkeit** fertility

Sterin/Sterol sterol
sterisch/räumlich steric, sterical, spacial
sterische Hinderung/sterische Behinderung steric hindrance
Sterkuliengewächse/ Kakaogewächse/Sterculiaceae cacao family, cocoa family
Stern (Pferd) snip (white spot)
Sternaktivität (veränderte Spezifität von Restriktionsenzymen) star activity
Sternanisgewächse/Illiciaceae star-anise family, illicium family
sternförmig star-shaped, stellate
Sternhaar stellate hair
Sternit (ventraler Sklerit/Teil des Sternum) sternite
Sternmine (Blattmine) *bot* star mine, asteronome
Sternparenchym stellate parenchyma
Sternum/Brustbein sternum, breastbone
Sternum/Brustplatte/Brustschild/ Bauchschild/Bauchteil (Insekten) sternum, ventral plate

Sternwürmer/Spritzwürmer/ Sipunculiden peanut worms, sipunculoids, sipunculans
Sternzelle stellate cell
sterzeln (Bienen) vibrating dance, dorsoventral abdominal vibrating dance (DVAV) (fanning with lifted abdomen: exposing scent organ)
Stetigkeit constancy, presence degree
Steuerfeder/Retrix (*pl* retrices) *orn* retrix (*pl* retrices)
steuern (in eine Richtung lenken) steer, steering; (regulieren) regulate, control
Steuerung (z. B. Stoffwechselvorgänge) control
Steuerung von Proteinen protein targeting
Steuerungsmechanismus regulatory mechanism
Stichkultur/Einstichkultur (Stichagar) stab culture
Stichlingsartige/ Stichlingverwandte/ Gasterosteiformes sticklebacks (and sea horses)
Stichprobe sample
- **Teilstichprobe** subsample
- **Zufallsstichprobe** random sample

Stichprobenerhebung sampling
Stichprobenfunktion *stat* sample function, sample statistic
Stichprobenumfang *stat* sample size
stickstoffenthaltend/Stickstoff... nitrogen-containing, nitrogenous
stickstofffixierende Bakterien nitrogen-fixing bacteria
Stickstofffixierung nitrogen fixation
stickstoffhaltige Base nitrogenous base
Stickstoffkreislauf nitrogen cycle
Stickstoffmangel nitrogen deficiency
Stickstoffverbindung nitrogenous compound, nitrogen-containing compound
Stickstoffzeiger nitrogen indicator
Stiel *bot/zool* stalk; pedicle, pedicel, peduncle
- **Blattstiel** leaf stalk, petiole
- **Blütenstiel** peduncle
- **Stiel eines Fiederblättchens** petiolule
- **Stiel einzelner Grasblüte** rachilla
- **Stiel einzelner Infloreszenzblüten** pedicel
- **kurzer Stiel** stipe
- **Moossporogon** seta
- **Pilzstiel** stipe

Stiel (Crinoide) *zool* stalk, stem, column, columna
Stielauge *entom* stalked eye
Stielchen/Caudicula/Kaudikula (Pollinienstielchen bei Gynostemium/Gynostegium) *bot* caudicle
Stielchen/Petiolus ("Taille"/ Hinterleibsstiel) *zool* waist, pedicel, petiole, podeon, podeum
Stielquallen/Becherquallen/ Stauromedusae stauromedusas
stielrund terete
Stielzelle/Dislokatorzelle/ Dislocatorzelle/Wandzelle (Cycadeenpollen) stalk cell
Stier (früh kastriert) steer (castrated early)
Stigma/Fleck stigma, spot
• **Tracheenstigma/Spiraculum** tracheal spiracle
Stigma/Spiraculum stigma, spiracle
Stigmasterin/Stigmasterol stigmasterol
Stilett stylet, stiletto
stilettförmig/griffelförmig styliform, prickle-shaped, bristle-shaped
Stillgewässer/Stehgewässer/ stehendes Gewässer/lenitisches Gewässer still water, stagnant water, standing water, still water, lenitic water, lentic water
Stillwasserzone *mar/limn* slack-water zone
Stimmband/Ligamentum vocale (*pl* Stimmbänder) vocal cord(s), vocal ligament
Stimmbandfortsatz/Processus vocalis vocal process
Stimmbildung/Lautbildung/ Phonation phonation
Stimmbruch change of voice, breaking of voice (at puberty)
Stimme voice
Stimmfalte/Stimmlippe/Plica vocalis vocal fold, true vocal cord
Stimmkopf/Syrinx *orn* syrinx (*pl* syringes/syrinxes)
Stimmritze/Rima glottidis (zw. Stimmlippen und Aryknorpeln des Kehlkopfs) rima glottidis (opening between the true vocal cords)
Stimmsack/Stimmbeutel vocal sac, vocal pouch, voice box
Stinkdrüse (*siehe:* Wehrdrüse) repugnatorial gland
Stinkmorcheln/Phallaceae stinkhorns, stinkhorn family
Stipel/Nebenblatt *bot* stipule
• **ohne Stipeln/nebenblattlos** exstipulate, astipulate, estipulate

Stipulardorn/Nebenblattdorn *bot* stipular spine
Stirn/Frons forehead, frons
Stirnauge/Scheitelauge (Stirn-Ocelle) dorsal ocellus
Stirnbein/Os frontale frontal bone
Stirndrüse/Frontaldrüse frontal gland
Stirnherz frontal heart, frontal sac
Stirnhöcker/Tuber frontale frontal tuber, frontal tuberosity
Stirnhöhle/Sinus frontalis frontal sinus
Stirnlappen/Frontallobus frontal lobe
Stirnleiste/Frontalleiste frontal carina
Stirnnaht/Sutura frontalis frontal suture
Stirnorgan frontal organ
Stirnplatte frontal plate
Stirnzapfen (Truthahn: Hautlappen an Schnabelbasis) snood
Stocherschnabel/Sondenschnabel *orn* probing bill
Stöchiometrie stoichiometry
stöchiometrisch stoichiometric(al)
Stock/Stecken stick, cane
• **Bienenstock** beehive
• **Grundstock/Grundlage (Fundament/Stammform)** foundation, base, stock
• **Tierstock (Korallenstock/ Bryozoenstock)** colony (corals/ bryozoans); cormus
• **Wurzelstock** rootstock, stock
Stockausschlag/Stockreis root bud, root sucker, tiller; sucker formation after coppicing
Stockbiene house bee
stockbildend forming a corm/ cormus; colonial
Stockwerk-Cambium/ etagiertes Cambium storied cambium, stratified cambium
stockwerkartig/etagiert/geschichtet storied, in tiers; stratified, layered
Stoffaustausch mass/substance exchange
Stofffluss material flow, chemical flow
Stoffkreislauf *ecol* nutrient cycle
• **Mineralstoffkreislauf** mineral cycle
• **Phosphorkreislauf** phosphorus cycle
• **Sauerstoffkreislauf** oxygen cycle
• **Schwefelkreislauf** sulfur cycle
• **Stickstoffkreislauf** nitrogen cycle
• **Wasserkreislauf** water cycle, hydrologic cycle

S Stoffübergang 696

**Stoffübergang/Massenübergang/
Stofftransport/Massentransport/
Massentransfer**
mass transfer
**Stoffübergangszahl/
Stofftransportkoeffizient/
Massentransferkoeffizient**
mass transfer coefficient
Stoffwechsel/Metabolismus
metabolism
● **Arbeitsstoffwechsel/
Leistungsstoffwechsel**
active metabolism
● **Betriebsstoffwechsel**
maintenance metabolism
● **Energiestoffwechsel**
energy metabolism
● **Grundstoffwechsel/
Ruhestoffwechsel**
basal metabolism
● **intermediärer Stoffwechsel/
Zwischenstoffwechsel**
intermediary metabolism
● **Leistungsstoffwechsel/
Arbeitsstoffwechsel**
active metabolism
● **Primärstoffwechsel**
primary metabolism
● **Sekundärstoffwechsel**
secondary metabolism
● **Synthesestoffwechsel/
Anabolismus**
synthetic reactions/metabolism,
anabolism
● **Zellstoffwechsel** cellular
metabolism
● **Zwischenstoffwechsel/
intermediärer Stoffwechsel**
intermediary metabolism
**Stoffwechselabbauprodukt/
Katabolit** catabolite
Stoffwechselmuster metabolic
pattern
Stoffwechselprodukt/Metabolit
metabolite
**Stoffwechselrate/
Stoffwechselintensität/
Stoffumsatz/Metabolismusrate**
rate of metabolism, metabolic rate
● **Arbeitsumsatz/Leistungsumsatz**
active metabolic rate
● **Grundstoffwechselrate/
Basalumsatz**
basal metabolic rate (BMR)
● **Standardstoffwechselrate**
standard metabolic rate
**Stoffwechselspektrum/
metabolisches Spektrum**
metabolic scope,
index of metabolic expansibility
Stoffwechselstörung metabolic
derangement, metabolic disturbance

**Stoffwechselsyntheseprodukt/
Anabolit** anabolite
Stoffwechselumsatz metabolic
turnover
Stoffwechselweg metabolic
pathway, metabolic shunt
stöhnen/ächzen moan
Stolon/Stolo//Ausläufer
(Gehölzausläufer) stolon;
(Hydrozoen) stolon (stalk-like
structure)
Stolperfaden *arach* trip-line,
tripping line, barrier thread
stolzieren strut (rooster),
prance (horse)
Stomiiformes (Tiefseefische)
deep-sea hatchetfishes and relatives
Stomochord stomochord, buccal
tube
Stopfen/Korken/Stöpsel *lab*
stopper, cork
**Stoppcodon/Terminationscodon/
Abbruchcodon** *gen* termination
codon, terminator codon, stop
codon, translational stop signal
● **PTC (vorzeitiges Stoppcodon)**
PTC (premature termination
codon)
Storaxgewächse/Styracaceae
storax family
**Storchennest/Abflachung der
Baumkrone** *bot/for* "stork's nest",
Storchennest (stunted treetop/crown)
**Storchschnabelgewächse/
Geraniengewächse/Geraniaceae**
geranium family, cranesbill family
**Störe & Löffelstöre/
Acipenseriformes** sturgeons &
sterlets & paddlefishes
Störgröße disturbance value,
interference factor
Stoß (mit Hörnern) butt
Stößel/Pistill (und Mörser)
pestle (and mortar)
**stoßen (Ziegen/Hirsche etc: mit
dem Kopf)** butting
Stoßfestigkeit (Holz) shock
resistance
Stoßtauchen *orn* power diving
(nose diving)
Stoßzahn tusk
Strahl ray; beam; jet
● **Baststrahl** bast ray
● **Branchiostegalstrahl/
Kiemenhautstrahl
(Radius branchiostegus)** *ichth*
branchiostegal ray
● **Federstrahl/Radius
(Bogenstrahl/Hakenstrahl)** *orn*
barbule (notched/hooked barbule)
● **Flossenstrahl (aus Hautknochen)/
Dermotrichium** dermotrichium

Strang S

697

- **Holzstrahl** wood ray
- **Hornstrahl (Huf)** horny frog
- **Hufstrahl/Cuneus ungulae** frog
- **Kiemenstrahl** gill ray
- **Lichtstrahl** beam of light
- **Markstrahl** *bot* pith ray, medullary ray
- **Polstrahl/Aster (meist** *pl*: **Asteren/Polstrahlen)** *cyt* aster (in mitosis)
- **Röntgenstrahl** X-ray
- **Sonnenstrahl** ray (of sunshine), sunbeam
- **Wasserstrahl** jet of water

Strahl *zool* (Pferde: Hufstrahl *siehe dort*) frog (triangular horny pad on underside of horse hoof)
Strahlbein/distale Sesambein/ Os sesamoideum distale navicular bone, distal sesamoid bone
strahlen shine; radiate
Strahlenbiologie radiation biology
Strahlenblüte (Zungenblüte) ray floret, ligulate flower
Strahlenbündel beam of rays
Strahlendiagramm *opt* ray diagram
Strahlenflosse ray fin
Strahlenflosser/Actinopterygii ray-finned bony fishes, actinopterygians
strahlenförmig/aktinomorph/ radiär/radiärsymmetrisch/zyklisch actinomorphic, radial, radially symmetrical, regular, cyclic
Strahlengang (Strahlendiagramm) path of light (ray diagram)
Strahlengriffelgewächse/ Actinidiaceae Chinese gooseberry family, actinidia family
Strahlenschäden radiation hazards, radiation injury
Strahlenschutz radiation control, radiation protection, protection from radiation
Strahlentherapie radiation therapy, radiotherapy
Strahlentierchen/Radiolarien radiolarians
Strahlkissen *siehe* Hufkissen
Strahlreaktor jet reactor
Strahlung radiation
- **Ausstrahlung/Emission/Ausstoss** emission
- **Bestrahlung** irradiation
- **Globalstrahlung** global radiation
- **ionisierende Strahlung** ionizing radiation
- **Kernstrahlung** nuclear radiation
- **photosynthetisch aktive Strahlung** photosynthetically active radiation (PAR)
- **radioaktive Strahlung** radioactive radiation

- **Sonneneinstrahlung** insolation
- **Sonnenstrahlung** solar radiation
- **Streustrahlung** scattered radiation, diffuse radiation
- **Wärmestrahlung** thermal radiation

Strahlungsenergie radiant energy
Strahlungsintensität radiation intensity
Strahlungsvermögen/ Emissionsvermögen (Wärmeabstrahlvermögen) emissivity
Strahlungswärme radiant heat
Strand beach, shore
- **Hochstrand/Sturmstrand (trockener Strand)** backshore
- **Vorstrand/Gezeitenstrand (nasser Strand)** foreshore

Strandbuhne/Seebuhne shore jetty
Stranddüne shore dune
Strandlinie/Küstenlinie shoreline, coastline
Strandmandelgewächse/ Combretaceae Indian almond family, white mangrove family
Strandpriel swash, tidal gully, tideway
Strandriff/Küstenriff fringing reef
Strang (*pl* **Stränge**) cord; *gen* strand
- **anticodierender Strang/Nicht-Sinnstrang/Matrizenstrang/ Mutterstrang/nichtcodierender Strang/Antisinn-Strang (transkribierter Strang)** anticoding strand, antisense strand, template strand
- **codierender Strang/ kodierender Strang/Sinnstrang (nicht-transkribierter Strang)** coding strand, sense strand
- **Doppelstrang** double strand
- **Einzelstrang** single strand
- **Folgestrang** lagging strand
- **Leitstrang** leading strand
- **Matrizenstrang/Mutterstrang/ anticodierender Strang/ nichtcodierender Strang/Nicht-Sinnstrang/Antisinn-Strang (transkribierter Strang)** template strand, antisense strand, anticoding strand
- **Minus-Strang/Negativ-Strang (nichtcodierender Strang)** minus strand (noncoding strand)
- **Nervenstrang** nerve strand, nerve cord
- **Plus-Strang/Positiv-Strang (codierender Strang)** plus strand (coding strand)
- **Sinnstrang** sense strand
- **Tochterstrang** daughter strand

Strang 698

Strang/Tractus (Nervenbahn) *anat/ neuro* tract
Strangassimilation *gen* strand assimilation
strangaufwärts *gen* upstream
Strangbruch (DNA) strand break
- **Doppelstrangbruch** double-strand break
- **Einzelstrangbruch** single-strand break
Strangmoor/Aapamoor string bog, aapa mire
Strangverdrängung *gen* strand displacement
Strasburger Zelle/Eiweißzelle Strasburger cell, albuminous cell
Strategie/Muster *ecol/evol* strategy, pattern
Stratifikation/Stratifizierung (Samenkeimung) stratification (seed germination)
Stratum germinativum/Keimschicht stratum germinativum, germinative layer
Strauch (pl Sträucher) shrub (*siehe*: Nanophanerophyt)
- **Dornstrauch/Dornenstrauch/ Dornbusch** thorn shrub, thorny thicket, thorn brush
- **Formbaum/Formstrauch (auch Zierschnitt)** topiary
- **Halbstrauch** half-shrub, semishrub, shrubby herb, suffrutescent plant
- **kleiner Strauch** shrublet
- **stacheliger Strauch** prickly shrub, bramble
- **Zierstrauch** ornamental shrub
- **Zwergstrauch/ holziger Chamaephyt** dwarf-shrub, woody chamaephyte
strauchartig shrub-like
Strauchbeeren/ Strauchbeerenobst bush fruit (*Ribes*: currents, gooseberries, etc.)
Strauchflechten fruticose lichens, shrublike lichens
strauchig shrubby, frutescent, fruticose
Strauchsavanne shrub savanna
Strauchschicht shrub layer (in lower canopy of forest)
Straußenvögel/Strauße/Laufvögel/ Struthioniformes ostriches
strecken (in die Länge ziehen) elongate, extend
Strecker/Extensor (Muskel) extensor
Streckspannung/Fließspannung yield stress

Streckung/Verlängerung elongation, extension
Streckungswachstum elongational growth, extension growth
Streckungszone region of elongation
Streifenanbau *agr* strip cropping
Streifenfarngewächse/Aspleniaceae spleenwort family
streifenförmig strap-shaped, ligulate
Streifenkörper/Basalkern/ Basalkörper/Corpus striatum striate body, corpus striatum
streifennervig/längsnervig striately veined, striate veined
streifig/gestreift/parallelgestreift striped, parallely striped
- **breitstreifig/breit gestreift/ gebändert** fasciate, broadely striped
- **feinstreifig/feingestreift** striate, finely striped
Streitkolbengewächse/ Casuarinaceae she-oak family, beefwood family
Streptoneurie/Chiastoneurie streptoneurous nerve pattern
Stress stress
stressen stress
Stressfaser stress fiber
stressig/anstrengend stressful
Streu litter
Streudiagramm scatter diagram (scattergram/scattergraph/ scatterplot)
streuen/verstreuen/ausstreuen/ verteilen scatter, spread, distribute
Streufrucht/Springfrucht/ Öffnungsfrucht dehiscent fruit
Streulichtsmessung/Nephelometrie nephelometry
Streuschicht/Streuhorizont/Förna (Wald) litter layer
Streustrahlung scattered radiation, diffuse radiation
Streutasche *ecol* litter bag
Streuung (Lichtstreuung) optical diffusion, dispersion, dissipation, scattering (light)
Streuung/Ausbreitung dispersal, dissemination
Streuung/Verstreuen/Verteilung scattering, spreading, distribution
Streuungstextur (Holz) irregular grain
Streuungsverhalten *stat* scedasticity, heterogeneity of variances
Streuwiese straw meadow
Strichdiagramm line diagram
Strichliste tally chart
Strichvogel bird of passage

Strickleiternervensystem
ladder-type nerve system,
double-chain nerve system
Stridulationsorgan/Schrillorgan/ Zirporgan stridulating organ
stridulieren/schrillen/zirpen
stridulate, chirp
Striegel (Kamm/Bürste: Pferdestriegel) currycomb
Striegel *entom* strigil, strigilis
(antennal comb/antennal cleaner; also file or scraper)
striegeln groom, brush; curry, currycomb (horses)
stringente Bedingungen/strenge Bedingungen stringent conditions
Stringenz (von Reaktionsbedingungen)
stringency (of reaction conditions)
Strobilation/Strobilisation
strobilization
Stroh straw
Strohblume strawflower
Strom (Flüssigkeit) stream, flow
Strom (großer Fluss) stream
Strom (Volumen pro Zeit) flow rate
Strom/Elektrizität *colloquial/ general* electricity, power, juice;
(Ladung/Zeit) current
stromaufwärts upstream
• **strangaufwärts** *gen* upstream
Strombrecher (z. B. an Rührer von Bioreaktoren) baffle
strömen stream, flow
Stromfluss *neuro* current flow
Stromquelle/Stromzufuhr *electr*
power supply
Stromschnelle rapids
Strömung *electr* flux
Strömung (Flüssigkeit)
current, flow
• **auf die Küste zufliessende Strömung** inshore current
• **Brandungslängsströmung/ Längsströmung (am Strand)**
longshore current
• **Brandungsrückströmung/ Rippstrom/Reißstrom** rip current
• **Gezeitenströmung/ Gezeitenstrom** tidal current
• **Konvektionsströmung/ Konvektionsstrom**
convection current
• **Konzentrationsströmung**
density current
• **laminare Strömung/ Schichtströmung** laminar flow
• **Meeresströmung** ocean current
• **Schichtströmung/laminare Strömung** laminar flow
• **Trübungsströmung/ Trübungsstrom** turbidity current

• **turbulente Strömung**
turbulent flow
• **Wirbelstrom (Vortex-Bewegung)**
eddy current
Strömungsmesser current meter
Strömungsmuster flow pattern
Strömungswiderstand flow
resistance, resistance to flow
Strophe (Vogelgesang)
verse (part of song)
STRPs (Polymorphismen von kurzen direkten Wiederholungen) *gen*
STRPs (short tandem repeat polymorphisms)
Strudel eddy, swirl
strudeln whirl, swirl, eddy
• **Nahrung herbei strudeln**
ciliary feeding
Strudelwürmer/Turbellarien
free-living flatworms, turbellarians
Strudler (Nahrungsstrudler)/ciliärer Suspensionsfresser ciliary feeder,
ciliary suspension feeder
Struktur structure
Struktur/Textur/Faser/ Fibrillenanordnung (Holz) grain
Strukturanalyse *chem*
structural analysis
Strukturaufklärung *chem*
structure elucidation
Strukturformel *chem*
structural formula
Strukturgen structural gene
Strukturprotein/Struktureiweiß
structural protein
Strunk/Blattstiel stipe
Strunk/Schaft/Stengel/Achse stalk,
stem, caudex
Strunk/Stumpf stump, stub, stool,
caudex
STS (sequenzmarkierte Stelle) *gen*
STS (sequence tagged site)
Stubbe(n)/Baumstubbe/ Baumstumpf tree stump
Studiergesang rehearsal song
Stufe level, stage
• **Entwicklungsstufe**
developmental level,
developmental stage
• **Höhenstufe**
altitudinal zone/region/belt
Stufenfolge/Rangordnung/ Rangfolge/Hierarchie order of
rank, ranking, hierarchy
Stuffer-DNA stuffer-DNA
Stufung zonation
• **vertikale Stufung**
altitudinal zonation
Stuhl/Fäzes/Kot (Mensch) stool,
feces
Stuhlgang/Darmentleerung/ Defäkation defecation, egestion

Stuhlgang haben/den Darm entleeren defecate, egest
Stuhlprobe stool sample
Stülpzitze crater teat
stumme Infektion/stille Feiung silent infection
stumme Mutation silent mutation
Stummel stump, stub
stummelartig stubby
Stummelbein/Stummelfuß stubby leg
Stummelfüßchen/Krüppelfüße/ Crepidotaceae Crepidotus family, crep fungus family
Stummelfüßer/Onychophoren velvet worms, onychophorans
Stummelschwanz (Pferde) dock
stummes Gen silent gene
stumpf obtuse, blunt
Stumpf/Strunk stump
• Stubbe/Stumpen/Baumstumpf tree stump, tree stub, "stool"
Stumpfaustrieb stump sprout, stump sucker, tiller
stumpfhöckrig/rundhöckrig/ bunodont (Zähne) with low crowns and cusps, bunodont
Sturm storm
• Schneesturm snowstorm; (heftiger S.) blizzard
Sturmbö/heftiger Sturmwind squall
Sturmwind gale, strong wind (51–101 km/h)
Sturmwurf/Windwurf wind fall
Sturzflug *orn* dive
Sturzpuppe/Pupa suspensa suspended pupa
Sturzquelle/Rheokrene flowing spring, rheocrene
Stutbuch/Gestütbuch/ Pferdestammbuch (Zuchtbuch für Pferde) studbook
Stute mare
• Zuchtstute broodmare
Stütze *hort/agr* prop; stake (zusätzliche Pfahlstütze)
stutzen/abschneiden (Fell) trim, crop
stützen support, prop up
Stützgewebe supporting tissue
Stützorgan fulcrum
Stützwurzel prop root, stilt root, brace root
Stützzelle supporting cell
Stygal (Grundwasser als Lebensraum) stygal
Stylidiumgewächse/ Säulenblumengewächse/ Stylidiaceae trigger plant family
stylopisieren/stylepisieren stylopize
Styracaceae/Storaxgewächse storax family

subalpin subalpine
Suberinsäure/Korksäure/ Octandisäure suberic acid, octanedioic acid
Suberinschicht suberin layer/lamella, suberized layer/lamella
suberisieren/verkorken suberize
Suberisierung/Verkorkung (Suberinanlagerung/ Suberinauflagerung) suberization; suberification
subgenomische Bibliothek/ subgenomische Genbank subgenomic library
Subgerminalhöhle subgerminal cavity
Subitanei/Jungfernei parthenogenetic egg
Subklima subclimate
Subklonierung subcloning
Subkultur/Subkultivierung/Passage (einer Zellkultur) subculture, passage (of cell culture)
subletal sublethal
Sublimation sublimation
sublimieren sublimate
Sublitoral (Zone des Kontinentalschelfs) sublittoral (continental shelf zone)
submental/unter dem Kinn submental, beneath the chin
Submerskultur submerged culture
Subsistenz subsistence
Subspezies/Unterart subspecies
Substanz P substance P
substituieren substitute
Substitution substitution
Substitutionsvektor replacement vector
Substrat substrate
• Folgesubstrat following substrate
• Leitsubstrat leading substrate
Substraterkennung substrate recognition
Substratfresser substrate feeder
Substrathemmung/ Substratüberschusshemmung substrate inhibition
Substratkettenphosphorylierung substrate-level phosphorylation
Substratkonstante (K_S) substrate constant
Substratsättigung substrate saturation
Substratspezifität substrate specificity
subtraktive Genbank/ Subtraktionsbank/ Subtraktionsbibliothek subtractive library
subtraktive Klonierung subtractive cloning

Subtypisierung subtyping
Succinylcholin succinylcholine
Sucht/Abhängigkeit addiction,
dependance
Suchtest *gen/med* screening,
screening test
süchtig/abhängig addicted,
dependant
süchtig machend/suchterzeugend
addictive
Süchtigkeit addiction
Suchtmittel/Droge drug
Suchtprophylaxe drug prevention
Süd-Huftiere/Notoungulata
notoungulates
südlich *biogeo* southern, austral
Suhle wallow
suhlen *vb* wallow
Suizidhemmung suicide inhibition
Sukkulente succulent
Sukkulenz/Dickfleischigkeit
succulence
Sukzession succession
● **primäre S./Erstbesiedlung**
primary succession
● **sekundäre S./Zweitbesiedlung**
secondary succession
Sukzessionsstufe/
Sukzessionsstadium
(ökologische S.) seral stage
Sulfat sulfate
Sulfurikanten sulfuricants
Sumachgewächse/Anacardiaceae
sumac family, cashew family
Summation (räumliche/zeitliche)
neuro (spatial/temporal) summation
Summe sum, total
summen (Insekten/Kolibri etc.)
hum, buzz
Summenformel *chem*
molecular formula
Summenhäufigkeit/
kumulative Häufigkeit *stat*
cumulative frequency
Summenpotential gross potential
Summenregel sum rule
Sumpf swamp (im Englischen:
vorwiegend bewaldeter Sumpf)
● **Küstensumpf** coastal swamp
● **Mangrovensumpf**
mangrove swamp
● **Riedsumpf** reed swamp
● **Salzsumpf** salt swamp
● **Tropensumpf/tropischer Sumpf**
tropical swamp
Sumpfblumengewächse/
Limnanthaceae false mermaid
family, meadowfoam family
Sumpferde muck
Sumpffarngewächse/Lappenfarne/
Thelypteridaceae
marsh fern family

Sumpffieber/Wechselfieber/Malaria
(*Plasmodium spp.*) malaria
sumpfig swampy, boggy
Sumpfland/Sumpflandschaft
swampland, moorland
Sumpfmoor muskeg
Sumpfpflanze (Moorpflanze)
helophyte, marsh plant (bog plant)
Sumpfschildkröten terrapins
Sumpfwald swamp forest
Sumpfwiese swamp meadow
Sumpfzypressengewächse/
Taxodiumgewächse/Taxodiaceae
swamp-cypress family,
redwood family, taxodium family
Supergenfamilie supergene family
Superhelix superhelix, supercoil
Superinfektion/Überinfektion
superinfection
Superovulation superovulation
Superpositionsauge
superposition eye
● **neurales S.**
neural superposition eye
● **optisches S.**
optical superposition eye,
clear-zone eye
superspiralisiert/superhelikal/
überspiralisiert supercoiled
Supinationsstellung
supinated position
Suppengrün herbs/vegetables for
soup making
Suppenkraut potherb
Suppression/Unterdrückung
suppression
Suppressorgen suppressor gene
Suppressormutation
suppressor mutation
supprimieren/unterdrücken/
zurückdrängen suppress
Surra (*Trypanosoma evansi*) surra
suspendieren (schwebende Teilchen
in Flüssigkeit) suspend
suspensionsfressend
suspension-feeding
Suspensionsfresser suspension feeder
Suspensor/Träger suspensor;
stalk (*Marchantia*)
süß sweet
Süße sweetness
Süßgräser/Gräser/echte Gräser/
Spelzenblütler/Gramineae/
Poaceae grasses, grass family
Süßstoff sweetener
Süßwasser freshwater
Süßwasserbryozoen/Armwirbler/
Lophopoda/Phylactolaemata
phylactolaemates,
"covered throat" bryozoans
Switchregion/Schalterregion
switch region

Syconium   702

Syconium syconium
Syllepsis syllepsis
sylleptischer Trieb sylleptic shoot
Sylvische Furche/Fissura lateralis cerebri lateral cerebral sulcus
Symbiont *allg* symbiont
Symbiont (in gegenseitiger Lebensgemeinschaft) mutualist
Symbiose *allg* symbiosis
Symbiose (gemeinnützige) mutualistic symbiosis, mutualism
● **Putzsymbiose** cleaning symbiosis
symbiotisch *allg* symbiotic
symbiotisch (gemeinnützig) mutualistic
Symmetrie symmetry
● **Bilateralsymmetrie** bilateral symmetry
● **Biradialsymmetrie** biradial symmetry
● **fünfstrahlige Symmetrie** pentamerous/pentameral symmetry, five-sided symmetry
● **Radiärsymmetrie/ Radialsymmetrie** radial symmetry
sympathisch (autonomes Nervensystem) sympathetic
Sympatrie sympatry
sympatrisch (in gleichen Arealen) sympatric
sympetal sympetalous
Symphile/echter Gast symphile
Symphilie/Gastpflege symphily
Symphorismus symphorism
Symphyse/Verwachsung symphysis, coalescence
"Symphyse"/Schambeinfuge/ Schamfuge/Symphysis pubica pubic symphysis
symplastes/symplastisches Wachstum symplastic growth
symplesiomorph symplesiomorphic
Symplocaceae/ Rechenblumengewächse sweetleaf family
sympodial sympodial, determinate
Sympodium/Scheinachse sympodium, pseudaxis
Symport symport
synandrisch synandrous
synanthrop synanthropic
synapomorph synapomorphic
Synapomorphie synapomorphy, cladistic homology
Synapse synapse
Synaptikel synapticle
synaptisch synaptic
synaptischer Spalt/Synapsenspalt synaptic cleft
synaptisches Potential synaptic potential

synaptonemaler Komplex synaptonemal complex
Synaptosom/synaptisches Vesikel/ synaptisches Bläschen synaptosome, synaptic vesicle
Synarthrose/Fuge/Haft synarthrosis, synarthrodial joint
Synchondrose/Knorpelhaft synchondrosis
Synchronkultur synchronous culture
syncytial/synzytial syncytial
Syncytium/Synzytium syncytium
Syndaktylie syndactylism
Syndrom/Symptomenkomplex syndrome, complex of symptoms
Synergismus/gegenseitige Förderung synergism
Synergist/Mitspieler/Förderer synergist
Syngamie/Gametogamie syngamy, gametogamy
syngen syngenic
Syngenese syngenesis
synkarp syncarpous
Synnema synnema
Synökologie synecology
Synostose/Knochenhaft synostosis, synosteosis
Synovialmembran synovial membrane
syntäne Gene (Gene auf *einem* Chromosom) syntenic genes
Syntänie synteny
Synthese synthesis
● **Biosynthese** biosynthesis
● **Chemosynthese** chemosynthesis
● **DNA-Synthese** DNA synthesis
● **Halbsynthese** semisynthesis
● **Neusynthese/*de-novo* Synthese** *de-novo*-synthesis
● **Photosynthese** photosynthesis
● **präbiotische Synthese** prebiotic synthesis
Synthesestoffwechsel/Anabolismus synthetic reactions/metabolism, anabolism
synthetisch synthetic(al)
● **biosynthetisch** biosynthetic(al)
● **halbsynthetisch** semisynthetic(al)
● **vollsynthetisch** totally synthetic(al)
synthetisieren synthesize
Syntypus/Syntyp syntype
Synusie/Synusia/Lebensverein synusia
Syrinx *orn* syrinx, voice box
System der amplifizierungsresistenten Mutation (ARMS) amplification refractory mutation system (ARMS)
Systemanalyse systems analysis

Systematik/Taxonomie systematics, taxonomy
Systematiker/Taxonom systematist, taxonomist
systematisch/taxonomisch systematic, taxonomic

systemisch systemic
Szintillationszähler ("Blitz"zähler) scintillation counter, scintillometer
szintillieren/funkeln/Funken sprühen/glänzen scintillate

T

T-Effektorzelle 704

T-Effektorzelle effector T cell
T-Zelle T cell, T lymphocyte
(T = thymic)
- **cytotoxische T-Zelle**
cytotoxic T cell, killer T cell,
T-killer cell (T_K or T_c)
- **T-Effektorzelle** effector T cell
- **T-Helferzelle/Helfer T-Zelle**
helper T cell, T-helper cell (T_H)
- **T-Suppressorzelle/Suppressor**
T-Zelle suppressor T cell,
T-suppressor cell (T_S),
regulator T-cell, regulatory T-cell
- **T-Vorläuferzelle** pre-T cell,
T-cell precursor
Tabakmosaik-Virus
tobacco mosaic virus
Taccaceae/Erdbrotgewächse
tacca family
Tachytelie tachytely
tachytelisch tachytelic
Tag-Nacht-Gleiche/Äquinotikum/
Tagundnachtgleiche
equinox
Tag-Nacht-Periodizität/Tag-Nacht-
Rhythmus (24-Stunden-Takt)
diel periodicity, diel pattern
tagaktiv day-active, diurnal
- **nachtaktiv** nocturnal
Tagblüher/Tagpflanze diurnal plant
Tageslänge day length
Tagesperiodizität/
Tag-Nacht-Periodizität
(24-Stunden-Takt)
diel periodicity, diel pattern
Tagesrhythmik/circadiane Rhythmik
circadian rhythm
Tagesrhythmus (Gegensatz zu:
Nachtrhythmus) diurnal rhythm
Tagliliengewächse/
Hemerocallidaceae daylily family
Tagma (*pl* **Tagmata**) tagma (fusion
of somites)
Tagmatisierung tagmatization,
tagmosis
tagneutrale Pflanze
day-neutral plant
Tagpflanze/Tagblüher diurnal plant
Taiga (Nadelwald der gemäßigten
Zone)
taiga (temperate coniferous forest)
Taille waist
"Taille"/Hinterleibsstiel/Stielchen/
Petiolus *entom* waist, podeon,
podeum, pedicel, petiole
Talg *med* sebaceous matter, sebum;
zool tallow (extracted from animals),
suet (from abdominal cavity of
ruminants)
Talg.../talgig sebaceous, tallowy
Talgdrüse sebaceous gland
Talsperre valley barrage (dam)

Tamariskengewächse/Tamaricaceae
tamarisk family, tamarix family
Tandemanordnung/
Tandemwiederholung *gen*
tandem duplication, tandem repeat
Tandemlauf *ethol* tandem running
Tandemwiederholungen *gen*
tandem repeats
- **variable Anzahl von**
Tandemwiederholungen (VNTR)
variable number of tandem repeats
(VNTR)
Tang/Seegras/Seetang seaweed
Tangentialschnitt tangential section
Tanggrasgewächse/Cymodoceaceae
manatee-grass family
Tannat (Gerbsäure)
tannate (tannic acid)
Tännelgewächse/Elatinaceae
waterwort family
Tannenfamilie/Kieferngewächse/
Föhrengewächse/Pinaceae
fir family, pine family
Tannenwedelgewächse/
Hippuridaceae marestail family,
mare's-tail family
Tannin (Gerbstoff) tannin (tanning
agent)
Tanz dance
- **Bienentanz (***siehe auch dort***)**
bee dance
Tanzsprache (Bienen) dance language
Tapetum tapetum
- **amöboides Tapetum**
ameboid tapetum
- **Periplasmodialtapetum** *bot*
plasmodial tapetum
- **Sekretionstapetum** *bot*
secretory tapetum
Taphonomie (Fossilisationslehre)
taphonomy
Taphozönose taphocenosis
Tara (Gewicht des Behälters/der
Verpackung) tare (weight of
container/packaging)
Tardigraden (*sg*** Tardigrad** *m***)/**
Bärentierchen/Bärtierchen
tardigrades, water bears
tarieren tare (determine weight of
container/packaging in order to
substract from gross weight)
Tarnfärbung/Schutzfärbung cryptic
coloration, concealing coloration
Tarntracht *orn* cryptic dress
(plumage/pelage/coat)
Tarnung camouflage
Tarpunähnliche/Elopiformes
tarpons
Tarsaldrüse tarsal gland
Tarsenspinner/Fußspinner/Embien/
Embioptera webspinners,
footspinners

Teilfrucht

Tasche (Enzym) pocket
Tasche/Beutel (Marsupialia) pouch
Tasche/Vertiefung (Elektrophorese-Gel) well, depression (at top of gel)
Taschenklappe (Herz)
semilunar valve
(consisting of three semilunar cusps)
Taste button, knob, key
tasten feel, touch, palpate
Taster/Tastfühler/Palpe labial feeler, palp
Tasterträger/Palpiger palpiger
Tasthaar/Spürhaar/Sinushaar/Vibrissa tactile hair, vibrissa
Tastkopf *micros* probe, probing head
Tastkörperchen tactile sensilla
Tastorgan tactile organ, touch sense organ
Tastsinn tactile sense, sense of touch
TATA-Box *gen* TATA box
Tatze/Pfote paw
Tau *meteo* dew
taub/gefühllos numb
taub/gehörlos deaf
taub/leer/hohl (Frucht/Same) empty, seedless
Taubenschlag dovecote, pigeonry
Taubenvögel/Columbiformes doves & pigeons and allies
Taubheit/Gefühllosigkeit numbness
Taubheit/Gehörlosigkeit deafness
Tauchflächenreaktor immersing surface reactor
Tauchglocke (Wasserspinne) *arach* diving bell
Tauchsieder *lab* immersion heater
Tauchstrahlreaktor plunging jet reactor, deep jet reactor, immersing jet reactor
taumeln (Bakterien) tumble
Taurin taurine
Täuschblume deceptive flower
Täuschung deception, delusion; illusion
Tausendblattgewächse/Seebeerengewächse/Haloragaceae water milfoil family
Tausendfüßler/Tausendfüßer/Myriapoden/Myriapoda millipedes ("thousand-leggers"), myriapodians
tautomere Umlagerung tautomeric shift
Tautropfen dewdrop
Taxaceae/Eibengewächse yew family
Taxis (*pl* Taxien) taxis (*pl* taxes)
Taxodiumgewächse/Sumpfzypressengewächse/Taxodiaceae redwood family, swamp-cypress family, taxodium family
Taxon/taxonomische Einheit taxon, taxonomic unit
Taxonom taxonomist
Taxonomie (biologische Klassifizierung) taxonomy
● **numerische Taxonomie/Phänetik** numerical taxonomy, phenetics, taxometrics
Technik (einzelnes Verfahren/Arbeitsweise) technique, technic
Technik/Technologie (Wissenschaft) technology
● **Umweltverfahrenstechnik** environmental process engineering
Technikfolgenabschätzung technology assessment
● **US-Büro für Technikfolgenabschätzung** OTA (Office of Technology Assessment)
technisch technic(al); (Laborchemikalie) lab grade
technischer Assistent (technische Assistentin)/Laborassistent (Laborassistentin)/Laborant (Laborantin) laboratory technician, lab technician, technical lab assistant
Technologie technology
technologisch technologic(al)
Teekräuter tea herbs
Teestrauchgewächse/Teegewächse/Kamel(l)iengewächse/Theaceae tea family, camellia family
Tegment/Knospenschuppe/Knospendecke tegmentum, protective bud scales
Tegula tegula (tile-shaped structure)
Teich pond
● **kleiner Teich/Tümpel** pool
Teichfadengewächse/Zannichelliaceae horned pondweed family
Teichonsäure teichoic acid
Teichuronsäure teichuronic acid
Teigreife (Getreide) dough stage (wax-ripe stage)
Teil (des Ganzen) moiety, part, section
Teil/Anteil/Hälfte moiety
Teilblume/Merianthium partial flower, merianthium
Teilblütenstand/Teilinfloreszenz partial inflorescence
Teilchen/Partikel particle
Teilchengröße (Bodenpartikel) particle size, soil texture
teilen divide, fission, separate
Teilerhebung *stat* partial survey
Teilfrucht/Karpid (ein ganzes Karpell) fruitlet

Teilfrucht

Teilfrucht/Merikarp (Teil eines Karpells) mericarp

teilig/geteilt (Blattrand) parted, partite

Teilkorrelationskoeffizient *stat* partial correlation coefficient

Teilmenge/Portion/Fraktion portion, fraction

Teilmengenauswahl *stat* subset selection

Teilstichprobe *stat* subsample

Teilung division, fission, separation
- **Äquatorialteilung** equatorial division
- **Furchungsteilung/Blastogenese** blastogenesis
- **Furchungsteilung/Eifurchung** cleavage; segmentation
- **Gabelteilung/Gabelung/ Dichotomie** forking, bifurcation, dichotomy
- **Kernteilung/Mitose** nuclear division, mitosis
- **Längsteilung** longitudinal division, fission
- **Reduktionsteilung/Reifeteilung/ Meiose** reduction division, meiosis
- **Vielfachteilung/Mehrfachteilung (Bakterien)** multiple fission
- **Zellteilung** cell division, cytokinesis
- **Zweiteilung/binäre Zellteilung** binary fission, bipartition

Teilungsphase division phase

Teilzieher *orn* partially migratory bird

Tektorialmembran/Deckmembran/ Membrana tectoris tectorial membrane

Teleutospore teleutospore, teliospore

Telma telma

Telmatophyt telmatophyte (wet meadow plant)

Telomer telomere

Telomtheorie telome theory

telozentrisches Chromosom telocentric chromosome

Temperatur temperature
- **obere kritische Temperatur (OKT)** upper critical temperature (UCT)
- **Phasenübergangstemperatur** phase transition temperature
- **Umgebungstemperatur** ambient temperature
- **untere kritische Temperatur (UKT)** lower critical temperature (LCT)

temperaturabhängig temperature-dependent

Temperaturgradient temperature gradient

Temperaturorgel *ecol* temperature-gradient apparatus

temperatursensitive Mutation temperature-sensitive mutation

temperenter Phage temperate phage

Tenaculum tenacle, tenaculum

Tentaculaten/Kranzfühler/ Fühlerkranztiere tentaculates (bryozoans/phoronids/brachiopods)

Tentaculiferen/tentakeltragende Rippenquallen (Ctenophora) tentaculiferans, "tentaculates"

Tentakel/Fanghaar tentacle

Tentakelarm tentacular arm

Tentakelebene tentacular plane

Tentakelgefäß (Ctenophoren) tentacular canal

Tentakelscheide (Ctenophoren) tentacle sheath

Tentorium tentorium

Tentoriumbrücke/Corpus tentorii/ Corpotentorium tentorial ridge

Tentoriumgrube tentorial pit

teratogen/Missbildungen verursachend teratogenic

Teratogenese/ Missbildungsentstehung teratogenesis, teratogeny

Teratologie (Lehre von Missbildungen) teratology

Teratom teratoma

Tergit/Rückenplatte (dorsale Sklerite) tergite

Tergum/Rückenschild tergum, back, roof, dorsal plate (consisting of tergites)

terminale Redundanz terminal redundancy

Terminalknospe/Endknospe terminal bud

Terminaltrieb/Endtrieb/Gipfeltrieb terminal shoot, apical shoot

Terminationscodon/Abbruchcodon/ Stoppcodon termination codon, terminator codon, stop codon

Terminus/Ende (Molekülende) terminus

Termiten/"Weiße Ameisen"/Isoptera termites

Terpen terpene

Terpentinharz pitch (resin from conifers)

Terrasierung terracing

terrestrisch/landlebend terrestrial, land-dwelling

Territorialität territoriality

Territorialverhalten territorial behavior

Territorium/Revier/Gebiet/ Wohnbezirk territory, range

Tertiär/Tertiärzeit/Braunkohlenzeit (erdgeschichtliche Periode) Tertiary, Tertiary Period
Tertiärfollikel/Graafscher Follikel/ Graaf-Follikel Graafian follicle, vesicular ovarian follicle
Tertiärstruktur (Proteine) tertiary structure
Test/Prüfung assay
Test auf verkürzte Proteine protein truncation test
testikuläre Feminisierung testicular feminization
Testis-Determinationsfaktor testis-determining factor (TDF)
Testkreuzung testcross
Testmedium/Prüfmedium (zur Diagnose) test medium
Testosteron testosterone
Testpartner *gen* tester
Testverfahren test procedure, testing procedure
Tetanus/Wundstarrkrampf (*Clostridium tetani*) tetanus
Tetrade tetrad
Tetradenanalyse tetrad analysis
tetraedrisch tetrahedral
tetraparental tetraparental
tetraploid tetraploid
Teuerlinge/Vogelnestpilze/ Nidulariaceae bird's-nest fungi, bird's-nest family
Teufelsklauengewächse/ Huperziaceae fir clubmoss family
Texasfieber (*Babesia ssp.*) Texas fever, red-water fever, hemoglobinuric fever (babesiosis)
Textur/Struktur/Faser/ Fibrillenanordnung (Dichte der Leitelemente in Jahresring) texture (*see* grain)
thallös membranous, foliose (body type/construction)
Thallus (*pl* Thalli)/Lager thallus (*pl* thalli/thalluses)
● **Fadenthallus** filamentous thallus
● **Prothallium/Vorkeim (Farne)** prothallus
Thanatose/Totstellen/Totstellung thanatosis, feigning death
Thanatozönose thanatocenosis
THE (transponierbares menschliches Element) *gen* THE (transposable human element)
Theaceae/Teegewächse/ Kamel(l)iengewächse/ Teestrauchgewächse tea family, camellia family
Thebain thebaine
Thein/Koffein theine, caffeine
Theka theca

Thekamöben/beschalte Amöben/ Testacea testate amebas
Theligonaceae/Hundskohlgewächse theliogonum family
Thelypteridaceae/ Lappenfarngewächse/ Sumpffarngewächse marsh fern family
Thelytokie thelytoky, thelyotoky
Theobromin theobromine
Theophrastaceae Joe-wood family
Theophyllin theophylline
theoretisch theoretic, theoretical
Theorie theory
Thermalquelle thermal spring
Thermiksegelflug/Thermiksegeln *orn* thermal soaring
Thermodynamik thermodynamics
● **1./2. Hauptsatz (der Thermodynamik)** first/second law of thermodynamics
Thermogenese thermogenesis
Thermokline/Sprungschicht thermocline
Thermometer thermometer
Thermoneutralzone thermal neutral zone
Thermoregulation thermoregulation
Therophyt/kurzlebige Pflanze/ Annuelle/Einjährige therophyte, annual plant
Thiamin (Vitamin B$_1$) thiamine (vitamin B$_1$)
Thigmotaxis (*pl* Thigmotaxien) thigmotaxis (*pl* thigmotaxes)
Thioharnstoff thiourea
Thorakalfuß/Thorakopode/ Thoracopod/Thoraxbein/ Rumpfbein thoracopod, thoracic leg
Thorakalrückenplatte/dorsales Thorakalschild/Notum notum (dorsal thoracic plate/thoracic tergum)
Thorakalschüppchen thoracic scale
Thorax/Brust/Brustkörper/ Brustkasten/Oberkörper thorax, breast, chest, pectus
Thorax/Mittelleib (Insekten) thorax
Threonin threonine
Thrombin thrombin
Thrombozyt/Thrombocyt/Plättchen/ Blutplättchen thrombocyte, platelet
Thylakoid thylakoid
Thylle tylosis, tylosis, tylose
Thyllenbildung tylosis formation, tylosis
Thymelaeaceae/ Spatzenzungengewächse/ Seidelbastgewächse daphne family, mezereum family

Thymin thymine
Thymindimer thymine dimer
**Thymus/Thymusdrüse/Bries
(Hals-/Brustthymus)**
thymus (gland)
**Thyreotropin/Tyrotropin/
thyreotropes Hormon (TSH)**
thyrotropin, thyroid-stimulating
hormone (TSH)
**Thyroliberin/Thyreotropin-
Freisetzungshormon (TRH/TRF)**
thyroliberin, thyrotropin releasing
hormone/factor (TRH/TRF)
Thyroxin (T_4) thyroxine (also:
thyroxin), tetraiodothyronine
**Thyrse/Thyrsus/Strauß
(Infloreszenz)** thyrse, thyrsus
**Thysanuren (Felsenspringer u.
Silberfischchen etc.)** thysanurans
Tide(n)/Gezeiten tide(s)
 • **Ebbe** low tide
 • **Flut/Tide** high tide, flood
 • **Nipptide** neap tide
 • **Springtide** spring tide
Tidebereich/Gezeitenzone
tidal zone, intertidal zone
Tidenhub tidal lift
**Tiedemannscher Körper/
schwammiger Körper**
Tiedemann's body
Tiefe (Meerestiefe) depth of the
ocean, profundal depth
Tiefebene lowland, low-lying plain
Tiefenätzung deep etching
Tiefenschärfe/Schärfentiefe opt
depth of focus, depth of field
Tiefland/Niederung bottomland,
lowland
Tiefsee deep sea
Tiefseebecken deep-sea basin
**Tiefseebereich, mittlerer/mittlere
Tiefseezone/Bathypelagial
(Wasser)** bathypelagic zone
**Tiefseebereich, unterster/unterste
Tiefseezone/Abyssopelagial
(Wasser)** abyssopelagic zone
Tiefseeberg seamount
Tiefseeboden/Abyssal (Boden)
abyssal zone
**Tiefseeboden bewohnend/
Abyssobenthal (Boden)**
abyssobenthic
Tiefseeebene abyssal plain
Tiefseeerhebung deep-sea rise,
oceanic rise
Tiefseegraben deep-sea trench
Tiefseegrabenbereich (Wasser)
hadopelagic zone
Tiefseegrabenzone/Hadal (Hänge)
hadal zone
**Tiefseequallen/Kranzquallen/
Coronatae** coronate medusas

Tiefseetafelberg/Tiefseekuppe
guyot, tablemount (flat-topped
seamount)
**Tiefseevampire/
Vampirtintenschnecken/
Vampyromorpha** vampire squids
tiefwurzelnde Pflanze
deep-rooted plant
Tiegelzange lab crucible tongs
Tier animal
Tierarzt/Veterinär veterinarian, vet
Tierasyl/Tierheim animal asylum,
animal shelter
Tierausbreitung zoochory,
animal-dispersal
tierblütig zoophilous
Tierblütigkeit/Zoophilie pollination
by animal vectors, zoophily
Tierchen animalcule,
little/small animal
Tierexkremente/Tierkot
animal feces, droppings
Tierfreund animal lover
Tiergarten/Zoo/Zoologischer Garten
zoo, zoological garden(s)
Tiergemeinschaft/Zoozönose
animal community, zoocoenosis
Tiergeographie/Zoogeographie
zoogeography
**Tiergesellschaft/essentielle
Vergesellschaftung/Sozietät**
(animal) society
**Tierheilkunde/Tiermedizin/
Veterinärmedizin** veterinary
science, veterinary medicine
Tierheim/Tierasyl animal shelter,
animal asylum
tierisch animal; (bestialisch/
animalisch) bestial; (brutal) brutal
tierisches Eiweiß animal protein
tierisches Fett animal fat
Tierkäfig animal cage
 • **kleiner Tierkäfig/Verschlag
(z.B. Geflügelstall)** hutch, coop
Tierklinik veterinary clinic,
animal hospital
Tierkunde/Zoologie zoology
**Tierläuse/Phthiraptera
(Mallophaga u. Anoplura)**
phthirapterans
tierlieb fond of animals,
animal-loving
**Tiermedizin/Tierheilkunde/
Veterinärmedizin** veterinary
medicine, veterinary science
Tiermodell animal model
Tierpension (Hunde/Katzen)
kennel (dogs/cats)
Tierpfleger/Tierwärter (animal)
keeper, warden
Tierphysiologie animal physiology,
zoophysiology

Torfmoose

Tierpsychologie animal psychology, zoopsychology
Tierquälerei animal abuse, cruelty to animals
Tierrechtler/Tierschützer animal rightist, animal rights advocate
Tierreich animal kingdom
Tierschutz (advocating of) animal rights, protection of animals
Tierschutzverein animal rights league
Tierseuche/Viehseuche epizooic disease, pest; livestock epidemic
Tierstaat animal society
Tierstock cormus; colony
Tiervirus animal virus
Tierwelt/Fauna fauna, animal life
Tierzucht (Nutztiere in der Landwirtschaft) animal husbandry, livestock breeding
Tierzucht/Tierzüchten *sensu lato* animal breeding
Tierzwinger (staatl. Verwahrung verwaister Tiere) pound
Tierzwinger (z. B. in Zoos) cage, enclosure
Tigrolyse tigrolysis
Tiliaceae/Lindengewächse lime tree family, linden family
Tinte ink
Tintenfische/Coleoidea/ Dibranchiata coleoids
Tintengang ink duct
Tintensack/Tintenbeutel/ Tintendrüse ink sac, ink gland
Tintenschnecken/eigentliche Tintenschnecken/Sepioidea (bzw.Sepiida) cuttlefish & sepiolas
• **achtarmige Tintenschnecken/ Kraken/Octopoda/Octobrachia** octopods, octopuses
• **zehnarmige Tintenschnecken/ Zehnarmer/Decabrachia/ Decapoda** cuttlefish & squids
Tintlinge/Tintenpilze/Coprinaceae inky cap family
Tisch, höhenverstellbarer *lab* laboratory jack
Tischveredelung *hort* bench grafting
Titer titer
Titin titin
Titrationskurve titration curve
titrieren titrate
Tittelpfropfung *hort* apical grafting (wedged scion inserted behind strip of bark) (after Tittel, 1916)
Tochter daughter
Tochtercyste exogenous cyst
Tochtergeneration/Filialgeneration (F1/F2) filial generation
Tochtergeschwulst/Metastase metastasis

Tochterstrang *gen* daughter strand
Tochterzelle daughter cell
Tochterzwiebeln bulblets
Tocopherol/Tokopherol (Vitamin E) tocopherol (vitamin E)
Tod *n* death
Todesforschung/Lehre vom Tod/ Thanatologie thanatology
Todeskampf death struggle, agony
Todestrieb death instinct
Todesursache cause of death
tödlich/letal deadly, lethal
Toleranzbereich tolerance range
Toleranzgrenze tolerance limit
Tolerogen tolerogen
Tollwut/Hundswut/Rabies/Lyssa (Lyssavirus) rabies
Tölt (Gangart bei Pferden) running-step
Tomographie tomography
Tömösvarysches Organ/ Postantennalorgan organ of Tömösvary
Ton *acust* tone, sound
Ton *geol* clay
Tönnchenlarve/Doliolaria (Crinoidea) doliolaria
Tönnchenlarve/Dotterlarve/ Vitellaria (Holothuroidea) vitellaria larva
Tönnchenpuppe/Pupa coarctata coarctate pupa
tonnenförmig barrel-shaped, dolioform
Tonnensalpen/Doliolida doliolids
Tonofilament tonofilament
Tonoplast tonoplast
Tonus tone
• **Muskeltonus** muscle tone
Topferde potting soil (potting mixture: soil & peat a.o.)
Topfpflanze potted plant
topogen topogenic, topogenous
Topotypus/Topostandard topotype
Torf peat
• **Bleichmoostorf/Sphagnumtorf** sphagnum peat
• **Schilftorf/Rohrgrastorf** reed peat
• **Schwarztorf** black peat
• **Seggentorf** sedge peat, carex peat
• **Weißtorf/Hochmoortorf** white peat
Torfhügelmoor/Palsenmoor palsa bog
Torfhumus peat humus
Torfmoor bog, peat bog, peat moor, muskeg
Torfmoos/Bleichmoostorf/ Sphagnumtorf sphagnum peat
Torfmoose (Sphagnidae) peat mosses

Torfmull granulated peat, garden peat
Torfstich/Torfgrube peat bank, peatery
tormogene Zelle/Balgzelle tormogen cell (socket-forming cell)
Tornaria-Larve tornaria larva
Torpor/Starre (Kältestarre/Winterstarre) torpor (hibernation)
Torsion/Drehung torsion
Torstrom (*pl* Torströme) *neuro* gating current
tot dead
tot geboren stillborn
Totenstarre/Leichenstarre rigor mortis
Totgeburt stillbirth
Totholz *for* wood litter
Totraum deadspace
Totreife (Getreide) dead ripeness
totstellen feign death, play dead
Totstellreflex/Sichtotstellen/Akinese akinesis
Totvolumen dead volume
Totwasser quiet water, still water (stationary eddies in river)
Toxikologie toxicology
Toxin/Gift toxin
toxisch/giftig toxic, poisonous
Toxizität/Giftigkeit toxicity, poisonousness
Toxophorium (Schmetterlingsraupen: Brenn- und Gifthaare) toxophore
Toxoplasmose (*Toxoplasma gondii*) toxoplasmosis
Trab/Trott (schnelle Gangart) trot (trotting gait)
Trabekel/Balken/Bälkchen trabecula
Traberkrankheit scrapie
Trachealdrüse/Luftröhrendrüse tracheal gland
Trachealring/Trachealknorpel/Knorpelspange der Luftröhre tracheal ring, tracheal cartilage
Trachee/Gefäß *bot* trachea, vessel
• **Leitertrachee** scalariform vessel
Trachee/Luftröhre/Atemröhre/Kehle *zool* trachea, windpipe, breathing tube
• **Fächertrachee/Buchlunge/Fächerlunge** book lung
• **Röhrentrachee** *arach* tube trachea, tubular trachea
• **Siebtrachee** *arach* sieve trachea
Tracheenglied vessel member, vessel element
Tracheenkieme tracheal gill
Tracheenöffnung/Tracheenstigma/Spirakulum/Ostium (tracheal) spiracle, ostium
Tracheentiere/Tracheata tracheates
Tracheide tracheid

Tracheole tracheole, tracheal capillary
Tracheophyt/Gefäßpflanze tracheophyte, vascular plant
Tracht (Ertrag an Honig) yield
Tracht (Jungtiere/Wurf) litter
Tracht/Kleid (Fell/Gefieder) coat, plumage
• **Schrecktracht/Schreckfärbung** fright coloration
• **Schutztracht/schützende Nachahmung/Mimikry** mimicry
• **Tarntracht** cryptic dress (plumage/pelage/coat)
• **Warntracht/Warnfärbung/Abschreckfärbung** warning coloration, aposematic coloration
Trachte (Huf) buttress of heel, angle of heel, angle of wall
Trachtenwand (Huf) wall of heel
trächtig/schwanger gravid, pregnant
Trächtigkeit/Schwangerschaft/Gravidität pregnancy, gravidity
Trachylina/Trachymedusae trachymedusas, trachyline medusas
Tractus olfactorius olfactory tract
träg/träge inert
Tragblatt/Deckblatt bract, subtending bract
Tragblatt zweiter Ordnung/Tragschuppe/Deckschuppe/Brakteole bract-scale, bracteole, bractlet, secondary bract
Träger carrier; *bot* suspensor; (*Marchantia*) stalk; *chromat* carrier
Trägerarm *micros* arm
Trägerelektrophorese carrier electrophoresis
Trägergas carrier gas
Trägerhyphe *fung* suspensor, zygosporophore
Trägermolekül carrier molecule
Trägerprotein/Schlepperprotein carrier protein
Trägersubstanz carrier
Trägheit inertia
Trägheitskraft inertial force
Tragschuppe/Deckschuppe/Tragblatt zweiter Ordnung/Brakteole bract-scale, bracteole, bractlet, secondary bract
Tragtier/Lasttier pack animal, beast of burden
Tragus (Ohr) tragus
Tragzeit/Tragezeit period of gestation
Trailer-Sequenz *gen* trailer segment
Trama (Lamellentrama) *fung* trama (of fungal gill), dissepiment
Tramaplatte tramal plate
Trampelklette traple burr

711   **Transplantat**

Tran/Fischöl train oil, fish oil (also from whales)
- **Lebertran** cod-liver oil

Träne tear
tränen tear
tränenartig/Tränen... lacrimal (lachrymal)
Tränenbein/Os lacrimale lacrimal bone
Tränendrüse lacrimal gland
Tränengang/Tränenkanal tear duct, lacrimal duct
Tränennasengang/ Tränennasenkanal/Ductus naso- lacrimalis nasolacrimal duct
Tränensack tear pouch
Tränke (für Tiere) watering place (for animals)
trans-Golgi-Netzwerk trans-Golgi network
Transadenylierung transadenylation
Transaminierung transamination
Transcytose transcytosis
Transduktion transduction
- **abortive T.** abortive transduction
- **spezielle T.** specialized transduction

Transduktionskartierung transduction mapping
Transekt transect
- **Gürteltransekt** belt transect
- **Linientransekt** line transect

Transfektion transfection
- **abortive T.** abortive transfection

Transfer-RNA (tRNA) transfer RNA (tRNA)
Transferöse transfer loop
Transformation transformation
- **Zelltransformation** cell transformation

Transformationskartierung transformation mapping
Transformationsreihe transformation series
transformieren transform
transformierendes (aktives) Prinzip transforming principle
transformierte Zelle transformed cell
transgen transgenic
transgene Pflanzen transgenic plants
transgenes Tier transgenic animal
Transhumanz transhumance, seasonal livestock movement
transienter Expressionsvektor transient expression vector
Transition transition
Transkript transcript
- **Primärtranskript** primary transcript

Transkript-Sequenzierung transcript sequencing

Transkription transcription
- **3′ → 5′ (drei Strich-fünf Strich/ drei Strich nach fünf Strich)** 3′ → 5′ (three prime five prime/ three prime to five prime)
- **divergente T.** divergent transcription
- **reverse T.** reverse transcription

Transkriptionsanalyse transcript analysis
Transkriptionsfaktor transcription factor
Transkriptionsfusion transcription fusion
Translation translation
- **hybridarretierte Translation** hybrid-arrested translation (HART)
- **Hybrid-Freisetzungstranslation** hybrid-release translation (HRT)
- **reverse Translation** reverse translation

Translationsbewegung translational motion
Translationsfusion translation fusion
Translator (Gynostegium) *bot* translator (caudicles + adhesive body)
Translokation translocation
- **balancierte Translokation** balanced translocation
- **reziproke Translokation/ interchromosomale Umordnung/ Austausch (zwischen Chromosomen)** interchange, interchromosomal rearrangement

transmembran *adj/adv* transmembrane
Transmembranprotein transmembrane protein
Transmission (horizontale/vertikale) *gen* (horizontal/vertical) transmission
Transphosphorylierung transphosphorylation
Transpiration transpiration
Transpirationssog/ Transpirationszug transpiration pull
Transpirationsstrom transpiration stream
Transpirationsweg transpiration pathway
Transplantat transplant, graft
- **Allotransplantat** allograft (allogeneic graft)
- **Autotransplantat** autograft (autologous graft)
- **Fremdtransplantat/ Xenotransplantat** xenograft (xenogeneic graft: from other species)

T Transplantat 712

- **Gewebetransplantat**
tissue graft, tissue transplant
- **Hauttransplantat**
skin graft, skin transplant
- **Homotransplantat/
Allotransplantat** homograft,
syngraft (syngeneic graft),
allograft

Transplantat-anti-Wirt-Reaktion
graft-versus-host reaction (GVH)
Transplantatabstoßung
graft rejection
transplantieren transplant
transponierbares Element *gen*
transposable element
- **transponierbares menschliches
Element (THE)** transposable
human element (THE)

Transport transport, transportation
- **aktiver Transport**
active transport, uphill transport
- **durch eine Membran hindurch**
membrane trafficking
- **erleichterter Transport**
facilitated transport
- **gekoppelter Transport**
coupled transport, co-transport
- **Membrantransport**
membrane transport

transportieren transport
Transportprotein transport protein
**Transportwirt/Sammelwirt/
Stapelwirt/paratenischer Wirt**
paratenic host
Transposition transposition
Transposon transposon
Transversalgefäß (Ctenophoren)
transverse canal
Transversalkanal (T-Kanal)
transverse tubule (T-tubule)
Transversion transversion
Trapaceae/Wassernussgewächse
water chestnut family
Traube/Botrys (Infloreszenz)
raceme, botrys
**traubenförmig/traubig/botryoid/
razemös/racemös/racemos**
grape-cluster-like, botryoid,
botryose, racemose
**Traubenzucker/Glukose/Glucose/
Dextrose** grape sugar, glucose,
dextrose
Traufe *zool* trough
Träufelspitze (Blatt) *bot* drip tip
**Träuschlinge/Schuppenpilze/
Strophariaceae**
Stropharia family
Treber/Biertreber brewers' grains
Treibhaus/Gewächshaus
forcing house, greenhouse
Treibhauseffekt greenhouse effect
Treibholz driftwood

Treibjagd drive
Treibmittel/Gärmittel/Gärstoff
leavening
Treibstoffalkohol/Gasohol gasohol
Trennart/Differentialart
differential species
trennen separate; divide
Trenngel separating gel
(running gel)
Trennkammer *chromat* **(DC)**
developing chamber (TLC)
Trennmethode separation method
Trennschärfe *chromat* resolution
Trennschicht *bot* (Blatt)
abscission layer
Trennstufe *chromat* (HPLC) plate
Trennstufenhöhe height equivalent
to theoretical plate (HETP)
Trennung separation
**Trennung/Verteilung/Disjunktion
(der Tochterchromosomen)**
disjunction
**Trennungsschicht/Trennschicht/
Ablösungsschicht/
Abszissionsschicht** *bot*
separation layer, abscission layer
Trennverfahren/Trennmethode
separation technique, separation
procedure, separation method
Trester/Treber (siehe auch dort)
(Fruchtpressrückstand/
Traubenpressrückstand) marc;
(Malzrückstand) draff
treten (begatten: Hahn) tread
treu/fest (Gesellschaftstreue)
exclusive
Treue (Gesellschaftstreue) fidelity
Triangulationszahl *vir* triangulation
number
Trias (erdgeschichtliche Periode)
Triassic, Triassic Period
Tribus (*pl*** Triben)/Sippe** tribe
**Tricarbonsäure-Zyklus/Citrat-Zyklus/
Citratcyclus/Zitronensäurezyklus/
Krebs-Cyclus** tricarboxylic acid
cycle (TCA cycle), citric acid cycle,
Krebs cycle
trichal (haarförmig)/fadenförmig
filamentous, filliform,
thread-shaped
Trichine (*Trichinella spiralis***)**
trichina worm
**Trichinose (Krankheit verursacht
durch:** *Trichinella spiralis***)**
trichinosis
trichogene Zelle trichogen cell
(seta-forming cell)
Trichogyne/Empfängnishyphe *fung*
trichogyne
Trichter funnel
Trichter/Sipho/Infundibulum
funnel, siphon, infundibulum

Trompetenbaumgewächse

Trichterblatt funnel-leaf
Trichterblüte funnel-shaped corolla, funnel-shaped flower
Trichterfalle/Reusenfalle *bot* (unidirectional) pitfall trap, funnel trap, weir basket trap
trichterförmig funnel-shaped, funnelform, infundibulate
Trichtergrube/Subgenitaltasche subgenital pit
Trichternetz/Röhrennetz *arach* funnel web, tube web
Trichterpflanze funnel-leaved plant, infundibulate plant
Trichterrispe/Spirre (Infloreszenz) anthela
tricolpat tricolpate
Trieb (Spross) *bot* shoot
Trieb/Drang/Impuls *ethol* urge, drive, compulsion, impulse
Triebkraft *phys/mech* (>Antrieb) propulsive force
Triebverhalten/Instinktverhalten instinctive behavior
Trift (Weide/Weidewiese) pasture, pasturage
Trift/Weg zur Weide (Vieh) cattle track
Triggerzone trigger zone
Triiodthyronin/Trijodthyronin (T3) triiodothyronine
trillern *orn* warble, trill
Trilliaceae/Einbeerengewächse trillium family
trimitisch trimitic
trimmen/abschneiden/ zurückschneiden *hort/zool* trim
Trimmschere trimming shears
Trinkwasser drinking water
Trinokularaufsatz/Tritubus *micros* trinocular head
Trinucleotid-Wiederholung trinucleotide repeat
Trinucleotidexpansion/ Trinukleotidexpansion/ Trinukleoidverlängerung trinucleotide expansion
triphasisch triphasic
Triplettbindungsversuch triplet binding assay
Triplettsequenzen *gen* triplet sequences
triploid triploid
Triploidie triploidy
triplostemon triplostemonous
trippeln (Bienen) spasmodic dance
trisom trisomic
Trisomie trisomy
tRNA/Transfer-RNA tRNA, transfer RNA

Trochodendraceae/ Radbaumgewächse trochodendron family, wheel-stamen tree family, yama-kuruma family
Trochus (vorderer Wimpernkranz) trochus (anterior circlet of cilia)
trocken dry, arid
Trockenbeet (Kläranlage) drying bed
Trockenblotten dry blotting
Trockenblume/Strohblume strawflower
Trockenfäule dry rot
Trockenfrucht dry fruit
Trockengebiet arid land, arid region, dryland
Trockengewicht (*sensu stricto*: Trockenmasse) dry weight (*sensu stricto*: dry mass)
Trockenheit/Dürre dryness, drought
Trockenheit ertragende Pflanze xerophyte, xeric/xerophilous plant, drought tolerator
Trockenkultur *agr* dry farming, dryland farming
Trockenlandwirtschaft/ Trockenkultur dry farming, dryland farming, dryland culture
trockenlegen (Sumpf) drain
Trockenlegung drainage
Trockenmasse/Trockensubstanz dry mass, dry matter
Trockenperiode dry spell, drought
Trockenpflanze xerophyte, xeric plant, xerophilous plant
Trockenrasen dry meadow, arid grassland
trockenresistent drought resistant, xerophytic
Trockenschlaf/Sommerschlaf/ Ästivation estivation
Trockenschrank *lab* drying cabinet (plant-drying cabinet)
Trockenschütte leaf cast (abscission of leaves) due to desiccation
Trockenstarre/Anhydrobiose anhydrobiosis
Trockensubstanz dry matter
trocknen dry
  • **austrocknen** desiccate
Trocknis/Austrocknungsschaden desiccation damage
Trog/Wanne trough
Trogons/Trogoniformes trogons
Trommelfell/Ohrtrommel/ Tympanalmembran/Tympanum eardrum, tympanic membrane, tympanum
Trompetenbaumgewächse/ Bignoniengewächse/Bignoniaceae trumpet-creeper family, trumpet-vine family, bignonia family

T Tropaeolaceae 714

Tropaeolaceae/
Kapuzinerkressengewächse
nasturtium family
Tropen tropics
Tropfberieselungsschlauch *agr*
soaker hose
Tropfbewässerung drip irrigation,
trickle irrigation
Tropfenfaden *arach* viscid line
Tropfflasche dropping bottle
Tropfglas/Tropfpipette dropper
Tropfkörper (Tropfkörperreaktor/
Rieselfilmreaktor) trickling filter
Tropftrichter *lab* dropping funnel
Trophamnion trophamnion
Trophie trophism
Trophieebene/Trophieniveau/
Trophiestufe/trophische Stufe/
trophische Ebene/
trophisches Niveau
trophic level, feeding level
Trophophase (Ernährungsphase)
trophophase
Tropismus tropism
trüb turbid
Trüffeln/Tuberales truffles
Trugdolde/Scheindolde/Cymus/
Zymus/Cyme (Infloreszenz)
cyme
● **eingablige T.** simple cyme,
monochasium
● **zweigablige T.** compound cyme,
dichasial cyme, dichasium
trugdoldig/cymös/zymös
cymoid, cymose
Trypsin trypsine
Tryptophan tryptophan
Tsetseseuche/Schlafkrankheit
(*Trypanosoma rhodesiense/
***gambiense*)** sleep sickness
tuberkular/knotig tubercular,
tuberculate, tuberculated
tuberkulös (>Tuberkulose)
tuberculous
Tuberkulose tuberculosis
tuberös tuberous, tuberal
Tubocurarin tubocurarine
tubulär tubular
Tubulin tubulin
Tubus *micros* tube, body tube
Tuff *geol* tuff, tufa
Tularämie/Hasenpest (*Pasteurella*
***tularensis*)** tularemia
Tullgren-Apparat *ecol*
Tullgren funnel
Tumor/Wucherung/Geschwulst
tumor
● **Wurzelhalstumor (Stamm- oder**
Wurzeltumor verursacht durch
A. tumefaciens) crown gall tumor
Tumornekrosefaktor (TNF)
tumor necrosis factor (TNF)

Tümpel pond, pool
● **Felstümpel** rockpool
● **Gezeitentümpel**
tidal pool, tide pool
Tümpelquelle/Limnokrene
limnocrene
Tundra tundra, polar grassland,
arctic grassland
Tundramoor muskeg, bog
tunikat/tunicat tunicate
Tunnelmikroskopie
tunneling microscopy
Tupelobaumgewächse/Nyssaceae
sourgum family
Tüpfel pit
● **einfacher Tüpfel** simple pit
● **Hoftüpfel** bordered pit
Tüpfelfarngewächse/Polypodiaceae
polypody family, fern family
Tüpfelfeld pit field
Tüpfelhof pit chamber
Tüpfelhöhle pit cavity
Tüpfelöffnung/Tüpfelporus/
Tüpfelapertur pit aperture
Tüpfelpaar pit-pair
● **behöftes T.** boardered pit-pair
● **verschlossenes/aspirates T.**
aspirated pit-pair
Tüpfelpfropfen (Rotalgen) pit plug
Tüpfelschließhaut pit membrane
Tüpfelung pitting
Tüpfelverbindung (Tüpfelkanal bei
Rotalgen) pit connection
Turbanauge (*Ephemeroptera*)
stalked compound eye
Turbellarien/Strudelwürmer
turbellarians, free-living flatworms
Turbidimetrie/Trübungsmessung
turbidimetry
Turbidostat turbidostat
turbulente Strömung
turbulent flow
Turbulenz turbulence
turgeszent/geschwollen/
angeschwollen/schwellend/prall
turgescent, swollen
Turgeszenz/Schwellung turgescence
Turgidität turgidity
Turgor/hydrostatischer Druck
turgor, hydrostatic pressure
Turgordruck turgor pressure
Turio (*pl* Turionen) turio, turion,
(detachable) winter bud,
hibernaculum
Türkensattel/Sella turcica
Turkish saddle, sella turcica
turmförmig gewunden conispiral
Turmreaktor column reactor
Turneraceae/
Safranmalvengewächse
turnera family
Tute/Ochrea ochrea, ocrea, mantle

Typusexemplar

Tympanalorgan/Gehörorgan
  tympanic organ
Typhaceae/Rohrkolbengewächse
  reedmace family, cattail family
Typhlosolis (Clitellata) typhlosole
**Typhus/Fleckfieber/Flecktyphus/
  Typhus exanthematicus
  (*Rickettsia spp.*)** typhus

**Typhus/Unterleibstyphus/
  Typhus abdominalis
  (*Salmonella typhi*)**
  typhoid fever, typhoid
Typus/Standard type
**Typusexemplar/Typusbeleg/
  Typbeleg**
  type specimen

Übelkeit

Übelkeit/Übelsein nausea, sickness, illness
übelriechend/stinkend fetid, smelly, smelling bad, stinking
Überart superspecies
Überaugenwulst/ Augenbrauenwulst brow ridge, brow crest
Überbevölkerung overpopulation
Überbeweidung/Überweidung overgrazing
Überbleibsel/Überrest relic
Überbrückung/Wundüberbrückung *hort* bridge grafting, repair grafting
Überbrückungskreuzung bridging cross
Überdauerung persistance, survival
Überdominanz overdominance
Überdosis overdose
Überdrehung overwinding
Überdüngung overfertilization
Überempfindlichkeit hypersensitivity
Überempfindlichkeitsreaktion hypersensitivity reaction
• **Soforttyp/anaphylaktischer Typ** immediate-type hypersensitivity reaction (TITH)
• **Spättyp/verzögerter Typ** delayed-type hypersensitivity reaction (TDTH)
Überexpression *gen* high-level expression, overexpression
Überfischung overfishing
Überfluss excess
Überfunktion overactivity, hyperactivity
Übergang/Entwicklungsübergang transition, developmental transition
Übergangs-Waldmoor carr
Übergangsblätter/Primärblätter/ Erstlingsblätter first foliage leaves
Übergangsepithel transitional epithelium
Übergangsform/Zwischenstufe intermediary form, transitory form, transient, intergrade
Übergangsfossil transitional fossil
Übergangsgesellschaft/Ökoton ecotone
Übergangsmoor/Zwischenmoor transitory bog
Übergangsphase transition phase
Übergangszone (Wurzel-Spross) transition(al) zone/region
Übergangszustand (Enzymkinetik) transition state
Übergipfelung *bot* overtopping (unilateral dominance)
übergreifen *med* spread (e.g. disease/epidemic)
Übergriff/Attacke attack

überhängendes Ende/klebriges Ende/kohäsives Ende cohesive end, protruding end, overhanging end, overhanging extension
Überinfektion/Superinfektion superinfection
überirdisch aboveground
Überkreuzungsaustausch/ Überkreuzungsstelle/Crossing over crossing over, crossover
Überkreuzvererbung criss-cross inheritance
überlappend overlapping
• **dachziegelartig/schuppenartig überlappend** imbricate
überlappende Gene overlapping genes
überleben *vb* survive
Überleben *n* survival
Überlebenskampf struggle for survival
Überlebenskurve survivorship curve
Überlebensrate survival rate
überlegen/vorherrschend/dominant superior, dominant
Überlegenheit/Dominanz superiority, dominance
Überordnung superorder
überragen protrude, project, stand/stick out, rise over
überragend/heraustretend (schlanker Wipfel/unten ausladend) excurrent
Überreplikation overreplication
überschießen *neuro/ecol* **(z.B. Kapazitätsgrenze)** overshoot
Überschuss *neuro/ecol* overshoot
Überschussproduktion surplus production
Überschwemmung flood, flooding, undulation
Überschwemmungswald (Flussaue) (river) floodplain forest
Überschwemmungswiese floodplain meadow
übersommern estivate, aestivate, pass summer in dormant stage
Übersommerung estivation, aestivation
überspiralisiert/superspiralisiert/ superhelikal supercoiled
Überspiralisierung supercoiling
Übersprungshandlung displacement activity
Überstand supernatant
übertragbar transmissible, communicable
übertragbare Krankheit transmissible disease, communicable disease
übertragen *med* transmit (e.g. a disease)

717 **Umordnung** **U**

Überträger/Überträgerstoff/ Transmitter transmitter
Überträger/Vektor vector
Übertragung (z. B. Krankheit) transmission
• **nicht-persistente Ü.** *vir* nonpersistent transmission
• **propagative Ü.** *vir* propagative transmission
• **venerische Ü.** venereal transmission
übervölkern (übervölkert) overpopulate(d)
Übervölkerung overpopulation
überwachen monitor, survey, supervise
Überwachung monitoring, surveillance, supervision
Überwallung/Wundüberwallung *bot* bulgy callus overgrowth of wound/other objects
überweiden overgraze
Überweidung overgrazing
überwintern overwinter, hibernate
Überwinterung/Winterschlaf hibernation
Überwinterungsknospe perennating bud, winter bud, hibernaculum
überwuchert overgrown
Überwuchs overstory growth
Ubichinon ubichinone
ubiqitär/weitverbreitet/überall verbreitet ubiquitous, widespread, existing everywhere
Ubiquinon/Coenzym Q ubiquinone, coenzyme Q
Ubiquist ubiquist
Ubiquitin ubiquitin
UESCE (ungleicher Austausch von Schwesterchromatiden) UESCE (unequal exchange of sister chromatids)
Ufer shore, coast, banks
• **Flussufer** riverbank, riverside
• **Meeresufer/Meeresküste** seashore, seaboard, seacoast
• **Seeufer** lakeshore, shore/banks of a lake
• **Steilufer/Felsufer (am Fluss)** (river) bluff
uferbewohnend/am Ufer lebend (Meeresküste) littoral; (Flussufer) riparious, riparial
Uferfiltration river-bed filtration
Uferfliegen/Steinfliegen/Plecoptera stoneflies
Uferlinie shoreline
Uferregion/Uferzone (Gewässer) littoral zone
Ufersaum/Küstensaum littoral fringe

Ufervegetation shore vegetation, shoreline vegetation
Uhr, biologische biological clock
Uhrglas *lab* watch glass
Ulmengewächse/Ulmaceae elm family
Ultradünnschnitt *micros* ultrathin section
Ultrafiltration ultrafiltration
Ultramikrotom ultramicrotome
Ultrasaprobität ultrasaproby
Ultraschall ultrasound, ultrasonics
Ultraschall.../den Ultraschall betreffend ultrasonic
Ultraschalldiagnose/Sonographie ultrasound, ultrasonography, sonography
Ultrastruktur ultrastructure
Ultrazentrifuge ultracentrifuge
Umdrehungszahl (DNA) twisting number
Umfang girth
Umgang (Schneckenschale) whorl
Umgebung surroundings, environs, environment, vicinity
Umgebungstemperatur ambient temperature
umgewendet (Samenanlage) anatropous
umgraben till, turn up (the soil)
umhüllen sheathe
Umkehrosmose/Reversosmose reverse osmosis
Umkehrphase/Reversphase reverse phase
Umkehrpotential reversal potential
Umkehrschleife (DNA) reverse turn
Umkehrtranskriptase/Revertase/ reverse Transkriptase reverse transcriptase
umkippen (Gewässer) turn over, become oxygen-deficient, turn anaerobic
umlagern/umordnen *chem* rearrange
Umlagerung/Umordnung *chem* rearrangement
• **tautomere Umlagerung** tautomeric shift
Umlaufdüsen-Reaktor/ Düsenumlaufreaktor nozzle loop reactor, circulating nozzle reactor
Umlaufreaktor/Umwälzreaktor/ Schlaufenreaktor loop reactor, circulating reactor, recycle reactor
Umordnung/Rearrangement (DNA/ Gene/Genom) rearrangement
• **interchromosomale Umordnung/ reziproke Translokation, Austausch (zwischen Chromosomen)** interchange, interchromosomal rearrangement

umpflanzen 718

umpflanzen/versetzen transplant, replant
Umriss contour, outline
Umrissfeder/Konturfeder *orn* contour feathers (pluma)
Umsatz turnover
Umsatzgeschwindigkeit/ Umsatzrate turnover rate, rate of turnover
Umsatzzeit turnover period
umsetzen *metabol* process, metabolize
umstimmen reorient, reorientate
Umstimmung reorientation
Umstimmungs-Effekt *physiol* primer effect
umtopfen repot
Umtrieb/Umtriebszeit (Forst) rotation, cutting cycle
Umtriebsbeweidung rotational grazing
Umtriebszeit rotation period
umwälzen *limno* overturn (lake water)
Umwandlung/Transformation transformation
Umwandlungsgestein metamorphic rock
Umwelt environment
Umweltanalyse environmental analysis
Umweltanalytik environmental analytics
Umweltansprüche environmental requirements
Umweltaudit/Öko-Audit environmental audit
Umweltbedingungen environmental conditions
Umweltbelastung environmental burden, environmental load
Umweltchemie environmental chemistry
Umweltfaktor environmental factor
umweltgerecht environmentally compatible
Umweltkapazität/Grenze der ökologischen Belastbarkeit carrying capacity
Umweltkriminalität environmental crime
Umweltmedizin environmental medicine
Umweltmesstechnik environmental monitoring technology
Umweltpolitik environmental politics
Umweltschutz environmental protection
Umweltschützer environmentalist
Umweltsünder person who litters or commits an environmental crime

Umweltvarianz environmental variance
Umweltverfahrenstechnik environmental process engineering
Umweltverhältnisse environmental conditions
Umweltverschmutzer polluter
Umweltverschmutzung environmental pollution
Umweltverträglichkeit environmental compatibility
Umweltverträglichkeitsprüfung (UVP) environmental impact assessment (EIA)
Umweltwiderstand environmental resistance
Umweltwissenschaft environmental science
Umweltzerstörung environmental degradation
Unabhängigkeitsregel/ Kombinationsregel (Mendel) law/principle of random (independent) assortment
unbefruchtet unfertilized
unbehaart/haarlos hairless, glabrous
unbelebt inanimate, lifeless, nonliving
unbeschränkt/unbegrenzt/ unbestimmt (Wachstum) indefinite, unrestricted
unbeweglich/bewegungslos/fixiert nonmotile, immotile, immobile, motionless, fixed
unbewohnbar uninhabitable
unbewusst unconscious
undulierende Membran undulating membrane
undurchlässig/impermeabel impervious, impermeable
Undurchlässigkeit/Impermeabilität imperviousness, impermeability
unersättlich insatiable
Unersättlichkeit insatiability
unfruchtbar/steril infertile, sterile
Unfruchtbarkeit/Sterilität infertility, sterility
ungeflügelt apterous, unwinged, exalate, wingless
ungefurcht/lissencephal *neuro* lissencephalous (no/few convolutions)
ungenießbar/nicht schmackhaft unpalatable
• **nicht essbar** uneatable, inedible
ungesättigt unsaturated
• **einfach u.** monounsaturated
• **mehrfach u.** polyunsaturated
ungeschlechtig/geschlechtslos/ ohne Geschlecht/sächlich agamous, neuter

ungeschlechtlich/asexuell/ nicht sexuell asexual
ungestielt/sitzend not stalked, sessile
Ungeziefer pest
ungleich/nicht identisch/anders unequal, different
Ungleichblättrigkeit/ Verschiedenblättrigkeit/ Anisophyllie/Heterophyllie anisophylly, heterophylly
ungleiches Crossing-over unequal crossing over
Ungleichgewicht imbalance, disequilibrium
ungleichmäßig irregular, non-uniform
ungleichzähnig/heterodont heterodont, anisodont
unifazial (ringsum gleiche Oberfläche) unifacial
Uniformismus uniformitarianism, principle of uniformity
Uniformitätsregel (*Mendel*) law/principle of uniformity (F1 of monohybrid cross)
unilokulär/einkammerig unilocular
unipar uniparous
uniparentale Mutation uniparental mutation
uniparentale Vererbung uniparental inheritance
Uniport uniport
uniseriat/einreihig uniseriate, uniserial, single rowed
Univarianzanalyse univariant analysis
Universalprimer *gen* universal primer
Univolitismus univolinism
univoltin/mit einer Jahresgeneration univoltine
Unkraut (*pl* Unkräuter) weed
Unkrautbekämpfung/ Unkrautvernichtung weed control
Unkrautbekämpfungsmittel/ Unkrautvernichtungsmittel/ Herbizid herbicide, weed killer
unlöslich insoluble
Unlöslichkeit insolubility
unnatürlich unnatural
unnütze DNA/überflüssige DNA/ wertlose DNA junk DNA
Unpaarhufer/Perssiodactyla odd-toed ungulates, perssiodactyls
unpolar apolar
unregelmäßig/irregulär/anomal irregular, anomalous
Unregelmäßigkeit/Anomalie irregularity, anomaly
unreif unripe, immature
unreife B-Zelle virgin B cell

Unreife immaturity, immatureness
unscharf *micro/photo* not in focus, out of focus, blurred
Unschärfe *micro/photo* blurredness, blur, obscurity, unsharpness
unspezifisch nonspecific
unsterblich immortal
Unsterblichkeit/Immortalität immortality
unteilbarer Faktor unit factor
Unterarm forearm
Unterarmschwungfedern/ Armschwingen secondaries, secondary feathers (secondary remiges)
Unterart/Subspezies subspecies
Unteraugendrüse suborbital gland
Unteraugenschild/Suboculare (Schlangen) subocular (scale)
Unterbewusstsein subconsciousness
Unterblatt underleaf, hypophyll; (Bauchblatt/Amphigastrium) underleaf, ventral leaf, amphigastrium
Unterboden (Einwaschungshorizont/ B-Horizont) subsoil (zone of accumulation/illuviation)
unterbrochene Paarung interrupted mating
Unterdominanz underdominance
unterdrückbar suppressible
Unterdrückung suppression
untereinanderliegend subtending
Untereinheit subunit
unterernährt undernourished (*siehe:* fehlernährt)
Unterernährung undernourishment
Unterfunktion/Insuffizienz hypofunction, insufficiency
• **Überfunktion** hyperfunction, hyperactivity
untergärig bottom fermenting
• **obergärig** top fermenting
Untergehölz/Unterholz understory
untergeordnet (Gruppenhierarchie) subordinate
Untergesellschaft/Subassoziation subassociation
untergetaucht/submers submerged, submersed
untergliedern (untergliedert) subdivide(d)
Untergliederung subdivision
Untergrund/Ausgangsgestein (C-Horizont) loose rock, boulder layer, parent material
Untergrund/Boden *allg* bottom, floor
• **fester Untergrund/festes Muttergestein/Grundgestein/ Ausgangsgestein** unmodified base, bedrock, parent rock

Untergrundbereich

Untergrundbereich (Binnensee) profundal zone
untergrundbewohnend (Ozean) benthic
Untergruppe subgroup
Unterhaar underhair
Unterhaar-Kleid underfur
Unterhaut/Unterhautbindegewebe/ Subcutis/Tela subcutanea subcutis
Unterholz/Untergehölz understory
unterirdisch underground, belowground, subterranean
• **oberirdisch** aboveground, overground, superterranean
Unterkategorie/Subkategorie subcategory
Unterkiefer/Mandibel/ Unterkieferknochen lower jaw, lower jawbone, submaxilla, submaxillary bone, mandible
Unterkiefer/1. Maxille/ Maxilla prima (Insekten) first maxilla
Unterkieferdrüse/Mandibulardrüse mandibular gland
Unterkühlung supercooling
Unterlage/Grundlage/ Untergrund/Substrat substrate
• **Pfropfunterlage** *hort* stock
unterlegen *adv/adj* inferior; defeated
Unterlegenheit inferiority; defeat
unterlegt von/mit/durch *bot* subtended by
Unterleib/Abdomen abdomen
unterliegen *bot* (ein Blatt dem anderen) subtending
Unterlippe/2. Maxille/ Maxilla secunda second maxilla
Unterlippe/Labium lower lip, labium
unterordnen subordinate, submit
Unterordnung (taxonomisch) suborder
Untersaat *agr* nurse crop
Unterscheidungsmerkmal differentiating characteristic
Unterschenkel shank
Unterschenkelknochen/Tibiotarsus (Vögel) tibiotarsus
unterschlächtig/succub undershot, succubous
Unterschlundganglion/ Subösophagealganglion subesophageal ganglion
Unterschlupf/Versteck hideout, hideaway, hiding place, retreat, refuge
Unterseite underside, undersurface
unterständig *bot* epigynous

untersuchen/prüfen/testen/ analysieren investigate, examine, test, assay, analyze
Untersuchung/Prüfung/Test/Probe/ Analyse investigation, examination (exam), study, search, test, trial, assay, analysis
• **Fruchtwasseruntersuchung** analysis of amniotic fluid (for prenatal diagnosis)
• **medizinische/ ärztliche Untersuchung** medical examination, medical exam, medical checkup, physical examination, physical
• **Wasseruntersuchung** water analysis
Untersuchungsmedium/ Prüfmedium/Testmedium assay medium
unterteilt/kompartimentiert septate, divided, compartmentalized
Unterwasservegetation underwater vegetation, submerged vegetation
unterwerfen subdue
Unterwerfung/Demut submission, yield
Unterwerfungsgebärde/ Unterwerfungshaltung/ Demutsgebärde/Demutshaltung submissive gesture, submissive posture
Unterwuchs undergrowth, understory
Unterzungendrüse sublingual gland
untranslatierte Region *gen* untranslated region (UTR)
unvermischbar immiscible
Unvermischbarkeit immiscibility
unverschmutzt uncontaminated
unverträglich/inkompatibel incompatible
Unverträglichkeit/Inkompatibilität incompatibility
Unverträglichkeitreaktion/ Inkompatibilitätreaktion incompatibility reaction
unverzerrt/unverfälscht *math/stat* unbiased
unverzweigt *bot/zool* unbranched, unramified
unverzweigt (Kette) *chem* unbranched (chain)
unvollständig gekoppelte Gene incompletely linked genes
unvollständige Blüte incomplete flower
unvollständige Penetranz incomplete penetrance
Up-Mutation up-mutation
üppige Vegetation lush vegetation

Ur-Leibeshöhlentiere/ Ur-Coelomaten/Archicoelomaten archicoelomates
Ur-Raubtiere/Creodonta creodonts
Uracil uracil
Urbanisierung urbanization
Urbanlandschaft urban landscape
Urbanökologie/Stadtökologie urban ecology
urbar/anbaufähig/nutzbar arable
Urbarmachung reclamation, making land suitable for cultivation
Urchromosom ancestral chromosome
Urdarm/Gastrocöl/Archenteron gastrocoel, archenteron (primitive gut of embryo)
ureotelisch/harnstoffausscheidend ureotelic, urea-excreting, excreting urea
Ureter/Harnleiter (sekundärer) ureter, urinary duct
Ureterknospe *embr* ureteric bud
Urethra/Harnröhre urethra
Urethraldrüse urethral gland
Urethralfalte urethral fold
Urethralplatte *embr* urethral plate
Urethralrinne urethral groove
Urethralsinus urethral sinus
URF (nicht zugeordnetes Leseraster) *gen* URF (unassigned reading frame)
Urfarne/Nacktfarne (Psilopsida) psilopsids
Urhirn/Archencephalon primitive brain, archencephalon
uricolytischer Weg/urikolytischer Weg uricolytic pathway
uricotelisch/urikotelisch/ harnsäureausscheidend uricotelic, excreting ureic acid
Uridin uridine
Uridintriphosphat (UTP) uridine triphosphate (UTP)
Uridylsäure uridylic acid
Urin/Harn urine
urinieren/harnlassen/harnen urinate
Urinsekten/Apterygota/Flügellose apterygotes
Urkaryot *m* urkaryote
Urkeimzelle primordial germ cell
Urlandschaft primeval landscape
Urmund/Blastoporus/Protostom blastopore, protostoma
Urmundtiere/Urmünder/ Erstmünder/Protostomia protostomes
Urmützenschnecken/ Monoplacophoren monoplacophorans
Urnenblatt (*Dischidia*) urn-shaped leaf, pouch leaf, "flower pot" leaf

urnenförmig urn-shaped, flask-shaped, urceolate
Urniere/Wolffscher Körper/ Mesonephros middle kidney, midkidney, Wolffian body, mesonephros
Urnierengang/Wolffscher Gang/ Ductus mesonephricus mesonephric duct, Wolffian duct
Urocaninsäure (Urocaninat)/ Imidazol-4-acrylsäure urocanic acid (urocaninate)
Urogenitalfalte *embr* urogenital fold
Urogenitalleiste *embr* urogenital ridge
Urogenitalplatte *embr* urogenital plate
Urogenitalsystem urogenital system
Uronsäure (Urat) uronic acid (urate)
Urophyse *ichth* urophysis
Uropod/Uropodium (*pl* Uropodien) uropod
Urpilze/Archimycetes/ Chytridiomycetes (Chytridiales) chytrids
Urraubsaurier/Pelycosauria pelycosaurs, pelycosaurians (early synapsids)
Ursamenzelle/Spermatogonium primordial male germ cell, spermatogonium
Urschleimpilze/haploide Schleimpilze/Protosteliomycetes protostelids
Urschuppensaurier/Eosuchia/ Younginiformes eosuchians (ancient two-arched reptiles)
Ursegment/Myotom myotome
Ursegment/Somit somite, somatome
Ursprung origin
ursprünglich/originär original, basic, simple, primitive
ursprünglich/urtümlich pristine
ursprungsgleich/homolog homologous
Ursuppe primordial soup
Urticaceae/Nesselgewächse nettle family
Urtierchen/Urtiere/"Einzeller"/ Protozoen first animals, protozoans
Urtyp archetype, prototype; stock
Urvögel/Archaeornithes ancestral birds, "lizard birds", archaeornithes
Urwald primeval forest, virgin forest, pristine forest, jungle
Urwurzelzähner/Thecodontia thecodonts, dinosaur ancestors
Urzeugungshypothese spontaneous generation hypothesis
Usninsäure usnic acid

Uterus

Uterus/Gebärmutter uterus, womb
Uterusepithel/Gebärmutterwand/ Endometrium
uteral epithelium, uteral lining, endometrium
Uterusglocke uterine bell
Uterushorn/Gebärmutterzipfel/ Cornu uteri
uterine horn, horn of uterus
Uterusmilch *ichth* uterine milk

UTR (untranslatierte Region) *gen*
UTR (untranslated region)
Utriculus/Bläschen/Fangblase
utriculus, utricle, small bladder
UTS (untranslatierte Sequenz) *gen*
UTS (untranslated sequence)
UV-Spektroskopie
ultraviolet spectroscopy,
UV spectroscopy

Vegetationskunde

vag/vage/indifferent (Bodentreue/ Gesellschaftstreue) indifferent
vagil/motil/frei beweglich vagile, freely motile
Vagilität/Motilität/Beweglichkeit/ aktive Ausbreitungsfähigkeit vagility, motility
Vagina/Scheide vagina
Vaginalträchtigkeit vaginal pregnancy
Vakuole vacuole
- **Gasvakuole** gas vacuole
- **Konkrementvakuole (Placozoa)** concrement vacuole
- **Nahrungsvakuole** food vacuole
- **pulsierende Vakuole/kontraktile Vakuole** contractile vacuole, water expulsion vesicle (WEV)
- **Verdauungsvakuole** digestive vacuole, secondary lysosome
- **Zentralvakuole/große Vakuole** *bot* central vacuole
vakuolisieren vacuolize, vacuolate
Vakuolisierung vacuolization, vacuolation
Vakuum vacuum
Vakuumpumpe vacuum pump
Vakzination/Vakzinierung/Impfung vaccination
Vakzine/Impfstoff vaccine (Impfstofftypen *siehe* Impfstoff)
Valenz valence, valency
Valerianaceae/Baldriangewächse valerian family
Valeriansäure/Baldriansäure/ Pentansäure (Valeriat/Pentanat) valeric acid, pentanoic acid (valeriate/pentanoate)
Valin valine
Valvifer valvifer
Valvula valve, valvula
Valvula uterina uterine valve
Vampirtintenschnecken/ Tiefseevampire/Vampyromorpha vampire squids
Vanillin vanillin
Vanillinsäure vanillic acid
Vannalfalte/Analfalte vannal fold, anal fold
Vannus/Analfeld/Analregion vannus, anal area
Variabilität/Veränderlichkeit/ Wandelbarkeit (*auch:* Verschiedenartigkeit) variability
Variabilitätsrückgang decay of variability
variable Anzahl von Tandemwiederholungen (VNTR) *gen* variable number of tandem repeats (VNTR)
variable Region (*Ig*) *immun* variable region

Variante variant
Varianz/mittlere quadratische Abweichung/mittleres Abweichungsquadrat *stat* variance, mean square deviation
- **additive genetische Varianz** additive genetic variance
- **Dominanzvarianz** dominance variance
- **Umweltvarianz** environmental variance
Varianzheterogenität/ Heteroskedastizität *stat* heteroscedasticity
Varianzhomogenität/ Varianzgleichheit/ Homoskedastizität *stat* homoscedasticity
Varianzquotientenverteilung/ F-Verteilung/Fisher-Verteilung *stat* variance ratio distribution, F-distribution, Fisher distribution
Variation variation
- **somaklonale Variation** somaclonal variation
Variationsbreite *stat* range of variation, range of distribution
Variationskoeffizient *stat* coefficient of variation
Variegation/Scheckung/ Buntblättrigkeit variegation
Varietät/Abart/Spielart variety; sport
Varikosität *neuro* varicosity
Variolation variolation
vasoaktives intestinales Peptid (VIP) vasoactive intestinal polypeptide (VIP)
Vasopressin/ antidiuretisches Hormon (ADH)/ Adiuretin vasopressin, antidiuretic hormone (ADH)
Vasotocin vasotocin
Vaterschaft paternity
Vaterschaftsbestimmung/ Vaterschaftstest paternity test
- **Ausschluss der Vaterschaft** paternity exclusion
Vegetarier vegetarian
- **Laktovegetarier** lactovegetarian
- **orthodoxer Vegetarier (streng vegetarisch ohne jegliche tierische Produkte)** vegan
- **Ovovegetarier** ovovegetarian
Vegetation vegetation, plant life
Vegetationsaufnahme relevé
Vegetationskegel vegetative cone, vegetative pole
Vegetationskunde/ Vegetationsökologie/ Pflanzenökologie plant ecology, phytoecology

Vegetationsperiode

**Vegetationsperiode/
Vegetationszeit** vegetation period
Vegetationsplan vegetation map
**Vegetationspunkt/Apicalmeristem
(Spross/Wurzel)** growing point,
apical meristem (shoot/root)
Vegetationsstufe/Höhenstufe
vegetation(al) zone/region/belt,
altitudinal zone/region/belt
Vegetationszone vegetational zone,
biome
vegetative Zelle vegetative cell,
somatic cell, body cell
vegetativer Pol vegetal pole,
vegetative pole
Veilchengewächse/Violoaceae
violet family
Vektor vector
- **Apportiervektor** eviction vector
- **bidirektionaler Vektor**
dual promoter vector,
bidirectional vector
- **bifunktionaler Vektor/
Schaukelvektor** bifunctional
vector, shuttle vector
- **Expressionsvektor**
expression vektor
- **multifunktioneller Vektor/
Vielzweckvektor** multifunctional
vector, multipurpose vector
- **Sicherheitsvektor**
containment vector
- **Substitutionsvektor**
replacement vector
- **transienter Expressionsvektor**
transient expression vector
Velamen velamen
Velichoncha/Pediveliger pediveliger
Veligerlarve/Segellarve
veliger larva
**Velloziagewächse/
Baumliliengewächse/Velloziaceae**
vellozia family
Velum velum
- **Apikalvelum** *fung* apical veil
- **Marginalvelum** *fung*
marginal veil
- **Schleier/Cortina (Rest des Velum
partiale/universale am Hutrand)**
fung cortina
Velum partiale/Velum hymeniale
fung partial veil
Velum universale *fung*
universal veil
Vene vein *(Venenbezeichnungen
finden Sie in einem Wörterbuch der
Human- bzw. Veterinärmedizin)*
- **Drosselvene/Vena jugularis**
jugular vein
- **Hohlvene/Vena cava** vena cava
- **Lungenvene/Vena pulmonis**
pulmonary vein

Venenwinkel/Angulus venosus
venous angle, Pirogoff's angle
venerische Übertragung
venereal transmission
Venerologie venereology
Venole venule
Ventil valve
Ventilationsvolumen ventilation
volume
**Ventiltrichter/Ventilkropf
(Vormagen/Proventriculus der
Honigbiene)** proventriculus
**Ventraldrüse (Nematoden/*siehe
auch:* H-Zelle)** ventral gland,
renette gland
**Ventralnaht/Bauchnaht (des
Fruchtblattes)** ventral suture,
ventral seam
Ventralschuppe/Bauchschuppe
ventral scale
Ventrikel ventricle
**Venushügel/Schamhügel/
Mons pubis**
pubic prominence
**Veränderlichkeit/Wandelbarkeit/
Variabilität** variability
Veränderung change, modification,
variation
verankern (befestigen)
anchor (fasten/attach)
Verankerung anchorage
**Veranlagung/Disposition/
Anfälligkeit** disposition;
(Prädisposition) predisposition
verarbeiten process, processing
Verarbeitung processing
verarmt/verkümmert depauperate,
starved, reduced, underdeveloped;
impoverished
**Verarmung/Reduktion/
Verringerung** reduction
verästeln, sich/sich verzweigen
ramify, branch; deliquescent
**Verästelung/Verzweigung/
Ramifikation** branching,
ramification
**Verband/Allianz/
Assoziationsgruppe** alliance
Verbänderung/Fasziation
fasciation
**Verbenengewächse/
Eisenkrautgewächse/Verbenaceae**
verbena family, vervain family
verbinden connect, bond, link
Verbindung *allg* connection, bond,
linkage; *chem* compound
- **chemische Verbindung**
(chemical) compound
- **energiereiche Verbindung**
high energy compound
Verbindungsstrang (Siebpore)
connecting strand

Vererbungslinienbestimmung

Verbiss/Wildverbiss (an Bäumen) damage caused by game, browsing damage

verborgene Spleißstelle *gen* cryptic splice site

verborgenfrüchtig cryptocarpous

Verbrauch consumption, use, usage

Verbraucher/Konsument consumer

Verbraucherschutz consumer protection

Verbreitung distribution, expansion; spread, spreading (dispersal *see* Ausbreitung)

● **disjunkte Verbreitung** disjunct distribution, discontinuous distribution

Verbreitungsgebiet/Areal geographic range, area of distribution

Verbreitungskarte/Arealkarte distribution map, range chart

verbrennen combust, incinerate, burn

Verbrennung combustion, incineration; *med* burn

Verbrennungswärme combustion heat, heat of combustion

Verdacht (auf eine Erkrankung) suspicion (of a disease)

Verdampfungswärme heat of vaporization

Verdau (enzymatischer) digest (enzymatic)

● **Doppelverdau** double digest

● **einfacher Verdau** single digest

● **Partialverdau** partial digest

verdauen digest

verdaulich digestible

Verdaulichkeit/Bekömmlichkeit digestibility

Verdauung digestion

Verdauungsdrüse digestive gland

Verdauungsenzym digestive enzyme

Verdauungshohlraum digestive cavity, gastrovascular cavity, enteron, coelenteron

Verdauungskanal/Verdauungstrakt alimentary canal/tract, digestive canal/tract

Verdauungssystem digestive system

verderblich perishable; (Früchte: leicht verderblich) highly perishable

verdickt enlarged, thickened

Verdickung thickening

Verdopplungszeit (Generationszeit) doubling time (generation time)

Verdrängung displacement

Verdrängungsreaktion *biochem* displacement reaction

● **doppelte/Doppel-Verdrängung (Pingpong-Reaktion)** double displacement reaction (ping-pong reaction)

● **einfache/Einzel-Verdrängung** single displacement reaction

● **zufällige/nicht-determinierte Verdrängung** random displacement reaction

● **geordnete Verdrängung** ordered displacement reaction

Verdrängungsschlaufe/ Verdrängungsschleife (DNA) *gen* displaced loop, displacement loop

verdreht/gedreht/verkrümmt/ eingewunden contorted

Verdriftung passive dispersal

verdünnen dilute, thin down

Verdünnung dilution, thinning down

Verdünnungs-Schüttelkultur dilution shake culture

Verdünnungsausstrich dilution streak, dilution streaking

verdunsten evaporate, vaporize

Verdunstung evaporation, vaporization

Verdunstungskälte/ Verdunstungsabkühlung evaporative cooling

Verdunstungswärme heat of vaporization

Veredelung/Pfropfung *hort* grafting

vereinigen unite, unify, combine

Vereinigung union, unification, combination

verenden (Tiere) die, perish

verengen/einschnüren constrict

Verengung/Enge/Einschnürung constriction

vererbbar transmissible, heritable

vererben transmit, pass on

Vererbung heredity, inheritance, transmission (of hereditary traits)

● **cytoplasmatische V.** cytoplasmic inheritance

● **maternale V.** maternal inheritance

● **matrokline V.** matroclinous inheritance

● **Mendelsche V. beim Menschen** Mendelian Inheritance in Man (MIM)

● **Mischvererbung** blending inheritance

● **mitochondriale V.** mitochondrial inheritance

● **partikuläre V.** particulate inheritance

● **Überkreuzvererbung** criss cross inheritance

● **uniparentale V.** uniparental inheritance

Vererbungslehre/Genetik genetics, transmission genetics

Vererbungslinienbestimmung gene tracking

Vererbungsmodus

Vererbungsmodus/Erbgang
mode of inheritance
verestern esterify
Veresterung esterification
Verfahrenstechnik
process engineering
- **biologische V./**
Bioingenieurwesen/Biotechnik
bioengineering

Verfall (körperlicher) decline
Verfallsdatum expiration date
verfaulen/zersetzen foul, rot,
decompose, decay
verfault/zersetzt foul, rotten,
decomposed, decayed
verfilzt matted, felted
verflochten interwoven, intertwined,
entangled
verflüssigen liquefy
Verflüssigung liquefaction
Verformung/Formänderung/
Deformation deformation
verfrüht precocious
Verfügbarkeit availability
vergären/fermentieren ferment
Vergärung/Fermentation
fermentation
Vergeilung/Etiolement etiolation
vergiften poison, intoxicate
vergiften (Tiergift) envenom
Vergiftung/Intoxikation poisoning,
intoxication
Vergiftung (Tiergift) envenomation,
envenomization
Vergiftungszentrale/
Entgiftungszentrale
poison control center
vergleichende
Genomhybridisierung (CGH)
comparative genome hybridization
(CGH)
vergleichende Morphologie
comparative morphology
Vergleichssubstanz
comparative substance
vergrößern magnify, enlarge
Vergrößerung magnification,
enlargement
- **x-fache Vergrößerung**
magnification at x diameters

Vergrößerungsglas
magnifying glass, magnifier, lens
Verhalten behavior
- **angeborenes Verhalten**
innate behavior
- **angepasstes Verhalten/**
beeinflusstes Verhalten
conditioned behavior
- **Appetenzverhalten**
appetitive behavior
- **Ausdrucksverhalten**
expressive behavior
- **Bettelverhalten** begging behavior
- **Brunstverhalten** rutting behavior
- **Drohverhalten** threat behavior
- **egoistisches Verhalten**
selfish behavior
- **epideiktisches Verhalten**
epideictic behavior
- **epimeletisches Verhalten**
epimeletic behavior
- **Erkundungsverhalten**
exploratory behavior
- **erlerntes Verhalten/**
Lernverhalten learned behavior,
acquired behavior
- **formkonstante**
Verhaltenselemente
fixed action pattern
- **Fortpflanzungsverhalten**
reproductive behavior
- **Geschlechtsverhalten/**
Sexualverhalten sexual behavior
- **Hassverhalten/Hassen**
mobbing behavior
- **Herden (Hüteverhalten)**
herding (guarding behavior)
- **Imponierverhalten/**
Imponiergehabe/
Imponiergebaren
display behavior
- **Instinktverhalten/Triebverhalten**
instinctive behavior,
instinct behavior
- **Kampfverhalten**
fighting behavior
- **Konsumverhalten**
consummatory behavior
- **Kontaktverhalten** huddling
- **Lernverhalten/erlerntes**
Verhalten learned behavior
- **Meideverhalten**
avoidance behavior
- **Nachfolgeverhalten**
following behavior
- **Neugierverhalten/Neugier**
curiosity, inquisitiveness
- **Orientierungsverhalten/**
Orientierung orientational
behavior, orientation
- **Paarungsverhalten**
mating behavior
- **Putzverhalten** preening behavior
- **Ritualverhalten**
ritualized behavior
- **Schreckverhalten** startle behavior
- **Sexualverhalten/**
Geschlechtsverhalten
sexual behavior
- **Sozialverhalten** social behavior
- **Spielverhalten** play,
play behavior
- **Spottverhalten**
mocking behavior

Verletzung

- **Territorialverhalten** territorial behavior
- **Triebverhalten/Instinktverhalten** instinctive behavior
- **Wanderverhalten** migratory behavior
- **Warnverhalten** warning behavior, alarm behavior
- **Weideverhalten** foraging behavior
- **Werbeverhalten** courting behavior

Verhaltensänderung behavioral change, change of behavior
Verhaltensbarriere behavioral barrier, ethological barrier, ethological isolation
Verhaltensforschung/ Verhaltensbiologie/Ethologie behavioral biology, ethology (study of animal behavior)
Verhaltensgenetik behavior genetics
Verhaltensmuster behavioral pattern
Verhaltensökologie/Ethökologie behavioral ecology
Verhaltensstörung/ Verhaltensanomalie behavioral disorder, behavioral anomaly, deviant behavior
Verhaltensweise behavior, mode of behavior
Verhältnis/Beziehung relationship
Verhältnis/Quotient/Proportion ratio, quotient, proportion
Verhältnisskala/Ratioskala stat ratio scale
verholzt/lignifiziert lignified
Verholzung/Lignifizierung lignification, sclerification
verhornen cornify (converting/changing into horn), keratinize
verhornt/keratinisiert horny, cornified, keratinized
Verhornung/Keratinisierung cornification, keratinization
Verhütung/Kontrazeption contraception
Verhütungsmittel/ empfängnisverhütendes Mittel/ Kontrazeptivum contraceptive
verjüngen/regenerieren rejuvenate, regenerate
verjüngen (spitz zulaufen) taper
verjüngt/spitz zulaufend (Blattspitze) attenuate, tapered, tapering
Verjüngung/Regeneration rejuvenation, regeneration
verkalken (verkalkt) calcify (calcified)
Verkalkung/Kalkeinlagerung/ Kalzifizierung/Calcifikation calcification

Verkarstung geol karstification
Verkehr/Geschlechtsverkehr (sexual) intercourse
verkehrt eiförmig obovate, inversely egg-shaped
verkehrt herzförmig obcordate, obcordiform, inversely heart-shaped
verkehrt lanzettförmig oblanceolate, inversely lanceolate
Verkernung medullation
verketten concatenate
Verkettung concatenation
Verkienung resinification, becoming resinous
Verklappung ocean dumping, discharge at sea
Verkleinerung photo (size) reduction
verknöchern ossify
Verknöcherung/Knochenbildung/ Ossifikation ossification
verkorkt/suberisiert corky, suberous
Verkorkung/Suberisierung suberification; suberization
verkrüppelt/krüppelig/krüppelhaft crippled, stunted
verkümmert/abortiv/rudimentär/ rückgebildet abortive
verkümmert/unterentwickelt vestigial (small and imperfectly developed), underdeveloped, stunted
verkümmert/verarmt depauperate, starved, reduced
verlagert/ektopisch (an unüblicher Stelle liegend) ectopic
Verlagerung (von Chromosomenabschnitten)/ Dislokation dislocation
Verlandung limn terrestrialization (lakes); geol silting up, filling up by sedimentation (rivers/lakes)
Verlangen/Bedürfnis desire; (starkes Verlangen) craving; (Sehnsucht) longing, yearning
Verlängerung elongation; (Ausdehnung) extension
Verlängerungszone/Streckungszone (Wurzel) zone/region of extension, zone/region of elongation
Verlangsamungsphase/ Bremsphase/Verzögerungsphase deceleration phase
Verlauf (einer Krankheit) course (of a disease), progress, development, trend
Verlauf (einer Kurve) path, course, trend
verletzen injure
verletzlich vulnerable
Verletzlichkeit vulnerability, vulnerableness
Verletzung injury

**Verlust der Heterozygotie/
Heterozygotieverlust**
loss of heterozygosity (LoH)
**vermehren/fortpflanzen/
reproduzieren** propagate,
reproduce
**Vermehrung/Amplifikation/
Vervielfältigung** amplification
**Vermehrung/Fortpflanzung/
Reproduktion** propagation,
reproduction
- **geschlechtliche/sexuelle V.**
sexual reproduction
- **ungeschlechtliche/vegetative V.**
asexual/vegetative reproduction
**Vermehrung/Vervielfältigung/
Multiplikation** multiplication
Vermehrungsorgan
reproductive organ
Vermehrungsrate reproductive rate
Vermeidestrategie
avoidance strategy
Vermeidung avoidance
**Vermeidungsreaktion/
Meidereaktion** *ethol* avoidance
reaction, avoiding reaction
Vermiculit *hort* vermiculite
vermischbar miscible
- **unvermischbar** immiscible
Vermischbarkeit miscibility
- **Unvermischbarkeit** immiscibility
vermischen mix
Vermischung mix, mixing
vermitteln mediate
Vermittler/Mediator mediator
vermodern/modern rot, decay,
decompose, putrefy
Vermutung/Annahme hunch, guess,
assumption
**Vernalisation/Kälteinduktion/
Keimstimmung** vernalization
Vernässung waterlogging
Vernation/Knospenlage vernation,
ptyxis, prefoliation
vernetzen/vernetzt netted, meshy,
reticulate
Vernetzung webbing
vernichten destroy, eliminate
Vernichtung destruction, elimination
veröden (Landschaft) become
desolate, become deserted,
obliterate; *med* obliterate
verödet (Landschaft) desolate(d),
deserted, obliterate(d);
med obliterate(d)
Verödung desolation, obliteration
**Verpackung (z. B. Virusnukleinsäure
mit Virusproteinen)** packaging
(e. g. viral nucleic acid by viral
proteins)
- ***in vitro*-Verpackung**
in vitro packaging

verpflanzen/transplantieren *zool*
transplant
**verpflanzen/umpflanzen/umsetzen/
versetzen** *bot* replant
Verpflanzung/Transplantation
transplantation
verpuffen *chem* deflagrate
Verpuffung *chem* deflagration
verpuppen pupate
Verpuppung pupation, pupating
Verrieselung (sprinkling)
purification of wastewater on
sewage fields
Versalzung (Boden) salinization
Versauerung acidification
verschachtelter Primer *gen*
nested primer
Verschiedenartigkeit/Variabilität
variability
**Verschiedenblättrigkeit/
Anisophyllie/Heterophyllie**
anisophylly, heterophylly
**Verschiedenheit/Mannigfaltigkeit/
Vielgestaltigkeit** diversity
**Verschleierung/Verkleidung
(Täuschung)** disguise
verschlingen devour
**Verschlusskontakt/Engkontakt/
Schlussleiste/Kittleiste/Tight
junction (Zonula occludens)**
tight junction
Verschlusskörper *vir* occlusion body
verschmelzen/fusionieren fuse
Verschmelzung/Fusion fusion
verschmutzen pollute, contaminate
verschmutzt polluted, contaminated
- **unverschmutzt** unpolluted,
uncontaminated
Verschmutzung pollution,
contamination
- **Lärmverschmutzung** noise
pollution
- **Luftverschmutzung** air pollution
- **Umweltverschmutzung**
environmental pollution
- **Wasserverschmutzung**
water pollution
Verschmutzungsgrad amount of
pollution, degree of contamination
verseifen saponify
Verseifung saponification
**Versengung/Brandfleck (Hitze/
Klima)** scorch
versetzen/umpflanzen transplant,
replant
Versickerung/Infiltration seepage,
infiltration
Verständigung/Kommunikation
communication
verstärken *techn* amplify; *metabol*
enhance; *neuro* (Reiz) reinforce,
amplify (stimulus)

Verstärker *techn* amplifier; *metabol* (Substanz) enhancer
Verstärker(sequenz)/Enhancer *gen* enhancer (sequence)
Verstärkerfolien (Autoradiographie) intensifying screens
Verstärkung *neuro* **(Reiz)** (stimulus) reinforcement, amplification
Verstärkungskaskade amplification cascade
Versteck hideout, hideaway, hiding place, retreat, refuge
verstecken hide, conceal
Verstecken hiding, concealment
versteift stiffened
versteinern petrify
Versteinerung petrification
Versteppung transformation into grassland
verstreuen/ausstreuen spread, scatter, disseminate
verstreut (liegen) intersperse, disperse
Versuch experiment, test, trial; (Ansatz) attempt
- **Doppelblindversuch** double blind assay, double-blind study
- **Feldversuch/ Freilanduntersuchung/ Freilandversuch** field study, field investigation, field trial
- **Isotopenversuch** isotope assay
- **Schutzversuch/ Schutzexperiment** protection assay, protection experiment
- **Triplettbindungsversuch** triplet binding assay
- **Vorversuch** pretrial, preliminary experiment
Versuchsanlage pilot plant
Versuchsanordnung experiment setup
Versuchsdurchführung performing an experiment, performance of an experiment
Versuchsreihe experimental series
Versuchsverfahren experimental procedure/protocol, experimental method
versumpfen paludify, become swampy
Versumpfung paludification
Verteidigung defense
Verteiler distributer
- **Gasverteiler (Düse in Reaktor)** *biot* sparger
Verteilung *chem/stat* distribution
Verteilung/Trennung/Disjunktion (der Tochterchromosomen) *gen* disjunction
- **alternierende Verteilung** alternate disjunction

Verteilung/Zerstreuung dispersion, spreading
- **Affinitätsverteilung** affinity partitioning
- **Altersverteilung** age distribution
- **bimodale Verteilung** bimodal distribution
- **Binomialverteilung** binomial distribution
- **F-Verteilung/Fisher-Verteilung/ Varianzquotientenverteilung** F-distribution, Fisher distribution, variance ratio distribution
- **freie/unabhängige Verteilung** *gen* independent assortment
- **Gauß-Verteilung/ Normalverteilung/ Gauß'sche Normalverteilung** Gaussian distribution (Gaussian curve/ normal probability curve)
- **Gegenstromverteilung** countercurrent distribution
- **Häufigkeitsverteilung** frequency distribution (FD)
- **Lognormalverteilung/ logarithmische Normalverteilung** lognormal distribution, logarithmic normal distribution
- **nicht-zufallsgemäße Verteilung** nonrandom disjunction
- **Normalverteilung** normal distribution
- **Poissonsche Verteilung/ Poisson Verteilung** Poisson distribution
- **Randverteilung** marginal distribution
- **statistische Verteilung** statistical distribution
- **Varianzquotientenverteilung/ F-Verteilung/Fisher-Verteilung** variance ratio distribution, F-distribution, Fisher distribution
Verteilungsfunktion *stat* distribution function
Verteilungsmuster distribution pattern
vertikale Luftführung (Vertikalflow-Biobench) vertical air flow (clean bench with vertical air curtain)
vertikale Transmission vertical transmission
Vertikalrotor *centrif* vertical rotor
Vertikalzonierung vertical zonation
vertilgen devour; (einverleiben) engulf
verträglich/kompatibel/tolerant compatible, tolerant
- **unverträglich/inkompatibel/ intolerant** incompatible, intolerant

Verträglichkeit 730

Verträglichkeit/Kompatibilität/ Toleranz compatibility, tolerance
- **Unverträglichkeit/ Inkompatibilität/Intoleranz** incompatibility, intolerance

Vertrauensintervall/ Konfidenzintervall *stat* confidence interval

Verunreinigung/Kontamination impurity, contamination

Vervielfältigung/Vermehrung/ Amplifikation amplification

verwachsen/angewachsen *allg* fused, coalescent

verwachsen (gleiche Organe) *bot/ zool* connate

verwachsen (ungleiche Organe) *bot/zool* adnate

verwachsenblättrig/ verwachsenblumenblättrig/ verwachsenkronblättrig/sympetal sympetalous; (Fruchtblätter) syncarpous

Verwachsenkiemer/Siebkiemer/ Septibranchia septibranch bivalves, septibranchs

Verwachsung *allg* fusion; coalescence, symphysis

Verwachsung *bot/zool* **(gleicher Organe)** connation, cohesion

Verwachsung *bot/zool* **(ungleiche Organe)** adnation

verwandeln/metamorphosieren/ die Gestalt ändern transform, metamorphose

Verwandlung/Metamorphose metamorphosis
- **unvollkommene/unvollständige V./M.** incomplete metamorphosis, gradual metamorphosis
- **vollkommene/vollständige V./M.** complete metamorphosis

verwandt akin, related

verwandt/zugehörig *gen* cognate

Verwandtenehe/konsanguine Ehe/ Ehe unter Blutsverwandten consanguineous marriage

Verwandtenselektion kin selection

Verwandter (ersten/zweiten Grades) (first-degree/second-degree) relative

Verwandtschaft (*generell*) relationship, relatedness, kinship; (*spezifisch*) relatives

verwandtschaftliche Beziehung hereditary relationship

Verwandtschaftskoeffizient coefficient of relatedness

Verwandtschaftsselektion kinship selection

Verwandtschaftstheorie kinship theory

Verweilzeit/Verweildauer/ Aufenthaltszeit/Verweildauer residence time; (Retentionszeit) retention time
- **mittlere V.** mean residence time

verwelken wither, wilt, fade (shrivel up)

verwelkend withering, wilting, fading, shrivelling, marcescent

verwerten *metabol/ecol* utilize

Verwertung *metabol/ecol* utilization

verwesen/zersetzen putrefy, rot, decompose

Verwesung/Zersetzung putrefaction, rotting, decomposition

verwildern *zool* become wild; *bot* overgrow, grow wild; (degenerieren) degenerate

verwildert *zool* feral; *bot* escaped (e.g. from cultivation)

verwittern *geol* weather; *bot* waste

Verwitterung *geol* weathering; *bot* wastage

Verwitterungsbeständigkeit durability

Verwitterungskruste/ Verwitterungsrinde *geol* weathering rind

verwüsten obliterate

Verwüstung obliteration

verzehren/verschlingen/ herunterschlingen devour, gulp down

verzerrt/verfälscht *math/stat* biased
- **unverzerrt/unverfälscht** *math/ stat* unbiased

verziehen (Holz) warp

verzögern delay, retard

Verzögerung delay, retardation

Verzögerungseffekt delayed effect

Verzögerungsphase/ Verlangsamungsphase/ Bremsphase (Wachstum) deceleration phase

verzuckern saccharify

Verzuckerung saccharification

verzweigen, sich branch out, ramify

verzweigt branched, ramified
- **unverzweigt** unbranched, unramified

verzweigtkettig *chem* branched-chained

Verzweigung/Verästelung/ Ramifikation branching, ramification; *chem* branching
- **gabelige Verzweigung** dichotomous branching

Verzweigungsstelle branch site

Verzweigungssystem (monopodiales/sympodiales) *bot* (monopodial/sympodial) branching system

Verzwergung/Nanismus nanism, dwarfishness
Vesikel *nt*/**Bläschen** vesicle
- **Amphiesmalvesikel/ Amphiesmalbläschen** amphiesmal vesicle
- **Endozytosevesikel/Endosom** endocytic vesicle, endosome
- **Golgi-Vesikel** *cyt* Golgi vesicle
- **Inside-out Vesikel (Vesikel mit der Innenseite nach außen)** inside-out vesicle
- **Korbvesikel/ Stachelsaumbläschen/ Stachelsaumvesikel/ coated vesicle** coated vesicle
- **Right side-out Vesikel (Vesikel mit der richtigen Seite nach außen)** right side-out vesicle
- **Stachelsaumvesikel/ Stachelsaumbläschen/ Korbvesikel/coated vesicle** coated vesicle
- **synaptisches Vesikel/ synaptisches Bläschen/ Synaptosom** synaptosome, synaptic vesicle
vesikulär/bläschenartig vesicular, bladderlike
Vesikulartransport/Zytopempsis/ Cytopempsis cytopempsis
Vestibül vestibule, vestibulum
Vestibulum labyrinthi (mit Utriculus und Sacculus) vestibule (with utricle and saccule)
Veterinärmedizin/Tiermedizin/ Tierheilkunde veterinary medicine, veterinary science
Vibracularie vibraculum
Vibrionen vibrios
Vieh/Tier animal
- **Federvieh/Geflügel** fowl
- **landwirtschaftlich genutztes Vieh** livestock
- **Rindvieh/Rinder** cattle
Viehfutter animal feed
Viehhaltung livestock keeping, animal husbandry
Viehstall cattle shed, barn
Viehwirtschaft animal husbandry, ranching
Viehzucht *sensu lato* livestock breeding; *allg* animal husbandry, ranging; (Rinderzucht) cattle production, cattle breeding
vielbeinig/polypod polypod
Vielborster/Borstenwürmer/ Polychaeten bristle worms, polychaetes, polychetes, polychete worms
Vieleckbein, großes/Os trapezium trapezium bone, greater multangular bone

Vieleckbein, kleines/ Os trapezoideum trapezoid bone, lesser multangular bone
Vielfachteilung/Mehrfachteilung (Bakterien) multiple fission
Vielfachzucker/Polysaccharid multiple sugar, polysaccharide
Vielfalt/Vielfältigkeit/ Vielgestaltigkeit/Mannigfaltigkeit diversity
- **biologische Vielfalt/ Lebensvielfalt** biodiversity, biological diversity, biological variability
Vielfraß polyphage
Vielgestaltigkeit/Vielfalt/ Vielfältigkeit/Mannigfaltigkeit diversity
vielkammerig/vielkämmrig/ polythalam/polythekal with many chambers, polythalamous, polythalamic, polythecal
vielkernig/mehrkernig multinucleate(d)
vielreihig/mehrreihig/multiseriat multiseriate, multiple rowed
vielschichtig/mehrschichtig multilayered
Vielweiberei/Polygynie (Polygamie) polygyny (polygamy)
Vielzeller *allg* multicellular lifeform
Vielzeller/Mitteltiere/Gewebetiere/ Metazoen/Metazoa metazoans
vielzellig multicellular
Vielzweckklonierungsstelle/ multiple Klonierungsstelle *gen* multiple cloning site (MCS)
Vielzweckvektor/multifunktioneller Vektor multifunctional vector, multipurpose vector
vierfächerig/vierfächrig *bot* (Fruchtknoten) four-locular
vierfüßig quadrupedal
Vierfüßigkeit quadrupedalism
viergeißlig tetraflagellate(d)
vierhändig four-handed, quadrumanous
Vierhügel/Corpora quadrigemina corpora quadrigemina
Vierhügelplatte/Lamina tecti/ Lamina quadrigemina quadrigeminal plate, lamina of roof of midbrain, tectal lamina of mesencephalon
vierkantig (Stengel) four-angled
vierlappig quadrilobate
Vierlinge quadruplets
vierspaltig quadrifid
vierteilig quadripartite
vierteilig-aufgesägt (Holzstamm) quartersawn
Viertelswert/Quartil *stat* quartile

vierwertig *chem* tetravalent
vierzählig/tetramer tetramerous, tetrameric, tetrameral
Vigreux-Kolonne *lab* vigreux column
Vikarianz vicariance
vikariieren vicariate
vikariierend vicarious
Vikariismus vicariism
Violdrüse (Schwanzwurzeldrüse des Fuchses) violet gland, supracaudal gland
Violoaceae/Veilchengewächse violet family
viral viral
Virämie viremia
Virion/Viruspartikel/Virusteilchen virion, viral particle
Virioplasma virioplasm
Viroid viroid
Virologie virology
Viropexis viropexis
Virose/Viruserkrankung virosis
Virostatikum virostatic
virtuelles Bild *micros* virtual image
virulent virulent
• **nicht virulent** avirulent
Virulenz/Infektionskraft virulence (disease-evoking power/ability of cause disease)
Virus (*pl* **Viren)** virus
• **amphotropes Virus** amphotropic virus
• **defektes Virus** defective virus
• **ecotropes Virus** ecotropic virus
• **ikosaedrisches Virus** icosahedral virus
• **Pflanzenvirus** plant virus
• **Satellitenvirus** satellite virus
• **Tiervirus** animal virus
• **xenotropes Virus** xenotropic virus
Viruserkrankung/Virose viral infection, virosis
Virushülle viral coat
Viruskern/Zentrum (zentrale Virionstruktur) core
Viruspartikel/Virusteilchen/Virion viral particle, virion
viruzid virucidal, viricidal
Viscaceae/Mistelgewächse christmas mistletoe family
viskos/viskös/zähflüssig/dickflüssig viscous, viscid (glutinous consistency)
Viskosität/Dickflüssigkeit/Zähflüssigkeit viscosity, viscousness
• **Viskositätskoeffizient** coefficient of viscosity
viszeral/Eingeweide.../zu den Eingeweiden gehörend visceral, splanchnic

Viszeralganglion visceral ganglion
Viszeralskelett visceral skeleton, visceroskeleton
Vitaceae/Weinrebengewächse vine family, grape family
Vitalfarbstoff vital dye, vital stain
Vitalfärbung/Lebendfärbung vital staining
Vitalität/Lebenskraft vitality
Vitalkapazität vital capacity
Vitamin(e) vitamin(s)
• **Ascorbinsäure (Vitamin C)** ascorbic acid
• **Biotin (Vitamin H)** biotin
• **Carnitin (Vitamin T)** carnitine (vitamin B_T)
• **Carotin/Caroten/Karotin (Vitamin A Vorläufer)** carotin, carotene (vitamin A precursor)
• **Cholecalciferol/Calciol (Vitamin D₃)** cholecalciferol
• **Citrin (Hesperidin) (Vitamin P)** citrin (hesperidin)
• **Cobalamin/Kobalamin (Vitamin B₁₂)** cobalamin
• **Ergocalciferol/Ergocalciol (Vitamin D₂)** ergocalciferol
• **Folsäure/Pteroylglutaminsäure (Vitamin B₂ Familie)** folic acid, folacin, pteroyl glutamic acid
• **Gadol/3-Dehydroretinol (Vitamin A₂)** gadol, 3-dehydroretinol
• **Menachinon (Vitamin K₂)** menaquinone
• **Menadion (Vitamin K₃)** menadione
• **Pantothensäure (Vitamin B₃)** pantothenic acid
• **Phyllochinon/Phytomenadion (Vitamin K₁)** phylloquinone, phytonadione
• **Pyridoxin/Pyridoxol/Adermin (Vitamin B₆)** pyridoxine, adermine
• **Retinol (Vitamin A)** retinol
• **Riboflavin/Lactoflavin (Vitamin B₂)** riboflavin, lactoflavin
• **Thiamin/Aneurin (Vitamin B₁)** thiamine, aneurin
• **Tocopherol/Tokopherol (Vitamin E)** tocopherol
Vitaminmangel vitamin deficiency
Vitellinmembran/primäre Eihülle/Dotterhaut/Dottermembran/Membrana vitellina vitelline layer, vitelline membrane
Vitellodukt yolk duct
Vitrine/Schaukasten showcase
Vittariaceae shoestring ferns

733   vorderseitig   **V**

vivipar/lebendgebärend viviparous, live-bearing
Viviparie/Lebendgebären vivipary, viviparity, live-bearing
Vochysiaceae/Rittersspornbäume vochysia family
Vögel/Aves birds
Vögel betreffend/Vogel... avian
Vogelaugenholz bird's eye (wood texture)
Vogelausbreitung/Ornithochorie bird-dispersal, ornithochory
Vogelbecken-Dinosaurier/ Ornithischia bird-hipped dinosaurs, ornithischian reptiles
Vogelblume bird flower, ornithophile, bird-pollinated flower
vogelblütig/ornithophil bird-pollinated, ornithophilous
Vogelblütigkeit/Vogelbestäubung/ Ornithophilie bird pollination, ornithophily
Vogelhaus/Voliere bird house, aviary
Vogelkäfig bird cage
"Vogelköpfchen"/Avicularie (Bryozoen) avicularium
Vogelkunde/Ornithologie ornithology, study of birds
Vogelkundler/Ornithologe ornithologist, birds specialist
Vogelnestpilze/Nestpilze/ Teuerlinge/Nidulariaceae bird's-nest fungi, bird's-nest family
Vogelwarte bird-watching station, bird-watching haunt, ornithological station
Vogelzucht aviculture
Vogelzüchter aviculturist
Vogelzug bird migration
Voliere/Vogelhaus aviary
Volk/Staat (Bienen/Ameisen) colony
Volkmannscher Kanal Volkmann canal, Volkmann's canal, canal of Volkmann (perforating canal)
volkstümlicher Name/ Vernakularname common name, vernacular name
Vollblut (Pferd) thoroughbred
Vollgesang *orn* full song
vollgesogen (mit Wasser) waterlogged
Vollinsekt/Vollkerf/Imago imago, adult insect
Vollkornmehl whole-grain flour
Vollmedium complete medium
Vollpipette/volumetrische Pipette transfer pipet, volumetric pipet
Vollplazenta/Placenta vera/Placenta deciduata deciduate placenta
Vollreife (z. B. Getreide) full ripeness

Vollschmarotzer/Vollparasit/ Holoparasit holoparasite, obligate parasite
vollständige Blüte complete flower
Vollzirkulation *limn* complete overturn
Volutinkörnchen/metachromatische Granula *pl* volutin granule, metachromatic granule
Volva/Velum universale/ becherförmige "Knolle" bei Agaricus volva, universal veil; cup, pouch
Von-Magnus-Partikel/DI-Partikel von Magnus particle, defective interfering particle (DI particle),
Vorauflaufbehandlung *agr* pre-emergence treatment
Voraugendrüse/Antorbitaldrüse preorbital gland, antorbital gland
Voraugenschild/Praeoculare preocular (scale)
Voraussage prediction
Voraussagemodell predictive model
voraussagend predictive
vorausschauende Ökologie/ voraussagende Ökologie predictive ecology
Vorblatt/Bracteola secondary bract, bracteole, bractlet
Vorblatt/Prophyll first leaf, prophyll
Vorbrunst/Proöstrus proestrus
Vorderbein foreleg
Vorderbrust/Prothorax prothorax
Vorderdarm/Stomodaeum foregut, stomodeum
Vorderextremität forelimb
Vorderflügel (Oberflügel/ Deckflügel/Flügeldecke) forewing, front wing, tegmina
Vorderflughaut/Propatagium propatagium
Vorderfuß forefoot (*pl* forefeet), front foot
Vorderhirn/Prosencephalon (Vertebraten) forebrain, prosencephalon (telencephalon + diencephalon)
Vorderhirn/Protocerebrum (Insekten) protocerebrum
Vorderkiemer/ Vorderkiemenschnecken/ Streptoneura/Prosobranchia prosobranch snails, prosobranchs
Vorderkörper/Vorderleib/Prosoma/ Cephalothorax/"Kopf" prosoma, proterosoma, cephalothorax
vorderseitig (bauchseitig) front side, ventral

Vorfahre

Vorfahre/Ahne ancestor, forebear, progenitor
Vorfluter receiving water
vorgeburtlich/pränatal antenatal, prenatal
vorgeschichtliche DNA ancient DNA
Vorhand (Pferd) forehand
Vorhaut/Präputium foreskin, preputium, prepuce
vorherrschen predominate
Vorhersage/Prognose prognosis
• **Wettervorhersage** weather forecast
vorhersagende Medizin predictive medicine
Vorhof/Atrium atrium
Vorhof/Vorraum/Vestibulum vestibule
Vorhofdrüse (Scheidenvorhof) vestibular gland
Vorhofgang/Scala vestibuli vestibular canal
Vorkeim (Prothallus: Farne) prothallus; (Protonema: Moose/ gewisse Algen) protonema
vorkeimen pregerminate
Vorkeimung pregermination
Vorkern/Pronukleus/Pronucleus pronucleus
vorkiefrig/prognath prognathous, prognathic
Vorkiefrigkeit/Prognathie prognathism, prognathy
Vorklärbecken primary settling tank
Vorkommen occurrence, presence
Vorkultur preculture
Vorlage *chem* destillation receiver
Vorlauf forerun
Vorläufer/Präkursor precursor
Vorläufer-mRNA/Prä-mRNA pre-mRNA (precursor mRNA)
Vorläufer-rRNA/Prä-rRNA pre-rRNA (precursor rRNA)
Vorläufer-tRNA/Prä-tRNA pre-tRNA (precursor tRNA)
Vorläuferzelle precursor cell
Vormagen/Blättermagen/Psalter/ Omasus (Wiederkäuer) third stomach, omasum, psalterium
Vormagen/Vorderdarm/Kropf/ Ingluvies (Insekten/Vögel) crop
Vormännlichkeit/Protandrie/ Proterandrie protandry
Vormilch/Biestmilch/Kolostralmilch/ Colostrum foremilk, colostrum

Vorniere/Pronephros fore-kidney, primitive kidney, primordial kidney, head kidney, pronephros
Vornierengang/primärer Nierengang/primärer Harnleiter/ Wolffscher Gang Wolffian duct, Leydig's duct, mesonephric duct, archinephric duct
Vorpuppe/Propupa/Präpupa/ Semipupa propupa, prepupa
Vorrat stock, supply (meist supplies), provisions, reserve
Vorratshaltung hoarding of food
Vorratskammer storage chamber
Vorratsschädling storage pest
Vorrichtung device
Vorriff fore reef
Vorschild/Präscutum prescutum
Vorsicht caution, cautiousness, care, carefulness, precaution
Vorsicht! caution! (careful!)
vorsichtig cautious, careful
Vorsichtsmaßnahme/ Vorsichtsmaßregel *lab* precaution, precautionary measure
Vorspann geben/Anhängen/ Ammenveredelung *hort* inarching
Vorspelze *bot* palea, palet, pale, glumella, inner glume
Vorsteherdrüse/Prostata/ Prostatadrüse prostate, prostate gland
Vorstoß/Attacke attack
Vorstoß *lab/chem* adapter
Vorstrand foreshore
"Vortex"/Mixer/Mixette/ Küchenmaschine vortex, mixer
Vortrieb/Anschub thrust
Vorversuch pretrial, preliminary experiment
Vorwald/Vorholz nurse crop, pioneer crop, pioneer forest
vorwärts gerichtet/ aufwärts gerichtet antrorse
Vorwärtsmutation forward mutation
Vorwärtszieher/Protraktor (Muskel) protractor
Vorweiblichkeit/Protogynie/ Proterogynie protogyny
Vorzieher/Protractor (Muskel) protractor muscle
Vorzugstemperatur cardinal temperature
Vulkan volcano
Vulkanasche volcanic ash
Vulkanausbruch volcanic eruption

Wachstumsphase

Waage scale (weight), balance (mass)
- **Analysenwaage** analytical balance
- **Feinwaage** precision balance
- **Laborwaage** laboratory balance

Wabe/Honigwabe honeycomb
wabenförmig/wabig honeycombed, alveolate, favose, faveolate
wabig honeycombed, favose, faveolate, alveolate
wach awake
Wachs wax
- **Plastilin** plasticine

wachsartig waxy, wax-like, ceraceous
wachsartig-weißlich reflektierend (Blattoberfläche) glaucous, "bloom"
Wachsbelag *bot* wax coating
Wachsblättler/Hygrophoraceae hygrophorus family
Wachsblume wax plant
Wachsdrüse wax gland, ceruminous gland
wachsen grow; thrive
Wachsfüßchen (Plastilinfüßchen an Deckgläschen) *micros* wax feet, plasticine supports on edges of coverslip
Wachshaut/Cera/Ceroma (am Schnabel) *orn* cere (on bill of birds)
Wachstum growth
- **akroplastes Wachstum** acroplastic growth
- **arithmetisches Wachstum** arithmetic growth
- **ausgewogenes Wachstum** balanced growth
- **basiplastes Wachstum** basiplastic growth
- **begrenztes/beschränktes Wachstum** determinate growth, restricted growth, localized growth
- **Dickenwachstum/Verdickung** thickening
- **Erstarkungswachstum** expansion growth (*Zimmermann/Tomlinson*), corroborative growth (*Troll*)
- **Erweiterungswachstum/ Dilatationswachstum** dilational growth, dilation, dilatation
- **intrusives Wachstum** intrusive growth
- **Jahreswachstum** annual growth
- **Kopfwachstum/kopfseitiges Wachstum** head growth
- **Längenwachstum** longitudinal growth
- **polares Wachstum** polar growth
- **Primärwachstum** primary growth
- **radiäres Wachstum** radial growth
- **Randwachstum** marginal growth
- **Schwanzwachstum/endständiges Wachstum** tail growth
- **Sekundärwachstum** secondary growth
- **Spitzenwachstum** apical growth
- **Streckungswachstum** extension growth, elongational growth
- **symplastes Wachstum** symplastic growth
- **unbegrenztes/unbeschränktes Wachstum** indeterminate growth, open growth, unrestricted growth, diffuse growth
- **Zellwachstum** cell growth

Wachstumsfaktor growth factor
- **epidermaler Wachstumsfaktor** epidermal growth factor (EGF)

wachstumsfördernd growth-stimulating
Wachstumsform growth form
Wachstumsgeschwindigkeit/ Wachstumsrate/Zuwachsrate growth rate
wachstumshemmend growth-retarding, growth-inhibiting
Wachstumshemmer/Wuchshemmer/ Wuchshemmstoff growth inhibitor
Wachstumshormon/Somatotropin/ somatotropes Hormon growth hormone (GH), somatotropin
- **menschliches W. (Somatotropin/ somatotropes Hormon)** human growth hormone (hGH), human somatotropin

Wachstumskegel (Axon) *neuro* growth cone
Wachstumskurve growth curve
Wachstumsleistung growth rate (vigor)
Wachstumsperiode growth period
Wachstumsphase growth phase
- **Absterbephase** decline phase, phase of decline, death phase
- **Adaptationsphase/Anlaufphase/ Latenzphase/Inkubationsphase/ lag-Phase** lag phase, latent phase, incubation phase, establishment phase
- **Beschleunigungsphase/ Anfahrphase** acceleration phase
- **Eingewöhnungsphase** establishment phase
- **exponentielle Wachstumsphase/ exponentielle Entwicklungs- phase** exponential growth phase
- **lag-Phase/Adaptationsphase/ Anlaufphase/Latenzphase/ Inkubationsphase** lag phase, incubation phase, latent phase, establishment phase

Wachstumspunkt

- **logarithmische Phase**
  logarithmic phase (log-phase)
- **Ruhephase/Ruheperiode**
  dormancy period
- **stationäre Phase** stationary
  phase, stabilization phase
- **Teilungsphase** division phase
- **Verlangsamungsphase/
  Bremsphase/Verzögerungsphase**
  deceleration phase,
  retardation phase

Wachstumspunkt growing point
(apical meristem)

**Wachstumsrate/Zuwachsrate/
Wachstumsgeschwindigkeit**
growth rate
- **spezifische W.**
  specific growth rate

Wachstumszone (Wurzel)
zone/region of cell division
(apical meristem)

Wächter guard

**Wackelgelenk/Amphiarthrosis
(straffes Gelenk)**
amphiarthrodial joint

Wade calf (of the leg)

Wadenbein/Fibula splint bone, fibula

Wägetisch weighing table

wahrnehmen/empfinden (Reiz)
perceive

**Wahrnehmung/Empfindung/
Perzeption (Reiz)** perception

Wahrscheinlichkeit probability

Wahrscheinlichkeitsfunktion
likelihood function

Wald forest (Forst/ausgedehnter
Wald), woods (Wald mittlerer
Größe), grove (Wäldchen)
- **Altbestand** old-growth (forest),
  mature forest
- **Ausschlagswald** coppice forest,
  sprout forest
- **Auwald/Auenwald**
  floodplain forest
- **Bannwald (in Austria)** protected
  forest for stabilizing slopes etc.
- **Bannwald/Naturwaldreservat
  (in S/W Germany)** protected
  forest (no commercial usage)
- **Bergregenwald**
  montane rain forest
- **Bergwald** *allg* mountain forest;
  (immergrüne Coniferenstufe)
  montane forest
- **Bewuchs, unterer (Waldschicht)**
  undergrowth
- **Blätterdach (Wald)** (forest) canopy
- **Bruchwald/Bruchwaldmoor/
  Bruchmoor/Sumpfwald/
  Waldmoor** carr (fen woodland),
  swamp woods/forest, wooded
  swamp, paludal forest
- **Buschwald** maquis
- **Dickung** young forest stand,
  young plantation
- **Dornwald** thorn woodland
- **Erholungswald** amenity forest,
  recreational forest
- **Fallaubwald** deciduous forest
- **Femelwald/Plenterwald**
  shelterwood: uneven-aged stand,
  uneven-aged plantation
  (with selective logging)
- **Forst/Kulturwald/
  Wirtschaftswald**
  cultivated forest, tree plantation
- **Galeriewald** gallery forest,
  fringing forest
- **Gebirgswald** mountain forest,
  montane forest
- **Hain/Gehölz/Waldung** grove
- **Hartlaubwald**
  sclerophyllous forest
- **Hegewald/Schutzwald/
  Schonwald** protected forest
  (limited/specified usage)
- **Heidewald** heath forest
- **Hochmoorwald**
  (upland) bog forest
- **Hochwald** high forest
- **Jungwald/junger Wald**
  young forest
- **Kulturwald/Forst**
  cultivated forest, tree plantation
- **Laubwald** deciduous forest,
  broadleaf forest
- **lichter Wald** low-density stand
- **Mischwald** mixed forest
- **Mittelwald** middle-aged forest
- **Monsunwald** monsoon forest
- **Nadelwald** coniferous forest
- **Nadelwaldstufe/
  hochmontane Stufe**
  upper montane/subalpine conifer
  forest zone
- **Nebelwald** cloud/fog forest,
  humid/perhumid forest,
  montane rainforest
- **Niederwald (durch Rückschnitt)**
  coppice
- **Parkwald** parkland
- **Plenterwald/Femelwald**
  shelterwood: selectively cut/
  uneven-aged stand,
  uneven-aged forest/plantation
- **Regenwald** rain forest
- **Saisonwald** seasonal forest
- **Schutzwald/Schonwald/
  Hegewald** protected forest
  (limited/specified usage)
- **Staatswald** state forest
- **Stadtwald/städtischer Wald/
  Kommunalwald/Gemeindewald**
  urban forest, community forest

Warmhaus

- **Streuschicht/Streuhorizont/ Förna (Wald)** litter layer
- **Sumpfwald** swamp forest
- **Urwald** primeval forest, virgin forest, pristine forest, jungle
- **Vorwald/Vorholz** nurse crop, pioneer crop, pioneer forest
- **Zwergnadelwald/ Zwergstrauchzone** pygmy conifer woodland
- **Zwergwald/Zwergwaldstufe** elfin forest, elfin woodland

Waldbau/Forstkultur silviculture
Waldboden forest floor
- **humusartiger W.** duff

Waldbrand forest fire
Wäldchen (kleines/niedriges) coppice, grove
Waldgesellschaft forest community
Waldgrenze forest line, timberline
Waldrand forest's edge
Waldschaden forest damage
Waldschlag clearing
- **Femelschlag/Femelhieb/ Plenterschlag/Plenterung/ Plenterbetrieb** sectional/uneven shelterwood method, femel coupe (selectively cut/uneven-aged stand, uneven-aged forest/plantation)
- **Kahlschlag** clear-cut, clearing, clearance
- **Plenterschlag/Plenterung/ Plenterbetrieb/Femelschlag/ Femelbetrieb** sectional/uneven shelterwood method, femel coupe (selectively cut/uneven-aged stand, uneven-aged forest/plantation)
- **Rückschnitt (bis auf den Stumpf/ für Neuaustrieb)** coppice, coppicing
- **Rückschnitt (Gehölzrückschnitt)** pruning, pruning back
- **Saumschlag/Saumhieb** aisle clearing, strip felling
- **Schirmschlag/Schirmhieb** shelterwood method, selective logging/cutting (even-aged stand, even-aged forest/plantation)

Waldsteppe woodland
Waldsterben "Waldsterben", forest deterioration, forest decline
Waldstreu forest litter
Waldstück woodlot
Waldung/Wäldchen/Hain/ Baumgruppe grove
Waldzerstörung deforestation
Wale und Delphine/Cetacea cetaceans: whales & porpoises & dolphins
Walfang whaling
Wallace-Linie *biogeo* Wallace's line
Wallach (Pferd) gelding

Wallriff/Barriereriff barrier reef
Walnussgewächse/Juglandaceae walnut family
Walrat spermaceti
Walratöl spermaceti oil, sperm oil
walzenförmig/zylindrisch/ cylindrisch cylindrical
Walzengelenk trochoidal joint
Walzenspinnen/Solifugae/ Solpugida sun spiders, false spiders, windscorpions, solifuges, solpugids
walzig/stielrund/länglich zylindrisch (an den Enden abgeflacht: z. B. Hülse) terete
Wamme/Triel/Brustlappen/Palear (Rind) dewlap, jowl
wandbrüchig/scheidewandbrüchig/ septifrag septifragal
Wanddruck *bot* wall pressure, turgor pressure
Wanderackerbau shifting agriculture/cultivation, swidden agriculture/cropping
Wanderdüne shifting dune
Wanderung/Migration *etholl chromatogrlelectrophor* migration
Wanderverhalten migratory behavior
Wandler transducer
wandspaltig/septizid *bot* septicidal
wandständig/wandbürtig/parietal *bot* parietal, borne on the wall
wandständige Plazentation/ Parietalplazentation *bot* parietal placentation
Wange/Gena cheek, gena
Wanne *electrophor* reservoir, tray
Wannenform (Cycloalkane) *chem* boat conformation
Wanzen/Heteroptera (Hemiptera) true bugs, heteropterans
Warburgsches Atmungsferment/ Cytochromoxydase Warburg's factor, cytochrome oxidase
warmblütig warm-blooded
Warmblütigkeit warm-bloodedness
Wärme/Hitze warmth, heat
- **Erwärmung** warming
- **globale Erwärmung** global warming

Wärmedurchgangszahl (C) thermal conductance
Wärmepumpe heat pump
Wärmeschrank incubator
Wärmestrahlung thermal radiation
wärmesuchend/thermophil thermophilic
Wärmetauscher *lab* heat exchanger
Wärmetransport heat transport
Wärmeübergang heat transfer
Warmhaus *hortlagr* hothouse

W Warnfärbung 738

**Warnfärbung/Warntracht/
Abschreckfärbung** *ethol* warning
coloration, aposematic coloration
Warnsignal/Alarmsignal
warning signal, alarm signal
Warnverhalten warning behavior,
alarm behavior
Warte *zool* observation point
(of a guard animal)
Warve/Jahresschicht *geol* varve
(one year's sediment deposit)
Warze (Höcker/Beule/Wölbung)
wart, tubercle, warty protuberance
warzenförmig wart-shaped,
verruciform
Warzenfortsatz (Schädel)
mastoid process
Warzenhof/Areola mammae (Brust)
areola
**Warzenschnecken/Sternschnecken/
Doridacea/Holohepatica**
doridacean snails, doridaceans
**Warzenschwämme/
Kellerschwämme/Coniophoraceae**
dry rot family
warzig/höckerig warty, verrucose,
tuberculate
**Washingtoner
Artenschutzabkommen/
Artenschutzübereinkommen**
Convention on International Trade
in Endangered Species (CITES),
Washington 1975
Wasser water
- **Abwasser** wastewater
- **Amnionwasser/Fruchtwasser/
Amnionflüssigkeit**
amniotic fluid, "water"
- **Binnengewässer** inland water,
inland waterbody
- **Brackwasser** brackish water
(somewhat salty)
- **Brunnenwasser** well water
- **destilliertes Wasser** distilled water
- **entionisiertes Wasser**
deionized water
- **Fließgewässer (Fluss/Strom)**
flowing water (river/stream)
- **gereinigtes Wasser/
aufgereinigtes Wasser/
aufbereitetes Wasser**
purified water
- **Gewässer** body of water,
water body
- **Grundwasser** ground water
- **Haftwasser** film water,
retained water
- **Hochwasser** high water, flood;
mar high tide
- **hartes Wasser** hard water
- **Kristallisationswasser**
water of crystallization

- **Küstengewässer** coastal waters
- **Leitungswasser** tap water
- **Meerwasser** seawater, saltwater
- **Niedrigwasser** low water
- **Oberflächenwasser**
surface water
- **Peptonwasser** peptone water
- **phreatisch/Grundwasser...**
phreatic (pertaining to
groundwater)
- **Quellwasser** springwater
- **salziges Wasser** saline water
- **Salzwasser** saltwater
- **Schmelzwasser** meltwater
- **Schwarzwasser (Fluss)**
black water (river)
- **Senkwasser/Sickerwasser**
gravitational water, seepage water
- **Sickerwasser** drainage water,
leachate/soakage, seepage/
gravitational water
- **Süßwasser** freshwater
- **trinkbares Wasser** potable water
- **Trinkwasser** drinking water
- **Vorfluter** receiving water
- **weiches Wasser** soft water
- **Weißwasser (Amazonas)**
white water
Wasser lassen/urinieren urinate
**Wasser speien/spritzen/abblasen
(Wale)** spout, blow (water)
wasserabweisend water-repellent,
water-resistant
**Wasserährengewächse/
Aponogetonaceae** cape-pondweed
family, water hawthorn family
Wasseraktivität/Hydratur
water activity
Wasseraufbereitung
water purification
Wasseraufbereitungsanlage
water purification plant/facility,
water treatment plant/facility
Wasseraufnahme water uptake
Wasserausbreitung/Hydrochorie
water-dispersal, hydrochory
Wasserbad *lab* water bath
Wasserbahne *limn* water track
Wasserbewegung water movement
**wasserbewohnend/im Wasser
lebend/aquatisch** aquatic
Wasserbilanz water balance
Wasserblatt submerged leaf
**Wasserblattgewächse/
Hydrophyllaceae** waterleaf family
Wasserblüte (meist Algen)
water bloom
Wasserblütigkeit/Hydrophilie
pollination by water, hydrophily
Wasserdampf water vapor
wasserdicht/wasserundurchlässig
waterproof

Wasserströmung

Wassereinlagerung/ Wasseranlagerung/Hydratation hydration

Wassereinzugsgebiet watershed, drainage basin/area/district, catchment area/basin

Wasserflöhe/Cladocera water fleas, cladocerans

wasserfrei free from water; moisture-free; anhydrous

Wasserfront *mar* waterfront

Wassergefäßsystem/ Ambulakralsystem/ Ambulakralgefäßsystem water-vascular system, ambulacral system

Wassergehalt water content

Wassergraben trench, ditch, moat

Wassergüte/Wasserqualität water quality

Wasserhärte water hardness

Wasserhaushalt/Wasserregime water regime

Wasserhülle/Hydrationsschale *chem* hydration shell

Wasserhülle/Hydrosphäre der Erde hydrosphere

Wasserkanal aqueous channel

Wasserkapazität moisture capacity, water-holding capacity of soil

Wasserkreislauf water cycle, hydrologic cycle

Wasserlauf (Fluss/Bach..) waterway, watercourse

Wasserleben aquatic life, life in the water

wasserlebend/im Wasser lebend/ wasserbewohnend/aquatisch aquatic

Wasserleitbahn/ Wasserleitungsbahn *lab* water-conducting element/pathway

wasserleitend water-conducting

Wasserleitung/Translokation (in Leitgewebe) water conductance/ conduction/translocation

Wasserlieschgewächse/ Schwanenblumengewächse/ Butomaceae flowering rush family

Wasserlinsengewächse/Lemnaceae duckweed family

wasserlöslich water-soluble

Wasserlunge (Holothurien) respiratory tree

Wasserlungenschnecken/ Basommatophora (Pulmonata) freshwater snails

Wassermohngewächse/ Limnocharitaceae water-poppy family

Wassernabelgewächse/ Hydrocotylaceae pennywort family

Wassernussgewächse/Trapaceae water chestnut family

Wassernymphe naiad

Wasserpflanze/Hydrophyt aquatic plant, water plant, hydrophyte

Wasserpotential/Hydratur/ Saugkraft water potential

Wasserprobe water sample

Wasserreis (Spross) *hort* sucker, coppice-shoot

Wasserröhrengewächse/ Hydrostachyaceae hydrostachys family

Wassersack/Ascus (Bryozoen) compensation sac, ascus

Wassersättigung water saturation

Wassersättigungsdefizit water saturation deficit (WSD)

Wasserscheide *geol* watershed, water parting, water divide

Wasserschimmel/Saprolegniales water molds

Wasserschlauchgewächse/ Lentibulariaceae bladderwort family, butterwort family

Wasserschutzgebiet water reserve

Wassersog water tension, water suction

Wasserspalte/Hydathode water stoma, water pore, hydathode

wasserspaltend/hydrolytisch hydrolytic

Wasserspaltung/Hydrolyse hydrolysis

Wasserspeicherung water storage

Wasserspiegel water level

Wasserstelle watering place

Wassersterngewächse/ Callitrichaceae water starwort family, starwort family

Wasserstoff hydrogen

Wasserstoff-Elektrode hydrogen electrode

Wasserstoffbakterien/ Knallgasbakterien hydrogen bacteria (aerobic hydrogen-oxidizing bacteria)

Wasserstoffbrücke/ Wasserstoffbrückenbindung hydrogen bond

Wasserstoffion (Proton) hydrogen ion (proton)

Wasserstoffperoxid hydrogen peroxide

Wasserstrahl jet of water

Wasserstrahlpumpe *lab* water pump, filter pump, vacuum filter pump

Wasserstress water stress

Wasserstrom, apoplastischer apoplast pathway

Wasserströmung water flow

W Wassertiere

Wassertiere aquatic animals, hydrocoles
Wassertransport (im Leitgewebe) water transport
Wassertransportweg (Wasser) water transport pathway
Wassertrieb *bot/hort* watershoot, water sprout, water sucker
wasserunlöslich insoluble in water
Wasseruntersuchung/ Wasseranalyse water analysis
Wasserverbrauch water consumption, water usage
Wasserverlust water loss
Wasserverschmutzung water pollution
Wasserversorgung water supply
Wasservögel waterfowl
wässrig aqueous
waten wade
watscheln waddle
Watt Wadden, coastal flat, tidal flat
• **Sandwatt** sandflat
• **Schlickwatt** mudflat
Watte absorbent cotton
Wattebausch/Tupfer cotton ball, cotton swab, swab
Wattenmeer intertidal flats
Wattestopfen cotton stopper
Wattrinne tidal channel
Watvögel & Möwenvögel & Alken/ Charadriiformes shorebirds & gulls & auks
Waugewächse/Resedagewächse/ Resedengewächse/Resedaceae mignonette family
weben/spinnen (Spinnenetz/ Kokon) spin
weben weave
Weberknechte/Opiliones/ Phalangida harvestmen, "daddy longlegs"
Webersche Linie/Weber-Linie Weber's line
Weberscher Apparat *ichth* Weberian apparatus
Webersches Knöchelchen Weberian ossicle, otolith
Webspinnen/Spinnen/Araneae spiders
Wechselbeziehung interrelation, interrelationship
Wechselfeld-Gelelektrophorese/ Puls-Feld-Gelelektrophorese pulsed field gel electrophoresis (PFGE)
wechselfeucht/poikilohydr/ poikilohydrisch poikilohydrous
Wechselfieber/Sumpffieber/Malaria (*Plasmodium spp.*) malaria
Wechselgesang antiphonal singing
Wechseljahre/Klimakterium climacteric, climacterium

wechselseitig mutual
wechselständig alternate
Wechseltierchen/Wurzeltierchen/ Rhizopoden/Amöben amoebas, amebas
wechselwarm/poikilotherm/ ektotherm poikilothermal, poikilothermous, cold-blooded, ectothermal, heterothermal
Wechselwirkung interaction
Wechselwirt alternate host
Wechselzahl kcat (katalytische Aktivität) turnover number
Weckamin analeptic amine
Weckreaktion *ethol* arousal reaction
Wedel *bot* frond
wedeln/wackeln (Schwanz/Kopf) waggle, wag
Weg/Pfad way, path, pathway, trail
Wegerichgewächse/Plantaginaceae plantain family
Wegfaden *arach* trail line
wegführend/ausführend/ableitend efferent
Wegweiserneuron guidepost neuron
Wegweiserzelle guidepost cell
Wegwerfgesellschaft throwaway society
Wehe (Gebärmutterkontraktion) uterine contraction
Wehen/in den Wehen liegen labor
Wehrdrüse (Schleimdrüse bei *Peripatus*: umgewandelte Schenkeldrüse) defensive gland (slime gland)
Wehrpolyp/Dactylozoid/ Dactylozooid stinging zooid, protective polyp, defensive polyp, dactylozooid
Weibchen female
weiblich *zool/bot* female
weiblich/pistillat *bot* pistillate, carpellate
Weichbast soft bast
Weichholz soft wood
Weichmacher/Plastifikator softener (esp. in foods), plasticizer (in plastics a.o.)
weichschalig soft-shelled, malacostracous
Weichtiere/Mollusken/Mollusca mollusks
Weichtierkunde/Malakologie malacology, study of mollusks
Weide (*Salix*) *siehe* Weidengewächse
Weide/Weidewiese (Grünland)/Trift (Heide) pasture
Weidefläche range, rangeland
Weidegänger/Abweider grazer
Weidekette/Weidenahrungskette grazing food chain

Werbeverhalten

Weideland rangeland, grazing land, pasture, pastureland, pasturage
weiden/abgrasen/abfressen graze (herbs), pasture, browse (twigs/leaves of shrubs)
Weidenahrungskette/Weidekette grazing food chain
Weidengewächse/Salicaceae willow family
Weiderichgewächse/ Blutweiderichgewächse/ Lythraceae loosestrife family
Weidespuren/Pascichnia *paleo* grazing traces
Weideverhalten foraging behavior
Weidevieh grazing animals
Weidewirtschaft *agr* pasture farming, pastoral economy, pastoralism, agropastoralism
Weiher (z. B. Fischweiher) small pond (e. g. fish pond)
Wein wine
Weinbau viticulture (viniculture)
Weinbaukunde/Önologie enology
Weinberg vineyard
Weinfass wine cask
Weingeist spirit of wine (rectified spirit: alcohol)
Weinrebe/Weinstock vine, grapevine
Weinrebengewächse/Vitaceae vine family, grape family
Weinsäure/Weinsteinsäure (Tartrat) tartaric acid (tartrate)
Weinstein/Tartarus (Kaliumsalz der Weinsäure) tartar
Weinstock/Weinrebe vine, grapevine
Weisel/Bienenkönigin queen bee
Weisheitszahn/dritter Molar wisdom tooth, third molar
weiße Linie/Zona alba (Huf) white line
weiße Substanz (Gehirn) white matter
Weißfäule/Korrosionsfäule white rot
Weißmoor open treeless bog
Weißrost/Albuginaceae *fung* white rusts
Weitbarkeit (Gefäßwand) compliance, capacitance
weiterleiten/fortleiten *neuro* propagate
Weiterleitung/Fortleitung *neuro* propagation
weiterverarbeiten/prozessieren process
weiterverarbeitetes Pseudogen processed pseudogene
Weiterverarbeitung/Prozessierung processing
weiterwachsend accrescent
Weithalsflasche *lab* wide-mouth flask, wide-neck bottle
Weitholz/Frühholz/Frühlingsholz earlywood, springwood
weitverbreitet/ubiquitär (überall verbreitet) widespread, ubiquitous (existing everywhere)
Weitwinkel *micros* widefield
welk/schlaff wilted, withered, faded, limp, flaccid
welken wilt, wither, fade
welkend wilting, withering, fading, flaccid, deficient in turgor
Welkepunkt wilting point
Welkungsgrad, permanenter permanent wilting percentage
Welkungskoeffizient wilting coefficient
Welle wave
Wellenauflauf *mar* uprush, swash
Wellenbewegung undulation
- **seitliche Schlängelbewegung/ horizontale Schlängelbewegung (Schlangen)** lateral undulation, lateral undulatory movement
Wellenexposition wave exposure
Wellenflug *orn* undulating flight
Wellenlänge wavelength
Wellenrücklauf/Wellenrückstrom/ Rücksog *mar* backwash
Wellenschlagzone *mar* breaker zone
wellig wavy, undulate, repand (slightly undulating)
Welpe/Jungtier (Fuchs/Wolf/ Schakal/Bär/Löwe) whelp, cub; (junger Hund) whelp, pup, puppy
Welse/Welsartige/Siluriformes catfishes
Weltbevölkerung global population
Weltmeere (sieben) oceans, seas (the seven seas)
weltweit verbreitet/kosmopolitisch occurring worldwide, cosmopolitan
Weltwirtschaftspflanze worldwide/global economic plant, world-trade plant/crop
Welwitschiagewächse/ Welwitschiaceae welwitschia
Wendeglied/Pedicellus (Antenne) antennal pedicel
Wendezehe *orn* **(z. B. Kuckuck/ Papageien)** zygodactylous toe
wenigbeinig/oligopod oligopod
Wenigborster/Oligochaeten oligochetes
Wenigfüßer/Pauropoden pauropods
Werbegesang *orn* mating song, courtship song
Werberuf *orn* mating call, courtship call
Werbeverhalten courting behavior

W werfen 742

werfen/verziehen, sich (Holz) warp
werfen/Junge werfen
litter, bear young
Werkbank (Labor-Werkbank)
bench (lab bench)
• **sterile Werkbank** sterile bench
Wertigkeit valency
• **dreiwertig** trivalent
• **einwertig** univalent
• **fünfwertig** pentavalent
• **vierwertig** tetravalent
• **zweiwertig** bivalent, divalent
Wesen/Kreatur being, creature
Wesen/Wesensart/Charakter/Natur
manner, character, nature
Western-Blot/Immunoblot
Western blot, immunoblot
Westwind(e)
westerly wind (westerlies)
• **Ostwind(e)**
easterly wind (easterlies)
Wettbewerb/Konkurrenz/
Existenzkampf competition
Wetter weather
• **Unwetter** storm, thunderstorm
Wetterbedingungen
weather conditions
Wetterkunde/Meteorologie
meteorology
Wettervorhersage weather forecast
Wickel (cymöse Infloreszenz)
cincinnus (scorpioid cyme)
Wickelkapsel/Volvent
(Nematocyste) volvent
Wicklung twist, coil, winding,
contortion
Widder/Schafbock/Rammler
(männliches Schaf) ram
Widerhaken barb
widerhakig provided with barbs,
glochidiate
Widerlagergewebe (Frucht)
resistance tissue
Widerrist withers
Widersachertum/Antibiose
antibiosis
Widerstand resistance
• **spezifischer Widerstand**
resistivity
widerstandsfähig resistive,
resistant, hardy
Widerstandsfähigkeit resistance,
resistivity, hardiness
Widerstandsthermometer
resistance thermometer
wiederaufforsten/aufforsten/
wiederbewalden/bestocken
reforest, reafforest
Wiederaufforstung/Aufforstung/
Wiederbewaldung/Bestockung
reforestation, reafforestation,
afforestation

Wiederaufnahme *physiol* re-uptake
wiederaufstoßen/regurgitieren/
hochwürgen regurgitate
Wiederaufstoßen/Regurgitation/
Hochwürgen regurgitation
Wiederbefall reinfestation
wiederbeleben revive, resuscitate
Wiederbelebung revival,
resuscitation
wiederbesieden reestablish, resettle,
recolonize
Wiederbesiedlung reestablishment,
resettlement
wiederergänzen/regenerieren
regenerate
Wiederergänzung/Regeneration
regeneration
Wiederfangmethode/
Rückfangmethode *ecol*
capture-recapture method
Wiederholung/
Sequenzwiederholung repeat,
repetition (of a sequence)
Wiederholungsrisiko
recurrence risk
wiederkäuen chew the cud
(regurgitate)
Wiederkäuer/Retroperistaltiker/
Ruminantier/Ruminantia
ruminant, "cud chewers"
• **Nichtwiederkäuer**
nonruminant
wiederverwenden reuse
Wiederverwendung reuse
wiederverwerten recycle
Wiederverwertung recycling
Wiederverwertungsreaktion
salvage pathway
Wiederverwertungs-
stoffwechselwege
salvage pathway
wiegen weigh
• **abwiegen (eine Teilmenge)**
weigh out
• **auswiegen (genau wiegen)**
weigh out precisely
• **einwiegen (nach Tara)**
weigh in (after setting tare)
wiehern neigh, whinny (low/gentle)
Wiese meadow
• **Auenwiese/Auwiese**
(Überschwemmungswiese)
riverine floodplain meadow,
bottomland meadow
• **Bergwiese/alpine Matte**
alpine meadow
• **Fettwiese** rich meadow,
rich pasture
• **Magerwiese** rough meadow,
rough pasture, poor grassland
• **Mähwiese** hay meadow,
mowed meadow

Wind **W**

- **Nasswiese** damp meadow, wet meadow (type of wetland)
- **Naturwiese** native meadow
- **Sumpfwiese** swamp meadow
- **Trift (Weide/Weidewiese)** pasture, pasturage
- **Überschwemmungswiese** floodplain meadow

Wiesenbach meadow creek

Wiesengrund lowlying meadow in a valley

Wiesenland meadowland

Wild/Großwild (jagbare Tiere) game

Wild/Wildbret/Wildfleisch venison

Wildallel wild-type allele

Wildbahn hunting ground
- **freie Wildbahn** in the wild, free-ranging

Wildbestand game population, stock of game

Wildbret/Wildfleisch venison

Wilderer/Wilddieb poacher

wildern poach

Wildform/Wildtyp wild type

Wildgeflügel wildfowl

Wildgehege game preserve, game reserve, game enclosure

wildlebend wild, living in the wild

Wildnis wilderness

Wildmanagement game management, wildlife management

Wildpflanze wildflower

Wildreservat/Wildtierpark/ Wildpark wildlife reserve, wildlife park, wild animal reserve, game reserve

Wildschutzgebiet wild animal sanctuary, wildlife sanctuary

Wildtyp/Wildform wild-type

Wildverbiss (z. B. an Baumrinde) damage caused by game, browsing damage

wildwachsend wild, growing in the wild

wildwachsende Pflanze wildflower

Wildwechsel deer path, run, runway

Willkür arbitrariness

willkürlich *generell* arbitrary, random; *med/psych* voluntary

Willkürmotorik voluntomotoricity, voluntary motility

Wimper/Augenwimper eyelash

Wimper/Zilie/Cilie/Flimmerhärchen cilium (*pl* cilia)

Wimperfeder (Ctenophoren) balancer, spring

Wimperflammenzelle flame cell

Wimperfurche (Ctenophoren) ciliated groove

Wimpergrube ciliated pit

Wimperkamm/Wimperplatte/ Wimperplättchen/Schwimmplatte/ Ruderplatte/Ruderplättchen/ "Kamm" (Ctenophoren) comb plate, swimming plate, ciliary comb, ctene

Wimperkölbchen/Wimperkolben flame bulb

Wimperkranzlarve/Trochophora trochophore larva

Wimperlarve ciliated larva

Wimpermeridian/Wimperrippe/ Rippe/Pleurostiche (Ctenophoren) comb row, costa

Wimperbüschel/Wimpernschopf tuft of cilia, ciliary tuft

Wimpernfarne/Woodsiaceae woodsia family

Wimpernkranz ciliary band

Wimpernschlag/Lidschlag (Auge) bat of an eye (lid)

Wimpernschlag/Zilienschlag/ Cilienschlag ciliary movement

Wimpernschopf ciliary tuft

Wimperntrichter/Flimmertrichter/ Eileitertrichter/Infundibulum (mit Ostium tubae) fimbriated funnel of oviduct, infundibulum

Wimpertierchen/Ciliaten/Ciliata ciliates

Wind wind
- **Abwind** downward wind; (Hang) katabatic wind
- **Antizyklon (Hochdruckgebiet)** anticyclone (rotating high-pressure wind system)
- **Aufwind** upwind, upcurrent; (Hang) anabatic wind
- **Bö** gust; (heftiger Windstoß/ Sturmbö) squall, sudden violent gusty wind
- **Brise** breeze
- **Gegenwind** head wind
- **geostrophischer Wind** geostrophic wind
- **Hangabwind** katabatic wind
- **Hangaufwind** anabatic wind
- **Hurrikan/Orkan (mittelamerik. Wirbelsturm) (>115 km/h)** hurricane
- **in Windrichtung** leeward
- **Landwind** offshore wind
- **Oberflächenwind** surface wind
- **Orkan/Hurrikan (mittelamerik. Wirbelsturm) (>115 km/h)** hurricane
- **Ostwinde** easterlies (easterly wind/current), eastern wind
- **Passatwinde** trade winds, trades
- **Rückenwind** tail wind
- **Sandteufel (Staubteufel)** sand devil (dust devil)

Windabrasion

- **Sandwirbel/Sandhose (Staubwirbel/Staubhose)** sand whirl (dust whirl)
- **Schneesturm (heftig)** blizzard
- **Seewind** onshore wind
- **Strahlströmung/Jetstream** jet stream
- **Sturmwind (51–101 km/h)** gale, strong wind
- **Taifun (tropischer Zyklon: Philippinen/Chinesisches Meer)** typhoon
- **Tornado (Nordamerik. kleinräumige Großtrombe/ Wirbelsturm)** tornado (North American whirlwind), "twister"
- **Trombe/Wirbelsturm** whirlwind
- **Wasserhose (eine Trombe)** waterspout
- **Westwinde** westerlies (westerly wind/current), western wind
- **Windhose (eine Trombe)** wind spout, vortex (of a tornado)
- **Wirbelsturm** whirlwind (violent windstorm)
- **Zyklon (trop. Wirbelsturm)** cyclone, tropical windstorm
- **Zyklone (Tiefdruckgebiet)** cyclone (rotating low-pressure wind system or storm)

Windabrasion wind abrasion
Windausbreitung (der Frucht)/ Windstreuung/Anemochorie wind-dispersal, anemochory
Windbestäubung/Windblütigkeit/ Anemophilie wind pollination, anemophily
windblütig/anemophil wind-pollinated, anemophilous
Windbö gust; (heftiger Windstoß/ Sturmbö) squall, sudden violent gusty wind
Windbruch windbreak (breaking of trees by wind), windfall
Windchill-Faktor/Windchill-Index (Abkühlungsgröße) windchill factor
winden wind, twist, coil
Windengewächse/Convolvulaceae bindweed family, morning glory family, convolvulus family
Windepflanze/Schlingpflanze winder, twiner; liana (woody)
Windfrost wind frost
Windgalle/Fesselgalle (Pferd) windgall
Windgeschwindigkeit wind speed, wind velocity
windgetragen airborne
Windmesser/Anemometer anemometer
Windmulde/Deflationskessel (im Sand) geol blowout, deflation basin

Windrichtung wind direction
- **in Windrichtung/Windseite/ Wetterseite** windward, windward side, luv
- **vorherrschende Windrichtung** prevailing direction of the wind

Windschatten/Windschattenseite lee, leeward, leeward side
Windschnappen/Luftkoppen (Pferd) wind sucking
Windschur wind shear, wind abrasion
Windschutz windbreak, shelterbelt
Windschutzbäume shelterwood, shelterbelt
Windstärke/Windintensität (siehe: Windgeschwindigkeit) wind force, wind strength, wind intensity
Windstille calm, windlessness
Windstoß/Bö gust (gust of wind)
Windstreuer bot anemochore
Windstreuung/Windausbreitung bot wind-dispersal, anemochory
Windung (Bewegung) spiral movement, spiral coiling
Windung/Gyros (Schneckenschale) whorl, spiral coil, gyre
Windung/Krümmung/Biegung winding, contortion, turn, bend
Windung/Spirale twist, coil, spiral (a series of loops)
Windungszahl (DNA) writhing number
Windwurf/Sturmwurf for windfall, windthrow, blowdown (of trees)
Winkel angle
Winkelrotor centrif angle rotor, angle head rotor
Winkelschleife/Umkehrschleife/ Haarnadelschleife/beta-Schleife/ β-Schleife hairpin loop, reverse turn, beta turn, β bend
Winteraceae/Winterrinden- gewächse wintera family, Winter's bark family, drimys family
Winterfell winter fur, winter coat
winterfest/winterhart hardy
Wintergrüngewächse/Pyrolaceae wintergreen family, shinleaf family
Winterhärte winter hardiness
Winterknospe/Hibernakel winter bud, hibernaculum, turio, turion
Winterquartier winter quarters
Winterrindengewächse/ Winteraceae wintera family, Winter's bark family, drimys family
Winterschlaf winter sleep, hibernation
Winterschläfer hibernating animal
Winterstagnation limn winter stagnation
Wipfel/Baumwipfel treetop

Wirbel *anat* vertebra
(auch bei **Ophiuroiden);**
(Umbo) umbo *(pl* umbones)
- **Brustwirbel/Thorakalwirbel**
  thoracic vertebra
- **Halswirbel/Cervikalwirbel**
  cervical vertebra
- **Kreuzbeinwirbel/Sakralwirbel**
  sacral vertebra
- **Lendenwirbel/Lumbarwirbel**
  lumbar vertebra
- **Schwanzwirbel/Kaudalwirbel**
  caudal vertebra
- **Steißwirbel/Steißbeinwirbel**
  coccygeal vertebra
Wirbel *meteo/hydro*
whirl, eddy
Wirbelbogen/Arcualium
arcualium
- **oberer Wirbelbogen/**
  Neuralbogen neural arch
Wirbelhöhle (Molluskenschale)
umbonal cavity
Wirbelkörper/Centrum centrum
Wirbellose/Wirbellose Tiere/
Invertebraten/Evertebraten
invertebrates
Wirbelsäule/Rückgrat
vertebral column, spinal column
Wirbelschichtreaktor/
Wirbelbettreaktor
fluidized bed reactor
Wirbelstrom eddy current
Wirbeltiere/Vertebraten/Vertebrata
vertebrates
wirken act, work, be effective,
causing an effect, take effect
Wirkstoff/Wirksubstanz
active ingredient, active principle,
active component
Wirkung effect, action
Wirkungsgrad efficiency
- **ökologischer W.**
  ecological efficiency
Wirkungsschlag/Kraftschlag
effective stroke, power stroke
Wirkungsspezifität
specificity of action
Wirkungsweise/Mechanismus
mode of action, mechanism
Wirt *allg* (Wirtsorganismus/Wirtstier)
host
- **Endwirt** final host
- **Fehlwirt/Irrwirt** accidental host,
  wrong host
- **Hauptwirt** main host,
  primary host, definitive host
- **Nebenwirt** secondary host
- **nicht-permissiver Wirt**
  non-permissive host
- **permissiver Wirt** permissive host
- **Reservoirwirt** reservoir host

- **Sammelwirt/Stapelwirt/**
  paratenischer Wirt/Transportwirt
  paratenic host, transfer host
- **Sicherheitswirt** containment host
- **Wechselwirt** alternate host
- **Zufallswirt** random host
- **Zwischenwirt** intermediary host
Wirtel/Quirl whorl, verticil
wirtelig/quirlig whorled
Wirtelung/Dekussation decussation
Wirtsbereich host range
Wirtschaftspflanze economic plant
Wirtsorganismus host organism
Wirtspflanze host plant
Wirtsrasse host race
Wirtsspektrum host range
Wirtsspezifität host specificity
Wirtstier host animal
Wirtswechsel alternation of hosts
Wirtszelle host cell
wittern scent, smell
Witterung/Geruchssinn *zool* scent
Witterung/Wetter *meteo* weather
Wobble-Base *gen* wobble base
Wobble-Hypothese *gen*
wobble hypothesis
Wohnbauten/Domichnia *paleo*
dwelling structures
Wohngebiet/Heimbereich/
Aktionsraum *ecol* home range
Wohnkammer (Schneckenschale)
body whorl
Wohnquartier/Behausung
dwelling
Wölbung (Höcker/Beule/Warze)
protuberance, tubercle, wart
Wölbung/Koeffizient der Wölbung
stat kurtosis
Wolffscher Gang/Vornierengang/
primärer Nierengang/primärer
Harnleiter Wolffian duct,
Leydig's duct, mesonephric duct
Wolfram tungsten
Wolfsmilchgewächse/
Euphorbiaceae spurge family
Wolkenbruch *meteo* downpour
Wolkenimpfung *meteo/ecol*
cloud seeding
Wollbaumgewächse/Bombacaceae
cotton-tree family, silk-cotton tree
family, kapok-tree family
Wolle wool
Wollfettdrüse wool fat gland
Wollhaar wooly hair
Wollhaarkleid undercoat
wollig wooly, lanate
Wollschildläuse (Coccoidea)
mealybugs
Woodsiaceae/Wimpernfarne
woodsia family
Woronin-Körper Woronin body
wuchern/proliferieren proliferate

W wuchernd

wuchernd/proliferierend
proliferative
Wucherung/Proliferation
proliferation
Wucherung/Tumor/Geschwulst
tumor
Wuchs growth, habit
Wuchsform/Habitus growth form,
appearance, habit
**Wuchshemmer/Wachstumshemmer/
Wuchshemmstoff** growth inhibitor
Wuchskraft growth vigor
Wuchsrichtung direction of growth
**Wuchsstoff (Pflanzenwuchsstoff)/
Phytohormon**
growth regulator, phytohormone,
growth substance
Wulst bulge, collar,
protuberant seam
**Wulstlinge/Freiblättler/
Amanitaceae** Amanita family
Wunde wound
**Wunderblumengewächse/
Nyctaginaceae** four-o'clock family
Wundernetz/Rete mirabile
rete mirabile
Wundfäule wound rot
**Wundgewebe/Wundcallus/
Wundholz** wound tissue, callus
Wundheilung wound healing
Wundkambium/Wundcambium
wound cambium
Wundparasit wound parasite
Wundparenchym traumatic
parenchyma
Wundpaste (beim Baumschnitt)
hort wound dressing
Wundüberbrückung/Überbrückung
hort repair grafting,
bridge grafting
Wurf *zool* litter; (Schweine) farrow;
vir burst
● **Zeitpunkt der Virusfreisetzung**
burst period
Würfelbein/Os cuboideum
cuboid bone
Würfelquallen/Cubozoa box jellies,
sea wasps, cubomedusas
Wurffaden/Angelfaden *arach*
casting line
**Wurfgeschwister/Geschwister eines
Wurfes** litter mate(s)
Wurfgröße *zool* (Anzahl Jungtiere/
Frischlinge) litter size; *vir* (Anzahl
freigesetzter Viren) burst size
Wurfnetz (Netzwerferspinnen)
arach casting web
würgen choke (an etwas würgen),
strangle (von etwas/jmd. gewürgt
werden)
Würger/Baumwürger strangler,
tree strangler

wurmartig wormlike, vermian,
vermicular
**Wurmfarngewächse/
Dryopteridaceae** male fern family,
dryopteris family
wurmförmig worm-shaped,
vermiform
**Wurmfortsatz des Blinddarms/
Appendix/Appendix vermiformis**
appendix, vermiform appendix
wurmig wormy
**Wurmmollusken/Wurmmolluscen/
Aplacophoren/Aplacophora**
aplacophorans
**Wurmschleichen/Doppelschleichen/
Amphisbaenia** worm lizards,
amphisb(a)enids, amphisbenians
Würze/Gewürz spice, seasoning
Würze (Bier) wort
**Würze/würzige Zutat/
Geschmacksverbesserer (kräftig)**
condiment
Wurzel *bot/zool* root;
(Crinoide) *zool* root, radix
● **Adventivwurzel** adventitious root
● **Ankerwurzel** anchorage root,
adhesion root
● **anwurzeln/anwachsen** take root
● **Assimilationswurzel**
assimilative root
● **Atemwurzel** pneumatophore,
air root, airial root, aerating root
● **Beiwurzel/Nebenwurzel/
Adventivwurzel** supplementary
root, adventitious root
● **bewurzeln** root
● **Bewurzelung** radication, rootage,
rooting
● **Brettwurzel** buttress root (*esp.*
tropical trees)
● **Büschelwurzelsystem (Gräser)**
fibrous root system
● **Dorsalwurzel** *neuro* dorsal root,
posterior root
● **Faserwurzel** fibrous root
● **Haarwurzel** hair root
● **Haftwurzel** holdfast root,
clinging root
● **Hauptwurzel/Primärwurzel**
main root, primary root
● **Keimwurzel (gesamte Anlage
innerhalb Samen)** seminal root
● **Keimwurzel/Radicula**
embryonic root, radicle
● **Luftwurzel** aerial root, air root
● **Nährwurzel** feeder root
● **Nebenwurzel/Beiwurzel/
Adventivwurzel** supplementary
root, adventitious root
● **Nebenwurzel/Seitenwurzel**
lateral root, secondary root
● **Pfahlwurzel** taproot

747   **Wüste** **W**

- **Pneumatophore/Atemwurzel**
  pneumatophore, aerating root
- **Primärwurzel/Hauptwurzel**
  primary root, main root
- **Saugwurzel** suction root, seeker
- **Seitentrieb** (am Wurzelhals)
  sucker; (kurz) offset
- **Seitenwurzel/Nebenwurzel**
  lateral root, secondary root
- **Speicherwurzel** storage root
- **sprossbürtige Wurzel**
  shoot-borne root
  (stem-borne adventitious root)
- **Streckungszone/
  Verlängerungszone** zone of
  expansion, region of elongation
- **Stützwurzel** prop root, stilt root,
  brace root
- **tiefwurzelnde Pflanze**
  deep-rooted plant
- **Übergangszone (Wurzel-Spross)**
  transition(al) zone/region
- **Ventralwurzel/motorische
  Wurzel** *neuro* ventral root,
  motor root, anterior root
- **Verlängerungszone/
  Streckungszone (Wurzel)**
  zone/region of extension,
  zone/region of elongation
- **Wachstumszone** zone/region of
  cell division (apical meristem)
- **Zugwurzel** contractile root
Wurzelanlage root primordium
- **Hauptwurzelanlage** radicula
**Wurzelanlauf/Stammanlauf/
Stammfuß** root butt, buttress
  (supportive ridge at base of tree
  trunk)
wurzelartig rootlike, rhizoid
Wurzelausläufer/Gehölzausläufer
  sobolifer, sobole
Wurzelausschlag/Wurzeltrieb
  root sucker
Wurzelbereich root system domain
Wurzelbulbille root bulbil
wurzelbürtiger Spross root sucker,
  offshoot; (kurz) offset
Würzelchen rootlet, radicle
Wurzeldruck root pressure
wurzelecht own-rooted
Wurzelepidermis/Rhizodermis
  rhizodermis, epiblem(a)
**Wurzelfäule (Krebsfäule der
Wurzeln)** root rot
wurzelförmig root-like, radiciform,
  rhizoid
Wurzelgalle (durch Nematoden)
  root gall, root knot
Wurzelgemüse root crop,
  root vegetable
Wurzelhaar/Wurzelhärchen
  root hair

Wurzelhaarzone root-hair zone,
  zone/region of maturation
Wurzelhals/Wurzelkrone
  root crown, root collar
Wurzelhalsfäule root-collar rot,
  collar rot
Wurzelhalsschössling root-collar
  shoot (sucker/offshoot)
**Wurzelhalstumor (Stamm- oder
Wurzeltumor verursacht durch
A. tumefaciens)** crown gall tumor
**Wurzelhaube/Wurzelhäubchen/
Kalyptra** root cap, calyptra
Wurzelkletterer/Wurzelklimmer
  root climber
Wurzelknie knee, root-knee
Wurzelknöllchen root nodule
Wurzelknolle
  tuber, root tuber, tuberous root,
  adventitious storage root
Wurzelkonkurrenz root competition
Wurzelkrebse/Rhizocephala
  rhizocephalans (parasitic
  "barnacles")
Wurzelkrone/Wurzelhals
  root crown
wurzellos rootless, arrhizous,
  arrhizal
Wurzelmundquallen/Rhizostomeae
  rhizostome medusas
Wurzelpfropf *hort* root graft
**Wurzelpfropfgrundlage/Wurzel-
pfröpfling** *hort* rootstock, stock
Wurzelpfropfung/Wurzelveredlung
  hort root grafting
Wurzelpol (Embryo) root apex
Wurzelranke root tendril
Wurzelreis *siehe* Wurzelspross
Wurzelsaugspannung root water
  tension
Wurzelscheide/Koleorhiza root
  sheath, radicle sheath, coleorhiza
Wurzelspitze root apex, root tip
**Wurzelspross/Wurzeltrieb/
Wurzelschössling/Wurzelreis/
Erdspross** root sucker; offset
  (short); (Gehölzausläufer) sobole
wurzelsprossbildend soboliferous
Wurzelspur root trace
Wurzelsteckling *hort* root cutting
Wurzelstele central cylinder of root
Wurzelstock/Rhizom/Erdausläufer
  rootstock, rhizome
Wurzelstock/Strunk/Caudex
  rootstock, caudex
Wurzeltrieb/Wurzelspross offshoot,
  offset, slip, sucker
Wurzelwerk/Wurzelsystem rootage,
  root system
Wüste desert
- **Halbwüste** semidesert
- **Kältewüste** cold desert

Wüste

- **Kieswüste/Geröllwüste/Serir**
  gravel desert, serir
- **Nebelwüste** fog desert
- **Sandwüste** sand desert
- **Steinwüste (Hamada)**
  stone desert, stony desert,
  rock desert (hamada/hammada)
- **Wärmewüste** hot desert

Wüstenausbreitung desertification,
desert expansion

Wüstenbewohner desert dweller
Wüstenbiom desert biome
Wüstenblüte desert bloom
Wüstengemeinschaft
desert community, eremium
Wüstenlack desert varnish
Wüstenpflanze/Eremiaphyt
desert plant, eremophyte, eremad
Wüstenpflaster/Steinpflaster
desert pavement, stone pavement

Xyrisgewächse

X-Chromosom X chromosome
- **fragiles X-Chromosom (Syndrom)** fragile X chromosome (syndrome)
- **verbundene X-Chromosomen/ verklebte X-Chromosomen** attached X chromosomes

X-Chromosom-Inaktivierung X-chromosome inactivation
X-Körper (Einschlusskörper) X body (inclusion body)
X-Organ/Bellonci-Organ X-organ
Xanthan xanthan
Xanthangummi xanthan gum
Xanthen/Methylene diphenylene oxid xanthene
Xanthin/2,6-Dioxopurin xanthine
Xanthismus xanthism
Xanthogensäure xanthogenic acid, xanthic acid, xanthonic acid, ethoxydithiocarbonic acid
xanthokarp/gelbfrüchtig xanthocarpous, having yellow fruits
Xanthophyll xanthophyll
Xanthorrhoeaceae/ Grasbaumgewächse grass tree family, blackboy family
Xenobiose xenobiosis
Xenobiotikum (*pl* **Xenobiotika)** xenobiotic (*pl* xenobiotics)
xenobiotisch xenobiotic
xenogen xenogenic
Xenoparasit xenoparasite
Xenospore xenospore (immediate germination)
Xenotransplantat/ Fremdtransplantat xenograft (xenogeneic graft: from other species)
xenotropes Virus xenotropic virus
Xeromorphismus xeromorphism
Xerophyt xerophyte, xeric plant, xerophilic plant
Xeroserie xerosere
xerotherm (trockenwarm) xerothermic
Xylem/Gefäßteil/Holzteil xylem
Xylemmutterzelle/ Tracheenmutterzelle/ Xylemprimane xylem mother cell
Xylemsaft xylem sap
Xylemsauger/Xylemsaftsauger xylem-sap feeders
Xylit xylitol/xylite
Xylol/Dimethylbenzol xylene, dimethylbenzene
Xylose xylose
Xylulose xylulose
Xyrisgewächse/Xyridaceae yellow-eyed grass family

Y-Organ/Carapaxdrüse
Y organ (molting gland)
Yamswurzelgewächse/ Schmerwurzgewächse/ Dioscoreaceae
yam family
YEp (episomales Hefeplasmid)
YEp (yeast episomal plasmid)

Zahn

Z-Form/Z-Konformation Z-form
Z-Linie/Z-Streifen (Z-Scheibe)
(Z = Zwischenscheibe)
Z line (Z disk)
Z-Schema (Zickzack-Schema:
Photosynthese) Z-scheme
Zacke indentation, projection, spike,
notch, serration
zackig/gekerbt crenate
zäh tough, rigid
zähflüssig/dickflüssig/viskos/viskös
viscous, viscid
Zähflüssigkeit/Dickflüssigkeit/
Viskosität viscosity, viscousness
Zählkammer counting chamber
Zählplatte counting plate
zahm tame
zähmen tame
Zähmung taming
Zahn (pl Zähne) tooth (pl teeth)
- **akrodont/auf der Kieferkante**
  stehend (Teleostei/Echsen)
  acrodont, attached to outer surface
  of bone/summit of jaws (teleosts/
  lizards)
- **Augenzahn (oberer Eckzahn)**
  eyetooth (canine tooth of upper
  jaw)
- **Backenzahn/Molar**
  molar (multicuspid tooth), grinder
- **>Prämolar/vorderer Backenzahn**
  premolar (bicuspid tooth)
- **>Weisheitszahn/dritter Molar**
  wisdom tooth, third molar
- **bleibende Zähne/zweite Zähne**
  (Dauergebiss) permanent teeth
  (permanent dentition)
- **brachyodont/niedrigkronig**
  brachyodont, brachyodont,
  with low crowns
- **bunodont/rundhöckrig/**
  stumpfhöckrig bunodont,
  with low crowns and cusps
- **diphyodont (einmaliger**
  Zahnwechsel) diphyodont
  (with two sets of teeth)
- **Eckzahn** canine
- **Eizahn/Eischwiele (Reptilien)**
  egg tooth
- **Fangzahn/Fang/Reißzahn**
  (Carnivora) fang, carnassial tooth
- **Giftzahn (Schlangen)**
  poison tooth, venom tooth, fang
- **gleichartig bezahnt/homodont**
  homodont, isodont
- **halbmondhöckrig/selenodont**
  crescentic, with crescent-shaped
  ridges, selenodont
- **Hauer (z. B. Eber)**
  tusk, fang (large teeth)
- **Hauptzahn (Bivalvia)**
  cardinal tooth

- **heterodont/ungleichzähnig**
  heterodont, anisodont
- **hochkronig/hypsodont/**
  hypselodont with high crowns,
  hypsodont, hypselodont
- **homodont/gleichartig bezahnt**
  homodont, isodont
- **hypsodont/hypselodont/**
  hochkronig
  hypsodont, hypselodont
  (high crowns/short roots)
- **labyrinthodont/mit komplexer**
  Struktur labyrinthodont
- **lophodont/mit Querjochen**
  lophodont, with transverse ridges
- **Mahlzahn** grinding tooth
- **Milchzahn** milk tooth, deciduous
  tooth (first teeth/primary teeth)
- **monophyodont (einfaches**
  Gebiss/ohne Zahnwechsel)
  monophyodont (only one set of
  teeth)
- **niedrigkronig/brachyodont**
  with low crowns, brachydont,
  brachyodont
- **pleurodont/an der**
  Kieferinnenseite pleurodont,
  attached to inside surface of jaws
- **plikodont/mit gefalteten**
  Höckern plicodont
- **polyphyodont (mehrfacher**
  Zahnwechsel) polyphyodont
- **Prämolaren/**
  vordere Backenzähne
  premolars, bicuspid teeth
- **Reißzahn/Fangzahn/Fang**
  (Carnivora) fang, carnassial tooth
- **rundhöckrig/stumpfhöckrig/**
  bunodont with low crowns and
  cusps, bunodont
- **Säbelzahn** sabre tooth,
  saber tooth
- **Schlosszähne (Muscheln)**
  hinge teeth
- **Schneidezahn/Vorderzahn/**
  Beißzahn incisor, front tooth
- **selenodont/halbmondhöckrig**
  (Zahnhöcker)
  selenodont, crescentic,
  with crescent-shaped ridges
- **stumpfhöckrig/rundhöckrig/**
  bunodont with low crowns and
  cusps, bunodont
- **tetralophodont/mit vier**
  Querjochen tetralophodont,
  with four transverse ridges
- **thekodont/in Zahnfächern**
  verankert
  thecodont, teeth in sockets
- **triconodont/dreihöckrig (in einer**
  Reihe) triconodont (three crown
  prominences in a row)

Z Zahnalveole   752

- **ungleichzähnig/heterodont**
  heterodont, anisodont
- **Weisheitszahn/dritter Molar**
  wisdom tooth, third molar
- **Wolfszähne** wolf teeth, remnant
  teeth (horse: first premolar)
Zahnalveole/Zahnfach tooth socket,
  alveolar cavity, alveolus
Zahnanlage/Zahnkeim tooth germ
**Zahnarme/Edentata/Xenarthra/
  Nebengelenktiere** "toothless"
  mammals, edentates, xenarthrans
Zahnbein/Dentin dentin, dentine,
  substantia eburnea
Zahnbeinbildner/Odontoblast
  odontoblast
Zahnbelag/Plaque dental plaque
Zähnchen denticle
**Zahndurchbruch/Duchbruch der
  Zähne/Dentition** dentition
  (development/cutting of teeth)
zahnen teethe, cut one's teeth,
  grow teeth
Zahnen/Zahnung teething,
  dentition
Zahnentwicklung
  tooth development
Zahnersatz tooth replacement
Zahnfach/Zahnalveole tooth socket,
  alveolar cavity, alveolus
Zahnfleisch gums, gingiva
Zahnformel/Gebissformel
  dental formula
**Zahngrube (am Schloss der
  Muschelschale)** hinge socket
Zahnhals tooth neck, dental neck
Zahnhöcker dental ridge,
  dental cusp, cusp
Zahnhöhle/Pulpahöhle
  dental cavity, pulp cavity
Zahnkeim/Zahnanlage tooth germ
Zahnkrone corona, dental crown
Zahnleiste dental lamina,
  dental lamella
zahnlos toothless, edentate
Zahnmark dental pulp, pulpa
Zahnreihe tooth row, arcade
Zahnschmelz dental enamel
Zahnschuppe/Placoidschuppe *ichth*
  dermal denticle, placoid scale
Zahnsystem/Zahnanordnung
  dentition
Zahntaucher/Hesperornithiformes
  western birds
Zahnwale/Odontoceti toothed
  whales & porpoises & dolphins
Zahnwechsel (second) dentition
Zahnwurzel dental root
Zange/Schere (Crustacea) forceps
  (seizing claws of crustaceans)
Zangensterne/Forcipulatida
  forcipulatids

**Zannichelliaceae/
  Teichfadengewächse**
  horned pondweed family
Zapfen *bot* cone, strobile, strobilus,
  "pine"
Zapfen/Zäpfchen/Zapfenzelle *opt*
  cone, cone cell
Zapfenbeere/Beerenzapfen *bot*
  fleshy cone, "berry"
Zapfenschuppe cone scale,
  cone bract
zappeln wiggle, struggle (fish on the
  hook)
**Zaubernussgewächse/
  Hamamelidaceae**
  witch-hazel family
Zecken/Ixodides ticks
Zedrachgewächse/Meliaceae
  mahogany family
Zeh/Zehe toe; (Rotatorien) toe
  (*see also:* Sporn>spur)
- **großer Zeh/Hallux**
  big toe, great toe, hallux
- **kleiner Zeh** small toe
Zeh/Zehe/Brutzwiebel bulbil, offset
  bulb; (Knoblauch) clove (of garlic)
Zehengang/Digitigradie
  digitigrade gait
Zehengänger/Digitigrade
  digitigrade
Zehenglied/Fingerglied/Phalanx
  phalanx (*pl* phalanges)
**Zehenspitzengang/Hufgang/
  Unguligradie** unguligrade gait
**Zehnarmer/zehnarmige
  Tintenschnecken/Decabrachia/
  Decapoda** cuttlefish & squids
Zehnfußkrebse/Decapoda
  decapods
**Zeichnung/Musterung (z. B. Tierfell/
  Haut/Flügel)** pattern
Zeichnung/Fladerung (Holz) *bot*
  figure
Zeigerart indicator species,
  index species
Zeigerokular *micros*
  pointer eyepiece
Zeigerpflanze/Leitpflanze
  indicator plant, index plant
Zeigerwerte indicator value
**Zeilandgewächse/
  Zwergölbaumgewächse/
  Cneoraceae** spurge olive family
**Zeitalter/Ära (erdgeschichtliches
  Zeitalter) (*siehe auch:*
  Äon/Epoche/Periode)** age,
  geological age, era, geological era
- **Eophytikum** Eophytic Era
- **Erdaltertum/Paläozoikum**
  Paleozoic, Paleozoic Era
- **Erdmittelalter/Mesozoikum**
  Mesozoic, Mesozoic Era

753 **Zellobiose** **Z**

- **Erdneuzeit/Neozoikum/ Känozoikum/Kaenozoikum** Neozoic, Neozoic Era, Cenozoic, Cenozoic Era (Cainozoic Era/ Caenozoic Era)
- **Känozoikum/Kaenozoikum/ Erdneuzeit/Neozoikum** Cenozoic, Cenozoic Era, Neozoic Era (Cainozoic Era/Caenozoic Era)
- **Mesophytikum** Mesophytic Era
- **Mesozoikum/Erdmittelalter** Mesozoic, Mesozoic Era
- **Neozoikum/Erdneuzeit/ Känozoikum/Kaenozoikum** Neozoic, Neozoic Era, Cenozoic, Cenozoic Era (Cainozoic Era/ Caenozoic Era)
- **Paläophytikum/Florenaltertum** Paleophytic Era
- **Paläozoikum/Erdaltertum** Paleozoic, Paleozoic Era
- **Präkambrium/Präcambrium** Precambrian, Precambrian Era

Zeitgeber Zeitgeber, synchronizer
zeitgenössisch extant, contemporary
Zeithorizont time horizon
zeitlebens during an entire life
Zeitlosengewächse/ Krokusgewächse/Colchicaceae crocus family
Zelladhäsionsproteine cell adhesion proteins
Zellafter/Zytopyge/Cytopyge/ Zytoproct/Cytoproct cell-anus, cytopyge, cytoproct
Zellatmung cellular respiration
Zellaufschluss/Öffnen der Zellmembran cell lysis
Zellaufschluss/Zellfraktionierung cell fractionation
Zellaufschluss/Zellhomogenisierung cell homogenization
Zellbiologie/Zytologie/Cytologie cell biology, cytology
Zellchemie/Zytochemie/Cytochemie cytochemistry
Zellcyclus/Zellzyklus cell cycle
Zelldichte cell density
Zelle cell
- **Dauerzelle** permanent cell
- **Feeder-Zelle** feeder cell
- **kernlose Zelle** enucleate cell, anucleate cell
- **nicht-permissive Zelle** non-permissive cell
- **permissive Zelle** permissive cell
- **somatische Zelle/Körperzelle** somatic cell, body cell
- **transformierte Zelle** transformed cell
- **vegetative Zelle** vegetative cell

Zelleinschluss (Inklusion) cell inclusion, cellular inclusion
Zellextrakt cell extract
Zellfaden/Filament chain of cells, filament
Zellfaden/Trichom (bei Algen/ Bakterien) trichome, trichoma
Zellfortsatz cell process
Zellfraktion cellular fraction
Zellfraktionierung cell fractionation
zellfrei cell-free
zellfreier Extrakt cell-free extract
zellfreies Proteinsynthesesystem cell-free protein synthesizing system
zellfreies System cell-free system
Zellfusion/Zellverschmelzung cell fusion
Zellgift/Zytotoxin/Cytotoxin cytotoxin
Zellhomogenisation/ Zellhomogenisierung cell homogenization
Zellhülle cell envelope
Zellhybridisierung cell hybridization
zellig cellular
- **nicht zellig/azellulär** acellular, noncellular

Zellinhalt cell content
Zellinie siehe Zelllinie
Zellkern/Nukleus nucleus, karyon
- **generativer Zellkern/ Mikronukleus** micronucleus
- **somatischer Zellkern/ Makronukleus** macronucleus, meganucleus

Zellkolonie cell colony
Zellkonstanz/Eutelie cell constancy, eutely
Zellkontakt cell junction
Zellkörper/Soma cell body, soma
Zellkultur cell culture
Zelllinie cell lineage, cell line, celline
- **etablierte Zelllinie** established cell line
- **kontinuierliche Zelllinie** continuous cell line

Zellmembran/Plasmamembran/ Plasmalemma (outer) cell membrane, biological membrane, unit membrane, plasmalemma
Zellmund/Zellmundöffnung/ Zytostom/Cytostom cell-mouth, cytostome, cytostoma
Zellmundhöhlung/Peristom (Protozoen) buccal cavity, peristome, peristomium
Zelloberfläche cell surface
Zelloberflächenmarker cell surface marker
Zellobiose/Cellobiose cellobiose

Z Zellplasma 754

Zellplasma/Zytoplasma/Cytoplasma cytoplasm
Zellproliferation cell proliferation
Zellsaft cell sap
zellschädigend/zytopathisch/ cytopathisch (zytotoxisch) cytopathic (cytotoxic)
Zellschicksal cell fate
Zellschlauch siphon
Zellschlund/Zytopharynx/ Cytopharynx gullet, cytopharynx
Zellskelett/Zytoskelett/Cytoskelett cytoskeleton
Zellsorter/Zellsortierer/ Zellsortiergerät (Zellfraktionator) cell sorter
• **fluoreszenzaktivierter Zellsorter** fluorescence-activated cell sorter (see: FACS)
Zellsortierung cell sorting
zellspezifisches Gen cell-specific gene
Zellstoff wood pulp
Zellstoffwatte wood wool
Zellstoffwechsel cellular metabolism
Zellteilung cell division, cytokinesis
Zelltheorie cell theory
Zelltod cell death
• **programmierter Zelltod (Apoptose)** programmed cell death (apoptosis)
zelltötend/zytozid cytocidal
Zelltransformation cell transformation
Zelltrennung/Zellseparation cell separation
• **fluoreszenzaktivierte Z.** fluorescence-activated cell sorting (FACS)
zellulär cellular
Zellulose/Cellulose cellulose
Zellverband cell aggregate
zellvermittelt cell-mediated
Zellverschmelzung/Zellfusion cell fusion
Zellwachstum cell growth
Zellwand cell wall
Zellzahl cell count
Zellzyklus/Zellcyclus cell cycle
Zementdrüse/Kittdrüse/Klebdrüse cement gland, adhesive gland, colleterial gland
Zenckersches Organ (Spermatozoenpumpe) Zencker's organ
Zentil/Perzentil/Prozentil *stat* centile, percentile
zentrales Dogma central dogma
Zentralkorn/Centroplast central granule, axoplast, centroplast

Zentralkörper (Insekten) central body
Zentralnervensystem (ZNS) central nervous system (CNS)
Zentralplazentation *bot* free central placentation
Zentralstrang central strand
zentralwinkelständig/axial axile
zentralwinkelständige Plazentation axile placentation
Zentralzylinder *bot* central cylinder, stele
zentrieren center
zentrifugal centrifugal
Zentrifugalkraft centrifugal force
Zentrifugation centrifugation
• **analytische Zentrifugation** analytical centrifugation
• **Dichtegradientenzentrifugation** density gradient centrifugation
• **Differentialzentrifugation/ differentielle Zentrifugation** differential centrifugation
• **isopyknische Zentrifugation** isopycnic centrifugation
• **präparative Zentrifugation** preparative centrifugation
• **Ultrazentrifugation** ultracentrifugation
• **Zonenzentrifugation** zonal centrifugation
Zentrifuge centrifuge
Zentrifugenröhrchen centrifuge tube
zentrifugieren centrifuge
Zentriol/Centriol centriole
zentripetal centripetal
Zentromer/Centromer centromere
Zentrum/Viruskern (zentrale Virionstruktur) core
• **aktives Zentrum (Enzym)** active site, catalytic site
Zerebralganglion cerebral ganglion
Zerfall/Abbau/Zusammenbruch breakdown
Zerfall/Zersetzung/Verrottung/ Verfaulen decay, disintegration, decomposition
• **radioaktiver Zerfall** radioactive decay, radioactive disintegration
zerfallen decay, disintegrate, decompose, fall apart
Zerfallfrucht fissile fruit
Zerfließen/Zerschmelzen/Zergehen deliquescence
zerfließend/zerschmelzend/ zergehend deliquescent
Zerkarie/Cercarie/Schwanzlarve cercaria
Zerkleinerer shredder (large-particle detritivore)
zermahlen/zerreiben (pulverisieren) grind (pulverize)

755 Zistrosengewächse Z

Zermahlen/Zerreibung (Pulverisierung) grinding, trituration (pulverization)
zermalmen (mit Zähnen/Kiefern) grind
zerreiben/zermahlen triturate
Zerreißfestigkeit/Reißfestigkeit/ Zugfestigkeit (Holz) tensile strength
zerschlitzt/geschlitzt (gleichmäßig) incised, cut; (ungleichmäßig) lacerate, torn
zerschnitten dissected
zersetzen disintegrate, decay, decompose, degrade
Zersetzer/Destruent/Reduzent decomposer
Zersetzung disintegration, decay, decomposition, degradation
zerstäuben atomize
Zerstäuber/Sprühgerät (Wasserzerstäuber) atomizer, humidifier, mist blower, sprayer
zerstreuen/dispergieren scatter, disperse
zerstreut/schraubig (Blattstellung) alternate
zerstreutporig (Holz) diffuse porous
Zerstreuung/Dispergierung scattering, dispersion
zeugen/fortpflanzen procreate, reproduce, propagate
Zeugung/Fortpflanzung procreation, reproduction, propagation
zeugungsfähig/fruchtbar capable of reproducing, fertile
zeugungsunfähig/steril incapable of reproducing, steril
Zicklein/Ziegenjunges kid
 • **frischgeborenes Zicklein/ Ziegenjunges** yeanling
Zicklein gebären kidding (parturition in goats)
Zickzackfaden _arach_ pendulum line
Ziegenbock billygoat, male goat
zielgerichtete "Konstruktion" neuer Medikamente am Computer drug design
Zielorgan target organ
Zielsequenz _gen_ target sequence
Zielzelle target cell
Zieralgen/Desmidiaceae desmids
Ziergarten ornamental/amenity garden
Ziergras ornamental grass
Zierpflanze ornamental plant
Zierpflanzenbau/Blumenzucht amenity horticulture, floriculture (flower growing)
Zierschnitt (Formbaum/ Formstrauch) topiary
Zierstrauch _hort_ ornamental shrub

Zikaden/Zirpen/Auchenorrhyncha (Homoptera) cicadas
Ziliarkörper/Ciliarkörper/Corpus ciliare ciliary body
Zilie/Cilie/Wimper/Flimmerhärchen cilium
zilientragend/cilientragend/ bewimpert bearing cilia, cilium-bearing, ciliated, ciliferous
Zimmerpflanze house plant
Zimtaldehyd cinnamic aldehyde, cinnamaldehyde
Zimtalkohol cinnamic alcohol, cinnamyl alcohol
Zimtsäure/Cinnamonsäure (Cinnamat) cinnamic acid
Zingiberaceae/Ingwergewächse ginger family
Zink zinc
Zinkfinger _gen_ zinc finger
Zinn tin
Zippering/Reißverschluss betätigen (Doppelstrangbildung: kooperativer Vorgang beim Bilden von Wasserstoffbrücken) _gen_ zippering
Zirbeldrüse/Pinealorgan/Epiphyse epiphysis (pineal body)
zirkular/zirkulär/kreisförmig/rund circular, round
Zirkularchromatographie circular chromatography
Zirkulardichroismus/ Circulardichroismus circular dichroism
Zirkularisierung/Ringschluss circularization
Zirkulation/Zirkulieren circulation
Zirkulationssystem/Kreislaufsystem (offenes/geschlossenes) circulatory system (open/closed)
zirkulieren circulate
zirkulierend/Zirkulations... circulating, circulatory
Zirkumanaldrüse circumanal gland
zirpen/schrillen/stridulieren chirp, stridulate
Zirpen/Schrillen/Stridulation chirping, stridulation
Zirporgan/Schrillorgan/ Stridulationsorgan stridulating organ
zischen (Schlange) hiss
Zisterne (ER) _cyt_ cisterna (_pl_ cisternae/cisternas)
 • **paarweise liegende Zisternen** paired cisternae
Zisterne (Wasserreservoir einiger Bromelien) _bot_ cistern, water tank
Zistrosengewächse/ Cistrosengewächse/Cistaceae rockrose family

Z Zitronensäure 756

Zitronensäure/Citronensäure (Zitrat/Citrat) citric acid (citrate)

Zitronensäurezyklus/ Citronensäurecyclus/Citrat-Zyklus/ Citratcyclus/Tricarbonsäure-Zyklus/Krebs-Cyclus citric acid cycle, tricarboxylic acid cycle (TCA cycle), Krebs cycle

Zitrullin/Citrullin citrulline

Zitrusfrucht/Citrusfrucht/ Hesperidium (eine Panzerbeere) hesperidium

zittern/schaudern shiver

Zittern/Schaudern shivering

zittern/vibrieren quiver, tremble, vibrate

Zittern/Vibration quivering, trembling, vibration

Zitterpilze/Gallertpilze/Tremellales jelly fungi

Zitterrochen/elektrische Rochen/ Torpediniformes electric rays

Zitze/Mamille nipple, mamilla, mammilla, teat
- **Afterzitze** accessory teat
- **Stülpzitze** crater teat

Zitzenfortsatz/Processus mamillaris (Wirbelkörper) mam(m)illary process, mam(m)illary tubercle

Zivilisation civilization

Zivilisationskrankheiten diseases of civilization, "affluent peoples' diseases"

Zivilisationslandschaft cultural landscape, anthropogenic landscape

ZNS (Zentralnervensystem) CNS (central nervous system)

Zoarium (Bryozoen-Kolonie) bryozoan colony

Zoëa (Decapoden-Larve) zoëa (decapod crustacean larva)

Zoecium (Bryozoen: Gehäuse des Einzeltiers) zoecium, zooecium

Zonensedimentation zone sedimentation, zonal sedimentation

Zonierung zonation

Zönobium/Cönobium/Coenobium (pl Coenobien) coenobium (pl coenobia), cell family

Zönogenese/Coenogenese cenogenesis

Zönose coenosis, cenosis

Zonoskelett (Extremitätengürtel) zonoskeleton

Zonulafasern zonule fibers

Zoocecidium Tiergalle

Zoogeographie/Tiergeographie zoogeography

Zoonose zoonosis

Zooparasit (Schmarotzer in/auf Tieren) zoophagous parasite (thriving in/on animals)

zoophag zoophagous

Zooplankton zooplankton

Zoozönose/Tiergemeinschaft zoocoenosis, zoocenosis, animal community

Zosteraceae/Seegrasgewächse eel-grass family

Zotte villus (pl villi)
- **Chorionzotten** chorionic villi
- **Darmzotten/Villi intestinales** intestinal villi
- **Gefäßzotten** vascular villi
- **Mikrovillus (pl Mikrovilli)** microvillus (pl microvilli)

Zottenglatze/Chorion laeve chorion laeve (nonvillous chorion)

Zottenhaut/Chorion frondosum chorion frondosum

Zottenplazenta/ Chorioallantoisplazenta chorioallantoic placenta

Zubehör accessories

Zucht bot cultivation, breeding, growing

Zucht/Züchterei (Farm) zool breeding farm

Zuchtbuch/Stammbuch/Herdbuch (Pferde: Stutbuch) studbook

Zuchtbulle/Zuchtstier studbull

züchten/kultivieren/aufziehen bot/ micb breed, cultivate, grow

züchten/kultivieren/aufziehen zool raise, rear

Züchterei breeding farm; (Zuchtanlage z.B. für Geflügel) hatchery; (Zuchtbetrieb/Gestüt für Pferde) studfarm

Zuchtform breed

Zuchthengst/Schälhengst/ Deckhengst/Beschäler studhorse, stud

Zuchtpferd stock horse

Zuchtstier/Zuchtbulle studbull

Zuchtstute broodmare

Züchtung/Kultivierung breed, breeding, cultivation, growing
- **aus der Kreuzung entfernt oder nicht verwandter Individuen gezüchtet** outbred

Züchtungsexperiment breeding experiment

Zuchtwahl selective breeding, breed selection
- **natürliche Zuchtwahl/ natürliche Selektion** natural selection

zucken (Muskel) twitch

Zucker sugar

zuckerbildend sacchariferous

Zuckerdrüse (Placophora) sugar gland, subradular organ

zuckerhaltig sugar-containing

757   Zungenmandel

Zuckerkrankheit/Diabetes mellitus
diabetes mellitus
Zuckerrohr sugar cane
Zuckersäure/Aldarsäure
saccharic acid, aldaric acid
Zuckerstoffwechsel
glycometabolism
Zuckreflex jerk
Zuckung twitch, twitching,
convulsion
Zufall chance
zufällig by chance, at random
zufällige Paarung/Panmixie
random mating
Zufallsabweichung *stat*
random deviation
Zufallsauslese random screening
Zufallsereignis random event
Zufallsfehler *stat* random error
"Zufallsknäuel" *gen* random coil
Zufallspaarung random mating
Zufallsstichprobe/Zufallsprobe *stat*
random sample
Zufallsvariable *stat* random variable
Zufallsverteilung *stat*
random distribution
Zufallswirt random host
Zufallszahl *stat* random number
Zufluss tributary
zuführend/hinführend/zuleitend
neuro afferent
Zug/Migration *orn* migration
Zug/Sog (Wasserleitung) tension,
suction, pull
zugehörig/verwandt (Nucleotid/
tRNA) cognate (nucleotide/tRNA)
Zügel/Stützleiste/Frenulum
frenulum
Zügel *orn* **(Bereich zwischen**
Schnabel und Augen) lore
Zügelschild/Loreale (Schlangen)
loreal (scale)
zugespitzt (z. B. Blatt) attenuate,
tapering, pointed
• **lang zugespitzt**
(konkav zulaufend) acuminate,
taper-pointed
zugewachsen overgrown
Zugfaser mantle fiber
Zugfestigkeit/Zerreißfestigkeit/
Reißfestigkeit (Holz)
tensile strength
Zugfisch migratory fish
Zuggeißel pulling flagellum
Zugholz tension wood
Zugspannung (Wasserkohäsion)
water tension
Zugstraße migratory route
Zugtier draft animal
Zugunruhe migratory restlessness
Zugvogel migratory bird
Zugwurzel contractile root

Zulauf (Eintrittsstelle einer
Flüssigkeit) inlet
Zulaufkultur/Fedbatch-Kultur
(semi-diskontinuierlich)
fed-batch culture
Zulaufverfahren/Fedbatch-
Verfahren (semi-diskontinuierlich)
fed-batch process,
fed-batch procedure
Zunahme gain, increase
• **Gewichtszunahme** weight gain,
gain in weight
Zündbarkeit ignitability
Zunder tinder
Zündflamme *lab* pilot flame
(from a pilot burner)
Zündfunke (ignition) spark
Zündung ignition
zunehmen gain, increase;
gain weight
Zunge/Glossa/Lingua tongue,
glossa, lingua
Zunge/Reibplatte/Radula radula
• **Balkenzunge/docoglosse Radula**
docoglossate radula
• **Bandzunge/taenioglosse Radula**
taenioglossate radula
• **Bürstenzunge/**
hystrichoglosse Radula
hystrichoglossate radula
• **Fächerzunge/**
rhipidoglosse Radula
rhipidoglossate radula
• **Federzunge/ptenoglosse Radula**
ptenoglossate radula
• **Pfeilzunge (hohl)/**
toxoglosse Radula
toxiglossate radula
(hollow radula teeth)
• **Schmalzunge/**
rhachiglosse Radula/
stenoglosse Radula
rachiglossate radula
züngeln tongue flickerng
Zungen.../die Zunge betreffend
lingual
Zungenbein/Hyoideum/
Os hyoideum hyoid bone,
lingual bone
Zungenbeinbogen/Schlundbogen/
Hyoidbogen *embr* hyoid arch
Zungenblüte (Strahlblüte)
ray flower, ray floret
Zungenbogen/Zwischenbogen
tongue bar, secondary bar
Zungenfarngewächse/
Elaphoglossaceae
elephant's-ear fern family
zungenförmig tongue-shaped,
linguiform, ligular, oblanceolate
Zungenmandel/Tonsilla lingualis
lingual tonsil

Z

Zungenpapille

Zungenpapille lingual papilla, gustatory papilla
- **Blattpapille/blättrige Papille** foliate papilla
- **fadenförmige Papille** threadlike papilla, filiform papilla
- **linsenförmige Papille** lentiform papilla
- **Pilzpapille** fungiform papilla
- **Wallpapille** vallate papilla

Zungenwürmer/Linguatuliden/ Pentastomitiden/Pentastomida tongue worms, linguatulids, pentastomids

Zünglein/Ligula ligule, ligula

zupacken/ergreifen grasping

zureiten (Pferd) break

zurückgeblieben/ in der Entwicklung gehemmt retarded, stunted

zurückgebogene DNA/ in sich gefaltete DNA snap-back DNA, fold-back DNA

zurückgerollt/nach hinten eingerollt rolled backward, revolute

zurückgesetztes Ende *gen* recessed end

zurückschneiden *hort* cut, prune, trim

Zusammenbau/Assemblierung *chem/gen* assembly

Zusammenbruch/Abbau/Zerfall breakdown; *ecol* (population) crash

zusammengesetzt (zusammengesetztes Blatt) compound (compound leaf)

Zusammenhang/Verhältnis/ Verbindung relation, correlation, interrelationship, connection

Zusatzbezeichnung/Epitheton epithet

zusätzliches Chromosom/ akzessorisches Chromosom accessory chromosome

Zusatzstoff/Zusatz/Additiv additive
- **Lebensmittelzusatzstoff** food additive

zuspitzen taper

Zustand state, condition
- **gleichbleibender/stationärer Zustand** steady state

zustöpseln stopper

zutage treten/zutage liegen *geo* outcrop

Zutageliegendes/Aufschluss *geo* outcrop

zuverlässig reliable

Zuverlässigkeit reliability

Zuwachs accretion, accrescence, additional growth, new growth, enlargement
- **Sprosszuwachs** shoot elongation

Zuwachs nach Blüte accrescence

zuwachsen/überwachsen overgrow

Zuwachsrate/Wachstumsrate/ Wachstumsgeschwindigkeit growth rate

Zuwachszone growth layer

Zuwanderung/Einwanderung/ Immigration immigration

zweiästig/biram two-branched, biramous

zweieiig/dizygot dizygous, dizygotic

zweifächerig (Fruchtknoten) bilocular

Zweiflügelfruchtgewächse/ Flügelnussgewächse/ Dipterocarpaceae meranti family, dipterocarpus family

Zweifüß(l)er biped

Zweifüßigkeit/Zweibeinigkeit/ Bipedie/Bipedität bipedalism, bipedality

Zweig twig, limb; (kleiner Zweig mit Blättern/Blüten) spray

zweigabelig/dichotom (einfach gegabelt) bifurcate, dichotomous

zweigeißlig biflagellated

zweigeschlechtig/zwittrig bisexual, hermaphroditic

zweigeschlechtige Blüte/ Zwitterblüte bisexual flower, hermaphroditic flower, perfect flower

zweigeteilt two-parted

Zweiglein (small) twig, sprig

zweigliedrige Benennung/ Bezeichnung binomial/binary nomenclature

Zweiglücke branch gap

Zweigspur branch trace

zweihäusig/getrenntgeschlechtig/ diözisch diecious, dioecious

Zweihäusigkeit/ Getrenntgeschlechtigkeit/Diözie dioecy, dioecism

zweihöckrig (Zahn) bicuspid

zweihörnig (z.B. Uterus) two-horned, bicornate, bicornuate, bicornuous

Zweihügel/Corpora bigemina corpora bigemina

zweijährig biennial

zweikammerig/zweikämmrig/ dithalam/dithekal with two chambers, dithalamous, dithalamic, dithecal

zweikammreihig bipectinate

zweikeimblättrig dicotyledonous

Zweikeimblättrige/Dikotyledone/ Dikotyle dicotyledon, dicot

zweiklappig/doppelklappig bivalve

Zwischenbild

zweireihig in two rows, two-row, biseriate
zweischeidig *zool* didelphic
zweischneidig (Initiale) *bot* with two cutting faces
Zweiseitentiere/Bilateria bilaterians
zweiseitig/bifazial/dorsiventral bifacial, dorsiventral
zweispaltig bifid
zweistachelig/zweidornig diacanthous
zweisträngig double-stranded, two-stranded
Zweisubstratreaktion/ Bisubstratreaktion bisubstrate reaction
• **Bi-Bi-Reaktion (zwei Substrate/ zwei Produkte)** Bi Bi reaction
• **doppelte Verdrängungsreaktion/ Doppel-Verdrängung (Pingpong-Reaktion)** double displacement reaction (ping-pong reaction)
• **einfache Verdrängungsreaktion/ Einzel-Verdrängung** single displacement reaction
• **geordnete Verdrängungsreaktion** ordered displacement reaction
• **zufällige Verdrängungsreaktion/ nicht-determinierte Verdrängungsreaktion** random displacement reaction
zweiteilig bipartite, dimeric, in two parts
Zweiteilung/binäre Zellteilung binary fission, bipartition
Zweitmünder/Neumundtiere/ Neumünder/Deuterostomia deuterostomes
zweiwertig/bivalent/divalent *chem* bivalent, divalent
Zweiwertigkeit *chem* bivalence, divalence
zweiwirtig/dixen dixenous, dixenic
zweizählig/dimer dimerous
zweizeilig distichous, two-ranked
Zwerchfell/Diaphragma diaphragm, diaphragma
Zwerchfellatmung/Bauchatmung diaphragmatic respiration, abdominal breathing
Zwerchfellnerv phrenic nerve
Zwergfadenwurm/*Strongyloides* spp. threadworms (causing strongyloidiasis)
Zwergfüßer/Symphyla symphylans
Zwerggeißelskorpione/Schizomida/ Schizopeltidia schizomids
Zwergmännchen/Nannandrium *bot/zool* dwarf male, nanander (e. g. male dwarf plant/ant)
Zwergmutante dwarf mutant

Zwergnadelwald/ Zwergstrauchzone pygmy conifer woodland
Zwergölbaumgewächse/ Zeilandgewächse/Cneoraceae spurge olive family
Zwergpfeffergewächse/ Peperomiaceae peperomia family
Zwergstrauch/holziger Chamaephyt dwarf-shrub, woody chamaephyte
Zwergvegetation dwarf vegetation
Zwergwald/Zwergwaldstufe elfin forest, elfin woodland
Zwergwuchs dwarfed growth, stunted appearance, dwarfism, nanism
zwergwüchsig stunt, dwarf
Zwicke/Zwitterrind (steriles Kuhkalb: Zwillings-Geschwister eines ♂ Kalbs) freemartin, martin heifer
Zwickel/Trigona fibrosa (Herz) fibrous trigone
Zwiebel/Zwiebelknolle/Bulbus bulb
zwiebelartig (Geruch/Geschmack) alliaceous
zwiebelförmig bulbous
Zwiebelgemüse bulb vegetable
Zwiebelgewächse/Alliaceae onion family
Zwiebelknolle bulb
Zwiebelkuchen bulb "plate" with short internodes, contracted disk-like axis of bulb
Zwiebelstaude bulbous perennial herb
zwieseln (sich)/sich gabeln fork, bifurcate
Zwieselung (Stammverzweigung nah am Boden) *bot* forking of trunk base/lower trunk
Zwilling(e) twin(s)
• **eineiige Zwillinge** monozygous/ monozygotic (identical) twins
• **zweieiige Zwillinge** dizygous/ dizygotic (fraternal) twins
Zwillingsflecken/Zwillingssektoren twin spots
Zwillingsstudien twin studies
Zwinger (staatl. Verwahrung verwaister Tiere) pound
Zwinger/Käfig (z. B. in Zoos) cage, enclosure
• **Hundezwinger (staatlich)** dog pound
• **Hundezwinger/Hundepension/ Hundeheim** dog kennel
zwischenartlich/interspezifisch interspecific
Zwischenbild *micros* intermediate image

Z Zwischenblatt 760

Zwischenblatt metaxyphyll
Zwischenbündelcambium
interfascicular cambium
Zwischenfeder/Mesoptile *orn*
mesoptile
zwischengeschlechtlich intersexual
Zwischenhirn diencephalon,
interbrain, betweenbrain
Zwischenkieferknochen
incisive bone
Zwischenkultur *agr* intercropping,
double cropping
Zwischenneuron/Interneuron
interneuron
Zwischenprodukt/Zwischenform
biochem intermediate (product),
intermediate form
• doppelköpfiges/janusköpfiges Z.
double-headed intermediate
• tetraedrisches Z.
tetrahedral intermediate
Zwischenrippenfeld/Interkostalfeld
intercostal field
Zwischenscheitelbein/
Os interparietale interparietal bone
Zwischensequenz/Spacer *gen*
spacer
Zwischenstadium/Zwischenstufe
intermediate state,
intermediate stage
Zwischenstoffwechsel/
intermediärer Stoffwechsel
intermediary metabolism
Zwischenstufe/Übergangsform
intergrade, intermediary form,
transitory form, transient
Zwischenveredlung/
Zwischenpropfung *hort* double-
working (grafting with interstock)
Zwischenwirbelkörper/
Intercentrum intercentrum,
hypocentrum
Zwischenwirbelscheibe/
Bandscheibe/
Discus intervertebralis
intervertebral disk
Zwischenwirt intermediary host
Zwischenzehendrüse/
interdigitale Drüse/
Interdigitaldrüse
interdigital gland
Zwischenzelle/Metula *fung* metula
Zwischenzellraum/Interzellulare
intercellular space
zwitschern twitter, chirp, (singen)
sing/warble
Zwitter/Hermaphrodit (siehe auch
dort) hermaphrodite
Zwitterblüte/zweigeschlechtige
Blüte bisexual flower,
hermaphroditic flower,
perfect flower

Zwitterdrüse/Zwittergonade/
Ovotestis hermaphroditic gland/
gonad, ovotestis
Zwittergang hermaphroditic duct
zwitterig/zwittrig/
hermaphroditisch
(zweigeschlechtlich)
hermaphroditic (bisexual)
Zwitterion zwitterion
(*not translated!*)
Zwitterrind/Zwicke (steriles
Kuhkalb: Zwillingsgeschwister
eines ♂ Kalbes) freemartin
Zwittertum/Zwittrigkeit/
Hermaphroditismus
hermaphroditism, hermaphrodism
• Scheinzwittertum/
Pseudohermaphroditismus
pseudohermaphrodism
zwittrig/zweigeschlechtlich
hermaphroditic, bisexual
Zwölffingerdarm/Duodenum
duodenum
Zyanthium/Cyanthium cyanthium
Zyathium/Cyathium cyathium
Zygapophyse zygapophysis
("yoking" process)
Zygokarp/Zygotenfrucht/
Cystozygote/Oospore
cystozygote, oospore
zygomorph zygomorphic,
monosymmetrical, irregular
Zygophyllaceae/Jochblattgewächse
caltrop family, creosote bush family
Zygospore/Hypnozygote zygospore
Zygosporocarp zygosporocarp
Zygotän (in meiotischer Prophase)
zygotene
Zygote zygote
Zygotenfrucht/Zygokarp/
Cystozygote/Oospore
(Dauerzygote) cystozygote,
oospore
Zygotenkern/Synkaryon
zygote nucleus, synkaryon
Zygotie zygosity
• Autozygotie autozygosity
• Dizygotie/Zweieiigkeit
dizygosity
• Hemizygotie hemizygosity
• Heterozygotie/Mischerbigkeit
heterozygosity
• Homozygotie/Reinerbigkeit/
Reinrassigkeit homozygosity
• Monozygotie/Eineiigkeit
monozygosity
zygotisch zygotic
zygotische Induktion
zygotic induction
zyklisch/cyclisch/ringförmig cyclic
Zyklisierung/Ringschluss *chem*
cyclization

Zytotoxizität

Zyklo-AMP/Cyclo-AMP/zyklisches AMP (cAMP) cyclic AMP (cAMP)
Zyklomorphose/Cyclomorphose/ Temporalvariation cyclomorphosis
Zyklus cycle
Zylinder cylinder
- **Leitbündelzylinder/Leitzylinder/ Leitbündelring** *bot* vascular cylinder
- **Linsenzylinder/Kristallkegel/ Kristallkörper/Conus** *zool* crystalline cone
- **Messzylinder** *lab* graduated cylinder
- **Mischzylinder** *lab* volumetric flask
- **Zentralzylinder** *bot* central cylinder, stele
Zylinderepithel/Säulenepithel columnar epithelium
- **hohes Zylinderepithel (hochprismatisches Epithel)** simple columnar epithelium
Zylinderglas/Becherglas *lab* beaker
Zylinderrosen/Ceriantharia tube anemones, cerianthids
Zylinderzelle/Axialzelle (*Dicyemida*) axial cell
Zylinderzelle/Wimperzelle (*Trichoplax*) cylinder cell, ciliated cell
zylindrisch/cylindrisch/ walzenförmig cylindric, cylindrical
Zyme/Zyma/Zymus/Cymus/Cyme/ cymöser Blütenstand cyme, cymose inflorescence
zymogen zymogenic
Zymogen/Proenzym (Enzymvorstufe) zymogen, proenzyme (enzyme precursor)
zymös/cymös/cymos/trugdoldig (sympodial verzweigt) cymose, cymoid (sympodially branched)
Zynthie/Cynthia cynthia
Zypressengewächse/Cupressaceae cypress family
Zyste/Cyste cyst

Zystid/Cystid cystidium
zystieren/enzystieren encyst
zystische Fibrose/Mukoviszidose/ Mucoviszidose cystic fibrosis, mucoviscidosis
Zytidin/Cytidin cytidine
Zytochemie/Cytochemie/Zellchemie cytochemistry
Zytochrom/Cytochrom cytochrome
Zytogenetik/Cytogenetik cytogenetics
Zytokeratin/Cytokeratin cytokeratin
Zytokin/Cytokin cytokine (biological response mediator)
Zytokinese/Zellteilung cytokinesis, cell division
Zytologie/Cytologie/Zellenlehre/ Zellbiologie cytology, cell biology
zytolytisch/cytolytisch cytolytic
Zytometrie/Cytometrie cytometry
zytopathischer Effekt/ zytopathogener Effekt cytopathic effect
Zytopempsis/Cytopempsis/ Vesikulartransport cytopempsis
Zytopenie/Cytopenie cytopenia
Zytopharynx/Cytopharynx/ Zellschlund cytopharynx, gullet
Zytoplasma/Cytoplasma/Zellplasma cytoplasm
Zytoplasmaströmung/ Cytoplasmaströmung cytoplasmic streaming
zytoplasmatische Vererbung/ cytoplasmatische Vererbung cytoplasmic inheritance
Zytoproct/Cytoproct/Zytopyge/ Cytopyge/Zellafter cytoproct, cytopyge, cell-anus
Zytoskelett/Cytoskelett/Zellskelett cytoskeleton
Zytosol/Cytosol cytosol
Zytostatikum/Cytostatikum (meist *pl* Zytostatika/Cytostatika) cytostatic agent, cytostatic
zytotoxisch/cytotoxisch cytotoxic
Zytotoxizität/Cytotoxizität cytotoxicity

Literatur / References

Ackermann HW, Berthiaume L, Tremblay M: *Virus Life in Diagrams*. CRC Press, Boca Raton, 1998

Alberts B, Bray D, Lewis J, Raff M, Roberts K, Watson JD: *Molecular Biology of the Cell*. 4th edn. Garland Science, New York, 2002. *Molekularbiologie der Zelle*, 4. Aufl., Wiley-VCH, Weinheim, 2004

Allaby M: *A Dictionary of Zoology*. 2nd edn. Oxford Univ. Press, Oxford/New York, 1999

Alsing I, Friesecke H, Guthy K: *Lexikon Landwirtschaft*. Ulmer, Stuttgart, 2002

Bahadir M, Parlar H, Spiteller M: *Springer Umwelt-Lexikon*, 2. Aufl. Springer, Berlin Heidelberg New York, 2000

Barker K: *At the Bench – A Laboratory Navigator*, 2nd edn. Cold Spring Harbor Laboratory Press, NY, 2004

Barnes RS, Calow P, Olive PJ: *The Invertebrates: A New Synthesis*, 2nd edn. Blackwell, Boston, 1993

Barrington EJW: *Invertebrate Structure and Function*, 2nd edn. Nelson, Middlesex, 1979

Barrows EM: *Animal Behavior Desk Reference*. CRC Press, Boca Raton, 1995

Bell AD: *Plant Form – An Illustrated Guide to Flowering Plant Morphology*. Oxford Univ. Press, Oxford/New York, 1993

Bender HF: *Das Gefahrstoffbuch*, 2. Aufl. Wiley-VCH, Weinheim, 2002

Benson L: *Plant Classification*. 2nd edn. Heath, Lexington MA, 1979

Berndt R, Meise W (Hrsg.): *Naturgeschichte der Vögel*, Bde. 1–3. Franckh'sche Verlagsbuchhandlung, Stuttgart, 1958–1966

Bezzel E, Prinzinger R: *Ornithologie*, 2. Aufl. Ulmer, Stuttgart, 1990

Böck P (Hrsg.): *Romeis – Mikroskopische Technik*. 17. Aufl. Urban & Schwarzenberg, München, 1989

Boden E: *Black's Veterinary Dictionary*, 19th edn. Rowman & Littlefield, London, 1998

Bold HC, Alexopoulos CJ, Delevoryas T: *Morphology of Plants and Fungi*. 5th edn. Harper-Collins, New York, 1987

Boolootian RA, Heyneman D: *An Illustrated Laboratory Text in Zoology*. 4th edn. Saunders, Philadelphia, 1980

Borror DJ, Triplehorn CA, Johnson NF: *An Introduction to the Study of Insects*. 6th edn. Saunders, Philadelphia, 1989

Brandis H, Eggers HJ, Köhler W, Pulverer G: *Lehrbuch der Medizinischen Mikrobiologie*, 7. Aufl. Fischer, Stuttgart, 1994

Braune W, Leman A, Taubert H: *Pflanzenanatomisches Praktikum*. Bd. I & II. jeweils 8./4. Aufl. Spektrum Akademischer Verlag, Heidelberg, 1999

Brenner S, Miller JH (eds) *Encyclopedia of Genetics*, 4 Vols. Academic Press, San Diego, 2002

Bresslau E, Ziegler HE: *Zoologisches Wörterbuch*, 4. Aufl. Fischer, Jena, 1927

Brockhaus Enzyklopädie, 19. Aufl. (24 Bde.) FA Brockhaus, Mannheim, 1986–1994

Brohmer – *Fauna von Deutschland*, 21. Aufl. (Schaefer M, Hrsg.) Quelle & Meyer, Wiebelsheim, 2002

Brooks GF, Butel JS: *Jawetz, Melnick & Adelberg's Medical Microbiology*, 23th edn. Appleton & Lange, East Norwalk, 2004

Brown TA: *Gene Cloning. An Introduction*, 3rd edn. Chapman & Hall, London/New York, 1995; *Gentechnologie für Einsteiger*, 3. Aufl. Spektrum Akademischer Verlag, Heidelberg, 2002

Brown TA: *Genetics – A Molecular Approach*, 3rd edn. Chapman & Hall, London/New York, 1998; *Moderne Genetik*, 2. Aufl. Spektrum Akademischer Verlag, Heidelberg, 1999

Brusca RC, Brusca GJ: *Invertebrates*. Sinauer, Sunderland, MA, 1990

Buddecke E: *Grundriss der Biochemie*, 9. Aufl. deGruyter, Berlin, 1994

Buhr H: *Bestimmungstabellen der Gallen (Zoo- und Phytocecidien) an Pflanzen Mittel- und Nordeuropas*. 2 Bde. Fischer, Jena, 1964

Campbell NA: *Biology*, 6th edn. Benjamin-Cummings/Pearson, San Francisco, CA, 2002; *Biologie*. Elsevier/Spektrum Akademischer Verlag, Heidelberg, 2003

Carlile MJ, Watkinson SC: *The Fungi*. Academic Press, London, 1994

Cheng TC: *General Parasitology*. 2nd edn. Academic Press, New York, 1986

Clark AN: *Dictionary of Geography*. Longman, Harlow, Essex, 1985

Cole TCH: *Taschenwörterbuch der Botanik/A Pocket Dictionary of Botany*. Thieme, Stuttgart, 1994

Cole TCH: *Taschenwörterbuch der Zoologie/A Pocket Dictionary of Zoology*. Thieme, Stuttgart, 1995

Cole TCH: *Wörterbuch der Tiernamen*. Spektrum Akademischer Verlag, Heidelberg, 2000

Cole TCH: *Wörterbuch Labor/Laboratory Dictionary*. Springer, Berlin Heidelberg New York, 2005

Coombs J: *Dictionary of Biotechnology*, 2nd edn. Macmillan, London, 1992

Cooper GM, Hausman RE: *The Cell – A Molecular Approach*, 3rd edn. ASM Press/Sinauer, Sunderland, 2004

Cronquist A: *An Integrated System of Classification of Flowering Plants*. Columbia Univ. Press, New York, 1981

Cruse JM, Lewis RE: *Illustrated Dictionary of Immunology*, 2nd edn. CRC Press, Boca Raton, 2003

Curtis H, Barnes NS: *Biology.* 5th edn. Worth, New York, 1989; *Invitation to Biology.* 5th edn. Worth, New York, 1994

Dahlgren G: *Systematische Botanik.* Springer, Berlin Heidelberg New York, 1987

Daly HV, Doyen JT, Ehrlich PR: *Introduction to Insect Biology and Diversity.* McGraw-Hill, New York, 1978

Daubenmire R: *Plant Communities. A Textbook of Plant Synecology.* Harper & Row, New York, 1968; *Plant Geography with Special Reference to North America.* Academic Press, New York, 1978

Davies RG: *Outline of Entomology,* 7th edn. Chapman & Hall, London/New York, 1988

Davis BD, Dulbecco R, Eisen HN, Ginsberg HS: *Microbiology,* 4th edn. Lippincott, Philadelphia, 1990

DeDuve C: *Blueprint for a Cell. The Nature and Origin of Life.* N Patterson Publ./Carolina Biological Supply Co, 1991; *Ursprung des Lebens. Präbiotische Evolution und die Entstehung der Zelle.* Spektrum Akademischer Verlag, Heidelberg, 1994

DeDuve C: *Die Zelle. Expeditionen in die Grundstruktur des Lebens.* Spektrum Akademischer Verlag, Heidelberg, 1989

Dellweg H: *Biotechnologie – Verständlich.* Springer, Berlin Heidelberg New York, 1995

Dellweg H, Schmid RD, Trommer WE: *Römpp Lexikon Biotechnologie.* Thieme, Stuttgart, 1992

Delvin TM: *Textbook of Biochemistry,* 5th edn. Wiley-Liss, New York, 2002

Dettner K, Peters W: *Lehrbuch der Entomologie,* 2. Aufl. Elsevier/Spektrum Akademischer Verlag, Heidelberg, 2003

Diekmann H, Metz H: *Grundlagen und Praxis der Biotechnologie.* Fischer, Stuttgart, 1991

Dörfelt H, Jetschke G: *Wörterbuch der Mykologie,* 2. Aufl. Spektrum Akademischer Verlag, Heidelberg, 2001

Dorit RL, Walker WF, Jr, Barnes RD: *Zoology.* Saunders, Philadelphia, 1991

Dressler D, Potter H: *Discovering Enzymes.* Scientific American Library, WH Freeman, New York, 1991; *Katalysatoren des Lebens – Struktur und Wirkung von Enzymen.* Spektrum Akademischer Verlag, Heidelberg, 1992

Drickamer LC, Vessey SH: *Animal Behavior.* 2nd edn. Prindle, Weber & Schmidt, Boston, 1986

Dyce KM, Sack WO, Wensing CJG: *Textbook of Veterinary Anatomy,* 2nd edn. Saunders, Philadelphia, 1996; *Anatomie der Haustiere,* 2. Aufl. Enke, Stuttgart, 1997

Eckhardt S, Gottwald W, Stieglitz B: *1×1 der Laborpraxis.* Wiley-VCH, Weinheim, 2002

Eisenreich G, Sube R: *Fachwörterbuch Mathematik.* Langenscheidt, München, 1996

Ellenberg H: *Vegetation Mitteleuropas mit den Alpen*, 4. Aufl. Ulmer, Stuttgart, 1986; *Vegetation Ecology of Central Europe*. Cambridge Univ. Press, Cambridge – New York, 1988

Esau K: *Plant Anatomy*. 2nd edn. Wiley, New York, 1965; *Pflanzenanatomie*. Fischer, Stuttgart, 1969

Fahn A: *Plant Anatomy*. 4th edn. Pergamon Press, New York, 1990

Fioroni P: *Allgemeine und vergleichende Embryologie der Tiere*. 2. Aufl. Springer, Berlin/Heidelberg, 1992

Flint SJ, Enquist LW, Racaniello VR, Skalka AM: *Principles of Virology*, 2nd edn. ASM Press, Wash. DC. 2004

Foelix RF: *Biologie der Spinnen*, 2. Aufl. Thieme, Stuttgart, 1992; *Biology of Spiders*, 2nd edn. Oxford Univ. Press, New York, 1996

Frandson RD, Spurgeon TL: *Anatomy and Physiology of Farm Animals*, 5th edn. Lea & Febiger, Philadelphia, 1992

Franke W: *Nutzpflanzenkunde*, 6. Aufl. Thieme, Stuttgart, 1997

Freifelder D, Malacinski GM: *Essentials of Molecular Biology*, 2nd edn. Jones & Bartlett, Boston, 1993

Freye HA, Kämpfe L, Biewald GA: *Zoologie*, 9. Aufl. Fischer, Jena, 1991

Furley PA, Newey WW: *Geography of the Biosphere*. Buttersworth, London, 1983

Futuyma DJ: *Evolutionary Biology*. 3rd edn. Sinauer, Sunderland, 1997; *Evolutionsbiologie*. Birkhäuser, Basel, 1990

Gardiner MS: *The Biology of Invertebrates*. McGraw-Hill, New York, 1972

Gattermann R: *Wörterbücher der Biologie: Verhaltensbiologie*. UTB/ Fischer, Jena – Stuttgart, 1993

Gemsa D, Kalden JR, Resch K: *Immunologie*, 4. Aufl. Thieme, Stuttgart, 1997

Gilbert SF: *Developmental Biology*. 7th edn. Sinauer, Sunderland, MA, 2003

Gilbert SF, Raunio AM (eds.) *Embryology. Constructing the Organism*. Sinauer, Sunderland, MA, 1997

Gill FB: *Ornithology*. WH Freeman, New York, 1990

Glick BR, Pasternak JJ: *Molecular Biotechnology – Principles and Applications of Recombinant DNA*. ASM Press, Washington, 2003; *Molekulare Biotechnologie*. Spektrum Akademischer Verlag, Heidelberg, 1996

Götting KJ: *Malakozoologie. Grundriß der Weichtierkunde*. Fischer, Stuttgart, 1974

Gould SJ: *Ontogeny and Phylogeny*. Belknap/Harvard Univ. Press, Cambridge MA, 1977

Grant R, Grant C: *Grant & Hackh's Chemical Dictionary*, 5th edn. McGraw-Hill, New York, 1987

Gray's Manual of Botany, 8th edn. (Fernald ML, ed.) Dioscorides, Portland OR, 1950

Griffiths AJF, Wessler SR, Lewontin RC, Gelbart WM, Suzuki DT, Miller JH: *An Introduction to Genetic Analysis.* 8th edn. WH Freeman, New York, 2005; *Genetik,* 3. Aufl. VCH, Weinheim/New York, 1991

Grzimek B: *Encyclopedia of Evolution.* Van Nostrand Reinhold, New York, 1976

Grzimek B: *Grzimeks Tierleben, Enzyklopädie des Tierreichs,* 13 Bde. Kindler, Zürich, 1971; *Grzimek's Animal Life Encyclopedia,* 13 Vols. Van Nostrand Reinhold, New York, 1972–1975

Grzimek B: *Grzimeks Enzyklopädie der Säugetiere,* Bde. 1–5. Kindler, München, 1988; *Encyclopaedia of Mammals.* McGraw-Hill, New York, 1989

Gullan PJ, Cranston PS: *The Insects,* 2nd edn. *An Outline of Entomology.* Blackwell Science, London, 2000

Harris JG, Harris MW: *Plant Identification Terminology. An Illustrated Glossary.* Spring Lake Publ., Spring Lake, UT, 1994

Hartmann HT, Davies FT, Geneve RL: *Plant Propagation – Principles and Practices,* 7th edn. Prentice-Hall/Pearson Englewood Cliffs, NJ, 2001

Hausmann K, Hülsmann N, Radek R: Protistology, 3rd edn. Schweizerbart'sche Verlagsbuchhandlung, Berlin Stuttgart, 2003

Hausmann K: *Protozoologie.* Thieme, Stuttgart, 1985; *Protozoology,* 2nd edn. Thieme, Stuttgart, 1996

Hawksworth DL, Kirk PM, Sutton BC, Pragler DN: Ainsworth & Bisby's – *Dictionary of the Fungi,* 8th edn. CAB Intl, Wallingford, Oxon, 1995

Hedrick PW: *Genetics of Populations.* Jones & Bartlett, Boston, 1983

Henderson's Dictionary of Biological Terms, 13th edn. (Lawrence E, ed.) Prentice Hall/Pearson, Harlow, 2005

Hennig W: *Genetik,* 3. Aufl. Springer, Heidelberg/Bertlin/New York, 2002

Herren RV, Donahue RL: *Delmar's Agriscience Dictionary.* Thomson Delmar Publ., Albany NY, 1998

Heymer A: *Ethologisches Wörterbuch,* D – E – F. Parey, Berlin/Hamburg, 1977

Heywood VH: *Flowering Plants of the World.* Oxford Univ. Press, Oxford, 1993; *Blütenpflanzen der Welt.* Birkhäuser, Basel – Boston, 1982

Hickey M, King C: *100 Families of Flowering Plants.* Cambridge Univ. Press, Cambridge – New York, 1981

Hickman CP et al.: *Integrated Principles of Zoology,* 12th edn. McGraw-Hill Higher Education, 2003

Hildebrand M, Goslow GE Jr.: *Analysis of Vertebrate Structure,* 5th edn. Wiley, NY, 2001; *Vergleichende und funktionelle Anatomie der Wirbeltiere.* Springer-Verlag, Berlin Heidelberg New York, 2004

Hill RW et al.: *Animal Physiology,* Sinauer, Sunderland, 2004

Hintermaier-Erhard G, Zech W: *Wörterbuch der Bodenkunde.* Enke, Stuttgart, 1997

Holmes S: *Outline of Plant Classification*. Longman, London, 1983

Hora B: *The Oxford Encyclopedia of Trees of the World*. Oxford Univ. Press, Oxford/New York, 1981

Horton HR, Moran LA, Ochs RS, Rawn JD, Scrimgeour KG: *Principles of Biochemistry*. Patterson/Prentice Hall, Englewood Cliffs, 1993

Hortus Third: *A Concise Dictionary of Plants Cultivated in the United States and Canada* (Bailey LH/Bailey EZ, eds.) Macmillan, New York, 1976

Hull R, Brown F, Pane C: *Virology – Directory and Dictionary of Animal, Bacterial, and Plant Viruses*. Macmillan, London/Stockton/New York, 1989

Huxley A (ed.) *The New Royal Horticultural Society: Dictionary of Gardening*. Macmillan, London/New York, 1992

Hyam R, Pankhurst R: *Plants and Their Names. A Concise Dictionary*. Oxford Univ. Press, Oxford, 1995

Hyman LH: *The Invertebrates*. Vols. 1–6. McGraw-Hill, New York, 1940–1967

Ibelgaufts H: *Gentechnologie von A bis Z*. VCH, Weinheim, 1990

Immelmann K, Beer C: *A Dictionary of Ethology*. Harvard Univ. Press, Cambridge, 1989

Immelmann K: *Wörterbuch der Verhaltensforschung*. Parey, Berlin/Hamburg, 1982

Jacks GV: *Multilingual Vocabulary of Soil Science*. Agriculture Division, FAO – United Nations, Rome, 1954

Jacobs W, Renner M: *Biologie und Ökologie der Insekten. Taschenlexikon*. 2. Aufl. Fischer, Stuttgart-Jena, 1988

Jaeger EC: *A Source-Book of Biological Names and Terms*, 3rd edn. Thomas Publ., Springfield, Illinois, 1978

Jäger EJ, Werner K: *Rothmaler – Exkursionsflora von Deutschland*. Band 2, 18. Aufl. Elsevier/Spektrum Akademischer Verlag, Heidelberg, 2002

Janeway CA, Travers P, Walport M, Shlomchik M: *Immunologie*, 5te Aufl. Elsevier/Spektrum Akademischer Verlag, Heidelberg, 2002

Jangi BS: *Economic Zoology. A Dictionary of Useful and Destructive Animals*. Balkema, Rotterdam, 1991

Judd WS, Campbell CS, Kellogg E: *Plant Systematics: A Phylogenetic Approach*, 2nd edn. Sinauer, Sunderland, 2002

Junqueira LC, Carneiro J: *Basic Histology*. Appleton/Lange, Stamford, CT, 1986

Junqueira LC, Carneiro J, Kelley RO: *Histologie* (Gratzl M, Hrsg.) Springer, Berlin Heidelberg New York, 2002

Kahl G: *The Dictionary of Gene Technology*, 3rd edn. Vols. 1 & 2. Wiley-VCH, Weinheim, 2004

Kämpfe L, Kittel R, Klapperstück J: *Leitfaden der Anatomie der Wirbeltiere*. 6. Aufl. Fischer, Jena, 1993

769

Kardong KV: *Vertebrates – Comparative Anatomy, Function, Evolution.* WmC Brown, Dubuque, IA, 1995

Karlson P, Doenecke D, Koolmann J: *Kurzes Lehrbuch der Biochemie,* 14. Aufl. Thieme, Stuttgart, 1994

Karp G: *Cell and Molecular Biology,* 4th edn. Wiley, New York, 2005

Kaufman PB: *Plants. Their Biology and Importance.* Harper & Row, New York, 1989

Kaussmann B, Schiewer U: *Funktionelle Morphologie und Anatomie der Pflanzen.* Fischer, Stuttgart, 1989

Kearey P: *The Encyclopedia of the Solid Earth Sciences.* Blackwell, Oxford, 1993

Kéler S, v: *Entomologisches Wörterbuch.* Akademie Verlag, Berlin, 1963

Kendrew, K Sir: *The Encyclopedia of Molecular Biology.* Blackwell, Oxford, 1994

Kindl H: *Biochemie der Pflanzen,* 4. Aufl. Springer, Berlin/Heidelberg, 1994

King RC, Stansfield WD: *A Dictionary of Genetics,* 6th edn. Oxford Univ. Press, Oxford/New York, 2002

Kleber HP, Schlee D: *Biochemie,* Bd. I & II. Fischer, Stuttgart, 1991/1992

Kleinig H, Maier U: *Zellbiologie.* 4. Aufl. Elsevier/Spektrum Akademischer Verlag, Heidelberg, 1999

Klemm M: *Wörterbuch Paläarktischer Tiere.* Deutsch – Latein – Russisch. Parey, Berlin, 1973

Knippers R: *Molekulare Genetik.* 8. Aufl., Thieme, Stuttgart, 2001

Kozloff EN: *Invertebrates.* Saunders Coll. Publ., Philadelphia, 1989

Kubitzki K, Rohwer JG, Bittrich V (Hrsg.): *The Families and Genera of Vascular Plants.* Vol. I-VI. Springer, Berlin Heidelberg New York, 1993–2004

Kuby J: *Immunology,* 3rd edn. Freemann, New York, 1997

Kükenthal W, Krumbach T (Hrsg.): *Handbuch der Zoologie/Handbook of Zoology.* Vol. I – VIII (Bde. 1–60). deGruyter, Berlin, 1923–1994

Lackie JM, Dow JAT: *The Dictionary of Cell Biology,* 2nd edn. Academic/ HBJ, London/San Diego, 1995

Landau SI (ed.) *International Dictionary of Medicine and Biology,* Vols. 1–3. Wiley, New York, 1986

Lee JJ, Leedale GF, Bradbury P: *An Illustrated Guide to the Protozoa.* 2nd edn. Society of Protozoologists, Lawrence, KA, 2000

Leeson CR, Leeson TS, Paparo AA: *Textbook of Histology,* 5th edn. Saunders, Philadelphia, 1985

Leftwich AW: *A Dictionary of Entomology.* Constable, London, 1977

Leftwich AW: *A Dictionary of Zoology.* Constable, London, 1975

Lehmann U: *Paläontologisches Wörterbuch,* 4. Aufl. Enke, Stuttgart, 1996

Lehrbuch der Speziellen Zoologie. (begr. A. Kaestner). Fischer, Jena/Spektrum Akad. Verlag, Heidelberg, 1954 bis heute

Lengeler JW, Drews G, Schlegel HG: *Biology of Prokaryotes.* Thieme, Stuttgart/New York, 1999

Leser H (Hrsg.) *Wörterbuch Allgemeine Geographie.* Westermann/DTV, München, 1997

Levy JA, Fraenkel-Conrat H, Owens RA: *Virology,* 3rd edn. Prentice-Hall, Englewood Cliffs, 1994

Lewin B: *Genes.* 8th edn. Prentice Hall, New York, 2004; *Molekularbiologie der Gene.* Spektrum Akademischer Verlag, Heidelberg, 2002

Lexikon der Biochemie und Molekularbiologie, 3 Bde. & Ergänzungsband. Elsevier/Spektrum Akademischer Verlag, Heidelberg, 2000

Lexikon der Biologie. 15 Bde. Elsevier/Spektrum Akademischer Verlag, 2004

Libbert E: *Lehrbuch der Pflanzenphysiologie,* 5. Aufl. Fischer, Stuttgart, 1993

Lincoln RJ, Boxshall GA: *The Cambridge Illustrated Dictionary of Natural History.* Cambridge Univ. Press, Cambridge/New York, 1990

Lincoln RJ, Boxshall GA, Clark PF: *A Dictionary of Ecology, Evolution, and Systematics,* 2nd edn. Cambridge Univ. Press, Cambridge/New York, 1998

Little RJ, Jones CE: *A Dictionary of Botany.* Van Nostrand Reinhold, New York, 1980

Lodish H, Berk A, Matsudaira P: *Molecular Cell Biology,* 5th edn. Freeman, New York, 2003

Lodish H, Berk A, Zipursky SL, Darnell J: *Molekulare Zellbiologie.* 4. Aufl. Spektrum Akademischer Verlag, 2001

Loewy AG, Siekevitz P, Menninger JR, Gallant JAN: *Cell Structure and Function,* 3rd edn. Saunders, Philadelphia, 1991

Lorenz RJ: *Grundbegriffe der Biometrie,* 3. Aufl. Fischer, Stuttgart, 1992

Lüttge U, Higinbotham N: *Transport in Plants.* Springer, New York/Berlin/Heidelberg, 1979

Mabberley DJ: *The Plant Book,* 2nd edn. Cambridge Iniv. Press, Cambridge/New York, 1997

Macura P: *Dictionary of Botany.* G – E – F – S – R, 2 Vols. Elsevier, New York, 1982

Mack R, Mikhail B, Mikhail M: Wörterbuch der Veterinärmedizin und Biowissenschaften, 2. Aufl., D – E/E – D. Blackwell, Berlin, 1996

Madigan MT, Martinko JM, Parker J: *Brock – Biology of Microorganisms,* 10th edn. Prentice-Hall/Pearson, NJ, 2003: *Brock – Mikrobiologie.* Spektrum Akademischer Verlag, Heidelberg, 2001

Magill RE: *Glossarium Polyglottum Bryologiae. A Multilingual Glossary for Bryology.* Missouri Botanical Garden, St. Louis, 1990

Margulis L, Corliss JO, Melkonian M, Chapman DJ: *Handbook of Protoctista*. Jones and Bartlett, Boston, 1990

Margulis L, McKhann HI, Olendzenski L: *Illustrated Glossary of Protoctista*. Jones and Bartlett, Boston, 1993

Marquardt WC, Demaree RS Jr: *Parasitology*. Macmillan, London/New York, 1985

Martin EA: *A Dictionary of Life Sciences*. Pan/Macmillan, London, 1976

Mathews CK, van Holde KE, Ahern KG: *Biochemistry*, 3rd edn. Benjamin/Cummings, San Francisco, 1999

Mauseth JB: *Introduction to Plant Biology*. Jones and Bartlett, Boston, 1997

Mauseth JB: *Plant Anatomy*. Benjamin-Cummings, Menlo Park, 1988

Mayr E, Ashlock PD: *Principles of Systematic Zoology*. 2nd edn. McGraw-Hill, 1991; *Grundlagen der zoologischen Systematik*. Parey, Hamburg, 1975

McFarland D: *Animal Behaviour*, 3rd edn. Addison Wesley/Longman Ltd., Essex, 1999; *Biologie des Verhaltens*, 2. Aufl. Spektrum Akademischer Verlag, Heidelberg, 1999

Meglitsch PA, Schram FR: *Invertebrate Zoology*, 3rd edn. Oxford Univ. Press, Oxford/New York, 1991

Mehlhorn H (ed) *Encyclopedic Reference of Parasitology*, 2nd. edn. 2 Vols., Springer, Berlin Heidelberg New York, 2001

Metzler DE: *Biochemistry*, 2nd edn. Vols. 1 & 2. Harcourt/Academic Press, Burlington, MA, 2001/2003

Michel G, Salomon FV, Gutte G: *Morphologie landwirtschaftlicher Nutztiere*. VEB Deutscher Landwirtschaftsverlag, Berlin, 1987

Mickols DA, Freyer GA, Crotty DA: *DNA Science*, 2nd edn. Cold Spring Harbor Laboratory Press, NY, 2003

Mückenhausen E: *Die Bodenkunde*, 4. Aufl. DLG, Frankfurt, 1993

Müller AH: *Lehrbuch der Paläozoologie*. Bde. I-III. Fischer, Jena, 1985–1994

Nelson DL, Cox MM, Lehninger AL: *Principles of Biochemistry*, 4th edn. Freeman, New York, 2005; *Lehninger – Biochemie*, 3. Aufl. Springer, Berlin Heidelberg New York, 2001

Nicholls J, Martin AR, Wallace BG, Fuchs PA: *From Neuron to Brain*, 4th edn. Sinauer, Sunderland, 2001; *Vom Neuron zum Gehirn*. Spektrum Akademischer Verlag, Heidelberg, 2002

Nichols SW: *The Torre-Bueno Glossary of Entomology*. The New York Entomological Society/American Museum of Natural History, New York, 1989

Nickel R, Schummer A, Seiferle E: *Lehrbuch der Anatomie der Haustiere*. Bde. I – V, Parey, Berlin, 1992–96

Nierenberg WA (ed.): *Encyclopedia of Environmental Biology.* Academic Press, San Diego, 1995

Nöhring FJ: *Fachwörterbuch Medizin*, 2 Bde. D-E/E-D. Langenscheidt/Urban Fischer, München, 2002/2003

Nowak RM (ed.) *Walker's Mammals of the World*, 5th. edn. Johns Hopkins Univ. Press, Baltimore, 1991

Nultsch W: *Allgemeine Botanik*, 11. Aufl. Thieme, Stuttgart, 2001

Ohman DE: *Experiments in Gene Manipulation.* Prentice Hall, Englewood Cliffs, New Jersey, 1988

Old RW, Primrose SB: *Principles of Gene Manipulation. An Introduction to Genetic Engineering*, 5th edn. Blackwell, Oxford, 1994; *Gentechnologie.* Thieme, Stuttgart, 1992

Orr RT: *Vertebrate Biology.* 5th edn. Saunders, Philadelphia, 1982

Ott J: *Meereskunde*. 2. Aufl. Ulmer, Stuttgart, 1995

Pagel M (ed) *Encyclopedia of Evolution*, Vols. 1 & 2. Oxford Univ. Press, Oxford/New York, 2002

Parslow TG, Stites DP, Terr AI, Imboden JB: *Medical Immunology*, 10th edn. McGraw-Hill/Appleton & Lange, New York, 2001

Passarge E: *Taschenatlas der Genetik.* Thieme, Stuttgart, 1994

Passarge E: *Color Atlas of Genetics*, 2nd edn. Thieme, New York/Stuttgart, 2001

Pearse V, Pearse J, Buchsbaum M, Buchsbaum R: *Living Invertebrates.* Blackwell, Boston, 1992

Pechenik JA: *Biology of the Invertebrates*, 2nd edn. WmC Brown, Dubuque, IA, 1991

Pennak RW: *Collegiate Dictionary of Zoology.* Ronald Press, 1964

Penzlin H: *Lehrbuch der Tierphysiologie*, 7. Aufl. Elsevier/Spektrum Akademischer Verlag, Heidelberg, 2005

Peters JA: *Dictionary of Herpetology.* Hafner, New York, 1964

Pettingill OS, Jr, Breckenridge WJ: *Ornithology in Laboratory and Field.* 5th edn. Academic Press, Orlando, FL, 1985

Pflumm W: *Biologie der Säugetiere*, 2. Aufl. Parey, Berlin, 1996

Pijl L van der: *Principles of Dispersal in Higher Plants*, 3rd edn. Springer, Berlin/Heidelberg, 1982

Plattner H, Zingsheim HP: *Elektronenmikroskopische Methodik in der Zell- und Molekularbiologie.* Fischer, Stuttgart, 1987

Präve P, Faust U, Sittig W, Sukatsch DA: *Handbuch der Biotechnologie*, 4. Aufl. Oldenbourg, München, 1994

Proctor NS, Lynch PJ: *Manual of Ornithology. Avian Structure and Function.* 2nd edn. Yale Univ. Press, New Haven/London, 1998

Pschyrembel Klinisches Wörterbuch (Hildebrandt H, Hrsg.) 260. Aufl. deGruyter, Berlin, 2004

Radford AE, Dickson WC, Massey JR, Bell CR: *Vascular Plant Systematics*. Harper & Row, New York, 1974

Randall D, Burggren W, French K: *Eckert Animal Physiology*, 5th edn. WH Freeman, New York, 2002; *Tierphysiologie*, 4. Aufl. Thieme, Stuttgart, 2002

Rapoport SM: *Medizinische Biochemie*, 9. Aufl. VEB Verlag Volk & Gesundheit, Berlin, 1987

Rasch D, Tiku ML, Sumpf D: *Dictionary of Biometry*. Elsevier, Amsterdam, 1994

Raven PH, Evert RF, Eichhorn SE: *Biology of Plants*, 6th edn. Freeman/ Worth, New York, 1999; *Biologie der Pflanzen*, 3. Aufl. deGruyter, Berlin/New York, 2000

Reuter P: *Springer Großwörterbuch Medizin – Medical Dictionary*. E-G/G-E. Springer, Berlin Heidelberg New York, 2001

Rieger R, Michaelis A, Green MM: *Glossary of Genetics. Classical and Molecular*, 5th edn. Springer, Berlin, 1991

Rimoin DL, Connor JM, Pyeritz R, Korf B: *Emery & Rimoin's Principles and Practice of Medical Genetics*. 4th edn. 3 Vols. Saunders, Philadelphia, 2001

Romer AS, Parsons TS: *The Vertebrate Body*, 6th edn. Saunders, Philadelphia, 1986; *Vergleichende Anatomie der Wirbeltiere*, 5. Aufl. Parey, Hamburg, 1983

Romoser WS, Stoffolano JG: *The Science of Entomology*, 3rd edn. WmC Brown, Dubuque, IA, 1994

Ross MH, Kaye GI, Pawlina W: *Histology – A Text and Atlas*, 4th edn. Lippincott/Williams & Wilkins, New York, 2002

Ruppert EE, Fox RS, Barnes RD: *Invertebrate Zoology – A Functional Evolutionary Approach*, 7th edn. Brooks/Cole, Pacific Groove, CA, 2003

Salisbury FB and Intl. Assoc. for Plant Physiology: *Units, Symbols, and Terminology for Plant Physiology*. Oxford Univ. Press, Oxford/New York, 1996

Salisbury FB, Ross CW: *Plant Physiology*, 4th edn. Wadsworth, Belmont, 1992

Schell T von, Mohr H: *Biotechnologie – Gentechnik*. Springer, Heidelberg/ New York, 1995

Schlegel HG: *Allgemeine Mikrobiologie*, 7. Aufl. Thieme, Stuttgart, 1992; *General Microbiology*, 7th edn. Cambridge Univ. Press, Cambridge, 1993

Schleif R: *Genetics and Molecular Biology*. John Hopkins Univ. Press, Baltimore/New York, 1993

Schmeil O, Fitschen J: *Flora von Deutschland*. 92. Aufl. Quelle & Meyer, Wiebelsheim, 2003

Schmidt RF, Lang F, Thews G: *Physiologie des Menschen*, 29. Aufl. Springer, Berlin Heidelberg New York, 2005

Schmidt-Nielsen K, Markl L: *Animal Physiology*, 4th edn. Cambridge Univ. Press, Cambridge MA, 1990; Physiologie der Tiere. Elsevier/Spektrum Akademischer Verlag, Heidelberg, 1999

Schneider CK: *Illustriertes Handwörterbuch der Botanik*. Engelmann, Leipzig, 1905

Schönborn W: *Lehrbuch der Limnologie*. Schweizerbart'sche Verlagsbuchhandlung, Stuttgart, 2003

Schubert R (Hrsg.): *Lehrbuch der Ökologie*, 3. Aufl. Fischer, Jena, 1991

Schubert R, Wagner G: *Botanisches Wörterbuch*. 12. Aufl. UTB/Ulmer, Stuttgart, 2000

Schütt P, Schuck HJ, Stimm B: *Lexikon der Forstbotanik*. Ecomed, Landsberg, 1992

Schwetlick K: *Organikum*, 22. Aufl. Wiley-VCH, Weinheim, 2004

Serré R: *Elsevier's Dictionary of Microscopes and Microtechnique*. Elsevier, Amsterdam, 1993

Seyffert W: *Lehrbuch der Genetik*, 2. Aufl. Elsevier/Spektrum Akademischer Verlag, Heidelberg, 2003

Shepherd GM: *Neurobiology*, 3rd edn. Oxford Univ. Press, New York/Oxford, 1994; *Neurobiologie*. Springer, Berlin, 1993

Shorthouse JD, Rohfritsch O (eds.): *Biology of Insect-Induced Galls*. Oxford Univ. Press, Oxford, 1992

Silbernagl S, Despopoulos A: *Taschenatlas der Physiologie*, 6. Aufl. Thieme, Stuttgart, 2003

Singer M, Berg P: *Genes and Genomes*. University Science Books, Mill Valley, California, 1991; *Gene und Genome*, Spektrum Akademischer Verlag, Heidelberg/New York, 1992

Singleton P, Sainsbury D: *Dictionary of Microbiology and Molecular Biology*, 3rd edn. Wiley, New York, 2001

Sitte P, Weiler EW, Kadereit JW, Bresinsky A, Körner C: *Strasburger – Lehrbuch der Botanik*. 35. Aufl. Elsevier/Spektrum Akademischer Verlag, Heidelberg, 2002

Skoog DA, Holler FJ, Nieman TA: *Principles of Instrumental Analysis*, 5th edn. Thomson Learning, New York, 1997

Skoog DA, Leary JJ: *Instrumentelle Analytik*. Springer, Berlin Heidelberg New York,1996

Smith AD (ed): *Oxford Dictionary of Biochemistry and Molecular Biology*, rev. edn. Oxford Univ. Press, Oxford/New York, 2000

Smith MM, Heemstra PC (eds.): *Smith's Sea Fishes*. Springer, Berlin Heidelberg New York, 1986

Springer O (Hrsg.) *Langenscheidts Großwörterbuch: „Der Große Muret-Sanders"* Deutsch – Englisch/Englisch – Deutsch, 4 Bde. Langenscheidt, München, 2000–2003

Stachowitsch M: *The Invertebrates. An Illustrated Glossary*. Wiley-Liss, New York, 1991

Stanier RY, Ingraham JL, Wheelis ML, Painter PR: *General Microbiology*, 5th edn. Macmillan, London, 1986

Starck D: *Vergleichende Anatomie der Wirbeltiere*, Bde. 1–3. Springer, Heidelberg/Berlin, 1978–1982

Starck D (Hrsg.): *Wirbeltiere (Lehrbuch der Speziellen Zoologie)*, Bd II, 5.Teil. Fischer, Jena/Stuttgart, 1995

Stearn WT: *Botanical Latin*. David & Charles, Newton Abbot, 2004

Stearn WT: *Stearn's Dictionary of Plant Names for Gardeners*. Cassell Publ., London, 2002

Steitz E, Stengel G: *Die Stämme und Klassen des Tierreichs*. VCH, Weinheim, 1984

Stenesh J: *Dictionary of Biochemistry and Molecular Biology*, 2nd edn. Wiley, New York, 1989

Storch V, Welsch U, Wink M: *Evolutionsbiologie*. Springer, Berlin Heidelberg New York, 2001

Storch V, Welsch U (Hrsg.): *Kükenthal – Zoologisches Praktikum*, 24. Aufl. Elsevier/Spektrum Akademischer Verlag, Heidelberg, 2002

Storch V, Welsch U: *Systematische Zoologie*, 6. Aufl. Elsevier/Spektrum Akademischer Verlag, Heidelberg, 2003

Storch V, Welsch U: *Kurzes Lehrbuch der Zoologie*, 8. Aufl. Elsevier/Spektrum Akademischer Verlag, Heidelberg, 2004

Storer TI, Usinger RL: *Laboratory Workbook for Zoology*. McGraw-Hill, New York, 1965

Storer TI, Usinger RL, Stebbins RC, Nybakken JW: *General Zoology*, 6th edn. McGraw-Hill, New York, 1979

Strasburger's Textbook of Botany (transl. of 30th Gerrman edn.) Longman, London/New York, 1976

Strauss JH, Strauss EG: *Viruses and Human Disease*. Academic Press, San Diego, 2002

Stryer L: *Biochemistry*, 5th edn. Freeman, New York, 2002; Berg JM, Tymoczko JL, Stryer L: *Biochemie*, 5. Aufl. Elsevier/Spektrum Akademischer Verlag, Heidelberg, 2003

Swartz D: *Collegiate Dictionary of Botany*. Ronald Press, New York, 1971

Taiz L, Zeiger E: *Plant Physiology*, 3rd edn. Sinauer, Sunderland, MA, 2002; *Physiologie der Pflanzen*. Elsevier/Spektrum Akademischer Verlag, Heidelberg, 2000

Tamarin RH: *Principles of Genetics*, 7th edn. McGraw/Hill, Boston, 2002

Tivy J: *Biogeography: A Study of Plants in the Ecosphere*, 2nd edn. Longman, London/New York, 1982

Tortora GJ, Grabowski SR: *Principles of Anatomy and Physiology*, 9th edn. Wiley, New York, 2000

Troll W, Höhn K: *Allgemeine Botanik*, 4. Aufl. Enke, Stuttgart, 1973

Urania-Tierreich. 6 Bde. Urania, Leipzig, 1991–1994

Vaughan TA: *Mammalogy*, 3rd edn. Saunders Coll. Publ., Philadelphia, 1986

Voet D, Voet JG: *Biochemistry*, 3rd edn. Wiley, New York, 2004

Voet D, Voet JG, Pratt CW: *Lehrbuch der Biochemie*, Wiley-VCH, Weinheim, 2002

Vogel F, Motulsky AG: *Human Genetics*, 3rd edn. Springer, Heidelberg/ New York, 1996

Walker JM, Cox M: *The Language of Biotechnology – A Dictionary of Terms*, 2nd edn. ACS, Wash. DC, 1995

Walker JM, Rapley R: *Molecular Biology and Biotechnology*, 4th edn. Royal Society of Chemistry, London, 2001

Walter H: *Die Vegetation der Erde*, Bd. 1 & 2. Fischer, Stuttgart, 1964/1968

Walter H, Breckle SW: *Ökologie der Erde*, Bde. 1–4. Elsevier/Spektrum Akademischer Verlag, Heidelberg, 1998–2004

Watson JD, Gilman M, Witkowski J, Zoller M: *Recombinant DNA*, 2nd edn. WH Freeman, New York, 1992; *Rekombinierte DNA*. Spektrum Akademischer Verlag, Heidelberg, 1993

Watson JD, Baker TA, Bell SP, Gann A, LevineM, Losick R: *Molecular Biology of the Gene*. 5th edn. Benjamin-Cummings/Pearson, San Francisco, CA, 2004

Watt IM: *The Principles and Practice of Electron Microscopy*. Cambridge Univ. Press, Cambridge, 1985

Weber H, Weidner H: *Grundriß der Insektenkunde*. 5. Aufl. Fischer, Stuttgart – Jena, 1974

Weberling F: *Morphologie der Blüten und der Blütenstände*. Ulmer, Stuttgart, 1981; *Morphology of Flowers and Inflorescences*. Cambridge Univ. Press, Cambridge/New York, 1989

Weberling F, Schwantes HO: *Pflanzensystematik*. 7. Aufl. Ulmer, Stuttgart, 2000

Webster's Twelfth New Collegiate Dictionary. Merriam-Webster, Springfield, MA, 2004

Webster's Third New International Dictionary. Merriam-Webster, Springfield, MA, 1986

Welsch U: *Lehrbuch Histologie*. Urban + Fischer, München, 2003

Westheide W, Rieger R: *Spezielle Zoologie*, Teil 1 & 2. Elsevier/Spektrum Akademischer Verlag, Heidelberg, 2003

Whittaker RH: *Classification of Plant Communities*. Junk Publ., The Hague/Boston, 1978

Winburne JN (ed.) *A Dictionary of Agricultural and Allied Terminology*. Michigan State Univ. Press, East Lansing, 1962

Wink M (Hrsg.) *Molekulare Biotechnologie*. Wiley-VCH, Weinheim, 2004

Wistreich GA: *Microbiology Laboratory. Fundamentals and Applications*, 2nd. edn. Prentice Hall, New Jersey, 2002

Wistreich GA, Lechtman MD: *Microbiology*, 5th edn. Macmillan, New York, 1988

Wood CE: *A Student's Atlas of Flowering Plants*. Harper & Row, New York, 1974

Woodland DW: *Contemporary Plant Systematics*. Prentice-Hall, N.J., 1991

Wurmbach H, Siewing R: *Lehrbuch der Zoologie. Allgemeine Zoologie/ Systematik*. Fischer, Stuttgart, 1980/1985

Zomlefer WB: *Guide to Flowering Plant Families*. Univ. of North Carolina Press, Chapel Hill/London, 1995

Zug GR, Vitt LJ, Caldwell JP: *Herpetology. An Introductory Biology of Amphibians and Reptiles*, 2nd. edn. Academic Press, New York/London, 2001

Kompaktlexikon der Biologie

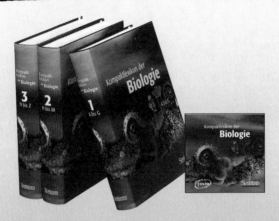

Das *Kompaktlexikon der Biologie* gibt in drei Bänden und auf CD-ROM einen umfassenden Überblick über das Spektrum der modernen Biologie. Dank der allgemein verständlich gehaltenen Darstellung ist das Lexikon für Studenten im Grundstudium sowie Leistungskurs-Schüler, insbesondere aber auch für biologisch interessierte Laien eine ausgezeichnete Hilfe zur Beantwortung von Fragen und ein Anreiz, tiefer in die faszinierende Welt der Biologie einzudringen: Von der Systematik der verschiedenen Organismengruppen, ihren Bauplänen und ihrer Lebensweise bis hin zu brandaktuellen Entwicklungen aus den Bereichen Bio- und Gentechnologie.

3 Bände mit jeweils ca. 500 S., geb. – insgesamt: mehr als 15.000 Stichwörter, 1.000 Abbildungen und Tabellen, 17 Essays und Methodenseiten.

Gesamtausgabe Buch: Früher € 387,-, jetzt € 99,-, ISBN 3-8274-0992-6
Gesamtausgabe CD-ROM: Früher € 387,-, jetzt € 99,-, ISBN 3-8274-1140-8
Gesamtausgabe Buch + CD-ROM: Früher € 580,50, jetzt € 149,-,
ISBN 3-8274-1141-6

 Mehr Information unter
www.elsevier.de

Stand 15.12.04 Preise unter Vorbehalt.

Weitere Wörterbücher

Erwin J. Hentschel/
Günther H. Wagner
■ **Wörterbuch der Zoologie**
Dieses *Wörterbuch der Zoologie* ist in der 7. Auflage umfassend überarbeitet und durchgängig aktualisiert worden. Im lexikalischen Hauptteil werden nunmehr 16.500 (!!) Stichwörter wissenschaftlich und etymologisch definiert. Vorangestellt ist eine gut fassliche "Einführung in die Terminologie und Nomenklatur", die mit der Geschichte und den Grundlagen der zoologischen Fachsprache vertraut macht. Einzigartig im deutschsprachigen Raum!!

7. Aufl. 2004, 604 S., kart.
€ 30,- / SFr 48,-
ISBN 3-8274-1479-2

Theodor C. Cole
■ **Wörterbuch der Tiernamen**
Dieses Buch enthält die Namen von mehr als 16.000 Tieren aus Europa und der ganzen Welt in lateinischer, deutscher und englischer Sprache. Seit "Grzimek" ist dies die erste aktuelle und umfassende Liste häufiger und gefährdeter Tiere. Coles Wörterbuch umfasst die wissenschaftlich korrekten lateinischen Bezeichnungen und international geläufigen englischen Begriffe ebenso wie Trivialnamen und Synonyme. Auch als CD-ROM erhältlich!

2000, 970 S., geb.
€ 82,- / sFr 132,-
ISBN 3-8274-0589-0

Gerhard Wagenitz
■ **Wörterbuch der Botanik**
In diesem Wörterbuch werden etwa 5700 Begriffe aus dem Gesamtgebiet der Botanik definiert. Herkunft und Geschichte wichtiger Termini werden kurz erläutert. Der Schwerpunkt liegt bei der Strukturforschung (Morphologie, Anatomie, Cytologie) und der Evolutionsbiologie (inkl. Taxonomie und Reproduktionsbiologie). Die erweiterten englisch-deutschen und französisch-deutschen Register sind eine wichtige Hilfe bei der Lektüre biologischer Fachliteratur.

2. Aufl. 2003, 552 S., 10 Abb., kart.
€ 25,- / SFr 40,-
ISBN 3-8274-1398-2

Mattias Schaefer
■ **Wörterbuch der Ökologie**
Das Wörterbuch definiert und erläutert Begriffe aus dem gesamten Gebiet der Ökologie. Viele Schlüsseldefinitionen und Begriffsfelder sind ausführlicher gehalten, um in Konzepte der Ökologie einzuführen. Die englischen Übersetzungen der Begriffe und das englisch-deutsche Register erleichtern den Zugang zur englischen Fachliteratur und sind eine Hilfe beim Verfassen eigener englischer Texte.

4. Aufl. 2003, 416 S., 45 Abb., kart.
€ 30,- / SFr 48,-
ISBN 3-8274-0167-4

 Mehr Information unter www.elsevier.de

Stand 15.12.04. Preise unter Vorbehalt.